KU-518-317

LIVERPOOL JMU LIBRARY
3 1111 01520 7853

Scholarpedia

Series editor

Eugene Izhikevich, San Diego, USA

More information about this series at http://www.springer.com/series/13574

Tony J. Prescott · Ehud Ahissar
Eugene Izhikevich
Editors

Scholarpedia of Touch

Editors
Tony J. Prescott
Department of Psychology
University of Sheffield
Sheffield
UK

Eugene Izhikevich
Brain Corporation
San Diego, CA
USA

Ehud Ahissar
Department of Neurobiology
Weizmann Institute of Science
Rehovot
Israel

Scholarpedia
ISBN 978-94-6239-132-1 ISBN 978-94-6239-133-8 (eBook)
DOI 10.2991/978-94-6239-133-8

Library of Congress Control Number: 2015948155

Printed on acid-free paper

Preface

Touch is the ability to understand the world through physical contact. The noun "touch" and the verb "to touch" derive from the Old French verb "tochier". Touch perception is also described by the adjectives *tactile*, from the Latin "tactilis", and *haptic*, from the Greek "haptós". Academic research concerned with touch is also often described as *haptics*.

The aim of *Scholarpedia of Touch*, first published by Scholarpedia (www.scholarpedia.org), is to provide a comprehensive set of articles, written by leading researchers and peer reviewed by fellow scientists, detailing the current scientific understanding of the sense of touch and of its neural substrates in animals including humans. It is hoped that the encyclopedia will encourage sharing of ideas and insights between researchers working on different aspects of touch in different species, including research in synthetic touch systems. In addition, it is hoped that the encyclopedia will raise awareness about research in tactile sensing and promote increased scientific and public interest in the field.

Our encyclopedia assembles a state-of-the-art understanding of the sense of touch across a broad range of species from invertebrates such as stick insects and spiders, terrestrial and marine mammals, through to humans. The different contributions show not only the varieties of touch—antennae, whiskers, fingertips—but also their commonalities. They explore how touch sensing has evolved in different animal lineages, how it serves to provide rapid and reliable cues for controlling ongoing behaviour, how it develops, and how it can disintegrate when our brains begin to fail. In addition to analysing natural touch, we also consider how engineering is beginning to exploit physical analogues of these biological systems so as to endow robots, and other engineered artefacts, with rich tactile sensing capabilities.

Scope and Structure

Following an introductory chapter—The World of Touch—our encyclopedia is structured into four parts:

- **Comparative Touch** There are a large number of specialist tactile sensory organs in the animal kingdom. This part contains articles on animal species that exhibit interesting or exceptional tactile sensing abilities. We particularly focus on antennal systems in insects, and on vibrissal systems in both terrestrial and marine mammals.
- **The Psychology of Touch** The study of human cutaneous touch has a rich and long history in psychology and psychophysics. The pioneering studies of Ernst Weber (1795–1878) distinguished different forms of touch—pressure, temperature, and pain—all of which are separately considered in our encyclopedia along with dynamical (effortful) touch, tactile perception of force, and relevant forms of interoception (internal sensing) and proprioception (sense of body position). A particular focus of recent research has been towards the combination of tactile sensing with manipulation and grasp in the human hand. Alongside the study of healthy touch, this part also considers research on touch disorders, loss of tactile acuity with ageing, and the phenomenon of phantom touch.
- **The Neuroscience of Touch** A scientific understanding of the biological substrates for tactile sensing is beginning to emerge at all levels from the sensory periphery through to the somatosensory and multimodal areas of cortex. Perhaps more than any other sensory modality, tactile sensing is critically dependent on the movement of the sensing apparatus, therefore touch is increasingly studied from an active perception perspective—understanding active touch (in contrast to passive touch) as an intentional, information-seeking activity that combines sensing with actuation. The investigation of the sensorimotor control loops involved in mammalian active touch has been significantly advanced by the availability of the rodent vibrissal sensory system as an animal model. This part therefore combines studies in both primates (including humans) and rodents to show how neurobiological research is beginning to demonstrate an in-depth understanding of tactile sensing systems in mammals.
- **Synthetic Touch** Touch sensing is giving rise to a range of exciting new technologies. This part highlights some of the most promising tactile sensors for robots and haptic displays for the visually-impaired, through to virtual touch systems that can allow the extension of touch, through telecommunication technologies, as a modality for communication.

Tony J. Prescott
Ehud Ahissar
Eugene Izhikevich

Acknowledgments

The Scholarpedia of Touch was sponsored in part by the European Union Framework 7 project BIOTACT (BIOmimetic Technology for vibrissal ACtive Touch) (FP7-ICT-215910), one of whose goals was to promote research in the fields of natural and synthetic tactile sensing. Tony J. Prescott's contribution to editing the Encyclopaedia was also assisted by support from the Framework 7 Co-ordination Action "The Convergent Science Network of Biomimetics and Neurotechnology" (FP7-ICT-601167). The collaboration among the editors was facilitated by support from a Weizmann-UK grant from the Weizmann Institute of Science.

All articles of this volume first appeared in Scholarpedia—the peer-reviewed open-access encyclopedia. They all are freely available at http://www.scholarpedia.org, and the readers are encouraged to revise and edit these articles to keep them up-to-date, similarly to how it is done in Wikipedia—the free encyclopedia. However, such revisions and edits are not published until they are approved by the articles' curators.

The following Scholarpedia Assistants provided substantial assistance to the specified chapters:

Swetamber Das (Tactile Temporal Order, Imaging Human Touch, Whisking Control by Motor Cortex), Javier Elkin (Touch in Aging), Serguei A. Mokhov (Models of Tactile Perception and Development, Systems Neuroscience of Touch), Shruti Muralidhar (S1 Microcircuits, Tactile Object Perception), Abdellatif Nemri (Dynamic Touch), B. Lungsi Sharma (A Spider's Tactile Hairs), and Juzar Thingna (Tactile Substitution for Vision, Whisking Pattern Generation).

Additionally, Nick Orbeck provided copy-editing for most of the articles, and Michelle Jones collated and edited the proof corrections.

Contents

Part II The Psychology of Touch

Contributors and Curators of Scholarpedia

Ehud Ahissar Department of Neurobiology, Weizmann Institute of Science, Rehovot, Israel

Kevin Alloway College of Medicine, Pennsylvania State University, State College, USA

Ehsan Arabzadeh Eccles Institute of Neuroscience, John Curtin School of Medical Research, the Australian National University, Canberra, ACT, Australia

Amos Arieli The Weizmann Institue of Science, Rehovot, Israel

Eldad Assa Department of Neurobiology, Weizmann Institute of Science, Rehovot, Israel

Soledad Ballesteros Universidad Nacional de Educación a Distancia, Madrid, Spain

Friedrich G. Barth Department of Neurobiology, University of Vienna, Vienna, Austria

Sliman Bensmaia Department of Organismal Biology and Anatomy, University of Chicago, Chicago, USA

Wouter M. Bergmann Tiest VU University, Amsterdam, The Netherlands

Bill Blessing Centre for Neuroscience, Flinders University, Adelaide, Australia

Yannick Bleyenheuft Université Catholique de Louvain, Louvain-la-Neuve, Belgium

David Brang Interdepartmental Neuroscience Program, Northwestern University, Evanstown, IL, USA

Michael Brecht Bernstein Center for Computational Neuroscience, Berlin, Germany

Claudia Carello Center for the Ecological Study of Perception and Action, University of Connecticut, Storrs, CT, USA

Manuel Castro-Alamancos Department of Neurobiology, Drexel University College of Medicine, Philadelphia, PA, USA

Shubhodeep Chakrabarti Werner Reichardt Center for Integrative Neuroscience; Hertie Institute for Clinical Brain Research, Tübingen, Germany

C. Elaine Chapman University of Montreal, Montreal, Canada

Adrian David Cheok Imagineering Institute, City University London, London, UK

Christopher Comer Division of Biological Sciences, University of Montana, Missoula, Montana, USA

Christine M. Crish Northeast Ohio Medical University, Rootstown, OH, USA

Samuel D. Crish Northeast Ohio Medical University, Rootstown, OH, USA

Guido Dehnhardt Institute for Biosciences, University of Rostock, Rostock, Germany

Stuart Derbyshire National University of Singapore, Singapore, Singapore

Martin Deschenes Université Laval Robert-Giffard, Québec, Canada

Mathew Diamond Tactile Perception and Learning Lab, International School for Advanced Studies, Trieste, Italy

Chris Dijkerman Utrecht University, Utrecht, The Netherlands

Volker Dürr Bielefeld University, Bielefeld, Germany

Valerie Ego-Stengel Unite de Neurosciences, Information et Complexite CNRS, Gif-sur-Yvette, France

Dirk Feldmeyer Research Centre Jülich, Institute of Neuroscience and Medicine, Jülich, Germany; Department of Psychiatry, Psychotherapy and Psychosomatics, RWTH Aachen University Hospital, Aachen, Germany

Ian Gibbins Centre for Neuroscience, Flinders University, Adelaide, SA, Australia

Goren Gordon Curiosity Lab, Department of Industrial Engineering, Tel-Aviv University, Tel-Aviv, Israel

Robyn A. Grant Conservation Evolution and Behaviour Research Group, Manchester Metropolitan University, Manchester, UK

Frank Grasso Department of Psychology, Brooklyn College, New York, USA

Sebastian Haidarliu Department of Neurobiology, Weizmann Institute of Science, Rehovot, Israel

Wolf Hanke Institute for Biosciences, University of Rostock, Rostock, Germany

Matthias Harders University of Innsbruck, Innsbruck, Austria

Mitra Hartmann Northwestern University, Evanston, USA

Vincent Hayward Université Pierre et Marie Curie, Paris, France

Moritz von Heimendahl Bernstein Center for Computational Neuroscience, Humboldt Universität zu Berlin, Berlin, Germany

Morton Heller Eastern Illinois University, Illinois, USA

Leif Johannsen Technische Universität München, Munich, Germany

Lynette Jones Department of Mechanical Engineering, MIT, Cambridge, USA

Jon H. Kaas Department of Psychology, Vanderbilt University, Nashville, TN, USA

Astrid M.L. Kappers VU University, Amsterdam, The Netherlands

Asaf Keller Department of Anatomy and Neurobiology, University of Maryland School of Medicine, Baltimore, USA

Yeongmi Kim The Department of Mechatronics, MCI, Innsbruck, Austria

Shigeru Kitazawa Graduate School of Frontier Biosciences, Osaka University, Suita, Japan

Roberta Klatzky Department of Psychology, Carnegie Mellon University, Pittsburgh, PA, USA

Per M. Knutsen UC San Diego, La Jolla, CA, USA

Simon Lacey Department of Neurology, Emory University, Atlanta, USA

Ilan Lampl Weizmann Institute of Science, Rehovot, Israel

Nathan Lepora University of Bristol, Bristol, UK

Gerald E. Loeb University of Southern California, Los Angeles, CA, USA

Ellen Lumpkin Columbia University, New York, NY, USA

Uriel Martinez-Hernandez University of Sheffield, Sheffield, UK

Sabah Master Hospital for Sick Children, Toronto, Canada

Milana Mileusnic Otto Bock Healthcare Products, Vienna, Austria

Ben Mitchinson Adaptive Behaviour Research Group, The University of Sheffield, Sheffield, South Yorkshire, UK

Yalda Moayedi Columbia University, New York, NY, USA

Chris Moore Department of Neuroscience, Brown University, Providence, RI, USA

Frank Moss Center for Neurodynamics, University of Missouri at St. Louis, St. Louis, USA

Masashi Nakatani Columbia University, New York, NY, USA

Guy Nelinger Department of Neurobiology, Weizmann Institute of Science, Rehovot, Israel

Jiro Okada Nagasaki University, Nagasaki, Japan

Michael Okun UCL, London, UK

Martin J. Pearson Bristol Robotics Laboratory, Bristol, UK

Tony Pipe Bristol Robotics Laboratory, Bristol, UK

Gilang Andi Pradana Imagineering Institute, City University London, London, UK

Tony J. Prescott Department of Psychology, University of Sheffield, Sheffield, UK

Vilayanur S. Ramachandran Center for Brain and Cognition, University of California, San Diego, CA, USA

Mark S. Redfern University of Pittsburgh, Pittsburgh, PA, USA

Catherine L. Reed Department of Psychology, Claremont McKenna College, Claremont, CA, USA

Roger Reep McKnight Brain Institute, University of Florida, Gainesville, USA

Claudia Roth-Alpermann Bernstein Center for Computational Neuroscience, Humboldt-University, Berlin, Germany

Diana K. Sarko Vanderbilt University, Nashville, TN, USA

K. Sathian Department of Neurology, Emory University, Atlanta, USA

Cornelius Schwarz Hertie Institute for Clinical Brain Research, University of Tübingen, Tübingen, Germany

Philip Servos Wilfrid Laurier University, Waterloo, ON, Canada

Namrata Shinde Department of Neurobiology, Weizmann Institute of Science, Rehovot, Israel

Daniel E. Shulz Unite de Neurosciences, Information et Complexite, CNRS, Gif-sur-Yvette, France

Jochen F. Staiger Center for Anatomy, Institute for Neuroanatomy, University Medicine Göttingen, Göttingen, Germany

Haike van Stralen Utrecht University, Utrecht, The Netherlands

Jean Louis Thonnard Université Catholique de Louvain, Brussels, Belgium

Francois Tremblay School of Rehabilitation Sciences, University of Ottawa, Ottawa, ON, Canada

Michael Turvey Center for the Ecological Study of Perception and Action, University of Connecticut, Storrs, CT, USA

Nadia Urbain EPFL, Lausanne, Switzerland

Jamie Ward University of Sussex, Sussex, UK

Martin Wells Department of Zoology, University of Cambridge, Cambridge, UK

Lon A. Wilkens Biology and Center for Neurodynamics, University of Missouri-St. Louis, St. Louis, USA

Stuart P. Wilson Dept. Psychology, The University of Sheffield, Sheffield, UK

Alan Wing Centre for Sensory Motor Neuroscience, University of Birmingham, Birmingham, UK

Shinya Yamamoto National Institute of Advanced Industrial Science and Technology (AIST), Tsukuba, Japan

Phil Zeigler Psychology Department, Hunter College of the City University of New York, New York, USA

Yael Zilbershtain-Kra Department of Neurobiology, Weizmann Institute of Science, Rehovot, Israel

Reviewers of *Scholarpedia of Touch*

Lazlo Acsady
Prof. Ehud Ahissar
Prof. Kevin Alloway
Dr. Gabriel Baud-Bovy
Prof. Sliman Bensmaia
Dr. Patrick A. Cabe
Corinna Darian-Smith
Prof. Guido Dehnhardt
Dr. Christiaan de Kock
Dr. David S. Deutsch
Prof. Mathew Diamond
Prof. Volker Dürr
Dr. Mathew Evans
Dr. John B. Furness
Dr. Daniel Goldreich
Dr. Goren Gordon
Dr. Robyn A. Grant
Dr. Wolf Hanke
Dr. Ealan Henis
Dr. John Horn
Dr. Thomas W. James
Dr. Lynette Jones
Prof. Astrid M.L. Kappers
Prof. Asaf Keller
Dr. Yeongmi Kim
Prof. Roberta Klatzky
Dr. Per M. Knutsen
Prof. James R. Lackner
Prof. Charles Lenay
Dr. Nathan Lepora
Dr. Mitchell Maltenfort

Dr. Ben Mitchinson
Dr. Shruti Muralidhar
Dr. Jiro Okada
Dr. Martin J. Pearson
Prof. Tony J. Prescott
Dr. Colleen Reichmuth
Dr. Robert Sachdev
Mr. Avraham Saig
Prof. Evren Samur
Prof. K. Sathian
Dr. Cornelius Schwarz
Dr. Jean Louis Thonnard
Dr. Avner Wallach
Lawrence M. Ward
Dr. Emma Wilson
Dr. Stuart P. Wilson
Dr. Jason Wolfe
Dr. Yael Zilbershtain-Kra

Introduction: The World of Touch

Tony J. Prescott and Volker Dürr

Despite its behavioural significance and omnipresence throughout the animal kingdom, the **sense of touch** is still one of the least studied and understood modalities. There are multiple forms of touch, and the mechanosensory basis underlying touch perception must be divided into several distinct sub-modalities (such as vibration or pressure), as will be made clear by the contributions elsewhere in this encyclopaedia. The commonality of all touch sensing systems is that touch experience is mediated by specialised receptors embedded in the integument—the outer protective layers of the animal such as the mammalian skin or the arthropod cuticle. Comparative research on touch, and its neuroethology, is only just beginning to provide a larger picture of the different forms of touch sensing within the animal kingdom. We begin our volume by reviewing works on several different invertebrate and vertebrate species, focusing on mechanosensation, each one with a specific requirement for tactile information. The aim of this introductory overview is to give selected examples of research on important model organisms from various classes of the animal kingdom, ranging from the skin of worms to the feelers of insects, and from the whiskers of a rat to the human hand. We conclude by discussing forms of human touch and the possibility of its future extension via synthetic systems.

Touch in Invertebrates

The Evolutionary Origins of Touch

Mechanical perturbation of the outer membrane of a ciliate such as *Paramecium* (Figure 1), will cause it to respond by moving away from the stimulus source (Naitoh and Eckert 1969). Thus, single-celled organisms already have a capacity for

T.J. Prescott (✉)
Department of Psychology, University of Sheffield, Sheffield, UK
e-mail: t.j.prescott@sheffield.ac.uk

V. Dürr
Bielefeld University, Bielefeld, Germany

T.J. Prescott et al. (eds.), *Scholarpedia of Touch*, Scholarpedia,
DOI 10.2991/978-94-6239-133-8_1

directional detection of tactile stimuli. The most primitive multicellular animals, the sponges, lack neurons, yet still show some capacity to respond to changes in water flow and pressure triggered by the deflection of non-motile cilia (Ludeman et al. 2014). Non-neural forms of sensitivity to tactile stimuli are also seen in many plants (Monshausen and Gilroy 2009; Coutand 2010). However, the evolution of neural conduction brings about a step-change in the capacity to respond rapidly and flexibly to tactile stimuli. Cnidarians, such as jellyfish and *Hydra*, despite having relatively simple nervous systems, can exhibit coordinated patterns of motor response to sensory stimuli and many have a rich capacity to respond to touch. For example, the nematocytes of *Hydra* are hair-like structures that respond to selective deflection and are thought to provide a good model for understanding the mechanoreceptors of more complex invertebrates (Thurm et al. 2004). In jellyfish such as *Aglantha digitale*, groups of hair cells, known as tactile combs, regulate complex behaviors including escape, feeding and locomotion (Arkett et al. 1988). The benefits of sensitivity to mechanical stimuli provided by hair-like structures may have encouraged their convergent evolution in multiple animal lineages. For instance, the hair cells of jellyfish appear to be sufficiently different from those in vertebrates that a common origin for both is unlikely (Arkett et al. 1988). Studies of the molecular basis of mechanosensation across different animal classes also suggest that cellular mechanisms to support tactile sensing may have evolved multiple times (Garcia-Anoveros and Corey 1997).

Lower Invertebrate Model Systems

All animals with a Central Nervous System (CNS) respond to touch. Even the tiny, un-segmented worm *Caenorhabditis elegans* (Figure 2), a nematode, shows touch-induced locomotion away from the stimulus. As in all higher animals, the corresponding mechanosensory cells are located in the integument, in this case beneath the cuticle. This worm's relevance to neuroscience stems from the fact that all of its 302 (somatic) neurons have been labelled and mapped. As a result, it has become the first animal system in which the entire network involved in touch-mediated behaviour—six sensory neurons, ten interneurons and 69 motoneurons—has been identified (Chalfie et al. 1985). Thirty years after the identification of all cellular components, many more details, including the biophysics of mechanosensory transduction (O'Hagan et al. 2005) and the molecular identity of several modulating signalling cascades (Chen and Chalfie 2014), have been unravelled.

In larger, arguably more advanced, animals such as the leech (*Hirudo medicinalis*, Figure 3), touch-induced behaviour becomes more versatile and complex. However, increased complexity generally means less complete understanding. In terms of complete mapping and understanding of touch-induced behaviour, the leech comes second to the champion nematode. Owing to the larger size of its neurons, the neurobiology of touch has been investigated primarily by means of electrophysiological recordings, which are almost impossible in tiny *Caenorhabditis*. As a

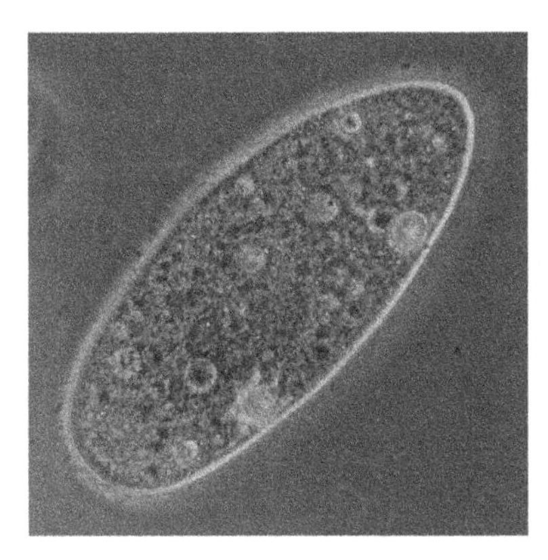

Figure 1 The ciliate *Paramecium* is a unicellular organism that responds to touch on either end of its oval-shaped cell body by active movement away from the stimulus. The corresponding change in the beat of the cilia is triggered by ion currents that are activated through touch (Photograph by Barfooz, CC BY-SA 3.0)

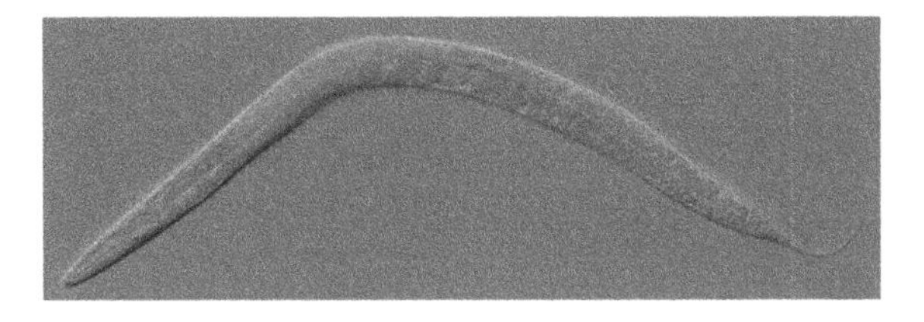

Figure 2 The nematode *Caenorhabditis elegans* is an important model organism in neuroscience. It has 16 sensory and interneurons involved in touch-mediated behaviour (CC BY-SA 2.5)

Figure 3 An important model organism for the electrophysiological study of touch-mediated movements is the medicinal leech (*Hirudo medicinalis*). The neurons of its ganglia have been mapped and many of them have been characterised individually (Photograph by Karl Ragnar Gjertsen, CC BY 2.5)

result, behaviourally relevant processing of touch-related information is perhaps best-studied in the leech (Muller et al. 1981). Because of its segmented body structure, the CNS of the leech has a chain of ganglia (one per segment), all of which contain the same, or at least very similar, sets of cellular elements. With regard to touch, three groups of mechanosensory neurons can be found in each ganglion, all of which respond to mechanical stimulation of the body, but with varying response thresholds: Touch cells (T-cells) are the most sensitive and respond to gentle touch of the body wall; Pressure cells (P cells) respond to stronger touch stimuli, and Nociceptive cells (N-cells) respond to very strong, potentially harmful, stimuli (Nicholls and Baylor 1968). After the original description of what has meanwhile become a textbook example of range fractionation of stimulus intensity, a number of general aspects of sensorimotor systems have been studied in leech: Notable examples are the recruitment of T-, P- and N- cells in crawling (Carlton and McVean 1995), the "mapping" of sensory input to distinct motor output by means of a population code (Lewis and Kristan 1998a), the encoding and decoding of touch (Thomson and Kristan 2006), and the modelling of the entire sensorimotor pathway underlying touch-induced directed movements (Lewis and Kristan 1998b).

Tactile Learning in Molluscs

Among the molluscs, a sister group of the segmented invertebrates, several species have become important model organisms in neuroscience, particularly with regard to research on learning and memory (see e.g. Brembs 2014). Most famous of these is probably the sea hare *Aplysia californica*, in which a number very fundamental cellular mechanisms underlying learning have been described for the first time. Although the historical paradigm of sensitisation and habituation of the siphon/gill withdrawal response can be induced by mechanical stimuli, the modality of touch has not been in the focus of these studies. Another fascinating mollusc model system is the **octopus**, that has long been known for its cognitive abilities. A number of behavioural studies on tactile discrimination and tactile shape recognition have been conducted, largely following an animal psychology approach, in combination with ablation studies (e.g., Young 1983). As a result, touch-related behaviour in octopus has been analysed at a different level of description than in other invertebrate groups. A review of the touch-related behavioural repertoire in octopus is given by Grasso and Wells (2013) in this volume.

The Arthropod Tactile Hair

Compared to the soft-bodied animals mentioned above, animals with a skeleton no longer work as hydrostats in which body deformation is best monitored by deformation of the surface, i.e., the integument. In contrast to a hydrostat, skeletons require the formation of joints, thus "focussing" any change in posture on a limited set of locations. As a consequence, skeletons impose a physical limit to the number of degrees of freedom of movement. With regard to touch, this is important because mechanoreceptors may be dedicated to "strategically relevant locations" across the body, i.e., locations where displacement is most likely to occur. In proprioception, such strategically relevant locations are the joints themselves (and/or the muscles and tendons actuating them). In touch, strategically relevant locations are surface areas where contact with external objects is most likely to occur. Arthropods (that is the group comprising spiders, Crustaceans, insects and their relatives) appear to exploit such strategically relevant locations in two ways: In proprioception, they often use patches of hairs, so-called *hair fields* or *hair plates*, to sense displacement of two adjoining body segments. In touch, they use very much the same kind of hairs, too, but at various locations on the body, with particularly high "hair density" at places where contacts are most likely to occur and/or most relevant to detect.

These arthropod hairs are not like mammalian hairs at all: The Arthropod tactile hair is a cone-shaped cuticular structure, filled with fluid and equipped with a number of cells as the base. At least one of these cells is a ciliated mechanoreceptor that encodes the deflection of the cuticular structure (Thurm 1964). Arthropods have a large variety of such sensory hairs (called *seta* or, more generally, *sensillum*; plural: *setae* and *sensilla*). Their information encoding properties have been described in different ways (e.g., see French 2009). Apart from various mechanoreceptors (the basic type being the *Sensillum trichoideum*), there are also variants with chemoreceptors, subserving gustation and olfaction.

Both spiders and insects use hair sensilla also in proprioception and, as a consequence, in active touch sensing, where the active movement of a limb needs to be monitored. To date, a number of such proprioceptive hair fields have been characterised functionally (e.g., French and Wong 1976). As part of this encyclopedia, tactile hairs are the key sensory elements in three chapters, one dealing with the functional properties and behavioural relevance of tactile hairs in **spiders** (Barth 2015), and two more dealing with tactile hairs on dedicated sensory organs in insects—the **antennae** (Okada 2009; Dürr 2014).

The Arthropod Antenna: Elaborate Touch in Invertebrates

Among the living Arthropods, only the Myriapods (millipedes, etc.), Crustaceans and insects carry antennae (singular: antenna). Antennae are commonly called feelers, which is appropriate given the fact that they are dedicated sensory limbs (Staudacher et al. 2005). As such, they are equipped with a particularly large number of sensilla, and the density of sensilla per unit surface area is much higher than on most other parts of the body. With regard to touch this means that the antennae are the major sensory organ of touch, although all other body parts that carry tactile hairs may contribute tactile information as well.

During evolution of the Arthropoda, several body segments have fused to form the head. For example, in insects, the common view is that the head has developed from six body segments, three of which carry the mouthparts, and another three that carry the main sensory organs of the head: the eyes and the antennae (or feelers). In Crustaceans, there are two pairs of antennae: the smaller, anterior pair—the *Antennules* (or 1st antennae)—are known to have important functions in chemoreception (as in an underwater sense of smell). The larger, posterior pair (the 2nd antennae) is known to serve the sense of touch.

Tactually mediated behaviour in Crustaceans has mostly been studied in lobsters, crayfish and other large, decapod Crustaceans (Staudacher et al. 2005). Given the fact that the *Decapoda* represent only a small fraction of the morphological, behavioural, and ecological diversity of Crustaceans, it is very likely that antennal touch works differently in different taxa. An example for a reasonably well-studied Crustacean with regard to touch is the Australian crayfish *Cherax destructor* in which both biomechanical and behavioural aspects have been studied (Sandeman 1985, 1989). *Cherax* actively explores the environment ahead with its 2nd antennae and shows directed attacks towards objects that it has localised tactually (e.g., Zeil et al. 1985).

Insects carry only one pair of antennae. The common view is that, during evolution, insects lost one pair of antennae. According to this view, insects must have lost the second pair of antennae, such that the insect antenna is homologous to the Crustacean antennule, i.e., both organs have a common evolutionary origin. Like the Crustacean antennule, the insect antenna is the most important sensory organ for olfaction. Other than the Crustacean antennule, the insect antenna is also the most important sensory organ for touch (Figure 4).

Tactually mediated behaviour in insects has been studied in great detail in a number of species, most notably in cockroaches, crickets, stick insects and bees (Staudacher et al. 2005). Antennal touch has been shown to be important for

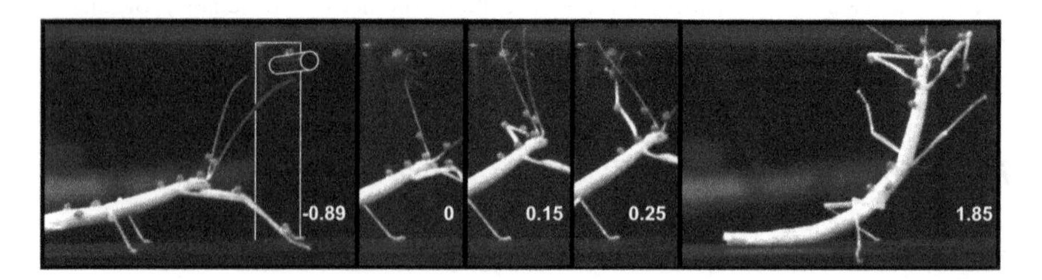

Figure 4 An Indian stick insect (*Carausius morosus*) climbs a rung that it has detected by tactile contact with its antennae. Location of the rung is highlighted in the leftmost frame. From left to right, frames show instants of (i) missing the rung with the antennae, (ii) touching the rung with the left antenna during upstroke, (iii) touching the rung again during the subsequent downstroke, (iv) first foot contact with the rung and (v) climbing the rung. *Numbers* indicate the time in [s] relative to the first antennal contact with the rung

near-range sensing during locomotion, for example in tactually mediated course control (Camhi and Johnson 1999). In many cases, the relevant mechanoreceptors have been identified, for example the antennal hair fields in touch-mediated turning towards an object (Okada and Toh 2000). In the past, sensors involved in course control have mostly been viewed as passive systems that do not require active movement for acquiring information. In some behavioural paradigms, such as tactile wall following, antennae are thought to be used passively. However, active movement of antennae is a key feature of active tactile exploration during locomotion. Antennal contact events may trigger decisions between mutually exclusive behavioural actions (e.g., Harley et al. 2009) or goal-directed motor actions such as tactually elicited reaching movements of the front legs (Schütz and Dürr 2011).

As a higher-order aspect to active tactile exploration, antennal movements are also involved in attentive behaviour such as visually guided pointing in crickets (e.g., Honegger 1981) and touch-related motor learning in the honeybee (Erber et al. 1997). In addition to non-associative motor learning, honeybees have also been demonstrated to show different associative learning behaviours (with regard to touch). For example, it is possible to condition them to different surface patterns (e.g., Erber et al. 1998) but also to condition antennal sampling movements as such in an operant conditioning paradigm (Erber et al. 2000).

The antennal mechanosensory system of insects has also been studied electrophysiologically, in particular with respect to the information transfer from the head to the ganglia of the ventral nerve cord. Owing to the fact that insect neurons can be identified and labelled individually, there is hope that some aspects of tactually mediated behaviour will be understood at the level of neuronal networks as research progresses. Today, the collection of identified antennal mechanosensitive neurons comprises descending interneurons in the brain (e.g., Gebhardt and Honegger 2001) and in the suboesophageal (or gnathal) ganglion (e.g., Ache et al. 2015).

As part of the present collection of articles, two insect model organisms are presented in more detail: the **cockroach** (Okada 2009) and the **stick insect** (Dürr 2014).

Touch in Vertebrates

The Lateral Line System of Fish and Amphibians

In fish and amphibians, vertebrate mechanoreceptor organs, known as *neuromasts*, are arranged in lines on the body, head and tail, either free-standing on the skin or inside fluid-filled canals. In many fish species, a line of neuromasts within a canal that lies on the center line of each flank forms what is known as the lateral line system (Bleckmann and Zelick 2009) (Figure 5). The number of neuromasts can vary between species from a few dozen to several thousand. The main sensory structure in each neuromast is a bundle of hair cells that respond to hydrodynamic stimuli resulting from water displacement or changes in water pressure. The lateral

Figure 5 The fish lateral line system provides a sensitive organ for detecting hydrodynamic stimuli. In the Rainbow Trout (*Oncorhynchus mykiss*), pictured here, the position of the lateral line is marked by the distinctive pink coloring. *Photograph* by Liquid Art (CC 4.0)

line system therefore provides a refined sensory system for detecting remote causes of water movement such as the behavior of predator or prey animals, water currents, and topographical features of the underwater environment. The lateral line, in effect, provides a form of remote or distal touch. A remnant of the lateral line system in mammals (including humans) are the liquid-filled (semicircular) canals of the inner ear, a structure informing us about rotational movements of the head.

Evolutionary Origins of Mammalian Hair and Skin

Skin serves many functions, including to house and protect the organs and internal body parts, and to act as the body's largest sensory organ sensitive to tactile, thermal, and chemical stimuli (Chuong et al. 2002). Mammalian skin has evolved from the integument of earlier vertebrates via a complex path that is still only partially charted (Maderson 1972, 2003). What is clear, however, is that hair evolved anew in early mammals, or in their therapsid reptilian ancestors, as a specialization of the outer epidermal layers of the integument (Sarko et al. 2011). Whereas dense or pelagic hair serves an obvious thermoregulatory function, the evolution of mammalian hair cannot be explained by the need to improve maintenance of body temperature. Rather, the first hairs almost certainly had a largely tactile function, and their subsequent proliferation allowed hair to gain a secondary role as an insulator. The primary sensory role of hair is retained in the tactile hair, or **vibrissae**, found on all therian mammals (marsupials and placentals) except humans (Prescott et al. 2011). In the **naked mole rat**, an animal that has lost all of its pelage, tactile hair has been retained across the whole body surface for its value in supporting the sense of touch (Crish et al. 2015) (Figure 6).

Figure 6 The naked mole rat (*Heterocephalus glaber*) has lost all of its pelagic hair but retains tactile hair across its body surface. *Photograph* by Trisha Shears

Many mammals have also evolved areas of non-hairy, glabrous skin. In humans, these include the skin areas on the lips, hands and fingertips and the soles of the feet. These are the parts of the body that are most important when physically interacting with the world and where accurate tactile discrimination is most critical; unsurprisingly, then, glabrous skin has a high density of **mechanosensory receptors** including Meissner corpuscles, Pacinian corpuscles, Merkel-cell neurite complexes and Ruffini endings (Moayedi et al. 2015). The same mechanoreceptor types are also found in hairy skin alongside hair follicles and a system of unmyelinated low threshold C-tactile (CT) mechanoreceptors (Loken et al. 2009). The CT fibers are particularly sensitive to light 'stroking' touch and thus are thought to underlie an affective, or social, touch capacity that may be unique to mammals (McGlone et al. 2014) (Figure 7).

Figure 7 The CT fiber system in mammalian hairy skin appears to be part of an affective system that encourages social touch and promotes bonding in mammals. *Photograph* by Jim Champion (CC 3.0)

Glabrous skin forms the basis for specialized sensory organs in a variety of animals including the electroreceptive bill of the platypus and the unusual tactile snout of the star-nosed mole (see below). A notable feature of mammalian tactile sensing systems is the presence of **somatotopic maps** of the skin surface in primary somatosensory cortex (S1) (Kaas 1997) (Wilson and Moore 2015). These maps are organized to match the topographic layout of the periphery but in a distorted manner, such that skin areas that have a higher density of receptors, greater receptor innervation, or that are functionally more important to the animal, have a proportionately larger representation in cortex. The human sensory "homunculus" described by Penfield and Boldrey (1937) is probably the best known of these maps, and the "barrel" field of rat and mice somatosensory cortex, first described by Woolsey and van der Loos (1970), the best studied. However, maps whose size, shape and organisation reflect the sensory specialization of the species have also now been described for a wide range of mammals (Krubitzer et al. 1995; Catania and Henry 2006).

Mechanosensation in Monotremes

Sensing in the monotremes, or egg-laying mammals, is perhaps most remarkable for the sensitive electroreceptive capability of the platypus (Scheich et al. 1986). However, all three monotreme species (the platypus and both species of spiny anteater—*Echidna*) also possess a distinctive tactile sensing system, quite different from the vibrissae of the therian mammals. Specifically, all monotremes have mechanoreceptive structures on the snout or bill known as "pushrods" (Proske et al. 1998). These are compact, rigid columns of cells embedded in the skin that are able to move relative to the surrounding tissue. Most of the rod structure is below the skin surface with a convex tip raised slightly above it. The tissues of the pushrods are associated with four types of nerve ending including Merkel cells, and the structure has been compared to the Eimer organs of moles (see below). In platypus, up to 50,000 pushrods are scattered across surface and along the edges of the bill. The bill also contains an extensive venous system that can be engorged with blood, possibly boosting the acuity of the pushrod system. In platypus (Figure 8) the electroreceptive system provides a strongly directional sense for detecting and orienting to prey animals; the pushrods might then assist targeting of prey in the final attack phase. In echidna the pushrod system might similarly be important when the animal probes the ground with its snout looking for insect prey.

Mammalian Vibrissal Systems

Long facial whiskers, or macrovibrissae, are found in many mammalian species, projecting outwards and forwards from the snout of the animal to form a tactile

Figure 8 Monotremes, such as the duck-billed platypus, combine mechanosensing through their unique "pushrod" system, with electroreception, to search for prey underwater or beneath the ground. *Photograph* from Brisbane City Council (CC 2.0)

sensory array that surrounds the head (Pocock 1914, Brecht et al. 1997; Mitchinson et al. 2011) (Figure 9). Pocock (1914) examined example specimens from all of the principal mammalian orders, concluding that facial vibrissae were present in at least some species in all orders except the monotremes, and that species that lacked whiskers, or in which the whiskers were less evident, were usually the more derived members of their order. He concluded that the possession of facial whiskers was a primitive mammalian trait. Each vibrissa, also known as a sinus hair, is composed of a flexible hair shaft that emerges from an intricate mechanosensory structure

Figure 9 Many terrestrial mammals have prominent vibrissae the evolution of which predates pelagic hair. Many species also actively control their whiskers using their facial musculature. *Top Left* The short-tailed Brazilian Opossum (*Monodelphis domestica*). *Bottom Left* South American river rat (*Myocastor coypus*). *Top Centre* Brazilian Porcupine (*Coendou prehensilis*). *Bottom Centre* European water vole (*Arvicola amphibius*). *Top Right* African hedgehog tenrec (*Echinops telfairi*). *Bottom Right* Eurasian water shrew (*Neomys fodiens*). *Photographs* by Tony Prescott (*top row*) and Sarah Prescott (*bottom row*) used with permission

known as a follicle-sinus complex (FSC), containing six distinct populations of receptors, and that differs in structure, and is more richly innervated, than the simpler follicle of a pelagic hair (Rice et al. 1986). Many terrestrial mammals, particularly ones that are nocturnal, inhabit poorly-lit spaces, or that are tree-climbing (scansorial), have evolved highly specialized vibrissal sensing systems. Research has focused on several such animals—rodents such the common rat, the common mouse, and the golden hamster, insectivores such as the **Etruscan shrew** (Roth-Alpermann and Brecht 2009) and the marsupial short-tailed Brazilian opossum (Grant et al. 2013).

The configuration of the facial whiskers varies substantially amongst rodents (Sokolov and Kulikov 1987) though the rat serves as a suitable example. In rats, the macrovibrissae form a two-dimensional grid of five rows on each side of the snout, each row containing between five and nine whiskers ranging between ~15 and ~50 mm in length (Brecht et al. 1997). During exploration and many other behaviors, these whiskers are swept back and forth at rates of up to 25 Hz, and in bouts that can last many seconds, in a behavior known as "whisking". The **kinematics** of each whisker is partly determined by its own intrinsic muscle [Berg and Kleinfeld (2003); see also (Knutsen 2015; Haidarliu 2015)]. These movements of the whiskers are also closely coordinated with those of the head and body, allowing the animal to locate interesting stimuli through whisker contact, then investigate them further using both the macrovibrissae and an array of shorter, non-actuated microvibrissae on the chin and lips.

Much of the neurobiological research on the rodent vibrissal system has focused on its accessibility, given the ease of use of rats and mice as laboratory animals, as a model of mammalian neural processing in general, and of cortical processing in particular, rather than as a target for understanding tactile sensing per se. Nevertheless, this research has revealed a great deal about neural processing for somatosensation, demonstrating, in particular, that it involves multiple closed sensorimotor loops at different levels of the neuraxis (Ahissar and Kleinfeld 2003; see also Ahissar, Shinde et al. 2015). Interest in the rodent vibrissal system as an active sensing system in its own right began with Vincent's 1912 monograph on "The function of the vibrissae in the behaviors of the white rat" and has seen significant growth in recent decades spurred by the availability of high-speed digital video recorders for accurate detection of vibrissal movement (for reviews see Hartmann (2001); Kleinfeld et al. (2006); Mitchinson et al. (2011).

The Refined Vibrissal Sense of Aquatic Mammals

Some of the most successful and intelligent aquatic animals are the **pinnipeds**—seals and walruses (Figure 10)—sea mammals that evolved when bear-like ancestors returned to the oceans more than twenty million years ago (Berta et al. 2006). A striking feature of these animals is that they have well-developed and

Figure 10 Aquatic mammals such as the Harbor seals have highly developed vibrissal sensing that they employ to detect hydrodynamic flows left by fish. *Photograph* by Barbara Walsh (CC 2.0)

sensitive vibrissal (whisker) sensory systems (Hanke and Dehnhardt 2015). In some species of seal, the vibrissae have adapted to be able to detect the disturbances left in the water by swimming fish. By measuring these hydrodynamic flow fields these animals are able to track and capture fast-moving prey in total darkness (Dehnhardt et al. 2001). Pinnipeds that find food at the bottom of the ocean, such as the bearded seal and the walrus, have evolved a very different form of vibrissal adaptation optimized for efficient foraging for clams and other invertebrates in the muddy substrates of the seafloor (Ray et al. 2006). The numerous, densely-packed, and moveable vibrissae of these animals form a sensitive tactile "rake" that allow them to recognize and consume prey animals at remarkably fast rates (up to 9 clams per minute has been estimated for the walrus).

In addition to the ocean-going pinnipeds, facial vibrissae are well-developed in many other aquatic mammalian species including **manatees**, otters, water shrews, water voles, and water rats. In manatees, vibrissae are found distributed across the entire body apparently as a compensation for the reduced availability of visual information in the natural habitat of these animals. The manatee's ability to navigate effectively in turgid water indicates a whole body capacity to detect hydrodynamic signals using vibrissal sensing that has interesting similarities to the fish lateral line (Sarko et al. 2007; Gaspard et al. 2013; Reep and Sarko 2009).

In small aquatic mammals whisker touch appears to be important in predation. For instance, the sensitivity and fast responsiveness of the vibrissae for underwater hunting has been demonstrated for the American water shrew using high-speed video data (Catania et al. 2008). These animals show no significant capacity for following hydrodynamic trials, but display a very rapid and precise attack response when they make direct contact with a prey animal or when there is nearby movement of the water generated by a possible prey. Attacks are initiated within 25 ms of first contact and usually completed with 1 s. The Etruscan shrew, the world's smallest terrestrial animal, has shown as similar facility for short-latency, high-precision attacks on insect prey in air, guided purely by vibrissal touch (Anjum et al. 2006; Roth-Alpermann and Brecht 2009).

The specialization of mammals to aquatic environments has been accompanied by adaption to the sensory innervation of the vibrissae. In a comparison of an aquatic (ringed seal), semi-aquatic (European otter) and terrestrial (pole cat) mammal species, the latter two being of the same mammalian family (Mustelidae), a marked increase was found in innervation by the deep vibrissal nerve in the semi-aquatic (4x terrestrial) and aquatic (10x terrestrial) animals (Hyvarinen et al. 2009). The form and structure of the whiskers also varies quite markedly between species. For instance, in terrestrial mammals, a typical vibrissal hair shaft is tapered and has a round cross-section, while those of eared seals and walruses are typically non-tapered and oval. These differences in hair structure are associated with a different response to tactile stimulation (Hanke et al. 2010), in particular their vibratory properties. These adaptations point to the increased importance of vibrissal touch sensing in the ecology of many aquatic mammals.

Active Sensing Strategies and the Unusual Case of the Star-Nosed Mole

Millions of years of tunneling in damp soil has given rise to the intriguing active-touch organ that is the "star" of the star-nosed mole (Catania and Kaas 1995; Catania 1999; Catania and Remple 2005; Catania 2011) (Figure 11). In this animal, the glabrous skin of the snout has expanded and diversified into twenty-two movable fleshy fingers, called rays, controlled by tendons connected to the facial musculature. Each ray is covered with thousands of Eimer organs (over 25,000 across the whole star), distributed in a honeycomb pattern. The Eimer organ is a

Figure 11 The nose of the star-nosed mole (*Condylura cristata*) has evolved twenty-two fleshy appendages that together constitute a unique active touch sensing system for rapid detection of invertebrate prey. *Photograph* from the US National Park Service

narrow column of skin tissue and nerve endings capped with a 30 μm dome, that sits atop one slowly adapting (Merkel) and one rapidly adapting (lamellated corpuscle) mechanoreceptor. Eimer organs are found in the snouts of all species of mole so far investigated and are known to be very sensitive to light touch.

The unusual structure of the star is thought to be a consequence of the mole's subterranean lifestyle (Catania and Remple 2005). Although many mole species tunnel in the ground and prey on small invertebrates, the star-nosed mole stands out for the speed at which it can detect and consume small prey animals—handling time, from detection to consumption, can be as little as 120 ms. The evolution of the star may thus have allowed the mole to specialize towards smaller and smaller prey sizes. To achieve this feat of speed foraging, touch is actively controlled. Specifically, as the animal searches for food, the rays of the star lightly palpate the surface of the substrate many times per second. If a potential prey item is detected the star is oriented to allow more extensive exploration with ray 11, a shorter ray at the center and bottom of the star. Ray 11 has been proposed to be a tactile fovea, with the orienting movements of the star likened to visual saccades (Catania 1999). Interestingly, ray 11 does not have higher Eimer organ density than other rays but it does have significantly higher innervation density (70 % higher than other rays), and a massively enlarged representation in somatosensory cortex (25 % of the S1 cortical area associated with the entire star). Enlarged cortical representation cannot be explained by increased innervation alone so likely reflects the central role of ray 11 in the mole's predatory behaviors. That is, it may have expanded to provide the mole with enhanced discrimination of tactile features relevant to prey identification and capture, whereas other rays are principally concerned with detection of possible prey.

In terms of its active sensing behavior, there appears to be an interesting convergence between control of its rays by the star-nosed mole, and of the vibrissae in terrestrial whisker specialists such as rats, mice, and hamsters. Each animal makes rapid palpations of a broad area around the snout with long protuberances, before homing in for a fine-grained investigation of regions of special interest with a tactile fovea, composed of the densely congregated microvibrissae, in the case of the rat, and ray 11 in the case of the mole. Note that although neurobiological investigations of S1 cortex in rats and mice have mainly focused on the prominent macrovibrissal "barrels", the microvibrissae on the lips and chin also map to individual barrel-like cell aggregates. Indeed, overall the area of S1 dedicated to representing the microvibrissae is larger than that for the longer macrovibrissae [see, e.g. Woolsey et al. (1975) and Figure 2 in Wilson and Moore (2015)], this is consistent with the notion of the microvibrissal array as a foveal region for the whisker sense (Brecht et al. 1997).

Varieties of Human Touch Sensing

In the womb, an unborn baby will be able to sense her surroundings through touch from as early as eight weeks. From sixteen weeks she will begin to move around inside the womb, exploring through touch—hand to face, one hand to the other, foot to other leg. In this way the unborn child begins to discover her own body. Indeed by the time she is born she will be able to distinguish self-touch from touch by someone else, a process that helps to establish the sense of self (Rochat and Hespos 1997).

Within the first few weeks of life, the mouth evolves from being a mechanism for gaining nutrition by sucking to being the infant's chief means for finding out about objects in her world. An infant will use her hands to transport an object to the mouth and then explore it by mouthing in a manner that is dependent on object properties (Rochat and Senders 1991). From the start of life, then, we engage in **active tactile perception** (Prescott, Diamond et al. 2011; Lepora 2015) to understand and interrogate our world. Whereas the mouth and lips are the initial focus for exploration, as we grow older our haptic attention shifts to our hands. Indeed, it is no surprise that humans have lost their facial whiskers during the course of evolution, as our hands more than compensate as devices for tactile exploration of nearby space whilst doubling up as manipulators for dextrous grasping and precise control. When tactile acuity is examined, for instance using Weber's (1834/1978) "two point discrimination" test, the lips, tongue, and fingertips stand out as the locations on the body with the greatest sensitivity to touch. In each place we are typically able to distinguish points that are as little as one or two millimeters apart (Weber 1834/1978; Weinstein 1968). At the other extreme, in the middle of the back, for instance, the two points can be centimeters apart yet you might still feel them as one, indicating a much lower concentration of tactile sensory afferents. Similar estimates can be made for other measures of sensitivity such as orientation, pressure, point localization and vibration detection. These differences in sensitivity at the periphery are also reflected in the proportions of the human somatosensory homunculus which has huge hands and lips and large feet.

Discriminative Touch as a Self-structuring Sensorimotor Activity

To understand active tactile sensing in humans, as in other animals, we need to think of touch as a co-ordinated sensorimotor activity with sensory feedback being a key determinant of future movement and sensory input. Crucially, the dynamic coupling of the sensorimotor system with the environment during active touch alters, or "self-structures", the flow of information between its different components (sensory, neural and motor) (Barlow 2001; Clark 2013). This can optimize the flow of information in measurable ways, for instance by reducing entropy (disorder) and by increasing mutual information (Lungarella and Sporns 2006; Pfeifer et al. 2007) making the challenge of understanding the world through touch significantly more tractable.

Self-structuring is evident in the **exploratory procedures** we engage during haptic tasks (Lederman and Klatzky 1993; Klatzky and Reed 2009). Each procedure is shaped by the goal (what it is we wish to find out), by the environmental structure, and by the morphology of the hand. The result is a pattern of behavior that is characterizable but not stereotyped, and that isolates and enhances (Gibson

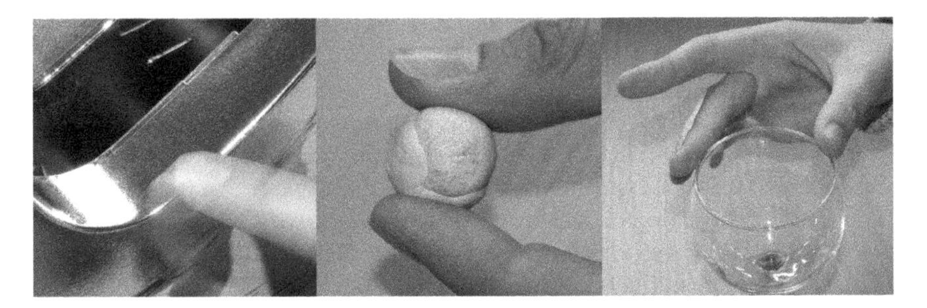

Figure 12 Examples of exploratory procedures in haptic touch for discrimination of texture, hardness, and shape. *Photographs* by Ben Mitchinson, reproduced with permission

1962) the object properties of interest. For instance, as illustrated in see Figure 12, you might stroke a surface to judge **texture** (Bensmaia 2009) by applying even pressure with your fingertip, your finger tracing a trajectory that naturally follows the contour of the surface. By contrast, you might squeeze an object between two fingers to assess hardness, or repeatedly enclose it within the hand to determine **shape** (Kappers and Bergmann Tiest 2014).

The notion of perception as inference due to Helmholtz (1866/1962) and Gregory (1980) (for more recent accounts see e.g. Kersten et al. 2004; Friston et al. 2012), is also important for understanding how the brain can make sense of the fragmentary and often fleeting signals we receive through touch. For instance, encountering a cool, hard surface in your pocket, you might manipulate the object with your fingers to determine the denomination of the coin. Encountering a straight-line edge, during this process, another hypothesis springs to mind—this could be a key. The ambiguity inherent in tactile signals requires the use of expectations (priors) to resolve a cloud of indeterminate stimuli into a specific and concrete percept. Fox et al. (2012) used an approach of this kind, employing strong top-down Bayesian priors, to construct spatial maps from sparse tactile sense data, Lepora et al. (2013) propose a similar Bayesian approach for simultaneous object localization and identification (SOLID) in fingertip touch.

Naturally, our understanding of tactile signals is also informed by information from **other modalities**, evidence of how and where this happens is shown in the activation of visual cortical areas during touch (Lacey and Sathian 2015). Being **visually-impaired**, on the other hand, does not always put you at a disadvantage. Indeed, congenitally blind individuals outperform those with intact vision on many haptic tasks as a consequence of "sensory compensation" thought to involve the recruitment of anatomically visual areas to enhance tactile processing (Heller and Ballesteros 2012).

The Gentler Side of Touch: Affect, Sociality and Sensuality

Although discrimination and the detection of affordances are critical tasks for tactile sensing, we should not overlook the central role of touch in human emotion. The slow CT fiber system found in other mammals is also present in humans and projects, not to the somatosensory cortices, but to the insula region (Olausson et al. 2002)—part of the Papez circuit for emotion and a gateway to the processing of reward and feeling in the frontal cortex. The soft stroking of the skin, to which the CT afferents are specifically sensitive, can promote release of endogenous opiates and the hormone oxytocin known to be important in pair bonding and in sexual behavior (Uvnäs-Moberg et al. 2005; Olausson et al. 2010).

The CT fiber system likely mediates the known benefits of gentle social touch which include reducing stress and blood pressure, and raising pain thresholds. The interpersonal effects of social touch may bypass conscious awareness. A light touch on the hand or the shoulder, can induce you to give a waiter a larger tip without realizing why (Crusco and Wetzel 1984)—the astute waiter is tapping into a subliminal mammalian channel for bonding via touch. Social touch also plays a notable role in communicating affect. For instance, Hertenstein et al. (2009) asked people to send emotional signals in the way they touched one another (no speaking or eye contact allowed). They found that people could communicate up to eight different emotions—fear, anger, disgust, sadness, gratitude, sympathy, happiness, and love —through touch alone, and with an accuracy of nearly eighty percent. Although CT fibers do not appear to penetrate glabrous skin, it is undoubtedly true that touch on surfaces such as the fingers and lips is often reported as pleasant. One possibility, suggested by McGlone et al. (2012) is that sensing in glabrous skin is primarily discriminatory, but that we learn to associate certain patterns of stimulation with positive affect.

Touch is also hugely important to human sexual behavior (see e.g. Gallace and Spence 2014), although research on touch is somewhat limited owing to social taboos. We know, however (and this is no surprise), that some of the most sensitive areas for touch on the human body are in the genitals, particularly, the male penis glans and the female clitoris (Soderquist 2002). Both sites have the same mechanoreceptors and sensory nerve endings found in other areas of the skin rather than having any specialized receptor types. Projections travel to at least two areas of S1 cortex, to secondary somatosensory cortex, and to the posterior insula, the latter presumably being key for processing the affective aspects of genital stimulation (Cazala et al. 2015).

Research on genital touch has important social and medical consequences. For instance, a number of studies have found reduced sensitivity in the circumcised penis, compared to uncircumcised (Bronselaer et al. 2013; Gallace and Spence 2014) raising questions about the use of circumcision as a surgical treatment and as a cultural practice. Research on female tactile sensitivity has failed to find support for a distinct erogenous zone on the vaginal wall, sometimes known as the G spot, helping to demystify our understanding of female sexual arousal (Hines 2001).

Whilst touch is clearly of great importance to human sexual response we should also note that its arousing effects strongly depend upon perception in other modalities, and on the broader psychological, social and cultural context. The flip-side of pleasurable touch is pain. Again specific neural pathways are involved (Derbyshire 2014), but as with pleasure, how and where in the wider brain these signals are processed, and what other contextual signals are present, will be critical to how there are subjectively experienced. Neither pleasure nor pain, is, in the end, simply the result of this or that kind of stimulus, or of activation in a particular kind of nerve fibre; both are, in a more holistic way, states of mind and body.

The benefits to personal well-being of affective touch are reflected in the universal interest in the therapeutic or healing potential of tactile stimulation. Beyond the immediately rewarding and restorative effects noted above for gentle stroking touch, there is evidence for some longer-term benefits, for instance in relieving chronic pain. A possible mechanism that could underlie such a lasting impact involves the reorganization of distorted somatotopic maps through gentle repeated stimulation (Kerr et al. 2007). The idea of using stimulation to unlearn a distorted brain map has also led to treatments such as mirror visual feedback for **phantom limb pain** (Ramachandran and Brang 2009), and could potentially lead to therapies for a wider range of **touch disorders** (van Stralen and Dijkerman 2011).

Synthetic Touch for Haptic Interfaces, Robots, Prostheses and Virtual Worlds

Touch sensing was a relatively slow starter in the realm of consumer electronics. Although we can think of the electronic keyboard as tactile input device, in its original manifestation this was essentially a bank of on-off switches. All that changed, however, with devices such as track pads for laptop computers and touch screens for phones and tablets. Since the turn of the century, haptic interfaces have become commonplace. Every smart-phone and tablet, and many computers now have a touch sensitive screen (Zhai et al. 2012), and laptops now replace the conventional mouse with an intelligent track-pad. These devices distinguish single from multiple touch, and track the direction and speed of motion, and possibly pressure. The remarkable speed with which haptic interfaces have been adopted is an indication of how natural it seems for people to interact with objects in this way—our electronic devices are at last catching up with the human predilection for fast and precise object manipulation and our ability to communicate intention through a simple touch (Vincent et al. 2004).

In robotics, touch was also, for a long time, the sensor system of last resort—a warning that something unexpected has happened. However, the latest generation of tactile sensing devices suggest that we are finally starting to catch up with nature which has always used touch as key modality for finding out about the world. Tactile sensing for robotics does not appear to be converging on a specific preferred transducer type, rather, as in animal integument, there are many different types of

Figure 13 Shadow Dexterous Robot Hand holding a lightbulb. Robots hands, such as the Shadow, are typically able to sense forces, micro-vibrations, and temperature, at multiple positions (taxels). *Photograph* by Richard Greenhill/Hugo Elias (CC 3.0)

sensor each one having benefits and limitations that make it suited to some sensing challenges and not others (Martinez-Hernandez 2015). A significant driver of technology development has been biomimetics—the goal to create an artificial fingertip or covering with the tactile sensitivity of human skin. Within the last decade many fingertip-like sensors have been developed for robot hands and grippers (see, e.g. Figure 13) although matching the human hand for the range of stimulus types it can recognise, and for its sensitivity and resolution is still some way off. Biomimetics is also looking to other species, and tactile systems found in nature. For instance, sensors that resemble insect antennae have been developed for a variety of wheeled and legged-robots (Russell 1985, Kaneko et al. 1998; Lewinger et al. 2005; Lee et al. 2008), and artificial vibrissae have been investigated both for **robots** operating in air, and those immersed in water (Prescott et al. 2009; Eberhardt et al. 2011; Pipe and Pearson 2015).

One of the most important roles for touch is likely to be in human-machine interfaces. A key technology will be tactile sensing for prosthetics (Clement et al. 2011). Whilst prosthetic hands are already being fitted with fingertip-like transducers, research to interface the outputs of these to the human nervous system, in a way that will allow users to experience a natural feeling of touch, is still in its infancy (but see Raspopovic et al. 2014; Tan et al. 2014). A non-invasive means to provide such an interface could be through the development of **haptic display** technologies that are evolving rapidly as a means to deliver complex stimulus patterns to areas of skin (Visell 2009; Kim and Harders 2015), taking advantage of the capacity of the human nervous system to discover meaning in novel patterns of peripheral stimulation. Haptic displays have long been a focus of research in sensory substitution, or sensory augmentation, beginning with the Polish ophthalmologist Kazimierz Noiszewski who, in the 1880s, invented a device to convert light energy into patterns of tactile stimulation called the *electroftalm*, or "artificial eye" (Figure 14). Other pioneering devices, created by Starkiewicz and Kuliszewski (1963) and by Bach-y-Rita (1972) also took visual images and displayed them as a pattern of vibration on the skin of the user. Research into sensory augmentation has discovered the importance of mapping into the sensing bandwidth, resolution, and psychophysics of the modality into which a signal is being converted. For instance, when going from vision to touch it is important to consider that tactile sensing in the human skin cannot match the spatial resolution of the eye, but that it does have good temporal resolution.

Some of the most effective augmentation or **substitution devices** operate by tapping into the active nature of human tactile sensing—allowing the human user to direct what might be a relatively low resolution device to explore areas of particular interest (Kra et al. 2015). A related idea, originating from the enactive perspective (e.g. Varela et al. 1993), is to seek to make the interface device "experientially transparent" (Froese et al. 2012) such that the goal-directed behavior of the user naturally incorporates properties of the artifact including its capacity to transform

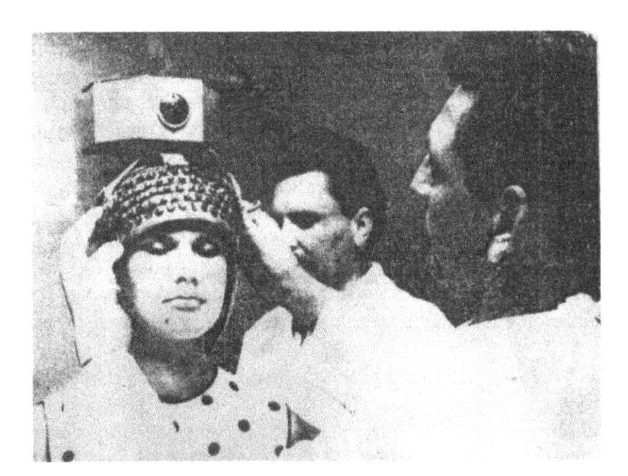

Figure 14 The *Electroftalm* developed by Starkiewicz and Kuliszewski in the 1960s was one of a number of pioneering attempts to create a sensory substitution device for the visually-impaired that mapped the visual scene to a pattern of vibration on the surface of the skin

from one sensory modality to another. Examples of effective devices adopting this approach include the *enactive torch* (Froese et al. 2012) that gives the user an augmented distance sense, and the *feelspace belt* (Nagel et al. 2005) that provides the wearer with a novel compass sense via touch.

Just as touch screens and trackpads have accelerated the developing of tactile sensors, telecommunication and virtual reality applications are beginning to drive the development of a new generation of haptic displays, that can enable, for instance, remote affective communication through the internet (Cheok and Pradana 2015). The possibility of immersive virtual reality in which tactile experience comes close to anything that can happen in the real world is far off, however, the importance of touch to our species is such that technologies that even crudely approximate a feeling of touch could have far-ranging applications, from touch-assisted key-hole surgery, to accurate manipulation of objects with robot hands, to a virtual handshake with a person in another country. The possibilities for synthetic touch are vast, and like much of the world of biological touch, we are only just beginning, so to speak, to feel our way.

Internal References

Ahissar, E; Shinde, N and Haidarliu, S (2015). Systems neuroscience of touch. *Scholarpedia* 10: 32785. http://www.scholarpedia.org/article/Systems_Neuroscience_of_Touch. (see also pages 401–405 of this book).

Barth, F G (2015). A spider's tactile hairs. *Scholarpedia* 10(3): 7267. http://www.scholarpedia.org/article/A_spider%27s_tactile_hairs. (see also pages 65–81 of this book).

Bensmaia, S (2009). Texture from touch. *Scholarpedia* 4(8): 7956. http://www.scholarpedia.org/article/Texture_from_touch. (see also pages 207–215 of this book).

Brembs, B (2014). Aplysia operant conditioning. *Scholarpedia* 9(1): 4097. http://www.scholarpedia.org/article/Aplysia_operant_conditioning.

Cheok, A and Pradana, G A (2015). Virtual touch. *Scholarpedia* 10: 32679. http://www.scholarpedia.org/article/Virtual_touch. (see also pages 837–849 of this book).

Crish, C M; Crish, S D and Comer, C M (2015). Tactile sensing in the naked mole rat. *Scholarpedia* 10(3): 7164. http://www.scholarpedia.org/article/Tactile_sensing_in_the_naked_mole_rat. (see also pages 95–101 of this book).

Derbyshire, S (2014). Painful touch. *Scholarpedia* 9(3): 7962. http://www.scholarpedia.org/article/Painful_touch. (see also pages 249–255 of this book).

Dürr, V (2014). Stick insect antennae. *Scholarpedia* 9(2): 6829. http://www.scholarpedia.org/article/Stick_Insect_Antennae. (see also pages 45–63 of this book).

Grasso, F and Wells, M (2013). Tactile sensing in the octopus. *Scholarpedia* 8(6): 7165. http://www.scholarpedia.org/article/Tactile_sensing_in_the_octopus. (see also pages 83–94 of this book).

Haidarliu, S (2015). Whisking musculature. *Scholarpedia* 10(4): 32331. http://www.scholarpedia.org/article/Whisking_musculature. (see also pages 627–639 of this book).

Hanke, W and Dehnhardt, G (2015). Vibrissal touch in pinnipeds. *Scholarpedia* 10(3): 6828. http://www.scholarpedia.org/article/Vibrissal_touch_in_pinnipeds. (see also pages 125–139 of this book).

Heller, M and Ballesteros, S (2012). Visually-impaired touch. *Scholarpedia* 7(11): 8240. http://www.scholarpedia.org/article/Visually-impaired_touch. (see also pages 387–397 of this book).

Kappers, A and Bergmann Tiest, W (2014). Shape from touch. *Scholarpedia* 9(1): 7945. http://www.scholarpedia.org/article/Shape_from_touch. (see also pages 197–206 of this book).
Kim, Y and Harders, M (2015). Haptic displays. *Scholarpedia* 10: 32376. http://www.scholarpedia.org/article/Haptic_Displays. (see also pages 817–827 of this book).
Klatzky, R and Reed, C L (2009). Haptic exploration. *Scholarpedia* 4(8): 7941. http://www.scholarpedia.org/article/Haptic_exploration. (see also pages 177–183 of this book).
Knutsen, P (2015). Whisking kinematics. *Scholarpedia* 10: 7280. http://www.scholarpedia.org/article/Whisking_kinematics. (see also pages 615–625 of this book).
Kra, Y; Arieli, A and Ahissar, E (2015). Tactile substitution for vision. *Scholarpedia* 10: 32457. http://www.scholarpedia.org/article/Tactile_Substitution_for_Vision. (see also pages 829–836 of this book).
Lacey, S and Sathian, K (2015). Crossmodal and multisensory interactions between vision and touch. *Scholarpedia* 10(3): 7957. http://www.scholarpedia.org/article/Crossmodal_and_multisensory_interactions_between_vision_and_touch. (see also pages 301–315 of this book).
Lepora, N (2015). Active tactile perception. *Scholarpedia* 10: 32364. http://www.scholarpedia.org/article/Active_tactile_perception. (see also pages 151–159 of this book).
Martinez-Hernandez, U (2015). Tactile sensors. *Scholarpedia* 10: 32398. http://www.scholarpedia.org/article/Tactile_Sensors. (see also pages 783–796 of this book).
Moayedi, Y; Nakatani, M and Lumpkin, E (2015). Mammalian mechanoreception. *Scholarpedia* 10: 7265. http://www.scholarpedia.org/article/Mammalian_mechanoreception. (see also pages 423–435 of this book).
Okada, J (2009). Cockroach antennae. *Scholarpedia* 4(10): 6842. http://www.scholarpedia.org/article/Cockroach_antennae. (see also pages 31–43 of this book).
Pipe, T and Pearson, M (2015). Whiskered robots. *Scholarpedia* 10: 6641. http://www.scholarpedia.org/article/Whiskered_robots. (see also pages 809–815 of this book).
Prescott, T J; Mitchinson, B and Grant, R A (2011). Vibrissal behavior and function. *Scholarpedia* 6(10): 6642. http://www.scholarpedia.org/article/Vibrissal_behavior_and_function. (see also pages 103–116 of this book).
Ramachandran, V S and Brang, D (2009). Phantom touch. *Scholarpedia* 4(10): 8244. http://www.scholarpedia.org/article/Phantom_touch. (see also pages 377–386 of this book).
Reep, R and Sarko, D K (2009). Tactile hair in manatees. *Scholarpedia* 4(4): 6831. http://www.scholarpedia.org/article/Tactile_hair_in_Manatees. (see also pages 141–148 of this book).
Roth-Alpermann, C and Brecht, M (2009). Vibrissal touch in the Etruscan shrew. *Scholarpedia* 4(11): 6830. http://www.scholarpedia.org/article/Vibrissal_touch_in_the_Etruscan_shrew. (see also pages 117–123 of this book).
van Stralen, H and Dijkerman, C (2011). Central touch disorders. *Scholarpedia* 6(10): 8243. http://www.scholarpedia.org/article/Central_touch_disorders. (see also pages 363–376 of this book).
Wilson, S and Moore, C (2015). S1 somatotopic maps. *Scholarpedia* 10: 8574. http://www.scholarpedia.org/article/S1_somatotopic_maps. (see also pages 565–576 of this book).

External References

Ache, J M; Haupt, S S and Dürr, V (2015). A direct descending pathway informing locomotor networks about tactile sensor movement. *The Journal of Neuroscience* 35: 4081–4091.
Ahissar, E and Kleinfeld, D (2003). Closed-loop neuronal computations: Focus on vibrissa somatosensation in rat. *Cerebral Cortex* 13(1): 53–62.
Anjum, F; Turni, H; Mulder, P G; van der Burg, J and Brecht, M (2006). Tactile guidance of prey capture in Etruscan shrews. *Proceedings of the National Academy of Sciences of the United States of America* 103(44): 16544–16549.
Arkett, S A; Mackie, G O and Meech, R W (1988). Hair cell mechanoreception in the jellyfish *Aglantha digitale*. *Journal of Experimental Biology* 135: 329–342.
Bach-y-Rita, P (1972). *Brain Mechanisms in Sensory Substitution*. New York: Academic Press.

Barlow, H B (2001). The exploitation of regularities in the environment by the brain. *Behavioral and Brain Sciences* 24: 602–660.

Berg, R W and Kleinfeld, D (2003). Rhythmic whisking by rat: Retraction as well as protraction of the vibrissae is under active muscular control. *Journal of Neurophysiology* 89(1): 104–117.

Berta, A; Sumich, J L and Kovacs, K M (2006). *Marine Mammals: Evolutionary Biology*. San Diego, CA: Academic Press.

Bleckmann, H and Zelick, R (2009). Lateral line system of fish. *Integrative Zoology* 4(1): 13–25.

Brecht, M; Preilowski, B and Merzenich, M M (1997). Functional architecture of the mystacial vibrissae. *Behavioural Brain Research* 84(1–2): 81-97.

Bronselaer, G A et al. (2013). Male circumcision decreases penile sensitivity as measured in a large cohort. *BJU International* 111(5): 820–827.

Camhi, J M and, Johnson (1999). High-frequency steering maneuvers mediated by tactile cues: Antennal wall-following in the cockroach. *Journal of Experimental Biology* 202: 631–643.

Carlton, T and McVean, A (1995). The role of touch, pressure and nociceptive mechanoreceptors of the leech in unrestrained behaviour. *Journal of Comparative Physiology A* 177: 781–791.

Catania, K C (1999). A nose that looks like a hand and acts like an eye: The unusual mechanosensory system of the star-nosed mole. *Journal of Comparative Physiology A* 185(4): 367–372.

Catania, K C (2011). The sense of touch in the star-nosed mole: from mechanoreceptors to the brain. *Philosophical Transactions of the Royal Society of London B: Biological Sciences* 366 (1581): 3016–3025.

Catania, K C; Hare, J F and Campbell, K L (2008). Water shrews detect movement, shape, and smell to find prey underwater. *Proceedings of the National Academy of Sciences of the United States of America* 105(2): 571–576.

Catania, K C and Henry, E C (2006). Touching on somatosensory specializations in mammals. *Current Opinion in Neurobiology* 16(4): 467–473.

Catania, K C and Kaas, J H (1995). Organization of the somatosensory cortex of the star-nosed mole. *Journal of Comparative Neurology* 351(4): 549–567.

Catania, K C and Remple, F E (2005). Asymptotic prey profitability drives star-nosed moles to the foraging speed limit. *Nature* 433(7025): 519–522.

Cazala, F; Vienney, N and Stoléru, S (2015). The cortical sensory representation of genitalia in women and men: A systematic review. *Socioaffective Neuroscience & Psychology* 5. 10.3402/snp.v3405.26428.

Chalfie, M et al. (1985).The neural circuit for touch sensitivity in *Caenorhabditis elegans*. *The Journal of Neuroscience* 5: 956–964.

Chen, X and Chalfie, M (2014). Modulation of *C. elegans* touch sensitivity is integrated at multiple levels. *The Journal of Neuroscience* 34: 6522–6536.

Chuong, C M et al. (2002). What is the 'true' function of skin? *Experimental Dermatology* 11(2): 159–187.

Clark, A (2013). Whatever next? Predictive brains, situated agents, and the future of cognitive science. *Behavioural and Brain Sciences* 36(3): 181–204.

Clement, R G E; Bugler, K E and Oliver, C W (2011). Bionic prosthetic hands: A review of present technology and future aspirations. *The Surgeon* 9(6): 336–340.

Coutand, C (2010). Mechanosensing and thigmomorphogenesis, a physiological and biomechanical point of view. *Plant Science* 179: 168–182.

Crusco, A H and Wetzel, C G (1984). The Midas Touch: The effects of interpersonal touch on restaurant tipping. *Personality and Social Psychology Bulletin* 10(4): 512–517.

Dehnhardt, G; Mauck, B; Hanke, W and Bleckmann, H (2001). Hydrodynamic trail-following in harbor seals (*Phoca vitulina*). *Science* 293(5527): 102–104.

Eberhardt, W C; Shakhsheer, Y A; Calhoun, B H; Paulus, J R and Appleby, M (2011). A bio-inspired artificial whisker for fluid motion sensing with increased sensitivity and reliability. In: *2011 IEEE Sensors*.

Erber, J; Kierzek, S; Sander, E and Grandy, K (1998). Tactile learning in the honeybee. *Journal of Comparative Physiology A* 183: 737–744.

Erber, J; Pribbenow, B; Grandy, K and Kierzek, S (1997). Tactile motor learning in the antennal system of the honeybee (*Apis mellifera* L.).*Journal of Comparative Physiology A* 181: 355–365.

Erber, J; Pribbenow, B; Kisch, J and Faensen, D (2000). Operant conditioning of antennal muscle activity in the honey bee (*Apis mellifera* L.). *Journal of Comparative Physiology A* 186: 557–565.

French, A S (2009). The systems analysis approach to mechanosensory coding. *Biological Cybernetics* 100: 417–426.

French, A S and Wong, R K S (1976). The responses of trochanteral hair plate sensilla in the cockroach to periodic and random displacements. *Biological Cybernetics* 22: 33–38.

Fox, C W; Evans, M H; Pearson, M J and Prescott, T J (2012). Towards hierarchical blackboard mapping on a whiskered robot. *Robotics and Autonomous Systems* 60(11): 1356–1366.

Friston, K; Adams, R A; Perrinet, L and Breakspear, M (2012). Perceptions as hypotheses: Saccades as experiments. *Frontiers in Psychology* 3: 151.

Froese, T; McGann, M; Bigge, W; Spiers, A and Seth, A K (2012). The enactive torch: A new tool for the science of perception. *IEEE Transactions on Haptics* 5(4): 363–375.

Gallace, A and Spence, C (2014). *In Touch with the Future: The Sense of Touch from Cognitive Neuroscience to Virtual Reality*. Oxford: OUP.

Garcia-Anoveros, J and Corey, D P (1997).The molecules of mechanosensation. *Annual Review of Neuroscience* 20: 567–594.

Gaspard, J C, 3rd et al. (2013). Detection of hydrodynamic stimuli by the Florida manatee (*Trichechus manatus latirostris*). *Journal of Comparative Physiology A. Neuroethology, Sensory, Neural, and Behavioral Physiology* 199(6): 441–450.

Gebhardt, M J and Honegger, H W (2001). Physiological characterisation of antennal mechanosensory descending interneurons in an insect (*Gryllus bimaculatus, Gryllus campestris*) brain. *Journal of Experimental Biology* 204: 2265–2275.

Gibson, J J (1962). Observations on active touch. *Psychological Review* 69: 477–491.

Grant, R A; Haidarliu, S; Kennerley, N J and Prescott, T J (2013). The evolution of active vibrissal sensing in mammals: Evidence from vibrissal musculature and function in the marsupial opossum *Monodelphis domestica*. *Journal of Experimental Biology* 216(Pt 18): 3483–3494.

Gregory, R L (1980). Perceptions as hypotheses. *Philosophical Transactions of the Royal Society of London B: Biological Sciences* 290: 181–197.

Hanke, W et al. (2010). Harbor seal vibrissa morphology suppresses vortex-induced vibrations. *Journal of Experimental Biology* 213(Pt 15): 2665–2672.

Harley, C M; English, B A and Ritzmann, R E (2009). Characterization of obstacle negotiation behaviors in the cockroach, *Blaberus discoidalis*. *Journal of Experimental Biology* 212: 1463–1476.

Hartmann, M J (2001). Active sensing capabilities of the rat whisker system. *Autonomous Robots* 11: 249–254.

Helmholtz, H (1866/1962). *Treatise on Physiological Optics*. New York: Dover.

Hertenstein, M J; Holmes, R; McCullough, M and Keltner, D (2009). The communication of emotion via touch. *Emotion* 9(4): 566–573.

Hines, T M (2001). The G-spot: A modern gynecologic myth. *American Journal of Obstetrics & Gynecology* 185(2): 359–362.

Honegger, H W (1981). A preliminary note on a new optomotor response in crickets: Antennal tracking of moving targets. *Journal of Comparative Physiology A* 142: 419–421.

Hyvarinen, H; Palviainen, A; Strandberg, U and Holopainen, I J (2009). Aquatic environment and differentiation of vibrissae: Comparison of sinus hair systems of ringed seal, otter and pole cat. *Brain, Behavior and Evolution* 74(4): 268–279.

Kaas, J H (1997). Topographic maps are fundamental to sensory processing. *Brain Research Bulletin* 44(2): 107–112.

Kaneko, M; Kanayama, N and Tsuji, T (1998). Active antenna for contact sensing. *IEEE Transactions on Robotics and Automation* 14(2): 278–291.

Kerr, C E; Wasserman, R H and Moore, C I (2007). Cortical dynamics as a therapeutic mechanism for touch healing. *The Journal of Alternative and Complementary Medicine* 13(1): 59–66.
Kersten, D; Mamassian, P and Yuille, A (2004). Object perception as Bayesian inference. *Annual Review of Psychology* 55: 271–304.
Kleinfeld, D; Ahissar, E and Diamond, M E (2006). Active sensation: Insights from the rodent vibrissa sensorimotor system. *Current Opinion in Neurobiology* 16(4): 435–444.
Krubitzer, L; Manger, P; Pettigrew, J and Calford, M (1995). Organization of somatosensory cortex in monotremes: In search of the prototypical plan. *Journal of Comparative Neurology* 351(2): 261–306.
Lederman, S J and Klatzky, R L (1993). Extracting object properties through haptic exploration. *Acta Psychologica (Amsterdam)* 84(1): 29–40.
Lee, J et al. (2008). Templates and anchors for antenna-based wall following in cockroaches and robots. *IEEE Transactions on Robotics* 24(1): 130–143.
Lepora, N; Martinez-Hernandez, U and Prescott, T J (2013). A SOLID case for active Bayesian perception in robot touch. In: *Biomimetic and Biohybrid Systems. Lecture Notes in Computer Science*, Vol. 8064 (pp. 154–166).
Lewis, J E and Kristan, W B (1998a). Representation of touch location by a population of leech sensory neurons. *Journal of Neurophysiology* 80: 2584–2592.
Lewis, J E and Kristan, W B (1998b). A neuronal network for computing population vectors in the leech. *Nature* 391: 76–79.
Lewinger, W A; Harley, C M; Ritzmann, R E; Branicky, M S and Quinn, R D (2005). Insect-like antennal sensing for climbing and tunneling behavior in a biologically-inspired mobile robot. In: *Proceedings of the IEEE International Conference on Robotics and Automation (ICRA).*
Loken, L S; Wessberg, J; Morrison, I; McGlone, F and Olausson, H (2009). Coding of pleasant touch by unmyelinated afferents in humans. *Nature Neuroscience* 12(5): 547–548.
Ludeman, D A; Farrar, N; Riesgo, A; Paps, J and Leys, S P (2014). Evolutionary origins of sensation in metazoans: Functional evidence for a new sensory organ in sponges. *BMC Evolutionary Biology* 14(3).
Lungarella, M and Sporns, O (2006). Mapping information flow in sensorimotor networks. *PLoS Computational Biology* 2(10): e144.
Maderson, P F A (1972). When? Why? and How?: Some speculations on the evolution of the vertebrate integument. *American Zoologist* 12(1): 159–171.
Maderson, P F A (2003). Mammalian skin evolution: A reevaluation. *Experimental Dermatology* 12(3): 233–236.
McGlone, F et al. (2012). Touching and feeling: Differences in pleasant touch processing between glabrous and hairy skin in humans. *European Journal of Neuroscience* 35(11): 1782–1788.
McGlone, F; Wessberg, J and Olausson, H (2014). Discriminative and affective touch: Sensing and feeling. *Neuron* 82(4): 737–755.
Mitchinson, B et al. (2011). Active vibrissal sensing in rodents and marsupials. *Philosophical Transactions of the Royal Society of London B: Biological Sciences* 366(1581): 3037–3048.
Monshausen, G B and Gilroy, S (2009). Feeling green: Mechanosensing in plants. *Trends in Cell Biology* 19: 228–235.
Muller, K J; Nicholls, J G and Stent, G S (1981). *Neurobiology of the Leech.* Cold Spring Harbor, New York: Cold Spring Harbor Publications.
Nagel, S K; Carl, C; Kringe, T; Märtin, R and König, P (2005). Beyond sensory substitution - learning the sixth sense. *Journal of Neural Engineering* 2: 4.
Naitoh, Y and Eckert, R (1969). Ionic mechanisms controlling behavioral responses of *Paramecium* to mechanical stimulation. *Science* 164: 963–965.
Nicholls, J G and Baylor, D A (1968). Specific modalities and receptive fields of sensory neurons in CNS of leech. *Journal of Neurophysiology* 31: 740–756.
O'Hagan, R; Chalfie, M and Goodman, M B (2005). The MEC-4 DEG/ENaC channel of *Caenorhabditis elegans* touch receptor neurons transduces mechanical signals. *Nature Neuroscience* 8: 43–50.

Olausson, H et al. (2002). Unmyelinated tactile afferents signal touch and project to insular cortex. *Nature Neuroscience* 5(9): 900–904.
Okada, J and Toh, Y (2000). The role of antennal hair plates in object-guided tactile orientation of the cockroach (*Periplaneta americana*). *Journal of Comparative Physiology A* 186: 849–857.
Olausson, H; Wessberg, J; Morrison, I; McGlone, F and Vallbo, A (2010). The neurophysiology of unmyelinated tactile afferents. *Neuroscience & Biobehavioral Reviews* 34(2): 185–191.
Penfield, W and Boldrey, E (1937). Somatic motor and sensory representation in the cerebral cortex of man as studies by electrical stimulation. *Brain* 60(4): 389–443.
Pfeifer, R; Lungarella, M; Sporns, O and Kuniyoshi, Y (2007). On the information theoretic implications of embodiment – Principles and methods. In: M Lungarella, F Iida, J Bongard and R Pfeifer (Eds.), *50 Years of Artificial Intelligence* Vol. 4850 (pp. 76–86). Berlin Heidelberg: Springer.
Pocock, R I (1914). On the facial vibrissae of mammalia. *Proceedings of the Zoological Society of London* 889–912.
Prescott, T J; Diamond, M E and Wing, A M (Eds.) (2011). *Active Touch Sensing. Philosophical Transactions of the Royal Society B: Biological Sciences.* London: Royal Society Publishing.
Prescott, T J; Pearson, M J; Mitchinson, B; Sullivan, J C W and Pipe, A G (2009). Whisking with robots: From rat vibrissae to biomimetic technology for active touch. *IEEE Robotics & Automation Magazine* 16(3): 42–50.
Proske, U; Gregory, J E and Iggo, A (1998). Sensory receptors in monotremes. *Philosophical Transactions of the Royal Society of London B: Biological Sciences* 353(1372): 1187–1198.
Raspopovic, S et al. (2014). Restoring natural sensory feedback in real-time bidirectional hand prostheses. *Science Translational Medicine* 6(222): 222ra219.
Ray, G C; McCormick-Ray, J; Berg, P and Epstein, H E (2006). Pacific walrus: Benthic bioturbator of Beringia. *Journal of Experimental Marine Biology and Ecology* 230(1): 403–419.
Rice, F L; Mance, A and Munger, B L (1986). A comparative light microscopic analysis of the sensory innervation of the mystacial pad. I. Innervation of vibrissal follicle-sinus complexes. *Journal of Comparative Neurology* 252(2): 154–174.
Rochat, P and Senders, S J (1991). Active touch in infancy: Action systems in development. In: M J S Weiss and P R Zelazo (Eds.), *Newborn Attention: Biological Constraints and the Influence of Experience* (pp. 412–442). Norwood, New Jersey: Ablex Publishing Corp.
Rochat, P and Hespos, S J (1997). Differential rooting response by neonates: Evidence for an early sense of self. *Early Development and Parenting* 6: 105–112.
Russell, R A (1985). Object recognition using articulated whisker probes. In: *Proceedings of the 15th International Symposium on Industrial Robots.*
Sandeman, D C (1985). Crayfish antennae as tactile organs: Their mobility and the responses of their proprioceptors to displacement. *Journal of Comparative Physiology A* 157: 363–374.
Sandeman, D C (1989). Physical properties, sensory receptors and tactile reflexes of the antenna of the Australian freshwater crayfish *Cherax destructor*. *Journal of Experimental Biology* 141: 197–218.
Sarko, D K; Reep, R L; Mazurkiewicz, J E and Rice, F L (2007). Adaptations in the structure and innervation of follicle-sinus complexes to an aquatic environment as seen in the Florida manatee (*Trichechus manatus latirostris*). *Journal of Comparative Neurology* 504(3): 217–237.
Sarko, D K; Rice, F L and Reep, R L (2011). Mammalian tactile hair: Divergence from a limited distribution. *Annals of the New York Academy of Sciences* 1225: 90–100.
Scheich, H; Langner, G; Tidemann, C; Coles, R B and Guppy, A (1986). Electroreception and electrolocation in platypus. *Nature* 319(6052): 401–402.
Schütz, C and Dürr, V (2011). Active tactile exploration for adaptive locomotion in the stick insect. *Philosophical Transactions of the Royal Society of London B* 366: 2996–3005.
Soderquist, D R (2002). *Sensory Processes*. Thousand Oaks, California: Sage.
Sokolov, V E and Kulikov, V F (1987). The structure and function of the vibrissal apparatus in some rodents. *Mammalia* 51(1): 1–15.

Starkiewicz, W and Kuliszewski, W (1963). The 80-channel elektroftalm. *International Congress on Technology and Blindness*. New York, USA: American Foundation for the Blindness.
Staudacher, E; Gebhardt, M J and Dürr, V (2005). Antennal movements and mechanoreception: Neurobiology of active tactile sensors. *Advances in Insect Physiology* 32: 49–205.
Tan, D W et al. (2014). A neural interface provides long-term stable natural touch perception. *Science Translational Medicine* 6(257): 257ra138.
Thomson, E E and Kristan, W B (2006). Encoding and decoding touch location in the leech CNS. *The Journal of Neuroscience* 26: 8009–8016.
Thurm, U (1964). Mechanoreceptors in the cuticle of the honey bee: Fine structure and stimulus mechanism. *Science* 145: 1063–1065.
Thurm, U et al. (2004). Mechanoreception and synaptic transmission of hydrozoan nematocytes. *Hydrobiologia* 530: 97–105.
Uvnäs-Moberg, K; Arn, I and Magnusson, D (2005). The psychobiology of emotion: The role of the oxytocinergic system. *International Journal of Behavioral Medicine* 12(2): 59–65.
Varela, F J; Thompson, E and Rosch, E (1993). *The Embodied Mind: Cognitive Science and Human Experience*. Cambridge, MA: MIT Press.
Vincent, H; Oliver, R A; Manuel, C H; Danny, G and Gabriel, R D L T (2004). Haptic interfaces and devices. *Sensor Review* 24(1): 16–29.
Vincent, S B (1912). The function of the vibrissae in the behaviour of the white rat. *Behavior Monographs* 1: 1–82.
Visell, Y (2009). Tactile sensory substitution: Models for enaction in HCI. *Interacting with Computers* 21(1–2): 38-53.
Weber, E H. (1834/1978). "The Sense of Touch" H E Ross (Trans. of "De Tactu") and D J Murray (Trans. of "Der Tastsinn"). New York: Academic Press.
Weinstein, S (1968). Intensive and extensive aspects of tactile sensitivity as a function of body part, sex, and laterality. In: D R Kenshalo (Ed.), *The Skin Senses: Proceedings*. Springfield, IL: Charles C. Thomas.
Woolsey, T A and Van der Loos, H (1970). The structural organization of layer IV in the somatosensory region (SI) of mouse cerebral cortex. The description of a cortical field composed of discrete cytoarchitectonic units. *Brain Research* 17(2): 205–242.
Woolsey, T A; Welker, C and Schwartz, R H (1975). Comparative anatomical studies of the SmL face cortex with special reference to the occurrence of "barrels" in layer IV. *Journal of Comparative Neurology* 164(1): 79–94.
Young, J Z (1983). The distributed tactile memory system of *Octopus*. *Proceedings of the Royal Society B: Biological Sciences* 218: 135–176.
Zeil, J; Sandeman, R and Sandeman, D C (1985). Tactile localisation: The function of active antennal movements in the crayfish *Cherax destructor*. *Journal of Comparative Physiology A* 157: 607–617.
Zhai, S; Kristensson, P O; Appert, C; Andersen, T H and Cao, X (2012). Foundational issues in touch-screen stroke gesture design - An integrative review. *Foundations and Trends in Human-Computer Interaction* 5(2): 97–205.

Part I
Comparative Touch

Cockroach Antennae

Jiro Okada

Cockroach antennae have been extensively used for studying the multifunctional sensory appendage that generates the olfactory, gustatory, tactile, thermal, and humidity senses. Of the variety of senses, the **tactile sense** is thought to play a key role for perceiving physical objects. Because most cockroach species are nocturnal, the tactile sense of the antenna would be essential to determine the position, shape, and texture of surrounding objects in the dark. Mechanoreceptors on the surface of the antenna are primarily responsible for the generation of tactile sense. In addition, the motor function of antenna also contributes to the **active tactile sense** (Staudacher et al. 2005; Comer and Baba 2011). The antennal movement is accompanied by the activation of **proprioceptors** at the antennal joints.

Cockroach Antenna as a Tactile Sense Organ

Cockroaches are insect species that are classified into the order Blattaria. Currently, over 4,000 species have been found in Blattaria, but of these, only a few are known to be pests. Because these species can tolerate the human environment, and are easily reared in laboratories, they are often used for various biological studies. *Periplaneta americana*, which is also known as the American cockroach, can be considered as the most representative species used for studying the antennal system.

Each antenna of adult *P. americana* is as long as its body length (≈40 mm), and consists of approximately 140 segments (Figures 1 and 2). The first and second proximal segments are called the **scape** and the **pedicel**, respectively, and the remaining distal segments are collectively referred to as the **flagellum**. Each segment is connected to the neighboring segments via flexible joints. However, only the head-scape and scape-pedicel joints can move actively with muscle contraction. The other joints connecting the flagellar segments are deflected only passively.

J. Okada (✉)
Nagasaki University, Nagasaki, Japan

T.J. Prescott et al. (eds.), *Scholarpedia of Touch*, Scholarpedia,
DOI 10.2991/978-94-6239-133-8_2

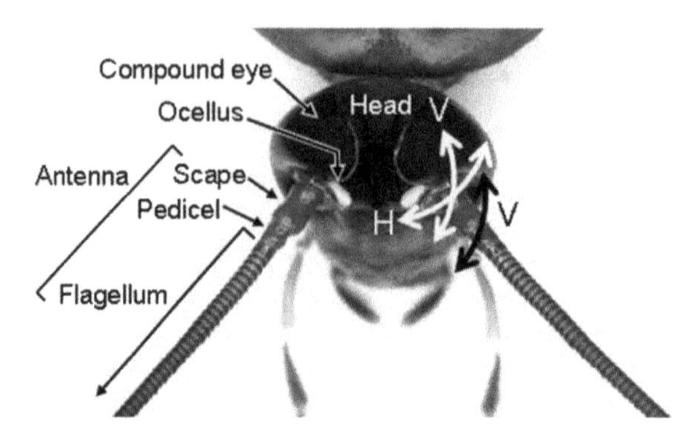

Figure 1 Head and antennae of the American cockroach *Periplaneta americana*. Antennae can move both horizontally (H) and vertically (V)

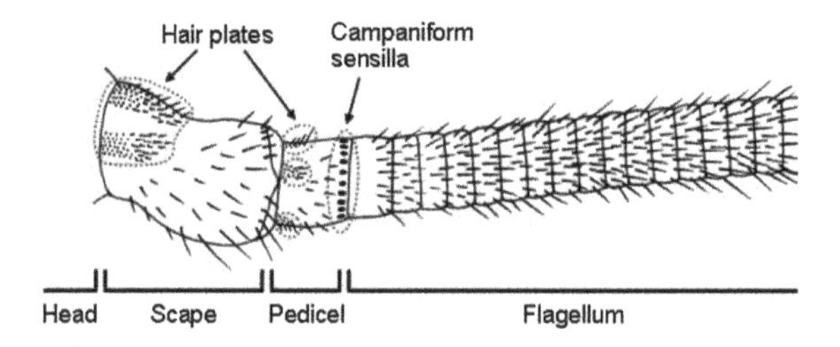

Figure 2 Lateral view of a cockroach's antenna

On the surface or beneath the cuticle of the antenna, there are numerous sensory units called **sensillum**. Some types of sensilla serve as mechanoreceptors that mediate the tactile sense.

Sensory System

Cockroach antennae have a variety of mechanoreceptors that differ in morphology, location, and function. Morphologically, the antennal mechanoreceptors are classified as follows: **hair sensillum** (or simply **hair**), **campaniform sensillum**, and **chordotonal sensillum**. The hair and campaniform sensilla are widely distributed on the surface of the antenna, whereas the chordotonal sensilla are located in the pedicel. Functionally, these receptors are classified as **exteroceptor** and **proprioceptor**, i.e. the external and internal mechanoreceptors, respectively. A recent histochemical study revealed that some mechanosensory neurons, including at least the cheatic (see below), chordotonal, and hair plate sensilla, are serotonergic (Watanabe et al. 2014).

Hair Sensillum

Hair sensilla are frequently found on the surface of the flagellum. Each of the mechanosensitive hairs in the flagellum contains a single mechanoreceptor and several chemoreceptor cells. Such multifunctional hairs with relatively thick and long shafts (>50 μm in length) are termed the sensilla cheatica or cheatic sensilla (the most prominent type of bristles, as shown in Figure 3). Each hair possesses a pore at the tip, which acts as a passage for molecules, and a flexible socket for passive deflection. The **cell body** (or perikaryon) of the mechanoreceptor beneath the socket is bipolar-shaped, and extends a **dendrite** distally to the base of the hair, and an **axon** proximally to the central nervous system. The dendrite contains longitudinally arranged sensory cilia (axoneme), and also a tightly packed microtubule structure (tubular body) at its tip. Deflection of the mechanosensory hair may deform the tubular body leading to the generation of electrical signals in the receptor cell via mechano-electric transduction. The **action potentials** appear only transiently upon the deflection of the hair, indicating the phasic (rapidly adapting) nature of the receptor (Hansen-Delkeskamp 1992).

Campaniform Sensillum

Campaniform sensilla are hairless mechanoreceptors located at each flagellar segment and the pedicel. They are characterized by an elliptical depression of the cuticles with a dome-like upheaval (Figure 4) or a pea-like protrusion (the latter is also called the marginal sensillum). Beneath the cuticle are a single mechanoreceptor cell and its accessory structures whose morphology is similar to the mechanosensitive hair. Each flagellar segment possesses several campaniform sensilla at both the distal and proximal margins. In *P. americana*, approximately 28 campaniform sensilla are arranged circularly around the distal margin of the pedicel (Toh 1981). The campaniform sensilla are presumably activated when the adjacent

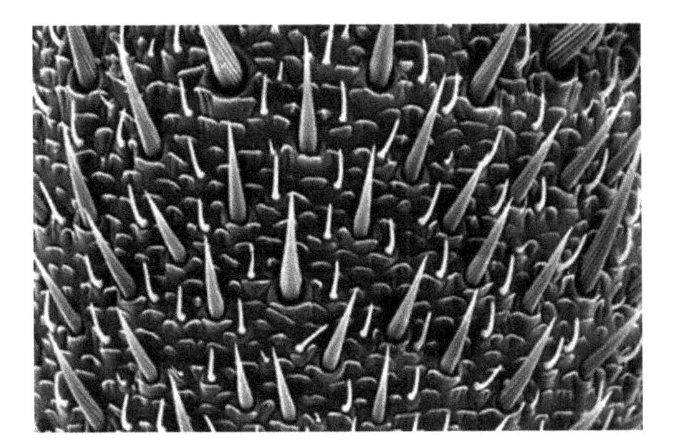

Figure 3 Scanning electron micrograph showing the surface view of the flagellum. Courtesy: Dr. Y. Toh

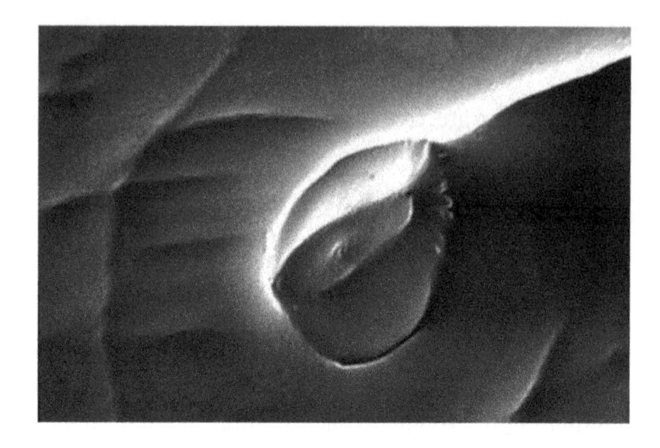

Figure 4 Magnified surface view of the campaniform sensillum. Courtesy: Dr. Y. Toh

cuticular structures are under stress. Therefore, the extent and direction of flexion at the intersegmental joints are speculated to be monitored by the campaniform sensilla. In this regard, they may function as proprioceptors.

Chordotonal Sensillum

Chordotonal sensilla are internal mechanoreceptors that serve as exteroceptors or proprioceptors. There are two types of chordotonal sensilla inside the pedicel of *P. americana*: the **connective chordotonal organ** (or simply **chordotonal organ**) (Figure 5) and **Johnston's organ**, which consist of 50 and 150 sensory units designated as the **scolopidia**, respectively (Toh 1981). The morphology of a scolopidium is characterized by the presence of a scolopale cell and an attachment cell that surrounds the dendrites of the receptor cell. The tip of the dendrite is covered by a cap of secreted matter from the scolopale cells. The attachment cell connects the distal part of the scolopidium to the inner surface of the cuticle. Each scolopidium consists of two and three bipolar receptor cells in the chordotonal and

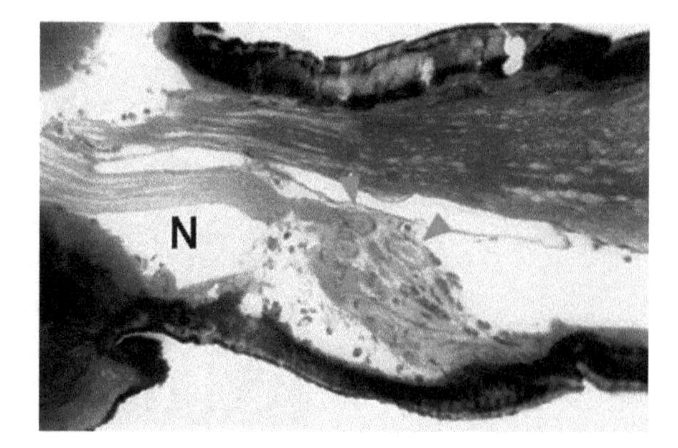

Figure 5 Internal morphology of the chordotonal organ. Arrowheads indicate cell bodies. N, afferent nerve. Courtesy: Dr. Y. Toh

Johnston's organs, respectively. The chordotonal organ may function as a proprioceptor for the scape-pedicel joint (Ikeda et al. 2004) or an exteroceptor for antennal contacts (Comer and Baba 2011). On the other hand, Johnston's organ is rather thought to be an exteroceptor for detecting sound or substrate vibration, considering its function in other insect species.

Hair Plate

Hair plates are clusters of the *pure* mechanosensory hair located at the base of an antenna. Several tens to over 100 hairs (15−60 μm in length) are arranged in the form of arrays adjacent to the head-scape and scape-pedicel joints (Okada and Toh 2000) (Figures 2 and 6). As the joint moves, a portion of the mechanosensory hairs is deflected by the joint membrane (Figure 6). At first, the hair plates seem to act as exteroceptors, but they actually function as proprioceptors for the active movements of the antennal basal joints. The receptor cells of the antennal hair plates in *P. americana* are known to be of a tonic-phasic type with a very slow adapting nature (Okada and Toh 2001).

Motor System

Antennal Muscles and Their Innervation

The cockroach antennae are controlled by five functionally different muscles located inside the head capsule and the scape (Figure 7, left side). The muscles in the head capsule span between the tentorium (an internal skeleton) and the proximal ends of the scape. The adductor muscle rotates the scape medially around the head-scape joint, while the abductor muscle rotates the scape laterally. The levator

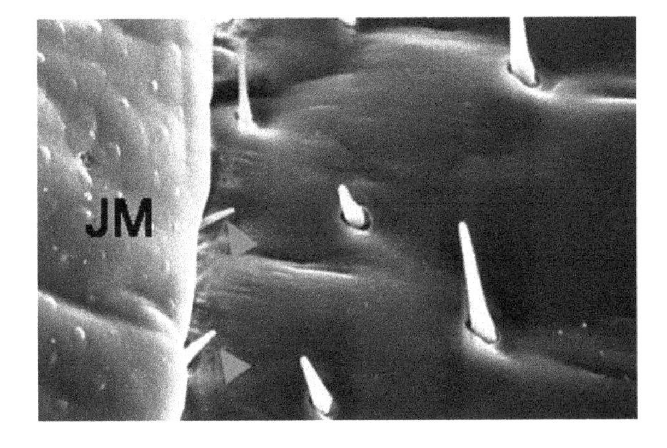

Figure 6 Magnified surface view of the hair plate. Some sensilla (*arrowheads*) are deflected by the joint membrane (JM). Courtesy: Dr. Y. Toh

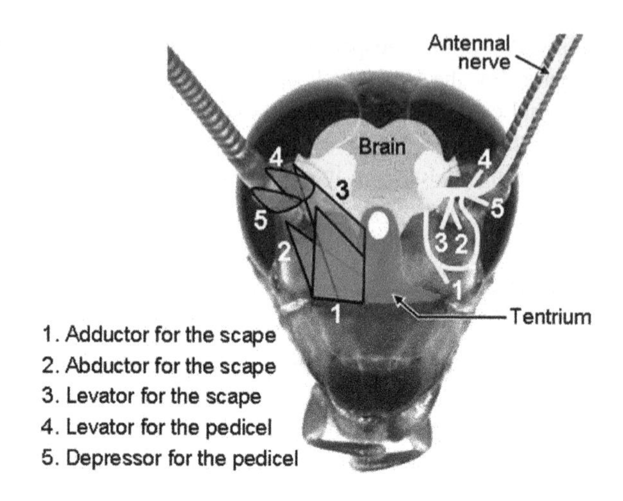

Figure 7 Frontal view of the five antennal muscles (*pink masses*) and the corresponding motor nerves (*yellow lines*)

muscle lifts the scape vertically. The other two muscles inside the scape span between the proximal ends of the scape and those of the pedicel: the levator and depressor muscles deflect the pedicel dorsally and ventrally, respectively, around the scape-pedicel joint. These five muscles are individually innervated by the antennal motor nerves arising from the brain (Figure 7, right side). Each antennal muscle may be innervated from 2–3 excitatory motor neurons (Baba and Comer 2008). It has also been suggested that some antennal motor nerves contain axons from a single "common" inhibitory motor neuron and dorsal unpaired median (DUM) neurons (Baba and Comer 2008). DUM neuron are considered responsible for releasing an excitatory modulator substance, possibly octopamine, to the antennal muscles.

Output Pattern

The output pattern of the antennal motor neurons has rarely been recorded for insect species probably because of the difficulty to generate their activities under the physiological experimental conditions. However, it has been reported in *P. americana* (Okada et al. 2009) that pilocarpine, a plant-derived muscarinic agonist, effectively induces the rhythmic bursting activities of antennal motor neurons even in isolated brain preparations. The drug-induced output pattern in cockroaches is well coordinated among the five antennal motor nerves. An agonistic pair of the motor nerves (3 vs. 4 in Figure 7) discharges bursting spikes with an in-phase relationship, whereas antagonistic pairs (1 vs. 2 and 4 vs. 5 in Figure 7) exhibit an anti-phase relationship with each other. These coordinated output patterns in an isolated brain preparation are comparable to the natural antennal movement.

Central System

Mechanosensory Center

The insect brain is composed of three distinct regions anteroposteriorly: the **protocerebrum**, **deutocerebrum**, and **tritocerebrum**. Sensory information received by the antennae is primarily conveyed to the deutocerebrum. Many anatomical studies clarified that the olfactory receptor afferents exclusively project to the primary olfactory center in the ventral deutocerebrum (**antennal lobe**), and non-olfactory (such as mechanosensory and gustatory) receptor afferents project to other regions from the deutocerebrum to the **subesophageal ganglion** (Figure 8). In contrast to the olfactory pathway, the mechanosensory pathways have not been adequately described in any insect species. An anatomical study in *P. americana* found that mechanoreceptor afferents in the basal segments project to the relatively dorsal part in the dorsal deutocerebrum (also known as the **dorsal lobe** or the antennal mechanosensory and motor center (**AMMC**)), while mechanoreceptor afferents in the flagellum project to the more ventroposterior areas in the dorsal lobe (Nishino et al. 2005). These antennal mechanosensory afferents also project up to the anterior region of the subesophageal ganglion. The central projections of the flagellar

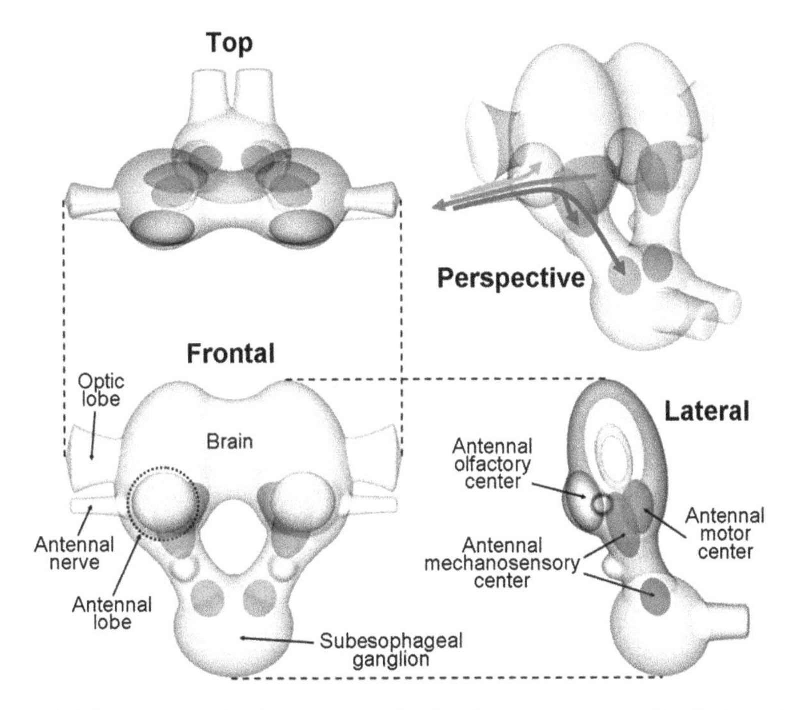

Figure 8 3-D cartoons showing antenna-related primary centers in the brain and the subesophageal ganglion. Arrows in the perspective view indicate the flow of chemosensory (*yellow*), mechanosensory (*blue*) and motor (*red*) information

mechanosensory afferents exhibit a topographical pattern, reflecting their peripheral locations, in the deutocerebrum (Nishino et al. 2005). Regarding the chordotonal sensilla, Comer and Baba (2011) showed that the central processes from the chordotonal organ are located from the deutocerebrum to the subesophageal ganglion. More recently, Watanabe et al. (2014) reported that serotonin-immunoreactive afferents from the chordotonal and Johnston's organs also project to the anteromedial region of the AMMC and subesophageal ganglion.

Motor Center

It is generally accepted that the primary motor center for the insect antenna is located in the dorsal lobe. Although the antennal motor center has been little known in the cockroach, an anatomical study elucidated its organization (Baba and Comer 2008). At least 17 antennal motor neurons were found in the deutocerebrum, and they were classified into five types according to their morphology. Although the positions of cell bodies differ according to their type, the dendritic processes from these neurons are colocalized at the dorsal area of the dorsal lobe (Figure 8). The dendritic zone is thought to be located dorsally to the antennal mechanosensory center. On the other hand, the central morphology of two dorsal unpaired median (DUM) neurons has been elucidated: both the cell bodies are located at the dorsomedial region of the subesophageal ganglion, and bilaterally symmetrical axons from the cell bodies run down to both sides of antennal muscles.

Interneurons

Tactile information from the antennal mechanoreceptors may be relayed to the following neurons for further sensory processing and/or expression of appropriate behavior. In this context, it would be essential to study individual antennal mechanosensory interneurons, particularly in the animals with a simpler central nervous system. Thus far, several physiological studies have identified such interneurons in cockroaches. Among the best known interneurons in *P. americana* are the "giant" descending mechanosensory interneurons (DMIs) (Burdohan and Comer 1996). Of the two DMIs identified, one (DMIa-1) has its cell body in the protocerebrum, and the other (DMIb-1) has its cell body in the subesophageal ganglion (Figure 9). The DMIa-1 axon extends fine branches into the ipsilateral dorsal deutocerebrum where the antennal mechanosensory and motor centers are thought to be located, runs to the contralateral hemisphere of the brain, and descends to the abdominal ganglia while extending fine processes into the thoracic ganglia. The DMIb-1 axon also descends in a similar morphological pattern. The DMIa-1 responds exclusively to the mechanical stimuli to the antenna. The DMIb-1 is responsive to the stimuli to the head and mouthparts as well as the antenna.

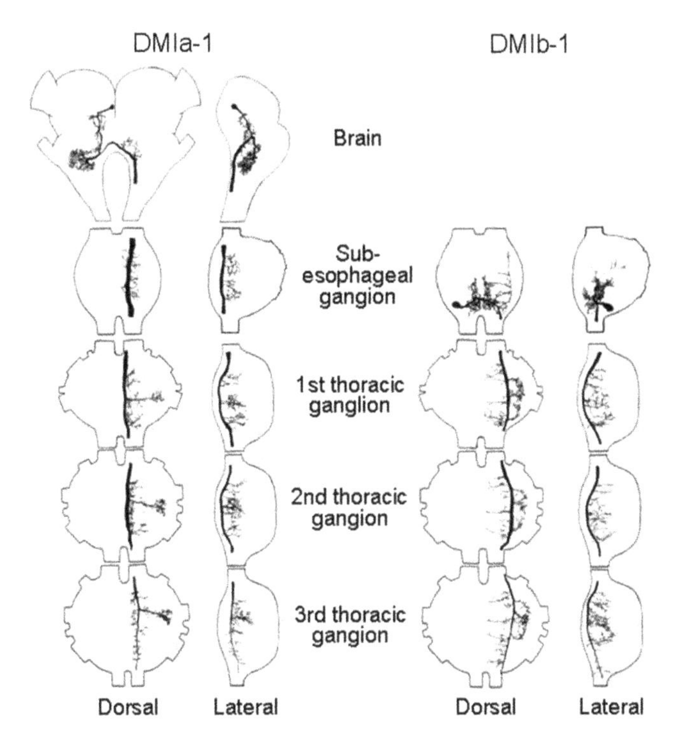

Figure 9 Central morphology of the antennal mechanosensory interneurons, DMIa-1 and DMIb-1. Modified from Burdohan and Comer (1996) with permission

The chordotonal organ is suspected to be a major mechanosensory input origin to the DMIs because its ablation leads to serious impairment in response to antennal mechanical stimulation, and the central processes of the chordotonal organ closely overlap the neurites of DMIs in the deutocerebrum (Comer and Baba 2011). The DMIs may be responsive for the control of an evasive locomotor behavior (see "Evasive behavior").

Behavior

Anemotactic Behavior

Anemotaxis, a locomotor response to the wind, is thought to be a fundamental strategy to find odor sources located upwind. In three cockroach species (*Blattella germanica*, *Periplaneta americana*, and *Blaberus craniifer*), the anemotactic behavior in the absence of olfactory cues was found to have different patterns depending on both the species and wind velocity (Bell and Kramer 1979). *B. craniifer* directed and ran upwind in a wide range of wind velocities, while

B. germanica oriented downwind. *P. americana* oriented upwind in low-wind velocities, and downwind in high-wind velocities. When both the scape-pedicel and pedicel-flagellar joints of *B. craniifer* were fixed to prevent movement at the pedicel, positive anemotaxis was considerably impaired. This implies that the putative mechanoreceptors adjacent to the pedicel are crucial to detect wind direction and velocity.

Evasive Behavior

The insect antenna has been considered rather non-responsive to escape behavior. However, the responsiveness seems to depend on the tactile feature of the stimulants. For instance, if a cockroach (*P. americana*) comes in contact with a potential predator (spider), as sensed by the antenna, it would immediately turn away and escape from the spider (Comer et al. 1994). Similarly, cockroaches may discriminate tactile features (probably texture) between the spider and the cockroach by means of antennal probing (antennation) (Comer et al. 2003). The mechanoreceptors in the flagellum and the antennal basal segments are thought to be essential for acquiring tactile information and initiating an escape response, respectively (Comer et al. 2003; Comer and Baba 2011).

Wall-Following

Since cockroaches exhibit a positive thigmotaxis, they tend to walk or run along the wall. During such wall-following, both the antennae extend forward to remain in contact with the adjacent wall. A behavioral study revealed that *P. americana* is capable of turning at 25 Hz during a rapid wall-following (Camhi and Johnson 1999). This high-frequency turn helps avoid collisions while maintaining body at an appropriate distance from the wall. Tactile cues from the antennae would be essential for such a high-performance wall-following behavior. Baba et al. (2010) found that the response type of collision avoidance (stop, climb, reverse, redirect, and so on) depends on the configuration of obstacles detected by the antennae. In particular, the stop response may occur when both antennae simultaneously contact an obstacle. On the other hand, the mechanical interaction between the flagellum and wall affects the state of the antenna: the tip of the wall-side flagellum tends to project forward when following a relatively smooth wall and backward when following a rougher wall (Mongeau et al. 2013). The distance of the body from the wall is significantly influenced by the antennal state, i.e. nearer in the forward state and farther in the backward state.

Obstacle Negotiation Behavior

Cockroaches with non-escape walking also encounter external objects with a wide variety of shapes in their path and negotiate them in proper ways by using tactile information from the antennae. Harley et al. (2009) investigated the behavioral choice between climbing over and tunneling under when a cockroach encounters an appropriate size of "shelf". They found that the choice depends on how the antenna hits the shelf. In the case of hitting from above the shelf, the cockroach may push up the body and try to climb, and in the case of hitting from below, it may tunnel. Furthermore, this behavioral choice is influenced by the ambient light conditions: under bright conditions, cockroaches prefer to tunnel. A line of studies has suggested that the central body complex (**CC**), a set of interconnected neuropils located along the midline of the protocerebrum, may play a key role in this behavioral choice because the focal lesioning of CC resulted in deficits in the relevant tactile behaviors (Ritzmann et al. 2012).

Tactile Orientation

When an antenna of a searching cockroach (*P. americana*) contacts a stable tactile object in an open space, the animal may stop and approach the object with antennation (see http://www.scholarpedia.org/). Similarly, a tethered cockroach mounted on a treadmill may attempt to approach a stable object presented to the antenna (Okada and Toh 2000). This antenna-derived tactile orientation behavior is probably due to the thigmotactic nature of cockroaches (see "Wall-following"). Because the extent of turn angle depends on the horizontal position of the presented object, cockroaches can discriminate the position of objects by antennation. The removal of hairs in the scapal hair plates resulted in significant deterioration in the performance of tactile orientation (Okada and Toh 2000), suggesting that the proprioceptors in the antennal joints are vital for the detection of an object's position.

Electrostatic Field Detection

It has been reported that cockroaches appear to avoid an artificially generated electrostatic field (Hunt et al. 2005; Newland et al. 2008; Jackson et al. 2011). This evasive response may help them avoid, for instance, friction-charged objects and high-voltage power lines. Newland et al. (2008) proposed a possible unique mechanism for electrostatic field detection in *P. americana*. After the antennal joints between the head and scape were fixed, the treated animal could no longer avoid electrostatic field. They proposed the following working hypothesis. As a cockroach approaches a charged field, it would produce biased charge distribution between the antennae and the cockroach's body. The antennae are then passively deflected by the Coulomb force in the electrostatic field. Finally, the antennal deflection may be detected by the hair plates at the head-scape joint.

Internal References

Llinas, R (2008). Neuron. *Scholarpedia* 3(8): 1490. http://www.scholarpedia.org/article/Neuron.
Pikovsky, A and Rosenblum, M (2007). Synchronization. *Scholarpedia* 2(12): 1459. http://www.scholarpedia.org/article/Synchronization.

External References

Baba, Y and Comer, C M (2008). Antennal motor system of the cockroach, "Periplaneta americana". *Cell and Tissue Research* 331: 751–762. [http://dx.doi.org/10.1007/s00441-007-0545-9 doi:10.1007/s00441-007-0545-9].
Baba, Y; Tsukada, A and Comer, C M (2010). Collision avoidance by running insects: Antennal guidance in cockroaches. *Journal of Experimental Biology* 213: 2294–2302. [http://dx.doi.org/10.1242/jeb.036996 doi:10.1242/jeb.036996].
Bell, W J and Kramer, E (1979). Search and anemotactic orientation of cockroaches. *Journal of Insect Physiology* 25: 631–640. [http://dx.doi.org/10.1016/0022-1910(79)90112-4 doi:10.1016/0022-1910(79)90112-4].
Burdohan, J A and Comer, C M (1996). Cellular organization of an antennal mechanosensory pathway in the cockroach, "Periplaneta americana". *Journal of Neuroscience* 16: 5830–5843.
Camhi, J M and Johnson, E N (1999). High-frequency steering maneuvers mediated by tactile cues: Antennal wall-following in the cockroach. *Journal of Experimental Biology* 202: 631–643.
Comer, C M; Mara, E; Murphy, K A; Getman, M and Mungy, M C (1994). Multisensory control of escape in the cockroach, "Periplaneta americana". II. Patterns of touch-evoked behavior. *Journal of Comparative Physiology A* 174: 13–26. [http://dx.doi.org/10.1007/bf00192002 doi:10.1007/bf00192002].
Comer, C M; Parks, L; Halvorsen, M B and Breese-Terteling, A (2003). The antennal system and cockroach evasive behavior. II. Stimulus identification and localization are separable antennal functions. *Journal of Comparative Physiology A* 189: 97–103.
Comer, C M and Baba, Y (2011). Active touch in orthopteroid insects: Behaviours, multisensory substrates and evolution. *Philosophical Transactions of the Royal Society B* 366: 3006–3015. [http://dx.doi.org/10.1098/rstb.2011.0149 doi:10.1098/rstb.2011.0149].
Hansen-Delkeskamp, E (1992). Functional characterization of antennal contact chemoreceptors in the cockroach, "Periplaneta americana": An electrophysiological investigation. *Journal of Insect Physiology* 38: 813–822. [http://dx.doi.org/10.1016/0022-1910(92)90034-b doi:10.1016/0022-1910(92)90034-b].
Harley, C M; English, B A and Ritzmann, R E (2009). Characterization of obstacle negotiation behaviors in the cockroach, "Blaberus discoidalis". *Journal of Experimental Biology* 212: 1463–1476. [http://dx.doi.org/10.1242/jeb.028381 doi:10.1242/jeb.028381].
Hunt, E P; Jackson, C W and Newland, P L (2005). "Electrorepellancy" behavior of "Periplaneta americana" exposed to friction charged dielectric surfaces. *Journal of Electrostatics* 63: 853–859. [http://dx.doi.org/10.1016/j.elstat.2005.03.081 doi:10.1016/j.elstat.2005.03.081].

Ikeda, S; Toh, Y; Okamura, J and Okada, J (2004). Intracellular responses of antennal chordotonal sensilla of the American cockroach. *Zoological Science* 21: 375–383. [http://dx.doi.org/10.2108/zsj.21.375 doi:10.2108/zsj.21.375].

Jackson, C W; Hunt, E; Sharkh, S and Newland, P L (2011). Static electric fields modify the locomotory behaviour of cockroaches. *Journal of Experimental Biology* 214: 2020–2026. [http://dx.doi.org/10.1242/jeb.053470 doi:10.1242/jeb.053470].

Mongeau, J M; Demir, A; Lee, J; Cowan, N J and Full, R J (2013). Locomotion- and mechanics-mediated tactile sensing: Antenna reconfiguration simplifies control during high-speed navigation in cockroaches. *Journal of Experimental Biology* 216: 4530–4541. [http://dx.doi.org/10.1242/jeb.083477 doi:10.1242/jeb.083477].

Newland, P L et al. (2008). Static electric field detection and behavioural avoidance in cockroaches. *Journal of Experimental Biology* 211: 3682–3690. [http://dx.doi.org/10.1242/jeb.019901 doi:10.1242/jeb.019901].

Nishino, H; Nishikawa, M; Yokohari, F and Mizunami, M (2005). Dual, multilayered somatosensory maps formed by antennal tactile and contact chemosensory afferents in an insect brain. *Journal of Comparative Neurology* 493: 291–308. [http://dx.doi.org/10.1002/cne.20757 doi:10.1002/cne.20757].

Okada, J and Toh, Y (2000). The role of antennal hair plates in object-guided tactile orientation of the cockroach ("Periplaneta americana"). *Journal of Comparative Physiology A* 186: 849–857. [http://dx.doi.org/10.1007/s003590000137 doi:10.1007/s003590000137].

Okada, J and Toh, Y (2001). Peripheral representation of antennal orientation by the scapal hair plate of the cockroach ("Periplaneta americana"). *Journal of Experimental Biology* 204: 4301–4309.

Okada, J; Morimoto, Y and Toh, Y (2009). Antennal motor activity induced by pilocarpine in the American cockroach. *Journal of Comparative Physiology A* 195: 351–363. [http://dx.doi.org/10.1007/s00359-008-0411-6 doi:10.1007/s00359-008-0411-6].

Ritzmann, R E et al. (2012). Deciding which way to go: How do insects alter movements to negotiate barriers? *Frontiers in Neuroscience* 6: 97. [http://dx.doi.org/10.3389/fnins.2012.00097 doi:10.3389/fnins.2012.00097].

Staudacher, E; Gebhardt, M J and Dürr, V (2005). Antennal movements and mechanoreception: Neurobiology of active tactile sensors. *Advances in Insect Physiology* 32: 49–205. [http://dx.doi.org/10.1016/s0065-2806(05)32002-9 doi:10.1016/s0065-2806(05)32002-9].

Toh, Y (1981). Fine structure of sense organs on the antennal pedicel and scape of the male cockroach, "Periplaneta americana". *Journal of Ultrastructure Research* 77: 119–132. [http://dx.doi.org/10.1016/s0022-5320(81)80036-6 doi:10.1016/s0022-5320(81)80036-6].

Watanabe, H; Shimohigashi, M and Yokohari, F (2014). Serotonin-immunoreactive sensory neurons in the antenna of the cockroach, "Periplaneta americana". *Journal of Comparative Neurology* 522: 414–434. [http://dx.doi.org/10.1002/cne.23419 doi:10.1002/cne.23419].

External Links

http://blattodea-culture-group.org/ Blattodea Culture Group website.
http://en.wikipedia.org/wiki/Cockroach Cockroach in Wikipedia.

Stick Insect Antennae

Volker Dürr

Insects have an elaborate sense of touch. Their most important source for tactile information is the pair of feelers on the head: the antennae (Figure 1; singular: antenna). The stick insect *Carausius morosus* is one of four major study organisms for the **insect tactile sense**. Accordingly, the **stick insect antenna** (or feeler), together with the antennae of the **cockroach** (Okada 2009), cricket and honeybee, belongs to the best-studied **insect antennae**. The aim of the present article is to provide an overview over the behavioural relevance, adaptive properties and sensory infrastructure of stick insect antennae, along with a complete bibliography thereof. For a more general treatment of the topic, please refer to the review by Staudacher et al. (2005).

As most stick insects are nocturnal, flightless insects, tactile sensing is of prime importance for exploration of the space immediately ahead of the animal. For example, obstacle detection and path-finding in the canopy at night are likely to be tasks where tactual near-range information is highly valuable to the animal.

Moreover, the stick insect is an important study organism in motor control of invertebrates and the **neuroethology of insect walking** (Ritzmann and Zill 2013). Nearly a century of research has led to a host of information about its walking behaviour and leg coordination (von Buddenbrock 1920; Wendler 1964; Bässler 1983; Cruse 1990), central and sensory contributions to motor pattern generation (Graham 1985; Bässler and Büschges 1998; Büschges and Gruhn 2007) the neural and muscular infrastructure (Jeziorski 1918; Marquardt 1939; Schmitz et al. 1991; Kittmann et al. 1991), and so on. On that background, the neuroethology and physiology the stick insect's tactile near-range sense can serve as a model for understanding the significance of the arthropod antenna in sensory-guided behaviour.

V. Dürr (✉)
Bielefeld University, Bielefeld, Germany

T.J. Prescott et al. (eds.), *Scholarpedia of Touch*, Scholarpedia,
DOI 10.2991/978-94-6239-133-8_3

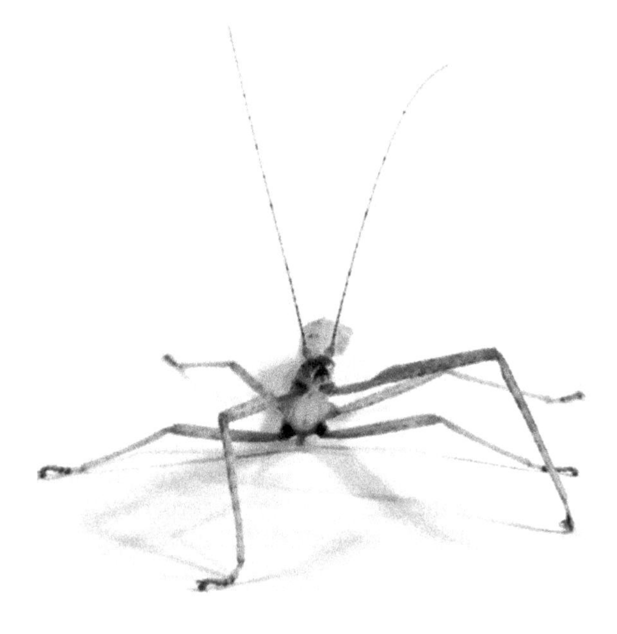

Figure 1 The stick insect *Carausius morosus* (de Sinéty 1901) carries a pair of long and straight feelers, or antennae. They are the main sensory organs for touch and smell

Antennae Are Active Near-Range Sensors

Stick insects continuously move their antennae during walking. Video 1 in the corresponding online article (http://www.scholarpedia.org) shows a blindfolded stick insect walking towards a block of wood. Note that the video was slowed down five-fold. Next to the walkway, there was a mirror mounted at an angle of 45 degrees, allowing synchronous recording of top and side views. If an antenna touches an obstacle, as the right antenna does in this video, then this touch event leads to an appropriate action of the front leg on the same side.

Like many other insects, *C. morosus* continuously moves its antennae during locomotion. By doing so, it actively raises the likelihood of tactile contacts with obstacles because each up-down or rear-to-front sweep of an antenna samples a volume of space immediately ahead. Accordingly, the antennal movement pattern can be considered an **active searching behaviour**. During this searching behaviour, antennal movement is generated by rhythmic movement of both antennal joints with the same frequency and a stable phase shift (Krause et al. 2013). Moreover, it is clear that this searching movement is generated within the brain, despite the fact that the brain requires additional activation, possibly through ascending neural input from the suboesophageal ganglion (Krause et al. 2013). Antennal searching movements have been shown to be adapted according to the behavioural context, e.g., when the insect steps across an edge and searches for foothold (Dürr 2001). Similarly, the beating field (or searched volume) of the antennae is shifted into the walking direction during turning (Dürr and Ebeling 2005).

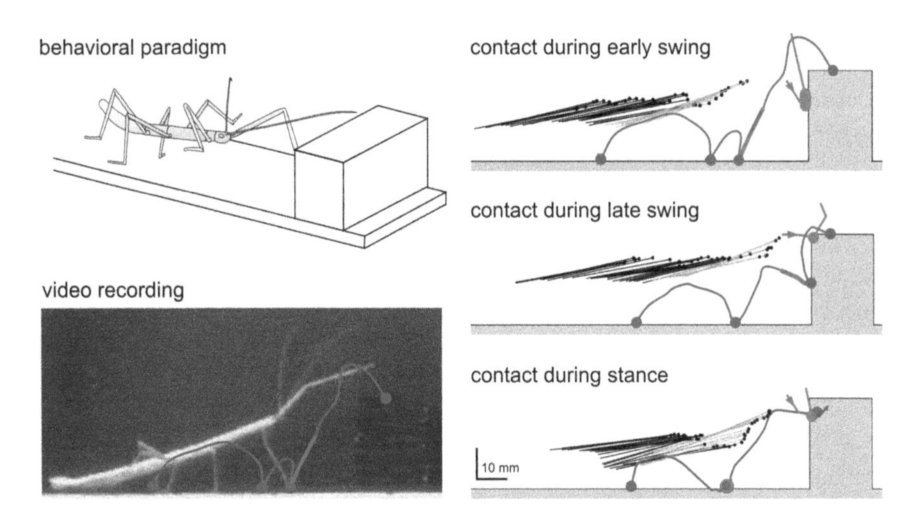

Figure 2 Stick insects use their antennae for obstacle detection during locomotion. When blindfolded, stick insects walk towards a block of wood (*top left*) they can readily climb obstacles that are much higher than the maximum foot height during a regular step. Following antennal contact with the obstacle, the next step is raised higher than normal (*bottom left* and *top right*: *blue line* indicates foot trajectory; *blue dots* mark contact sites; *red line* segment labels the period of concurrent antennal contact). Antennal contact during early swing often leads to re-targeting (*top right*). Contact during late swing typically ensues a correcting step (*right middle*). Antennal contact during stance leads to a higher step than regular (*lower right*). *Red lines* and *dots* show trajectories and contact sites of the antennal tip (first contact only). *Black lines* and *dots* show the body axis and head every 40 ms. Following antennal contact, body *axis lines* are drawn in *grey*

Experiments have shown that the movement pattern of the antennae is rhythmical and may be coordinated with the stepping pattern of the legs (Dürr et al. 2001). This can be seen in the video in the corresponding online article (http://www.scholarpedia.org), showing a blindfolded stick insect that walks towards a block of wood. During approach of the obstacle, the top view shows how both antennae are moved rhythmically from side to side. Soon after a brief contact of the right antenna, the insect lifts its ipsilateral leg higher than normal and steps onto the top of the block (same trial as in Figure 2, *top right*).

Examples like this show that stick insects respond to antennal tactile contact with appropriate adaptation of their stepping pattern and leg movements. Depending on the timing of the antennal contact with respect to the step cycle of the ipsilateral leg, three kinds of behavioural responses are easily distinguished:

- The first kind is the one shown in the video in the corresponding online article (http://www.scholarpedia.org). If the antennal contact occurs during early swing, the foot trajectory often reveals a distinct upward kink, indicating re-targeting of an ongoing swing movement (Figure 2, top right). Essentially this means that antennal touch information can interfere with the cyclic execution of the normal stepping pattern of the front legs, such that it can trigger re-targeting of an ongoing swing movement.

- The second kind occurs when the antennal contact happens during late swing. In this case, reaction time appears to be too short for re-targeting and the foot hits the obstacle, only to be raised in a second, correcting step (Figure 2, middle right).
- Finally, in the third kind of response, antennal contact occurs during stance. Then, stance movement is completed and the following swing movement is higher than a regular step (Figure 2, lower right).

The analysis of adaptive motor behaviour in response to antennal touch can be simplified if the touched obstacle is nearly one-dimensional, e.g., a vertical rod (Schütz and Dürr 2011). In this case, it can be demonstrated that antennal touch information is not only used for rapid adaptation of goal-directed leg movements, but also triggers distinct changes of the antennal tactile sampling pattern, including an increase in cycle frequency and a switch in inter-joint coupling. Therefore, **tactile sampling behaviour** is clearly distinct from searching behaviour.

Tactile sampling of other obstacles, e.g., stairs of different height, bears very similar characteristics (Krause and Dürr 2012). During crossing of large gaps, antennal contacts of the species *Aretaon asperrimus* (Brunner von Wattenwyl 1907) have been shown to 'inform' the animal about the presence of an object in reach (Bläsing and Cruse 2004a, b), though antennal movements have not been analysed yet in detail in this animal. In both *Carausius* and *Aretaon*, antennal contact information ensues a change in leg movement, as necessary during climbing. Similar tactually mediated climbing behaviour has been described in the cockroach (Okada 2009).

In summary, tactile information from the antenna induces an adjustment of leg movements in a context-dependent manner. Moreover it induces changes in the antennal movement itself.

Antennae Are Dedicated Sensory Limbs

The antenna of the Arthropoda (crustacea, millipedes and insects carry antennae, arachnids do not) is considered to be a true limb that evolved from a standard locomotory limb into a dedicated sensory limb. Various lines of evidence suggest that the segmented body structure of the arthropods once carried leg-like motor appendages on each segment. As some body segments adopted specialised functions during the course of evolution, their motor appendages changed their function accordingly. For example, the insect head is thought to have evolved by fusion of the six most frontal body segments, with three pairs of limbs having turned into mouthparts that specialised for feeding, two pairs having been lost, and one pair having turned into dedicated, multimodal sensory organs—the antennae. This common theory is supported by palaeontological, morphological, genetic and developmental evidence.

For example, a simple experiment on stick insects illustrates the close relationship between walking legs and antennae: if a young stick insect larva loses an

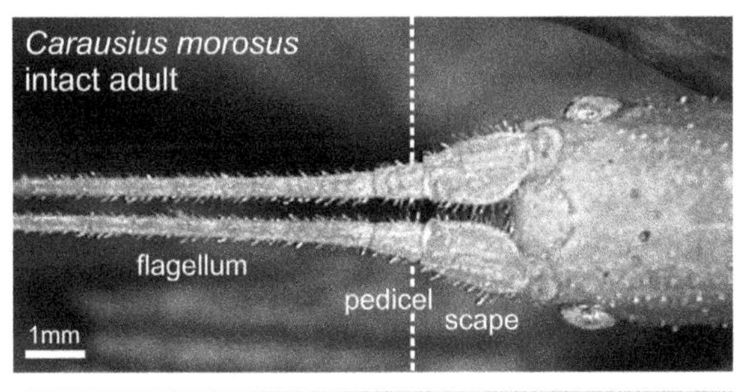

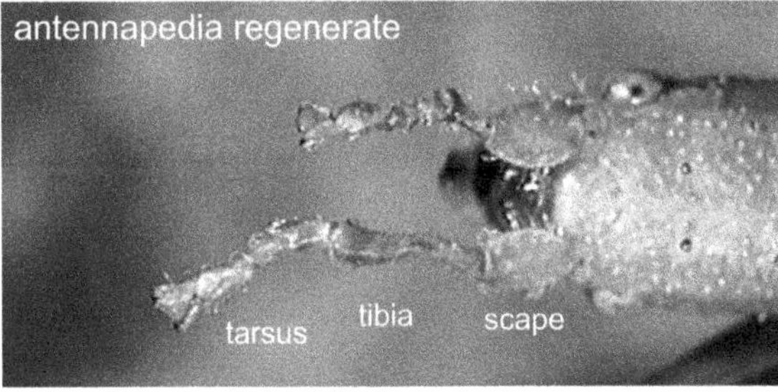

Figure 3 Stick insects can regenerate a leg in place of an antenna. The *top image* shows the head and first half of the antennae of an adult female stick insect. Each antenna has three functional segments: scape, pedicel and flagellum. The larvae of stick insects look very similar to the adult, except that they are smaller. If an antenna of a larva is cut at the level of the proximal pedicel (*dashed line*), the animal often regenerates a leg instead of an antenna during the next moult. The *bottom image* shows the head and antennapedia regenerates of a stick insect whose antennae were cut in the third or fourth instar (same scale as above)

antenna, it often regenerates a walking leg instead of an antenna (Schmidt-Jensen 1914). This "faulty" regeneration is called an antennapedia regenerate (literally: "antennal feet") It happens during the next moult and becomes more pronounced with each successive moult. Antennapedia regenerates can be induced experimentally by cutting an antenna of a stick insect larva (e.g., 3rd instar) at the second segment, the pedicel (Figure 3). The site of the cut determines the outcome of the regeneration: distal cuts lead to a "correct" antenna regenerate, proximal cuts lead to complete failure of regeneration. Only cuts through a narrow region of the pedicel reliably induce an antennapedia regenerate (Cuénot 1921; Brecher 1924; Borchardt 1927). Note that there is also a *Drosophila* gene called Antennapedia (Casares and Mann 1998).

Morphological similarity between walking legs and antennae concerns the structure of joints, musculature, innervation and most types of mechanoreceptors and contact chemoreceptors (Staudacher et al. 2005). Differences concern the number of functional segments (five in legs, three in antennae), cuticle properties (see below) and sensory infrastructure (e.g., olfactory receptors on antennae, only).

The antenna has three functional segments: they are called scape, pedicel and flagellum, from base to tip. In the stick insect, only the scape contains muscles (Dürr et al. 2001). This is the same in all higher insects (the Ectognatha; see Imms

1939). In other words, true joints that are capable of active, muscle-driven movement occur only between head and scape (HS-joint) and between scape and pedicel (SP-joint). The HS-joint is moved by three muscles inside the head capsule (so-called extrinsic muscles, because they are outside the antenna), the SP-joint is moved by two muscles inside the scape (so-called intrinsic muscles, because they are located inside the antenna).

Four Adaptations Improve Tactile Efficiency

Several morphological, biomechanical and physiological properties of the stick insect antenna are beneficial for its function in tactually guided behaviour. Four of such adaptations are:

- Matching lengths of antennae and front legs (in all developmental stages)
- Co-ordinated movement of antennae and legs
- Non-orthogonal, slanted joint axes
- Special biomechanical properties, such as damping

Matching Lengths of Antennae and Front Legs

As active tactile sensors are particularly well-suited for exploration of the near-range environment, it is reasonable to assume that their action range tells us something about the animal's behavioural requirements to react to near-range information. Since stick insects generally are nocturnal animals, their antennae are likely to be their main "look-ahead sense" (rather than the eyes). Moreover, as many stick insect species are obligatory walkers, tactile exploration is likely to serve obstacle detection and orientation during terrestrial locomotion. In *C. morosus*, the length of the antenna matches that of a front leg such that the action radii of the antennal tip and the end of the foot are nearly the same (See Figure 4). Thus, potentially, anything the antenna touches is located within reach of the front leg. As Figure 4 shows, this is the case throughout development. Owing to this match, the stick insect is able to adapt its locomotory behaviour to a touch event within the action distance of one length of a leg, and with a look-ahead time of up to one step cycle period. In fact, reaction times of a front leg reacting to an antennal touch event is in the range of 40 ms (Schütz and Dürr 2011), and several classes of descending interneurons have been described that convey antennal mechanosensory information from the antenna to the motor centres of the front leg (Ache and Dürr 2013). Some of these descending interneurons have been characterized individually to quite some detail. For example, a set of three motion-sensitive descending interneurons encodes information about antennal movement velocity in a complementary manner: two of them respond by increases in spike rate, the third one

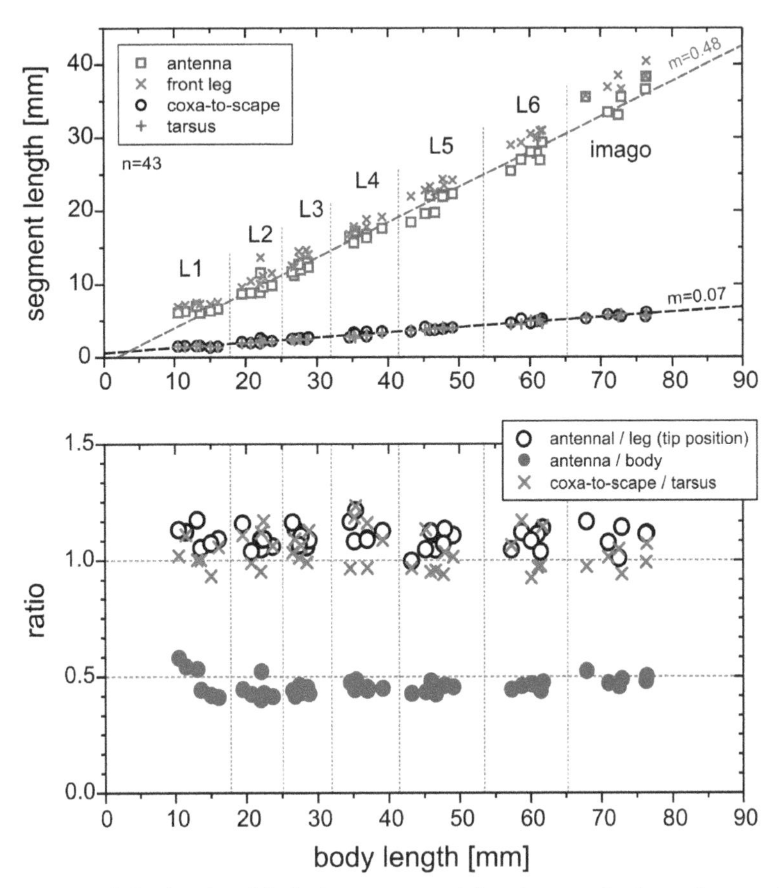

Figure 4 The relative lengths of the body, antennae and front legs remain almost the same in all seven developmental stages (larval stages L1 to L6, and imago). *Top* A front leg is slightly longer than an antenna. However, since the attachment site of the front leg is more posterior than that of the antenna, the tip of the antenna reaches slightly further than the front leg tarsi (feet). The distance coxa-to-scape (between the attachment sites) is about equal to the tarsus length. All four measures increase linearly with body length (see *dashed lines* and *slopes* for antenna length and coxa-to-scape). Values are from at least 6 female animals per developmental stage. *Bottom* length ratios for absolute values above. For the ratio antenna/leg, the offset coxa-to-scape was subtracted from leg length, in order to relate the workspaces of antennal tip and tarsus tip. *Vertical dotted lines* separate developmental stages

responds by a decrease in spike rate (Ache et al. 2015). Because these movement-induced changes in spike rate occur with very short latency (approx. 15 ms at the entry of the prothoracic ganglion) they are suitable candidates for mediating the fast, tactually induced reach-to-grasp reaction described above.

Note that the length of the antennae does not match the length of the front legs in all stick insect species. In some species, like *Medauroidea extradentata* (Redtenbacher 1906), the antennae are much shorter than the front legs, indicating that their antennae are not suited for tactile near-range searching (because the feet of the front leg will nearly always lead the antennal tips). In the case of *M. extradentata* (= *Cuniculina impigra*), it has been shown that the front legs execute very high swing movements during walking and climbing (Theunissen et al. 2015). Thus, in these animals the front legs appear to take on the function of tactile near-range searching.

Coordinated Movement of Antennae and Legs

Antennae and front legs not only match in length, their movement may be coordinated in space and time, too (Dürr et al. 2001). Temporal coordination is revealed by the gait pattern of a straight walking stick insect in Figure 5 (*left*), where the rhythmical protraction phases of the antenna (red) are shifted relative to the rhythmical retraction phases of the legs (blue). The pattern is the same for left and right limbs. In both cases, coordination is well-described by a rear-to-front wave travelling along the body axis (indicated by grey arrows). As if activated by such a wave, the middle legs follow the hind leg rhythm, the front legs follow the middle legs, and the antennae follow the front legs. Note that this pattern depends on walking conditions and behavioural context.

Antennae and front legs are also coordinated spatially (Figure 5, right) such that the antennal tip leads the foot of the same body side (in the top view plot, grey lines

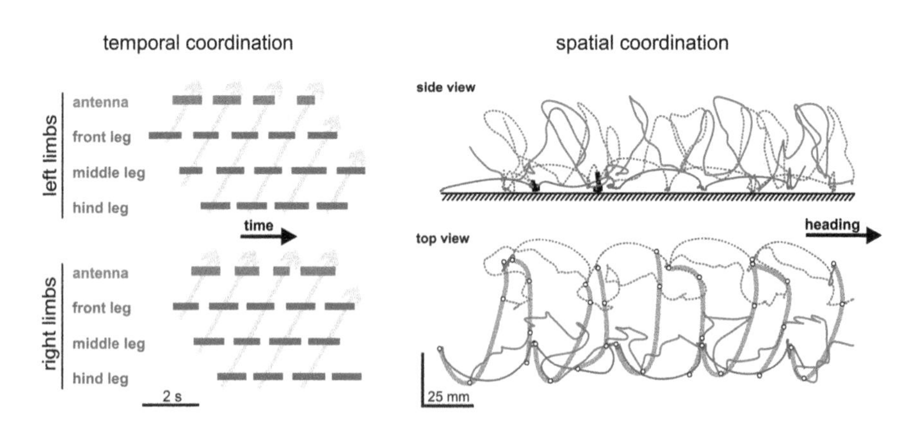

Figure 5 During walking, antennal movements are often coordinated with the stepping movements of the legs. *Left* Temporal coordination of antennal adduction/protraction phases (*red line segments*) and stance/retraction phases of the legs (*blue line segments*). *Light gray arrows* indicate a back-to-front wave that describes the coordination pattern of left limbs (*top*) and right limbs (*bottom*). *Right* Spatial trajectories of antennal tips (*red*) and front leg tarsi (*blue*) during a walking sequence of four steps. *Side view* (*top*) and *top view* (*bottom*) of *left* (*broken lines*) and *right* (*solid lines*) limbs. *Bold gray lines* connect coincident points. Note how these lines always curve in the same direction. Therefore, antennal tips tend to lead the tarsus movement [modified from Dürr et al. (2001)]

connect coincident points). Moreover, the lateral turning points of the antennal tip's trajectory are always very close to the most lateral position of the foot during the next step. Both, temporal and spatial coordination support the hypothesis that the antenna actively explores the near-range space for objects which require the ipsilateral leg to adapt its movement. Indeed, it has been shown that the likelihood of the antenna to detect an obstacle before the leg gets there, i.e., in time for the leg to react, increases with the height of the obstacle, and gets very large for obstacles that are so high that climbing them requires an adaptation of leg movements (Dürr et al. 2001).

Non-orthogonal, Slanted Joint Axes

As mentioned above, all higher insects have only two true joints per antenna (Imms 1939), meaning that only two joints are actively moved by muscles. In stick insects, both of these joints are revolute joints (hinge joints) with a single, fixed rotation axis. Because of the fixed joint axes, a single joint angle accurately describes the movement of each joint (they have a single) degree of freedom (DOF). Therefore, the two DOF associated with the two revolute joints of a stick insect antenna, fully describe its posture.

The spatial arrangement of the two joint axes determines the action range of the antenna (Krause and Dürr 2004). For example, if both joint axes were orthogonal to each other, and if the length of the scape was zero, the antennal joints would work like a universal joint (or Cardan joint): the flagellum could point into any direction. Because the scape is short relative to the total length of the antenna, its length has virtually no effect on the action range of the antenna. However, the joint axes of sick insect antennae are not orthogonal to each other (Dürr et al. 2001; Mujagic et al. 2007), which has the effect that the antenna can not point into all directions (Figure 6). Instead, there is an out-of reach zone, the size of which directly depends on the angle between the joint axes (Krause and Dürr 2004).

This is different than in other insect groups, for example crickets, locusts and cockroaches. Mujagic et al. (2007) argued that this non-orthogonal arrangement of the joint axes might be an adaptation to efficient tactile exploration. Their argument goes as follows: For a given minimal change in joint angle, the resulting change in pointing direction will be smaller in an arrangement with out-of-reach zones than in an arrangement without (Dürr and Krause 2001). Essentially, this is because the surface of the area that can be reached by the antennal tip decreases (due to the out-of-reach zone), but the number of possible joint angle combinations stays the same. Therefore, theoretically, an arrangement with out-of-reach zones has improved positioning accuracy. Since the order of the stick insects (Phasmatodea) is thought to have evolved as a primarily wingless group of insects (Whiting et al. 2003), and

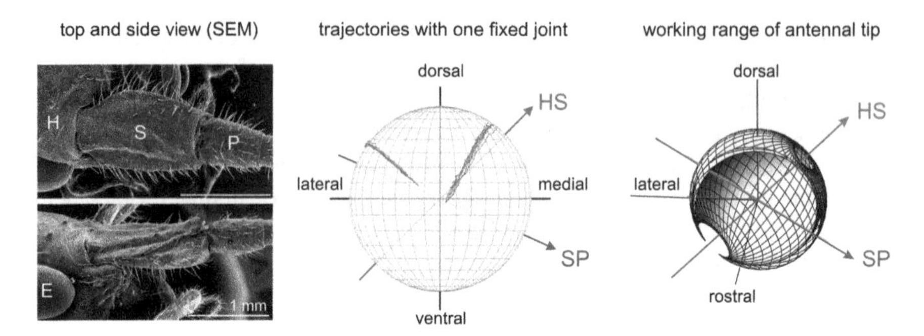

Figure 6 Both antennal joints of the stick insect are simple rotary joints (hinge joints). *Left* One joint is located between head (H) and scape (S), the other between scape and pedicel (P). *Middle* When one joint is immobilised, movement of the other joint causes the antennal tip to move along a circular line, i.e., a cross-section of a sphere. The axis perpendicular to this cross-section is the joint axis. In stick insects, the joint axes are non-orthogonal and slanted with respect to the body coordinate frame. *Right* The non-orthogonal axis orientation results in out-of-reach zones. By assuming unrestrained rotation around both joint axes, the antennal workspace and its out-of-reach zones can be visualised as a torus with holes. Theoretically, increased positioning accuracy can be traded off by increasingly large out-of-reach-zones. [modified from Mujagic et al. (2007)]

because all Phasmatodea are nocturnal, tactile exploration must be expected to be their prime source of spatial information. The finding that slanted, non-orthogonal joint axes are an autapomorphy of the Phasmatodea suggests that this insect order has evolved an antennal morphology that is efficient for tactile near-range sensing (Mujagic et al. 2007; strictly, is has been shown for the so-called Euphasmatodea, only).

Special Biomechanical Properties

Last but not least, there are particular biomechanical features that support the sensory function of the antenna. In particular, this concerns the function of the delicate, long and thin flagellum that carries thousands of sensory hairs. In *Carausius morosus*, the flagellum is about 100 times as long as its diameter at the base. If this structure was totally stiff, it would break very easily. In the other extreme, if it was too flexible, it would be very inappropriate for spatial sampling, simply because its shape would change all the time. As a consequence, much of the sensory resources would have to go into monitoring the own curvature, at least if contact locations in space were to be encoded. In *C. morosus*, the skeletal properties of the antenna are such that the flagellum remains stiff during self-generated movement (e.g., during searching) but is very compliant when in contact with obstacles (e.g., during tactile sampling). For example, the flagellum frequently bends very much as the stick insect samples an obstacle during climbing (Figure 2).

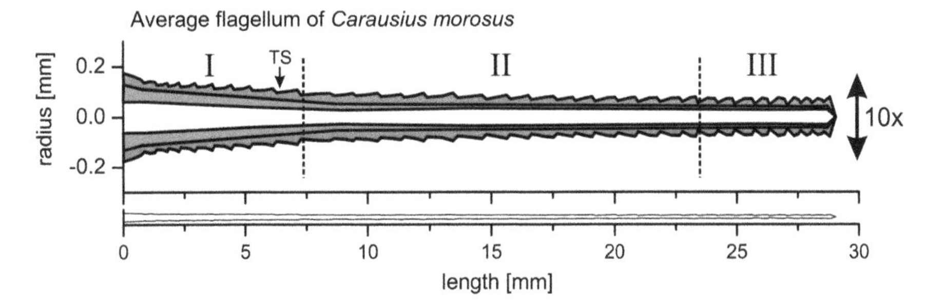

Figure 7 Schematic longitudinal section through the flagellum of the stick insects antenna (*Carausius morosus*). *Top* The flagellum has a two-layered cuticle, the outer exocuticle (*orange*) and the inner endocuticle (*blue*). As the diameter of the flagellum tapers, the soft endocuticle gets thinner and thinner. As a result, the fraction of cross-sectional area filled by endocuticle gets less from base to tip, much like the damping properties change from strongly overdamped at the base to slightly underdamped near the tip. Note that the diameter is enlarged 10-fold relative to the length. *Bottom* True proportions of the Flagellum. TS: Location of the temperature sensor. [adapted after Dirks and Dürr (2011)]

The structure of the cuticle, i.e. the layered material of the external skeleton, suggests that this combination of features is caused by the balanced function of a stiff outer cone (of exocuticle) that is lined by a soft inner wedge (of endocuticle, see Figure 7). Owing to the different material properties of the inner and outer cuticle layers, the water-rich inner endocuticle supposedly acts like a damping system that prevents oscillations that would be caused by the stiff material alone. As a result, a bent antenna can snap back into its resting posture without over-shoot, a sign of over-critical damping. Indeed, experimental desiccation of the flagellum (a method known to reduce water content and, therefore, damping of the endocuticle) strongly changes the damping regime of the antenna, to under-critical damping (Dirks and Dürr 2011).

There Are Specialised Sensors for Pointing Direction, Contact Site, Bending, Vibration, and More

The antenna is the most complex sensory organ of an insect (Staudacher et al. 2005). It carries thousands of sensory hairs, the sensilla, that subserve the modalities of **smell** (olfaction: Tichy and Loftus 1983), **taste** (gustation), **temperature** (thermoreception: Cappe de Baillon 1936; Tichy and Loftus 1987; Tichy 2007), **humidity** (hygroreception: Tichy 1979, 1987; Tichy and Loftus 1990), **gravity** (graviception: Wendler 1965; Bässler 1971), and **touch**.

With regard to touch, there are at least four mechanosensory submodalities that contribute to touch perception in insects. Each one of them is encoded by different types of sensory hairs (sensilla) or modifications thereof (for review, see Staudacher et al. 2005). At least one of these sensory structures, the **hair fields**, must be considered as proprioceptors, as they encode the (actively controlled) antennal posture. On the other hand, there are also genuine exteroceptive sensory structures, the **tactile hairs**, that encode touch location along the antenna. As yet, there are also sensory structures that convey both proprioceptive and exteroceptive information: **Campaniform sensilla** and a **chordotonal organ** are thought to encode the bending and vibration of the flagellum and, therefore, postural changes. The energy required to bring about the postural change may be of external cause (e.g., wind) or self-generated during locomotion and/or active searching (e.g., self-induced contact with an obstacle). As a consequence, the distinction between exteroception and proprioception vanishes for these sensory structures for as long as there is no additional information about whether the flagellum had been moved or relocated by self-generated movement.

Pointing Direction (Hair Fields)

Hair fields are patches of sensory hairs that are located near the joints. Essentially, they serve as joint angle sensors in which individual hairs get deflected if it is pressed against the juxtaposing joint membrane or segment. The number of deflected hairs and the degree of deflection causes mechanosensory activity that encodes the joint angle and, possibly, joint angle velocity. The encoding properties of antennal hair fields have been nicely demonstrated for the cockroach antenna.

Hair fields at the antennal joints are sometimes referred to as *Böhm's bristles* or *Böhm's organ*. This is a relict from a time when sensory organs were named after the person who first described them. In this case, Böhm (1911) described tactile hairs on the antenna of the moth *Macroglossum stellatarum*, although not all of these hairs belong to what we call *hair field* today.

Hair fields come in two kinds: **hair plates** are patches and **hair rows** are rows of hairs. It is not known whether these subtypes differ functionally. In the stick insect *C. morosus* there are seven antennal hair fields (Figure 8), three on the scape (monitoring the HS-joint) and four on the pedicel (monitoring the SP-joint). For a detailed description and revision see Urvoy et al. (1984) and Krause et al. (2013). Ablation of all antennal hair fields severely impairs the inter-joint coordination of antennal movement and leads to an enlargement of the working-range. In particular, the hair plate on the dorsal side of the scape appears to be involved in the control of the upper turning point of antennal movement (Krause et al. 2013).

The properties of hair field afferents are unknown for the stick insect antenna, but it is not unlikely that their encoding properties are similar to those of cockroach hair plate afferents (Okada and Toh 2001). This is supported by a recent characterisation of descending interneurons, where some groups were shown to convey antennal mechanoreceptive information reminiscent of what has been described for the cockroach hair plates (Ache and Dürr 2013). Indeed, the terminals of hair field afferents arborize in close vicinity to the dendrites of two motion-sensitive

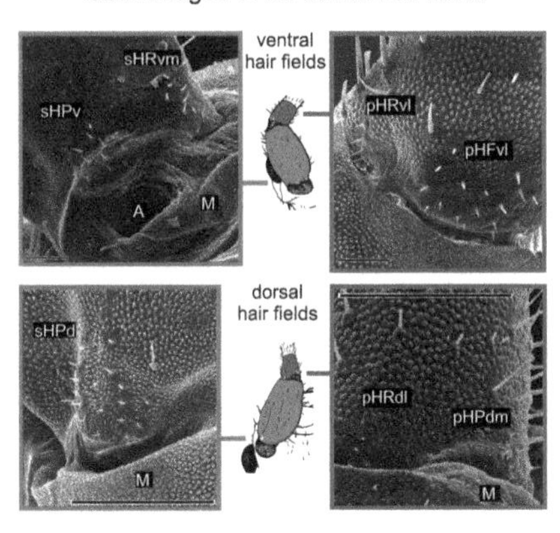

Figure 8 Hair fields near the antennal joints serve as external proprioceptors. In stick insects, there are seven hair fields, four on the scape (HS-joint) and three on the pedicel (SP-joint angle). The *lower left* drawing shows the location of these hair fields on the ventral (*right antenna*) and dorsal (*left antenna*) surfaces. Acronyms code for scape (s) or pedicel (p), hair plate (HP) or hair row (HR), and dorsal (d), ventral (v), medial (m) or lateral (l). For example, pHRvl is the ventrolateral hair row of the pedicel. Scale bars on the electron micrographs are 100 microns. [adapted from Krause et al. (2013).]

descending interneurons in the suboesophageal ganglion (Ache et al. 2015). Assuming that hair field afferents drive at least part of the observed activity in descending interneurons in stick insects, they may provide an important input to the coordinate transfer that is necessary for tactually induced reaching movements as observed by Schütz and Dürr (2011). In order to execute such aimed leg movements, the motor centres in the thorax that control leg movement need to be informed about the spatial coordinates of the antennal contact site, and antennal pointing direction encoded by descending interneurons may be part of that.

Contact Site (Tactile Hairs)

The flagellum of an adult stick insect carries thousands of sensilla or sensory hairs (Weide 1960), and several different sensillum types have been described (Slifer 1966), some of which can only be distinguished electronmicroscopically. According to Monteforti et al. (2002), the flagellum carries seven types of hair-shaped sensilla. Two types are innervated by a mechanosensory neuron, the other five are purely chemoreceptive sensilla. As a consequence, there are two kinds of mechanoreceptive tactile hairs on the flagellum that may encode touch location. Assuming that the occurrence of spikes in any given mechanoreceptor afferent would code for a certain distance along the flagellum, the identity of active afferents would encode the size, location and, potentially, the structure of the contacted

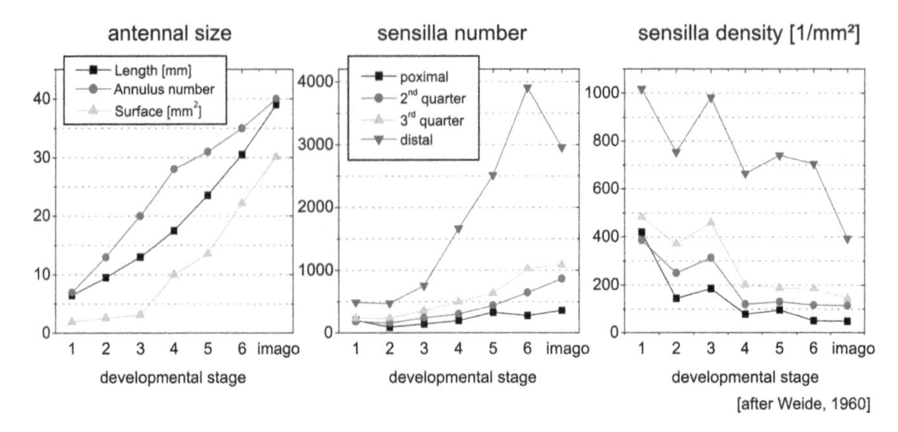

Figure 9 Development of antennal contact mechanoreception. *Left* Change of antennal size and annulus number between the seven developmental stages of *Carausius morosus*. *Middle* Change of sensilla number during development. *Right* Change of sensilla density during development. [graphed data from Weide (1960)]

surface. As yet, to date, nobody has recorded the afferent activity of a population of antennal touch receptors to verify this idea.

What is known is that sensilla density (irrespective of their morphological type) increases towards the tip. Moreover, sensilla density decreases during development, because sensilla number increases less than the surface area of the flagellum (Weide 1960; for a graphical presentation of his data, see Figure 9).

Bending (Campaniform Sensilla)

Campaniform sensilla are coffee-bean-shaped structures that are embedded within the cuticle. Essentially, they serve as strain sensors. Depending on their location in the exoskeleton they may encode different kind of information. Quite often, they are found at the base of a long body segment, an ideal location for sensing shear forces induced by bending of the segment: If the tip of the segment is deflected, the long segment acts like a lever that increases the torque acting on the base of the structure. On the insect antenna, a prominent site for campaniform sensilla is the distal pedicel. Given the long flagellum that is "held" by the pedicel, bending of the flagellum exerts a torque at the pedicel-flagellum junction that can be picked up by the strain sensors embedded in the cuticle of the distal pedicel (remember that this junction is not a true joint as it is not actuated and has little slack).

The ring of campaniform sensilla at the distal pedicel is sometimes called *Hicks' organ*, a similar relict as mentioned above for the hair fields: Hicks described this structure for the locust antenna (Hicks 1857). More than a century later, Heinzel and Gewecke (1979) recorded the activity of Hicks' organ afferents in the antenna of the locust *Locusta migratoria*. They found a strong directional selectivity of the

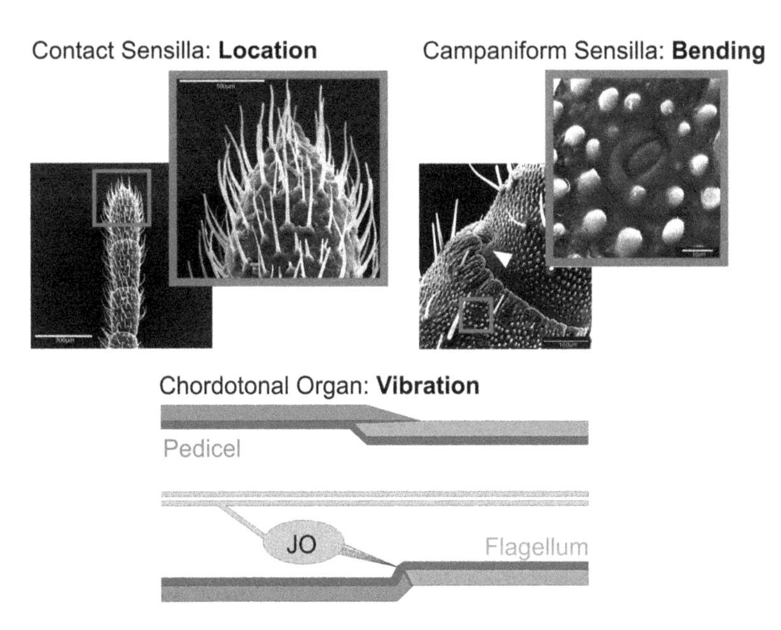

Figure 10 Three kinds of mechanoreceptors are involved in tactile sensing. Contact sensilla encode contact location, campaniform sensilla encode bending of the flagellum and a chordotonal organ encodes vibration of the flagellum. *Top left* The flagellum carries many sensilla, some of which are innervated by a mechanoreceptor (Scale bars: left 300 microns; right 100 microns). *Top right* At the pedicel-flagellum junction, the base of the flagellum 'sits' in a shaft formed by the distal pedicel, such that their cuticles overlap. The distal pedicel contains a number of campaniform sensilla, oval-shaped mechanoreceptors that are embedded in the cuticle (scale bars: left 100 microns; enlargement 10 microns). *Bottom* The pedicel contains a chordotonal organ which is often referred to as Johnston's organ (JO). In the Stick insect, it attaches at a single cuticular dent on the ventral side of the pedicel-flagellum junction (labelled by the arrowhead in top right image)

sensilla, depending on their location within the ring around the pedicel. This property makes Hicks' organ very suitable for encoding bending direction of the flagellum. In locusts this is thought to be important for measuring air flow direction and speed during flight.

In stick insects, campaniform sensilla can be found on the distal pedicel, too (see enlargement in Figure 10, *top right*), but there they are much less pronounced than in the locust. Moreover, campaniform sensilla are sparsely distributed along the flagellum of *C. morosus* (Urvoy et al. 1984).

Vibration (Johnston's Organ)

Finally, there is at least one **chordotonal organ** in the pedicel of an insect antenna. As in the cases mentioned above, a historical relict determines the naming of this organ until today: *Johnston's organ* originally referred to a conspicuous swelling at the base of the mosquito antenna (Johnston 1855) which was later described in more detail by Child (1894). Eggers (1924) then correctly placed Johnston's organ in the group of sensory organs containing so-called scolopidia, small rod-like structures that are characteristic of all chordotonal organs.

In several insect species, most importantly in mosquitoes, fruit flies and bees, Johnston's organ is known to be an auditory organ that can pick up vibrations generated by the wing beat (in mosquitoes and flies) or whole-body movements (in bees) of a conspecific (Staudacher et al. 2005). Therefore, they serve intraspecific communication by means of sound.

In the stick insect *C. morosus*, there is a large chordotonal organ in the ventral part of the pedicel. Other than the 'classical' Johnston's organ of the mosquito, it does not attach to the entire ring of the pedicel-flagellum junction. Instead, all scolopidial cells appear to attach at a single cuticular indentation. This dent is formed by the sclerotised ventral cuticle of the pedicel. It can be seen from the outside, as on the SEM of the pedicel-flagellum junction in Figure 10, *top right* (there, the ventral side is pointing upwards). The flagellar cuticle forms two sclerotised brace-like structures that surround the dent. Both the dent and the brace are embedded in soft endocuticle. The function of this structure is unknown but the focal attachment of the chordotonal organ at a single point (see schematic in Figure 10, *bottom*) suggests that it is not equally sensitive in all directions.

Internal References

Okada, J (2009). Cockroach antennae. *Scholarpedia* 4(10): 6842. http://www.scholarpedia.org/article/Cockroach_antennae. (see also pages 31–43 of this book).

Ritzmann, R and Zill, S N (2013). Neuroethology of insect walking. *Scholarpedia* 8(9): 30879. http://www.scholarpedia.org/article/Neuroethology_of_Insect_Walking.

External References

Ache, J M and Dürr, V (2013). Encoding of near-range spatial information by descending interneurons in the stick insect antennal mechanosensory pathway. *Journal of Neurophysiology* 110: 2099–2112. [http://dx.doi.org/10.1152/jn.00281.2013 doi:10.1152/jn.00281.2013].

Ache, J M; Haupt, S S and Dürr, V (2015). A direct descending pathway informing locomotor networks about tactile sensor movement. *The Journal of Neuroscience* 35: 4081–4091. [http://dx.doi.org/10.1523/JNEUROSCI.3350-14.2015 doi:10.1523/JNEUROSCI.3350-14.2015].

Bässler, U (1971). Zur Bedeutung der Antennen für die Wahrnehmung der Schwerkraftrichtung bei der Stabheuschrecke *Carausius morosus*. *Kybernetik* 9: 31–34. [http://dx.doi.org/10.1007/bf00272558 doi:10.1007/bf00272558].

Bässler, U (1983). Neural Basis of Elementary Behavior in Stick Insects. *Berlin Heidelberg New York:* Springer. doi:10.1007/978-3-642-68813-3.

Böhm, L (1911). Die antennalen Sinnesorgane der Lepidopteren. *Arbeiten aus dem Zoologischen Institut der Universität Wien und der Zoologischen Station in Triest* 19: 219–246.

Bässler, U and Büschges, A (1998). Pattern generation for stick insect walking movements - Multisensory control of a locomotor program. *Brain Research Reviews* 27: 65–88. [http://dx.doi.org/10.1016/s0165-0173(98)00006-x doi:10.1016/s0165-0173(98)00006-x].

Bläsing, B and Cruse, H (2004a). Mechanisms of stick insect locomotion in a gap crossing paradigm. *Journal of Comparative Physiology A* 190: 173–183. [http://dx.doi.org/10.1007/s00359-003-0482-3 doi:10.1007/s00359-003-0482-3].

Bläsing, B and Cruse, H (2004b). Stick insect locomotion in a complex environment: Climbing over large gaps. *The Journal of Experimental Biology* 207: 1273–1286. [http://dx.doi.org/10.1242/jeb.00888 doi:10.1242/jeb.00888].

Borchardt, E (1927). Beitrag zur heteromorphen Regeneration bei *Dixippus morosus*. *Arch Entw Mech* 110: 366–394. [http://dx.doi.org/10.1007/bf02252442 doi:10.1007/bf02252442].

Brecher, L (1924). Die Bedingungen für Fühlerfüße bei *Dixippus morosus*. *Arch Mikr Anat Entw Mech* 102: 549–572. [http://dx.doi.org/10.1007/bf02292960 doi:10.1007/bf02292960].

Büschges, A and Gruhn, M (2007). Mechanosensory feedback in walking: From joint control to locomotor patterns. *Advances in Insect Physiology* 34: 193–230. [http://dx.doi.org/10.1016/s0065-2806(07)34004-6 doi:10.1016/s0065-2806(07)34004-6].

Cappe de Baillon, P (1936). L'organe antennaire des Phasmes. *Bulletin biologique de la France et de la Belgique* 70: 1–35.

Casares, F and Mann, R S (1998). Control of antennal versus leg development in *Drosophila*. *Nature* 392: 723–726.

Child, C M (1894). Ein bisher wenig beachtetes antennales Sinnesorgan der Insekten mit besonderer Berücksichtigung der Culiciden und Chironomiden. *Zeitschrift für wissenschaftliche Zoologie* 58: 475–528.

Cruse, H (1990). What mechanisms coordinate leg movement in walking arthropods? *Trends in Neurosciences* 13: 15–21. [http://dx.doi.org/10.1016/0166-2236(90)90057-h doi:10.1016/0166-2236(90)90057-h].

Cuénot, L (1921). Regeneration de pattes á la place d'antennes sectionnées chez un Phasme. *Comptes Rendus de l'Académie des Sciences* 172: 949–952.

Dirks, J-H and Dürr, V (2011). Biomechanics of the stick insect antenna: Damping properties and structural correlates of the cuticle. *Journal of the Mechanical Behavior of Biomedical Materials* 4: 2031–2042. [http://dx.doi.org/10.1016/j.jmbbm.2011.07.002 doi:10.1016/j.jmbbm.2011.07.002].

Dürr, V (2001). Stereotypic leg searching-movements in the stick insect: Kinematic analysis, behavioural context and simulation. *The Journal of Experimental Biology* 204: 1589–1604.

Dürr, V and Ebeling, W (2005). The behavioural transition from straight to curve walking: Kinetics of leg movement parameters and the initiation of turning. *The Journal of Experimental Biology* 208: 2237–2252. [http://dx.doi.org/10.1242/jeb.01637 doi:10.1242/jeb.01637].

Dürr, V; König, Y and Kittmann, R (2001). The antennal motor system of the stick insect *Carausius morosus*: Anatomy and antennal movement pattern during walking. *Journal of Comparative Physiology A* 187: 131–144. [http://dx.doi.org/10.1007/s003590100183 doi:10.1007/s003590100183].

Dürr, V and Krause, A F (2001). The stick insect antenna as a biological paragon for an actively moved tactile probe for obstacle detection. In: K Berns and R Dillmann (Eds.), *Climbing and Walking Robots - From Biology to Industrial Applications. Proceedings of the 4th International Conference on Climbing and Walking Robots (CLAWAR 2001, Karlsruhe)* (pp 87–96). Bury St. Edmunds, London: Professional Engineering Publishing.

Eggers, F (1924). Zur Kenntnis der antennalen stiftführenden Sinnesorgane der Insekten. *Z Morph Ökol Tiere* 2: 259–349. [http://dx.doi.org/10.1007/bf01254870 doi:10.1007/bf01254870].

Graham, D (1985). Pattern and control of walking in insects. *Advances in Insect Physiology* 18: 31–140. [http://dx.doi.org/10.1016/s0065-2806(08)60039-9 doi:10.1016/s0065-2806(08)60039-9].

Heinzel, H-G and Gewecke, M (1979). Directional sensitivity of the antennal campaniform sensilla in locusts. *Naturwissenschaften* 66: 212–213. [http://dx.doi.org/10.1007/bf00366034 doi:10.1007/bf00366034].

Hicks, J B (1857). On a new organ in insects. *Journal of the Proceedings of the Linnean Society (London)* 1: 136–140. [http://dx.doi.org/10.1111/j.1096-3642.1856.tb00965.x doi:10.1111/j.1096-3642.1856.tb00965.x].

Imms, A D (1939). On the antennal musculature in insects and other arthropods. *Quarterly Journal of Microscopical Science* 81: 273–320.

Johnston, C (1855). Auditory apparatus of the "Culex" mosquito. *Quarterly Journal of Microscopical Science* 3: 97–102.

Jeziorski, L (1918). Der Thorax von *Dixippus morosus* (*Carausius*). *Zeitschrift für wissenschaftliche Zoologie* 117: 727–815.

Kittmann, R; Dean, J and Schmitz, J (1991). An atlas of the thoracic ganglia in the stick insect, *Carausius morosus*. *Philosophical Transactions of the Royal Society of London B* 331: 101–121. [http://dx.doi.org/10.1098/rstb.1991.0002 doi:10.1098/rstb.1991.0002].

Krause, A F and Dürr, V (2004). Tactile efficiency of insect antennae with two hinge joints. *Biological Cybernetics* 91: 168–181. [http://dx.doi.org/10.1007/s00422-004-0490-6 doi:10.1007/s00422-004-0490-6].

Krause, A F and Dürr, V (2012). Active tactile sampling by an insect in a step-climbing paradigm. *Frontiers in Behavioural Neuroscience* 6: 1–17. [http://dx.doi.org/10.3389/fnbeh.2012.00030 doi:10.3389/fnbeh.2012.00030].

Krause, A F; Winkler, A and Dürr, V (2013). Central drive and proprioceptive control of antennal movements in the walking stick insect. *Journal of Physiology Paris* 107: 116–129. [http://dx.doi.org/10.1016/j.jphysparis.2012.06.001 doi:10.1016/j.jphysparis.2012.06.001].

Marquardt, F (1939). Beiträge zur Anatomie der Muskulatur und der peripheren Nerven von "Carausius" ("Dixippus") "morosus". *Zoologische Jahrbücher. Abteilung für Anatomie und Ontogenie der Tiere* 66: 63–128.

Monteforti, G; Angeli, S; Petacchi, R and Minnocci, A (2002). Ultrastructural characterization of antennal sensilla and immunocytochemical localization of a chemosensory protein in *Carausius morosus* Brunner (Phasmida: Phasmatidae). *Arthropod Structure & Development* 30: 195–205. [http://dx.doi.org/10.1016/s1467-8039(01)00036-6 doi:10.1016/s1467-8039(01)00036-6].

Mujagic, S; Krause, A F and Dürr, V (2007). Slanted joint axes of the stick insect antenna: An adaptation to tactile acuity. *Naturwiss* 94: 313–318. [http://dx.doi.org/10.1007/s00114-006-0191-1 doi:10.1007/s00114-006-0191-1].

Okada, J and Toh, Y (2001). Peripheral representation of antennal orientation by the scapal hair plate of the cockroach *Periplaneta americana*. *The Journal of Experimental Biology* 204: 4301–4309.

Schmidt-Jensen, H O (1914). Homoetic regeneration of the antennae in a phasmid or walking-stick. *Smithsonian Report* 1914: 523–536.

Schmitz, J; Dean, J and Kittmann, R (1991). Central projections of leg sense organs in *Carausius morosus* (Insecta, Phasmida). *Zoomorphology* 111: 19–33. [http://dx.doi.org/10.1007/bf01632707 doi:10.1007/bf01632707].

Schütz, C and Dürr, V (2011). Active tactile exploration for adaptive locomotion in the stick insect. *Philosophical Transactions of the Royal Society of London B* 366: 2996–3005. [http://dx.doi.org/10.1098/rstb.2011.0126 doi:10.1098/rstb.2011.0126].

Slifer, E H (1966). Sense organs on the antennal flagellum of a walkingstick *Carausius morosus* Brunner (Phasmida). *Journal of Morphology* 120: 189–202. [http://dx.doi.org/10.1002/jmor.1051200205 doi:10.1002/jmor.1051200205].

Staudacher, E; Gebhardt, M J and Dürr, V (2005). Antennal movements and mechanoreception: Neurobiology of active tactile sensors. *Advances in Insect Physiology* 32: 49–205. [http://dx.doi.org/10.1016/s0065-2806(05)32002-9 doi:10.1016/s0065-2806(05)32002-9].

Theunissen, L M; Bekemeier, H H and Dürr, V (2015). Comparative whole-body kinematics of closely related insect species with different body morphology. *The Journal of Experimental Biology* 218: 340–352. [http://dx.doi.org/10.1242/jeb.114173 doi:10.1242/jeb.114173].

Tichy, H (1979). Hygro- and thermoreceptive triad in the antennal sensillum of the stick insect, *Carausius morosus*. *Journal of Comparative Physiology A* 132: 149–152. [http://dx.doi.org/10.1007/bf00610718 doi:10.1007/bf00610718].

Tichy, H (1987). Hygroreceptor identification and response characteristics in the stick insect *Carausius morosus*. *Journal of Comparative Physiology A* 160: 43–53. [http://dx.doi.org/10.1007/bf00613440 doi:10.1007/bf00613440].

Tichy, H (2007). Humidity-dependent cold cells on the antenna of the stick insect. *Journal of Neurophysiology* 97: 3851–3858. [http://dx.doi.org/10.1152/jn.00097.2007 doi:10.1152/jn.00097.2007].

Tichy, H and Loftus, R (1983). Relative excitability of antennal olfactory receptors in the stick insect, *Carausius morosus* L.: in search of a simple, concentration-independent odor-coding parameter. *Journal of Comparative Physiology A* 152: 459–473. [http://dx.doi.org/10.1007/bf00606436 doi:10.1007/bf00606436].

Tichy, H and Loftus, R (1987). Response characteristics of a cold receptor in the stick insect *Carausius morosus*. *Journal of Comparative Physiology A* 160: 33–42. [http://dx.doi.org/10.1007/bf00613439 doi:10.1007/bf00613439].

Tichy, H and Loftus, R (1990). Response of moist-air receptor on antenna of the stick insect *Carausius morosus* to step changes in temperature. *Journal of Comparative Physiology A* 166: 507–516. [http://dx.doi.org/10.1007/bf00192021 doi:10.1007/bf00192021].

Urvoy, J; Fudalewicz-Niemczyk, W; Petryszak, A and Rosciszewska, M (1984). Etude des organs sensoriels externes de l'antenne de Phasme *Carausius morosus* B. (Phasmatodea). *Acta Biologica Cracoviensia Series: Zoologia* XXVI: 57–68.

von Buddenbrock, W (1920). Der Rhythmus der Schreitbewegungen der Stabheuschrecke Dyxippus. *Biologisches Zentralblatt* 41: 41–48.

Weide, W (1960). Einige Bemerkungen über die antennalen Sensillen sowie über das Fühlerwachstum der Stabheuschrecke *Carausius* (*Dixippus*) *morosus* Br. (Insecta: Phasmida). *Wiss Z Univ Halle Math.-Nat.* IX/2: 247–250.

Wendler, G (1964). Laufen und Stehen der Stabheuschrecke: Sinnesborsten in den Beingelenken als Glieder von Regelkreisen. *Zeitschrift für vergleichende Physiologie* 48: 198–250. [http://dx.doi.org/10.1007/bf00297860 doi:10.1007/bf00297860].

Wendler, G (1965). Über den Anteil der Antennen an der Schwererezeption der Stabheuschrecke *Carausius morosus* Br. *Zeitschrift für vergleichende Physiologie* 51: 60–66.

Whiting, M F; Bradler, S and Maxwell, T (2003). Loss and recovery of wings in stick insects. *Nature* 421: 264–267. [http://dx.doi.org/10.1038/nature01313 doi:10.1038/nature01313].

A Spider's Tactile Hairs

Friedrich G. Barth

With more than 43,000 known extant species, some 500 new species currently described per year and an educated estimate of 150,000 or more existing species (Penney et al. 2012; Platnick 2013) **spiders** form a large group of animals, much larger than other groups traditionally receiving much more attention (e.g. birds with ca. 10,000 species, amphibians ca. 6,500 and mammals with ca. 5,500 species). They have existed at least since the carboniferous and their age of at least 320 mya is about twice that of mammals. From an evolutionary and ecological point of view spiders have been very successful to this very day. They have conquered practically all terrestrial biotopes and impress with their sheer number of individuals and their role as the main predators of insects. Undoubtedly, the spiders' success to a large extent is due to their rich repertoire of highly developed **sensory systems**. Among these the cuticular hairs are the most obvious ones. They also provide the input stage for a spider's tactile sense. As shown by studies in *Cupiennius salei* (Ctenidae), a Central American wandering spider (Barth 2002, 2008) (Figure 1), which has attained particular importance as an exemplary spider in a number of international laboratories, these tactile hairs are surprisingly well "designed" to serve their particular sensory purpose. This will be outlined in the following.

The Exoskeleton and Its Sensors

No other structure dominates the life of arthropods more than their **exoskeleton**. It not only protects them and provides the lever system needed for locomotion, but also carries a multitude of sensors at the interface between the organism and its environment. In addition to being the base for sense organs the exoskeleton is an auxiliary structure propagating the stimuli for a unique skeletal sense alien to us humans. Spiders in particular have an elaborate skeletal sense with up to some 3,500 sensory slits embedded into it. These represent sites of enhanced mechanical compliance responding to the slightest strain and deformation in the nanometer

F.G. Barth (✉)
Department of Neurobiology, University of Vienna, Vienna, Austria

T.J. Prescott et al. (eds.), *Scholarpedia of Touch*, Scholarpedia,
DOI 10.2991/978-94-6239-133-8_4

Figure 1 *Cupiennius salei* (Ctenidae), a much investigated wandering spider from Central America; adult female with a leg span of ca. 12 cm [*Photograph*: *FG Barth*]

range, going along with stresses in the exoskeleton due to muscular activity, hemolymph pressure or substrate vibrations (Barth 2002, 2012a, b; Fratzl and Barth 2009). Supported by many studies on the functional properties and behavioral role of its strain sensors we have to conclude that the spider is informed about the mechanical status of its exoskeleton in remarkable detail.

The sense of touch, however, is served by the other group of sensors, the hair sensilla. These are particularly numerous in hunting or wandering spiders which do not build webs to catch prey but roam around and much more rely on their tactile sense than do for instance orb weavers. Although counter to the intuition of most people (associating spiders with a web for prey capture) not only more than half of the known spider species, including *Cupiennius*, are such hunters, but there also seems to be an evolutionary tendency to abandon web building for the sake of free roaming hunting (Jackson 1986; Jocqué et al. 2013).

Mechano-Sensitive Hair Sensilla

Cupiennius belongs to the hairy type of wandering spiders with several 100 thousands of hairs on its exoskeleton (Figure 2). The vast majority of these hairs is innervated and serving a tactile function, the major exception being the non-innervated short yellowish plume hairs of adult spiders. There are also some chemoreceptive hairs among them, however, which are mostly found on the distal sections of walking legs and pedipalps and recognized by the steeper angle to the cuticular surface and often a slightly S-shaped hair shaft. Among the 21 sensory cells of each of these hairs there are two mechanosensitive ones, responding to the deflection of the hair shaft (Harris and Mill 1973, 1977). The most interesting known function of pedipalpal chemosensory hairs is their response in *Cupiennius* to a pheromone (S-dimethyl ester of citric acid) attached to the female dragline which stimulates the male to start his vibratory courtship (Tichy and Barth 1992; Barth 1997, 2002; Gingl 1998; Papke et al. 2000). Different from chemoreceptive hairs

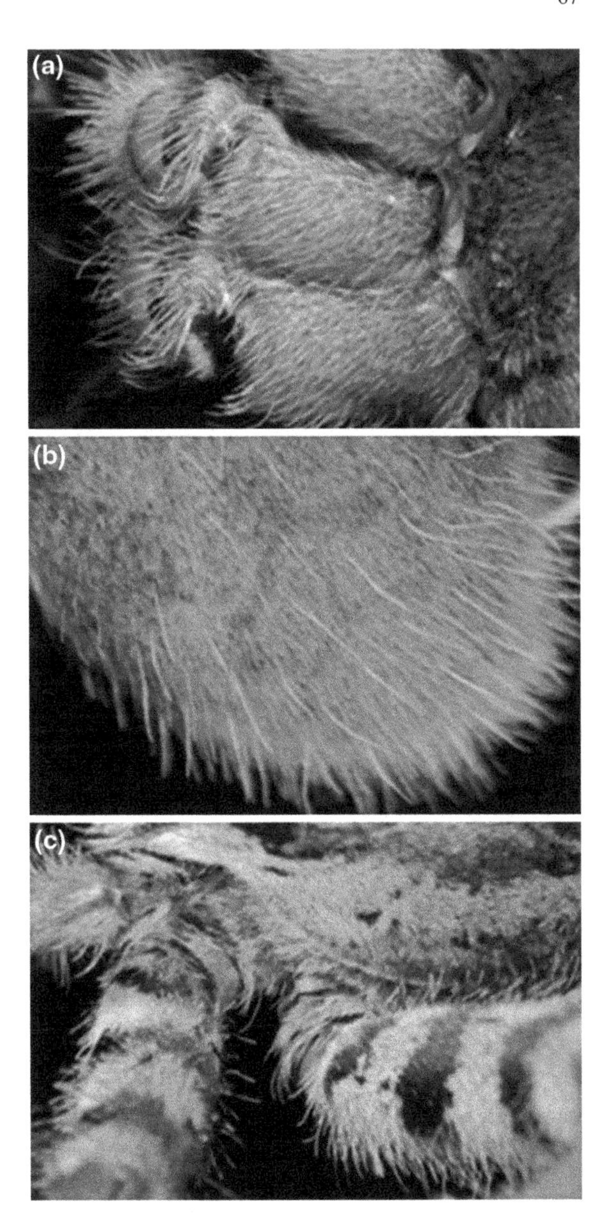

Figure 2 The exoskeleton of *Cupiennius salei* is densely covered by cuticular hairs, the majority of which is innervated. **a** Ventral view of first two segments of walking legs (coxa, trochanter, trochanter/femur joint), **b** dorsal side of opisthosoma, and **c** dorsolateral aspect of proximal leg segments and prosomal tergum [*Photographs*: *FG Barth*]

the majority of the other hairs are unimodal, responding to mechanical forces deflecting the hair shaft by direct contact (Foelix 1985). The density of these hairs is up to about 400/mm^2 (Eckweiler 1983; Friedrich 1998). This is remarkable, in particular when compared to our human skin, where the densely innervated glabrous skin of the fingertip has only 1.5 Meissner afferents per mm^2, to give an example (Johnson et al. 2000). A lot still has to be learned to understand whether the spider uses the high spatial resolution potentially provided by such a dense arrangement and what for (Barth 2015).

At a closer look a large **diversity of mechanosensitive hair sensilla** can be seen. And it is to a large extent this diversity of the mechanosensory hairs which enriches the small sensory space of spiders which lack long distance senses like our vision and hearing. The tactile sense is a close-range sense par excellence. Its hairs all form first-order lever systems and follow the same basic Bauplan. Their outer hair shaft always protrudes from a cuticular socket and they are innervated by 1–3 (most typically three) sensory cells with dendritic terminals close to the end of the inner lever arm. Tactile hairs all have a so-called tubular body in their dendritic ends indicating their mechanoreceptive function. The primary afferent fibers into the central nervous system show a somatotopic organization in the longitudinal tracts of the subesophageal ganglionic mass. There are also projections into the brain proper, the supraesophageal central nervous system (Figure 3).

Types of Morphology and Arrangement

The most obvious difference among sensory hairs is that between hairs adequately deflected by direct contact and those deflected by medium flow,that is by the frictional forces contained in the slightest flow of air (Figure 4). The latter type are the so-called **trichobothria**, of which *Cupiennius* has about 100 on each of its legs. The trichobothria are easy to identify: Due to their outstanding mechanical sensitivity they can easily be seen moving under the dissecting microscope following the ever present air movement. Quite differently, the deflection of tactile hairs needs much larger forces (see below). Trichobothria have been the subject of numerous studies, including physical-mathematical modeling, physiological and behavioral experiments, analyzing their dominant role in the spider's capture of flying insect prey (Humphrey and Barth 2007; Klopsch et al. 2012, 2013; Barth 2014). According to studies in several laboratories and including the analog sensors (filiform hairs) of insects they are working down to energy levels close to that contained in thermal noise and for that reason mainly have received a lot of attention from physicists and engineers as well. For the biologist the trichobothria are an impressive demonstration of the functional changes achieved by varying and properly adjusting just a few parameters contained in the Bauplan common to all hairs. Thus the spring stiffness contained in the articulation of the airflow sensing hair and resisting its deflection is smaller by up to about four powers of ten than in a typical tactile hair, where it measures between 10^{-8} and 10^{-9} Nm/rad (Dechant et al. 2001; McConney et al. 2008; Barth 2014). Similarly, low mass and the effect of hair length on the frequency tuning are used in a very "clever" way closely related to fluid mechanics in trichobothria. Details are found in reviews by Humphrey and Barth (2007) and Barth (2014).

A simple way to classify spider tactile hairs is the **number of sensory cells** attached to them. The exceptions to the rule of 3 bipolar cells mentioned above (Foelix and Chu-Wang 1973) are hairs supplied by only 1 sensory cell, as shown

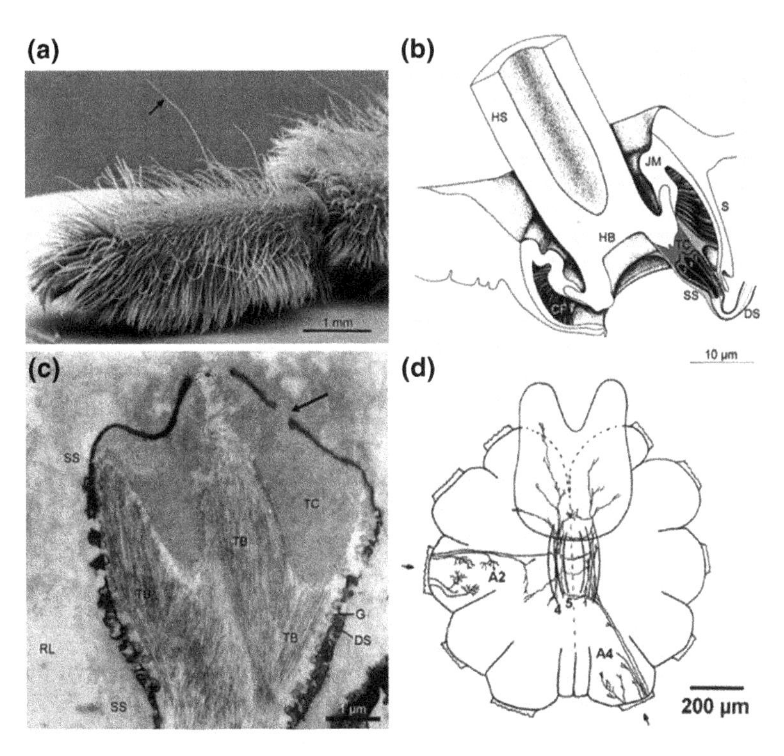

Figure 3 Structural details of the long tactile hairs sticking out of the carpet of other hairs. **a** Scanning electron micrograph of a walking leg tarsus. Arrow points to one of the long tactile hairs dorsally. **b** Basal part of tactile hair indicated in (**a**) according to electron micrographs. CF connecting fibrils, DS dendrite sheath, HB hair base, HS hair shaft, JM joint membrane, S socket, SS socket septum, TC terminal connecting material. **c** Dendrite terminals in longitudinal section (transmission electron micrograph). TB tubular bodies, RL receptor lymph. **d** Example of projections of tactile hairs on the femur of the second left (A2) and fourth right walking leg (A4) into the prosomal (thoracic) ganglionic mass; dorsal view. Note longitudinal sensory tracts 4 and 5 [(**a**)–(**c**) from (Barth et al. 2004), (**d**) from Ullrich (2000), unpublished]

for the short and stout hairs of the coxal hair plates (Seyfarth et al. 1990) and most likely applying to the two hair plates more recently found on the chelicerae. Another type of tactile hairs with one sensory cell only are the so called long smooth hairs on the coxae as described by Eckweiler et al. (1989) (Figure 5).

When going into more **morphological detail** (like shape of hair socket, hair length, shape of hair shaft), a further classification of tactile hairs turns out to be difficult and its value is still doubtful. Intriguingly, the distribution of hair types characterized by a certain combination of these parameters is conservative, possibly indicating a relation to particular stimulus patterns, being kind of templates of them (Friedrich 1998; Barth 2015). However, in order to turn this speculation into reliable data it still needs a lot of research. Hair sockets measure between 3 and 15 µm in diameter and vary in shape and their degree of openness, which strongly affects the mechanical directional characteristics of hair deflection.

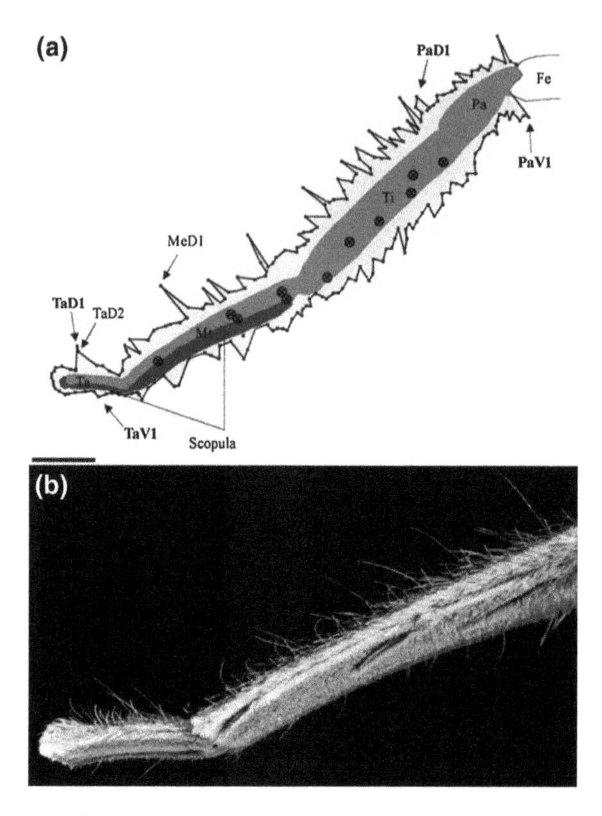

Figure 4 Distal part of *C. salei* walking leg. **a** The tactile space of the spider's walking leg (not considering active leg movement) determined by indicating the tips of the tactile hairs forming the outer borderline. Individually labeled hairs all belong to the particularly long type primarily exposed to tactile stimuli. Ta tarsus, Mt metatarsus, Ti tibia, Pa patella, Fe femur; encircled crosses mark the position of spines, scale bar 5 mm. **b** Tarsus and metatarsus, showing the most intensively studied long tactile hairs. Note presence of such tactile hairs on the ventral side as well [(**a**) adapted from (Friedrich 2001), (**b**) *photograph: FG Barth*]

Proprioreceptive versus Exteroreceptive

Tactile spider hairs represent both proprioreceptors and exteroreceptors, the difference being that the first ones are stimulated by self-generated stimuli whereas the latter respond to stimuli from an outside source. Proprioreceptive stimulation amply occurs during locomotion when hairs located at joints are deflected by joint movement or when a joint membrane rolls over a field of coxal hair plate sensilla. The long smooth hairs are stimulated when two neighboring coxae are approaching each other, most likely measuring the distance between them (Eckweiler et al. 1989; Seyfarth et al. 1990; Schaber and Barth 2014).

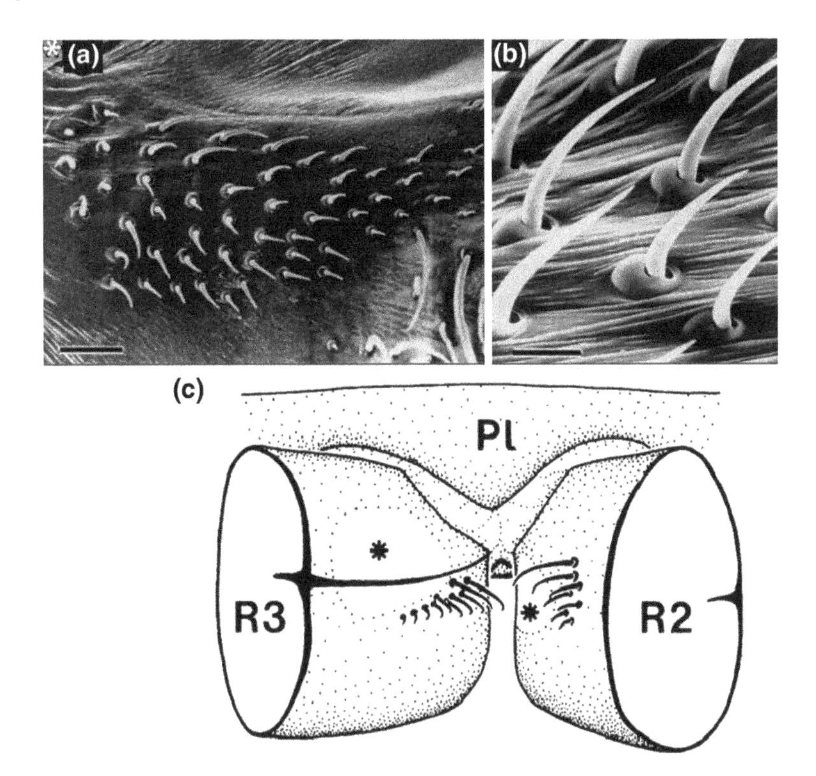

Figure 5 Tactile hairs with one sensory cell only. **a** and **b** hair plate on the anterior side of a walking leg coxa of *C. salei*; scale bars 100 and 20 μm, respectively [from Seyfarth et al. (1990)]. **c** Long smooth hairs (length from 40 to 1000 μm) on the coxae of two neighbouring walking legs (R2 and R3); Pl pleural sclerite of prosoma; asterisks indicate areas of smooth hairless cuticle opposing both groups of hairs [from Eckweiler et al. (1989)]

Well Matched Micromechanics

A number of electrophysiological experiments as well as computational studies of their **micromechanical properties** have revealed surprisingly "clever" details of tactile spider hair properties, reflecting the properties of the stimuli they have to cope with. Examples are the following.

Hair Shaft or Outer Lever Arm

1. The tactile hairs studied in detail are the long hairs (length up to ca. 3.2 mm) dorsally on the walking leg tarsus and metatarsus, forming the outer boundary of the spider's tactile space (Figure 4) (Albert et al. 2001). They are stimulated from above when the spider is moving around in small spaces and hitting

obstacles while wandering at night (Schmid 1997; Barth 2015). As already mentioned the stiffness of the hairs' articulation is larger by up to four powers of ten than that of the trichobothria. As a consequence the forces (which are in the μN range) needed to deflect these and similar tactile hairs on other body parts (Figure 2) not only deflect but at the same time **bend the hair shaft**. Accordingly, Young's modulus E and the second moment of inertia J along the bending hair shaft dominate the hair's mechanical behavior during stimulation. Inertial forces due to the mass of the hair shaft and highly relevant in case of the trichobothria may be neglected in case of the tactile hairs. An important consequence of the bending of the hair shaft, its cross sectional heterogeneity and the increase of J by roughly 4 powers of ten from the tip of the hair towards the hair base is that the point of contact of the stimulus moves closer and closer towards the hair base with increasing stimulus force from above. From this it follows that at the same time the effective lever arm decreases. The bending moment therefore increases more and more slowly until it saturates at ca. 4×10^{-9} Nm (Dechant et al. 2001). This in turn implies protection against breaking, an increased working range (as compared to a non-bending rod) and higher mechanical sensitivity for small deflections (forces ca. 5×10^{-4} N/°) and the initial phase of a stimulus than for large stimuli (forces ca. 1×10^{-5} N/°).

2. According to Finite Element Analysis the hair shaft may be considered a structure of equal maximum strength underlining its mechanical robustness. Axial stresses do not exceed ca. 3.2×10^5 N/m^2 (Dechant et al. 2001).
3. A seemingly perfect match between stimulus and hair micromechanics is in addition found in what was identified as a "second joint" within the socket. There the deflected hair shaft bends even before it contacts the socket (Barth et al. 2004).

Inner Lever Arm and Dendrites

As it seems the hair base and the inner lever arm of the tactile hairs (on the inner side of the axis of rotation) likewise are structures favoring a combination of high sensitivity and mechanical protection of the dendritic terminals from being overloaded and damaged (Barth 2004). The inner lever arm is only ca. 3.5 μm long, that is at least ca. 750 times shorter than the outer one. This implies a considerable scaling down of the hair tip movement and a corresponding scaling up of the force close to the dendrites. When the hair is deflected by 10°, which is close to the maximum deflection under biological conditions, the torque counteracting the stimulus measures ca. 10^{-8}–10^{-9} Nm and the displacement close to the dendrites ca.0.5 μm. At the physiologically determined threshold stimulus of 1° ("slow cell", see below) the latter value is only 0.05 μm (Albert et al. 2001).

Directionality

Apart from effects of the asymmetry of the socket structure (as clearly seen from above in some tactile spider hairs) there are directional dependencies of the torques resisting hair deflection before any contact with the socket. The most pronounced case so far known of such a mechanical directionality are the hairs at the joint between walking leg tibia and metatarsus. For the natural direction of stimulation the torsional restoring constant S is smaller by a factor of about 100 as compared to all other directions (Schaber and Barth 2014). Dechant et al. (2006) provide a quantitative mathematical description of any cuticular hair, based on the stiffness of its articulation in the preferred direction and transversal to it.

Physiological Responses of Sensory Cell

The tactile hairs sticking out of the carpet of hairs dorsally on the walking legs of *Cupiennius* have also been subjects of electrophysiological experiments (Albert et al. 2001). Like in *Ciniflo* (Harris and Mill 1977) for unknown reasons extracellular recordings were only possible from two of the three sensory cells, one of which is much larger than the others. Tactile hair sensory cells consistently show phasic response characteristics, answering to the dynamic stimulus phase, that is to the velocity of hair deflection. Using biologically relevant stimulus velocities the maximum response is seen with a latency of 1 to 2 ms only, a very short time typical of many mechanoreceptors and implying high temporal resolution. Adaptation time to static deflection varies; consistently, a **"slow" and a "fast" cell were found**, the latter being much less sensitive in terms of the deflection velocity threshold than the former (ca. 30°/s vs. <0.1°/s) (Figure 6). When exposed to ramp-and-hold stimuli the action potential frequency follows a simple power function $y(t) = a \times \mathrm{d} \times \mathrm{t}^{-\mathrm{k}}$ in both cases. Here y is the impulse rate, t is the time, a a constant representing the amplification, d stimulus amplitude and k a receptor constant describing how quickly the response to a maintained stimulus declines. In the present case k-values are around around 0.5, implying properties in between that of a pure displacement receiver ($k = 0$; response independent of frequency) and that of a velocity sensor ($k = 1$; differentiator of first order). Lowest threshold deflection angles are in the range of 1°. The characteristic curves (impulse rate vs. angular velocity of hair shaft deflection) are saturation curves for both the slow and the fast cell. However, the slow cell saturates at much lower velocities than the fast cell (Figure 6). The corresponding values for the tarsal tactile hairs (TaD1 and TaD2; Figure 4) are 250 and 650° s^{-1}, respectively. For the slow cell threshold deflection angles are independent of deflection velocity, whereas they highly depend on it for the fast cell for which the minima occur at ca. ≥100° s^{-1} (Albert et al. 2001). Importantly, the cells do not provide information about the exact time course of hair deflection but only about its presence and onset. Whereas the "fast" cell is working like a mere quasi-digital "event detector", the "slow" cell is suggested to serve the

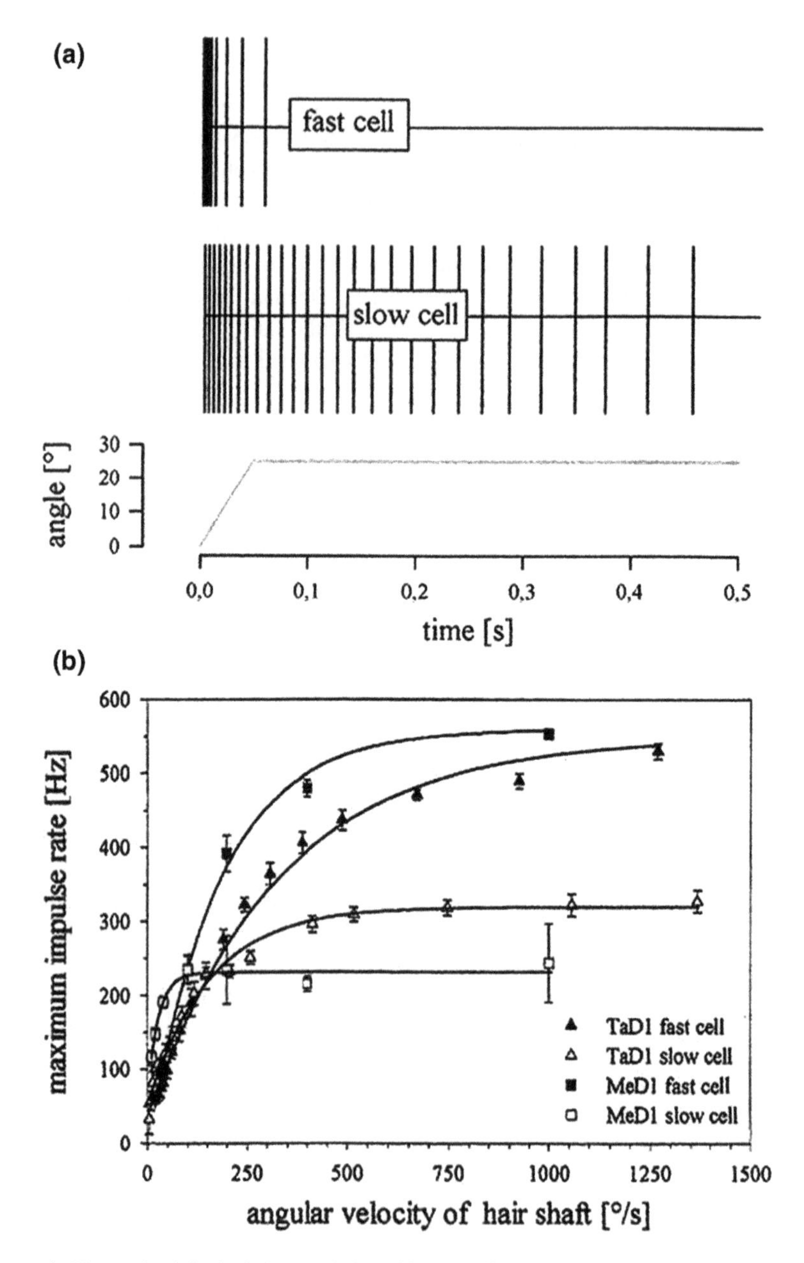

Figure 6 Electrophysiological characteristics of long tactile hairs. **a** Responses of the fast and the slow cell to a ramp and hold stimulus (hair deflection). **b** Characteristic curves of slow and fast cells of tarsal and metatarsal tactile hairs [from Albert et al. (2001)]

analysis of the texture and shape of surfaces actively scanned by the spider and to be well adapted to that function, similar to the SAI tactile units in the vertebrate glabrous skin (Albert et al. 2001; Johnson 2001; Barth 2015).

The reader interested in insect mechanoreceptive hairs is referred to a review article by Keil (1997) and to a paper by Theiß (1979) who found a spring stiffness *S* in fly macrochaetae very similar to that described here for the spider case. Of particular interest may be the remarkable lack of a standing transepithelial potential in spiders. Such a potential was found to be fundamental for the primary processes (transduction) in insect sensilla but is absent from spiders according to all knowledge currently available (Thurm and Wessel 1979; Thurm and Küppers 1980; Barth 2002).

Behavioral Roles

A lot still needs to be done to better understand why *Cupiennius* and other wandering spiders are so extremely well equipped with tactile hairs. A particular research requirement is the analysis of complex **stimulus patterns** like those seen during active tactile probing in the dark and courtship and copulation (Barth 2002). Upon simple stimulation like the experimental deflection of a few neighboring or individual tactile hairs the spider withdraws the stimulated body part or turns away from the source of stimulation. Often this behavior follows the deflection of a single hair only, like raising the opisthosoma ("abdomen"), lowering the prosoma ("cephalothorax") or withdrawing the spinnerets (Seyfarth and Pflüger 1984; Friedrich 1998; Barth 2015). Interestingly, the long tactile hairs forming the outposts of the sense of touch are not only found dorsally on the leg, but also ventrally on the tarsus (Figure 4) and among all tactile hairs so far studied they were the ones most easily deflected (smallest values for elastic restoring constant S; 5.9×10^{-11} Nm/° for distad deflection). They are assumed to provide sensory feedback information during locomotion.

A full neuroethological analysis of a tactile behavior comes from the work of E-A Seyfarth and his associates who examined "**body raising**" behavior in *Cupiennius*, which is also found in salticid and theraphosid and probably many other spiders (Figure 7) (see review by Seyfarth 2000). This behavior must be very helpful when the spiders walk around on structurally complex terrain and have to avoid and cope with all sorts of mechanical obstacles. Seyfarth and co-workers successfully traced the information flow from the sensory receptors to the central nervous system and the motor output. The deflection of long tactile hairs on the sternum or ventrally on the proximal leg first activates the coxa levator muscle of the stimulated leg. Thus a primary local response pulls the coxa against the prosoma while the distal leg joints are extended hydraulically. Coxal muscle activity in turn leads to the stimulation of internal joint receptors at the tergo-coxal joint, which triggers a second, the pluri-segmental response: The muscles of the remaining seven legs contract almost simultaneously and the legs are extended. By intracellular

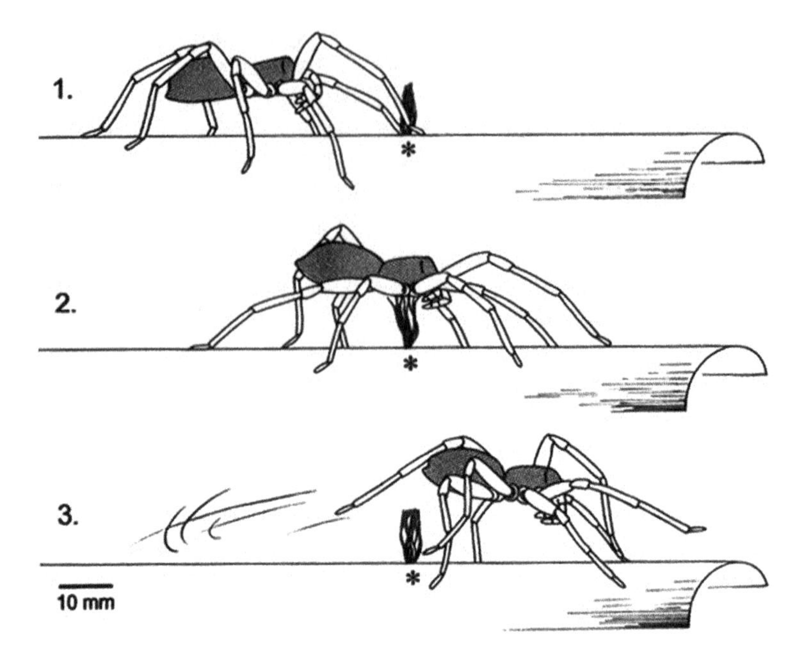

Figure 7 The body raising behavior of *C. salei*. *1* The spider approaches a 10 mm high wire obstacle from the left. *2* It raises its body as soon as tactile hairs ventrally on the proximal leg and the sternum touch it. *3* Having passed the obstacle the spider returns to its undisturbed walking position [from Seyfarth (2000)]

recording and staining the neuronal correlates of "body raising" behavior could be identified from the stimulated tactile hair and its sensory afferents to inter- and motor neurons (Figures 8 and 9).

Another, more recent analysis, asked how well adapted tactile hairs ventrally at the tibia-metatarsus joint might be to a proprioreceptive function, monitoring the **movement of the joint** (Schaber and Barth 2014). The results are much in favor of such an adaptedness.

1. Hairs opposing each other on the tibia and metatarsus side of the joint, respectively, deflect each other by some 30° (mean) during locomotion, with the microtrichs covering the hair shafts interlocked.
2. For both hairs the torque resisting deflection during locomotion (ca. 10^{-10} Nm rad^{-1}) is smaller by up to two orders of magnitude than that in the opposite direction.
3. Action potential frequencies recorded from individual tactile hair sensory cells follow the velocity of hair deflection within the naturally occurring range of step frequencies between 0.3 and 3 Hz. Obviously then, these joint hairs are well suited for a proprioreceptive function.

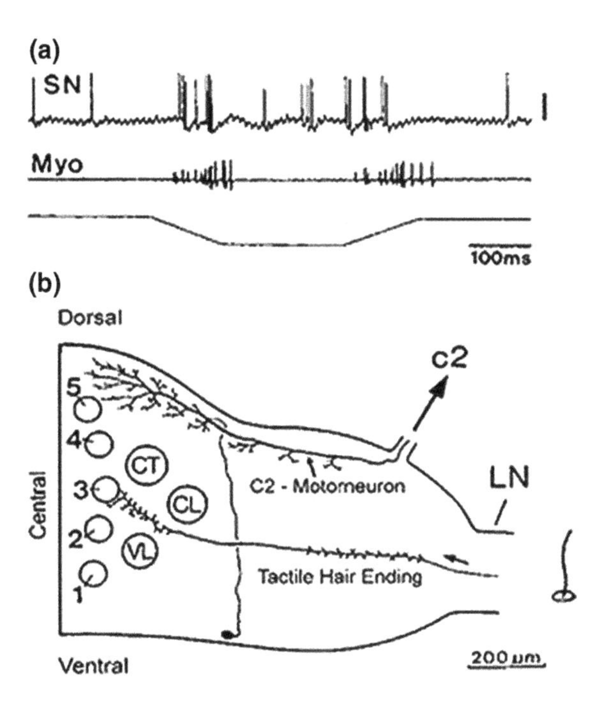

Figure 8 a Action potentials of tactile hair sensory cell (SN) and coxal muscle c2 (Myo) responding to the deflection of 10 tactile hairs on the hind leg coxa. **b** Cross sectioned neuromer of leg 4 showing the incoming tactile hair ending and the motor neuron supplying coxal muscle c2. There are local and plurisegmental interneurons which are not shown here. Numbers 1–5 indicate the centrally located sensory longitudinal tracts; CT, CL, and VL represent the central, centro-lateral and ventro-lateral tract, respectively [from Milde and Seyfarth (1988)]

There is much left for future research on the tactile sense of spiders. Do wandering spiders use tactile information on form, size and texture of an object's surface? When watching their smooth and elegant way of moving around on geometrically complex structures like bromeliads or other plants and watching the females producing and manipulating their egg sac one is strongly inclined to assume that they do use it. The same applies to the capture and handling of their prey. The next question then is: How are they doing it? The active tactile probing of their immediate environment (Schmid 1997) should be examined in more detail as should be the handling of prey and the sexual partner during precopulatory and copulatory communication (Barth 2002, 2015). Such knowledge would now allow us to predict the respective stimulation patterns of the sensory periphery and to hypothesize on the information theoretically available to the central nervous system. What the central nervous system is doing with it still largely is in the dark but the neuroethological analysis of the body raising behavior of *Cupiennius* (Seyfarth 2000) nicely shows what can be done and hoped for.

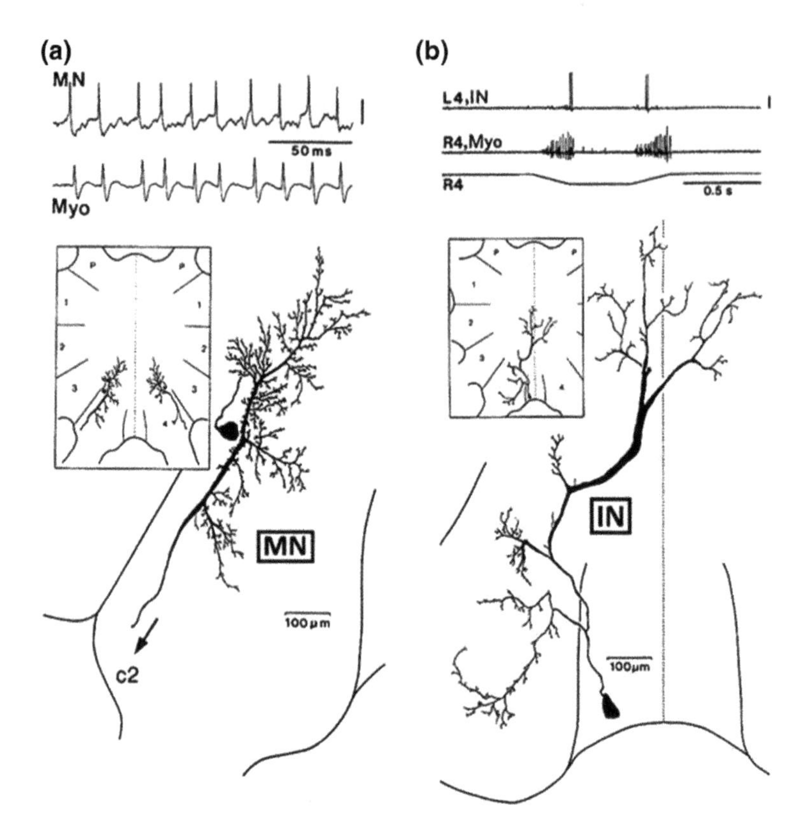

Figure 9 **a** Excitatory motor neurons innervating coxal muscle c2. Above: Motor neuron activity (MN) and the corresponding muscle response (Myo) following tactile hair stimulation. **b** Plurisegmental interneuron involved in body raising behavior of *C. salei*. Above: Stimulation of right hind leg (R4) induces reflex muscle activity in the same leg (R4, Myo) and also activity of an interneuron on the contralateral side (L4, IN) [from Milde and Seyfarth (1988)]

Internal References

Dürr, V (2014). Stick insect antennae. *Scholarpedia* 9(2): 6829. http://www.scholarpedia.org/article/Stick_Insect_Antennae. (see also pages 45–63 of this book).

Reep, R and Sarko, D K (2009). Tactile hair in manatees. *Scholarpedia* 4(4): 6831. http://www.scholarpedia.org/article/Tactile_hair_in_Manatees. (see also pages 141–148 of this book).

External References

Albert, J T; Friedrich, O C; Dechant, H-E and Barth, F G (2001). Arthropod touch reception: Spider hair sensilla as rapid touch detectors. *Journal of Comparative Physiology A* 187(4): 303–312. doi:10.1007/s003590100202.

Barth, F G (1997). Vibratory communication in spiders: Adaptation and compromise at many levels. In: M Lehrer Birkhäuser (Ed.), *Orientation and Communication in Arthropods* (pp. 247–272). Basel. ISBN: 9783764356934.

Barth, F G (2002). *A Spider's World: Senses and Behavior* (pp. 394). Berlin Heidelberg New York: Springer. ISBN: 9783540420460.

Barth, F G (2004). Spider mechanoreceptors. *Current Opinion in Neurobiology* 14(4): 415–422. doi:10.1016/j.conb.2004.07.005.

Barth, F G (2008). *Cupiennius* (Araneae, Ctenidae): Biology and sensory ecology of a model spider. In: A Weissenhofer (Ed.), *Natural and Cultural History of the Golfo Dulce Region, Costa Rica*, Vol. 88 (pp. 267–276). Stapfia.

Barth, F G (2012a). Spider strain detection. In: F G Barth, J A C Humphrey and M V Srinivasan (Eds.), *Frontiers in Sensing: From Biology to Engineering* (pp. 251–273). New York, USA and Wien, Austria: Springer. ISBN: 9783211997482.

Barth, F G (2012b). Arthropod strain sensors. In: B Bharat (Ed.), *Encyclopedia of Nanotechnology* (pp. 127–136). New York: Springer. ISBN: 9789048197521.

Barth, F G (2014). The slightest whiff of air: Airflow sensing in arthropods. In: H Bleckmann, J Mogdans and S L Coombs, *Flow Sensing in Air and Water: Behavioral, Neural and Engineering Principles of Operation* (pp. 169–196). Berlin Heidelberg New York: Springer. ISBN: 9783642414466.

Barth, F G (2015). A spider's sense of touch: What to do with myriads of tactile hairs? In: G von der Emde and E Warrant (Eds.), *The Ecology of Animal Senses. Matched Filters for Economical Sensing*. Berlin Heidelberg New York: Springer. IN PRESS.

Barth, F G; Nemeth, S S and Friedrich, O C (2004). Arthropod touch reception: Structure and mechanics of the basal part of a spider tactile hair. *Journal of Comparative Physiology A* 190 (7): 523–530. doi:10.1007/s00359-004-0497-4.

Dechant, H-E; Rammerstorfer, F G and Barth, F G (2001). Arthropod touch reception: Stimulus transformation and finite element model of spider tactile hairs. *Journal of Comparative Physiology A* 187(10): 851. doi:10.1007/s00359–001-0255-9.

Dechant, H-E; Hößl, B; Rammerstorfer, F G and Barth, F G (2006). Arthropod mechanoreceptive hairs: Modeling the directionality of the joint. *Journal of Comparative Physiology A* 192(12): 1271–1278. doi:10.1007/s00359-006-0155-0.

Eckweiler, W (1983). Topographie von Propriorezeptoren, Muskeln und Nerven im Patella-Tibia- und Metatarsus-Tarsus-Gelenk des Spinnenbeins, Diploma thesis. Fachbereich Biologie, University of Frankfurt am Main.

Eckweiler, W; Hammer, K and Seyfarth, E-A (1989). Long smooth hair sensilla on the spider leg coxa: Sensory physiology, central projection pattern, and proprioceptive function. *Zoomorphology* 109(2): 97–102. doi:10.1007/bf00312315.

Foelix, R and Chu-Wang, I-W (1973). The morphology of spider sensilla I. Mechanoreceptors. *Tissue and Cell* 5(3): 451–460. doi:10.1016/s0040-8166(73)80037-0.

Foelix, R F (1985). Mechano- and chemoreceptive sensilla. In: F G Barth (Ed.), *Neurobiology of Arachnids* (pp. 118–137). Berlin, Heidelberg, New York, Tokyo: Springer. ISBN:3-540-15303-9 and 0-387-15303-9.

Fratzl, P and Barth, F G (2009). Biomaterial systems for mechanosensing and actuation. *Nature* 462(7272): 442–448. doi:10.1038/nature08603.

Friedrich, O C (1998). Tasthaare bei Spinnen: Zur äußeren Morphologie, Biomechanik und Innervierung mechanoreceptiver Haarsensillen bei der Jagdspinne *Cupiennius salei* Keys. (Ctenidae). Diploma thesis. Faculty of Life Sciences, University of Vienna.

Friedrich, O C (2001). Zum Berührungssinn von Spinnen. Doctoral thesis. Faculty of Life Sciences, University of Vienna.

Gingl, E (1998). Nachweis der Rezeptoren für das Kontaktpheromon an der Seide der weiblichen Jagdspinne. Diploma thesis. Faculty of Life Sciences, University of Vienna.

Harris, D J and Mill, P J (1973). The ultrastructure of chemoreceptor sensilla in *Ciniflo* (Arachnida, Araneida). *Tissue and Cell* 5(4): 679–689. doi:10.1016/s0040-8166(73)80053-9.

Harris, D J and Mill, P J (1977). Observations on the leg receptors of *Ciniflo* (Araneidae, Dictynidae) I external mechanoreceptors. *Journal of Comparative Physiology* 119(1): 37–54. doi:10.1007/bf00655870.

Humphrey, J A C and Barth, F G (2007). Medium-flow sensing hairs: Biomechanics and models. In: J Casas and S J Simpson (Eds.), *Advances in Insect Physiology. Insect Mechanics and Control*, Vol. 34 (pp. 1–80). Elsevier Academic Press. ISBN: 9780123737144.

Jackson, R R (1986). Web building, predatory versatility and the evolution of the *Salticidae*. In: W Shear (Ed.), *Spiders: Webs, Behavior and Evolution* (pp. 232–268). Stanford: Stanford University Press. ISBN: 978-0804712033.

Jocqué, R; Alderweireldt, M and Dippenaar-Schoeman, A (2013). Biodiversity: An African perspective. In: D Penney (Ed.), *Spider Research in the 21st Century*. Manchester: Siri Scientific Press. ISBN: 9780957453012.

Johnson, K O (2001). The roles and functions of cutaneous mechanoreceptors. *Current Opinion in Neurobiology* 11(4): 455–461. doi:10.1016/s0959-4388(00)00234-8.

Johnson, K O; Yoshioka, T and Vega-Bermudez, F (2000). Tactile functions of mechanoreceptive afferents innervating the hand. *Journal of Clinical Neurophysiology* 17(6): 539–558. doi:10.1097/00004691-200011000-00002.

Keil, T A (1997). Functional morphology of insect mechanoreceptors. *Microscopy Research and Technique* 39(6): 506–531.

Klopsch, C; Kuhlmann, H C and Barth, F G (2012). Airflow elicits a spider's jump towards airborne prey. I. Airflow around a flying blowfly. *Journal of The Royal Society Interface* 9(75): 2591–2602. doi:10.1098/rsif.2012.0186.

Klopsch, C; Kuhlmann, H C and Barth, F G (2013). Airflow elicits a spider's jump towards airborne prey. II. Flow characteristics guiding behaviour. *Journal of The Royal Society Interface* 10(82): 20120820. doi:10.1098/rsif.2012.0820.

McConney, M E et al. (2008). Surface force spectroscopic point load measurements and viscoelastic modelling of the micromechanical properties of air flow sensitive hairs of a spider (*Cupiennius salei*). *Journal of The Royal Society Interface* 6(37): 681–694. doi:10.1098/rsif.2008.0463.

Milde, J J and Seyfarth, E-A (1988). Tactile hairs and leg reflexes in wandering spiders: Physiological and anatomical correlates of reflex activity in the leg ganglia. *Journal of Comparative Physiology A* 162(5): 623–631. doi:10.1007/bf01342637.

Papke, M; Schulz, S; Tichy, H; Gingl, E and Ehn, R (2000). Identification of a new sex pheromone from silk dragline of the tropical hunting spider *Cupiennius salei*. *Angewandte Chemie* 112: 4517–4518.

Penney, D; Dunlop, J A and Marusik, Y M (2012). Summary statistics for fossil spider species taxonomy. *Zoo Keys* 192: 1–13. doi:10.3897/zookeys.192.3093.

Platnick, N I (2013). In: P Merrett and H D Cameron (Eds.), *The World Spider Catalog 13.5*. American Museum of Natural History. http://research.amnh.org/iz/spiders/catalog.

Schaber, C F and Barth, F G (2014). Spider joint hairs: Adaptation to proprioreceptive stimulation. *Journal of Comparative Physiology A* 201(2): 235–248. doi:10.1007/s00359-014-0965-4.

Schmid, A (1997). A visually induced switch in mode of locomotion of a spider. *Zeitschrift für Naturforschung* 52c: 124–128.

Seyfarth, E-A (2000). Tactile body raising: Neuronal correlates of a 'simple' behavior in spiders. In: S Toft and N Scharff, *European Arachnology 2000. Proceedings of the 19th European College of Arachnology* (pp. 19–32). Aarhus: Aarhus University Press.

Seyfarth, E-A and Pflüger, H-J (1984). Proprioreceptor distribution and control of a muscle reflex in the tibia of spider legs. *Journal of Neurobiology* 15(5): 365–374. doi:10.1002/neu.480150506.

Seyfarth, E-A; Gnatzy, W and Hammer, K (1990). Coxal hair plates in spiders: Physiology, fine structure, and specific central projections. *Journal of Comparative Physiology A* 166(5): 633–642. doi:10.1007/bf00240013.

Theiß, J (1979). Mechanoreceptive bristles on the head of the blowfly: Mechanics and electrophysiology of the macrochaetae. *Journal of Comparative Physiology A* 132(1): 55–68. doi:10.1007/bf00617732.

Thurm, U and Wessel, G (1979). Metabolism-dependent transepithelial potential differences at epidermal receptors of arthropods. *Journal of Comparative Physiology A* 134(2): 119–130. doi:10.1007/bf00610470.

Thurm, U and Küppers, J (1980). Epithelial physiology of insect sensilla. In: M Locke and D S Smith (Eds.), *Insect Biology in the Future VBW80* (pp. 735–763). New York: Academic Press. ISBN: 9780124543409.

Tichy, H and Barth, F G (1992). Fine structure of olfactory sensilla in myriapods and arachnids. *Microscopy Research and Technique* 22(4): 372–391. doi:10.1002/jemt.1070220406.

Ullrich, N D (2000). Zum Berührungssinn von Spinnen: Feinstruktur und zentrale Projektion von Tasthaaren bei *Cupiennius salei* Keys. (Ctenidae). Diploma Thesis. Faculty of Life Sciences, University of Vienna.

Tactile Sensing in the Octopus

Frank Grasso and Martin Wells

All animals must have a sense of touch, if only to avoid damage. The problem is to discover how much information about their environment the animals can derive from this sense. *Octopus vulgaris* has proved to be a useful tool for the investigation of how much a soft-bodied invertebrate animal can derive from contacts with its environment because it learns rapidly in the laboratory. The limits of its ability to discriminate can be deduced from the results of training experiments (Figure 1).

Observations of the animal in the laboratory and in its natural habitat show that it is sensitive to contact all over its body surface. It uses its long flexible arms to investigate objects and to grope below and around surfaces as it moves about. It gathers much of its food in this way. The hundreds of suckers in particular are used to move across and to grasp objects that it encounters. Males use the tip of the third right arm, the so-called *hectocotylus*, to insert packets of sperm into the oviducts of females.

Sense Organs in the Skin

Graziadei and Gagne (1976) have reviewed light and EM studies of sensory cells in the suckers. There are enormous numbers of these, a single sucker of 3 mm diameter (from halfway along the arm of an animal of 250 g) will have several tens of thousands of receptors with single axons running inwards from the single-cell thick columnar epithelium of the rim of the sucker. Similar but more widely spaced receptors are found all over the skin surface. In addition, see below, there are receptors within the muscles. The eight arms together would carry an estimated 2.4×10^8 receptors (Graziadei 1971). Of these the overwhelming number are long narrow cells with a ciliated tip at the surface, assumed to be chemoreceptors. A second category is formed by rounded cells, near the base of the epithelium,

F. Grasso (✉)
Department of Psychology, Brooklyn College, New York, USA

M. Wells
Department of Zoology, University of Cambridge, Cambridge, UK

T.J. Prescott et al. (eds.), *Scholarpedia of Touch*, Scholarpedia,
DOI 10.2991/978-94-6239-133-8_5

Figure 1 A blind octopus grasps a test object [object P1, see Figure 5a] with its suckers prior to making a decision to take or reject it

presumed mechanoreceptors. Two further sorts of receptors are found, one is fusiform, like the presumed chemoreceptors, but terminating in a pore with stiff immobile cilia, again possibly a form of chemoreceptor. Finally there are flattened cells spread close to the base of the epithelium, which seem likely to be a further category of contact receptor (see Figure 2).

As well as receptors in the epithelium there are stellate cells among the muscles of the arms and suckers (see Figures 3 and 4). These are presumed to be proprioceptors, signalling muscular stretch. Degeneration experiments show that at least some of the primary receptors send axons direct to the ganglia of the nerve cord running down each arm. Others run to the acetabular ganglion at the base of each sucker. The majority of the axons from the presumed chemoreceptors run to encapsulated neurones in the connective tissue between the sucker epithelium and sucker musculature (Figure 2). This is the first stage in an integrative process that terminates in a mere 30,000 neurones running up into the brain from the tens of millions of receptors found in each arm. The tiny axons from the receptors in the suckers have proved too small for electrical recordings to be made. Rowell (1966) managed to record from nerves running into the arm nerve cords. Recordings were dominated by rapidly adapting neurones signalling mechanical distortion of the suckers. Surprisingly, no response to chemical stimulation (acetic acid, to which the animals were known to react behaviourally) was found. Recordings from the arm nerve cords showed rapid habituation to repeated peripheral stimulation with apparent *novelty units* responding when stimulation was changed. It appears that some learning occurs even at this level in the octopus nervous system.

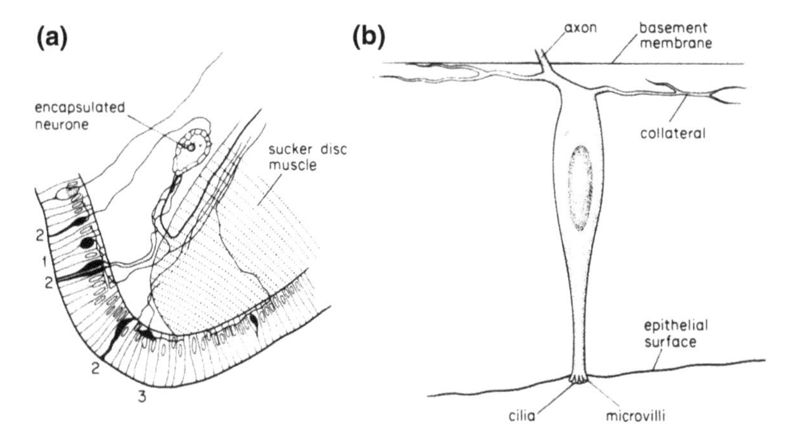

Figure 2 **a** The position of sense organs and an encapsulated neurone in the rim of a sucker. Sense organs of types 1 and 3 are presumed to be mechanoreceptors while the more numerous type 2, which reach to the surface and sometimes occur in clusters seem likely to be chemoreceptors. **b** shows more detail of a type 2 receptor. **a** After Graziadei (1964). **b** after Graziadei (1964)

Training Experiments

Octopus vulgaris adapts very quickly to life in aquaria, rapidly becoming tame enough to associate people with food. There is a large literature on the results of experiments in which the animals were shown pairs of cut-out shapes in a semi-random sequence, rewarded for attacking one with a fragment of fish and punished with a small electric shock for attacking the other. Under these conditions Octopus learns simple discriminations in ten or twenty trials. There is even some evidence that the animals can learn by observing the performance in training of other individuals (Fiorito and Scotto 1992). octopus' visual acuity is similar to ours. They cannot see colours, but can detect the plane of polarisation of light (for reviews of visual learning see Wells 1978; Budelmann 1996; Hanlon and Messenger 1996).

Tactile discrimination can be investigated by similar means. It is convenient to use animals with the optic nerves cut so that they cannot see the trainer or the objects to be distinguished. Such animals tend to sit on the sides of their aquaria with the arms extended along the walls around them. Objects can be presented by touching them gently against the side of an arm. The almost invariable response is for the arm to twist and grasp the object with three or four of the nearest suckers. If the octopus considers that the object is or might be edible the arm bends and passes it under the interbrachial web towards the mouth. If the object is of no interest or distasteful it is dropped or thrust away. These behaviours can be accentuated by giving a small piece of fish impaled on the end of a fine wire to the suckers if the 'positive' object is taken, and a small [12 V.dc] electric shock from a probe if the 'negative' is passed below the web. Because the trainer usually wants to present a succession of objects it is convenient to suspend each on a monofilament line so

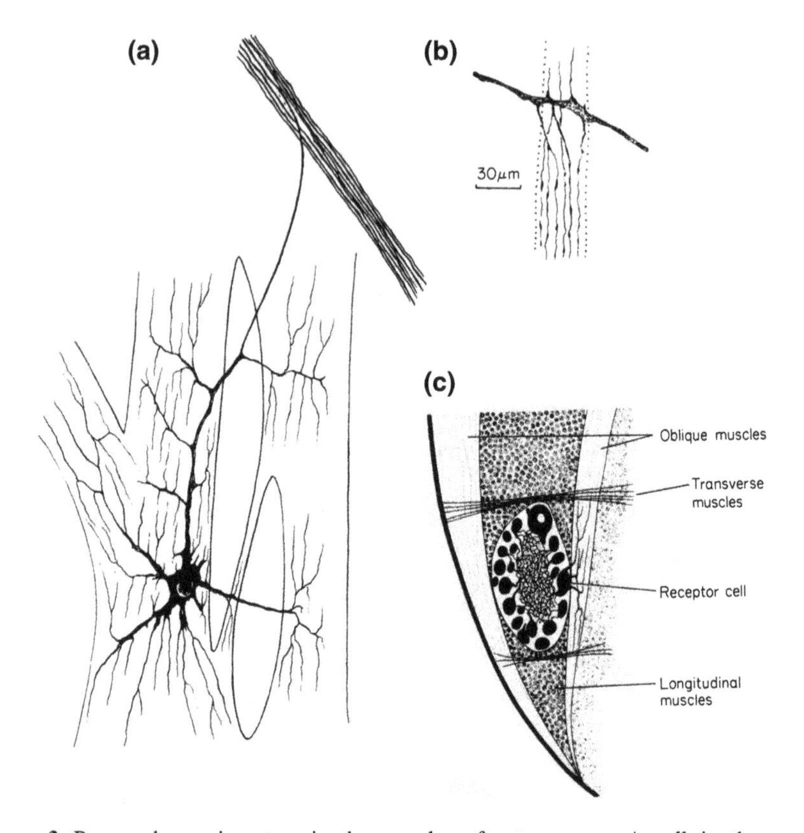

Figure 3 Presumed proprioceptors in the muscles of octopuses. **a** A cell in the mantle musculature of Eledone. **b** Fine branches from a similar cell in a small strip of muscle (*outlined by dots*). **c** shows a receptor cell in one of the intramuscular nerve cords in the arm of octopus with dendrites extending into an oblique muscle. The monopolar elements in the cord are assumed to be motor neurones. The relationship between the intramuscular cords and other structures in the arm are shown in this figure. **a** and **b** after Alexandrowicz (1960), **c** after Graziadei (1965a, b)

that it can be recovered. This is easy enough if the object has been rejected, more difficult if the object has been passed below the web. In the latter case it can generally be recovered by a quick jerk on the line; if the pull is developed slowly the animal will tighten its grip and refuse to let go. See Figures 5 and 6 show the results of a typical series of training experiments with Perspex cylinders differing in the number and orientation of right-angled grooves cut into their sides. Individual performances vary, but it is immediately obvious that the orientation of the grooves in relation to the shape of the cylinders is irrelevant; what matters is the proportion of each surface cut away. Grooves in the surface will distort the rims of the suckers in contact. One would expect maximum stimulation of mechanoreceptors buried deeply in the rims to occur when the surface of the object grasped is 50 % grooved. In contrast a smooth cylinder will only slightly distort the suckers in contact, the degree of distortion depending on the diameter of the cylinder. A very narrow cylinder should develop signals approaching those caused by an abrupt corner.

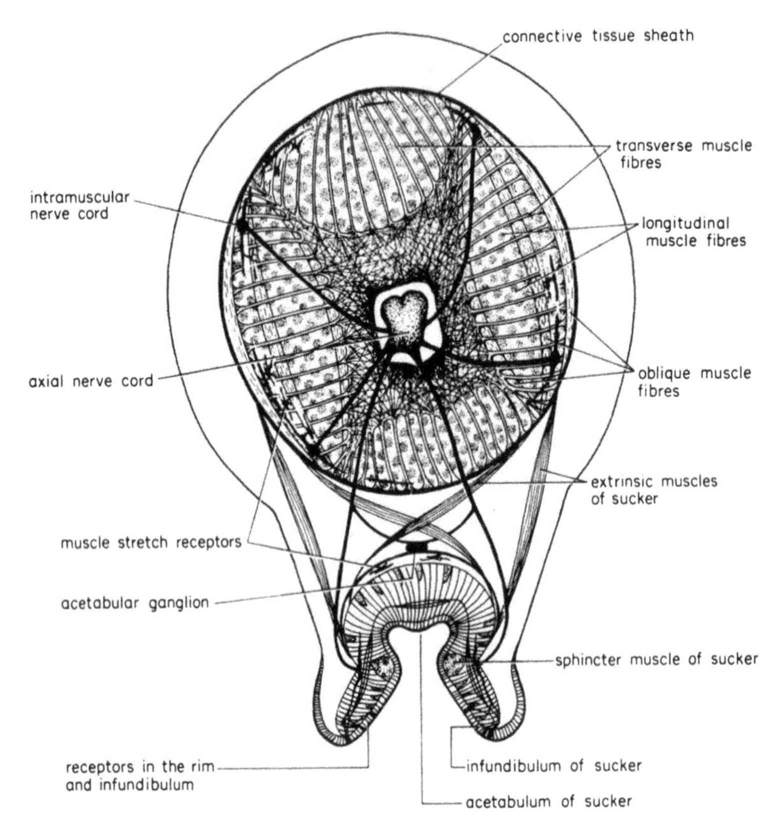

Figure 4 Nerves and muscles in the arm of an octopus. Nerves and muscles in the skin have been omitted outside the rim and infundibulum of a sucker. After Graziadei (1965a, b)

To see if these predictions were correct, a series of objects was developed as shown in Figure 7. If the distortion of mechanoreceptors in the rims of the suckers is all that the octopus recognises, a smooth sphere should yield the same signals as a flat surface. A cube is mostly flat surface but will differ from a sphere because any sucker that happens to bend around a corner will produce a distortion signal. Objects with the same ratio of flat surface to corner [such as objects PC and PSL in Figure 7] should appear identical to the octopus.

The hypothesis that Octopus does not take into account the relative position of suckers in contact appears to be upheld by the experimental results (Figures 8, 9 and 10). This means that the animal cannot be taking into account any input from proprioceptors in the muscles. This observation is confirmed by experiments in which attempts were made to train animals to distinguish between objects with the same surface texture, but different weight. The arms quite obviously react when a heavy object is grasped and the trainer relaxes the line on which the object is suspended; The muscles at once tighten up to take the weight (Figure 11) but the

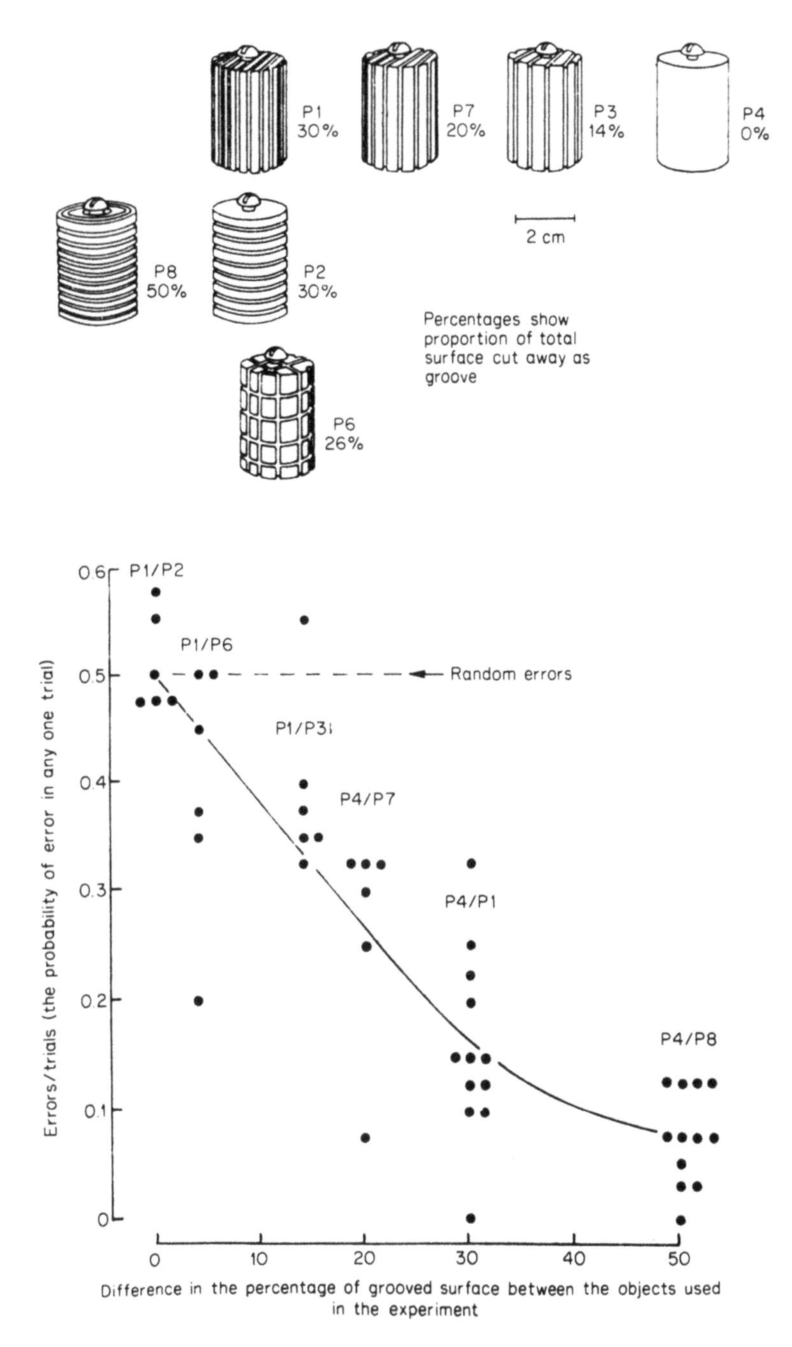

Figure 5 **a** Perspex cylinders used in tactile training experiments with octopus. In each case the percentage of the surface cut away in grooves is shown. The screw on top of each was for the attachment of a monofilament line, so that the trainer could recover the object after acceptance or rejection. **b** Each point shows the probability of error by one animal in 40 trials on days 8–12 of training at 8 trials (4+, 4−) per day. From Wells (1978)

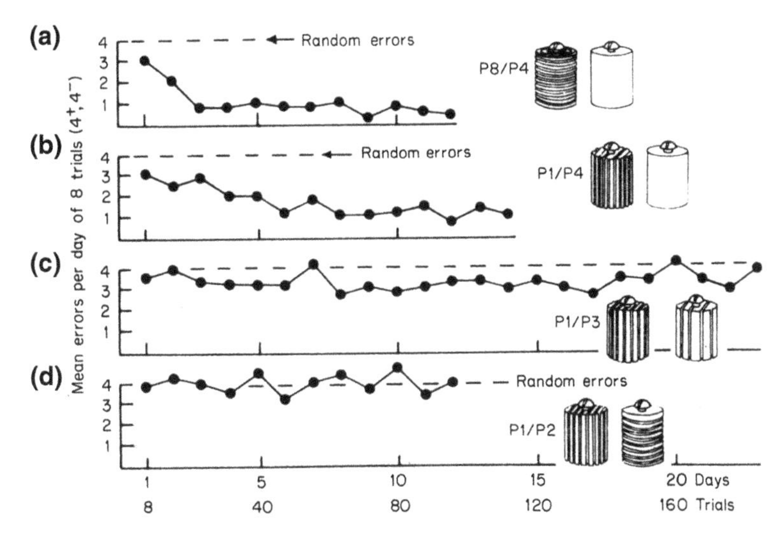

Figure 6 The progress of some typical training experiments using the objects shown in Figure 5a. From Wells and Wells (1957)

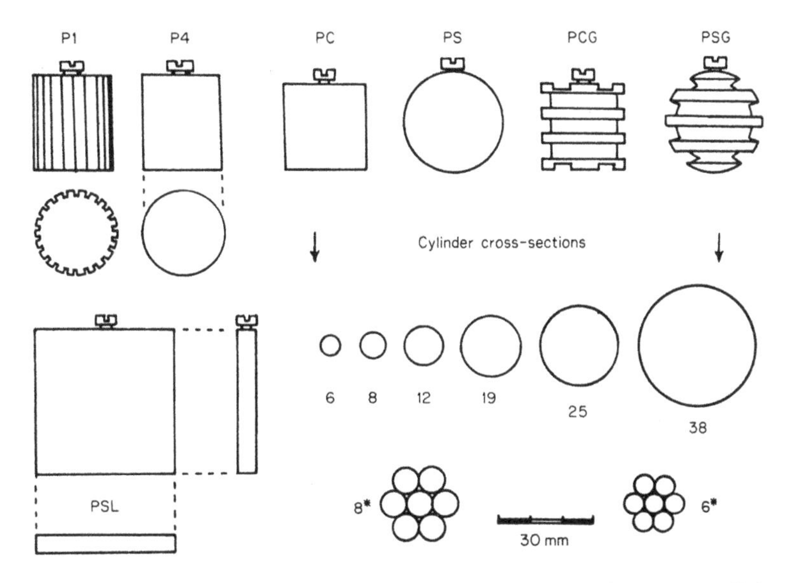

Figure 7 Further Perspex objects used in tactile training experiments with octopus. All drawn to the same scale. Simple and compound cylinders are shown in cross-section only, all were 50 mm long. Octopuses in the size range used in the experiments have suckers from 10 mm downwards in diameter as measured when applied to a flat glass surface. From Wells (1964a, b)

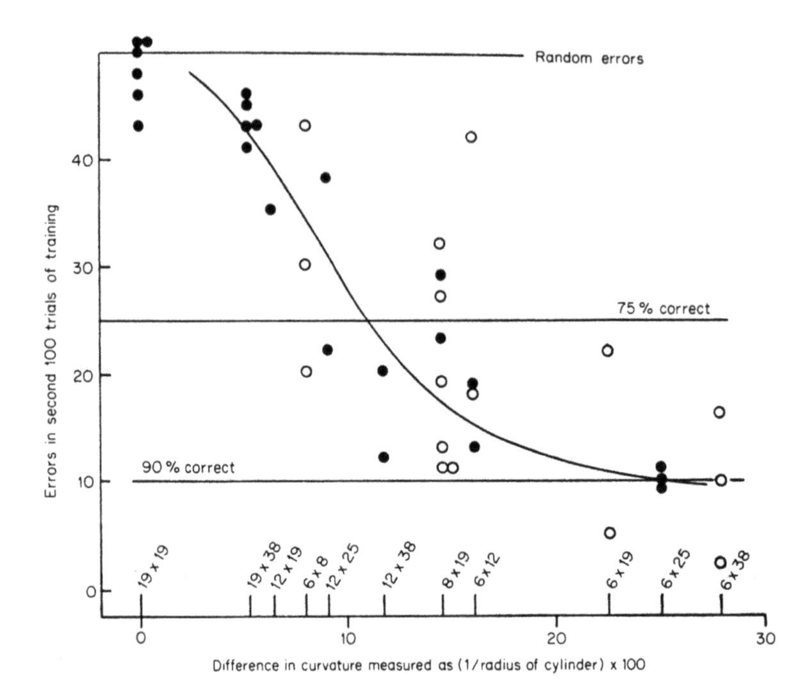

Figure 8 The proportion of errors made in training varies with the difference in surface curvature of the objects to be distinguished. Points plotted 0 and 0 show results with simple and compound cylinders (Figure 7) respectively, the stated diameter being that of the units not the overall diameter of the compound bundle, which appears to be unrecognised by the animals. From Wells (1964a)

animal never learns that such an object is in any way different from its *light* alternative. The overall conclusion from all of the experiments outlined above must be that octopuses cannot tell shapes by touch.

Taste by Touch

The presence of so many chemoreceptors in the rims of the suckers implies that it should be possible to train octopuses to make chemotactile discriminations. This is easily shown to be so. Objects with spongy surfaces soaked in solutions of presumed different taste can be used in training just as readily as *tasteless* objects. Using the crude human classification of tastes as Sweet (sucrose), Sour (hydrochloric acid) and Bitter (quinine sulphate) it emerges that the octopus chemotactile sense, for these substances in seawater is at least 100 times as sensitive as our own tongue for tastes of the same substances in distilled water (Wells 1963). The animal can detect small differences in concentration and differing dilutions of seawater. Acids appear to be distinguished on a basis of pH, with acetic acid apparently tasting more *acid* than hydrochloric or sulphuric acids at the same pH. (Wells et al. 1965).

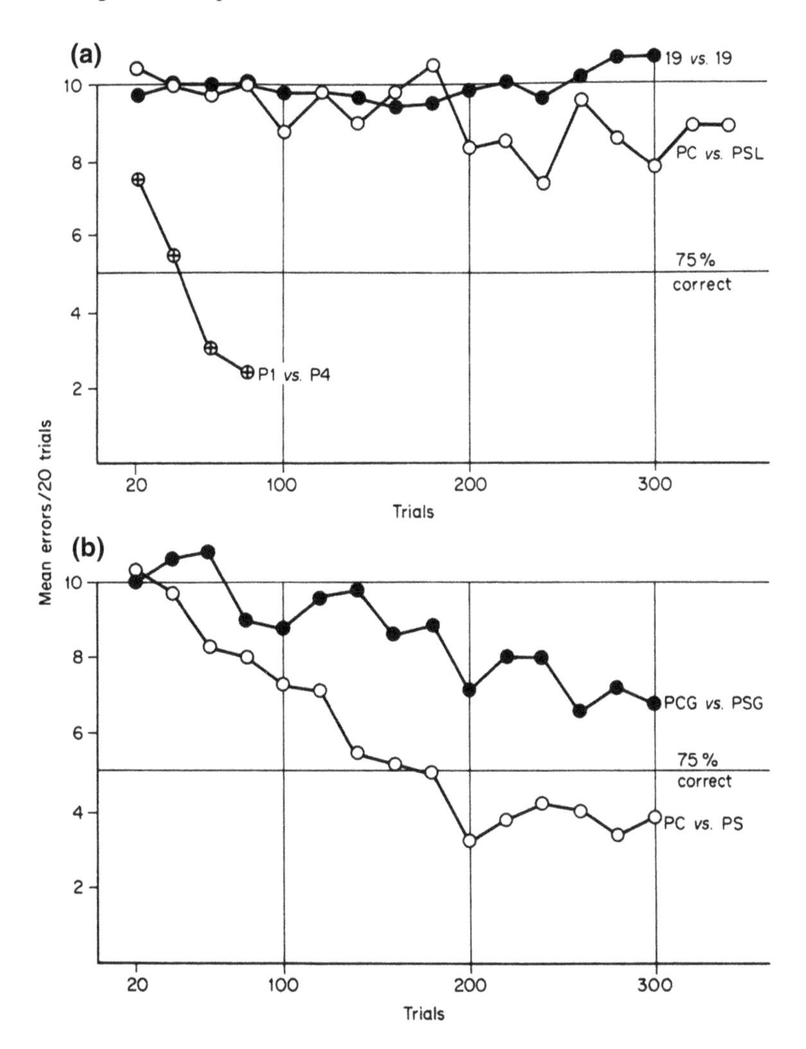

Figure 9 Plots show the course of four series of experiments using the Objects shown in Figure 7. P1 versus P4 is a rough-smooth discrimination, PC a cube and PS a sphere. PC and PSL are a cube and a slab with the same ratio of flat surface to length of right-angled corners. 19 versus 19 is an *impossible* discrimination between cylinders of the same diameter. From Wells (1964a)

Further investigation of the chemotactile sense is hampered by our own inability to imagine suitable experiments. The few results available, however, make it quite clear that one must be very wary of interpreting responses to naturally occurring objects such as shellfish—an apparent shape or surface textural discrimination is just as likely to be being made on a basis of taste. In this respect it is encouraging to find that some of the Perspex objects used in the experiments outlined above were in fact indistinguishable to the octopus; there was always the possibility that the *positive* object would become contaminated and identified on a basis of the taste of the fish given as rewards.

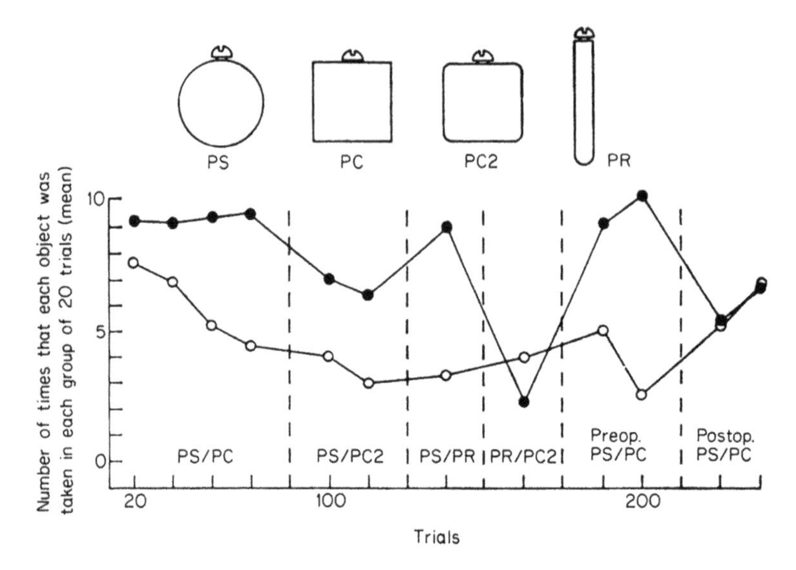

Figure 10 Summarises the results of the last 80 trials of training to distinguish between a Cube and a Sphere. Training was then continued with the cube replaced by a similar cube with rounded corners, evidently more difficult to distinguish from the sphere. Substitution of the narrow rod PR for PC2 restored performance while replacing the sphere with PR revealed that PR was in fact more cube-like than a cube. Training was continued with PS and PC. Finally the animals were operated upon, lesions being made in the Inferior Frontal system of the supraesophageal brain. After this the ability to discriminate broke down completely. From Wells (1964b)

Learning to Make Skilled Movements

Octopus arms are flexible and very strong. Each can extend and contract, twist and bend, independently of the others. Each sucker can move independently, grasp, extend and contract, and exert suction. In total a very remarkable apparatus potentially capable of performing very intricate tasks. But the animal never builds a shelter more elaborate than walls of debris pulled towards it to block the entrance to the hole in which it is living. It never assembles traps for its prey, makes or dismantles anything. Even Aristotle's contention that Octopus would carefully place a stone between the valves of shellfish to prevent their closing seems never to happen in the laboratory. It has sometimes been said that octopuses will learn to open screw-top jars to get at crabs seen inside, which would suggest learning to repeat a relatively complex series of movements, but again there is little more than anecdotal evidence of results that could be achieved by random pulling.

The problem is that muscular stretch receptors cannot signal position achieved independently of contractive force exerted. To do that an animal needs joints and further receptors that can signal the angles of joints. Vertebrates and Arthropods have joints and can repeat movements, hence honey combs and spiders webs, bird's

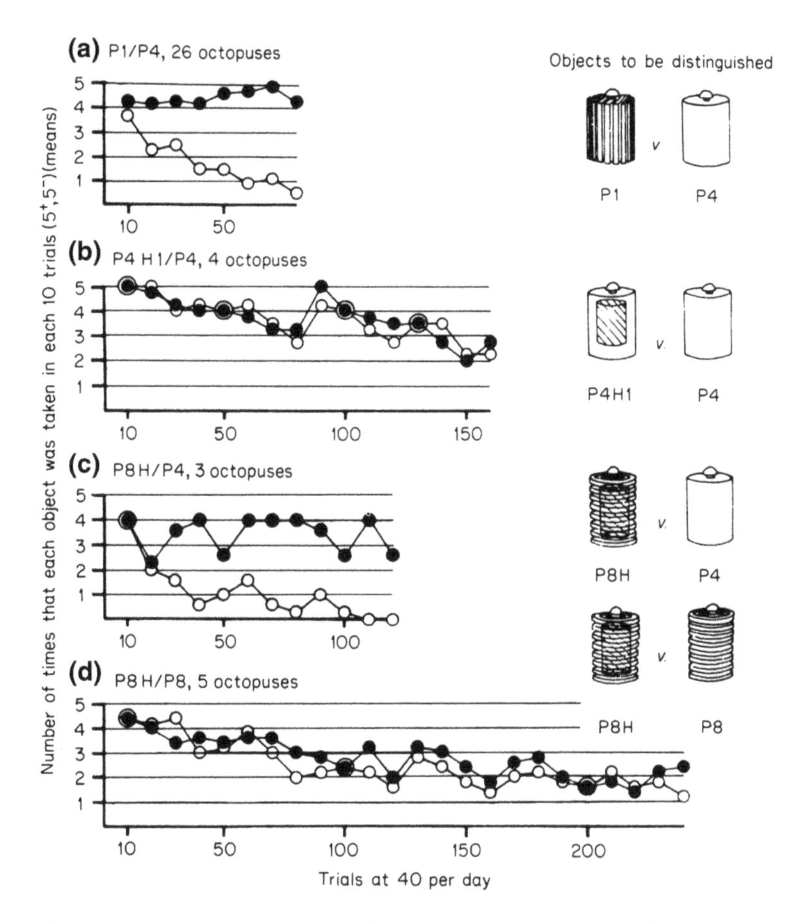

Figure 11 Attempts to teach octopuses to distinguish between objects of different weight failed unless there was also a difference in surface texture. *Light* objects weighed 5 g. *Heavy* objects drilled out and filled with lead weighted 25 (P8H) or 45 g (P4H1). From Wells (1961)

nests and bicycles. The natural world is sharply divided into two categories, with the Octopus clearly demonstrating that whatever a soft-bodied animal's intelligence and muscular equipment, there are things that it will never be able to do.

References

Alexandrowicz, J S (1960). A muscle receptor organ in Eledone cirrhosa. *Journal of the Marine Biological Association of the United Kingdom* 39: 419–431.

Budelmann, B U (1996). Active marine predators: The sensory world of cephalopods. *Marine and Freshwater Behaviour and Physiology* 27: 59.

Fiorito, G and Scotto, P (1992). Observational learning in *Octopus vulgaris*. *Science* 256: 545–547.

Graziadei, P (1964). Electron microscopy of some primary receptors in the sucker of *Octopus vulgaris*. *Zeitschrift fur Zellforschung und mikroskopische Anatomie* 64: 510–522.
Graziadei, P (1965a). Electron microscope observations of some peripheral synapses in the sensory pathway of the sucker of *Octopus vulgaris*. *Zeitschrift fur Zellforschung und mikroskopische Anatomie* 65: 363–379.
Graziadei, P (1965b). Muscle receptors in cephalopods. *Proceedings of the Royal Society of London B* 161: 392–402.
Graziadei, P (1971). The nervous system of the arms. In: J Z Young (Ed.), *The Anatomy of the Nervous System of Octopus vulgaris* (pp. 44–61). Oxford: Clarendon Press.
Graziadei, P P C and Gagne, H T (1976). Sensory innervation of the rim of the octopus sucker. *Journal of Morphology* 150: 639–680.
Hanlon, R T and Messenger, J B (1996). *Cephalopod Behaviour Cambridge*: Cambridge University Press.
Rowell, C H F (1966). Activity of interneurones in the arm of Octopus in response to tactile stimulation. *Journal of Experimental Biology* 44: 589–605.
Wells, M J (1961). Weight discrimination by Octopus. *Journal of Experimental Biology* 38: 127–133.
Wells, M J (1963). Taste by touch: Some experiments with Octopus. *Journal of Experimental Biology* 40: 187–193.
Wells, M J (1964a). Tactile discrimination of surface curvature and shape by the octopus. *Journal of Experimental Biology* 41: 433–445.
Wells, M J (1964b). Tactile discrimination of shape by Octopus. *Quarterly Journal of Experimental Psychology* 15(2): 156–162.
Wells, M J (1978). In: *Octopus: Physiology and Behaviour of an Advanced Invertebrate*, (p. 417). Chapman and Hall.
Wells, M J; Freeman, N H and Ashburner, M (1965). Some experiments of the chemotactile sense of octopuses. *Journal of Experimental Biology* 43: 553–563.
Wells, M J and Wells, J (1957). The function of the brain of Octopus in tactile discrimination. *Journal of Experimental Biology*, 34: 131–142.

Recommended Readings

Hanlon, R T and Messenger, J B (1996). *Cephalopod Behaviour Cambridge*: Cambridge University Press.
Wells, M J (1978). *Octopus: Physiology and Behaviour of an Advanced Invertebrate*. Chapman and Hall.
Young, J Z (1971). *The Anatomy of the Nervous System of Octopus vulgaris Oxford*: Clarendon Press.

Tactile Sensing in the Naked Mole Rat

Christine M. Crish, Samuel D. Crish and Christopher Comer

Tactile sensing in the naked mole rat refers to the ability of this naturally blind species to respond to, and localize, stimuli that deflect facial vibrissae, but also an array of somatic vibrissae.

Naked Mole-Rat Sensory World

Naked mole-rats (*Heterocephalus glaber*) are eusocial rodents that spend their entire lives in extensive subterranean burrows where visual and auditory cues are poor. As such, they have reduced their dependence on these systems; they exhibit microphthalmia and have small external ears (Figure 1; Hetling et al. 2005). Form vision is essentially lost, with related atrophy of brain structures dedicated to processing vision (Catania and Remple 2002; Crish et al. 2006a). However, structures mediating circadian rhythms in the naked mole-rat brain appear relatively normal (Crish et al. 2006a), the retina does respond to light (Hetling et al. 2005), and they can be entrained to light cues (Riccio and Goldman 2000a, b). This is a similar degree of highly limited visual processing to that present in other subterranean mammals, such as the blind mole rat (Spalax ehrenbergi; Cooper et al. 1993)[1]. Naked mole-rats also have poor high frequency hearing and an inability to localize sound effectively (Heffner and Heffner 1993). Little is known about their

[1]Naked mole-rats are Bathyergid rodents, a family of more than a dozen species found in sub-Saharan Africa. Blind mole rats are taxonomically quite distinct. They are in the Family Spalicidae (notice lack of hyphen for rodents in this taxon) and inhabit the eastern Mediterranean region.

C.M. Crish (✉) · S.D. Crish
Northeast Ohio Medical University, Rootstown, OH, USA

C. Comer
Division of Biological Sciences, University of Montana, Missoula, Montana, USA

T.J. Prescott et al. (eds.), *Scholarpedia of Touch*, Scholarpedia,
DOI 10.2991/978-94-6239-133-8_6

Figure 1 Photograph of two adult naked mole-rats. Note the small eyes, reduced auditory pinnae, and prominent sensory body hairs that line the body and tail

chemical senses. They display what appears to be scent marking, but unlike other rodents, they have an underdeveloped vomeronasal organ necessary for pheromone sensation (Smith et al. 2007).

Somatosensation

Like many other underground mammals such as the star-nosed mole, naked mole-rats have a highly derived somatosensory system. This system consists of increased capabilities compared with "standard" rodent features, addition of new features, and loss of functions (such as certain types of pain).

Skin

As their name indicates, naked mole rats lack fur that characterizes nearly all mammals except for humans, cetaceans, and manatees. They are not completely devoid of dermal appendages however. In addition to a dense array of facial vibrissae characteristic of rodents, they have a sparse arrangement of stiff sensory hairs (body vibrissae) on the rest of their head and along the entire body, including the tail. These "vibrissae" lack the blood sinus characteristic of true vibrissae but are heavily innervated. Thus they resemble guard hairs that are found interspersed among the fur of many mammals. The mechanism by which naked mole-rats have lost insulating fur but retained these dermal specializations is unknown.

The body vibrissae are not randomly arranged on the body but occur in a more or less regular array of about 40–50 receptors on each side of the body (Crish et al. 2003). In this sense, they share some general organizational similarities with arrays of somatic receptors in a few other mammals including manatees and rock hyraxes (Reep et al. 1998, 2007).

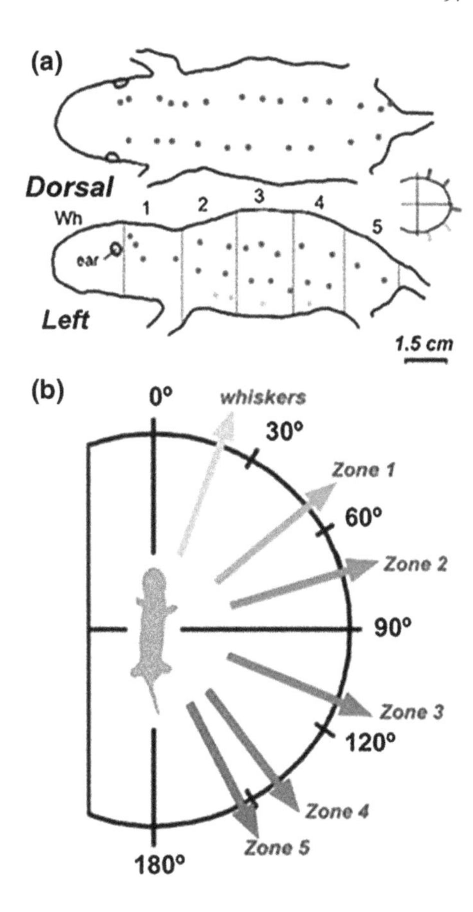

Figure 2 *Panel a.* Schematic depicting distribution of sensory body hairs on the dorsal and lateral surfaces of a naked mole-rat. Tactile stimulation of hairs in the numbered rostral-caudal zones shown on the schematic produce a topographic turning response that orients the mole-rat's snout to the locations depicted by the *arrows* in *Panel b.* For example, stimulation of the extreme caudal hairs in zone 5 produce the greatest magnitude turns. Figure adapted from Crish et al. (2003). Copyright 2003 with permission from Karger Publishers

These mole-rat body vibrissae are not vestigial remnants left after evolutionary loss of fur. These receptors mediate a well-organized topographic orienting behavior (Figure 2; Crish et al. 2003). This indicates that spatial information from these hairs is processed centrally in a distinct, systematic way. This non-visual orienting behavior has been shown by lesion studies to be mediated, at least in part, by the superior colliculus (Crish et al. 2006b). This midbrain area is well known for its involvement in visual orientation and multisensory integration in mammals. Mole-rats seem to provide an intriguing case where the midbrain tectum has reduced or lost its circuitry for visual orientation but retained components that process non-visual spatial information in parallel to the more typical visual circuitry.

The skin between hairs is also unusual when compared to other mammals. It is heavily innervated by a-delta fibers which transmit certain types of pain and temperature information to the central nervous system. C-fibers, which carry other types of painful stimuli, are absent, and the CNS shows a lack of CGRP-positive cells in early somatosensory processing areas (Park et al. 2008). Because of this absence naked mole-rats lack the ability to sensitize to chemicals such as capsaicin and are relatively unperturbed by acid and formalin challenge (Park et al. 2008).

Naked-mole rat burrow systems present a very atypical mammalian environment. The temperature in mole-rat burrows remains fairly constant and the concentration of carbon dioxide is unusually high. Not surprisingly, naked mole-rats are poikilothermic (unable to physiologically regulate their own body temperature) and huddle together for warmth as well as move to warmer areas when their core temperature is low. Additionally, there is speculation that the loss of CGRP-positive components of the pain system is related to the fact that these components may play a role in signaling pain toward high CO_2 concentrations (Park et al. 2008).

Teeth

As evident in Figure 1, naked mole-rats have prominent upper and lower incisors that are used for a range of tasks that include digging through harsh substrates, harvesting food, manipulating objects, and delicately carrying pups. The mandibular symphysis in the naked mole-rat jaw is unusually flexible, enabling them to move their contralateral lower incisors independently. Both upper and lower incisors are also highly sensitive to tactile stimulation, with each incisor having its own receptive field in primary somatosensory cortex (Catania and Remple 2002; Henry et al. 2006). In fact, these 4 upper and lower teeth have a hypertrophied representation in the brain—occupying 10 % of the entire neocortex (Catania and Remple 2002).

Brain Structures Involved in Sensation and Perception

Naked mole-rats have a large primary somatosensory cortex (S1) that significantly extends into the extreme caudal-medial cortices that are usually dominated by primary visual cortex (V1) and other visual areas. Somatosensory cortex accounts for 47 % of all sensory-related cortices in the naked mole-rat, contrasting with 27 % of cortical sensory tissue in other rodents with well-developed somatosensory systems such as the rat (Figure 3; Catania and Remple 2002). The incisors occupy approximately 1/3 of naked mole-rat S1 (Catania and Remple 2002), suggesting a specialized role for teeth in the uniquely enhanced somatosensory system of this animal.

Naked mole-rat subcortex also has some striking features. Areas responsible for visual processing such as the superficial superior colliculus, lateral geniculate nucleus, and central zone of the cerebellum are greatly reduced (Crish et al. 2006a, b; Marzban et al. 2011; Sarko et al. 2013). It is unknown whether these areas do not form normally or develop and then degenerate; recent investigations indicate that the superficial SC and LGN are reduced at birth (Dengler-Crish et al. 2013).

Like cortex, areas of the brainstem and cerebellum responsible for processing somatosensation are expanded (Henry et al. 2008; Marzban et al. 2011; Sarko et al. 2013). The deep superior colliculus, which contains visual, auditory, and

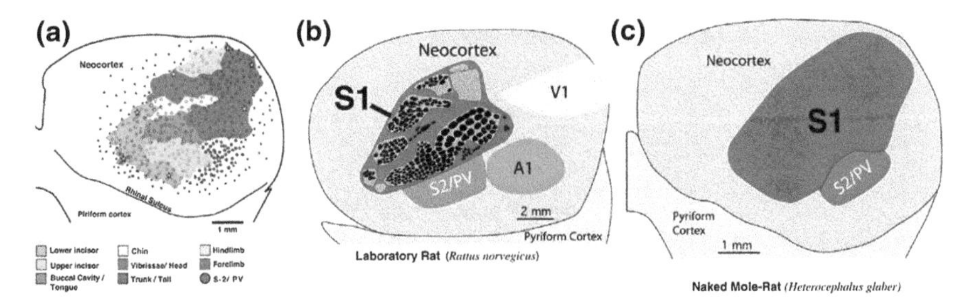

Figure 3 *Panel a.* Map of flattened naked mole-rat neocortex depicting electrode penetrations from electrophysiological recording experiments where neurons responded to tactile stimulation of the body surface. Color-coded regions indicate the extent of the representation of each body part. Note the large representation of the teeth as indicated in *green*. *Panel b.* Map of sensory cortices in the rat neocortex can contrasted with the same map (*Panel c*) in naked mole-rats. Note the expansion of primary somatosensory cortex (S1) in naked-mole rat and relative absence of primary visual (V1) and auditory (A1) cortices compared to the laboratory rat. Figures reprinted from Catania and Remple (2002), Figures 2 and 4. Copyright 2002, The National Academy of Sciences

somatosensory information in other animals, exhibits normal organization. Lesions of this structure abolish the tactile orienting behavior to stimulation of body vibrissae (Crish et al. 2003).

Computational Properties of Somatosensory Function

There are very few instances in which the computational abilities of touch-sensory processing in mammals have been amenable to investigation. However naked-mole-rats have yielded some very clear information on this processing. Studies of the tactile orienting behavior of naked mole-rats in response to deflection of their sensory body hairs have shown that the brains of these animals use computational strategies similar to that shown in other mammals' midbrain-mediated visual orienting responses (i.e. saccades). Deflecting two unilateral sensory hairs in located in rostral and caudal trunk regions of naked mole-rats will produce an orienting response to a location intermediate of the two regions (Figure 4). This outcome suggests that neural activity is averaged across the active populations of neurons in one hemisphere of the superior colliculus to produce this movement (Crish et al. 2006a, b). Other computational strategies are employed by collicular neurons when naked mole-rat sensory hairs in the same rostral-caudal plane are stimulated on contralateral sides of the body—producing a turn to one side or the other but no intermediate response (Crish et al. 2006a, b). This strategy known as "winner-take-all" has been demonstrated previously in the visual orienting responses of primates (Groh 1998).

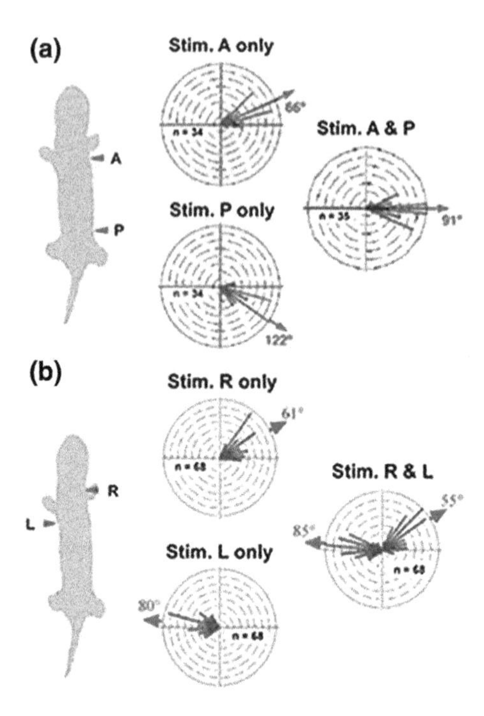

Figure 4 Data from behavioral experiments investigating neuronal population coding strategies in the tactile orienting response of naked mole-rats. Two vibrissa were stimulated either individually or simultaneously and the schematic shows the approximate location of the vibrissae tested (*red arrowheads*) on a gray drawing of an animal. Circular histograms display the distribution of turns elicited by each stimulus condition (*blue bars*) and the mean turn vector (*red arrow*) is also shown. *Panel a* shows orienting turns produced when deflecting sensory hairs in unilateral rostral and caudal regions. Simultaneous deflection produces an intermediate response indicative of a neuronal averaging computation. *Panel b* shows orienting turns elicited by stimulating sensory hairs bilaterally in the rostral plane. Simultaneous deflection produces a turn towards one side or the other, indicative of a winner-take-all neuronal computational strategy. Figure adapted from Crish et al. (2006b). Copyright 2006 with permission of Elsevier

References

Catania, K C and Remple, M S (2002). Somatosensory cortex dominated by the representation of teeth in the naked mole-rat brain. *Proceedings of the National Academy of Sciences of the United States of America* 99(8): 5692–5697. doi:10.1073/pnas.072097999.

Cooper, H M; Herbin, M and Nevo, E (1993). Visual system of a naturally microphthalmic mammal: The blind mole rat, *Spalax ehrenbergi*. *Journal of Comparative Neurology* 328(3): 313–350.

Crish, S D; Rice, F L; Park, T J and Comer C M (2003). Somatosensory organization and behavior in naked mole-rats I: Vibrissa-like body hairs comprise a sensory array that mediates orientation to tactile stimuli. *Brain, Behavior and Evolution* 62(3): 141–151.

Crish, S D; Dengler-Crish, C M and Catania, K C (2006a). Central visual system of the naked mole-rat (*Heterocephalus glaber*). *The Anatomical Record. Part A, Discoveries in Molecular, Cellular, and Evolutionary Biology* 288(2): 205–212.

Crish, S D; Dengler-Crish, C M and Comer, C M (2006b). Population coding strategies and involvement of the superior colliculus in the tactile orienting behavior of naked mole-rats. *Neuroscience* 139(4): 1461–1466.

Dengler-Crish, C M and Crish, S D (2013). Postnatal development of the subcortical visual system in naked mole-rats. Presented at the Society for Neuroscience Annual Meeting.

Groh, J M (1998). Reading neural representations. *Neuron* 21: 661–664.

Heffner, R S and Heffner, H E (1993). Degenerate hearing and sound localization in naked mole rats (*Heterocephalus glaber*), with an overview of central auditory structures. *Journal of Comparative Neurology* 331(3): 418–433.

Henry, E C; Remple, M S; O'Riain, M J and Catania, K C (2006). Organization of somatosensory cortical areas in the naked mole-rat (*Heterocephalus glaber*). *Journal of Comparative Neurology* 495(4): 434–452.

Henry, E C; Sarko, D K and Catania, K C (2008). Central projections of trigeminal afferents innervating the face in naked mole-rats (*Heterocephalus glaber*). *Anatomical Record (Hoboken, N.J. : 2007)* 291(8): 988–998. doi:10.1002/ar.20714.

Hetling, J R et al. (2005). Features of visual function in the naked mole-rat *Heterocephalus glaber*. *Journal of Comparative Physiology. A, Neuroethology, Sensory, Neural, and Behavioral Physiology* 191(4): 317–330.

Marzban, H et al. (2011). Compartmentation of the cerebellar cortex in the naked mole-rat (*Heterocephalus glaber*). *Cerebellum* 10(3): 435–448. doi:10.1007/s12311-011-0251-8.

Park, T J et al. (2008). Selective inflammatory pain insensitivity in the African naked mole-rat (*Heterocephalus glaber*). *PLOS Biology* 6(1): e13. doi:10.1371/journal.pbio.0060013.

Reep, R L; Marshall, C D; Stoll, M L and Whitaker, D M (1998). Distribution and innervation of facial bristles and hairs in the Florida manatee (*Trichechus manatus latirostris*). *Marine Mammal Science* 14: 257–273.

Reep, R L; Sarko, D K and Rice, F L (2007). Rock hyraxes (*Procavia capensis*) possess vibrissae over the entire postfacial body. Poster presented at Society for Neuroscience Annual Meeting.

Riccio, A P and Goldman, B D (2000a). Circadian rhythms of locomotor activity in naked mole-rats (*Heterocephalus glaber*). *Physiology & Behavior* 71(1–2): 1–13.

Riccio, A P and Goldman, B D (2000b). Circadian rhythms of body temperature and metabolic rate in naked mole-rats. *Physiology & Behavior* 71(1–2): 15–22.

Sarko, D K; Leitch, D B and Catania, K C (2013). Cutaneous and periodontal inputs to the cerebellum of the naked mole-rat (*Heterocephalus glaber*). *Frontiers in Neuroanatomy* 7: 39. doi:10.3389/fnana.2013.00039.

Smith, T D; Bhatnagar, K P; Dennis, J C; Morrison, E E and Park, T J (2007). *Growth-deficient vomeronasal organs in the naked mole-rat* (Heterocephalus glaber). *Brain Research* 1132(1): 78–83.

Vibrissal Behavior and Function

Tony J. Prescott, Ben Mitchinson and Robyn A. Grant

Tactile hair, or *vibrissae*, are a mammalian characteristic found on many mammals (Ahl 1986). Vibrissae differ from ordinary (pelagic) hair by being longer and thicker, having large follicles containing blood-filled sinus tissues, and by having an identifiable representation in the somatosensory cortex. Vibrissae are found on various parts of the body, but those most frequently studied are the facial or mystacial vibrissae, also called whiskers. Long facial whiskers, or macrovibrissae, are found in many mammalian species, projecting outwards and forwards from the snout of the animal to form a tactile sensory array that surrounds the head. For example, in rats, the macrovibrissae form a two-dimensional grid of five rows on each side of the snout, each row containing between five and nine whiskers ranging between ~ 15 and ~ 50 mm in length (see Figure 1 for illustration). The macrovibrissae of many rodents and some other species can move back and forth at high-speed thus explaining the term "vibrissa" which derives from the Latin "vibrio" meaning to vibrate.

The study of any behavior involves identifying the circumstances under which it arises, and then characterizing its nature, as precisely as possible, in all of the relevant contexts. However, in addition to describing what animals do, behavioral science also seeks to understand the function of behavior, both 'proximally' in terms of its immediate consequences for the animal, and 'ultimately', in terms of its adaptive significance and contribution to the evolutionary fitness of the species. In the context of the vibrissal system, we are only just beginning to piece together descriptions of how, and in what contexts, animals use their whiskers. Even less is known about the function of vibrissae, beyond the obvious intuition that whiskers

T.J. Prescott (✉)
Department of Psychology, University of Sheffield, Sheffield, UK

B. Mitchinson
Adaptive Behaviour Research Group, University of Sheffield, Sheffield, South Yorkshire, UK

R.A. Grant
Conservation Evolution and Behaviour Research Group, Manchester Metropolitan University, Manchester, UK

T.J. Prescott et al. (eds.), *Scholarpedia of Touch*, Scholarpedia,
DOI 10.2991/978-94-6239-133-8_7

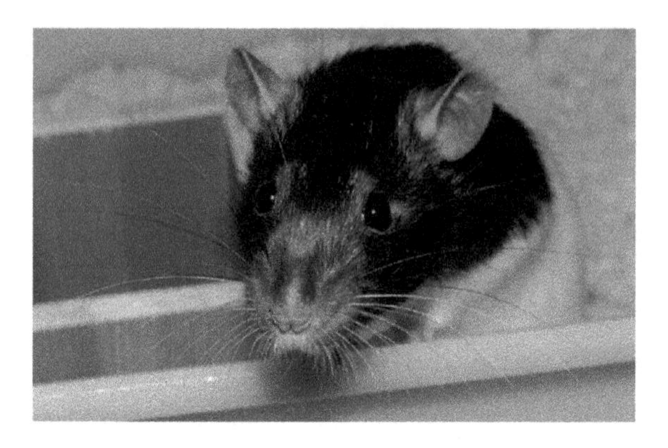

Figure 1 Facial vibrissae of the common rat

are 'for touch' just as the eyes are 'for sight'. We would like to understand much more about the specific contribution of the vibrissae to the life of the animal, both in order to explain the emergence, through natural selection, of this important mammalian sense, and also to be able to frame better functional hypotheses for physiologists investigating its biological and neural substrates. Here, we provide a brief comparative and ethological overview of vibrissal behavior and function.

Cross-Species Comparisons of Vibrissal Touch

Comparative studies of vibrissal sensing are important as they indicate the various roles of vibrissal touch in different environmental settings (e.g. in air or under water) and can reveal evolutionary convergences and divergences that allow us to draw conclusions about the generality of observed solutions. Vibrissal sensing is most often studied in rodents, aquatic mammals, and most recently in shrews (insectivores). Brecht (1997) has provided a morphological analysis of vibrissae systems in ten mammalian species including examples from marsupials, rodents, insectivores, pinnipeds, and primates, concluding that the presence of multiple rows of macrovibrissae increasing in length along the rostrocaudal axis is a shared feature of mammalian vibrissal sensing systems. Ahl (1986, 1987) has also reviewed comparative data from a wide range of mammalian groups, concluding that there is great variation between species but low variation within species, and arguing that these differences in vibrissal morphology could provide useful clues to function. For instance, a study of Old World field mice (genus *Apodemus*) found that a smaller facial vibrissal field was associated with a burrowing lifestyle and a larger field with a more arboreal one. Vibrissal function in a number of tactile sensing specialists is explored in the articles on **pinnipeds** (Hanke and Dehnhardt 2015), **manatees** (Reep and Sarko 2009), and **Etruscan shrews** (Roth-Alpermann and Brecht 2009). These studies provide islands of illumination, in a predominantly dark comparative landscape; much further research is needed to characterize the natural variation in vibrissal sensing systems across the different mammalian orders. By far the largest amount of research has concerned the facial vibrissae of mice and rats, which therefore form the main focus of the remainder of this article.

Active Control of Vibrissal Movement

In rats and mice the facial whiskers are repetitively and rapidly swept back and forth, during many behaviors including locomotion and exploration. These "whisking" movements generally occur in bouts of variable duration, and at rates between 3 to about twenty-five "whisks" per second (with rates generally being somewhat faster in mice than in rats). Movements of the whiskers are also closely coordinated with those of the head and body allowing the animal to locate interesting stimuli through whisker contact, then investigate them further using both the macrovibrissae and an array of shorter, non-actuated microvibrissae on the chin and lips (see Figure 2). Movement of the vibrissae and its measurement is discussed at length in Zeigler and Keller (2009) and Knutsen (2015).

Whisking is observed in only a sub-set of animals possessing prominent macrovibrissae. In addition to some species of rats and mice, it has also been reported in flying squirrels, gerbils, chinchillas, hamsters, shrews, the African porcupine, and in two species of marsupial—the Brazilian short-tailed opossum *Monodelphis domestica*, and the Australian brush-tailed possum (Rice 1995; Welker 1964; Wineski 1985; Woolsey et al. 1975; Mitchinson et al. 2011). Although whisking appears to be most prominent in rodents, there are several rodent genera, such as capybara and gophers, that do not appear to whisk, and others, such as guinea pigs, that display only irregular and relatively short whisking bouts (Jin et al. 2004). Whisking behavior has not been observed in carnivores (e.g. cats, dogs, raccoons, pandas), although some species, such as pinnipeds have well-developed sinus muscles making the whiskers highly mobile (Ahl 1986). In animals such as rats and mice, that are capable of whisking at high frequencies, the whisking musculature contains a high proportion of type 2B muscle fibers that can support faster contractions than normal skeletal muscles (Jin et al. 2004). A comparative study of whisker movement in rats, mice, and the marsupial *M. domestica* found similar patterns of whisker movement, and of active whisker control (see below), in all three species, with mouse whisking having the most complex movement patterns. The presence of whisking in both rodents and marsupials implies that early mammals may also have exhibited this behavior, with evidence of similarities between the whisking musculature of rats and the marsupial opossum adding further support for the idea of a whisking common ancestor for modern mammals (Grant et al. 2013) (see also Haidarliu 2015).

Since rapid movement of the vibrissae consumes energy, and has required the evolution of specialised musculature, it can be assumed that whisking must convey some sensory advantages to the animal. Likely benefits are that it provides more

Figure 2 The mobile macrovibrissae are often used to locate objects that are then investigated further with the shorter, non-actuated microvibrissae on the chin and lips. In these example high-speed videoframes a rat locates a coin with its macrovibrissae (*top frame*) and, in the next whisk cycle (*bottom frame*), brushes the microvibrissae against the coin surface. From Fox et al. 2009

degrees of freedom for sensor positioning, that it allows the animal to sample a larger volume of space with a given density of whiskers, and that it allows control over the velocity with which the whiskers contact surfaces. In addition, the ability to employ alternative whisking strategies in different contexts, may constitute an important gain. In other words, vibrissal specialists may whisk for the same reason that humans carefully and repeatedly adjust the position of their fingertips whilst exploring objects with their hands, and adopt different exploratory strategies depending on the type of tactile judgement they are seeking to make (see Klatzky and Reed 2009). In both vibrissal and fingertip touch, better quality and more appropriate tactile information may be obtained by exerting precise control over how the sensory apparatus interacts with the world.

Evidence to support this "active sensing" view (see Prescott et al. 2011; Lepora 2015) of whisking behavior comes from a number of sources. First, conditioning studies have shown that rats can be trained to vary some of the key parameters of whisking, such as amplitude and frequency in a stimulus-dependent manner (Bermejo et al. 1996; Gao et al. 2003). Second, recordings of whisker movements in freely-moving animals, show that this often diverges from the regular, bilaterally-symmetric and synchronous motor pattern that is usually observed in head-restrained animals. In the freely-moving animal, asymmetries (see, e.g. Figure 3), asynchronies, and changes in whisker protraction angle and timing have been noted some of which seem likely to boost the amount of useful sensory information obtained by the animal. For instance, it has been shown that head movements tend to direct the whiskers in the direction in which the animal is turning (Towal and Hartmann 2006), suggesting increased exploration of space in the direction in which the animal is moving. Unilateral contact with a nearby surface tends to reduce whisker protraction on the side of the snout ipsilateral to that contact and increase protraction on the contralateral side (Mitchinson et al. 2007). Such a strategy would tend to increase the number of contacts made with a surface of interest whilst ensuring that the whiskers do not press hard against the contacted object. Other work has shown that the frequency of whisking, the starting position of the whiskers (the minimum protraction angle), or the angular spread between the whiskers, could each be controlled in a context- or behavior-specific way (Berg and Kleinfeld 2003; Carvell and Simons 1990; Sachdev et al. 2003; Sellien et al. 2005;

Figure 3 Frame from a video recording (with *side-on* and *top-down views*) showing bilateral asymmetry of whisker movements during contact with a wall. The whiskers on the side of the snout closest to the wall protract less, which may serve to reduce bending against the surface; the whiskers on the opposite side protract more, which may serve to increase the number of whisker-surface contacts. From Mitchinson et al. 2007

Grant et al. 2009). Perhaps the strongest evidence that whisking control is purposive, in an active perception sense, comes from a study of sightless rats trained to run up a corridor for food (Arkley et al. 2014). Animals that were trained to expect obstacles at unpredictable locations in the corridor tended to run more slowly, and push their whiskers further forward, than animals trained to expect an empty corridor. This evidence suggests that the configuration of the whiskers is dependent on the animal's expectations about what it may encounter in the environment. The growing evidence for active control of whisker movement also implies that the vibrissal system can provide an accessible model for studying purposive behavior in mammals.

Behavioral Evidence of Vibrissal Function

Experiments in adult rats and shrews involving whisker clipping, cauterization of the whisker follicle, section of the peripheral nerves, or lesion of critical structures in the vibrissal pathway, using appropriate controls for other sensory modalities such as vision and olfaction, have found significant deficits in exploration, thigmotaxis, locomotion, maintenance of equilibrium, maze learning, swimming, locating food pellets, and fighting (for reviews see Ahl 1982; Gustafson and Felbain-Keramidas 1977). Adults in which whiskers have been removed also show alterations of posture that have the effect of bringing the snout (and whisker stubs) into contact with surfaces of interest (Meyer and Meyer 1992; Vincent 1912), whilst animals in which whiskers are plucked shortly after birth and repeatedly thereafter exhibit pronounced behavioral compensations as adults (Gustafson and Felbain-Keramidas 1977; Volgyi et al. 1993), and show altered behavior even when whiskers are allowed to regrow (Carvell and Simons 1996). Whisker clipping in infants disrupts early (post-natal days 3–5) nipple attachment and huddling (Sullivan et al. 2003). One of the most interesting and demanding uses of the vibrissal sense is in predation. For instance, pygmy shrews are known to prey on insects such as crickets that are themselves highly agile and almost as large as the shrew itself. Anjum et al. (2006) have demonstrated that tactile shape cues are both necessary and sufficient for evoking and controlling these attacks, whereas visual and olfactory cues are not needed (see Roth-Alpermann and Brecht 2009). Rats are also efficient predators that can detect, track, and immobilise prey animals without vision (see, e.g. Gregoire and Smith 1975).

Disruptions of predation, following whisker removal or damage, or of other behaviors such as maze learning, food finding, gap crossing, and fighting, are presumably the consequence of the loss of fundamental tactile sensory abilities that use vibrissal signals to (i) localise, orient to, and track objects/surfaces in space, and (ii) discriminate between objects/surfaces based on their tactile properties. These vibrissal perceptual functions are considered in more detail next.

Localising, Orienting, and Tracking

Object localization and distance, orientation detection. Rats have been shown to use vibrissal information in the following tasks: gap measurement (Krupa et al. 2001), gap jumping (Hutson and Masterton 1986; Jenkinson and Glickstein 2000; Richardson 1909), measuring angular position along the sweep of the whisker (Knutsen et al. 2006; Mehta et al. 2007), and distinguishing horizontal versus vertical orientation of bars (Polley et al. 2005). Schiffman et al. (1970) demonstrated that rats confronted with a visual cliff do not show cliff-avoidance behavior unless their whiskers have been trimmed. This finding illustrates the primacy of vibrissal tactile sensing over vision in these animals with regard to depth perception. Ahissar and Knutsen (2008) have proposed that whisker identity, activation intensity, and timing of contact relative to the whisk cycle, together provide sufficient information to localise the position of a contact in 3-dimensional space (elevation, azimuth, and radial distance, respectively) (See Ahissar and Knutsen 2011).

Orienting. Contact of the macrovibrissae with a surface often brings about orientation towards that surface which is then further explored using both the macro- and micro- vibrissae (Brecht 1997; Hartmann 2001). Grant et al. (2012b) showed that rats reliably orient to the nearest macrovibrissal contact on an unexpected object, progressively homing in on the nearest contact point on the object in each subsequent whisk. When exploring an object, the touching or brushing of the microvibrissae against the surface is largely synchronised with maximal macrovibrissal protraction and with a special pattern or rapid inhalation termed "sniffing" (Welker 1964; Hartmann 2001). Long-term bilateral removal of the whiskers has been shown to reduce orienting towards a tactile stimulus on the snout (Symons and Tees 1990). When the various modulations of active whisking control are considered together it is possible to consider them as part of a general orienting system that both moves the tactile "fovea", at the tip of the snout, towards points of interest, and, at the same time modifies the positions of the whiskers in the wider array in a way that should increase the number of whisker contacts within an attentional zone (Mitchinson and Prescott 2013; Mitchinson 2015).

Detection of movement. The importance of vibrissae to prey-capture in shrews (Anjum et al. 2006), and in mouse-killing in rats (Gregoire and Smith 1975) suggests that these animals may be able to estimate some of movement parameters of their targets using vibrissal information. In the case of the Etruscan shrew, effective use of the whiskers in prey-capture appears to be experience-dependent (Anjum and Brecht 2012).

Tactile Discrimination

Texture discrimination. Initial studies of vibrissal texture discrimination (Carvell and Simons 1990; Guic-robles et al. 1989, 1992) showed that rats can be trained to reliably discriminate multiple textures using only their macrovibrissae, and can make fine distinctions with similar sensory thresholds to humans. Movement of an individual whisker across a texture has been shown to give rise to a characteristic whisker vibration or "kinetic signature" which is thought to form the basis on which sensory discriminations are made (Diamond et al. 2008; Hipp et al. 2006; Wolfe et al. 2008; Arabzadeh et al. 2009).

Air flow/water currents/vibrations. Air, or water, currents are effective stimuli for generating responses to vibrissal signals (e.g. in rats, Hutson and Masterton 1986; in seals, Dehnhardt et al. 2001). Rat vibrissae should also be capable of discriminating vibrations of appropriate frequencies either applied directly to the vibrissal shaft or transmitted through the air as sound waves (Shatz and Christensen 2008).

Shape discrimination. Compared to texture (microgeometry), shape and other macrogeometric properties are relatively poorly studied in rats. Brecht et al. (1997) have shown that rats can distinguish between triangular and square cookies (each 6 mm per side) using their whiskers, although the evidence suggests that the microvibrissae are used in this task alongside the macrovibrissae. Experiments in shrews, using artificial prey replica, suggest that these animals may respond to some macrogeometric properties of prey animals (Anjum et al. 2006). Dehnhardt and co-workers have shown that seals can discriminate size and shape with their vibrissae (Dehnhardt and Ducker 1996). Experiments with artificial vibrissal systems show that 3D shape of complex surfaces can, in principle, be recovered from vibrissal data (Kim and Moller 2007; Solomon and Hartmann 2006).

Numerical discrimination. Davis et al. (1989) have demonstrated that rats can be trained to discriminate between 2, 3, and 4 anterior-posterior strokes on their vibrissae, but not on their flanks. This result is indicative of vibrissal short-term memory and of some enumerative capacity in the vibrissal system.

Locomotion

The emergence of active whisker control alongside locomotion during development (Grant et al. 2012a) suggests a potentially tight relationship between vibrissal sensing and the control of locomotion. However, it appears that the vibrissal sense may be used in different modes in support of locomotion depending on speed of travel. Specifically, in rats, walking appears to be accompanied by broad whisker sweeps, whereas running is accompanied by the protraction of whiskers in front of the animal with much less forward-and-back oscillatory movement (Arkley et al. 2014) (see Figure 4). This pattern suggests a switch from using the whiskers to explore the floor surface, perhaps to find good locations for footfalls, to one where the whiskers are primarily used for obstacle detection and collision avoidance. The value of whiskers in complex locomotor tasks, such as climbing, is also indicated

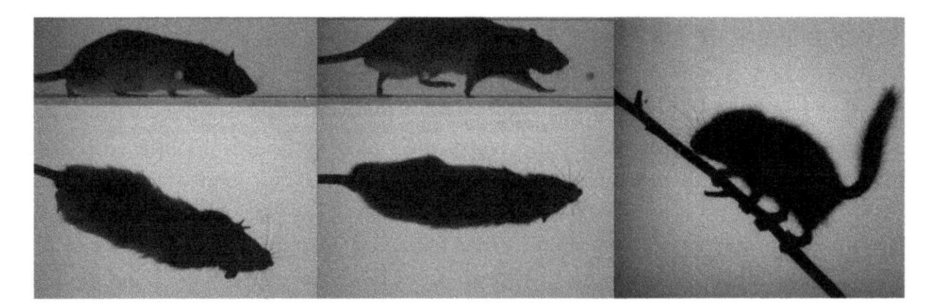

Figure 4 The role of vibrissae in locomotion. *Left* when exploring a novel environment a rat typically keeps its snout close to the ground making many whiskers contacts and performing broad whisking sweeps. *Centre* in a familiar environment, when motivated to move quickly, the head is held higher with fewer whiskers contacting the ground; the vibrissae are pushed forward, making smaller sweeps, and suggesting a role in obstacle detection and collision avoidance (see Arkley et al. 2014). *Right* scansorial (tree-living) and nocturnal rodents, such as this dormouse, have longer whiskers, implying a role for the vibrissae in climbing, particularly in darkness

by evidence that small arboreal mammals, particularly nocturnal ones, have longer macrovibrissae than similar ground-dwelling species (Ahl 1986; Sokolov and Kulikov 1987). In rats, a three-whisker array on the underside of the chin, called the "trident" whiskers, has been shown to drag along the ground during exploratory locomotion (Thé et al. 2013), suggesting that these whiskers could provide information to the animal about its velocity and heading direction. Niederschuh et al. (2015) studied rhythmic whisker movement alongside locomotor gait in rats running on a treadmill with or without intact vibrissae. Whilst arguing against a close coupling between whisking and running, they suggest a likely role for the mystacial vibrissae in foot placement, in addition, they found that the carpal (wrist) vibrissae may assist the animal in monitoring its speed of movement. Experiments with movement on more complex surfaces, and in the absence of light, will be needed to better understand the importance of the vibrissae to locomotion control in rodents.

Social Behavior

In rodents, the vibrissae are also used during social behaviors. Typically, during encounters between resident and intruder rats, physical contact is often initiated by whisker-to-whisker contact, followed by potential aggressive behavior. Such nose-off's, along with other behaviors such as teeth-chattering and threat postures, may be used to convey social signals such as submissiveness or dominance. A study of whisker-to-whisker interaction in rats found that the social context of the interaction modulates whisker-related activity in primary somatosensory cortex (Bobrov et al. 2014).

Development of Vibrissal Function

Vibrissal system development has been studied most extensively in rats. Rat pups are born with an intact whisker field constituted of very fine, immobile vibrissae, that are invisible to the naked eye. The whiskers grow to their adult size in the first month of life, however, rats sustaining denervation of the whisker pad grow whiskers that are thinner and smaller than those of normal adults (Landers et al. 2006). Whisking begins around postnatal day 11–13 (Welker 1964; Landers and Ziegler 2006; Grant et al. 2012a), prior to the opening of the eyes, and achieves adult amplitudes and characteristics by the end of the third post-natal week. The onset of active whisking control emerges a few days after the initial appearance of synchronised bilateral whisking (Grant et al. 2012a). Prior to the onset of whisking, neonatal rats show behavioral activation in response to whisker stimulation, and tactile learning in a classical conditioning avoidance paradigm (Landers and Sullivan 1999; Sullivan et al. 2003). Micro-movements of the vibrissae in the first week of life have also been observed (Grant et al. 2012a). Rat pups are able to orient to contacts with nearby conspecifics before their eyes open implying an important role for the macrovibrissae in maintaining contact with conspecifics (Grant et al. 2012b). The emergence of vibrissal tactile sensing in rats may parallel the gradually increasing motor capacities of these animals that allow adult upright locomotion to occur around the same time as whisking onset. Tactile experience may be very different for neonatal rats than for adults both because of the small size and relative immobility of the whiskers, and because rat pups spend much of their time in "huddles" with littermates which is likely to produce stimulation of the whiskers from many different directions (Figure 5).

Figure 5 Huddling behavior of 6-day old rat pups

Conclusion

Interest in the rodent vibrissal system has often stemmed from its accessibility as a model of mammalian sensory processing, rather than from the perspective of trying to understand its role in the life of the animal. For this reason, an accurate characterization of the contribution of vibrissal touch to rat or mouse behavior is some way off. We contend that such an understanding will be important for understanding the processing of vibrissal signals throughout the brain; after all to understand how a system works it should certainly help to know how it is used.

Internal References

Ahissar, E and Knutsen, P (2011). Vibrissal location coding. *Scholarpedia* 6: 6639. http://www.scholarpedia.org/article/Vibrissal_location_coding. (see also pages 725–735 of this book).

Arabzadeh, E; von Heimendahl, M and Diamond, M (2009). Vibrissal texture decoding. *Scholarpedia* 4(4): 6640. http://www.scholarpedia.org/article/Vibrissal_texture_decoding. (see also pages 737–749 of this book).

Haidarliu, S (2015). Whisking musculature. *Scholarpedia* 10(4): 32331. http://www.scholarpedia.org/article/Whisking_musculature. (see also pages 627–639 of this book).

Hanke, W and Dehnhardt, G (2015). Vibrissal touch in pinnipeds. *Scholarpedia* 10(3): 6828. http://www.scholarpedia.org/article/Vibrissal_touch_in_pinnipeds. (see also pages 125–139 of this book).

Klatzky, R and Reed, C L (2009). Haptic exploration. *Scholarpedia* 4(8): 7941. http://www.scholarpedia.org/article/Haptic_exploration. (see also pages 177–183 of this book).

Knutsen, P (2015). Whisking kinematics. *Scholarpedia* 10: 7280. http://www.scholarpedia.org/article/Whisking_kinematics. (see also pages 615–625 of this book).

Lepora, N (2015). Active tactile perception. *Scholarpedia* 10: 32364. http://www.scholarpedia.org/article/Active_tactile_perception. (see also pages 151–159 of this book).

Mitchinson, B (2015). Tactile attention in the vibrissal system. *Scholarpedia* 10: 32361. http://www.scholarpedia.org/article/Tactile_Attention_in_the_Vibrissal_System. (see also pages 771–779 of this book).

Reep, R and Sarko, D K (2009). Tactile hair in manatees. *Scholarpedia* 4(4): 6831. http://www.scholarpedia.org/article/Tactile_hair_in_Manatees. (see also pages 141–148 of this book).

Roth-Alpermann, C and Brecht, M (2009). Vibrissal touch in the Etruscan shrew. *Scholarpedia* 4 (11): 6830. http://www.scholarpedia.org/article/Vibrissal_touch_in_the_Etruscan_shrew. (see also pages 117–123 of this book).

Zeigler, P and Keller, A (2009). Whisking pattern generation. *Scholarpedia* 4(12): 7271. http://www.scholarpedia.org/article/Whisking_pattern_generation. (see also pages 641–656 of this book).

External References

Ahissar, E and Knutsen, P M (2008). Object localization with whiskers. *Biological Cybernetics* 98 (6): 449–458.

Ahl, A S (1982). Evidence of use of vibrissae in swimming in Sigmo don fulviventer. *Animal Behaviour* 30: 1203–1206.

Ahl, A S (1986). The role of vibrissae in behavior - A status review. *Veterinary Research Communications* 10(4): 245–268. http://www.springerlink.com/content/k438303108l38260/.

Ahl, A S (1987). Relationship of vibrissal length and habits in the Sciuridae. *Journal of Mammalogy* 68(4): 848–853.

Anjum, F; Turni, H; Mulder, P G; van der Burg, J and Brecht, M (2006). Tactile guidance of prey capture in Etruscan shrews. *Proceedings of the National Academy of Sciences of the United States of America* 103(44): 16544–16549.

Anjum, F and Brecht, M (2012). Tactile experience shapes prey-capture behavior in Etruscan shrews. *Frontiers in Behavioral Neuroscience* 6: 28.

Arkley, K; Grant, R A; Mitchinson, B and Prescott, T J (2014). Strategy change in vibrissal active sensing during rat locomotion. *Current Biology* 24(13): 1507–1512.

Berg, R W and Kleinfeld, D (2003). Rhythmic whisking by rat: Retraction as well as protraction of the vibrissae is under active muscular control. *Journal of Neurophysiology* 89(1): 104–117.

Bermejo, R; Harvey, M; Gao, P and Zeigler, H P (1996). Conditioned whisking in the rat. *Somatosensory & Motor Research* 13(3–4): 225–233.

Bobrov, E; Wolfe, J; Rao, R P and Brecht, M (2014). The representation of social facial touch in rat barrel cortex. *Current Biology* 24(1): 109–115.

Brecht, M; Preilowski, B and Merzenich, M M (1997). Functional architecture of the mystacial vibrissae. *Behavioural Brain Research* 84(1–2): 81–97.

Carvell, G E and Simons, D J (1990). Biometric analyses of vibrissal tactile discrimination in the rat. *The Journal of Neuroscience* 10(8): 2638–2648.

Carvell, G E and Simons, D J (1996). Abnormal tactile experience early in life disrupts active touch. *The Journal of Neuroscience* 16(8): 2750–2757.

Davis, H; Mackenzie, K A and Morrison, S (1989). Numerical discrimination by rats (*Rattus norvegicus*) using body and vibrissal touch. *Journal of Comparative Psychology* 103(1): 45–53.

Dehnhardt, G and Ducker, G (1996). Tactual discrimination of size and shape by a California sea lion (*Zalophus californianus*). *Animal Learning & Behavior* 24(4): 366–374.

Dehnhardt, G; Mauck, B; Hanke, W and Bleckmann, H (2001). Hydrodynamic trail-following in harbor seals (*Phoca vitulina*). *Science* 293(5527): 102–104.

Diamond, M E; von Heimendahl, M and Arabzadeh, E (2008). Whisker-mediated texture discrimination. *PLoS Biology* 6(8): e220.

Fox, C W; Mitchinson, B; Pearson, M J; Pipe, A G and Prescott, T J (2009). Contact type dependency of texture classification in a whiskered mobile robot. *Autonomous Robots* 26(4): 223–239. doi:10.1007/s10514-009-9109-z.

Gao, P; Ploog, B O and Zeigler, H P (2003). Whisking as a "voluntary" response: Operant control of whisking parameters and effects of whisker denervation. *Somatosensory & Motor Research* 20(3–4): 179–189.

Grant, R A; Mitchinson, B; Fox, C and Prescott, T J (2009). Active touch sensing in the rat: Anticipatory and regulatory control of whisker movements during surface exploration. *Journal of Neurophysiology* 101(2): 862–874.

Grant, R A; Mitchinson, B and Prescott, T J (2012a). The development of whisker control in rats in relation to locomotion. *Developmental Psychobiology* 54(2): 151–168.

Grant, R A; Sperber, A L and Prescott, T J (2012b). The role of orienting in vibrissal touch sensing. *Frontiers in Behavioral Neuroscience* 6: 39.

Grant, R A; Haidarliu, S; Kennerley, N J and Prescott, T J (2013). The evolution of active vibrissal sensing in mammals: Evidence from vibrissal musculature and function in the marsupial opossum *Monodelphis domestica*. *The Journal of Experimental Biology* 216: 3483–3494.

Gregoire, S E and Smith, D E (1975). Mouse-killing in the rat: Effects of sensory deficits on attack behaviour and stereotyped biting. *Animal Behaviour* 23(Part 1): 186–191.

Guic-robles, E; Guajardo, G and Valdivieso, C (1989). Rats can learn a roughness discrimination using only their vibrissal system. *Behavioural Brain Research* 31: 285–289.

Guic-robles, E; Jenkins, W M and Bravo, H (1992). Vibrissal roughness discrimination is barrelcortex-dependent. *Behavioural Brain Research* 48(2): 145–152.

Gustafson, J W and Felbain-Keramidas, S L (1977). Behavioral and neural approaches to the function of the mystacial vibrissae. *Psychological Bulletin* 84(3): 477–488.

Hartmann, M J (2001). Active sensing capabilities of the rat whisker system. *Autonomous Robots* 11: 249–254.
Hipp, J et al. (2006). Texture signals in whisker vibrations. *Journal of Neurophysiology* 95(3): 1792–1799.
Hutson, K A and Masterton, R B (1986). The sensory contribution of a single vibrissa's cortical barrel. *Journal of Neurophysiology* 56(4): 1196–1223.
Jenkinson, E W and Glickstein, M (2000). Whiskers, barrels, and cortical efferent pathways in gap crossing by rats. *Journal of Neurophysiology* 84(4): 1781–1789.
Jin, T-E; Witzemann, V and Brecht, M (2004). Fiber types of the intrinsic whisker muscle and whisking behavior. *The Journal of Neuroscience* 24(13): 3386–3393.
Kim, D and Moller, R (2007). Biomimetic whiskers for shape recognition. *Robotics and Autonomous Systems* 55(3): 229–243.
Knutsen, P M; Pietr, M and Ahissar, E (2006). Haptic object localization in the vibrissal system: Behavior and performance. *The Journal of Neuroscience* 26(33): 8451–8464.
Krupa, D J; Matell, M S; Brisben, A J; Oliveira, L M and Nicolelis, M A (2001). Behavioral properties of the trigeminal somatosensory system in rats performing whisker-dependent tactile discriminations. *The Journal of Neuroscience* 21(15): 5752–5763.
Landers, M; Haidarliu, S and Philip Zeigler, H (2006). Development of rodent macrovibrissae: Effects of neonatal whisker denervation and bilateral neonatal enucleation. *Somatosensory & Motor Research* 23(1–2): 11–17.
Landers, M S and Sullivan, R M (1999). Virissae-evoked behavior and conditioning before functional ontogeny of the somatosensory vibrissae cortex. *The Journal of Neuroscience* 19 (12): 5131–5137.
Landers, M and Zeigler, P H (2006). Development of rodent whisking: Trigeminal input and central pattern generation. *Somatosensory & Motor Research* 23(1–2): 1–10.
Mehta, S B; Whitmer, D; Figueroa, R; Williams, B A and Kleinfeld, D (2007). Active spatial perception in the vibrissa scanning sensorimotor system. *PLoS Biology* 5(2): 309–322.
Meyer, M E and Meyer, M E (1992). The effects of bilateral and unilateral vibrissotomy on behavior within aquatic and terrestrial environments. *Physiology & Behavior* 51(4): 877–880.
Mitchinson, B; Martin, C J; Grant, R A and Prescott, T J (2007). Feedback control in active sensing: Rat exploratory whisking is modulated by environmental contact. *Proceedings of the Royal Society B: Biological Sciences* 274(1613): 1035–1041.
Mitchinson, B et al. (2011). Active vibrissal sensing in rodents and marsupials. *Philosophical Transactions of the Royal Society B: Biological Sciences* 366(1581): 3037–3048.
Mitchinson, B and Prescott, T J (2013). Whisker movements reveal spatial attention: A unified computational model of active sensing control in the rat. *PLOS Computational Biology* 9(9): e1003236.
Niederschuh, S J; Witte, H and Schmidt, M (2015). The role of vibrissal sensing in forelimb position control during travelling locomotion in the rat (*Rattus norvegicus, Rodentia*). *Zoology* 118(1): 51–62.
Polley, D B; Rickert, J L and Frostig, R D (2005). Whisker-based discrimination of object orientation determined with a rapid training paradigm. *Neurobiology of Learning and Memory* 83(2): 134–142.
Prescott, T J; Wing, A and Diamond, M E (2011). Active touch sensing. *Philosophical Transactions of Royal Society B: Biological Sciences*. 366(1581): 2989–2995.
Rice, F L (1995). Comparative aspects of barrel structure and development. In: E G Jones and I T Diamond (Eds.), *Cerebral Cortex Volume II: The Barrel Cortex of Rodents*. New York: Plenum Press.
Richardson, F (1909). A study of sensory control in the rat. *Psychological Monographs Supplement* 12(1): 1–124.
Sachdev, R N; Berg, R W; Champney, G; Kleinfeld, D and Ebner, F F (2003). Unilateral vibrissa contact: Changes in amplitude but not timing of rhythmic whisking. *Somatosensory & Motor Research* 20: 163–169.

Schiffman, H R; Lore, R; Passafiume, J and Neeb, R (1970). Role of vibrissae for depth perception in the rat (*Rattus norvegicus*). *Animal Behaviour* 18(Part 2): 290–292.

Sellien, H; Eshenroder, D S and Ebner, F F (2005). Comparison of bilateral whisker movement in freely exploring and head-fixed adult rats. *Somatosensory & Motor Research* 22: 97–114.

Shatz, L F and Christensen, C W (2008). The frequency response of rat vibrissae to sound. *The Journal of the Acoustical Society of America* 123(5): 2918–2927.

Sokolov, V E and Kulikov, V F (1987). The structure and function of the vibrissal apparatus in some rodents. *Mammalia* 51: 125–138.

Solomon, J H and Hartmann, M J (2006). Biomechanics: Robotic whiskers used to sense features. *Nature* 443(7111): 525.

Sullivan, R M et al. (2003). Characterizing the functional significance of the neonatal rat vibrissae prior to the onset of whisking. *Somatosensory & Motor Research* 20(2): 157–162.

Symons, L A and Tees, R C (1990). An examination of the intramodal and intermodal behavioral consequences of long-term vibrissae removal in rats. *Developmental Psychobiology* 23(8): 849–867.

Thé, L; Wallace, M L; Chen, C H; Chorev, E and Brecht, M (2013). Structure, function, and cortical representation of the rat submandibular whisker trident. *The Journal of Neuroscience* 33: 4815–4824.

Towal, R B and Hartmann, M J (2006). Right-left asymmetries in whisking behavior of rats anticipate head movements. *The Journal of Neuroscience* 26(34): 8838–8846.

Vincent, S B (1912). The function of the vibrissae in the behaviour of the white rat. *Behavior Monographs* 1: 1–82.

Volgyi, B; Farkas, T and Toldi, J (1993). Compensation of a sensory deficit inflicted upon newborn and adult animals - A behavioral study. *Neuroreport* 4(6): 827–829.

Welker, C I (1964). Analysis of sniffing in the albino rat. *Behaviour* 22: 223–244.

Wineski, L E (1985). Facial morphology and vibrissal movement in the Golden Hamster. *Journal of Morphology* 183(2): 199–217.

Wolfe, J et al. (2008). Texture coding in the rat whisker system: Slip-stick versus differential resonance. *PLoS Biology* 6(8): e215.

Woolsey, T A; Welker, C and Schwartz, R H (1975). Comparative anatomical studies of the SmL face cortex with special reference to the occurrence of *barrels* in layer IV. *Journal of Comparative Neurology* 164(1): 79–94.

Vibrissal Touch in the Etruscan Shrew

Claudia Roth-Alpermann and Michael Brecht

The **Etruscan shrew** *Suncus etruscus* (also known as white-toothed pygmy shrew) is the smallest terrestrial mammal with a body weight of 2 g and a body length of around 4 cm without tail (Figure 1a). Shrews feed on insects and they use the sense of touch to detect and hunt prey. The elongated rostrum of the shrew has long whiskers referred to as macrovibrissae (Figure 1a, b); the shrew's mouth is surrounded by a dense array of short whiskers, the so-called microvibrissae (Figure 1c).

Behavioral Ecology

Etruscan shrews belong to the family of *Soricidae* (shrews) and therein to the subfamily of *Crocidurinae* (white-toothed shrews) (Wilson and Reeder 2005). It is widely believed that among recent mammals, shrews represent the closest relative to the ancestor of all placental mammals and the earliest shrew-like fossils date back approximately 70–100 million years ago (Archibald et al. 2001)

Etruscan shrews can be found from the Mediterranean to Southeast Asia in a belt extending between 10° and 30°N latitude (Figure 2). Their habitat includes scrub, open forest and grassland environments. In the Mediterranean region it prefers abandoned olive groves, vineyards, and other cultivated areas overrun by shrubs (Aulagnier et al. 2008). Being hunted by predatory birds and owls, shrews try to avoid moving uncovered in the open field, but rather seek shelter in old dry stone walls, under leaves or pieces of bark on the ground or in loose soil. They are specialized for a life in slits found in stone walls or piles of rock and they are able to enter and capture prey in slits as thin as 7 mm. Etruscan shrews successfully hunt

C. Roth-Alpermann (✉)
Bernstein Center for Computational Neuroscience, Humboldt-University, Berlin, Germany

M. Brecht
Bernstein Center for Computational Neuroscience, Berlin, Germany

T.J. Prescott et al. (eds.), *Scholarpedia of Touch*, Scholarpedia,
DOI 10.2991/978-94-6239-133-8_8

Figure 1 The Etruscan shrew. **a** Etruscan shrew with whisker fan. **b** Frontal view of the shrew's head. **c** High magnification view of the microvibrissae surrounding the shrew's mouth

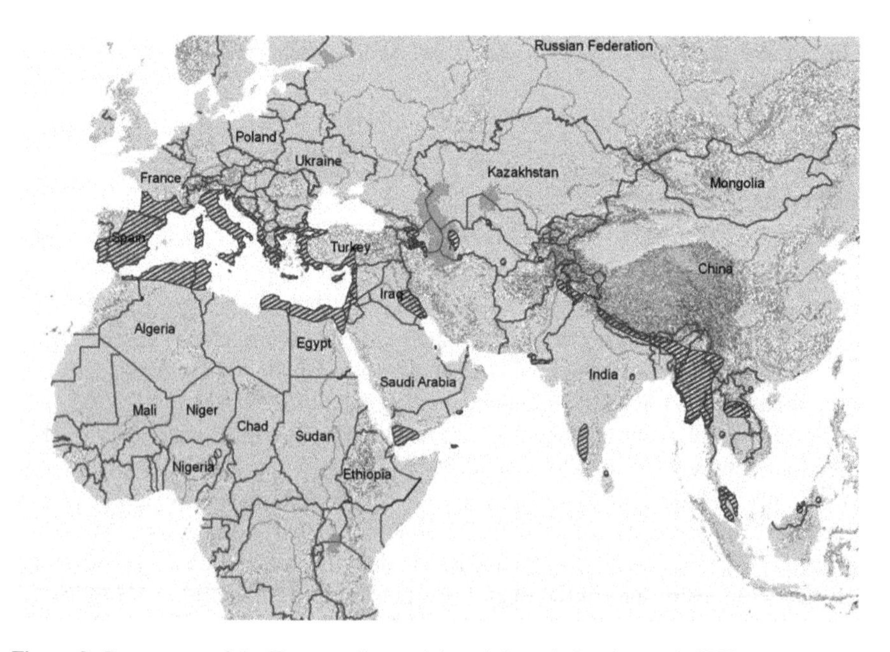

Figure 2 Range map of the Etruscan shrew. Adapted from Aulagnier et al. 2008

and feed on insects that have almost the same body size as themselves and crickets are amongst their preferred food.

The small body size of Etruscan shrews goes along with an extraordinarily high energy turnover. They feed up to 25 times a day and consume more than their own body weight in food. Heart, respiratory system and skeletal muscles are functionally and structurally adapted to meet the enormous metabolic needs (Jürgens 2002). In case of food restriction and at low ambient temperature, Etruscan shrews can reduce their body temperature and enter a resting state called "torpor" to cut down their energy expenditure (Fons et al. 1997). Traditionally, shrews (including *Suncus etruscus*) have been regarded as nocturnal animals. However, probably due to their constant food requirement, they may actually have a polyphasic circadian activity pattern with frequent activity bouts distributed over a period of 24 h. This means that shrews have to be able to successfully hunt in twilight as well as in darkness, i.e., under conditions where vision is of limited use and, indeed, sight only seems to play a minor role. Work on relative shrew species from the [http://en.wikipedia.org/wiki/Crocidura Crocidura] genus implicated vibrissae in navigation, but argued against the presence of echolocation in these animals (Grünwald 1969).

A synopsis of sensory ecology of Etruscan shrews suggests the following picture: They live and hunt in slits inaccessible to larger animals. Here they rely on touch rather than on long-distance sensory modalities such as vision. Thus, shrews can be regarded as short-range/high-speed animals.

Tactile Prey Capture Behavior

Etruscan shrew prey capture is guided by tactile cues (Anjum et al. 2006). In a laboratory setting, hunting was filmed in total darkness under infrared illumination while crickets were offered as prey. These experiments demonstrated that Etruscan shrews attack in a precise and fast manner and that they need their whiskers to hunt successfully.

Spatio-Temporal Analysis of Attacks on Crickets

Spatial attack characteristics. Etruscan shrews place their attacks selectively on the cricket's thorax (Figure 3a, b) and manage to keep this precision regardless of the size of the prey. They attack crickets from the side with a narrow distribution of attack angles around 90° relative to the cricket's body axis. Although most attacks are directed straight ahead, some shrews show a lateralization in their hunting: they preferentially attack with their head turned to their right rather than to their left side.

Temporal attack characteristics. Prey capture occurs very fast, in 80–200 ms per attack, with short inter-attack intervals of around 200 ms (Figure 3c, d).

Attack dynamics. While first attacks are distributed relatively broadly over the cricket's body, subsequent attacks are directed more and more precisely to the thorax with the help of corrective head turns that the shrews perform. Thus, shrews can use contact information from a distant body part of the cricket to guide attacks towards their preferred location.

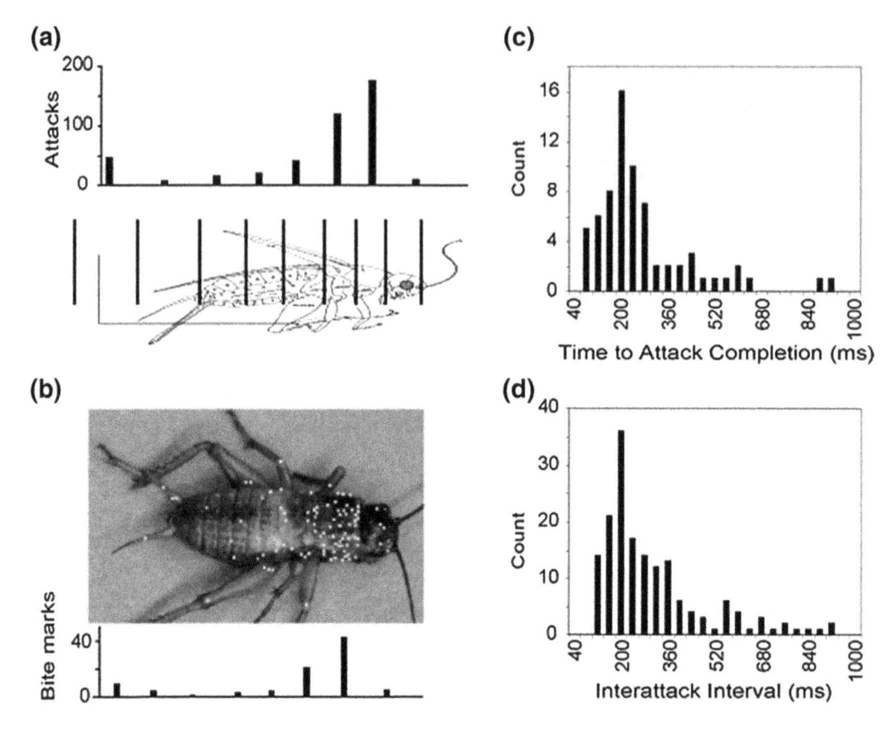

Figure 3 Spatial and temporal analysis of shrew attacks. Shrew attacks are selectively placed on the thorax of crickets. **a** Attack histogram showing the location of bites. **b** Bite mark positions (*yellow squares* superimposed on a cricket photograph) and bite mark histogram. **c** Histogram of attack latencies (time to completion of first attacks) showing the time from encounter to the end point of an attack on a cricket. **d** Histogram of inter-attack intervals (time from attack end point to the beginning point of the successive attack

Whisker Dependence and Tactile Shape Recognition

Both macro- and microvibrissae are required for hunting, with the macrovibrissae being particularly relevant for attack targeting. Shrews attack a plastic replica of a cricket but not other objects of similar size. Altering the shape of crickets by gluing on additional body parts from donor animals reveals that the jumping legs but not the head are key features in prey recognition. Addition of such "ectopic" jumping legs is highly confusing for shrews and leads to dramatic changes in attack pattern. The anterior thorax, a preferred target in normal crickets, is not attacked at all. However, the number of attacks targeted to the legs greatly increases. In summary, these experiments show that tactile shape cues are both necessary and sufficient for evoking attacks.

Shrews distinguish and memorize prey features and their prey representation is motion and size invariant. Shrew behavior appears to be based on the "Gestaltwahrnehmung" of crickets: they form a global construct of a cricket rather than only recognizing local elements. Thus, tactile object recognition in Etruscan shrews shares characteristics of human visual object recognition, but it proceeds faster and occurs in a 20,000-times-smaller brain.

Experience Shapes Tactile Behavior in Shrews

Little is known about the development of behavioral capacities in the somatosensory system. Three lines of evidence suggest that shrew tactile behaviors are not hard-wired but modified by tactile experience. The hunting behavior of young animals differs in subtle but significant ways from the hunting behavior of adults, whisker deprivation in young shrews disrupt the acquisition of normal hunting skills and shrews can acquire new hunting strategies in response to novel prey (Anjum and Brecht, in prep.)

The Etruscan Shrew Brain

Histological analyses as well as physiological techniques have been applied to assess the organization of the Etruscan shrew cortex. Based on Nissl and NeuN stains of coronal brain sections (Figure 4, left), the number of neurons is estimated to be ~1 million per cortical hemisphere, the surface area 11 mm^2 per hemisphere and the volume 4.5 mm^3 (Naumann 2008). The cortex is between 300 and 600 μm thick.

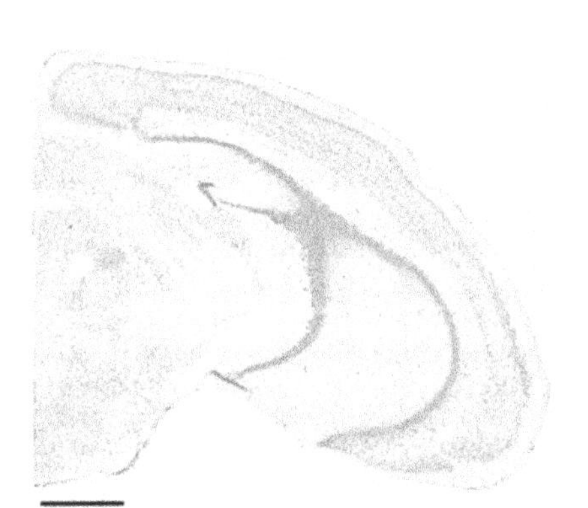

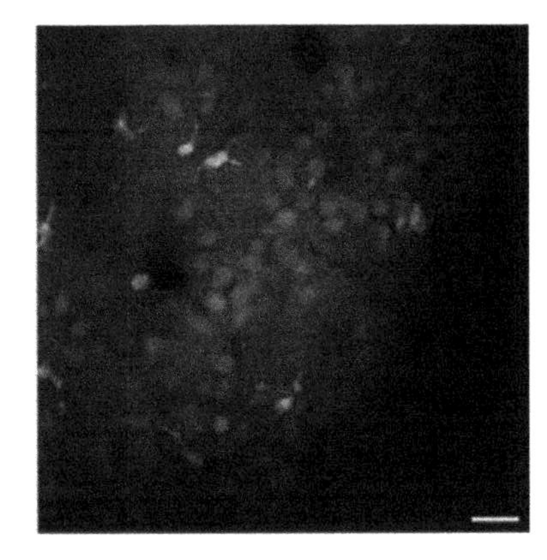

Figure 4 The Etruscan shrew brain. *Left* Coronal section through a shrew brain (Nissl stain), *scale bar* 500 μm. *Right* Two-photon microscopy of shrew cortex layer 2, *scale bar* 20 μm. Neurons are labeled with the green Calcium indicator dye OregonGreenBapta1-AM, which is also taken up by astrocytes. In addition, astrocytes are labeled with the *red* dye sulforhodamine and thus appear *yellow* in this overlay

The small brain size of the Etruscan shrew and its thin cortex makes this animal uniquely accessible for imaging based approaches to study brain function. In particular, using [http://en.wikipedia.org/wiki/Two-photon_excitation_microscopy two-photon-microscopy] might allow visualizing the structure and function of cortical networks in unprecedented completeness (Figure 4, right; Roth-Alpermann, Houweling and Brecht, unpublished).

Cortical sensory areas have been delineated using multi-unit electrophysiological mapping of sensory responses (Brecht et al. submitted). Large parts of Etruscan shrew cortex (i.e. 60 % of the total neocortical surface) respond to sensory stimuli. A small visual and a small auditory area have been identified. The majority of recording sites respond to tactile stimuli and more than half of these sites respond to macrovibrissae stimulation.

These findings demonstrate a remarkable degree of tactile specialization in the Etruscan shrew cortex. In comparison to other mammals studied so far, it is clear that the Etruscan shrew is one of the most extreme tactile specialists studied to date.

Internal References

Braitenberg, V (2007). Brain. *Scholarpedia* 2(11): 2918. http://www.scholarpedia.org/article/Brain.

Llinas, R (2008). Neuron. *Scholarpedia* 3(8): 1490. http://www.scholarpedia.org/article/Neuron.

External References

Anjum, F and Brecht, M (in preparation). Tactile experience shapes prey-capture behavior in Etruscan shrews.

Anjum, F; Turni, H; Mulder, P G H; van der Burg, J and Brecht, M (2006). Tactile guidance of prey capture in Etruscan shrews. *Proceedings of the National Academy of Sciences of the United States of America* 103(44): 16544–16549.

Archibald, J D; Averianov, A O and Ekdale, E G (2001). Late Cretaceous relatives of rabbits, rodents, and other extant eutherian mammals. *Nature* 414(6859): 62–65.

Aulagnier, S et al (2008). *Suncus etruscus*. In: IUCN 2009. IUCN Red List of Threatened Species, Version 2009.2. http://www.iucnredlist.org. Downloaded on 10 November 2009.

Brecht, M; Anjum, F; Naumann, R and Roth-Alpermann, C (submitted). Cortical organization in the Etruscan shrew (*Suncus etruscus*).

Fons, R; Sender, S; Peters, T and Jürgens, K D (1997). Rates of rewarming, heart and respiratory rates and their significance for oxygen transport during arousal from torpor in the smallest mammal, the Etruscan shrew *Suncus etruscus*. *The Journal of Experimental Biology* 200(Pt10): 1451–1458.

Grünwald, A (1969). Untersuchungen zur Orientierung der Weisszahnspitzmäuse (Soricidae – Crocidurinae). *Zeitschrift für vergleichende Physiologie* 65: 91–217.
Jürgens, K D (2002). Etruscan shrew muscle: The consequences of being small. *The Journal of Experimental Biology* 205(Pt15): 2161–2166.
Naumann, R K (2008). *Neuroanatomy of the Etruscan Shrew*. Diploma thesis. Humboldt-University, Berlin, Germany.
Wilson, D E and Reeder, D A M (2005). *Mammal Species of the World: A Taxonomic and Geographic Reference*. Baltimore: Johns Hopkins University Press.

External links

http://www.activetouch.de/index.php?id=7&L=1 Movies of Etruscan shrew prey capture.
http://www.activetouch.de The authors' website.
http://www.biotact.org Research project developing novel tactile sensory technologies inspired by the vibrissal sensory systems of mammals.
http://www.activetouch.org Active touch community website.

Vibrissal Touch in Pinnipeds

Wolf Hanke and Guido Dehnhardt

Vibrissal touch is the sensory modality that enables animals to detect and analyse objects and to orient themselves using their **vibrissae** (whiskers). **Pinnipeds** (**Pinnipedia**) are aquatic carnivores of the families **Phocidae** (**true seals**), **Otariidae** (**eared seals**), and **Odobenidae** (**walruses**). All pinnipeds possess prominent vibrissae (whiskers) in the facial region (Figure 1). Pinnipeds use their whiskers for orientation by directly touching objects (Dehnhardt 1990, 1994; Dehnhardt and Dücker 1996; Dehnhardt and Kaminski 1995; Dehnhardt and Mauck 2008; Kastelein et al. 1990) and by perceiving and analyzing water movements (Dehnhardt et al. 1998a; Dehnhardt et al. 2001; Gläser et al. 2011; Hanke et al. 2012, 2013; Wieskotten et al. 2010a, b, 2011).

Morphology and Anatomy

The vibrissae of pinnipeds are mostly located on the snout (mystacial vibrissae), but also above the eyes (supraorbital or superciliary vibrissae) and above the nares (rhinal vibrissae, only in true seals). Numbers of mystacial vibrissae range from approximately 15 (in the Ross seal *Ommatophoca*) to 350 (in the walrus *Odobenus*) on each side (Ling 1977), with considerable intraspecific as well as interspecific variability. True seals and eared seals have approximately 3 to 7 supraorbital vibrissae above each eye (Ling 1977), while supraorbital vibrissae in the walrus are rarely noticeable in animals more than a few weeks old (Fay 1982). True seals have one (rarely two) additional vibrissa above each nare (rhinal vibrissae) (Ling 1977). In the harbour seal (*Phoca vitulina*), which is the true seal species best studied behaviourally to date, there are approximately 44 mystacial vibrissae (Dehnhardt and Kaminski 1995), exactly one rhinal vibrissa (Ling 1977) and usually 5 supraorbital vibrissae (pers. obs.) per side. The California sea lion (*Zalophus*

W. Hanke (✉)
Institute for Biosciences, University of Rostock, Rostock, Germany

G. Dehnhardt
Institute for Biosciences, University of Rostock, Rostock, Germany

T.J. Prescott et al. (eds.), *Scholarpedia of Touch*, Scholarpedia,
DOI 10.2991/978-94-6239-133-8_9

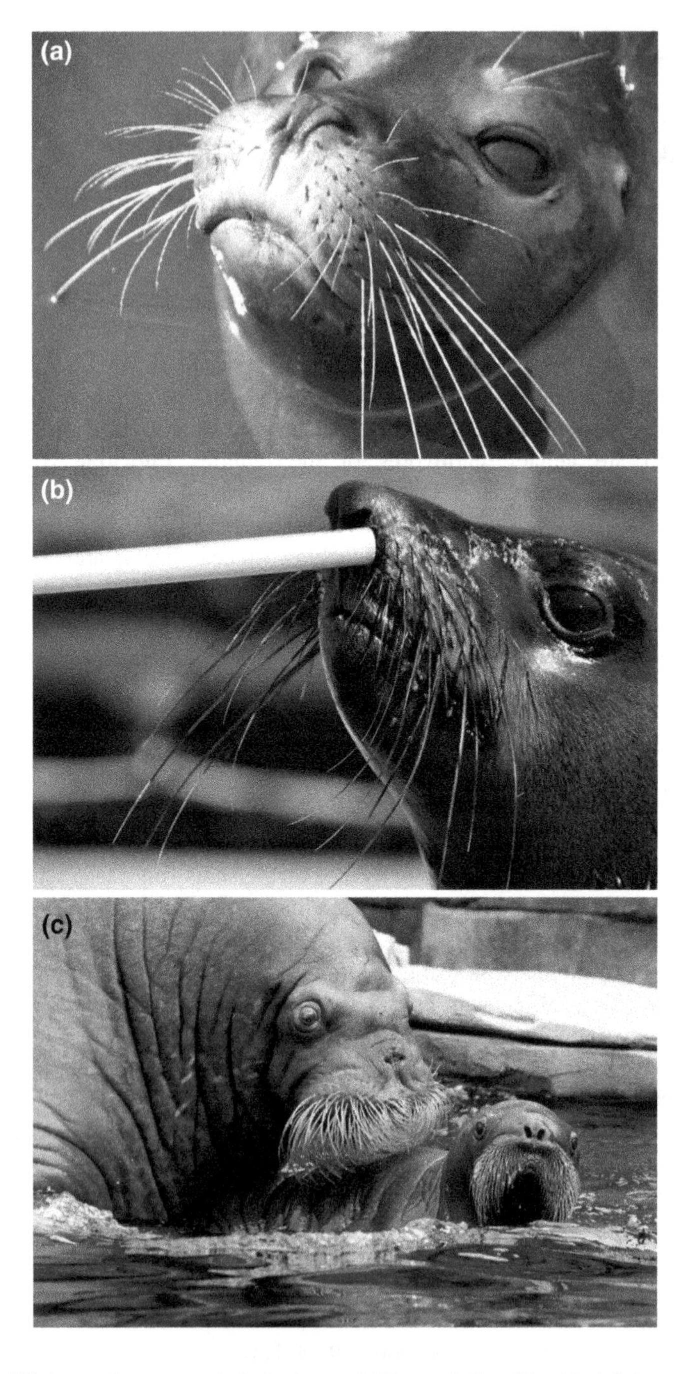

Figure 1 Vibrissae **a** in a true seal, the harbor seal (*Phoca vitulina*, Phocidae); **b** in an eared seal, the South African sea lion (*Arctocephalus pusillus*, Otariidae); **c** in the walrus (*Odobenus rosmarus*, Odobenidae). **a, b** Marine Science Center Rostock, reproduced from Hanke et al. (2013), with kind permission of Springer Science+Business Media; **c** photographed at Tierpark Hagenbeck, Hamburg, Germany; Copyright: Götz Berlik

californianus), which is the eared seal species best studied behaviourally to date, has 38 mystacial vibrissae on each side (counted in 4 animals) (Dehnhardt 1994), no rhinal vibrissae (Ling 1977), but supraorbital vibrissae, which are less prominent than in true seals (pers. obs.).

Each vibrissa emerges from a vibrissal **follicle-sinus complex** (F-SC). F-SCs are found at the base of the vibrissae of all mammals and constitute complex and richly innervated sensory organs. They consist of large blood-filled sinuses in a dermal capsule, and the hair follicle around the hair shaft. The blood sinuses supply nutrients, but must also crucially influence the biomechanics of vibrissal sensing, as they form the bearing where the vibrissa is supported. Pinnipeds have particularly large F-SCs that can reach a length of more than 2 cm (Marshall 2006). F-SCs are richly endowed with receptor cells and nerve endings.

Figure 2 shows the F-SC of a ringed seal (*Phoca hispida*) based on information and presentations in Dehnhardt et al. (1998b), Hyvärinen and Katajisto (1984), and Hyvärinen et al. (2009). F-SCs of California sea lions (*Zalophus californianus*)

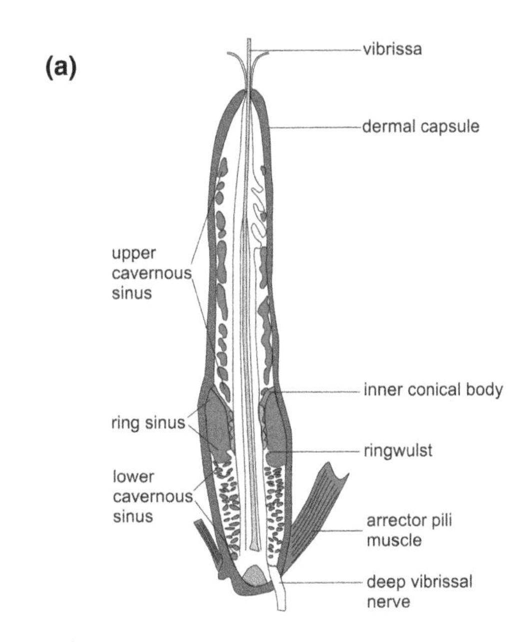

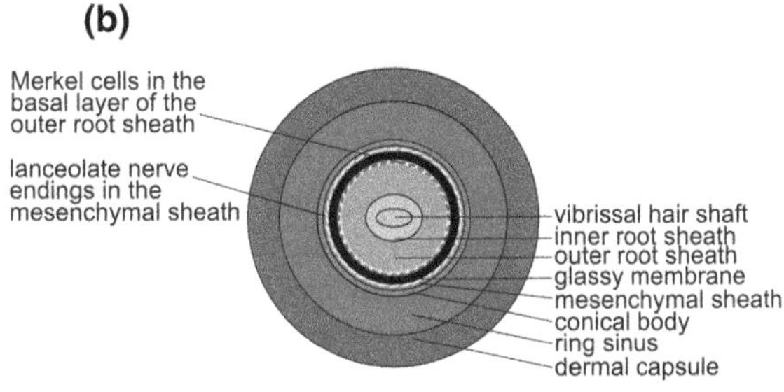

Figure 2 Vibrissal follicle-sinus complex (F-SC) of a ringed seal (Phoca hispida, Phocidae).
a Longitudinal section of the F-SC of a ringed seal (Phoca hispida, Phocidae).
b transverse section of the same F-SC at the level of the ring sinus above the ringwulst, showing the different layers surrounding the hair, and the position of Merkel cells and lanceolate nerve endings on either side of the glassy membrane. Authors' own work based on information in Dehnhardt et al. (1998b), Hyvärinen (1989), Hyvärinen and Katajisto (1984), and Hyvärinen et al. (2009)

(Stephens et al. 1973), bearded seals (*Erignathus barbatus*) (Marshall et al. 2006) and Northern elephant seals (*Mirounga angustirostris*) (McGovern et al. 2014) are generally similar to the F-SCs of ringed seals. The F-SCs of all pinnipeds are particularly large and possess a feature that distinguishes them from the F-SCs of almost all other mammals except the sea otter (*Enhydra lutris*): the blood sinus is not bipartite, but tripartite. In addition to a (lower) cavernous sinus and a ring sinus found in all mammalian F-SCs, pinnipeds (Stephens 1973; Hyvärinen and Katajisto 1984) and sea otters (Marshall et al. 2014) have an upper cavernous sinus (Figure 2a). The upper cavernous sinus, possibly in addition to biomechanical functions that are not yet fully understood, serves to protect the receptors from cooling by the surrounding water (Erdsack et al. 2014). The ring sinus surrounds the vibrissa basally to the upper cavernous sinus, separating it from the lower cavernous sinus at the basal end of the F-SC. The ring sinus is strongly believed to play a crucial role in the biomechanics of the vibrissa. The ring sinus is separated from the upper cavernous sinus by the inner conical body, a collagenous structure. At the level of the ring sinus, a ringwulst (also called ring bulge) surrounds the outer root sheath of the hair. The ringwulst is an asymmetrical bulge of connective tissue attached to the glassy membrane by strong collagenous bundles. The F-SC is equipped with mechanoreceptors at the level of the lower cavernous sinus and especially the ring sinus, but not the upper cavernous sinus. Figure 2b shows a transverse section of the F-SC of a ringed seal at the lower part of the ring sinus, above the ringwulst. The vibrissal hair shaft is surrounded subsequently by an inner root sheath, an outer root sheath, a glassy membrane, and a mesenchymal root sheath (these are general features of a hair follicle), followed by the conical body, the ring sinus, and a dermal capsule. Nerve endings are located on each side of the glassy membrane. On the inside of the glassy membrane, within the basal layer of the outer root sheath, there are Merkel cells in Merkel cell-neurite complexes; on the outside of the glassy membrane, there are lanceolate nerve endings within the outer root sheath. The number of Merkel cell-neurite complexes is up to 20 000, and the number of lanceolate endings is approximately 1000–8000 (Hyvärinen et al. 2009). The F-SC also contains 100–400 lamellated endings and numerous free nerve endings in the ring sinus and lower cavernous sinus area (Dehnhardt and Mauck 2008).

The F-SC is innervated by a deep vibrissal nerve (DVN) that penetrates the outer capsule at the base of the F-SC. A superficial vibrissal nerve, as found in other mammals, is lacking in pinnipeds. The DVN is part of the trigeminal nerve and can contain approximately 1000–1600 nerve fibers in harp seals (Hyvärinen and Katajisto 1984), bearded seals (Marshall et al. 2006) and Northern elephant seals (McGovern et al. 2014), that is about ten times as many as in the rat, the cat and most other terrestrial mammals (Hyvärinen et al. 2009).

Little is known about the central neuroanatomy of the vibrissal system of pinnipeds (see below, neurophysiology).

The vibrissal hair shaft has an ellipsoidal cross section (Watkins and Wartzok 1985; Hyvärinen 1989; Ginter et al. 2010), contrary to the vibrissal hair shafts of terrestrial mammals, which are round in cross section. The vibrissal hair shaft (often

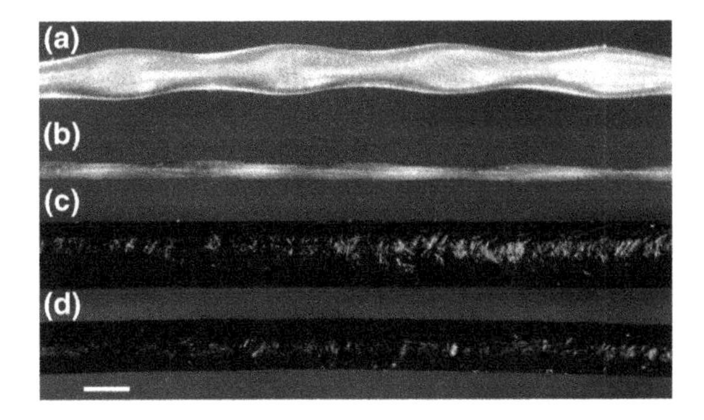

Figure 3 Structure of typical true seal and eared seal vibrissae. **a, b** The vibrissal hair shaft of a harbour seal (true seals) in *dorsal* and *frontal view*. **c, d** The vibrissal hair shaft of a California sea lion (eared seals) in *frontal* and *dorsal view*. The vibrissa of the true seal is undulated, while the vibrissa of the eared seal is smooth. Vibrissae of walruses (not shown) are smooth. From Hanke et al. (2010), under the permission for authors of original work granted by the Journal of Experimental Biology

called the vibrissa for short) is strongly flattened in most of the true seals (Dehnhardt and Mauck 2008), and slightly flattened in eared seals (Dehnhardt and Mauck 2008) and walruses (Fay 1982). In most true seals, the vibrissa is undulated (synonymously called "waved" or "beaded") (King 1983; Ling 1977, Ginter et al. 2010, 2012), that means that the diameter changes along the length sinusoidally with a period of a few millimeters, reminding of a pearl necklace at first glance (Figures 1a, 3a, b). However, the greater and the smaller diameter of the cross section change out of phase by approximately 180 deg. along the length of the vibrissa (Dehnhardt and Mauck 2008; Hanke et al. 2010). The shape of the undulated vibrissae shows interspecific variation (Ginter et al. 2012). Within the true seals, the monk seals (*Monachus* spp.) and the bearded seal (*Erignathus barbatus*) are exceptions, their vibrissae are smooth in outline. Eared seals (Figures 1b, 3 c, d) and the walrus (Figure 1c) have smooth vibrissae.

Neurophysiology

Dykes (1975) recorded action potentials from the infraorbital nerve (a branch of the trigeminal nerve) in grey seals (*Halichoerus grypus*) and harbour seals (*Phoca vitulina concolor*). The nerve consisted of about 45 bundles. In most bundles, most of the nerve fibers were associated with vibrissae. Among these, slowly adapting (SA) and rapidly adapting (RA) fibers could be distinguished; two thirds of these fibers were RA. SA fibers were defined by ongoing neural activity after a vibrissa had been deflected and held steady for about 1 s; in RA fibers, neural activity had ceased at that point.

Ladygina et al. (1985) mapped somatic projections in the cerebral cortex of northern fur seals (*Callorhinus ursinus*). They recorded neuronal activity (multiple spike activity) in a depth range of 600–1000 μm and monitored it acoustically (by listening to the action potentials fed into a loudspeaker). Neuronal activity upon touching the body surface was found in a region on the dorso-rostral part of the cerebral cortex, bounded caudally by the anterior suprasylvian sulcus, dorsomedially by the ansate sulcus, rostrally by the postcruciate and coronal sulci. Only contralateral touch of the body elicited responses in this part of the cortex. Receptive fields (body surface areas that on touch elicited response at a given electrode position in the brain) were smallest on the tip of the nose and anterior regions of the upper and lower lips. Within the somatosensory cortex, vibrissae were overrepresented with each vibrissa corresponding to a considerable portion of the cortex, up to 3–4 mm in diameter. Representations of the vibrissae were arranged somatotopically in the cortex. Intensive neuronal responses were recorded from all vibrissae regardless of position or length. The region mapped was believed to correspond to the primary somatosensory cortex (SI) studied in other animals. The method of acoustical monitoring did not allow for further conclusions beyond the somatotopic organization of the cortex and the approximate size of receptive fields and the approximate size of the cortex representing a vibrissa.

In summary, neurophysiological data on the vibrissal system of pinnipeds is scarce.

Biomechanics

The biomechanical properties of the vibrissal hair shaft are largely influenced by its moment of inertia (that is, by its cross-sectional shape). The modulus of elasticity (a measure of how the material itself deforms elastically under stress) of a harbour seal vibrissa is similar to other keratinous structures and decreases after immersion in water (Marine Science Center Rostock, unpublished data). Contrary to the vibrissae of rats, the modulus of elasticity of harbour seal vibrissae changes along the length of the vibrissa (Hans et al. 2014).

The undulated shape of harbour seal vibrissae reduces vortex-induced vibrations as the animal swims (Hanke et al. 2010). Vortex-induced vibrations are vibrations of cylindroid objects in a perpendicular flow that are caused by vortices detaching from the object on alternating sides. Figure 4 represents a numerical simulation of the flow behind a harbour seal vibrissa compared with the flow behind cylinders with an elliptical cross section and a round cross section, respectively. Behind the vibrissa, vortices form at a greater distance, they are weaker, and they constitute a complex three-dimensional pattern that makes forces on the vibrissa more symmetrical (Hanke et al. 2010; Witte et al. 2012).

By reducing vortex-induced vibrations at the vibrissa, the signal-to-noise ratio at the receptors in the F-SC is enhanced.

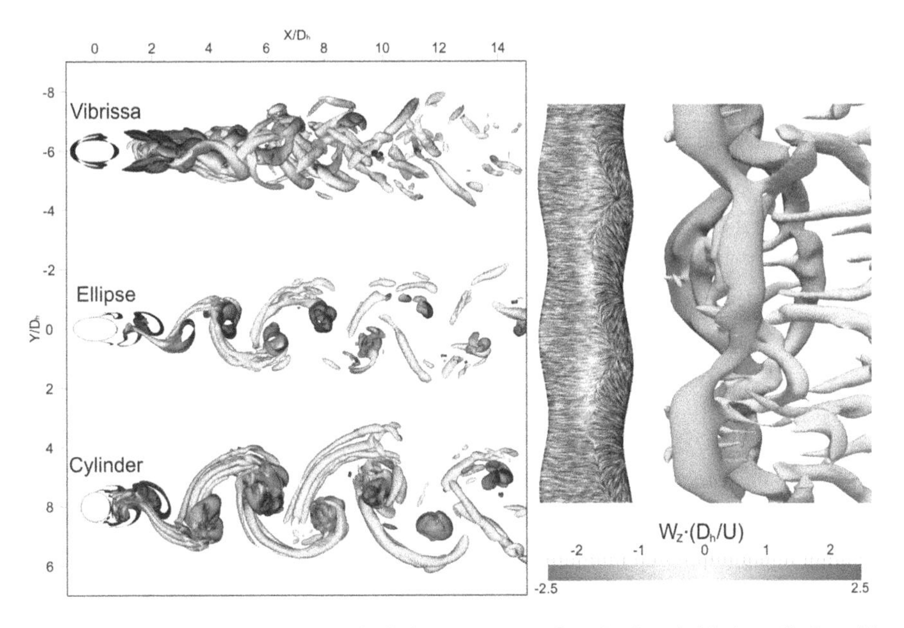

Figure 4 Flow behind a harbour seal vibrissa as compared to the flow behind a cylinder with elliptical and round cross section. Numerical simulation of the wake-flow behind different cylinder bodies at a Reynolds number of 500, vortex cores depicted as isosurfaces. Color: cross-stream vorticity. *Left panel* (*top* to *bottom*): wake behind a vibrissa of a harbor seal, behind a cylinder with elliptic cross-section and behind a circular cylinder. The radius ratio of the elliptic cylinder corresponds to the mean radius ratio along the vibrissa. *Right panel*: *side view* of the wake behind a vibrissa of a harbor seal. The *surface flow pattern* clearly indicates a wavy separation line along the axis of the vibrissa. From Hanke et al. (2010), under the permission for authors of original work granted by the *Journal of Experimental Biology*

Not surprisingly, the orientation of the flattened vibrissal cross section towards the flow influences fluid mechanical properties including vortex-induced vibrations (Murphy et al. 2013). Phocid vibrissae appear specifically designed to reduce vortex-induced vibrations during forward swimming.

Behaviour

General Observations

On shore, use of the vibrissae appears to be limited to social interactions. Pinnipeds use their vibrissae predominantly under water, making it difficult to observe this behaviour in the natural habitat. Seals have been observed in the wild that were blind but nevertheless well nourished (Hyvärinen 1989). In the light of the fact that pinnipeds have highly developed vibrissal systems while they do not possess any noticeably advanced form of biosonar (Schusterman et al. 2000), it was hypothesized that the vibrissae play a most important role in prey capture and underwater feeding.

Use of the vibrissae in underwater orientation tasks other that feeding has been investigated by Wartzok et al. (1992) and Oliver (1978). Wartzok and coauthors found that ringed seals and Weddell seals navigating under ice towards breathing holes can use their vibrissae to center within the hole, but not to find the hole. Oliver found that a blindfolded grey seal in an artificial pool could navigate through a maze of vertical rods without displacing the rods. It was suggested that the vibrissae played a role in this task.

Experiments on Direct Touch

Pinnipeds have been investigated for their ability to discriminate objects by touching them with their vibrissae in behavioural experiments (reviewed by Dehnhardt 2002; Dehnhardt and Mauck 2008). Psychophysical experiments have shown that all three groups of pinnipeds (investigated in harbour seals as representatives of the true seals, California sea lions as representatives of the eared seals, and in the walrus) can discriminate objects with high accuracy by direct touch with their vibrissae. This direct vibrissal touch is an example of a haptic sense, i.e. information from cutaneous mechanosensation and kinesthetic mechanosensation is integrated. Pinnipeds applied adequate exploratory strategies by performing head movements while whiskers were in contact with the objects. Contrary to terrestrial mammals such as rats, pinnipeds did not apply movement of whiskers relative to the head (whisking), although their mystacial vibrissae are mobile.

The experimental setup for a size discrimination experiment is exemplified in Figure 5a, b. A sea lion was trained to wait near two test objects (Perspex discs of different size) while blindfolded (Figure 5a). On an acoustic start signal, the sea lion began investigating the objects with its vibrissae (Figure 5b). The animal protracted its vibrissae as it actively investigated the objects (Figure 5b). Similar behaviour is observed in pinnipeds in experiments on shape discrimination (Dehnhardt 1990; Dehnhardt and Dücker 1996; Dehnhardt and Kaminski 1995) or surface structure discrimination (Dehnhardt et al. 1998b). Alternatively, in a discrimination task where size differences between objects were large, harbour seals appeared to discriminate sizes by the number of vibrissae that were in touch with the objects (Grant et al. 2013). Sea lions performing the complex sensorimotor task of balancing balls on their snouts moved their vibrissae independently of head movements to some degree (Milne and Grant 2014).

Tactile shape recognition in California sea lions matches their visual shape recognition in speed and reliability. Harbour seals tested in similar experiments detected size differences of objects tactually with a resolution similar to the haptic resolution of some primates. Behavioural experiments also showed that tactile

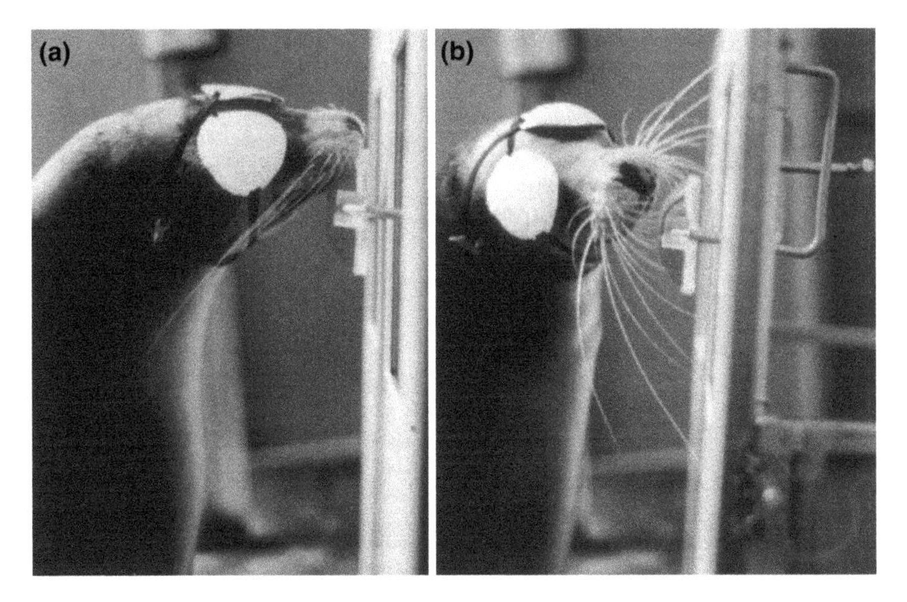

Figure 5 Behavioural experiment on size discrimination by vibrissal touch in a California sea lion. **a** The sea lion waits near the test objects with its vibrissae adducted. **b** Upon a signal that indicates the start of the experiment, the sea lion protracts its vibrissae, holds them in a stable position and investigates the objects using head movements. Reproduced from Dehnhardt (1994), with kind permission of Springer Science+Business Media

sensitivity in harbour seals is not affected by ambient temperature (Dehnhardt et al. 1998b), contrary to human tactile sensitivity. Temperature measurements on vibrissal pads and inside F-SCs (Erdsack et al. 2014) confirmed the hypothesis that extensive heating of the F-SCs is responsible for this performance.

Experiments on Hydrodynamic Perception

True seals and eared seals can detect water movements with high sensitivity using their whiskers (Dehnhardt et al. 1998a, 2001; Hanke et al. 2013). Harbour seals (*Phoca vitulina*) detect the water movements caused by a sinusoidally vibrating sphere (hydrodynamic dipole stimuli) at water velocities down to 245 µm/s when presented with 3 s long stimuli in a go/no-go experiment (Dehnhardt et al. 1998a). Figure 6 depicts the experimental setup and results from the experiment on hydrodynamic dipole detection in harbor seals. California sea lions are even more sensitive to this kind of stimuli (Dehnhardt and Mauck 2008). Harbour seals can not only detect hydrodynamic dipole stimuli, but also discriminate between stimuli of different amplitude (Dehnhardt and Mauck 2008).

Harbour seals and California sea lions are also able to detect and follow hydrodynamic trails (Dehnhardt et al. 2001; Gläser et al. 2011). Hydrodynamic

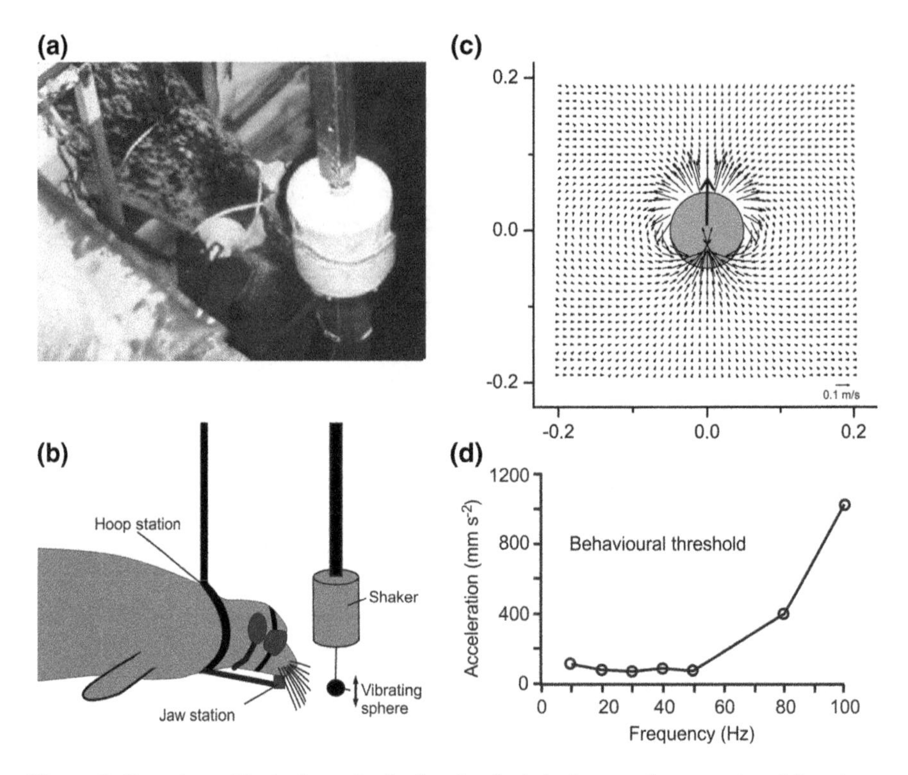

Figure 6 Detection of hydrodynamic dipole stimuli. A harbour seal was stationed in a hoop station (**a, b**). The seal was trained to respond to hydrodynamic dipole stimuli and to remain in station in the absence of hydrodynamic dipole stimuli in a go/no go psychophysical experiment. Dipole stimuli were calculated using potential theory; the flow pattern is shown in (**c**). Behavioural sensory thresholds (here in terms of water acceleration) are shown in (**d**). At 50 Hz, the behavioural threshold was at 77 mm/s^2 water acceleration, or 245 μm/s water velocity, or 0,8 μm water displacement. After the experiment performed by Dehnhardt et al. (1998a, b), reproduced from Hanke et al. (2013), with kind permission of Springer Science+Business Media

trails (Hanke et al. 2000) are the water movements left behind by swimming objects such as prey fish. They can last up to several minutes after a fish swam by (Hanke 2014; Hanke and Bleckmann 2004; Hanke et al. 2000; Niesterok and Hanke 2013) and thus form a trail tens of meters in length that leads to a prey fish. Using this kind of stimulus, seals extend their range of hydrodynamic perception, which would otherwise be limited to a few tens of centimeters. Behavioural experiments with both species led to the conclusion that harbour seals are more accomplished hydrodynamic trail followers than California sea lions. This is in accordance with the effect of the undulated whisker structure in harbour seals that reduces vortex-induced vibrations (see above, Biomechanics).

Figure 7 exemplifies an experiment where a harbour seal followed the hydrodynamic trail generated by a miniature submarine. The submarine (Figure 7a) was remote controlled to swim arbitrary paths. The harbour seal was blindfolded and

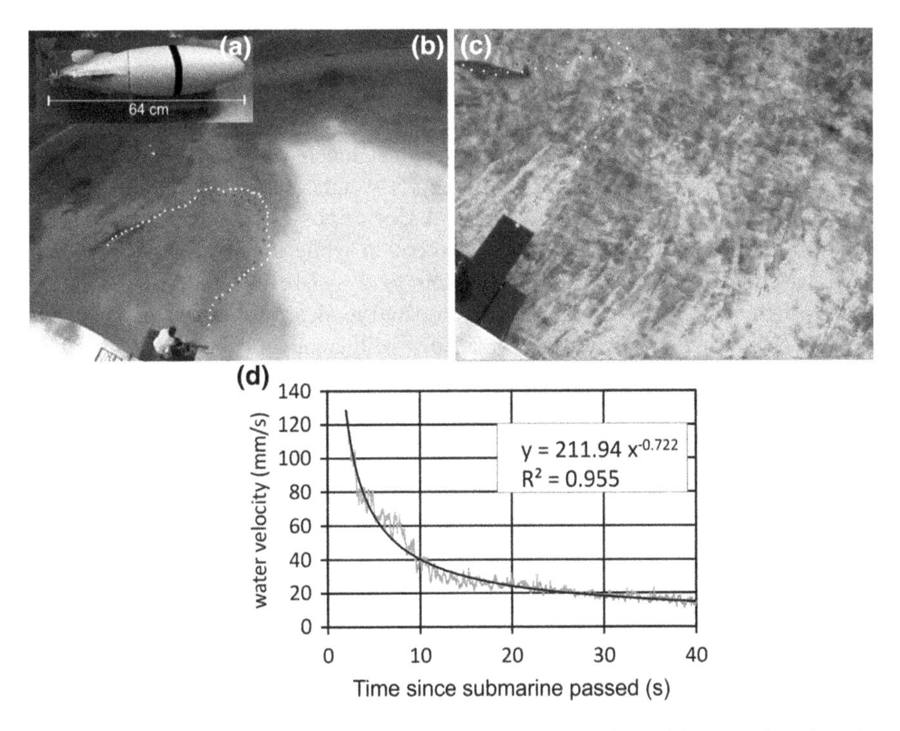

Figure 7 Hydrodynamic Trail Following: Swimming paths of a miniature submarine that generated an experimental hydrodynamic trail, and a harbour seal that followed that trail after a time delay while blindfolded using only its vibrissae. **a** The miniature submarine. **b, c** Curved swimming paths of the miniature submarine (*red dots*) and the harbour seal (*white dots*). **d** Water velocities in the hydrodynamic trail of the miniature submarine. Reproduced from Hanke et al. (2013), with kind permission of Springer Science+Business Media

used only its vibrissae. Harbour seals can follow the swimming path of the submarine when the submarine and the animal both start at the same position (Figure 7b) or when the animal intercepts the submarine's trail (Figure 7c) (Dehnhardt et al. 2001).

Harbour seals can extract more information from a hydrodynamic trail beyond the mere presence of an object. They can discriminate different moving directions, sizes and shapes of the objects that generated the trail (Wieskotten et al. 2010a, 2011). Harbour seals also cope with the reduced hydrodynamic trail that passively gliding objects generate as compared to propelled objects, in line with the hypothesis that they need to pursue fish that swim in burst-and-coast mode (Wieskotten et al. 2010b). Hydrodynamic trail following can also be used to follow other seals instead of fish (Schulte-Pelkum et al. 2007). This may serve in the interaction of pups with their mothers and many other intraspecific interactions, as living in groups offers a variety of benefits (Krause and Ruxton 2002).

Feeding Ecology

Pinnipeds not only live in different climate zones, but also have different feeding habits. Accordingly, the roles of vibrissal direct touch versus vibrissal hydrodynamic perception are expected to be differently weighed across species. This is in line with the observation that most true seal species possess the undulated whisker structure that improves hydrodynamic perception while the animal swims, while eared seals and walruses do not. True seals tend to live in temperate and arctic waters where vision is more limited as compared to clear tropical waters; catching actively swimming fish when vision is low will greatly benefit from improved hydrodynamic perception. Exceptions among the true seals are the bearded seal (*Erignathus barbatus*) and the monk seals (*Monachus* spp.) that do not possess the specialized whisker structure. The former feeds on benthic prey (Marshall et al. 2006, 2008 and references therein), while the latter live in clear tropical or subtropical waters.

Different true seal species (Phocidae) possess similar numbers of nerve fibers per vibrissa (Hyvärinen 1989; Marshall et al. 2006; McGovern et al. 2014) (note that fiber counts for eared seals or walruses are not yet available). The picture emerges that numbers of axons per F-SC are similar in different pinniped species. Another marine carnivore, the sea otter (*Enhydra lutris*), possesses a similar number of nerve fibers per F-SC as pinnipeds (Marshall et al. 2014), adding to this picture for marine carnivores. Differences between species appear to manifest themselves mainly in the number of F-SC of the mystacial array, their arrangement on the snout, and the morphology of the hair shafts.

True seals (Phocidae) are mostly opportunistic feeders with a broad prey spectrum including pelagic and benthic species, and their vibrissae are adapted to both feeding situations. The two pinniped species that are most specialized for benthic feeding are the bearded seal (*Erignathus barbatus*, Phocidae) and the walrus (*Odobenus rosmarus*, Odobenidae). They possess the best endowed vibrissal systems in terms of number of vibrissae and therefore also total nerve fibers (Marshall et al. 2006), indicating that vibrissal systems designed for direct touch may generally have higher counts of vibrissae and F-SCs than vibrissal systems designed for hydrodynamic perception.

References

Dehnhardt, G (1990). Preliminary results from psychophysical studies on the tactile sensitivity in marine mammals. In: J A Thomas and R A Kastelein (Eds.), *Sensory abilities of cetaceans* (pp. 435–446). New York: Plenum Press.

Dehnhardt, G (1994). Tactile size discrimination by a California sea lion (*Zalophus californianus*) using its mystacial vibrissae. *Journal of Comparative Physiology A - Neuroethology, Sensory, Neural, and Behavioral Physiology* 175: 791–800.

Dehnhardt, G (2002). Sensory systems. In: A R Hoelzel (Ed.), *Marine Mammal Biology* (pp. 116-141). Oxford: Blackwell Publishing.

Dehnhardt, G and Dücker, G (1996). Tactual discrimination of size and shape by a California sea lion (*Zalophus californianus*). *Animal Learning and Behavior* 24: 366–374.

Dehnhardt, G and Kaminski, A (1995). Sensitivity of the mystacial vibrissae of harbour seals (*Phoca vitulina*) for size differences of actively touched objects. *Journal of Experimental Biology* 198: 2317–2323.

Dehnhardt, G and Mauck, B (2008). Mechanoreception in secondarily aquatic vertebrates. In: J G M Thewissen and S Nummela (Eds.), *Sensory Evolution on the Threshold: Adaptations in Secondarily Aquatic Vertebrates* (pp. 295–314). Berkeley: University of California Press.

Dehnhardt, G; Mauck, B and Bleckmann, H (1998a). Seal whiskers detect water movements. *Nature* 394: 235–236.

Dehnhardt, G; Mauck, B; Hanke, W and Bleckmann, H (2001). Hydrodynamic trail following in harbor seals (*Phoca vitulina*). *Science* 293: 102–104.

Dehnhardt, G; Mauck, B and Hyvärinen, H (1998b). Ambient temperature does not affect the tactile sensitivity of mystacial vibrissae of harbour seals. *Journal of Experimental Biology* 201: 3023–3029.

Dykes, R W (1975). Afferent fibers from mystacial vibrissae of cats and seals. *Journal of Neurophysiology* 38: 650–662.

Erdsack, N; Dehnhardt, G and Hanke, W (2014). Thermoregulation of the vibrissal system in harbor seals (*Phoca vitulina*) and Cape fur seals (*Arctocephalus pusillus pusillus*). *Journal of Experimental Marine Biology and Ecology* 452: 111–118.

Fay, F H (1982). Ecology and biology of the Pacific walrus, *Odobenus rosmarus divergens* Illiger. *North American Faun* 74: 1–279.

Ginter, C C; Fish, F E and Marshall, C D (2010). Morphological analysis of the bumpy profile of phocid vibrissae. *Marine Mammal Science* 26: 733–743.

Ginter, C C; DeWitt, T J; Fish, F E and Marshall, C D (2012). Fused traditional and geometric morphometrics demonstrate pinniped whisker diversity. *Plos One* 7: 1–10.

Gläser, N; Wieskotten, S; Otter, C; Dehnhardt, G and Hanke, W (2011). Hydrodynamic trail following in a California sea lion (*Zalophus californianus*). *Journal of Comparative Physiology A - Neuroethology, Sensory, Neural, and Behavioral Physiology* 197: 141–151.

Grant, R A; Wieskotten, S; Wengst, N; Prescott, T and Dehnhardt, G (2013). Vibrissal touch sensing in the harbor seal (*Phoca vitulina*): How do seals judge size? *Journal of Comparative Physiology A - Neuroethology, Sensory, Neural, and Behavioral Physiology* 199: 521–533.

Hanke, W (2014). Natural hydrodynamic stimuli. In: H Bleckmann, J Mogdans and S Coombs (Eds.), *Flow Sensing in Air and Water - Behavioural, Neural and Engineering Principles of Operation* (pp. 3–29). Berlin: Springer.

Hanke, W and Bleckmann, H (2004). The hydrodynamic trails of *Lepomis gibbosus* (Centrarchidae), *Colomesus psittacus* (Tetraodontidae) and *Thysochromis ansorgii* (Cichlidae) measured with Scanning Particle Image Velocimetry. *Journal of Experimental Biology* 207: 1585–1596.

Hanke, W; Brücker, C and Bleckmann, H (2000). The ageing of the low-frequency water disturbances caused by swimming goldfish and its possible relevance to prey detection. *Journal of Experimental Biology* 203: 1193–1200.

Hanke, W; Wieskotten, S; Marshall, C D and Dehnhardt, G (2013). Hydrodynamic perception in true seals (Phocidae) and eared seals (Otariidae). *Journal of Comparative Physiology A - Neuroethology, Sensory, Neural, and Behavioral Physiology* 199: 421–440.

Hanke, W et al (2012). Hydrodynamic perception in pinnipeds. In: C Tropea and H Bleckmann (Eds.), *Nature-inspired fluid mechanics* (pp. 225–240). Berlin: Springer.

Hanke, W et al (2010). Harbor seal vibrissa morphology suppresses vortex-induced vibrations. *Journal of Experimental Biology* 213: 2665–2672.

Hans, H; Miao, J M and Triantafyllou, M S (2014). Mechanical characteristics of harbor seal (*Phoca vitulina*) vibrissae under different circumstances and their implications on its sensing methodology. *Bioinspiration & Biomimetics* 9(3): 036013.

Hyvärinen, H (1989). Diving in darkness: Whiskers are sense organs of the ringed seal (*Phoca hispida saimensis*). *Journal of Zoology* 218: 663–678.

Hyvärinen, H and Katajisto, H (1984). Functional structure of the vibrissae of the ringed seal (*Phoca hispida*). *Acta Zoologica Fennica* 171: 27–30.

Hyvärinen, H; Palviainen, A; Strandberg, U and Holopainen, I J (2009). Aquatic environment and differentiation of vibrissae: Comparison of sinus hair systems of ringed seal, otter and pole cat. *Brain, Behavior and Evolution* 74: 268–279.

Kastelein, R A; Stevens, S and Mosterd, P (1990). The sensitivity of the vibrissae of a Pacific walrus (*Odobenus rosmarus divergens*). Part 2: Masking. *Aquatic Mammals* 16: 78–87.

King, J E (1983). *Seals of the world*. Ithaca: Cornell University Press.

Krause, J and Ruxton, G D (2002). *Living in Groups*. Oxford: Oxford University Press.

Ladygina, T F; Popov, V V and Supin, A Y (1985). Topical organization of somatic projections in the fur seal cerebral cortex. *Neurophysiology* 17: 246–252.

Ling, J K (1977). Vibrissae of marine mammals. In: R J Harrison (Ed.), *Functional Anatomy of Marine Mammals* (pp. 387–415). London: Academic Press.

Marshall, C D; Amin, H; Kovacs, K M and Lydersen, C (2006). Microstructure and innervation of the mystacial vibrissal follicle-sinus complex in bearded seals, *Erignathus barbatus* (Pinnipedia: Phocidae). *The Anatomical Record Part A: Discoveries in Molecular, Cellular, and Evolutionary Biology* 288: 13–25.

Marshall, C D; Kovacs, K M and Lydersen, C (2008). Feeding kinematics, suction and hydraulic jetting capabilities in bearded seals (*Erignathus barbatus*). *Journal of Experimental Biology* 211: 699–708.

Marshall, C D; Rozas, K; Kot, B and Gill, V (2014). Innervation patterns of sea otter (*Enhydra lutris*) mystacial follicle-sinus complexes: Support for a specialized somatosensory system. *Frontiers of Neuroanatomy*. doi:10.3389/fnana.2014.00121.

McGovern, K A; Marshall, C D and Davis, R W (2014). Are vibrissae of northern elephant seals viable sensory structures for prey capture? *The Anatomical Record*. doi:10.1002/ar.2306100.

Milne, A O and Grant, R A (2014). Characterisation of whisker control in the California sea lion (*Zalophus californianus*) during a complex, dynamic sensorimotor task. *Journal of Comparative Physiology A - Neuroethology, Sensory, Neural, and Behavioral Physiology* 200: 871–879. doi:10.1007/s00359-014-0931-1.

Murphy, C T; Eberhardt, W D; Calhoun, B H; Mann, K and Mann, D A (2013). Effect of angle on flow-induced vibrations of pinniped vibrissae. *PLOS One*. doi:10.1371/journal.pone.0069872.

Niesterok, B and Hanke, W (2013). Hydrodynamic patterns from fast-starts in teleost fish and their possible relevance to predator–prey interactions. *Journal of Comparative Physiology A* 199: 139–149.

Oliver, G W (1978). Navigation in mazes by a grey seal, *Halichoerus grypus* (Fabricius). *Behaviour* 67: 97–114.

Schulte-Pelkum, N; Wieskotten, S; Hanke, W; Dehnhardt, G and Mauck, B (2007). Tracking of biogenic hydrodynamic trails in a harbor seal (*Phoca vitulina*). *The Journal of Experimental Biology* 210: 781–787.

Schusterman, R J; Levenson, D H; Reichmuth, C J and Southall, B L (2000). Why pinnipeds don't echolocate. *Journal of the Acoustical Society of America* 107: 2256–2264.

Stephens, R J; Beebe, I J and Poulter, T C (1973). Innervation of the vibrissae of the California sea lion, *Zalophus californianus*. *Anatomical Record* 176: 421–441.

Wartzok, D; Elsner, R; Stone, H; Kelly, B P and Davis, R W (1992). Under-ice movement and the sensory basis of hole finding by ringed and Weddell seals. *Canadian Journal of Zoology-Revue Canadienne De Zoologie* 70: 1712–1722.

Watkins, W A and Wartzok, D (1985). Sensory biophysics of marine mammals. *Marine Mammal Science* 1(3): 219–260.

Wieskotten, S; Dehnhardt, G; Mauck, B; Miersch, L and Hanke, W (2010a). Hydrodynamic determination of the moving direction of an artificial fin by a harbour seal (*Phoca vitulina*). *The Journal of Experimental Biology* 213: 2194–2200.

Wieskotten, S; Dehnhardt, G; Mauck, B; Miersch, L and Hanke, W (2010b). The impact of glide phases on the trackability of hydrodynamic trails in harbour seals (*Phoca vitulina*). *The Journal of Experimental Biology* 213: 3734–3740.
Wieskotten, S; Mauck, B; Miersch, L; Dehnhardt, G and Hanke, W (2011). Hydrodynamic discrimination of wakes caused by objects of different size or shape in a harbour seal (*Phoca vitulina*). *The Journal of Experimental Biology* 214: 1922–1930.
Witte, M (2012). On the wake flow dynamics behind harbor seal vibrissae – A fluid mechanical explanation for an extraordinary capability. In: C Tropea and H Bleckmann (Eds.), *Nature-inspired fluid mechanics* (pp. 241–260). Berlin: Springer.

Recommended Readings

Dehnhardt, G (2002). Sensory systems. In: A R Hoelzel, *Marine Mammal Biology* (pp. 116–141). Oxford: Blackwell Publishing.
Hanke, W (2014). Natural hydrodynamic stimuli. In: H Bleckmann, J Mogdans and S Coombs, *Flow Sensing in Air and Water - Behavioural, Neural and Engineering Principles of Operation* (pp. 3–29). Berlin: Springer.
Thewissen, J G M and Nummela, S (2008). *Sensory Evolution on the Threshold - Adaptations in Secondarily Aquatic Vertebrates*. Berkeley: University of California Press.

Tactile Hair in Manatees

Roger Reep and Diana K. Sarko

Tactile hairs function to detect mechanosensory stimuli rather than for warmth or protection, which is the main function of pelage hair. Vibrissae, which have a specific set of structural features, are the main class of tactile hair.

Manatee Hair

Florida manatees (Figure 1) have ~5300 vibrissae scattered over the entire body, with ~2000 on the face and ~3300 on the postfacial body (Reep et al. 1998). This is unusual because in other mammals vibrissae are generally confined to the face area (whiskers), with some taxa exhibiting a few vibrissae on the limbs or ventrum (Sokolov and Kulikov 1987). The follicles associated with vibrissae differ from those of other hairs in being associated with a circumferential blood sinus, dense connective tissue capsule, and a variety of mechanoreceptors that are extensively innervated. Because of these distinctions they are referred to as Follicle-Sinus Complexes (FSCs). Vibrissae are specialized for detecting movement rather than for providing protection or warmth. Manatees and dugongs are unique among mammals in having only vibrissae and no other type of hair on their bodies. Why is this the case?

The rock hyrax, one of the closest living relatives of manatees, has vibrissae interspersed among its fur (or pelage) over the entire body (Reep et al. 2007). Since vibrissae are part of highly specialized and well-innervated FSCs, the elaboration of their distribution to the entire body in both manatees and hyraxes indicates a dedication of neural resources (and by extension, perceptual capabilities) to somatosensation. Naked mole-rats have only sparsely scattered hairs that are sensory but these do not have the structural features of vibrissae, possibly representing an example of convergent evolution. For each of these species, elaboration of the

R. Reep (✉)
McKnight Brain Institute, University of Florida, Gainesville, USA

D.K. Sarko
Vanderbilt University, Nashville, TN, USA

T.J. Prescott et al. (eds.), *Scholarpedia of Touch*, Scholarpedia,
DOI 10.2991/978-94-6239-133-8_10

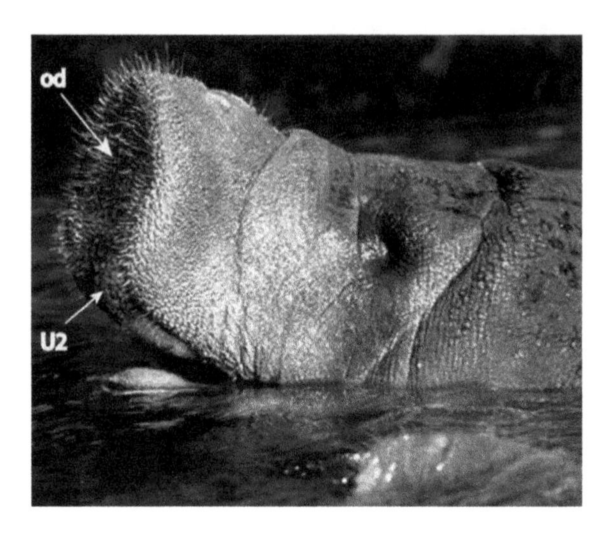

Figure 1 Manatee face. The oral disk (*od*) is flared and its bristle-like hairs are everted. The large perioral bristles of the upper lip (*U2*) are withdrawn into their fleshy pad. Reproduced from Reep and Sarko (2006) with permission from Elsevier Press

tactile system appears to compensate for a reduced reliance on visual cues. This raises an interesting evolutionary question: is the condition of having vibrissae distributed over the entire body a primitive or derived mammalian condition? Furthermore, what developmental mechanisms may have been modified to produce an arrangement of vibrissae on the entire body as opposed to the distribution limited to restricted regions that occurs in most species?

Facial Vibrissae

Two prominent features of the manatee facial region are a broad oral disk (analogous to the mystacial region in other mammals) containing ~600 bristle-like hairs (thin vibrissae) and six distinct paired fields of perioral bristles (thick vibrissae), each containing 8–30 vibrissae per side. Vibrissae in each of these regions of the face vary with respect to length, diameter, and innervation density (Reep et al. 1998, 2001). The largest, stoutest vibrissae are found on the lateral margin of each upper lip. Vibrissae of the oral disk are much thinner than the perioral bristles (diameter/length ratio of ~0.03 vs. 0.30). The number of axons innervating perioral bristles ranges from 70–225 per FSC; the number innervating vibrissae of the oral disk is ~54 per FSC. It is estimated that there is a total of ~110,000 axons innervating the ~2000 facial FSCs. This is comparable to the ~100,00 axons that innervate the nasal appendages in the star-nosed mole (Catania and Kaas 1997), another tactile specialist.

Vibrissae of the oral disk are used for tactile investigation of objects. In the relaxed state the oral disk vibrissae are withdrawn into the fleshy folds of skin of the oral disk. As a manatee approaches an object to be investigated, it engages in a 'flare response', in which orofacial muscles contract to flatten and expand the oral disk (Marshall et al. 1998). This causes the vibrissae of the oral disk to extend outward as a collective sensory array that is used to perform tactile investigations during direct contact with objects. These investigations can take the form of brief touches or sweeping scans.

Perioral bristles are used to grasp objects, including plants that are ingested during feeding (Marshall et al. 1998). Because this grasping involves the face rather than the hand, it is referred to as 'oripulation'. Often, tactile scanning by the oral disk is followed by vigorous, repetitive oripulation by the large perioral vibrissae on the upper lips. This occurs not only with plants during feeding, but also during investigation of novel objects including anchor lines, bathing suits, and human legs. Such a behavioral specialization may be indicative of extensive sensorimotor integration within the central nervous system representations for the face. Psychophysical tests of a few captive subjects have demonstrated that tactile acuity by manatees using the vibrissae of the face is at least as good as that of an elephant using the tip of its trunk (Weber fraction of 3–14 %; Bachteler and Dehnhardt 1999; Bauer et al. 2012). Interestingly, the eyes often close during feeding, which may heighten tactile acuity. Manatees also detect hydrodynamic stimuli of 15–150 Hz with high sensitivity, using the facial vibrissae (Gaspard et al. 2013).

Postfacial Vibrissae

The postfacial vibrissae are not actively involved in tactile exploration. It has been hypothesized that they function to detect water movements, thus functioning like a mammalian version of the lateral line system found in fishes and amphibians (Reep et al. 2002). This possibility is currently being examined through behavioral psychophysics experiments involving two captive manatees at Mote Marine Laboratory. Such an ability would be useful for detecting water movements associated with conspecifics, tides, and currents. It may also function to detect energy reflected from objects as a manatee moves underwater, thereby assisting in navigating the murky water environment in which manatees spend much of their time.

Postfacial vibrissae are smaller and thinner (diameter/length ratio of ~0.01) than those on the face, and their FSCs are less extensively innervated (~30 axons/FSC) (Reep et al. 2002). However, innervation density appears to scale with size of FSC (Reep et al. 2001), perhaps preserving a relatively constant innervation density per unit area. Considering that there are ~3300 postfacial FSCs innervated by ~30 axons/FSC, this represents a total investment of ~100,000 axons.

Innervation of FSCs

Immunolabeling studies have revealed that all manatee FSCs (Figure 2) have large Aß fibers terminating as club endings, longitudinal lanceolate endings, and Merkel endings (Sarko et al. 2007a). The particularly dense distribution of Merkel endings

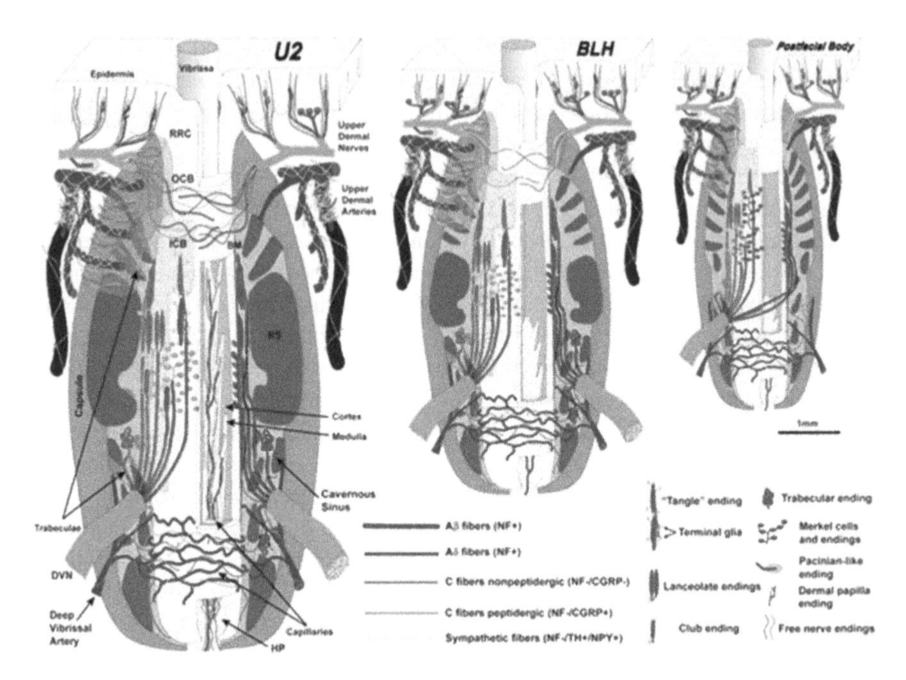

Figure 2 Schematic drawings of the structure and innervation of manatee tactile hair follicle-sinus complexes. These correspond externally to the large bristles on the upper lip (*U2*), intermediate bristle-like hairs on the oral disk (*BLH*), and thin hairs on the postfacial body. Several types of axons and nerve endings are found, resulting in a complex receptor anatomy. Reproduced from Sarko et al. (2007a, b) with permission from Wiley-Liss

indicates that manatees may be especially attuned to detecting the directionality of follicle deflection. Other types of endings generally seen in mammalian FSCs at a deeper level of the FSC known as the cavernous sinus include reticular and spiny endings; these mechanoreceptors were notably absent in manatees and may have been functionally replaced by a novel type of nerve ending thought to be adaptive to the aquatic environment. This novel ending, termed a "trabecular ending," was found only in facial FSCs and terminates along the connective tissue traversing between the FSC capsule and the basement membrane at the cavernous sinus level. A second novel mechanoreceptor type discovered in the same study consisted of gigantic, spindle-shaped "tangle" endings abutting the mesenchymal sheath at the lower extent of the inner conical body level of all FSCs studied. Both novel types of nerve endings, the trabecular and tangle endings, appear to be low-threshold mechanoreceptors (as indicated by positive BNaC labeling) and may confer additional directionality detection at the lower and upper levels of each FSC, respectively. The stout perioral follicles involved in oripulation displayed an additional specialization—the medullary core of the hair papilla exhibited dense, superficially extensive, small-caliber innervation. This adaptation is thought to allow these specialized vibrissae to detect force applied to the follicle with extremely little displacement of the follicle, much like what occurs in tooth pulp.

Vibrissal Information and the Brain

A total of >200,000 axons is devoted to conveying neural signals from the manatee vibrissal FSCs to the central nervous system. This is a sizable investment, and implies that rather elaborate CNS machinery is required to analyze this information. Regions of the brainstem, thalamus, and cerebral cortex dedicated to somatic sensation are disproportionately large, and exhibit lobulations or specialized cell aggregations indicative of functional compartmentalization.

Brainstem

Histochemical assessments of the manatee brainstem (Figure 3), combined with analysis of cytoarchitecture, have shown that the brainstem nuclei dedicated to somatosensation are well-developed (Sarko et al. 2007b). Subdivisions of the trigeminal nucleus, receiving sensory inputs from the face, revealed dense cytochrome oxidase staining and parcellation potentially related to the oral disk and perioral divisions of facial vibrissae. The cuneate-gracile complex, which represents sensory input from the upper and lower body, also stained densely for cytochrome oxidase and appeared lobulated. Bischoff's nucleus, which has been noted in animals with functionally important tails, is large in manatees and is thought to represent the

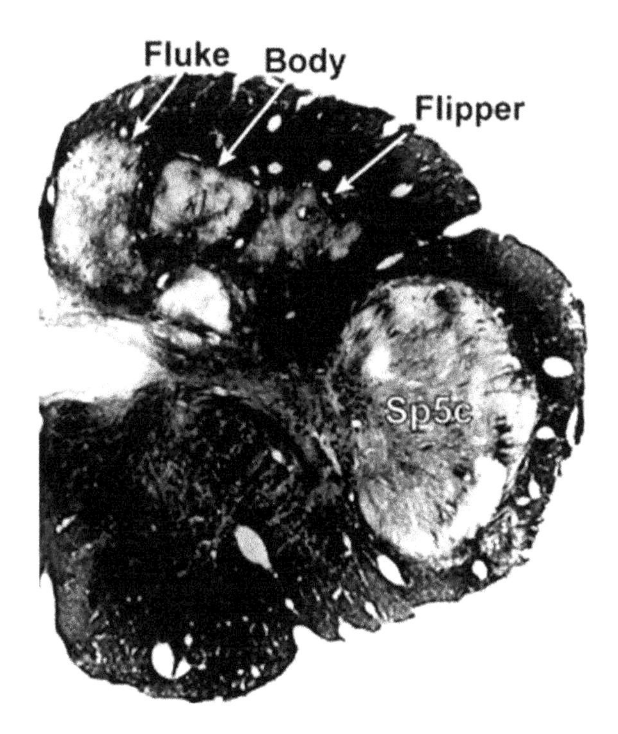

Figure 3 Within the cuneate-gracile complex of the brainstem are separate zones hypothesized to correspond to representations of tactile information coming from the fluke, body, and flipper. Tactile information from the face reaches the trigeminal complex (*Sp5c*). Reproduced from Sarko et al. (2007a, b) with permission from Wiley-Liss

fluke. The facial motor nucleus, which innervates the facial musculature involved in the flare response and oripulation, is also large and parcellated (Marshall et al. 2007). Although the manatee cerebral cortex contains cellular aggregates similar to “barrels” (see below), which are the functional representations of vibrissae in other taxa, no definitive counterparts (“barrelettes”) were seen in the brainstem.

Thalamus

Within the thalamus, the principal nucleus devoted to somatosensation (the ventroposterior nucleus, or VP) was disproportionately large compared to nuclei receiving inputs predominantly from other sensory modalities. In fact the lateral geniculate nucleus, devoted to visual inputs, appears greatly diminished in size and displaced by somatosensory and auditory nuclei (Sarko et al. 2007b). As in the brainstem, no definitive counterparts to cortical barrels (“barreloids” in the thalamus) were seen in the manatee.

Cerebral Cortex

Based on flattened cortex preparations using cytochrome oxidase staining to delineate primary sensory areas, the presumptive primary somatosensory cortex (SI) (see Figure 4) occupies approximately 25 % of total cortical area (comparable

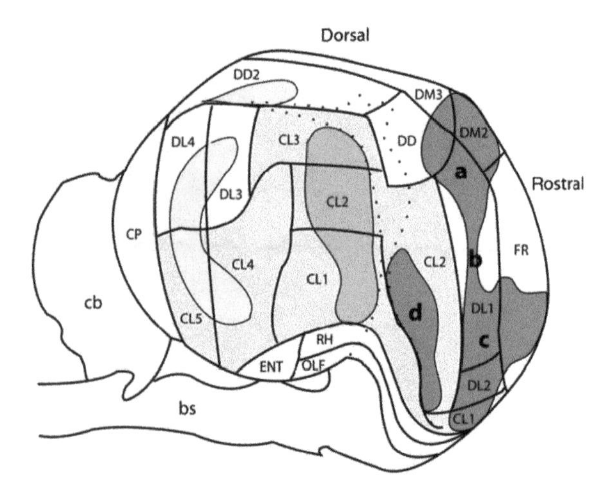

Figure 4 Sensory areas of cerebral cortex as revealed by cytochrome oxidase staining. *Yellow* areas are presumed visual cortex, *green* areas are presumed auditory cortex. *Red* areas are presumed somatic sensory cortex, which includes regions thought to represent the fluke (**a**), flipper (**b**), tactile hairs of the perioral face (**c**), and tactile hairs of the oral disk and postfacial body (**d**). Reproduced from Reep and Sarko (2006) with permission from Elsevier Press

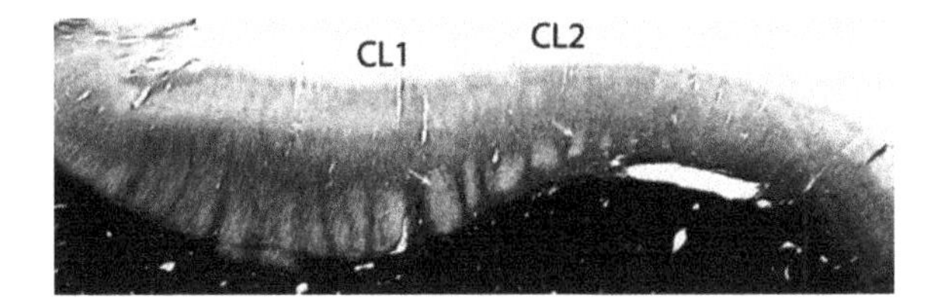

Figure 5 Rindenkerne are clusters of neurons hypothesized to process information from individual tactile hairs. Here they appear as *light oval* regions (*arrows*) in layer VI of areas CL1 and CL2 within presumed somatic sensory cerebral cortex. Reproduced from Reep and Sarko (2006) with permission from Elsevier Press

to 31 % of neocortex in naked mole-rats; Catania and Remple 2002). In contrast, presumptive primary visual and auditory cortices combined occupied 4–17 % of total cortex in the specimens examined (Sarko and Reep 2007).

One of the most definitive and intriguing specializations within the sirenian CNS is the Rindenkerne ('cortical nuclei'; Figure 5), first described by Dexler in 1912. Rindenkerne are aggregates of neurons found in layer VI of the presumptive somatosensory cortex (Reep et al. 1989; Marshall and Reep 1995). The largest Rindenkerne are found in the presumed face area. Rindenkerne stain positively for cytochrome oxidase and acetylcholinesterase. These features, together with their location, suggest that Rindenkerne may be the sirenian counterpart to the neuron aggregates known as barrels, which are found in layer IV of the face area in rodents and marsupials. Barrels are known to constitute a topographic array in which each barrel processes information related to one vibrissa. Thus, it may be that Rindenkerne represent another case of convergent evolution. Some Rindenkerne are found in presumed auditory cortex. This may represent a region of overlap between somatosensory and auditory processing, perhaps reflecting the ability of low frequency energy to stimulate the postfacial vibrissae, and the inherent continuity between these two modes of detecting vibratory stimuli.

References

Bachteler, D and Dehnhardt, G (1999). Active touch performance in the Antillean manatee: Evidence for a functional differentiation of facial tactile hairs. *Zoology* 102: 61–69.

Bauer, G et al. (2012). Tactile discrimination of textures by Florida manatees (*Trichechus manatus latirostris*). *Marine Mammal Science* 28: E456–471.

Catania, K C and Kaas, J H (1997). Somatosensory fovea in the star-nosed mole: Behavioral use of the star in relation to innervation patterns and cortical representation. *Journal of Comparative Neurology* 387: 215–233.

Catania, K C and Remple, M S (2002). Somatosensory cortex dominated by the representation of teeth in the naked mole-rat brain. *Proceedings of the National Academy of Sciences of the United States of America* 99: 5692–5697.

Gaspard, J C III et al. (2013). Detection of hydrodynamic stimuli by the Florida manatee (*Trichechus manatus latirostris*). *Journal of Comparative Physiology A* 199: 441–450.

Marshall, C D; Vaughn, S D; Sarko, D K and Reep, R L (2007). Topographical organization of the facial motor nucleus in Florida manatees (Trichechus manatus latirostris). *Brain, Behavior and Evolution* 70: 164–173.

Marshall, C D and Reep, R L (1995). Manatee cerebral cortex: Cytoarchitecture of the caudal region in Trichechus manatus latirostris. *Brain, Behavior and Evolution* 45: 1–18.

Marshall, C D; Huth, G D; Edmonds, V M; Halin, D L and Reep, R L (1998). Prehensile use of perioral bristles during feeding and associated behaviors of the Florida manatee (Trichechus manatus latirostris). *Marine Mammal Science* 14: 274–289.

Reep, R L and Sarko, D K (2006). Somatosensory specializations in the nervous systems of manatees. In: J H Kaas (Ed.), *Evolution of Nervous Systems*, Vol. 3 (pp. 207–214). Oxford: Elsevier.

Reep, R L; Sarko, D K and Rice, F L (2007). Rock hyraxes (*Procavia capensis*) possess vibrissae over the entire postfacial body. Poster presented at Society for Neuroscience.

Reep, R L; Marshall, C D and Stoll, M L (2002). Tactile hairs on the postcranial body in Florida manatees: A mammalian lateral line? *Brain, Behavior and Evolution* 59: 141–154.

Reep, R L; Marshall, C D; Stoll, M L; Homer, B L and Samuelson, D A (2001). Microanatomy of facial vibrissae in the Florida manatee: The basis for specialized sensory function and oripulation. *Brain, Behavior and Evolution* 58: 1–14.

Reep, R L; Marshall, C D; Stoll, M L and Whitaker, D M (1998). Distribution and innervation of facial bristles and hairs in the Florida manatee (Trichechus manatus latirostris). *Marine Mammal Science* 14: 257–273.

Reep, R L; Johnson, J I; Switzer, R C and Welker, W I (1989). Manatee cerebral cortex: Cytoarchitecture of the frontal region in Trichechus manatus latirostris. *Brain, Behavior and Evolution* 34: 365–386.

Sarko, D K; Rice, F L; Reep, R L and Mazurkiewicz, J E (2007a). Adaptations in the structure and innervation of follicle-sinus complexes to an aquatic environment as seen in the Florida manatee (Trichechus manatus latirostris). *Journal of Comparative Neurology* 504: 217–237.

Sarko, D K; Johnson, J I; Switzer, R C; Welker, W I and Reep, R L (2007b). Somatosensory nuclei of the brainstem and thalamus in Florida manatees. *The Anatomical Record* 290: 1138–1165.

Sarko, D K and Reep, R L (2007). Somatosensory areas of manatee cerebral cortex: Histochemical characterization and functional implications. *Brain, Behavior and Evolution* 69: 20–36.

Sokolov, V E and Kulikov, V F (1987). The structure and function of the vibrissal apparatus in some rodents. *Mammalia* 51: 125–138.

Internal References

Llinas, R (2008). Neuron. *Scholarpedia* 3(8): 1490. http://www.scholarpedia.org/article/Neuron.

Sherman, S M (2006). Thalamus. *Scholarpedia* 1(9): 1583. http://www.scholarpedia.org/article/Thalamus.

Recommended Reading

Hughes, H C (2001). *Sensory Exotica* Cambridge, MA: MIT Press.

Part II
The Psychology of Touch

Active Tactile Perception

Nathan Lepora

We do not just touch, we feel (Bajcsy 1988). Our tactile sense is not merely a passive receiver of information, but actively selects and refines sensations according to our present goals and perceptions (Gibson 1962). Our fingers, hands and bodies are not external from the world, but direct actions within it to access the information that we need. Thus, tactile sensation, perception and action cannot be considered simply as a forward process, but instead form a closed 'active perception' loop.

Active Perception Versus Active Sensing

A potential confusion lies in that some researchers refer to *active sensing* and others to *active perception*, when describing the same process. There is also a distinction between *teleceptive* and *contact* sensory systems.

Sensation and perception are considered distinct stages in the processing of the senses in humans and animals. Tactile sensation refers to the first stages in the functioning of the senses, related to the effect of a physical stimulus on touch receptors in the skin and their transduction and transmittal from the peripheral nervous system to the sensory areas of the brain; tactile perception refers to later stages where the sensation is processed, organized and interpreted so that the organism may use the information to guide its behaviour based on understanding its environment.

Therefore active sensing could refer to controlling the movements of the sensory apparatus while contacting a stimulus; for example, brushing our fingertips across a surface to feel texture. However, those movements are themselves guided in response to perceiving other sensory information; for example, we may control the force exerted by our fingertips against the surface to best feel the texture. Therefore the active process can refer to the entire sensation-perception-action loop, rather than just the sensory or perceptual parts of it. A further subtlety is that some of the movements involved in active sensing may be purely reflexive, in that they been

N. Lepora (✉)
University of Bristol, Bristol, UK

T.J. Prescott et al. (eds.), *Scholarpedia of Touch*, Scholarpedia,
DOI 10.2991/978-94-6239-133-8_11

hardwired by evolution as successful strategies to take during sensing; for example, reflexive orienting movements are made by the superior colliculus in the midbrain without involvement of higher brain areas such as sensory cortex. It is then debateable whether the movement was truly from 'perceiving' the stimulus, so the process could then just be described as *active sensing*, which is the term commonly used in the biological literature.

Another source of confusion is that the term *active sensing* is commonly used to refer to sensory receptors that are activated by probing the environment with self-generated energy. Examples include the echolocation of bats and dolphins, and the electrosensory detection of electric fish. Thus the term *active perception* can remove ambiguity by emphasizing that the process referred to is the control of the sensor to select and refine perceptions according to the present goals and perceptions of the organism. Thus even if the movement was reflexive, if the movement was to aid perception, then the process could be referred to as active perception. For this reason, in this article we will refer to *active tactile perception*, but we acknowledge that some authors prefer the term *active tactile sensing* and in many ways the terms can be treated synonymously.

Active sensory systems, in the sense of using self-generated energy, are also further distinguished into teleceptive systems that propagate energy (e.g. acoustic or electromagnetic) and contact sensory systems that use physical contact between the stimulus and the sensor. One can therefore view tactile sensing as mediated by contact active sensory systems, which can use active perception to select, refine and interpret the sensory information to understand the tactile environment.

Views of Active Perception

Concepts related to active tactile perception and active touch have been defined by several scientists over the last 50 years (Prescott et al. 2011). Their terminology does change. For example Gibson (1962) refers to active touch, (Bajcsy et al. 1987) sometimes refers to active sensing and O'Regan and Noe (2001) refer to sensory contingencies. However, all definitions can be seen as applying to active tactile perception, albeit possibly in a more limited manner than that intended originally by the authors. This is particularly the case with *touch*, which encompasses all tactile phenomena not just perception; also there are differences in the meaning of *perception* versus *exploration*, and the sensory contingency theory of O'Regan and Noe is intended as a broader treatment of conscious phenomena rather than just active perception.

Active Touching

In his 1962 article on ‘Observations on Active Touch’, the Psychologist James J. Gibson makes a number of observations about the relation between bodily movements and the sense of touch that would underlie later work on active tactile perception.

He begins by defining that: ‘Active touch refers to what is ordinarily called touching. This ought to be distinguished from passive touch, or being touched. In one case the impression on the skin is brought about by the perceiver himself and in the other case by some outside agency.’ (Gibson 1962). Here he is making a distinction between whether the agent is itself controlling its body to sense, or instead whether a tactile sensation is exerted on the agent. It would therefore be wrong, in Gibson’s view, to think of passive touch as simply the absence of movement during tactile perception; immobile touch may in some circumstances be appropriate to best sense a tactile percept, such as holding a fingertip against a vibrating surface to best perceive the strength of vibration. Instead, passive touch is when a tactile event occurs that is unanticipated or not of concern, which could then become active touch if the organism responds to that passive sensation to further select or refine the sensory information.

Gibson clarifies his position further by criticizing a view of active touch as merely a combination of kinesthesis (the feeling of bodily movement) and touch proper (the feeling of contact), in that ‘it fails to take account of the purposive character of touching.’ He emphasizes further that movements (or lack thereof) underlying active touch are purposive: ‘the act of touching or feeling is a search for stimulation or, more exactly, an effort to obtain the kind of stimulation which yields a perception of what is being touched. When one explores anything with his hand, the movements of the fingers are purposive. An organ of the body is being adjusted for the registering of information.’ Here he is using the purposiveness of movements in active touch, or for active tactile perception, to emphasize that organisms use perception for a reason: to achieve their goals and needs. Active tactile perception seeks to help achieve those goals by actively selecting and refining the sensations to give the appropriate perceptual information.

Active Perception

In her 1988 paper on ‘Active Perception’, the Engineering Scientist Ruzena Bajcsy used the term *active* to denote ‘purposefully changing the sensor’s state parameters according to sensing strategies,’ such that these controlling strategies are ‘applied to the data acquisition process which will depend on the current state of the data interpretation and the goal or the task of the process’ (Bajcsy 1988). As such, her definition of *active perception* is in essence equivalent to Gibson’s definition of *active touch*, although her terminology and nomenclature are adapted to

Engineering. In particular, her phrase 'changing the sensor's state parameters' is equivalent to 'moving' or 'adjusting' the sensor, assuming all state parameters correspond to physical changes of the sensor brought about by movement; 'data acquisition' means the same as 'sensing'; and 'data interpretation' the same as 'perceiving'.

Since Bajcsy brought her definition within Engineering terminology, one could perhaps interpret active perception as an application of control theory, by interpreting active perception as a closed-loop system that uses sensory feedback to control the state parameters of the sensor. However, Bajcsy emphasis that this interpretation is too simplistic, because: 'the feedback is performed not only on sensory data but on complex processed sensory data' and 'the feedback is dependent on a priori knowledge.' Instead she views active perception as 'an application of intelligent control theory which includes reasoning, decision making and control.' Here she is indicating the limitations of classical control theory for addressing active perception, which conventionally applies to systems with a simple relation between the sensory data and the variables to be controlled. Modern techniques in intelligent control have helped in utilizing complex processed sensory data, although there remain significant challenges, as discussed below in the section on *Active Tactile Perception in Robotics*.

Similar views were also brought forwards by other authors around the same time, focussing on the modality of vision. Dana H. Ballard in his work on 'Animate Vision' argues that vision is understood 'in the context of the visual behaviours that the system is engaged in, and that these behaviours may not require elaborate categorical representations of the 3-D world' (Ballard 1991). Meanwhile, John Aloimonos and colleagues in 'Active vision' take a similar approach to Bajcsy in defining 'an observer is called active when engaged in some kind of activity wholse purpose is to control the geometric parameters of the sensor sensory apparatus.' (Aloimonos et al. 1990), with 'The purpose of the activity is to manipulate the constraints underlying the observed phenomena in order to improve the quality of the perceptual results.'

Haptic Exploration

Closely related to active tactile perception, Haptic exploration was introduced in Psychologists Susan J. Lederman and Roberta L. Klatzky's (1987) article on 'Hand Movements: A Window into Haptic Object Recognition' (Lederman and Klatzky 1987) They propose that 'the hand (more accurately, the hand and brain) is an intelligent device, in that it uses motor capabilities to greatly extend its sensory functions,' consistent with observations on active touch by Gibson and others. They then extended these ideas by offering a 'taxonomy for purposive hand movements that achieve object apprehension. Specific exploratory procedures appear to be linked with specific object dimensions.' Thus, like Gibson, they emphasise the purposive nature of the movements guiding the hand and fingers to recognize objects, but they go further in proposing the motor mechanisms—a set of *exploratory procedures*—that constitute those purposive movements.

Although Lederman and Klatzky's proposals on Haptic exploration and Bajcsy's definition of Active Perception were presented independently, there are various ways they can be related. For example, Active Haptic Exploration can also be contrasted with passive touch (Loomis and Lederman 1986), as discussed in the article (Klatzky and Reed 2009). Moreover, Haptic Exploration can be active in the sense of Bajcsy's formulation as feedback control, as is apparent from considering specific exploratory procedures; for example, contour following requires that the orientation of the contour be perceived at each instance and used to guide the motor commands to keep the finger in contact with the object. Another relation between their proposals concerns the optimality of the purposive movements. Lederman and Klatzky observe that in most cases the 'exploratory procedures... optimize the speed or accuracy with which the readings of the object along the named dimensions are obtained.' Meanwhile, Bajcsy also discusses optimality, but within the context that 'A control sequence is to be evaluated relative to its expected utility... of two parts: the performance of the estimation procedure for that choice of strategy and the cost of implementing that strategy.' If 'strategy' is read as 'exploratory procedure' then there is confluence between the two bodies of work.

Sensorimotor Contingencies

Although formulated in the modality of vision, the concept of sensorimotor contingencies also formalises aspects of active perception and active exploration, and has general applicability across all sensory modalities. Psychologist Kevin O'Regan and Philosopher Alva Noë in 'A sensorimotor account of vision and visual consciousness' propose that 'seeing is a way of acting' so that 'vision is a mode of exploration of the world that is mediated by knowledge of ... sensorimotor contingencies' (O'Regan and Noë 2001), with sensorimotor contingencies being the 'rules governing the sensory changes produced by various motor actions.' Their sensorimotor contingency theory is intended to go further than active perception by offering an explanation of visual consciousness: 'the basic thing people do when they see is that they exercise mastery of the sensorimotor contingencies governing visual exploration. Thus visual sensation and visual perception are different aspects of a person's skillful exploratory activity... Visual awareness depends further on the person's integration of these patterns of skillful exercise into ongoing planning, reasoning, decision making, and linguistic activities.' Hence they conclude that 'these ingredients are sufficient to explain the otherwise elusive character of visual consciousness.

Active Tactile Perception in Humans and Animals

A large body of literature has documented aspects of active tactile sensing and perception with the human hand and fingertips. Notable early works include observations that the sensation of the roughness of a surface is altered by the changes in the manner of feeling that surface (Lederman and Taylor 1972), a comparison of active and passive touch on the perception of surface roughness (Lederman 1983), and a psychophysical comparison of active and passive touch in perceiving texture (Lamb 1983). A comprehensive review of foundational research into touch and haptics is documented in the article 'tactual perception' (Loomis and Lederman 1986).

Complementary observation on active tactile perception have been made from studies of how rodents deploy their whisker system, which involves modulating how they rhythmically 'whisk' their facial vibrissae during perception and exploration of their environment. An early study observed that better performers appeared to optimize whisking frequency bandwidth and the extent to which the vibrissae would be bent by object control (Carvell and Simons 1995). A large number of studies have since been made on active tactile perception in the rodent vibrissal system, including on the neural encoding of vibrissal active touch (Szwed et al. 2003) and the role of feedback in active tactile sensing (Mitchinson et al. 2007; Grant et al. 2009). A review of the earlier work in this area is documented in the article 'Active sensation: insights from the rodent vibrissae sensorimotor system' (Kleinfeld et al. 2006). A more recent review article (Diamond and Arabzadeh 2013) further clarifies whisker-mediated active perception by arguing for two distinct modes of operation: generative and receptive modes. In the generative mode, the rodent whisks to actively seek and explore objects; in the receptive mode, the rodent immobilizes its whiskers to optimize information gathering from a self-moving object. Both the generative and receptive modes are strategies for active perception, with the task and stimulus determining which mode is more suited for a given situation.

Studies of active touch in other whiskered mammals and with insect antennae have also been made, including cockroach antennae, stick insect antennae, and tactile sensing in the naked mole rat, Etruscan shrew and seals. Sensorimotor contingencies were also found to be essential for coding vibrissal active touch; where sensory cues alone convey ambiguous information about object location, the relationships between motor and sensory cues convey unique information (Bagdasarian et al. 2013).

Active Tactile Perception in Robots

Around the time of the work on active perception (Bajcsy 1988) and haptic exploration (Lederman and Klatzky 1987), there were several initial proposals that robot touch could be based around related principles (Bajcsy et al. 1987; Roberts

1990; Allen 1990). Early examples of active perception with tactile fingertips came soon after, including motion control of a tactile fingertip for profile delineation of an unseen object (Maekawa et al. 1992) and controlling the speed, and hence spatial filtering, of a tactile fingertip to measure surface roughness (Shimojo and Ishikawa 1993). Implementations of haptic exploration of unknown objects with dexterous robot hands followed a few years later; both by using exploratory procedures where some fingers grasp and manipulate the object while others feel the object surface (Okamura et al. 1997) and to use active perception to compute contact localization for grasping (Kaneko and Tanie 1994). A surveys of early work in these areas can be found in (Cutkosky and Howe 2008); the review ‘tactile sensing—from humans to humanoids’ also usefully contrasts *perception for action*, such as grasp control and dexterous manipulation, with *action for perception*, such as haptic exploration and active perception (Dahiya et al. 2010).

Active perception has also been demonstrated on whiskered robots. Early work included using an *active antenna* to detect contact distance (Kaneko et al. 1998) and an *active artificial whisker array* for texture discrimination (Fend et al. 2003). These early studies used *active* to mean the sensor is moving, but did not consider a feedback loop to modulate the whisking. Later whiskered robots implemented sensorimotor feedback between the tactile sensing and motor control (Pearson et al. 2007) enabling demonstration of the contact dependency of texture perception with a whiskered robot (Fox et al. 2009) and that texture perception was improved when the whisking motion was controlled using a sensory feedback mechanism (Sullivan et al. 2012).

Recent work on active touch with robot fingertips has used probabilistic approaches for active perception, where the statistical hypotheses correspond to discrete percepts. In one body of work, *active Bayesian perception* is implemented with a control policy that takes intermediate estimates of the percept probabilities during perception to guide movements of the sensor (Lepora et al. 2013a, b). This approach can be used to improve the quality of the perception, for example to attain tactile superresolution sensing (Lepora et al. 2015), gives robust sensing under positional uncertainty (Lepora et al. 2013a, b) and can implement exploratory procedures such as contour following (Martinez-Hernandez et al. 2013). A related method of *Bayesian exploration* selects movements that best disambiguate a percept from a set of plausible candidates, and on a large database of textures with control of contact force and movement speed ‘was found to exceed human capabilities’ (Fishel and Loeb 2012).

Conclusions

Although there has been an evolving definition of active perception and active touch, an underlying theme is that active tactile perception utilizes sensor movement control to aid interpretation of the tactile data. A large body of research has documented that active tactile perception is central to understanding the function of the human hand and fingertips and also vibrissal sensing systems in rodents and other animals. This psychological and neuroscientific research has inspired the implementation of active tactile perception with robot fingertips and whiskers, with potential to lead to artificial tactile devices with comparable or superior performance to human and animal touch.

References

Allen, P K (1990). Mapping haptic exploratory procedures to multiple shape representations. In: *ICRA 1990* (pp. 1679–1684).

Aloimonos, J (1990). Purposive and qualitative active vision. In: *10th International Conference on Pattern Recognition, 1990, Proceedings* (vol. 1, pp. 346–360).

Bagdasarian, K et al. (2013). Pre-neuronal morphological processing of object location by individual whiskers. *Nature Neuroscience* 16: 622–631.

Bajcsy, R. (1988). Active perception. *Proceedings of the IEEE* 76(8): 966–1005.

Bajcsy, R; Lederman, S J and Klatzky, R (1987). Object exploration in one and two fingered robots. *Proceedings of the 1987 IEEE International Conference on Robotics and Automation* 3: 1806–1810.

Ballard, D H (1991). Animate vision. *Artificial intelligence* 48(1): 57–86.

Carvell, G E and Simons, D J (1995). Task-and subject-related differences in sensorimotor behavior during active touch. *Somatosensory and Motor Research* 12(1): 1–9.

Cutkosky, M R and Howe, R D (2008). Force and Tactile Sensors. *Springer Handbook of Robotics*: 455–476.

Dahiya, R S; Metta, G; Valle, M and Sandini, G (2010). Tactile sensing—from humans to humanoids. *IEEE Transactions on Robotics* 26(1): 1–20.

Diamond, M and Arabzadeh, E (2013). Whisker sensory system—from receptor to decision. *Progress in Neurobiology* 103: 28–40.

Fend, M; Bovet, S; Yokoi, H and Pfeifer, R (2003). An active artificial whisker array for texture discrimination. In: *2003 IEEE/RSJ International Conference on Intelligent Robots and Systems (IROS 2003), Proceedings* 2: 1044–1049.

Fishel, J A and Loeb, G E (2012). Bayesian exploration for intelligent identification of textures. *Frontiers in Neurorobotics* 6.

Fox, C W; Mitchinson, B; Pearson, M J; Pipe, A G and Prescott, T J (2009). Contact type dependency of texture classification in a whiskered mobile robot. *Autonomous Robots* 26(4): 223–239.

Gibson, J J (1962). Observations on active touch. *Psychological review* 69(6).

Grant, R A; Mitchinson, B; Fox, C W and Prescott, T J (2009). Active touch sensing in the rat: Anticipatory and regulatory control of whisker movements during surface exploration. *Journal of Neurophysiology* 101(2): 862–874.

Kaneko, M; Kanayama, N and Tsuji, T (1998). Active antenna for contact sensing. *IEEE Transactions on Robotics and Automation* 14(2): 278–291.

Kaneko, M and Kazuo, T (1994). Contact point detection for grasping an unknown object using self-posture changeability. *IEEE Transactions on Robotics and Automation* 10(3): 355–367.

Klatzky, R and Reed, C (2009). Haptic exploration. *Scholarpedia* 4(8): 7941.

Kleinfeld, D; Ahissar, E and Diamond, M E (2006). Active sensation: Insights from the rodent vibrissa sensorimotor system. *Current Opinion in Neurobiology* 16: 435–444.

Lamb, G D (1983). Tactile discrimination of textured surfaces: Psychophysical performance measurements in humans. *Journal of Physiology (London)* 338: 551–565.

Lederman, S J (1983). Tactual roughness perception: Spatial and temporal determinants. *Canadian Journal of Psychology/Revue Canadienne de Psychologie* 37(4).

Lederman, S J and Klatzky, R L (1987). Hand movements: A window into haptic object recognition. *Cognitive Psychology* 19(3): 342–368.

Lederman, S J and Taylor, M M (1972). Fingertip force, surface geometry, and the perception of roughness by active touch. *Perception & Psychophysics* 12(5): 401–408.

Lepora, N F; Martinez-Hernandez, U and Prescott, T J (2013). Active Bayesian perception for simultaneous object localization and identification. In: *Robotics: Science and Systems.*

Lepora, N F; Martinez-Hernandez, U and Prescott, T J (2013). Active touch for robust perception under position uncertainty. In: *2013 IEEE International Conference on Robotics and Automation (ICRA)* (pp. 3020–3025)

Lepora, N F; Martinez-Hernandez, U and Prescott, T J (2015). Tactile superresolution and biomimetic hyperacuity. 31(3): 605–618.

Loomis, J M and Lederman, S J (1986). Tactual perception. *Handbook of Perception and Human Performances* 2(2).

Maekawa, H; Tanie, K; Komoriya, K; Kaneko, M; Horiguchi, C and Sugawara, T (1992). Development of a finger-shaped tactile sensor and its evaluation by active touch. In: *1992 IEEE International Conference on Robotics and Automation, Proceedings* (pp. 1327–1334).

Martinez-Hernandez, U; Metta, G; Dodd, T J; Prescott, T J and Lepora, N F (2013). Active contour following to explore object shape with robot touch. In: *World Haptics Conference (WHC)* (pp. 341–346).

Mitchinson, B; Martin, C J; Grant, R A and Prescott, T J (2007). Feedback control in active sensing: Rat exploratory whisking is modulated by environmental contact. *Proceedings of the Royal Society, B* 274: 1035–1041.

Okamura, A M; Turner, M L and Cutkosky, M R (1997). Haptic exploration of objects with rolling and sliding. In: *1997 IEEE International Conference on Robotics and Automation, Proceedings* (pp. 2485–2490).

O'Regan, J K and Noë, A (2001). A sensorimotor account of vision and visual consciousness. *Behavioral and Brain Sciences* 24(5): 939–973.

Pearson, M J; Pipe, A G; Melhuish, C; Mitchinson, B and Prescott, T J (2007). Whiskerbot: A robotic active touch system modelled on the rat whisker sensory system. *Adaptive Behavior* 15(3): 223–240.

Prescott, T J; Diamond, M E and Wing, A M (2011). Active touch sensing. *Philosophical Transactions of the Royal Society B: Biological Sciences* 366(1581): 2989–2995.

Roberts, K S (1990). Robot active touch exploration: Constraints and strategies. In: *1990 IEEE International Conference on Robotics and Automation, Proceedings* (pp. 980–985).

Shimojo, M and Ishikawa, M (1993). An active touch sensing method using a spatial filtering tactile sensor. In: *1993 IEEE International Conference on Robotics and Automation, Proceedings* (pp. 948–954).

Sullivan, J C; Mitchinson, B; Pearson, M J; Evans, M; Lepora, N F and Prescott; T J (2012). Tactile discrimination using active whisker sensors. *IEEE Sensors* 12(2): 350–362.

Szwed, M; Bagdasarian, K and Ahissar, E (2003). Encoding of vibrissal active touch. *Neuron* 40 (3): 621–630.

Tactile Object Perception

Guy Nelinger, Eldad Assa and Ehud Ahissar

It is commonly assumed that object perception is the combination of sensory features into unified perceptual entities. Tactile object perception may therefore be defined as the perception of objects whose feature information is acquired via touch. Consequently, research relevant to the topic of tactile objects has focused on exploring the primitives of the tactile system, their interrelation, and how they may be bound together. The current discussion does not explicitly rule out, nor does it address, kinesthetic sensation. As such, tactile perception is used here interchangeably with haptic perception (Lederman and Klatzky 2009).

The Concept of an Object in Different Modalities

The question of objects has received a great deal of attention in philosophical treatises spanning many centuries. Perhaps most accurate in representing modern science's stance was Kant (1781/1999). Kant argued that the existence and nature of objects as a self-contained entity, unperceived, is unknowable to us. Therefore, we should not attempt to know the 'object in itself' (i.e., noumenon), but reformulate our topic of interest altogether. The question at issue should be what determines the objects of the senses (i.e., phenomenon). More specifically, how do sensory and cognitive faculties constrain the way that sensory information is obtained, processed and combined into compounded entities? Accepting the Kantian perspective, it may be appropriate to preface the discussion of tactile objects with a more general theoretical enquiry. Explicitly, is the concept of an object different for separate modalities, and if so—how?

To answer this question, it is useful to consider the cases of olfaction and audition, in which the idea of a modality-specific object is still a matter of debate. For audition, conjunction of features into an object has been suggested to occur in a

G. Nelinger (✉) · E. Assa · E. Ahissar
Department of Neurobiology, Weizmann Institute of Science, Rehovot, Israel

T.J. Prescott et al. (eds.), *Scholarpedia of Touch*, Scholarpedia,
DOI 10.2991/978-94-6239-133-8_12

hierarchical fashion. Processing in the cochlea is based on putative two-dimensional time-frequency primitives. Features are then extracted, compared to templates, and integrated with other modalities, as one ascends the primary and secondary auditory cortices (Griffiths and Warren 2004). For olfactory objects, it has been suggested that the molecular structure of the entire olfactory 'landscape' is essentially projected to a single dimension, and this dimension corresponds to the inherent pleasantness of the odor. The percept of an odor object is then given by integration of this pleasantness value with the perceiver's internal state (Yeshurun and Sobel 2010).

The cases of audition and olfaction clearly demonstrate two contrasting views of object perception. Audition suggests that feature conjunction into objects is achieved by hierarchical convergent processing of isolated feature information, a framework widely accepted in vision. Studies in olfaction appear to suggest a common representation across all levels of processing, questioning the involvement of a hierarchical processing framework. This chapter examines the perception of tactile objects in light of this theoretical divide, through comparisons to visual object perception, concluding by reviewing the suggestion that objects in touch are best understood not through comparison with vision, but within a unique framework of active sensing which takes into account both sensory and motor variables.

Co-processing of Tactile Features

In order to characterize the ways in which features are combined by a perceptual system, it is imperative to first probe the fundamental units, or "primitive units", of feature processing by that system. It is quite possible that these primitive units do not correspond to the units of description used by humans when referring to the aforementioned features. In the following we thus refer to description-derived units, such as texture, length or softness, as 'features' and to the, yet unknown, fundamental units of processing as 'processing-primitive units'. A specific individual feature can be considered as corresponding to a specific individual primitive unit if the processing of that feature is not affected by simultaneous processing ("co-processing") of other features. Behavioral studies that have examined the degree to which co-processing of different tactile features occurs may provide the first clues for the identities of such processing-primitive units.

Both examples of co-processing and independent processing have been presented. For example, Corsini and Pick (1969) demonstrated co-processing of texture and length. They showed that sheets shorter in length than a reference were better discriminated when they had a coarse texture, and those longer than the reference—when they had a fine texture (for similar examples of tactile illusions, readers are encouraged to turn to Hayward 2008; Lederman and Jones 2011). Sinclair et al. (2000), on the other hand, demonstrated that texture and vibration frequency are perceptually separable from duration, even when processed simultaneously.

Klatzky, Lederman and Reed examined similar questions in a series of studies. They presented subjects with objects which could differ from one another by one or more tactile features, and taught them to classify these objects into groups. The classification rule was based on the values of a single feature (e.g., groups A, B, and C contain small, medium and large objects, respectively). However, for some groups of subjects an additional feature covaried with the group-determining feature (e.g., small, medium and large objects had fine, semi-rough and rough texture). They found that this feature covariation significantly improved response times in the classification task (Klatzky et al. 1989; Reed et al. 1990). These redundancy gains in performance appeared to result from covariation in practically any two features (Lederman et al. 1993). However, covariation of an additional third feature (i.e., covariation in texture, shape and hardness) did not yield significant response time improvement compared to covariation of any two features (Klatzky et al. 1989).

Task improvement achieved by co-processing of covarying features is, in itself, not a trivial finding. However, the opposite question may be more informative. That is, can subjects ignore information from one tactile feature while processing another, when such information adversely affects performance? To examine this question, Klatzky, Lederman and Reed used a paradigm they termed 'withdrawal'. Here, as before, subjects learned to classify objects by a sorting rule which relied only on one tactile feature as a determinant. Again, the values of a second tactile feature initially covaried with those of the first. This time, however, at some point during testing, the covariation of the second feature was stopped and it was fixed to a given value (i.e., it was 'withdrawn'). If response times were lengthened by this manipulation, one can assume that the two features were co-processed despite explicit motivation to avoid such integration.

When the covarying features used in this method were size and shape, withdrawing either significantly worsened performance times (symmetrical deficits), indicating these features tend to be co-processed. However, withdrawing shape damaged performance more than withdrawing size (Reed et al. 1990). When hardness and texture were the covarying features, withdrawing either yielded comparable, significant response time deficits (Klatzky et al. 1989; Lederman et al. 1993). As a putative rule of thumb, it appears that when features are related (i.e., both properties of either structure or material), co-processing is more likely to occur. Cases in which features were non-related (e.g., size and hardness, shape and texture, etc.) produced only asymmetrical deficits, if at all, after withdrawal. This is not in accord, of course, with the hypothesis that these features were being co-processed by default. These findings were virtually unaffected by whether subjects were explicitly told that classification depends only on one feature (Klatzky et al. 1989; Reed et al. 1990) or not (Klatzky et al. 1989).

However, one final research in this line of studies had quite different results (Lederman et al. 1993). This study found that covariation of shape (a structural property) and texture (a material property), followed by withdrawal of either, significantly worsened reaction times. That is, deficits were symmetrical, implying co-processing. These results were also replicated using a method called 'orthogonal insertion', which is sort of a mirror-image of withdrawal. In 'orthogonal insertion', the non-focal feature was first fixed to a given value and then varied orthogonally compared to the feature of interest (Lederman et al. 1993). These results do not fit with those of a previously mentioned study, in which only the withdrawal of shape significantly damaged performance after a period in which shape and texture covaried (unidirectional deficits, Klatzky et al. 1989). This discrepancy can be accounted for by the fact that evidence of co-processing was found using 3D ellipsoids as shapes. With these objects it was possible to extract both texture and shape information (in terms of curvature) from any local region; exploratory procedures for shape and texture extraction (contour following and lateral motion, respectively) could be performed together in a more compatible manner. Section "Feature Processing from an Active Sensing Perspective" expands on this point. Other explanations are, of course, possible. For example, it was recently shown that different cognitive styles (namely, the tendency to employ different types of imagery) may affect the co-processing of texture and shape, suggesting a meaningful role of individual differences in these tasks (Lacey et al. 2011).

Taken together, the studies cited in this section support the idea that individual features corresponding to everyday experience may not necessarily correspond to individual processing primitive units. Yet, it appears that simple behavioral paradigms for testing co-processing of features should be further developed to achieve more meaningful results. This can be performed in two ways: either by analyzing more carefully the effect that task demands and specific parameter values have on the convergent results of these studies, or by approaching their results from the perspective of a guiding theoretical framework. As explained in section "The Concept of an Object in Different Modalities", the following sections attempt to do the latter: section "Tactile Objects and Visual Objects" discusses results in line with existing models in vision, and section "Feature Processing from an Active Sensing Perspective" discusses results in line with a unique theory of object perception in touch as an active-sensing process.

Tactile Objects and Visual Objects

'Feature Integration Theory' and Tactile Objects

Arguably the most widely accepted model of feature binding developed in vision is Treisman's 'Feature Integration Theory' (FIT). According to FIT, features that are processed by primitive units are pre-attentively and automatically registered in separate feature maps (Treisman and Souther 1985; Treisman 1998a, b). An additional master-map of locations is formed, containing information of the boundaries of all these features (where features are), but lacking the information of feature identity (which features are where; Treisman and Souther 1985; Treisman 1998a, b). A scalable window (Treisman 1998a), or 'spotlight' (Treisman 1998b), of attention may then define a specific location, exclude features outside of it, and thereby bind all features falling within its range to a given location (Treisman 1998a, b). A priori, FIT does not intuitively apply to touch, mainly because touch is a proximal sense and at any given moment information is available only from a small set of locations in the surrounding world. Nonetheless, this does not preclude FIT, or a variant of it, from adequately describing feature binding in touch.

Much of the appeal of FIT stems from its clearly defined experimental paradigms and extremely specific predictions. A particularly relevant example is FIT's predictions regarding search time when subjects are required to determine whether a target is present or absent within an array of distractors. If the target is defined by the presence of a unique single feature which is absent from distractors (e.g., a red 'Q' in an array of red O's; 'disjunctive search'), it should 'pop-out' immediately, independent of the number of distractors (up to ~3–4 ms per distractor; Treisman and Souther 1985; Treisman 1998b). This is because information about the target is immediately present within the activity of the dedicated feature map (or the lack of such activity; Treisman 1998b). Therefore, if the target is present, it will 'pop-out', and if not its absence will also be effortlessly detected. In neuronal terms, this simply requires perceivers to examine the pooled response of feature detectors and determine if feature detection has taken place. However, if the target differs from distractors by a conjunction of features (e.g., a red 'Q' among red 'O's and blue 'Q's'; 'conjunctive search'), a serial self-terminating approach is required, in which the features of each element are bound and only then is the element accepted or rejected as the target (Treisman 1998b).

To summarize these hypotheses in quantitative terms, FIT predicts that slopes of search time in a disjunctive task will be virtually flat, indicating parallel search, while slopes in a conjunctive task will be steeply linear with the number of distractors, indicating serial self-terminating search. Moreover, in the latter task, the ratio of search time slopes between target-absent and target-present trials is expected to be 2:1 (Treisman and Souther 1985; Treisman 1998b). This is because the former condition requires scanning of the entire array for a response to be made, while in the latter condition the target will be found after scanning half of the array, on average.

An additional finding of FIT is that when the target is defined by the absence of a unique feature (e.g., a red 'O' among red 'Q's) search will also be serial (Treisman and Souther 1985; Treisman 1998b). This is presumably because the overall activity in any single feature map is highly similar for arrays in which the target is absent and those where it is present, hampering pop-out detection (Treisman and Souther 1985). This phenomenon, in which a target defined by the presence of a feature pops out from among distractors lacking the same feature, while the opposite arrangement produces serial search, is one example of what is called a 'search asymmetry'. A search asymmetry of this type strongly suggests that the feature in question is represented by a dedicated map, or equivalently—that it is processed by a unique primitive unit. Many of the studies described below have used search asymmetry to characterize tactile primitives.

Two studies using a tactile search paradigm demonstrated results very close to those obtained by Treisman in vision (Sathian and Burton 1991; Whang et al. 1991). In the tactile paradigm, the individual fingertips of subjects simultaneously received vibrotactile stimulation, in two consecutive intervals (Whang et al. 1991). Each fingertip received its own stimulation, making their combination an array. On one interval all fingertips were presented with distractors, while on another, one of the fingertips was presented with a target. Subjects were asked to report the interval in which a target appeared. In one condition, the target had a segment of altered frequency and distractors had constant frequency ('presence' search condition), while in another these roles were reversed (that is, all stimuli but one had a segment of altered frequency; 'absence' search condition). Results showed that pre-cuing the targeted fingertip benefited task accuracy in the 'absence' condition, but not in the 'presence' condition. The fact that frequency change detection does not benefit from cuing might suggest, in the terms of FIT, that it is processed by a specific primitive unit. The same principal finding was obtained, using the same paradigm, for textured stimuli (Sathian and Burton 1991).

Plaisier et al. (2008) found that texture occurrence (as opposed to texture change, reported above) also produces a search asymmetry. Their subjects performed a free hand sweep over a surface in which elements of texture were embedded. One element served as a target, and had either a rougher or a smoother texture than the other elements. Subjects were required only to report whether a target was present in the array. Rough targets popped-out easily from among smooth distractors; search time in this condition had little dependence on the number of distractors, yielding a close-to-flat search-time slope. Additionally, this condition typically required only a single sweep over the array. Conversely, smooth targets among rough distractors required more complex hand-motions, and the search-time slope of this condition depended on the number of distractors more steeply. This asymmetry was given a mechanical explanation, according to which rough-textured stimuli elicit more friction in the course of hand motion, and therefore pop-out during a single sweep.

An additional asymmetry, one of edge presence, was examined by Plaisier et al. (2009). They had subjects cup (in their hands) an array of objects which either did or did not contain a single target whose shape differed from distractors. Subjects were to report if a target was present, and could release the array and re-examine it in almost any way they wished. If information was not available in the initial grasp, search times lengthened, trivially. Results showed that cubes tended to somewhat pop-out from among spheres (that is, they could be sensed with minimal release of the stimuli) while the opposite did not hold (supporting a search asymmetry). A follow-up experiment tested for search of ellipsoid, cylinder or tetrahedron targets among cube or sphere distractors; it demonstrated that search times for a target were heavily affected by the identity of distractors. To quantify the effect of features on search times, correlation was computed for the slope of search-time from the target-present trials with a variety of geometrical features. The geometrical feature most correlated with search-time was the difference in edge acuteness values for each shape and its distractor in absolute value (with edge acuteness defined as the smallest angle between two planes of a shape). This result suggests that, contrary to the initial results obtained with cubes and spheres, edge acuteness affects pop-out in a symmetrical manner (whether edges are a feature of the target or the distractors).

As such, it appears that the differences for search time of a cube among spheres and vice versa, found in the initial experiment, reflect search asymmetry in the domain of objects (cubes vs. spheres) but not necessarily in that of tactile primitive units.

In a similar behavioral paradigm and setup (grasp-and-release), objects colder than body temperature were found to pop-out from among objects warmer than body temperature (Plaisier and Kappers 2010). The slope of search-time for cold objects was similar to the one found for a tetrahedron target among spheres in the experiment reported in the previous paragraph. However, here, the reverse correspondence of a warm target among cold distractors was not examined, making talk of an asymmetry impossible.

Perhaps the most detailed test of FIT's hypothesis of conjunction using tactile features was performed by Lederman et al. (1988). Subjects placed their hand in a passive manner, with each finger receiving stimulation individually. In a 'disjunctive search' condition, subjects were instructed to search for a target, either rough or vertical, among smooth horizontal distractors. In a 'conjunctive search' condition, the target was both rough and vertical, and it appeared among rough-and-horizontal as well as smooth-and-vertical distractors. In both conditions, 50 % of trials contained no target at all. Subjects were only asked to indicate if a target was present in each trial. The ratio of slopes between the target-absent and target-present conditions was approximately 2:1 in the conjunctive search task, as predicted by findings in vision, indicating a serial self-terminating search. For the disjunctive search task, slopes were not completely flat as predicted. Thus, while disjunctive tactile search may have a meaningful parallel-search component, it does not precisely fit the quantitative predictions derived from search tasks in vision (different interpretations are therefore possible for these findings). In the absence of a clear 'pop-out' effect in this study, neither orientation nor roughness can be confidently considered to be processed by tactile primitive units (this is further complicated by the fact that the slopes show some profound qualitative differences from equivalent slopes in vision).

In general, thus, FIT-inspired methodology rarely produced unequivocal results in touch. Still, several pieces of data seem to be valuable for tracing the fundamental components of tactile object perception. The tactile perceptual system seems to be quite sensitive to presence over absence, whether the present element is an edge, or roughness of texture (Plaisier et al. 2008). It is also sensitive to change over constancy, whether the change is in texture (Sathian and Burton 1991), or vibration frequency (Whang et al. 1991). Cold items were shown to pop-out from among warm ones (Plaisier and Kappers 2010). Also, the tactile system was demonstrated to more quickly find targets defined by either orientation or texture than their conjunction (Lederman et al. 1988). While these studies cannot indubitably indicate tactile primitives, their collective summary suggests that edges, texture, vibration and temperature are all strong candidates. Perhaps the most important finding supporting FIT in touch was that search for a conjunction of texture and orientation was slower and, presumably, more effortful than search for either component (Lederman et al. 1988).

Neuronal Conjunction: Convergence, Population Firing and Synchrony

Classical electrophysiological findings have demonstrated that, at least in the passive mode, some neurons in the visual cortex respond selectively to specific features of a given sensory stimulation (Hubel and Wiesel 1962). Furthermore, as one ascends the cortical hierarchy, specific cells tend to respond to complex combinations of such features (Barlow 1972; Hubel and Wiesel 1962; Kobatake and Tanaka 1994). Convergence of feature information to a single neuron was thus suggested to occur in a hierarchical manner. Simultaneous appearance of features should lead to simultaneous activation of several 'feature detector' neurons. Input to a single higher-level neuron from several such feature detectors would allow it to act as a coincidence detector and represent the combination of features. In the extreme case, such a cell could correspond perfectly with a specific compounded percept of an object (Figure 1a; Barlow 1972). Indeed, the existence of such extremely selective cells, colloquially termed 'grandmother cells', has been documented in vision (Gelbard-Sagiv et al. 2008; Quiroga et al. 2005, 2008).

Similarly, single neurons displaying highly selective responses to complex tactile features have also been reported. Jörntell and collaborators have found that when the glabrous pads of decerebrate cats were stimulated in different fashions and velocities, cuneate neurons showed very different selectivity in their response patterns (Jörntell et al. 2014). Iwamura and collaborators have also documented specific neurons with strikingly selective responses, located in or around the post-central gyrus of monkeys. For example, some neurons, which had rather complex receptive fields spanning large parts of the hand, did not respond to passive stimulation with a probe object nor did they respond to positioning of the joints in a form matching the object's shape. They did respond, however, to a combination of the two, when the object was actually grasped (Iwamura and Tanaka 1978). Other neurons in this area appeared to be highly tuned to specific features such as softness, mobility, and even familiarity, reflecting varying degrees of sensory integration and abstraction (Iwamura et al. 1995). Saal and Bensmaia (2014) argued further that when the relevant literature is inspected more closely, with an emphasis on appropriate and naturalistic stimulation paradigms, it appears likely that a typical cortical tactile neuron receives inputs from various mechanoreceptor types and can potentially encode information from a diversity of features.

With time, however, the idea that feature conjunction is represented via the activity of a single cell has mostly fallen from grace. This is mainly because such a representation would clearly be noise-sensitive and vulnerable, and the system as a whole would require a superfluous number of such explicit elements (Barlow 1972; Quiroga et al. 2005; Treisman 1996; Von Der Malsburg 1995). This does not mean that convergent anatomical hierarchy cannot represent or compute feature conjunction. It only suggests that such a representation would be implemented at the level of a neuronal population response (Figure 1b).

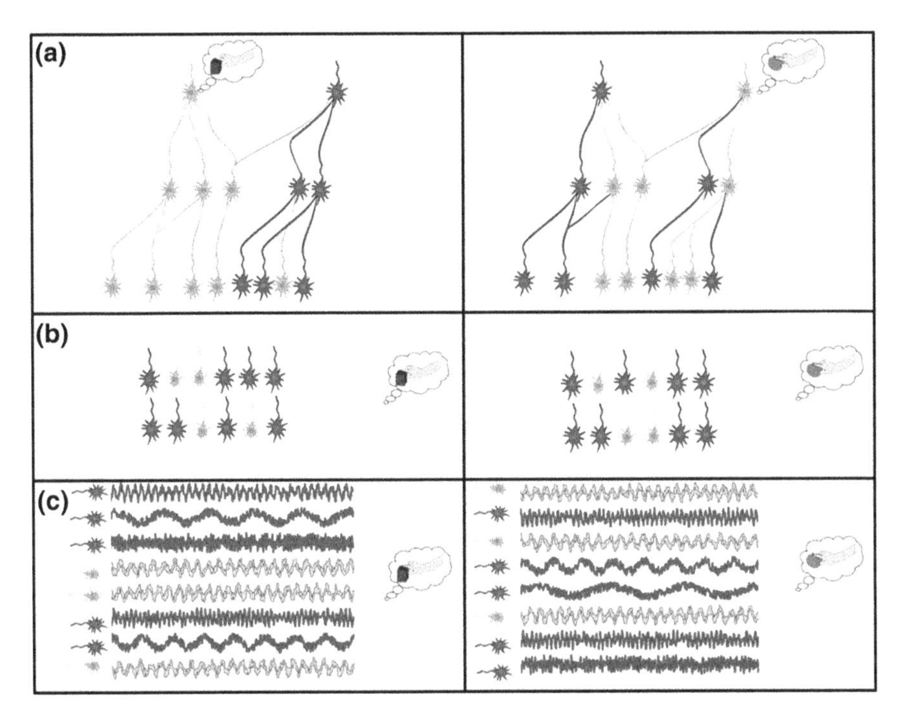

Figure 1 Three suggested schemes by which neurons can represent the existence of external objects via feature binding. Each row shows one scheme. The *left* and *right panels* show how two different objects (a cube and an apple, respectively) can be differently encoded within each scheme. **a** "Grandmother cells": neurons send their outputs in a converging manner, such that each layer has more selective response properties than its predecessor. The activity of specific cells in the topmost layer represents the existence of a specific external object (active neurons appear in *yellow*, inactive ones in *blue*). **b** Population coding: each specific pattern of activation across the entire population of neurons represents the existence of a specific external object. **c** Binding by synchrony: the firing rate of each single neuron may hold some information about the object. The complete bound representation, however, is achieved by compounding those neurons whose firings are synchronized temporally (marked here in *yellow*).

Some studies have specifically examined the existence of anatomical areas co-processing tactile features by using imaging techniques. Roland et al. (1998) used PET imaging on 3 groups of subjects who were asked to discriminate two objects differing either by length, shape, or roughness. Results showed that the lateral parietal operculum (LPO) was more significantly activated during roughness discrimination compared to both shape and length discrimination (using a Boolean intersect of contrasts). This suggests that the LPO is more sensitive to properties of structure than those of material. Indeed, recent studies have found further support for the involvement of the parietal opercular cortex in haptic texture processing (Stilla and Sathian 2008; Sathian et al. 2011). The intraparietal sulcus showed significant activation in the exact opposite circumstances. That is, it was significantly less activated by roughness discrimination than by either length or shape discrimination. Hence, the intraparietal sulcus may actually encode relatively abstract features, such as the combination of length and shape (i.e., representation of structure, and not of material). This, of course, would support the idea that conjunction of features is coded by population activity of higher processing areas.

However, a different study using a slightly more relevant paradigm found no support of the existence of such convergent areas (Burton et al. 1999). In this second study, subjects were asked to discriminate which of two stimuli had either greater roughness or was longer. Additionally, subjects were either cued to notice a specific tactile feature (selective attention condition) or were not cued (divided attention condition). A PET scan was performed during the task, allowing researchers to examine if any areas were activated when subjects attended both features (which was the case during the non-cued, divided attention condition). An activation contrast showed that one area was indeed more significantly activated during divided attention than during selective attention: the orbitofrontal region. However, in light of existing literature, the authors suggest that this activation should be attributed to memory demands. Indeed, when subjects in the divided attention condition awaited the appearance of the second stimulus in each trial, the task required them to keep both features of the first stimulus in memory. Subjects in the selective attention condition only needed to memorize one feature.

Taken together, the findings outlined above do not strongly support the idea that convergence along anatomical areas is a correlate of perceptual compounding of tactile features in the somatosensory cortex. One intriguing suggestion has been that some of these features may be processed by non-somatosensory cortices, such as the visual cortex (Lacey and Sathian 2015). Another suggestion raised by another group of studies is that binding of features into compounded representations may be achieved through synchronized or temporally correlated responses of population firing (Figure 1c; Damasio 1989; Treisman 1996; Von Der Malsburg 1994; Von Der Malsburg 1995). This may be achieved by phase-coherent activity of different areas processing fragments of the entire object (Damasio 1989). It has been further postulated that it is specifically neuronal discharges at the band of gamma oscillations that yield the perceptual binding of co-processed features (Fries et al. 2007; Tallon-baudry and Bertrand 1999).

In the framework of synchronization (or correlation), it is quite possible that no single brain area would selectively and consistently respond to a complex combination of features, which would explain why no such area was found by Burton et al. (1999) in the study cited above. This would be expected when the processing of individual features has spatial signatures (i.e., via the firing of dedicated neuronal circuits), and object binding has only temporal signatures (i.e., via synchronization).

At least one study is in line with the involvement of neuronal synchrony in tactile perception (Steinmetz et al. 2000). In this study, neurons were recorded in the secondary somatosensory cortex of monkeys that performed both a tactile and a visual task and were cued to alternate attention between them. Synchronization analyses were performed on pairs of neurons. Of the 427 neuron pairs which had significant cross-correlogram peaks during either task, 17 % (79/427) demonstrated changes in firing synchrony between the two conditions. 80 % of these pairs (59/74) displayed significantly more synchronous firing when attention was focused on the tactile task compared to the visual one. Tactile stimulation was continued whether monkeys were attending and performing the tactile task or not (i.e., were performing the visual task). Hence, the rise in synchrony does not result purely from stimulation. Rather, it corresponds to attention and to perception of the tactile stimuli. This would of course be in line with the relation between attention and feature binding proposed by FIT, as reviewed in the previous subsection.

As an ending remark to this subsection, we note that the assumption of distributed representations, either over space or time, does not satisfy a solution to binding per se; the questions of how correlations in neuronal firing are computed by the brain, or how semantics emerge from such a correlation, remain open (Roskies 1999).

Feature Processing from an Active Sensing Perspective

In everyday experience, the importance of action in touch is surely quite prominent. While 'foveating' objects in touch (i.e., exploring them with the fingertips) interaction often affects both the object and the perceiver. This is to be expected, as the tactile fovea and the body's main instrument of fine-tuned external manipulation are essentially one and the same: the hand. Given this consideration, it stands to reason that specifically in the tactile modality, the way in which sensors are moved and the perception derived from sensory input should be tightly coupled. Indeed, the way in which one chooses to explore a tactile stimulus has been shown to have a meaningful effect on resulting percepts (Lederman and Klatzky 1987; Locher 1986). While free tactile exploration of everyday stimuli leads to virtually perfect recognition (Klatzky et al. 1985), constrained exploration under short exposure time hampers recognition severely (Klatzky and Lederman 1995).

A major theoretical contribution to the question of action in touch is Lederman and Klatzky's definition of exploratory procedures (EPs; Lederman and Klatzky 1987). Noting that haptic exploration typically involves a stereotypical set of hand gestures, these researchers hypothesized that each such procedure specifically relates to the extraction of a given tactile property (Figure 2). Of the identified procedures, five were most prominently explored in later works: lateral motion (for texture), pressure (for compliance), static contact (for temperature), enclosure (for global shape or volume) and contour following (for exact shape or volume). Preliminary validation showed that the profile of EP frequencies in a trial could generally be predicted by the feature required for discrimination, and vice versa (classification of the trial type by EP profile; Lederman and Klatzky 1987).

EPs form a meaningful explanatory variable. For example, such is the case with the withdrawal studies mentioned above (section "Co-processing of Tactile Features"). When two features covaried redundantly, and one was then withdrawn, EPs were found to mostly correspond with performance time. Specifically, Klatzky et al. (1989) found that for most pairs of features (e.g. shape and hardness),

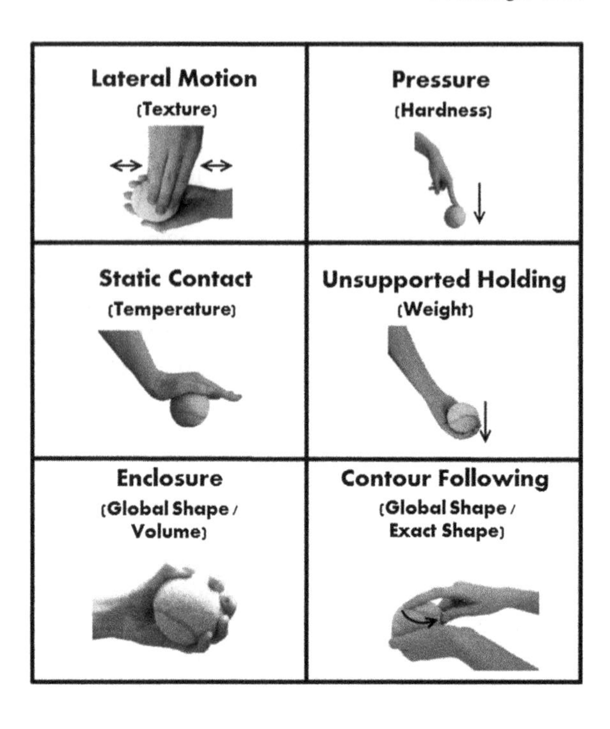

Figure 2 Illustration of exploratory procedures, as described by Lederman and Klatzky (1987). Out of 8 procedures described in the original article, 6 prominently explored ones are presented here. For each such procedure, there exists a haptic feature which it extracts in an optimal manner, appearing in parentheses

performance after withdrawal suffered only slightly and EPs relevant to the withdrawn features were shown to be minimized. However, when the pair of covarying features was texture and hardness, not only was performance hampered by withdrawal, but persistence of EPs relevant for the withdrawn feature was evident. Recall also the fact that covariation in one property of material and one of structure (e.g., texture and shape, respectively) inconsistently led to symmetrical deficits after withdrawal, discussed in section "Co-processing of Tactile Features". Symmetrical deficits were found in Lederman et al. (1993), but not in Klatzky et al. (1989). Under the framework of EPs, this difference is readily understood. In the former experiment all shapes were 3D ellipsoids. As such, information of shape, just like information of material, could be assessed at any regional section of the object. The same cannot be said of stimuli in the latter experiment. Thus, in the former experiment, EPs for extraction of shape and texture (i.e., contour following and lateral motion, respectively) were more compatible with each other. That is, in this experiment both procedures could be optimally performed on interior homogenous sections of the object. This is presumed to have led to stronger co-processing of relevant features. Taken together, these results suggest that feature co-processing (and as a direct result, conjunction) is highly affected by EPs.

EPs are, despite their major contribution, quite a crude description of tactile exploration. Over the last decades, behavioral studies have shifted to using more elaborate analyses, which are designed to help uncover the complex dynamic nature of many phenomena. In an influential line of works, Turvey has suggested that during object hefting and wielding, the tactile perceptual system can extract information by exploiting a variety of rotational factors. Particularly, the inertia tensor about a point (i.e., the object's resistance to rotation in different axes) is an invariant property of the object, which corresponds with subjects' perception of objects magnitudes (e.g., length, width and weight) as well as orientation (summarized in Turvey 1996). Klatzky and Lederman (1992) have found that exploration often starts with a generalized sequence (grasp-and-lift) followed by more detailed exploration. Taking these observations together, it is expected that exploration in a given trial will develop over time, affected by slight variations in movement, object properties, and their interaction. This is because even when correspondence between physical and perceptual properties is strong, information is gathered in a piecemeal manner.

To address the actual complex motions of exploration, methods that explicitly model development over time are required. Indeed, methods for fine-grained modeling of exploratory procedures are gradually being adapted, as discussed in Klatzky and Reed (2009). One such method is called recurrence quantification analysis (RQA), which enables the detection of recurring patterns in a time series. Riley and colleagues (Riley et al. 2002) found that dynamics of wielding under perceptual intent were more complex and more stable (in terms of response to perturbations) than those of free wielding. Another task examined dynamics of wielding for identifying either height or width. The objects used did not have a very large ratio of height to width, but the magnitude of the principal moment of inertia corresponding to height was at least five times greater than that corresponding to width. Consequently, height was predicted to be a much more salient property for perceivers than width. Results showed that discriminations of width led to wielding dynamics more deterministic, complex, and stable than those of height, suggesting fine-tuning of exploratory manipulations. These results affirm the idea outlined above. That is, they extend the concept suggested by Lederman and Klatzky (1987) to show that exploration is anticipatory not only at higher levels of description, but also over very short timescales, presumably adapting itself in an iterative manner.

This last property is in line with the behavior expected of closed-loop systems, systems in which signals affect their sources via feedback. Indeed, studies in humans and rats seeking out features have shown that gradually convergent changes of motor and sensory variables occur during perceptual tasks (Horev et al. 2011; Mitchinson et al. 2007; Saig et al. 2012). It is quite plausible, then, that EP compatibility has an even more profound role in tactile object perception then originally proposed. Importantly, EP dynamics likely vary greatly over short time intervals, given the complex interactions between intent, exploratory performance, and sensory information received. We thus predict that works extensively examining co-processing of features from the perspective of dynamical systems are imminently the next step within the field of tactile object perception.

Conclusion

In sum, the research of tactile object perception is a multi-faceted field of study, with various sub-fields and paradigms contributing unique insights into its functions and mechanisms. In some cases these contributions are synergistic. For example, attention's role in tactile perception is emphasized both by behavioral studies and by electrophysiological recordings of firing synchrony in neurons (e.g., Steinmetz et al. 2000). The dependency of object perception on motor patterns is demonstrated in primate and rodent by behavioral studies (e.g., Horev et al. 2011; Lederman and Klatzky 1987; Mitchinson et al. 2007; Riley et al. 2002; Turvey 1996) and supported by electrophysiology studies (e.g., Iwamura and Tanaka 1978). Action, in general, appears to be indispensable in interpreting results of behavioral paradigms, including those manipulating attention (Riley et al. 2002), as predicted decades ago by Gibson (Gibson 1962). The accumulated evidence suggest that action, sensation and neuronal activity form a loop of variables affecting each other, and by doing so determining tactile object perception.

We hence conclude by suggesting that study of tactile objects from the perspective of a dynamical system may help complete the unifying framework that the field is currently lacking. When research questions are reformulated under this approach theoretical difficulties may simply dissolve; Questions related to attention may find their resolution in kinematic descriptions, and the ambiguity of primitives might be resolved by considering motor-sensory contingencies (Bagdasarian et al. 2013; O'Regan and Noë 2002). Importantly, experiments addressing these questions must allow natural-like employment of motor patterns.

References

Bagdasarian, K et al. (2013). Pre-neuronal morphological processing of object location by individual whiskers. *Nature Neuroscience* 16(5): 622–631.

Barlow, H B (1972). Single units and sensation: A neuron doctrine for perceptual psychology. *Perception* 1(4): 371–394.

Burton, H; Abend, N S; Macleod, A M; Sinclair, R J; Snyder, A Z and Raichle, M E (1999). Tactile attention tasks enhance activation in somatosensory regions of parietal cortex: A positron emission tomography study. *Cerebral Cortex* 9(7): 662–674.

Corsini, D A and Pick, H L (1969). The effect of texture on tactually perceived length. *Perception & Psychophysics* 5(6): 352–356.

Damasio A R (1989). The brain binds entities and events by multiregional activation from convergence zones. *Neural Computation* 1(1): 123–132.

Fries, P; Nikolić, D and Singer, W (2007). The gamma cycle. *Trends in Neurosciences* 30(7): 309–316.

Gelbard-Sagiv, H; Mukamel, R; Harel, M; Malach, R and Fried, I (2008). Internally generated reactivation of single neurons in human hippocampus during free recall. *Science* 322(5898): 96–101.

Gibson, J J (1962). Observations on active touch. *Psychological Review* 69(6): 477–491.

Griffiths, T D and Warren, J D (2004). What is an auditory object? *Nature Reviews Neuroscience* 5(11): 887–892.
Hayward, V (2008). A brief taxonomy of tactile illusions and demonstrations that can be done in a hardware store. *Brain Research Bulletin* 75(6): 742–752.
Horev, G; Saig, A; Knutsen, P M; Pietr, M; Yu, C and Ahissar, E (2011). Motor-sensory convergence in object localization: A comparative study in rats and humans. *Philosophical Transactions of the Royal Society of London. Series B, Biological Sciences* 366(1581): 3070–3076.
Hubel, D H and Wiesel, T N (1962). Receptive fields, binocular interaction and functional architecture in the cat's visual cortex. *The Journal of Physiology* 160(1): 106–154.
Iwamura, Y and Tanaka, M (1978). Postcentral neurons in hand region of area 2: Their possible role in the form discrimination of tactile objects. *Brain Research* 150(3): 662–666.
Iwamura, Y; Tanaka, M; Hikosaka, O and Sakamoto, M (1995). Postcentral neurons of alert monkeys activated by the contact of the hand with objects other than the monkey's own body. *Neuroscience Letters* 186(2): 127–130.
Jörntell, H; Bengtsson, F; Geborek, P; Spanne, A; Terekhov, A V and Hayward, V (2014). Segregation of tactile input features in neurons of the cuneate nucleus. *Neuron* 83(6); 1444–1452.
Kant, I (n.d.). "Critique of pure reason." (P Guyer and A W Wood, Eds.). Cambridge, MA: Cambridge University Press.
Klatzky, R L and Lederman, S J (1992). Stages of manual exploration in haptic object identification. *Perception* 52(6): 661–670.
Klatzky, R L; Lederman, S J and Metzger, V A (1985). Identifying objects by touch: An "expert system." *Perception & Psychophysics* 37(4): 299–302.
Klatzky, R L; Lederman, S J and Reed, C L (1989). Haptic integration of object properties: Texture, hardness, and planar contour. *Journal of Experimental Psychology: Human Perception and Performance* 15(1): 45–57.
Klatzky, R L and Lederman, S J (1995). Identifying objects from a haptic glance. *Perception & Psychophysics* 57(8): 1111–1123.
Klatzky, R L and Reed, C L (2009). Haptic exploration. *Scholarpedia* 4(8): 7941.
Kobatake, E and Tanaka, K (1994). Neuronal selectivities to complex object features in the ventral visual pathway of the macaque cerebral cortex. *Journal of Neurophysiology* 71(3): 856–867.
Lacey, S; Lin, J B and Sathian, K (2011). Object and spatial imagery dimensions in visuo-haptic representations. *Experimental Brain Research* 213(2–3): 267–273.
Lacey, S and Sathian, K (2015). Crossmodal and multisensory interactions between vision and touch. *Scholarpedia* 10(3): 7957. http://www.scholarpedia.org/article/Crossmodal_and_multisensory_interactions_between_vision_and_touch. (see also pages 301–315 of this book).
Lederman, S J; Browse, R A and Klatzky, R L (1988). Haptic processing of spatially distributed information. *Perception & Psychophysics* 44(3): 222–232.
Lederman, S J and Jones, L A (2011). Tactile and haptic illusions. *IEEE Transactions on Haptics* 4(4): 273–294.
Lederman, S J and Klatzky, R L (1987). Hand movements: A window into haptic object recognition. *Cognitive Psychology* 19(3): 342–368.
Lederman, S J and Klatzky, R L (2009). Haptic perception: A tutorial. *Attention, Perception, & Psychophysics* 71(7): 1439–1459.
Lederman, S J; Klatzky, R L and Reed, C L (1993). Constraints on haptic integration of spatially shared object dimensions. *Perception* 22(6): 723–743.
Locher, P J (1986). Aspects of haptic perceptual information processing. *Bulletin of the Psychonomic Society*, 24(3), 197–200.
Mitchinson, B; Martin, C J; Grant, R A and Prescott, T J (2007). Feedback control in active sensing: Rat exploratory whisking is modulated by environmental contact. *Proceedings of the Royal Society B: Biological Sciences* 274(1613): 1035–1041.
O'Regan, J K and Noë, A (2002). A sensorimotor account of vision and visual consciousness. *Behavioral and Brain Sciences* 24(5): 939–973.

Plaisier, M A; Bergmann Tiest, W M and Kappers, A M L (2008). Haptic pop-out in a hand sweep. *Acta Psychologica* 128(2): 368–377.

Plaisier, M A; Bergmann Tiest, W M and Kappers, A M L (2009). Salient features in 3-D haptic shape perception. *Attention, Perception, & Psychophysics* 71(2): 421–430.

Plaisier, M A and Kappers, A M L (2010). Cold objects pop out! *Lecture Notes in Computer Science* 6192: 219–224.

Quiroga, R Q; Mukamel, R; Isham, E A; Malach, R and Fried, I (2008). Human single-neuron responses at the threshold of conscious recognition. *Proceedings of the National Academy of Sciences* 105(9): 3599–3604.

Quiroga, R Q; Reddy, L; Kreiman, G; Koch, C and Fried, I (2005). Invariant visual representation by single neurons in the human brain. *Nature* 435(7045): 1102–1107.

Reed, C L; Lederman, S J and Klatzky, R L (1990). Haptic integration of planar size with hardness, texture, and planar contour. *Canadian Journal of Psychology* 44(4): 522–545.

Riley, M A; Wagman, J B; Santana, M V; Carello, C and Turvey, M T (2002). Perceptual behavior: Recurrence analysis of a haptic exploratory procedure. *Perception* 31(4): 481–510.

Roland, P E; O'Sullivan, B and Kawashima, R (1998). Shape and roughness activate different somatosensory areas in the human brain. *Proceedings of the National Academy of Sciences* 95(6): 3295–3300.

Roskies, A L (1999). The Binding Problem. *Neuron* 24(1): 7–9.

Saig, A; Gordon, G; Assa, E; Arieli, A and Ahissar, E (2012). Motor-sensory confluence in tactile perception. *The Journal of Neuroscience* 32(40): 14022–14032.

Saal, H P and Bensmaia, S J (2014). Touch is a team effort: Interplay of submodalities in cutaneous sensibility, *Trends in Neuroscience* 37(12): 689–697.

Sathian, K and Burton, H (1991). The role of spatially selective attention in the tactile perception of texture. *Perception & Psychophysics* 50(3): 237–248.

Sathian, K et al. (2011). Dual pathways for haptic and visual perception of spatial and texture information. *Neuroimage* 57(2): 462–475.

Sinclair, R J; Kuo, J J and Burton, H (2000). Effects on discrimination performance of selective attention to tactile features. *Somatosensory & Motor Research* 17(2): 145–157.

Steinmetz, P N; Roy, A; Fitzgerald, P J; Hsiao, S S and Johnson, K O (2000). Attention modulates synchronized neuronal firing in primate somatosensory cortex. *Nature* 404(6774): 187–190.

Stilla, R and Sathian, K (2008). Selective visuo-haptic processing of shape and texture. *Human Brain Mapping* 29(10): 1123–1138.

Tallon-Baudry, C and Bertrand, O (1999). Oscillatory gamma activity in humans and its role in object representation. *Trends in Cognitive Science* 3(4): 151–162.

Treisman, A (1996). The binding problem. *Current Opinion in Neurobiology* 6(2): 171–178.

Treisman, A (1998a). Feature binding, attention and object perception. *Philosophical Transactions of the Royal Society of London. Series B: Biological Sciences* 353(1373): 1295–1306.

Treisman, A (1998b). Features and objects: The fourteenth Bartlett memorial lecture. *The Quarterly Journal of Experimental Psychology Section A* 40(2): 201–237.

Treisman, A and Souther, J (1985). Search asymmetry: A diagnostic for preattentive processing of separable features. *Journal of Experimental Psychology: General* 114(3): 285–310.

Turvey, M T (1996). Dynamic touch. *American Psychologist* 51(11): 1134–1152.

Von Der Malsburg, C (1994). *The Correlation Theory of Brain Function* (pp. 95–119). New York: Springer.

Von Der Malsburg, C (1995). Binding in models of perception and brain function. *Current Opinion in Neurobiology* 5(4): 520–526.

Whang, K C; Burton, H and Shulman, G L (1991). Selective attention in vibrotactile tasks : Detecting the presence and absence of amplitude change. *Perception & Psychophysics* 50(2): 157–165.

Yeshurun, Y and Sobel, N (2010). An odor is not worth a thousand words: From multidimensional odors to unidimensional odor objects. *Annual Review of Psychology*, 61: 219–241.

Haptic Exploration

Roberta Klatzky and Catherine L. Reed

Haptic exploration refers to purposive action patterns that perceivers execute in order to encode properties of surfaces and objects, patterns that are executed spontaneously and also appear to optimize information uptake.

Exploratory Procedures

In 1987, Lederman and Klatzky described a set of specialized patterns of exploration, called exploratory procedures (EPs). These exploratory patterns are linked to specific object properties, in two respects: The EP associated with an object property is (a) executed spontaneously when information about that property is desired, and (b) appears to optimize information uptake about that property. A basic set of EPs, along with their associated properties and behavioral invariants, is as follows:

- **Exploratory Procedure**: Lateral motion
- Associated Property: Surface texture
- Behavior: The skin is passed laterally across a surface, producing shear force.
- **Exploratory Procedure**: Pressure
- Associated Property: Compliance or hardness
- Behavior: Force is exerted on the object against a resisting force; for example, by pressing into the surface, bending the object, or twisting.
- **Exploratory Procedure**: Static contact
- Associated Property: Apparent temperature
- Behavior: The skin surface is held in contact with the object surface, without motion; typically a large surface (like the whole hand) is applied. This EP gives rise to heat flow between the skin and the object.

R. Klatzky (✉)
Department of Psychology, Carnegie Mellon University, Pittsburgh, PA, USA

C.L. Reed
Department of Psychology, Claremont McKenna College, Claremont, CA, USA

T.J. Prescott et al. (eds.), *Scholarpedia of Touch*, Scholarpedia,
DOI 10.2991/978-94-6239-133-8_13

- **Exploratory Procedure**: Unsupported holding
- Associated Property: Weight
- Behavior: The object is held while the hand is not externally supported; typically this EP involves lifting, hefting or wielding the object.
- **Exploratory Procedure**: Enclosure
- Associated Property: Volume; Global shape
- Behavior: The fingers (or other exploring effector) are molded closely to the object surface.
- **Exploratory Procedure**: Contour following
- Associated Property: Exact shape
- Behavior: Skin contact follows the gradient of the object's surface or is maintained along edges when they are present.

People can acquire accurate information about the touched environment not only by directed exploration, but by dynamic touch in the act of manipulating tools and other objects. Turvey and colleagues focused on information that can be obtained from grasping and wielding (e.g., raising, lowering, pushing, turning, transporting: (Turvey and Carello 1995)). Information obtained about an object from wielding includes length, weight, width, shape of the object tip, and orientations of hands in relation to the object. This information is obtained from the sensitivity of body tissues to values of rotational dynamics under rotational forces (torques) and motions.

Optimization of Exploratory Procedures

For purposes of haptic perception, people tend to explore objects using EPs that optimize information apprehension (Lederman and Klatzky 1987). An EP that is optimal for one property may also deliver information about another; for example, contour following on a surface will be informative about its texture, since the contour following EP, like lateral motion, produces shear forces. Conversely, although a specialized EP maximizes information intake about an associated object property, it has the further consequence of limiting access to some other properties. For example, use of the static hand to perceive temperature is incompatible with executing lateral motion to perceive texture. Some EPs can be executed together, allowing simultaneous access to multiple object properties (Klatzky et al. 1987). For example, people tend to exert pressure while they are rubbing an object to obtain simultaneous compliance and texture information. Also, when people grasp and lift an object, they obtain information about its shape, volume, and weight.

Fine-grained analysis of exploration has found that the parameterization of EPS is optimized to the local context (Riley et al. 2002; Smith et al. 2002). When judging roughness, for example, people vary contact force more with smooth than rough exemplars, and they scan more rapidly when discriminating surfaces than identifying them (Tanaka et al 2014). When exploring to determine compliance, people use greater force when expecting a rigid object than a compliant one. The optimality of this approach is indicated by the finding that when fine compliance discriminations are called for, enforcing unnaturally low forces impairs performance (Kaim and Drewing 2011).

The presence of vision changes the nature of optimal haptic exploration. When objects can be viewed as well as touched, specialized EPs tend to be executed only when people wish to perceive material properties (e.g., compliance, texture), and then only when relatively precise discrimination is demanded (Klatzky et al. 1993). For example, the rough texture of coarse sandpaper is salient to vision, and hence the object is unlikely to elicit haptic exploration. In contrast, a person attempting to determine whether a surface is free from grit (a fine discrimination) will be likely to explore using a specialized EP, lateral motion.

Haptic Exploration in Nonhumans and Young Humans

A variety of species exhibit specialized patterns of exploration similar to those of humans. These include squirrel and capuchin monkeys (Hille et al. 2001; Lacreuse and Fragaszy 1997). Whisker sweeps of seals during shape discrimination may serve a similar function to following contours of an object by humans (Denhardt and Dücker 1996).

In humans, systematic exploration similar to that of adults is observed in young children and follows a developmental progression. Early grasping and fingering may be precursors of the EPs of enclosure and lateral motion (Ruff 1989). The occurrence of these EPs has been found to be predicated on which properties are most salient; for example, textured objects promote fingering. The fact that infants tend to explore with the mouth may reflect early motor control of the oral musculature as well as the density of sensory input (Rochat 1989).

Pre-school aged children execute adult-like EPs when given similar perceptual goals (Kalagher and Jones 2011). Pre-schoolers spontaneously show dedicated exploratory patterns not only when a target dimension such as hardness is explicitly mentioned, but when its appropriateness arises in tool use. For example, they use pressure to test a stirring stick to ensure that it is sufficiently rigid for the substance that must be stirred (Klatzky et al. 2005). Appropriate exploration is also found when blind children match objects on designated dimensions (Withagen et al. 2013).

Haptic Exploration from a Neurophysiological Perspective

To understand why there is specialization of exploration during haptic perception, one must consider two perspectives: the physical interaction between the perceiver and the object, and the neural consequences of that interaction. In general, EPs put into place physical interactions that then optimize the signal to sensory receptors and higher-order neural computations. For example, the EP called Static Contact, whereby a large skin surface is held without motion against a surface in order to perceive its temperature, provides an opportunity for heat flow between the skin and surface. The resulting change of temperature at the skin surface is sensed by neurons within the skin that specialize in the detection of coolness and warmth, and initiate signals of thermal change to the brain (Jones and Ho 2008). In contrast, dynamic contact, such as grasping and releasing a warming or cooling surface, actually diminishes the thermal sensation relative to the static EP (Green 2009).

As was noted above, fine-grained parameters of exploration are tuned to the perceptual and task environment. Such tuning appears to be the result of pre-cortical as well as cortical neural interactions. When Weiss and Flanders (2011) asked participants to follow, with their fingertip, the contour of a spherical surface rendered in a virtual environment, an unexpected change in surface curvature led to a compensatory adjustment in contact force with a latency associated with spinal control (approximately 50 ms). Other control mechanisms appeared to be regulated at cortical levels.

Specialized exploration has implications for the brain's ability to recognize objects by touch. For tactile object recognition, perceptual input information must be transformed through a series of processing stages before object identification. The first stage of processing involves the extraction and processing of the properties and features of the felt object. For example, to recognize a quarter one takes in its small size, cool temperature, round shape, and rough edges. The perceptual information provided by EPs is combined or integrated into a common modality-specific object description, which is then used to access information about object identity and function. When people are allowed to either use specialized EPs or a more general grasp for recognizing real, multidimensional objects, brain regions specific to touch are activated in addition to brain regions activated for visual object recognition. These touch-specific brain regions include primary somatosensory cortex (SI), secondary somatosensory cortex (SII), parietal operculum, and the insula. Some activated brain regions are similar to object recognition using vision, such as the lateral occipitotemporal cortex, medial temporal lobe, and prefrontal areas, all of which support cross-modal information integration leading to object recognition (Reed et al. 2004). However, when the touched stimuli are primarily two-dimensional spatial patterns that do not permit EPs, a greater reliance on visual cortical regions (e.g., V1, V2, etc.) is observed (Zangaladze et al. 1999).

To what extent are EPs necessary for object recognition? Patients with brain damage provide some insight into the relative contributions from sensory and motor inputs. To examine the extent to which somatosensory and motoric inputs influence the tactile object recognition process, researchers have demonstrated that patients who have damage to their hands or peripheral nervous system, as well as patients with lesions in SI, do not always show tactile object recognition deficits (Valenza et al. 2001). Likewise, patients with hand paralysis do not necessarily have significant deficits in tactile object recognition (Caselli 1991). Because the hand uses both tactile inputs and hand movements, usually in the form of EPs, to extract somatosensory information, these patients may be using the motions of object parts or hand movement cues to extract relevant object information that they cannot obtain through purely tactile perception. Finally, patients with "tactile agnosia" have a deficit in recognizing common objects by touch following brain damage. Although rare, these patients tend to have lesions in the left inferior parietal lobe (e.g., Reed et al. 1996). Their disorder tends to be at a higher cognitive level because their EPs are normal and they tend to have relatively intact tactile sensation, memory, spatial processing, and general intellectual function. Conversely tactile object recognition deficits are also observed in patients who have right hemisphere parietal lobe damage and demonstrate disorganized or random EPs. In sum, the inability to execute EPs does contribute to tactile agnosia and make object recognition more difficult, but the object recognition process can be aided by previous knowledge of object parts and functions that can assist sensory and motoric limitations.

Haptic Exploration and Tool Use

Finally, people explore objects not only with their hands or other parts of the body, but also with tools. Although tools limit what can be perceived, they can nevertheless provide at least coarse information about some object properties. For example, when a rigid tool is used to explore a rigid object, the resulting vibrations can enable people to perceive its surface texture (Katz 1925). Any dentist knows that the deformation and resistance of an object under the force of a probing tool can allow the perception of compliance. The same EP that is executed with the bare hand to extract an object property may not be observed when a tool is used, given the changes in the physical interaction that arise from tool use.

Conclusion

In conclusion, haptic exploration involves exploratory procedures, active touch patterns that are specific to the demands of the task in that they optimize the extraction of information the perceiver needs to obtain. Purpose haptic exploration allows people and animals to extract specific types of tactile information from the environment and provides information about the material or substance properties of objects that cannot be achieved by vision alone. Further, exploratory procedures expand the basic sensory functions of animate bodies by allowing them to perceive the world through tools.

Internal References

Bensmaia, S (2009). Texture from touch. *Scholarpedia* 4(8): 7956. http://www.scholarpedia.org/article/Texture_from_touch. (see also pages 207–215 of this book)

Braitenberg, V (2007). Brain. *Scholarpedia* 2(11): 2918. http://www.scholarpedia.org/article/Brain.

Eichenbaum, H (2008). Memory. *Scholarpedia* 3(3): 1747. http://www.scholarpedia.org/article/Memory.

Llinas, R (2008). Neuron. *Scholarpedia* 3(8): 1490. http://www.scholarpedia.org/article/Neuron.

External References

Caselli, R J (1991). Rediscovering tactile agnosia. *Mayo Clinic Proceedings* 66(2): 129–142. doi:10.1016/s0025-6196(12)60484-4 (http://dx.doi.org/10.1016/s0025-6196(12)60484-4).

Denhardt, G and Dücker, G (1996). Tactual discrimination of size and shape by a California sea lion (*Zalophus californianus*). *Animal Learning & Behavior* 24(4): 366–374. doi:10.3758/bf03199008 (http://dx.doi.org/10.3758/bf03199008).

Green, B G (2009). Temperature perception on the hand during static versus dynamic contact with a surface. *Attention, Perception, & Psychophysics* 71(5): 1185–1196. doi:10.3758/app.71.5.1185 (http://dx.doi.org/10.3758/app.71.5.1185).

Hille, P; Becker-Carus, C; Dücker, G and Dehnhardt, G (2001). Haptic discrimination of size and texture in squirrel monkeys (*Saimiri sciureus*). *Somatosensory & Motor Research* 18(1): 50–61. doi:10.1080/08990220020021348 (http://dx.doi.org/10.1080/08990220020021348).

Jones, L A and Ho, H (2008). Warm or cool, large or small? The challenge of thermal displays. *IEEE Transactions on Haptics* 1(1): 53–70. doi:10.1109/toh.2008.2 (http://dx.doi.org/10.1109/toh.2008.2).

Kaim, L and Drewing, K (2011). Exploratory strategies in haptic softness discrimination are tuned to achieve high levels of task performance. *IEEE Transactions on Haptics* 4(4): 242–252. doi:10.1109/toh.2011.19 (http://dx.doi.org/10.1109/toh.2011.19).

Kalagher, H and Jones, S S (2011). Young children's haptic exploratory procedures. *Journal of Experimental Child Psychology* 110(4): 592–602. doi:10.1016/j.jecp.2011.06.007 (http://dx.doi.org/10.1016/j.jecp.2011.06.007).

Katz, D (1925). *Der aufbau der tastwelt (The world of touch)*. L. Krueger and L. Erlbaum (Eds.). Mahwah, NJ. ISBN: 9780805805291.

Klatzky, R L; Lederman, S J and Mankinen, J M (2005). Visual and haptic exploratory procedures in children's judgments about tool function. *Infant Behavior and Development* 28(3): 240–249. doi:10.1016/j.infbeh.2005.05.002 (http://dx.doi.org/10.1016/j.infbeh.2005.05.002).

Klatzky, R L; Lederman, S J and Matula, D E (1993). Haptic exploration in the presence of vision. *Journal of Experimental Psychology: Human Perception and Performance* 19(4): 726–743. doi:10.1037//0096-1523.19.4.726 (http://dx.doi.org/10.1037//0096-1523.19.4.726).

Klatzky, R L; Lederman, S J and Reed, C L (1987). There's more to touch than meets the eye: the salience of object attributes for haptics with and without vision. *Journal of Experimental Psychology: General* 116: 356–369.

Lacreuse, A and Fragaszy, D M (1997). Manual exploratory procedures and asymmetries for a haptic search task: A comparison between capuchins (*Cebus apella*) and humans. *Laterality* 2 (3): 247–266. doi:10.1080/135765097397477 (http://dx.doi.org/10.1080/135765097397477).

Lederman, S J and Klatzky, R L (1987). Hand movements: a window into haptic object recognition. *Cognitive Psychology* 19(3): 342–368. doi:10.1016/0010-0285(87)90008-9 (http://dx.doi.org/10.1016/0010-0285(87)90008-9).

Reed, C L; Caselli, R J and Farah, M J (1996). Tactile agnosia: Underlying impairment and implications for normal tactile object recognition. *Brain* 119(3): 875–888. doi:10.1093/brain/119.3.875 (http://dx.doi.org/10.1093/brain/119.3.875).

Reed, C L; Shoham, S and Halgren, E (2004). Neural substrates of tactile object recognition: An fMRI study. *Human Brain Mapping* 21(4): 236–246. doi:10.1002/hbm.10162 (http://dx.doi.org/10.1002/hbm.10162).

Riley, M A; Wagman, J B; Santana, M-V; Carello, C and Turvey, M T (2002). Perceptual behavior: Recurrence analysis of a haptic exploratory procedure. *Perception* 31(4): 481–510. doi:10.1068/p3176 (http://dx.doi.org/10.1068/p3176).

Rochat, P (1989). Object manipulation and exploration in 2- to 5-month-old infants. *Developmental Psychology* 25(6): 871–884. doi:10.1037//0012-1649.25.6.871 (http://dx.doi.org/10.1037//0012-1649.25.6.871).

Ruff, H A (1989). The infant's use of visual and haptic information in the perception and recognition of objects. *Canadian Journal of Psychology/Revue Canadienne De Psychologie* 43 (2): 302–319. doi:10.1037/h0084222 (http://dx.doi.org/10.1037/h0084222).

Smith, A; Gosselin, G and Houde, B (2002). Deployment of fingertip forces in tactile exploration. *Experimental Brain Research* 147(2): 209–218. doi:10.1007/s00221-002-1240-4 (http://dx.doi.org/10.1007/s00221-002-1240-4).

Tanaka, Y; Bergmann Tiest, W M; Kappers, A M L and Sano, A (2014). Contact force and scanning velocity during active roughness perception. *PLoS ONE* 9(3): e93363. doi:10.1371/journal.pone.0093363 (http://dx.doi.org/10.1371/journal.pone.0093363).

Turvey, M T and Carello, C (1995). Dynamic touch. In: W Epstein and S Rogers (Eds.), *Handbook of perception and cognition, Vol. 5, Perception of space and motion* (pp. 401–490). San Diego: Academic Press.

Valenza, N et al. (2001). Dissociated active and passive tactile shape recognition: A case study of pure tactile apraxia. *Brain* 124(11): 2287–2298. doi:10.1093/brain/124.11.2287 (http://dx.doi.org/10.1093/brain/124.11.2287).

Weiss, E J and Flanders, M (2011). Somatosensory comparison during haptic tracing. *Cerebral Cortex* 21(2): 425–434. doi:10.1093/cercor/bhq110 (http://dx.doi.org/10.1093/cercor/bhq110).

Withagen, A; Kappers, A M L; Vervloed, M P J; Knoors, H and Verhoeven, L (2013). The use of exploratory procedures by blind and sighted adults and children. *Attention, Perception, & Psychophysics* 75(7): 1451–1464. doi:10.3758/s13414-013-0479-0 (http://dx.doi.org/10.3758/s13414-013-0479-0).

Zangaladze, A; Epstein, C M; Grafton, S T and Sathian, K (1999). Involvement of visual cortex in tactile discrimination of orientation. *Nature* 401(6753): 587–590. doi:10.1038/44139 (http://dx.doi.org/10.1038/44139).

External Links

http://www.psy.cmu.edu/faculty/klatzky/ Roberta Klatzky's academic webpage
http://www.claremontmckenna.edu/psych/creed/ Catherine L. Reed's academic webpage

Haptic Saliency

Astrid M.L. Kappers and Wouter M. Bergmann Tiest

Humans can distinguish many object and surface properties by touch. Some of these properties are highly salient: they are immediately perceived after just a brief touch. Since the processing of such features is apparently highly efficient, it is of interest to investigate which features are salient to touch and which are not. In vision, there already exists a large body of literature concerning salient features (e.g., Itti 2007). A typical way to investigate visual saliency is by means of a search task (e.g., Treisman and Gelade 1980; Treisman and Gormican 1988; Wolfe and Horowitz 2008). Here, we will describe this search task, its adaptations to haptic research and give an overview of the features that are salient to touch.

Visual Search

The idea behind the search tasks is that items with a salient feature stand out among distractor items. In Figure 1a, b, examples are given of a red disk among blue disks. In both examples, the red disk is seen immediately; finding the red disk does not depend on the number of blue distractor items in the direct neighbourhood. On the other hand, finding the letter "o" among the distracting letters "c" as in Figure 1d, e, is much harder and clearly depends on the number of distractors. In a typical visual search task, participants are confronted with a series of displays like in this figure (with one type of stimuli) and they have to decide as quickly as possible whether or not the target item is present, without making too many errors. A measure for the saliency of the target item property (in these examples "red" or "opening", respectively) is the so-called search slope, which is the slope of the line fitted through the response times plotted against the total number of items. If the property is salient, the search slope will be rather flat, as the response times hardly depend on number of distractors (see Figure 1c); the target item is said to "pop-out" and the search is termed "parallel". In case of a less salient target, the response times

A.M.L. Kappers (✉) · W.M. Bergmann Tiest
VU University, Amsterdam, The Netherlands
e-mail: a.m.l.kappers@vu.nl

T.J. Prescott et al. (eds.), *Scholarpedia of Touch*, Scholarpedia,
DOI 10.2991/978-94-6239-133-8_14

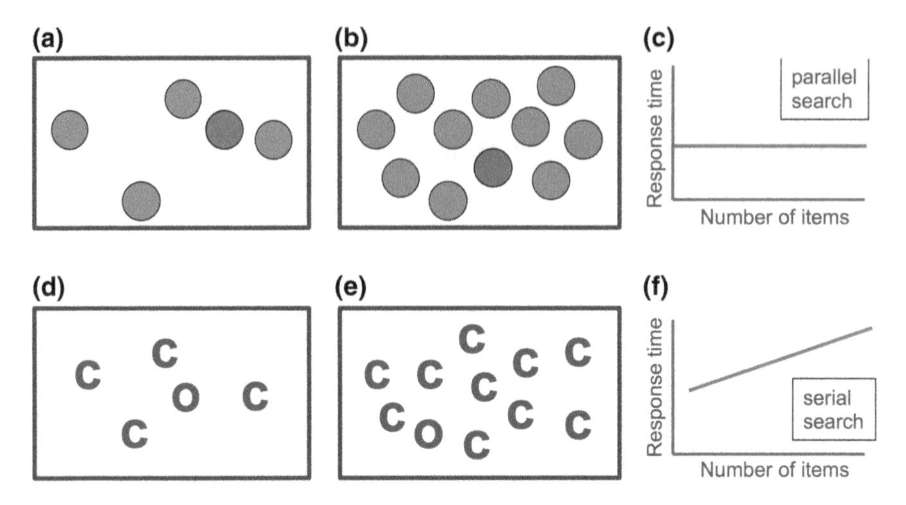

Figure 1 Schematic illustration of a visual search task. Observers have to decide as fast as possible whether or not a target is present. In (**a**) and (**b**), the target is a *red disk*, in (**d**) and (**e**), it is a letter "o". In parallel search, the response time does not depend on the number of items (**c**); in serial search, the response time increases linearly with the number of items (**f**)

increase with number of distractors (see Figure 1f) and the search is termed "serial"; participants have to go over each item one by one until the target is found, or, in case the target is not present, all items have been inspected.

Although the search slope gives a good indication of the saliency of the target property, it should be noted that the division between parallel and serial search is not clear-cut (Wolfe 1998): the (relative) saliency of a property depends on the context. For example, the property "red colour" might be salient among blue items, it will be much less so among orange items. Therefore, search slopes can be anything between very flat and very steep. The important message is that in some conditions, colour can be salient and thus colour must be processed very fast or very early on in the visual system.

Haptic Search for Saliency

In this section, we will first describe the various methods that have been introduced to make the visual search paradigm suitable for touch experiments. In the subsections that follow, we will discuss the results obtained with these methods for the various stimulus properties that have been studied.

Lederman and colleagues (1988) were the first to adapt the visual search paradigm to the sense of touch. Stimuli could be presented to the index, middle and ring fingers of both hands. Like in the visual search tasks, blindfolded participants were required to decide as quickly as possible whether a stimulus with a certain property or combination of properties was present or not. In an extensive follow-up study using a sophisticated device (see Moore et al. (1991) for a description), Lederman and Klatzky (1997) investigated a large number of stimulus properties, such as roughness, hardness, temperature, surface discontinuity and surface orientation. From trial to trial the number of stimulated fingers varied from 1 to 6, with at most one target present. The fingers could make small movements, but basically remained in place. They used the search slopes as outcome measure.

A key feature of haptic perception is that it is active and that may be essential for optimal performance. Therefore, Plaisier et al. (2008a) created a dynamic version of the search task, first in two dimensions and subsequently also in three. In two dimensions, displays like those in vision, were produced with smooth and rough patches of sandpaper as target and distractor (in either way). Participants were asked to sweep their hand over the display in search for a target and again decide as fast as possible whether the target was present. Depending on the difference between the roughnesses of target and distractors, movement patterns were either simple (see movie http://www.scholarpedia.org/article/File:OneSweep.gif) or complex (see movie http://www.scholarpedia.org/article/File:Serialsearch2.gif), and this was used as an additional measure for distinguishing parallel from serial searches.

In a three-dimensional version of the active haptic search task, Plaisier et al. (2008b) asked participants to grasp a bunch of items, such as spheres, cubes or ellipsoids. Again, a target could either be present or not. With easily discriminable features, a single grasp sufficed, but in harder discrimination conditions, the items had to be explored one by one and occasionally be thrown out of the hand (see movie http://www.scholarpedia.org/article/File:Grasp.gif). As release of items is a clear indication of serial search, the percentage of trials with released items was taken as an additional measure for saliency.

Another way to address the haptic saliency of features is to investigate what can be observed in a very brief touch. In vision, salient properties pop-out and can be observed in just a glance. Klatzky and Lederman (1995) investigated the recognition of objects in what they termed a "haptic glance", a brief touch of about 200 ms. Features that can be observed within such a short period pop-out and need to be processed early on in the haptic system.

Roughness

In the experiments of Lederman and Klatzky (1997), the target could be either rough or smooth and the difference between the roughnesses of target and distractor could be either small or large. In the easy condition with the rough target, they found very shallow search slopes. These were slightly higher when the target was smooth. In the difficult condition, all search slopes were very steep, indicating that participants had to examine the items one at a time. The important finding was that it was possible to create a condition with very flat search slopes, indicating that roughness could be processed fast. The saliency of roughness depends, as always, on the relative difference with the distractors.

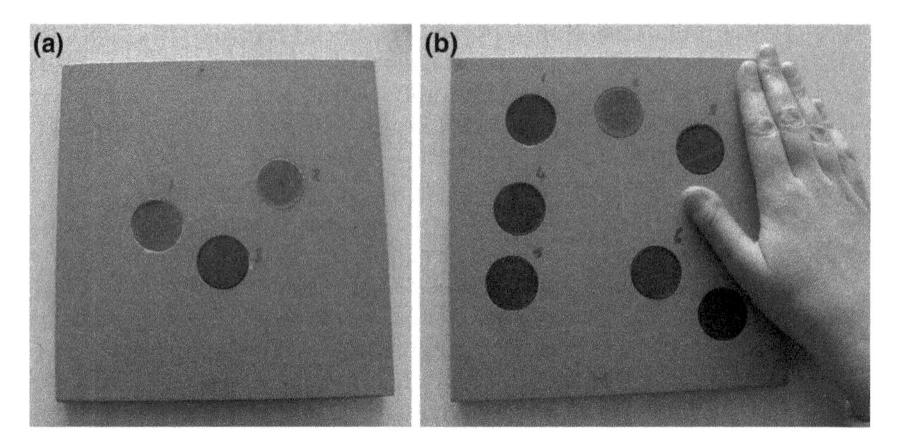

Figure 2 Examples of displays with patches of sandpaper used in haptic search tasks. The *dark patch* is the target, the other patches the distractors. The target was only present in half of the trials

Examples of the displays used by Plaisier et al. (2008a) are shown in Figure 2. Although their experimental set-up and procedure were quite different from those of Lederman and Klatzky (1997), they obtained similar results: very flat search slopes for the rough target among smooth distractors, somewhat higher slopes for the smooth target among rough distractors and much steeper slopes if the difference between target and distractor was smaller.

They also measured and analysed the movement patterns of the participants by placing an infra-red LED on the index finger of the exploring hand and recording its position with an NDI Optotrak Certus system. Schematic examples of such movement tracks can be seen in Figure 3. In easy searches, a single hand sweep over the display was sufficient to determine whether or not a target was present. Note that in the example of Figure 3a the tip of the index finger did not even touch the target; touching the target item with the palm of the hand was sufficient to make the decision. In difficult searches, the items had to be explored one by one by the tip of the fingers, as can clearly be seen in Figure 3b. As a consequence, the length of the trajectory was shorter in easy searches and movement speed was higher. Based on these findings, flat search slopes and simple movement tracks, Plaisier and colleagues concluded that roughness is an object property that might "pop-out" and thus is a salient feature for touch.

Allowing participants just a haptic glance, Klatzky and Lederman (1995) investigated whether familiar objects could be recognized within 200 ms. When participants did not get any prior information about the object that would be presented, they were still able to recognize 15 % of the objects. Performance increased when participants were given prior information about the kind of object (47 % correct) or if they only had to decide whether or not the stimulus was a certain named object (74 % correct). One of their conclusions was that texture, of which roughness was one of the aspects, could be extracted within a haptic glance.

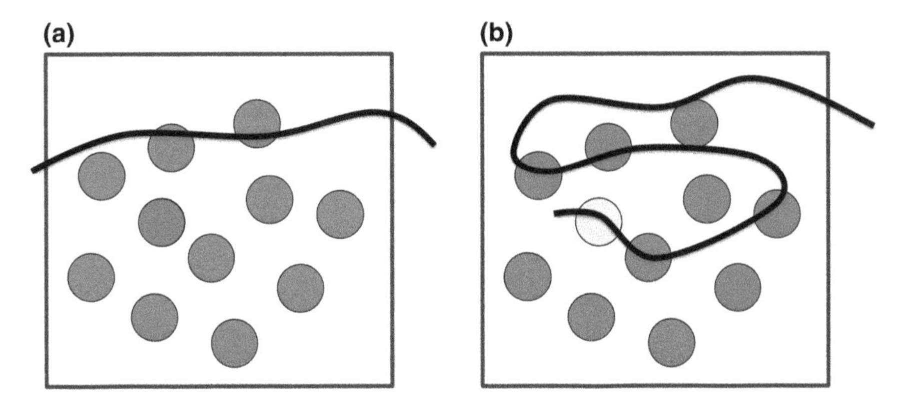

Figure 3 Schematic illustration of hand movements made in a parallel search (**a**) and in a serial search (**b**). The trajectory shows the path of the index finger. All movements start from the *right side* of the display. The colors *red* and *light blue* indicate the difference in roughness of the target with the distractors (*blue*)

Edges

In the search experiments of Lederman and Klatzky (1997) where stimuli were pressed to the fingers, a small raised bar was used as edge. Its orientation could be either horizontal or vertical (that is, perpendicular or aligned with the finger, respectively). If the presence of such an edge had to be detected among flat surfaces or vice versa, search slopes were relatively shallow. Also other types of edges, like a cylindrical hole in a flat surface, led to good performance. Although these response times were somewhat higher than in their roughness experiments, the conclusion still had to be that abrupt-surface discontinuities or edges are processed early on in the haptic system. However, if a specific orientation of an edge (horizontal or vertical) served as target among distractors with a perpendicular orientation, performance was much slower. Therefore, relative orientation must be processed much later in the system.

In the experiments of Plaisier et al. (2009a) participants had to grasp bunches of items with or without edges, such as cubes and spheres (see Figure 4). Especially for the detection of a cube or a tetrahedron among spheres, search slopes were very flat. Also the reversed situation of a sphere among cubes led to fast performance. In these experiments, participants were allowed to remove items from their hand if they thought that this would speed up their performance. Release of items is a clear indication that the search could not be performed in parallel, and therefore the percentage of trials in which at least one item is removed, is another measure for the search to be parallel or serial. In the case of the search for a cube among spheres hardly any trials occurred in which an item was released, and when the target was a tetrahedron even no items were released. A single grasp followed by a quick release of all items was sufficient, showing indeed strong evidence for a parallel search. Prominent features of a cube and a tetrahedron in comparison to a sphere are their

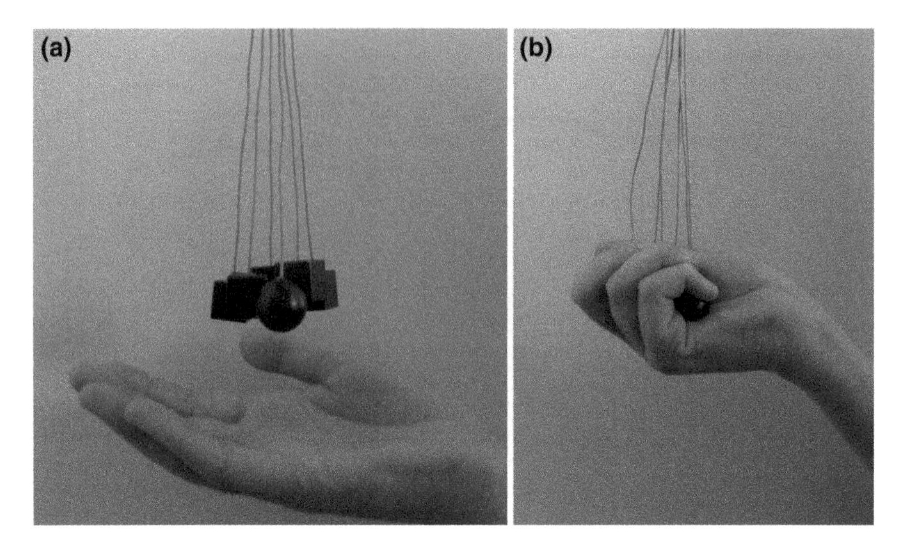

Figure 4 Stimuli in the three-dimensional haptic search task. **a** A bunch of *cubes* with one *sphere* as target. **b** A participant grasps the bunch and has to decide as fast as possible whether the *sphere* is present

edges and vertices. These results indicate that edges and vertices are very salient features for touch.

The fact that edges and vertices are salient to touch can also be shown by their disrupting influence on other tasks. Using a search paradigm similar to that shown in Figure 4, Van Polanen et al. (2013) showed that it was much harder to decide whether a rough target was present among smooth distractors if the shapes were cubes instead of spheres. Thus although the shape was irrelevant to the task, the saliency of the edges and vertices distracted from the saliency of the roughness and performance deteriorated. In a completely different experimental paradigm, Panday et al. (2012) found that the presence of edges disrupted the perception of overall shape.

Movability

Pawluk and colleagues (2010, 2011) studied whether participants were able to decide within a haptic glance whether or not an object was present, that is, to segregate figure from ground. In their studies, they found that if an object moves upon touch, this greatly adds to the perception of indeed touching an object and not just the background.

Van Polanen and colleagues (2012a) investigated movability using a search task. They created displays with a varying number of ball transfer units (for examples, see Figure 5). By gluing some of the balls to their casing, they created stationary

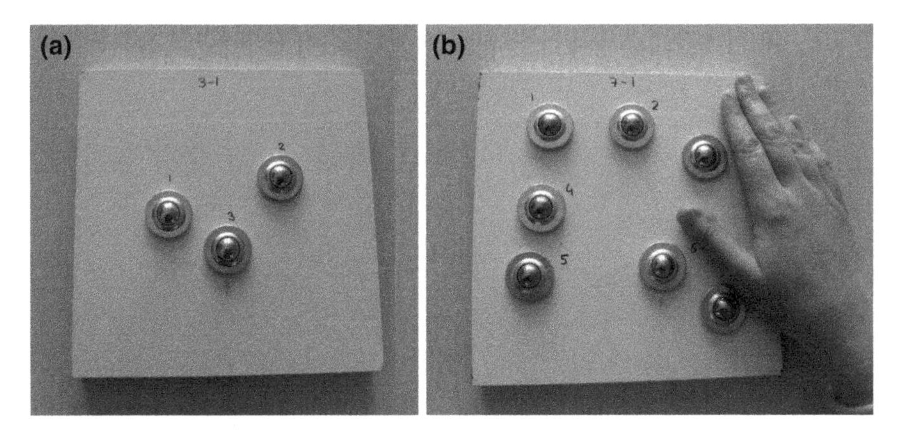

Figure 5 Examples of displays with *ball* transfer units. The *ball* of the target is fixed (this is, of course, not visible). The hand is added to give information about the scale

stimuli. Participants were instructed to move their hand over the display and decide as fast as possible whether the target (either a movable or a stationary) stimulus was present. Again, search slopes were one of the measures for saliency. They found that if the target was the movable stimulus, search slopes were very flat and the target "popped-out" from the distractors. In the reverse case, the response times were higher and indicated a more serial search. They also measured and analysed the movement patterns and these confirmed their conclusions from the search slopes: a movable target is salient and can be found in a single hand sweep; a stationary target among moving distractors is harder to find and requires serial exploration. This asymmetry is again an indication of the saliency of movability: if many items move, this disrupts performance.

Plaisier et al. (2009b) investigated the role of fixation of the items in a search task. Instead of stimuli attached to flexible wires as shown in Figure 4, they pulled the wires to which the cubes and spheres were fixed through small metal tubes, thereby restricting the movability of the shapes. By lowering the wire 0.5 cm, they could also create a condition which allowed the stimuli a limited amount of movability. They found the fastest response times and thus the lowest slopes in the conditions where all items were fixed. Compared to their earlier research with the flexible wires (Plaisier et al. 2009a), the slopes in the completely fixed condition were even lower (but care should be taken with this finding, as the experimental conditions were somewhat different).

Hardness

Lederman and Klatzky (1997) used two types of rubber (hard and soft) to investigate the saliency of hardness. A typical movement their participants made was

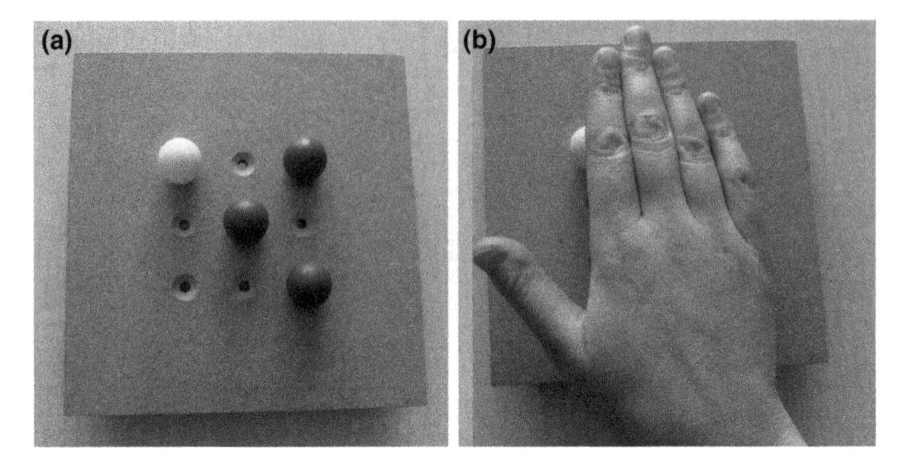

Figure 6 Search for hard or soft target. **a** Display with hard (*red*) and soft (*white*) stimuli. **b** A participant pressed her hand on the stimulus display

pressing their fingers on the surfaces. They found relatively fast performance, both with the hard and the soft stimuli as target. In the overview of their experiment, the saliency of hardness reached the second place, directly after roughness.

Van Polanen and colleagues (2012b) created a set of spheres that were made of three types of silicon rubber and therefore differed in hardness; the "hard" and "middle-soft" spheres were solid, the "soft" ones were hollow. These spheres were used in both two-dimensional (see Figure 6) and three-dimensional search tasks. In the three-dimensional task, the spheres were suspended from wires and participants had to grasp the bunch, like in the experiments described above. Especially the conditions with the hard and soft spheres as target and distractors (in either order) led to very flat search slopes. Also the very low number of items that were released from the hand in these conditions, indicated a parallel search. Conditions with the middle-soft stimuli as either target or distractor were more difficult and led to steeper search slopes.

Although these results indicated that hardness could be a salient feature, the possibility existed that the fast performance was not (only) due to the saliency of hardness, but (also) due to the weight difference; hard spheres were heavier than the soft spheres and thus a bunch of, for example, four items would be somewhat heavier if the target were present. Therefore, a two-dimensional version of this experiment was designed in which the participants had to press their hand on a static display (see Figure 6b). In this experiment, the weight of the stimuli could not play a role. Again, a hard target popped-out among soft distractors. The opposite was not the case and this was explained by the fact that due to the presence of the surrounding hard distractors, the hand was blocked from compressing the soft target, which thus could not be identified. The combination of the results of these two experiments led to the conclusion that both hardness and softness could be salient features to touch.

Temperature

Lederman and Klatzky (1997) did not actually vary the temperature of their stimuli, but they used different materials that due to their different heat conductance were perceived as either warm (soft pine) or cool (copper). They found search slopes that were only slightly steeper than those for roughness and hardness. However, overall response times were higher, which should not be surprising as it takes some time before the difference in heat flow creates a temperature difference that can be observed.

Plaisier and Kappers (2010) used the brass spheres from the earlier experiments (Plaisier et al. 2009a) of which they manipulated the temperature. The target was always a sphere at room temperature (about 22 °C), while the distractors had a temperature of about 38 °C. The distractor spheres were heated by placing them on a plastic layer on a water bath of about 41 °C. The hand temperature of the participants was kept between 28 and 33 °C by having them place their hand on a plastic layer on another water bath before and in between trials. The set-up and experimental paradigm were otherwise the same as in the previous experiments. They found search slopes similar to that of tetrahedra between spheres and as participants did not release any items, this provided strong evidence for parallel search. Therefore, they concluded that coldness pops out and thus temperature can be a salient feature to touch.

Bergmann Tiest and colleagues (2012) investigated the influence of the thermal conductivity on the haptic perception of volume. Participants had to compare the size of a cube made out of a synthetic material to that of a brass cube. The results showed that at room temperature, the brass cube was perceived as significantly larger than the synthetic cube. Cooling or heating the brass cube made no difference and thus these results are independent of and therefore not caused by temperature. The brass cube has a higher thermal conductivity than the synthetic one, so the heat flow is higher in the brass cube; it was concluded that the salience of this heat flow resulted in a larger perceived size.

Conclusion

In their research, Lederman and Klatzky (1997) were especially interested in the order in which stimulus features were processed by the haptic system. In their set-up, stimuli were pressed to the almost static fingers. They found that material properties like roughness and hardness are almost immediately available upon contact. Abrupt surface discontinuities like edges were also processed fast, though slightly slower. All these features are most likely processed in parallel by the haptic system. Orientations of edges or surfaces require much longer exploration times and there is no indication that such information can be processed in parallel. These findings are in agreement with results obtained in more active exploration conditions: roughness (Plaisier et al. 2008a; van Polanen et al. 2013), hardness (van Polanen et al. 2012b), movability (van Polanen et al. 2012a), edges (Plaisier et al. 2008a) and temperature (Plaisier and Kappers 2010) were all found to be salient features to touch. These features can be processed fast and early by the haptic system and by their presence they may disrupt the perception of other features.

Applications

Salient features stand out among their surroundings. Knowledge about which features are salient and which are not is therefore essential in situations where haptic stimulation is used for conveyance of information. One can think of very diverse situations. In gaming and in virtual worlds, haptic sensations may add to the experience of immersion; in navigation and communication, haptic stimulation provides information instead of or in addition to visual and auditory information; in teleoperations, complicated actions have to be performed either with or without vision. For the purpose of multisensory navigation and communication, Elliot and colleagues (2013) developed a model of tactile salience, which depends on user, environment, technology and tactile parameters. They developed a prototype system for the guidance of soldiers which received positive feedback from potential users. Also for the further development of robotic hands and hand prosthetics, understanding haptic perception is of eminent importance. Not only is the human sense of touch so sophisticated that it may inspire technology, it may also guide the direction of research as obviously salient features will be more important than non-salient features.

Internal References

Itti, L (2007). Visual salience. *Scholarpedia* 2(9): 3327. http://www.scholarpedia.org/article/Visual_salience.

Wolfe, J M and Horowitz, T S (2008). Visual search. *Scholarpedia* 3(7): 3325. http://www.scholarpedia.org/article/Visual_search.

External References

Bergmann Tiest, W M; Kahrimanovic, M; Niemantsverdriet, I; Bogale, K and Kappers, A M L (2012). Salient material properties and haptic volume perception: The influences of surface texture, thermal conductivity, and compliance. *Attention, Perception & Psychophysics* 74(8): 1810–1818.

Elliott, L R et al. (2013). Development of dual tactor capability for a soldier multisensory navigation and communication system. In: S Yamamoto (Ed.), *HIMI/HCII 2013, Part II, Volume 8017 of Lecture Notes in Computer Science* (pp. 46–55). Berlin: Springer-Verlag.

Klatzky, R L and Lederman, S J (1995). Identifying objects from a haptic glance. *Perception & Psychophysics* 57(8): 1111–1123.
Lederman, S J; Browse, R A and Klatzky, R L (1988). Haptic processing of spatially distributed information. *Perception & Psychophysics* 44(3): 222–232.
Lederman, S J and Klatzky, R L (1997). Relative availability of surface and object properties during early haptic processing. *Journal of Experimental Psychology: Human Perception and Performance* 23(6): 1680–1707.
Moore, T; Broekhoven, M; Lederman, S J and Wug, S (1991). Q'Hand: A fully automated apparatus for studying haptic processing of spatially distributed inputs. *Behaviour Research Methods, Instruments & Computers* 23: 27–35.
Ni, B et al. (2014). Touch saliency: Characteristics and prediction. *IEEE Transactions on Multimedia* 16(6): 1779–1791.
Panday, V; Bergmann Tiest, W M and Kappers, A M L (2012). Influence of local properties on haptic perception of global object orientation. *IEEE Transactions on Haptics* 5(1): 58–65.
Pawluk, D; Kitada, R; Abramowicz, A; Hamilton, C and Lederman, S J (2010). Haptic figure-ground differentiation via a haptic glance. In: *IEEE Haptics Symposium* (pp. 63–66).
Pawluk, D; Kitada, R; Abramowicz, A; Hamilton, C and Lederman, S J (2011). Figure/ground segmentation via a haptic glance: Attributing initial finger contacts to objects or their supporting surfaces. *IEEE Transactions on Haptics* 4(1): 2–13.
Plaisier, M A; Bergmann Tiest, W M and Kappers, A M L (2008a). Haptic pop-out in a hand sweep. *Acta Psychologica* 128: 368–377.
Plaisier, M A; Bergmann Tiest, W M and Kappers, A M L (2008b). Haptic search for spheres and cubes. In: M Ferre (Ed.), *Haptics: Perception, Devices and Scenarios, Volume 5024 of Lecture Notes on Computer Science* (pp. 275–282). Berlin/Heidelberg: Springer.
Plaisier, M A; Bergmann Tiest, W M and Kappers, A M L (2009a). Salient features in three-dimensional haptic shape perception. *Attention, Perception & Psychophysics* 71(2): 421–430.
Plaisier, M A and Kappers, A M L (2010). Cold objects pop out! In: A M L Kappers, J B F van Erp, W M Bergmann Tiest and F C T van der Helm (Eds.), *Haptics: Generating and Perceiving Tangible Sensations. Part II, Volume 6192 of Lecture Notes in Computer Science* (pp. 219–224). Berlin/Heidelberg: Springer.
Plaisier, M A; Kuling, I A; Bergmann Tiest, W M and Kappers, A M L (2009b). The role of item fixation in haptic search. In: J Hollerbach (Ed.), *Proceedings 3rd Joint EuroHaptics Conference and Symposium on Haptic Interfaces for Virtual Environment and Teleoperator Systems* (pp. 417–421). Salt Lake City, UT, USA: IEEE.
Treisman, A M and Gelade, G (1980). A feature-integration theory of attention. *Cognitive Psychology* 12(1): 97–136.
Treisman, A and Gormican, S (1988). Feature analysis in early vision: evidence from search asymmetries. *Psychological Review* 95(1): 15–48.
Van Polanen, V; Bergmann Tiest, W M and Kappers, A M L (2012a). Haptic pop-out of movable stimuli. *Attention, Perception & Psychophysics* 74(1): 204–215.
Van Polanen, V; Bergmann Tiest, W M and Kappers, A M L (2012b). Haptic search for hard and soft spheres. *PLoS ONE* 7(10): e45298.
Van Polanen, V; Bergmann Tiest, W M and Kappers, A M L (2013). Integration and disruption effects of shape and texture in haptic search. *PLoS ONE* 8(7): e70255.
Wolfe, J M (1998). What can 1 million trials tell us about visual search? *Psychological Science* 9 (1): 33–39.
Xu, M et al. (2012). Touch saliency. In: *Proceedings of the 20th ACM International Conference on Multimedia* (pp. 1041–1044).

Note on Terminology

One final note about "touch saliency". Ni, Xu and colleagues investigated the characteristics of what they call "touch saliency" (Xu et al. 2012; Ni et al. 2014). Although this term seems very relevant in the present context, it is not. What they mean by touch saliency are the finger fixation maps on touch screens, that might provide useful information about the regions of interest on the screen. As these regions can only be perceived and identified visually, this does not tell us anything about touch perception. A more proper term would be "touch map".

Shape from Touch

Astrid M.L. Kappers and Wouter M. Bergmann Tiest

The shape of objects cannot only be recognized by vision, but also by touch. Vision has the advantage that shapes can be seen at a distance, but touch has the advantage that during exploration many additional object properties become available, such as temperature (Jones 2009), texture (Bensmaia 2009), and weight (Jones 1986). Moreover, also the invisible backside of the objects can provide shape information (Newell et al. 2001). In active touch, both the cutaneous sense (input from skin receptors) and the kinesthetic sense (input from receptors located in muscles, tendons, and joints) play a role. For such active tactual perception, the term "haptic perception" is used (Loomis and Lederman 1986). Typical exploratory movements to determine shape by touch are "enclosure" for global shape and size, and "contour following" for exact shape (Lederman and Klatzky 1987; Klatzky and Reed 2009). By means of touch, both three-dimensional and two-dimensional shapes can be recognized, although the latter is much harder.

Three-Dimensional Shape from Touch

Humans are well able to recognize familiar objects by means of touch (Klatzky et al. 1985). In an extensive study, blindfolded students were asked to haptically identify as fast as possible objects such as a comb, a carrot, a tennis racket, a padlock, a knife, and many other daily life objects. Most objects were identified correctly (96 %) and were recognized within 3 s. Thus, this study established that human performance with such objects is fast and accurate. Although these authors were the first to investigate response times and accuracy in great detail, a similar test was already in use for clinical studies of sensory function (e.g., Wynn Parry and Salter 1976).

Other studies investigating shape from touch used better-defined stimuli than just objects from daily life. Gibson (1963) asked an artist to create a set of stimuli that

A.M.L. Kappers (✉) · W.M. Bergmann Tiest
VU University, Amsterdam, The Netherlands

T.J. Prescott et al. (eds.), *Scholarpedia of Touch*, Scholarpedia,
DOI 10.2991/978-94-6239-133-8_15

were "equally different" from one another, were about hand-sized, were smooth and had a regular convex backside. He concluded that his participants were able to distinguish such objects by just haptic exploration. The same stimuli were used by Norman and colleagues (2012). They showed that visual and haptic shape discrimination performance is similar as long as manipulation of the stimuli was similar to that in daily life.

Roland and Mortensen (1987) were the first who used a set of parametrically defined stimuli, such as spheres, ellipsoids and rectangular blocks of different sizes. By keeping the surface properties and the weights constant (the spheres were hollow), shape and size were the only factors distinguishing the stimuli. They were interested in the way the objects were manipulated during a size or shape discrimination task and in discrimination performance. They described a model that could predict human performance in the size discrimination task, but the results for shape discrimination were not so good. Kappers and colleagues (1994) created a set a doubly curved surfaces such as convex and concave elliptic and hyperbolic paraboloids by means of a computer-controlled milling machine. They found that hyperbolic stimuli were somewhat harder to identify than elliptic stimuli. Moreover, if the curvature of the stimuli was more pronounced, identification performance improved.

More recently, Norman et al. (2004) made plastic copies of 12 bell peppers in order to have a set of "natural" stimuli. They compared discrimination and matching performance in several conditions: using just touch, using just vision, and using both touch and vision. Performance in all conditions was very similar, which led them to conclude that the 3D representations of shape in vision and in touch are functionally overlapping. Also Gaissert and Wallraven (2012) used natural stimuli, namely a set of seashells. They compared the visual and haptic perceptual spaces and found that these are nearly identical. They suggest that haptic and visual similarity perception are linked by the same cognitive processes.

Curvature from Touch

An important and often investigated aspect of shape is curvature, as many smooth objects contain curved parts. Gordon and Morrison (1982) were interested in how well humans can distinguish flat from curved stimuli using active touch. Using small curved stimuli of just a few centimeters, they showed that the threshold for curvature is determined by the overall gradient of the stimulus. This gradient is the ratio between the base-to-peak height of the stimulus and half the stimulus length (see Figure 1). Goodwin and colleagues (1991) did similar experiments using passive touch, as they wanted to limit the role of sensory and motor afferents. They found that a difference in curvature of about 10 % could be distinguished. In a sequel to this study (Goodwin and Wheat 1992), the contact area between the finger and the curved stimulus was kept constant. As the thresholds remained the same, they could conclude that contact area could not be the determining factor in

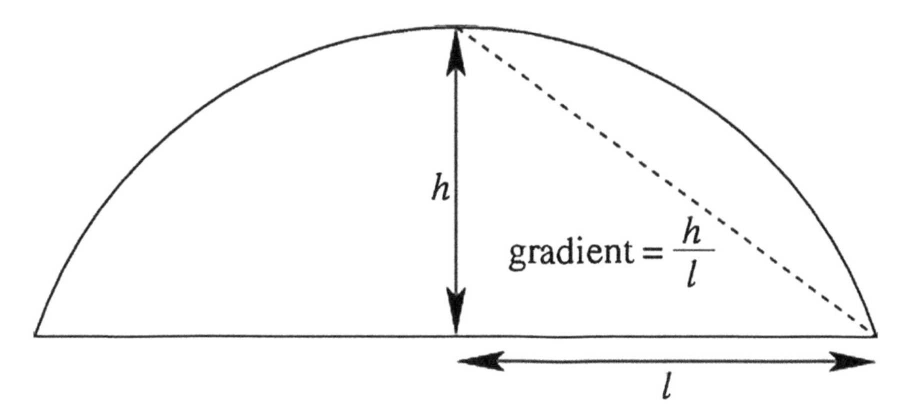

Figure 1 Illustration of how the gradient of a curved stimulus is defined as the ratio between its height h and half the stimulus length l

curvature discrimination. As larger contact areas led to smaller thresholds, their finding was consistent with that of Gordon and Morrison (1982) for active touch.

Pont et al. (1997) did similar experiments with hand-sized stimuli (and much less curved stimuli) and they also came to a similar conclusion: the difference in local slopes between two stimuli of the same length determines curvature discrimination performance. They went a step further in proving this conclusion, by creating two new sets of stimuli that either contained only height differences (so no slope and no curvature information) or contained both height and slope differences (but no curvature information) (Pont et al. 1999). Stimuli with only height information were much harder to discriminate than those with also slope and curvature information, but the sets with or without curvature indeed led to similar thresholds as long as slope information was present. Wijntjes and colleagues (2009) used a device so that similar experiments could be run in active instead of passive touch and also their experiments showed that slope information was the determining factor for curvature discrimination. An overview of many of the curvature discrimination experiments can be found in Kappers (2011).

Influence of Curvature on Shape from Touch

Haptic curvature discrimination gets better if the length of the stimuli is larger (Pont et al. 1999). In addition, the same study showed that perception of the curvature of a stimulus is influenced by its length: if the curvature is the same, the longer of two stimuli will be perceived as more curved. This led to an intriguing prediction: as the hands are usually longer than wide, a spherical object should be perceived as an ellipsoid if judged by placing the hand on it (see Figure 2). An experiment with blindfolded participants confirmed that this is indeed the case (Pont et al. 1998).

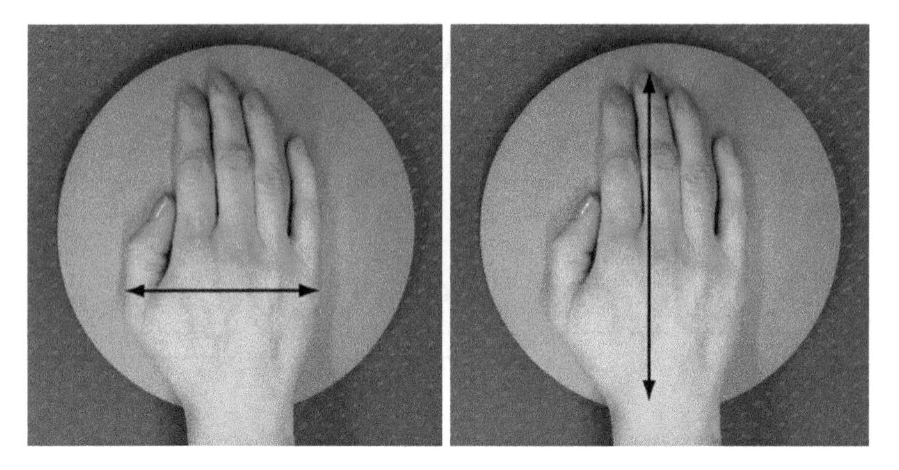

Figure 2 Hand positioned on a curved surface. As the hand is longer than wide, the perceived curvature along the hand will be larger than across the hand

Van der Horst and Kappers (2008) performed an experiment in which participants had to distinguish cylinders with circular cross-section from those with an elliptical cross-section. In addition, they did a similar experiment with rectangular blocks, in which a square cross-section had to be distinguished from a rectangular one. In both experiments, the aspect ratio (the ratio between the longer and the shorter axes) was varied systematically. Whereas the task was the same, performance with the cylinders was much better than with the blocks: an aspect ratio of 1.03 was sufficient for the cylinders, whereas 1.11 was necessary for the rectangular blocks (see Figure 3). They concluded that the curvature information in the cylindrical stimuli aided in discrimination, although the aspect ratio itself was less readily available in these stimuli as compared to the blocks. In a follow-up study with similar stimuli, Panday et al. (2012) showed that indeed the availability of curvature and curvature changes improves human perception of such stimuli and that the edges of the rectangular blocks probably disturbed the shape perception by touch.

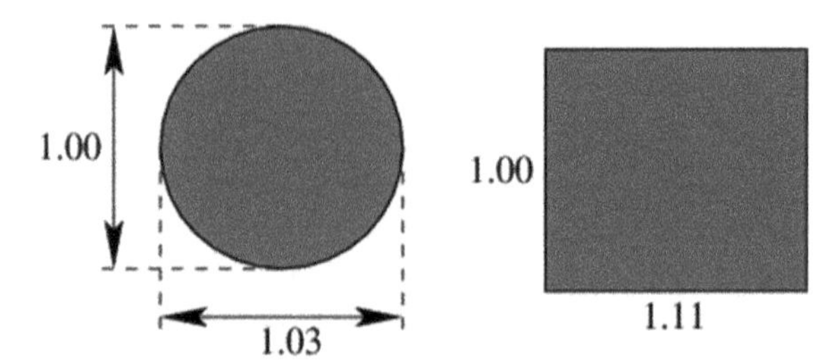

Figure 3 Cross-sections of a cylindrical object (*left*) and a rectangular block (*right*). Illustration of the difference in aspect ratio needed for correct discrimination from *circular* (1.03 in the case of *cylinders*) and *square* (1.11 in the case of the *rectangular blocks*)

Influence of Aftereffects on Shape from Touch

Touching a shape for only a few seconds influences the perceived curvature of the subsequently touched shape (Vogels et al. 1996): after touching a convex shape, a flat surface will be perceived as concave and a concave shape will be perceived as even more concave; a slightly convex shape will be perceived as flat. Although the built-up of this so-called aftereffect is quite fast (a few seconds and saturation after 10–15 s), the decay is quite slow: after a delay of 40 s before touching the next shape, an aftereffect can still be measured, even if participants stretch and bend their fingers during this delay (Vogels et al. 1997). Van der Horst et al. (2008a) showed that also the finger tips are sensitive to a curvature aftereffect. They also showed that this aftereffect partially transferred to other fingers, so after touching a curved surface with a finger, the curvature perceived by another finger is somewhat changed. Interestingly, if curved surfaces were explored dynamically, a full transfer to the other hand occurred (van der Horst et al. 2008b).

Uznadze (1966) let observers repeatedly grasp spheres of different sizes, but always the small one with one hand and the large one with the other hand. After a sequence of about 10–15 of such grasps, he presented both hands with identical spheres of intermediate size. He observed that this intermediate sphere felt small to the hand used to the large sphere and large to the other hand. Kappers and Bergmann Tiest (2013) repeated and confirmed this experiment in a more quantitative way. In a subsequent study (Kappers and Bergmann Tiest 2014), they varied the shape of the objects and showed that this size aftereffect is at least partly a shape aftereffect: only after testing with an object of similar shape (either a sphere or a tetrahedron), there occurs a large aftereffect.

Influence of Blindness on Shape from Touch

Norman and Bartholomew (2011) used the plastic bell pepper stimuli also in a study comparing discrimination performance of congenitally, early and late blind observers and blindfolded sighted observers. As the early and late blind observers performed better than the blindfolded sighted observers, they suggested that early visual experience plays a role in haptic shape perception. Withagen et al. (2012) performed a haptic matching experiment in which participants had to find the best match to the standard from a set of three test items. They found that blind and blindfolded sighted participants were equally accurate. However, the response times were much higher for the blindfolded sighted (adult) observers as compared to the blind observers if they had to match the exact shape of the stimuli.

Hunter (1954) measured how well blind and sighted participants could classify straight, concave and convex shapes. He found that both groups of participants had the tendency to judge a convex shape as straight, but this was more pronounced in the sighted group. Davidson (1972) also compared the performance of blind and blindfolded sighted participants in distinguishing whether a curved shape was convex, straight or concave. In his first experiment, he found that the blind performed better than the sighted. However, he also noticed that the blind used different strategies than the sighted. In his second experiment, he instructed the blindfolded sighted participants to use the strategies of the blind and then their performance improved.

Influence of Age on Shape from Touch

Norman et al. (2013) showed that depending on the exact task, shape discrimination might or might not depend on age. In a curvature discrimination task, participants were required to explore the stimuli by using either static touch or dynamic touch. The "young" participant group consisted of individuals aged between 20 and 29 years, the "old" group participants were between 66 and 85 years. As long as the participants were allowed to use dynamic touch, there was no difference in discrimination performance between the two age groups. However, when using static touch, performance of the old group was significantly worse than that of the young group. This suggests that the cutaneous system is more sensitive to degradation with age than the kinesthetic system. Withagen et al. (2012) showed that in a haptic shape matching task, adults performed significantly better than children (mean age 9.3 years). However, sighted adults were much slower than sighted and blind children, indicating a possible speed-accuracy trade-off; blind adults were as fast as both groups of children.

Influence of Shape on Size Perception

The shape of an object might influence its perceived size. In vision, there exists the well- known elongation bias: of two cylindrical objects with the same volume (i.e. size), the higher one is perceived as being the larger (e.g., Piaget 1968). Krishna (2006) investigated whether a similar bias could also be observed in touch and in bimodal conditions (both vision and touch). She found that as long as vision was present (either vision alone or in combination with touch) indeed this elongation bias was found. However, if only touch was used to judge the size of the cylinders, the bias reversed: the wider of the two objects of a pair was perceived as being the larger. She suggested that in grasping a glass without looking, the diameter and thus the width of the glass is a much more salient feature than the height of the glass. In contrast, when looking at a glass, the height is more prominent.

Kahrimanovic et al. (2010) compared the perceived sizes of spheres, cubes and tetrahedra (see Figure 4), both in a condition with weight information present, and one without. They found very substantial differences: if no weight information was available, a sphere had on average to be 60 % larger in volume than a tetrahedron to

Figure 4 Some examples of *tetrahedra*, *spheres* and *cubes* as used in the experiment by Kahrimanovic et al. (2010)

be perceived as equal in size. When weight information was added (without explicitly telling participants that they could as well base their judgement on weight), the difference became somewhat smaller, but they were still around 30 %. Also the perceived differences between the sizes of spheres and cubes, and those between cubes and tetrahedra were quite large. With a theoretical analysis and a second experiment, they could show that participants do not actually compare volumes, but instead use the surface area of the objects (note that spheres, cubes and tetrahedra with the same volume have different surface areas).

Two-Dimensional Shape from Touch

Recognizing two-dimensional common objects, usually in the form of raised line drawings, by just using touch is quite poor as compared to recognizing them by sight or as compared to recognizing the three-dimensional objects by touch (e.g., Ikeda and Uchikawa 1978; Klatzky et al. 1993). Whereas in vision such drawings might be recognized almost instantaneously, in touch it may require minutes! In such tasks, it helps if categorical information about the object is given (e.g., Heller et al. 1996; Heller 2002). Also guidance of the exploring finger helps in identifying the depicted object (Magee and Kennedy 1980). Kennedy (1993) showed that also congenitally blind individuals who had never been exposed to such drawings before, were able to make use of them. Wijntjes et al. (2008) showed that blind-folded sighted observers who did not recognize a certain raised line drawing, were often able to draw what they felt and then recognize their drawing visually. This shows that apparently it is very hard to integrate the serially acquired information about the line drawing into a percept of an object (See Figure 5).

An overview of studies investigating the ease or difficulty of recognizing raised line drawings by touch is given by Picard and Lebaz (2012). More details about picture recognition performance of blind individuals can be found in (Visually-impaired touch; Heller and Ballesteros (2012)).

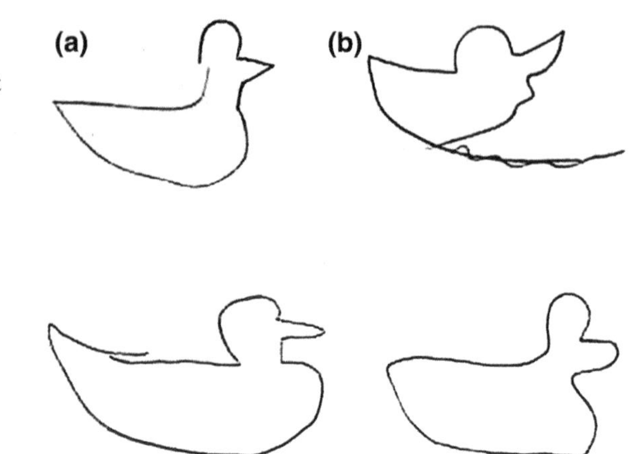

Figure 5 Drawings of a stimulus after blindfolded exploring. In **a** examples that were already haptically recognized; in **b** examples that were recognized *after* drawing

Internal References

Bensmaia, S (2009). Texture from touch. *Scholarpedia* 4(8): 7956. http://www.scholarpedia.org/article/Texture_from_touch. (see also pages 207–215 of this book)

Heller, M and Ballesteros, S (2012). Visually-impaired touch. *Scholarpedia* 7(11): 8240. http://www.scholarpedia.org/article/Visually-impaired_touch. (see also pages 387–397 of this book)

Jones, L (2009). Thermal touch. *Scholarpedia* 4(5): 7955. http://www.scholarpedia.org/article/Thermal_touch. (see also pages 257–262 of this book)

Klatzky, R and Reed, C (2009). Haptic exploration. *Scholarpedia* 4(8): 7941. http://www.scholarpedia.org/article/Haptic_exploration. (see also pages 177–183 of this book)

External References

Davidson, P W (1972). Haptic judgments of curvature by blind and sighted humans. *Journal of Experimental Psychology* 93(1): 43–55.

Gaissert, N and Wallraven, C (2012). Categorizing natural objects: A comparison of the visual and the haptic modalities. *Experimental Brain Research* 216(1): 123–134.

Gibson, J J (1963). The useful dimensions of sensitivity. *American Psychologist* 18: 1–15.

Goodwin, A W; John, K T and Marceglia, A H (1991). Tactile discrimination of curvature by humans using only cutaneous information from the fingerpads. *Experimental Brain Research* 86(3): 663–672.

Goodwin, A W and Wheat, H E (1992). Human tactile discrimination of curvature when contact area with the skin remains constant. *Experimental Brain Research* 88(2): 447–450.

Gordon, I A and Morison, V (1982). The haptic perception of curvature. *Perception & Psychophysics* 31: 446–450.

Heller, M A (2002). Tactile picture perception in sighted and blind people. *Behavior Brain Research* 135(1–2): 65–68.

Heller, M A; Calcaterra, J A; Burson, L L and Tyler, L A (1996). Tactual picture identification by blind and sighted people: Effects of providing categorical information. *Perception & Psychophysics* 58(2): 310–323.

Hunter, I M L (1954). Tactile-kinaesthetic perception of straightness in blind and sighted humans. *Quarterly Journal of Experimental Psychology* 6: 149–154.

Ikeda, M and Uchikawa, K (1978). Integrating time for visual pattern perception and a comparison with the tactile mode. *Vision Research* 18(11): 1565–71.

Jones, L A (1986). Perception of force and weight: Theory and research. *Psychological Bulletin* 100(1): 29–42.

Kahrimanovic, M; Bergmann Tiest, W M and Kappers, A M L (2010). Haptic perception of volume and surface area of 3-D objects. *Attention, Perception & Psychophysics* 72(2): 517–527.

Kappers, A M L (2011). Human perception of shape from touch. *Philosophical Transactions of the Royal Society of London B* 366: 3106–3114.

Kappers, A M L and Bergmann Tiest, W M (2013). Haptic size aftereffects revisited. In: *World Haptics Conference (WHC)* (pp. 335–339).

Kappers, A M L and Bergmann Tiest, W M (2014). Influence of shape on the haptic size aftereffect. *PLoS One* 9(2): e88729.

Kappers, A M L; Koenderink, J J and Lichtenegger, I (1994). Haptic identification of curved surfaces. *Perception & Psychophysics* 56(1): 53–61.

Kennedy, J R (1993). *Drawing & the blind: Pictures to touch.* New Haven, CT: Yale University Press.

Klatzky, R L; Lederman, S J and Metzger, V A (1985). Identifying objects by touch: an "expert system". *Perception & Psychophysics* 37(4): 299–302.

Klatzky, R L; Loomis, J M; Lederman, S J; Wake, H and Fujita, N (1993). Haptic identification of objects and their depictions. *Perception & Psychophysics* 54(2): 170–8.

Krishna, A (2006). Interaction of senses: The effect of vision versus touch on the elongation bias. *The Journal of Consumer Research* 32(4): 557–566.

Lederman, S J and Klatzky, R L (1987). Hand movements: A window into haptic object recognition. *Cognitive Psychology* 19(3): 342–368.

Loomis, J M and Lederman, S J (1986). Tactual perception. In: K R Boff, L Kaufman and J P Thomas (Eds.), *Cognitive Processes and Performance, volume 2 of Handbook of Perception and Human Performance* (pp. 31.1–31.41). New York: John Wiley & Sons.

Magee, L E and Kennedy, J M (1980). Exploring pictures tactually. *Nature* 283: 287–288.

Newell, F N; Ernst, M O; Tjan, B S and Bülthoff, H H (2001). Viewpoint dependence in visual and haptic object recognition. *Psychological Science* 12(1): 37–42.

Norman, J F and Bartholomew, A N (2011). Blindness enhances tactile acuity and haptic 3-D shape discrimination. *Attention, Perception & Psychophysics* 73(7): 2323–2331.

Norman, J F et al. (2013). Aging and curvature discrimination from static and dynamic touch. *PLoS One* 8(7): e68577.

Norman, J F; Norman, H F; Clayton, A M; Lianekhammy, J and Zielke, G (2004). The visual and haptic perception of natural object shape. *Perception & Psychophysics* 66(2): 342–351.

Norman, J F et al. (2012). Solid shape discrimination from vision and haptics: natural objects (*Capsicum annuum*) and Gibson's "feelies"'. *Experimental Brain Research* 222(3): 321–332.

Panday, V; Bergmann Tiest, W M and Kappers, A M L (2012). Influence of local properties on haptic perception of global object orientation. *IEEE Transactions on Haptics* 5: 58–65.

Piaget, J (1968). Quantification, conservation, and nativism. Quantitative evaluations of children aged two to three years are examined. *Science* 162(3857): 976–981.

Picard, D and Lebaz, S (2012). Identifying raised-line drawings by touch: A hard but not impossible task. *Journal of Visual Impairment & Blindness* 106(7): 427–431.

Pont, S C; Kappers, A M L and Koenderink, J J (1997). Haptic curvature discrimination at several regions of the hand. *Perception & Psychophysics* 59(8): 1225–1240.

Pont, S C; Kappers, A M L and Koenderink, J J (1998). Anisotropy in haptic curvature and shape perception. *Perception* 27(5): 573–589.

Pont, S C; Kappers, A M L and Koenderink, J J (1999). Similar mechanisms underlie curvature comparison by static and dynamic touch. *Perception & Psychophysics* 61(5): 874–894.

Roland, P E and Mortensen, E (1987). Somatosensory detection of microgeometry, macrogeometry and kinesthesia in man. *Brain Research* 434(1): 1–42.

Uznadze, D N (1966). *The psychology of set.* New York: Consultants Bureau.

Van der Horst, B J et al. (2008a). Intramanual and intermanual transfer of the curvature aftereffect. *Experimental Brain Research* 187(3): 491–496.

Van der Horst, B J and Kappers, A M L (2008). Using curvature information in haptic shape perception of 3D objects. *Experimental Brain Research* 190(3): 361–367.

Van der Horst, B J; Willebrands, W P and Kappers, A M L (2008b). Transfer of the curvature aftereffect in dynamic touch. *Neuropsychologia* 46(12): 2966–2972.

Vogels, I M L C; Kappers, A M L and Koenderink, J J (1996). Haptic aftereffect of curved surfaces. *Perception* 25(1): 109–119.

Vogels, I M L C; Kappers, A M L and Koenderink, J J (1997). Investigation into the origin of the haptic after-effect of curved surfaces. *Perception* 26: 101–107.

Wijntjes, M W A; Sato, A; Hayward, V and Kappers, A M L (2009). Local surface orientation dominates haptic curvature discrimination. *IEEE Transactions on Haptics* 2(2): 94–102.

Wijntjes, M W A; van Lienen, T; Verstijnen, I M and Kappers, A M L (2008). Look what I have felt: Unidentified haptic line drawings are identified after sketching. *Acta Psychologica* 128(2): 255–263.

Withagen, A; Kappers, A M L; Vervloed, M P J; Knoors, H and Verhoeven, L (2012). Haptic object matching by blind and sighted adults and children. *Acta Psychologica* 139(2): 261–271.

Wynn Parry, C B and Salter, M (1976). Sensory re-education after median nerve lesions. *Hand* 8 (3): 250–257.

Texture from Touch

Sliman Bensmaia

Texture from touch refers to the processing of information about surface material and microgeometry obtained from tactile exploration. Though textural information can be obtained both visually (Heller 1989) and auditorily (Lederman 1979), touch yields much finer and more complex textural information than do the other sensory modalities. Humans can distinguish textures whose elements differ in size by tens of nanometers or whose spatial periods differ by hundreds of nanometers (Skedung et al. 2013). When we run our fingers across a surface, we may perceive the surface as being rough, like sandpaper, or smooth, like glass; the surface may also vary along other sensory continua, such as hardness (e.g., stone) versus softness (e.g. moist sponge), stickiness (e.g., tape) versus slipperiness (e.g., soap). Also, whether a texture is thermally isolating (e.g., leather) or thermally conductive (like metal) contributes to the textural percept (Hollins et al. 1993, 2000; Bensmaia and Hollins 2005). Tactile texture perception plays a role in the tactile recognition of objects (Klatzky et al. 1987) as most natural objects differ not only in shape but in texture as well. Furthermore, certain types of texture information are essential in order to properly manipulate objects (Johansson and Flanagan 2009). Texture is represented at the somatosensory periphery in the spatiotemporal pattern of activity in populations of receptors embedded in the skin and in cortex by different populations of neurons with different response properties.

Neural Basis of Texture Perception in the Nerve

Three types of mechanoreceptive afferents that innervate the glabrous skin contribute to the tactile perception of texture, namely slowly adapting type 1 (SA1), rapidly adapting (RA), and Pacinian (PC) afferents, which innervate Merkel cells, Meissner corpuscles, and Pacinian corpuscles, respectively. Coarse textural features, with element sizes on the order of millimeters, are encoded in the responses of SA1 afferents. Specifically, the spatial configuration of coarse surface elements is

S. Bensmaia (✉)
Department of Organismal Biology and Anatomy, University of Chicago, Chicago, USA

T.J. Prescott et al. (eds.), *Scholarpedia of Touch*, Scholarpedia,
DOI 10.2991/978-94-6239-133-8_16

reflected in the spatial pattern of SA1 activation. However, this encoding mechanism is limited by the innervation density of these receptors (which are spaced about 1 mm apart; cf. Vallbo and Johansson 1979). To perceive fine textural features, motion between skin and surface is required: When the exploring finger scans a fine texture, small vibrations are produced in the skin, which reflect both the microgeometry of the surface and that of the fingerprint (Hollins et al. 1998, 2001, 2002; Bensmaia and Hollins 2003, 2005; Manfredi et al. 2014). These texture-elicited vibrations are transduced and processed by two other populations of vibration-sensitive fibers, namely RA and PC afferents (Weber et al. 2013). These fibers respond to textures by producing highly repeatable temporal spiking patterns, which are sufficiently informative about texture identify to mediate our ability to identify and discriminate tactile textures (Weber et al. 2013). This vibrotactile coding scheme is analogous to that is observed in the vibrissal system of rodents (see http://www.scholarpedia.org/article/Vibrissal_texture_decoding).

Neural Basis of Texture Perception in Cortex

As discussed above, tactile texture perception relies on two distinct mechanisms: a spatial mechanism mediated by SA1 fibers, and a temporal mechanism mediated by RA and PC fibers. In cortex, a subpopulation of neurons exhibit receptive field properties that are well suited to analyze the spatial image of a texture conveyed by SA1 afferents. To investigate the receptive field properties of these neurons, random dot patterns (DiCarlo et al. 1998; DiCarlo and Johnson 1999, 2000) or spatiotemporal white noise stimuli (Sripati et al. 2006) were presented to the glabrous skin of the distal finger pads while neuronal responses in primary somatosensory cortex (SI) were recorded. The spatial and spatiotemporal receptive fields (SRFs and STRFs) of SI neurons were computed from these measurements using reverse correlation (DiCarlo et al. 1998; DiCarlo and Johnson 1999, 2000; Sripati et al. 2006). A subpopulation of neurons, whose SRFs or STRFs comprise excitatory and inhibitory sub-regions and are thus highly analogous to neurons in primary visual cortex, are ideally suited to analyze the spatial image conveyed by SA1 neurons. The texture signal carried in the responses of RA and PC afferents likely drives the responses of another population of SI neurons that exhibits highly patterned responses to skin vibrations (Harvey et al. 2013). Indeed, these neurons receive input from both RA and PC afferents (Saal et al. 2014) and produce entrained responses to simple and complex (texture-like) vibrations across the range of frequencies that are experienced during tactile texture exploration. How texture information from these two streams—spatial and temporal—is combined to culminate in a holistic percept of texture remains to be elucidated.

The Dimensions of Texture

The tactile exploration of a surface has been shown to yield a multidimensional textural percept that includes sensations of roughness/smoothness, hardness/softness, stickiness/slipperiness, and warmth/coolness (Hollins et al. 1993, 2000; Bensmaia and Hollins 2005). The overall textural percept of a surface is strongly determined by three of these texture properties, namely roughness, hardness and stickiness. Of all textural continua, the study of roughness has been the most extensive and has yielded important insights into neural coding in the somatosensory system.

Roughness

The subjective sense of roughness seems to vary along a single dimension and has been shown to vary predictably with surface properties. In psychophysical studies, the perceived roughness of sandpapers increases as a power function of particle size (exponent $\approx$ 1.5), the roughness of gratings increases linearly with spatial period (Lederman and Taylor 1972; Chapman et al. 2002), and that of embossed dots increases monotonically with inter-element spacing up to a spatial period of about 2 mm, then decreases with further increases in spatial period (Morley et al. 1983; Connor et al. 1990). For gratings, however, the spatial period does not seem to be the relevant stimulus property. For instance, changing the groove width and ridge width of gratings has opposite effects on perceived roughness (Lederman and Taylor 1972; Sathian et al. 1989). The main determinant of perceived roughness seems to be the spatial pattern of deformation of the skin (Taylor and Lederman 1975) although temporal cues (Cascio and Sathian 2001; Gamzu and Ahissar 2001) and tangential forces (Smith et al. 2002) also play a role.

In a series of studies pairing human psychophysics with macaque neurophysiology, Johnson and colleagues set out to establish the peripheral neural code underlying roughness perception (Connor et al. 1990; Connor and Johnson 1992; Blake et al. 1997; Yoshioka et al. 2001). Their approach consisted of devising and testing a set of hypotheses linking the activity evoked by textured surfaces in populations of mechanoreceptive afferents, measured in macaque monkeys, to estimates of their perceived roughness, measured in human observers. The stimuli consisted of embossed dot patterns, varying in their spatial properties, presented passively to the skin using a rotating drum stimulator (Johnson and Phillips 1988). The roughness estimates, obtained for a variety of dot patterns, were plotted against predictions derived from each putative neural code. A hypothesis was eliminated if it failed to account for roughness estimates under any single experimental condition. The putative neural codes for roughness included (1) the mean firing rate elicited in a given population of mechanoreceptive afferents fibers; (2) the temporal variability in the firing of a given population of mechanoreceptive afferent fibers; (3) the spatial variability in the firing of a given population of mechanoreceptive afferent fibers. The spatial variability in the responses of slowly adapting type 1 afferents was found to account for perceived roughness of all the textures tested.

In the aforementioned studies, however, textured surfaces tended to be relatively coarse, ranging in inter-element spacing from 0.5 mm to 5 mm (with the exception one or two stimuli in Yoshioka et al. 2001). A first hint that SA1 afferents were not solely responsible for roughness coding was provided in a study that showed that low-frequency vibratory adaptation (at 10 Hz), which would primarily desensitize RA and SA1 fibers (Bensmaia et al. 2005), had no effect on fine texture discrimination (with elements sized in the tens of microns), whereas high-frequency adaptation, targeting PC fibers, abolished subjects ability to discriminate fine textures (Hollins et al. 2001). When roughness coding was investigated with a wide range of textures, it was found that the spatial code mediated by SA1 fibers could not account for the perceived roughness over the range of tangible textures. Rather, roughness judgments could be accounted for based on a combination of spatial variation in SA1 (following Johnson and colleagues) and temporal variation in RA and PC fibers (Weber et al. 2013). That is, a texture is rough to the extent that the response it evokes in SA1 afferents is spatially inhomogeneous and the response it evokes in RA and PC afferents is temporally inhomogeneous.

Information about roughness is encoded in SI as evidenced by the fact that the responses of neurons in this brain area are sensitive to changes in surface properties that determine perceived roughness, namely the spatial period of embossed dot patterns and the groove width of tactile gratings (Darian-Smith et al. 1982, 1984; Sinclair and Burton 1991; Tremblay et al. 1996; Sinclair et al. 1996; Chapman et al. 2002). Lesions in SI, particularly in areas 3b and 1, lead to severe impairments in roughness discrimination (Randolph and Semmes 1974). The second somatosensory cortex (SII) has also been implicated in the processing of surface roughness as it contains neurons that are sensitive to the relevant surface properties (Jiang et al. 1997; Pruett, Jr. et al. 2000) and lesions in SII cause impairments in roughness discrimination (Murray and Mishkin 1984). Finally, the lateral parietal opercular cortex (Roland et al. 1998; Stilla and Sathian 2008) is selectively activated when human subjects perform a roughness discrimination task, as are the posterior insula and the medial occipital cortex (Stilla and Sathian 2008), indicating that they too are involved in the cortical processing of roughness information. Spatial variability in the peripheral response—which drives the perception of roughness for coarse textures—is likely computed by the subpopulation of neurons, described above, whose receptive fields comprise excitatory and inhibitory sub-regions. In fact, a subset of these neurons exhibit responses to embossed dot patterns that match their perceived roughness: Their response increases for inter-element spacings up to about 2 mm, then decreases (Arun Sripati, personal communication); this population of neurons is thus ideally suited to extract roughness information from spatial patterns of SA1 activation. The roughness signal carried in the responses of RA and PC afferents likely drives responses in another population of SI neurons, also described above, that are highly sensitive to skin vibrations over a wide range of frequencies (Harvey et al. 2013) and receive input from both RA and PC fibers (Saal et al. 2014).

Hardness

Hardness/softness is the subjective continuum associated with the compliance of an object: hardness ratings have been shown to be inversely proportional to softness ratings; these ratings are, in turn, related to surface compliance following Stevens's power law (Harper and Stevens 1964). Softness perception has been shown to rely primarily on cutaneous cues: eliminating kinesthetic information has no effect on subjects' ability to discriminate softness (Srinivasan and LaMotte 1995). As the hand is pressed up against a compliant object, it conforms to the contour of the hand in proportion to the contact force. The compliance (and the softness) of the object may be signaled by the growth of the area over which the skin contacts the object as the contact force increases, as well as the increase in the force exerted by the object on the skin across the contact area. Softness perception likely relies on signals from SA1 fibers (Srinivasan and LaMotte 1996): First, PC fibers are too sparse and their RFs too large to play a significant role in softness perception. Second, the response of RA fibers to a surface indented into the skin is not modulated by the compliance of the surface whereas the response of SA1 fibers is (Srinivasan and LaMotte 1996). Although the evidence suggests that SA1 fibers are implicated in softness perception, the neural code for softness is unclear: the rate of indentation (in addition to surface compliance) modulates the discharge rate in individual SA1 afferents whereas softness perception is independent of the rate with which a surface is (passively) indented into the skin. Thus, the firing rate in a single SA1 fiber does not unambiguously encode the compliance of an object. One possibility is that the average pressure exerted across the contact area is predictive of compliance and invariant with respect to indentation velocity; it may then be this quantity—average pressure—that is encoded in the population response of SA1 fibers.

Stickiness

Stickiness/slipperiness is the sensory continuum associated with the friction between skin and surface. Indeed, magnitude estimates of stickiness have been shown to closely match the measured kinetic friction between skin and surface, i.e. the ratio between the force exerted normal to the surface to that exerted parallel to the plane of the surface (Smith and Scott 1996). Furthermore, when judging stickiness, subjects do not substantially vary the normal forces they apply on the surface, but the applied tangential forces tend to vary across surfaces, suggesting that tangential forces are critical in the perception of stickiness. As slowly adapting type 2 fibers are sensitive to skin stretch (Witt and Hensel 1959; Iggo 1966; Knibestöl 1975), this population of mechanoreceptive afferent fibers may provide the peripheral signals underlying stickiness perception, although recent evidence suggests that other mechanoreceptive afferents also convey information about tangential forces exerted on the skin (Birznieks et al. 2001).

Thermal Conductivity

Because ambient temperatures are generally cooler than the temperature of the skin, objects in the environment tend to conduct heat out of the skin when contacted. The perceived warmth or coolness of a surface is determined by how slowly or rapidly heat is conducted out of the skin (Ho and Jones 2006, 2008). The perception of the thermal quality of a surface is likely mediated by thermoreceptors in the skin (Darian-Smith et al. 1973, 1979; Johnson et al. 1973, 1979).

References

Bensmaia, S J; Hollins, M (2003). The vibrations of texture. *Somatosensory & Motor Research* 20: 33–43.

Bensmaia, S J; Hollins, M (2005). Pacinian representations of fine surface texture. *Perception & Psychophysics* 67: 842–854.

Bensmaia, S J; Leung, Y Y; Hsiao, S S and Johnson, K O (2005). Vibratory adaptation of cutaneous mechanoreceptive afferents. *Journal of Neurophysiology* 94: 3023–3036.

Birznieks, I; Jenmalm, P; Goodwin, A W and Johansson, R S (2001). Encoding of direction of fingertip forces by human tactile afferents. *The Journal of Neuroscience* 21: 8222–8237.

Blake, D T; Hsiao, S S and Johnson, K O (1997). Neural coding mechanisms in tactile pattern recognition: the relative contributions of slowly and rapidly adapting mechanoreceptors to perceived roughness. *The Journal of Neuroscience* 17: 7480–7489.

Cascio, C J and Sathian, K (2001). Temporal cues contribute to tactile perception of roughness. *The Journal of Neuroscience* 21: 5289–5296.

Chapman, C E; Tremblay, F; Jiang, W; Belingard, L and Meftah, E M (2002). Central neural mechanisms contributing to the perception of tactile roughness. *Behavioural Brain Research* 135: 225–233.

Connor, C E; Hsiao, S S; Phillips, J R and Johnson, K O (1990). Tactile roughness: neural codes that account for psychophysical magnitude estimates. *The Journal of Neuroscience* 10: 3823–3836.

Connor, C E and Johnson, K O (1992). Neural coding of tactile texture: comparisons of spatial and temporal mechanisms for roughness perception. *The Journal of Neuroscience* 12: 3414–3426.

Darian-Smith, I; Goodwin, A W; Sugitani, M and Heywood, J (1984). The tangible features of textured surfaces: their representation in the monkey's somatosensory cortex. In: G M Edelman, W E Gall and W M Cowan (Eds.), *Dynamic Aspects of Neocortical Function* (pp. 475–500). New York: Wiley.

Darian-Smith, I; Johnson, K O and Dykes, R W (1973). The 'cold' fiber population innervating palmar and digital skin of the monkey: responses to cooling pulses. *Journal of Neurophysiology* 36: 325–346.

Darian-Smith, I et al. (1979). Warm fibers innervating palmar and digital skin of the monkey: responses to thermal stimuli. *Journal of Neurophysiology* 42: 1297–1315.

Darian-Smith, I; Sugitani, M; Heywood, J; Karita, K and Goodwin, A W (1982). Touching textured surfaces: Cells in somatosensory cortex respond both to finger movements and to surface features. *Science* 218: 906–909.

DiCarlo, J J and Johnson, K O (1999). Velocity invariance of receptive field structure in somatosensory cortical area 3b of the alert monkey. *The Journal of Neuroscience* 19: 401–419.

DiCarlo, J J and Johnson, K O (2000). Spatial and temporal structure of receptive fields in primate somatosensory area 3b: effects of stimulus scanning direction and orientation. *The Journal of Neuroscience* 20: 495–510.

DiCarlo, J J; Johnson, K O and Hsiao, S S (1998). Structure of receptive fields in area 3b of primary somatosensory cortex in the alert monkey. *The Journal of Neuroscience* 18: 2626–2645.

Gamzu, E and Ahissar, E (2001). Importance of temporal cues for tactile spatial-frequency discrimination. *The Journal of Neuroscience* 21: 7416–7427.

Harper, R and Stevens, S S (1964). Subjective hardness of compliant materials. *The Quarterly Journal of Experimental Psychology* 16: 204–215.

Harvey, M; Saal, H P; Dammann, J F and Bensmaia, S J (2013). Multiplexing stimulus information through rate and temporal codes in primate somatosensory cortex. *PLoS Biology* 11: e1001558.

Heller, M A (1989). Texture perception in sighted and blind observers. *Perception & Psychophysics* 45: 49–54.

Ho, H N and Jones, L A (2006). Contribution of thermal cues to material discrimination and localization. *Perception & Psychophysics* 68: 118–128.

Ho, H N and Jones, L A (2008). Modeling the thermal responses of the skin surface during hand-object interactions. *Journal of Biomechanical Engineering* 130: 021005.

Hollins, M; Bensmaïa, S J; Karlof, K and Young, F (2000). Individual differences in perceptual space for tactile textures: Evidence from multidimensional scaling. *Perception & Psychophysics* 62: 1534–1544.

Hollins, M; Bensmaïa, S J and Risner, R (1998). The duplex theory of tactile texture perception. In: S Grondin and Y Lacouture (Eds.), *Fechner Day 98: Proceedings of the Fourteenth Annual Meeting of the International Society for Psychophysics* (pp. 115–120). Quebec, Canada: The International Society for Psychophysics.

Hollins, M; Bensmaia, S J and Roy, E A (2002). Vibrotaction and texture perception. *Behavioural Brain Research* 135: 51–56.

Hollins, M; Bensmaia, S J and Washburn, S (2001). Vibrotactile adaptation impairs discrimination of fine, but not coarse, textures. *Somatosensory & Motor Research* 18: 253–262.

Hollins, M; Faldowski, R; Rao, S and Young, F (1993). Perceptual dimensions of tactile surface texture: A multidimensional-scaling analysis. *Perception & Psychophysics* 54: 697–705.

Iggo, A (1966). Cutaneous receptors with a high sensitivity to mechanical displacement. In: A V de Reuck and J Knight (Eds.), *Touch, Heat and Pain* (pp. 237–260). Boston: Little, Brown and Company.

Jiang, W; Tremblay, F and Chapman, C E (1997). Neuronal encoding of texture changes in the primary and the secondary somatosensory cortical areas of monkeys during passive texture discrimination. *Journal of Neurophysiology* 77: 1656–1662.

Johansson, R S and Flanagan, J R (2009). Coding and use of tactile signals from the fingertips in object manipulation tasks. *Nature Reviews Neuroscience* 10: 345–359.

Johnson, K O; Darian-Smith, I and LaMotte, C C (1973). Peripheral neural determinants of temperature discrimination in man: a correlative study of responses to cooling skin. *Journal of Neurophysiology* 36: 347–370.

Johnson, K O; Darian-Smith, I; LaMotte, C C; Johnson, B and Oldfield, S R (1979). Coding of incremental changes in skin temperature by a population of warm fibers in the monkey: Correlation with intensity discrimination in man. *Journal of Neurophysiology* 42: 1332–1353.

Johnson, K O and Phillips, J R (1988). A rotating drum stimulator for scanning embossed patterns and textures across the skin. *Journal of Neuroscience Methods* 22: 221–231.

Klatzky, R L; Lederman, S J and Reed, C L (1987). There's more to touch than meets the eye: The salience of object attributes for haptics with and without vision. *Journal of Experimental Psychology: General* 116: 356–369.

Knibestöl, M (1975). Stimulus-response functions of slowly adapting mechanoreceptors in the human glabrous skin area. *The Journal of Physiology* 245: 63–80.

Lederman, S J (1979). Auditory texture perception. *Perception* 8: 93–103.

Lederman, S J and Taylor, M M (1972). Fingertip force, surface geometry, and the perception of roughness by active touch. *Perception & Psychophysics* 12: 401–408.
Manfredi, L R et al. (2014). Natural scenes in tactile texture. *Journal of Neurophysiology* 111(9): 1792–1802.
Morley, J W; Goodwin, A W and Darian-Smith, I (1983). Tactile discrimination of gratings. *Experimental Brain Research* 49: 291–299.
Murray, E A and Mishkin, M (1984). Relative contributions of SII and area 5 to tactile discrimination in monkeys. *Behavioural Brain Research* 11: 67–85.
Pruett, J R, Jr.; Sinclair, R J and Burton, H (2000). Response patterns in second somatosensory cortex (SII) of awake monkeys to passively applied tactile gratings. *Journal of Neurophysiology* 84: 780–797.
Randolph, M and Semmes, J (1974). Behavioral consequences of selective ablations in the postcentral gyrus of Macaca mulatta. *Brain Research* 70: 55–70.
Roland, P E; O'Sullivan, B and Kawashima, R (1998). Shape and roughness activate different somatosensory areas in the human brain. *Proceedings of the National Academy of Sciences of the United States of America* 95: 3295–3300.
Saal, H P; Harvey, M A and Bensmaia, S J (2014). Integration of cutaneous submodalities in primate somatosensory cortex. *SfN abstracts* 44: 339.07.
Sathian, K; Goodwin, A W; John, K T and Darian-Smith, I (1989). Perceived roughness of a grating: correlation with responses of mechanoreceptive afferents innervating the monkey's fingerpad. *The Journal of Neuroscience* 9: 1273–1279.
Sinclair, R J and Burton, H (1991). Neuronal activity in the primary somatosensory cortex in monkeys (*Macaca mulatta*) during active touch of textured surface gratings: Responses to groove width, applied force, and velocity of motion. *Journal of Neurophysiology* 66: 153–169.
Sinclair, R J; Pruett, J R and Burton, H (1996). Responses in primary somatosensory cortex of rhesus monkey to controlled application of embossed grating and bar patterns. *Somatosensory & Motor Research* 13: 287–306.
Skedung, L et al. (2013). Feeling small: exploring the tactile perception limits. *Scientific Reports* 3: 2617.
Smith, A M; Chapman, C E; Deslandes, M; Langlais, J S and Thibodeau, M P (2002). Role of friction and tangential force variation in the subjective scaling of tactile roughness. *Experimental Brain Research* 144: 211–223.
Smith, A M and Scott, S H (1996). Subjective scaling of smooth surface friction. *Journal of Neurophysiology* 75: 1957–1962.
Srinivasan, M A and LaMotte, R H (1995). Tactual discrimination of softness. *Journal of Neurophysiology* 73: 88–101.
Srinivasan, M A and LaMotte, R H (1996). Tactual discrimination of softness: Abilities and mechanisms. In: O Franzén, R S Johansson, L Terenius (Eds.), *Somesthesis and the Neurobiology of the Somatosensory Cortex* (pp. 123–135). Basel: Birkhäuser.
Sripati, A P; Yoshioka, T; Denchev, P; Hsiao, S S and Johnson, K O (2006). Spatiotemporal receptive fields of peripheral afferents and cortical area 3b and 1 neurons in the primate somatosensory system. *The Journal of Neuroscience* 26: 2101–2114.
Stilla, R and Sathian, K (2008). Selective visuo-haptic processing of shape and texture. *Human Brain Mapping* 29: 1123–1193.
Taylor, M M and Lederman, S J (1975). Tactile roughness of grooved surfaces: A model and the effect of friction. *Perception & Psychophysics* 17: 23–36.
Tremblay, F; Ageranioti-Bélanger, S A and Chapman, C E (1996). Cortical mechanisms underlying tactile discrimination in the monkey. I. role of primary somatosensory cortex in passive texture discrimination. *Journal of Neurophysiology* 79: 3382–3403.
Vallbo, A B and Johansson, R S (1979). Tactile sensibility in the human hand: relative and absolute densities of four types of mechanoreceptive units in glabrous skin. *The Journal of Physiology* 286: 283–300.
Weber, A I et al. (2013). Spatial and temporal codes mediate the tactile perception of textures. *Proceedings of the National Academy of Sciences* 110: 18279–18284.

Witt, I and Hensel, H (1959). Afferente Impulse aus der Extremitätenhaut der Katze bei thermischer und mechanischer Reizung. *Pflügers Archiv: European Journal of Physiology* 268: 582–596.

Yoshioka, T; Gibb, B; Dorsch, A K; Hsiao, S S and Johnson, K O (2001). Neural coding mechanisms underlying perceived roughness of finely textured surfaces. *The Journal of Neuroscience* 21: 6905–6916.

Haptic Perception of Force

Wouter M. Bergmann Tiest and Astrid M.L. Kappers

This chapter reviews research that deals with measuring the precision and/or the accuracy of haptic force perception, using psychophysical methods. Other aspects of force perception might include physiological aspects or neuronal processing, but these are beyond the scope of the present chapter. Furthermore, force sensation plays a role in the perception of other physical aspects, such as friction or stiffness. These aspects are reviewed in Chap. 16 of this book.

Force perception relates to two aspects of force: its magnitude and its direction. Perception of these aspects is discussed separately below. Concerning force magnitude, many early studies have focused on the perception of force in the direction of gravity, i.e. the weight of objects. This is discussed first. Then, force magnitude perception in other directions is discussed. Finally, some applications of this research are discussed.

Different psychophysical techniques have been used to study force perception. To study the relationship between physical and perceived force magnitude, the technique of magnitude estimation has been used. To measure the smallest difference in force magnitude or direction that can be perceived, discrimination experiments have been performed. Finally, to study the relationship between force magnitudes perceived in different ways, or between physical and perceived force direction, matching experiments were done. This subdivision is used to categorise the different studies.

Perception of Force Magnitude

An overview of perception of force magnitude is given by Jones (1986). The most important findings, as well as some more recent research, will be discussed below. First, the focus will be on force magnitude perception in the direction of gravity, and then, in other directions.

W.M. Bergmann Tiest (✉) · A.M.L. Kappers
VU University, Amsterdam, The Netherlands

T.J. Prescott et al. (eds.), *Scholarpedia of Touch*, Scholarpedia,
DOI 10.2991/978-94-6239-133-8_17

In the Direction of Gravity (Weight Perception)

Discrimination Formal investigation of force perception started with the weight discrimination experiments of E. H. Weber, published in 1834 (Weber 1834/1996). By placing weights on the hands of subjects, he was able to measure the smallest weight difference that could be perceived. With the subjects' hands lying on the table, he found Weber fractions (the ratio between the smallest detectable difference and the reference weight) of about 0.3. However, with the hands and weights lifted in the air, the Weber fractions were in the order of 0.09, indicating a much smaller difference. This shows that force perception is much more precise when both the cutaneous sense (skin receptors) and kinaesthetic sense (receptors in limbs and muscles) are involved, compared to just the cutaneous sense. The relative importance of the cutaneous and kinaesthetic senses probably depends on the magnitude of the force itself, with cutaneous inputs playing a larger role in the perception of weaker forces. The experiment was performed with weights of about 1.2 and 10 N, with similar results in terms of Weber fractions. Thus, although in *absolute* terms the discrimination threshold depends on the force used, the *relative* discrimination threshold seems fairly constant. Note that for very small weights, the relative discrimination threshold is expected to go up, as there also exists a minimum discrimination threshold in absolute terms.

Weight discrimination appears to depend on object size: in experiments on weight discrimination with objects of different densities (the ratio between mass and volume), it was found that weight perception was most precise for the density that corresponded to the expected density of the object's material (Ross and Gregory 1970). With more or less dense (i.e. smaller or larger) objects, discrimination was poorer. This suggests that there is more to weight discrimination than just force perception. This is also illustrated by the fact that people can use inertial cues to improve their weight discrimination (Brodie and Ross 1985; Ross and Brodie 1987). When not only the static force of earth's gravity on the object is available, but also the dynamic inertial force of an object that is moved about, then extra information is available to determine the object's mass when combined with an estimate of the object's acceleration. In fact, mass discrimination based just on inertial force is possible in the absence of gravitational force, e.g. in orbit in space (Ross and Reschke 1982; Ross et al. 1984, 1986).

Magnitude estimation Many studies have used magnitude estimation to characterise the relationship between physical and perceived heaviness of objects, which corresponds to the magnitude of the force that gravity exerts on these objects. It has been found that this relationship can be described by a power function with an exponent of about 1.45 (Stevens and Galanter 1957). In the papers reviewed by Jones (1986), a range of exponents of 0.8–2.0 have been reported, depending on the specific conditions of the experiments. Indeed, perceived heaviness is subject to the influence of numerous factors. For example, in a matching task between forces exerted on the left and the right wrists, grasping an object firmly in the hand made the force feel smaller, whereas anaesthetising the hand (not the muscles involved in lifting) made the force feel larger (Gandevia and McCloskey 1976).

Furthermore, it has been found that with forces exerted on the skin of the hand palm, the area of stimulation plays a role in the perceived magnitude of the force: with a larger area of stimulation, the same force felt smaller (Bergmann Tiest et al. 2012). This occurred mainly with the hand lying flat on a table, so that only cutaneous force information was available. With the unsupported hand, when also proprioceptive force information was available, this effect was reduced significantly. This is in accordance with the idea that proprioception of force is unaffected by the area of stimulation on the skin.

Lastly, it has been long known that perceived heaviness is influenced by object size: a smaller object of the same weight feels heavier, an effect known as the *size-weight illusion* (Charpentier 1891, as discussed in Murray et al. 1999; Stevens and Rubin 1970; see also Chap. 27 of this book). This is the topic of a large body of research, but is beyond the scope of the present article.

In Other Directions

Discrimination Using an electromechanical apparatus, Pang et al. (1991) let subjects squeeze two plates between thumb and forefinger, which offered an adjustable resistive force. In this way, different forces could be displayed to the subjects in a horizontal direction, and force discrimination thresholds could be measured. In this dynamic scenario (involving movement of the fingers), the Weber fraction for force discrimination was found to be about 0.07 for a wide range of reference forces, initial finger spans, and squeezing distances, as long as the kinaesthetic sense was involved. A Weber fraction of about 0.1 for force magnitude discrimination was found in an experiment using a hand-held stylus to which forces were applied (Pongrac et al. 2006). These findings are consistent with the earlier weight discrimination experiments (Weber 1834/1996; Brodie and Ross 1985; Ross and Brodie 1987), indicating that the direction of the presented force is not critical to force discrimination. However, when force magnitude had to be discriminated while the whole hand was moving, discrimination thresholds were found to be higher, with Weber fractions around 0.45 (Yang et al. 2008b). In this experiment, subjects moved a hand-held stylus from left to right while forces were applied at five different angles with respect to the movement direction: either in the same or opposite direction, perpendicular, or at an angle of 45° or 135°. In particular, subjects found it difficult to discriminate the magnitude of forces applied at 45°, with a Weber fraction as high as 0.6.

Such an angular dependence was not observed in an experiment where subjects had to discriminate the magnitude of a horizontally applied force (to the right) of 2.5 N from forces applied in all six cardinal directions of the 3D space (Dorjgotov et al. 2008). Weber fractions for all directions were around 0.33. This suggests that for the stationary hand, and a relatively low magnitude, force discrimination is quite isotropic. However, it should be noted that in this experiment, the reference force was always in the same direction. This was different in the experiment by Vicentini et al. (2010), which used five reference force magnitudes in each of the six cardinal directions, but with the test and reference forces always in matching directions. They found differences in discrimination thresholds between the different directions, with the highest Weber fractions (~0.14) in the horizontal-tangential direction, and the lowest (~0.11) in the radial direction. These numbers are asymptotic Weber fractions for high force magnitudes; with lower forces (<5 N), a significant increase in Weber fractions was found. Thus, we can conclude that with these higher forces, force discrimination thresholds are in accordance with the earlier weight discrimination experiments, but that there are anisotropies depending on the direction of the force.

Magnitude estimation Forces on the fingers can be applied either normal or tangential to the skin. In the former case, the skin is indented, whereas in the latter case, the skin is stretched sideways. In a magnitude estimation experiment by Paré et al. (2002), both methods were used with a range of low forces (0.15–0.70 N). For both methods, a linear relationship between physical and perceived force was found, with correlation coefficients ranging from 0.48–0.96, and 0.70–0.97 for the tangential and normal force conditions, respectively. There was no significant difference in slope between the two conditions. This suggests that, at least for these small forces, the human sensitivity for tangential force is comparable to that for normal force. However, an anisotropy in perceived force magnitude was found for forces applied to the hand, both for relatively high forces (5–30 N; Tanaka and Tsuji 2008) and for medium forces (2–6 N; Van Beek et al. 2013). In both these experiments, forces in different directions in the horizontal plane were applied to a handle held in the subject's hand. The perceived magnitude was found to be the greatest in the directions of the line connecting the shoulder to the hand and vice versa, and smallest in the perpendicular directions. Thus, when applied to the whole hand, the same force might feel different, depending on the direction.

Matching If forces from different directions on the same body part feel different, how do forces applied to different body parts compare? This was investigated in a force matching experiment where subjects had to produce a force with one body part, based on a visual display of the required force, and then reproduce that force with another body part, without visual feedback (Jones 2003). The results showed that forces were reproduced as larger by the elbow than by the hand, and as larger by the hand than by the finger. Thus, the perceived magnitude of a force varies as a function of the muscle group generating the force. When the reproduced forces were expressed as a percentage of the maximum force that can be generated by each body part, they showed a much better correspondence. It seems that the forces are matched based on their relative magnitude with respect to the maximum.

Furthermore, the perceived magnitude of a force has been found to depend on whether it is passively perceived or actively generated. In an experiment where pairs of subjects were asked to reproduce a force to the other person that this other person just applied to their index finger, the magnitude of the force going back and forth quickly escalated (Shergill et al. 2003). On average, subjects generated a force

38 % higher than was applied to them. The authors concluded that the perceived magnitude of self-generated forces is attenuated as a result of a mechanism that removes some of the predictable sensory consequences of a movement, in order for externally generated sensations to become relatively more salient. This might also explain an effect observed by Bergmann Tiest and Kappers (2010), in an experiment where mass perceived through gravitational force (weight) and inertial force (resistance to acceleration or deceleration) had to be matched. When subjects accelerated a mass by pushing it, it had to be twice as heavy as a mass resting on the hand in order to feel equal. However, when subjects decelerated a moving mass by stopping it, it was matched veridically to a mass resting on the hand. Thus, the force used to actively push a mass is perceived as smaller than the force necessary to passively stop it. Taken together, this illustrates that the perceived magnitude of a force depends on many factors, such as the direction, the body part involved, and whether or not the force is self-generated.

Perception of Force Direction

Let us now turn our attention from force magnitude to force direction perception. The precision of this can be assessed using discrimination experiments, which will be discussed first. Afterwards, matching experiments will be discussed, which are used to measure the accuracy of force direction perception.

Discrimination Using a force-feedback joystick, Elhajj et al. (2006) determined that a 83 % correct force direction discrimination level was attained for an angle difference of 15°. This can be interpreted as the discrimination threshold for direction of force applied to the whole hand. However, the research does not tell us how this threshold depends on the magnitude or direction of the force. The latter question was investigated by Tan et al. (2006) for five directions in the frontoparallel plane and a 2 N force applied to the finger tip. They found an average discrimination threshold of 33°, but no effect of direction. It is unknown whether the considerably higher threshold compared to Elhajj et al. (2006) should be ascribed to a difference in force magnitude or to whether the force is applied to the whole hand or just the index finger. It should be noted that these thresholds might be quite variable: using the same setup and paradigm, the authors found a threshold of 26° in another experiment (Barbagli et al. 2006). In that experiment, they also included conditions in which (congruent or incongruent) visual information was present. This led to a decrease or increase, respectively, of the discrimination threshold, even though the subjects were instructed to base their judgements on the haptic information only. This prompted the authors to conclude that visual information can modulate haptic force perception (Ho et al. 2006).

The question whether the force direction discrimination threshold depends on the magnitude of the force was investigated by Pongrac et al. (2006). They applied forces to a hand-held stylus, either straight or with a perturbation in the magnitude or direction. When the perturbation was perpendicular to the reference force, the test force differed only in direction from the reference force. In those cases, they found discrimination thresholds corresponding to a perpendicular force magnitude of 25 and 20 % of the reference force for 1 and 2 N, respectively. Using the arctangent, this corresponds to angle differences of 14° and 11°, respectively. This difference was statistically significant, so it seems that with a higher force magnitude, the force direction is perceived with better precision, at least for forces in this range.

Furthermore, the force direction discrimination threshold has been investigated in the case of a moving hand, using a hand-held stylus and a 1.5 N force (Yang et al. 2008a). Averaged over five different reference directions and two speeds, the discrimination threshold was found to be 32°. However, no effect was found of reference direction or movement speed. To sum up, it seems that the precision of haptic force perception is fairly isotropic, with a discrimination threshold of around 30° for low forces applied to the fingers or during movement, and about 10–15° for higher forces or forces applied to the whole hand.

Finally, a discrimination experiment has been used to determine the smallest magnitude of force necessary to differentiate between two diametrically opposing directions (Baud-Bovy and Gatti 2010). Subjects had hold the handle of a robotic device, to which either a left- or rightward force was applied, and then select the correct direction. When they had to keep their hand still, the minimum required force magnitude was 0.1 N, but when they were allowed to move their hand left and right, this was only 0.05 N, suggesting that movement can actually help in determining the direction of a force.

Matching In this last paragraph, we look at the accuracy of force direction perception; that is, how well does the perceived force direction correspond to the physical one? This type of question is usually investigated using matching experiments, in which the subject has to adjust some parameter until it matches his/her perception. In one such task, subjects had to hold a joystick and resist a 5 N force that was applied in a certain direction (Toffin et al. 2003). Then, they had to reproduce it, that is, apply a force to the joystick in the exact opposite direction. The difference between the presented force direction and the reproduced one is the bias. The magnitude and the sign of this bias were found to depend on the force direction, indicating an anisotropy in either the perception or the reproduction of force direction. However, since every trial consisted of both a perception and a reproduction phase, it was unclear from which phase the anisotropy originates. This problem was circumvented in the experiment by Van Beek et al. (2013), in which only a perception phase was present, and the perceived direction had to be matched by turning a visual arrow in the correct direction. Subjects showed substantial biases, up to 30° in both directions. Within subjects, repeated measurements (using different force levels (2–6 N), different hands, and on different days) showed consistent bias patterns, but these differed between subjects. These intra-subject similarities and inter-subject differences were further explored in a follow-up study (Van Beek et al. 2014). In one experiment, subjects performed the same task three or four times on different days, with a month between the third and the fourth session. Six out of eight subjects showed high correlations between the first three sessions, and three of the five subjects that completed a fourth session showed high correlations between the first three and the fourth session. The second experiment revealed that only 7 % of the bias patterns showed a high correlation with patterns of other subjects. These findings indicate that the bias pattern of a particular subject is not random, but a unique feature of that person. This knowledge could be used to compensate for biases in force direction perception on a person-by-person basis, after a simple characterisation procedure.

Applications

In many situations in which an operator remotely controls a vehicle or robotic arm, force feedback can be used to provide the operator with information about the task or the situation. Especially when visual feedback is limited (for example under water), haptic force feedback becomes more important and is relied upon more. In order to be able to correctly provide this force feedback, it is important to know how forces are perceived by humans. Using the thresholds reported above, the minimum resolution can be calculated that a system needs to have so that no perceptible degradation occurs. Furthermore, the discussed research illustrates that there are systematic deviations in force perception that should be taken into account when designing such a system for a situation with reduced or absent visual information.

It should be noted that most studies discussed above deal with perception of forces that are constant in time, or do not change when the subject moves. However, these are rare in everyday interactions with the world around us. We can assume that the knowledge about perception of these constant forces generalises to the perception of dynamic forces, but future research should confirm whether this assumption is justified.

References

Barbagli, F; Salisbury, K; Ho, C; Spence, C and Tan, H Z (2006). Haptic discrimination of force direction and the influence of visual information. *ACM Transactions on Applied Perception* 3: 125–135.

Baud-Bovy, G and Gatti, E (2010). Hand-held object force direction identification thresholds at rest and during movement. In: A M L Kappers, J B F van Erp, W M Bergmann Tiest and F C T van der Helm (Eds.), *Haptics: Generating and perceiving tangible sensations. Part II,* Vol. 6192 of *Lecture Notes in Computer Science* (pp. 231–236). Berlin/Heidelberg: Springer-Verlag.

Bergmann Tiest, W M and Kappers, A M L (2010). Haptic perception of gravitational and inertial mass. *Attention, Perception & Psychophysics* 72(4): 1144–1154.

Bergmann Tiest, W M; Lyklema, C and Kappers, A M L (2012). Investigating the effect of area of stimulation on cutaneous and proprioceptive weight perception. In: P Isokoski and J Springare

(Eds.), *Haptics: Perception, devices, mobility, and communication. Part II*, Vol. 7283 of *Lecture Notes in Computer Science* (pp. 7–12). Berlin/Heidelberg: Springer-Verlag.

Brodie, E E and Ross, H E (1985). Jiggling a lifted weight does aid discrimination. *American Journal of Psychology* 98(3): 469–471.

Dorjgotov, E; Bertoline, G R; Arns, L; Pizlo, Z and Dunlop, S R (2008). Force amplitude perception in six orthogonal directions. In: *Proc. Symposium on Haptic Interfaces for Virtual Environments and Teleoperator Systems* (pp. 121–127). Reno, Nevada, USA: IEEE.

Elhajj, I; Weerasinghe, H; Dika, A and Hansen, R (2006). Human perception of haptic force direction. In: *Proc. IEEE/RSJ International Conference on Intelligent Robots and Systems* (pp. 989–993).

Gandevia, S C and McCloskey, D I (1976). Perceived heaviness of lifted objects and effects of sensory inputs from related, non-lifting parts. *Brain Research* 109: 399–401.

Ho, C; Tan, H Z; Barbagli, F; Salisbury, K and Spence, C (2006). Isotropy and visual modulation of haptic force direction discrimination on the human finger. In: *Proceedings of EuroHaptics* (pp. 483–486).

Jones, L A (1986). Perception of force and weight: Theory and research. *Psychological Bulletin* 100(1): 29–42.

Jones, L A (2003). Perceptual constancy and the perceived magnitude of muscle forces. *Experimental Brain Research* 151(2): 197–203.

Murray, D; Ellis, R; Bandomir, C and Ross, H (1999). Charpentier (1891) on the size-weight illusion. *Perception & Psychophysics* 61: 1681–1685.

Pang, X D; Tan, H Z and Durlach, N I (1991). Manual discrimination of force using active finger motion. *Perception & Psychophysics* 49(6): 531–540.

Paré, M; Carnahan, H and Smith, A M (2002). Magnitude estimation of tangential force applied to the fingerpad. *Experimental Brain Research* 142(3): 342–348.

Pongrac, H; Hinterseer, P; Kammerl, J; Steinbach, E and Färber, B (2006). Limitations of human 3d-force discrimination. In: *Proc. Second International Workshop on Human-Centered Robotics Systems* (pp. 109–114).

Ross, H E and Brodie, E E (1987). Weber fractions for weight and mass as a function of stimulus intensity. *Quarterly Journal of Experimental Psychology* 39A: 77–88.

Ross, H E; Brodie, E E and Benson, A J (1984). Mass discrimination during prolonged weightlessness. *Science* 225: 219–221.

Ross, H E; Brodie, E E and Benson, A J (1986). Mass-discrimination in weightlessness and readaptation to earth's gravity. *Experimental Brain Research* 64(2): 358–366.

Ross, H E and Gregory, R L (1970). Weight illusions and weight discrimination—a revised hypothesis. *Quarterly Journal of Experimental Psychology* 22(2): 318–328.

Ross, H E and Reschke, M F (1982). Mass estimation and discrimination during brief periods of zero gravity. *Perception & Psychophysics* 31(5): 429–436.

Shergill, S S; Bays, P H; Frith, C D and Wolpert, D M (2003). Two eyes for an eye: The neuroscience of force escalation. *Science* 301(5630): 187.

Stevens, J C and Rubin, L L (1970). Psychophysical scales of apparent heaviness and the size-weight illusion. *Perception & Psychophysics* 8(4): 225–230.

Stevens, S S and Galanter, E H (1957). Ratio scales and category scales for a dozen perceptual continua. *Journal of Experimental Psychology* 54(6): 377–411.

Tan, H Z; Barbagli, F; Salisbury, K; Ho, C and Spence, C (2006). Force direction discrimination is not influenced by reference force direction. *Haptics-e* 4: 1–6.

Tanaka, Y and Tsuji, T (2008). Directional properties of human hand force perception in the maintenance of arm posture. In: M Ishikawa et al. (Eds.), *Proc. International Conference on Neural Information Processing*. Part I, Vol. 4984 of *Lecture Notes in Computer Science* (pp. 933–942). Berlin/Heidelberg: Springer-Verlag.

Toffin, D; McIntyre, J; Droulez, J; Kemeny, A and Berthoz, A (2003). Perception and reproduction of force direction in the horizontal plane. *Journal of Neurophysiology* 90(5): 3040–3053.

Van Beek, F E et al. (2014). Subject-specific distortions in haptic perception of force direction. In: M Auvray and C Duriez (Eds.), *Haptics: Neuroscience, Devices, Modeling, and Applications.*

Part I, Vol. 8618 of *Lecture Notes in Computer Science* (pp. 48–54). Berlin/Heidelberg: Springer-Verlag.

Van Beek, F E; Bergmann Tiest, W M and Kappers, A M L (2013). Anisotropy in the haptic perception of force direction and magnitude. *IEEE Transactions on Haptics* 6(4): 399–407.

Vicentini, M; Galvan, S; Botturi, D and Fiorini, P (2010). Evaluation of force and torque magnitude discrimination thresholds on the human hand-arm system. *ACM Transactions on Applied Perception* 8(1): article 1.

Weber, E H (1834/1996). De tactu. In: H E Ross & D J Murray (Eds.), *E. H. Weber on the Tactile Senses*. Hove: Erlbaum (UK) Taylor & Francis.

Yang, X-D; Bischof, W F and Boulanger, P (2008a). The effects of hand motion on haptic perception of force direction. In: M Ferre (Ed.), *Haptics: Perception, devices and scenarios,* Vol. 5024 of *Lecture Notes in Computer Science* (pp. 355–360). Berlin/Heidelberg: Springer-Verlag.

Yang, X-D; Bischof, W F and Boulanger, P (2008b). Perception of haptic force magnitude during hand movements. In: *Proc. IEEE International Conference on Robotics and Automation* (pp. 2061–2066). Pasadena, CA, USA.

Dynamic (Effortful) Touch

Claudia Carello and Michael Turvey

Dynamic or **effortful touch** is a subsystem of the haptic perceptual system (Gibson 1966). It is the most common form of touch, hardly noticed as such, and rarely studied (until recently). Its commonplace nature is conveyed by the multiplicity of words in the lexicon needed to capture its variant forms. For example, *perceiving x by*: supporting, shaking, lifting, hefting, wielding, pushing, pulling, probing, chewing, prodding, groping, bending, stretching, striking, tugging, folding, twisting, squeezing, turning, rolling, vibrating, and so on.

Dynamic (effortful) touch addresses the nature of touch as expressed in Figures 1 and 2. Figure 1a identifies the four kinds of perceiving encompassing properties of self, environment, and self-environment relations (Turvey and Fonseca 2014). Figure 2 identifies the moments of the mass distributions integral to the "effort of touching" in the manners of Figure 1b, c and the attendant deformation of the body's muscular and connective tissues. A working presumption of the theory and research summarized is that the information in support of the perceptual capabilities of Figure 1 is defined over the patterns of tissue deformation.

Methods for Studying Dynamic Touch

Research directed at dynamic (effortful) touch most commonly addresses how a person knows, without benefit of vision, properties of an object held in the hand. In the paradigm reviewed here, the object is grasped at one place and either wielded freely or held still. It is not spanned between the hands or traced along its length. (In the terminology of exploratory procedures (Klatzky & Reed, this volume), the focus here, for example, is on objects held without the hand itself being supported, so-called unsupported holding and not contour following or enclosure.)

A number of response methods are employed to ascertain whether a property has been perceived successfully. The most common is *magnitude production*, which

C. Carello (✉) · M. Turvey
Center for the Ecological Study of Perception and Action,
University of Connecticut, Storrs, CT, USA

T.J. Prescott et al. (eds.), *Scholarpedia of Touch*, Scholarpedia,
DOI 10.2991/978-94-6239-133-8_18

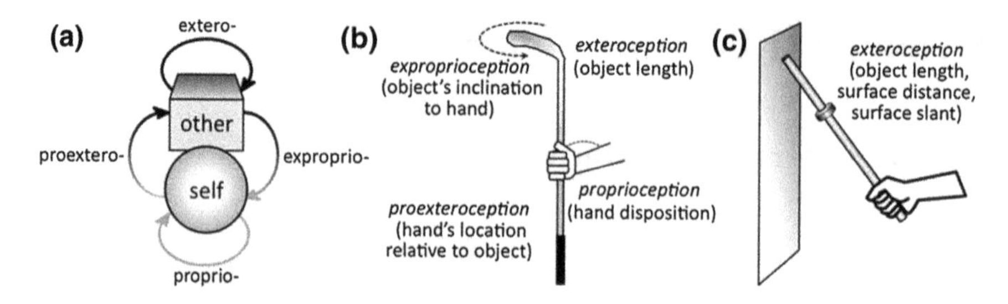

Figure 1 **a** Perception is of environmental properties (exteroception), body properties (proprio-ception), the environment relative to the body (exproprioception), and the body relative to the environment (proexteroception). **b** Dynamic touch examples for a hand-held object. **c** Probing the environment with an implement reveals properties of the probe and the surface probed

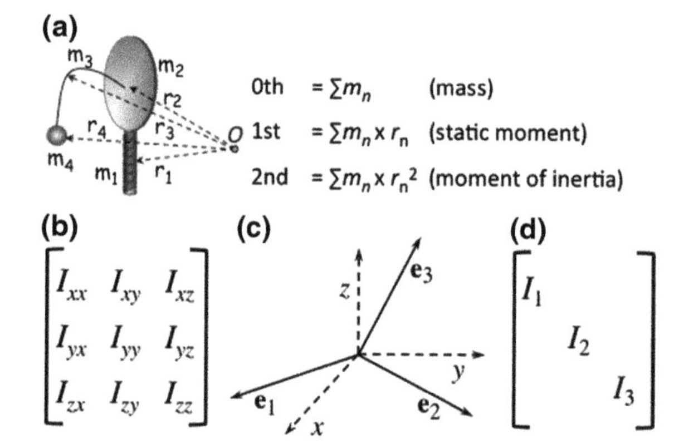

Figure 2 Mass-moments are illustrated for a paddleball attached by an elastic band to a paddle. **a** Calculating the mass-moments about an origin O (taken to be in the wrist for hand-held objects). **b** When an inertia tensor, defined in an xyz-coordinate system is referred to **c** the symmetry axes or eigenvectors e_k, **d** only the principal moments of inertia or eigenvalues remain

requires positioning a marker to coincide with the perceived property (e.g., length, height, width, orientation). The advantage of this method is that the quantitative response can be compared against the actual object dimension. In *magnitude estimation*, a standard object is assigned a value of 100 and target objects are assigned numbers scaled to it (e.g., an object that feels twice as large as the standard would receive a 200; an object that felt just a little smaller could receive a 90). Magnitude estimation can be used for any geometric property, but it is especially useful for something like weight that doesn't have a straightforward magnitude production implementation. A *rating scale* is typically used for more functional properties. An object is rated on a scale from 1 to 7 for how well it would serve a particular purpose. A formal test of *perceptual independence* evaluates the extent to which perception of one property does or does not depend on perception of another property.

In a representative experiment, a rod-like object is grasped and held at just one place along its length, as in Figure 1b. The mechanoreceptors embedded in the body's tissues cannot, therefore, be affected by the object's length as such but they can be affected by the object's mass moments identified in Figure 2. Rod-like objects,

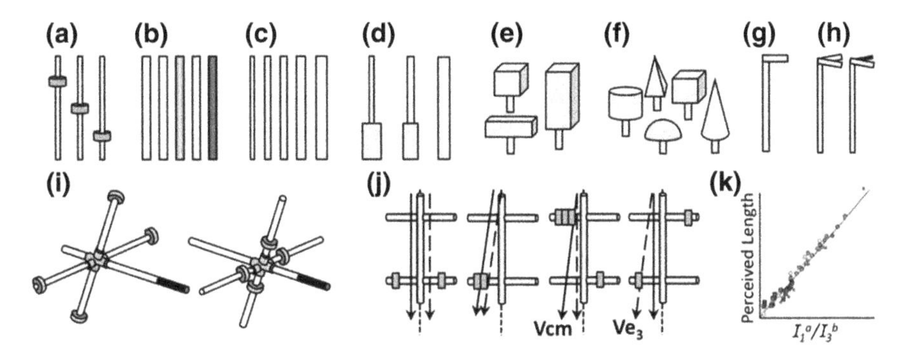

Figure 3 **a–j** Dynamic touch experiments use object sets that vary the moments of the mass distribution. **k** Data from experiments using objects from *b*, *c*, *d*, and *f*

of course, are but one kind of object that can be readily grasped and wielded. Other kinds are shown in Figure 3. Differences among these objects (e.g., their spatial magnitudes, material composition, geometry) have consequences for the mass moments and, perforce, consequences for perception by dynamic effortful touch.

Indifference to Force Magnitude, Significance of Force Variation

A telling feature of dynamic touch is its dependence upon the variation of muscular forces and its independence from the magnitude of muscular forces. As Gibson (1966, p. 127) observed:

> The mass of an object can be judged, in fact, by wielding it in any of a variety of ways, such as tossing and catching, or shaking it from side to side. One can only conclude that the judgment is based on information, not on the sensations. The stimulus information from wielding can only be an invariant of the changing flux of stimulation in the muscles and tendons, an exterospecific invariant in this play of forces. Whatever specifies the mass of the object presumably can be isolated from the change, and the wielding of the object has a function of separating off the permanent component from the changes.

In substantiation of this observation are experiments that examined perceiving the lengths of handheld rods wielded freely about (a) the wrist, (b) the elbow (with wrist firm), (c) the shoulder (with wrist and elbow firm) and (d) all three joints simultaneously (Figure 6 in Turvey 1996). Perceived length was the same for all four conditions, suggestive of invariance detection rather than inference (from torques and movements). Allied observations are that perceiving rod length is invariant over imposed levels of acceleration and over imposed levels of drag (wielding rods in water versus in air, Pagano and Cabe 2003).

The partnership of change and nonchange, variance and invariance, suggests, however, that a perceiver is mandated to modulate force during a bout of dynamic touching. Such modulation and its benefits have been observed and may be allied

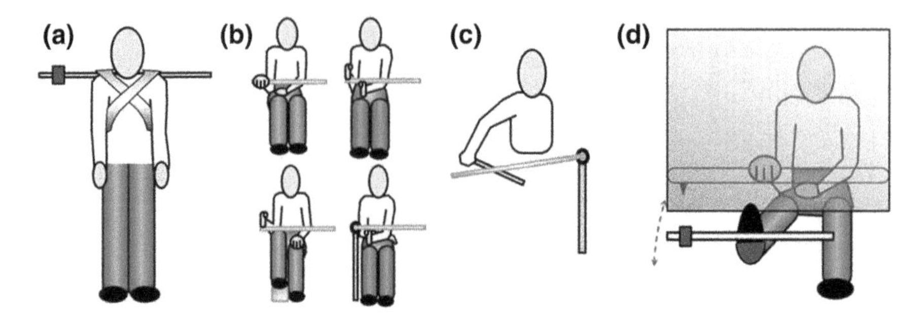

Figure 4 Length can be perceived by **a** wielding by the torso, **b** holding with one hand, propped by two hands, the hand and knee, or the hand and an environmental support; **c** wielding one rod with another; or **d** wielding with rotations about the ankle

with the observation that the microstructure of wielding differs for different object properties.

An apparent challenge to the preceding is the fact that rod lengths can be perceived without intended movement (without the explicit and intended torque variations of wielding, hefting, etc.), that is, by simply holding the rod still. As an example, for rods of lengths 45, 61, 76, 91, 107, and 122 cm placed directly into the hand in the posture in which the judgments were to be made, mean perceptions were 41, 58, 77, 91, 109, and 111 cm. That absence of intended movement does not mean no-movement is key to addressing this capability. In respect to Figure 4a, the selective perception of the whole rod versus a part can be achieved in "quiet standing," that is, without intended movement. Multifractal analysis reveals different spatiotemporal structure in the fluctuations of the body's center of pressure at the mm/ms scale for whole and part perceiving (Palatinus et al. 2014). Force modulation could be very, very subtle.

Intention, Attention

For a rod held at an intermediate position 1/4, 1/2, and 3/4 along its length, one can wield to determine "rod length if held at an end" and wield to determine *length of the rod part forward of grasp*. For the set of rods used in the experiments, the mean actual length was 76 cm and the mean partial length was 37 cm. The respective mean perceptions were 76 and 38 cm. This apparent ability to "fractionate" objects on instruction emphasizes investigations in terms of "to intend perception of *x* requires attending to information about *x*." That rod fractions can be perceived when rods are simply held (no wielding) at different locations along their lengths underscores the theoretical challenge. The challenge is amplified if emphasis is given to 0th and 1st moments. The 0th is the same value for all hand positions. The 1st is zero at the 1/2 position.

Experiments on perceiving the sweet spot of wielded occluded rackets and rods invite similar considerations. The "sweet spot" of an implement refers to the best location along an object's length (its center of percussion) to strike something such as a tennis ball. Perception given the "sweet spot intent" is distinguished from perception given the "length intent" both in magnitude production and in dependence on mass moments. Perceived sweet spot follows closely the value of I_1/static moment, both for tennis rackets of different lengths (from junior through stretch rackets) and for wooden rods with attached masses used to manipulate the moments (reviewed in Carello and Wagman 2009).

Definite Scaling and the Issue of Colinearity

For rods of a fixed material and fixed diameter, magnitude productions of perceived length are ordered appropriately and in the size ranges of the experimental objects. Perception that approximates values measured by a ruler suggests the availability of information that is more definite than the information supporting merely relative scaling (in which values are correctly ordered but arbitrary). This definite scaling occurs despite apparatus-allowable responses ranging from 0 to at least twice object size, for objects as small as a cocktail stirrer, as large as a pool cue, when the mix of sizes is distributed unevenly and adaptation levels are within either a narrow or wide range and when the data points are drawn, point for point, from different participants. Figure 3k suggests that there might be a definable physical basis for this definite scaling (reviewed in Turvey and Carello 1995). The *y*-axis is the mean perceived length of each of 48 objects differing in material heterogeneity, geometric shape, and density (Figure 3b–f), with most affixed to a uniform handle eliminating tactile information about the objects' width, material, shape, and other characteristics. Of note is the fact that definite scaling is evident when one rod is used to wield another (Figure 4c) and regressions are conducted such that different participants contribute the means for randomly selected rod configurations (reviewed in Carello and Turvey 2000).

Revealing the physical basis for definite scaling is confronted by the fact that the 0th, 1st, and 2nd moments over a set of experimental objects will necessarily exhibit some degree of covariation. To confront the consequent colinearity (a) stimulus objects must be inclusive of the hypothesized candidate moments and (b) multiple regressions must be used to reveal what matters (e.g., Kingma et al. 2004). Citing complications with the preceding strategy, Cabe (2010) has advanced a radically different experimental strategy: *eliminate access to all but the parameter of interest*. He demonstrated length perception in a situation (rolling cylindrical objects of different lengths and radii about their longitudinal axes) where the only variable was I_3. Cabe's strategy had been anticipated in respect to heaviness (Shockley et al. 2004, Experiment 1). The tensor objects depicted by Figure 3i are equal in mass. By situating the equally weighted crossbars at the same coordinates of the central rod they are rendered equal in static moment. They are unequal, however, in rotational inertia (their inertia tensors are different). With the objects wielded individually at their marked ends, the left variant of tensor object 3i is perceived as heavier than the right variant of tensor object 3i.

Shape Perception by Wielding

Figure 3e, f are solid objects wieldable by their handles. Experimenters provided participants with a visible array of objects and had them simply point to a match for the wielded object. Perception was well above chance and the confusions—cylinders with rectangular parallelepipeds, cones with pyramids—were predictable on the basis of I_1/I_3. The eigenvalue ratio for hemispheres was quite distinct and this shape was not confused with other shapes (reviewed in Turvey and Carello 1995).

An Object's Mass, an Object's Heaviness

What makes something held or moved feel heavy? Convention would say "its mass," "how much it weighs in kg," but investigations of dynamic (effortful) touch suggest it could be otherwise, that the answer is closer to moveable-ness or maneuverable-ness or controllable-ness.

Charpentier's size-weight illusion has been addressed in many ways. The illusion makes clear that one's perception of an object's heaviness does not refer to the object's weight. In the ecological approach to dynamic touch the so-called illusion is a point of entry into the haptic perception of what a hand-held object affords by way of neuromuscular control. Tensor objects of the kind shown in Figure 3i were introduced to study heaviness perception's dependence on the inertial eigenvalues (Figure 2d; reviewed in Turvey and Carello 1995). Experiments have varied the mass M of the tensor objects and, independently, two scalar variables, symmetry $S = 2I_3/(I_1 + I_2)$ and volume $V{=}4\pi/3(Det\, I_{ij})^{-1/2}$ of their inertia ellipsoids (reviewed in Turvey and Carello 2011). The latter are physical characterizations of an object's resistance to rotational acceleration taken in reference to the movement system. Arguably, they are the right degrees of freedom (the inertia tensor per se is not). They bear, respectively, on the patterning and level of muscular forces needed to move a handheld object in a controlled fashion. The experiment revealed additive effects of mass and S and V. Using single rods that could be systematically varied in 0th, 1st, and 2nd moments, Kingma et al. (2002) linked heaviness perception (on grounds of multiple regression analysis) to mass (0th) and static moment (1st) with no contribution from moments of inertia (2nd). In three experiments (one of which was noted above), Shockley et al. (2004) varied only the 2nd moment (in the S and V forms). The experiments revealed that variation of the 2nd moment is sufficient for heaviness perception and variation in the 0th and 1st moments is unnecessary.

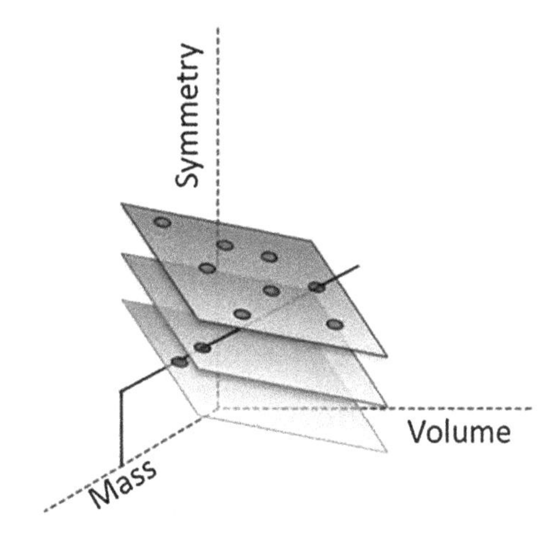

Figure 5 The line for a single mass intersects three different planes yielding three levels of perceived heaviness. All points on a given plane feel equally heavy despite being different masses

The perennial question of "What feels heavier, a pound of feathers or a pound of lead?" has been addressed in the foregoing terms (Wagman et al. 2007).

The observations of a change in perceived heaviness for a fixed mass have a complement in *weight metamers*: objects of different mass that have the same perceived heaviness (reviewed in Turvey and Carello 2011). As shown in Figure 5, combinations of mass, S and V yield metameric planes. Length perception by wielding a given object is different from heaviness perception (Amazeen 1997, 1999). It can be expected therefore that metamers for heaviness are not metamers for length, an expectation that was confirmed (Shockley et al. 2004, Experiment 5).

Orienting Objects and Limbs

Perception of a nonvisible object's orientation in the hand can be related to the orientation of the object's eigenvectors e_i (Figure 2c; reviewed in Turvey and Carello 1995). The perception of limb orientation might be similarly based. The arm's inertial quantities can be manipulated through a hand-held rod extending along the underside of the arm with a cross bar to which masses can be added (reviewed in Turvey and Carello 2011). Masses evenly distributed on the left and right keeps the ellipsoid aligned with the long axis of the arm. Uneven distribution of those masses diverts the ellipsoid from the arm's axis. In such a case, an individual asked to point at a visible target with the occluded arm points with e_i. Matching the positions of the left and right arms when splints are held in the two hands also result in matching the ellipsoids of the two limbs rather than matching the angles of the joints. The most recent experiments reveal, however, that these phenomena are based in the center of mass vector, V_{cm}, of the arm and not in its eigenvector, V_{ei}. The two cross bars in Figure 3j allow manipulations of added masses that disentangle V_{cm} and V_{ei} and show the former to be the constraining invariant in matching joint angles (van de Langenberg et al. 2007), pointing (van de Langenberg et al. 2008), and phase in interlimb rhythmic coordination (Silva and Turvey 2012). These outcomes challenge a general inertia-tensor based theory of perceiving limb orientation.

The Challenge of Selectivity

An object grasped other than at the end has, in effect, two extents, one on either side of the grasp. These two extents can be perceived reliably. If a mass is attached to the left of the grasp, that rod segment will feel longer than the rod segment on the right. If the configuration is flipped, the partial length perceptions of the segments will be flipped (reviewed in Carello and Turvey 2000; Turvey and Carello 1995). The working hypothesis from the inertia tensor perspective has been that distinct consequences for the mass distribution are found in the orientation of e_i. There is, however, just one inertia tensor for a given object. To distinguish reliably left from right requires something that entails a sign change of e_i. One conjecture is that the "spinor" theory of mechanical rotations, as a supplement to the inertia tensor, provides a solution within rotational dynamics. Whereas the tensor quantifies an object's resistance to being rotated, the spinor quantifies an object's orientation relative to some reference frame (e.g., the hand). The spinor entails two oppositely oriented orientations relative to the grasp. In this context, selective perception requires selection of one of the two distinctly oriented rotations (reviewed in Turvey 1996).

Selective perception is also at work when someone perceives different properties of the same object. An individual can successfully perceive by wielding an object's whole length, the partial length in front of the grasp, where the hand is on the object (reviewed in Turvey and Carello 1995), the object's orientation relative to the hand, and the object's heaviness. All of the perceived properties have been shown to scale to moments of the mass distribution. Experiments by van de Langenberg et al. (2006) suggest that we should expect the involvement of any given moment in these cases of selective perception to be conditional on its salience in the act of wielding.

Mass moments are affected by whole length, grasp location, object orientation, and mass. Perhaps perceiving x derives from perceiving y and z (e.g., perceiving partial length depends on perceiving whole length and grasp position). Formal procedures for evaluating perceptual independence require (a) that the properties of interest be obtained for each experimental object, (b) that the properties be manipulated categorically (short, middle, long length; near, middle, far grasp; light, medium, heavy mass), and (c) that the focal properties vary orthogonally (Amazeen 1999). For example, small, medium and heavy mass should each accompany short, middle, and long whole length. Building orthogonality into the experiment can reveal independence otherwise obscured by the shared underlying mass-distribution constraints.

Complementary analytic procedures evaluate the extent to which perception of property A depends on the value of property A, the value of property B, and perception of property B (e.g., the extent to which perception of partial length depends on actual partial length, where the grip really is, and perception of where the grip is). These procedures have revealed the following to be perceptually independent: whole length and partial length; partial length and grip position; whole length and grip position; heaviness and whole length (e.g., Cooper et al. 2000). These observations suggest that selective perception is reliable and constrained in a principled way by particular invariants manifest in the act of wielding.

Functional Properties: Perceiving Affordances

One can look at rotational inertia from the perspective of its relevance for controlling movement (reviewed in Carello and Wagman 2009). With $S = 2I_3/(I_1 + I_2) \approx 1$, a hand held object affords, in Gibson's (1986) sense, being moved as easily in one direction as in any other. Experiments on weight perception can be conducted with the alternate question of how movable is a given object— tapping into distinctions of how moving is to be accomplished (e.g., pushed or wielded, using one hand or two). Objects rated with respect to their affordance for striking with power or aiming with precision differ in their particular inertial configurations. They can be tailored to these tasks by allowing an individual to vary grasp position, in effect, changing a "hammer" into a "poker." Moreover, distinct relations among the mass moments functions can produce metamers for hammer-with-ability that are different from metamers for poke-with-ability (Wagman and Shockley 2011).

Independence of Local Anatomy

The hand can be considered an expert wielder. It is how objects are usually grasped and maneuvered. The logic of tissue deformation, however, does not convey a special advantage on any part of the body. An object attached to the body will deform tissues in a way that reflects the object's mass distribution. Rod length perception is invariant over the four conditions of Figure 4b. Perception of lengths of rods in Figure 4d, whole and partial (e.g., that leftward of foot), is comparable to such perception for handheld wielded rods (e.g., Hajnal et al. 2007). It is similarly the case for postural wielding depicted in Figure 4a (Palatinus et al. 2011).

Mechanoreceptor changes with age have consequences for spatial acuity and vibration sensitivity. The ability to perceive properties by dynamic touch (specifically, length and sweet spot), however, does not differ dramatically between the old and young (reviewed in Carello and Wagman 2009). Their judgments are single-valued functions of the same invariants and their definite scaling is in the same range. Moreover, old-young plots (regressions of length judgments by the old on length judgments by the young) are linear. The field of geriatric research takes linear old-young plots as an indication that the process underlying an achievement is the same. In this case, such an interpretation means that differences in sensory machinery between old and young do not enforce differences in how they perceive properties by dynamic touch.

Case studies of certain clinical conditions that bring about sensory neuropathy—a loss of feeling, typically in the extremities—also show that dynamic touch is preserved. A participant with a complete loss of sensitivity in one arm due to lesions on one side of the dorsal column system successfully perceived length by dynamic touch with that arm (Carello et al. 2009). On a stereognosis test with that arm (identifying three-dimensional objects placed in the hand), the participant's score was 0.

Dynamically Touching Things at a Distance

The centrality of dynamic touch to everyday activity is obscured by the fact that it is usually an aspect of a coordinated enterprise among perceptual systems. Exploratory or goal-directed contacts with an occluded surface by means of an occluded handheld probe provide a minimal example (reviewed in Turvey and Carello 1995). Investigations of probing a horizontal surface at variable depth used a probe (Figure 6a) that varied in moment of inertia, center of percussion and angle of inclination to the surface. Perceived depth was one single-valued function of the variables, perceived probe length was another. Perception of a vertical surface's distance and perception of a handheld probe's length are summarized in Figure 6b. Extending tangible affordances, probing with a rod (without benefit of vision or sound) reveals whether an inclined surface affords upright standing (Figure 6c) and whether a gap in a surface affords crossing (Figure 6d).

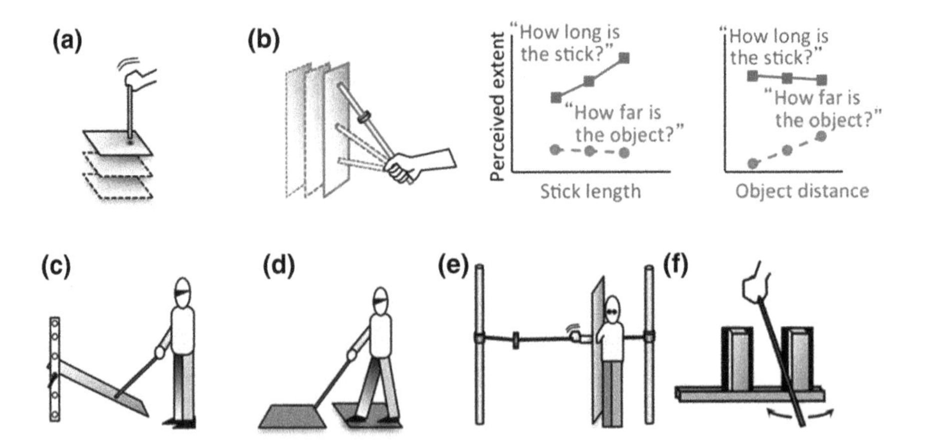

Figure 6 Experimental investigations of dynamic touch at a distance

A canonical case of dynamical touching at a distance is that of a spider in its web. A minimal simulation is possible. It can be shown that for an object on a single taut strand (Figure 6e) its distance from the participant's hand is perceptible whether this minimal haptic web is put into vibration by the participant or by the object (via the experimenter). A theoretically special case is dynamical touching of an aperture at a distance by means of a probe with the intent to perceive the aperture's size (Figure 6f). A single scalar quantity connects the muscle forces imposed upon the probe to the reactive forces impressed upon the tissues of the body. Perception proves to be a single-valued function of this quantity but it is not of "aperture size." It seems to be of "size-at-a-distance-contacted-with-a-particular-implement"!

The Medium of Dynamic (Effortful) Touch

Air and water, the media for the other perceptual systems, share the features of being homogeneous (physical properties are place invariant) and isotropic (physical properties are direction invariant) making them appropriate to being reliably structured by events that tie the perceiver to the environment. The body interpreted as a *tensegrity architecture*—continuous tension elements and intermittent compression elements at all of the body's scales—promises the requisite homogeneity and isotropy for the medium of haptic perception. Turvey and Fonseca (2014) suggest that this tension array is information about, in the sense of specificity to, the layout of the body, its transformations, and its attachments. By hypothesis, the haptic medium grounds the phenomena of dynamic (effortful) touch.

Learning and Expertise

Even though the dominant hand is more expert at wielding objects than the non-dominant hand, and the hand is more expert than the foot or torso, these differences do not alter the invariants these effectors extract. The role of specific expertise seems task-dependent. While experts and novices do not differ in inertial constraints on perceived length or sweet spot of tennis rackets or in rating whether weighted hockey sticks are suitable for intercepting a puck, they do differ in preferences for sticks suitable for hitting (Hove et al. 2006). However, novices' preferences become "more expert" after minimal opportunities to use hockey sticks to hit pucks and intercept projectiles.

In addition to attunement, learning involves changes in calibration to environmental properties (reviewed in Wagman and Chemero 2014). Although initial responses are already definitely scaled, responses are recalibrated after training: The slope of the perceived length/actual length regression approaches 1. Training may be visual or auditory. A magnitude estimation of length might be followed by the experimenter's indicating actual length with a visible marker or with sounds from the object dropping onto a surface. Haptics has also been shown to "train itself" through changes in grasp position, a recalibration that transfers across properties (e.g., calibrating perceived length incidentally calibrates perceived partial length).

A theoretical context for learning in dynamic touch settings conjectures an "information space" as a navigable low-dimensional manifold (e.g., Jacobs et al. 2009; Michaels and Isenhower 2011a, b), built from moments of the mass distribution that might be relevant to a property. Learning entails moving through the information space toward the optimal locus for that task as the perceiver becomes attuned to the invariants specific to their intentions (e.g., from I_1 towards I_1/I_3 for perceived length). But in order for attunement to happen, tissue must be deformed reliably by distinct properties in the first place, allowing learning to be an information-guided process rather than a trial-and-error search (Michaels et al. 2008).

Controversies in Perception by Dynamic Touch

It is noncontroversial that dynamic touch allows awareness of a wide variety of extero-, proprio-, exproprio- and proextero-specific properties (Figure 1). But controversies arise with respect to:

- *how* those properties are perceived
- *what* kind of variable matters (higher-order invariant or collection of cues)
- *how* a variable that matters is used (detection or cue weighting and inference)
- *whether* scaling is just relative or definite (If scaling is definite, is it just in the approximate ballpark or is it consistent as well? Can manipulations make scaling more absolute?)

The preceding issues are addressed within two primary perspectives on dynamical (effortful) touch, the Cognitive and the Ecological. Both are challenged by the phenomena, but in different ways.

The cognitive perspective is commonplace. It is typically in the tradition of Helmholtz. It assumes abductive inference: the making of inferences from effects to cause. Assumed but unexplained is knowledge of the causes. It is quite likely that, as computational problems, those posed by dynamical (effortful) touch are insoluble. They are NP Hard problems, meaning they are noncomputable (N) in polynomial (P) time. Instead, computation time increases exponentially with the size of the input (e.g., time-varying tissue deformation values).

The ecological perspective is not commonplace. It is in the newer tradition of Gibson. It assumes lawfulness, namely, specificity of the deformation of the body's tissues to dynamically touched objects and specificity of the concordant perception to the dynamically touched objects contributing causally to that deformation. Assumed but undefined is the nature of the lawfulness. The physical and mathematical problems posed by dynamical (effortful) touch are intractable in the dictionary sense of not easily solved. They cross multiple length and time scales, are nonlinear, and entail mixtures of thermodynamic and mechanical principles.

References

Amazeen, E (1997). The effects of volume on perceived heaviness by dynamic touch: With and without vision. *Ecological Psychology* 9: 245–263.

Amazeen, E (1999). Perceptual independence of size and weight by dynamic touch. *Journal of Experimental Psychology: Human Perception & Performance* 25: 102–119.

Cabe, P A (2010). Sufficiency of longitudinal moment of inertia for haptic cylinder length judgments. *Journal of Experimental Psychology: Human Perception & Performance*. 36: 373–394.

Carello, C and Turvey, M T (2000). Rotational invariants and dynamic touch. In: M Heller (Ed.), *Touch, Representation and Blindness* (pp. 27–66). Oxford: Oxford University Press.

Carello, C and Wagman, J B (2009). Mutuality in the perception of affordances and the control of movement. In: D Sternad (Ed.), *Progress in Motor Control* (pp. 273–292). New York: Springer Verlag.

Carello, C; Silva, P L; Kinsella-Shaw, J M and Turvey, M T (2009). Sensory and motor challenges to muscle-based perception. *Brazilian Journal of Physical Therapy* 12: 339–350.

Cooper, M; Carello, C and Turvey, M T (2000). Perceptual independence of whole length, partial length, and hand position in wielding a rod. *Journal of Experimental Psychology: Human Perception and Performance* 26: 74–85.

Gibson, J J (1966). *The Senses Considered as Perceptual Systems*. Boston: Houghton Mifflin.

Gibson, J J (1986). *The Ecological Approach to Visual Perception*. Hillsdale, NJ: Lawrence Erlbaum Associates, Inc. (Original work published in 1979.).

Hajnal, A; Fonseca, S T; Harrison, S; Kinsella-Shaw, J M and Carello, C (2007). Comparison of dynamic (effortful) touch by hand and foot. *Journal of Motor Behavior* 39: 82–88.

Hove, P; Riley, M A and Shockley, K (2006). Perceiving affordances of hockey sticks by dynamic touch. *Ecological Psychology* 18: 163–189.

Jacobs, D M; Silva, P L and Calvo, J (2009). An empirical formalization of the theory of direct learning: The muscle-based perception of kinetic properties. *Ecological Psychology* 21: 245–289.

Kingma, I; Beek, P J and van Dieën, J H (2002). The inertia tensor versus static moment and mass in perceiving length and heaviness of hand-wielded rods. *Journal of Experimental Psychology: Human Perception and Performance* 28: 180–191.

Kingma, I; van de Langenberg, R and Beek, P J (2004). Which mechanical invariants are associated with the perception of length and heaviness of a nonvisible handheld rod? Testing the inertia tensor hypothesis. *Journal of Experimental Psychology: Human Perception and Performance* 30: 346–354.

Michaels, C F and Isenhower, R W (2011a). An information space for partial length perception in dynamic touch. *Ecological Psychology* 23: 37–57.

Michaels, C F and Isenhower, R W (2011b). Information space is action space: Perceiving the partial lengths of rods rotated on an axle. *Attention, Perception, & Psychophysics* 73: 160–171.

Michaels, C F; Arzamarski, R; Isenhower, R and Jacobs, D M (2008). Direct learning in dynamic touch. *Journal of Experimental Psychology: Human Perception and Performance* 34: 944–957.

Pagano, C C and Cabe, P A (2003). Constancy in dynamic touch: Length perceived by dynamic touch is invariant over changes in media. *Ecological Psychology* 15: 1–17.

Palatinus, Z; Carello, C and Turvey, M T (2011). Principles of selective perception by dynamic touch extend to the body. *Journal of Motor Behavior* 43: 87–93.

Palatinus, Z; Kelty-Stephen, D; Kinsella-Shaw, J; Carello, C and Turvey, M T (2014). Haptic intent in quiet standing affects multifractal scaling of postural fluctuations. *Journal of Experimental Psychology: Human Perception and Performance* 40: 1808–1818.

Shockley, K; Carello, C and Turvey, M T (2004). Metamers in the haptic perception of heaviness and moveable–ness. *Perception & Psychophysics* 66: 731–742.

Silva, P and Turvey, M T (2012). The role of haptic information in shaping coordination dynamics: Inertial frame of reference hypothesis. *Human Movement Science* 31: 1014–1036.

Turvey, M T (1996). Dynamic touch. *American Psychologist* 51: 1134–1152.

Turvey, M T and Carello, C (1995). Dynamic touch. In: W Epstein and S Rogers (Eds.), *Handbook of Perception and Cognition, Vol. V. Perception of Space and Motion* (pp. 401–490). San Diego: Academic Press.

Turvey, M T and Carello, C (2011). Obtaining information by dynamic (effortful) touching. *Philosophical Transactions of the Royal Society B: Biological Sciences* 366: 3123–3132.

Turvey, M T and Fonseca, S T (2014). The medium of haptic perception: A tensegrity hypothesis. *Journal of Motor Behavior* 46: 143–187.

van de Langenberg, R; Kingma, I and Beek, P J (2006). Mechanical invariants are implicated in dynamic touch as a function of their salience in the stimulus flow. *Journal of Experimental Psychology: Human Perception and Performance* 32: 1093–1106.

van de Langenberg, R; Kingma, I and Beek, P J (2007). Perception of limb orientation in the vertical plane depends on center of mass rather than inertial eigenvectors. *Experimental Brain Research* 180: 595–607.

van de Langenberg, R; Kingma, I and Beek, P J (2008). The perception of limb orientation depends on the center of mass. *Journal of Experimental Psychology: Human Perception and Performance* 34: 624–639.

Wagman, J B and Chemero, A (2014). The end of the debate over extended cognition. In: T Solymosi and J Shook (Eds.), *Neuroscience, Neurophilosophy, and Pragmatism: Brains at Work in the World. New Directions in Philosophy and Cognitive Science* (pp. 105–124). Hampshire, UK: Palgrave Macmillen.

Wagman, J B and Shockley, K (2011). Metamers for hammer-with-ability are not metamers for poke-with-ability. *Ecological Psychology* 23: 76–92.

Wagman, J B; Zimmerman, C and Sorric, C (2007). Which feels heavier—a pound of lead or a pound of feathers? A potential perceptual basis of a cognitive riddle. *Perception* 36: 1709–1711.

Suggested Readings

Summaries of much of the cited research and theorizing are to be found in the following articles.

Carello, C and Turvey, M T (2000). Rotational dynamics and dynamic touch. In: M Heller (Ed.), *Touch, Representation and Blindness* (pp. 27–66). Oxford: Oxford University Press.

Carello, C and Turvey, M T (2004). Physics and psychology of the muscle sense. *Current Directions in Psychological Science* 13,: 25–28.

Carello, C and Wagman, J B (2009). Mutuality in the perception of affordances and the control of movement. In: D Sternad (Ed.), *Progress in Motor Control: A Multidisciplinary Perspective* (pp. 273–292). New York: Springer Verlag.

Carello, C; Silva, P L; Kinsella-Shaw, J M and Turvey, M T (2009). Sensory and motor challenges to muscle-based perception. *Brazilian Journal of Physical Therapy* 12: 339–350.

Turvey, M T and Carello, C (1995). Dynamic touch. In: W Epstein and S Rogers (Eds.), *Handbook of Perception and Cognition, Vol. V. Perception of Space and Motion* (pp. 401–490). San Diego: Academic Press.

Turvey, M T (1996). Dynamic touch. *American Psychologist* 51: 1134–1152.

Turvey, M T and Carello, C (2011). Obtaining information by dynamic (effortful) touching. *Philosophical Transactions of the Royal Society B: Biological Sciences* 366: 3123–3132.

Turvey, M T and Fonseca, S T (2014). The medium of haptic perception: A tensegrity hypothesis. *Journal of Motor Behavior* 46: 143–187.

Observed Touch

Jamie Ward

Observed touch refers simply to the process of watching touch to an inanimate or animate object. In many situations, this is a purely visual event. For example, when watching inanimate objects or other people being touched. However, just because the stimulus is visual in nature does not mean that the brain processes it in purely visual terms. There is now convincing evidence that observed touch, at least in humans, involves cognitive and neural processes that are traditionally considered part of the somatosensory system. In other situations, observed touch may consist of both a visual and a tactile event. For example, when observing our own body being touched or when observing touch to somebody/something else whilst our own body is being touched. Studies on observed touch can therefore provide important insights into how these two senses (vision and touch) interact with each other.

Observed Touch Influences Felt Touch

Observing touch can enhance the ability to detect real tactile stimuli. For example, watching a movie of a right hand being stroked increases subsequent tactile sensitivity in the participant's own (unseen) right hand but not his/her left hand (Schaefer et al. 2005). Similarly watching a mirror reflection of one's own right hand being stroked (i.e. so that it looks like the left hand is stroked) can increase subsequent tactile sensitivity in the participant's left hand, although merely looking at the left hand (without the mirror reflection of the touched right hand) does not (Ro et al. 2004). In some cases, merely observing a body part, without observing the touch itself, can increase tactile acuity (Kennett et al. 2001). This effect can be enhanced by making the body part seem larger (using a magnifying glass) and reduced by placing an inanimate object in the same apparent location.

The studies mentioned above involve observing touch (or observing the body part to be touched) followed later in time by unseen tactile stimuli. An alternative approach is to observe touch at exactly the same time as real touch is applied to see

J. Ward (✉)
University of Sussex, Sussex, UK

T.J. Prescott et al. (eds.), *Scholarpedia of Touch*, Scholarpedia,
DOI 10.2991/978-94-6239-133-8_19

if the two interact. In such experiments, it is important that the tactile stimulator cannot be seen otherwise it could provide trivial information about when, how and where the participant was touched. One can circumvent this problem by touching the participants hands or face (or other body part) whilst they observe touch to a different person, dummy body parts, or images of bodies or objects on a computer screen (Maravita et al. 2002; Pavani et al. 2000; Spence et al. 2004; Tipper et al. 1998, 2001). For example, one study applied weak tactile stimuli to the participant's left or right cheek whilst they observed touch either to an image of themselves, an image of another person, or an image of a house (Serino et al. 2008). Their task was to report the location of the felt touch (left, right or both) whilst ignoring the location of the observed touch (also left, right, or both). Accuracy of detection was enhanced when the observed touch matched the felt touch (e.g. observed touch to both cheeks and felt touch to both cheeks) and was greatest for the self-face relative to other-face, and for other-face relative to the house images. This suggests that the interplay between observed touch and felt touch depends on self-other similarity. This extends beyond physical similarity to also include similarity in political attitudes (Serino et al. 2009). Thus, the system linking observed touch with felt touch is sensitive to top-down biases.

In some cases, when observed touch is incongruent with felt touch it can give rise to bizarre illusions. Perhaps the most commonly studied illusion involving observed touch is the so-called 'rubber hand illusion' (Figure 1) (Botvinick and Cohen 1998). In this illusion, the participants own hand is hidden from view whilst a rubber hand is placed on the table near their real hand. The experimenter then strokes both the rubber hand and the real hand in the same place at the same time. As such, the participant observes touch to the rubber hand but feels touch in their own hand. Over time, participants report the following statements to be true: "I felt as if the rubber hand were my hand", and "It seemed as though the touch I felt was caused by the paintbrush touching the rubber hand" (Botvinick and Cohen 1998). A more objective correlate of the illusion is to ask participants to report the felt location of the finger that is being touched by placing a ruler on the seen surface where the rubber hand lies. Participants in the illusion are less accurate at doing this (the position of the felt finger gravitates towards the rubber hand). The presence of the illusion, measured both subjectively and objectively, depends on a number of factors. It disappears when the observed and actual stroking are out of synchrony; if the observed hand is the 'wrong' hand, for example a left hand instead of a right hand; if the hand is replaced by a stick; and if the observed hand is in the 'wrong' orientation, for example at 90 degrees to the real hand (Tsakiris and Haggard 2005). The explanation for this illusion is that the sense of proprioception is fooled. Our proprioceptive sense conveys information about the current position of the muscles and joints. It provides information about where the hand is in space, e.g. relative to the body, at a given location in time (note: the sense of touch carries the information 'my finger is being stroked' but not where the finger is located in space). In the rubber hand illusion, there is a spatial conflict between information from proprioception and vision but the discrepancy is resolved in vision's favour (so-called visual capture) because vision tends to be spatially more accurate. This illusion has recently been

Figure 1 The rubber hand illusion (from Ward 2008, *The Frog Who Croaked Blue: Synesthesia and the Mixing of the Senses*)

extended using virtual reality to create something akin to an out-of-body experience (Ehrsson 2007; Lenggenhager et al. 2007).

Observed Touch Activates the Somatosensory System

Several functional imaging studies, using fMRI, have measured brain responses to watching humans being touched. These studies show activation of the primary and/or secondary somatosensory cortex either when touch is compared to no touch or when touch to a human is compared to touch to an object (Keysers et al. 2010). As such a purely visual stimulus can, under some situations, reliably activate the somatosensory system. These same regions respond when the real touch to the appropriate body part is felt, which suggests that there is a mirror system for touch (i.e. a system that responds both to our own touch and that of other people and possibly objects). This is perhaps remarkable given the general belief that our tactile sensory experiences are private rather than shared.

The question of why somatosensory activity when observing touch is not accompanied by a conscious experience of touch is interesting. It may reflect sub-threshold activity and/or the absence of activity in other parts of the brain outside of the primary and secondary somatosensory cortex. Nonetheless, this modulation of somatosensory processing by observed touch may be sufficient to alter behaviour (e.g. improve acuity) or lead to certain illusions. There is evidence that activity in the somatosensory cortex is reduced during the rubber hand illusion presumably as a result of the mismatch between the location of real touch and the illusion that one is feeling touch to the rubber hand (Tsakiris et al. 2007).

It is unlikely that these behavioural and neurophysiological effects are mediated by direct connections between early visual areas and primary/secondary somatosensory regions. The effect of observed touch depends on far more than vision per se. It depends, to some degree, on whether the touch is to humans or objects, and the perspective and location of the observed body part relative to the observer's own body. There are neurons within the primate parietal lobe and frontal lobe (premotor region) that respond to both touch and vision (Graziano et al. 2000; Graziano 1999). Observed touch may affect the somatosensory regions via feedback from these multi-sensory regions. The visual and tactile receptive fields of these neurons tend to be aligned such that the visual receptive field will respond to visual events on or near the arm, irrespective of where the arm is. In some situations the visual receptive fields can extend beyond the body. For example, when using tools (Iriki et al. 1996) or watching one's own body projected onto a computer screen (Iriki et al. 2001). There is a class of neurons termed 'body matching neurons' discovered in the macaque parietal lobe that responds to both felt touch and observed touch on the same body part of another person (Ishida et al. 2009).

Observed Touch in Atypical Populations

Amputees

After loss of a limb, the majority of patients experience (at some point in time) a vivid sense that the missing limb is still present. It may itch, gesticulate, or be painful. Several studies report that these phantom limb experiences can be modified by visual feedback, for example, by placing a mirror down the midline so that the intact limb is seen to be reflected into the space where the phantom is felt (Ramachandran and Rogers-Ramachandran 1996; Ramachandran et al. 1995). Touching the normal hand whilst watching its mirror reflection in the space where the phantom is felt can create a sensation of touch in the phantom hand as well as the real hand (Ramachandran et al. 1995). It has subsequently been found that some amputees experience touch when observing touch, particularly painful touch, on another person (Goller et al. 2013).

Brain-Damaged Patients

Some brain damaged patients with impaired ability to detect touch (whilst blindfolded) nevertheless report tactile sensations when observing the touch applied to their arm or a live/recorded movie of touch to their arm (Halligan et al. 1996, 1997). Vision of an appropriate body part (e.g. watching an arm whilst an arm is touched) can enhance tactile acuity in brain damaged patients with poor somatosensation even if the act of touching is blacked out (Serino et al. 2007). The same is not found if an inanimate object or the wrong body part (e.g. a foot) is observed.

Mirror-Touch Synaesthesia

Synaesthesia (also spelled synesthesia) is the involuntary experience of a perceptual quality that is triggered by a stimulus that does not normally evoke such a quality (for examples, letters evoking colours, sounds evoking tastes). Observed touch can sometimes trigger involuntary experiences of felt touch and this constitutes a form of synaesthesia. This can be acquired after amputation or brain damage (Goller et al. 2013; Halligan et al. 1996), or it could be developmental in origin with no known external cause (Banissy and Ward 2007; Holle et al. 2013). In an experimental study involving congruent or incongruent observed and felt touch, synaesthetic touch (induced by observed touch) tended to be confused with real touch (Banissy and Ward 2007). There were other intriguing aspects of this study. There appears to be two spatial mechanisms for linking observed touch with ones own body. In most synaesthetes, observed touch on the left of someone's face triggers a tactile experience on the synaesthete's right cheek (i.e. like a reflection) but in others it is felt on their left cheek (i.e. like a rotation). A functional imaging study using fMRI of developmental mirror-touch synaesthesia showed hyper-activity within certain regions of the same network that responds when non-synaesthetic people observe touch (Holle et al. 2013). This included the primary and secondary somatosensory cortex. The same study found, using structural MRI, increased grey matter density in a region of secondary somatosensory cortex (amongst other regions). Activity beyond some threshold may be linked to conscious experiences of touch (in synaesthetes) but below the threshold may have unconscious behavioural correlates (in controls) such as those documented previously. This was assumed to reflect a difficulty in a self-other gating mechanism that enables ones own body to be processed as separate from other bodies (Figure 2).

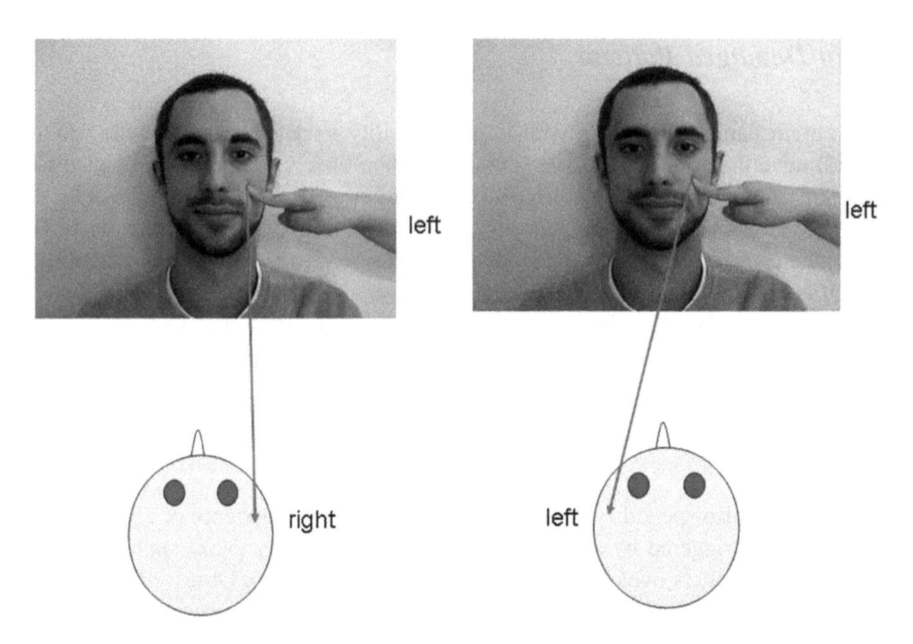

Figure 2 In this example, the stimulus consists of observed touch to the left cheek. Some synaesthetes experience a touch on their right cheek (as if the other person is a mirror-reflection of themselves) whereas other synaesthetes experience touch on their left cheek (as if they are rotated into the other persons perspective)

Internal References

Alonso, J-M and Chen, Y (2009). Receptive field. *Scholarpedia* 4(1): 5393. http://www.scholarpedia.org/article/Receptive_field.

Braitenberg, V (2007). Brain. *Scholarpedia* 2(11): 2918. http://www.scholarpedia.org/article/Brain.

Llinas, R (2008). Neuron. *Scholarpedia* 3(8): 1490. http://www.scholarpedia.org/article/Neuron.

Ogawa, S and Sung, Y-W (2007). Functional magnetic resonance imaging. *Scholarpedia* 2(10): 3105. http://www.scholarpedia.org/article/Functional_magnetic_resonance_imaging.

Penny, W D and Friston, K J (2007). Functional imaging. *Scholarpedia* 2(5): 1478. http://www.scholarpedia.org/article/Functional_imaging.

Pikovsky, A and Rosenblum, M (2007). Synchronization. *Scholarpedia* 2(12): 1459. http://www.scholarpedia.org/article/Synchronization.

Ramachandran, V S and Brang, D (2008). Synesthesia. *Scholarpedia* 3(6): 3981. http://www.scholarpedia.org/article/Synaesthesia.

Rizzolatti, G and Fabbri-Destro, M (2008). Mirror neurons. *Scholarpedia* 3(1): 2055. http://www.scholarpedia.org/article/Mirror_neurons.

External References

Banissy, M and Ward, J (2007). Mirror touch synaesthesia is linked with empathy. *Nature Neuroscience* 10: 815–816.
Botvinick, M and Cohen, J (1998). Rubber hands 'feel' touch that eyes see. *Nature* 391: 756.
Ehrsson, H H (2007). The experimental induction of out-of-body experiences. *Science* 317: 1048.
Goller, A I et al. (2013). Mirror-touch synaesthesia in the phantom limb of amputees. *Cortex* 49 (1): 243–251.
Graziano, M S; Cooke, D F and Taylor, C S R (2000). Coding the location of the arm by sight. *Science* 290: 1782–1786.
Graziano, M S A (1999). Where is my arm? The relative role of vision and proprioception in the neuronal representation of limb position. *Proceedings of the National Academy of Sciences of the United States of America* 96: 10418–10421.
Halligan, P W et al. (1996). When seeing is feeling: Acquired synaesthesia or phantom touch? *Neurocase* 2: 21–29.
Halligan, P W et al. (1997). Somatosensory assessment: Can seeing produce feeling? *Journal of Neurology* 244: 199–203.
Holle, H et al. (2013). Functional and structural brain correlates of mirror-touch synaesthesia. *NeuroImage* 83: 1041–1050.
Iriki, A et al. (2001). Self-images in the video monitor coded by monkey intraparietal neurons. *Neuroscience Research* 40: 163–173.
Iriki, A; Tanaka, M and Iwamura, Y (1996). Coding of modified body schema during tool use by macaque postcentral neurons. *NeuroReport* 7: 2325–2330.
Ishida, H et al. (2009). Shared Mapping of Own and Others' Bodies in Visuotactile Bimodal Area of Monkey Parietal Cortex. *Journal of Cognitive Neuroscience* 22(1): 83–96.
Kennett, S; Taylor-Clarke, M and Haggard, P (2001). Noninformative vision improves the spatial resolution of touch in humans. *Current Biology* 11: 1188–1191.
Keysers, C et al. (2010). Somatosensation in social perception. *Nature Reviews Neuroscience* 11: 417–428.
Lenggenhager, B et al. (2007). Video ergo sum: Manipulating bodily self-consciousness. *Science* 317: 1096–1099.
Maravita, A et al. (2002). Seeing your own touched hands in a mirror modulates cross-modal interactions. *Psychological Science* 13: 350–355.
Pavani, F; Spence, C and Driver, J (2000). Visual capture of touch: Out-of-the-body experiences with rubber gloves. *Psychological Science* 11: 353–359.
Ramachandran, V S and Rogers-Ramachandran, D (1996). Synaesthesia in phantom limbs induced with mirrors. *Proceedings of the Royal Society of London B* 263: 377–386.
Ramachandran, V S; Rogers-Ramachandran, D and Cobb, S (1995). Touching the phantom limb. *Nature* 377: 489–490.
Ro, T et al. (2004). Visual enhancing of tactile perception in the posterior parietal cortex. *Journal of Cognitive Neuroscience* 16: 24–30.
Schaefer, M; Heinze, H-J and Rotte, M (2005). Viewing touch improves tactile sensory threshold. *NeuroReport* 16: 367–370.
Serino, A et al. (2007). Can vision of the body ameliorate impaired somatosensory function? *Neuropsychologia* 45: 1101–1107.
Serino, A et al. (2009). I Feel what You Feel if You Are Similar to Me. *PLoS One* 4(3): 7.
Serino, A; Pizzoferrato, F and Ladavas, E (2008). Viewing a face (especially one's own face) being touched enhances tactile perception on the face. *Psychological Science* 19: 434–438.
Spence, C; Pavani, F and Driver, J (2004). Spatial constraints on visual-tactile cross-modal distractor congruency effects. *Cognitive, Affective and Behavioral Neuroscience* 4: 148–169.
Tipper, S P et al. (1998). Vision influences tactile perception without proprioceptive orienting. *NeuroReport* 9: 1741–1744.

Tipper, S P et al. (2001). Vision influences tactile perception at body sites that cannot be viewed directly. *Experimental Brain Research* 139: 160–167.

Tsakiris, M and Haggard, P (2005). The Rubber Hand Illusion revisited: Visuotactile integration and self-attribution. *Journal of Experimental Psychology: Human Perception and Performance* 31: 80–91.

Tsakiris, M et al. (2007). Neural signatures of body ownership: A sensory network for bodily self-consciousness. *Cerebral Cortex* 17: 2235–2244.

Ward, J (2008). *The Frog who croaked blue: Synesthesia and the mixing of the senses*. London: Routledge.

Painful Touch

Stuart Derbyshire

Defining pain is not as straightforward as many might assume. Intuitively, it seems reasonable to define pain as the response to tissue damage or disease. Simply put, it hurts when we hit our hand with a hammer and it hurts because we cause damage to the hand. Unfortunately that definition is problematic for at least two reasons. First, it tends towards tautology and circularity because it reduces to the statement that pain is caused by painful stimuli. In other words, pain is caused by pain. Second, it encourages a focus on a stimulus, which does not feel pain, rather than on a person, who does feel pain. Pain cannot be defined by stimuli, pain must be defined by the content of painful experience. Most pain researchers adopt a definition of pain that emphasises the sensory, cognitive and affective content of pain experience that typically follows a noxious stimulus. This understanding of pain is supported by the International Association for the Study of Pain (IASP) who define pain as "an unpleasant sensory and emotional experience associated with actual or potential tissue damage, or described in terms of such damage… pain is always subjective. Each individual learns the application of the word through experiences related to injury in early life" (Merskey 1991). By this definition pain is not part of an objective stimulus but is a part of subjective experience.

Much pain research has been motivated by efforts to combine the objective and subjective or to understand subjective experience in terms of objective responses to stimulation. Many of the non-painful tactile perceptions we feel, for example, originate from the activation of mechanoreceptors that have a distinct structure. The Pacinian corpuscle, one major kind of tactile sensory receptor for example, is shaped like a bean and layered like an onion. This structure enables the Pacinian corpuscle to detect changes in pressure and the fibres Pacinian corpuscles give rise to rapidly adapt to facilitate responses to changing pressure. **Painful touch** typically involves a mechanical stimulus that exceeds a noxious threshold or includes a noxious component such as excessive heat. Unlike the mechanoreceptors, nerve endings that transmit noxious information are naked and lie free in the skin. Thus they are known as free nerve endings and they arise mostly from the peripheral termination of Aδ (pronounced: A-delta) and C fibres. The free nerve endings are

S. Derbyshire (✉)
National University of Singapore, Singapore, Singapore

T.J. Prescott et al. (eds.), *Scholarpedia of Touch*, Scholarpedia,
DOI 10.2991/978-94-6239-133-8_20

polymodal and can respond to non-noxious and noxious temperatures or mechanical stimuli.

When activity in Aδ and C fibres gives rise to pain or behaviour associated with pain then they are labelled as nociceptors. Fibres that only respond in the noxious range are labelled as nociceptive-specific while those that respond across the noxious and non-noxious range are labelled as wide dynamic range. There is considerable controversy as to whether nociceptive pathways code a single modality of pain or even if nociceptive pathways can be specifically associated with pain (Craig et al. 1995; Han et al. 1998; Wall 1995).

Pain Pathways

The primary afferent Aδ and C fibres terminate on neurons in the superficial dorsal horn of the spinal cord. Ascending projections to the thalamus originate from the most superficial layer, known as lamina I, and project contralaterally in the spinothalamic tract (STT). Intracellular recordings from lamina I neurons revealed neurons with seemingly modality-specific responses [add reference]. One class of neurons was nociceptive specific, responsive only to noxious pinch, heat or both. Another class was thermoreceptive-specific, responding only to non-noxious cooling. A final class was polymodal, responding to heat, pinch and cooling (HPC). Provocatively, the existence of lamina I neurons, with specific responses and distinct morphology, motivated the suggestion that there are dedicated pathways for pain and temperature detection (Han et al. 1998).

In a particularly elegant study, Craig and colleagues used the 'thermal grill illusion' to illustrate the potential influence of these pathways (Craig and Bushnell 1994). The thermal grill illusion is generated by interleaved bars of warm and cool placed against the skin. The subject should experience interleaved warm and cool consistent with the stimulus actually delivered, but typically the subject will report a burning cold that is painful.

Craig explains this illusion as the consequence of integration across multiply activated pathways. The cool bars activate both thermoreceptive cells, which respond to non-noxious cooling, and polymodal cells, which respond to heat, pinch and cooling. Under conditions of normal, non-noxious cooling, activity in the thermoreceptive pathway dominates and the HPC pathway is inhibited. The presence of the warm bars, however, complicates the situation. The warm bars do not affect activity in the HPC pathway but do inhibit the cool pathway. Consequently, the interleaving of warm and cool bars causes relative excitation in the HPC pathway and generates a painful burning cold. When Craig and colleagues used the thermal grill in conjunction with brain imaging they generated an experience of painful burning cold and demonstrated activity in the anterior cingulate, insula and secondary somatosensory cortex (Craig and Bushnell 1994). The proposed mechanism is illustrated in Figure 1.

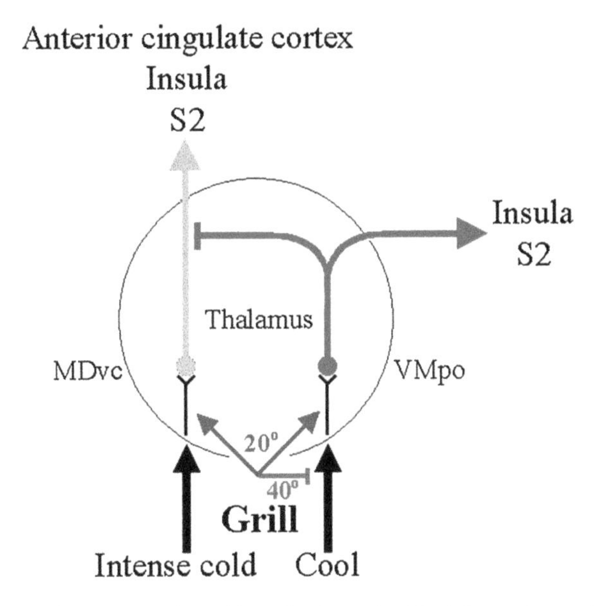

Figure 1 A schematic representation of how the thermal grill might produce an illusion of pain and illustrates the pathways described by Craig et al. Cool stimuli excite the spinothalamic pathway through the VMpo (*purple*), which subsequently inhibits the pathway through the MDvc (*turquoise*). In the thermal grill experiment both pathways are excited by the 20 °C stimulation (*blue*). The warm, 40 °C, bars (*red*), however, also inhibit the VMpo, but not the MDvc pathway, resulting in less excitation entering the inhibitory pathway. The net output is through the MDvc pathway to the anterior cingulate cortex resulting in the experience of a burning pain. Developed from Derbyshire (2002)

The proposal of dedicated pathways for specific experiences, including pain, implies that there ought to be a one-to-one relationship between specific noxious stimuli and pain experience. This proposal is controversial because pain occurs quite frequently in the absence of specific noxious stimuli and pain experience is easily influenced by activation in other sensory modalities, especially touch.

Pain Without Noxious Stimulation

A significant number of patients have chronic unremitting pain in the absence of any identifiable injury or diagnostic marker. Such patients include those suffering from low back pain, fibromyalgia, irritable bowel syndrome, atypical facial pain and a number of other disorders that can be loosely grouped under the category of 'functional disorders' (Barsky and Borus 1999; Wessely et al. 1999). These disorders are considered 'functional' because they can only be diagnosed following the report of symptoms rather than via an objective diagnostic test. Despite the lack of objective indicators, functional pain reduces quality of life, generates considerable health care costs, decreases work-related productivity and may increase mortality (Inadomi et al. 2003; Maetzel and Li 2002; McFarlane et al. 2001).

The persistence, intractability and apparent absence of peripheral disease to account for functional pain has led to an increasing interest in the neuropsychological mechanisms that might underpin functional pain. A series of studies have now demonstrated that pain can be generated in normal volunteers without the presence of a stimulus that could cause tissue damage. For example, participants report pain in the absence of a stimulus more readily after viewing images of other people in pain and a significant minority report pain directly associated with the image (Kirwilliam and Derbyshire 2008; Osborn and Derbyshire 2010). Participants reporting pain when viewing an image of someone else in pain activate the anterior cingulate, insular and somatosensory cortices (Osborn and Derbyshire 2010). Highly suggestible participants can also be induced to experience pain without a noxious stimulus under hypnosis and also activate the anterior cingulate, insular and somatosensory cortices (Derbyshire et al. 2004).

Pain Interacts with Touch

Anyone who has ever had a mild injury will know that rubbing the area tends to reduce the experience of pain. This simple observation demonstrates that pain and touch must sometimes interact. Over 40 years ago, Melzack and Wall proposed that the interaction of pain and touch begins in the dorsal horn where touch fibre terminations presumably connect with cells transmitting noxious information and damp the flow of noxious activity (Melzack and Wall 1965). Melzack and Wall reasoned that the cells of the spinal cord, which first receive incoming information, must select and compute outputs based on the combination of signals received. A descending influence upon these cells from the brain was also included, which in turn was influenced by input from the spinal cord cells, thus forming a spinal cord-brain loop.

Although Melzack and Wall's gate control theory is no longer considered correct in detail it is still widely quoted for its heuristic value (Sufka and Price 2002). Importantly, gate control theory shifted attention away from the peripheral source of injury and towards the spinal cord and brain. It also made it difficult to argue that pain is a direct consequence of activity in a dedicated 'pain pathway'. The gate control theory also stimulated the treatment of pain as a symptom or problem in itself, independent of any clinical diagnosis. After all, if pain cannot be reliably judged following an objective measure of injury or receptor activation, then assessment of pain must fall to subjective factors.

The Definition of Pain

The importance of subjectivity in pain is captured by the IASP definition of pain referenced at the beginning. Although heavily criticized and debated (Anand and Craig 1996; Derbyshire 1999), the IASP definition remains largely accepted by those that research and treat pain.

The IASP definition is important because it shifts attention away from the stimulus and towards the experience of pain. We cannot rely on the stimulus to provide a clear indication of experience because subjectivity does not exist in the stimulus; subjectivity exists in the person responding to the stimulus. The IASP definition reminds us that pain is subjective and that the content of pain subjectivity includes both sensory and emotional factors.

Pain Activates Sensory and Emotional Areas of the Brain

Consistent with the multidimensional nature of pain, brain imaging studies demonstrate that pain experience correlates with activity in both sensory and emotional regions of the brain (Apkarian et al. 2005; Derbyshire 2000). Noxious stimuli consistently activate primary and secondary somatosensory cortices, which presumably code the intensity, location and other sensory factors associated with the stimulus. Noxious stimuli also consistently activate the anterior cingulate, insular and prefrontal cortices, which presumably code the unpleasantness and cognitive components associated with the stimulus. Rainville and colleagues used hypnosis to dissociate the sensory and emotional components of pain and demonstrated greater activation of the anterior cingulate cortex when pain unpleasantness was increased via hypnotic suggestion (Rainville 1997). Activation in primary sensory cortex (S1) was unaffected. A further study demonstrated increased S1 activity when pain intensity was increased via hypnotic suggestion (Hofbauer 2001).

Closing the Cartesian Divide

The fact that both sensory and emotional components of pain can be influenced by hypnosis, and that pain can be generated directly by hypnosis and other types of suggestion (Osborn and Derbyshire 2010; Derbyshire et al. 2004), highlights the fact that pain experience involves more than stimulation of neurons in the skin. Pain is something consciously active; a product of minds rather than physics. Descartes explained long ago that while the mind is clearly exposed to sensory information, the mind is not drowned or dissolved by the senses (Descartes 1997). Human beings are self-located within sensory experience but we are not sensorily immersed; our intuition of ourselves as particular things with particular location and experience is opened up by, rather than collapsed into, our senses.

There is tension between pain as a physical, or objective, reaction and pain as a subjective experience (Grahek 2007). There is a temptation to reduce pain to its physical, raw, nature by stripping away the layers of conceptualisation that we associate with knowledge. The challenge is to understand pain as something apprehended rather than comprehended. Pain, at some primitive, raw level just is.

Pain that just is, immediate and without further elaboration, might presumably be the product of activity in dedicated pathways that process stimuli in the noxious range. Once activity in those pathways reaches a critical threshold then pain will be registered for all sentient beings in roughly comparable fashion. In more complex, conscious, beings such as humans, that initial immediate experience will be taken up and elaborated through a nexus of knowledge. Understanding pain as a part of knowledge means taking account the whole complex of traits by which we are characterised (Mannheim 1935). The very fact that pain is experienced as something multidimensional but separate from other experiences implies the existence of a conceptual apparatus that can marshal the dimensions of pain into a coherent experience. It is no longer possible to experience pain as a raw, pure, apprehension, pain can only be experienced as an elaborated comprehension. There is no return to innocence once pain is marshalled into knowledge. And even though dedicated pathways for noxious information remain a part of the story, they cannot be isolated from the broader neural and informational influences over pain. Pain does not have primacy over subjectivity, existing before and in addition to subjectivity, but is now experienced through subjectivity. That is what makes painful touch so interesting and maddeningly capricious.

References

Anand, K J S and Craig, K D (1996). New perspectives on the definition of pain. *PAIN* 67: 3–6.

Apkarian, A V; Bushnell, M C; Treede, R-D and Zubieta, J-K (2005). Human brain mechanisms of pain perception and regulation in health and disease. *European Journal of Pain* 9: 463–484.

Barsky, A J and Borus, J F (1999). Functional Somatic Syndromes. *Annals of Internal Medicine* 130: 910–921.

Craig, A D and Bushnell, M C (1994). The thermal grill illusion: unmasking the burn of cold pain. *Science* 265: 252–255.

Craig, A D; Zhang, E T; Bushnell, M C and Blomqvist, A (1995). Reply to P.D. Wall. *Pain* 62: 391-393.

Derbyshire, S W G (1999). The IASP definition captures the essence of pain experience. *Pain Forum* 8: 106–109.

Derbyshire, S W G (2000). Exploring the pain "neuromatrix". *Current Review of Pain* 6: 467–477.

Derbyshire, S W G (2002). Measuring our natural painkiller. *Trends in Neurosciences* 25: 65–66.

Derbyshire, S W G; Whalley, M G; Stenger, V A and Oakley, D A (2004). Cerebral activation during hypnotically induced and imagined pain. *NeuroImage* 23: 392–401.

Descartes, R (1997). Meditations on First Philosophy (1641). In: E S Haldane and G R T Ross (Trls.), *Descartes Key Philosophical Writings* (pp. 139–146). Wordsworth Classics.

Grahek, N (2007). *Feeling Pain and Being in Pain.* Bradford Books.

Han, Z S; Zhang, E T and Craig, A D (1998). Nociceptive and thermoreceptive lamina I neurons are anatomically distinct. *Nature Neuroscience* 1: 218–225.

Hofbauer, R K; Rainville, P; Duncan, G H and Bushnell, M C (2001). Cortical representation of the sensory dimension of pain. *Journal of Neurophysiology* 86: 402–411.

Inadomi, J M; Fennerty, M B and Bjorkman, D (2003). The economic impact of irritable bowel syndrome. *Alimentary Pharmacology & Therapeutics* 18: 671–682.

Kirwilliam, S S and Derbyshire, S W G (2008). Increased bias to report heat or pain following emotional priming of pain-related fear. *Pain* 137: 60–65.

Maetzel, A and Li, L (2002). The economic burden of low back pain: a review of studies published between 1996 and 2001. *Best Practice & Research: Clinical Rheumatology* 16: 23–30.

Mannheim, K (1935). *Ideology and Utopia*. New York: Harcourt, Brace and Company.

Macfarlane, G J; McBeth, J and Silman, A J (2001). Widespread body pain and mortality: prospective population based study. *British Medical Journal* 323: 1–4.

Melzack, R and Wall, P D (1965). Pain mechanisms: a new theory. *Science* 150: 971–979.

Merskey, H (1991). The definition of pain. *European Psychiatry* 6: 153–159.

Osborn, J and Derbyshire, S W G (2010). Pain sensation evoked by observing injury in others. *PAIN* 148: 268–274.

Rainville, P; Duncan, G H and Price, D D (1997). Pain affect encoded in human anterior cingulate but not somatosensory cortex. *Science* 277: 968–971.

Sufka, K J and Price, D D (2002). Gate control theory reconsidered. *Brain and Mind* 3: 277–290.

Wall, P D (1995). Pain in the brain and lower parts of the anatomy. *PAIN* 62: 389–391.

Wessely, S; Nimnuan, C and Sharpe, M (1999). Functional somatic syndromes: one or many? *Lancet* 354: 936–939.

Thermal Touch

Lynette Jones

Thermal touch refers to the perception of temperature of objects in contact with the skin. When the hand makes contact with an object, the temperatures of the object and the skin change at a rate that is determined by the thermal properties of the object and skin and their initial temperatures. On the basis of these changes in temperature, people can identify the material composition of objects, for example, whether the object is made from copper or wood.

Perception of Object Temperature

When the hand grasps an object, changes in skin temperature can assist in identifying the object and discriminating between different types of objects. These cues become especially important when objects must be identified without visual feedback, such as when reaching for objects in the dark. The thermal cues that assist in identifying an object arise from the changes in skin temperature that occur when the object and hand are in contact. The thermal properties of the object, such as its conductivity and specific heat capacity, the initial temperatures of the skin and object, the thermal contact resistance between the skin and object, and the object's size and shape all determine the rate at which heat is conducted out of the skin or object during contact. Because of the differences in the thermal properties of materials, when an object made from plastic is held in the hand, skin temperature changes much more slowly than when the hand grasps an object made from stainless steel or copper. This means that the metal object feels "cooler" than one made from plastic, even though both objects are at the same temperature. After 10 s of contact with a copper object at room temperature, skin temperature can decrease by as much as 5 °C, whereas it changes by less than 2 °C after 10 s of contact with a plastic object (Ho and Jones 2006). The temperature of the skin is usually higher than the temperature of objects encountered in the environment, and so it is the decrease in skin temperature on contact that is used to identify whether an object is

L. Jones (✉)
Department of Mechanical Engineering, MIT, Cambridge, USA

T.J. Prescott et al. (eds.), *Scholarpedia of Touch*, Scholarpedia,
DOI 10.2991/978-94-6239-133-8_21

made from metal, wood or plastic. For objects warmer than the hand, such as the water in a shower or the handle of a pot on a stove, increases in skin temperature are typically used to evaluate the temperature of the object, and not to identify it.

Temperature Receptors

The sensory system involved in perceiving the changes in skin temperature begins with free nerve endings found in the dermal and epidermal layers of skin that can be functionally classified as cold and warm thermoreceptors. Warm and cold receptors respond similarly to radiant and conducted thermal energy and are involved in the perception of innocuous (harmless) temperatures. The molecular mechanisms underlying temperature sensation have been extensively studied over the past decade with the result that several temperature-sensitive ion channels of the transient receptor potential (TRP) family have been identified as candidate temperature sensors. These thermoTRP channels are expressed in sensory nerve endings and are active at specific temperatures ranging from noxious cold to burning heat (Dhaka et al. 2006). In addition to responding to changes in temperature, these thermoTRPs are involved in chemesthesis, and so mediate the pungent qualities of stimuli such as capsaicin, the "hot" ingredient in chili peppers and menthol, the "cooling" compound from mint.

Thermoreceptors in the glabrous skin on the palm of the hand are mainly used to assist in identifying objects in contact with the hand, whereas thermoreceptors in hairy skin are particularly important in thermoregulation. Cold receptors respond to decreases in skin temperature over a range of 5–43 °C, and discharge most vigorously at skin temperatures around 25 °C. In contrast, warm receptors signal that skin temperature has increased and are most responsive at approximately 45 °C (Darian-Smith and Johnson 1977). When the temperature of the skin is between 30–36 °C, which is the normal range for skin temperature, both types of receptor are spontaneously active, but there is generally no awareness of cold or warmth. This is called the neutral thermal region; at higher or lower temperatures, there is an enduring sensation of warmth or coolness, respectively. In contrast to body temperature which varies by less than 1 °C across healthy individuals, skin temperature can vary by as much as 12 °C in normal individuals, particularly on the hands and feet (Parsons 2003).

In addition to sensing the temperature of objects in contact with the skin, afferent signals arising from cold thermoreceptors have been shown to play a role in the perception of wetness. It appears that thermal cues are used in conjunction with tactile inputs to perceive the wetness experienced when the skin is in contact with a wet surface (Filingeri et al. 2014). These interactions between thermal and tactile inputs presumably account for the illusion of skin wetness that can occur when the skin is exposed to cold-dry stimuli which result in cooling rates similar to those that occur during evaporation of water from the skin surface.

The number and density of thermoreceptors in the skin has been measured by placing small warm and cold stimulators on the skin and recording the sites at which a person detects a change in temperature. The locations at which a thermal stimulus is detected are known as warm and cold spots and are assumed to mark the receptive fields of underlying thermoreceptors. Warm and cold spots are only a few millimeters in diameter, and are distributed independently. There are more cold spots than warm spots, and the density of spots varies across the body. For example, on the forearm it is estimated that there are approximately 7 cold spots and 0.24 warm spots per 100 mm^2.

In addition to differences in the distribution of cold and warm thermoreceptors across the skin surface, the two types of receptor differ with respect to the conduction velocities of the afferent fibers that convey information from the receptor to the central nervous system. Cold afferent fibers are myelinated and so are much faster than unmyelinated warm afferent fibers with conduction velocities of 10–20 m/s as compared to 1–2 m/s for warm fibers. As would be expected from these differences in conduction velocities, the time to respond to a cold stimulus is significantly shorter than that for a warm stimulus.

The skin also contains thermally sensitive receptors leading to pain sensation known as thermal nociceptors that respond to noxious or harmful temperatures. Nociceptors that are responsive to temperature signal to the central nervous system that tissue damage is imminent and that the affected body part should be withdrawn immediately from the thermal source (e.g. a finger on a hot plate). These receptors are active when the temperature of the skin falls below 15–18 °C or rises above 45 °C. When they are activated, the sensation is one of pain. Although the thresholds for activating heat and cold-sensitive nociceptors are usually described as being greater than 45 °C and less than 15 °C, in some individuals mild cooling (25–31 °C) and warming (34–40 °C) of the skin can evoke sensations of burning and stinging as well as innocuous sensations of cold and warmth (Green 2002). Changes in skin temperature also affect the responses of mechanoreceptors in the skin that signal mechanical deformation, such as pressure or vibration, which is why the hands often seem clumsy when they are cold. However, it is generally accepted that mechanoreceptors do not have sufficient encoding capacity to account for thermal sensations.

Thermal Thresholds

The ability to perceive changes in skin temperature depends on a number of variables including the location on the body stimulated, the amplitude and rate of temperature change, and the baseline temperature of the skin. There is a 100-fold variation in sensitivity to changes in skin temperature across the body, with the cheeks and the lips being the most sensitive area, and the feet being the least sensitive region. For the hand, cold and warm thresholds are lower on the thenar eminence at the base of the thumb as compared to the forearm and fingertips (Stevens and Choo 1998). Thermal sensitivity maps of the body are therefore quite different from homologous maps of spatial tactile acuity in which the exquisite sensitivity of the fingertips is immediately apparent. A common finding in many studies of thermal thresholds is that despite the variability in thresholds across the body, all regions are more sensitive to cold than to warmth. In general, the threshold for detecting a decrease in temperature (cold) is half that of detecting an increase in skin temperature (warmth), and the better a site is at detecting cold, the better it is at detecting warmth (Stevens and Choo 1998).

The thermal sensory system is extremely sensitive to very small changes in temperature and on the hairless skin at the base of the thumb, people can perceive a difference of 0.02–0.07 °C in the amplitudes of two cooling pulses or 0.03–0.09 °C of two warming pulses delivered to the hand. The threshold for detecting a change in skin temperature is larger than the threshold for discriminating between two cooling or warming pulses delivered to the skin. When the skin at the base of the thumb is at 33 °C, the threshold for detecting an increase in temperature is 0.20 °C and is 0.11 °C for detecting a decrease in temperature.

The rate that skin temperature changes influences how readily people can detect the change in temperature. If the temperature changes very slowly, for example at a rate of less than 0.5 °C per minute, then a person can be unaware of a 4–5 °C change in temperature, provided that the temperature of the skin remains within the neutral thermal region of 30–36 °C. If the temperature changes more rapidly, such as at 0.1 °C/s, then small decreases and increases in skin temperature are detected. However, warm and cold thresholds do not decrease any further if the rate at which temperature changes is faster than 0.1 °C/s.

Spatial Aspects of Thermal Perception

In contrast to the visual, auditory and haptic modalities, the range of sensations evoked by changes in skin temperature is rather limited. In response to thermal stimulation, people report that the skin has been cooled or warmed, and perceive the intensity and duration of the stimulus. They may also note the hedonic nature of the stimulus, that is, whether it is pleasant (e.g. standing near a warm fire when cold) or unpleasant (e.g. standing by an open door on a cold day), and at extreme temperatures, sensations of pain predominate (Jones and Ho 2008). The spatial features of the thermal stimulus, such as its area and shape, or changes in intensity within an area of stimulation are barely resolved. Yang et al. (2009) showed that when two thermal stimuli were presented on the same fingertip, participants were unable to discriminate between them even though they could discriminate between them quite reliably when they were presented to two fingers on opposite hands. Spatial resolution is poor because the sensory system involved in processing information from thermoreceptors in the skin summates intensity over the area of stimulation, which means that changes in stimulus area are often indistinguishable from changes in stimulus intensity. This property increases the detection of small changes in temperature that occur over a large surface area, which is important to thermoregulation, that is maintaining core temperature constant. For the thermal senses, the spatial extent of stimulation affects the perceived intensity of the stimulus with the result that as the area of stimulation increases the stimulus is perceived to be more intense, rather than just larger. This characteristic also means that a warm or cold stimulus becomes more detectable (i.e. the threshold decreases) if the area of stimulation increases. Spatial summation does, however, decline as the temperature approaches the threshold of pain.

One of the interesting properties of thermal spatial summation is that the areas stimulated do not have to be contiguous for summation to occur. When two symmetrical sites on the body (e.g. both forearms) are stimulated simultaneously, the thermal stimulus at one site is perceived to be more intense than when only a single site is stimulated. As would be expected from this result, both warm and cold thresholds are lower when stimuli are presented bilaterally. However, there is no change in thresholds if the two sites are asymmetric, such as the forehead and the contralateral hand.

A further dimension of spatial processing that is used to characterize sensory systems is spatial acuity, which in the context of the tactile modality refers to the spatial resolution of the skin. As would be expected for a sensory system that displays pervasive spatial summation, the thermal senses are poor at localizing the site of thermal stimulation on the body and at differentiating spatially two thermal stimuli delivered in close proximity. If tactile cues are eliminated by using non-contact thermal stimulation such as radiant heat, the ability to localize the site of stimulation is very poor, particularly if the stimuli are not very intense. The errors of localization are such that warm stimuli delivered to the front or back of the torso can be misperceived as being presented to the other side of the torso. Such errors of localization are never found with mechanical stimulation to the torso and do not occur for thermal stimuli near the pain threshold. Information from tactile and thermal receptors in the skin is conveyed to the brain via different anatomical pathways, and the spatial properties of the tactile and thermal sensory systems reflect this distinction.

Temporal Aspects of Thermal Sensation

The duration of a thermal stimulus and the rate with which it changes can have a marked effect on perception. It is a frequently experienced phenomenon, that the sensation of warmth that is aroused when one steps in the shower gradually diminishes with time. With continuous exposure to a thermal stimulus there is a decrease in neural responsiveness, a process referred to as adaptation. The skin adapts to both warm and cold stimuli over time, and for temperatures close to that of the skin, the rate at which adaptation occurs is rapid, in the order of 60 s for changes of ±1 °C in skin temperature. It takes much longer for the skin to adapt to more extreme temperatures, and for the forearm complete adaptation occurs within about 25 min for temperatures between 28 and 37.5 °C (Kenshalo and Scott 1966).

The time required to respond to a thermal stimulus depends on the intensity of the stimulus and the response required. Temperatures close to the thermal pain thresholds are responded to rapidly due to the possibility of tissue damage, but the response to more moderate temperatures is sluggish, when compared to other sensory systems.

Internal References

Alonso, J-M and Chen, Y (2009). Receptive field. *Scholarpedia* 4(1): 5393. http://www.scholarpedia.org/article/Receptive_field.

Braitenberg, V (2007). Brain. *Scholarpedia* 2(11): 2918. http://www.scholarpedia.org/article/Brain.

Llinas, R (2008). Neuron. *Scholarpedia* 3(8): 1490. http://www.scholarpedia.org/article/Neuron.

External References

Darian-Smith, I and Johnson, K O (1977). Thermal sensibility and thermal receptors. *Journal of Investigative Dermatology* 69: 146–153.

Dhaka, A; Viswanath, V; and Patapoutian, A (2006). TRP ion channels and temperature sensation. *Annual Review of Neuroscience* 29: 135–161.

Filingeri, D; Fournet, D; Hodder, S and Havenith, G (2014). Why wet feels wet? A neurophysiological model of human cutaneous wetness sensitivity. *Journal of Neurophysiology* 112: 1457–1469.

Green, B G (2002). Synthetic heat at mild temperatures. *Somatosensory & Motor Research* 19: 130–138.

Ho, H and Jones, L A (2006). Contribution of thermal cues to material discrimination and localization. *Perception & Psychophysics* 68: 118–128.

Jones, L A and Ho, H-N (2008). Warm or cool, large or small? The challenge of thermal displays. *IEEE Transactions on Haptics* 1: 53–70.

Kenshalo, D R and Scott, H A (1966). Temporal course of thermal adaptation. *Science* 152: 1095–1096.

Parsons, K (2003). *Human Thermal Environments*, 2nd ed. Taylor & Francis.

Stevens, J C and Choo, K K (1998). Temperature sensitivity of the body surface over the life span. *Somatosensory & Motor Research* 15: 13–28.

Yang, G-H; Kwon, D-S and Jones, L A (2009). Spatial acuity and summation on the hand: The role of thermal cues in material identification. *Attention, Perception & Psychophysics* 71: 156–163.

Tactile Control of Balance

Leif Johannsen, Alan Wing and Mark S. Redfern

Introduction

Light tactile contact to the fingertips or other body segments improves the control of body balance in quiet stance as well as locomotion through augmented self-motion feedback.

The stability of a mechanical system may be defined as the time taken to return to its initial state when perturbed by external forces or torques. Thus a stable chair is one which, when tipped so that two legs rise of the ground, returns to its initial upright position with all four legs in contact. A chair that returns to this state immediately may be said to be more stable than one that alternately rocks onto opposing pairs of legs several times before settling on all four legs. Correspondingly, the stability of someone standing upright may be defined as the speed with which the initial quiet standing state is restored after a horizontal push at waist level (Gilles et al. 1999). For example, a 10 N (1 kg) horizontal forward push at the level of the pelvis will accelerate body Centre of Mass (CoM) forward relative to the base of support. Restoration of the initial state involves a postural reflex including toe-down ankle torque generated by the calf plantar flexors (gastrocnemius, soleus). The speed of that response will depend on a number of factors including the level of sensory input, the amplitude of the muscle response, intra- and inter-segmental coordination of muscle activation. Sensory motor disorder, such as hemiparetic stroke with asymmetrically reduced sensory input, slowed, weakened and incoordinated muscle activation, can profoundly affect the standing stability assessed by a horizontal

L. Johannsen (✉)
Technische Universität München, Munich, Germany

A. Wing
Centre for Sensory Motor Neuroscience, University of Birmingham, Birmingham, UK

M.S. Redfern
University of Pittsburgh, Pittsburgh, PA, USA

T.J. Prescott et al. (eds.), *Scholarpedia of Touch*, Scholarpedia,
DOI 10.2991/978-94-6239-133-8_22

forward push. And such impairment would also be expected to signal that difficulties would arise in practical contexts such as maintaining stable standing on transport systems where the base of support is subject to horizontal accelerations e.g. on a bus or train.

Quiet standing without external perturbations is also a challenge for the Central Nervous System (CNS). The CoM is located somewhat in front of the ankle so that upright posture must be maintained by the ankle muscles. Passive stiffness of the ankle muscles and tendons can prevent around 90 % of the tendency to dorsiflexion (Lakie et al. 2003; Loram and Lakie 2002a, b). However, this is not sufficient to prevent forward sway so active muscle contraction is indicated. Loram and Lakie (2002a, b) used very small perturbations to the ankle and show no phasic activation of calf muscles. Instead they argued for intermittent anticipatory adjustments of the muscle resting length to vary stiffness and hence change the forward lean (Loram et al. 2005).

In order to perceive both self-motion and motion of a dynamic environment when maintaining body balance, the human CNS has at its disposal multiple sensory channels (Horak and Macpherson 1996) including vision, vestibular sensation, leg muscle proprioception and tactile sensations in the soles of the feet. Deprivation of any one of these senses or the inability to resolve conflict between sensory cues, potentially caused by disorders of the peripheral or central nervous system, invariably causes conspicuous instability and increases in body sway (Dichgans and Diener 1989; Diener and Dichgans 1988). In fact, the detection of body sway by means of somatosensory input channels (muscle proprioception and cutaneous plantar sensation) plays a predominate role during quiet stance (Fitzpatrick and McCloskey 1994). Body sway in a static posture means that the CoM moves relative to the feet which causes not only stretch in lower limb muscles but also changes the pressure distribution under the feet. Modulating plantar tactile sensation and perceived length of distal leg muscles by local vibratory stimulation directly affects body sway as much as perceived postural orientation and body tilt (Kavounoudias et al. 1998, 2001; Roll et al. 2002). Therefore, Roll and Roll and colleagues (Kavounoudias et al. 2001) suggested that both modalities are combined additively into one representation of body orientation for sway control.

The versatility of the human postural control system, however, also enables us to utilize other sensory channels, such as non-plantar skin receptors, for the control of body sway. Of course, a prerequisite of this feat is that these afferences convey body sway-related information. For example, upper limb tactile feedback, for example from the fingertips, can be recruited as a sensory source about body sway. In the following review we will present the current state of knowledge concerning mechanisms involved in the processing of tactile afferent information for the optimization of body balance in quiet stance and locomotion. We will begin with a discussion of the traditional approach investigating sway feedback by means of light touch during quiet standing. It follows a summary of light touch effects on sway in the aging population and patients with sensorimotor impairments and increased fall risk. In this context, consideration of object-mediated light contact such as sticks and canes is also relevant. We will then discuss evidence that high-order anticipatory processes are also likely to be involved in sway control when contacting an external reference location. From the static contact situation, we will move on to more dynamic postural situations in which the contact is kept with a reference showing own motion dynamics or where the participant him- or herself is actively involved in locomotion. We will then tackle current knowledge about central integration mechanisms for light touch control of balance. We will conclude our review with an overview of the role of tactile feedback during postural and locomotor development in early infancy.

Basic Findings in Quiet Stance

Strong interactions between multiple sensory modalities occur with respect to the representation of the orientation of our body and its segments, such as the eyes, head or limbs, relative to its environment. Uncertainty about the state of the body or a limb can give rise to sensorimotor illusions. Is additional tactile and proprioceptive information available, however, state estimates can be improved in their accuracy which suppresses any illusory states. For example, when prolonged fixation of a visual target is required involuntary ocular drift is observed that correlates with perceived illusory self-motion ("autokinesis") (Barlow 1952). Extra somatosensory information about the location of the fixation target, however, for example by grasping the target's mount, improves the stability of fixations and reduces the self-motion illusion (Lackner and Zabkar 1977). Further, illusory arm movement induced by muscle tendon vibration will be suppressed if the other arm touches the stimulated arm after onset of vibration, but not if touch is established before onset (Lackner and Taublieb 1983). Given the benefit of somatosensory information for spatial limb orientation estimates, one can ask whether this effect generalizes to more complex postural tasks such as upright standing balance, perhaps in terms of improved representation of the gravitational vertical or better localization relative to the environment? Indeed, this seems to be the case.

Lackner, Jeka and colleagues (Holden et al. 1994) showed in healthy participants that mechanically non-supportive fingertip contact (<1 N) with an earth-fixed reference while the eyes are closed results in reductions in sway, which may be even more efficient than vision alone (Jeka and Lackner 1994; Lackner et al. 1999). Cross-correlation time lags indicated that the force signal at the fingertip precedes postural adjustments, which implies feedback processing of touch for sway control (Jeka and Lackner 1994). The processing of the touch signal is affected by the relative position of the touch location to the body and the contacting limb posture. Depending on the current body posture and the relative position of the contact, a "radial" touch signal, meaning a touch vector directed parallel to the dominant direction of sway, is more effective than a "laminar" touch vector orthogonal to the sway plane (Jeka et al. 1998b). Nevertheless, the touch effect seems to be independent of the degrees of freedom of the contacting limb in contrast to the precision with which contact is being kept (Rabin et al. 2008).

Skin feedback can also be utilized when the tactile contact is received passively (Krishnamoorthy et al. 2002; Rogers et al. 2001). Cutaneous receptors sensitive to skin stretch detecting differences in shear forces at the contact location but also muscle proprioception of the contacting upper limb provide feedback about the direction and amplitude of body sway relative to the contact location (Rabin et al. 1999). Fingertip contact modulates the H reflex of the soleus muscle (Huang et al. 2009). Simultaneous contact of both hands to reference locations induces greater sway reduction than a single hand contact (Dickstein 2005), which indicates that sensory summation works in this modality too. Withdrawal of the differential by keeping the relative position of the contact constant to the body (Reginella et al. 1999) or removal of finger tactile feedback by tourniquet ischemia removes the attenuation of body sway with light touch (Kouzaki and Masani 2008). Finally, it has been shown that the contacting location does not need to be earth-fixed since haptic force feedback from holding a weight via a non-rigid link such as a cable or string reduces body sway as well (Krishnamoorthy et al. 2002; Mauerberg-deCastro et al. 2010, 2012).

Compensation for Sensorimotor Impairments

Baldan et al. (2014) reviewed the literature on the effect of light touch on postural sway in individuals with balance problems due to aging, brain lesions or other motor or sensory deficits. The effect of light touch on body sway was found irrespective of the impaired balance. They suggested that the maintenance of the fingertip lightly touching an external reference point could provide somatosensory information for individuals with poor balance universally. In healthy individuals, light touch has been demonstrated to counter the disruptive effects (spatial disorientation by proprioceptive distortion and abnormal motor commands) of neck muscle vibration as well as peroneal muscle tendon and Achilles tendon vibration on body sway (Lackner et al. 2000; Bove et al. 2006; Caudron et al. 2010). Older adults retain the ability to use fingertip contact for augmentation of body sway feedback despite reductions in their tactile sensitivity (Tremblay et al. 2004). In fact, older adults tend to show even greater efficacy of touch feedback for sway reduction than younger adults possibly due to greater sensory loss in distal parts of the lower extremities (Baccini et al. 2007).

Interestingly, blind individuals appear to have an advantage in terms of the tactile integration latency compared to sighted individuals (Schieppati et al. 2014), which implies that in sighted individuals intermodal integration involving both the visual and tactile sensory systems (even when visual feedback is currently not available) makes up a significant portion of processing time. A more detailed discussion of the factors affecting integration of tactile as well as visual information for body balance can be found in a recent review by Honeine and Schieppati (2014).

Cunha et al. (2012) demonstrated that individuals who suffered a stroke are able to use fingertip light contact to reduce body sway. Participants with peripheral neuropathy demonstrated reduced body sway with fingertip contact (Dickstein et al. 2001, 2003) as well as with passive stimulation applied to the skin of the leg (Menz et al. 2006). Without visual feedback, individuals with bilateral vestibular loss demonstrated greater reductions in body sway with haptic contact than normal controls (Lackner et al. 1999). In Parkinson's disease (Franzen et al. 2012; Rabin et al. 2013) and multiple sclerosis (Kanekar et al. 2013) light haptic contact at the fingertip has been reported to improve postural stability in quiet standing.

Object-Mediated Contact Effects

Jeka (1997) suggested that augmented haptic feedback supplied by handles of canes and sticks might lead to the development of mobility aids for balance-impaired populations. A cane as a hand-held tool allows a much greater number of contacting degrees of freedom and therefore variation of haptic force feedback compared to any earth-fixed stand. Indeed, cane-mediated ground contact benefits sway if force feedback is sufficiently strong and correlated to body sway (Albertsen et al. 2010, 2012). Jeka and colleagues (Jeka et al. 1996) showed that blind individuals' sway benefits significantly from holding a cane, especially when contacting the ground in a slanted orientation. In addition, Albertsen and colleagues (Albertsen et al. 2012) demonstrated that holding a long stick contacting the ground increases postural stability in young and older adults irrespective of whether the contact location of the stick was earth-fixed or spatially unconstrained due to a low-friction surface over which the contact could slide as a result of participants' body motion. These effects also generalize to an unstable seated posture when holding a pen in younger and older adults (Albertsen et al. 2014).

Light Touch as a Suprapostural Task

When fixating an external visual target at varying distances, postural sway is modulated by the demands of the oculomotor task in terms of minimizing retinal slip of the target (Stoffregen et al. 1999). Thus, looking at the target can be considered a "suprapostural" task to the control of body sway. Similarly, the requirement of lightly keeping contact during standing and walking, for example precise control of the contacting forces in terms of the perceived tactile variability, may represent explicit and implicit goals of a suprapostural task with an external focus of attention (McNevin et al. 2000, 2013; McNevin and Wulf 2002; Riley et al. 1999). Riley and co-workers (Riley et al. 1999) demonstrated that the light touch effect on sway is dependent on the salience of the contact within the current postural context. In other words, participants, for whom finger contact occurred only coincidentally, did not show any reductions in sway. That the light touch effect is not only a consequence of tactile feedback processing is also indicated by evidence in favour of proactive sway control. It is remarkable that merely the intention to establish light contact with an earth-fixed reference soon (<5 s) can result in effects on body sway similar to actual contact (Bove et al. 2006). Schieppati and colleagues (Bove et al. 2006) proposed that transient anticipatory processes are involved in the preparation of the central postural set to the context of stance control with light contact.

Contact with a Reference Possessing Own Motion Dynamics

Contact with a non-biological reference location that demonstrates its own oscillatory motion causes involuntary postural sway entrainment as well as increases in body sway compared to a static contact for oscillation frequencies less than 0.8 Hz (Jeka et al. 1997, 1998a). The default (mis-)attribution of any tactile sensation to own body motion as the origin results in postural adjustments coordinated with the motion of the external source. Thus, body sway oscillations will become synchronized to the velocity as well as the position of the contact point (Jeka et al. 1998a). Interestingly, this effect of spontaneous postural entrainment to a moving contact may be used for closed-loop driving postural sway of an individual (Verite et al. 2014). It is surprising in this respect that contact to a dynamic reference such as another human (light interpersonal touch; IPT) actually reduces postural sway and leads to interpersonal sway synchronization potentially, but not necessarily, due to similar entrainment mechanisms (Johannsen et al. 2009, 2012). This reduction of sway during contact with another individual may also be caused by mechanisms involved in the reduction of sway by low-noise vibratory stimulation of the fingertip or the foot soles ("stochastic resonance" resulting in enhanced somatosensory feedback; (Collins et al. 2003; Kimura et al. 2012; Magalhaes and Kohn 2011; Magalhaes and Kohn 2012; Priplata et al. 2003)). Passive exposure to the recorded sway dynamics of another individual via a haptic force feedback device, however, does not result in the sway reductions seen during contact with an actual human partner (Wing et al. 2011). This implies that sway reduction with IPT reflects a mutually adaptive process between two contacting individuals. This interaction does not seem to be the result of a mechanical coupling between both individuals but instead represents the effect of mutually shared sensory information (Reynolds and Osler 2014). The IPT phenomenon has implications for clinical assistance with the balance of neurological patients. For example, sway reduction with passively received IPT has also been reported in patients with chronic stroke and Parkinson's disease as a function of the contact location (Johannsen et al. 2014b).

Perturbed Stance and Dynamic Postural Activities

Light finger contact with an earth-fixed reference point modulates compensatory postural adjustments following a support surface perturbation (Dickstein et al. 2003). Also, light shoulder contact shortens postural compensation of both self-imposed and externally imposed perturbations at hip level (Johannsen et al. 2007). Fingertip also reduces gastrocnemius activity and sway following a hip perturbation, possibly by modulating the postural response strategy (Martinelli et al. 2015).

While most studies demonstrated the benefits of haptic feedback exclusively in quiet standing, a few have reported increases in postural stability in a more dynamic context such as overground and treadmill walking (Dickstein and Laufer 2004; Forero and Misiaszek 2013; Fung and Perez 2011; Kodesh et al. 2015) and staircase negotiation (Reeves et al. 2008; Reid et al. 2011). For example, Dickstein and colleagues (Kodesh et al. 2015) reported that fingertip contact with an elastic thera-band reduces the variability of step width presumably by stabilizing trunk excursions. In older adults, light use of handrails during stair negotiation changes ankle and knee coordination during ascent and improves postural stability in terms of the separation distance between centre-of-mass and centre-of-pressure (Reeves et al. 2008). During treadmill walking, handrail use induces specific changes in reflex patterns all over the body that might facilitate postural stability during gait (Lamont and Zehr 2007).

Effects of interpersonal touch have also been reported during overground walking. Hausdorff and colleagues (Zivotofsky and Hausdorff 2007; Zivotofsky et al. 2012) reported that when two individuals walk next to each other strong and spontaneous synchronization can be observed, especially when interpersonal tactile feedback is present, for example by holding hands. In this context, tactile and auditory cues appear more effective than visual feedback alone (Zivotofsky et al. 2012) Zivotofsky and Hausdorff (2007) suggested that this interpersonal entrainment, perhaps enhanced by interpersonal touch, may be used to facilitate the locomotor pattern in individuals with impaired gait.

Central Processing for Tactile Integration

Tactile feedback delays of about 300 ms duration as reported in several studies suggest supraspinal pathways affording complex processing (Clapp and Wing 1999; Jeka and Lackner 1994; Rabin and Gordon 2006). Utilizing light touch for sway control demands attention (Vuillerme et al. 2006). This may be related to high-order processing of the tactile feedback or representation of the postural context. For example, Franzen and colleagues (Franzen et al. 2011) suggested that the postural control system switches between reference frames (from a global to a local trunk-centred) during the integration of light touch. Studies of the time course of sway before, during and after periods of intermittent touch indicate that sway stabilization with light tactile contact is a time-consuming integrative process (Rabin et al. 2006; Sozzi et al. 2012). Initial integration of touch information into the postural control loop seems to happen in a third of the time (100–300 ms) required for the integration of visual information (Lackner and DiZio 2005; Rabin et al. 2006). In other situations, however, the processing of touch for postural control may be more complex than the processing of visual feedback and therefore may require longer integration latencies due to increased computational load (Sozzi et al. 2012). Nevertheless, sway reduction following integration of fingertip afferent information into the postural control loop occurs between 1.5 to 2 s after contact onset (Rabin et al. 2006). Compared to periods of 2 s (or shorter) intermittent fingertip contact, intermittent contact of just 5 s duration results in an after-effect on sway in terms of a reduced return rate to no contact levels (Johannsen et al. 2014a). This after-effect may indicate that the integration of fingertip contact requires no less than about 2 s of computation and is likely to involve not only bottom-up sensory processing but also top-down, "intentional" control of body sway and tactile attention.

An indication of the involvement of cortical processes in the light touch effect on sway is the observation that the disruption of the right-hemisphere prefrontal cortex of the brain by continuous theta-burst transcranial magnetic stimulation (cTBS) modulates the processing of somatosensory evoked potentials during standing with contact to an earth-fixed reference location (Bolton et al. 2011, 2012). In addition, Johannsen et al. (2015) showed that disruption of left inferior parietal gyrus by 1 Hz repetitive TMS (rTMS) alters the time course of sway following unpredictable contact removal. This finding is another indication that top-down tactile attentional processes may have a prominent influence on the control of body sway during postural state transitions.

Developmental Aspects

Interpersonal support as well as external earth-fixed balance support plays a prominent role in the early development of motor capabilities such as standing and walking. Lacquaniti and colleagues (Ivanenko et al. 2005) demonstrated in toddlers who had just begun to walk independently how unilateral single hand support with a parent improves postural stability in terms of reduced trunk sway, sideways hip motion as well as step width. Older children, 2 years and above, did not show any similar effects. Ivanenko et al. (2005) interpreted these benefits as an indication of increased toddlers' stepping confidence.

On the other hand, there is evidence that early stage toddlers are quite susceptible to tactile input that conveys self-motion information gained from contact with an environmental reference. Metcalfe, Clark and colleagues (Chen et al. 2007, 2008; Metcalfe et al. 2005a, b) conducted a longitudinal study on the development of upright standing and walking in infants and toddlers. They assessed body sway in a quiet posture, sitting on a saddle-styled support and standing, at regular intervals from one month before to nine months after the onset of independent walking. In addition, body sway in the respective posture was assessed with and without static external light touch. When standing, variability of sway did not change a lot across the observation period. The sway dynamics, however, indicated a progression towards a lower sway frequency. The availability of hand light touch reduced the variability and the frequency of sway even further. Metcalfe, Clark and colleagues (Chen et al. 2008; Metcalfe et al. 2005b) interpreted this as an indication for a progressively refined internal representation of own sway dynamics during standing and walking in toddlers. In the semi-sitting posture in contrast to upright standing, hand light touch reduced sway only during the transition period when the toddlers began to take their first independent steps (Chen et al. 2007).

It can easily be imagined that tactile feedback from a contacting hand represents a highly salient sensory signal during the transition to independent walking. When contacting a support, every voluntary step as well as every unintended sway excursion will cause correlated tactile feedback. Thus, it seems reasonable to assume that the tactile feedback resembles an important error signal in the course of learning to walk independently. This notion is confirmed by Barela et al. (1999), who demonstrated that the utilization of tactile information in the context of postural control changes its qualitative character after the onset of independent walking. Cross-correlation analysis indicated that before this event, body sway had a very short lead of about 45 ms on the contact forces, expressing mechanical support functions of the contact, while afterwards sway was following the contacting forces by about 140 ms. The authors suggested that the tactile contact from that point on facilitates anticipatory postural control in the toddlers (Barela et al. 1999). Metcalfe and Clark (2000) elaborated on this conclusion by suggesting that the tactile signal enhances the exploration of perception-action relationship for the generation of internal models for the control of body balance during standing and walking.

Conclusions

In this chapter we have reviewed how light tactile contact to the fingertips or other body segments improves the control of body balance in quiet stance as well as locomotion. We interpret the effects, which extend to light contact with moving targets, such as another individual, as representing augmented self-motion feedback. The context of keeping light touch, however, imposes task demands which are likely to influence the selection of any concurrent postural strategy. Therefore, keeping light touch requires both proactive as well as reactive balance control. Interpretation of any tactile signal in the current postural context involves high-order representations, which are still understood poorly. Nevertheless, the beneficial effect of tactile information for the control of body balance is quite robust and has been demonstrated in a number of neurological disorders. Thus, the findings reviewed have implications for rehabilitation and handling of neurological patients.

Acknowledgments We acknowledge the financial support by the Biotechnology and Biological Sciences Research Council of the United Kingdom (BBSRC; BBI0260491) and the Federal Ministry of Education and Research of Germany (BMBF; 01EO1401).

References

Albertsen, I M; Temprado, J J and Berton, E (2010). Effect of haptic supplementation on postural stabilization: A comparison of fixed and mobile support conditions. *Human Movement Science* 29(6): 999–1010. doi:10.1016/j.humov.2010. 07.013.

Albertsen, I M; Temprado, J J and Berton, E (2012). Effect of haptic supplementation provided by a fixed or mobile stick on postural stabilization in elderly people. *Gerontology* 58(5): 419–429. doi:10.1159/000337495.

Albertsen, I M; Temprado, J J; Berton, E and Heuer, H (2014). Effect of haptic supplementation on postural control of younger and older adults in an unstable sitting task. *Journal of Electromyography and Kinesiology* 24(4): 572–578. doi:10.1016/j.jelekin.2014.03.005.

Baccini, M et al. (2007). Effectiveness of fingertip light contact in reducing postural sway in older people. *Age and Ageing* 36(1): 30–35. doi:10.1093/ageing/afl072.

Baldan, A M; Alouche, S R; Araujo, I M and Freitas, S M (2014). Effect of light touch on postural sway in individuals with balance problems: A systematic review. *Gait Posture* 40(1): 1–10. doi:10.1016/j.gaitpost.2013.12.028.

Barela, J A; Jeka, J J and Clark, J E (1999). The use of somatosensory information during the acquisition of independent upright stance. *Infant Behavior and Development* 22: 87–102. doi:10.1016/s0163-6383(99)80007-1.

Barlow, H B (1952). Eye movements during fixation. *The Journal of Physiology* 116(3): 290–306. doi:10.1113/jphysiol.1952.sp004706.

Bolton, D A; Brown, K E; McIlroy, W E and Staines, W R (2012). Transient inhibition of the dorsolateral prefrontal cortex disrupts somatosensory modulation during standing balance as measured by electroencephalography. *Neuroreport* 23(6): 369–372. doi:10.1097/WNR. 0b013e328352027c.

Bolton, D A; McIlroy, W E and Staines, W R (2011). The impact of light fingertip touch on haptic cortical processing during a standing balance task. *Experimental Brain Research* 212(2): 279–291. doi:10.1007/s00221-011-2728-6.

Bove, M; Bonzano, L; Trompetto, C; Abbruzzese, G and Schieppati, M (2006). The postural disorientation induced by neck muscle vibration subsides on lightly touching a stationary surface or aiming at it. *Neuroscience* 143(4): 1095–1103. doi:10.1016/j.neuroscience.2006.08. 038.

Caudron, S; Nougier, V and Guerraz, M (2010). Postural challenge and adaptation to vibration-induced disturbances. *Experimental Brain Research* 202(4): 935–941. doi:10.1007/ s00221-010-2194-6.

Chen, L C; Metcalfe, J S; Chang, T Y; Jeka, J J and Clark, J E (2008). The development of infant upright posture: sway less or sway differently? *Experimental Brain Research.* 186(2): 293–303. doi:10.1007/s00221-007-1236-1

Chen, L C; Metcalfe, J S; Jeka, J J and Clark, J E (2007). Two steps forward and one back: Learning to walk affects infants' sitting posture. *Infant Behavior and Development* 30(1): 16–25. doi:10.1016/j.infbeh.2006.07.005.

Clapp, S and Wing, A M (1999). Light touch contribution to balance in normal bipedal stance. *Experimental Brain Research* 125(4): 521–524. doi:10.1007/s002210050711.

Collins, J J et al. (2003). Noise-enhanced human sensorimotor function. *IEEE Engineering in Medicine and Biology Magazine* 22(2): 76–83. doi:10.1109/memb.2003.1195700.

Cunha, B P; Alouche, S R; Araujo, I M and Freitas, S M (2012). Individuals with post-stroke hemiparesis are able to use additional sensory information to reduce postural sway. *Neuroscience Letters* 513(1): 6–11. doi:10.1016/j.neulet.2012.01.053.

Dichgans, J and Diener, H C (1989). The contribution of vestibulo-spinal mechanisms to the maintenance of human upright posture. *Acta Oto-laryngologica* 107(5–6): 338-345. doi:10.3109/00016488909127518.

Dickstein, R (2005). Stance stability with unilateral and bilateral light touch of an external stationary object. *Somatosensory & Motor Research* 22(4): 319–325. doi:10.1080/08990220500420640.

Dickstein, R and Laufer, Y (2004). Light touch and center of mass stability during treadmill locomotion. *Gait & Posture* 20(1): 41–47. doi:10.1016/s0966-6362(03)00091-2.

Dickstein, R; Peterka, R J and Horak, F B (2003). Effects of light fingertip touch on postural responses in subjects with diabetic neuropathy. *Journal of Neurology, Neurosurgery & Psychiatry* 74(5): 620–626. doi:10.1136/jnnp.74.5. 620.

Dickstein, R; Shupert, C L and Horak, F B (2001). Fingertip touch improves postural stability in patients with peripheral neuropathy. *Gait & Posture* 14(3): 238–247. doi:10.1016/s0966-6362(01)00161-8.

Diener, H C and Dichgans, J (1988). On the role of vestibular, visual and somatosensory information for dynamic postural control in humans. *Progress in Brain Research* 76: 253–262. doi:10.1016/s0079-6123(08)64512-4.

Fitzpatrick, R and McCloskey, D I (1994). Proprioceptive, visual and vestibular thresholds for the perception of sway during standing in humans. *The Journal of Physiology* 478(1): 173–186. doi:10.1113/jphysiol.1994.sp020240.

Forero, J and Misiaszek, J E (2013). The contribution of light touch sensory cues to corrective reactions during treadmill locomotion. *Experimental Brain Research* 226(4): 575–584. doi:10.1007/s00221-013-3470-z.

Franzen, E; Gurfinkel, V S; Wright, W G; Cordo, P J and Horak, F B (2011). Haptic touch reduces sway by increasing axial tone. *Neuroscience* 174: 216–223. doi:10.1016/j.neuroscience.2010.11.017.

Franzen, E; Paquette, C; Gurfinkel, V and Horak, F (2012). Light and heavy touch reduces postural sway and modifies axial tone in Parkinson's disease. *Neurorehabilitation and Neural Repair* 26(8): 1007–1014. doi:10.1177/1545968312437942.

Fung, J and Perez, C F (2011). Sensorimotor enhancement with a mixed reality system for balance and mobility rehabilitation. *Annual International Conference of the IEEE Engineering in Medicine and Biology Society* 2011: 6753–6757. doi:10.1109/iembs.2011.6091666.

Gilles, M; Wing, A M and Kirker, S G (1999). Lateral balance organisation in human stance in response to a random or predictable perturbation. *Experimental Brain Research* 124(2): 137–144. doi:10.1007/s002210050607.

Holden, M; Ventura, J and Lackner, J R (1994). Stabilization of posture by precision contact of the index finger. *Journal of Vestibular Research* 4(4): 285–301.

Honeine, J L and Schieppati, M (2014). Time-interval for integration of stabilizing haptic and visual information in subjects balancing under static and dynamic conditions. *Frontiers in Systems Neuroscience* 8: 190. doi:10.3389/fnsys.2014.00190.

Horak, F B and Macpherson, J M (1996). Postural orientation and equilibrium. In: L B Rowell and J T Shepherd (Eds.), "Handbook of Physiology, section 12, Exercise: Regulation and Integration of Multiple Systems" (pp. 255–292). New York: Oxford University Press. 9780195091748.

Huang, C Y; Cherng, R J; Yang, Z R; Chen, Y T and Hwang, I S (2009). Modulation of soleus H reflex due to stance pattern and haptic stabilization of posture. *Journal of Electromyography and Kinesiology* 19(3): 492–499. doi:10.1016/ j.jelekin.2007.07.014.

Ivanenko, Y P; Dominici, N; Cappellini, G and Lacquaniti, F (2005). Kinematics in newly walking toddlers does not depend upon postural stability. *Journal of Neurophysiology* 94(1): 754–763. doi:10.1152/jn.00088.2005.

Jeka, J J; Oie, K; Schoner, G; Dijkstra, T and Henson, E (1998a). Position and velocity coupling of postural sway to somatosensory drive. *Journal of Neurophysiology* 79(4): 1661–1674.

Jeka, J J (1997). Light touch contact as a balance aid. *Physical Therapy* 77(5): 476–487.

Jeka, J J; Easton, R D; Bentzen, B L and Lackner, J R (1996). Haptic cues for orientation and postural control in sighted and blind individuals. *Perception & Psychophysics* 58(3): 409–423.
Jeka, J J and Lackner, J R (1994). Fingertip contact influences human postural control. *Experimental Brain Research* 100(3): 495–502. doi:10.1007/bf00229188.
Jeka, J J; Ribeiro, P; Oie, K and Lackner, J R (1998b). The structure of somatosensory information for human postural control. *Motor Control* 2(1): 13–33.
Jeka, J J; Schoner, G; Dijkstra, T; Ribeiro, P and Lackner, J R (1997). Coupling of fingertip somatosensory information to head and body sway. *Experimental Brain Research* 113(3): 475–483. doi:10.1007/pl00005600.
Johannsen, L; Guzman-Garcia, A and Wing, A M (2009). Interpersonal light touch assists balance in the elderly. *Journal of Motor Behavior* 41(5): 397–399. doi:10.3200/35-09-001.
Johannsen, L; Hirschauer, F; Stadler, W and Hermsdorfer, J (2015). Disruption of contralateral inferior parietal cortex by 1 Hz repetitive TMS modulates body sway following unpredictable removal of sway-related fingertip feedback. *Neuroscience Letters* 586: 13–18. doi:10.1016/j.neulet.2014.11.048.
Johannsen, L; Lou, S Z and Chen, H Y (2014). Effects and after-effects of voluntary intermittent light finger touch on body sway. *Gait & Posture* 40(4): 575–580. doi:10.1016/j.gaitpost.2014.06.017.
Johannsen, L; McKenzie, E; Brown, M; Redfern, M S and Wing, A M (2014). Deliberately light interpersonal touch as an aid to balance control in neurologic conditions. "Rehabilitation Nursing".
Johannsen, L; Wing, A M and Hatzitaki, V (2007). Effects of maintaining touch contact on predictive and reactive balance. *Journal of Neurophysiology* 97(4): 2686–2695. doi:10.1152/jn.00038.2007.
Johannsen, L; Wing, A M and Hatzitaki, V (2012). Contrasting effects of finger and shoulder interpersonal light touch on standing balance. *Journal of Neurophysiology* 107(1): 216–225. doi:10.1152/jn.00149.2011.
Kanekar, N; Lee, Y J and Aruin, A S (2013). Effect of light finger touch in balance control of individuals with multiple sclerosis. *Gait & Posture* 38(4): 643–647. doi:10.1016/j.gaitpost.2013.02.017.
Kavounoudias, A; Roll, R and Roll, J P (1998). The plantar sole is a 'dynamometric map' for human balance control. *Neuroreport* 9(14): 3247–3252. doi:10.1097/00001756-199810050-00021.
Kavounoudias, A; Roll, R and Roll, J P (2001). Foot sole and ankle muscle inputs contribute jointly to human erect posture regulation. *The Journal of Physiology* 532(3): 869–878. doi:10.1111/j.1469-7793.2001.0869e.x.
Kimura, T; Kouzaki, M; Masani, K and Moritani, T (2012). Unperceivable noise to active light touch effects on fast postural sway. *Neuroscience Letters* 506(1): 100–103. doi:10.1016/j.neulet.2011.10.058.
Kodesh, E; Falash, F; Sprecher, E and Dickstein, R (2015). Light touch and medio-lateral postural stability during short distance gait. *Neuroscience Letters* 584: 378–381. doi:10.1016/j.neulet.2014.10.048.
Kouzaki, M and Masani, K (2008). Reduced postural sway during quiet standing by light touch is due to finger tactile feedback but not mechanical support. *Experimental Brain Research* 188(1): 153–158. doi:10.1007/s00221-008-1426-5.
Krishnamoorthy, V; Slijper, H and Latash, M L (2002). Effects of different types of light touch on postural sway. *Experimental Brain Research* 147(1): 71–79. doi:10.1007/s00221-002-1206-6.
Lackner, J R and DiZio, P (2005). Vestibular, proprioceptive, and haptic contributions to spatial orientation. *Annual Review of Psychology* 56: 115–147. doi:10.1146/annurev.psych.55.090902.142023.
Lackner, J R et al. (1999). Precision contact of the fingertip reduces postural sway of individuals with bilateral vestibular loss. *Experimental Brain Research* 126(4): 459–466. doi:10.1007/s002210050753.

Lackner, J R and Taublieb, A B (1983). Reciprocal interactions between the position sense representations of the two forearms. *The Journal of Neuroscience* 3(11): 2280–2285.
Lackner, J R; Rabin, E and DiZio, P (2000). Fingertip contact suppresses the destabilizing influence of leg muscle vibration. *Journal of Neurophysiology* 84(5): 2217–2224.
Lackner, J R and Zabkar, J J (1977). Proprioceptive information about target location suppresses autokinesis. *Vision Research* 17(10): 1225–1229. doi:10.1016/0042-6989(77)90158-4.
Lakie, M; Caplan, N and Loram, I D (2003). Human balancing of an inverted pendulum with a compliant linkage: neural control by anticipatory intermittent bias. *The Journal of Physiology* 551(1): 357–370. doi:10.1113/jphysiol.2002.036939.
Lamont, E V and Zehr, E P (2007). Earth-referenced handrail contact facilitates interlimb cutaneous reflexes during locomotion. *Journal of Neurophysiology* 98(1): 433–442. doi:10.1152/jn.00002.2007.
Loram, I D and Lakie, M (2002a). Direct measurement of human ankle stiffness during quiet standing: the intrinsic mechanical stiffness is insufficient for stability. *The Journal of Physiology* 545(3): 1041–1053. doi:10.1113/jphysiol.2002.025049.
Loram, I D and Lakie, M (2002b). Human balancing of an inverted pendulum: position control by small, ballistic-like, throw and catch movements. *The Journal of Physiology* 540(3): 1111–1124. doi:10.1111/j.1469-7793.2002. 01111.x.
Loram, I D; Maganaris, C N and Lakie, M (2005). Active, non-spring-like muscle movements in human postural sway: how might paradoxical changes in muscle length be produced? *The Journal of Physiology* 564(1): 281–293. doi:10.1113/jphysiol.2004.073437.
Magalhaes, F H and Kohn, A F (2011). Vibratory noise to the fingertip enhances balance improvement associated with light touch. *Experimental Brain Research* 209(1): 139–151. doi:10.1007/s00221-010-2529-3.
Magalhaes, F H and Kohn, A F (2012). Imperceptible electrical noise attenuates isometric plantar flexion force fluctuations with correlated reductions in postural sway. *Experimental Brain Research* 217(2): 175–186. doi:10.1007/ s00221-011-2983-6.
Martinelli, A R; Coelho, D B; Magalhães, F H; Kohn, A F and Teixeira, L A (2015). Light touch modulates balance recovery following perturbation: from fast response to stance restabilization. *Experimental Brain Research* 233(5): 1399–1408. doi:10.1007/s00221-015-4214-z.
Mauerberg-deCastro, E et al. (2010). Haptic stabilization of posture in adults with intellectual disabilities using a nonrigid tool. *Adapted Physical Activity Quarterly* 27(3): 208–225.
Mauerberg-deCastro, E; Moraes, R and Campbell, D F (2012). Short-term effects of the use of non-rigid tools for postural control by adults with intellectual disabilities. *Motor Control* 16(2): 131–143.
McNevin, N H; Weir, P and Quinn, T (2013). Effects of attentional focus and age on suprapostural task performance and postural control. *Research Quarterly for Exercise and Sport* 84(1): 96–103. doi:10.1080/02701367.2013.762321.
McNevin, N H and Wulf, G (2002). Attentional focus on supra-postural tasks affects postural control. *Human Movement Science* 21(2): 187–202. doi:10.1016/s0167-9457(02)00095-7.
McNevin, N H; Wulf, G and Carlson, C (2000). Effects of attentional focus, self-control, and dyad training on motor learning: implications for physical rehabilitation. *Physical Therapy* 80(4): 373–385.
Menz, H B; Lord, S R and Fitzpatrick, R C (2006). A tactile stimulus applied to the leg improves postural stability in young, old and neuropathic subjects. *Neuroscience Letters* 406(1–2): 23-26. doi:10.1016/j.neulet.2006.07.014.
Metcalfe, J S et al. (2005). The temporal organization of posture changes during the first year of independent walking. *Experimental Brain Research* 161(4): 405–416. doi:10.1007/s00221-004-2082-z.
Metcalfe, J S and Clark, J E (2000). Sensory information affords exploration of posture in newly walking infants and toddlers. *Infant Behavior and Development* 23: 391–405. doi:10.1016/s0163-6383(01)00054-6.

Metcalfe, J S et al. (2005). Development of somatosensory-motor integration: an event-related analysis of infant posture in the first year of independent walking. *Developmental Psychobiology* 46(1): 19–35. doi:10.1002/dev.20037.

Priplata, A A; Niemi, J B; Harry, J D; Lipsitz, L A and Collins, J J (2003). Vibrating insoles and balance control in elderly people. *Lancet* 362(9390): 1123–1124. doi:10.1016/S0140-6736(03) 14470-4.

Rabin, E; Bortolami, S B; DiZio, P and Lackner, J R (1999). Haptic stabilization of posture: changes in arm proprioception and cutaneous feedback for different arm orientations. *Journal of Neurophysiology* 82(6): 3541–3549.

Rabin, E; Chen, J; Muratori, L; DiFrancisco-Donoghue, J and Werner, W G (2013). Haptic feedback from manual contact improves balance control in people with Parkinson's disease. *Gait & Posture* 38(3): 373–379. doi:10.1016/ j.gaitpost.2012.12.008.

Rabin, E; DiZio, P and Lackner, J R (2006). Time course of haptic stabilization of posture. *Experimental Brain Research* 170(1): 122–126. doi:10.1007/s00221-006-0348-3.

Rabin, E; DiZio, P; Ventura, J and Lackner, J R (2008). Influences of arm proprioception and degrees of freedom on postural control with light touch feedback. *Journal of Neurophysiology* 99(2): 595–604. doi:10.1152/jn.00504.2007.

Rabin, E and Gordon, A M (2006). Prior experience and current goals affect muscle-spindle and tactile integration. *Experimental Brain Research* 169(3): 407–416. doi:10.1007/s00221-005-0154-3.

Reeves, N D; Spanjaard, M; Mohagheghi, A A; Baltzopoulos, V and Maganaris, C N (2008). Influence of light handrail use on the biomechanics of stair negotiation in old age. *Gait & Posture* 28(2): 327–336. doi:10.1016/j. gaitpost.2008.01.014.

Reginella, R L; Redfern, M S and Furman, J M (1999). Postural sway with earth-fixed and body-referenced finger contact in young and older adults. *Journal of Vestibular Research* 9(2): 103–109.

Reid, S M; Novak, A C; Brouwer, B and Costigan, P A (2011). Relationship between stair ambulation with and without a handrail and centre of pressure velocities during stair ascent and descent. *Gait & Posture* 34(4): 529–532. doi:10.1016/j.gaitpost.2011.07.008.

Reynolds, R F and Osler, C J (2014). Mechanisms of interpersonal sway synchrony and stability. Journal of The Royal Society Interface 11(101). doi:10.1098/rsif.2014.0751.

Riley, M A; Stoffregen, T A; Grocki, M J and Turvey, M T (1999). Postural stabilization for the control of touching. *Human Movement Science* 18(6): 795–817. doi:10.1016/s0167-9457(99) 00041-x.

Rogers, M W; Wardman, D L; Lord, S R and Fitzpatrick, R C (2001). Passive tactile sensory input improves stability during standing. *Experimental Brain Research* 136(4): 514–522. doi:10. 1007/s002210000615.

Roll, R; Kavounoudias, A and Roll, J P (2002). Cutaneous afferents from human plantar sole contribute to body posture awareness. *Neuroreport* 13(15): 1957–1961. doi:10.1097/ 00001756-200210280-00025.

Schieppati, M; Schmid, M and Sozzi, S (2014). Rapid processing of haptic cues for postural control in blind subjects. *Clinical Neurophysiology* 125(7): 1427–1439. doi:10.1016/j.clinph. 2013.11.011.

Sozzi, S; Do, M C; Monti, A and Schieppati, M (2012). Sensorimotor integration during stance: processing time of active or passive addition or withdrawal of visual or haptic information. *Neuroscience* 212: 59–76. doi:10.1016/ j.neuroscience.2012.03.044.

Stoffregen, T A; Smart, L J; Bardy, B G and Pagulayan, R J (1999). Postural stabilization of looking. *Journal of Experimental Psychology: Human Perception and Performance* 25(6): 1641–1658. doi:10.1037/0096-1523.25.6.1641.

Tremblay, F; Mireault, A C; Dessureault, L; Manning, H and Sveistrup, H (2004). Postural stabilization from fingertip contact: I. Variations in sway attenuation, perceived stability and contact forces with aging. *Experimental Brain Research* 157(3): 275–285. doi:10.1007/ s00221-004-1830-4.

Verite, F; Bachta, W and Morel, G (2014). Closed loop kinesthetic feedback for postural control rehabilitation. *IEEE Transactions on Haptics* 7(2): 150–160. doi:10.1109/TOH.2013.64.

Vuillerme, N; Isableu, B and Nougier, V (2006). Attentional demands associated with the use of a light fingertip touch for postural control during quiet standing. *Experimental Brain Research* 169(2): 232–236. doi:10.1007/s00221-005- 0142-7.

Wing, A M; Johannsen, L and Endo, S (2011). Light touch for balance: influence of a time-varying external driving signal. *Philosophical Transactions of the Royal Society of London B: Biological Sciences* 366(1581): 3133–3141. doi:10.1098/rstb.2011.0169.

Zivotofsky, A Z; Gruendlinger, L and Hausdorff, J M (2012). Modality-specific communication enabling gait synchronization during over-ground side-by-side walking. *Human Movement Science* 31(5): 1268–1285. doi:10.1016/j. humov.2012.01.003.

Zivotofsky, A Z and Hausdorff, J M (2007). The sensory feedback mechanisms enabling couples to walk synchronously: an initial investigation. *Journal of NeuroEngineering and Rehabilitation* 4(1): 28. doi:10.1186/1743-0003-4-28.

Tactile Temporal Order

Shinya Yamamoto and Shigeru Kitazawa

The perception of **tactile temporal order**, the order of multiple touches to the skin in time, has long been a topic of research in psychophysics but has drawn new attention since the discovery that it depends not only on the stimulation interval but also on body postures in space. When two stimuli are delivered in succession one to each hand, 75 % correct judgment is achieved with a Stimulus Onset Asynchrony (SOA) of approximately 20–60 ms (as long as the hands are not crossed). However, when the arms are crossed, judgments are often inverted at longer SOAs (100–200 ms), and correct judgments are recovered at much longer SOAs (0.5–1 s) yielding an N-shaped psychometric function in some individuals. The inverted judgment and the recovery are now considered to result from an initial erroneous remapping of the touch to the spatial location of the other hand and further remapping of the signal to the correct hand thereafter. Thus, tactile temporal order depends critically on the process of localizing tactile stimuli in space, which develops over time. Multiple regions are now implicated in the judgment of tactile temporal order, including the posterior parietal cortices for remapping tactile signals in space, and the perisylvian areas that are implicated in the temporal order judgment of visual stimuli as well.

What Is Unique About Tactile Temporal Order?

The perception of temporal order between two events has been an important issue in psychophysics for many years (Hirsh and Sherrick 1961; Efron 1963; Sternberg and Knoll 1973; Pöppel 1997). However, tactile temporal order, the temporal order between two tactile stimuli, had been only one issue of many combinations of the three sensory modalities (visual, auditory, tactile, audio-visual, visuo-tactile, audio-tactile). Indeed, it was generally accepted that the brain can judge the order of

---

S. Yamamoto (✉)
National Institute of Advanced Industrial Science and Technology (AIST), Tsukuba, Japan

S. Kitazawa
Graduate School of Frontier Biosciences, Osaka University, Suita, Japan

T.J. Prescott et al. (eds.), *Scholarpedia of Touch*, Scholarpedia,
DOI 10.2991/978-94-6239-133-8_23

two stimuli that are separated in time by 20–30 ms independent of the sensory modalities of these stimuli (Hirsh and Sherrick 1961; Pöppel 1997).

In the 2000s, tactile temporal order began to attract particular attention in two contexts. First, it was discovered that tactile temporal order is dramatically altered just by crossing the hands (Yamamoto and Kitazawa 2001a; Shore et al. 2002). This was a finding unique to the tactile modality, because we can cross the hands (tactile sense organs) in space but not the eyes or ears. Second, calibration in tactile temporal order, after the repeated presentation of a pair of stimuli in a particular order, occurred in the opposite direction (Miyazaki et al. 2006) compared to lag adaptation that had been discovered in audio-visual modalities (Fujisaki et al. 2004; Vroomen et al. 2004). Sound and light with fixed intervals are likely to originate from a single event. By contrast, two tactile signals with a fixed interval are most likely to originate from two different events.

Here, we review these recent findings on tactile temporal order and discuss their general implications.

Essential Contribution of Spatial Coordinates and Body Postures to Tactile Temporal Order

Temporal Resolution in Normal Postures

When two touches are delivered one to each hand with the arms in a normal posture (uncrossed), 75 % correct judgment can be achieved with a stimulation interval of 20–60 ms (Shore et al. 2002; Röder et al. 2004; Wada et al. 2004; Shore et al. 2005; Craig and Belser 2006; Kóbor et al. 2006; Schicke and Röder 2006; Azañón and Soto-Faraco 2007; Roberts and Humphreys 2008; Fujisaki and Nishida 2009; Heed et al. 2012). To be more precise, the stimulation interval may be replaced with the stimulus onset asynchrony (SOA), which refers to the interval between the onsets of the two stimuli. The just noticeable difference (JND) is a measure of temporal resolution, which is defined as half of the difference between two SOAs that yield the "right hand first" judgment in 75 and 25 % of the trials (Figure 1b).

Hirsh and Sherrick (1961) reported that a JND of approximately 20 ms was shared across 6 combinations of the 3 different sensory modalities (visual, auditory, tactile, audio-tactile, visuo-tactile, and audio-visual temporal order judgment, TOJ). However, a recent study (Fujisaki and Nishida 2009) reported that the JND was smallest in tactile TOJ (19 ms) followed by the JNDs in audio-tactile (23 ms) and visuo-tactile (30 ms) TOJs and was largest in audio-visual TOJs (37 ms).

Sambo et al. (2013) recently reported that the JND with nociceptive stimuli increased to 80 ms as compared to 49 ms with tactile stimuli. This may be attributed to a larger variance in conduction latency with Aδ fibers that convey nociception than that with Aβ fibers that convey the sense of touch.

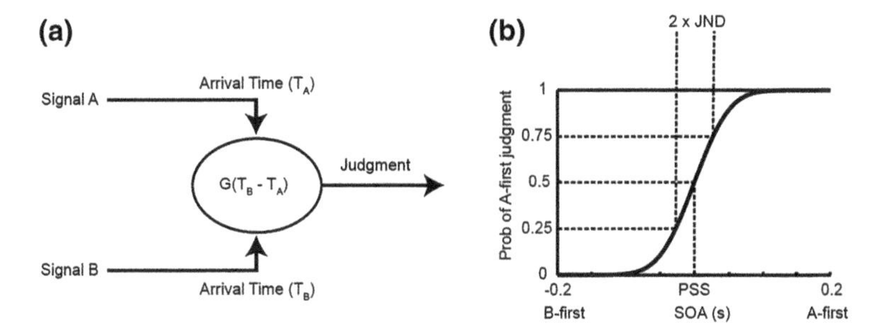

Figure 1 A general model (**a**) and an ordinary psychometric function (**b**) of temporal order judgment. Sternberg and Knoll (1973) hypothesized that there is a decision mechanism that receives signals A and B through independent channels and yields an "A-first-then-B" judgment according to the difference in the arrival times (T_B–T_A). The probability function G was hypothesized to be a non-decreasing (monotonous) function of the time difference. SOA: stimulus onset asynchrony, JND: just noticeable difference, PSS: point of subjective simultaneity

Judgment Reversal in Tactile Temporal Order Due to Arm Crossing

Tactile temporal order is dramatically altered simply by crossing the hands (Yamamoto and Kitazawa 2001a; Shore et al. 2002). When participants judged the order of two tactile stimuli with their arms crossed, the JND increased to 124 ms from 34 ms (Shore et al. 2002). The effect was beyond the mere increase in the JND (Yamamoto and Kitazawa 2001a). In some participants, the subjective temporal order was inverted at SOAs of 100–200 ms but was corrected at longer SOAs (1500 ms). As a result, the psychometric function was no longer monotonic but N-shaped (Figure 2a, b). The effect of arm crossing on tactile TOJ has now been replicated in many studies (Yamamoto and Kitazawa 2001a; Shore et al. 2002; Röder et al. 2004; Wada et al. 2004; Craig and Belser 2006; Kóbor et al. 2006; Schicke and Röder 2006; Azañón and Soto-Faraco 2007; Roberts and Humphreys 2008; Pagel et al. 2009; Heed et al. 2012; Cadieux and Shore 2013), and the mean JND in the arms-uncrossed posture was in the range of 18–58 ms but increased to 91–447 ms. It is worth noting that the JND, which implicitly assumes a monotonic function, is not useful for quantifying N-shaped responses because it yields a negative value. Instead, Yamamoto and Kitazawa (2001a) introduced a quantitative model with a probability of judgment reversal from the right-hand first to the left-hand first or vice versa, which can be fit to N-shaped or S-shaped responses. The peak probability of judgment reversal varied across participants and ranged from 0.1 to 1 with an average of ~0.3 (Yamamoto and Kitazawa 2001a; Wada et al. 2004).

A reversal of tactile temporal order also occurs when two legs are crossed (with foot stimulation) or when one arm and the contralateral leg are crossed (Schicke and Röder 2006). It even occurs without crossing the arms when crossing two drumsticks (one held in each hand) (Figure 2c) when participants are required to judge

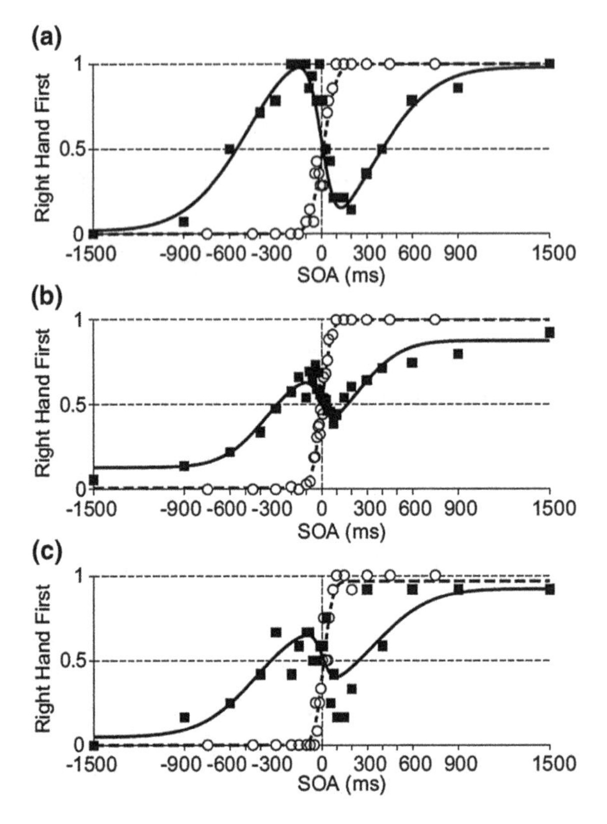

Figure 2 Reversal of subjective temporal order due to arm (**a**, **b**) and stick (**c**) crossing. **a** The judgment probability that the *right hand* was stimulated earlier than the *left hand* is plotted against the stimulation interval. A positive SOA indicates that the *right hand* was stimulated first. The response *curve* was a classical sigmoid when the arms were uncrossed (*black circles* and *curves*), whereas it became N-shaped in the exemplified subject when the arms were crossed (*red dots* and *curves*). **b** Pooled data from eight naive subjects who crossed the *left hand* over the *right hand* in the crossed-arm condition (*red dots* and *curves*). The N-shape is still apparent. **c** Data from a subject with an apparent judgment reversal when the sticks were crossed without crossing the arms (*red dots* and *curve*). Stimulation was delivered to the tip of each stick, and the subject was required to judge which of the two tips was stimulated first. **Panel A was reproduced from Yamamoto and Kitazawa** (2001a)**. Panels a–c were reproduced from Kitazawa et al.** (2008) **with permission from Oxford University Press (In: Sensorimotor Foundations of Higher Cognition, page 76)**

the order of stimuli that were delivered to the tips of the sticks with their eyes closed (Yamamoto and Kitazawa 2001b). Straight sticks can be replaced with L-shaped (Yamamoto et al. 2005) or virtual tools in virtual reality (Moizumi et al. 2007).

These reversals in tactile temporal order occur when participants respond with their hand with their eyes closed. Under these conditions, it is reasonable to base their judgments simply on the locations of the stimuli on the body surface. Nonetheless, tactile temporal order was reversed. This indicates that tactile signals are not ordered in time while the signals are represented somatotopically. Tactile temporal order is determined only after the tactile signals are referred to a relevant location in space, which could be the hand itself or the tip of a tool in the hand.

Other Evidence for the Involvement of Spatial Coordinates

The perceived hand positions in space affect tactile temporal order even when the arms are not crossed. The JND decreases as the distance of two hands increase (Shore et al. 2005; Kuroki et al. 2010). The JND increases when the two hands are virtually placed close together with a mirror (Gallace and Spence 2005). Hermosillo et al. (2011) demonstrated that the planning of arm movements from uncrossed to crossed impairs tactile TOJ, even when the stimuli are delivered prior to the onset of actual arm-crossing movements. The subjective temporal order is affected by future (planned) arm configuration.

Involvement of Local Motion and Global "Apparent Motion"

Craig (2003) and Craig and Busey (2003) examined TOJs of two tactile stimuli presented to two finger pads when each of the two stimuli simulated a local motion of the skin. They found that the judgments were affected by the local tactile motions. The judgments were biased toward the correct judgment, (or toward the incorrect judgment) according to whether the direction of the local motions were consistent (or inconsistent) with the global "apparent motion" defined by the two successive stimuli across the finger pads. These results clearly demonstrate that local tactile motions, which should be represented in the somatotopic coordinates in the initial stage, interact with the direction of the global motion vector between the two hands defined in the external coordinates. Another study (Kitazawa et al. 2008) reported that tactile TOJ was affected by visual distractors that defined a global motion vector between the two hands. By using visual distractors that participants were told to ignore, some participants yielded N-shaped response curves even when the arms were not crossed. Takahashi et al. (2013) reported that the participants felt a sense of "motion" in >70 % of trials when the two stimuli were delivered one to each hand with SOAs of 50–200 ms. The sense of "motion" occurred irrespective of whether the arms were crossed or uncrossed even though the participants closed their eyes.

Altogether, we can speculate that tactile TOJ involves signals regarding a global motion vector defined between the spatial locations of tactile stimuli in the external coordinates.

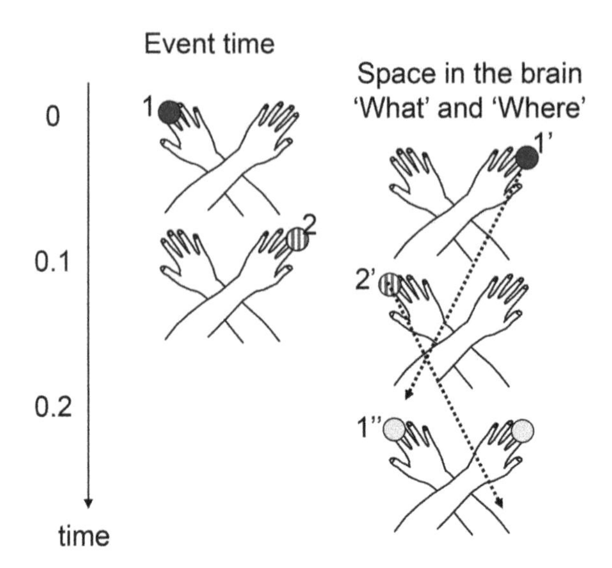

Figure 3 Two steps in spatial remapping. It is now generally accepted that a stimulus to a crossed hand (1) is initially remapped to the wrong hand (1') but remapped further to the correct hand thereafter (1"). When the second stimulus (2) is delivered before the first stimulus is correctly remapped, the second stimulus is also remapped to the wrong hand (2'). Tactile temporal order is then inverted in the space coordinates in the brain. **The figure was reproduced from Kitazawa et al.** (2008) **with permission from Oxford University Press (In: Sensorimotor Foundations of Higher Cognition, page 91)**

Spatial Remapping of Tactile Stimuli Implicated for Judgment Reversal

It is now generally accepted that the judgment reversal due to arm crossing occurs because of an initial erroneous remapping of the tactile signal to the spatial position of the wrong (contralateral) hand (Yamamoto and Kitazawa 2001a; Kitazawa 2002; Fujisaki et al. 2012; Heed and Azañón 2014). The tactile signal, initially remapped to the wrong position (1' in Figure 3), is then remapped to the correct hand thereafter (1" in Figure 3). The dynamic process of remapping is supported by several findings as follows. Groh and Sparks (1996) reported that some trajectories of saccadic eye movements in response to a touch to the crossed hand began in the direction of the wrong hand, curved toward the correct hand, and reached the correct hand approximately 400 ms after the delivery of the stimulus. A recent study examined the finding in more detail (Overvliet et al. 2011). By using a cross-modal cueing paradigm, Azañón and Soto-Faraco (Azañón and Soto-Faraco 2008; Azañón et al. 2010b) demonstrated that a touch to a crossed hand initially facilitated a response to a visual stimulus to the wrong hand but then facilitated a response to a visual stimulus to the correct hand 180–360 ms after the touch.

What happens when two stimuli are delivered one to each of the crossed hands (1 and 2 in Figure 3) before the first stimulus is remapped to the correct hand?

We may speculate that both stimuli are mapped to the wrong hand (1' and 2' in Figure 3). That is, tactile temporal order is inverted in the space represented in the brain. By contrast, when the second stimulus is delivered with an interval greater than 500 ms, after the first stimulus is remapped to the correct hand, the second stimulus is also remapped to the correct hand after an interval, and the correct judgment can be recovered.

Recent studies using functional imaging (Wada et al. 2012), electroencephalography (Heed and Röder 2010; Soto-Faraco and Azañón 2013) and transcranial magnetic stimulation (Azañón et al. 2010a) suggested that the posterior parietal cortex is involved in the process of tactile remapping.

Congenitally (early) blind people do not demonstrate reversals in tactile TOJ even when their arms are crossed but late blind participants (aged 12 years and above) do (Röder et al. 2004; Collignon et al. 2009). Pagel et al. (2009) demonstrated that the crossing effect on tactile TOJ was not observed before the age of 5 years but is observed after the age of 5 1/2 years. Those data suggest that the use of external coordinates for tactile localization is acquired by the age of 5 years and requires normal vision. Notably, the crossing effect is smaller when the normal sighted cross the hands in the back than when they cross their hands in front (Kóbor et al. 2006). These findings further suggest that the development of the crossing effect requires experience to direct the eyes to the position of touch on the skin.

A Model of Tactile TOJ

A model of tactile TOJ (a motion projection hypothesis) that accounts for all of the findings explained above has been proposed (Figure 4) (Kitazawa et al. 2008; Fujisaki et al. 2012; Takahashi et al. 2013). In the model, two successive tactile signals to both hands are initially represented in somatotopic coordinates (such as in the primary sensory cortex) but are then remapped to spatial coordinates (such as in the parietal cortex). These signals also evoke a global motion signal in the perisylvian areas. The separate information regarding "what happened where" and the motion vector is finally integrated to construct a perception regarding "what happened in which order".

However, it is worth noting that the model does not apply to all cases in tactile TOJ. For example, Roberts and Humphrey (2008) reported that crossing the arms had no effect on tactile TOJ when order was judged by the frequency or duration of the tactile stimuli. The underlying mechanisms should be different in those tasks that do not require the discrimination of one hand from the other.

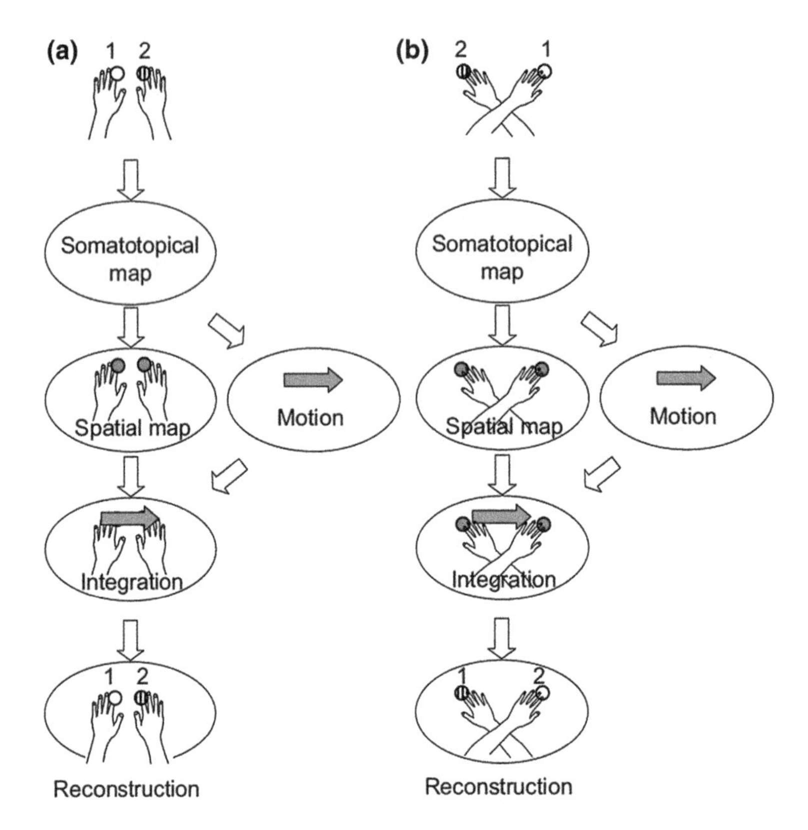

Figure 4 Motion-projection hypothesis. Tactile events are hypothesized to be ordered in time by combining information regarding "what happened where" represented in external spatial coordinates and information regarding "when". The "when" information is hypothesized to be captured in a motion vector. Illustrations show the process of correct judgment in the arms-uncrossed condition (**a**) and inverted judgment in the arms-crossed condition (**b**). The same networks are hypothesized to be activated in both cases. **Adapted from Takahashi et al.** (2013) **with permission from Oxford University Press**

Neuroimaging

A functional imaging study (Takahashi et al. 2013) examined the brain regions involved in judgments of the temporal order of successive taps delivered to both hands. A numerosity judgment task of similar difficulty with identical tactile stimulation was used as a control. In both arm postures (arms crossed and uncrossed) the following regions were activated: the bilateral premotor cortices, the bilateral middle frontal gyri, the bilateral inferior parietal cortices and supramarginal gyri, and the bilateral posterior part of the superior and middle temporal gyri.

Activation in the premotor and inferior parietal areas agrees with the involvement of spatial coordinates. Stronger activation was found in the parietal region, which is implicated in the process of remapping tactile stimuli to spatial coordinates when the participants crossed their arms.

The activation in the perisylvian areas was close to (and partially overlapped with) the visual-motion–sensitive areas. This was also in good agreement with the suggested involvement of the motion vector defined between the two hands. Activation close to the perisylvian area was also reported during a visual TOJ task (Davis et al. 2009).

Temporal Order Judgments and Simultaneity Judgments

Simultaneity judgment (SJ) refers to a judgment of whether two signals are simultaneous or not. The probability of the judgment that two signals are simultaneous is generally approximated by a Gaussian function of the SOA with a peak at approximately zero (e.g., Fujisaki and Nishida 2009). The SOA that yields the peak probability of simultaneity judgment is generally close to the point of subjective simultaneity (PSS) during TOJ, which is defined as an SOA where the order judgment probability is 0.5 (Figure 1b). This may yield the impression that the two stimuli are perceived as simultaneous at the PSS during TOJ. However, this is not necessarily true (Weiss and Scharlau 2011). Even if we are totally uncertain about the temporal order of two stimuli, we can still perceive them as not being simultaneous. This point is clearly demonstrated in the participant illustrated in Figure 2a. The participant's N-shaped response curve cut the level of 0.5 at 3 points; one near zero; and the others at −600 and +400 ms. At the two points with SOAs of ~500 ms, the stimuli cannot be perceived as simultaneous, though the order was totally ambiguous for the participant. Indeed, it has been repeatedly shown that crossing the arms does not impair SJ at all (Axelrod et al. 1968; Geffen et al. 2000; Fujisaki and Nishida 2009). We are not necessarily able to judge the temporal order even if we are certain that they are not simultaneous. That is, some mechanisms for TOJ do not overlap with those for SJ (Hirsh and Sherrick 1961; Ulrich 1987; Jaśkowski 1991; Yamamoto and Kitazawa 2001a; Shore et al. 2005; Fujisaki and Nishida 2009).

Patients with Parkinson's disease (PD), a neurodegenerative disease strongly associated with basal ganglia dysfunction, demonstrate deficits in temporal processing when performing temporal reproduction tasks (Pastor et al. 1992a; O'Boyle et al. 1996; Harrington et al. 1998), perceptual timing tasks (Pastor et al. 1992b; Malapani et al. 1998), SJs, and temporal discrimination between two stimuli (Artieda et al. 1992; Fiorio et al. 2008; Conte et al. 2010). A neuroimaging study of healthy participants confirmed that the basal ganglia and the substantia nigra pars compacta are involved in the reproduction of short and long time intervals (Jahanshahi et al. 2006). These previous findings suggest that dopaminergic transmission in the basal ganglia and related cortical structures play an important role in interval timing tasks, including SJ (for review see Buhusi and Meck 2005; Merchant et al. 2013). However, patients with PD do not necessarily demonstrate deficits in TOJ compared with their age-matched controls (Nelson et al. 2012; Nishikawa et al. 2015). It is possible that SJ may depend more critically on dopaminergic transmission in the basal ganglia than TOJ.

Taken together, the mechanisms underlying tactile TOJ are not identical to SJ, though some overlap may exist.

Effects of Past Experiences on Tactile Temporal Order

After repeated exposures to a constant time lag of audiovisual stimulus pairs, the lag tends to be ignored to make participants perceive the audiovisual stimuli as being simultaneous (Figure 5b) (Fujisaki et al. 2004; Vroomen et al. 2004). This is called lag adaptation, which is considered to be helpful for binding two signals that have arrived in the brain with a certain delay but actually originated from a single event. Due to the difference between the conduction velocity of light (3 × 1010 m/s) and sound (3 × 102 m/s), the light and sound that occurred simultaneously from a single

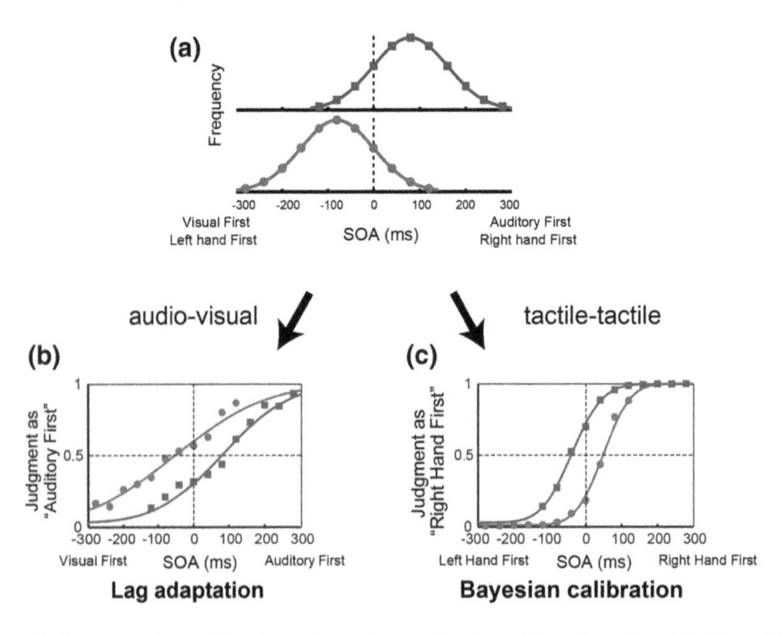

Figure 5 Two opposing calibrations observed in audiovisual (**b**) and tactile (**c**) TOJs. **a** Biased distributions of SOAs with positive (*blue*) and negative (*red*) peaks. Positive SOA indicates "auditory first" in audiovisual TOJ and "*right hand* first" in tactile TOJ. **b–c** Opposite types of temporal calibration: lag adaptation (**b**) and Bayesian calibration (**c**). The probability of "auditory first" (**b**) and "right hand first" (**c**) judgments (ordinate) is plotted against the SOAs. The psychometric function shifts toward the peak of each distribution of SOAs in audiovisual TOJs (**b**, lag adaptation), while it shifts away from the peak in tactile TOJs (**c**, Bayesian calibration). **The figure was reproduced from Miyazaki et al.** (2006)

event arrive at the sensory organs with some delay (light then sound) depending on the distance between the event location and the head. Thus, it is reasonable to ignore the constant lag. However, it is less likely that two tactile stimuli to both hands originated from a single event because the spatial locations of the two hands are not always identical. In fact, lag adaptation is not observed in tactile temporal order judgment when stimuli are delivered one to each hand. Instead, the opposite phenomenon is observed. When the right (left) hand is stimulated earlier than the left (right) hand, subjects tend to have more judgments of right-hand-first (left-hand-first) (Figure 5c). Since it conforms to Bayesian integration theory, the phenomenon is referred to as “Bayesian calibration” (Miyazaki et al. 2006). More specifically, this type of perceptual change can be considered as a learning of “prior” probability in the Bayesian terminology. By contrast, the lag adaptation can be considered as a learning of “likelihood” (Sato and Aihara 2011; Fujisaki et al. 2012). How these two types of perceptual changes are implemented in the brain should be the focus of future studies.

Future Directions

N-shaped response curves (Figure 2a), discovered by crossing the arms, cannot be explained by the long-held decision center model with two independent channels (Figure 1a). The discovery unique to tactile temporal order revealed that the process of TOJ is more complex than had once been believed. A proposed model hypothesizes that spatial information (what happened where) and temporal information (direction of motion vector) are combined, and the most likely scenario on what happened where in which order is constructed (Figure 4). The validity of the hypothesis in other sensory modalities and its neural correlates warrant further investigation.

Another line of studies in tactile temporal order have revealed that repeated exposure to stimuli with a certain order results in a calibration of PSS in the opposite direction (Bayesian calibration) compared to the lag adaptation discovered during audiovisual TOJ (Figure 5). The new finding in tactile temporal order has already been tested with audiovisual TOJ, and it was confirmed that Bayesian calibration occurs even in the audiovisual TOJ (Yamamoto et al. 2012). Generalizing the discoveries in tactile temporal order to other modalities will provide important hints to uncover the mechanisms of temporal order information processing in general.

References

Artieda, J; Pastor, M A; Lacruz, F and Obeso, J A (1992). Temporal discrimination is abnormal in Parkinson's disease. *Brain* 115 (Pt. 1): 199–210.

Axelrod, S; Thompson, L W and Cohen, L D (1968). Effects of senescence on the temporal resolution of somesthetic stimuli presented to one hand or both. *Journal of Gerontology* 23: 191–195.

Azañón, E and Soto-Faraco, S (2007). Alleviating the 'crossed-hands' deficit by seeing uncrossed rubber hands. *Experimental Brain Research* 182: 537–548.

Azañón, E and Soto-Faraco, S (2008). Changing reference frames during the encoding of tactile events. *Current Biology* 18: 1044–1049.

Azañón, E; Longo, M R; Soto-Faraco, S and Haggard, P (2010a). The posterior parietal cortex remaps touch into external space. *Current Biology* 20: 1304–1309.

Azañón, E; Camacho, K and Soto-Faraco, S (2010b). Tactile remapping beyond space. *European Journal of Neuroscience* 31: 1858–1867.

Buhusi, C V and Meck, W H (2005). What makes us tick? Functional and neural mechanisms of interval timing. *Nature Reviews Neuroscience* 6: 755–765.

Cadieux, M L and Shore, D I (2013). Response demands and blindfolding in the crossed-hands deficit: An exploration of reference frame conflict. *Multisensory Research* 26: 465–482.

Collignon, O; Charbonneau, G; Lassonde, M and Lepore, F (2009). Early visual deprivation alters multisensory processing in peripersonal space. *Neuropsychologia* 47: 3236–3243.

Conte, A et al. (2010). Subthalamic nucleus stimulation and somatosensory temporal discrimination in Parkinson's disease. *Brain* 133: 2656–2663.

Craig, J C and Busey, T A (2003). The effect of motion on tactile and visual temporal order judgments. *Perception & Psychophysics* 65: 81–94.

Craig, J C (2003). The effect of hand position and pattern motion on temporal order judgments. *Perception & Psychophysics* 65: 779–788.

Craig, J C and Belser, A N (2006). The crossed-hands deficit in tactile temporal-order judgments: The effect of training. *Perception* 35: 1561–1572.

Davis, B; Christie, J and Rorden, C (2009). Temporal order judgments activate temporal parietal junction. *The Journal of Neuroscience* 29: 3182–3188.

Efron, R (1963). The effect of handedness on the perception of simultaneity and temporal order. *Brain* 86: 261–284.

Fiorio, M (2008). Subclinical sensory abnormalities in unaffected PINK1 heterozygotes. *Journal of Neurology* 255: 1372–1377.

Fujisaki, W; Shimojo, S; Kashino, M and Nishida, S (2004). Recalibration of audiovisual simultaneity. *Nature Neuroscience* 7: 773–778.

Fujisaki, W and Nishida, S (2009). Audio-tactile superiority over visuo-tactile and audio-visual combinations in the temporal resolution of synchrony perception. *Experimental Brain Research* 198: 245–259.

Fujisaki, W; Kitazawa, S and Nishida, S (2012). Multisensory timing. In: B Stein (Ed.), *The New Handbook of Multisensory Processes* (pp. 301–318). Cambridge: MIT Press.

Gallace, A and Spence, C (2005). Visual capture of apparent limb position influences tactile temporal order judgments. *Neuroscience Letters* 379: 63–68.

Geffen, G; Rosa, V and Luciano, M (2000). Effects of preferred hand and sex on the perception of tactile simultaneity. *Journal of Clinical and Experimental Neuropsychology* 22: 219–231.

Groh, J M and Sparks, D L (1996). Saccades to somatosensory targets. I. Behavioral characteristics. *Journal of Neurophysiology* 75: 412–427.

Harrington, D L; Haaland, K Y and Hermanowicz, N (1998). Temporal processing in the basal ganglia. *Neuropsychology* 12: 3–12.

Heed, T and Röder, B (2010). Common anatomical and external coding for hands and feet in tactile attention: evidence from event-related potentials. *Journal of Cognitive Neuroscience* 22: 184–202.

Heed, T; Backhaus, J and Röder, B (2012). Integration of hand and finger location in external spatial coordinates for tactile localization. *Journal of Experimental Psychology: Human Perception and Performance* 38: 386–401.
Heed, T and Azañón, E (2014). Using time to investigate space: A review of tactile temporal order judgments as a window onto spatial processing in touch. *Frontiers in Psychology* 5: 76.
Hermosillo, R; Ritterband-Rosenbaum, A and van Donkelaar, P (2011). Predicting future sensorimotor states influences current temporal decision making. *The Journal of Neuroscience* 31: 10019–10022.
Hirsh, I J and Sherrick, C E, Jr. (1961). Perceived order in different sense modalities. *Journal of Experimental Psychology* 62: 423–432.
Jahanshahi, M; Jones, C R; Dirnberger, G and Frith, C D (2006). The substantia nigra pars compacta and temporal processing. *The Journal of Neuroscience* 26: 12266–12273.
Jaśkowski, P (1991). Two-stage model for order discrimination. *Perception & Psychophysics* 50: 76–82.
Kitazawa, S (2002). Where conscious sensation takes place. *Consciousness and Cognition* 11: 475–477.
Kitazawa, S et al. (2008). Reversal of subjective temporal order due to sensory and motor integrations. In: P Haggard, M Kawato, Y Rossetti (Eds.), *Attention and Performance XXII* (pp. 73–97). New York: Oxford University Press.
Kóbor, I; Furedi, L; Kovacs, G; Spence, C and Vidnyanszky, Z (2006). Back-to-front: Improved tactile discrimination performance in the space you cannot see. *Neuroscience Lettters* 400: 163–167.
Kuroki, S; Watanabe, J; Kawakami, N; Tachi, S and Nishida, S (2010). Somatotopic dominance in tactile temporal processing. *Experimental Brain Research* 203: 51–62.
Malapani, C et al. (1998). Coupled temporal memories in Parkinson's disease: A dopamine-related dysfunction. *Journal of Cognitive Neuroscience* 10: 316–331.
Merchant, H; Harrington, D L and Meck, W H (2013). Neural basis of the perception and estimation of time. *Annual Review of Neuroscience* 36: 313–336.
Miyazaki, M; Yamamoto, S; Uchida, S and Kitazawa, S (2006). Bayesian calibration of simultaneity in tactile temporal order judgment. *Nature Neuroscience* 9: 875–877.
Moizumi, S; Yamamoto, S and Kitazawa, S (2007). Referral of tactile stimuli to action points in virtual reality with reaction force. *Neuroscience Research* 59: 60–67.
Nelson, A J et al. (2012). Dopamine alters tactile perception in Parkinson's disease. *The Canadian Journal of Neurological Sciences. Le Journal Canadien des Sciences Neurologiques* 39: 52–57.
Nishikawa, N; Shimo, Y; Wada, M; Hattori, N and Kitazawa, S (2015). Effects of aging and idiopathic Parkinson's disease on tactile temporal order judgment. *PLoS ONE* 10(3): e0118331.
O'Boyle, D J; Freeman, J S and Cody, F W (1996). The accuracy and precision of timing of self-paced, repetitive movements in subjects with Parkinson's disease. *Brain* 119 (Pt 1): 51–70.
Overvliet, K E; Azañón, E and Soto-Faraco, S (2011). Somatosensory saccades reveal the timing of tactile spatial remapping. *Neuropsychologia* 49: 3046–3052.
Pöppel, E (1997). A hierarchical model of temporal perception. *Trends in Cognitive Sciences* 1: 56–61.
Pagel, B; Heed, T and Röder, B (2009). Change of reference frame for tactile localization during child development. *Developmental Science* 12: 929–937.
Pastor, M A; Jahanshahi, M; Artieda, J and Obeso, J A (1992a). Performance of repetitive wrist movements in Parkinson's disease. *Brain* 115(Pt 3): 875–891.
Pastor, M A; Artieda, J; Jahanshahi, M and Obeso, J A (1992b). Time estimation and reproduction is abnormal in Parkinson's disease. *Brain* 115(Pt 1): 211–225.
Roberts, R D and Humphreys, G W (2008). Task effects on tactile temporal order judgments: when space does and does not matter. *Journal of Experimental Psychology: Human Perception and Performance* 34: 592–604.
Röder, B; Rosler, F and Spence, C (2004). Early vision impairs tactile perception in the blind. *Current Biology* 14: 121–124.

Sambo, C F et al. (2013). The temporal order judgement of tactile and nociceptive stimuli is impaired by crossing the hands over the body midline. *Pain* 154: 242–247.

Sato, Y and Aihara, K (2011). A bayesian model of sensory adaptation. *PloS ONE* 6: e19377.

Schicke, T and Röder, B (2006). Spatial remapping of touch: confusion of perceived stimulus order across hand and foot. *Proceedings of the National Academy of Sciences of the United States of America* 103: 11808–11813.

Shore, D I; Spry, E and Spence, C (2002). Confusing the mind by crossing the hands. *Brain Research. Cognitive Brain Research* 14: 153–163.

Shore, D I; Gray, K; Spry, E and Spence, C (2005). Spatial modulation of tactile temporal-order judgments. *Perception* 34: 1251–1262.

Soto-Faraco, S and Azañón, E (2013). Electrophysiological correlates of tactile remapping. *Neuropsychologia* 51: 1584–1594.

Sternberg, S and Knoll, R (1973). The perception of temporal order: fundamental issues and a general model. In: S Kornblum (Ed.), *Attention and Performance IV* (pp. 629–685). New York: Academic Press.

Takahashi, T; Kansaku, K; Wada, M; Shibuya, S and Kitazawa, S (2013). Neural correlates of tactile temporal-order judgment in humans: An fMRI study. *Cerebral Cortex* 23: 1952–1964.

Ulrich, R (1987). Threshold models of temporal-order judgments evaluated by a ternary response task. *Perception & Psychophysics* 42: 224–239.

Vroomen, J; Keetels, M; de Gelder, B and Bertelson, P (2004). Recalibration of temporal order perception by exposure to audio-visual asynchrony. *Brain Research. Cognitive Brain Research* 22: 32–35.

Wada, M; Yamamoto, S and Kitazawa, S (2004). Effects of handedness on tactile temporal order judgment. *Neuropsychologia* 42: 1887–1895.

Wada, M et al. (2012). Spatio-temporal updating in the left posterior parietal cortex. *PLoS ONE* 7: e39800.

Weiss, K and Scharlau, I (2011). Simultaneity and temporal order perception: Different sides of the same coin? Evidence from a visual prior-entry study. *Quarterly Journal of Experimental Psychology (Hove)* 64: 394–416.

Yamamoto, S and Kitazawa, S (2001a). Reversal of subjective temporal order due to arm crossing. *Nature Neuroscience* 4: 759–765.

Yamamoto, S and Kitazawa, S (2001b). Sensation at the tips of invisible tools. *Nature Neuroscience* 4: 979–980.

Yamamoto, S; Moizumi, S and Kitazawa, S (2005). Referral of tactile sensation to the tips of L-shaped sticks. *Journal of Neurophysiology* 93: 2856–2863.

Yamamoto, S; Miyazaki, M; Iwano, T and Kitazawa, S (2012). Bayesian calibration of simultaneity in audiovisual temporal order judgments. *PloS ONE* 7: e40379.

Tactile Suppression

C. Elaine Chapman and François Tremblay

Tactile suppression commonly refers to the reduction in tactile perception that occurs during movement, or what is also called movement-related gating. The function of tactile suppression is most likely to suppress redundant movement-related feedback that can be predicted from the motor command so that the perception of unexpected or novel inputs is enhanced. The central motor command plays a key role in generating tactile suppression. Peripheral feedback from the moving limb also contributes because tactile suppression is seen during passive movement, i.e. in the absence of a motor command.

Introduction

The senses provide continuous information about the extra- and intra-personal environment that we inhabit and, together with the motor systems, allow us to both interpret and interact with the environment. The somatosensory system occupies a particularly privileged position in this schema since primary somatosensory cortex, S1, is the only primary sensory receiving area to have direct, and reciprocal, connections with primary motor cortex, M1, and so direct access to the motor system. One challenge for the Central Nervous System (CNS) is, however, to process the vast amount of somatosensory input to the CNS at any moment of time since the receptor sheet for discriminative touch covers the entire surface of the body ($\sim 2\ m^2$). Not too surprisingly, the brain has developed mechanisms to control the flow of sensory inputs to the central processing regions, with tactile suppression during movement (also referred to as movement-related gating) being an example of one such control mechanism. While the following text refers mainly to results

C.E. Chapman (✉)
University of Montreal, Montreal, Canada

F. Tremblay
School of Rehabilitation Sciences, University of Ottawa, Ottawa, ON, Canada

T.J. Prescott et al. (eds.), *Scholarpedia of Touch*, Scholarpedia,
DOI 10.2991/978-94-6239-133-8_24

obtained in primates, tactile suppression is not limited to higher mammals but has also been intensively studied in rodents specifically in relation to whisking movements of the vibrissae (for a review, see Kleinfeld et al. 2006).

Tactile Suppression of Perception

During movement, near threshold tactile stimuli (intensities that can be detected at rest) are not detected, i.e. they are totally suppressed (Chapman et al. 1987; Post et al. 1994). Similar results are obtained with mechanical (e.g. vibration) and electrical stimulation. The decrease in perceptual sensitivity begins well before the onset of movement (Figure 1B(a)). Tactile suppression is maximal (no stimuli detected) at about the same time as the limb begins to move. When the timing of the decrease is examined relative to the onset of electrical activity in the muscles involved in the movement (EMG, electromyographic activity), i.e. relative to the time that the motor command arrives at the periphery (Figure 1B(b)), it can be seen that detection concomitantly begins to decrease prior to EMG onset (Williams et al. 1998). Such observations suggest that the motor command contributes to generating tactile suppression.

As the intensity of the test stimulus is increased (not illustrated) the degree of suppression declines so that stronger tactile stimuli are all perceived, but their subjective intensity is diminished (Williams and Chapman 2000). If the test stimulus is sufficiently strong, then there may be no drop in subjective intensity during movement. Finally, the ability to discriminate differences in the intensity of pairs of

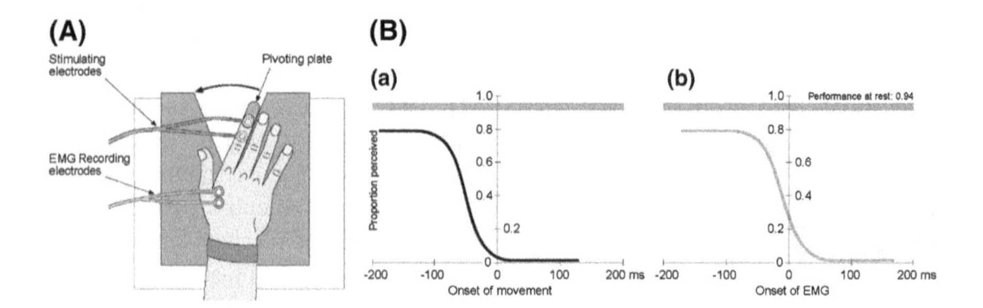

Figure 1 Suppression of tactile detection before voluntary movement. **A** *Motor task*, abduction of the index finger in response to a visual go cue. *Perceptual task*, detect the presence of a near-threshold electrical stimulus applied to the index finger either during the motor task or at rest. **B** Time-course for the modulation of tactile detection in relation to the onset of movement (**a**) and electromyographic (EMG) activity (**b**). The *shaded bar* (*top*) shows performance at rest; the latter is better than baseline performance during the motor task (200 ms before movement onset), an effect attributed to divided attention (dual task situation). Adapted from Williams et al. (1998)

tactile stimuli (just noticeable difference) is unaffected during movement, indicating that the relative differences between tactile inputs are preserved during movement even though their subjective intensity may be decreased (Chapman et al. 1987; Post et al. 1994).

The degree of tactile suppression with movement is modulated by several factors. The speed of the movement is important (Cybulska-Klosowicz et al. 2011): there is no attenuation of tactile detection for very slow movements, <200 mm/s or 25°/s, corresponding to speeds often used in tactile exploration (Smith et al. 2002). Beyond this, faster movements result in greater suppression. The physical relation between the movement and the stimulus is also critical, with greater and earlier suppression occurring with closer physical proximity between the active muscles and the tactile stimulus (Williams et al. 1998). Moreover, there is an anatomical limit to the suppressive effects with the effects being largely, if not exclusively, restricted to the moving limb. The critical factor for tactile suppression is motor activity and not movement per se, since similar sensory attenuation is seen with isometric contractions, when no overt movement is produced (Williams and Chapman 2002). Tactile suppression can be elicited in the absence of descending motor commands when the limb is passively moved, indicating that reafferent feedback also has a role in modulating sensory perception (Williams and Chapman 2002; Chapman and Beauchamp 2006).

Sensory inputs that are self-induced also show evidence of a modest reduction in their subjective intensity during the movement that generates the sensation (e.g. Bays et al. 2005, 2006). Under certain conditions, this modulation appears to be entirely central in origin, dependent on the predicted sensory consequences of the movement. This mechanism differs from the modulation seen for externally applied stimuli, however, since the effects are not limited to the moving limb but extend to include the contralateral limb. This predictive mechanism, triggered by the motor command, helps to explain why we cannot tickle ourselves (Weiskrantz et al. 1971).

To summarize, the signals responsible for suppressing or modulating tactile perception arise from at least two different sources—the motor command (efference copy or corollary discharge) and the sensory feedback associated with the movement or motor activity (reafference). Finally, peripheral feedback alone can suppress the detection of threshold-level stimuli "prior" to the onset of movement (Williams and Chapman 2002; Chapman and Beauchamp 2006), an effect attributed to backward masking whereby the strong signal from movement reafference suppresses the perception of an earlier, but weak, test stimulus. Interestingly, similar effects are also reported in the visual system (reviewed in Wurtz 2008), with both the motor command and visual feedback contributing to visual suppression. Furthermore, there is evidence for a suppression of proprioception ("muscular sense") as well (Collins et al. 1998).

Neurophysiological Substrates

Tactile inputs from cutaneous mechanoreceptors of, for example, the hand are relayed through the dorsal column-medial lemniscal (ML) pathway to sensory thalamus (ventral posterior lateral nucleus, VPL) and thence to S1. Inputs to the dorsal column nuclei are both direct (collaterals from primary afferents) and indirect (after first synapsing in the spinal dorsal horn). The central neural mechanisms underlying tactile suppression have been studied by recording the amplitude of short-latency somatosensory evoked responses (SEPs) to tactile stimulation in the alert monkey, with the test stimulus being applied at various times prior to and during movement. Figure 2 shows representative recordings from S1 taken with the animal at rest (left, inter trial interval) and while performing a simple reaction-time task (right). In this example, the tactile stimulus, an air puff directed to the centre of the receptive field on the forearm, was applied shortly after the onset of movement (elbow flexion), i.e. after the onset of electromyographic (EMG) activity in the biceps, the main agonist. There was a pronounced decrease in the amplitude of the S1 cortical SEP during movement (Chapman et al. 1988), consistent with a reduction in the transmission of tactile inputs to S1. Short-latency single unit responses to natural tactile stimuli (air puff) in S1 are likewise decreased prior to and during voluntary movement, consistent with a profound decrease in S1 responsiveness to tactile inputs before and during movement (Jiang et al. 1991).

The time-course for movement-related SEP modulation in S1 is shown in Figure 3a. The SEP begins to decrease at about the onset of EMG activity (first vertical line), with the maximal decrease occurring after the onset of movement

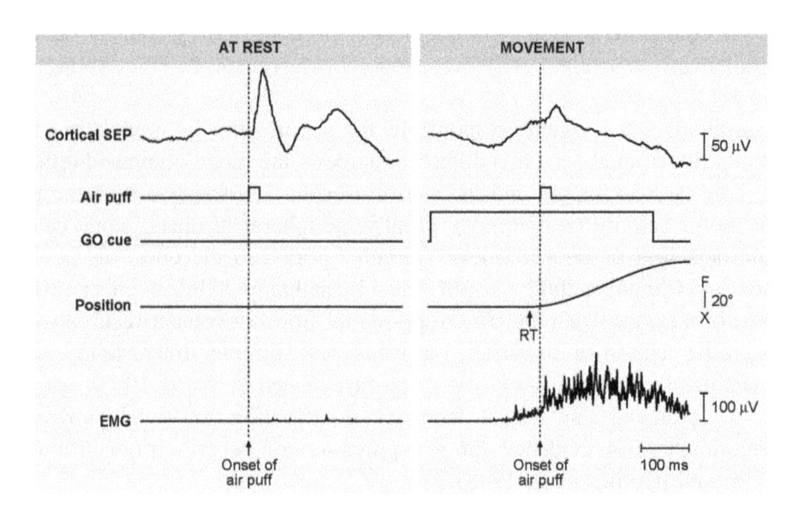

Figure 2 Intracortical somatosensory evoked potentials (SEPs) recorded from S1 in the macaque are decreased in amplitude when the stimulus (air puff applied to the receptive field on the forearm) is applied during movement (elbow flexion in response to a visual go cue). Response at rest (no movement) also shown. Reprinted, with permission, from Jiang et al. (1990)

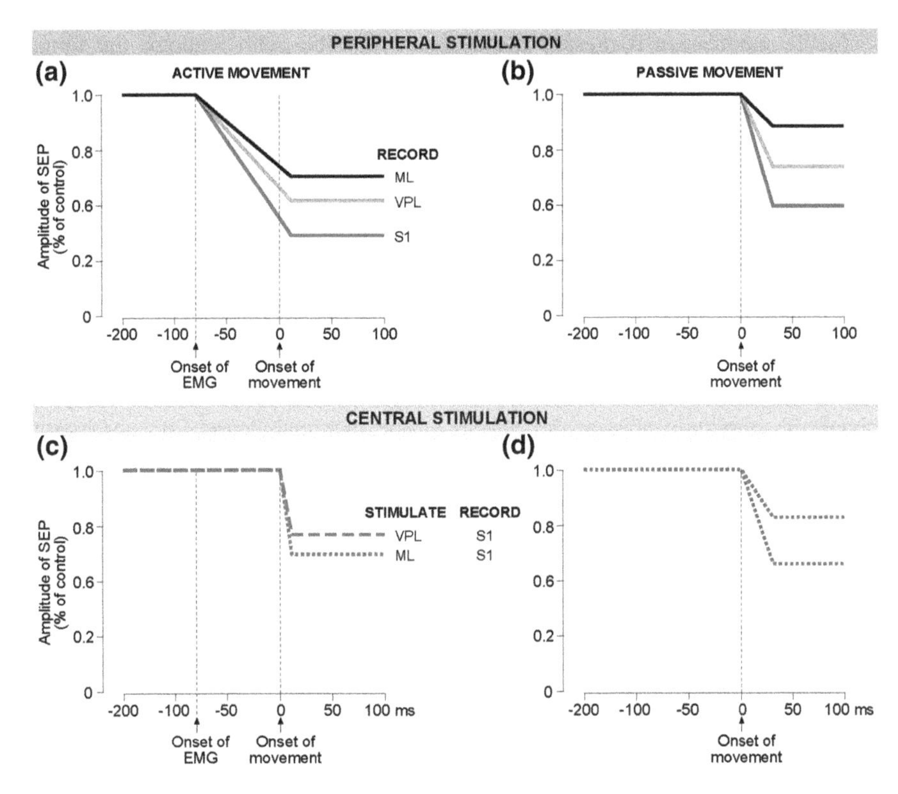

Figure 3 **a-b**: Time course for modulation of the amplitude of SEPs elicited by peripheral stimulation recorded from the medial lemniscus (ML), sensory thalamus (VPL) and S1 in relation to active (**a**) and passive (**b**) elbow movements. **c-d**: time course for modulation of S1 SEPs elicited by stimulating the ML and VPL (recording sites from **a-b**) in relation to active (**c**) and passive (**d**) movements. Data are all aligned on the onset of movement; the time of the approximate onset of EMG activity is shown for active movements (*left*). Reproduced, with permission, from Chapman (1994)

(second vertical line). Similar effects are seen with (isotonic) and without (isometric) movement about the joint (Jiang et al. 1990). The effects are, moreover, independent of the direction of the movement (flexion versus extension of the elbow).The degree of modulation covaries with the speed of movement—greater gating with faster movements (not shown). As in the perceptual studies, the gating effects seen with active movement are also observed with passive movements (Figure 3b), confirming the importance of peripheral reafference to tactile gating. The motor command clearly also contribute: the onset of the decrease in the S1 SEP during passive movement is delayed by about 80 ms compared to active movement, beginning at the onset of passive movement. Thus, the signals contributing to tactile suppression are *both peripheral and central in origin.*

Tactile suppression is present at the earliest possible relay stations, the spinal cord (Seki et al. 2003) and the dorsal column nuclei (Ghez and Lenzi 1971), and precedes the onset of movement. Pioneering studies by Ghez and Lenzi showed that the underlying synaptic mechanism involves presynaptic inhibition (primary afferent depolarization, PAD) of primary cutaneous afferents at the level of the dorsal column nuclei, thus reducing transmission at the first main relay from the periphery to S1 cortex. More recently Fetz and colleagues (Seki et al. 2003) have extended these results to show that PAD is also present at the level of the dorsal horn of the spinal cord. Their recordings, made in awake, behaving primates, showed that the inhibition of cutaneous afferent input to spinal cord interneurones begins well before the onset of EMG activity and voluntary movement.

The suppressive effects seen at the earliest relay stations are, however, quite modest in amplitude (Figure 3a, b—ML). The maximal degree of suppression during movement gradually increases on ascending to higher levels of the pathway (VPL, S1), with the time-course for this modulation being almost identical at all levels (Chapman et al. 1988). Most importantly, the time-course for the decrease in SEP amplitude is almost identical to the time-course for tactile suppression of detection in humans (Figure 1). Peripheral feedback from the moving limb (passive movements) decreases the amplitude of SEPs at all levels of the pathway (Figure 3b), with greater decreases occurring at higher levels (S1 versus ML). In contrast, the motor command appears to particularly act at lower levels of the pathway (spinal cord, dorsal column nuclei) since S1 cortical responses to stimulation of either the medial lemniscus or VPL thalamus are only modulated *after the onset of movement*, active or passive (Figure 3c–d).

While tactile suppression is present at all levels of transmission of tactile inputs, there is little or no evidence for visual suppression during saccades in, for example, primary visual cortex, V1 (Wurtz 1968) although it is found in the extrastriate areas (see Wurtz 2008). This suggests that the neuronal mechanisms underlying visual suppression are different from those underlying tactile suppression.

Evidence that the motor commands contribute to generating tactile suppression has been obtained in studies carried out in awake monkeys. Jiang et al. (1990) showed that intracortical microstimulation applied to the arm representation of M1, essentially generating a weak artificial motor command, leads to a decrease in the amplitude of SEPs in S1 elicited by air puff stimuli applied to the arm. Similar results are obtained even when the stimulation intensity is reduced below the threshold for eliciting EMG activity. Moreover, the effects show a spatial gradient, limited to gating inputs from skin areas overlying, or distal to, the motor output. Thus there is direct evidence that the motor command can diminish the transmission of cutaneous inputs to S1 cortex. Similar observations have since been made in humans using transcranial magnetic stimulation to activate M1 (Voss et al. 2006).

Finally, there is evidence that these suppressive effects extend beyond the pathway relaying tactile inputs to S1. Seki and Fetz (2012) have shown that cutaneous inputs to M1 and premotor cortex are also diminished during active, as well as passive, movement.

Function of Tactile Suppression

The function of tactile suppression, or movement-related gating of tactile stimuli, is most likely to suppress inputs that can be predicted from the motor command, so as to enhance the detection of other novel stimuli (Coulter 1974; Chapman et al. 1988). These controls act to limit the quantity of afferent feedback that is processed at higher levels of the neuraxis.

References

Bays, P M; Flanagan, J R and Wolpert, D M (2006). Attenuation of self-generated tactile sensations is predictive, not postdictive. *PLoS Biology* 4(2): 281–284. doi:10.1371/journal.pbio.0040028.

Bays, P M; Wolpert, D M and Flanagan, J R (2005). Perception of the consequences of self-action is temporally tuned and event driven. *Current Biology* 15(12): 1125–1128. doi:10.1016/j.cub.2005.05.023.

Chapman, C E (1994). Active versus passive touch: factors influencing the transmission of somatosensory signals to primary somatosensory cortex. *Canadian Journal of Physiology and Pharmacology* 72(5): 558–570. doi:10.1139/y94-080.

Chapman, C E and Beauchamp, E (2006). Differential controls over tactile detection in humans by motor commands and peripheral reafference. *Journal of Neurophysiology* 96(3): 1664–1675. doi:10.1152/jn.00214.2006.

Chapman, C E; Bushnell, M C; Miron, D; Duncan, G H and Lund, J P (1987). Sensory perception during movement in man. *Experimental Brain Research* 68(3): 516–524. doi:10.1007/bf00249795.

Chapman, C E; Jiang, W and Lamarre, Y (1988). Modulation of lemniscal input during conditioned arm movements in the monkey. *Experimental Brain Research* 72(2): 316–334. doi:10.1007/bf00250254.

Collins, D F; Cameron, T; Gillard, D M and Prochazka, A (1998). Muscular sense is attenuated when humans move. *Journal of Physiology (London)* 508(2): 635–643. doi:10.1111/j.1469-7793.1998.00635.x.

Coulter, J D (1974). Sensory transmission through lemniscal pathway during voluntary movement in cat. *Journal of Neurophysiology* 37(5): 831–845.

Cybulska-Klosowicz, A; Meftah, E M; Raby, M; Lemieux, M-L and Chapman, C E (2011). A critical speed for gating of tactile detection during voluntary movement. *Experimental Brain Research* 210(2): 291–301. doi:10.1007/s00221-011-2632-0.

Ghez, C and Lenzi, G L (1971). Modulation of sensory transmission in cat lemnsical system during voluntary movements. *Pflügers Archiv European Journal of Physiology* 323(3): 273–278. doi:10.1007/bf00586390.

Jiang, W; Chapman, C E and Lamarre, Y (1990). Modulation of somatosensory evoked responses in the primary somatosensory cortex produced by intracortical microstimulation of the motor cortex in the monkey. *Experimental Brain Research* 80(2): 333–344. doi:10.1007/bf00228160.

Jiang, W; Lamarre, Y and Chapman, C E (1990). Modulation of cutaneous cortical evoked potentials during isometric and isotonic contractions in the monkey. *Brain Research* 536(1–2): 69–78. doi:10.1016/0006-8993(90)90010-9.

Jiang, W; Chapman, C E and Lamarre, Y (1991). Modulation of the cutaneous responsiveness of neurones in the primary somatosensory cortex during conditioned arm movements in the monkey. *Experimental Brain Research* 84(2): 342–354. doi:10.1007/bf00231455.

Kleinfeld, D; Ahissar, E and Diamond, M E (2006). Active sensation: insights from the rodent vibrissa sensorimotor system. *Current Opinion in Neurobiology* 16(4): 435–444. doi:10.1016/j.conb.2006.06.009.

Post, L J; Zompa, I C and Chapman, C E (1994). Perception of vibrotactile stimuli during voluntary motor activity in human subjects. *Experimental Brain Research* 100(1): 107–120. doi:10.1007/bf00227283.

Seki, K and Fetz, E E (2012). Gating of sensory input at spinal and cortical levels during preparation and execution of voluntary movement. *Journal of Neuroscience* 32(3): 890–902. doi:10.1523/jneurosci.4958-11.2012.

Seki, K; Perlmutter, S I and Fetz, E E (2003). Sensory input to primate spinal cord is presynaptically inhibited during voluntary movement. *Nature Neuroscience* 6(12): 1309–1316. doi:10.1038/nn1154.

Smith, A M; Chapman, C E; Deslandes, M; Langlais, J-S and Thibodeau, M-P (2002). The role of friction and tangential force in the subjective scaling of tactile roughness. *Experimental Brain Research* 144(2): 211–223. doi:10.1007/s00221-002-1015-y.

Voss, M; Ingram, J N; Haggard, P and Wolpert, D M (2006). Sensorimotor attenuation by central motor command signals in the absence of movement. *Nature Neuroscience* 9(1): 26–27. doi:10.1038/nn1592.

Weiskrantz, L; Elliott, J and Darlington, C (1971). Preliminary observations on tickling oneself. *Nature* 230(5296): 598–599. doi:10.1038/230598a0.

Williams, S R and Chapman, C E (2000). Time-course and magnitude of movement-related gating of tactile detection in humans. II. Importance of stimulus intensity. *Journal of Neurophysiology* 84(2): 863–875.

Williams, S R and Chapman, C E (2002). Time-course and magnitude of movement-related gating of tactile detection in humans. III. Importance of the motor task. *Journal of Neurophysiology* 88(4): 1968–1979.

Williams, S R; Shenasa, J and Chapman, C E (1998). Time-course and magnitude of movement-related gating of tactile detection in humans. I. Importance of stimulus location. *Journal of Neurophysiology* 79(2): 947–963.

Wurtz, R H (1968). Visual cortex neurons: Response to stimuli during rapid eye movements. *Science* 162(3858): 1148–1150. doi:10.1126/science.162.3858.1148.

Wurtz, R H (2008). Neuronal mechanisms of visual stability. *Vision Research* 48(20): 2070–2089. doi:10.1016/j.visres.2008.03.021.

Crossmodal and Multisensory Interactions Between Vision and Touch

Simon Lacey and K. Sathian

Over the past two decades, there has been growing appreciation of the multisensory nature of perception and its neural basis. Consequently, the concept has arisen that the brain is "metamodal", with a task-based rather than strictly modality-based organization (Pascual-Leone and Hamilton 2001; Lacey et al. 2009b; James et al. 2011). Here we focus on interactions between vision and touch in humans, including crossmodal interactions where tactile inputs evoke activity in neocortical regions traditionally considered visual, and multisensory integrative interactions. It is now established that cortical areas in both the ventral and dorsal pathways, previously identified as specialized for various aspects of visual processing, are also routinely recruited during the corresponding aspects of touch (for reviews see Amedi et al. 2005; Sathian and Lacey 2007; Lacey and Sathian 2011, 2014). When these regions are in classical visual cortex so that they would traditionally be regarded as unisensory, their engagement is referred to as crossmodal, whereas other regions lie in classically multisensory sectors of the association neocortex. Much of the relevant work concerns haptic perception (active sensing using the hand) of shape; this work is therefore considered in detail. We consider how vision and touch might be integrated in various situations and address the role of mental imagery in visual cortical activity during haptic perception. Finally, we present a model of haptic object recognition and its relationship with mental imagery (Lacey et al. 2014).

Activation of Visually Responsive Cortical Regions During Touch

The first demonstration that a visual cortical area is active during normal tactile perception came from a positron emission tomographic (PET) study in humans (Sathian et al. 1997). In this study, tactile discrimination of the orientation of

S. Lacey (✉) · K. Sathian
Department of Neurology, Emory University, Atlanta, USA

T.J. Prescott et al. (eds.), *Scholarpedia of Touch*, Scholarpedia,
DOI 10.2991/978-94-6239-133-8_25

gratings applied to the immobilized fingerpad, relative to a control task requiring tactile discrimination of grating groove width, activated a focus in extrastriate visual cortex, close to the parieto-occipital fissure. This focus, located in the region of the human V6 complex of visual areas (Pitzalis et al. 2006), is also active during visual discrimination of grating orientation (Sergent et al. 1992). Similarly, other neocortical regions known to selectively process particular aspects of vision are activated by analogous non-visual tasks: The human MT complex (hMT), a region well-known to be responsive to visual motion, is also active during tactile motion perception (Hagen et al. 2002; Blake et al. 2004; Summers et al. 2009). This region is sensitive to auditory motion as well (Poirier et al. 2005), but not to arbitrary cues for auditory motion (Blake et al. 2004). Together, these findings suggest that hMT functions as a modality-independent motion processor. Parts of early visual cortex and a focus in the lingual gyrus are texture-selective in both vision and touch (Stilla and Sathian 2008; Sathian et al. 2011; Eck et al. 2013), although one group found that haptically and visually texture-selective regions in medial occipitotemporal cortex were adjacent but non-overlapping (Podrebarac et al. 2014). Further, the early visual regions are sensitive to the congruence of texture information across the visual and haptic modalities (Eck et al. 2013), and information about haptic texture flows from somatosensory regions into these early visual cortical areas (Sathian et al. 2011). Both visual and haptic location judgments involve a dorsally directed pathway comprising cortex along the intraparietal sulcus (IPS) and that constituting the frontal eye fields (FEFs) bilaterally: the IPS is classically considered multisensory while the FEF is now recognized to be so. For both texture and location, several of these bisensory areas show correlations of activation magnitude between the visual and haptic tasks, indicating some commonality of cortical processing across modalities (Sathian et al. 2011).

Most research on visuo-haptic processing of object shape has concentrated on the lateral occipital complex (LOC), an object-selective region in the ventral visual pathway (Malach et al. 1995) that is also object- or shape-selective in touch (Amedi et al. 2001, 2002; James et al. 2002; Stilla and Sathian 2008). The LOC responds to both haptic 3-D (Amedi et al. 2001, 2002; Stilla and Sathian 2008) and tactile 2-D stimuli (Stoesz et al. 2003; Prather et al. 2004) but does not respond during auditory object recognition cued by object-specific sounds (Amedi et al. 2002). However, this region is activated when participants listen to the impact sounds made by metal or wood objects and categorize these sounds by the shape of the associated object (James et al. 2011). Auditory shape information can be conveyed by a visual-auditory sensory substitution device using a specific algorithm to convert visual information into an auditory stream or ‘soundscape’. Both sighted and blind humans can learn to recognize objects by extracting shape information from such soundscapes, albeit after extensive training; interestingly, the LOC responds to soundscapes after such training, but not when participants simply learn to arbitrarily associate soundscapes with particular objects (Amedi et al. 2007). Thus, the LOC can be regarded as processing geometric shape information independently of the sensory modality used to acquire it, similar to the view of hMT+ as processing modality-independent motion information (see above).

Apart from the LOC, visuo-haptic responses have also been observed in several, classically multisensory, parietal regions, including multiple loci along the IPS (Grefkes et al. 2002; Saito et al. 2003; Stilla and Sathian 2008). Given that many of these IPS regions are involved in discrimination of both visual and haptic location of object features, their responsiveness during shape perception may be concerned with reconstruction of global shape representations from object parts (Sathian et al. 2011). The postcentral sulcus (PCS), which corresponds to Brodmann's area 2 of primary somatosensory cortex (S1) (Grefkes et al. 2001), also shows visuo-haptic shape-selectivity (Stilla and Sathian 2008).

It is critical to determine whether haptic or tactile recruitment of visual cortical areas is functionally relevant, i.e. whether these regions are actually necessary for task performance. Although research along these lines remains sparse, there is some evidence in support of this idea. Firstly, case studies indicate that the LOC is necessary for both haptic and visual shape perception: A lesion to the left occipito-temporal cortex, which likely included the LOC, resulted in both tactile and visual agnosia even though somatosensory cortex and basic somatosensory function were intact (Feinberg et al. 1986). Another patient with bilateral LOC lesions was unable to learn new objects either visually or haptically (James et al. 2006, 2007). Transcranial magnetic stimulation (TMS) is a technique used to temporarily deactivate specific, functionally defined, cortical areas, i.e. to create 'virtual lesions' (Sack 2006). TMS over the parieto-occipital region active during tactile grating orientation discrimination (Sathian et al. 1997) interfered with performance of this task (Zangaladze et al. 1999) indicating that this area is functionally, rather than epiphenomenally, involved in the task. More work is necessary to fully test the dependence of haptic perception on classic visual cortical areas.

Why Are Visual Cortical Regions Active During Touch?

Activation of the LOC and other visual cortical areas during touch might arise from direct, "bottom-up" or "feedforward" somatosensory input. Human electrophysiological studies are consistent with this possibility: activity in somatosensory cortex propagates to the LOC as early as 150 ms after stimulus onset (Lucan et al. 2010; Adhikari et al. 2014) in a beta-band oscillatory network (Adhikari et al. 2014). This might reflect cortical pathways between primary somatosensory and visual cortices demonstrated in the macaque (Négyessy et al. 2006). A case study is also illuminating: a patient with bilateral ventral occipito-temporal lesions, with sparing of the dorsal part of the LOC that likely included the multisensory sub-region, showed visual agnosia but intact haptic object recognition associated with activation of the intact dorsal part of the LOC, suggesting that somatosensory input could directly activate this region (Allen and Humphreys 2009). Thus, neocortical regions classically considered to engage in unisensory visual processing may in actuality integrate multisensory inputs.

Studies of congenitally or early blind individuals are consistent with the notion that many classical visual cortical areas are modality-independent but task-specific. Thus, non-visual stimuli in the early blind, and visual stimuli in the sighted, activate the same extrastriate cortical regions on comparable tasks (reviewed by Sathian 2014). For instance, an area known as the visual word-form area in the left fusiform gyrus, which responds selectively to visually presented words in the sighted, is also sensitive to Braille words (Reich et al. 2011). Congenitally blind people also engage visual cortical regions that are not active during corresponding tasks in the normally sighted population, most interestingly, visual cortical areas located at the site of primary (and adjacent non-primary) visual cortex of the sighted, i.e. medial occipital cortex (reviewed in Pascual-Leone et al. 2005; Sathian 2014, 2005; Sathian and Lacey 2007; Pavani and Röder 2012). As pointed out earlier (Sathian 2014), it cannot be assumed that primary visual cortex occupies exactly the same anatomical extent in those who are born without vision as in normally sighted people. Across a host of studies, a wide range of non-visual tasks has been reported to recruit medial occipital cortex in the congenitally blind but not the sighted—these tasks include somatosensory, auditory and language tasks (Pavani and Röder 2012; Sathian 2014). However, each study has typically focused on just one or a few tasks, so how these different functional domains are organized in visual cortex of the congenitally blind remains essentially unknown. Further, the idea that blind people are superior to their sighted counterparts on non-visual tasks is not a universal truth; rather, the evidence pooled over many studies suggests that their superiority reflects the specifics of their experience in the absence of vision (Sathian and Lacey 2007; Pavani and Röder 2012; Sathian 2014).

An alternative to feedforward activation of visual cortex by tactile inputs is that haptic perception might evoke visual imagery of the felt object resulting in "top-down" activation of the LOC by "feedback" connections from higher-order areas (Sathian and Zangaladze 2001). In keeping with this hypothesis, many studies demonstrate LOC activity during visual imagery: During auditorily-cued mental imagery of the shape of familiar objects, both blind and sighted participants show left LOC activation, where shape information would arise mainly from haptic experience for the blind and mainly from visual experience for the sighted (De Volder et al. 2001). The left LOC is also active when geometric and material object properties are retrieved from memory (Newman et al. 2005) and haptic shape-selective activation magnitudes in the right LOC were highly correlated with ratings of visual imagery vividness (Zhang et al. 2004). Even early visual cortical areas have been reported to respond to changes in haptic shape (Snow et al. 2014); however, as with other studies of crossmodal recruitment of visual cortex, it is not possible to exclude visual imagery as an explanation. A counter-argument is that visual imagery cannot explain haptically-evoked LOC activity. In support of this, LOC activity was found to be substantially lower during visual imagery compared to haptic shape perception (Amedi et al. 2001); however, this study did not verify that participants actually engaged in imagery throughout the scan. Another argument against the role of visual imagery is based on the observations that early- as well as late-blind individuals show haptic shape-related LOC activation (Pietrini et al. 2004). While the early blind clearly do not experience visual imagery, these findings do not necessarily rule out a visual imagery explanation in the sighted, given the extensive consequences of visual deprivation on neocortical organization (see above).

Recently, multivariate pattern analyses of voxel-wise activity have been used to demonstrate that activity patterns in primary sensory cortices can differentiate stimuli presented in modalities other than the canonical one. Thus, S1 activity could distinguish between objects in video clips that were being haptically explored, although there was only visual and no somatosensory input (Meyer et al. 2011). Along similar lines, stimulus modality (tactile, pain, auditory or visual) could be decoded in primary sensory cortices (S1, primary visual cortex or primary auditory cortex) regardless of their canonical associations (Liang et al. 2013). These studies provide further evidence of widely distributed multisensory processing in the neocortex; however, it remains uncertain whether the observed non-canonical activity arises from feedforward or feedback projections.

Integration of Visual and Tactile Inputs

Vision and touch share many similarities in the way stimuli are processed. Ahissar and Arieli (2001) proposed that visual and tactile systems use analogous spatiotemporal coding schemes. Consistent with this idea, single neurons in macaques are similarly tuned for curvature direction at intermediate levels of the processing hierarchy in both visual and somatosensory cortex (areas V4 and S2) (Yau et al. 2009). Such similarities in coding lend themselves to multisensory integration. Evidence that visual and tactile inputs are indeed integrated comes from studies showing that the orientation of a tactile grating can disambiguate binocular rivalry (Lunghi et al. 2010), and that tactile motion can bias visually perceived bistable alternations of motion direction (James and Blake 2004). Ernst and Banks (2002) demonstrated that humans integrate visual and haptic inputs in a manner that turns out to be statistically optimal, with the dominant modality being the one associated with lower variance in its estimates. Thus, vision tends to be dominant when assessing object shape but not surface texture (Klatzky et al. 1987). Statistically optimal integration is probably learnt during development, since it is not apparent in the first decade of life (Gori et al. 2008). A dramatic illustration of the importance of multisensory integration in body ownership is offered by the rubber-hand illusion (RHI), in which a viewed rubber hand is aligned with one's own hand screened from view; when both the rubber hand and the real hand are tapped synchronously, the rubber hand is rapidly incorporated into the body image and perceived as one's own (Botvinick and Cohen 1998). The RHI can be induced in sighted people in the absence of vision, if an experimenter taps the subject on the real hand and synchronously moves the subject's other hand to tap the rubber hand, suggesting that it is multisensory congruence of body-related information (in this case between tactile and proprioceptive inputs) that is critical, rather than visuo-tactile congruence specifically (Ehrsson et al. 2005). However, this version of the RHI is absent in early blind people, pointing to potential differences in multisensory integrative processes as a consequence of early visual deprivation (Petkova et al. 2012). In a variant of the RHI, one can be induced to perceive a third arm (Guterstam et al. 2011), and visuotactile conflicts can disrupt the feeling of ownership of one's own limb (Gentile et al. 2013). The illusion has even been extended to the entire body using head-mounted virtual reality displays to create an out-of-body experience (Ehrsson 2007; Lenggenhager et al. 2007).

There has been some study of the neural processes underlying visuo-tactile integration, although a comprehensive account is not yet feasible. Most neurons in the ventral intraparietal areas (VIP) of macaques exhibit modulation of their responses by bisensory stimulation, even when their overt responses to unisensory stimuli are limited to either vision or touch (Avillac et al. 2007). Similarly, bisensory stimuli in rats augment the somatosensory evoked response and reset the phase of induced network oscillations (Sieben et al. 2013). In the putative human homolog of VIP, topographic maps of tactile and proximal visual stimuli are aligned (Sereno and Huang 2006), although at a single neuron level in macaques, the reference frame for tactile stimulation is head-centered whereas that for visual stimuli varies between head-centered, eye-centered or intermediate (Avillac et al. 2005). In one study, visuo-haptic responses were enhanced in the LOC and IPS by stimulus salience (Kim and James 2010), although a subsequent study by the same group showed that when spatiotemporal congruence was maximized across modalities, the inverse effectiveness pattern characteristic of classic neurophysiologic studies of multisensory integration emerged in the LOC and IPS as well as parietal opercular cortex, all on the left (Kim et al. 2012). Visuo-haptic responses in perirhinal cortex are also sensitive to the congruence of stimuli across modalities (Holdstock et al. 2009). The RHI appears to have a different neural substrate, being associated with activity in ventral premotor cortex (Ehrsson et al. 2004), although IPS and cerebellar activity is also reported (Ehrsson et al. 2004, 2005). Repetitive TMS (rTMS) over the left anterior IPS impairs visual-haptic, but not haptic-visual, shape matching using the right hand (Buelte et al. 2008), while rTMS over occipitotemporal cortex affects the Müller-Lyer illusion regardless of whether it is induced visually, haptically or cross-modally (Mancini et al. 2011)—these studies imply that the multisensory convergence reported in the preceding studies is functionally relevant. Further study of visuo-tactile integration and its neural substrates is warranted.

Individual Differences

Two kinds of visual imagery have been described: "object imagery" (involving pictorial images that are vivid and detailed, dealing with the literal appearance of objects in terms of shape, color, brightness, etc.) and "spatial imagery" (involving schematic images more concerned with the spatial relations of objects, their component parts, and spatial transformations) (Kozhevnikov et al. 2002, 2005; Blajenkova et al. 2006). An experimentally important difference is that object imagery includes surface property information while spatial imagery does not. To establish whether object and spatial imagery differences occur in touch as well as vision, we required participants to discriminate shape across changes in texture, and texture across changes in shape (Figure 1), in both visual and haptic within-modal conditions. We found that spatial imagers could discriminate shape despite changes in texture but not vice versa, presumably because their images tend not to encode surface properties. By contrast, object imagers could discriminate texture despite changes in shape, but not the reverse (Lacey et al. 2011), indicating that texture, a surface property, is integrated into their shape representations. In addition, greater preference for object imagery was associated with greater impairment by texture changes (Lacey et al. 2011). Thus, the object-spatial imagery continuum characterizes haptics as well as vision, and individual differences in imagery preference along this continuum affect the extent to which surface properties are integrated into object representations (Lacey et al. 2011).

Cross-modal visuo-haptic object recognition, while fairly accurate, comes at a cost compared to within-modal recognition (Bushnell and Baxt 1999; Casey and Newell 2007; Lacey et al. 2007). Importantly, while within-modal recognition of a set of previously unfamiliar and highly similar objects is view-dependent in both vision and touch, cross-modal recognition of these objects turns out to be view-independent (Lacey et al. 2007). Moreover, training in either the visual or the haptic modality to induce view-independence in the trained modality automatically

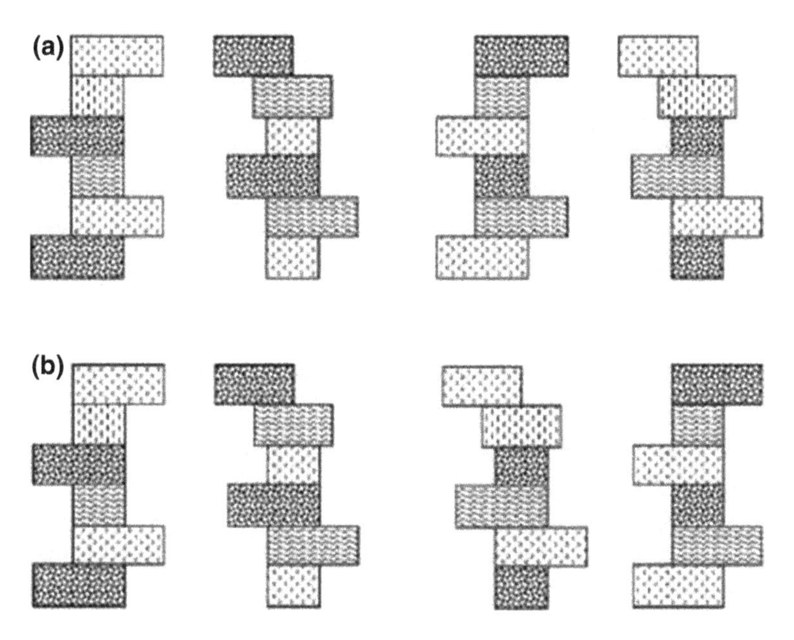

Figure 1 **a** Schematic example of Shapes 1 and 2 with original texture schemes (*left pair*) and the texture schemes exchanged (*right pair*). **b** Example of Textures 1 and 2 with original shapes (*left pair*) and the shapes exchanged (*right pair*). (From Lacey et al. 2011)

confers view-independence in the untrained modality, and cross-modal training yields view-independence in each modality, implying that unisensory, view-dependent representations converge onto a bisensory, view-independent representation, possibly in the LOC (Lacey et al. 2009b). Further, spatial imagery preference correlates with the accuracy of cross-modal object recognition (Lacey et al. 2007). It appears, then, that the multisensory representation has some features that are stable across individuals, like view-independence, and some that vary across individuals, such as integration of surface property information and individual differences in imagery preference.

A Model of Visuo-Haptic Multisensory Object Representation

We now describe a model of visuo-haptic multisensory object representation (Lacey et al. 2009a, 2014) and review the evidence for this model from studies designed to explicitly test the visual imagery hypothesis discussed above (Lacey et al. 2010, 2014). In this model, object representations in the LOC can be flexibly accessed either bottom-up or top-down, independently of the input modality, and object familiarity plays a modulatory role. There is no stored representation for unfamiliar objects so that during haptic recognition, an unfamiliar object has to be explored in its entirety in order to compute global shape and to relate component parts to one another. We proposed (Lacey et al. 2009a) that this occurs in a bottom-up pathway from somatosensory cortex to the LOC, with involvement of the IPS in computing part relationships and thence global shape, facilitated by spatial imagery processes. For familiar objects, global shape can be inferred more easily, perhaps from distinctive features or one diagnostic part, and we suggested (Lacey et al. 2009a) that haptic exploration rapidly acquires enough information to trigger a stored visual image and generate a hypothesis about its identity, involving primarily object imagery processes via a top-down pathway from prefrontal cortex to LOC, as has been proposed for vision (e.g., Bar 2007).

We tested this model by directly comparing activations and effective connectivity during haptic shape perception and both visual object imagery and spatial imagery (Lacey et al. 2010, 2014), reasoning that reliance on similar processes across tasks would lead to correlations of activation magnitude across participants, as well as similar patterns of effective connectivity across tasks. In contrast to previous studies, we ensured that participants engaged in the desired kind of imagery throughout each scan by using appropriate tasks and recording responses. Participants also performed haptic shape discrimination using either familiar or unfamiliar objects. We found that object familiarity modulated inter-task correlations: Of eleven regions common to visual object imagery and haptic perception of familiar shape, six (including bilateral LOC) showed inter-task correlations of activation magnitude. By contrast, object imagery and haptic perception of unfamiliar shape shared only four regions, only one of which (an IPS region) showed an inter-task correlation (Lacey et al. 2010). Relatively few regions showed inter-task correlations between spatial imagery and haptic perception of either familiar or unfamiliar shape, with parietal foci featuring in both sets of correlations (Lacey et al. 2014). This suggests that spatial imagery is relevant to haptic shape perception regardless of object familiarity (contrary to the initial model), whereas object imagery is more strongly associated with haptic perception of familiar, than unfamiliar, shape (in agreement with the initial model). However, it remains possible that the parietal foci showing inter-task correlations between spatial imagery and haptic shape perception reflect spatial processing more generally, rather than spatial imagery per se (Jäncke et al. 2001; Lacey et al. 2014), or generic imagery processes, e.g., image generation, common to both object and spatial imagery (Lacey et al. 2014; Mechelli et al. 2004).

We also conducted effective connectivity analyses, based on the inferred neuronal activity derived from deconvolving the hemodynamic response out of the observed BOLD signals (Lacey et al. 2014). These analyses supported the broad architecture of the model, showing that the spatial imagery network shared much more commonality with the network associated with unfamiliar, compared to familiar, shape perception, while the object imagery network shared much more commonality with familiar, than unfamiliar, shape perception (Lacey et al. 2014). More specifically, the model proposes that the component parts of an unfamiliar object are explored in their entirety and assembled into a representation of global shape via spatial imagery processes (Lacey et al. 2009a). Consistent with this, in the parts of the network common to spatial imagery and unfamiliar haptic shape perception, the LOC is driven by parietal foci, with complex cross-talk between posterior parietal and somatosensory foci. These findings fit with the notion of bottom-up pathways from somatosensory cortex and a role for cortex in and around the IPS in spatial imagery (Lacey et al. 2014). The IPS and somatosensory interactions were absent from the sparse network shared by spatial imagery and haptic perception of familiar shape. By contrast, the relationship between object imagery and familiar shape perception is characterized by top-down pathways from prefrontal areas reflecting the involvement of object imagery (Lacey et al. 2009a). Supporting this, the LOC was driven bilaterally by the left inferior frontal gyrus in the network shared by object imagery and haptic perception of familiar shape, while these pathways were absent from the sparse network common to object imagery and unfamiliar haptic shape perception (Lacey et al. 2014).

Figure 2 shows the current version of our model for haptic shape perception in which the LOC is driven bottom-up from primary somatosensory cortex as well as top-down via object imagery processes from prefrontal cortex, with additional input from the IPS involving spatial imagery processes (Lacey et al. 2014). We propose that the bottom-up route is more important for haptic perception of unfamiliar than familiar objects, whereas the converse is true of the top-down route, which is more important for haptic perception of familiar than unfamiliar objects. It will be interesting to explore the impact of individual preferences for object vs. spatial imagery on these processes and paths.

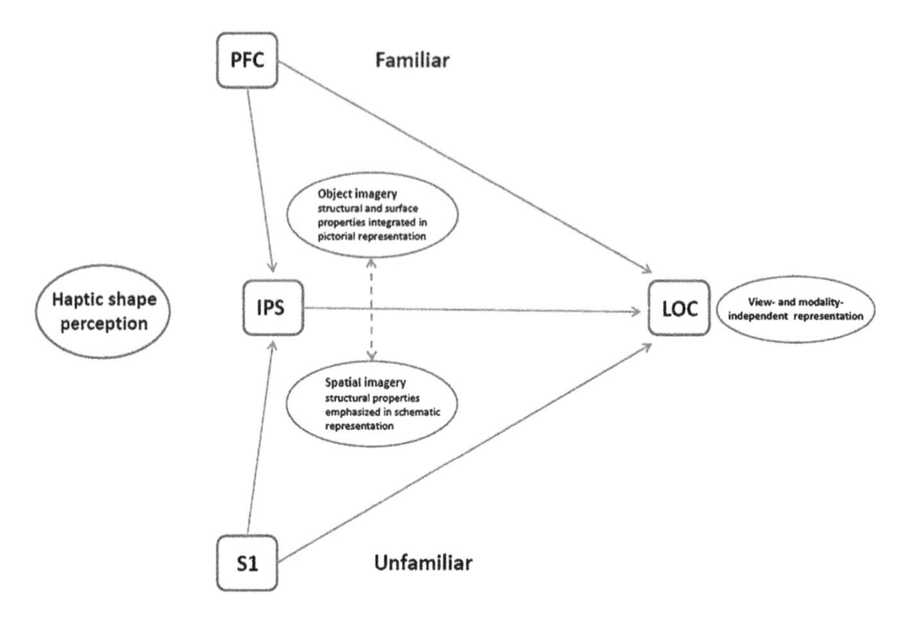

Figure 2 Schematic model of haptic object representation in LOC modulated by object familiarity and imagery type. For unfamiliar more than familiar objects, the LOC is driven bottom-up from somatosensory cortex (S1) with support from spatial imagery processes in the IPS. For familiar more than unfamiliar objects, the LOC is driven top-down from prefrontal cortex (PFC) via object imagery processes. The LOC thus houses an object representation that is flexibly accessible, both bottom-up and top-down, and which is modality- and possibly view-independent (From Lacey et al. 2014)

Acknowledgments Support to KS from the National Eye Institute at the NIH, the National Science Foundation, and the Veterans Administration is gratefully acknowledged.

References

Adhikari, B M; Sathian, K; Epstein, C M; Lamichhane, B and Dhamala, M (2014). Oscillatory activity in neocortical networks during tactile discrimination near the limit of spatial acuity. *NeuroImage* 91: 300–310. doi:10.1016/j.neuroimage.2014.01.007.

Ahissar, E and Arieli, A (2001). Figuring space by time. *Neuron* 32(2): 185–201. doi:10.1016/s0896-6273(01)00466-4.

Allen, H A and Humphreys, G W (2009). Direct tactile stimulation of dorsal occipito-temporal cortex in a visual agnosic. *Current Biology* 19(12): 1044–1049. doi:10.1016/j.cub.2009.04.057.

Amedi, A; Jacobson, G; Hendler, T; Malach, R and Zohary, E (2002). Convergence of visual and tactile shape processing in the human lateral occipital complex. *Cerebral Cortex* 12(11): 1202–1212. doi:10.1093/cercor/12.11.1202.

Amedi, A; Malach, R; Hendler, T; Peled, S and Zohary, E (2001). Visuo-haptic object-related activation in the ventral visual pathway. *Nature Neuroscience* 4(3): 324–330. doi:10.1038/85201.

Amedi, A et al. (2007). Shape conveyed by visual-to-auditory sensory substitution activates the lateral occipital complex. *Nature Neuroscience* 10(6): 687–689. doi:10.1038/nn1912.

Amedi, A; von Kriegstein, K; van Atteveldt, N M; Beauchamp, M S and Naumer, M J (2005). Functional imaging of human crossmodal identification and object recognition. *Experimental Brain Research* 166(3–4): 559-571. doi:10.1007/s00221-005-2396-5.

Avillac, M; Ben Hamed, S and Duhamel, J-R (2007). Multisensory integration in the ventral intraparietal area of the macaque monkey. *Journal of Neuroscience* 27(8): 1922–1932. doi:10.1523/jneurosci.2646-06.2007.

Avillac, M; Denève, S; Olivier, E; Pouget, A and Duhamel, J-R (2005). Reference frames for representing visual and tactile locations in parietal cortex. *Nature Neuroscience* 8(7): 941–949. doi:10.1038/nn1480.

Bar, M (2007). The proactive brain: using analogies and associations to generate predictions. *Trends in Cognitive Sciences* 11(7): 280–289. doi:10.1016/j.tics.2007.05.005.

Blajenkova, O; Kozhevnikov, M and Motes, M A (2006). Object-spatial imagery: A new self-report imagery questionnaire. *Applied Cognitive Psychology* 20(2): 239–263. doi:10.1002/acp.1182.

Blake, R; Sobel, K V and James, T W (2004). Neural synergy between kinetic vision and touch. *Psychological Science* 15(6): 397–402. doi:10.1111/j.0956-7976.2004.00691.x.

Botvinick, M and Cohen, J (1998). Rubber hands 'feel' touch that eyes see. *Nature* 391: 756. doi:10.1038/35784.

Buelte, D; Meister, I G and Staedtgen, M (2008). The role of the anterior intraparietal sulcus in crossmodal processing of object features in humans: An rTMS study. *Brain Research* 1217: 110–118. doi:10.1016/j.brainres.2008.03.075.

Bushnell, E W and Baxt, C (1999). Children's haptic and cross-modal recognition with familiar and unfamiliar objects. *Journal of Experimental Psychology: Human Perception & Performance* 25(6): 1867–1881. doi:10.1037//0096-1523.25.6.1867.

Casey, S J and Newell, F N (2007). Are representations of unfamiliar faces independent of encoding modality? *Neuropsychologia* 45(3): 506–513. doi:10.1016/j.neuropsychologia.2006.02.011.

De Volder, A G et al. (2001). Auditory triggered mental imagery of shape involves visual association areas in early blind humans. *NeuroImage* 14(1): 129–139. doi:10.1006/nimg.2001.0782.

Eck, J; Kaas, A L and Goebel, R (2013). Crossmodal interactions of haptic and visual texture information in early sensory cortex. *NeuroImage* 75: 123–135. doi:10.1016/j.neuroimage.2013.02.075.

Ehrsson, H H (2007). The experimental induction of out-of-body experiences. *Science* 317(5841): 1048. doi:10.1126/science.1142175.

Ehrsson, H H; Holmes, N P and Passingham, R E (2005). Touching a rubber hand: Feeling of body ownership is associated with activity in multisensory brain areas. *Journal of Neuroscience* 25(45): 10564–10573. doi:10.1523/jneurosci.0800-05.2005.

Ehrsson, H H; Spence, C and Passingham, R E (2004). That's my hand! Activity in premotor cortex reflects feeling of ownership of a limb. *Science* 305(5685): 875–877. doi:10.1126/science.1097011.

Ernst, M O and Banks, M S (2002). Humans integrate visual and haptic information in a statistically optimal fashion. *Nature* 415: 429–433. doi:10.1038/415429a.

Feinberg, T E; Rothi, L J and Heilman, K M (1986). Multimodal agnosia after unilateral left hemisphere lesion. *Neurology* 36(6): 864–867. doi:10.1212/wnl.36.6.864.

Gentile, G; Guterstam, A; Brozzoli, C and Ehrsson, H H (2013). Disintegration of multisensory signals from the real hand reduces default limb self-attribution: An fMRI study. *Journal of Neuroscience* 33(33): 13350–13366. doi:10.1523/jneurosci.1363-13.2013.

Gori, M; Del Viva, M; Sandini, G and Burr, D C (2008). Young children do not integrate visual and haptic form information. *Current Biology* 18(9): 694–698. doi:10.1016/j.cub.2008.04.036.

Grefkes, C; Geyer, S; Schormann, T; Roland, P and Zilles, K (2001). Human somatosensory area 2: Observer-independent cytoarchitectonic mapping, interindividual variability, and population map. *NeuroImage* 14(3): 617–631. doi:10.1006/nimg.2001.0858.

Grefkes, C; Weiss, P H; Zilles, K and Fink, G R (2002). Crossmodal processing of object features in human anterior intraparietal cortex: An fMRI study implies equivalencies between humans and monkeys. *Neuron* 35(1): 173–184. doi:10.1016/s0896-6273(02)00741-9.

Guterstam, A; Petkova, V I and Ehrsson, H H (2011). The illusion of owning a third arm. *PLoS ONE* 6: e17208. doi:10.1371/journal.pone.0017208.

Hagen, M C et al. (2002). Tactile motion activates the human middle temporal/V5 (MT/V5) complex. *European Journal of Neuroscience* 16(5): 957–964. doi:10.1046/j.1460-9568.2002.02139.x.

Holdstock, J S; Hocking, J; Notley, P; Devlin, J T and Price, C J (2009). Integrating visual and tactile information in the perirhinal cortex. *Cerebral Cortex* 19(12): 2993–3000. doi:10.1093/cercor/bhp073.

James, T W et al. (2002). Haptic study of three-dimensional objects activates extrastriate visual areas. *Neuropsychologia* 40(10): 1706–1714. doi:10.1016/S0028-3932(02)00017-9.

James, T W and Blake, R (2004). Perceiving object motion using vision and touch. *Cognitive, Affective, Behavioral Neuroscience* 4(2): 201–207. doi:10.3758/cabn.4.2.201.

James, T W; James, K H; Humphrey, G K and Goodale, M A (2006). Do visual and tactile object representations share the same neural substrate? In: M A Heller and S Ballesteros (Eds.), "Touch and Blindness, Psychology and Neuroscience" (pp. 139–155). Mahwah, New Jersey: Lawrence Erlbaum Associates. ISBN: 9780805847253.

James, T W; Kim, S and Fisher, J S (2007). The neural basis of haptic object processing. *Canadian Journal of Experimental Psychology* 61(3): 219–229. doi:10.1037/cjep2007023.

James, T W; Stevenson, R W; Kim, S; VanDerKlok, R M and James, K H (2011). Shape from sound: Evidence for a shape operator in the lateral occipital cortex. *Neuropsychologia* 49(7): 1807–1815. doi:10.1016/j.neuropsychologia.2011.03.004.

Jäncke, L; Kleinschmidt, A; Mirzazade, S; Shah, N J and Freund, H-J (2001). The role of the inferior parietal cortex in linking the tactile perception and manual construction of object shapes. *Cerebral Cortex* 11(2): 114–121. doi:10.1093/cercor/11.2.114.

Kim, S and James, T W (2010). Enhanced effectiveness in visuo-haptic object-selective brain regions with increasing stimulus salience. *Human Brain Mapping* 31: 678–693. doi:10.1002/hbm.20897.

Kim, S; Stevenson, R A and James, T W (2012). Visuo-haptic neuronal convergence demonstrated by an inversely effective pattern of BOLD activation. *Journal of Cognitive Neuroscience* 24(4): 830–842. doi:10.1162/jocn_a_00176.

Klatzky, R L; Lederman, S J and Reed, C (1987). There's more to touch than meets the eye: The salience of object attributes for haptics with and without vision. *Journal of Experimental Psychology* 116(4): 356–369. doi:10.1037/0096-3445.116.4.356.

Kozhevnikov, M; Hegarty, M and Mayer, R E (2002). Revising the visualiser-verbaliser dimension: Evidence for two types of visualisers. *Cognition & Instruction* 20(1): 47–77. doi:10.1207/S1532690XCI2001_3.

Kozhevnikov, M; Kosslyn, S M and Shephard, J (2005). Spatial versus object visualisers: A new characterisation of cognitive style. *Memory & Cognition* 33: 710–726. doi:10.3758/bf03195337.

Lacey, S; Flueckiger, P; Stilla, R; Lava, M and Sathian, K (2010). Object familiarity modulates the relationship between visual object imagery and haptic shape perception. *NeuroImage* 49(3): 1977–1990. doi:10.1016/j.neuroimage.2009.10.081.

Lacey, S; Lin, J B and Sathian, K (2011). Object and spatial imagery dimensions in visuo-haptic representations. *Experimental Brain Research* 213(2–3): 267-273. doi:10.1007/s00221-011-2623-1.

Lacey, S; Peters, A and Sathian, K (2007). Cross-modal object representation is viewpoint-independent. *PLoS ONE* 2(9): e890. doi:10.1371/journal.pone.0000890.

Lacey, S and Sathian, K (2011). Multisensory object representation: Insights from studies of vision and touch. *Progress in Brain Research* 191: 165–176. doi:10.1016/b978-0-444-53752-2.00006-0.

Lacey, S and Sathian, K (2014). Visuo-haptic multisensory object recognition, categorization and representation. *Frontiers in Psychology* 5: 730. doi:10.3389/fpsyg.2014.00730.

Lacey, S; Tal, N; Amedi, A and Sathian, K (2009a). A putative model of multisensory object representation. *Brain Topography* 21(3–4): 269-274. doi:10.1007/s10548-009-0087-4.

Lacey, S; Pappas, M; Kreps, A; Lee, K and Sathian, K (2009b). Perceptual learning of view-independence in visuo-haptic object representations. *Experimental Brain Research* 198 (2–3): 329-337. doi:10.1007/s00221-009-1856-8.

Lacey, S; Stilla, R; Sreenivasan, K; Deshpande, G and Sathian, K (2014). Spatial imagery in haptic shape perception. *Neuropsychologia* 60: 144–158. doi:10.1016/j.neuropsychologia.2014.05.008.

Lenggenhager, B; Tadi, T; Metzinger, T and Blanke, O (2007). Video ergo sum: manipulating bodily self-consciousness. *Science* 317(5841): 1096–1099. doi:10.1126/science.1143439.

Liang, M; Mouraux, A; Hu, L and Iannetti, G D (2013). Primary sensory cortices contain distinguishable spatial patterns of activity for each sense. *Nature Communications* 4: 1979. doi:10.1038/ncomms2979.

Lucan, J N; Foxe, J J; Gomez-Ramirez, M; Sathian, K and Molholm, S (2010). Tactile shape discrimination recruits human lateral occipital complex during early perceptual processing. *Human Brain Mapping* 31: 1813–1821. doi:10.1002/hbm.20983.

Lunghi, C; Binda, P and Morrone, M (2010). Touch disambiguates rivalrous perception at early stages of visual analysis. *Current Biology* 20(4): R143-R144. doi:10.1016/j.cub.2009.12.015.

Malach, R et al. (1995). Object-related activity revealed by functional magnetic resonance imaging in human occipital cortex. *Proceedings of the National Academy of Sciences of the United States of America* 92(18): 8135–8139. doi:10.1073/pnas.92.18.8135.

Mancini, F; Bolognini, N; Bricolo, E and Vallar, G (2011). Cross-modal processing in the occipito-temporal cortex: A TMS study of the Müller-Lyer illusion. *Journal of Cognitive Neuroscience* 23(8): 1987–1997. doi:10.1162/jocn.2010.21561.

Mechelli, A; Price, C J; Friston, K J and Ishai, A (2004). Where bottom-up meets top-down: Neuronal interactions during perception and imagery. *Cerebral Cortex* 14(11): 1256–1265. doi:10.1093/cercor/bhh087.

Meyer, K; Kaplan, J T; Essex, R; Damasio, H and Damasio, A (2011). Seeing touch is correlated with content-specific activity in primary somatosensory cortex. *Cerebral Cortex* 21(9): 2113–2121. doi:10.1093/cercor/bhq289.

Négyessy, L; Nepusz, T; Kocsis, L and Bazsó, F (2006). Prediction of the main cortical areas and connections involved in the tactile function of the visual cortex by network analysis. *European Journal of Neuroscience* 23(7): 1919–1930. doi:10.1111/j.1460-9568.2006.04678.x.

Newman, S D; Klatzky, R L; Lederman, S J and Just, M A (2005). Imagining material versus geometric properties of objects: An fMRI study. *Cognitive Brain Research* 23(2–3): 235-246. doi:10.1016/j.cogbrainres.2004.10.020.

Pascual-Leone, A; Amedi, A; Fregni, F and Merabet, L B (2005). The plastic human brain cortex. *Annual Review of Neuroscience* 28(1): 377–401. doi:10.1146/annurev.neuro.27.070203.144216.

Pascual-Leone, A and Hamilton, R H (2001). The metamodal organization of the brain. *Progress in Brain Research* 134: 427–445. doi:10.1016/s0079-6123(01)34028-1.

Pavani, F and Röder, B (2012). Crossmodal plasticity as a consequence of sensory loss: Insights from blindness and deafness. In: B E Stein (Ed.), "The New Handbook of Multisensory Processes" (pp. 737–759). Cambridge, MA: MIT Press.

Petkova, V I; Zetetrberg, H and Ehrsson, H H (2012). Rubber hands feel touch, but not in blind individuals. *PLoS ONE* 7: e335912. doi:10.1371/journal.pone.0035912.

Pietrini, P et al. (2004). Beyond sensory images: Object-based representation in the human ventral pathway. *Proceedings of the National Academy of Sciences of the United States of America* 101(15): 5658–5663. doi:10.1073/pnas.0400707101.

Pitzalis, S et al. (2006). Wide-field retinotopy defines human cortical visual area V6. *Journal of Neuroscience* 26(30): 7962–7973. doi:10.1523/jneurosci.0178-06.2006.

Podrebarac, S K; Goodale, M A and Snow, J C (2014). Are visual texture-selective areas recruited during haptic texture discrimination? *NeuroImage* 94: 129–137. doi:10.1016/j.neuroimage.2014.03.013.

Poirier, C et al. (2005). Specific activation of the V5 brain area by auditory motion processing: An fMRI study. *Cognitive Brain Research* 25(3): 650–658. doi:10.1016/j.cogbrainres.2005.08.015.

Prather, S C; Votaw, J R and Sathian, K (2004). Task-specific recruitment of dorsal and ventral visual areas during tactile perception. *Neuropsychologia* 42(8): 1079–1087. doi:10.1016/j.neuropsychologia.2003.12.013.

Reich, L; Szwed, M; Cohen, L and Amedi, A (2011). A ventral visual stream reading center independent of visual experience. *Current Biology* 21(5): 363–368. doi:10.1016/j.cub.2011.01.040.

Sack, A T (2006). Transcranial magnetic stimulation, causal structure-function mapping and networks of functional relevance. *Current Opinion in Neurobiology* 16(5): 593–599. doi:10.1016/j.conb.2006.06.016.

Saito, D N; Okada, T; Morita, Y; Yonekura, Y and Sadato, N (2003). Tactile-visual cross-modal shape matching: A functional MRI study. *Cognitive Brain Research* 17(1): 14–25. doi:10.1016/s0926-6410(03)00076-4.

Sathian, K (2005). Visual cortical activity during tactile perception in the sighted and visually deprived. *Developmental Psychobiology* 46(3): 279–286. doi:10.1002/dev.20056.

Sathian, K (2014). Cross-modal plasticity in the visual system. In: M E Selzer, S Clarke, L Cohen, G Kwakkel and R Miller (Eds.), "Textbook of Neural Repair and Rehabilitation, 2nd Ed." (pp. 140–153). Cambridge, UK: Cambridge University Press. ISBN: 9781107010475.

Sathian, K and Lacey, S (2007). Journeying beyond classical somatosensory cortex. *Canadian Journal of Experimental Psychology* 61(3): 254–264. doi:10.1037/cjep2007026.

Sathian, K et al. (2011). Dual pathways for somatosensory information processing. *NeuroImage* 57: 462–475.

Sathian, K and Zangaladze, A (2001). Feeling with the mind's eye: the role of visual imagery in tactile perception. *Optometry & Vision Science* 78(5): 276–281. doi:10.1097/00006324-200105000-00010.

Sathian, K; Zangaladze, A; Hoffman, J M and Grafton, S T (1997). Feeling with the mind's eye. *NeuroReport* 8(18): 3877–3881.

Sereno, M I and Huang, R-S (2006). A human parietal face area contains aligned head-centered visual and tactile maps. *Nature Neuroscience* 9(10): 1337–1343. doi:10.1038/nn1777.

Sergent, J; Ohta, S and MacDonald, B (1992). Functional neuroanatomy of face and object processing. A positron emission tomography study. *Brain* 115(1): 15–36. doi:10.1093/brain/115.1.15.

Sieben, K; Röder, B and Hanganu-Opatz, I L (2013). Oscillatory entrainment of primary somatosensory cortex encodes visual control of tactile processing. *Journal of Neuroscience* 33 (13): 5736–5749. doi:10.1523/jneurosci.4432–12.2013.

Snow, J C; Strother, L and Humphreys, G W (2014). Haptic shape processing in visual cortex. *Journal of Cognitive Neuroscience* 26(5): 1154–1167. doi:10.1162/jocn_a_00548.

Stilla, R and Sathian, K (2008). Selective visuo-haptic processing of shape and texture. *Human Brain Mapping* 29(10): 1123–1138. doi:10.1002/hbm.20456.

Stoesz, M et al. (2003). Neural networks active during tactile form perception: Common and differential activity during macrospatial and microspatial tasks. *International Journal of Psychophysiology* 50(1–2): 41–49. doi:10.1016/s0167-8760(03)00123-5.

Summers, I R; Francis, S T; Bowtell, R W; McGlone, F P and Clemence, M (2009). A functional magnetic resonance imaging investigation of cortical activation from moving vibrotactile stimuli on the fingertip. *The Journal of the Acoustical Society of America* 125(2): 1033–1039. doi:10.1121/1.3056399.

Yau, J M; Pasupathy, A; Fitzgerald, P J; Hsiao, S S and Connor, C E (2009). Analogous intermediate shape coding in vision and touch. *Proceedings of the National Academy of Sciences of the United States of America* 106(38): 16457–16462. doi:10.1073/pnas.0904186106.

Zangaladze, A; Epstein, C M; Grafton, S T and Sathian, K (1999).Involvement of visual cortex in tactile discrimination of orientation. *Nature* 401(6753): 587–590. doi:10.1038/44139.

Zhang, M; Weisser, V D; Stilla, R; Prather, S C and Sathian, K (2004). Multisensory cortical processing of object shape and its relation to mental imagery. *Cognitive, Affective, & Behavioral Neuroscience* 4(2): 251–259. doi:10.3758/cabn.4.2.251.

Aftereffects in Touch

Astrid M.L. Kappers and Wouter M. Bergmann Tiest

An aftereffect is the change in the perception of a (test) stimulus after prolonged stimulation with an (adaptation) stimulus. Usually, this change is in the negative direction, that is, in a direction opposite to that of the adaptation stimulus. Aftereffects are often fast and strong. A well-known example in vision is the waterfall illusion: when looking at trees after staring at a waterfall for a minute or more, the subsequently viewed trees seem to move upwards (Addams 1834; Swanston and Wade 1994). Also touch is susceptible to strong aftereffects: temperature, roughness, shape, curvature, motion and size of an object all give rise to aftereffects in touch.

Temperature Aftereffect

One of the first reports of an aftereffect came from the 17th-century philosopher John Locke (1690/1975). In his essay on human understanding, he describes how water of a certain temperature can be perceived as different by the two hands. Although he did not perform an actual experiment, the idea is that if one hand is placed for some time in a bucket of cold water and the other in a bucket of warm water, the lukewarm water in a third bucket will feel as warm to the hand coming from the cold water and as cold to the other hand. This is an aftereffect of temperature. Arnold and colleagues (1982) actually performed this experiment with adults and children. Both groups of participants experienced this aftereffect. They also report that about half of the children responded that the water temperature in the third bucket was really different for the two hands, instead of just feeling differently. This was confirmed by the answer to the follow-up question about what would be the perceived temperature if the bucket were rotated by 180°. The majority of the children expected that the perceived temperature would switch after the rotation. Most of the adults responded to both questions correctly.

A.M.L. Kappers (✉) · W.M. Bergmann Tiest
VU University, Amsterdam, The Netherlands

T.J. Prescott et al. (eds.), *Scholarpedia of Touch*, Scholarpedia,
DOI 10.2991/978-94-6239-133-8_26

Size Aftereffect

Walker and Shea (1974) placed a narrow bar to the right of a participant and a wide bar to the left (or vice versa). The participant was asked to repeatedly grasp the left and right bars with one of his/her hands and judge their width. After an exploration period of 2 min, the two bars were replaced with two identical bars of intermediate width and the participant had to judge which of the two bars was the wider. A significant aftereffect was found: the bar that was placed at the side where during the exploration phase the wider bar had been, felt narrower than the bar on the other side. Exploring with one hand and testing with the other hand did not result in an aftereffect. In a subsequent study (Walker 1978), the participants had to explore and judge the length instead of the width of the bars and this led to similar results. Both these aftereffects were termed contingent (i.e. dependent) aftereffects, as they depended on the location of the stimulus. A similar experiment was performed using two hands: one hand repeatedly explored the length of a long bar, the other hand explored a short bar. This also led to a significant aftereffect: when the two hands were subsequently presented with bars of intermediate length, the bar that was touched with the hand that previously touched the longer one felt shorter than the other.

Uznadze (1966) did a similar experiment with three-dimensional shapes, namely spheres. His participants were asked to repeatedly grasp spheres with their hands, always a small one on one side and a large one on the other side. After about 15 such grasps, the spheres were replaced with two identical spheres with a size in between that of the small and large ones. As in the width and length aftereffects, the size of the sphere that was grasped with the hand that previously grasped the large one was perceived as smaller than that grasped by the other hand. Kappers and Bergmann Tiest (2013) repeated this experiment in a more formal way. They showed that the strength of this aftereffect, expressed as the difference in perceived size of the spheres, was 24 % of the size of the test sphere. They also showed that the grasping tetrahedra instead of spheres led to similar aftereffects (Kappers and Bergmann Tiest 2014). Interestingly, Maravita (1997) showed that the same size aftereffect existed in a right-brain damaged patient. Although this patient did not consciously perceive the sphere presented to his left hand, the size of this sphere influenced the perception of the size of the sphere presented to his right hand. The occurrence of the aftereffect showed that somatosensory information that is not consciously perceived is still processed by the brain.

Curvature Aftereffect

In 1933, Gibson published a study on the visual and kinesthetic perception of curved lines. He observed that in both modalities, the perception of curvature was strongly biased by previous exposure to a curved stimulus. In his kinesthetic experiment, blindfolded participants were instructed to run their fingers over a convexly curved piece of cardboard for 3 min. After this adaptation period, they were asked to move their fingers in the same manner along a straight edge. All the participants reported that the straight edge felt concave. Moreover, the participants also reported a gradual decrease of the perceived curvature during the 3 min exposure period.

Vogels and colleagues (1996) used solid three-dimensional shapes to study the aftereffect of touching a curved surface. Participants seated behind a curtain were required to place one of their hands on a curved conditioning shape. After a fixed time (usually just 5 s), participants lifted their hand and the experimenter presented them with a differently curved test surface. Subsequently, they had to respond as quickly as possible whether this test surface was convex or concave. Such trials were repeated for many different combinations of adaptation and test curvatures. Even after such a short adaptation phase, significant aftereffects were found: the perceived curvature of the test stimulus was strongly shifted in a direction opposite to that of the conditioning curvature. In other words, after adapting to a concave stimulus, a flat surface feels convex, and vice versa. In a second experiment, the length of the conditioning period was varied. Interestingly, after adapting for only 2 s to a curvature, the perception of the subsequently touched curved surface was changed. The strength of the aftereffect reached its maximum after about 10 s; longer adaptation periods did not increase the strength of the effect. In the third experiment, the duration of the aftereffect was tested. Instead of touching the test surface immediately after the adaptation phase, participants had to wait for up to 80 s. Although the strength of the aftereffect diminished with time, a significant aftereffect was still present after 40 s. Thus, touching a curved surface only briefly has a relatively long effect on the perception of objects that are touched next. Indeed, when touching three curved surfaces in a row, the perceived curvature of the third is influenced by the curvatures of both the first and the second touched surfaces (Vogels et al. 2001).

In another study, Vogels et al. (1997) investigated the origin of this curvature aftereffect. In the first experiment, they varied the hand activity of the participant in the period of 5, 20 or 40 s between the adaptation and the test phases: participants either held their hand passively in the air, made a fist, or repeatedly bent and stretched their fingers. The idea behind these conditions was that the stimulation of cutaneous receptors would be quite different in the three conditions. Therefore, if the cutaneous receptors played an important role in the aftereffect, the rate of decay of the aftereffect would strongly depend on the intervening activity. However, the decay rates in the various conditions were quite similar, and so the conclusion was that the role of the cutaneous receptors (i.e. the peripheral level) in the aftereffect is at most minor. Thus the origin of the aftereffect had to be at a higher processing level. In the second experiment, they tested whether the aftereffect transferred to the other hand. If that were the case, the origin of the aftereffect would be highly central. In this experiment, participants adapted to a curved surface with one hand and the test was presented to the other hand. Only two participants took part in this study, but neither of them showed a transfer of the aftereffect to the other hand. Thus, the origin of the aftereffect could not be highly central, either.

Van der Horst and colleagues (2008a) showed that a curvature aftereffect also occurs when the adaption surface is only touched with a static finger. They also showed a partial but significant transfer of the aftereffect between fingers, not only between those of the same hand, but also across hands. This was again strong evidence that the aftereffect does not originate at the peripheral level of the cutaneous receptors in the skin. Moreover, because the partial transfer also occurred between the fingers of different hands, the origin of this particular curvature aftereffect necessarily lies at a high central level, probably at the level of the somatosensory cortex. In an other study (Van der Horst et al. 2008b), they investigated the curvature aftereffect and its possible transfer when fingers either actively or passively touched curved stimuli. In the active dynamic condition, participants moved their fingers over a curved surface. This condition is reminiscent of that of Gibson (1933), except that the stimuli were presented vertically instead of horizontally and the adaptation phase was much shorter (three back-and-forth movements instead of 3 min of exploration). In the passive dynamic condition, the stimulus moved repeatedly from left to right and back underneath the static finger. In both conditions, a full transfer to the other hand was found. Thus a dynamically obtained aftereffect has to originate at a high level. It was also found that the strength of the aftereffect was substantially larger in the active condition. From this it followed that the aftereffect and thus also the representation of curvature, depends on the exploration mode, and that kinesthetic information plays an additional role in the aftereffect. A detailed overview of the various experimental conditions in the above-cited papers can be found in Kappers (2011).

Denisova and colleagues (2014) investigated whether an aftereffect also occurred when the curved surface was touched with a tool instead of directly with the hand or fingers. Their virtual stimuli were simulated versions of the real ones used by Van der Horst et al. (2008b). They found both an aftereffect and a partial transfer of the aftereffect to the other hand, which confirms that in the dynamic case, kinesthetic information plays a major role in the origin of the aftereffect.

Shape Aftereffect

As mentioned above, Kappers and Bergmann Tiest (2014) showed that size aftereffects occurred for both spheres and tetrahedra. They also tested "mixed" conditions: adaptation to one type of shape, sphere or tetrahedron, and testing with the other type. Interestingly, in these conditions the aftereffects were much smaller and for many participants they did not even exist. They concluded that the "size aftereffect" for three-dimensional shapes, should more properly be termed "size-shape aftereffect". As shape information is not processed at a peripheral level, this finding suggests that higher cortical areas are involved in this aftereffect.

Inspired by findings in vision, a rather different shape aftereffect was investigated by Matsumiya (2012). His participants had to haptically explore face masks with different expressions (sad, happy or neutral). He ran two conditions. In the non-adaptation condition, the participants had to explore one of the three test faces and subsequently make a decision about whether the face had a sad or happy expression. In the adaptation condition, participants first explored either the happy or the sad face mask for 20 s and subsequently had to judge the expression of the neutral face as either happy or sad. He then compared the percentages "sad" responses to the neutral face mask in the different conditions. Without adaptation, the percentage of "sad" faces was 50 %. With adaptation to the happy face, the percentage of "sad" responses to the neutral face increased to about 90 %, whereas after adaptation to the sad face, this percentage decreased to about 35 % (percentages estimated from his Figure 2). The main conclusion that he drew from these results is that probably the face aftereffect originates at higher cortical areas, such as the inferior frontal gyrus, inferior parietal lobe, and/or superior temporal sulcus, as that is where haptic facial information is processed (Matsumiya 2012; Kitada et al. 2010).

Tactile Motion Aftereffect

Motion of a stimulus over the skin may give rise to a tactile motion aftereffect, which is the impression that after removing the adaptation stimulus, a stimulus still moves over the skin. However, both the occurrence and the direction of this motion aftereffect critically depends on the precise experimental conditions, as not all studies have found effects (Watanabe et al. 2007). One of the first studies was rather informal, with participants (the author and colleagues) describing their introspective perceptions (Thalman 1922). In some of the conditions, some or all of the participants experienced an aftereffect. Hazlewood (1971), using a rather sophisticated set-up, failed to find a tactile motion aftereffect. Only one of her 80 participants reported an aftereffect. Hollins and Favorov (1994) report that a surface with a smooth microtexture elicits the best aftereffects. Their five participants experienced a mostly negative (that is, in the opposite direction as in most other aftereffects) tactile motion aftereffect in most of the trials. The strength and the duration of the aftereffect increased with the duration of the adaptation phase (between 30 and 120 s). However, in a replication of this experiment, Lerner and Craig (2002) found that only about half of their 50 participants reported an aftereffect, which were mostly in positive or other directions. Using a different set-up, again about 50 % reported an aftereffect, but this time mostly in a negative direction.

Watanabe and colleagues (2007) argued that in order to elicit a reliable tactile motion aftereffect, appropriate stimuli have to be used to stimulate the same mechanoreceptors in the adaptation and test phases. They used a set of three pins spaced 5 mm apart, that vibrated with a frequency of 30 Hz. By giving the pins a different phase, participants perceived apparent motion over their fingers. Such stimulation was used during an adaptation phase of 10 s. This resulted in robust tactile motion aftereffects in the opposite direction. In a number of sophisticated experimental conditions, Kuroki et al. (2012) found that the direction of the tactile motion aftereffect was determined by the environmental direction and not by the somatotopic direction. Moreover, they concluded that stimulation of peripheral receptors is essential for the occurrence of the aftereffect.

Planetta and Servos (2008) investigated, among others, the influence of the speed of the adapting motion. They found that duration, frequency and vividness of the aftereffect increased with speed. They found a tactile motion aftereffect in only about 50 % of their trials of which the direction could be positive, negative or "other". In a subsequent study (Planetta and Servos 2010), they investigated which type(s) of mechanoreceptors are involved in the tactile motion aftereffect. They compared different sites of stimulation, namely the hand, the cheek and the forearm. A tactile motion aftereffect was most often reported after stimulation of the hand. From their findings they concluded that most likely the fast adapting type I receptors and the hair follicles are involved in the tactile motion aftereffect. The main conclusion from most of the above-mentioned studies is that tactile motion aftereffects are much harder to induce than visual motion aftereffects (for example, the above-mentioned waterfall illusion).

A different aftereffect of tactile motion is the tactile speed aftereffect: after an adaptation phase of exposure to a moving stimulus, a moving test stimulus is perceived as moving slower as compared to the same test stimulus presented to a non-adapted hand (McIntyre et al. 2012). This effect was independent of the direction of the adapting stimulus relative to the test direction. As peripheral afferents are direction sensitive, this independence of direction suggests that adaptation of these afferents cannot be the cause of the aftereffect: adaptation in their preferred direction should cause stronger adaptation and as a consequence, a direction sensitive aftereffect. As this was clearly not the case, the authors conclude that this aftereffect has to be of central origin.

Vibration Aftereffect

Lederman and colleagues (1982) asked participants to place their finger for 10 min on a stimulus vibrating at either 20 or 250 Hz. After this adaptation period, the participants had to make magnitude estimates of vibrating stimuli with different intensities. In most cases, the magnitude estimates of the intensity of vibrating test stimuli decreased in comparison to estimates made without adaptation. The effects were stronger for the test stimuli with vibration frequencies equal to the adaptation vibration frequency. In the case of the 250 Hz adaptation and test vibration frequencies, the test stimuli with the lowest intensities were no longer detected. The authors also showed that such an adaptation to vibration does not influence the perceived roughness of stimuli. From this finding, they concluded that the sense of vibration is not the underlying mechanism for the perception of roughness. However, more recently, Hollins et al. (2001) found that adaptation to vibration does affect roughness discrimination performance of very fine textures (spatial periods below 100 μm), but not that of rougher textures. They conclude that in the former case, the peripheral Pacinian receptors play a major role.

Hahn (1966) showed that longer adaptation periods lead to stronger aftereffects, both with a method of absolute thresholds and a matching experiment. Even after an adaptation period of 25 min, the increase in the strength of the aftereffect was not yet saturated. After such a long adaptation period, a recovery period of 10 min was needed to return to values before adaptation.

Roughness Aftereffect

Kahrimanovic and colleagues (2009) let participants adapt to a rough surface by making a back-and-forth scanning movement with a finger. The perceived roughness of the next surface was decreased in comparison to the roughness perceived with a non-adapted finger. Also the opposite occurred: after adaptation to a smooth surface, the perceived roughness of the next surface was increased. The authors argue that from these findings that this aftereffect is not caused at a peripheral level. If the aftereffect were due to overstimulation of certain types of mechanoreceptors, the aftereffect after adaptation to smooth and rough stimuli should be in the same direction and not in opposite directions, with a smaller effect for smooth stimuli as compared to rough stimuli. They hypothesize that the adaptation originates in the somatosensory cortex.

Weight Aftereffect

De Mendoza (1979) asked participants in a series of trials to compare weights presented to the two hands. These weights were either equal or differed substantially. He found a significant aftereffect of weight: if two identical weights were presented after substantially different ones, the stimulus presented to the hand that in the previous trial held the heavier one, was perceived as lighter than the identical weight presented to the other hand. The author proposed to term this aftereffect the "gravimetric aftereffect", as it was unclear whether it originated in proprioception, kinesthesis, tactile sensitivity or combinations thereof.

Cross-Modal Aftereffect

Many of the aftereffects mentioned above exist in both vision and touch. It is therefore of interest to investigate whether these aftereffects have a similar origin. One approach to this question is to test whether an aftereffect transfers from one modality to the other. Konkle and colleagues (2009) tested transfer of the motion aftereffect. They found that after adaptation to a visually moving stimulus, a tactile motion aftereffect could be measured. The illusion of motion was in a direction opposite to that of the inducing visual motion. They also found a visual motion aftereffect after adapting to tactile motion. This study provides strong evidence that processing of visual and tactile motion needs at least partially to take place in a common brain area. In a subsequent study, Konkle and Moore (2009) argue that the origin of aftereffects can lie in many different brain areas, depending on the type of processing.

Matsumiya (2013) found face aftereffects, the changed perception of emotional expressions (see above), in both touch and vision. He also found that these transferred both from vision to touch and vice versa. Again the conclusion had to be that face processing in vision and touch share representations in the brain.

Conclusion

Touch is susceptible to strong aftereffects: curvature, size, shape, roughness, vibrations, temperature and weight all lead to aftereffects. These aftereffects are always negative, that is, in a direction opposite to that of the adapting stimulus. Also motion sometimes induces aftereffects in touch, but the effects are usually less strong, not always found in all participants, and the direction can be positive, negative or even in other directions. For most of these aftereffects there exists evidence that they originate in higher brain areas and not by overstimulation of peripheral receptors.

References

Addams, R (1834). An account of a peculiar optical phenomenon seen after having looked at a moving body. *London and Edinburgh Philosophical Magazine and Journal of Science* 5: 373–374.

Arnold, K D; Winer, G A and Wickens, D D (1982). Veridical and nonveridical interpretations to perceived temperature differences by children and adults. *Bulletin of the Psychonomic Society* 20(5): 237–238.

de Mendoza, J-L J (1979). Demonstration of an aftereffect occurring in the tactile-kinesthetic domain - The gravimetric aftereffect. *Psychological Research* 40: 415–422.

Denisova, K; Kibbe, M M; Cholewiak, S A and Kim, S-H (2014). Intra- and intermanual curvature aftereffect can be obtained via tool-touch. *IEEE Transactions on Haptics* 7(1): 61–66.

Gibson, J J (1933). Adaptation, after-effect and contrast in the perception of curved lines. *Journal of Experimental Psychology* 16(1): 1–31.

Hahn, J F (1966). Vibrotactile adaptation and recovery measured by two methods. *Journal of Experimental Psychology* 71(5): 655–658.

Hazlewood, V (1971). A note on failure to find a tactile motion aftereffect. *Australian Journal of Psychology* 23(1): 59–61.

Hollins, M; Bensmaïa, S J and Washburn, S (2001). Vibrotactile adaptation impairs discrimination of fine, but not coarse, textures. *Somatosensory & Motor Research* 18(4): 253–262.

Hollins, M and Favorov, O (1994). The tactile movement aftereffect. *Somatosensory & Motor Research* 11(2): 153–162.

Kahrimanovic, M; Bergmann Tiest, W M and Kappers, A M L (2009). Context effects in haptic perception of roughness. *Experimental Brain Research* 194(2): 287–297.

Kappers, A M L (2011). Human perception of shape from touch. *Philosophical Transactions of the Royal Society B* 366: 3106–3114.

Kappers, A M L and Bergmann Tiest, W M (2013). Haptic size aftereffects revisited. *World Haptics Conference (WHC)* (pp. 335–339).

Kappers, A M L and Bergmann Tiest, W M (2014). Influence of shape on the haptic size aftereffect. *PLoS One* 9(2): e88729.

Kitada, R; Johnsrude, I S; Kochiyama, T and Lederman, S J (2010). Brain networks involved in haptic and visual identification of facial expressions of emotion: An fMRI study. *NeuroImage* 49(2): 1677–1689.

Konkle, T and Moore, C I (2009). What can crossmodal aftereffects reveal about neural representation and dynamics? *Communicative & Integrative Biology* 2(6): 479–481.

Konkle, T; Wang, Q; Hayward, V and Moore, C I (2009). Motion aftereffects transfer between touch and vision. *Current Biology* 19(9): 745–750.

Kuroki, S; Watanabe, J; Mabuchi, K; Tachi, S and Nishida, S (2012). Directional remapping in tactile inter-finger apparent motion: A motion aftereffect study. *Experimental Brain Research* 216(2): 311–320.

Lederman, S J; Loomis, J M and Williams, D A (1982). The role of vibration in the tactual perception of roughness. *Perception & Psychophysics* 32(2): 109–116.

Lerner, E A and Craig, J C (2002). The prevalence of tactile motion aftereffects. *Somatosensory & Motor Research* 19(1): 24–29.

Locke, J (1690/1975). In: P H Nidditch (Ed.), *An essay concerning human understanding*, Vol. II, Ch. 8, §21. Oxford: Clarendon Press.

Maravita, A (1997). Implicit processing of somatosensory stimuli disclosed by a perceptual after-effect. *NeuroReport* 8(7): 1671–1674.

Matsumiya, K (2012). Haptic face aftereffect. *i-Perception* 3(2): 97–100.

Matsumiya, K (2013). Seeing a haptically explored face: visual facial-expression aftereffect from haptic adaptation to a face. *Psychological Science* 24(10): 2088–2098.

McIntyre, S; Holcombe, A O; Birznieks, I and Seizova-Cajic, T (2012). Tactile motion adaptation reduces perceived speed but shows no evidence of direction sensitivity. *PLoS One* 7(9): e45438.

Planetta, P J and Servos, P (2008). The tactile motion aftereffect revisited. *Somatosensory & Motor Research* 25(2): 93–99.

Planetta, P J and Servos, P (2010). Site of stimulation effects on the prevalence of the tactile motion aftereffect. *Experimental Brain Research* 202(2): 377–383.

Swanston, M and Wade, N (1994). Editorial - A peculiar optical phenomenon. *Perception* 23: 1107–1110.

Thalman, W A (1922). The after-effect of movement in the sense of touch. *The American Journal of Psychology* 33(2): 268–276.

Uznadze, D N (1966). *The psychology of set*. New York: Consultants Bureau.

Van der Horst, B J et al. (2008a). Intramanual and intermanual transfer of the curvature aftereffect. *Experimental Brain Research* 187(3): 491–496.

Van der Horst, B J; Willebrands, W P and Kappers, A M L (2008b). Transfer of the curvature aftereffect in dynamic touch. *Neuropsychologia* 46(12): 2966–2972.

Vogels, I M L C; Kappers, A M L and Koenderink, J J (1996). Haptic aftereffect of curved surfaces. *Perception* 25(1): 109–119.

Vogels, I M L C; Kappers, A M L and Koenderink, J J (1997). Investigation into the origin of the haptic after-effect of curved surfaces. *Perception* 26: 101–107.

Vogels, I M L C; Kappers, A M L and Koenderink, J J (2001). Haptic after-effect of successively touched curved surfaces. *Acta Psychologica* 106(3): 247–263.

Walker, J T (1978). Simple and contingent aftereffects in the kinesthetic perception of length. *Journal of Experimental Psychology: Human Perception and Performance* 4(2): 294–301.

Walker, J T and Shea, K S (1974). A tactual size aftereffect contingent on hand position. *Journal of Experimental Psychology* 103(4): 668–674.

Watanabe, J; Hayashi, S; Kajimoto, H; Tachi, S and Nishida, S (2007). Tactile motion aftereffects produced by appropriate presentation for mechanoreceptors. *Experimental Brain Research* 180(3): 577–582.

Tactile Illusions

Vincent Hayward

Tactile illusions are found when the perception of a quality of an object through the sense of touch does not seem to be in agreement with the physical stimulus. They can arise in numerous circumstances and can provide insights into the mechanisms subserving haptic sensations. Many of them can be exploited, or avoided, in order to create efficient haptic display systems or to study the nervous system.

All Senses, Including Touch, Are Subject to Illusions

It is sometimes assumed that vision is the main source of perceptual illusions and that, in contrast, touch is not subject to surprising perceptual phenomena. This belief is ancient. George Berkeley (1685–1753), Étienne Bonnot de Condillac (1714–1780), and others of this era frequently referred to touch as the provider of the 'truth' to the other senses. While this observation appears to be borne out frequently in everyday life, touch is similarly subject to ambiguous or conflicting sources of information, which, as in vision, audition, and other sensory inputs, provide circumstances in which touch-based illusions can arise. It could be that tactile illusions simply go unnoticed more frequently.

Like the systems subserving audition, vision, vestibular inputs, and taste/olfaction, (all of which are subject to illusions) the somatosensory system has evolved to solve perceptual problems quickly and reliably, subject to constraints that are physical (i.e., skin mechanoreceptors cannot be at a small distance from the surface comparatively to their size), physiological (i.e., neural computation is powerful but limited; mechanoreceptors must have a refractory period), and metabolic (i.e., only a proportion of afferent fibers from the periphery to the brain can be myelinated) in origin.

What is an illusion, visual, tactile, or otherwise? Gregory (1997) wrote that illusions are difficult to define. The commonly adopted definition of an illusion is that it is a discrepancy between perception and reality. This definition rapidly leads

V. Hayward (✉)
Université Pierre et Marie Curie, Paris, France

T.J. Prescott et al. (eds.), *Scholarpedia of Touch*, Scholarpedia,
DOI 10.2991/978-94-6239-133-8_27

to the unsatisfactory conclusion that all percepts are illusions since a percept—which is a brain state—is always discrepant with a stimulus—which is a physical object. In fact these two notions cannot even be compared. For this reason, a simple operational definition which has the advantage of comparing things of the same nature was proposed (Hayward 2008a). An illusion is a percept that arises from a stimulus combining two separable components. One component is fixed and the observer attends to it. What makes it an illusion is that the perception of this component is strongly contingent on the variation of a second component, perplexing the person made aware of the unchanging component of the stimulus. For instance, the moon disk appears larger when viewed close to the horizon than up in the sky. The fixed component is its angular size and the variable component is its elevation. The variable component can also be an internal state of the brain. The Necker cube illusion and other rivalry-based illusions are percepts that change through time according to brain states variations, although the stimulus is invariant.

Several surveys on tactile and haptic illusions were published recently (Hayward 2008a; Bresciani et al. 2008; Lederman and Jones 2011), describing several dozens of categories. The word 'haptic' is often used to refer to touch sensations that involve a motor component. According to Grunwald and John (2008) the term 'haptic' was coined by Max Dessoir (1867–1947), who was in need of a counterpart term for the terms 'optic' and 'acoustic'. In the past few years, the rate of discovery of new tactile and haptic illusions has increased greatly, indicating renewed interest in the subject, with more and better informed web-based resources than in the recent past, although much catching up remains to be done.

A particular aspect of haptic perception in humans, like in most animals, is that the whole body is a mechanically sensitive system. Of course, some species have developed specialized sensing organs: whiskers for rodents, pinnipeds, and other mammalians; dextrous fingers for primates, procyonids, and other families; scales for reptiles, crocodilians, and others; antennae, cuticles for insects and arthropods; spines in echinoids; and so on. In many cases these sensory organs are appendages with motor capabilities. Nevertheless, the whole body of an animal is, to some extent, mechanosensitive. Superficial mechanoreceptors are found in hair follicles, skin, scales, lips, cuticles; and deep receptors are found in muscles, tendons, ligaments, as well as other connective tissues, providing a great diversity of overlapping sensing options. What these sensing options have in common is that they all inform the brain of the mechanical state of the tissues in which they are embedded, but never provide direct information about the contacting objects. This information is always mediated by the laws of mechanics that govern the change of the mechanical state of tissues subject to internal and external loads.

Haptics and tactile sensing is thus the province of mechanics where the assumption of rigidity, which for simplification we are so easily inclined to adopt, is misleading even if we assume for the sake of analysis that the stimulated tissues take a quasi-static state (Hayward 2011). The consequence is that we instinctively adopt finite-dimensional notions such as 'force' (which has meaning only for ideal 'point masses') or 'pressure' (which can be sensed only by compressible organs; since tissues are incompressible, the somatosensory system cannot sense pressure).

These abstract notions are certainly convenient for helping our understanding of the tactile functions but have probably no significance for the brain. If we abandon these simplifications then the occurrence of haptic and tactile illusions—or surprising perceptual behaviors as defined earlier—can be expected. For example, a given load, which we normally express as a force, or a given displacement of a solid, which we normally express as a distance, do not correspond univocally to mechanical states of the body.

What follows is a selection of tactile perceptual phenomena that undoubtedly merit the status of being an illusion and that teach something specific about how the mechanical properties of objects are perceived by humans coarsely organized according to their likely attribution from more central to more peripheral haptic neural processing.

Haptic Perception Interacts with Other Senses

When discussing haptic and tactile illusions it must always be kept in mind that senses rarely operate in isolation and that they all interact with each other in the formation of perceptual estimates and judgements. Many studies have shown that touch interacts with taste, vision, and audition, and thus specific perceptual effects are elicited when these interactions are strong. Here is a very classic and powerful example of such interactions which can easily be demonstrated in a classroom or elsewhere. Procure two graspable boxes of similar appearance but of different sizes as illustrated in Figure 1 and arrange them so they have the same mass. When asked to judge the relative heaviness of the two blocks, most people will be convinced that the smaller is heavier than the larger block. This effect has been known for more than a century and is frequently termed the Charpentier illusion (Charpentier 1891), or more commonly the 'size-weight illusion'. The effect is not small (it can be of 20 % or more of difference in judgement of heaviness) and has been, and continues

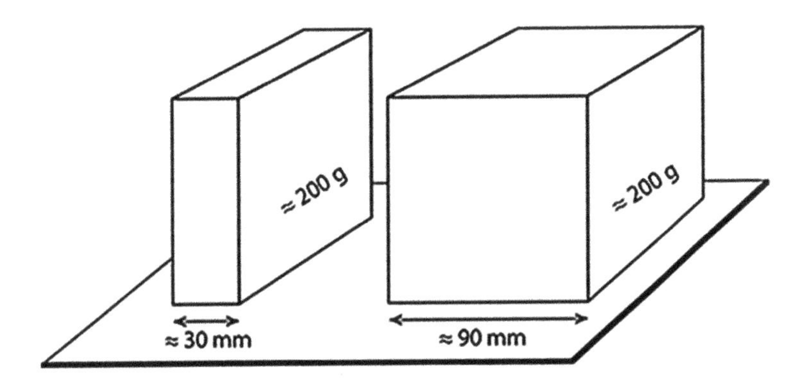

Figure 1 A convenient setting to demonstrate the Charpentier Illusion

to be, the subject of a very large number of studies that appear at a rate that does not seem to subside. Despite numerous attempts, a principled explanatory mechanism for its occurrence remains to be found.

Figure 1 shows two boxes or wood blocks that are easily graspable are arranged to have the same mass, viz. 200 g. Here they differ by one dimension only (say, 30 mm versus 90 mm) which make it possible to show, using the same blocks in the dark, that the two objects do cause a similar sensation of heaviness if the grasp is carefully executed in order to conceal information about their size difference. Conversely, the same blocks can be used to show that, in the absence of vision, the haptic finger-span size estimation method gives similar information to the brain as vision, which results in a similar illusion.

An uncontroversial aspect of the 'size-weight illusion' is the role played by prior experience. The majority of the available, numerous explanations that have been discussed for now a century give a central role to expectation based on prior information (Ross 1966; Buckingham 2014), an hypothesis that has received strong support with the demonstration that the effect can be inverted after sufficiently long practice (Flanagan et al. 2008).

There are numerous other haptic illusions that can arise from interactions between vision and touch, and this is also true of touch and audition. It is worth describing a representative demonstration of such interactions, as one example among many. Because frictional interactions between solids are generally accompanied by acoustic emissions that can be heard and because the vibrations of the source of emission can also be felt, audition and touch are in a position to collaboratively determine the mechanical characteristics of surfaces sliding against one another. Specifically, the glabrous skin of our hands—the skin inside the hands that we use to interact with objects—is covered by a layer of keratin, a material that has strong affinity with water. The mechanical properties of keratin change profoundly with hydration and so do its frictional properties (Johnson et al. 1993; Adams et al. 2013). As a result, the frictional sound made by rubbing hands is a direct function of their moisture content. If the sound emitted by rubbing hands is artificially modified, then the sensation of hand dryness is also modified (Jousmäki and Hari 1998).

The set-up shown in Figure 2 can be used to demonstrate interactions between audition and touch. One needs a microphone (directional) to pick up an auditory scene, such as rubbing hands; a frequency equalizer (analog or digital); headphones (closed) to reproduce the modified scene. The high frequencies characteristic of

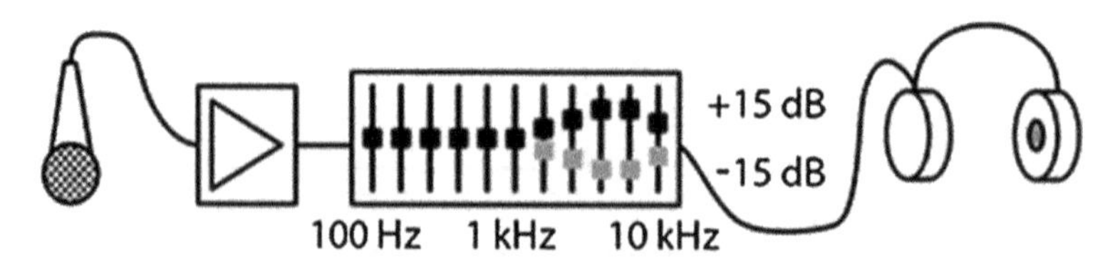

Figure 2 Equipment needed to observe audio-tactile interactions

frictional sounds can be enhanced or attenuated, affecting tactile perception. The perception of other frictional interactions will be affected similarly (Guest et al. 2002), most notably chalk against a blackboard, etc.

In this subsection, we have seen two examples, selected from many, where sensory information supplied by different senses interfered sufficiently to give the resulting percept an illusory quality, suggesting that a fundamental type of brain mechanism is the fusion of sensory information to extract a single object property such as weight, size, distance, numerosity, movement, mobility, wetness, softness, smoothness, and so on.

Similarities of Certain Illusions Across Senses

In some cases of illusory perceptual phenomena there is a remarkable analogy between perceptual effects across senses, suggesting that certain brain mechanisms, even neural circuits, are shared by the senses, sometimes in surprising ways (Konkle et al. 2009). In vision there are many well-known effects arising from viewing certain line drawings (e.g. Delboeuf, Bourdon, Ebbinghaus, Müller-Lyer, Poggendorff, or Ponzo illusions). Interestingly, most of these visual illusions also operate in haptics when the figure represented as a raised drawing is explored with the finger (Suzuki and Arashida 1992).

The interpretations of these visual illusions frequently appeal to brain mechanisms engaged in resolving ambiguities introduced by optical projections (Howe and Purves 2005, Wolfe et al. 2005). It is therefore surprising that these illusions also operate in touch (albeit not always as stably), since visual projections arise from the laws of optics and haptic projections come from self-generated movements (Hartcher-O'Brien et al. 2014). In contrast, explanations based on the anisotropy of fundamental sensory discrimination thresholds could apply in the two modalities (Heller et al. 1997; Mamassian and de Montalembert 2010).

The so-called visual vertical-horizontal illusion exemplified in the Figure 3 is a good representative example. For most people, the vertical segment appears to be longer than the horizontal one. They have the same length. Next, procure a page-size

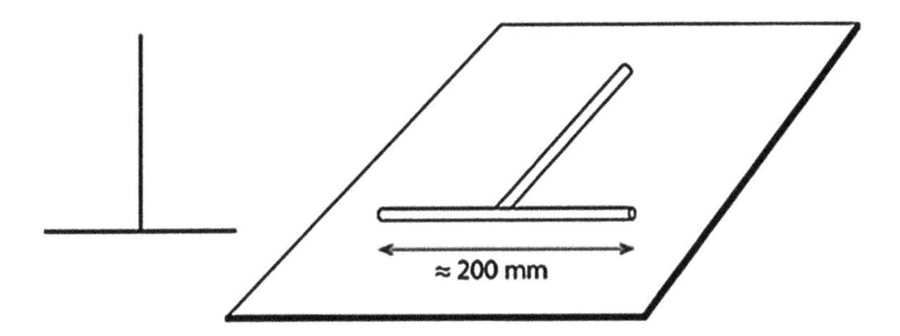

Figure 3 Geometrical visual illusions operate with touch

cardboard sheet and glue two 200 mm sticks on it, as indicated (chopsticks cut at length will do). Blindfolded exploration of the sticks will cause most people to feel, similarly to vision, that the vertical stick is longer than the horizontal one.

To mention another class of illusions that is common to all three non-chemical modalities, the so-called "tau effect" stands out. If two stimuli localized in time and in space are attended to, in all modalities: in visual space (Benussi 1913), in auditory tonal space (Cohen et al. 1954), in auditory physical space (Sarrazin et al. 2007), on the skin (Gelb 1914; Helson 1930), the perceived distance between those stimuli depends on their temporal separation. A shorter time separation corresponds to a smaller perceived spatial separation. The reverse is also true and is called the "kappa effect" (Cohen et al. 1953). Numerous studies have been conducted about these and related phenomena, and the most commonly adopted approach to explain them is to evoke brain mechanisms aimed at coping with moving sources of stimulation in the presence of uncertainty (Goldreich 2007). If the reader is interested in replicating any of these effects with electronically controlled stimuli, it is strongly advised to avoid employing the type of vibrator employed in consumer devices, particularly those based on eccentric motors, because their poor temporal resolution precludes the production of sufficiently brief stimuli.

Lateral inhibition is another neural computational principle that is shared by all senses and that can be invoked to explain universal interactions between intensity and proximity (von Békésy 1959). Thus, apparent motion, which is tightly connected to the latter interaction, operates in touch as in other sensory modalities (Wertheimer 1912; Bregman 1990; Gjerdingen 1994) by modulating the relative intensity of simultaneous stimuli that are separated in space (von Békésy 1959). In the same vein, the permutability of amplitude and duration of short stimuli seems to be a general phenomenon (Bochereau et al. 2014). Perceptual rivalry can likewise be demonstrated in all three sensory modalities (Carter et al. 2008), so does the phenomenon of capture where the localization of a stimulus in space by one sensory modality is modified by synchronous inputs from other sensory modalities (Caclin et al. 2002), as well as the family of attentional and change blindness phenomena (Gallace et al. 2006).

The types of tactile and haptic illusions discussed so far (namely, interactions between sensory modalities, geometrical illusions, or space time interactions) share the quality of being classical in the sense that they have been known for a century or so. In the foregoing, haptic illusions that have been described more recently are described.

Order of Differences: The Particular Multi-scale Nature of Touch

Sensory processes must deal with scale differences because auditory, visual, and haptic scenes can be examined at different spatial and temporal scales. For example, when looking at a tree, the details of the venation of its leaves need not to be considered in assessing the shape of the whole tree. Visual information also frequently has a self-similar character when the scale varies. For example the fundamental process of the extraction of illumination discontinuities in an image is similar when examining leaf venation or the tree branch patterns. Visual objects are also self-similar when viewed from different distances. In audition, a musical melody exists independently from the timber of the sounds of each note. Sounds also often have a self-similar character in their spectral characteristics (Voss and Clark 1975). The situation is more complex in touch because, unlike the other senses, the physics at play differs fundamentally according to the scale at which haptic interaction is considered, even though certain self-similarity characteristics can also be observed (Wiertlewski et al. 2011). Tactile mechanics begin at the molecular scale since touch clearly depends on friction-related phenomena that depend on microscopic-scale physics, and it ends at the scales covered during ambulation.

At the macroscopic scale the multi-scale character of haptic perception can be demonstrated by the following illusion. If a flat plate is made to roll on the fingertip, that is, if the observer is provided with no other information than the orientation of the direction of the normal to a solid object while exploring it as depicted by Figure 4a, then the resulting percept is comparable to that of exploring a real slippery object where the observer is given displacement, orientation, and curvature information as shown in Figure 4b (Dostmohamed and Hayward 2005). Provided that appropriate precautions are taken such as averting vision and ensuring that the observer is not aware of the mechanical details of the stimulation, then observers feel as if they were touching a curved object.

Figure 4c shows a cam mechanism capable of generating the sensation of exploring a virtual object with two fingers obtained by combining two stimuli as in Figure 4a. This effect can be achieved by assembling two of the mechanisms described in (Hayward 2008a) in mirror opposition as in Figure 4d. During exploration, the two fingers remain at a constant distance from each other, as

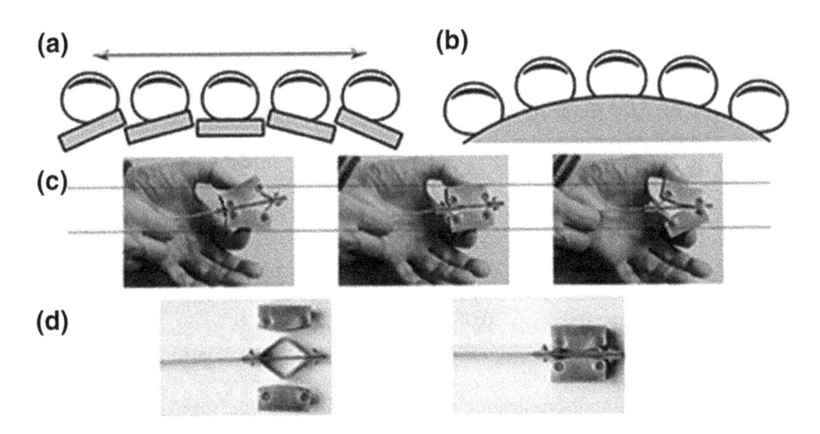

Figure 4 Bent plate illusion

indicated in Figure 4c by the two thin lines, but the sensation is that of exploring a round object.

The relationship of this illusion with the notion of scale can be established assuming that one of the fundamental haptic perceptual tasks is to assess the local curvature of solid objects. It may be accepted without proof that in the simplified case of a profile of constant curvature the measurement of three points on this profile is the minimum information required to determine its curvature. Figure 5a illustrates this necessity. Measurements are necessarily corrupted by errors which translate to discrimination thresholds. It can be intuitively seen that, ceteris paribus, the greater is the portion of the profile that is considered, parametrized by the length of the cord, d, the more accurate is the measurement of curvature (Wijntjes et al. 2009). Assuming the existence of an osculating circle to a shape, the estimation of its curvature requires the measurement of the relative position of at least three points (circles). For a given scale, d, the displacement, h, the slope, ϕ, or the curvature, c, are all potential sensory cues. Measurement errors can be represented either by the relative change in height, Δh, by the relative change in slope, $\Delta\phi$, at its opposite ends, or by the relative change in curvature, Δc, everywhere. Figure 5a is

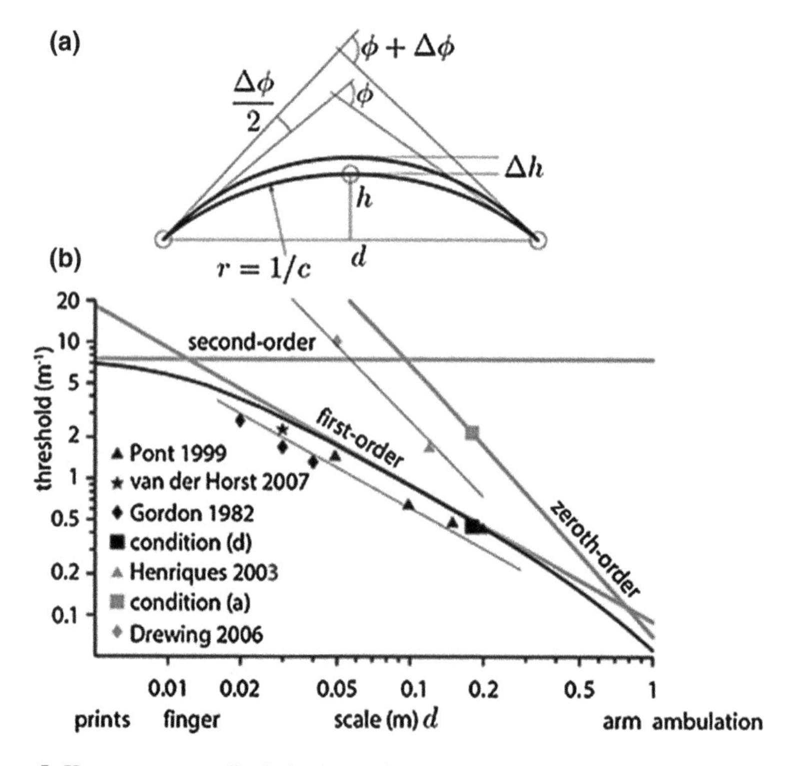

Figure 5 Human curvature discrimination performance model (with permission of the IEEE)

an abstraction of the curvature sensing problem. Zero-order error (Δh), first-order error ($\Delta \phi$), and second order error (Δc) can be related to each other with simple algebra. Figure 5b shows the results of a weak fusion cue combination model where the weights attributed to each sensory cue increase according to the reliability of the corresponding cue (Wijntjes et al. 2009). Given the known discrimination thresholds for these quantities, the model predicts that in the small scales (approx. $d < 1.0$ cm) curvature is the most reliable quantity to be sensed, in the intermediate scales (approx. $1.0 < d < 75$ cm), slope has this role, and in the large scales (approx. $d > 75$ cm) it is displacement, as corroborated by numerous psychophysical studies.

It was thus found that in the range of scales comprised between the size of a finger and the size of an arm, first-order information—that is, orientation—dominates over the other sources. These numbers suggest that the anatomical sizes of the human haptic appendages impose strict limits on the type of features that can be felt. These quantities correspond to orders of differences of displacement: zero, one, and second order; reflecting physiological constraints which in turn reflect the scale at which processing is performed. Of course one could speculate that higher derivatives could be leveraged to discriminate smaller scale features. The change of curvature over space would then be characteristic of a surface with asperities where curvature changes over very small length scales, viz. 1.0 mm and less.

On Contact Mechanics

One source of tactile illusions is clearly derived from contact mechanics effects. As alluded to earlier, extracting the attributes of a touched object from partial knowledge of one's own tissue deformation, is a noisy and ambiguous process. It occurs under the influence of internal and external loads, and is at the root of all effects described thus far. Contact mechanics, or the analysis of the deformation of solids in contact, is thus of immediate relevance in the perception of small-scale attributes such as surface details. Nakatani et al. (2006) described an intriguing effect where strips with different small-scale mechanical properties are juxtaposed to form a flush surface. When explored actively, such surfaces cause the sensation that they have raised or recessed geometries.

In its original form, Figure 6a, the stimulus is a rigid surface textured as shown. A raised pattern (0.1 mm thick) has a 3 mm wide central spine with orthogonal processes extending on each side with a 2 mm spatial period. When rubbing the finger on the spine, it is perceived as a recessed feature compared to the sides. Variants of this stimulus can be realized by juxtaposing strips of different materials having different roughnesses, different frictional properties (such as metal and rubber), or even different mobilities (Nakatani et al. 2008). Figure 6b shows a variant that can be easily realized by drilling holes in a plastic or metal plate.

A rough explanation for this illusion involves the observation that, during sliding, surfaces with different frictional or mobility properties create different boundary conditions that cause a complex tissue deformation field to propagate

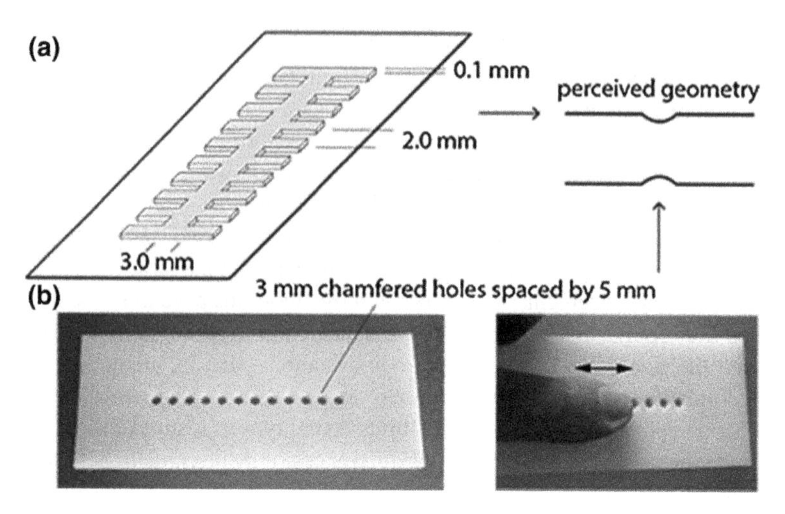

Figure 6 Fishbone illusion and variants

inside the finger. Since the tactile system is by necessity capable of reporting a highly simplified version of the actual deformation field of the finger tissues, then peripheral or central neural processes provide their best guess of what the boundary condition could be. The difficulty of the inverse problem involved has been recognized by roboticists who noticed the inherent ambiguous nature of the corresponding computational problem (Ricker and Ellis 1993; de Rossi et al. 1991).

In vision, it was found that the brain had preferences for certain solutions to ambiguous perceptual problems. As one instance among many, it is well known that the visual system prefers to accept motion over deformation to explain the raw visual inputs (Wallach and O'Connell 1953). So we could conclude that the tactile system prefers to assign the possible cause of an effect to variations of geometry over variations of surface frictional properties or other factors that could affect an unknown boundary condition. This conclusion is supported by a number of related effects that are only briefly mentioned here (Wang and Hayward 2008; Kikuuwe et al. 2005; Hayward and Cruz-Hernandez 2000; Smith et al. 2009; Robles-De-La-Torre and Hayward 2001) but which all point to the same conclusion.

Mechanical Regularities

It may be surmised that the mechanical world is considerably more complicated than the optical or the acoustic world. This argument rests on the observation that the diversity of mechanical phenomena that can take place is truly great for the reason, as alluded to earlier, that different physics apply at different scales. Moreover, a variety of nonlinear and complex mechanical behaviors take place when objects come into contact, slide on each other, are compressed, are collided with, and so on. Only a small subset of objects we interact with are simple, smooth, solid objects. Most other solid objects are aggregations of small scale structures like fabrics, soil, wood, or have multi-stable mechanics like retractable ball pens or keyboards, and so-on, multiplying the possible mechanical behaviors at infinitum. Yet, universal, environmentally driven regularities must exist that the brain can initially extract and later rely upon. In vision, instances of such regularities include the celebrated convexity, light-from-above, or object rigidity assumptions (Ramachandran 1988, Gregory 1980, Ullman 1979). Surely, similar notions must exist in touch and haptics.

Crushing things. Many surfaces on which one steps are made of complex, inhomogeneous, aggregated materials. These include carpets, gravel, soils, underbrush, snow, which have a broadband mechanical response due to the nonlinear mechanics at play. Despite their variety, these materials all share the property of a stronger response when they are crushed faster. If this regularity is artificially reproduced by vibrating a rigid tile with a random signal modulated in amplitude, one experiences the strong sensation that the tile gives under the foot, as shown by Visell et al. (2011). A related effect was demonstrated by Kildal et al. (2010) when pressing on a rigid surface with a vibrating pen.

Gravity. A omnipresent regularity that the brain should have internalized is the movement of objects under the influence of gravity (McIntyre et al. 2001). Balls rolling down a slope of inclination, α, accelerate according to 0.7 $\mathrm{Sin}(\alpha)$, no matter what is their size and what is the substance they are made of. (This regularity was discovered by Galileo circa 1638 in one of the most far-reaching experiments in the history of science (Settle 1961)). If one holds a stick made to vibrate with an amplitude $f(t) \propto g\left[7.0 \iint \mathrm{Sin}(\alpha(t))\mathrm{dt},\right]$, where g is a periodic function and α is the stick inclination angle, then the person holding the stick spontaneously experiences the irrepressible sensation that a ball is rolling inside the stick (Yao and Hayward 2006). The coefficient 7.0 is the corrected acceleration of gravity to account for the rolling movement of a ball. Different functions g give different levels of realism but the effect is highly robust. The perceptual problem is to determine the ball displacement, $x(t)$, knowing $f(t)$, a type of inverse problem that the brain solves effortlessly despite the fact that g is unknown but periodic.

Contact mechanics. Another example of a regularity which is linked to what our body experiences when pushing against a stiff surface. Almost all solid objects in contact obey to a Hertzian law which states that the area of the surface of contact between the bodies increases with the load. The rate of increase is a function of the relative geometry of the two bodies but also of their material properties (Hayward 2008b). Thus, softer materials correspond to a lower rate of increase of the contact area. If an apparatus is constructed to modify the finger contact surface as a function of the pressing force independently of the finger displacement, then the modification of the rate of increase of the area of the contact surface can induce an illusory sensation of finger motion (Moscatelli et al. 2014). A related effect appealing to similar principles is the sensation of heaviness induced by the lateral deformation of the fingertips in the absence of net loading (Minamizawa et al. 2007).

Absence of slip. The notion of mechanical regularity can be exploited in the opposite manner. What would the brain make of stimuli which, precisely, do not contain the regularities that can normally be relied upon? Here is an example of an illusory effect that could be interpreted in this light. The so-called 'velvet hand illusion' (Mochiyama et al. 2005) occurs when one moves the two hands in contact with each other without slip but with an interposed network of wires or thin rods in-between. It is a conflicting stimulus since, normally, moving the hands together in mutual does not generate any significant tactile sensation, but here, the thin objects sliding between the two hands do cause a powerful tactile input. To the violation of the aforementioned regularity, the brain responds by 'feeling' a film interposed between the two hands (Kawabe et al. 2010).

The nonlinear nature of small scale mechanics. There are very few natural mechanical phenomena of relevance to touch that could be said to have a "linear character". Moreover, there is no indication that linearity is a useful concept in the mechano-transduction to tactile inputs (see for instance Lamoré et al. (1986). Thus it comes as no surprise that complex signals used to drive somatosensation may create surprising effects if they deviate in specific ways from the natural signals that the somatosensory system has evolved to process. The somatosensory system has been shown to have evolved to optimize the detection of fast rate stimuli differently from slow rate stimuli (Iggo and Ogawa 1977, Edin and Valbo 1990). It is otherwise known that if a signal detection system exhibits this property, then periodic excitatory signals having an odd symmetry will cause the output of the detector to undergo a DC drift (a ratcheting behavior). There are many examples in biology of such behaviors including, for instance, the pupillary reflex or heart rate regulation (Clynes 1962). In touch, odd-symmetrical stimuli do cause a sensation of a persisting external load on the limb or the finger (Amemiya et al. 2005; Amemiya and Gomi 2014).

Conclusion

In this note, only a small subset of known tactile and haptic illusions was discussed. They were used to point out the similarities and the differences of the putative perceptual mechanisms in other sensory modalities. In sum, touch exhibits a number of similarities to other perceptual systems, but touch has idiosyncrasies which can be understood from the observation that certain perceptual problems that touch faces cannot be related to those faced by other modalities.

It would be natural to ask whether tactile illusions are the expression of imperfections of the somatosensory system or if illusions are a necessity. In the opening paragraphs, the impossibility for the brain to gain perfect knowledge of the mechanical state of the body that it inhabits, let alone of the external objects that perturb its state, was made clear. Evolution has found methods able to expedite the resolution of these problems at speeds and accuracies that are compatible with the survival of the organism, such as quickly grabbing and evaluating the mass of an object, whether it is a 20 kg suitcase or a flimsy paper cup. These solutions are sometimes surprising and we call them illusions. So the answer to the question of the imperfection of the somatosensory system is rather a question of whether it could be improved. The answer is emphatically yes, through perceptual learning and other skill-based mechanisms.

References

Adams, M et al. (2013). Finger pad friction and its role in grip and touch. *Journal of the Royal Society Interface* 10(80): 20120467.

Amemiya, T and Gomi, H (2014). Distinct pseudo-attraction force sensation by a thumb-sized vibrator that oscillates asymmetrically. In: M Auvray and C Duriez (Eds.), *Haptics: Neuroscience, Devices, Modeling, and Applications*," Part-II (pp. 88–95).

Amemiya, T; Ando, H and Maeda, T (2005). Virtual force display: Direction guidance using asymmetric acceleration via periodic translational motion. *Proceedings of the World Haptics Conference* (pp. 619–622).

Benussi, V (1913). *Psychologie der Zeitauffassung*. Heidelberg: Carl Winter's Universitätsbuchhandlung.

Bochereau, S; Terekhov, A V and Hayward, V (2014). Amplitude and duration interdependence in the perceived intensity of complex tactile signals. In: M Auvray and C Duriez (Eds.), *Haptics: Neuroscience, Devices, Modeling, and Applications*, Part I (pp. 93–100).

Bregman, A S (1990). *Auditory scene analysis*. Cambridge, MA: The MIT Press.

Bresciani, J P; Drewing, K and Ernst, M O (2008). Human haptic perception and the design of haptic-enhanced virtual environments. In: *The Sense of Touch and its Rendering* (pp. 61–106). Berlin Heidelberg: Springer.

Buckingham, G (2014). Getting a grip on heaviness perception: a review of weight illusions and their probable causes. *Experimental Brain Research* 232: 1623–1629.

Caclin, A; Soto-Faraco, S; Kingstone, A and Spence, C (2002). Tactile *capture* of audition. *Perception & Psychophysics*, 64(4): 616–630.

Carter, O; Konkle, T; Wang, Q; Hayward, V and Moore, C I (2008). Tactile rivalry demonstrated with an ambiguous apparent-motion quartet. *Current Biology* 18(14): 1050–1054.

Charpentier, A (1891). Analyse expérimentale de quelques éléments de la sensation de poids. *Archives de Physiologie Normale et Pathologique* 3: 122–135.

Clynes, M (1962). The non-linear biological dynamics of unidirectional rate sensitivity illustrated by analog computer analysis, pupillary reflex to light and sound, and heart rate behavior. *Annals of the New York Academy of Sciences* 98(4): 806–845.

Cohen, J; Hansel, C E M and Sylvester, J D (1953). A new phenomenon in time judgment. *Nature* 172: 901.

Cohen, J; Hansel, C E M and Sylvester, J D (1954). Interdependence of temporal and auditory judgments. *Nature* 174: 642–644.

De Rossi, D; Caiti, A; Bianchi, R and Canepa, G (1991). Fine-form tactile discrimination through inversion of data from a skin-like sensor. In: *Proceedings of the IEEE International Conference on Robotics and Automation* (pp. 398–403).

Dostmohamed, H and Hayward, V (2005). Trajectory of contact region on the fingerpad gives the illusion of haptic shape. *Experimental Brain Research* 164(3): 387–394.

Edin, B B and Vallbo, A B (1990). Dynamic response of human muscle spindle afferents to stretch. *Journal of Neurophysiology* 63(6): 1297–1306.

Flanagan, J R; Bittner, J P and Johansson, R S (2008). Experience can change distinct size-weight priors engaged in lifting objects and judging their weights. *Current Biology* 18: 1742–1747.

Gallace, A; Tan, H Z and Spence, C (2006). The failure to detect tactile change: A tactile analogue of visual change blindness. *Journal Psychonomic Bulletin & Review* 13(2): 300–303.

Gelb, A (1914). Versuche auf dem Gebiete der Zeit- und Raumanschauung, Bericht uber der VI. *Kongress fur Experimentelle Psychologie* (pp. 36–42).

Gjerdingen, R O (1994). Apparent motion in music? *Music Perception* 11(4): 335–370.

Goldreich, D (2007). A bayesian perceptual model replicates the cutaneous rabbit and other spatiotemporal illusions. *PLoS ONE* 2(3): e333.

Gregory, R L (1980). Perceptions as hypotheses. *Philosophical Transactions of the Royal Society of London B: Biological Sciences* 290(1038): 181–197.

Gregory, R L (1997). Knowledge in perception and illusion. *Philosophical Transactions of the Royal Society of London B: Biological Sciences* 352(1358): 1121–1127.

Grunwald, M and John, M (2008). German pioneers of research into human haptic perception. In: *Human Haptic Perception: Basics and Applications* (pp. 15–39). Basel: Birkhäuser.

Guest, S; Catmur, C; Lloyd, D and Spence, C (2002). Audiotactile interactions in roughness perception. *Experimental Brain Research* 146: 161–171.

Hartcher-O'Brien, J; Terekhov, A V; Auvray, M and Hayward, V (2014). Haptic shape constancy across distance. In: M Auvray and C Duriez (Eds.), *Haptics: Neuroscience, Devices, Modeling, and Applications*, Part I (pp. 77–84).

Hayward, V (2008a). A brief taxonomy of tactile illusions and demonstrations that can be done in a hardware store. *Brain Research Bulletin* 75(6): 742–752.

Hayward, V (2008b). Haptic shape cues, invariants, priors, and interface design. In: M Grunwald (Ed.), *Human Haptic Perception—Basics and Applications* (pp. 381–392). Birkhauser Verlag.

Hayward, V (2011). Is there a 'plenhaptic' function? *Philosophical Transactions of the Royal Society B: Biological Sciences* 366(1581): 3115–3122.

Hayward, V and Cruz-Hernandez, M (2000). Tactile display device using distributed lateral skin stretch. In: *Proceedings of the Haptic Interfaces for Virtual Environment and Teleoperator Systems Symposium, DSC-69-2* (pp. 1309–1314).

Heller, M A; Calcaterra, J A; Burson, L L and Green, S L (1997). The tactual horizontal-vertical illusion depends on radial motion of the entire arm. *Perception & Psychophysics* 59: 1297–1311.

Helson, H (1930). The tau effect—An example of psychological relativity. *Science* 71(1847): 536–537.

Howe, C Q and Purves, D (2005). *Perceiving geometry: Geometrical illusions explained by natural scene statistics*. New York: Springer.

Iggo, A and Ogawa, H (1977). Correlative physiological and morphological studies of rapidly adapting mechanoreceptors in cat's glabrous skin. *Journal of Physiology* 266(2): 275–296.

Johnson, S A; Gorman, D M; Adams, M J and Briscoe, B J (1993). The friction and lubrication of human stratum corneum. *Tribology Series* 25: 663–672.

Jousmäki, V and Hari, R (1998). Parchment-skin illusion: sound-biased touch. *Current Biology* 8 (6): 190–191.

Kawabe, Y; Chami, A; Ohka, M and Miyaoka, T (2010). A basic study on tactile displays using velvet hand illusion. In: *IEEE Haptics Symposium* (pp. 101–104).

Kikuuwe, R; Sano, A; Mochiyama, H; Takasue, N and Fujimoto, H (2005). Enhancing haptic detection of surface undulation. *ACM Transactions on Applied Perception* 2(1): 46–67.

Kildal, J (2010). 3d-press: haptic illusion of compliance when pressing on a rigid surface. In: *Proceedings of the International Conference on Multimodal Interfaces and the Workshop on Machine Learning for Multimodal Interaction, ICMI-MLMI'10* (pp. 21:1–21:8).

Konkle, T; Wang, Q; Hayward, V and Moore, C I (2009). Motion aftereffects transfer between touch and vision. *Current Biology* 19(9): 745–750.

Lamoré, P J J; Muijser, H and Keemink, C J (1986). Envelope detection of amplitude-modulated high-frequency sinusoidal signals-by skin mechanoreceptors. *Journal of the Acoustical Society of America* 79(4): 1082–1085.

Lederman, S J and Jones, L A (2011). Tactile and haptic illusions. *IEEE Transactions on Haptics* 4 (4): 273–294.
Mamassian, P and de Montalembert, M (2010). A simple model of the vertical-horizontal illusion. *Vision Research* 50(10): 956–962.
McIntyre, J; Zago, M; Berthoz, A and Lacquantini, F (2001). Does the brain model Newton's laws? *Nature Neuroscience* 4(7): 693–694.
Minamizawa, K; Kajimoto, H; Kawakami, N and Tachi, S (2007). A wearable haptic display to present the gravity sensation—Preliminary observations and device design. In: *World Haptics Conference* (pp. 133–138).
Mochiyama, H et al. (2005). Haptic illusions induced by moving line stimuli. In: *World Haptics Conference* (pp. 645–648).
Moscatelli, A et al. (2014). A change in the fingertip contact area induces an illusory displacement of the finger. In: M Auvray and C Duriez (Eds.), *Haptics: Neuroscience, Devices, Modeling, and Applications*, Part II (pp. 72–79).
Nakatani, M; Sato, A; Tachi, S and Hayward, V (2008). Tactile illusion caused by tangential skin srain and analysis in terms of skin deformation. In: *Proceedings of Eurohaptics, LNCS 5024* (pp. 229–237). Springer-Verlag.
Nakatani, M; Howe, R D and Tachi, S (2006). The fishbone tactile illusion. In: *Proceedings of EuroHaptics* (pp. 69–73).
Ramachandran, V S (1988). Perception of shape-from-shading. *Nature* 331: 163–166.
Ricker, S L and Ellis, R E (1993). 2-D finite-element models of tactile sensors. In: *Proceedings of the IEEE International Conference on Robotics and Automation*, Vol. 1 (pp. 941–947).
Robles-De-La-Torre, G and Hayward, V (2001). Force can overcome object geometry in the perception of shape through active touch. *Nature* 412: 445–448.
Ross, H E (1966). Sensory information necessary for the size-weight illusion. *Nature* 212: 650.
Sarrazin, J C; Giraudo, M D and Pittenger, J B (2007). Tau and kappa effects in physical space: The case of audition. *Psychological Research* 71(2): 201–218.
Settle, T B (1961). An experiment in the history of science: with a simple but ingenious device Galileo could obtain relatively precise time measurements. *Science* 133: 19–23.
Smith, A M; Chapman, C E; Donati, F; Fortier-Poisson, P and Hayward, V (2009). Perception of simulated local shapes using active and passive touch. *Journal of Neurophysiology* 102: 3519–3529.
Suzuki, K and Arashida, R (1992). Geometrical haptic illusions revisited: Haptic illusions compared with visual illusions. *Perception & Psychophysics* 52(3): 329–335.
Ullman, S (1979). The interpretation of structure from motion. *Proceedings of the Royal Society of London B: Biological Sciences* 203(1153): 405–426.
Visell, Y; Giordano, B L; Millet, G and Cooperstock, J R (2011). Vibration influences haptic perception of surface compliance during walking. *PLoS ONE* 6(3): e17697.
von Békésy, G (1959). Neural funneling along the skin and between the inner and outer cells of the cochlea. *Journal of the Acoustical Society of America* 31(9): 1236–1249.
Voss, R F and Clarke, J (1975). 1/f noise in music and speech. *Nature* 258: 317–318.
Wallach, H and O'Connell, D N (1953). The kinetic depth effect. *Journal of Experimental Psychology* 45: 205–217.
Wang, Q and Hayward, V (2008). Tactile synthesis and perceptual inverse problems seen from the view point of contact mechanics. *ACM Transactions on Applied Perception* 5(2): 1–19.
Wertheimer, M (1912). Experimentelle Studien über das Sehen von Bewegung. *Zeitschrift für Psychologie* 61: 161–278. [Excerpted and translated in: T Shipley (Ed.), *Classics in Psychology*. New York: Philos. Lib. 1961].
Wiertlewski, M; Hudin, C and Hayward, V (2011). On the 1/f noise and non-integer harmonic decay of the interaction of a finger sliding on flat and sinusoidal surfaces. In: *Proceedings of World Haptics Conference* (pp. 25–30).
Wijntjes, M W A; Sato, A; Hayward, V and Kappers, A M L (2009). Local surface orientation dominates haptic curvature discrimination. *IEEE Transactions on Haptics* 2(2): 94–102.

Wolfe, U; Maloney, L T and Tam, M (2005). Distortions of plane: Tests of perspective theories. *Perception & psychophysics* 67(6): 967–979.
Yao, H-Y and Hayward, V (2006). An experiment on length perception with a virtual rolling stone. In: *Proceedings of Eurohaptics* (pp. 325–330).

Recommended Reading

Grunwald, M (Ed.) (2008). *Human haptic perception: Basics and applications*. Springer Science & Business Media.

Development of Touch

Yannick Bleyenheuft and Jean Louis Thonnard

Among the four somesthetic qualities of touch, warmth, coolness, and pain described by Mountcastle (2005), touch is the most difficult to define due to its multimodality. The sense of touch is enabled by "afferents sensitive to mechanical stimulation of the skin; they provide signals to the brain concerning the form, texture, location, intensity, movement, direction, and temporal cadence of mechanical stimuli, forms of somesthesis highly developed in the hand" (Mountcastle 2005, p. 72). The sensations included in the sense of touch are also categorized as epicritic. In the glabrous skin, the sense of touch is mediated by four types of classically described cutaneous receptors (Merkel, Ruffini, Pacini, Meissner). Tactile information from the body travels through large myelinated axons in the peripheral nerves to the dorsal root ganglia. From there, the information ascends to the medulla via the ipsilateral dorsal columns (gracilis and cuneatus tracts). In the dorsal column nuclei, the second-order neurons send projections that cross the mid-line, where they form the medial lemniscus, which further ascends in the pons and mid-brain to terminate in the ventral posterior lateral nucleus of the thalamus. From there, third-order neurons send their axons to the primary somatosensory cortex in the post-central gyrus (Kandel et al. 2000). The "**development of touch**" sensation is dependent upon maturational processes affecting mechanoreceptor populations, cortical neurons and myelinated fibers.

Receptors

Four types of receptors have been identified in the human glabrous skin (Vallbo and Johansson 1984).

Y. Bleyenheuft (✉)
Université Catholique de Louvain, Louvain-la-Neuve, Belgium

J.L. Thonnard
Université Catholique de Louvain, Brussels, Belgium

T.J. Prescott et al. (eds.), *Scholarpedia of Touch*, Scholarpedia,
DOI 10.2991/978-94-6239-133-8_28

- Merkel discs: slow adapting type I (SAI) receptors that are dynamically sensitive and exhibit a response linked to the strength of maintained skin deformation. They have small and well-defined cutaneous receptive fields.
- Meissner corpuscles: fast adapting type I (FAI) receptors that have small and well-defined cutaneous receptive fields. They only respond to changes in skin deformation.
- Ruffini receptors: slow adapting type II (SAII) receptors that are dynamically sensitive and exhibit a response linked to the strength of maintained skin deformation. They have receptive fields that are larger and less well defined than those of type I receptors.However, the role of Ruffini receptors has been put into question due to the few number of such receptors in the glabrous skin of the human hand (Paré et al. 2003).
- Pacini corpuscles: fast adapting type II (FAII) receptors that respond to changes in skin deformation. They have receptive fields that are larger and less well defined than those of type I receptors.

Additional contributions to the sense of touch are made by hair follicles in non-glabrous skin areas, as well as by muscular and joint receptors. From embryogenesis to adulthood, these receptors encounter developmental changes. Merkel receptors originate from migratory neural crest cells (Szeder et al. 2003). They appear in the epidermis of the palms of the hands and the soles of the feet between 8 and 12 weeks of gestation (Standring 2005). By that time, cutaneous plexi are already functioning. By the fourth gestational month, the dermal plexi are very well developed, and Meissner and Pacinian receptors have emerged (Standring 2005). The number of Merkel cells begins to decrease during the last part of gestation (Kim and Holbrook 1995). Evidence suggests that the number of Merkel cells continues to decrease throughout life (Besné et al. 2002). Meissner as well as Pacinian corpuscles also decrease in number throughout life (Bruce 1980). This is consistent with the psychophysical findings for age-related PC channel decline (Cauna 1965; Verillo 1979, 1982)

Cortical Maturation and System Myelination

The cortical areas dedicated to touch and the encoding of sensory information are the primary somatosensory cortex (SI, including Brodmann's areas 3a, 3b, 1 and 2); the secondary somatosensory cortex (SII) and the insular (retro and posterior) cortex. The insular cortex, which receives projections from SII, is thought to be important for tactile learning and tactile memory (Kandel et al. 2000). In addition, associative functions of the posterior parietal cortex play an important role in the sense of touch (Kandel et al. 2000). These cortical areas, along with the entire cortex, develop the first synapses from the 23rd week of gestation. The cortex continues to develop until birth. Afterward, changes persist; cortical thickness and size continue to increase until 4 years of age. Dendritic connections, synaptic stabilization, myelination and maturation of associative pathways are developed post-natally following a predetermined sequence. The primary areas develop early, followed by the secondary association areas and finally by the terminal zones, i.e. the long association fibers that may only become operative in the second decade of life (Connolly and Forssberg 1997).

The myelination of both central and peripheral pathways may also play a role in the development of the sense of touch. The myelination follows a defined order: peripheral nerves are myelinated first, followed by the spinal cord, the brainstem, the cerebellum, the basal ganglia and the thalamus. Cortical myelination begins last (Yakovlev and Lecours 1967). Whereas peripheral nerves, such as the ischiatic nerve, are already myelinated at 12 weeks of gestation, short association fibers in the cortex are not fully myelinated before the age of 16 years (Connolly and Forssberg 1997).

Measuring the Sense of Touch Throughout Life

The description of how the sense of touch evolves throughout life is a complex challenge due to both the multimodal aspect of tactile perception and the lack of systematic investigations in the field. Tactile sensations can be roughly categorized (Jones and Lederman 2006) into simple stimuli, such as touch detection or vibration, or complex stimuli (texture, spatial acuity/orientation, size/shape/form, manual exploration).

Simple Stimuli

Touch detection, sensing pressure and vibration, is provided primarily by the Meissner and the Pacinian receptors, respectively. No publication has systematically investigated the changes in sensitivity to pressure and vibration throughout life; however, a decline in both has been observed in aging adults (Thornbury and Mistretta 1981; Bruce 1980; Kenshalo 1986; Gescheider et al. 1994; Goble et al. 1996; Verrillo et al. 2002).The decline of the vibrotactile sensitivity has been observed by studying detection thresholds of vibrotactile signals (Gescheider et al. 1994 I and II), as well as by measuring absolute difference limens (Gescheider et al. 1996). The investigation of the subjective intensity of vibration provided also higher thresholds in older subjects (Verrillo et al. 2002). Touch detection measured with Semmes-Weinstein aesthesiometer filaments also showed increased thresholds with age (Thornbury and Mistretta 1981; Bruce 1980).

Complex Stimuli

Texture discrimination, spatial acuity and orientation, and size/shape/form and manual exploration have been described as complex stimuli (Jones and Lederman 2006). The Merkel-SA1 afferents are selectively sensitive to particular components of local stress-strain fields, which makes them sensitive to edges, points and curvatures (Johnson 2001). These receptors are believed to be determinant for size/shape/form perception, spatial acuity and orientation. Texture discrimination, or a simplified conception thereof, is also commonly attributed to the Merkel-SA1 receptors.

- Texture
  It is estimated that from 4 to 6 months of age, infants are only able to discriminate coarse differences in surface texture (Bushnell and Boudreau 1991; Moranje-Majoux et al. 1997).
- Orientation and spatial acuity
  The performance of children in a grating orientation task (GOT) was tested in a study showing that the spatial acuity of children from 6–10 years of age was less accurate than that of older children (Bleyenheuft et al. 2006). From 10 until 16 years of age, children performed at a level equal to that of young adults. In older adults, spatial acuity performance measured with GOT is decreased (Tremblay et al. 2003). Previous studies of tactile acuity during childhood were based on a gap detection task or on a two-point discrimination (TPD) test (Gellis and Pool 1977; Stevens and Choo 1996). Gellis and Pool (1977) reported that children (0 to 19 years old) had worse TPD when compared with young adults (20–29 years old), and performance peaked at 30 years of age. Stevens and Choo (1996) observed similar results when comparing 8–14 years-old children to adults.
- Size/shape/form
  Size perception abilities have been observed in infants as young as 6–9 months (Palmer 1989). It has been suggested that this ability may actually develop even earlier, around 2–4 months of age (Bushnell and Boudreau 1991). Thus, infants are probably able to detect different object size before they are able to discern shape, texture or compliance variations of an object (Bushnell and Boudreau 1991). The perception of shape begins at about 6 months. At this age, infants are able to differentiate shapes on the basis of features like sharp angles versus smooth curves (Bryant et al. 1972; Brown and Gottfried 1986). However, it is only at 15 months of age that children differentiate shapes using overall spatial configuration (Bushnell and Weinberger 1987). The identification of geometric forms develops later, at around 4–4.5 years of age (Bushnell and Boudreau 1991). The shape/form perception performance continues to increase with age, because children from 8–14 years of age showed improved performance in a two-dimensional form identification task (Benton et al. 1983). At 14 years of age, the performance reached that of 15–50 year-old adults, and a slight decline in accuracy was observed in 51–70 year-olds (Benton et al. 1983).

- Manual exploration
  Manual exploration is probably dependent on the development of object shape and form perception (Jones and Lederman 2006). It has been shown that infants as young as 4 months of age use manual exploration to recognize the boundaries and unity of an object (Streri and Spelke 1988). However, a study in 3–8 year-olds demonstrated that older children perform object recognition with greater speed and accuracy (Morrongiello et al. 1994). The authors also showed that the older children explored common objects with more thoroughness.

Conclusion

Whereas many of the multi-modal capabilities related to tactile perception can be observed in infants aged less than one year, the refinement of touch requires at least the first decade of development to reach young adult values. These observations could have implications for children's manual abilities.

Internal References

Alonso, J-M and Chen, Y (2009). Receptive field. *Scholarpedia* 4(1): 5393. http://www.scholarpedia.org/article/Receptive_field.
Bensmaia, S (2009). Texture from touch. *Scholarpedia* 4(8): 7956. http://www.scholarpedia.org/article/Texture_from_touch. (see also pages 207–215 of this book).
Braitenberg, V (2007). Brain. *Scholarpedia* 2(11): 2918. http://www.scholarpedia.org/article/Brain.
Burke, R E (2008). Spinal cord. *Scholarpedia* 3(4): 1925. http://www.scholarpedia.org/article/Spinal_cord.
Eichenbaum, H (2008). Memory. *Scholarpedia* 3(3): 1747. http://www.scholarpedia.org/article/Memory.
Llinas, R (2008). Neuron. *Scholarpedia* 3(8): 1490. http://www.scholarpedia.org/article/Neuron.
Redgrave, P (2007). Basal ganglia. *Scholarpedia* 2(6): 1825. http://www.scholarpedia.org/article/Basal_ganglia.
Sherman, S M (2006). Thalamus. *Scholarpedia* 1(9): 1583. http://www.scholarpedia.org/article/Thalamus.

External References

Benton, A L; Hamsher, K D; Varney, N R and Spreen, O (1983). Contributions to neuropsychological assessment: A clinical manual. Oxford university press.
Besné, I; Descombes, C and Breton, L (2002). Effect of age and anatomical site on density of sensory innervation in human epidermis. *Archives of Dermatology* 138: 1445–1450.
Bleyenheuft, Y; Cols, C; Arnould, C and Thonnard, J L (2006). Age-related changes in tactile spatial resolution from 6 to 16 years old. *Somatosensory & Motor Research* 23(3): 83–87.
Brown, K W and Gottfried, A W (1986). Cross-modal transfer of shape in early infancy: Is there a reliable evidence? In: L P Lipsitt and C Rovee-Collier (Eds.), *Advances in Infancy Research*, Vol. 4. Norwood, NJ: Albex.

Bruce, M F (1980). The relation of tactile thresholds to histology in the fingers of elderly people. *Journal of Neurology, Neurosurgery & Psychiatry* 43(8): 730–734.

Bryant, P E; Jones, P; Claxton, V and Perkins, G M (1972). Recognition of shapes across modalities by infants. *Nature* 240(5379): 303–304.

Bushnell, E W and Boudreau, P (1991). The development of haptic perception during infancy. In: M Heller and W Schiff (Eds.), *The psychology of touch*. Hillsdale, NJ: Erlbaum.

Bushnell, E W and Weinberger, N (1987). Infants' detection of visual-tactual discrepancies: asymmetries that indicate a directive role of visual information. *Journal of Experimental Psychology: Human Perception and Performance* 13(4): 601–608.

Cauna, N (1965). The effects of aging on the receptors organs of the human dermis. In: W Montagna (Ed.), *Advances in Biology of Skin* (pp. 63–96). Pergamon Press.

Connolly, K J and Forssberg, H (1997). *Neurophysiology and Neuropsychology of Motor Development*. Mac Keith Press.

Gellis, M and Pool, R (1977). Two point discrimination distances in the normal hand and forearm. *Plastic and Reconstructive Surgery* 59: 57–63.

Gescheider, G A; Beiles, E J; Checkosky, C M; Bolanowski, S J and Verrillo, R T (1994). The effects of aging on information-processing channels in the sense of touch: II. Temporal summation in the P channel. *Somatosensory & Motor Research* 11(4): 359–365.

Gescheider, G A; Edwards, R R; Lackner, E A; Bolanowski, S J and Verrillo, R T (1996). The effects of aging on information-processing channels in the sense of touch: III. Differential sensitivity to changes in stimulus intensity. *Somatosensory & Motor Research* 13(1): 73–80.

Goble, A K; Collins, A A and Cholewiak, R W (1996). Vibrotactile threshold in young and old observers: the effects of spatial summation and the presence of a rigid surround. *The Journal of the Acoustical Society of America* 99(4 Pt 1): 2256–2269.

Jones, L and Lederman, S (2006). *Human Hand Function*. Oxford university Press.

Johnson, K O (2001). The roles and functions of cutaneous mechanoreceptors. Current Opinion in Neurobiology 11: 455–461.

Kandel, R; Schwartz, J and Jessell, T (2000). *Principles of Neural Science*, 4th ed. McGraw-Hill companies.

Kenshalo, D R, Sr. (1986). Somesthetic sensitivity in young and elderly humans. *Journal of Gerontology* 41(6): 732–742.

Kim, D K and Holbrook, A (1995). The appearance, density, and distribution of merkel cells in human embryonic and fetal skin: their relation to sweat gland and hair follicle development. *The Journal of Investigative Dermatology* 104: 411–416.

Moranje-Majoux, E; Cougnot, P and Bloch, H (1997). Hand tactual exploration of textures in infants from 4 to 6 months. *Early Development and Parenting* 6: 127–135.

Morrongiello, B A; Humphrey, G K; Timney, B; Choi, J and Rocca, P T (1994). Tactual object exploration and recognition in blind and sighted children. *Perception* 23(7): 833–848.

Mountcastle, V B (2005). *The sensory hand: Neural mechanisms of somatic sensation*. Cambridge, MA: Harvard University Press.

Palmer, C F (1989). The discriminating nature of infant's exploratory actions. *Developmental Psychology* 25: 885–893.

Paré, M; Behets, C and Cornu, O (2003). Paucity of presumptive ruffini corpuscles in the index finger pad of humans. *Journal of Comparative Neurology* 456(3): 260–266.

Standring, S (2005). *Gray's Anatomy: The Anatomical Basis of Clinical Practice*, 39th ed. Edinburgh: Elsevier Churchill Livingstone.

Streri, A and Spelke, E (1988). Haptic perception of objects in infancy. *Cognitive Psychology* 20: 1–23.

Stevens, J C and Choo, K K (1996). Spatial acuity of the body surface over the life span. *Somatosensory & Motor Research* 13(2): 153–166.

Szeder, V; Grim, M; Halata, Z and Sieber-Blum, M (2003). Neural crest origin of the mammalian Merkel cells. *Developmental Biology* 253(2): 258–263.

Thornbury, J M and Mistretta, C M (1981). Tactile sensitivity as a function of age. *Journal of Gerontology* 36(1): 34–39.

Tremblay, F; Wong, K; Sanderson, R and Cote, L (2003). Tactile spatial acuity in elderly persons: Assessment with grating domes and relationship with manual dexterity. *Somatosensory & Motor Research* 20(2): 127–132.
Vallbo, A B and Johansson, R S (1984). Properties of cutaneous mechanoreceptors in the human hand related to touch sensation. *Human Neurobiology* 3(1): 3–14.
Verrillo, R T (1979). Change in vibrotactile thresholds as a function of age. *Sens Processes* 3(1): 49–59.
Verrillo, R T (1982). Effects of aging on the suprathreshold responses to vibration. *Perception & Psychophysics* 32: 61–68.
Verrillo, R T; Bolanowski, S J and Gescheider, G A (2002). Effect of aging on the subjective magnitude of vibration. *Somatosensory & Motor Research* 19(3): 238–244.
Yakovlev, P I and Lecours, A R (1967). The myelogenetic cycles of regional maturation of the brain. In: A Minkowski (Ed.), *Regional Development of the Brain in Early Life* (pp. 3–65). Philadelphia: Davis.
Authors, please check this list and remove any references that are irrelevant. This list is generated automatically to reflect the links from your article to other accepted articles in Scholarpedia.

Further Readings

Connolly, K J (1998). *The Psychobiology of the Hand*. London: Mac Keith Press.
Jones, L and Lederman, S (2006). *Human Hand Function*. Oxford, UK: Oxford University Press.
Mountcastle, V B (2005). *The Sensory Hand: Neural Mechanisms of Somatic Sensation*. Cambridge, MA: Harvard University Press.

Touch in Aging

Francois Tremblay and Sabah Master

While hearing loss and decreased eyesight are things we all expect with aging, we are less aware of the changes that also affect other sensory systems. For instance, people often experience substantial decline in their ability to detect and discriminate touch stimuli as they age and yet, these changes often go unnoticed for years. **Touch in Aging** provides an overview of results from past and current studies that have examined the impact of age on tactile performance in human observers. We also address briefly the reasons as to why some tactile abilities are more affected than others by age and why tactile experience might be an important factor in modulating age effects in senior individuals.

Changes in Tactile Performance with Age

Given its primary importance in the sense of touch, this section will focus on the changes affecting the sensory function of the hand. In addition, following the terminology introduced by Jones and Lederman (2006) to distinguish between passive and active modes of touch, we will first address changes affecting tactile sensing abilities, i.e. when stimuli are applied to the passive hand. This mode of sensing allows for control of stimulus delivery and thus provides an ideal context to examine changes affecting the hand's sensitivity and spatial resolution. Afterwards, we will turn our attention to changes affecting haptic sensing abilities, i.e. when the hand is actively engaged in tactile exploration of objects' material and spatial properties.

F. Tremblay (✉)
School of Rehabilitation Sciences, University of Ottawa, Ottawa, ON, Canada

S. Master
Hospital for Sick Children, Toronto, Canada

T.J. Prescott et al. (eds.), *Scholarpedia of Touch*, Scholarpedia,
DOI 10.2991/978-94-6239-133-8_29

Changes in Touch Sensitivity and Spatial Resolution with Age

One of the early noticeable signs of aging in sensory perceptual systems is the decline of various forms of sensitivity, i.e. a decrease in the ability to detect near-threshold stimuli. Two forms of sensitivity that have been particularly investigated in the context of aging are pressure sensitivity and vibrotactile detection.

Pressure Sensitivity

The ability to detect light pressure is usually assessed using von Frey filaments of different diameters, such as those introduced by Semmes and Weinstein (Weinstein 1993). Each filament is calibrated to bend to a specific buckling force which translates into pressure in g/mm^2. For instance, the 2.83 filament, which is often taken as the norm for screening changes in pressure detection, corresponds to 0.07 g of force (Bell-Krotoski et al. 1993, 1995). In healthy seniors, pressure sensitivity in the hand, as assessed with filaments, is usually decreased relative to norms (Desrosiers et al. 1999) or when compared to a younger population (Bruce and Sinclair 1980; Tremblay et al. 2005). There is also a general consensus that these changes become more noticeable during the sixth decade and then, tend to increase as people advance in age (Thornbury and Mistretta 1981; Bowden and McNulty 2013). The degree of impaired sensitivity, however, varies considerably between individuals. The importance of this variability was highlighted in a report by Thornbury and Mistretta (1981), who noted that despite the fact that most seniors exhibited higher detection thresholds after 60 years, a substantial proportion ($\sim$40 %) still retain a sensitivity comparable to that of younger groups of participants. In this study, the mean detection threshold at the fingertip was 2.74 in the group of seniors aged over 60 years, whereas it was 2.18 in the participants aged less than 30 years, which translates into a 3-fold increase with age when expressed in mg of force. Using a similar methodological approach (i.e., forced choice paradigm), Tremblay et al. (2005) reached a similar conclusion with regard to both individual variability among seniors and the relative increase with age in detection threshold for light pressure at the fingertip (young, 2.50, senior 3.21).

Vibrotactile Sensitivity

Much like pressure, sensitivity to vibration also declines with age. In this regard, the work of Verrillo and colleagues (Verrillo 1979, 1980) has been particularly enlightening. From the early 80s to mid-90s, these investigators ran a series of experiments examining vibrotactile sensitivity in relation to age linked with factors such as frequency, duration of stimulation and contact area. In general, these studies have revealed that sensitivity to vibration starts to decline substantially during the sixth decade and then deteriorates further with more advancing age (Gescheider 1965).

In addition, the loss in sensitivity was far more pronounced for detection of high frequencies (e.g. 250–300 Hz) than for low frequencies (e.g. 35–40 Hz) (Verrillo 1979, 1980; Gescheider et al. 1996). For example, detection thresholds for 300 Hz stimuli increased by 30 dB over the age range 10–89 years (Gescheider et al. 1996). Along the same line, the perceived intensity of supra-threshold vibratory stimuli (250 Hz) was greatly decreased with age, with a 16 dB loss between 20 and 60 years of age (Verrillo et al. 2002). Collectively, these observations (e.g., decreased sensitivity for high frequencies) were interpreted as evidence for an age-related sensory loss particularly in the Pacinian channel (i.e., Pacini receptors). As for the Non-Pacinian channel (e.g., Meissner receptors) mediating detection of low frequencies, although the loss seems less pronounced than in the PC channel, it is still significant with age (Gescheider et al. 1994). Subsequent works have largely confirmed the observations of Verrillo and colleagues as to the primary importance of age in influencing vibrotactile sensitivity for both low and high frequencies (Goble et al. 1996; Stevens et al. 1998; Lin et al. 2005; Bhattacherjee et al. 2010).

Spatial Acuity

Humans possess a very refined ability to discriminate spatial details, especially at the fingertip, where the high density innervation allows for fine spatial resolution. Various tasks and stimuli have been employed to examine spatial acuity in the resting hand, including "the classical" two-point limen test. Although still widely used clinically, the two-point test and its variants remain problematic for their interpretation is blurred by the presence of intensive cues affecting the perception of single vs. double point stimuli (Johnson and Persinger 1994). In fact, careful investigations by Johnson and colleagues have shown that observers can actually discriminate between single and double-point stimuli even when there is no physical separation between the two probes, thus invalidating the two-point limen as a test for spatial acuity (for more recent development on this issue see Tong et al. 2013). Instead, after the seminal work of Johnson and colleagues (Johnson and Lamb 1981), investigators turn to grooves and gaps as spatial stimuli to test spatial acuity. With such stimuli, the limit of spatial resolution at the fingertip in young adults has been shown to be ~ 1 mm (Van Boven et al. 1991; Van Boven and Johnson 1994; Sathian and Zangaladze 1996), i.e. close to the theoretical limit imposed by the estimated innervation density in this area (Johansson and Vallbo 1979). As expected, with age there is marked decline in spatial resolution. In an elegant series of experiments led in the 90s, Stevens and colleagues (Stevens 1992; Stevens et al. 1995; Stevens and Patterson 1995; Stevens and Choo 1996; Stevens and Cruz 1996) examined age-related changes in spatial acuity across different body areas. These experiments showed that age-related alterations in spatial acuity were ubiquitous across the body surface, although the decline was more pronounced in the distal extremities. For instance, changes in resolution thresholds with advancing age averaged 400 % in the foot area whereas it averaged 130 % at the fingertip. The fact that spatial acuity was particularly affected at the fingers was further corroborated by Tremblay and colleagues (Tremblay et al. 2000, 2003) who used the grating orientation task developed by Van Boven and colleagues (Van Boven et al. 1991) to assess spatial resolution threshold. Their observations revealed a substantial decline in acuity with age, older adults exhibiting almost a threefold increase in grating resolution threshold (mean 2.7 mm) at the index finger. In addition, the decline in spatial acuity observed in seniors correlated strongly with decreased manual dexterity, suggesting a link between changes in tactile sensation at the fingertip and declining manual dexterity with advancing age. Subsequent investigations by Goldreich and colleagues (2003), Wong et al. (2011) examining grating orientation acuity using a precision-controlled stimulus system in large groups of sighted as well as blind participants showed significant age-related decline among both groups. Interestingly, while at any given age blind participants had on average better performance, spatial acuity in the two groups declined at an equal rate with age (see Figure 4 of Goldreich and Kanics 2003), lending support to the hypothesis that the age-related decline in spatial acuity is due to loss of peripheral receptors (e.g. SA1 afferents or Merkel cells, see below). Still, it seems clear that blind people somehow process tactile input more effectively than sighted people as their performance always outweighed that of their sighted peers.

Haptic Sensing

While there are numerous reports describing age-related changes in tactile sensing, the range of observations for haptic sensing is comparatively smaller. In the next section we will highlight the results of experiments examining how age affects haptic performance in recognizing two-dimensional (2-D) patterns, shapes and texture by touch. We will conclude by examining performance for three-dimensional (3-D) objects.

Tactile Recognition of 2-D Patterns

Recognition of 2D patterns is a common task that we do almost every day (e.g., determine which button to press using patterns imprinted on your car's key). Yet, pattern recognition is an inherently difficult task for its performance relies on the integration of sparse inputs generated from multiple contacts as the finger explores the contour (Lederman and Klatzky 1997). In spite of this difficulty, young adults can achieve very high levels of accuracy even after minimal training (Loomis 1982; Vega-Bermudez et al. 1991; Manning and Tremblay 2006). This level of performance, however, is greatly compromised with age. For instance, Manning and Tremblay (2006) found a 38 % decline in performance when comparing young and older adults' ability to recognize Roman letters by haptic exploration (mean hit rate, 88 ± 8 % vs. 55 ± 18 %, respectively). Interestingly, individual variations in letter recognition were highly correlated with spatial acuity thresholds determined at rest, stressing the critical link between spatial resolution and pattern recognition. In a subsequent study, Tremblay and colleagues (Master et al. 2010) compared tactile letter recognition across three age groups, youth, young adults and seniors. The performance was monitored not only in terms of accuracy but also in terms of response time. The comparison revealed a major age effect on performance with youth and young adults largely outperforming the seniors performance both in terms of accuracy (20 % decline) and response time. The latter outcome was particularly affected by age, seniors showing a 2–3 × increase in response time. Thus, pattern recognition appears to become much less efficient with age, with decreasing accuracy and longer processing time.

Texture Perception

Perception of texture by touch has two main dimensions: the roughness-smoothness continuum and the hardness-softness continuum. Only the former has been the subject of systematic investigations. Psychophysical studies conducted by Lederman et al. in the 70s (Lederman and Taylor 1972; Lederman 1974, 1978) using mechanical gratings have contributed significantly in our understanding of texture perception. These studies revealed that spacing between repeated elements (e.g. groove width in a grating) was a major determinant of perceived roughness (Lederman 1982). This work and subsequent investigations have also shown that the ability to perceive relative roughness was quite refined, observers being able to reliably detect increase in spatial period of 1–3 % between gratings (Lamb 1983). Only a handful of studies have actually examined changes in texture perception with age. Sathian et al. (1997) reported on the relatively good performance exhibited by healthy seniors in discriminating grating roughness when compared to that of age-matched senior patients affected by Parkinson's disease (difference threshold, 4 vs. 14 %, respectively). Along the same line, Tremblay et al. (2002), in investigating the impact of computer usage on tactile perception, showed that roughness discrimination of gratings was largely unaffected by the degree of exposure or by age. More recently, Bowden and McNulty (2013) compared texture discrimination in different age groups and concluded that performance of seniors was largely comparable to that of younger participants. Thus, it seems that, unlike other forms of tactile abilities, the perception of surface texture remains relatively impervious to the effect of age.

Tactile Recognition of 3-D Objects

Tactile gnosis or stereognosis refers to the ability to recognize common objects by touch. Such recognition involves a complex process whereby material (e.g., texture, compliance) and geometrical properties (e.g., shape) are first extracted by manual exploration and then integrated centrally to allow for proper identification. Performance is usually assessed by asking participants to recognize a series of familiar objects placed in their hand and by recording the response time and overall accuracy (Jones 1989). In general, healthy people can recognize familiar objects placed in their hand within 3 s (Jones 1989). Another form of test to assess tactile gnosis is to ask participants to explore an object and then match it with a sample of objects having similar shapes. Although we rely on it almost every day, tactile recognition of 3-D objects has not been extensively studied, especially with regard to the effect of aging. Ballesteros and colleagues (Ballesteros and Reales 2004; Ballesteros et al. 2008) investigated the influence of priming in the ability of participants (healthy young, senior adults and patients with dementia) to recognize common 3-D objects (e.g. tools) via haptic touch. Participants were first allowed to tactually explore a sample of different objects for 10 s. Then, in a second testing session, some of the previously explored objects were presented along with new "unexplored" objects. Their results showed a clear priming effect for faster recognition of previously "explored" objects when compared to "unexplored" objects. Interestingly, this priming effect was present in both young and older participants and of similar magnitude, indicating that healthy seniors were as susceptible as younger participants to haptic priming. In the same vein, Norman et al. (2006) compared the ability of younger and older participants to discriminate 3-D objects with familiar shape (Bell peppers) and found no difference in performance between the two groups. In a subsequent study (Norman et al. 2011), the same investigators examined whether age affects the ability to perceive large and small 3D objects by haptic touch. The large objects were explored using the whole hand whereas the small objects were explored with a single finger. In both tasks, the performance in recognizing shapes of the older group was as accurate and precise as that of the younger group. From this limited set of observations, it appears that the ability to recognize 3-D objects by haptic exploration is largely preserved with advancing age.

Why Are Some Tactile Abilities More Affected Than Others by Advancing Age?

As we saw in the preceding sections, some forms of touch perception seem to be particularly affected by advancing age (e.g. pressure sensitivity) while others seem largely preserved (e.g. texture discrimination). Such observations reflect differences in the effects of advancing age on the various neurophysiological mechanisms underlying tactile perception.

For instance, at the peripheral level, age-dependent reductions in the density of receptors supplying the glabrous skin (Cauna 1965; Bruce 1980) might account for the substantial decline observed in the ability to resolve spatial details, irrespective of the mode of touch (tactile or haptic sensing). Similarly, decreased sensitivity to mechanical pressure or vibration may also reflect age-related change in both the number and in the transduction properties of peripheral receptors such as Merkel' discs, Meissner's and Pacini corpuscles (Kenshalo 1986; Gescheider et al. 1996). Interestingly, changes in the skin itself seem to have only a negligible contribution to age-related loss in tactile sensibility (Woodward 1992; Vega-Bermudez and Johnson 2004).

In addition to receptors in the periphery, there are also changes with age occurring centrally affecting the processing of tactile information. For example, early EEG studies in the 1970s and 80s described changes in somatosensory evoked potentials with age including longer latencies and smaller amplitudes (Lüders 1970; Allison et al. 1984). More recent work using advanced imaging techniques has corroborated these early neurophysiological findings (e.g., Sebastian et al. 2011) and suggests that such changes are indicative of aging effects inducing both slower tactile sensory processing possibly due to age-related demyelination of axons as well as somatosensory cortical atrophy resulting in fewer neurons available for sensory processing at the cortical level (Raz et al. 1997; Gunning-Dixon et al. 1998; Abe et al. 2002; Good et al. 2002; Raz et al. 2003, 2004; Salat et al. 2004; Hsu et al. 2008). Indirect evidence for impaired central processing with age was obtained by Master et al. (2010) when examining the performance of a group of healthy seniors in recognizing raised letters by touch. While seniors exhibited relatively good accuracy, their response times were substantially longer than those observed in young adults, indicating slower processing possibly linked with impaired tactile working memory.

Thus both changes at the peripheral and central level might contribute to impaired performance with age in the ability to detect near-threshold stimuli, or to process fine spatial details. Conversely, when tasks are based on above threshold stimulation and do not involve fine spatial discrimination, then seniors can show performance levels comparable to those seen in the younger population. This is likely the case for roughness perception, where discrimination performance relies largely on intensity signaling at the cortical level (Sinclair and Burton 1991; Jiang et al. 1997). The same applies for the recognition of common 3-D objects, where the multiple sources of information available (e.g., shape, temperature, consistence) might ease the identification even if one source is not reliable.

On a final note, one may ask whether expertise in manual tasks could potentially counteract age-related changes in tactile performance. Two recent studies have examined this question. In one study, Reuter et al. (2012) examined whether work-related expertise in manual tasks influenced performance in a variety of tasks assessing tactile and haptic sensing abilities. Their results showed that only the older workers (54–65 years) exhibited signs of declining performance when compared to young and middle-aged workers. In addition, their results showed no influence of tactile expertise on age-related changes in performance. In investigating the same question, Guest, Mehrabyan et al. (2014) reached a similar conclusion. Their results showed that degrees of tactile expertise had no relationship with performance levels measured in different tests evaluating tactile and haptic sensing abilities. While manual expertise and tactile experience might not drastically change the course of age-related decline in most seniors, the story might be different for those seniors who had to adjust to a loss in functional vision at some point in their life. In fact, a study by Legge and colleagues (2008) showed that blind braille readers, unlike sighted subjects, did not experience decline with age on a 2D haptic test, which suggests that extensive tactile experience (or some other feature related blindness) may be able to overcome the impact of age on tactile performance.

Conclusion

As we have seen through this chapter, much like the other senses, the sense of touch tends to decline as we age. This decline, however, is not uniform across individuals, some being more affected than others. The decline is also more noticeable in certain forms of touch perception (e.g. spatial discrimination), leaving other forms relatively preserved (e.g., 3-D object recognition). Interestingly, factors such as mechanical changes in the skin seem to play a minor role in mediating age-related changes; pointing to the importance of alterations in the somatosensory system both at the level of peripheral receptors and centrally. Further evidence for the importance of central factors in influencing age-related changes in tactile performance, comes from observations on blind people showing that extensive practice over a lifetime can somehow help to counteract the decline in sensitivity associated with advancing age.

References

Abe, O et al. (2002). Normal aging in the central nervous system: quantitative MR diffusion-tensor analysis. *Neurobiology of Aging* 23(3): 433–441.

Allison, T; Hume, A L; Wood, C C and Goff, W R (1984). Developmental and aging changes in somatosensory, auditory and visual evoked potentials. *Electroencephalography and Clinical Neurophysiology* 58(1): 14–24.

Ballesteros, S and Reales, J M (2004). Intact haptic priming in normal aging and Alzheimer's disease: Evidence for dissociable memory systems. *Neuropsychologia* 42(8): 1063–1070.

Ballesteros, S; Reales, J M; Mayas, J and Heller, M A (2008). Selective attention modulates visual and haptic repetition priming: Effects in aging and Alzheimer's disease. *Experimental Brain Research* 189(4): 473–483.

Bhattacherjee, A et al. (2010) Vibrotactile masking experiments reveal accelerated somatosensory processing in congenitally blind Braille readers. *The Journal of Neuroscience* 30(43): 14288–14298.

Bell-Krotoski, J; Weinstein, S and Weinstein, C (1993). Testing sensibility, including touch-pressure, two-point discrimination, point localization, and vibration. *Journal of Hand Therapy* 6(2): 114–123.

Bell-Krotoski, J A; Fess, E E; Figarola, J H and Hiltz, D (1995). Threshold detection and Semmes-Weinstein monofilaments. *Journal of Hand Therapy* 8(2): 155–162.

Bowden, J L and McNulty, P A (2013). Age-related changes in cutaneous sensation in the healthy human hand. *Age (Dordrecht, Netherlands)* 35(4): 1077–1089.

Bruce, M F (1980). The relation of tactile thresholds to histology in the fingers of elderly people. *Journal of Neurology, Neurosurgery & Psychiatry* 43(8): 730–734.

Bruce, M F and Sinclair, D C (1980). The relationship between tactile thresholds and histology in the human finger. *Journal of Neurology, Neurosurgery & Psychiatry* 43(3): 235–242.

Cauna, N (1965). The effects of aging on the receptors organs of the human dermis. In: W Montagna (Ed.), *Advances in Biology of Skin*,Vol. 6 (pp. 63–96). Elmsford, NY: Pergamon Press.

Desrosiers, J; Hebert, R; Bravo, G and Rochette, A (1999). Age-related changes in upper extremity performance of elderly people: A longitudinal study. *Experimental Gerontology* 34(3): 393–405.

Gescheider, G A (1965). Cutaneous sound localization. *Journal of Experimental Psychology* 70 (6): 617–625.

Gescheider, G A; Bolanowski, S J; Hall, K L; Hoffman, K E and Verrillo, R T (1994). The effects of aging on information-processing channels in the sense of touch: I. Absolute sensitivity. *Somatosensory & Motor Research* 11(4): 345–357.

Gescheider, G A; Edwards, R R; Lackner, E A; Bolanowski, S J and Verrillo, R T (1996). The effects of aging on information-processing channels in the sense of touch: III. Differential sensitivity to changes in stimulus intensity. *Somatosensory & Motor Research* 13(1): 73–80.

Goble, A K; Collins, A A and Cholewiak, R W (1996). Vibrotactile threshold in young and old observers: The effects of spatial summation and the presence of a rigid surround. *The Journal of the Acoustical Society of America* 99(4 Pt 1): 2256–2269.

Goldreich, D and Kanics, I M (2003). Tactile acuity is enhanced in blindness. *The Journal of Neuroscience* 23(8): 3439–3445.

Good, C D et al. (2002). A voxel-based morphometric study of ageing in 465 normal adult human brains. In: *5th IEEE EMBS International Summer School on Biomedical Imaging, 2002.*

Guest, S et al. (2014). Tactile experience does not ameliorate age-related reductions in sensory function. *Experimental Aging Research* 40(1): 81–106.

Gunning-Dixon, F M; Head, D; McQuain, J; Acker, J D and Raz, N (1998). Differential aging of the human striatum: A prospective MR imaging study. *American Journal of Neuroradiology* 19(8): 1501–1507.

Hsu, J-L et al. (2008). Gender differences and age-related white matter changes of the human brain: a diffusion tensor imaging study. *NeuroImage* 39(2): 566–577.

Jiang, W; Tremblay, F and Chapman, C E (1997). Neuronal encoding of texture changes in the primary and the secondary somatosensory cortical areas of monkeys during passive texture discrimination. *Journal of Neurophysiology* 77(3): 1656–1662.

Johansson, R S and Vallbo, A B (1979). Tactile sensibility in the human hand: relative and absolute densities of four types of mechanoreceptive units in glabrous skin. *Journal of Physiology* 286: 283–300.

Johnson, C P and Persinger, M A (1994). The sensed presence may be facilitated by interhemispheric intercalation: Relative efficacy of the Mind's Eye, Hemi-Sync Tape, and bilateral temporal magnetic field stimulation. *Perceptual and Motor Skills* 79(1 Pt 1): 351–354.

Johnson, K O and Lamb, G D (1981). Neural mechanisms of spatial tactile discrimination: neural patterns evoked by Braille dot patterns in the monkey. *Journal of Physiology* (*London*) 310: 117–144.

Jones, L A (1989). The assessment of hand function: A critical review of techniques. *Journal of Hand Surgery* 14a: 221–228.

Jones, L A and Lederman, S J (2006). *Human Hand Function.* New York: Oxford University Press.

Kenshalo, D R, Sr. (1986). Somesthetic sensitivity in young and elderly humans. *Journal of Gerontology* 41(6): 732–742.

Lamb, G D (1983). Tactile discrimination of textured surfaces: Psychophysical performance measurements in humans. *Journal of Physiology* 338: 551–565.

Lederman, S J (1974). Tactile roughness of grooved surfaces: The touching process and effects of macro- and microsurface structure. *Perception & Psychophysics* 16: 385–395.

Lederman, S J (1978). Heightening tactile impressions of surface texture. In: *Active Touch: the Mechanisms of Recognition of Objects by Manipulation. G. Gordon* (pp. 205–214). Oxford: Pergamon.

Lederman, S J (1982). The perception of texture by touch. In: W Schiff and E Foulke (Ed.), *Tactual Perception: A Source Book* (pp. 130–167). Melbourne, Australia: Cambridge University Press.

Lederman, S J and Klatzky, R L (1997). Relative availability of surface and object properties during early haptic processing. *Journal of Experimental Psychology: Human Perception and Performance* 23(6): 1680–1707.

Lederman, S J and Taylor, M M (1972). Fingertip forces, surface microgeometry, and the perception of roughness by active touch. *Perception & Psychophysics* 12: 401–408.

Legge, G E; Madison, C; Vaughn, B N; Cheong, A M Y and Miller, J C (2008). Retention of high tactile acuity throughout the life span in blindness. *Perception & Psychophysics* 70(8): 1471–1488.

Lin, Y H; Hsieh, S C; Chao, C C; Chang, Y C and Hsieh, S T (2005). Influence of aging on thermal and vibratory thresholds of quantitative sensory testing. *Journal of the Peripheral Nervous System* 10(3): 269–281.

Loomis, J M (1982). Analysis of tactile and visual confusion matrices. *Perception & Psychophysics* 31(1): 41–52.

Lüders, H (1970). The effects of aging on the wave form of the somatosensory cortical evoked potential. *Electroencephalography and Clinical Neurophysiology* 29(5): 450–460.

Manning, H and Tremblay, F (2006). Age differences in tactile pattern recognition at the fingertip. *Somatosensory & Motor Research* 23(3–4): 147-155.

Master, S; Larue, M and Tremblay, F (2010). Characterization of human tactile pattern recognition performance at different ages. *Somatosensory & Motor Research* 27: 60–67.

Norman, J F et al. (2006). Aging and the visual, haptic, and cross-modal perception of natural object shape. *Perception* 35(10): 1383–1395.

Norman, J F et al. (2011). Aging and the haptic perception of 3D surface shape. *Attention, Perception, & Psychophysics* 73(3): 908–918.

Raz, N et al. (2004). Aging, sexual dimorphism, and hemispheric asymmetry of the cerebral cortex: Replicability of regional differences in volume. *Neurobiology of Aging* 25(3): 377–396.

Raz, N et al. (1997). Selective aging of the human cerebral cortex observed in vivo: Differential vulnerability of the prefrontal gray matter. *Cerebral Cortex* 7(3): 268–282.

Raz, N et al. (2003). Differential aging of the human striatum: Longitudinal evidence. *American Journal of Neuroradiology* 24(9): 1849–1856.

Reuter, E-M; Voelcker-Rehage, C; Vieluf, S and Godde, B (2012). Touch perception throughout working life: Effects of age and expertise. *Experimental Brain Research* 216(2): 287–297.

Salat, D H et al. (2004). Thinning of the cerebral cortex in aging. *Cerebral Cortex* 14(7): 721–730.

Sathian, K and Zangaladze, A (1996). Tactile spatial acuity at the human fingertip and lip: Bilateral symmetry and interdigit variability. *Neurology* 46(5): 1464–1466.

Sathian, K; Zangaladze, A; Green, J; Vitek, J L and DeLong, M R (1997). Tactile spatial acuity and roughness discrimination: Impairments due to aging and Parkinson's disease. *Neurology* 49(1): 168–177.

Sebastian, M; Reales, J M and Ballesteros, S (2011). Ageing affects event-related potentials and brain oscillations: A behavioral and electrophysiological study using a haptic recognition memory task. *Neuropsychologia* 49(14): 3967–3980.

Sinclair, R J and Burton, H (1991). Neuronal activity in the primary somatosensory cortex in monkeys (Macaca mulatta) during active touch of textured surface gratings: responses to groove width, applied force, and velocity of motion. *Journal of Neurophysiology* 66(1): 153–169.

Stevens, J C (1992). Aging and spatial acuity of touch. *Journal of Gerontology* 47(1): 35–40.

Stevens, J C and Choo, K K (1996). Spatial acuity of the body surface over the life span. *Somatosensory & Motor Research* 13(2): 153–166.

Stevens, J C and Cruz, L A (1996). Spatial acuity of touch: ubiquitous decline with aging revealed by repeated threshold testing. *Somatosensory & Motor Research* 13(1): 1–10.

Stevens, J C; Cruz, L A; Hoffman, J M and Patterson, M Q (1995). Taste sensitivity and aging: High incidence of decline revealed by repeated threshold measures. *Chemical Senses* 20(4): 451–459.

Stevens, J C; Cruz, L A; Marks, L E and Lakatos, S (1998). A multimodal assessment of sensory thresholds in aging. *The Journals of Gerontology Series B: Psychological Sciences and Social Sciences* 53(4): 263–272.

Stevens, J C and Patterson, M Q (1995). Dimensions of spatial acuity in the touch sense: Changes over the life span. *Somatosensory & Motor Research* 12(1): 29–47.

Thornbury, J M and Mistretta, C M (1981). Tactile sensitivity as a function of age. *Journal of Gerontology* 36(1): 34–39.

Tong, J; Mao, O and Goldreich, D (2013). Two-point orientation discrimination versus the traditional two-point test for tactile spatial acuity assessment. *Frontiers in Human Neuroscience* 7: 579. doi:10.3389/fnhum.2013.00579.

Tremblay, F; Backman, A; Cuenco, A; Vant, K and Wassef, M A (2000). Assessment of spatial acuity at the fingertip with grating (JVP) domes: Validity for use in an elderly population. *Somatosensory & Motor Research* 17(1): 61–66.

Tremblay, F; Mireault, A C; Dessureault, L; Manning, H and Sveistrup, H (2005). Postural stabilization from fingertip contact II. Relationships between age, tactile sensibility and magnitude of contact forces. *Experimental Brain Research* 164(2): 155–164.

Tremblay, F; Mireault, A C; Letourneau, J; Pierrat, A and Bourrassa, S (2002). Tactile perception and manual dexterity in computer users. *Somatosensory & Motor Research* 19(2): 101–108.

Tremblay, F; Wong, K; Sanderson, R and Cote, L (2003). Tactile spatial acuity in elderly persons: Assessment with grating domes and relationship with manual dexterity. *Somatosensory & Motor Research* 20(2): 127–132.

Van Boven, R W and Johnson, K O (1994). The limit of tactile spatial resolution in humans: Grating orientation discrimination at the lip, tongue, and finger. *Neurology* 44(12): 2361–2366.

Van Boven, R W; Johnson, K O and Tilghman, D M (1991). A new clinical test for quantifying somatosensory impairment. *Journal of Oral and Maxillofacial Surgery* 49(8): 141.

Vega-Bermudez, F and Johnson, K O (2004). Fingertip skin conformance accounts, in part, for differences in tactile spatial acuity in young subjects, but not for the decline in spatial acuity with aging. *Perception & Psychophysics* 66(1): 60–67.

Vega-Bermudez, F; Johnson, K O and Hsiao, S S (1991). Human tactile pattern recognition: Active versus passive touch, velocity effect and patterns of confusion. *Journal of Neurophysiology* 65: 531–546.

Verrillo, R T (1979). Change in vibrotactile thresholds as a function of age. *Sens Processes* 3(1): 49–59.

Verrillo, R T (1980). Age related changes in the sensitivity to vibration. *Journal of Gerontology* 35 (2): 185–193.

Verrillo, R T; Bolanowski, S J and Gescheider, G A (2002). Effect of aging on the subjective magnitude of vibration. *Somatosensory & Motor Research* 19(3): 238–244.

Weinstein, S (1993). Fifty years of somatosensory research: from the Semmes-Weinstein monofilaments to the Weinstein Enhanced Sensory Test. *Journal of Hand Therapy* 6(1), 11–22.

Wong, M; Gnanakumaran, V and Goldreich, D (2011). Tactile spatial acuity enhancement in blindness: Evidence for experience-dependent mechanisms. *The Journal of Neuroscience* 31 (19): 7028–7037. doi:10.1523/jneurosci.6461-10.2011.

Woodward, K L (1992). The relationship between skin compliance, age, gender and tactile discriminative thresholds in humans. *Somatosensory & Motor Research* 10: 63–67.

Central Touch Disorders

Haike van Stralen and Chris Dijkerman

Central touch disorders comprise a wide range of deficits in somatosensory perception than can occur after damage to the central nervous system. They vary from deficits in the detection of a touch to complex cognitive deficits such as the inability to recognize objects through touch or the experience of having an additional body part such as a third arm. To understand these disorders, first the neural pathways involved in tactile information processing in the central nervous system will be summarized. This is followed by an overview of the touch disorders ranging from primary-, to higher order deficits.

Pathways of Tactile Information Processing in the Central Nervous System

Tactile information is processed within the somatosensory system. Somatosensory input is derived from a variety of receptors in the skin, muscles and joints which convey information about different elementary sensory modalities such as (i) discriminative touch (pressure, vibration), (ii) proprioception which concerns information about the position and movement of one's own body and limbs, (iii) pain and sensitivity to hot and cold and (iv) affective touch (induced by slow stroking with a soft brush) (Loken et al. 2009). Two ascending systems are responsible for conveying somatosensory input to the brain (Figure 1). The medial lemniscal system is involved in discriminative touch and proprioception, while the spinothalamic tract mediates pain, thermal and affective tactile information. The medial lemniscal system projects contralaterally to the thalamus after which most somatosensory input is relayed to the primary somatosensory cortex (SI), located in the anterior parietal cortex (Figure 2).

The SI of each hemisphere contains somatotopic maps of the contralateral side of the body (Figure 3). In these somatotopic maps, each body part is represented according to the degree of innervation density, e.g. body parts with higher receptor

H. van Stralen (✉) · C. Dijkerman
Utrecht University, Utrecht, The Netherlands

T.J. Prescott et al. (eds.), *Scholarpedia of Touch*, Scholarpedia,
DOI 10.2991/978-94-6239-133-8_30

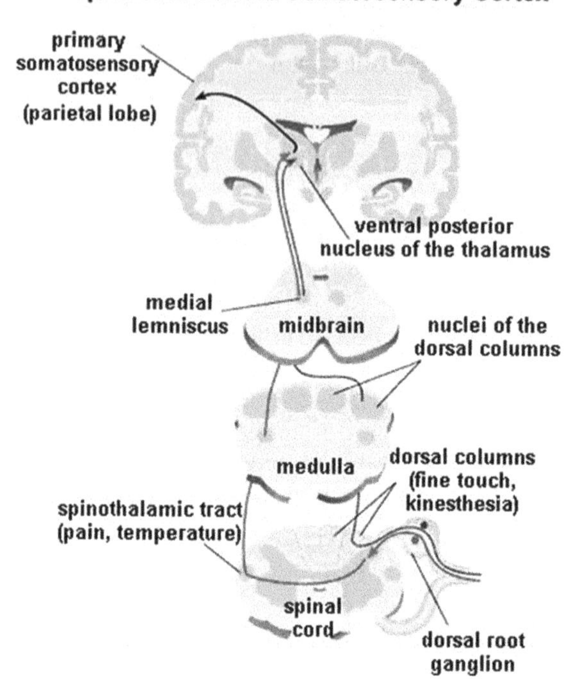

Figure 1 The somatosensory ascending pathways responsible for conveying somatosensory input to the brain

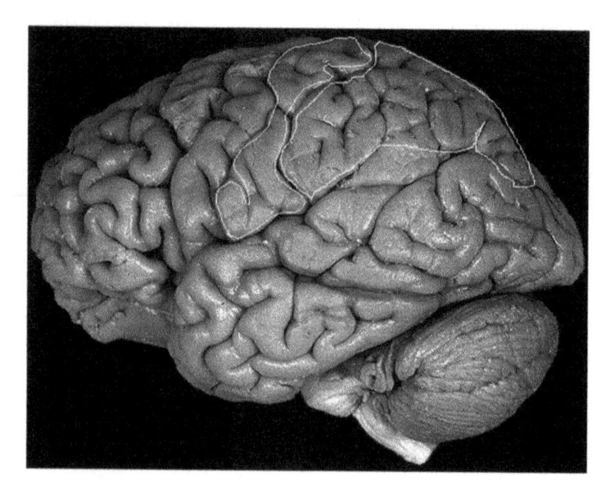

Figure 2 The primary somatosensory area (SI) is located in the anterior parietal cortex (left delineated area). The posterior parietal cortex (right delineated area) is involved in various aspects of higher order somatosensory processing. *From: the digital anatomist project in the department of biological structure at the University of Washington*

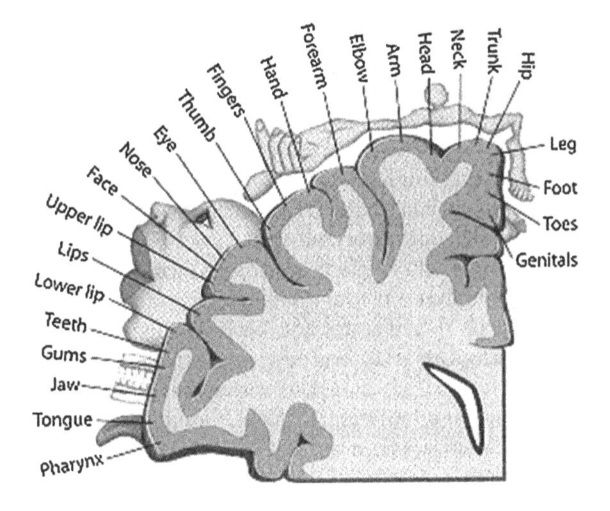

Figure 3 The representation of body parts in the somatosensory homunculus

density occupy larger areas in SI (Penfield et al. 1950). Damage to SI is associated with primary discriminative tactile processing disorders, i.e. impairments in processing the physical, elementary characteristics of tactile stimuli. The posterior insular cortex also contains somatotopic maps for pain, temperature sensitivity and affective touch (Bjornsdotter et al. 2010) (Figure 4).

Higher order somatosensory processes involve wider more distributed networks, including the secondary somatosensory cortex (SII), the posterior parietal cortex (Figure 2) and the anterior insula. The higher order processes range from extracting the features of an object, to the recognition of an object and to body-perception related processes. In contrast to the contralateral involvement for primary tactile

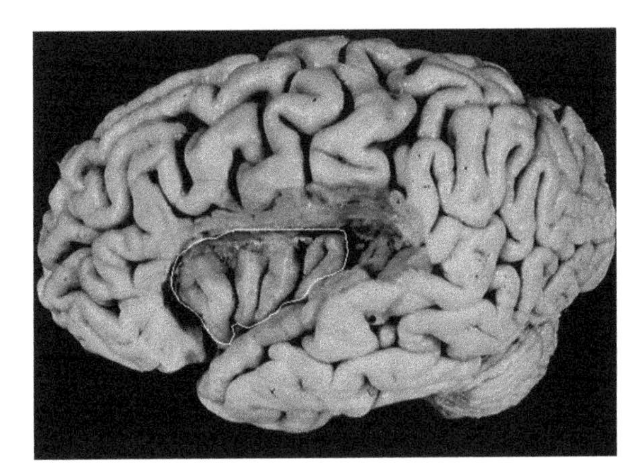

Figure 4 The insula is located beneath the frontal, parietal and temporal opercula and is involved in affective touch, pain and temperature sensitivity and in higher order tactile processing. *From: the digital anatomist project in the department of biological structure at the University of Washington*

information, higher order processes can be bilaterally disturbed after a unilateral lesion. For example, a right hemispheric lesion can cause problems in object recognition in the left hand as well as in the right hand. In addition, there is growing evidence for hemispheric specialization in higher order tactile processes, where right-sided brain lesions result in more severe spatial defects (Heilman et al. 2003) or impairments in body awareness (Baier and Karnath 2008) compared to left-sided lesions. No hemispheric lateralization for primary (elementary) tactile function appears to exist.

Overview of Central Touch Disorders

Primary Somatosensory Disorders

Primary tactile disorders consist of an inability to detect elementary somatosensory aspects, including impaired sensitivity to pressure applied to the skin, elevated two-point discrimination thresholds (i.e. impaired spatial acuity), loss of vibratory sense, or deficits in proprioception. Primary tactile impairments have been reported usually after damage to the contralateral SI, the thalamus, or the subcortical ascending somatosensory pathways. These deficits can selectively affect one somatosensory submodality while others remain functionally intact (Corkin 1978). For example, some patients are able to feel hot and cold while they have no sense of where their limbs are when they have their eyes closed. This is consistent with the idea that these features are processed in parallel. Obviously, primary tactile disorders can lead to problems in higher order touch disorders such as an inability to recognize objects by touch. However, higher order tactile disorders can be present in the absence of primary elementary defects (Wiebers et al. 1998).

Higher Order Touch Disorders

Discrimination of Features

A next hierarchical level in the somatosensory processing of stimuli, is the discrimination of haptic features. The term haptic is used here to show that it involves more than just passive tactile input, but a combination of tactile and proprioceptive information gained through active exploratory hand movements (see below under Object exploration). Haptic features include texture, substance, size, shape, weight and the hardness of a stimulus. Evidence from studies with healthy participants suggests that haptic object feature discrimination can be designated into two categories, i.e. pertaining to the micro- and macrogeometrical properties of an object (Morley et al. 1983). Texture, density or thermal properties are regarded as the microgeometrical aspects, whereas size and shape are regarded as macrogeometrical properties. Evidence for this segregation stems also from reports of selective dissociated impairments within tactile feature discrimination (Delay 1935). That is, hylognosia is an impairment that is characterized by the inability to discriminate texture, density or the thermal properties of an object (microgeometrical). Conversely, suffering from morphognosia means that the patient has an inability to discriminate the size or shape of an object (macrogeometrical). Is has been found that discriminating microgeometrical properties of an object is associated with activation in the parietal operculum (Roland and Mortensen 1987; Roland et al. 1998; Binkofski et al. 1999; O'Sullivan et al. 1994), whereas the anterior part of the intraparietal sulcus is predominantly associated with processing macrogeometrical properties, suggesting that these two functions are segregated at a neuroanatomical level as well (Caselli 1991; Knecht et al. 1996; Homke et al. 2009). However, other reports have disputed the theory of two separate feature processing disorders. It has been argued that impairments in perceiving macrogeomatrical properties of an object is a consequence of impaired spatial abilities. For example, perceiving the size or shape of an object requires for an analysis of the direction and extension of the movement, the sense of limb position in space and tactile localisation (Saetti 1998). However, some reported patients with morhphognosia showed no spatial deficits in other (visual) modalities (Reed et al. 1996) or only mild spatial deficits (Delay 1935).

Object Exploration

Discrimination of features and recognizing an object through touch is not a passive process and requires hand movements to interact with the object. These are stereotypical hand movements that are elicited spontaneously through interaction with an object by touch (Lederman and Klatzky 1987). The type of hand movements depends on the object characteristics we want to extract. Deficits in making these hand movements at this level are called **tactile apraxia**, in which difficulties arise in attuning hand movements to the characteristics of an object in the presence of preserved elementary motor or sensory abilities (Figure 5). Tactile apraxia is usually associated with damage to superior posterior parietal areas (Binkofski et al. 2001). Not surprisingly, difficulties in the exploration of an object can lead to problems in object recognition (Valenza et al. 2001), although this is not obligatory (Caselli 1991). In the case of problems in object recognition, different causes can underlie this deficit. In the next paragraph, the haptic recognition of objects and their associated disorders are discussed.

Object Recognition

Besides intact somatosensory processing at lower levels and purposeful exploratory hand movements, multiple somatosensory signals have to be combined to form a

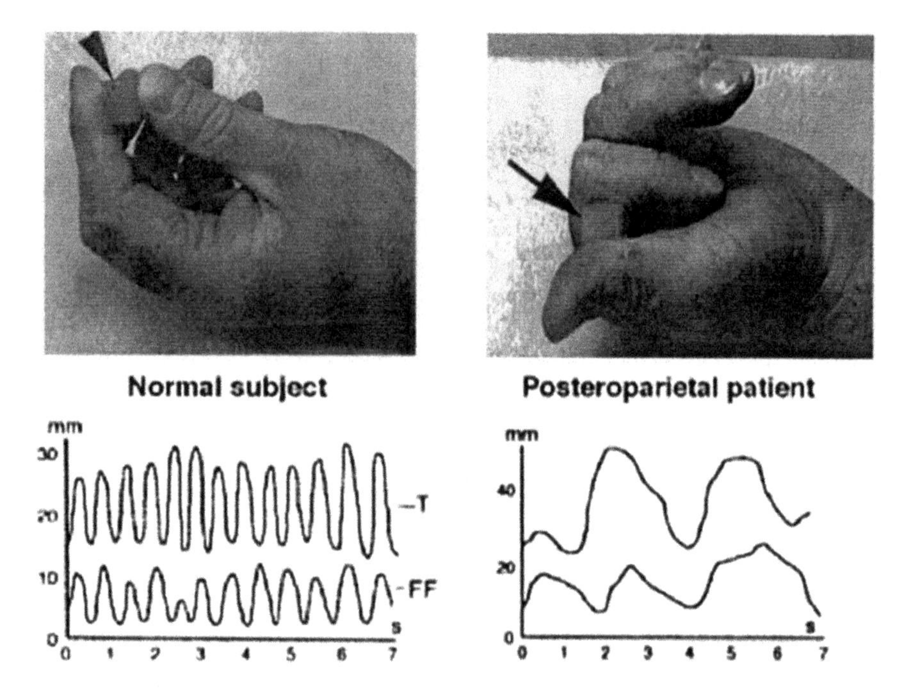

Figure 5 Difficulty in attuning hand movements to the characteristics of an object in a case of tactile apraxia. *From*: Binkofski et al. (2001)

representation of an object in order to recognize it. An example is when you try to grasp your keys from your pocket. This requires purposeful hand movements to search for the expected object features, for example a flat, hard, cool object with a circular shape on one end and ridges on that other end. The information then needs to be integrated into a coherent representation of an object. Subsequently, the semantic properties (its use and function) are retrieved. A deficit in building the object representation or in accessing the semantic properties of the object is called 'tactile agnosia' (Delay 1935; Caselli 1991; Endo et al. 1992; Reed and Caselli 1994; Reed et al. 1996; Caselli 1997). In tactile agnosia, primary somatosensory processing as well as object recognition through other modalities are usually preserved. The level at which an abnormality in information processing occurs in tactile agnosia can vary. First, the integration of the micro- and/or macrogeometrical properties into a coherent representation of the object can be impaired, which is called tactile apperceptive agnosia or astereognosis. These patients are unable to draw the object they have explored through touch. Clinical reports of apperceptive agnosia without primary somatosensory or motor deficits are rare and are often linked to right hemispheric damage. Since the right hemisphere is associated with spatial perception, some authors have suggested that higher order tactile disorders are merely a consequence of impairments in spatial skills (Semmes 1965; Sterzi et al. 1993). Indeed, somatosensory impairments often occur together with deficits in higher order spatial processing such as neglect. More recent studies, however, reported that tactile agnosia can exist without spatial deficits (Reed et al. 1996).

The second type of haptic object recognition deficit is tactile associative agnosia and occurs when a representation of object is achieved (i.e. the patient can make a drawing of the object), but when access to semantic knowledge of the object is is lost, therefore preventing recognition. Recently, a case of pure associative agnosia of the left hand has been described (Veronelli et al. 2014). A right haemorrhagic lesion limited to the post-central and supra-marginal gyri resulted in an inability to recognize objects in only the left hand, with a preserved tactile discrimination or visuo-tactile matching of objects. Thus, patients with associative tactile agnosia can describe the object (e.g. a metal object with an irregular side in case of a key) but are not able to indicate either the use nor name the object. To access this semantic knowledge, input from memory storage about this object is needed (Mesulam 1998). Furthermore, prior semantic knowledge about an object improves tactile recognition performance, suggesting that top-down mechanisms are involved in tactile processing (Bohlhalter et al. 2002).

Another haptic object recognition disorder is **tactile aphasia** (anomia), where the patient is unable to name the object when perceived by touch. Interestingly, the patient is capable of naming the object when it is perceived through another modality. In addition, patients can pantomime the use of a tactile presented object or are able to categorize objects by their meaning, indicating that the semantic knowledge of the object is accessible. Whether the semantic knowledge of the object is completely intact remains controversial but it is clear that semantic problems do not fully account for the problems in naming the object. Thus, a patient with tactile aphasia would be able to successfully discriminate his key from the coins in his pocket. He is capable to describe that this is the object for opening his front door, although he is unable to come up with the word ‘key’. When he can see the key (other modality), he immediately is able to name it. Although not many case studies on tactile aphasia have been described, there is evidence that tactile aphasia and tactile agnosia are different on symptomatology as well as the neuroanatomical substrate (Endo et al. 1992).

The somatosensory system is not only important for recognizing external stimuli such as objects, but primarily provides information and a conscious experience about the body of the observer. A wide range of disorders in bodily experience after damage to the central nervous system has been reported. In the next section an overview of these deficits is given.

Body Related Disorders

Information about our body is based on an integration of visual, vestibular, proprioceptive and tactile input. Several authors have proposed that different representations of our body exist and a common distinction is that between body image and body schema (Gallagher 2005; Paillard 1999; but see De Vignemont 2010 for a critical appraisal of this idea). The body image represents a conscious 'perceptual identification of body features'. It may be more visually based and is influenced by stored knowledge about the body structure and semantics as well as by bottom-up incoming sensory input. In contrast, the body schema codes the position of body parts in space for the guidance of action and is mainly based on tactile input combined with proprioceptive information. The body schema is continuously updated as our body moves or changes. The cerebral basis of the body schema is still unclear, though a central role for the superior part of the posterior parietal cortex has frequently been suggested (Dijkerman and de Haan 2007). Disorders in the body representations may include features of both types.

An alternative organization with respect to the role of the somatosensory system to body representations is proposed by Longo et al. (2010). They described different levels of processing with somatosensation being the lowest (linked to a primary somatosensory processing in SI). Somatoperception concerns the process leading to bodily perception especially related to achieving perceptual constancy. Finally, somatorepresentation refers to more cognitive processes that result in building semantic, configural and emotional knowledge and attitudes about the body. Both somatoperception and somatorepresentation seem to be linked more to the concept of body image rather than the action related body schema. In the next two paragraphs, we will describe a few examples from the huge variety of disorders that exists and we will link them to some of the more cognitive concepts described above.

Structural Body Representation Disorders

Structural body representation concerns the knowledge about the arrangement and form of body parts, crucial to form a sense of body awareness. In "autotopagnosia", patients are unable to point to their own body parts on a visual scheme (Poeck and Orgass 1971; Semenza and Goodglass 1985), whereas in **heterotopagnosia** problems arise in pointing to somebody else's body parts. These disorders have been associated with middle-temporal or parietal lesions of the dominant hemisphere (Schwoebel and Coslett 2005). Structural body representation disorders not necessarily affect the whole body, but can be selectively impaired for the fingers, in the case of "finger agnosia" in which patients are unable to identify the fingers despite a preserved ability to use them (Gerstmann 1940; Kinsbourne and Warrington 1962). It usually affects the middle three fingers of both hands (Frederiks 1985). Although finger agnosia was initially regarded as a form of autotopagnosia (Gerstmann 1940), the disorders appeared to be dissociated (De Renzi and Scotti 1970). Finger gnosis has been repeatedly associated with bilateral parietal activation (Rusconi 2005). The bilateral parietal lobe has been repeatedly associated with finger gnosis. A recent study on the neuroanatomical correlates of finger gnosis specified that left anteromedial parietal lobule plays an important role in finger identification. (Rusconi et al. 2014). Finger agnosia can be considered to be a body image deficit, as tactile input to individual fingers can be used correctly to guide movements (Anema et al. 2008).

Traditionally, finger agnosia was not regarded a unitary phenomenon, but has been described a part of a cluster of impairments, known as the Gerstmann syndrome. Gerstmann syndrome is characterized by four core symptoms, i.e. finger agnosia, dyscalculia, dysgraphia and left-right orientation. The latter is also regarded as a body representation disorder and concerns the impairment in the identification of the left and right side of one's own, but also someone else's body.

In addition to deficits in structural and spatial aspects of body representation, disturbances in body size perception have also been reported. Macrosomatognosia refers to the perception of a body part being larger than it's actual size, while in microsomatognosia patients experience their body (part) as being smaller (Frederiks 1985). These deficits have been associated with a range of paroxysmal disorders such migraine or seizures and often occur temporarily (Rode et al. 2012). They also have been reported for the affected hand in patients with complex regional pain syndrome. Moreover, the perception of a smaller or larger body part can also be induced in healthy participants through proprioceptive illusions (De Vignemont et al. 2005) or through temporary peripheral proprioceptive deafferentation, which results in the affected body part feeling larger (Gandevia and Phegan 1999). Damage to the central nervous also can affect body size perception. Macrosomatognosia is reported more frequently than microsomatognosia and is usually associated with parietal lesions (Frederiks 1985). However, it has also been reported after a frontal lesion (Weijers et al. 2013) or in Parkinson's patients (Sandyk 1998).

Body Awareness Disorders

Disorders in body awareness can arise at different levels. First, being aware of a physical deficit such as a paralysis can be disturbed. Different gradations exist, patients with anosognosia for hemiplegia reject the idea of physical impairment, while other patients admit the existence of their deficit but underestimate the severity and the implications of their physical impairment (**anosodiaphoria**). Anosognosia can exist for both motor and somatosensory impairments. The first is related to lesions to the posterior insula (Karnath et al. 2005), the second is a consequence of lesions to the basal ganglia (i.e. putamen) (Pia et al. 2014). Anosognosia for hemiplegia is relatively common with 18 % of the first-ever stroke patients and 32 % of the right hemisphere stroke patients suffering from this disorder (Appelros et al. 2002; Vocat et al. 2010). A second group of disorders concerns impairments in awareness about (part of) the body itself. Patients with **asomatognosia** experience that a body part is 'missing' or has disappeared from corporal awareness, for example the loss of awareness of one body-half (which may or may not be paralyzed). Another disorder related to disturbances in body ownership is **somatoparaphrenia** in which patients suffer from asomatognosia plus extensive delusions, misidentifications, and confabulations regarding the limb. For example, patients might give an alternative explanation for the limb disownership, for example by believing that the affected limb belongs to someone else, that it is an animal or that it is part of a rotting corpse. A reverse interpretation has been observed as well, with patients identifying body parts of another person as their own (Garcin et al. 1938; Gerstmann 1942). Different theories on the etiology of asomatognosia and somatoparaphrenia exist, also because patients often suffer from other neuropsychological impairments. Hemispatial neglect, hemianesthesia, proprioceptive and attentional impairments have been commonly associated with asomatognosia and somatoparaphrenia (Feinberg et al. 2010; Gandola et al. 2012; Vallar and Ronchi 2009). Furthermore, it has been suggested that multisensory integration and building a spatial representation of the body is hampered (Vallar and Ronchi 2009). Interestingly, these problems are only present for a first-person perspective, but diminishes when a patients views him or herself in the mirror (Fotopoulou et al. 2011). The duration of symptoms of asomatognosia and somatoparaphrenia varies from minutes to months. Providing evidence to the patient that contradicts the delusion only temporarily reduces the denial, after which the asomatognosia or somatoparaphrenia returns. **Misoplegia** is a more affective form of body awareness disorder and can be defined as a hatred for the affected limb, with offensive behaviours toward the limb as a result.

With respect to the underlying neural substrate, disorders of body awareness and ownership are often a consequence of large right hemispheric lesions, involving premotor, parietal and posterior insular areas (for review see Vallar and Ronchi 2009). In addition, cases have been described in which somatoparaphrenic symptoms were induced through vestibular stimulation suggesting that body awareness disorders might be a consequence of functional rather that structural deficits (Ronchi et al. 2013).

Problems concerning body ownership and awareness can also occur when a limb is no longer a physical part of the body, as in the case of phantom limb phenomenon (Figure 6). This can be defined as "the persistent experience of the postural and motor aspects of a limb after its physical loss" (Brugger 2006). Phantom limb experiences a present in approximately 95 % of patients who undergo amputation of a limb (Melzack 1990).

A similar phenomenon (i.e. experience of a limb that is not physically present) can also occur after brain damage, but is less frequently observed. This is referred to as **supernumerary phantom limb (SPL)**, and is defined as "the awareness of having an "extra limb" in addition to the regular set of two arms and two legs." Multimodal experience of the extra limb has been reported including tactile (feel objects with their phantom arm), visual (visually perceive their phantom limb) and motor components (generate action). Neural representations of this extra limb have been reported in the brain areas that represent these modalities, particularly in the left hemisphere (Khateb 2009), although the phenomenon has been described after right-sided lesions as well (Miyazawa et al. 2004).

Except for anosognosia, disorders of body awareness after brain damage appear to occur relatively infrequently, although it's prevalence so far has not been well-documented. They also tend to recover over time. These disorders are

Figure 6 An example of the phantom limb phenomenon. A patient experiences a phantom hand but no forearm after amputation. *From: Wright Halligan and Kew, Wellcome SciArt Project 1997*

nevertheless of great interest as they can further our understanding of the functional and neural mechanisms underlying bodily awareness, body ownership and self-other distinctions.

Conclusion

Central touch disorders can occur on multiple levels ranging from primary somatosensory perception disorders (e.g. a deficit in spatial acuity) to higher order disorders (e.g. shape detection, object recognition or body related disorders). These disorders can be present in absence of other deficits, although it is more common that they influence, and are influenced by other tactile and/or cognitive deficiencies. Compared to visual disorders, touch disorders receive less attention both in research as well as in clinical practice and their presence can therefore be underestimated. Nevertheless, they are linked to limitations in functional independence and therefore deserve more attention.

References

Anema, H A et al. (2008). Differences in finger localisation performance in patients with finger agnosia. *Neuroreport* 19: 1429–1433.

Appelros, P; Karlsson, G M; Seiger, A and Nydevik, I (2002). Neglect and anosognosia after first-ever stroke: Incidence and relationship to disability. *Journal of Rehabilitation Medicine* 34: 215–220.

Baier, B and Karnath, H-O (2008). Tight link between our sense of limb ownership and self-awareness of actions. *Stroke* 39: 486–488.

Binkofski, F et al. (1999) A parieto-premotor network for object manipulation: Evidence from neuroimaging. *Experimental Brain Research* 128: 210–213.

Binkofski, F; Kunesch, E; Classen, J; Seitz, R J and Freund, H J (2001). Tactile apraxia. *Brain* 124: 132–144.

Bohlhalter, S; Fretz, C and Weder, B (2002). Hierarchical versus parallel processing in tactile object recognition: A behavioural-neuroanatomical study of aperceptive tactile agnosia. *Brain* 125: 2537–2548.

Björnsdotter, M; Morrison, I and Olausson, H (2010). Feeling good: On the role of C fiber mediated touch in interoception. *Experimental Brain Research* 207: 149–155.

Brugger, P (2006). From phantom limb to phantom body: varieties of extracorporeal awareness. In: G Knoblich, I M Thornton, M Grosjean and M Shiffrar (Eds.), *Human Body Perception from the Inside Out* (pp. 171–210). New York: Oxford University Press.

Caselli, R J (1991). Rediscovering tactile agnosia (review). *Mayo Clinical Proceedings* 66: 129–142.

Caselli, R J (1997). Tactile agnosia and disorders of tactile perception. In: T E Feinberg and M J Farah (Eds.), *Behavioral Neurology and Neuropsychology* (pp. 277–288). New York: McGraw-Hill.

Corkin, S (1978). The role of different cerebral structures in somesthetic perception. In: E C Carterette and M P Friedman (Eds.), *Handbook of Perception*, Vol. 6B (pp. 105–155). New York: Academic press.

Delay, J P (1935). *Les Astereognosies: Pathologie du Toucher*. Paris: Masson.

De Renzi, E and Scotti, G (1970). Autotopagnosia: fiction or reality? Report of a case. *Archives of Neurology* 23: 221–227.

De Vignemont, F; Ehrsson, H H and Haggard, P (2005). Bodily illusions modulate tactile perception. *Current Biology* 15: 1286–1290.

De Vignemont, F (2010). Body schema and body image–Pros and cons. *Neuropsychologia* 48: 669–680.

Dijkerman, H C and de Haan, E H (2007). Somatosensory processing subserving perception and action. *Behavioral and Brain Sciences* 30: 189–201.

Endo, K et al. (1992). Tactile agnosia and tactile aphasia: Symptomatological and anatomical differences. *Cortex* 28: 445–469.

Feinberg, T E; Venneri, A; Simone, A M; Fan, Y and Northoff, G. The neuroanatomy of asomatognosia and somatoparaphrenia. *Journal of Neurology*, Neurosurgery, and Psychiatry. 2010. 81: 276–281.

Frederiks, J A M (1985). Disorders of the body schema. In: F J Vinken and G W Bruyn (Eds.), *Handbook of Clinical Neurology* 4: 207–240. Amsterdam, North Holland.

Fotopoulou, A et al. (2011). Mirror-view reverses somatoparaphrenia: Dissociation between first- and third-person perspectives on body ownership. *Neuropsychologia* 49: 3946–3955.

Gallagher, S (2005). *How the Body Shapes the Mind*. Oxford: Oxford University Press.

Gandevia, S C and Phegan, C M (1999). Perceptual distortions of the human body image produced by local anaesthesia, pain and cutaneous stimulation. *Journal of Physiology* 514: 609–616.

Gandola, M et al. (2012). An anatomical account of somatoparaphrenia. *Cortex* 48: 1165–1178.

Garcin, R; Varay, A and Hadji-Dimo (1938). Document pour servir à l'étude des troubles du schéma corporel (sur quelques phénomènes moteurs, gnosiques et quelques troubles de l'utilisation des membres du côté gauche au cours d'un syndrome temporo-pariétal par tumeur,

envisagés dans leurs rapports avec l'anosognosie et les troubles du schéma corporel). Revue Neurologique, Paris 69: 498–510.
Gerstmann, J (1940). Syndrome of finger agnosia, disorientation for right and left, agraphia and aculculia. *Archives of Neurology & Psychiatry* 44: 398–407.
Gerstmann, J (1942). Problem of imperception of disease and of impaired body territories with organic lesions. *Archives of Neurology and Psychiatry* 48: 890–913.
Heilman, K (2003). *Clinical Neuropsychology*, 4th ed. USA: Oxford University Press.
Homke, L et al. (2009). Analysis of lesions in patients with unilateral tactile agnosia using cytoarchitectonic probabilistic maps. *Human Brain Mapping* 30: 1444–1456.
Karnath, H O; Baier, B and Nagele, T (2005). Awareness of the functioning of one's own limbs mediated by the insular cortex? *Journal of Neuroscience* 25: 7134–7138.
Kinsbourne, M and Warrington, E K (1962). A study of finger agnosia. *Brain* 85: 47–66.
Khateb, A (2009). Seeing the phantom: a functional magnetic resonance imaging study of a supernumerary phantom limb. *Annals of Neurology* 65: 698–705.
Knecht, S; Kunesch, E and Schnitzler, A (1996). Parallel and serial processing of haptic information in man: Effects of parietal lesions on sensorimotor hand function. *Neuropsychologia* 37: 669–687.
Lederman, S J and Klatzky, R L (1987). Hand movements: A window into haptic object recognition. *Cognitive Psychology* 19: 342–368.
Loken, L S; Wessberg, J; Morrison, I; McGlone, F and Olausson, H (2009). Coding of pleasant touch by unmyelinated afferents in humans. *Nature Neuroscience* 12: 547–548.
Longo, M R; Azañón, E and Haggard, P (2010). More than skin deep: Body representation beyond primary somatosensory cortex. *Neuropsychologia* 48: 655–668.
Melzack, R (1990). Phantom limbs and the concept of a neuromatrix. *Trends in Neuroscience* 13: 88–92.
Mesulam, M (1998). From sensation to cognition. *Brain* 121: 1013–1152.
Miyazawa, N; Hayashi, M; Komiya, K and Akiyama, I (2004). Supernumerary phantom limbs associated with left hemisphere stroke: Case report and review of the literature. *Neurosurgery* 54: 228–231.
Morley, J W; Goodwin, A W and Darian-Smith, I (1983). Tactile discrimination of gratings. *Experimental Brain Research* 49: 291–299.
O'Sullivan, B T; Roland, P E and Kawashima, R (1994). A PET study of somatosensory discrimination in man. Microgeometry versus macrogeometry. *European Journal of Neuroscience* 6: 137–148.
Paillard, J (1999). Body schema and body image - A double dissociation in deafferented patients. In: G N Gantchev, S Mori and J Massion (Eds.), *Motor Control, Today and Tomorrow* (pp. 197–214). Sofia: Akademicno Izdatelstvo.
Penfield, W (1950). *The Cerebral Cortex of Man*. New York: McMillan.
Pia, L et al. (2014). Anosognosia for hemianaesthesia: A voxel-based lesion-symptom mapping study. *Cortex* 61: 158–166.
Poeck, K and Orgass, B (1971). The concept of the body schema: A critical review and some experimental results. *Cortex* 7: 254–277.
Reed, C L and Caselli, R J (1994). The nature of tactile agnosia: A case study. *Neuropsychologia* 32: 527–539.
Reed, C L; Caselli, R J and Farah, M J (1996). Tactile agnosia. Underlying impairment and implications for normal tactile object recognition. *Brain* 119: 875–888.
Rode, G et al. (2012). Facial macrosomatognosia and pain in a case of Wallenberg's syndrome: Selective effects of vestibular and transcutaneous stimulations. *Neuropsychologia* 50: 245–253.
Roland, P E and Mortensen, E (1987). Somatosensory detection of microgeometry, macrogeometry and kinesthesia in man. *Brain Research* 434: 1–42.
Roland, P E; O'Sullivan, B and Kawashima, R (1998). Shape and roughness activate different somatosensory areas in the human brain. *Proceedings of the National Academy of Sciences of the United States of America* 95: 3295–3300.

Ronchi, R et al. (2013). Remission of anosognosia for right hemiplegia and neglect after caloric vestibular stimulation. *Restorative Neurology and Neuroscience* 31: 19–24.
Rusconi, E; Walsh, V and Butterworth, B (2005). Dexterity with numbers: rTMS over left angular gyrus disrupts finger gnosis and number processing. *Neuropsychologia* 43: 1609–1624.
Rusconi, E et al. (2014). Neural correlates of finger gnosis. *The Journal of Neuroscience* 34: 9012–9023.
Saetti, M C and De Renzi, E (1999). Tactile morphagnosia secondary to spatial deficit. *Neuropsychologia* 37: 1087–1100.
Sandyk, R (1998). Reversal of a body image disorder (macrosomatognosia) in Parkinson's disease by treatment with AC pulsed electromagnetic fields. *International The Journal of Neuroscience* 93: 43–54.
Schwoebel, J and Coslett, H B (2005). Evidence for multiple, distinct representations of the human body. *Journal of Cognitive Neuroscience* 17: 543–553.
Semenza, C and Goodglass, H (1985). Localization of body parts in brain injured subjects. *Neuropsychologia* 23: 161–175.
Semmes, J (1965). A non-tactual factor in astereognosis. *Neuropsychologia* 3: 295–315.
Sterzi, R et al. (1993). Hemianopia, hemianaesthesia, and hemiplegia after left and right hemisphere damage: A hemispheric difference. *Journal of Neurology, Neurosurgery, and Psychiatry* 56: 308–310.
Valenza, N et al. (2001). Dissociated active and passive tactile shape recognition: A case study of pure tactile apraxia. *Brain* 124: 2287–2298.
Vallar, G and Ronchi, R (2009). Somatoparaphrenia: A body delusion. A review of the neuropsychological literature. *Experimental Brain Research* 192: 533–551.
Veronelli, L; Ginex, V; Dinacci, D; Cappa, S F and Corbo, M (2014). Pure associative tactile agnosia for the left hand: Clinical and anatomo-functional correlations. *Cortex* 58: 206–216.
Vocat, R; Staub, F; Stroppini, T and Vuilleumier, P (2010). Anosognosia for hemiplegia: a clinical-anatomical prospective study. *Brain* 33: 3578–3597.
Weijers, N R; Rietveld, A; Meijer, F J and de Leeuw, F E (2013). Macrosomatognosia in frontal lobe infarct–A case report. *Journal of Neurology* 260: 925–926.
Wiebers, D O et al. (1998). The sensory examination. In: *Mayo Clinic Examinations in Neurology*, 7th ed. (pp. 255–274). St Louis, MO: Mosby.

Phantom Touch

Vilayanur S. Ramachandran and David Brang

Phantom Touch is continued experience of sensations and presence of a missing limb often occurring after amputation; it is frequently referred to as a phantom limb.

Introduction

About 95 % of amputees experience phantoms which often emerge immediately after amputation but sometimes after weeks or months. In roughly two thirds of patients the phantom is extremely painful. Phantoms are most commonly seen after limb amputation but can occur also occur for other body parts (e.g. phantom breasts, phantom uterus, and phantom appendix). After amputation of the penis, many patients even experience a phantom penis and phantom erections (intriguingly some otherwise intact individuals also report having mainly phantom erections; S.M. Anstis, personal communication).

Although known since antiquity, phantom limbs were first described scientifically by Silas Weir-Mitchell (1872). Since then there have been hundreds of case studies reported in the medical literature (Sunderland 1972; Riddoch 1941; Melzack 1992) but very few systematic experiments. The currentera of experimental work on human patients was inspired, in part, by animal experiments (Kaas and Florence 1996; Jenkins et al. 1990). The combined use of systematic psychophysics and brain imaging has allowed researchers to link neurophysiological experiments in animals with perceptual phenomenology in humans (Ramachandran et al. 1992; Ramachandran and Hirstein 1998) (Figure 1).

V.S. Ramachandran (✉)
Center for Brain and Cognition, University of California, San Diego, CA, USA

D. Brang
Interdepartmental Neuroscience Program, Northwestern University, Evanstown, IL, USA

T.J. Prescott et al. (eds.), *Scholarpedia of Touch*, Scholarpedia,
DOI 10.2991/978-94-6239-133-8_31

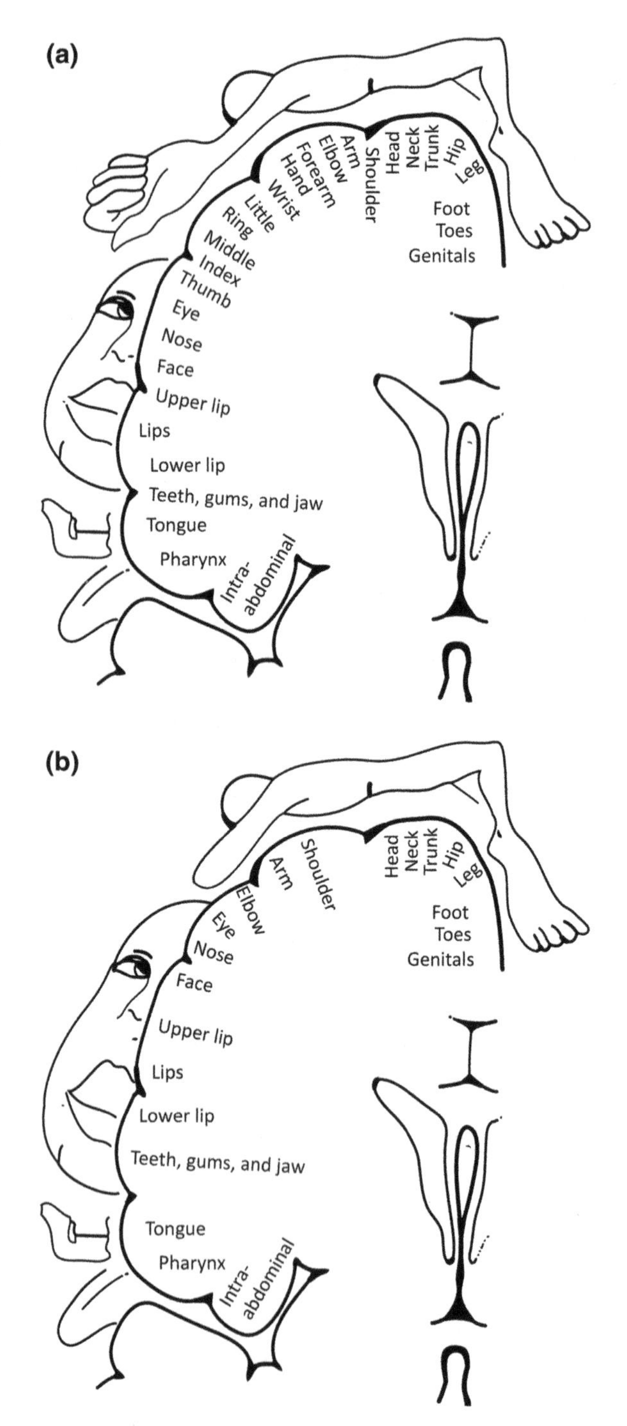

Figure 1 **a** Normal 'homunculus' map showing physical representation of primary somatosensory cortex. **b** After amputation of the hand, the 'face' area of the map invades the former 'hand' territory

Remapping/Referral of Sensations

A few weeks after amputation of an arm, sensory stimuli applied to the ipsilateral face are experienced by the patient as arising from the missing (phantom) arm. There is often a highly specific topographically organized map of the hand on the face (Figure 2) with clearly delineated digits.

This referral of sensations is possibly caused by reorganization of somatosensory maps in the brain (Figure 1). The entire right side of the body is mapped onto the postcentral gyrus of the left hemisphere, known since Penfield, and the map is systematic except for the face being directly below the hand rather than near the neck. After arm amputation the sensory input from the face, which normally projects only to the face area, now "invades" the vacated territory corresponding to the denervated hand territory. As a result, stimuli applied to the face now activate

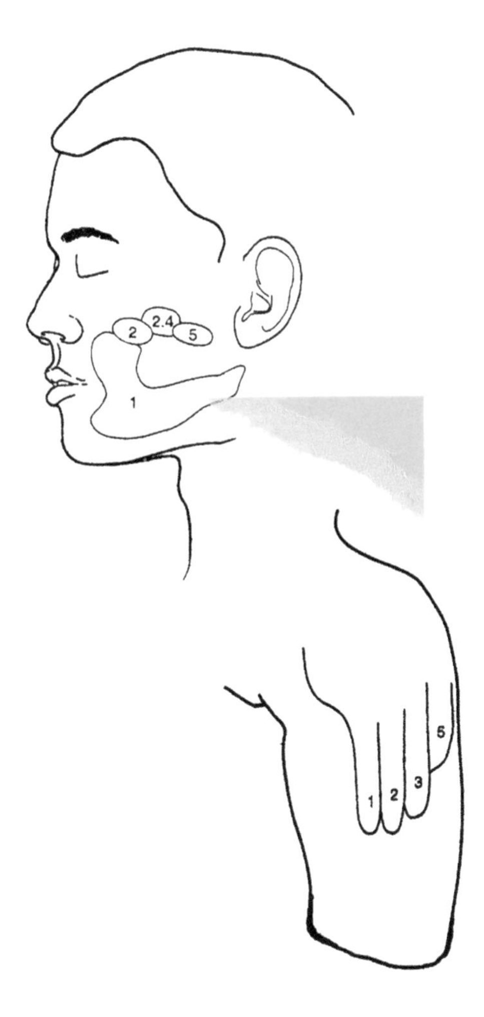

Figure 2 Topographically organized map of sensations referred from face to arm patient DS; numbers indicate digits

the hand region of the brain and are therefore interpreted by higher brain centers as arising from the missing phantom hand (the "remapping hypothesis"). A second map of referred sensations is often seen on the arm proximal to the amputation. This is probably caused by cross-activation of the hand area of cortex by afferents from the upper arm which normally project only to the upper arm region of cortex (Ramachandran and Hirstein 1998; Merzenich et al. 1984). We confirmed these conjectures using MEG; a noninvasive brain imaging technique (Yang et al. 1994).

The map of referred sensations in phantom touch is modality specific; warmth on the face elicits warmth in the phantom thumb, and the same follows for cold and vibration. As touch, warmth, cold, and vibration are separately remapped in separate brain regions, the remapping must then be modality specific. While it is possible that some of this occurs in the thalamus, strokes damaging the touch fibers from the thalamic hand representation to the cortical hand map can result in sensations that are referred from face to hand, suggesting that cortical remapping is sufficient to cause the phenomenon.

Other predictions from the hypothesis were also confirmed. If the trigeminal nerve innervating the face is cut, touching the hand evokes referred sensations in the face in a topographically organized manner (Clarke et al. 1996). Amputation of a finger results in referral of sensations from adjacent fingers (Ramachandran and Hirstein 1998) and sometimes a representation of a single finger is seen on the face (Aglioti et al. 1997). Intensity of phantom pain correlates well with the degree of remapping as explored with brain imaging, suggesting that the remapping is one of the main causes of phantom pain (Flor et al. 1995).

The remapping of hand to face possibly involves the sprouting of new axon terminals as well as the "unmasking" of pre-existing connections (Florence et al. 1998; for a review see Buonomano and Merzenich 1998). Recent evidence for the latter hypothesis comes from the result of left ear cold caloric irrigation which is known to stimulate the vestibular cortex- adjacent to posterior insula—and probably superior parietal lobule (SPL) as well. After caloric irrigation, the patient reports the temporary reduction in the magnitude of referred sensations (Ramachandran and Azoulai 2007), demonstrating the connections responsible for referred sensations can be inhibited or unmasked on small time scales (within a matter of hours). These findings further highlight labile nature of connections within the somatosensory cortex.

The phantom usually has a "habitual" position e.g. supinated. Remarkably a pronation of the phantom (in some subjects) also leads to a small but systematic shift in the topography of the map on the face (and proximal to the stump) along the same direction as the movement of the phantom (Ramachandran and Hirstein 1998; Ramachandran, Brang, and McGeoch, in review), attesting to the extraordinary malleability of neural connections that determine topography in the brain. These shifts are probably caused by changes in body image in the superior temporal sulcus (STS) produced by reafference from motor commands to the phantom. It is conceivable that the signals also feedback to S1/S2 affecting topography.

In a recent study a normal volunteer placed her intact arm near the patient's phantom limb (but not overlapping the phantom). If the patient watched the volunteers hand being stroked or rubbed vigorously, he felt the sensations in his own hand (Ramachandran and Rogers-Ramachandran 2008). This curious observation can be explained in terms of the activity of mirror neurons (Rizzolatti et al. 2006). When you are touched, sensory neurons are activated in your brain's somatosensory cortex. It has been discovered that a subset of these neurons called "mirror neurons" will fire even if you watch someone else being touched—as if the neuron was "putting you in the other person's shoes" or 'empathizing' with the touch delivered to the other person. But if your sensory mirror neurons fire when you watch someone else being touched why don't you literally feel her touch? This is presumably because the absence of touch signals from your skin sends a null signal that vetoes one of the outputs of the mirror neurons. If the arm were to be amputated then you would indeed quite literally feel the other person's sensations in your phantom. This hypothesis would explain why touching another person elicits phantom sensations in the patient. Given their ability to dissolve the barrier between self and others we have dubbed these neurons "Gandhi neurons".

In one case, suddenly "stabbing" the student volunteer's hand, the patient winced in pain and "withdrew" the phantom. This was instantly relieved by the patient simply watching the students intact hand being rubbed; an observation that might have therapeutic implications and give new meaning to the word "empathy".

Genetic Template for Body Image

Some subjects report a phantom arm even if their arm has been missing from birth; suggesting that in spite of its extreme malleability there must also be a genetic scaffolding for body image (La Croix et al. 1992). The same might be true for transgender female to male; they often report having had a phantom penis from early childhood (Ramachandran and McGeoch 2008). We postulate that this genetically specified scaffolding is present in the right SPL.

A "mirror image" of phantom limbs may be the curious syndrome known as apotemnophilia. In this condition a person who is otherwise completely normal experiences an intense desire to have a specific arm or leg amputated; a desire that begins in early childhood. Using MEG (magnetoencephalography) we were able to show that these patients have a body part missing from the body representation in SPL (McGeoch, Brang, and Ramachandran, in review).

Visual Feedback

Amputees usually report feeling movements in the phantom ("It's patting my brother's shoulder" etc.). When you send a command to move your arm, a copy of the command originating in the motor and pre-motor cortex is sent to the parietal cortex where you monitor these commands, combining them with visual and proprioceptive feedback to construct your body image. The monitoring of commands continues to occur even after amputation, "fooling" the brain into thinking that the phantom is moving.

In some patients the arm had been paralyzed and painful for a few months due to peripheral nerve injury. If the arm is then amputated the paralysis is "carried over" into the phantom; a phenomenon that we dubbed "learned paralysis." We speculated that the continued absence of visual feedback signals that the motor commands are being obeyed causes the brain to learn that the arm is paralyzed and this "learned paralysis" persists in the phantom.

The phantom is often fixed in a very painful position. You can prop up a mirror vertically (parasagittally) in front of the patient, and have him look into the mirror so that the reflection of his normal hand is superimposed optically on the felt position of the phantom (Figure 3). This creates the illusion that the phantom arm has been resurrected and if the patient sends motor commands to make bilaterally symmetrical hand movements, the phantom appears to obey the commands. This restores the visuo-motor loop and alleviates pain (Ramachandran and Hirstein 1998; Tsao et al. 2008) by eliminating the discrepancy that is thought to cause phantom pain (Harris 1999). In some cases the entire phantom limb itself disappears—along with the pain (Ramachandran and Hirstein 1998).

The efficacy of MVF for stroke rehabilitation in many patients has also now been demonstrated in a number of placebo controlled trials, following a pilot study we published in the Lancet (Altschuler 1999; Figure 4).

These sorts of experiments suggested that there may also be a "learned paralysis" component (in addition to permanent damage to motor pathways) to the paralysis

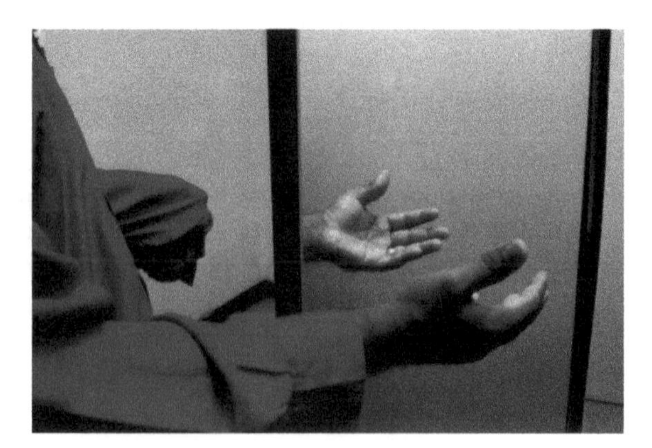

Figure 3 Mirror visual feedback (MVF) for phantom pain was the first demonstration that "real" somatic pain caused by cerebral changes can be modulated rapidly or eliminated by visual feedback. Such mirror feedback can also cause shrinkage of the phantom and corresponding shrinkage of pain. Similar reduction of pain has also been shown using other optical techniques and virtual reality

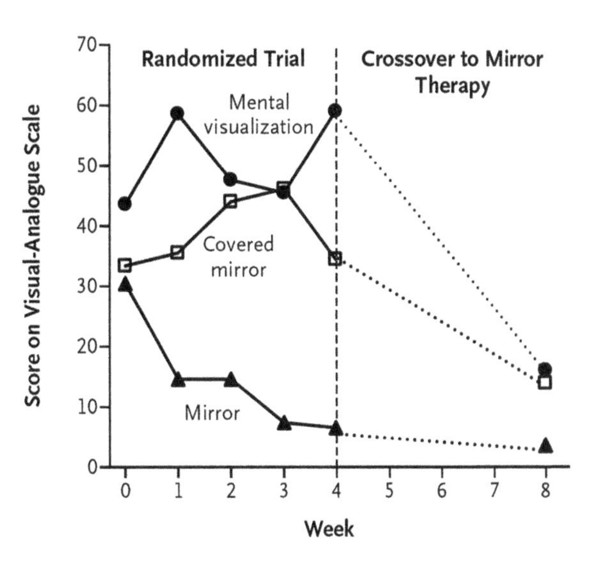

Figure 4 Blind placebo controlled cross-over study demonstrating the stroking efficacy of MVF on phantom pain. The score on the visual-analogue scale ranges from 0 to 100, with higher scores indicating a greater severity of pain. Reproduced from Chan el al. 2007

following stroke and that this could be overcome with visual feedback provided by mirrors or virtual reality (Ramachandran 1992). There is considerable clinical evidence that this is the case (Altschuler et al. 1999; Yavuzer et al. 2008). Other neurological disorders such as complex regional pain syndrome (RSD) have also been treated successfully using the procedure (McCabe et al. 2003).

McCabe et al.'s observations are of special interest. They had patients viewing the reflection of the normal limb superimposed optically on the dystrophic limb. The mirror visual feedback procedure not only reduced pain but produced measurable changes in the temperature of the abnormal limb. This finding, that illusory visual feedback from the location of the dystrophic arm can produce instant physiological changes, cannot be "faked" and is a convincing example of genuine "mind—body" interactions.

Ordinarily our sense of having an arm arises from at least four sources (1) monitoring of feedforward or corollary discharge of commands sent from motor centers to the arm (2) Proprioceptive feedback arising from muscles and joint (3) visual feedback (4) a genetic scaffolding of ones body image. The interplay of the same sources of signals (or lack thereof) must be involved in the genesis of phantom limbs and phantom pain. In addition to pain signals arising from stump neuromas, the discrepancies between these various sources of information may itself be contributing to phantom pain and removing the discrepancies (as with the mirror) seems to relieve the pain.

Illusions of body image can also be produced viewing yourself in a "non-reversing" mirror constructed out of two mirrors with their reflecting surfaces facing each other at right angles. If you wink or move your hands randomly you can get a tingling sensation in your arms and (rarely) a momentary out-of-body experience; especially if you wear heavy makeup (or a mask) to disguise your appearance. We tested one naïve subject who noted, without prompting, that for ten minutes after she left the lab she continued to feel eerily disembodied as if she had "left her body behind" in the experimental set up. Similar experiences occur if subjects look through a half-silvered mirror to superimpose their reflection on a mask seen through the mirror. A momentary "fusion" of selves can occur (Ramachandran and Hirstein 1998).

An observation made by Gawande (2008) is noteworthy. His patient had long suffered from a painfully "swollen" phantom. Looking at the reflection of the normal hand superposed on the phantom caused it to shrink instantly causing the pain to "shrink" as well.

Conclusion

During the last decade the study of phantom limbs has moved from the obscurantism of clinical phenomenology to the era of experimental science. In addition to providing insights into how the brain constructs body image, the study of phantom limbs has broader theoretical and clinical implications The idea that the adult brain consists of independent modules that are largely autonomous and fixed by genes has been replaced by the idea that the so-called modules are in a state of constant dynamic equilibrium with each other and with the sensory input. Neurological dysfunction is caused just as often by shifts in these equilibria as by permanent anatomical damage and if so pushing a reset button can restore the equilibrium; a radical concept in the rehabilitation of neurological patients. These new approaches to treatment should, however, be regarded as complimenting rather than replacing conventional therapies.

Internal References

Braitenberg, V (2007). Brain. *Scholarpedia* 2(11): 2918. http://www.scholarpedia.org/article/Brain.

Cullen, K and Sadeghi, S (2008). Vestibular system. *Scholarpedia* 3(1): 3013. http://www.scholarpedia.org/article/Vestibular_system.

Izhikevich, E M (2007). Equilibrium. *Scholarpedia* 2(10): 2014. http://www.scholarpedia.org/article/Equilibrium.

Llinas, R (2008). Neuron. *Scholarpedia* 3(8): 1490. http://www.scholarpedia.org/article/Neuron.

Penny, W D and Friston, K J (2007). Functional imaging. *Scholarpedia* 2(5): 1478. http://www.scholarpedia.org/article/Functional_imaging.

Rizzolatti, G and Fabbri Destro, M (2008). Mirror neurons. *Scholarpedia* 3(1): 2055. http://www.scholarpedia.org/article/Mirror_neurons.

Sherman, S M (2006). Thalamus. *Scholarpedia* 1(9): 1583. http://www.scholarpedia.org/article/Thalamus.

External References

Aglioti, S; Smania, N; Atzei, A and Berlucchi, G (1997). Spatiotemporal properties of the pattern of evoked sensations in a left index finger amputee patient. *Behavioral Neurology* 111(5): 867–872.

Altschuler, E; Wisdom, S; Stone, L; Foster, C and Ramachandran, V S (1999). Rehabilitation of hemiparesis after stroke with a mirror. *Lancet* 353: 2035–2036.

Buonomano, D V and Merzenich, M M (1998). Cortical plasticity: From synapses to maps. *Annual Review of Neuroscience* 21(1): 149–186.

Chan, B L et al. (2007). Mirror therapy for phantom limb pain. *New England Journal of Medicine* 22: 2206–2207.

Clarke, S; Regali, L; Janser, R C; Assal, G and De Tribolet, N (1996). Phantom face. *Neuroreport* 7: 2853–2857.

Flor, H et al. (1995). Phantom-limb pain as a perceptual correlate of cortical reorganization following arm amputation. *Nature* 375: 482–484.

Florence, S L; Taub, H B and Kaas, J H (1998). Large-scale sprouting of cortical connections after peripheral injury in adult macaque monkeys. *Science* 282(5391): 1117–1121.

Gawande, A (2008, June 30). Annals of medicine: The itch. *New Yorker* 58–64.

Harris, A J (1999). Cortical origin of pathological pain. *Lancet* 354: 1464–1466.

Jenkins, W M; Merzenich, M; Och, M T; Allard, T and Guic-Robles, E (1990). Functional reorganization of primary somatosensory cortex in adult owl monkeys alters behaviorally controlled tactile stimulation. *Journal of Neurophysiology* 63: 82–104.

Kaas, J H and Florence, S L (1996). Brain reorganization and experience. *Peabody Journal of Education* 71(4): 152–167.

La Croix, R; Melack, D and Mitchell, N (1992). Multiple phantom limbs in a child. *Cortex* 28: 503–507.

McCabe, C S et al. (2003). A controlled pilot study of the utility of mirror visual feedback in the treatment of complex regional pain syndrome (type 1). *Rheumatology* 42: 97–101.

Melzack, R (1992). Phantom limbs. *Scientific American* 266: 120–126.

Merzenich, M M et al. (1984). Somatosensory map changes following digit amputation in adult monkeys. *Journal of Comparative Neurology* 224: 591–605.

Ramachandran, V S; Brang, D and McGeoch, P M (in review). Dynamic reorganization of referred sensations caused by volitional movements of phantom limbs.

Ramachandran, V S; Brang, D and McGeoch, P M (in review). Optical shrinkage using mirror visual feedback (MVF) reduces phantom pain.

Ramachandran, V S and McGeoch, P (2008). Phantom penises in female to male transsexuals. *Journal of Consciousness Studies* 15.

Ramachandran, V S; Rogers-Ramachandran, D and Stewart, M (1992). Perceptual correlates of massive cortical reorganization. *Science* 258: 1159–1160.

Ramachandran, V S and Hirstein, W (1998). Perception of phantom limbs. *Brain* 121: 1603–1630.

Ramachandran, V S and Azoulai, S (2007). Caloric stimulation modulates Phantom Limbs. Annual meeting of the Psychonomics Society.

Ramachandran, V S and Rogers-Ramachandran, D (2008). Sensations referred to a patient's phantom arm from another subjects intact arm: Perceptual correlates of mirror neurons. *Medical Hypotheses* 70: 1233–1234.

Ramachandran, V S; McGeoch, P M and Brang, D (2008). Society for Neuroscience abstracts.

Riddoch, G (1941). Phantom limbs and body shape. *Brain* 64: 197.

Rizzolatti, G; Fogassi, L and Gallese, V (2006). Mirrors in the mind. *Scientific American* 295(5): 54–61.

Sunderland, S (1972). *Nerves and Nerve Injury*. Edinburgh: Churchill Livingstone.
Weir MS (1872) Injuries of nerves and their consequences. Lippin-cott: Philadelphia, Pa.
Yang, T T et al. (1994). Sensory maps in the human brain. *Nature* 368: 592–593.
Yavuzer, G et al. (2008). Mirror therapy improves hand function in subacute stroke: A randomized controlled trial. *Archives of Physical Medicine and Rehabilitation* 89: 393–398.

Visually-Impaired Touch

Morton Heller and Soledad Ballesteros

Blindness

Blind individuals rely on their sense of touch for pattern perception, much as the rest of us depend on vision. If a blind person has extra training in the use of touch for tasks such as Braille or spatial orientation, then we might expect increased skill as a consequence. This is the **sensory compensation hypothesis**, and there is evidence that practice can aid touch (Sathian 2000).

Most of us have excellent vision. However, if we are fortunate to live long enough, we are likely to suffer some degree of visual impairment, including, perhaps, blindness. A number of eye diseases increase in frequency with aging, notably cataracts, glaucoma and Age related Macular Degeneration (AMD). It is fortunate, however, that new very successful treatments now exist for cataracts. Diseases that inevitably led to blindness in the past, for example, wet AMD, are now treatable. Unfortunately, this is not the case for all eye disorders, and some individuals have to contend with blindness. The age of onset of blindness has implications for how well people use their sense of touch. For example, if people are **Congenitally Blind** (CB), that is, born without sight or lose it soon after birth, they will not benefit from visual experience or visual imagery. However, they may have an advantage conveyed by their education in mobility skills. People who lose sight later on in life, the **Late Blind** (LB) or **adventitiously blind**, retain the influence of visual experience and are likely to retain memory of visual imagery. Low vision may increase reliance on the sense of touch, but sight of the hand and crude large object perception may aid mobility and touch perception.

M. Heller (✉)
Eastern Illinois University, Illinois, USA

S. Ballesteros
Universidad Nacional de Educación a Distancia, Madrid, Spain

T.J. Prescott et al. (eds.), *Scholarpedia of Touch*, Scholarpedia,
DOI 10.2991/978-94-6239-133-8_32

Several studies supported the sensory compensation hypothesis, suggesting that the superior tactile spatial acuity in the blind may be a form of adaptation (e.g., Sathian 2000; Van Boven et al. 2000). Short-term visual deprivation does not improve passive spatial acuity but long-term visual deprivation enhances tactile acuity (Wong et al. 2011a, b). Blind participants are more accurate with the stationary fingertips than sighted participants, but the two groups show equivalent accuracy with the lips. Braille reading is related to enhanced fingertip acuity, suggesting that tactile experience drives tactile acuity enhancement in blindness (Voss 2011; Wong et al. 2011a, b).

Pictures and Pattern Perception by Blind People

This research area has been controversial, with some researchers emphasizing difficulties in the production and interpretation of tangible pictures. Many researchers have pointed out the difficulties involved in translating 3D information to a 2D display (Jansson and Holmes 2003), while others have noted that it may be easy to name visual pictures, but not so easy to name **haptic** counterparts. **Haptics** involves the use of active touch to perceive objects and forms. The empirical data are variable, with some studies showing lower performance by the CB (Heller 1989a; Lederman et al. 1990; Heller and Schiff 1991), while others have found better performance with raised-line pictures if they are larger (Wijntjes et al. 2008). Kennedy and Juricevic (2006) have argued that touch is suited for the understanding of pictures, even when they involve linear perspective. **Perspective** is a sort of illusory distortion that is found in vision. When we view railroad tracks receding in the distance, it looks as if they converge. Of course they do not. Kennedy has proposed that since perspective involves direction, it should be accessible to CB individuals and the sense of touch (see Heller et al. 1996, 2009; Heller and Kennedy 1990).

Failures to name pictures could derive, at least in part, from lack of familiarity with the rules of depiction in the sense of touch. Thus, most blind people have had little experience with raised-line pictures, and if CB, then they won't have seen pictures when younger. Experience with drawing, combined with increased tactile skill, probably aided the LB in a number of picture perception studies (e.g., Heller 1989a; Heller and Ballesteros 2006). There are reports of excellent performance in a variety of picture perception experiments that required the understanding of depth relations and perspective (Heller et al. 2009). Some experiments indicate that CB individuals do not spontaneously follow the rules of perspective in their drawings, but may come to quickly understand aspects of perspective. Note that failures to name a picture could involve lack of access to categorical information, rather than a perceptual problem. If a young child calls a bus a "train," that does not mean that the child cannot see the bus. Note that visual and haptic experience can alter whether responses to tangible patterns are global or local in blind individuals (Heller and Clyburn 1993).

Maps are useful for blind people, but can be very difficult to use successfully. If a tangible map is too small, there will be problems with resolving fine detail. If a map is too large, there can be difficulties getting an overview of the map. Scale is often a problem for touch, just as in vision. Moreover, if straight ahead on the map does not conform to straight ahead in the world, people can be confused. For example, upward (straight ahead) on a path in the map needs to conform to the direction of locomotion, or people will have difficulty making directional judgments. There is little doubt about the possibility of CB individuals adopting a number of different vantage points, but maps can be difficult to interpret. Of course, individual differences play an important role for blind individuals, just as in the sighted.

Braille

The Braille system of embossed dots was developed because of difficulties with embossed print. Print must be much larger than Braille, before it can be comprehensible for touch. The Braille code is a two by three set of coordinate locations where dot patterns correspond to letters in the alphabet. Braille characters are just over 6 mm in length (vertical dimension), with the dots themselves about 1.44 mm in diameter. The spacing between adjacent centers of the dots is 2.34 mm. While there is little doubt about the utility of this reading scheme for blind people, most blind people do not read Braille, and reading speed is slower than for reading print in most sighted individuals. Skilled readers use both hands for reading Braille, with a variety of methods in evidence (see Davidson et al. 1992; Millar 1997). Some blind people use their right index finger to smoothly and rapidly scan lines of text and the left index finger for finding the beginning of the next line. Poor readers tend to make frequent pauses to attempt to identify individual patterns, and may read with a single finger. High-proficiency Braille readers read with both hands and manage to read at a faster rate with their left index finger, scanning almost twice as many Braille cells as low-proficiency readers (Davidson et al. 1992).

The Braille reading rate using touch is variable and dependent upon a number of factors, including the difficulty of the text (Davidson et al. 1992). According to Foulke (1982) the average Braille reading rate is about 104 words per minute (wpm) for experienced adults, but there are reports of rates as high as 250–300 wpm for excellent Braille readers. The rate of 100 wpm for touch is much slower than visual rates for typical high school students reading print.

A major source of the difficulty in acquiring Braille reading skills derives from the late age of onset of blindness in most instances. Visual impairment is much more common in the aged, and older individuals are much less likely than young children to learn to read with their fingers. Tactile acuity declines with age, along with a number of other functions (Stevens et al. 1996).

It is easier to identify Braille patterns than corresponding letters by touch, if stimulus size is the same. Loomis (1981) demonstrated the superior performance for

Braille. He argued that low-pass spatial filtering by the skin is a problem for letters, but Braille characters are far more distinctive.

Illusions in Touch and Blindness

Illusions occur in touch and in blind people, and are not solely visual. For example, the Mueller-Lyer illusion has been found in CB individuals (Heller et al. 2005). This indicates that explanations of the illusion in terms of size/constancy scaling may not suffice. Thus, one explanation of the illusion has been that the wings-out and wings-in versions of the illusion prompt perceptual differences in depth. If one sees two edges that are judged to be at different distances, but the same size, the "distant" one will be perceived as greater in extent. While this may contribute to the illusion in vision, it is an unlikely explanation for the CB participants. It was proposed that in touch, blind people may have difficulty noting where the straight line ends and the arrows/wings begin. This would lead to overestimation (d) or underestimation of the line (c), depending upon the endings (Figure 1).

The horizontal-vertical illusion (Figure 2) also occurs in CB touch, just as in sight. In the horizontal-vertical illusion, vertical lines are judged as longer than horizontals that are of equal length, when in the form of an inverted T shape. The illusion also occurs with curved shapes, as in the Saint Louis Arch (Heller et al. 2008, 2010, 2013). Heller et al. had blind and sighted participants make judgments about the height and width of curves, and found overestimation of vertical extents. This overestimation also occurs in touch when the height and width are equal, as with the St. Louis Arch. However, it is important to note that while some causal factors are similar in vision and touch, others may be different. For example, the horizontal-vertical illusion is affected by bisection, but it is also influenced by radial/tangential scanning in touch. If one makes radial movements towards the body, these movements are overestimated compared with tangential movements that do not converge on the body (Heller et al. 2010, 2013).

Heller et al. (2013) examined the horizontal-vertical curvature illusion with raised-lines, as well as solid objects. The illusion was found with both types of stimuli, indicating that the haptic illusion is not the result of the use of line drawings. In addition, the illusion was stronger when the curved stimuli were frontally placed, as on a computer screen or wall. These results indicate that the horizontal-curvature illusion is not entirely dependent upon radial-tangential scanning, and can occur in their absence. Radial scanning is not possible with frontal placement.

It is interesting that the Ponzo illusion does not occur in CB individuals or in blindfolded sighted participants. The Ponzo illusion involves making size estimates about two lines that are equal in extent, but appear between converging lines. The converging lines mimic railroad tracks that seem to converge in the distance. The failure to see the illusion in blind people using touch was explained in terms of a failure of the CB to make use of perspective cues. In vision, the higher line is perceived as further away than the second line, so it is judged as larger.

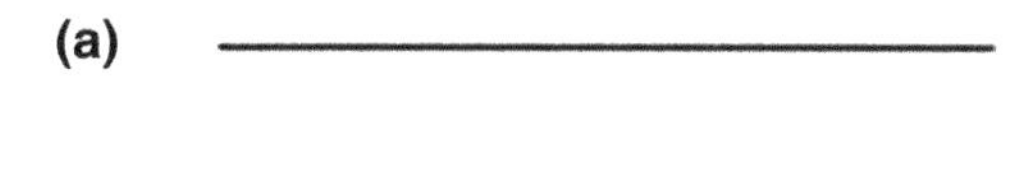

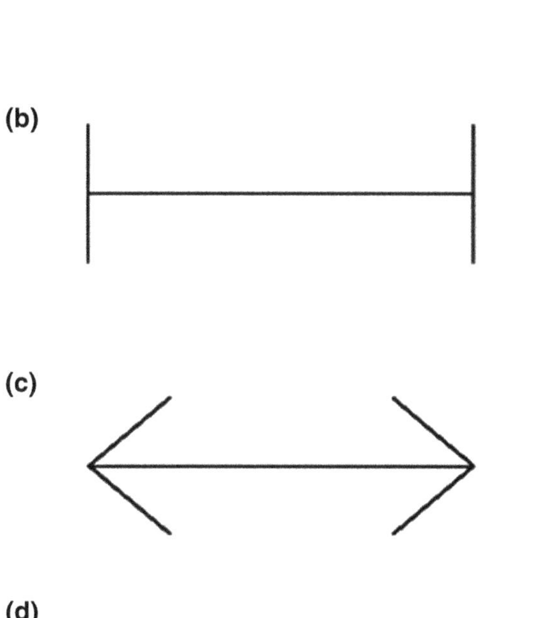

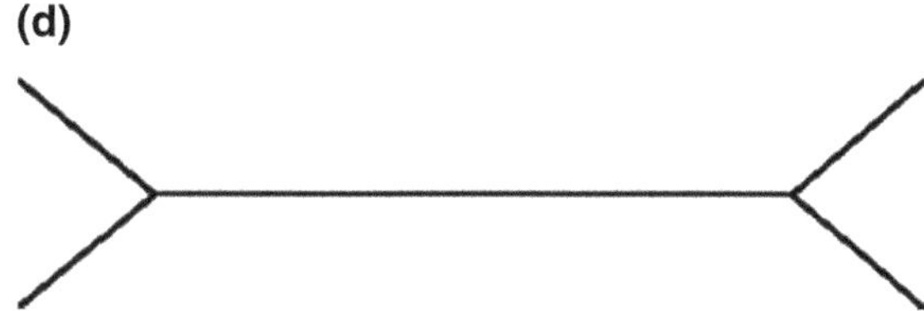

Figure 1 This figure shows the Mueller-Lyer stimuli [adapted from Heller et al. (2005), Perception, 34, 1475–1500, Pion Press, Ltd.]

Aging and Developmental Issues

Do blind children perform better than sighted children of the same age in haptic spatial and memory tasks? In a study with 119 participants (59 blind) from 3 to 16 years of age, blind children performed significantly better than age-matched sighted children in a number of haptic tasks. These tasks were involved in different aspects of shape and spatial perception and cognition, including dimensional structure, spatial orientation, symmetry detection in raised-line and in raised surfaces, and dot spans (Ballesteros 2005). Testing dimensional structure involved a matching-to-sample task to assess whether the child can use different haptic dimensions (shape, size and texture) concomitantly. Spatial orientation measured the ability to recognize the spatial orientation of a shape in tabletop space. Symmetry detection in raised-line and in raised surfaces measured the accuracy of detecting bilateral symmetry (Ballesteros et al. 1997). The 3-D stimuli were

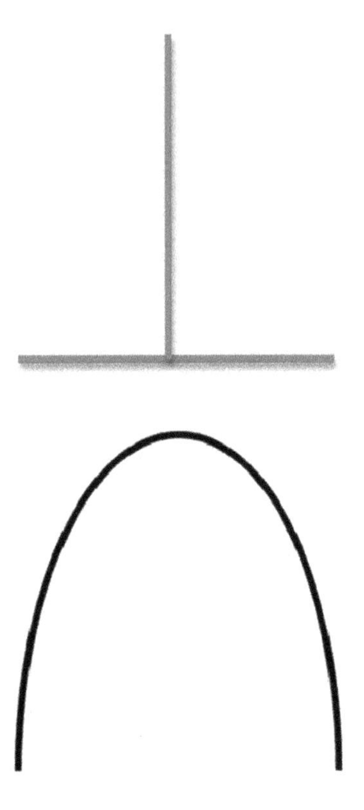

Figure 2 This figure illustrates two forms of the horizontal-vertical illusion. The inverted T shape is above, and the illusion with a curve appears below. Note that in both instances, the horizontal extent is equal to the vertical

constructed by extending the third dimension of figures (Ballesteros and Reales 2004a). Finally, dot span is a short term memory task consisting of a series of items that the child had to repeat correctly in the same order.

Considering older adults, declines in many cognitive domains during the aging process are well documented. Aging negatively affects cognitive processing and brain activity (Park et al. 2002; Park and Reuter-Lorenz 2009). Declines in tactile perception with age could be related to the loss of tactile receptors, in addition to other factors (Bolton et al. 1966; Bruce 1980; Cauna 1965). However, declines are not uniform across cognitive functions but follow different patterns of decline and stability across the lifespan (see Ballesteros et al. 2009a, b). An ability that does not decline with age is implicit memory, an unconscious memory assessed by showing faster and/or more accurate responses for repeated stimuli compared to new ones (repetition priming effects). Implicit memory for familiar objects that were presented to touch without vision was similar in young adults, older adults and in Alzheimer's patients (Ballesteros and Reales 2004b; Ballesteros et al. 2008). Not only within-modal haptic priming, but also cross-modal priming (vision to touch and touch to vision) are spared with aging. Young and older adults show similar perceptual facilitation when the stimuli are presented twice in the same tactual modality (e.g., touch) than when the stimuli are presented first to touch and then to vision (Ballesteros et al. 2009a, b).

When blindfolded participants explored raise-line convex curves with one finger and two fingers and judged the sizes of the curves (horizontal/vertical), it was found that the haptic curvature illusion is not only experienced in young adults but across adulthood and even into old age (Ballesteros et al. 2012). Young and older haptic explorers overestimated the vertical but adolescents did not show the haptic illusion. However, when adolescents performed the task visually, they showed a strong horizontal-vertical illusion.

In the elderly blind, extra training in the use of touch acts as a protective factor against the decline in tactile acuity. Until quite recently it was believed that tactile detection and discrimination are similar in blind and sighted older adults (Hollins 1989). However, the use of modern and more precise psychophysical methods and passive touch have shown that tactile acuity is better in blind participants than in age-matched older adults (Stevens et al. 1996). Stevens et al. (1996) as well as Goldreich and Kanics (2003, 2006) found that, in passive spatial tasks, the acuity of blind individuals declined with age. Tactile acuity was better in the blind at any age but acuity of blind participants declined with age, in parallel with that of sighted participants (but see Heller and Gentaz 2014).

In a recent study, using tactile-acuity charts that required active exploration Legge et al. (2008) found that sighted subjects showed an age-related decrease in tactile acuity of nearly 1 % per year. In contrast, blind individuals did not show an age-related decline using active touch. What can account for the superiority of blind individuals in tasks involving shape and spatial perception and cognition and the enhanced tactile acuity in old age? Enriched tactile experience may explain these differences between blind and sighted people across the lifespan. However, there are reports that with passive spatial tasks, the acuity of blind individuals also declines with age, but acuity remains higher than that for sighted individuals (Goldreich and Kanics 2003, 2006).

The findings discussed above agree with the sensory compensation hypothesis (Sathian 2000). Practice can aid touch. However, recruitment of the visual cortex in blind individuals might explain their better tactile and haptic perception (Cohen et al. 1997). However, when sighted individuals wore blindfolds for 2 h, this resulted in significant deactivation in intermediate regions in visual shape processing, i.e. V3A and ventral intraparietal sulcus—vIPS. (Weisser et al. 2005). This was not found in controls. These results suggest that short-term blindfolding induces changes in the neural processing of tactile form perception and may reflect short-term neural plasticity. Merabet et al. (2006) investigated the involvement of early visual cortical areas in normally sighted, briefly blindfolded subjects as they tactually explored and rated raised-dot patterns using fMRI. They found that tactile form exploration produced activation in the primary visual cortex (V1) and deactivation of extrastriate cortical regions V2, V3, V3A, and hV4, with greater deactivation in dorsal subregions and higher visual areas. These researchers interpreted the findings as suggesting that tactile processing affects the occipital cortex via a suppressive top-down pathway descending through the visual cortical hierarchy, and an excitatory pathway arising from outside the visual cortical hierarchy that drives area V1 directly.

These findings from transcranial magnetic stimulation (TMS), structural and functional imaging studies suggest that the human brain reacts dynamically in response to visual deprivation. Brain regions usually involved in visual processing are involved in processing inputs from other sensorial modalities (for a review see Noppeney 2007). The brain clearly can benefit from reorganization as a function of experience, and this cortical plasticity could mediate sensory compensation (Amedi et al. 2005). In addition, there is evidence that there is plasticity within the somatosensory cortex. Tactile experience could alter the organization and sensitivity of these regions of the cortex (Pascual-Leone and Torres 1993; Sterr et al. 1998).

Passive Tactile Perception in Blindness

In passive touch, the individual is immobile and stimulation is imposed on the skin (Gibson 1966). Gibson argued that passive touch is atypical and prompts subjective sensations; however, many useful sensory aids for blind people make good use of passive touch. Also, people who are both deaf and blind may communicate via **Print-On-Palm** (POP), where letters and words are printed on the skin of the palm. Performance can be high if the rate of presentation is slow enough to avoid the interfering effects of **after-sensations**, where people continue to feel a stimulus after it has disappeared (see Heller 1980, 1989b). Performance in congenitally blind individuals is lowered when patterns are tilted, but this could be a consequence of lack of familiarity and experience with letter and number shapes (Heller 1989b). Congenitally blind individuals may not be familiar with number patterns, since they are taught Braille from an early age. Circumstances are very different, of course, for people who are both deaf and blind or LB. The deaf-blind are generally taught to read POP and achieve high rates of speed and accuracy.

It is possible to present Braille patterns in a passive mode via a tachistometer, allowing precise control over stimulus timing. Legibility thresholds are less than 100 ms using passive touch with skilled readers of Braille (Foulke 1982).

Conclusions

Many studies demonstrate advantages in pattern perception in the late blind and the beneficial role of visual experience and imagery, but this is context dependent. If a task involves reading Braille, one might expect that age of onset matters, with much later onset of blindness a negative predictor of skill. If tasks involve situations that control for differential familiarity, one may see comparable or lower performance by CB and sighted individuals compared to the LB, when using their sense of touch. Late blind persons may have the combined benefits of haptic and visual experience. Again, there are large individual differences and the effects of visual experience can be negative, when one is blinded very late in life. Under typical circumstances in the sighted, vision is used to guide touch, but this advantage is lacking in the totally blind, or blindfolded individual. Blind participants tend to be much faster than blindfolded sighted individuals using their touch for pattern perception (Heller and Ballesteros 2006; Heller and Gentaz 2014).

References

Amedi, A; Merabet, L B; Bermpohl, F and Pascual-Leone, A (2005). The occipital cortex in the blind. Lessons about plasticity and vision. *Current Directions in Psychological Science* 14(6): 306–311. doi:10.1111/j.0963-7214.2005.00387.x.

Ballesteros, S (2005). The haptic test battery: A new instrument to test tactual abilities in blind and visually impaired and sighted children. *British Journal of Visual Impairment* 23(1): 11–24. doi:10.1177/0264619605051717.

Ballesteros, S; González, M; Mayas, J; García-Rodríguez, B and Reales, J M (2009a). Cross-modal repetition priming in young and old adults. *European Journal of Cognitive Psychology* 21(2–3): 366-387. doi:10.1080/09541440802311956.

Ballesteros, S; Manga, D and Reales, J M (1997). Haptic discrimination of bilateral symmetry in two-dimensional and three-dimensional unfamiliar displays. Perception & Psychophysics 59: 37–50.

Ballesteros, S; Mayas, J; Reales, J M and Heller, M A (2012). The effect of age on the haptic horizontal-vertical curvature illusion with raised-line shapes. *Developmental Neuropsychology* 37: 653–667.

Ballesteros, S; Nilsson, L-G and Lemaire, P (2009b). Ageing, cognition, and neuroscience: An introduction. *European Journal of Cognitive Psychology* 21(2–3): 161-175. doi:10.1080/09541440802598339.

Ballesteros, S and Reales, J M (2004a). Visual and haptic discrimination of symmetry in unfamiliar displays extended in the z-axis. *Perception* 33(3): 315–327. doi:10.1068/p5017.

Ballesteros, S and Reales, J M (2004b). Intact haptic priming in normal aging and Alzheimer's disease: Evidence for dissociable memory systems. *Neuropsychologia* 42(8): 1063–1070. doi:10.1016/j.neuropsychologia.2003.12.008.

Ballesteros, S; Reales, J M; Mayas, J and Heller, M A (2008). Selective attention modulates visual and haptic repetition priming: Effects in aging and Alzheimer's disease. *Experimental Brain Research* 189(4): 473–483. doi:10.1007/s00221-008-1441-6.

Bolton, C F; Winkelmann, R K and Dyck, P J (1966). A quantitative study of Meissner's corpuscles in man. *Neurology* 16(1): 1. doi:10.1212/wnl.16.1.1.

Boven, R W V; Hamilton, R H; Kauffman, T; Keenan, J P and Pascual-Leone, A (2000). Tactile spatial resolution in blind Braille readers. *Neurology* 54(12): 2230–2236. doi:10.1212/wnl.54.12.2230.

Bruce, M F (1980). The relation of tactile thresholds to histology in the fingers of elderly people. *Journal of Neurology, Neurosurgery & Psychiatry* 43(8): 730–734. doi:10.1136/jnnp.43.8.730.

Cauna, N (1965). The effects of aging on the receptor organs of the human dermis. In: W Montagna (Ed.), *Advances in Biology of Skin* (pp. 63–96). Pergamon Press.

Cohen, L G et al. (1997). Functional relevance of cross-modal plasticity in blind humans. *Nature* 389: 180–183.

Davidson, P; Appelle, S and Haber, R (1992). Haptic scanning of braille cells by low- and high-proficiency blind readers. *Research Developmental Disabilities* 13: 99–111.

Foulke, E (1982). Reading Braille. In: W Schiff and E Foulke (Eds.), *Tactual Perception: A Sourcebook*. New York: Cambridge University Press.

Gibson, J J (1966). *The Senses Considered as Perceptual Systems*. Boston: Houghton Mifflin.

Goldreich, D and Kanics, I M (2003). Tactile acuity is enhanced in blindness. *The Journal of Neuroscience* 23: 3439–3445.

Goldreich, D and Kanics, Ingrid M (2006). Performance of blind and sighted humans on a tactile grating detection task. *Perception & Psychophysics* 68(8): 1363–1371. doi:10.3758/bf03193735.

Heller, M A and Clyburn, S (1993). Global versus local processing in haptic perception of form. *Bulletin of the Psychonomic Society* 31: 574–576.

Heller, M A (1989a). Picture and pattern perception in the sighted and the blind: The advantage of the late blind. *Perception* 18(3): 379–389. doi:10.1068/p180379.

Heller, M A (1989b). Tactile memory in sighted and blind observers: The influence of orientation and rate of presentation. *Perception* 18(1): 121–133. doi:10.1068/p180121.

Heller, M A; Calcaterra, J A; Tyler, L A and Burson, L L (1996). Production and interpretation of perspective drawings by blind and sighted people. *Perception* 25(3): 321–334. doi:10.1068/p250321.

Heller, M A and Gentaz, E (2014). *Psychology of Touch and Blindness*. New York: Taylor & Francis.

Heller, M A and Kennedy, J M (1990). Perspective taking, pictures and the blind. *Perception & Psychophysics* 48: 459–466.

Heller, M A and Schiff, W (Eds.) (1991). *The Psychology of Touch*. Hillsdale, NJ: Lawrence Erlbaum Associates.

Heller, M A et al. (2005). The influence of exploration mode, orientation and configuration on the haptic Mueller-Lyer illusion. *Perception* 34: 1475–1500.

Heller, M A and Ballesteros, S (Eds.) (2006). *Touch and Blindness: Psychology and Neuroscience*. Mahwah, NJ: Lawrence Erlbaum Associates.

Heller, M A et al. (2008). The effects of curvature on haptic judgments of extent in sighted and blind people. *Perception* 37(6): 816–840. doi:10.1068/p5497.

Heller, M A et al. (2009). The influence of viewpoint and object detail in blind people when matching pictures to complex objects. *Perception* 38(8): 1234–1250. doi:10.1068/p5596.

Heller, M A; Smith, A; Schnarr, R; Larson, J and Ballesteros, S (2013). The horizontal-vertical curvature illusion in touch is present in three-dimensional objects and raised-lines. *American Journal of Psychology* 126: 67–81.

Heller, M A. et al. (2010). Attenuating the haptic horizontal—vertical curvature illusion. *Attention, Perception, & Psychophysics* 72(6): 1626–1641. doi:10.3758/app.72.6.1626.

Hollins, M (1989). *Understanding Blindness: An Integrative Approach*. Hillsdale, NJ: Lawrence Erlbaum Associates.

Jansson, G and Holmes, E (2003). Can we read depth in tactile pictures? In: E Axel and N Levent (Eds.), *Art Beyond Sight: A Resource Guide to Art, Creativity and Visual Impairment* (pp. 1146–1156). New York: American Foundation for the Blind.

Kennedy, J M and Juricevic, I (2006). Form, projection and pictures for the blind. In: M A Heller and S Ballesteros (Eds.), *Touch and Blindness: Psychology and Neuroscience* (pp. 73–93). Mahwah, NJ: Lawrence Erlbaum Associates.

Lederman, S J; Klatzky, R L; Chataway, C and Summers, C D. (1990). Visual mediation and the haptic recognition of two-dimensional pictures of common objects. *Perception & Psychophysics* 47(1): 54–64. doi:10.3758/bf03208164.

Legge, G E; Madison, C; Vaughn, B N; Cheong, A M Y and Miller, J C (2008). Retention of high tactile acuity throughout the life span in blindness. *Perception & Psychophysics* 70(8): 1471–1488. doi:10.3758/pp.70.8.1471.

Loomis, J M (1981). On the tangibility of letters and braille. *Perception & Psychophysics* 29(1): 37–46. doi:10.3758/bf03198838.

Merabet, L B et al. (2006). Combined activation and deactivation of visual cortex during tactile sensory processing. *Journal of Neurophysiology* 97(2): 1633–1641. doi:10.1152/jn.00806.2006.

Millar, S (1997). *Reading by Touch*. London: Routledge.

Noppeney, U (2007). The effects of visual deprivation on functional and structural organization of the human brain. *Neuroscience & Biobehavioral Reviews* 31(8): 1169–1180. doi:10.1016/j.neubiorev.2007.04.012.

Park, D C and Reuter-Lorenz, P (2009). The adaptive brain: Aging and neurocognitive scaffolding. *Annual Review of Psychology* 60(1): 173–196. doi:10.1146/annurev.psych.59.103006.093656.

Park, D C et al. (2002). Models of visuospatial and verbal memory across the adult life span. *Psychology and Aging* 17(2): 299–320. doi:10.1037//0882-7974.17.2.299.

Pascual-Leone, A and Torres, F (1993). Sensorimotor cortex representation of plasticity of the reading finger in Braille readers. *Brain* 116: 39–52.

Sathian, K (2000). Practice makes perfect: Sharper tactile perception in the blind. *Neurology* 54 (12): 2203–2204. doi:10.1212/wnl.54.12.2203.

Sterr, A et al. (1998). Perceptual correlates of changes in cortical representation of fingers in blind multifinger Braille readers. *The Journal of Neuroscience* 18(11): 4417–4423.

Stevens, J C; Foulke, E and Patterson, M Q (1996). Tactile acuity, aging, and braille reading in long-term blindness. *Journal of Experimental Psychology: Applied* 2(2): 91–106. doi:10.1037/1076-898X.2.2.91.

Voss, P (2011). Superior tactile abilities in the blind: Is blindness required? *The Journal of Neuroscience* 31(33): 11745–11747. doi:10.1523/JNEUROSCI.2624-11.2011.

Weisser, V; Stilla, R; Peltier, S; Hu, X and Sathian, K (2005). Short-term visual deprivation alters neural processing of tactile form. *Experimental Brain Research* 166(3–4): 572-582. doi:10.1007/s00221-005-2397-4.

Wijntjes, M W A; van Lienen, T; Verstijnen, I M and Kappers, A M L (2008). The influence of picture size on recognition and exploratory behaviour in raised-line drawings. *Perception* 37 (4): 602–614. doi:10.1068/p5714.

Wong, M; Gnanakumaran, V and Goldreich, D (2011). Tactile spatial acuity enhancement in blindness: Evidence for experience-dependent mechanisms. *The Journal of Neuroscience* 31 (19): 7028–7037. doi:10.1523/jneurosci.6461-10.2011.

Wong, M; Hackeman, E; Hurd, C and Goldreich, D (2011). Short-term visual deprivation does not enhance passive tactile spatial acuity. *PLoS ONE* 6(9): e25277. doi:10.1371/journal.pone.0025277.t001.

Recommended Readings

Grunwald, M (Ed.) (2008). *Human Haptic Perception: Basics and Applications*. Boston: Birkhäser Verlag.

Hertenstein, M J and Weiss, S J (2011). *The Handbook of Touch*. New York, NY: Springer.

Part III
The Neuroscience of Touch

Systems Neuroscience of Touch

Ehud Ahissar, Namrata Shinde and Sebastian Haidarliu

Systems neuroscience attempts to understand the realization of brain-related functions at the level of organization that can be captured by logico-mathematical models (Von Bertalanffy 1950), such as those describing man-made systems. In the case of perceptual systems, like that of touch, this level of organization is captured by schemes such as the one in Figure 1. Research groups addressing vibrissal touch are trying to map the vibrissal system anatomically, reveal the physiological properties of its components and the interactions between the components, and understand the function of each component and sub-system. As a result, an understanding of the emergence of tactile perception in this system is expected to arise.

The Rodent Vibrissal-Touch System

The rodent vibrissal system is arguably the leading choice for systems neuroscientists addressing touch. This system offers outstanding experimental traceability at all relevant levels. At the level of whisker-object interactions the accuracy of tracking of whisker motion is at the level of the movement evoked by a single motor spike (Herfst and Brecht 2008; Simony et al. 2010), accuracy unparalleled by any other mammalian sensory system. Clear anatomical layout allows tracing and manipulating activity in every major pathway at all levels between the sensory organ and the cortex. Importantly, rodents can perceive their environment via single whiskers, and the circuits related to single whiskers can be delineated with high confidence in many of the major stations. This combination addresses one of the largest obstacles of systems neuroscience, the sampling problem: It makes possible the tracing of a substantial portion of the neurons that are relevant to the accomplishment of a specific tactile task.

Every whisker is controlled by several intrinsic and extrinsic muscles (Hill et al. 2008; Haidarliu et al. 2010; Simony et al. 2010), and is equipped with several thousand mechano-receptors that sense the whisker's motion as well as its contacts

E. Ahissar (✉) · N. Shinde · S. Haidarliu
Department of Neurobiology, Weizmann Institute of Science, Rehovot, Israel

T.J. Prescott et al. (eds.), *Scholarpedia of Touch*, Scholarpedia,
DOI 10.2991/978-94-6239-133-8_33

Figure 1 Major motor-sensory-motor connections in the rodent vibrissal system. Sensory information from follicle/whisker complex is transmitted via multiple pathways through the sensory regions of the brain, feeding motor subcircuits of the neocortex, cerebellum, midbrain and brainstem, which in turn feed back to the muscles that move the whiskers. Oval circles indicate the brain regions involved in vibrissal sensory-motor processing (TG, trigeminal ganglion; TN, trigeminal brainstem nuclei; RN, red nucleus; VPMdm and VPMvl, ventroposteromedial complex of the thalamus; SI, primary (barrel) somatosensory cortex; SII, secondary somatosensory cortex; ZI, zona incerta; VL, ventrolateral thalamic nucleus; MCx, motor cortex; BG, basal ganglia; Cer, cerebellum; Pn, pontine nuclei; IO, inferior olive; SC, superior colliculus; POm, posteromedial thalamic nucleus; RF, brainstem reticular formation (primarily, gigantocellular reticular nucleus, raphe magnus nucleus, supratrigeminal region); FN, facial nucleus). Black lines connecting brain regions indicate anatomical connections. Arrows indicate the direction of information flow between brain regions. Connections not labeled with arrows are reciprocal. (We thank Naama Rubin for her help in preparing the figures.)

with external objects (Ebara et al. 2002). This motor-sensory organ is equipped with mechanisms enabling it to sense specific features of the environment (Vibrissa mechanical properties). While the scope and specificity of these features is not yet known, specific mechanisms enabling the sensation of specific features have been mapped out. Salient examples are the coordinates of object location (Vibrissal location coding) and the texture of objects' surface (Vibrissal texture decoding). A common denominator of these mechanisms is their motor-sensory nature, or, more accurately, motor-object-sensory nature. That is, sensation is based on active probing of the environment by sensor motion, the result of which is received by the follicle's mechanoreceptors. Tactile perception, thus, must be based on the interpretation of the relationships between motor and sensory signals, relationships that are determined by the touched object and thus represent it (Bagdasarian et al. 2013).

The Primate Finger-Touch System

Tactile perception in primates is typically mediated via hand and finger movements (Lederman and Klatzky 1987; Sathian 1989; Turvey 1996; Cascio and Sathian 2001; Gamzu and Ahissar 2001). The anatomical structure of the related sensory and motor pathways resembles that of the vibrissal system (Jones and Powell

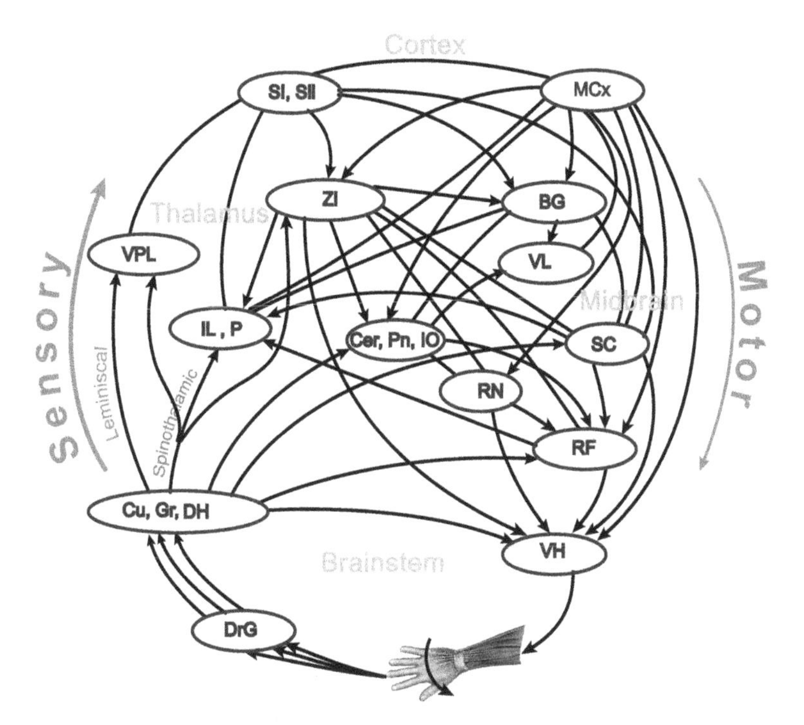

Figure 2 Major motor-sensory-motor connections in the primate hand tactile system. Sensory information from the skin of the primate fingers is transmitted via multiple pathways through the sensory regions of the brain, feeding motor subcircuits of the neocortex, cerebellum, midbrain and spinal cord, which in turn feed back to the muscles that move the fingers. Oval circles indicate the brain regions involved in the hand tactile sensory-motor processing (DrG, Dorsal root ganglion; DH, dorsal horn of the spinal cord; Cu, nucleus cuneatus; Gr, nucleus gracilis; SC, superior colliculus; ZI, zona incerta; VPL, ventral posterolateral thalamic nucleus; IL, intralaminar nuclei of thalamus; P, posterior nuclear group of thalamus; VL, ventrolateral thalamic nucleus; SI, primary somatosensory cortex; SII, secondary somatosensory cortex; MCx, motor cortex; BG, basal ganglia; RN, red nucleus; Cer, cerebellum; IO, inferior olive; RF, reticular formation; VH, ventral horn of the spinal cord). Other details are similar to those in Figure 1

1969a, b; Loeb et al. 1990; Kaas 2004) (Figure 2). Interestingly, whenever compared primates and rodents were found to exhibit comparable tactile acuities (Carvell and Simons 1995; Horev et al. 2011; Saig et al. 2012). Although it is widely agreed that perceiving objects via touch is an active process, there is no agreement about the way in which active touch is implemented. Mainly, it is not agreed whether scanning movements are there merely to activate receptors or that the exact pattern of movements is an inherent component of perception (Connor and Johnson 1992; Johnson and Hsiao 1994; Gamzu and Ahissar 2001). The latter scenario entails that touch-related movements are controlled in a closed-loop form via the tactile system (Saig et al. 2012). In contrast, as receptor activation can be achieved via a wide range of movements, open-loop activation of pre-determined movements would suffice for the former scenario.

Two (Among Many) Challenges for Systems Neuroscience of Touch

The question whether tactile perception emerges via closed- or open-loop processing forms one of the major current challenges facing systems neuroscience of touch. Addressing this question is tricky and the key for a successful resolution probably lies in the development of accurate testable discriminatory predictions. Another major challenge facing this field is the integration of insights obtained from rodent vibrissal touch and primate finger touch into one coherent theory of mammalian tactile perception. As a future goal one should also aspire to integrate insights from mammalian and non-mammalian tactile studies into a coherent theory of touch—given the striking similarities observed across animals (Introduction: The World of Touch) this is not an impossible task. It seems that evidently, addressing these challenges is essential for allowing common-ground interpretations of results obtained in different experimental settings.

References

Bagdasarian, K et al. (2013). Pre-neuronal morphological processing of object location by individual whiskers. *Nature Neuroscience* 16: 622–631.

Carvell, G E and Simons, D J (1995). Task- and subject-related differences in sensorimotor behavior during active touch. *Somatosensory & Motor Research* 12: 1–9.

Cascio, C and Sathian, K (2001). Temporal cues contribute to tactile perception of roughness. *The Journal of Neuroscience* 21: 5289–5296.

Connor, C E and Johnson, K O (1992). Neural coding of tactile texture: Comparison of spatial and temporal mechanisms for roughness perception. *The Journal of Neuroscience* 12: 3414–3426.

Ebara, S; Kumamoto, K; Matsuura, T; Mazurkiewicz, J E and Rice, F L (2002). Similarities and differences in the innervation of mystacial vibrissal follicle-sinus complexes in the rat and cat: A confocal microscopic study. *Journal of Comparative Neurology* 449: 103–119.

Gamzu, E and Ahissar, E (2001). Importance of temporal cues for tactile spatial-frequency discrimination. *The Journal of Neuroscience* 21: 7416–7427.

Haidarliu, S; Simony, E; Golomb, D and Ahissar, E (2010). Muscle architecture in the mystacial pad of the rat. *Anatomical Record (Hoboken)* 293: 1192–1206.

Herfst, L J and Brecht, M (2008). Whisker movements evoked by stimulation of single motor neurons in the facial nucleus of the rat. *Journal of Neurophysiology* 99: 2821–2832.

Hill, D N; Bermejo, R; Zeigler, H P and Kleinfeld, D (2008). Biomechanics of the vibrissa motor plant in rat: Rhythmic whisking consists of triphasic neuromuscular activity. *The Journal of Neuroscience* 28: 3438–3455.

Horev, G et al. (2011). Motor-sensory convergence in object localization: A comparative study in rats and humans. *Philosophical Transactions of the Royal Society of London B: Biological Sciences* 366: 3070–3076.

Johnson, K O and Hsiao, S S (1994). Evaluation of the relative roles of slowly and rapidly adapting afferent fibers in roughness perception. *Canadian Journal of Physiology and Pharmacology* 72: 488–497.

Jones, E and Powell, T (1969a). Connexions of the somatic sensory cortex of the rhesus monkey i. — Ipsilateral cortical connexions. *Brain* 92: 477–502.

Jones, E and Powell, T (1969b) Connexions of the somatic sensory cortex of the rhesus monkey ii. — Contralateral cortical connexions. *Brain* 92: 717–730.

Kaas, J H (2004). Evolution of somatosensory and motor cortex in primates. *The Anatomical Record. Part A, Discoveries in Molecular, Cellular, and Evolutionary Biology* 281: 1148–1156.

Lederman, S J and Klatzky, R L (1987). Hand movements: A window into haptic object recognition. *Cognitive Psychology* 19: 342–368.

Loeb, G; Levine, W and He, J (1990). Understanding sensorimotor feedback through optimal control. In: *Cold Spring Harbor Symposia on Quantitative Biology* (pp. 791–803). Cold Spring Harbor Laboratory Press.

Saig, A; Gordon, G; Assa, E; Arieli, A and Ahissar, E (2012). Motor-sensory confluence in tactile perception. *The Journal of Neuroscience* 32: 14022–14032.

Sathian, K (1989). Tactile sensing of surface features. *Trends in Neuroscience* 12: 513–519.

Simony, E et al. (2010). Temporal and spatial characteristics of vibrissa responses to motor commands. *The Journal of Neuroscience* 30: 8935–8952.

Turvey, M T (1996). Dynamic touch. *American Psychologist* 51: 1134–1152.

Von Bertalanffy, L (1950). An outline of general system theory. *British Journal for the Philosophy of Science* 1: 134–165.

Mechanoreceptors and Stochastic Resonance

Lon A. Wilkens and Frank Moss

Stochastic Resonance (SR) is a counterintuitive phenomenon whereby noise under appropriate conditions can enhance the detection of weak signals rather than interfering with signal transmission. Originally described in nonlinear physical systems, SR is applicable as well for biological processes that are also frequently nonlinear. The first demonstration of SR in biology utilized the mechanoreceptive hairs of crayfish. Since mechanoreceptors are ubiquitous in the biological world, enabling animals and humans to sense their environment, we introduce the topic of mechanoreception as fundamental to a broad range of sensory modalities followed by SR and its application in the crayfish.

Introduction to Mechanoreception

The term mechanoreceptor applies to any sensory receptor that transduces mechanical energy into an electrical current, the universal signal currency of the nervous system. Although mechanoreception to the non-specialist is generally associated with skin sensations such as touch, vibration, and pressure, mechanoreceptors are also the primary receptors underlying many other senses such as, in vertebrates, audition, muscle and joint position, and the vestibular senses essential to balance and posture and contributing to the perception of self-motion and orientation relative to gravity. In lower (aquatic) vertebrates mechanoreceptors are employed for detecting water motion and pressure waves, e.g., the lateral line. In less familiar involuntary roles, mechanosensitivity contributes to such health related phenomena as regulation of blood pressure, cardiac overload and enlargement of the heart, osmotic homeostasis, and polycystic kidney disease. In addition to electrical signaling, mechanotransduction is now linked directly to biochemical

L.A. Wilkens (✉)
Biology and Center for Neurodynamics, University of Missouri-St. Louis,
St. Louis, USA

F. Moss
Center for Neurodynamics, University of Missouri at St. Louis, St. Louis, USA

T.J. Prescott et al. (eds.), *Scholarpedia of Touch*, Scholarpedia,
DOI 10.2991/978-94-6239-133-8_34

signaling, e.g., where forces transmitted by or involving transport via the cytoskeleton affect changes including gene expression (Huang et al. 2004; Singla and Reuter 2006).

Sensory Mechanoreceptors

For the skin senses, the primary cutaneous receptors include free nerve endings but more so nerve fibers variously encapsulated to afford specificity and/or amplification. The most common receptor, the Meissner corpuscle that lies just beneath the epidermis, is an elongated fiber encapsulated in connective tissue that responds to low-frequency vibrations but adapts rapidly with a decrease in firing rate following stimulus onset. Pacinian corpuscles, located in dermal layers of the skin as well as skeletal joints, are unmyelinated nerve terminals encapsulated in onion-like layers of cells that act as high-pass filters and contribute to a rapidly adapting, high frequency vibration sensitivity. At the other end of the spectrum, the slowly adapting Merkel's disk is a branching pressure-sensitive nerve ending contacting the epidermis. Below the skin are mechanoreceptors of the musculoskeletal system, proprioceptors that provide information on body position. Embedded within and parallel to the somatic muscles are muscle spindle organs, specialized muscle fibers with a central non-contractile nuclear bag encircled by sensory neurons that monitor stretch. Sensitivity is maintained by efferent (motor) feedback to the muscle spindle to take up the slack during passive shortening as the main muscle contracts. Sensory nerve terminals embedded serially in tendons, the Golgi tendon organs, monitor muscle tension. The tendon organs are low-threshold, rapidly-adapting mechanoreceptors important for initiating spinal reflexes that also control muscle contraction.

The primary sensory function in skin and proprioceptive receptors involves mechanoelectric transduction mechanisms resident in the dendritic terminals of afferent nerve fibers that project directly to the central nervous system. In contrast, mechanotransduction in the auditory, vestibular, and lateral line systems resides in epithelial receptor cells that relay signals synaptically to afferent fibers. The receptor cell common to these systems is the sensory hair cell, so named for the fine ciliary processes that arise from the apical surface of the cell membrane (Figure 1, adapted from Holt et al. 1997). A stepwise series of parallel stereocilia, increasing in length until they meet a single, longer kinocilium, form an intricate, polarized transducing system. Displacement of these hairs toward the kinocilium results in an excitatory depolarization of the hair cell whereas motion away from the kinocilium results in hyperpolarization. These voltage changes modulate the release of excitatory neurotransmitter onto the postsynaptic afferent fiber. The structural and molecular mechanisms that underlie mechanotransduction have been studied extensively in hair cells beginning with the frog auditory and vestibular systems, providing the following gating-spring model (Hudspeth 1985; see also Eatock et al. 2006). The tips of adjacent kinocilia are connected by a filamentous 'tip link' such that

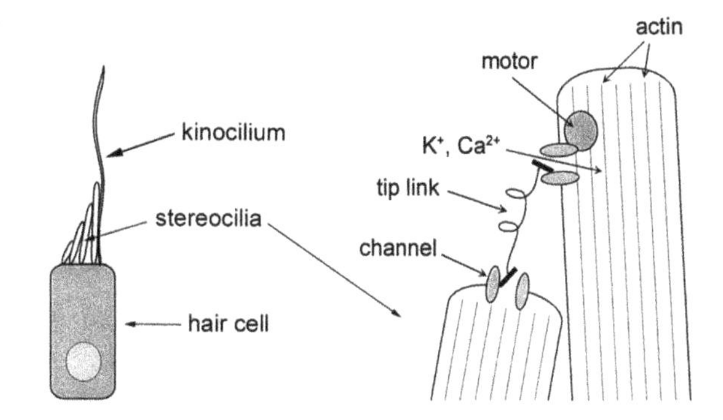

Figure 1 A model of the hair cell illustrating the mechanical linkage between stereo and kinocilia that results in polarizing channel currents in response to ciliary translation

translation of the ciliary bundle toward the stereocilium imparts a mechanical force that activates cation channels near their tips (Figure 1).

Since the apical ciliary surface of both auditory and vestibular hair cells is bathed in a high-potassium concentration endolymph, inward potassium current depolarizes the hair cell and in turn activates voltage-gated calcium channels in the cell soma. The resulting influx of calcium ions facilitates the release of neurotransmitter. This mechanism imparts great sensitivity, with threshold movements of ~ 0.3 nm, and response times on the order of tens of microseconds for receptor potentials. Such rapid responses are characteristic of mechanoelectric transduction in contrast to the slower response of photo- and chemotransduction whose receptors rely heavily on indirect G-protein activated second messenger systems.

The mechanism of hair cell transduction as the basis of multiple sensory organs is therefore a function of mechanical coupling to environmental forces. In the cochlea of the inner ear, hair cells are supported by the basilar membrane with their ciliary bundles bathed in a viscous endolymphatic fluid or in contact with the tectorial membrane. The differential movement of the cochlear membranes, or fluid motions, mediated by the middle ear bones results in shearing forces that bend the cilia. Similarly, within the vestibular labyrinth (see Highstein et al. 2004), hair cells of the utricle and saccule macula (the sensory epithelium) are overlain by an otolithic membrane embedded with calcium carbonate crystals. The greater density of these otoliths imparts a similar shearing force onto the ciliary bundles whenever the head alters its position with respect to gravity or experiences acceleration. In the semicircular canals the hair bundles of the sensory epithelium extend into a compliant gelatinous cupula that occludes the canal. Thus, movements of the endolymphatic fluids deflect the cupula in response to angular acceleration of the head. An equivalent system is found in the lateral line of fish and some amphibians, a fluid-filled skin canal with numerous pores opening to the outside. Here, neuromasts lining the canals of the trunk and head feature ciliary tufts embedded in a gelatinous cupula that extends into the canal. Fluid movements inside the canals deflect the cupula as a result of pressure differences alongside the fish. Additional, superficial neuromasts are scattered over the skin surface of fish and amphibians, as

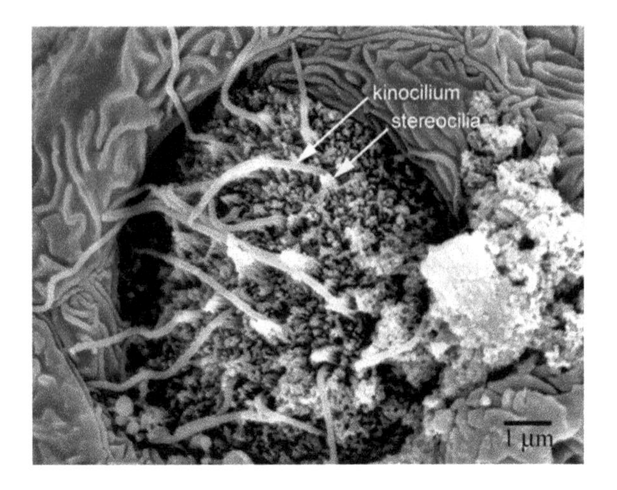

Figure 2 A superficial neuromast in the skin of the catfish *Ancistrus* showing the ciliary apparatus modeled in Figure 1. Neuromasts constitute the mechanoreceptors of the lateral line pressure receptors in fish

illustrated in a superficial neuromast of the bristlenose catfish, *Ancistrus* (Figure 2, M.H. Hofmann, unpublished; k, kinocilia; s, stereocilia).

Hair cell morphology highlights a conservative evolutionary feature of cell biology as it relates to sensory systems. Although the mechanically-gated ion channels are located in the kinocilia, somewhat of a misnomer since they lack the complete axoneme of 9 + 2 microtubules characteristic of cilia and flagella, the presence of a kinocilium with its ciliary axoneme and polarizing role in directional sensitivity invites comparison with other sensory receptors that feature non-motile cilia. For example, vertebrate rods and cones arise during development as extensions of a ciliary membrane whose invaginations and vesicles form the outer segments containing photopigments. Olfactory cells also feature cilia, the membranes of which contain the odorant binding receptors and G-proteins that initiate sensory transduction. Likewise, modified ciliary structures are characteristic of many primary mechanosensory neurons where dendrites contain the 9 pairs of ciliary fibrils extending from a ciliary base. Indeed, the cilium arose early on in eukaryotic organisms as a cellular compartment for sensing the environment, as in ciliate protozoans where it functions simultaneously as a sensory antenna and locomotory organelle. For example, a mechanical stimulus applied to the anterior cilia of *Paramecium* triggers a spike-like membrane potential and calcium current that reverses the effective stroke of ciliary beating resulting in reversed swimming.

Mechanoreception in Invertebrates

Cilia aside, mechanoreception is poorly understood in invertebrate animals, with the notable exception of arthropods. Many of these organisms (cnidarians, various phyla of worms, mollusks, echinoderms, and protochordates) feature a soft flexible integument (~skin) without specialized mechanosensory organs. Rather, branching free nerve endings in the epithelial tissues, about which little is known other than their morphological description, appear to mediate tactile responses. Various invertebrates feature a sensory epithelium that extends into bristles or papillae that function as sensory hairs. Statocysts, invaginations lined with sensory hairs deflected by a statolith, are recognizable in animals as diverse as jellyfish, mollusks, and a few worms and echinoderms, but corresponding physiological analyses are largely unavailable. Exceptions include the body wall stretch receptors of annelid worms, the tactile sense organs richly endowed in cephalopod suckers, and arthropods. Nevertheless, fruit fly (*Drosophila*) and soil nematode (*C. elegans*) invertebrates now serve as important models for the genetic analysis of mechanoreceptor ion channels, including ciliary proteins homologous with those of vertebrates.

Mechanoreceptors in Arthropods

The study of arthropods mechanoreceptors rivals that of vertebrates and the diversity of internal and external receptors is enormous. Like vertebrates, arthropods feature a rigid (exo) skeleton, which also serves as the external integument. With a relatively non compliant exterior surface devoid of epithelium and undifferentiated free nerve endings, crustaceans, insects, spiders and their allies feature a wide variety of receptors specialized to detect both the outside world as well as body structures internally. Many are analogous to those of the skeletonized vertebrates, *e.g.*, proprioceptors. Here, stretch sensitive sensory cells are embedded within strands of muscle or elastic elements that span the joints of body and appendage segments. The muscle receptor organ (MRO) of crayfish and lobster is a clear analog of the vertebrate muscle spindle organ. Here, the dendrites of two sensory neurons, one phasic and one tonic, are embedded within muscle strands stretched as the abdomen flexes. Sensitivity is modulated by efferent feedback to the muscle fibers that adjusts length and tension in the muscle over a range of postures.

As in vertebrates, arthropods mechanoreceptors underlie the function of multiple sensory modalities, including the tactile senses, audition, and sensitivity to gravity. The scolopidial organ is a common element of many of these systems and the site of sensory transduction where one or more bipolar receptor cells with a ciliary distal process are ensheathed and capped by various cuticular end organs. For example, the chordotonal organs are scolopidia embedded in connective tissue attached to and stretched by joint movements. Campaniform organs, oval patches of flexible cuticle, cap receptors that monitor cuticular stress. Often located in the leg bases, these receptors detect loading relative to orientation and gravity. Thin membranous cuticle forms the insect tympanum, the end organ that caps the ciliary scolopidium of auditory organs located in the thorax or forelegs.

Most visible are the external, cuticular setal hairs flexibly articulated with the exoskeleton that are sensitive to touch or fluid motion in the environment. These hairs exhibit a wide variety of shapes and articulations consistent with their

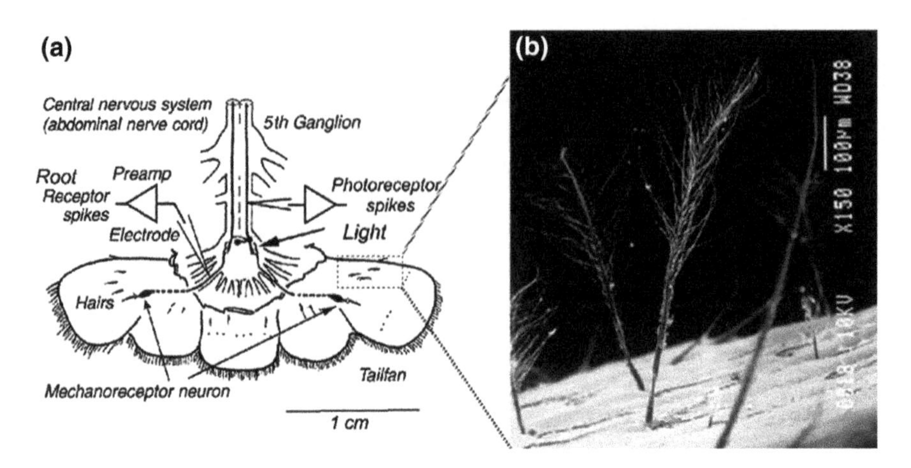

Figure 3 (**a**) Diagram of crayfish tailfan cuticular hairs whose two mechanosensory neurons make synaptic connections with interneurons in the abdominal nerve cord. (**b**) Scanning electron micrograph of long-feathered filiform hairs on the crayfish tailfan lateral uropod

sensitivity and function. For example, the thin filiform hairs on the cercal appendages of insects (crickets, cockroaches) easily detect subtle wind currents and mediate effective escape responses (Levin and Miller 1996). These hairs exhibit a range of sensitivities to wind velocity and direction and are perhaps the most sensitive of all mechanoreceptors, responding theoretically at the level of thermal noise (Shimozawa et al. 2003). Chelicerates (spiders, etc.) also exhibit a range of low-threshold wind sensitive setae. In crustaceans, hairs range from short, stiff tactile setae to long feathered seta deflected by the weakest of water currents. Stretch sensitive receptor cell dendrites extend to various extents into the hair shaft, or chords relay hair movements to the more proximal transducing scolopidia. In a specialized function, setal hairs also form gravitational sensory organs, the statocysts. These cuticular invaginations, prominent within crustaceans, contain a statolith supported by a dense population of fine hairs that are deflected according to the position of the animal.

The crayfish mechanoreceptors, chosen for the study of random noise effects (Section "Stochastic Resonance in Biology"), are within a graded population of well-studied setal hairs on the surface of the tailfan appendages (Figure 3). The afferent fibers of short stiff hairs, phasic and rapidly adapting, activate the system of giant fibers that mediate tailflip escape responses. Our experiments sampled long feathered setal hairs articulated flexibly and responding tonically to gentle water motion (Figure 3b). Each hair is innervated by the distal, ciliated dendrites of two bipolar sensory cells (Mellon 1963). These hairs are directionally selective, bending in a planar motion defined by the geometry of their articulating socket. Directionality is further encoded in the response properties of the innervating receptors, each receptor cell firing a burst of spikes when the hair is deflected in the direction opposite to the other (Douglass and Wilkens 1998). Figure 4 is a polar plot in the x-y plane of the sensitivity of a selected hair (Douglass and

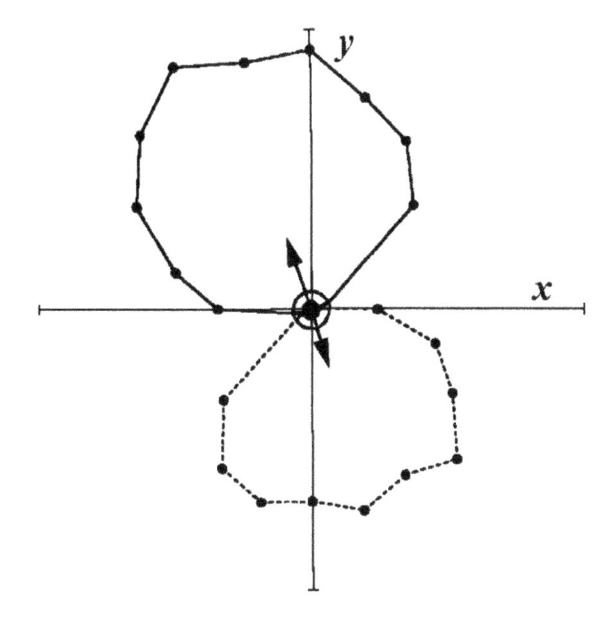

Figure 4 A bidirectional sensitivity plot for an identified mechanosensory L-F hair to water motion, with maximum sensitivity in the rostral-caudal direction

Wilkens 1998). Arrows indicate the mean vectors of the paired anterior-sensitive (solid line) and posterior-sensitive (dashed line) sensory neurons.

The response function of a generalized sensory receptor as a function of stimulus intensity is shown in Figure 5, where the symbols represent experimental data. In the next section we discuss "threshold" stochastic resonance. For calculations of the characteristics of stochastic resonance we use an idealized rectangular function for the threshold as shown by the blue lines and the right hand scale in Figure 5 (Adapted from Mellon 1968).

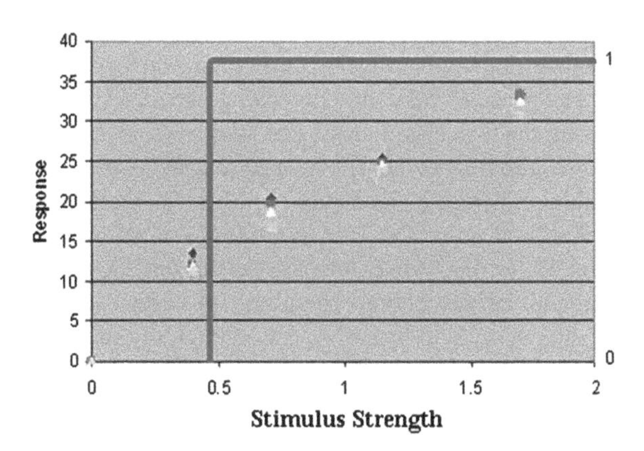

Figure 5 Stimulus-response plot for a generalized crayfish mechanosensory receptor based on the response of four cells (symbols in black, red, yellow, light blue). An idealized threshold function is shown as a blue line

Stochastic Resonance

Stochastic resonance (SR) is a counterintuitive phenomenon occurring in some nonlinear systems, whereby the addition of random noise to a weak signal causes it to become more detectable or enhances the transmission of the information in the signal through the system. The origins of SR, its early demonstrations in dynamical systems and its important applications to global climate modeling have been amply reviewed elsewhere in Scholarpedia: (Nicolis and Rouvas-Nicolis, Stochastic Resonance). We will therefore concentrate in this section on applications of SR in biology, and we will only briefly mention its migration into chemistry and medical science.

Dynamical SR

For many years after its discovery SR was thought to exist only in stochastic dynamical systems (Gammaitoni et al. 1998). A paradigmatic and often studied system is the overdamped Brownian motion of particles in a one-dimensional double well potential,

$$U(x) = -x^2/2 + x^2/4.$$

The dynamics of a single particle in one or the other well subject to random noise, $\xi(t)$, plus a weak, information carrying signal, ε, is governed by a stochastic Langevin equation,

$$\dot{x} = x - x^3 + \xi(t) + \varepsilon \cos \omega t,$$

where ε is small compared with the height of the barrier in the potential well. An example is shown in Figure 6, where (a) shows the potential, (b) shows a time series of the "particle" randomly switching between the left and right wells (the noise on top of the waveform has been removed so that only the barrier crossings are shown), and (c) is the power spectrum ***P(f)*** versus the frequency (3 harmonics are visible), where the signal strength S is the sharp peak (fundamental) on top of the background noise, N. The signal-to-noise ratio, SNR = S/N. Now consider what happens as the noise intensity is systematically increased. The SNR, in this example, or some other measure of order, information, detectability, or more recently success in natural selection (Garcia et al. 2007) passes through a maximum at an optimal value of the noise intensity, $D = \sqrt{<\xi^2> /2}$. This characteristic *noise optimization* is the *defining signature of SR*, as discussed further below.

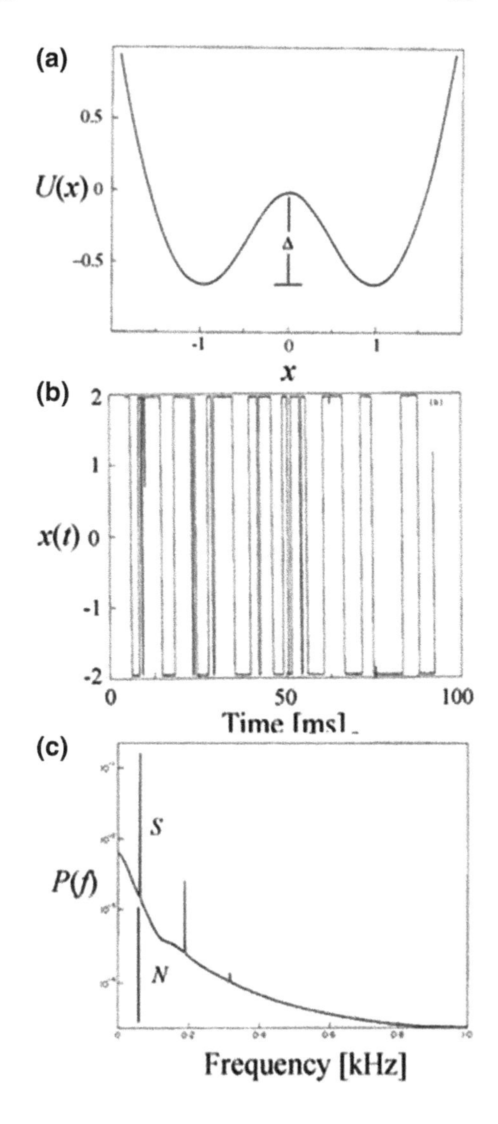

Figure 6 Dynamical stochastic resonance (SR) based on Brownian movement in a double-well potential (a) where random particle switching is illustrated over time (b). A power spectrum versus frequency (c) of signal strength relative to background noise (SNR) shows optimums at 3 harmonics, a fundamental illustration of SR

Threshold or Non Dynamical SR

In 1995 a quite different view of SR emerged (Gingl et al. 1995; Pierson et al. 1995). The new picture was not based on dynamics at all. Instead, the description became purely statistical, so that stochastic differential equations as descriptors were unnecessary. This perception came to be called *threshold* or *non dynamical SR*. The process can be described quite simply as depicted in Figure 7. First one must focus on (b). The *threshold* is shown by the dashed black line. Note the sub threshold *signal*, indicated by the sinusoidal black line. The mean of this signal (blue line) lies a distance Δ_0 below the threshold. The *noise* (red trace) is added to the signal and

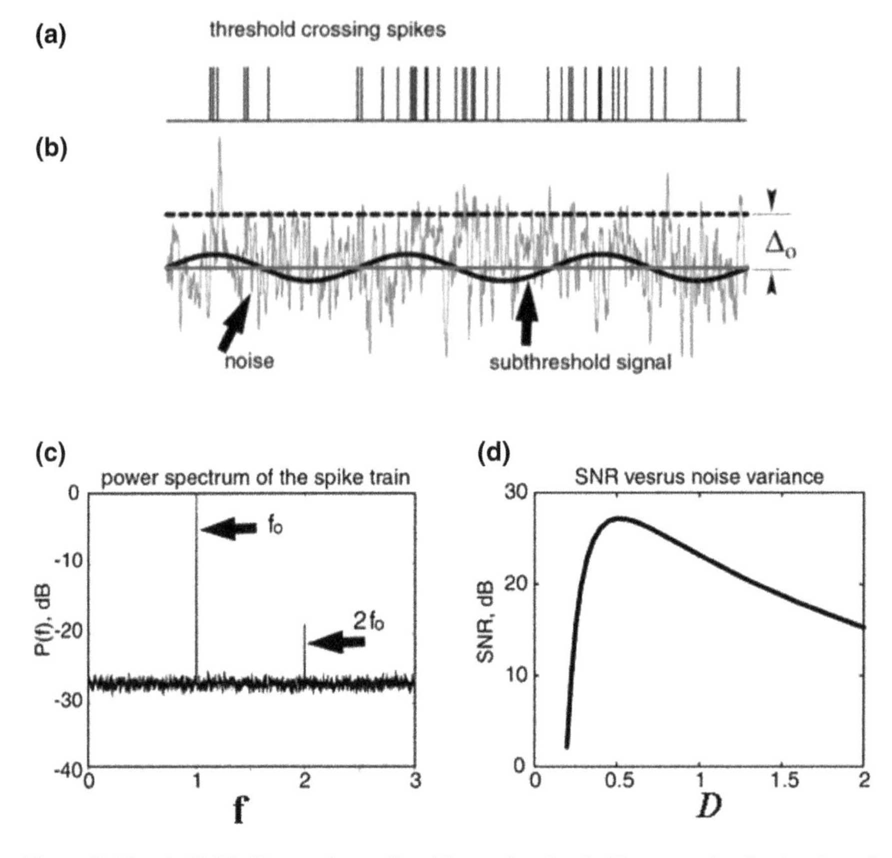

Figure 7 Threshold SR illustrated as spikes (a) crossing threshold as a result of noise (in red) added to a subthreshold sinusoidal signal in a statistical model (b). A power spectrum of the spike train (c) shows peak SNR at the fundamental frequency. SNR as a function of noise (d) is nonlinear peaking at optimum noise intensity

causes the time sequence of positive-going threshold crossing events (spikes) shown in (a). The power spectrum of the spike train, shown in (c) is a bit different from that of the dynamical example discussed above. The spectrum of the noise, N, is flat as shown by the "noisy" line at about—29 db in this case. The signal features are the two sharp peaks at $\boldsymbol{f_0}$ and its 2nd harmonic at $\boldsymbol{2f_0}$ As before, we calculate the SNR as the ratio of the amplitude of the signal feature above the noise to the noise level. The SNR (in decibels) versus the noise intensity, $\boldsymbol{D}$ is shown in (d). We note that the SNR enhancement is maximal at an optimal noise intensity, $\boldsymbol{D_o} \approx \mathbf{0.53}$. A simple, approximate theory gives for the SNR (in dB) (Pierson et al. 1995):

$$SNR = 10\log_{10}\left[\frac{2\sigma_0\Delta_0^2B^2}{\sqrt{3}D^4}\right]\exp\left[-\frac{\Delta_0^2}{D^2}\right]$$

where $\boldsymbol{B}$ is the peak amplitude of the subthreshold signal, and σ_0 is the bandwidth of the noise. [The noise in the dynamical case is usually assumed to be of infinite bandwidth, since that makes certain calculations exact. Here the noise must be of limited bandwidth, but that is no restriction, since *all* noise in the natural world as well as in designed instruments is in fact band limited.] The maximum of SNR in the formula above is accounted for by the factors $\boldsymbol{D}^{-4}$ in the prefactor and $\boldsymbol{D}^{-2}$ in the exponent. Figure 7 reveals an interesting fact. Only *three ingredients are necessary for threshold SR*: a threshold, a subthreshold signal and noise. Moreover, since the process is only statistical, theoretical calculations for specific examples are much simpler than is the case for dynamical SR. Since the three ingredients are ubiquitous in nature, and because the threshold view is more intuitive, SR has migrated into many diverse fields, so that by now there is an enormous literature on the subject in virtually all areas of science and engineering: (909,000 hits on Google as of January 2007). As examples of this migration we very briefly describe and cite work in only two fields outside of physics and biology: chemistry and medical science. In chemistry, the Belousov-Zhabotinsky reaction (BZ) has long been used to demonstrate nonlinear dynamical phenomena, particularly the nucleation and propagation of spiral waves. A light-sensitive, two-dimensional realization of the BZ reaction has been used to demonstrate spatiotemporal SR (Kádár et al. 1998). Noise, applied over the surface in independently fluctuating cells, has been shown to enhance the propagation of BZ traveling waves, and the enhancement is maximal at optimal noise intensity. In medical science, two demonstrations (of many) can be cited here. First, noise enhanced propagation of electromyographic (EMG) pulses was observed in the human median nerve (Chiou-Tan et al. 1996). In this experiment, the noise was generated internally by the incoherent firings of many compound motor neurons. The firing rate (noise intensity) was controlled externally by contractions by the subject of the abductor pollicis muscle against a calibrated external force gauge. Standard EMG stimuli were applied transcutaneously above the median nerve on the right upper arm. In the absence of muscle contractions the stimulus amplitudes were adjusted to be just below or near the threshold of detection by electrodes on the middle finger of the right hand. As subjects contracted their brevis muscles against the force gauge, the subthreshold EMG signal emerged from the noise with increased amplitude. A different version of this same experiment using human muscle spindles was also carried out (Cordo et al. 1996) as well as the ground-breaking use of SR to enhance human balance in healthy and pathological subjects (Priplata et al. 2003).

Stochastic Resonance in Biology

The introduction of SR into experimental biology, based on theoretical predictions by Longtin et al. (1991), was its first demonstration in any field other than physics (Douglass et al. 1993). Above we learned that mechanoreceptors are widespread among sensory modalities in biology. A particularly accessible and simple mechanoreceptor exists on the tail fan of the crayfish in the form of an array of hydrodynamically sensitive hairs innervated by sensory neurons. There are hairs of both long and short length, but here we focus on the long (about 250 μm) hairs that sense water disturbances caused by the approach of a predator (among other things). Figure 3 shows this arrangement (a) with an inset (b) that shows a close-up view of several hairs. Glass pipette suction electrodes were attached to a sensory neuron in the root (a) and connected to the amplifier on the left. The entire tailfan was moved sinusoidally through fluid at about 10 Hz. Thus the hairs were exposed to a periodic, hydrodynamic stimulus. The amplitude of the stimulus was adjusted to be near subthreshold, and spike trains were recorded.

Figure 8 shows three power spectra of spike trains recorded at near optimal noise intensity using a neuron model (Fitzhugh-Nagumo). Stimulated at 55 Hz, the signal

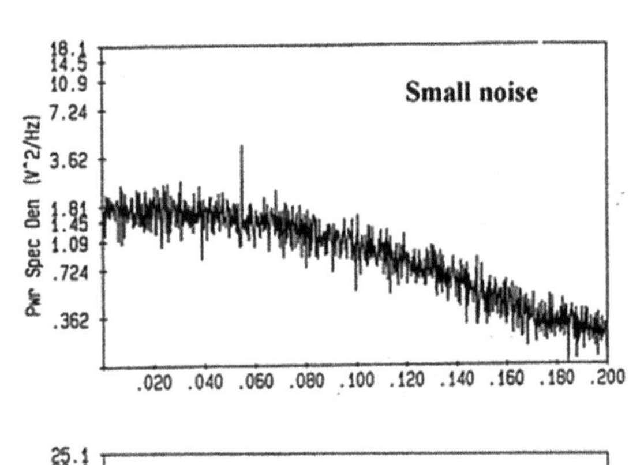

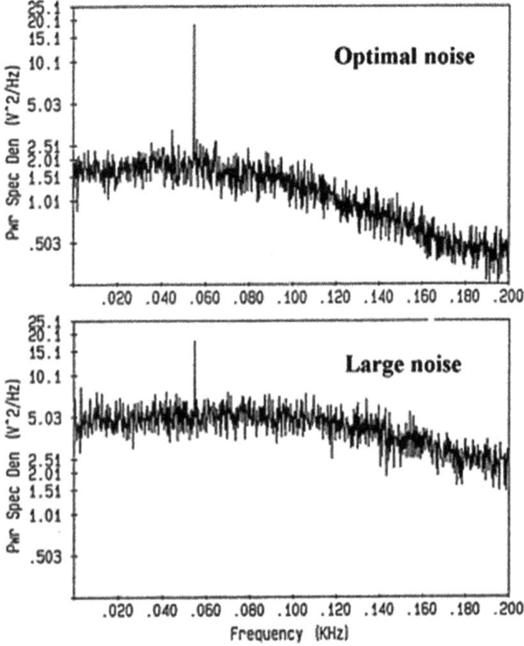

Figure 8 Power spectra are shown for three noise levels in a Fitzhugh-Nagumo model neuron added to a 55 Hz stimulus signal. The second plot shows optimal SNR with a sharp peak at the fundamental frequency

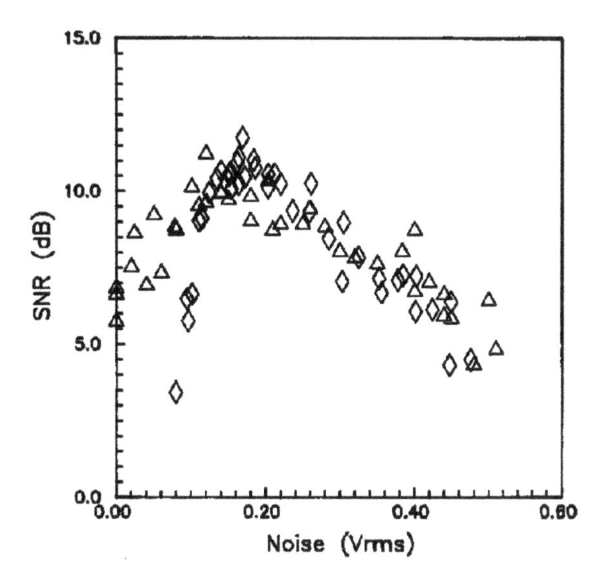

Figure 9 A plot of SNR versus noise shown for the model neuron in Figure 8 (diamonds) superimposed on results obtained from a mechanosensory neuron (triangles) in a crayfish sensillum. A nonlinear SNR demonstrates stochastic resonance in both sets of data

features (sharp peaks at about 55 Hz) are clearly visible. It is obvious that their amplitude changes with noise intensity with the maximum amplitude obtained at the optimal value of noise. The lower panel also shows clearly that noise larger than optimum raises the noise floor and reduces the relative amplitude of the signal feature. The SNRs measured from these (and additional) power spectra versus the noise intensity are shown by the diamonds in Figure 9. The diamonds are from the simulation using the Fitzhugh-Nagumo neuron model to represent the crayfish mechanoreceptor and its associated sensory neuron. The actual crayfish mechanoreceptor data, stimulated at 55 Hz, are shown by the triangles. These data in Figure 9 show that the SNR is maximized at an optimal value of the noise intensity, clearly demonstrating SR in a biological experiment for the first time.

The threshold view of SR also stimulated experiments in animal (Russell et al. 1999; Freund et al. 2001) and human (Ward 2002, Ward et al. 2002) behavior based on perceptive thresholds that are well known in both. SR applied to perception in behavioral experiments is called Behavioral SR.

Internal References

Cullen, K and Sadeghi, S (2008). Vestibular system. Scholarpedia 3(1): 3013. http://www.scholarpedia.org/article/Vestibular_system.

Izhikevich, E M (2006). Bursting. Scholarpedia 1(3): 1300. http://www.scholarpedia.org/article/Bursting.

Johnson, D H (2006). Signal-to-noise ratio. Scholarpedia 1(12): 2088. http://www.scholarpedia.org/article/Signal-to-noise_ratio.

Meiss, J (2007). Dynamical systems. Scholarpedia 2(2): 1629. http://www.scholarpedia.org/article/Dynamical_systems.

Rouvas-Nicolis, C and Nicolis, G (2007). Stochastic resonance. Scholarpedia 2(11): 1474. http://www.scholarpedia.org/article/Stochastic_resonance.
Zhabotinsky, A M (2007). Belousov-Zhabotinsky reaction. Scholarpedia 2(9): 1435. http://www.scholarpedia.org/article/Belousov-Zhabotinsky_reaction.

External References

Chiou-Tan, F Y et al. (1996). Enhancement of subthreshold sensory nerve action potentials during muscle tension mediated noise. *International Journal of Bifurcation and Chaos* 6: 1389–1396.
Cordo, P et al. (1996). Noise in human muscle spindles. *Nature* 383: 769–770.
Douglass, J K and Wilkens, L A (1998). Directional selectivities of near-field filiform mechanoreceptors on the crayfish tailfan (Crustacean: Decapoda). *Journal of Comparative Physiology A* 183: 23–34.
Douglass, J; Wilkens, L; Pantazelou, E and Moss, F (1993). Noise enhancement of information transfer in crayfish mechanoreceptors by stochastic resonance. *Nature* 365: 337–340.
Eatock, R A; Fay, R R and Popper, A N (Eds.) (2006). *Vertebrate Hair Cells. Springer Handbook of Auditory Research*, Vol. 27.
Freund, J A et al. (2001). Behavioral stochastic resonance: How a noisy army betrays its outpost. *Physical Review E* 63: 031910.
Gammaitoni, L; Hanggi, P; Jung, P and Marchesoni, F (1998). Stochastic resonance. *Reviews of Modern Physics* 7: 223–288.
Garcia, R et al. (2007). Optimal foraging by zooplankton within patches: The case of Daphnia. *Mathematical Biosciences* 207: 165–188.
Gingl, Z; Kiss, L and Moss, F (1995). Non-dynamical stochastic resonance: theory and experiments with white and arbitrarily coloured noises. *Europhysics Letters* 29: 191.
Highstein, S M; Fay, R R and Popper, A N (Eds.) (2004). *The Vestibular System. Springer Handbook of Auditory Research*, Vol. 19.
Holt, J R; Corey, D P and Eatock, R A (1997). Mechanoelectrical transduction and adaptation in hair cells of the mouse utricle, a low-frequency vestibular organ. *The Journal of Neuroscience* 17: 8739–8748.
Huang, H; Kamm, R D and Lee, R T (2004). Cell mechanics and mechanotransduction: Pathways, probes, and physiology. *American Journal of Physiology - Cell Physiology* 287: C1-C11.
Hudspeth, A J (1985). The cellular basis of hearing: The biophysics of hair cells. *Science* 230: 745–752.
Kádár, S; Wang, J and Showalter, K (1998). Noise-supported traveling waves in sub-excitable media. *Nature* 391: 770–772.
Levin, J E and Miller, J P (1996). Broadband neural encoding in the cricket cercal sensory system enhanced by stochastic resonance. *Nature* 380: 165–168.
Longtin, A; Bulsara, A and Moss, F (1991). Time-interval sequences in bistable systems and the noise-induced transmission of information by sensory neurons. *Physical Review Letters* 67: 656–659.
Mellon, D (1963). Electrical responses from dually innervated tactile receptors on the thorax of the crayfish. *The Journal of Experimental Biology* 40: 127–148.
Mellon, D (1968). *The Physiology of Sense Organs* (p. 14). San Francisco: Freeman.
Pierson, D; O'Gorman, D and Moss, F (1995). Stochastic resonance: Tutorial and update. *International Journal of Bifurcation and Chaos* 4: 1–15.
Priplata, A; Niemi, J; Harry, J; Lipsitz, L and Collins, J J (2003). Vibrating insoles and balance control in elderly people. *Lancet* 362: 1123–1124.
Russell, D; Wilkens, L and Moss, F (1999). Use of behavioral stochastic resonance by paddlefish for feeding. *Nature* 402: 219–223.

Shimozawa, T; Murakami, J and Kumagai, T (2003). Cricket wind receptors: Thermal noise for the highest sensitivity known. In: F G Barth, J A C Humphrey, T Secomb (Eds.), *Sensors and Sensing in Biology and Engineering* (pp. 145–157). Berlin: Springer-Verlag.
Singla, V and Reiter, J F (2006). The primary cilium as the cell's antenna: Signaling at a sensory organelle. *Science* 313: 629–633.
Ward, L M (2002a). *Dynamical Cognitive Science.* Cambridge, MA: MIT Press.
Ward, L M; Neiman, A and Moss, F (2002b). Stochastic resonance in psychophysics and animal behavior. *Biological Cybernetics* 87: 91–101.

External Links

http://www.umsl.edu/~biology/faculty/wilkens.html Lon A. Wilkens web page.
http://www.media.mit.edu/people/bio_fmoss.html Frank Moss web page.

Mammalian Mechanoreception

Yalda Moayedi, Masashi Nakatani and Ellen Lumpkin

Mechanotransduction, the conversion of mechanical stimuli into biochemical or electrical signals within cells, is necessary for many aspects of development, homeostasis and sensation (Chalfie 2009). For example, in vascular development transduction of shear forces is needed for development and alignment of endothelial cells (Li et al. 2014; Ranade et al. 2014a; Tzima et al. 2005). In the vertebrate ear, movement of specialized stereocilia on mechanosensory hair cells transduces sound waves to initiate hearing (Kazmierczak and Muller 2012). In mammalian skin, an array of somatosensory neurons transduces distinct components of cutaneous sensations, including pressure, stretch, flutter, vibration and pain.

In touch receptors and other mechanosensory cells, transduction is thought to be mediated by mechanically activated ion channels that are embedded in the cell's plasma membrane. These ion channels open in response to cellular movement or plasma membrane deformation, allowing cations to enter or exit the cell, resulting in a change in membrane potential. This change in membrane voltage triggers a cascade of neural signaling that results in perception of discriminative, affective or painful touch. Here, we review current knowledge on mammalian touch receptors and mechanically activated ion channels in the skin.

Anatomy of Touch Receptors in Mammalian Skin

Skin, which is the largest sensory organ of the body, is equipped for many sensory functions. Non-hairy, or glabrous, skin that covers our palms and soles has distinctive receptors to facilitate high acuity (or discriminative) touch, allowing identification of shape and texture of objects. Hairy skin, which predominates on mammals, performs discriminative roles as well as encodes affective aspects of touch linked to "pleasant" emotional reactions. Classically, studies on discriminative touch reception have focused on the glabrous skin of non-human primates (Johnson 2001). In recent years, there has been an expanded interest in the innervation of hairy skin in

Y. Moayedi · M. Nakatani · E. Lumpkin (✉)
Columbia University, New York, NY, USA

T.J. Prescott et al. (eds.), *Scholarpedia of Touch*, Scholarpedia,
DOI 10.2991/978-94-6239-133-8_35

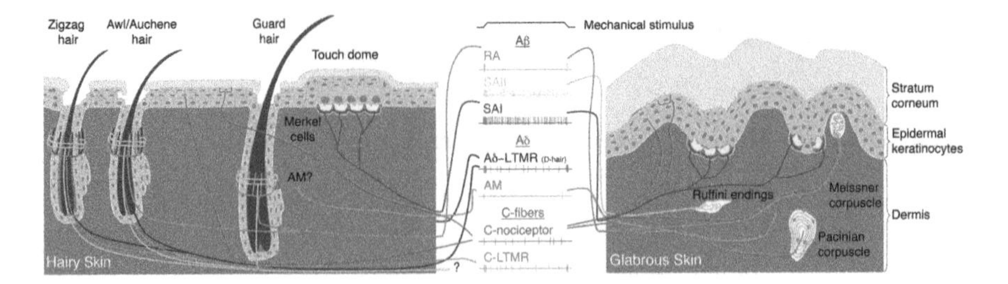

Figure 1 Mechanoreceptors in the skin. Sensory afferents innervating mammalian skin display distinct morphologies and response patterns to mechanical stimulation. Cartoons depict end organs in hairy skin (*left*) and glabrous skin (*right*). Modified from (Bautista and Lumpkin 2011; Li and Ginty 2014; Li et al. 2011; Rutlin et al. 2014)

murine models (Lechner and Lewin 2013). Somatosensory neurons and their anatomically specialized peripheral terminals, termed end organs, reside within different skin structures and facilitate the unique receptive abilities of each area (Figure 1). Four end-organ types reside in mammalian glabrous skin: Meissner corpuscles, Pacinian corpuscles, Merkel-cell neurite complexes and Ruffini endings. In the coat of hair that covers most of the mammalian body, called the pelage, four different types of hair follicles are richly innervated by neurons that help transduce different aspects of touch (Brown and Iggo 1967; Li et al. 2011). In addition to these hair follicle types, mammals have vibrissae, or whiskers, that are organs of discriminative touch in some species. The anatomy and location of tactile afferents, specialized end organs and associated cells help to define their functional roles in touch reception.

Somatosensory Neurons

Somatosensory neurons, with cell bodies in dorsal root ganglia (DRG) or trigeminal ganglia, innervate the skin and transmit mechanosensory information to the spinal cord and hindbrain. Anatomically, DRG neurons are pseudounipolar and are classified based on degree of myelination and conduction velocity. These include heavily myelinated Aβ afferents, lightly myelinated Aδ afferents, and unmyelinated C-fibers. Aβ neurons have the fastest conduction velocity and the largest axonal diameter. Aδ fibers have a slightly smaller diameter than Aβ fibers and a slower conduction velocity due to thinner myelination. C-fibers are the slowest and smallest diameter class of somatosensory neurons and account for the majority of neurons innervating the skin (Smith and Lewin 2009). The stimulus thresholds of somatosensory neurons in the skin help to define their function: neurons underlying discriminative and affective touch have low-threshold responses to mechanical stimuli, and are thus called low threshold mechanoreceptors (LTMRs). Nociceptive stimuli tend to have much higher thresholds for mechanical activation than LTMRs. Many nociceptive neurons are also activated by thermal and chemical stimulation.

Pacinian Corpuscles

Pacinian corpuscles lie deep within the dermis of glabrous skin, as well as in other organs such as pancreas, gut, joints, tendons and interosseous membranes in different species (Bell et al. 1994). Pacinian corpuscles are large, oval shaped end organs that resemble onions in cross sections due to stacks of lamellar Schwann cells that surround the sensory afferents along a parallel axis. The end organ is further encased in layers of perineural cells. Within each Pacinian corpuscle, a single Aβ-LTMR axon follows a linear path through the center of the corpuscle. Interestingly, Pacinian corpuscles grow throughout life and are found to be up to 4 mm long in adult humans (Cauna and Mannan 1958).

Meissner Corpuscles

Meissner corpuscles are situated in the dermis of glabrous skin directly below the epithelium of rete ridges. The Meissner corpuscle itself is an oval shaped structure of stacked lamellar Schwann cells encapsulated in a layer of fibroblasts. By contrast with the Pacinian corpuscle's concentric layers, the Meissner corpuscle's lamellae are stacked perpendicular to the neuron's entrance site. Aβ-axons course through the corpuscle following a tortuous route through lamellae. An Aβ fiber can innervate many Meissner corpuscles, and a single corpuscle can be innervated by multiple axons (Abraira and Ginty 2013; Cauna 1956; Halata 1975). Meissner corpuscles also receive C-fiber innervation (Pare et al. 2001) but the functional significance of this innervation is unclear.

Merkel Cell-Neurite Complexes

Merkel cells are derived from epithelial precursors and are positioned in the basal layer of the epidermis (Morrison et al. 2009; Van Keymeulen et al. 2009). Merkel cells associate with slowly adapting type I (SAI) afferents (described below) to form Merkel cell-neurite complexes. In glabrous skin, Merkel cells reside at the base of rete ridges. In hairy skin, Merkel cell-neurite complexes are found in touch domes, which are epidermal thickenings that form a crescent around guard hairs. SAI neurons form elaborate arborizations, innervating Merkel cells in one or more touch domes (Iggo and Muir 1969; Lesniak et al. 2014; Tapper 1965; Woodbury and Koerber 2007). Merkel cells are mechanosensitive cells that are capable of exciting SAI afferents (Ikeda et al. 2014; Maksimovic et al. 2014); however the basis for synaptic transmission has not been identified. Although Merkel cells express synaptic machinery, they have not been shown to contain clear-core vesicles associated with fast synaptic transmission (Maksimovic et al. 2013). Interestingly,

Merkel cells also express several biogenic amines and neuropeptides, leading them to also be classified as neuroendocrine cells of the skin (Halata et al. 2003), however, neuroendocrine functions have yet to be identified.

Ruffini Endings

Ruffini endings are structures of dense spindle shaped neurite networks in the deep layers of the dermis of both hairy and glabrous skin. Ruffini afferent endings are encased in layers of perineural cells and filled with Schwann cells and a network of endings from a single neuron. The presence and distribution of Ruffini endings in different species is debated, as they are elusive in histology (Johnson 2001).

Hair Types and Innervation

The mammalian coat contains four hair follicle types that fall into three classes: guard, awl/auchene and zigzag hairs. Hair types are distinguished by size, number of columns of cells in the medulla, and the number of kinks in the hair shaft (Duverger and Morasso 2009). Guard hairs, also called tylotrich hairs, are the largest hair type, have two medulla columns, no kinks and compose between 1–3 % of the mouse coat. In rodent species, touch domes are found adjacent to guard hairs. Awl and auchene hairs are of medium size, have 2–4 columns in the medulla and make up about 25–30 % of the mouse coat. Awl and auchene hairs are indistinguishable except for a single kink in the latter. Zigzag hairs are small, contain one row of medulla cells, have several kinks and comprise roughly 65–70 % of the mouse coat (Duverger and Morasso 2009; Li et al. 2011). Hair types are not only morphologically distinct, but are also innervated by a unique array of neuronal types.

Four types of somatosensory neurons with two distinct histological profiles innervate mammalian hair follicles. Circumferential neurons form rings while lanceolate endings form interdigitated, fence-like structures associated with terminal Schwann cells around the hair follicles. All hair follicles have circumferential endings and a unique cadre of lanceolate endings, which fall into Aβ, Aδ and C-LTMR classes (Abraira and Ginty 2013; Li et al. 2011). Guard hairs receive Aβ-rapidly adapting (RA) LTMR innervation; awl/auchene receive Aβ-RA LTMR, Aδ-LTMR and C-LTMR; and zigzag hairs receive Aδ and C-LTMR innervation. The response properties of circumferential endings have yet to be identified (Li and Ginty 2014; Li et al. 2011). As in other hair types, vibrissae are innervated by circumferential and lanceolate endings, but are also associated with Merkel cells, Ruffini-like endings, encapsulated endings and other specialized endings (Ebara et al. 2002).

Functions of Cutaneous Mechanoreceptors

In addition to their end-organ structures, somatosensory neurons can be functionally classified based on physiological properties, including the stimulus modality to which they best respond, conduction velocity and adaptation properties (Figure 1). Aβ and Aδ LTMRs transmit information about gentle touch and are categorized based on physiological responses: rapidly adapting (RA) afferents fire phasically at the beginning and the end of a stimulus. RA afferents are particularly sensitive to vibrations and slip. By contrast, slowly adapting (SA) afferents display sustained firing during a prolonged stimulus, such as constant pressure. In addition to these classes, C-fibers that respond to innocuous touch have been identified in hairy skin. These are known as C-LTMRs in mice and C-tactile afferents in humans. The significance of and the relationship between C-LTMRs and C-tactile afferents are unclear at this time (Liu et al. 2007; Vrontou et al. 2013; Wessberg et al. 2003).

In glabrous skin, RA afferents are associated with Meissner corpuscles and Pacinian corpuscles. Meissner afferents respond best to stimuli moving across their receptive field, allowing them to encode texture. Meissner corpuscles have small receptive fields and respond well to moving and low frequency vibratory stimuli, triggering the sensation of flutter. Pacinian corpuscles respond best to sudden changes in skin pressure and high frequency vibrations, and are thought to contribute information about digit and joint positions. Pacinian corpuscles have the properties of having very low thresholds for activation and very little spatial resolution (Brisben et al. 1999; Johnson 2001). Unlike Meissner corpuscles, Pacinian corpuscles respond to distant stimuli conducted through bone (Macefield 2005). This feature is due to the exquisite sensitivity of Pacinian corpuscles and to their deep position within the dermis. These traits also cause Pacinian corpuscles to facilitate the transmission of information about attributes of distant objects during tool use, such as the texture of an object that is being manipulated with forceps or the soil below a shovel (Johnson 2001). Meissner and Pacinian corpuscles also differ in their frequency response properties. Meissner corpuscles respond best to low frequency vibrations with optimal responses in the 40–60 Hz range whereas Pacinian corpuscles respond to higher frequency ranges with optimal responses at the frequency range of 200–300 Hz (Johnson et al. 2000).

In the hairy skin, RA afferents and C-LTMRs associate with hair follicles to detect hair movement. As described above, three distinct lanceolate types innervate hairs including Aβ-RA LTMRs, Aδ-LTMRs and C-LTMRs. While Aβ-RA LTMRs and C-LTMRs encircle the entire hair follicle, Aδ-LTMRs have recently been found to polarize around the caudal side of zigzag and awl/auchene hairs. These afferents are exquisitely sensitive to hair movement in the rostral direction (Rutlin et al. 2014). C-LTMRs also form lanceolate endings around coat hairs (Abraira and Ginty 2013; Li et al. 2011). Stimulation of these fibers is hypothesized to be associated with affective touch, such as in maternal grooming (Vrontou et al. 2013).

At least two types of SA afferents innervate mammalian skin. SAI afferents innervate Merkel cells in both hairy and glabrous skin. SAI afferents display the highest spatial acuity amongst mechanosensory afferents and are thought to encode information about object edges and curvature. SAII afferents respond to skin stretch, and are believed to provide information about finger positions and handgrip. SAII afferents are hypothesized to terminate in Ruffini endings. Microneurography studies in humans have identified SAII responses; however, the presence of SAII responses are contested in other species, causing this to be the least studied cutaneous mechanoreceptor type (Chambers et al. 1972; Johnson 2001; Koltzenburg et al. 1997; Wellnitz et al. 2010).

Many nociceptors are polymodal sensory neurons that can respond to multiple modalities including mechanical, thermal and chemical stimuli (Cain et al. 2001). Nociceptors have high stimulus activation thresholds and slow inactivation kinetics. The majority of nociceptive neurons are unmyelinated C-fibers, however a subset of Aδ fibers, also known as a mechanonociceptor (AM) fibers, respond to noxious mechanical and thermal stimuli. The end organs of AM fibers are unknown.

Central Projections of Peripheral Mechanosensitive Neurons

How information about touch is transmitted to and integrated in the central nervous system is an intriguing open question. A labeled line code where sensory organs relay singular aspects of touch is likely too simplistic to account for the richness of cutaneous sensations. Much research has focused on single unit recordings of sensory fibers to determine response properties to stimuli, but it is unknown precisely how the central nervous system processes these inputs.

In the classical view of touch transduction, LTMR axons branch directly into the dorsal columns via the direct path and terminate in dorsal column nuclei in the hindbrain. From there, information feeds forward via second order neurons to the medial lemniscus of the thalamus where tertiary neurons send signals to the somatosensory cortex. LTMR axons also send collaterals to deep layers of the spinal cord dorsal horn. Little is known about the molecular identities of the spinal cord interneurons with which LTMRs synapse, or whether initial processing of touch information occurs within the spinal cord. Other projection pathways have been identified in the spinal cord, suggesting that the classic model may be too simplistic (Abraira and Ginty 2013). The first is a "postsynaptic dorsal column pathway" where LTMRs first synapse in the dorsal horn of the spinal cord and second order neurons send projections to the hindbrain via the dorsal columns. The second pathway is the spinocervical tract, in which dorsal horn neurons send axons to the lateral cervical nuclei and then to the ventral posterior lateral nucleus of the thalamus via the medial lemniscus. The significance and relative contributions of these alternate pathways has yet to be determined, however, these open up the possibility of initial touch processing occurring within the spinal cord.

Mechanotransduction Channels

How are somatosensory afferents capable of detecting mechanical signals and converting these into action potentials? This question has been challenging since the discovery of mechanoreceptive neurons in the nineteen century. Recent advances in the molecular and biophysical study of touch reception has begun to address this long-standing mystery.

Mechanically activated ion channels can initiate the process of mechanical signal detection. Ion channels are macromolecular pores in the cell membrane. Mechanical stimulation of the plasma membrane opens mechanotransduction channels, leading to an increase in ion conductance and depolarizing the neuron's membrane potential. This series of processes converts mechanical stimulation into electrical signals in the cell membrane.

Mechanotransduction channels in vertebrates have remained elusive for decades. In invertebrates, transient receptor potential (TRP) channels and degenerin and epithelial Na^+ channel (DEG/ENaC) are bona fide mechanotransduction channels that are necessary and sufficient to confer mechanosensitivity in cells. The acid sensing ion channels (ASICs), which are DEG/ENaC homologues in vertebrates, also gathered attention as candidate mechanotransduction channels. ASIC and TRP channels have been examined extensively in mammalian touch receptors, but do not appear to be essential for mechanosensory transduction. Recent work identified proteins of the Piezo family as mechanically activated ion channels (Coste et al. 2010). *Piezo* genes are broadly expressed in non-mammalian cells and in a variety of mechanosensitive tissues. Three lines of evidence illustrate that Piezo channels, particularly Piezo2, are important for mammalian touch reception.

First, Piezo2 channels are expressed in touch receptors, including Merkel cells and a subset of somatosensory neurons. In Merkel cell-neurite complexes, both the sensory terminals and Merkel cells express Piezo2 (Ranade et al. 2014b; Woo et al. 2014). The sensory terminal of Meissner corpuscles also expresses Piezo2. In hairy skin, Aβ-RA-LTMR lanceolate endings and circumferential fibers express this molecule (Ranade et al. 2014b).

Second, Piezo2 is necessary for the mechanosensitivity of touch receptors. Selective genetic deletion of Piezo2 in mouse Merkel cells abolishes mechanically activated currents in vitro (Woo et al. 2014), and blocking Piezo2 activity in Merkel cells of rat whisker follicles exhibited similar results in situ (Ikeda et al. 2014). Similarly, genetic deletion of Piezo2 in DRG neurons dramatically reduces the proportion of neurons that display rapidly inactivating mechanotransduction currents, a hallmark of LTMRs (Ranade et al. 2014b). By contrast, slowly and intermediately inactivating currents, which are found in nociceptors, are unchanged in conditional Piezo2 knockout mice. These findings suggest that Piezo2 is required for mechanotransduction in LTMRs but that Piezo2-independent mechanisms mediate mechanically activated currents in other DRG neurons.

Finally, Piezo2 is required in mechanosensory function. In Merkel cell-neurite complexes, SAI afferents whose Merkel cells lack Piezo2 still exhibit action potentials; however, they show intermediate adaptation characteristics compared with control genotypes (Maksimovic et al. 2014; Woo et al. 2014). Deletion of Piezo2 in both Merkel cells and LTMRs completely abolishes mechanically evoked responses in many LTMRs, and decreases the touch-evoked activity in those afferents that retain some mechanosensitivity (Ranade et al. 2014b). Moreover, mechanically evoked responses in Aδ nociceptors are also decreased by eliminating Piezo2 (Ranade et al. 2014b). By contrast, mechanically evoked firings are normal in C fibers of Piezo2 knockout mice. These findings support the notion that Piezo2 is particularly important for gentle touch reception rather than nociception.

The contribution of Piezo2 in touch sensation is also supported by behavioral experiment in rodents. Mice that selectively lack Piezo2 in epidermal Merkel cells display higher thresholds to gentle touch stimuli (1.0–1.5 g), but the sensitivity to suprathreshold mechanical stimuli remains intact (Woo et al. 2014). The disruption of Piezo2 in both sensory neurons and Merkel cells causes reduced mechanical sensitivity up to 3 g (Ranade et al. 2014b). In both cases, withdrawal responses to larger forces (>4 g) remain intact (Figure 2). Knocking down Piezo2 in rat whisker follicles by injecting shRNA lentiviral particles also decreases behavioral avoidance to mechanical stimulation (Ikeda et al. 2014).

Collectively, these studies indicate that Piezo2 plays a critical role in detecting touch in several types of mammalian touch receptors. A recent study also found a physical interaction between Piezo2 and a protein tether that amplifies mechanical displacement (Poole et al. 2014). Further studies are needed to understand how Piezo2 channels are gated in mammalian touch receptors.

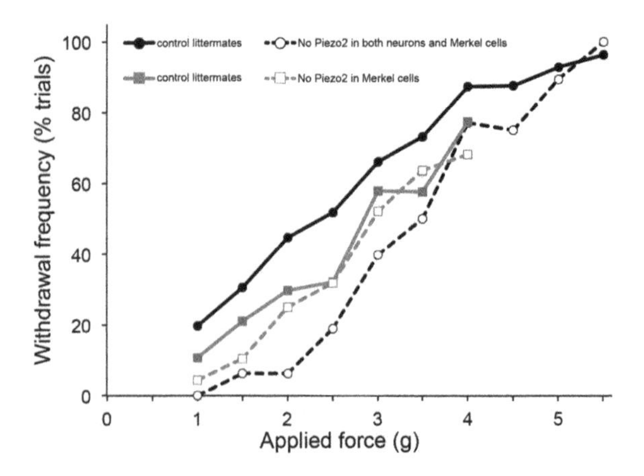

Figure 2 Comparison of behaviors between mutants that lack Piezo2 protein in sensory neurons and in Merkel cells. Data taken and modified from (Ranade et al. 2014b; Woo et al. 2014)

Neural Versus Perceptual Responses to Touch Stimuli

Although electrophysiological studies have been instrumental for analyzing the function of cell types and molecules in mammalian touch reception, behavioral assays are required to assess the importance of these cells and molecules to sensory signaling in the intact nervous system. Subjective perception is important for guiding appropriate behavioral outputs in response to environmental stimuli. For example, a painful stimulus causes a typical avoidance behavior like a flinch. Based on this notion, an experimenter quantifies the number of avoidance behaviors and uses it as an index to evaluate the intensity of pain. By modifying behavioral assays used in pain studies, researchers can examine subjective gentle touch perception. As described below, several behavioral assays have been developed to assess tactile responsiveness in rodents.

The Von Frey test is a widely utilized behavioral assay to measure perceptual mechanical thresholds. In this test, an experimenter uses a variety of filaments of specific diameters to mechanically stimulate test animals. Generally speaking, a fiber-like filament buckles at a mechanical load that is determined by its diameter. By using a variety of filaments with different diameters, an experimenter can exert different magnitudes of mechanical stimulation to body sites. In rodent models, the most frequently used region for testing touch sensitivity is the hind paw. An experimenter counts the number of withdrawals preceded by the mechanical stimulus. The magnitude of applied force that produces 50 % paw withdrawal is a measure of perceptual threshold (Dixon 1980). The Von Frey test is commonly used in recent literature (Garrison et al. 2012; Zheng et al. 2012).

A cotton swab test is an alternative method to measure the sensitivity to mechanical stimulus. The tip of cotton swab is puffed out until it has a light furry appearance that can uniformly stimulate the paw (Garrison et al. 2012). The experimenter gently touches or strokes the hind paw of animals and quantifies the number of withdrawals. This stimulus is more naturalistic rather than giving a localized stimulus as with the Von Frey filament. The cotton swab has been employed in human experiments to examine hypersensitivity to innocuous mechanical stimulations [known as mechanical allodynia (Treede and Cole 1993)].

The adhesive tape removal test measures how quickly an animal can detect a continuous mechanical stimulus. The experimenter cuts adhesive tape into tiny pieces [for example, 30 × 40 mm for the paw (Bouet et al. 2009)] and attaches the tape strip onto the paw or the back (Ranade et al. 2014b). The experimenter then observes the animal for the presence of paw shakes (Bradbury et al. 2002) or counts the number of attempts to remove the tape (Ranade et al. 2014b).

The texture preference test examines the sensitivity to surface roughness. An experimenter prepares sandpaper of two different grits and carpets them onto the floor of the testing arena. Test animals are placed in the testing arena and the experimenter quantifies the duration spent on each texture. Based on the conditioned place preference paradigms, test subjects tend to spend more time in rewarding places [in this case, comfortable textures are assumed to be a reward

(Wetzel et al. 2007)]. If test subjects cannot discriminate surface textures, they spend approximately equal time on both textures. In mouse models, it has been reported that female mice prefer rough textures, while male mice show no preference (Maricich et al. 2012).

A recently developed vibration test examines the perceptual sensitivity to vibratory stimulation. This test also employs a two-choice preference paradigm. An experimenter prepares two pairs of platforms and vibrators. The vibrator is attached to the bottom of the platform via a plate spring. In the experiment, one of the vibrators is actuated at a set frequency and the other one is left static. Recent studies showed that mice tend to avoid vibratory stimulus at 150 Hz (Ranade et al. 2014b), but mice that have Piezo2 deleted in sensory neurons do not show this avoidance behavior.

The advantage of behavioral assays to physiology experiments is that they can estimate the neural correlates of touch sensitivity and perceptual modalities. Recordings from sensory afferent responses by employing an ex vivo preparation [e.g., skin-nerve preparation (Zimmermann et al. 2009)] or in vivo preparation [in the dorsal root ganglion (Boada and Woodbury 2007)] are capable of providing a better understanding of the relationship between the subtype of sensory afferents and characteristics of mechanical stimuli. By combining behavioral responses with neural responses from the same animals, we may answer whether touch coding is based on a labeled line theory or pattern theory.

The current disadvantage of behavioral assays in rodent models is that it is not possible to give a localized stimulus to a single receptive field. For example, the size of a receptive field in mice is generally less than $<0.3\ mm^2$ (Wellnitz et al. 2010), which is one-thirtieth the size of receptive field of human SAI afferents in hairy skin [about $10\ mm^2$ (Johansson 1978)]. This makes it difficult to directly compare neural responses to focused stimulus with behavioral responses to blunt mechanical stimulus.

Closing Remarks

Recent findings have revealed details about the mechanisms underlying the sense of touch. Peripheral somatosensory neurons and their end organs have anatomically distinct morphology and responses to mechanical stimuli. They encode various kinds of mechanical modalities into distinct neural responses. The recent identification of Piezo2 as required for mechanically activated currents in touch receptors has propelled the field forward by promoting an understanding of the underlying encoding mechanisms of mechanoreceptors; however, the data indicate that additional mechanosensitive channels remain to be identified. Behavioral assays work as a window to clarify the causal relationship between neural responses and behaviors. Currently, we understand the detailed classification of mechanoreceptive cells and the response characteristics of associated sensory neurons. One of the next major questions is how somatosensory information is integrated and utilized in the spinal cord and the central nervous system for feeding, rearing, avoiding or emotionally bonding with others.

References

Abraira, V E and Ginty, D D (2013). The sensory neurons of touch. *Neuron* 79: 618–639.

Bautista, D M and Lumpkin, E A (2011). Perspectives on: Information and coding in mammalian sensory physiology: Probing mammalian touch transduction. *The Journal of General Physiology* 138: 291–301.

Bell, J; Bolanowski, S and Holmes, M H (1994). The structure and function of Pacinian corpuscles: A review. *Progress in Neurobiology* 42: 79–128.

Boada, M D and Woodbury, C J (2007). Physiological properties of mouse skin sensory neurons recorded intracellularly in vivo: Temperature effects on somal membrane properties. *Journal of Neurophysiology* 98: 668–680.

Bouet, V et al. (2009). The adhesive removal test: a sensitive method to assess sensorimotor deficits in mice. *Nature Protocols* 4: 1560–1564.

Bradbury, E J et al. (2002). Chondroitinase ABC promotes functional recovery after spinal cord injury. *Nature* 416: 636–640.

Brisben, A J; Hsiao, S S and Johnson, K O (1999). Detection of vibration transmitted through an object grasped in the hand. *Journal of Neurophysiology* 81: 1548–1558.

Brown, A G and Iggo, A (1967). A quantitative study of cutaneous receptors and afferent fibres in the cat and rabbit. *Journal of Physiology* 193: 707–733.

Cain, D M; Khasabov, S G and Simone, D A (2001). Response properties of mechanoreceptors and nociceptors in mouse glabrous skin: An in vivo study. *Journal of Neurophysiology* 85: 1561–1574.

Cauna, N (1956). Nerve supply and nerve endings in Meissner's corpuscles. *American Journal of Anatomy* 99: 315–350.

Cauna, N and Mannan, G (1958). The structure of human digital pacinian corpuscles (corpuscula lamellosa) and its functional significance. *Journal of Anatomy* 92: 1–20.

Chalfie, M (2009). Neurosensory mechanotransduction. *Nature Reviews Molecular Cell Biology* 10: 44–52.

Chambers, M R; Andres, K H; von Duering, M and Iggo, A (1972). The structure and function of the slowly adapting type II mechanoreceptor in hairy skin. *Quarterly Journal of Experimental Physiology and Cognate Medical Sciences* 57: 417–445.

Coste, B et al. (2010). Piezo1 and Piezo2 are essential components of distinct mechanically activated cation channels. *Science* 330: 55–60.

Dixon, W J (1980). Efficient analysis of experimental observations. *Annual Review of Pharmacology and Toxicology* 20: 441–462.

Duverger, O and Morasso, M I (2009). Epidermal patterning and induction of different hair types during mouse embryonic development. *Birth Defects Research Part C: Embryo Today* 87: 263–272.

Ebara, S; Kumamoto, K; Matsuura, T; Mazurkiewicz, J E and Rice, F L (2002). Similarities and differences in the innervation of mystacial vibrissal follicle-sinus complexes in the rat and cat: A confocal microscopic study. *Journal of Comparative Neurology* 449: 103–119.

Garrison, S R; Dietrich, A and Stucky, C L (2012). TRPC1 contributes to light-touch sensation and mechanical responses in low-threshold cutaneous sensory neurons. *Journal of Neurophysiology* 107: 913–922.

Halata, Z (1975). The mechanoreceptors of the mammalian skin ultrastructure and morphological classification. *Advances in Anatomy, Embryology and Cell Biology* 50: 3–77.

Halata, Z; Grim, M and Bauman, K I (2003). Friedrich Sigmund Merkel and his "Merkel cell," morphology, development, and physiology: Review and new results. *The Anatomical Record. Part A, Discoveries in Molecular, Cellular, and Evolutionary Biology* 271: 225–239.

Iggo, A and Muir, A R (1969). The structure and function of a slowly adapting touch corpuscle in hairy skin. *Journal of Physiology* 200: 763–796.

Ikeda, R et al. 2014. Merkel cells transduce and encode tactile stimuli to drive Abeta-afferent impulses. *Cell* 157: 664–675.

Johansson, R S (1978). Tactile sensibility in the human hand: receptive field characteristics of mechanoreceptive units in the glabrous skin area. *Journal of Physiology* 281: 101–125.

Johnson, K O (2001). The roles and functions of cutaneous mechanoreceptors. *Current Opinion in Neurobiology* 11: 455–461.

Johnson, K O; Yoshioka, T and Vega-Bermudez, F (2000). Tactile functions of mechanoreceptive afferents innervating the hand. *Journal of Clinical Neurophysiology* 17: 539–558.

Kazmierczak, P and Muller, U (2012). Sensing sound: Molecules that orchestrate mechanotransduction by hair cells. *Trends in Neuroscience* 35: 220–229.

Koltzenburg, M; Stucky, C L and Lewin, G R (1997). Receptive properties of mouse sensory neurons innervating hairy skin. *Journal of Neurophysiology* 78: 1841–1850.

Lechner, S G and Lewin, G R (2013). Hairy sensation. *Physiology (Bethesda)* 28: 142–150.

Lesniak, D R et al. (2014). Computation identifies structural features that govern neuronal firing properties in slowly adapting touch receptors. *eLife* 3: e01488.

Li, J et al. (2014). Piezo1 integration of vascular architecture with physiological force. *Nature* 515: 279–282.

Li, L and Ginty, D D (2014). The structure and organization of lanceolate mechanosensory complexes at mouse hair follicles. *eLife* 3: e01901.

Li, L et al. (2011). The functional organization of cutaneous low-threshold mechanosensory neurons. *Cell* 147: 1615–1627.

Liu, Q et al. (2007). Molecular genetic visualization of a rare subset of unmyelinated sensory neurons that may detect gentle touch. *Nature Neuroscience* 10: 946–948.

Macefield, V G (2005). Physiological characteristics of low-threshold mechanoreceptors in joints, muscle and skin in human subjects. *Clinical and Experimental Pharmacology and Physiology* 32: 135–144.

Maksimovic, S; Baba, Y and Lumpkin, E A (2013). Neurotransmitters and synaptic components in the Merkel cell-neurite complex, a gentle-touch receptor. *Annals of the New York Academy of Sciences* 1279: 13–21.

Maksimovic, S et al. (2014). Epidermal Merkel cells are mechanosensory cells that tune mammalian touch receptors. *Nature* 509: 617–621.

Maricich, S M; Morrison, K M; Mathes, E L and Brewer, B M (2012). Rodents rely on Merkel cells for texture discrimination tasks. *The Journal of Neuroscience* 32: 3296–3300.

Morrison, K M; Miesegaes, G R; Lumpkin, E A and Maricich, S M (2009). Mammalian Merkel cells are descended from the epidermal lineage. *Developmental Biology* 336: 76–83.

Pare, M; Elde, R; Mazurkiewicz, J E; Smith, A M and Rice, F L (2001). The Meissner corpuscle revised: A multiafferented mechanoreceptor with nociceptor immunochemical properties. *The Journal of Neuroscience* 21: 7236–7246.

Poole, K; Herget, R; Lapatsina, L; Ngo, H D and Lewin, G R (2014). Tuning Piezo ion channels to detect molecular-scale movements relevant for fine touch. *Nature Communications* 5: 3520.

Ranade, S S et al. (2014a). Piezo1, a mechanically activated ion channel, is required for vascular development in mice. *Proceedings of the National Academy of Sciences of the United States of America* 111: 10347–10352.

Ranade, S S et al. (2014b). Piezo2 is the major transducer of mechanical forces for touch sensation in mice. *Nature* 516: 121–125.

Rutlin, M et al. (2014). The cellular and molecular basis of direction selectivity of a delta-LTMRs. *Cell* 159: 1640–1651.

Smith, E S and Lewin, G R (2009). Nociceptors: A phylogenetic view. *Journal of Comparative Physiology. A, Neuroethology, Sensory, Neural, and Behavioral Physiology* 195: 1089–1106.

Tapper, D N (1965). Stimulus-response relationships in the cutaneous slowly-adapting mechanoreceptor in hairy skin of the cat. *Experimental Neurology* 13: 364–385.
Treede, R D and Cole, J D (1993). Dissociated secondary hyperalgesia in a subject with a large-fibre sensory neuropathy. *Pain* 53: 169–174.
Tzima, E et al. (2005). A mechanosensory complex that mediates the endothelial cell response to fluid shear stress. *Nature* 437: 426–431.
van Keymeulen, A (2009). Epidermal progenitors give rise to Merkel cells during embryonic development and adult homeostasis. *Journal of Cell Biology* 187: 91–100.
Vrontou, S; Wong, A M; Rau, K K; Koerber, H R and Anderson, D J (2013). Genetic identification of C fibres that detect massage-like stroking of hairy skin in vivo. *Nature* 493: 669–673.
Wellnitz, S A; Lesniak, D R; Gerling, G J and Lumpkin, E A (2010). The regularity of sustained firing reveals two populations of slowly adapting touch receptors in mouse hairy skin. *Journal of Neurophysiology* 103: 3378–3388.
Wessberg, J; Olausson, H; Fernstrom, K W and Vallbo, A B (2003). Receptive field properties of unmyelinated tactile afferents in the human skin. *Journal of Neurophysiology* 89: 1567–1575.
Wetzel, C et al. (2007). A stomatin-domain protein essential for touch sensation in the mouse. *Nature* 445: 206–209.
Woo, S H et al. (2014). Piezo2 is required for Merkel-cell mechanotransduction. *Nature* 509: 622–626.
Woodbury, C J and Koerber, H R (2007). Central and peripheral anatomy of slowly adapting type I low-threshold mechanoreceptors innervating trunk skin of neonatal mice. *Journal of Comparative Neurology* 505: 547–561.
Zheng, Q et al. (2012). Enhanced excitability of small dorsal root ganglion neurons in rats with bone cancer pain. *Molecular Pain* 8: 24.
Zimmermann, K et al. (2009). Phenotyping sensory nerve endings in vitro in the mouse. *Nature Protocols* 4: 174–196.

Proprioceptors and Models of Transduction

Gerald E. Loeb and Milana Mileusnic

Even when deprived of exteroceptive sensory information such as vision and touch, we are aware of the posture and motion of our bodies (kinaesthesia) and the amount of effort being exerted by our muscles. Most of this information comes from large diameter, myelinated, afferent nerve fibers innervating specialized mechanoreceptors called muscle spindles and Golgi tendon organs, whose structure and function have been modelled in detail as reviewed herein. Muscles, tendons, ligaments and joint capsules contain a host of other sensory endings serviced by smaller diameter and more slowly conducting nerve fibers, some of which are also sensitive to mechanical stimuli such as strain (Paintal 1960; Rymer et al. 1979; Grigg et al. 1986; Martin et al. 2006). Their structure is more heterogeneous and their roles in sensorimotor control are less well understood.

Muscle Spindle

The muscle spindle is a sense organ found in most vertebrate skeletal muscles. In a typical mammalian lower limb muscle, 20–500 muscle spindles can be found lying in parallel with extrafusal muscle fibers and experiencing length changes representative of muscle length changes (Chin et al. 1962; Voss 1971). Their sensory transducers (primary and secondary afferents) provide the central nervous system with information about the length and velocity of the muscle in which the spindle is embedded (Figure 1). Their sensitivity is dynamically modulated by a complex fusimotor system of tiny intrafusal muscle fibers under separate control by the central nervous system. The spindles provide the main source of proprioceptive feedback for kinesthetic perception and spinal coordination.

Early spindle literature described very complex firing patterns from spindle afferents when subject to various kinematic and fusimotor conditions. Various

G.E. Loeb (✉)
University of Southern California, Los Angeles, CA, USA

M. Mileusnic
Otto Bock Healthcare Products, Vienna, Austria

T.J. Prescott et al. (eds.), *Scholarpedia of Touch*, Scholarpedia,
DOI 10.2991/978-94-6239-133-8_36

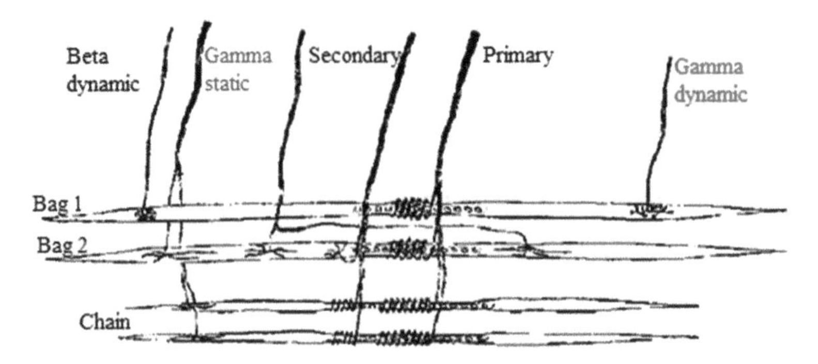

Figure 1 A typical muscle spindle consists of 3 types of intrafusal fibers that receive several fusimotor inputs (gamma static and dynamic) while giving rise to primary (Ia) and secondary (II) afferent (modified from Bakker 1980)

stretches (ramp, triangular, sinusoidal) at different speeds, starting from different fiber lengths or during different fusimotor activations were utilized to help elucidate the mechanisms responsible for the observed afferent behaviour (transducer properties, mechanical properties, etc.). Early models of muscle spindles usually addressed only specific afferent behavior such as small amplitude sinusoidal stretches that were hypothesized to be adequately described by linear system (Matthews and Stein 1969; Poppele and Bowman 1970; Chen and Poppele 1978). They proposed transfer functions to describe the system and in some cases also discussed potential ways to integrate fusimotor activity. While accurately capturing the specific range of behaviour, such models failed to capture the nonlinear afferent behaviour during larger amplitudes and higher velocities of stretch (Houk et al. 1981). With time, the understanding of spindle structure and function became more complete and the models became more detailed and complex. We first review models of spindle transduction that are based on their anatomical structure, followed by a black box model by Maltenfort and Burke (2003).

Spindle Components

The structural models rely on anatomy and physiology to explain spindle afferent behaviour during different kinematic and fusimotor conditions. The structural spindle models usually use the extensive information available about the internal spindle components and their properties as measured directly in intrafusal fibers, or as inferred indirectly from spindle activity during controlled experiments, or as extrapolated from general properties of striated muscle fibers. Similarly to the biological spindle, the spindle models are driven by a combination of length changes and fusimotor activation (where modelled) in order to generate primary and/or secondary afferent activity. Most models use the extrafusal fiber length as the kinematic input, assuming that it is representative of the length changes being experienced by a spindle lying in parallel to extrafusal fibers. Because of series elasticity in tendons, the velocities of the spindles and extrafusal muscle fibers may differ transiently from the length of the overall musculotendon unit from origin to insertion, which more directly reflects the skeletal posture (Hoffer et al. 1992). The models address either both the primary and the secondary afferents activity (Lin and Crago 2002; Mileusnic et al. 2006) or concentrate only on the primary afferent (Rudjord 1970a, b; Hasan 1983; Schaafsma et al. 1991).

Intrafusal Fiber Models

The structural spindle models include explicit models of intrafusal fibers. All the spindle models include a model representing a bag1 intrafusal fiber, while a bag2 and chain fibers are either lumped together in a single model (Schaafsma et al. 1991; Rudjord 1970a, b; and partly Mileusnic et al. 2006) or modelled separately (Lin and Crago 2002; Hasan 1983; partly Mileusnic et al. 2006). In a particular model system, all the intrafusal fiber models typically share the same structure, but their parameters and fusimotor innervation (when included) depend on fiber type. The intrafusal fiber is generally modelled as consisting of a polar and sensory region in series.

In a biological spindle, the centrally located sensory (transduction) region contains afferent endings wrapped around it. This region is commonly modeled as an elastic element (spring). The stretch in this region is related to the distortion of afferent endings and therefore depolarization of the membranes and generation or contribution to generation of the action potential firing (depending on the model).

On each side of the sensory region are the polar regions which share many properties with well-studied striated skeletal (extrafusal) muscle fibers. When modelling the polar region, most of the models take advantage of this similarity and utilize Hill-type muscle models or typical force-length and force-velocity curves obtained for extrafusal fibers to model the polar region's behavior (Hill 1968). Therefore, the polar region is usually modeled as a passive spring and a contractile element consisting of an active force generator and a damping element. In models where fusimotor effects are included, the polar region receives fusimotor activation, dynamically changing its mechanical properties. Dynamic fusimotor activation innervating the bag1 fibers has its major effect on increasing the stiffness of the myofilaments, thereby increasing the velocity sensitivity of the primary afferent. Static fusimotor activation of the bag2 and chain fibers induces polar region contraction, effectively introducing a positive bias to the activity of primary and secondary afferents.

Several models also use the polar region to model the experimentally observed effect of an initial burst ('stiction') in afferent firing at the start of small stretching movements from rest (Schaafsma et al. 1991; Lin and Crago 2002; Rudjord 1970a, b; Hasan 1983; Matthews and Stein 1969; Chen and Poppele 1978). This seems likely to be caused by a very high viscosity in the quiescent bag1 fiber that exists for a very small range of motion, as a consequence of a small number of residual cross-bridges between myofilaments that form in the absence of background fusimotor drive

(Hill 1968; Chen and Poppele 1978). In addition to being responsible for the initial burst, this phenomenon could be also a factor responsible for the observed elevated afferent activity during very slow and small-amplitude sinusoidal stretches. The effect is modelled differently depending on the model. The model by Schaafsma et al. (1991), for example, assumes the existence of one hundred cross-bridge regions that are initially fixed and thereby make the muscular part rigid. When a stretch is introduced, all elongation is initially transmitted to the sensory region resulting in intrafusal fiber force. As the force increases, the cross-bridges gradually release when their individual breaking forces are reached. The model by Lin and Crago (2002) represented this effect as a potentiation effect (tx) that is a negative (decreasing) exponential function of the activation level. Although several models omit this "stiction" effect, it is debatable whether that represents a significant limitation because the effect seems unlikely to be a factor during most natural motor behavior, in which spindles usually operate over longer stretches and with continuously modulated fusimotor input, both of which eliminate stiction.

Like all muscle fibers, intrafusal fibers exhibit a substantial delay between fusimotor excitation and the generation of contractile output in the myofilaments. This is particularly relevant given that some theories of fusimotor function propose rapid modulation of fusimotor activation such as during a specific phase of the locomotor cycle. In the two models addressing this aspect (Lin and Crago 2002; Mileusnic et al. 2006), different types of fusimotor fibers were given different time-constants for a low-pass filter between excitation and mechanical output, as reported by (Boyd 1976; Boyd et al. 1977). While both models assume short time constant for the rapidly contracting chain fibers and longer for bag1 and bag2 fibers, they disagree regarding the time constants for the bag fibers. Boyd et al. (1977) reported a very long time-constant for bag1 fibers based on video observations of visible contractions, but this may under-estimate the rise-time for the viscosity increase in these tonic-like muscle fibers that underlies the velocity sensitivity of the primary afferents.

Primary and Secondary Afferent

Because the primary afferent endings spiral around all three intrafusal fibers, most of the models assume that the primary afferent ending reflects stretches in the sensory zones of all intrafusal fibers, except for a model by Lin and Crago (2002) where only the contributions from bag1 and bag2 fibers are recognized.

The few spindle models that address the secondary afferent model its activity as resulting from the bag2 and chain fibers where its spiral endings are located (Hasan 1983; Lin and Crago 2002; Mileusnic et al. 2006). In these models, the fiber's potential to contribute to the afferent firing is obtained either by calculating the stretch in the sensory region (Lin and Crago 2002; Hasan 1983) or the stretch in both sensory and polar regions (Mileusnic et al. 2006) to account for the juxtaequatorial location of the secondary afferent endings straddling both sensory and polar regions of bag2 and chain intrafusal fibers. While assuming that the sensory endings are located only on the central sensory region might be acceptable during muscle lengthening where the primary and secondary afferents behave similarly, it might introduce errors during shortening where very different behaviors have been observed in primary and secondary afferents (especially during low or absent static fusimotor stimulation, when the primary afferent becomes silent, whereas the secondary afferent continues to fire (Hulliger et al. 1977)).

Early spindle models computed the resulting primary or secondary afferent activity as a simple summation of the relevant intrafusal fiber contributions (Rudjord 1970a, b; Hasan 1983), while the more recent models assume the existence of multiple spike-generating sites (Schaafsma et al. 1991; Lin and Crago 2002; Mileusnic et al. 2006). Multiple, competing spike-generation sites appear to be necessary to account for the experimentally observed effect of occlusion, in which primary afferent firing that results from the simultaneous activation of static and dynamic fusimotor input innervating the same spindle is less than the sum of the rates produced by either fusimotor input individually (Banks et al. 1997; Carr et al. 1998; Fallon et al. 2001). The number of the sites and the exact interaction between them, however, varies across the models. Some suggest two spike-generating sites, one on the bag1 and the other on the bag2 (and chain) (Mileusnic et al. 2006; Schaafsma et al. 1991), while the others suggest three sites, each related to individual intrafusal fiber (Lin and Crago 2002). The secondary afferent activity results from either pure summation of bag2 and chain activities (Mileusnic et al. 2006) or the competition between them (Lin and Crago 2002). Some models permit complete occlusion (Lin and Crago 2002; Schaafsma et al. 1991), while a recent model assumes that the spread of transduction current takes place between the impulse generating sites prior to the competition and suppression, therefore resulting in increased impulse generation at the dominant site (partial occlusion) (Mileusnic et al. 2006). Partial occlusion between two or more spike initiation regions may give rise to irregular firing patterns as a result of collisions between action potentials (Loeb and Marks 1985) but this has not been modeled explicitly; see Ensemble spindle activity below.

Spindle Motoneuron Innervation

All the spindle models addressing the fusimotor innervation assume the existence of single dynamic and single static fusimotor input. While not being widely accepted, there exists limited evidence supporting the existence of two different sources of static fusimotor drives. The direct fusimotor recordings of two static fusimotor efferents firing during decerebrate locomotion in the cat suggest that the two efferent firing profiles are somewhat out of phase; type 2 drive (presumed to be innervating bag2 fiber) leads the type 1 drive (innervating chain fibers) by about 0.17 s (Taylor et al. 2000). A recent spindle model (Mileusnic et al. 2006) incorporating the temporal properties of bag2 and chain fibers divided the static fusimotor input into two different static drives. The model more accurately predicted experimentally recorded secondary afferent activity when incorporating the different temporal properties of the different intrafusal fibers. Furthermore, the model findings suggested that the experimentally observed phase differences between the two static fusimotor drives would compensate for the differences in contractile dynamics of their respective intrafusal fibers, resulting in synchronous static fusimotor modulation of the receptors. This provides support for the simplification of including only one type of static fusimotor fiber and drive in spindle models, but it seems likely that they are actually used differentially in some behaviors.

In addition to incorporating fusimotor effects, two spindle models also addressed the effect of beta (β) motoneurons (Maltenfort and Burke 2003; Lin and Crago 2002). The beta (or also called skeleto-fusimotor) motoneurons sometimes innervate both intra- and extrafusal muscle fibers. Various evidence suggests that beta motoneurons receive monosynaptic group Ia excitation comparable to that in motoneurons, which would then result in a positive feedback loop (Burke et al. 1973). While the model of Lin and Crago (2002) suggests that the beta input could be received by the model, this is somewhat unclear from their description. In contrast, the development of Maltenfort's black box spindle model was largely inspired by the author's desire to study the effects of beta motoneurons. After designing a spindle model that responds to fusimotor input, the β feedback was simulated as a spindle model receiving a fusimotor drive that is proportional to the instantaneous firing rate of that afferent. The model assumes that the effect of β motoneuron activity on spindle afferent firing has the same strength and time course as the effect for a fusimotor neuron of the same type (static or dynamic), but it is not clear whether or how to account for the general tendency of alpha (extrafusal) motoneurons to fire much slower than gamma (fusimotor) neurons.

Model Parameters and Performance

The number of model parameters varies across the models depending on their complexity. When possible, many parameters were derived directly from spindle or extrafusal fiber literature. The remaining free parameters were typically optimized to account for various afferent records under experimentally controlled conditions of kinematics and fusimotor stimulation, mostly from feline hindlimb muscles. Usually no independent data sets were available to validate the models. Their strengths and weaknesses in fitting the available data are discussed for each model below.

Spindle Models

The structural models rely on anatomy and physiology to explain spindle afferent behaviour during different kinematic and fusimotor conditions. The structural spindle models usually use the extensive information available about the internal spindle components and their properties as measured directly in intrafusal fibers, or as inferred indirectly from spindle activity during controlled experiments, or as extrapolated from general properties of striated muscle fibers. Similarly to the biological spindle, the spindle models are driven by a combination of length changes and fusimotor activation (where modelled) in order to generate primary and/or secondary afferent activity. Most models use the extrafusal fiber length as the kinematic input, assuming that it is representative of the length changes being experienced by a spindle lying in parallel to extrafusal fibers. Because of series elasticity in tendons, the velocities of the spindles and extrafusal muscle fibers may differ transiently from the length of the overall musculotendon unit from origin to insertion, which more directly reflects the skeletal posture (Hoffer et al. 1988). Some models address the primary and secondary afferents separately (Mileusnic et al. 2006; Lin and Crago 2002); others concentrate only on the primary afferent (Rudjord 1970b; Schaafsma et al. 1991; Hasan 1983).

Structural Model of Rudjord (1970a, b)

Contrary to the initial modelling attempts that used linear transfer functions to describe spindle afferent behaviour, Rudjord pointed out various non-linear transfer properties and developed a second-order model of primary afferents (Figure 2). The model was created to describe the behaviour of the de-efferented spindle, but it included two types of intrafusal muscle fibers (bag and chain) having different mechanical properties. The mechanical model consisted of two parallel mechano-elastic transducers (that simulate the effect of the terminations on the bag and chain fibers) that are in series with the polar region. Therefore, when the length changes are imposed on the entire system, the length changes of the two elastic elements ($k\lambda$—chain and $k\alpha$—bag) are considered to activate two mechano-electric transducers. An amplitude dependent gain of the α-branch transducer is included to account for nonlinear transfer properties prior to superimposing two output components. The two transducers were assumed to have different gains.

The equations of movement for the entire system were developed and a transfer function derived that completely characterized the static and dynamic behaviour of the model. Rather than optimizing the free parameters for specific experimental data, the parameters were adjusted to capture qualitatively various types of published data. By introducing a much higher gain for bag than for chain transducers, the model demonstrated also the ability to capture the stiction overshoot at the initiation of ramp stretches. For sinusoidal stretches of varying frequency, both the amplitude ratio and the phase advance showed only moderate agreement with experimental data (Matthews and Stein 1969) (Figure 2).

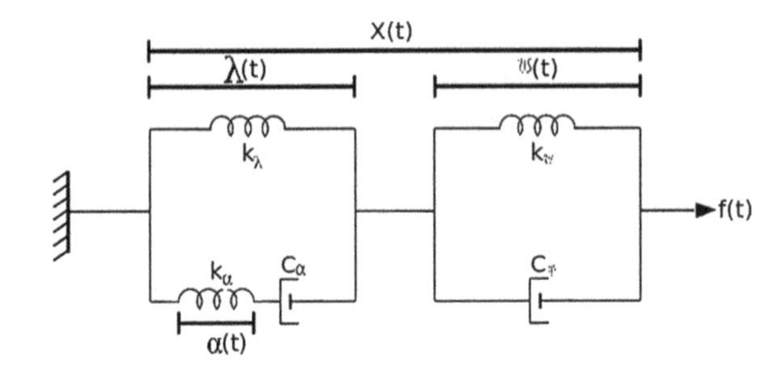

Figure 2 Primary afferent model from (Rudjord 1970a, b)

Structural Model of Hasan (1983)

Hasan (1983) presented one of the first models consisting of mathematical elements representing the anatomical components found in the biological spindle. Hasan's model separates the mechanical filtering process from the later transducing and encoding processes. It describes the mechanical filtering that converts the change in muscular length to change in extension of the nerve endings as a nonlinear process. This is supported by the observation in isolated spindles that increasing amplitudes of sinusoidal stretch produce less than proportional increases in the amplitudes of modulation of spindle tension or receptor potentials of both primary and secondary endings. Transduction and spike encoding were linear, on the assumption that any nonlinearities could be incorporated into the mechanical filtering.

The model differentiates between the sensory and polar region of the intrafusal fibers. The sensory region is modelled as a spring. The polar region is assumed to share many similarities with mechanical features observed in extrafusal fibers and included features of nonlinear velocity dependency and a property akin to friction. Each of the regions (sensory, polar) is described by an equation and the two equations are then summarized in a differential equation. By solving the equation, the length of the sensory region was obtained. Once the equation for length in the sensory region was calculated, the instantaneous discharge rate (afferent activity) was assumed to be a function of the rate and extent of sensory region elongation.

The model also addresses the fusimotor activation but in a qualitative rather than quantitative way. Particularly when integrating fusimotor activation, the model demonstrates a very rigid structure that requires changing its parameters to account for each fusimotor state. It cannot account for the modulated fusimotor activation usually found in cyclic tasks such as locomotion. Like Rudjord's model, it can account for the initial burst in afferent activity at the start of each stretch, whereas in the biological spindle this occurs only during the first stretch after a period of rest (Matthews 1972). Nevertheless, the model captures qualitatively many aspects of afferent activity. Because of both its innovative nature and its easy implementation, Hasan's model attracted the attention of many researchers.

Structural Model of Schaafsma et al. (1991)

This structural model of the primary afferent incorporates fusimotor effects that arise from two intrafusal fiber submodels (Figure 3). One submodel represents the bag1 intrafusal fiber and receives dynamic fusimotor input, while the other represents bag2 and chain fibers together and receives static fusimotor input (called bag2 submodel). Both submodels contain a sensory part (the central nuclear bag around which primary afferents terminate) and a muscular part (the contractile portion of a nuclear bag, containing sarcomeres). The sensory parts are modeled as identical linear elastic elements. The muscular part of the bag1 submodel is modelled as a slow twitch muscle fiber, and that of the bag2 as a fast twitch fiber. The forces generated by this part depend on fiber activation, fiber length and contraction velocity (similar to extrafusal muscle fibers). The two submodels differ in terms of the parameters that control the maximal isometric force, the passive elasticity, the maximal force for an eccentric contraction relative to the maximal isometric force, and the passive damping coefficient. The fusimotor input is linearly related to the active force, and scaled with respect to the fusimotor excitation for maximal effect. The model furthermore includes the force enhancement in response to stretch that has been observed in extrafusal muscle fibers.

Each bag submodel generates a depolarization (transducer process) that depends linearly on the length of its sensory region. The firing rate of the bag submodel is linearly dependent on the receptor potential but also on the rate of change of the receptor potential. The final primary afferent firing is obtained by competition between the two action potential trains. The model assumes complete occlusion—the final primary afferent discharge is driven by the intrafusal fiber providing the highest discharge rate.

The bag1 submodel's muscular part also includes the cross-bridge fixation (or stiction) in order to account for the high sensitivity of muscle spindles to small amplitudes of stretching (e.g. initial burst at the onset of ramp-and-hold stretches). The model assumes one hundred cross-bridge regions that are initially fixed and thereby make the muscular part rigid. If a stretch is introduced, all elongation is initially transmitted to the sensory region resulting in intrafusal fiber force. As the force increases, the cross-bridges gradually release one after the other when their individual breaking forces are reached.

While many parameters used in the model are based on spindle or extrafusal fiber literature, it was necessary to define the remaining ten parameters. The parameters were optimized on experimentally recorded primary afferent activity during 36 ramp-and-hold stretches under 6 different conditions of fusimotor activity (constant static or dynamic fusimotor input, not simultaneous) and 6 different velocities of stretching. The resulting model accounted well for the data used for the fitting process.

The output of the model during small amplitude, 1 Hz sinusoidal stretches was compared to experimental data in terms of amplitude, modulation depth and phase during constant static or dynamic fusimotor input. The model underestimated the

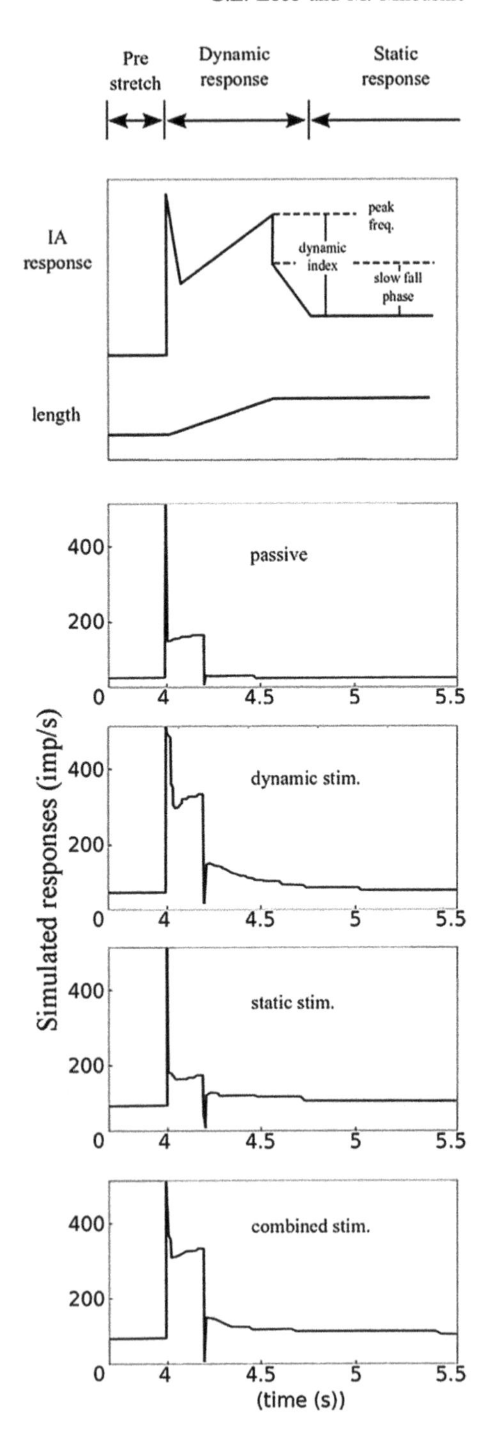

Figure 3 The structural model of primary afferent from Schaafsma et al. 1991). The figure presents primary afferent response during passive stretch without fusimotor stimulation, with constant dynamic, constant static and combined static and dynamic fusimotor stimulation

afferent activity, generating about one-half the discharge modulation reported in experimental studies. The phase advances predicted by the model tended to be slightly lower than those observed for real spindles.

Structural Model of Lin and Crago (2002)

A combined model of extrafusal muscle and spindles was designed in order to study their interaction (Figure 4). A Hill-type muscle model was used to obtain the extrafusal fiber length, which was assumed to be proportional to the intrafusal fiber length. The spindle model is a structural one consisting of three intrafusal fiber models (bag1, bag 2 and chain). The three intrafusal fiber models share the same structure (normalized Hill-type muscle model) while having different parameters (maximal muscle force, optimal muscle fiber length, tendon slack length, muscle mass, etc.). Each intrafusal fiber has a sensory region in series with a polar region consisting of intrafusal contractile element (CEi), intrafusal fiber mass (Mi), parallel viscous element (Ri) and nonlinear spring corresponding to series elastic element (Kt,i) and passive elastic element (Kp,i).

Dynamic fusimotor activity excites the bag1 fiber's polar region, while static fusimotor activity excites bag2 and chain fibers' polar regions. The model calculates the intrafusal muscle force by multiplication of the activation level, the output of the force–length curve and the output of the force–velocity curve, similarly to extrafusal fibers. The initial burst is modeled as a potentiation of stretch sensitivity (tx) that is a negative (decreasing) exponential function of the activation level. The model assumes that each intrafusal fiber has an independent impulse generator site. The stretch in each intrafusal fiber's sensory region is calculated to obtain its activity. The primary afferent firing rate is modelled as resulting from the complete occlusion between the bag1 and bag2 fiber activities.

The secondary afferent endings are assumed to be located only on the sensory region rather than juxtaequatorially. The secondary afferent activity results from the complete occlusion among the activities in bag2 and chain fibers. Individual intrafusal fibers also include time constants of activation dynamics (low-pass filter properties). The chain fiber is modelled as having a significantly shorter time constant than slow bag fibers. Overall, 89 parameters, out of which 24 were free parameters, were used in this model.

The model correctly predicted the fractional power dependence on stretch velocity of the primary and secondary ending responses. In the case of the primary afferent, the model performed well during ramp-and-hold and sinusoidal stretches, especially during the stimulation of individual fusimotor inputs, while demonstrating certain limitations in the absence of such inputs. In the case of the ramp-and-hold stretch, the model underestimated the afferent decay time after completion of the stretch, especially in the absence of any fusimotor stimulation (0.2 vs. 0.5 s). At the completion of the stretching phase of a ramp-and hold stretch, the model predicted an abrupt and brief cessation of firing, which might be a

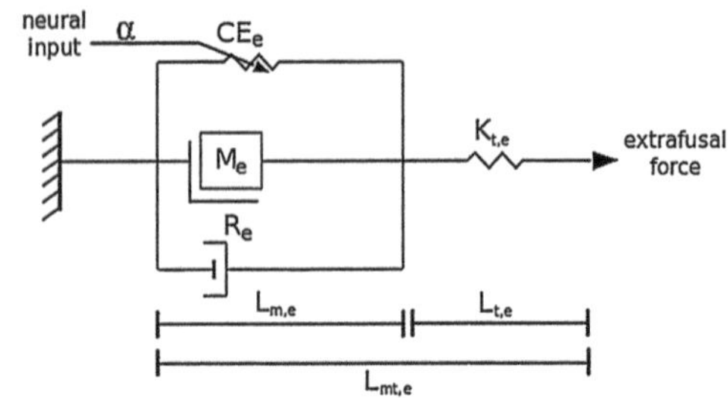

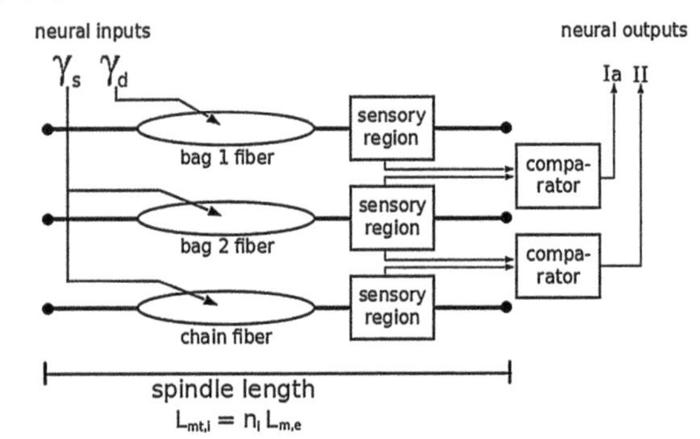

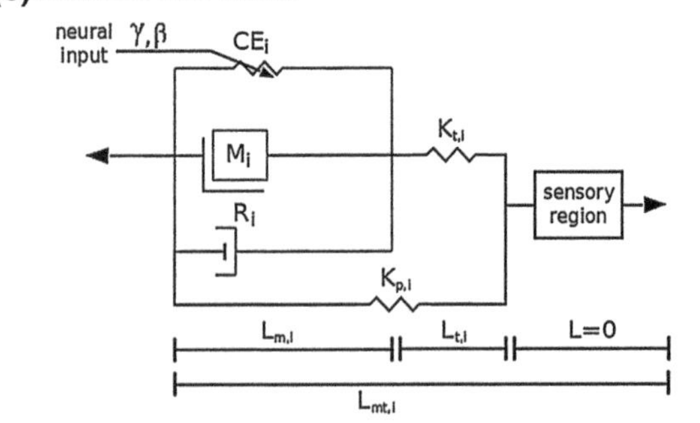

Figure 4 The structural spindle model by Lin and Crago (2002). A. A combined model of extrafusal muscle and spindles assumes intrafusal length to be proportional to extrafusal fiber length. B. Spindlle model consists of three intrafusal fiber models. C. Intrafusal fiber models share the same structure while having different parameters. Each intrafusal fiber has a sensory region in series with a polar region consisting of intrafusal contractile element (CEi), intrafusal fiber mass (Mi), parallel viscous element (Ri) and nonlinear spring corresponding to series elastic element (Kt,i) and passive elastic element (Kp,i)

troubling artefact if the model were to be incorporated into a model of segmental regulation of motor output. Finally, the performance of the model during combined fusimotor stimulation is not described, except for a single ramp-and-hold example presented without matching experimental data. Because the model uses complete rather than partial occlusion, it seems likely that the model will underestimate afferent activity during conditions of simultaneous fusimotor stimulation.

Structural Model of Mileusnic et al. (2006)

This model consists of mathematical elements closely related to the anatomical components found in the biological spindle (Figure 5). It is composed of three nonlinear intrafusal fiber models (bag1, bag2, and chain). Each intrafusal fiber model responds to two inputs: fascicle length and the relevant fusimotor drive (dyanmic fusimotor drive to the bag1 and static fusimotor drive to the bag2 and chain fibers). The spindle model generates two outputs: primary (Ia) and secondary (II) afferent activity. The primary afferent response results from the contributions of all three intrafusal fiber models on which the primary afferent receptor has transduction endings. The secondary afferent receives inputs from only the bag2 and chain intrafusal fiber models.

All the intrafusal fiber models share the same general structure, a modified version of McMahon's spindle model (McMahon 1984). The relative importance of model parameters differs for three intrafusal fiber models to account for the different properties of three fiber types. The intrafusal fiber model consists of two regions, sensory and polar. The sensory (transduction) region contains afferent endings wrapped around it. Stretch of this region results in distortion of afferent endings, depolarization of their membranes, and increase in the rate of action potential firing

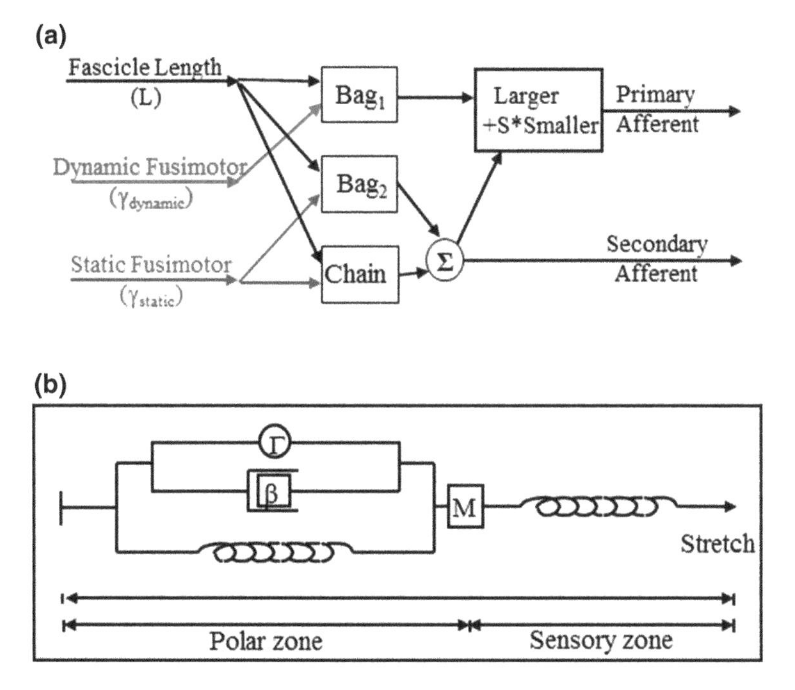

Figure 5 The structural spindle model by Mileusnic et al. (2006). **a** Spindle model consisting of 3 intrafusal fiber models, **b** intrafusal fiber model consisting of sensory and polar region. Three intrafusal fibers share same structure while having different parameter values

(Boyd and Smith 1984). The sensory region of each intrafusal fiber model is modelled as a linear elastic element. The model calculates the stretch in the sensory region to obtain the intrafusal fiber's contribution to the activity of the afferent (prior to taking into account the partial occlusion described below). The polar region, constituting essentially a striated muscle fiber innervated by fusimotor endings, is modelled as a spring and a parallel contractile element that consists of an active force generator and a damping element, both of whose properties are modulated by fusimotor input. The polar region's damping term is primarily modified by dynamic fusimotor stimulation and to a smaller extent by static fusimotor stimulation. Increases in damping result in increases in the velocity sensitivity of the primary afferent, which plays an important role in modulating the spindle's behaviour during dynamic fusimotor excitation of the bag1 fiber. By contrast, static fusimotor activation produces a small decrease in damping. Active force generation is strongly modulated by the static fusimotor input, which causes a sustained, strong contraction within the bag2 and chain polar regions, producing a stretch in the sensory region and a bias in the afferent activity. Dynamic fusimotor input produces a similar but much weaker effect in the bag1 fiber. The model also incorporates the experimentally observed nonlinear velocity dependency of the spindle's afferent response (Houk et al. 1981). Furthermore, it incorporates the experimentally observed asymmetric effect of velocity on force production during lengthening and shortening observed in extrafusal striated muscle. The intrafusal fiber models include a biochemical Hill-type equation to capture the saturation effects that take place at high fusimotor stimulation frequencies. Additionally, the models incorporate low-pass filters when converting the fusimotor input into model activation to account for the experimentally measured temporal properties of individual intrafusal fibers, particularly for slow bag intrafusal fibers.

The output of the primary afferent is captured by nonlinear summation between the bag1 and combined bag2 plus chain intrafusal fiber outputs, to account for the effect of partial occlusion that has been observed in primary afferents during simultaneous static and dynamic fusimotor stimulation (Banks et al. 1997; Carr et al. 1998; Fallon et al. 2001). The driving potentials produced by bag1 and combined bag2 and chain intrafusal fibers are compared and the larger of the two plus a fraction (S) of the smaller are summed to obtain the primary afferent firing. The secondary afferent output (which is not influenced by bag1 receptors) is obtained from the simple summation of the outputs of bag2 and chain intrafusal fiber models.

The model behaviour is described by a second order differential equation which uses muscle fascicle length and fusimotor inputs (static and dynamic) to calculate the primary and secondary afferent firing. Numerous model parameters were based on the experimental literature for the intrafusal or extrafusal muscle fibers while the remaining parameters were fit using standard least-squares fitting algorithms. A single set of model parameters was optimized on a number of data sets collected from feline soleus muscle, accounting accurately for afferent activity during a variety of ramp, triangular, and sinusoidal stretches. The primary and secondary afferent firing during ramp and triangual stretches at different velocities and during different fusimotor stimulation conditions on which model data were optimized is accurately captured (Figure 6). Secondary afferent response is shown for 3 different whole muscle stretch velocities in absence of fusimotor stimulation (Figure 7). During simultaneous static and dynamic fusimotor stimulation the model correctly captured the primary afferent activity as influenced by the partial occlusion effect (Figure 8).

The model was validated on an independent data set recorded in the decerebrate cat preparation where two types of naturally occurring static fusimotor drives were recorded (Taylor et al. 2000). Although the existence of two types of static fusimotor drives is still not universally accepted, the researchers suggested that type-1 static fusimotor drive innervates chain or bag2 and chain fibers together, whereas type-2 static fusimotor drive innervates solely the bag2 fiber. Although originally planned to receive one type of static fusimotor drive, the spindle model was modified to account for two static drives. The experimentally recorded kinematic and fusimotor

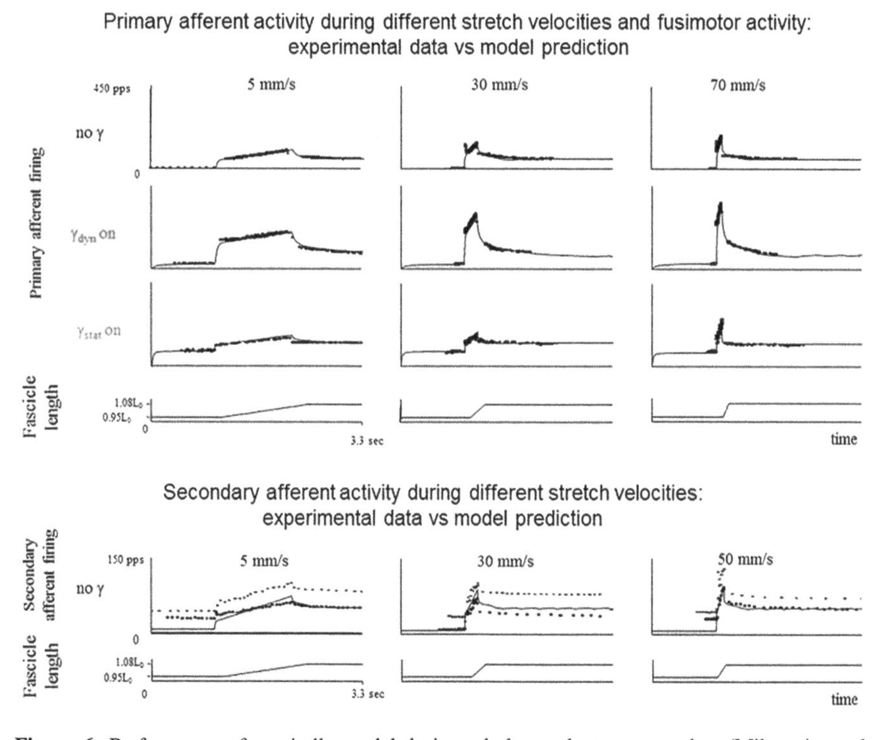

Figure 6 Performance of a spindle model during whole mucle ramp stretches (Mileusnic et al. 2006). Primary afferent response is shown for 3 different whole muscle stretch velocities and 3 different fusimotor conditions (no fusimotor stimulation, constant dynamic fusimotor stimulation, constant static fusimotor stimulation)

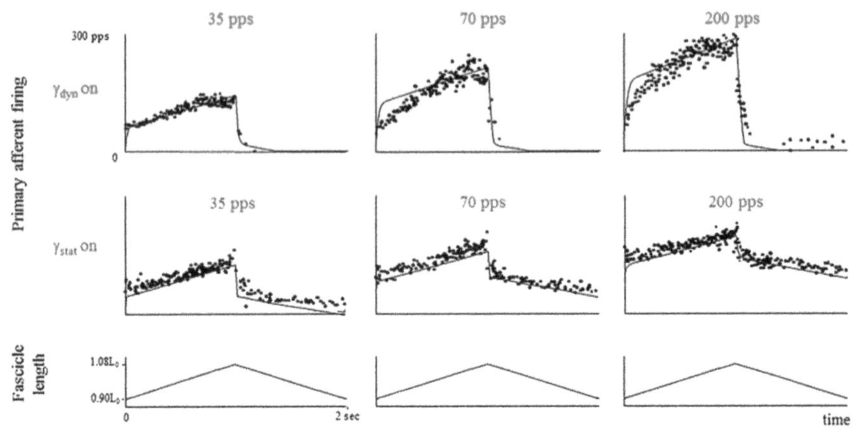

Figure 7 Performance of a spindle model during whole muscle triangular stretches (8 mm/s) (Mileusnic et al. 2006). Primary afferent activity was recorded during constant static or dynamic fusimotor stimulation

profiles were used to run the modified model and calculate secondary afferent activity. The model simulations were repeated twice, with and without the inclusion of the low-pass filter for bag2 intrafusal fiber model. When subject to two static fusimotor drives, the model more accurately predicted experimentally recorded secondary afferent activity when incorporating the known different temporal properties of the two fiber types. The model suggests that the experimentally observed phase differences between two static fusimotor drives compensate for the differences in contractile dynamics of their respective intrafusal fibers, resulting in synchronous static fusimotor modulation of the receptors, at least in this preparation.

Black Box Model of Maltenfort and Burke (2003)

The development of the model was largely inspired by the desire to study the effects of beta (β) (or also called skeleto-fusimotor) motoneurons which innervate both intra- and extrafusal muscle fibers (Laporte and Emonet-Dénand 1976). Some evidence suggests that β motoneurons receive monosynaptic group primary afferent excitation comparable to that in motoneurons, potentially resulting in a positive feedback loop (Burke and Tsairis 1977).

As proposed previously by several researchers, the model describes the cat's primary afferent (Ia) as a sum of four components: a pure velocity sensitivity, a pure position sensitivity, a mixed velocity and position sensitivity, and baseline afferent activity at the initial length of the muscle (Lennerstrand and Thoden 1968a, b;

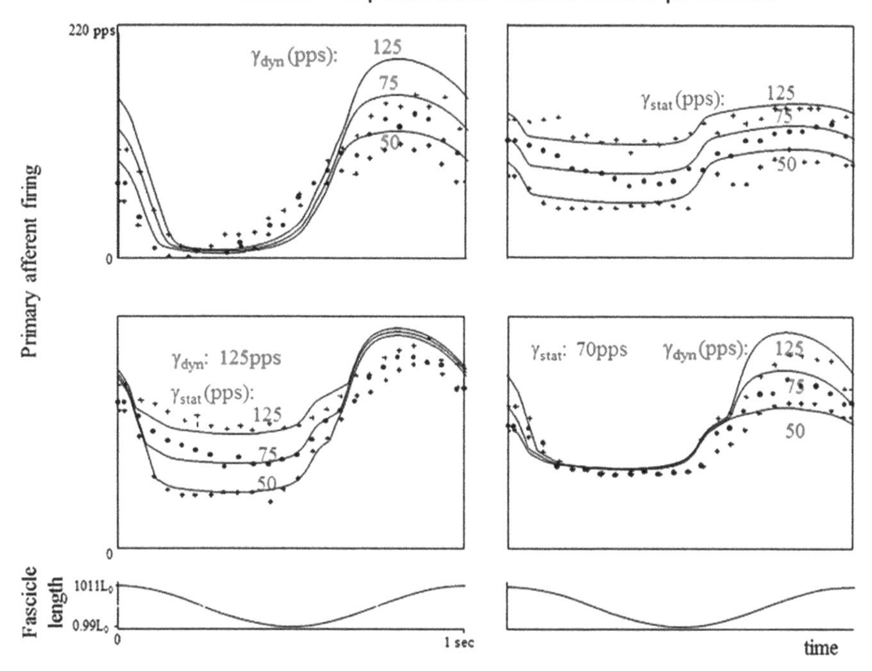

Figure 8 Performance of a spindle model during sinusoidal stretched and fusimotor stimulation (Mileusnic et al. 2006). Two upper graphs show the primary afferent activity during constant dynamic or static fusimotor stimulation. Two lower graphs show the activity during constant, stimultaneous static and dynamic fusimotor stimulation

Hasan 1983; Prochazka and Gorassini 1998) (Figure 9). Spindle afferent response R to muscle length (x) and velocity (v) is summarized by the following equation

$$R = [S_v(\gamma, v) + S_{SS}(\gamma)]x + Q(\gamma, v) + B(\gamma)$$

where γ denotes fusimotor drive, R instantaneous primary afferent firing rate, $S_{ss}(\gamma)$ a velocity-independent position sensitivity, $S_v(\gamma, v)$ a velocity-dependent position sensitivity, $Q(\gamma, v)$ a pure velocity sensitivity, and $B(\gamma)$ the baseline firing of the spindle at the initial length.

In their primary afferent spindle model Maltenfort and Burke extended the above mentioned theory and subdivided B, S_{ss}, Q and S_v into the response of the passive spindle to stretch ($R_{passive}$; no γ input) and additive responses to stretch modulated by dynamic (γ_d) and static (γ_s) fusimotor inputs ($\Delta R_{dynamic}, \Delta R_{static}$):

$$R(t) = R_{passive}(t) + f_{occlusion}(\Delta R_{dynamic}, \Delta R_{static}).$$

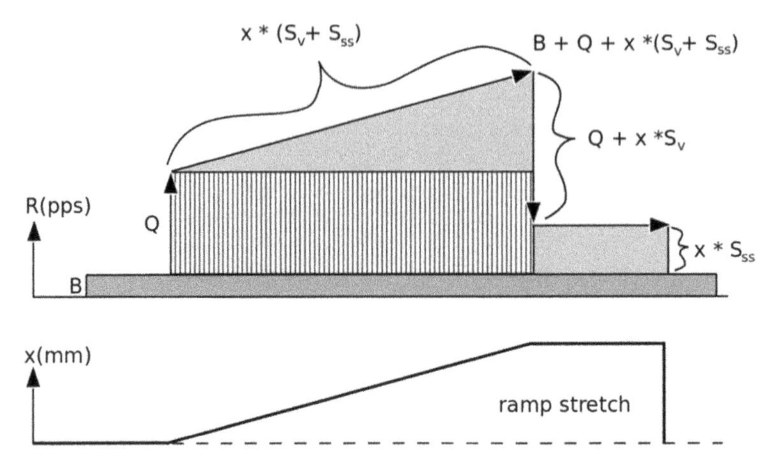

Figure 9 The primary afferent spindle model from Maltenfort and Burke (2003)

The passive response is given by

$$R(t) = R_{passive}(t) + f_{occlusion}(\Delta R_{dynamic}, \Delta R_{static}).$$

The portion of spindle response to stretch that is modulated by fusimotor drive is

$$\Delta R_g(t) = [S_{v,g}(\gamma_g, v_f) + S_{ss,g}(\gamma_g)] * x + Q_g(\gamma_g, v_f) + B_g(\gamma_g),$$

where subscript g represents static or dynamic drive, x is muscle mass, v_f estimate of velocity.

The model also includes occlusion:

$$f_{occlusion}(\Delta R_{g1}, \Delta R_{g2}) = \Delta R_{g1} + \Delta R_{g2} - \frac{1}{\frac{1}{\Delta R_{g1}} + \frac{1}{\Delta R_{g2}}},$$

where ΔR_{g1} and ΔR_{g2} are the increments in primary firing rate (over passive) caused by fusimotor inputs γ_1 and γ_2 individually.

Finally, baseline firing (B), position sensitivity (S_{ss}) and velocity-dependent terms (Q and S_v) during lengthening and shortening were estimated based on the published experimental records of spindle activity.

The model fairly accurately captured primary afferent activity during ramp stretches at different velocities without fusimotor stimulation as well as with either constant static or dynamic fusimotor stimulation present. During small sinusoidal stretches without fusimotor stimulation as well as during individual or simultaneous static and/or dynamic fusimotor stimulations, the model also performed relatively well for most of the conditions.

In terms of the role of β feedback that inspired the development of the spindle model in the first place, this was simulated as a single spindle receiving a fusimotor drive that was proportional to the instantaneous firing rate of that afferent. The model also assumes that β input has the same influence on primary afferent firing as γ motoneurons. In other words, the effect of β motoneuron activity on spindle afferent firing had the same strength and time course as the effect of a γ motoneuron of the same type (static or dynamic). During triangular stretching, the β feedback enhanced the primary afferent firing during shortening, while the closure of the β dynamic loop increased primary afferent firing during lengthening. In general, the loop gain increased with velocity and amplitude of stretch but decreased with increased superimposed γ rates. The effect of β feedback was most noticeable when β loop and γ bias were of different type.

Ensemble Spindle Activity

Computing kinaesthesia from spindle afferent activity is not trivial. The firing rates of individual afferents tend to be noisy (Prochazka et al. 1977; Loeb and Duysens 1979) as a result of transduction noise and mechanical ripple from fusimotor activity. Combining signals from many asynchronously firing afferents results in a Poisson distribution for which signal-to-noise ratio improves with the square root of spike events (Loeb and Marks 1985). Accurate information about the length and velocity of each muscle can only be extracted by combining activity from all available pindle afferents and only after accounting for fusimotor drive and tendon stretch, as noted above.

Most joints have more than one axis of motion and most muscles cross more than one joint, so information about the length and velocity of many muscles must be combined to build up an accurate sense of posture and motion. The numbers of spindles in almost all human skeletal muscles have been counted (Voss 1971) and have been used in large-scale musculoskeletal models to test hypotheses about how these signals are combined to account for psychophysical measurements of kinaesthetic resolution at each joint (Scott and Loeb 1994).

Golgi Tendon Organs

Golgi tendon organs (GTO) are tension-sensitive mechanoreceptors found in mammalian skeletal muscles that supply the central nervous system with information regarding active muscle tension by their Ib afferents. The GTOs are most commonly located at the junctions between the muscle fibers and the collagen strands composing tendons and aponeuroses (Figure 10) (Golgi 1878; Golgi 1880). The GTO receptor consists of bundles of collagen fibers that connect small fascicles of muscle to the whole muscle tendon or aponeurosis (Schoultz and Swett 1972). In other words, the GTO is placed in series between muscle fibers ("muscle end") and tendon and aponeurosis ("tendon end"), the oppostie of the muscle spindle which

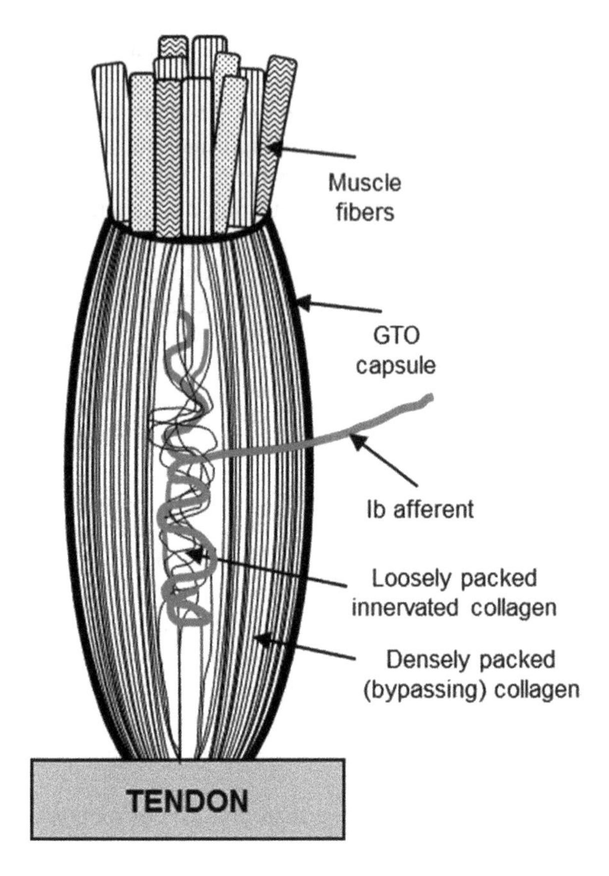

Figure 10 The structure of Golgi tendon organ. The GTO receptor is located in series between the tendon and muscle fibers that insert into it. A single GTO afferent axon innervates GTO's complexly interwoven collagen strands. (Figure adapted from Mileusnic and Loeb 2006)

lies in parallel with extrafusal muscle fibers. Each GTO receptor is usually innervated by a single large, myelinated Ib afferent that enters the GTO capsule near its equator. The activation of a muscle fiber that inserts onto tendon strands that pass through the GTO straightens some of the collagen strands, compressing and depolarizing the pressure-sensitive afferent endings and eventually resulting in membrane depolarization and initiation of action potentials in the Ib axon (Fukami and Wilkinson 1977).

GTO Models

To date, only two models of GTO activity have been published (Houk and Henneman 1967; Mileusnic and Loeb 2006). The later model was subsequently used to study the ability of the ensemble firing of all GTO receptors in a muscle to accurately encode whole muscle force (Mileusnic and Loeb 2009).

Model of Houk and Henneman (1967)

Although the main objective of the published work by (Houk and Henneman 1967) was to test whether active contraction rather than muscle stretching was more effective in activating the GTO, they also offered a linear model of GTO response. In the studies, a laminectomy was performed on adult cats and the dorsal and ventral roots subdivided to record individual GTO activity during stimulation of individual motor units.

The firing rate of the Ib afferent $r(t)$ to any input whose time course is specified by $f(t)$ is calculated from the convolution or superposition integral

$$r(t) = \int_0^t f(t - t')h(t')dt'$$

where h represents a model response to a unit impulse. In other words, the convolution integral computes the output by adding up the responses to an infinite series of infinitesimal impulses. The dynamic properties are defined by the means of h, where h is defined as

$$h(t) = K(1 + B + C]u_0(t) - K[bBe^{-bt} + cCe^{-ct}]u_{-1}(t)$$

and characterizes the tendon organ. B, b, C, c, K are parameters, $u_0(t)$ is the unit impulse function, and $u_{-1}(t - t')$ is the unit step function the effect of exponential parameters can be best appreciated by looking at the response to a step input

$$r(t) = K[1 + Be^{-bt} + Ce^{-ct}], \quad t \geq 0$$

The dynamic properties of tendon organs to stimulations of single motor units at different velocities were recorded. Subsequently, these different responses were fitted by the simulated responses of the mathematical model of the receptor where model parameters were adjusted for each individual GTO response. The exponential parameters of the model indicate the magnitude of overshoot (B and C) and the decay rate (b and c), while K is a gain factor whose physiological correlate is the sensitivity of the receptor (Figure 11).

In addition to GTO responses to individual motor unit stimulations, the researchers recorded also GTO activity during multiple motor unit stimulation, which results in non-linear summation. If two motor units are stimulated simultaneously, the resultant response is greater than that which could be obtained from either individually but smaller than that expected from simple addition. While providing an experimental example of non-linear summation, the researchers do not report on model parameters that would capture it.

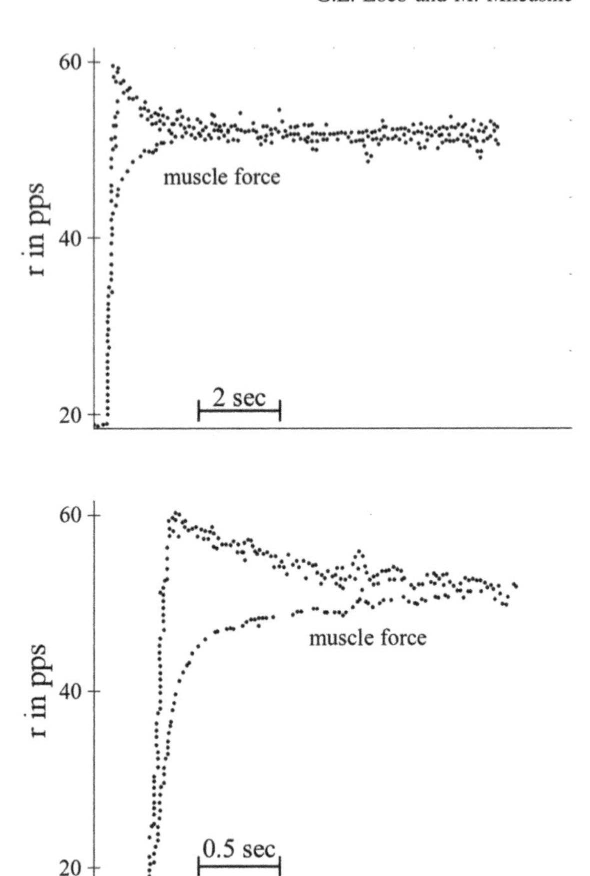

Figure 11 GTO response to a tetanic contraction of a single motor unit (upper curve-slow contraction, lower curve-fast contraction). The input force is represented by solid line while modelled GTO rate of discharge 'r' in pulses per second as +. (From Houk and Henneman 1967)

Model of Mileusnic and Loeb (2006)

A recently developed, anatomically based GTO models attempts to explain the experimentally recorded GTO firing by interactions of various histological components of the GTO. It is based on feline medial gastrocnemius (MG) muscle in which the GTO has been most extensively studied.

A biological GTO consists of bundles of collagen fibers that connect small fascicles of muscle fibers to the whole muscle tendon or aponeurosis (Figure 10). Similarly to a biological GTO, the GTO model receives the tension input from multiple muscle fibers inserting into its capsule. A typical GTO in feline MG is attached to 20 muscle fibers belonging to 13 different motor units (Gregory 1990). The structure of the GTO model is also strongly influenced by the published literature which suggests that the collagen within the GTO capsule is unevenly packed. The marginal areas of the GTO are typically occupied by densely packed collagen whose fibers run in parallel with one another and only rarely make contact with the GTO afferent axon. The central area of the capsule is occupied largely by loosely packed collagen containing many afferent branches among the complexly interwoven collagen strands (Schoultz and Swett 1972; Nitatori 1988).

The GTO model assumes that a single muscle fiber inserting into the capsule interacts with both types of collagen (Figure 12). Because at least two large myelinated branches are typically found within the GTO capsule, each innervating separate portions of the GTO's loosely packed collagen network, the model assumes the existence of two separate transduction zones, each having its own impulse-generating site. The bypassing collagen attached to an inserting muscle fiber is modelled as uninnervated nonlinear spring, while the remaining innervated

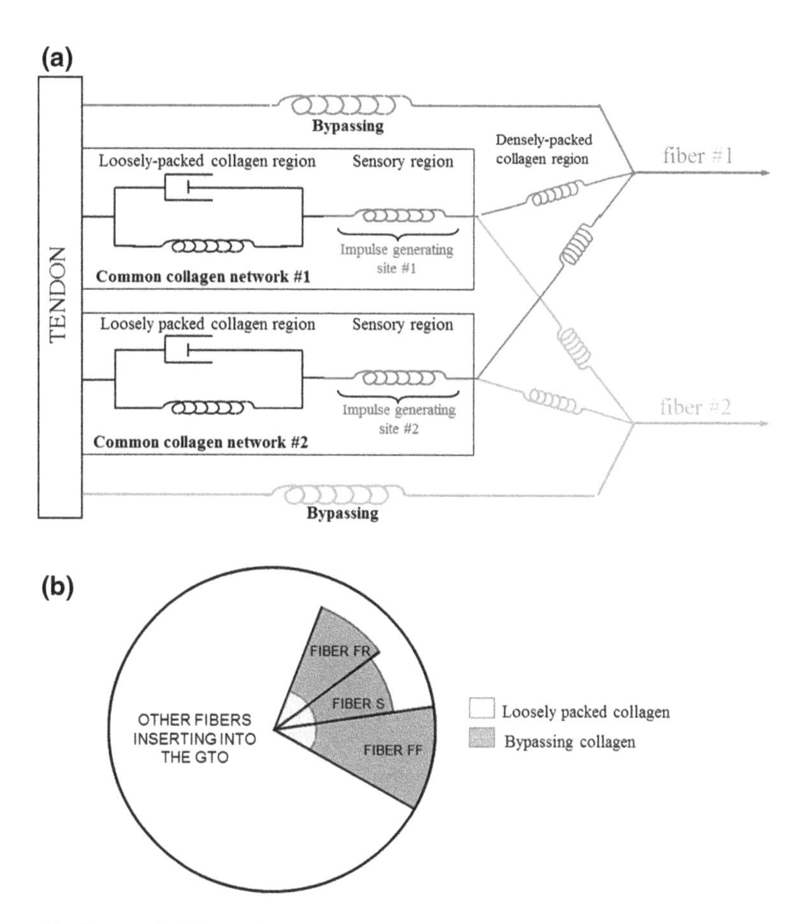

Figure 12 Structural GTO model (Mileusnic and Loeb 2006). **a** The structure of the GTO model. For simplicity the figure represents only two of the many muscle fibers that insert into the GTO capsule. Each fiber has its collagen divided among bypassing and innervated collagen which is then divided between two common collagen networks having separate impulse generating sites. **b** The flower-shaped cross sectional area of collagen within the GTO. The muscle fibers' collagen was assumed to be arranged in the manner of the flower petals where the fiber's collagen that occupies the central area of flower-shaped structure belongs to the innervated collagen type (yellow) while the more distally located collagen is of bypassing type (dark grey)

collagen is distributed between two common transduction zones. The loosely packed collagen within each transduction zone consists of a nonlinear sensory spring (collagen in direct contact with sensory endings) in series with an element consisting of a nonlinear spring in parallel with a damper that reflects the tendency of collagen strands within the network to gradually rearrange in response to stretch.

The amount of innervated and bypassing collagen that is attached to a single muscle fiber is determined by using a conceptual geometrical arrangement of fibers inserting into the GTO where the GTO's cross-sectional area is assumed to be flower shaped and the fibers are arranged in the manner of flower petals. The cumulative innervated collagen belonging to all the muscle fibers inserting into the GTO is assumed to occupy an inner zone (10 % of the total GTO collagen) of the flower-shaped cross-sectional area. Therefore 90 % of the cross-sectional area is assumed to be of the bypassing type. The various extrafusal fiber types (S, FF, FR) have different cross-sectional areas and therefore different amounts of the collagen attached to them, accounting for their different abilities to deform innervated collagen and excite the nerve endings.

The sensory region is modelled as a spring in series with loosely packed collagen and represents the collagen that is in direct contact with the sensory endings. The model assumes that transduction in the sensory endings reflects the amount of stretch of the innervated collagen attached to a given muscle fiber weighted by the cross-sectional area of that collagen. The two separate transduction zones each have their own impulse-generating sites and the net activity of the Ib afferent reflects the dominant site (complete occlusion).

The equations of tensions in various model regions were derived and combined. The model parameters were manually tuned rather than optimized because the available experimental data are sparse and highly variable, particularly for the more complex aspects of GTO behaviour.

The model captures GTO activity well as reported in the literature, including the following complexities:

Dynamic and static responses: After a sudden step activation of the MU, the GTO response consists of a burst (dynamic response) that gradually decays to a constant afferent firing (static response) as a result of the damping term as well as the time-dependencies of force generation in various fiber types. Figure 13 demonstrates the model's response to slow, non-fatiguable, and fast fatiguable motor unit activation.

Self- and cross-adaptation: This refers to a phenomenon whereby a GTO's dynamic response during motor unit activation is decreased after prior activation of the same motor unit (self-adaptation) or a different motor unit (cross-adaptation) (Gregory and Proske 1979; Gregory et al. 1985). In Figure 13, the second responses of the modeled GTO to the same tetanic activation of same (or different) motor unit shows a typical decrease confined to the dynamic response. The experimental literature describes large variability in the amount of cross-adaptation that exists between different MUs (Gregory et al. 1985). Two FF muscle fibers were modeled as sharing either 90 or 10 % of the innervated collagen. Figure 4 shows a large or small amount of cross-adaptation, respectively, when the units are activated in succession.

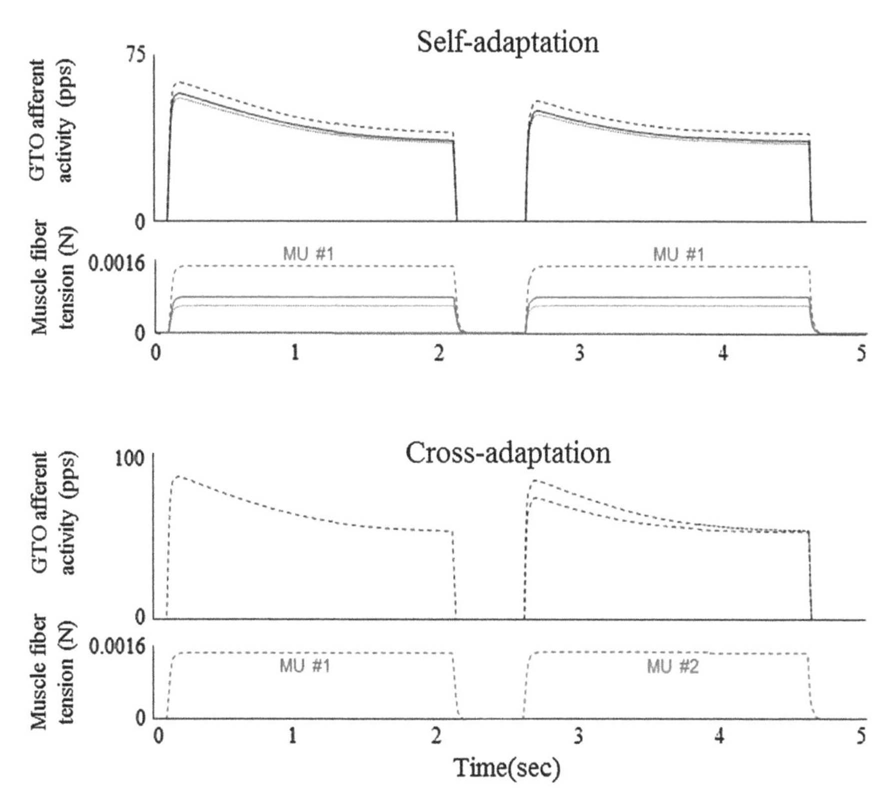

Figure 13 The GTO model's ability to capture self- and cross-adaptation (Mileusnic and Loeb 2006).The self-adaptation is given for three types of MUs (S: ··, FR: — and FF:–). Different amounts of cross-adaptation were obtained by varying the amount of innervated collagen that 2 MUs share. (Mileusnic and Loeb 2006)

Nonlinear summation: During the activation of multiple MUs the GTO demonstrates nonlinear summation, where two MUs when stimulated simultaneously produce afferent activity that is smaller than the linear summation of the GTO firing rates that individual MUs produce when stimulated independently (Gregory and Proske 1979) (Figure 14).

Ensemble GTO Activtity

Typically, tens of GTOs are distributed unevenly across the myotendinous junction, where each of them responds vigorously to active tension produced by those muscle fibers that insert into its capsules. The ensemble firing of all GTO receptors in the muscle has been hypothesized to represent a reliable measure of the whole muscle force during normal, size-ordered recruitment of motor units (Jami 1992) but the

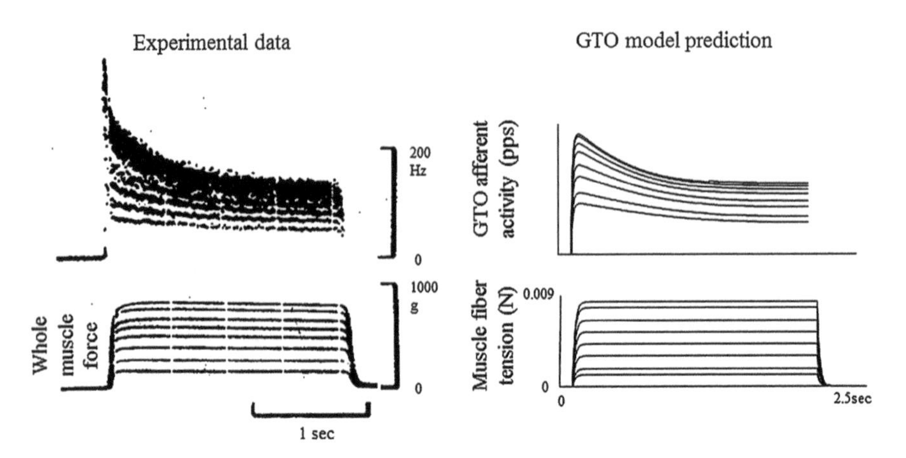

Figure 14 The GTO model's ability to capture nonlinear summation (Mileusnic and Loeb 2006). The GTO afferent activity is obtained in response to progressive recruitment of 8 MUs

precision and accuracy of that information are largely unknown because it is impossible to record activity simultaneously from all GTOs in a muscle.

The GTO model developed by Mileusnic and Loeb (2006) was used subsequently in a larger model of ensemble GTO firing to test Jami's hypothesis. The model of ensemble GTO firing in a muscle was designed to examine the reliability of the ensemble sensory information (to study the relationship between the aggregate GTO activity and total muscle force). To derive the ensemble GTO activity, the model utilizes several other models. The previously described GTO model (Mileusnic and Loeb 2006) was employed to obtain afferent activity of each individual GTO receptor in response to tensions of muscle fibers inserting into its capsule. A model of the cross-sectional area of the muscle-tendon junction (including the distributions of various types of muscle fibers, motor units and GTOs) was designed in order to obtain a realistic sample of muscles fibers that insert into each GTO receptor. Finally, the realistic muscle model was used to calculate the tensions of the fibers inserting into individual GTOs.

The relationship between the aggregate GTO activity and whole muscle tension was investigated under conditions of natural muscle recruitment and natural distribution of GTOs and muscle fibers as well as in pathological muscles (for example, reinnervated muscles, artificial recruitment by functional electrical stimulation). The ensemble activity was studied for all conditions during very slow ramp muscle activations as well as during phasic excitation similar to that associated with cyclical behavior such as locomotion.

The model suggests that in the intact muscle where MUs are recruited based on the size principle (Henneman and Mendell 1981), the ensemble GTO activity accurately encodes force information according to a nonlinear, monotonic relationship that has its steepest slope for low force levels and tends to saturate at the

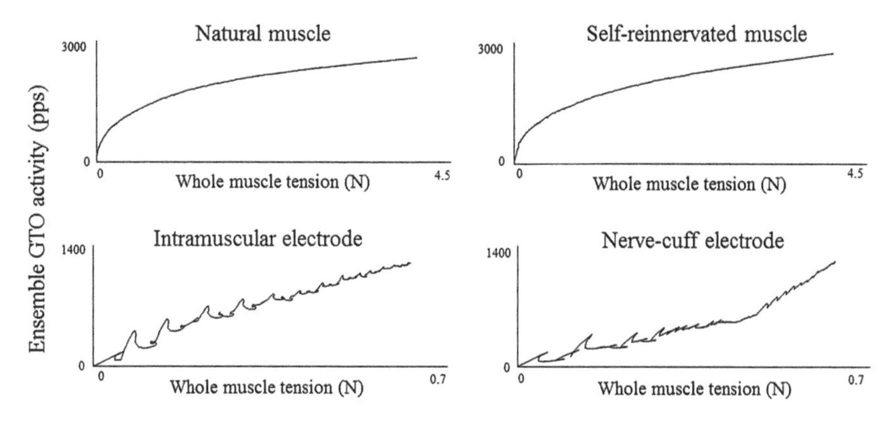

Figure 15 The relationship between the ensemble Golgi tendon organ activity and whole muscle tension (as predicted by ensemble GTO model from Mileusnic and Loeb 2009). The relationship is shown for four conditions: intact muscle, self-reinnervated muscle, intramuscular electrode stimulation and nerve-cuff electrode stimulation

highest force levels. Various types of neuromuscular pathology and electrical recruitment produced different and sometimes non-monotonic input-output relationships, as illustrated in Figure 15.

References

Bakker, G J (1980). Histochemical characteristics of muscle spindles in cat dorsal neck muscles. In: *Physiology*, Vol. M.Sc. Kingston, Ontario, Canada: Queen's University.

Banks, R; Hulliger, M; Scheepstra, K and Otten, E (1997). Pacemaker activity in a sensory ending with multiple encoding sites: The cat muscle spindle primary ending. *The Journal of Physiology* 498: 177–199.

Boyd, I; Gladden, M H; McWilliam, P and Ward, J (1977). Control of dynamic and static nuclear bag fibres and nuclear chain fibres by gamma and beta axons in isolated cat muscle spindels. *The Journal of Physiology* 265: 133–162.

Boyd, I A (1976). The response of fast and slow nuclear bag fibres and nuclear chain fibres in isolated cat muscle spindles to fusimotor stimulation, and the effect of intrafusal contraction on the sensory endings. *Quarterly Journal of Experimental Physiology and Cognate Medical Sciences* 61: 203–254.

Boyd, I A and Smith, R S (1984). The muscle spindle. In: P J Dyck et al. (Eds.), *Peripheral Neuropathy*, Vol. 2 (pp. 171–202). W.B. Saunders company.

Burke, R; Levine, D; Tsairis, P and Zajac, F, III (1973). Physiological types and histochemical profiles in motor units of the cat gastrocnemius. *The Journal of Physiology* 234: 723.

Burke, R E and Tsairis, P (1977). Histochemical and physiological profile of a skeletofusimotor (β) unit in cat soleus muscle. *Brain Research* 129: 341–345.

Carr, R W; Gregory, J E and Proske, U (1998). Summation of responses of cat muscle spindles to combined static and dynamic fusimotor stimulation. *Brain Research* 800: 97–104.

Chen, W and Poppele, R (1978). Small-signal analysis of response of mammalian muscle spindles with fusimotor stimulation and a comparison with large-signal responses. *Journal of Neurophysiology* 41: 15–27.

Chin, N K; Cope, M and Pang, M (1962). Number and distribution of spindle capsules in seven hindlimb muscles of the cat. In: D Barker (Ed.), *Symposium on Muscle Receptors* (pp. 241–248). Hong Kong: Hong Kong University Press.

Fallon, J B; Carr, R W; Gregory, J E and Proske, U (2001). Summing responses of cat soleus muscle spindles to combined static and dynamic fusimotor stimulation. *Brain Research* 888: 348–355.

Fukami, Y and Wilkinson, R S (1977). Responses of isolated golgi tendon organs of the cat. *The Journal of Physiology* 265: 673–689.

Golgi, C (1878). Intorno alla distribuzione e terminazione dei nervi nei tendini dell'uomo e di altri vertebrati. *Rend R Ist Lomb Sci Lett* B11: 445–453.

Golgi, C (1880). Sui nervi dei tendini dell'uomo et di altri vertebrati e di un nuovo organo nervosa terminale muscolo-tendineo. *Mem R Acad Sci Torino* 32: 359–385.

Gregory, J E (1990). Relations between identified tendon organs and motor units in the medial gastrocnemius muscle of the cat. *Experimental Brain Research* 81: 602–608.

Gregory, J E; Morgan, D and Proske, U (1985). Site of impulse initiation in tendon organs of cat soleus muscle. *Journal of Neurophysiology* 54: 1383–1395.

Gregory, J E and Proske, U (1979). The responses of Golgi tendon organs to stimulation of different combinations of motor units. *The Journal of Physiology* 295: 251–262.

Grigg, P; Schaible, H G and Schmidt, R F (1986). Mechanical sensitivity of group III and IV afferents from posterior articular nerve in normal and inflamed cat knee. *Journal of Neurophysiology* 55: 635–643.

Hasan, Z (1983). A model of spindle afferent response to muscle stretch. *Journal of Neurophysiology* 49: 989–1006.

Henneman, E and Mendell, L M (1981). Functional organization of the motoneuron pool and its inputs. In: V B Brooks (Ed.), *Handbook of Physiology Sect I The Nervous System*, Vol II, Part 1 (pp. 423–507). Washington, DC: American Physiological Society.

Hill, D (1968). Tension due to interaction between the sliding filaments in resting striated muscle. The effect of stimulation. *The Journal of Physiology* 199: 637–684.

Hoffer, J A; Caputi, A A and Pose, I E (1988). Role of muscle activity on the relationship between muscle spindle length and whole muscle length in the freely walking cat. *Proceedings of the Conference on Afferent Control of Posture and Locomotion* 1–4: 32.

Hoffer, J A; Caputi, A A and Pose, I E (1992). Activity of muscle proprioceptors in cat posture and locomotion: Relation to EMG, tendon force and the movement of fibres and aponeurotic segments. *IBRO Symposium Series* 1–6.

Houk, J and Henneman, E (1967). Responses of Golgi tendon organs to active contractions of the soleus muscle of the cat. *Journal of Neurophysiology* 30: 466–481.

Houk, J C; Rymer, W Z and Crago, P E (1981). Dependence of dynamic response of spindle receptors on muscle length and velocity. *Journal of Neurophysiology* 46: 143–166.

Hulliger, M; Matthews, P B C and Norht, J (1977). Static and dynamic fusimotor action on the response of Ia fibers to low frequency sinusoidal stretching of widely ranging amplitude. *Journal of Physiology* 267: 811–838.

Jami, L (1992). Golgi tendon organs in mammalian skeletal muscle: Functional properties and central actions. *Physiological Reviews* 72: 623–666.

Laporte, Y and Emonet-Dénand, F (1976). The skeleto-fusimotor innervation of cat muscle spindle. In: *Progress in Brain Research Understanding the Stretch Reflex*, Vol. 44 (pp. 99–105).

Lennerstrand, G and Thoden, U (1968a). Position and velocity sensitivity of muscle spindles in the cat. II. Dynamic fusimotor single-fibre activation of primary endings. *Acta Physiologica Scandinavica* 74: 16–29.

Lennerstrand, G and Thoden, U (1968b). Position and velocity sensitivity of muscle spindles in the cat. III. Static fusimotor single-fibre activation of primary and secondary endings. *Acta Physiologica Scandinavica* 74: 30–49.

Lin, C-C K and Crago, P E (2002). Structural model of the muscle spindle. *Annals of Biomedical Engineering* 30: 68–83.

Loeb, G E and Duysens, J (1979). Activity patterns in individual hindlimb primary and secondary muscle spindle afferents during normal movements in unrestrained cats. *Journal of Neurophysiology* 42: 420–440.

Loeb, G E and Marks, W B (1985). Optimal control principles for sensory transducers. In: I A Boyd and M H Gladden (Eds.), *Proceedings of the International Symposium: The Muscle Spindle* (pp. 409–415). London: MacMillan Ltd.

Maltenfort, M G and Burke, R E (2003). Spindle model responsive to mixed fusimotor inputs and testable predictions of β feedback effects. *Journal of Neurophysiology* 89: 2797–2809.

Martin, P G; Smith, J L; Butler, J E; Gandevia, S C and Taylor, J L (2006). Fatigue-sensitive afferents inhibit extensor but not flexor motoneurons in humans. *The Journal of Neuroscience* 26: 4796–4802.

Matthews, P B C (1972). *Mammalian Muscle Receptors and Their Central Actions*. London: Edward Arnold Publishing Ltd.

Matthews, P B C and Stein, R B (1969). The sensitivity of muscle spindle afferents to small sinusoidal changes of length. *Journal of Physiology* 200: 723–743.

McMahon, T A (1984). *Muscles, Reflexes and Locomotion*. Princeton, NJ: Princeton University Press.

Mileusnic, M and Loeb, G (2009). Force estimation from ensembles of Golgi tendon organs. *Journal of Neural Engineering* 6: 036001.

Mileusnic, M P; Brown, I E; Lan, N and Loeb, G E (2006). Mathematical models of proprioceptors. I. Control and transduction in the muscle spindle. *Journal of Neurophysiology* 96: 1772–1788.

Mileusnic, M P and Loeb, G E (2006). Mathematical models of proprioceptors. II. Structure and function of the Golgi tendon organ. *Journal of Neurophysiology* 96: 1789–1802.

Nitatori, T (1988). The fine structure of human Golgi tendon organs as studied by three-dimensional reconstruction. *Journal of Neurocytology* 17: 27–41.

Paintal, A S (1960). Functional analysis of group III, afferent fibres of mammalian muscles. *Journal of Physiology* 152: 250–270.

Poppele, R E and Bowman, R J (1970). Quantitative description of linear behavior of mammalian muscle spindles. *Journal of Neurophysiology* 33: 59–72.

Prochazka, A and Gorassini, M (1998). Models of ensemble firing of muscle spindle afferents recorded during normal locomotion in cats. *Journal of Physiology (London)* 507(Pt. 1): 277–291.

Prochazka, A; Westerman, R A and Ziccone, S P (1977). Ia afferent activity during a variety of voluntary movements in the cat. *Journal of Physiology (London)* 268: 423–448.

Rudjord, T (1970a). A mechanical model of the secondary endings of mammalian muscle spindles. *Kybernetik* 7: 122–128.

Rudjord, T (1970b). A second order mechanical model of muscle spindle primary endings. *Kybernetik* 6: 205–213.

Rymer, W Z; Houk, J C and Crago, P E (1979). Mechanisms of the clasp-knife reflex studied in an animal model. *Experimental Brain Research* 37: 93–113.

Schaafsma, A; Otten, E and Van Willigen, J D (1991). A muscle spindle model for primary afferent firing based on a simulation of intrafusal mechanical events. *Journal of Neurophysiology* 65: 1297–1312.

Schoultz, T W and Swett, J E (1972). The fine structure of the Golgi tendon organ. *Journal of Neurocytology* 1: 1–25.

Scott, S H and Loeb, G E (1994). The computation of position sense from spindles in mono-and multiarticular muscles. *The Journal of Neuroscience* 14: 7529–7540.

Taylor, A; Durbaba, R; Ellaway, P H and Rawlinson, S (2000). Patterns of fusimotor activity during locomotion in the decerebrate cat deduced from recordings from hindlimb muscle spindles. *Journal of Physiology* 522: 515–532.

Voss, H (1971). Tabelle der absoluten und relativen muskelspindelzahlen der menschlichen skelettmuskulatur. *Anatomischer Anzeiger* 129: 562–572.

Autonomic Nervous System

Bill Blessing and Ian Gibbins

The term **autonomic nervous system** (ANS) refers to collections of motor neurons (ganglia) situated in the head, neck, thorax, abdomen, and pelvis, and to the axonal connections of these neurons (Figure 1). Autonomic pathways, together with somatic motor pathways to skeletal muscle and neuroendocrine pathways, are the means whereby the central nervous system (CNS) sends commands to the rest of the body. There are also CNS components of the ANS, including brainstem and spinal autonomic preganglionic neurons that project to the autonomic motor neurons in the peripheral ganglia. In this respect preganglionic autonomic motor neurons are clearly distinguished from somatic motor neurons that project from the CNS directly to the innervated tissue (skeletal muscle), without any intervening ganglia.

Post-ganglionic axonal processes of motor neurons in the autonomic ganglia innervate organs and tissues throughout the body (eyes, salivary glands, heart, stomach, urinary bladder, blood vessels, etc). The motor neurons in the autonomic ganglia are sometimes referred to as "postganglionic neurons". This traditional terminology is confusing and we use the term "autonomic motoneurons" or "final motoneurons" for the ganglionic cells.

Complex autonomic ganglia in the walls of the stomach and small intestine are separately classified as the enteric nervous system. Most of the neural pathways in the enteric plexuses lack direct preganglionic inputs and can operate independently of central control. Indeed, uniquely within the ANS, the enteric plexuses contain primary sensory neurons that connect to extensive networks of interneurons as well as excitatory and inhibitory enteric motor neurons.

B. Blessing
Centre for Neuroscience, Flinders University, Adelaide, Australia

I. Gibbins (✉)
Centre for Neuroscience, Flinders University, Adelaide, SA, Australia

T.J. Prescott et al. (eds.), *Scholarpedia of Touch*, Scholarpedia,
DOI 10.2991/978-94-6239-133-8_37

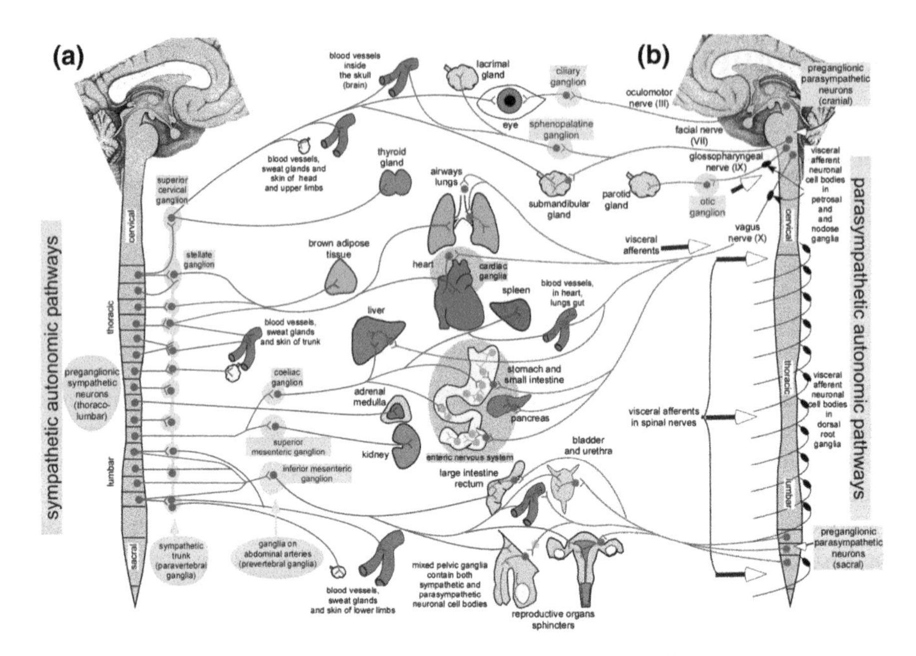

Figure 1 Summary of sympathetic (A) and parasympathetic (B) autonomic neural outflows from the central nervous system. Figure drawn by the authors, incorporating material from *Gray's Anatomy* 31st Edition 1954, and from Cannon and Rosenblueth *Physiology of the Autonomic Nervous System*, 1937

History of the Definition and Functional Conception of the ANS

Emotional feeling has traditionally been seen as distinct from rational thought (Ackerknecht 1974). The brain, locked away in its bony case, was conceived as responsible for rational thought and for ideas that direct behavioral interactions with the external environment. Emotions, visceral rather than rational, were linked with the functions of the internal bodily organs. We have "gut feelings", the heart is the "seat of love" and we "vent our spleen". Bichat (1771–1802) divided life into two distinct forms, one (relational life) governed by the brain, and the other (organic, vegetative, life) by the abdominal ganglia. Vegetative life was seen as connected with the passions and independent of education, governed by independently functioning abdominal ganglia, a chain of "little brains". Phillipe Pinel, one of the founders of psychiatry, and Bichat's teacher, even considered mental disease to be caused by abnormal function of these ganglia, and modern psychiatry still refers to "vegetative functions".

Langley (1852–1925) coined the term autonomic nervous system. Langley noted the absence of sensory (afferent) nerve cell bodies in autonomic ganglia and defined the ANS as a purely motor system. He nevertheless continued the tradition whereby the ANS is seen as functioning in its own right, with independence from the CNS. It should be noted that Langley did not completely adhere to this simplification. In his introduction to the ANS (1903) he wrote that it is possible to "consider as afferent autonomic fibres those which give rise to reflexes in autonomic tissues, and which are incapable of directly giving rise to sensation". Moreover, the discovery of primary afferent neurons that are part of the ANS, but lie entirely outside the CNS, and make no direct connection with the CNS, make it difficult to conceive of the ANS as an entirely efferent system (Furness 2006a, b; see further below).

Modern experiments have shown that neurons in autonomic ganglia do not have inbuilt discharge patterns sufficiently integrated to regulate physiological functions, with the possible exception of neurons within the enteric nervous system of the small and large intestines. The classic description of hexamethonium man summarizes the state of an individual after drug-mediated separation of the ANS from functional control by the brain. Similarly, when brain control of spinal autonomic preganglionic neurons is removed (as in quadriplegia), cardiovascular, bowel and bladder functions are profoundly impaired. Thus the ANS is best seen as one of the outflows whereby the CNS controls bodily organs, so that "peripheral autonomic pathways" is a better term, but "autonomic nervous system" is well-established.

ANS pathways are divided into sympathetic and parasympathetic (around the sympathetic) divisions and enteric plexuses (Gibbins 2012). Preganglionic cell bodies for the sympathetic outflow are in the thoracic spinal cord. Preganglionic cell bodies for the parasympathetic outflow are in the brainstem (cranial) and in the sacral spinal cord (sacral). The idea that the divisions oppose each other is a misleading simplification. Neither division is ever activated in its entirety. Rather, each division consists of a series of discrete functional pathways that may be activated from the CNS either independently or in patterns, according to the particular requirement of the particular daily activity that is contributing to bodily homeostasis (Jänig 2006). The primacy of integrative brain control of all bodily functions was recognized by Walter Cannon, but his idea that the brain activates sympathetic nerves diffusely and non-specifically during bodily emergencies ("fight or flight reaction") is an over simplification. Different emergency states require different patterns of autonomic activity, and normal daily life (apart from emergencies) also requires patterned autonomic activity (Loewy and Spyer 1990). The individual functions as a whole: there is just one nervous system (Blessing 1997).

Sensory information (visceral afferent information) relevant to autonomic control (eg degree of bladder distention or level of blood pressure) travels in visceral afferent nerves and enters the CNS via spinal afferent pathways, or via vagal or glossopharyngeal afferents that project into the lower brainstem (see white-filled black arrows in Figure 1).

Autonomic Neurotransmitters

All preganglionic autonomic neurons, both sympathetic and parasympathetic use acetylcholine (ACh) as their fast excitatory transmitter. In the ganglia, ACh acts on a subclass of nicotinic receptors, distinct from nicotinic receptors at the skeletal muscle neuromuscular junction. Many preganglionic autonomic neurons also contain neuropeptides, usually acting as co-transmitters that mediate slow excitatory post-synaptic potentials, facilitating cholinergic transmission.

Most sympathetic final motor neurons utilise noradrenaline (norepinephrine) as their primary transmitter, together with co-transmitters such as adenosine triphosphate (ATP) and peptides, including neuropeptide Y (NPY), galanin, somatostatin or opioid peptides. Some sympathetic final motor neurons (especially those innervating sweat glands) use ACh as their main non-peptide transmitter. Parasympathetic final motor neurons pathways usually use ACh, nitric oxide, or both as non-peptide transmitters, as well as a wide range of co-transmitter peptides including vasoactive intestinal peptide (VIP), calcitonin gene-related peptide (CGRP), somatostatin and opioid peptides. No parasympathetic neurons use noradrenaline as a transmitter. ACh is also a major excitatory transmitter utilised by enteric neurons. Other enteric neurotransmitters include nitric oxide (probably the main inhibitory transmitter to gut muscle), substance P, VIP, enkephalin, serotonin (5-hydroxytryptamine, 5-HT) and ATP.

Axons of final motor neurons ramify throughout their target tissues, typically smooth muscle, secretory tissue or cardiac muscle. Axon terminals are specialized for neurotransmission, but they usually lack the structures characteristic of conventional synaptic contacts. Many target tissues are innervated by both sympathetic and parasympathetic nerves (eg the heart, the iris muscle, some salivary glands, the gastrointestinal tract and pelvic organs).

Cranial Parasympathetic Pathways

The cranial parasympathetic pathways project to a wide variety of targets in the head, neck, thorax and abdomen (Figure 1). The pathways are associated with four of the cranial nerves: the oculomotor (III), facial (VII), glossopharyngeal (IX) and the vagus (X). Most final motor neurons in these cranial autonomic pathways are in four pairs of major ganglia: the ciliary ganglia (III), sphenopalatine or pterygopalatine ganglia (VII), submandibular ganglia (VII), and otic ganglia (IX). The final motor neurons of the vagal autonomic pathways lie mostly in microganglia located near or within the target organs.

The major target of cranial parasympathetic pathways are secretory glands associated with the eye (tears), mouth (saliva) and nose (mucus). They stimulate the secretion of watery fluid, often with a concomitant vasodilation. Parasympathetic pathways also have a critical role in focusing the eye and regulating pupil diameter.

Blood vessels in the brain also receive a parasympathetic vasodilator innervation, but the actual physiological function of these nerves is not well understood.

The vagus nerve innervates microganglia in the neck, thorax and abdomen, including the airways, heart, thyroid, pancreas, gall bladder and the upper gastrointestinal tract. Consequently, the vagus nerve has a vast array of actions. It alters resistance to airflow and increases mucus secretion from the upper respiratory tract; it slows the heart; it stimulates secretion of digestive enzymes and bicarbonate from the pancreas; it either increases or decreases both secretory activity and smooth muscle contractility in the stomach.

Some parasympathetic pathways tend to be tonically active (eg vagal pathways that keep heart rate low when we are not exercising) whereas others are activated only when required, eg salivary secretion during eating; relaxation of gastric smooth muscle; or near focus of the eyes when reading.

Sympathetic Pathways

Neurons of the sympathetic division of the autonomic nervous system are aggregated into two main collections of ganglia: the paravertebral ganglia, which form the sympathetic chain each side of the vertebral column, and the prevertebral ganglia lying around the origins of the coeliac and mesenteric arteries (Figure 1). Sympathetic neurons project to most tissues of the body, commonly reaching them by traveling with major nerves containing predominantly sensory and somatic motor nerve fibers.

Sympathetic pathways have a diverse range of activities. Many are active nearly all the time, eg, vasoconstrictor pathways to the muscles that maintain central blood pressure, vasoconstrictor pathways to the skin that help prevent excessive heat loss, or prevertebral pathways to the gastrointestinal tract that help prevent excessive water loss from the gut. Other sympathetic pathways are activated only on demand, eg those to that increase heart rate during exercise; sudomotor neurons stimulating sweating during high body temperature; or those stimulating ejaculation during sexual activity. In some circumstances, sympathetic and parasympathetic pathways to a target tissue are co-activated eg sympathetic pathways to the salivary glands are co-activated with parasympathetic pathways when we eat something potentially noxious, such as hot chillies. The sympathetic co-activation results in the production of a thicker, more viscous saliva.

Sympathetic pathways normally are never activated all at once. Despite the widespread belief that they are only activated during stressful situations, on-going activity of specific sympathetic pathways are essential for our day-to-day health and well-being. Even when we are faced with extreme stress, only a subset of sympathetic pathways will be involved.

Pelvic Autonomic Pathways

Regulation of the activity of many pelvic organs requires coordinated control via both sympathetic and sacral parasympathetic pathways, often in association with the relevant somatic motor pathways. Indeed, many of the ganglia in pelvic pathways contain mixtures of neurons, some of which receive preganglionic inputs from lumbar spinal levels (by definition, sympathetic) and others of which receive preganglionic input from sacral spinal levels (by definition, parasympathetic). Some individual neurons receive convergent inputs from both lumbar and sacral preganglionic neurons, and there may be considered to lie in both sympathetic and parasympathetic pathways.

Control of bladder function requires sympathetic activity to relax the bladder wall and combined sympathetic and somatic motor activity to keep sphincters closed during continence. In contrast, micturition (urination) involves parasympathetic activation to contract the bladder wall and relax the sphincters, along with somatic motor pathways to increase intra-abdominal pressure. During sexual activity, erection requires coordinated activity of parasympathetic and somatic pathways, whilst ejaculation is the result of coordinated sympathetic and somatic motor activity.

Brain and Spinal Cord Pathways Regulating Autonomic Outflow

Preganglionic neurons for parasympathetic and sympathetic autonomic outflow are located in the brainstem and in thoracic, upper lumbar and sacral regions of the spinal cord (Figure 1). Several different brain centres control these preganglionic neurons. For the sympathetic outflow, brain regions containing premotor neurons include medulla oblongata, pons and hypothalamus. Many of these premotor neurons synthesize a monoamine (noradrenaline, adrenaline, dopamine or serotonin). For parasympathetic outflows, premotor neurons occur mainly in the brainstem and hypothalamus. The premotor neurons themselves are controlled by inputs from diverse regions of the brain, including other regions of the brainstem and hypothalamus, the amygdala, basal ganglia, anterior cingulate cortex, insular cortex, visual centres, and pre-frontal cortical centres involved in emotional processing, for example.

Afferent Inputs to Autonomic Pathways

Nearly all neural communication from one viscera to another (eg from the gut or the lung to the heart) are mediated via afferent neurons with cell bodies in the dorsal root ganglia (near the spinal cord) or in the nodose and petrosal ganglia of the lower cranial nerves (located in the neck), as shown in Figure 1. These visceral afferent neurons have a central process that projects into the dorsal horn of the spinal cord or into afferent nuclei in the brainstem (eg the nucleus tractus solitarius in the dorsal medulla oblongata).

Langley initially expected to find afferent cell bodies in autonomic ganglia, with projections to other ganglia. He believed that activation of these "autonomic afferents" should lead to purely autonomic responses. However Langley's own careful work demonstrated that there were no such neurons.

Complex neuronal networks within and closely associated with the gastrointestinal tract regulate digestive, absorptive and excretory functions. This enteric nervous system is structurally and functionally organized into afferent neurons, interneurons and motoneurons, with characteristic projections and neurochemical profiles. There are some projections from afferent cell bodies within the enteric nervous system to neurons in autonomic ganglia that project back to the gut, but projections to other parts of the autonomic nervous system are sparse or absent.

Thus, rather than "autonomic afferents" (or sympathetic or parasympathetic afferents) we prefer the term "visceral afferents". The fundamentally important point is that integrative processes responsible for the organization of visceral function occur principally within the central nervous system (brain and/or spinal cord). Both somatic and visceral afferents result in complex, brain mediated, responses that include somatic and visceral function. Autonomic motor activity can be generated by both somatic and visceral inputs to the CNS, and visceral inputs to the CNS initiate responses that are both somatic and autonomic. Natural bodily functioning does not include "purely autonomic" or "purely somatic" responses, just as it does not include 'purely sympathetic" or "purely parasympathetic" responses. The best way to illustrate this idea is by examples.

Nociceptive Visceral Afferents (Pain from Internal Organs)

Probably all the viscera are innervated by the unmyelinated axons of dorsal root ganglia neurons that respond to a range of noxious stimuli, such as tissue inflammation, low pH, or ischaemia. When activated, these pain afferents produce a conscious perception of pain reasonably localized to the organ. These visceral afferent neurons can result in sympathetically-mediated responses (eg increased blood pressure), but they also activate somatic motor activity, such as spasm of the facial muscles (grimacing), as well as the abdominal ("doubling over with pain") and the respiratory muscles (rapid breathing).

Baroreceptors and Chemoreceptors

The baroreceptors measure blood pressure via specialised sensory endings in the carotid arteries, just before they enter the skull. Changes in baroreceptor activity, via afferents in IX (glossopharyngeal) and X (vagal) cranial nerves, activate brain centres that lead to altered sympathetic motor outflow to the heart and blood vessels. This response helps to maintain blood flow to the brain under a wide range of circumstances. We have little conscious awareness of these actions unless they fail to work properly, as when we feel lightheaded after standing up too quickly.

Other specialized receptors (chemoreceptors) in the carotid sinus signal alterations in blood oxygen levels to the brain. As well as changes in blood pressure and heart rate, responses to low blood oxygen levels include increased breathing, and moving the head and face to clear the airway. Thus medical and nursing staff caring for babies have a rule: "the restless infant is hypoxic until proven otherwise".

Control of Accommodation and Pupil Diameter

Accommodation refers to the ability of the eye to focus on nearby objects by changing the shape of the lens. This is a parasympathetic motor function that is largely under conscious control, with sensory input arising from the visual system. Changes in pupil diameter regulate the amount of light reaching the retina and allow the eye to adapt to varying levels of ambient light. Pupil diameter is regulated by a combination of parasympathetic and sympathetic innervation of smooth muscle in the iris, in response to the global level of incident light. The overall level of illumination is detected by a special set of photosensitive ganglion cells in the retina. Thus a "somatic" stimulus causes an "autonomic response". If the light is extra bright we may also screw up our eyelids (squint), and this is a "somatic" response.

Tears in the Eyes

If we are sad or upset or, perhaps, incredibly relieved or deliriously happy, we may cry. Lacrimation, the production of tears, is mediated by purely parasympathetic motor activity. Normally there is a low level of tear production that lubricates the eye when we blink. Lacrimation also occurs in response to mechanical irritation of the eye (eg a grain of sand) or chemical irritation (eg a squirt of lemon juice). We also may "cry" following noxious mechanical stimulation of the face (eg a whack across the bridge of the nose). A psychological visual stimulus, for example a sad scene in a movie, may also initiate crying. In a heightened emotional state, or after particular kinds of strokes, we may cry in the absence of any immediate external stimulus. In all of these situations, the increased parasympathetic activity may be accompanied by characteristic patterns of somatic motor activity such as vocalisations (eg wailing) and facial expressions.

Auditory System Input to Cardiovascular System and Cutaneous Thermoregulators

Many types of auditory input can activate sympathetic output to the heart and blood vessels. A sudden unexpected sound may cause an increase in heart rate and vasoconstriction in the skin (we go pale with fright). Alternatively, music with special emotional resonance may "send shivers down our spines" and give us "goosebumps". Goosebumps are generated by sympathetic activation of special smooth muscles associated with each hair follicle, an evolutionary remnant from a time when we presumably possessed a much more luxuriant pelage.

Indeed, if we really do need to raise our body temperature, either because the environment is cold (detected by cutaneous thermoreceptors) or because we have a fever, generated from the thermoregulatory areas of the hypothalamus, we will shiver (a somatic motor response) and reduce blood flow to the skin (a sympathetic response).

Sexual Activity

Sexual activity requires coordinated motor activity of parasympathetic, sympathetic and somatic motor pathways. In males, erection is maintained mostly by parasympathetic activity, whilst ejaculation is controlled mostly by sympathetic activity. In both these components, somatic motor activity is required to control muscles of the pelvic floor and the external sphincters, for example, as well as all the various body movements involved in intercourse. As is well known, erection can be elicited either by appropriate cutaneous mechanical stimulation, which activates a special set of cutaneous mechanoreceptors in genital skin, or by psychogenic means.

Feeling Nervous

One of the best known but most misinterpreted autonomic motor patterns is the response to stress. Typically this involves an increase in sympathetic activity in selected pathways, such as those to the cardiovascular system, producing increased heart rate, skin blanching and perhaps high blood pressure, as well as an increased sympathetic output to the sweat glands, of the face, armpits and hands. This pattern of autonomic output is psychogenic (i.e. "brainogenic") in origin, even if triggered by visual, auditory or tactile somatic inputs: is that a spider crawling up the back of my neck?

Feeling Sick

The archetypal "visceral afferents" are those arising from the gastro-intestinal tract. Different functional classes of these afferent nerves respond to distention of the gut; or to changes in the contents of the gut, Yet others respond to inflammation or damage to the gut wall. Motor outputs from the brain to the gut utilize parasympathetic or sympathetic pathways. With food poisoning, activation of gut afferents generates autonomic motor activity in addition to coordinated somatic motor activity. Vomiting involves activation of somatic motor pathways to the pharyngeal and abdominal muscles. Autonomic pathways include those regulating contraction and relaxation of the stomach and oesophagus, saliva secretion from the main salivary glands, and probably the cardiovascular system as well. Intriguingly, we can generate the same coordinated set of responses entirely from central pathways, such as when we view and emotionally disgusting event that literally "makes us sick", or if we are "sick with worry".

Internal References

Braitenberg, V (2007). Brain. *Scholarpedia* 2(11): 2918. http://www.scholarpedia.org/article/Brain
.
Burke, R E (2008). Spinal cord. *Scholarpedia* 3(4): 1925. http://www.scholarpedia.org/article/Spinal_cord.
Corringer, P-J and Changeux, J-P (2008). Nicotinic acetylcholine receptors. *Scholarpedia* 3(1): 3468. http://www.scholarpedia.org/article/Nicotinic_acetylcholine_receptors.
Dowling, J (2007). Retina. *Scholarpedia* 2(12): 3487. http://www.scholarpedia.org/article/Retina.
Furness, J B (2007). Enteric nervous system. *Scholarpedia* 2(10): 4064. http://www.scholarpedia.org/article/Enteric_nervous_system.
Myers, D (2007). Psychology of happiness. *Scholarpedia* 2(8): 3149. http://www.scholarpedia.org/article/Psychology_of_happiness.
Schüz, A (2008). Neuroanatomy. *Scholarpedia* 3(3): 3158. http://www.scholarpedia.org/article/Neuroanatomy.

External References

Ackerknecht, E H (1974). The history of the discovery of the vegetative (autonomic) nervous system. *Medical History* 18: 1–8.
Blessing, W W (1997). *The Lower Brainstem and Bodily Homeostasis*. New York: Oxford University Press.
Furness, J B (2006a). *Enteric Nervous System*. Oxford: Blackwell Publishing.

Furness, J B (2006b). The organisation of the autonomic nervous system: Peripheral connections. *Autonomic Neuroscience* 130: 1–5.
Gibbins, I L (2012). Peripheral autonomic pathways. In: G Paxinos and J K Mai (Eds.), *The Human Nervous System, Third edition* (pp. 141–185). Amsterdam: Academic Press.
Jänig, W W (2006). *The Integrative Action of the Autonomic Nervous System: Neurobiology of Homeostasis*. Cambridge: Cambridge University Press.
Langley, J N (1903). The autonomic nervous system. *Brain* 26: 1–26.
Loewy, A D and Spyer, K M (1990). *Central Regulation of Autonomic Function*. New York: Oxford University Press.

Imaging Human Touch

Philip Servos

Since the early 20th century researchers have used techniques to visualize the neural organization of the human touch system. **Imaging human touch** covers research from the earliest known work involving electrical stimulation mapping to more recent techniques such as event-related potentials, magnetoencephalography, positron emission tomography, and functional magnetic resonance imaging. Topics include somatotopy, the cortical processing of perceptual attributes such as shape, texture, and hardness, the neural bases of tactile illusions as well as the time course of somatosensory processing.

Early Work Localizing Touch to the Postcentral Gyrus

As with other sensory modalities early work in the 1870s investigating the cortical representation of human touch focused on the effects of brain damage (typically tumours and vascular accidents). Much debate centered on the degree to which the pre- and post-Rolandic gyri (respectively, the precentral gyrus and postcentral gyrus) were involved in the cutaneous senses (Dana 1888).

Electrical Stimulation Mapping

The Work of Harvey Cushing

Subsequent work (in particular that of Harvey Cushing) at the turn of the 20th century localized the cutaneous senses to the postcentral gyrus. The electrical stimulation studies in humans by Harvey Cushing ushered in the modern era of mapping out the human touch system (Cushing 1909) (Figure 1).

P. Servos (✉)
Wilfrid Laurier University, Waterloo, ON, Canada

T.J. Prescott et al. (eds.), *Scholarpedia of Touch*, Scholarpedia,
DOI 10.2991/978-94-6239-133-8_38

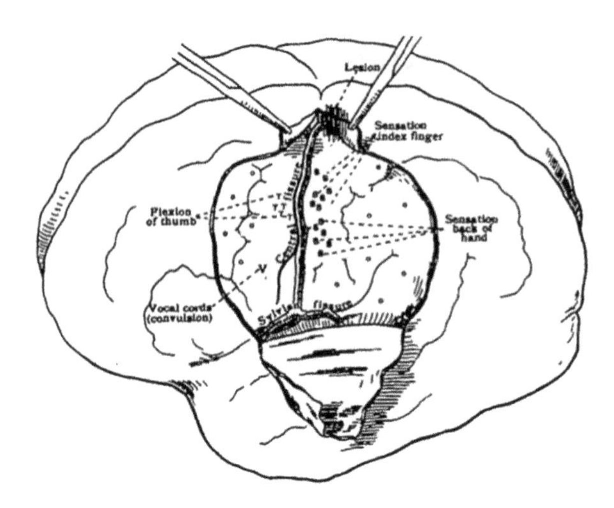

Figure 1 Diagram of stimulation sites along the central sulcus that resulted in cutaneous sensations published by Cushing in 1909 (Figure 2). Region to the *right* of the 'central fissure' represents the postcentral gyrus

The Work of Wilder Penfield

Cushing's methods were refined in the 1930s at the Montreal Neurological Institute by Wilder Penfield (Penfield and Boldrey 1937). By this point Brodmann's areas were commonly used to describe cortical regions of the brain and Brodmann's areas 3b, 1, and 2—all within the postcentral gyrus—were collectively known as primary somatosensory cortex (S1). Penfield did not attempt to differentiate between these subregions within S1. Working mainly with epileptic patients in whom pre-surgical explorations were being made Penfield used the electrical stimulation mapping technique to map out many aspects of brain function including the touch system. Small sterile tags were used to keep track of the stimulation points on the surface of the brain (see Figures 2, 3). Penfield's work on the cortical representation of human touch is widely known, in part for his much reproduced somatosensory homunculus.

An additional insight from Penfield's work is that there might be multiple somatotopic maps for certain regions of the body. For example, he found evidence for two separate cortical representations of the hand (Penfield and Rasmussen 1950). One of these, area S1, was located in the postcentral gyrus whereas the other representation, was located in the upper bank of the Sylvian fissure. Much work since then has confirmed the role in touch of the secondary somatosensory cortex (S2) located in the parietal operculum (Brodmann's areas 40 and 43).

There were, however, limitations to Penfield's methods. He examined primarily epileptic patients—it is possible the cortical organization of such patients may differ from that of neurologically intact research subjects (Maegaki et al. 1995; Maldjian et al. 1996; Weiller et al. 1993). Consider too the possibility of the spread of charge in the electrical stimulation mapping method and that the method was restricted to the cortical surface. A large proportion, probably 30–40 % (Geyer et al. 1999, 2000) of primary somatosensory cortex, including the entire extent of area 3b

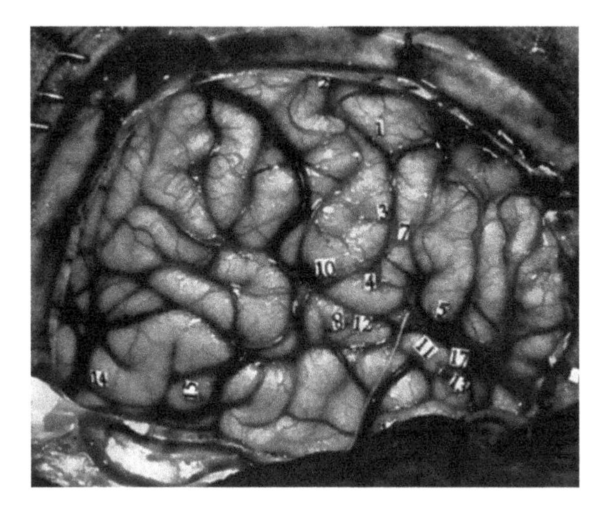

Figure 2 Lateral view of exposed cortex—anterior regions to *right* of image—posterior regions to *left*. The superior portion of the postcentral gyrus is indicated by tag '2' and tag '3' corresponds to a relatively inferior region of the postcentral gyrus. Tag '1' corresponds to a superior portion of the precentral gyrus and tag '7' corresponds to a relatively inferior region of the precentral gyrus. Figure 11 from Penfield (1958)

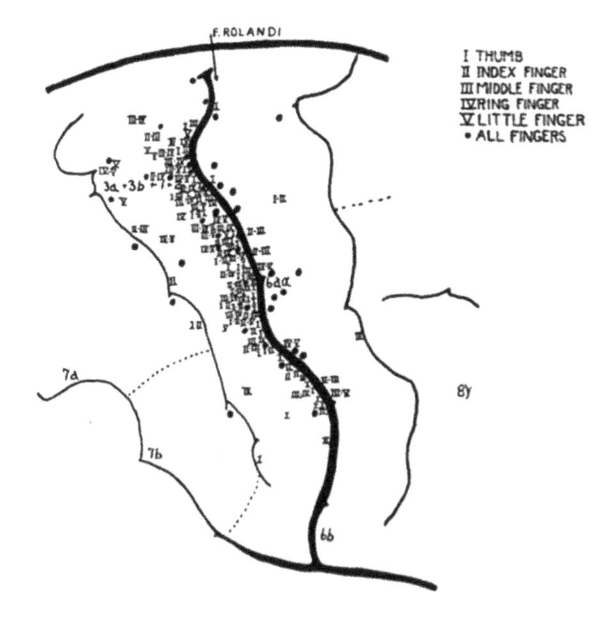

Figure 3 Diagram from Penfield and Boldrey (1937) depicting the overwhelming number of stimulation sites located within the postcentral gyrus (as opposed to the precentral gyrus) that led to cutaneous sensations. Note the relative overlap of the digits at the various stimulation sites

(Geyer et al. 1999, 2000; White et al. 1997) is located in the fundus and posterior bank of the central sulcus. The presence of blood vessels within the most superior aspects of the central sulcus would not have allowed Penfield to stimulate these regions of the postcentral gyrus. Nevertheless, much has been learned from this work and Woolsey and colleagues have subsequently confirmed these findings and extended them by not only using electrical stimulation mapping techniques on exposed cortex of humans but also recording the electrical activity from these regions (Woolsey et al. 1979).

Noninvasive Methods of Somatotopic Mapping

Positron Emission Tomography (PET)

The advent of non-invasive imaging techniques in humans in the 1980s—positron emission tomography (PET) and the 1990s—functional magnetic resonance imaging (fMRI)—allowed for further advances in mapping human touch. Much of the human neuroimaging work has been inspired by discoveries of the neural bases of touch in the non-human primate brain by Jon Kaas and his colleagues (Kaas et al. 1979). Perhaps the most telling finding from this work is that four different cortical maps of the body surface exist in the primate—in a rostral-caudal gradient (from areas 3a, 3b, 1, and 2). At least two of these maps (3b and 1) involve cutaneous receptors and would play a critical role in human touch.

The earliest PET work imaging human touch was published in 1978 (Reivich et al. 1978). Early PET studies involved stroking of the hand with a brush. Increased PET signal occurred in what appeared to be the contralateral post-central gyrus (Greenberg et al. 1981; Reivich et al. 1979). Subsequent work in the 1980s refined this work and showed in greater detail the activation produced by contralateral hand stimulation (Fox et al. 1986, 1987; Hagen and Pardo 2002) and when subjects haptically identified textured tiles (Ginsberg et al. 1987). Distinct regions within the contralateral postcentral gyrus were also identified for the lips, fingers, and toes (Fox et al. 1987). Stimulation of the fingers also produced activation within posterior parietal regions (Hagen and Pardo 2002) as well as within S2 both contralaterally and ipsilaterally (Burton et al. 1993; Hagen and Pardo 2002).

Functional Magnetic Resonance Imaging (FMRI)

Finger Somatotopy

On/Off Paradigm

With the increasing popularity of fMRI in the 1990s, due in part to its higher spatial resolution (PET voxels typically in the 1–1.5 cubic cm range as opposed to fMRI voxels in the 5 cubic mm range), the field shifted away from PET studies of touch.

fMRI work has focused primarily on the somatotopic organization of the arm (Servos et al. 1995, 1998), face (Eickoff et al. 2007, 2008; Huang and Sereno 2007; Iannetti et al. 2003; Kopietz et al. 2009; Moulton et al. 2009; Servos et al. 1999) and with a particular emphasis on the fingers (Blankenburg et al. 2003; Kurth et al. 1998; Martuzzi et al. 2014; Overduin and Servos 2004, 2008; Polonara et al. 1999; Puce et al. 1995) or the tips of the fingers (Francis et al. 2000; Gelnar et al. 1998; Maldjian et al. 1999; McGlone et al. 2002; Nelson et al. 2008; Sakai et al. 1995; Schweizer et al. 2008; Stippich et al. 1999).

The bulk of these studies involving the fingers have consistently shown activation within contralateral S1 (Eickoff et al. 2008; Francis et al. 2000; Gelnar et al. 1998; Jang et al. 2013; Kurth et al. 1998; Maldjian et al. 1999; Martuzzi et al. 2014; McGlone et al. 2002; Moore et al. 2000; Nelson et al. 2008; Overduin and Servos 2004, 2008; Polonara et al. 1999; Puce et al. 1995; Sakai et al. 1995; Schweizer et al. 2008; Stippich et al. 1999), and S2 (Burton et al. 2008; Eickoff et al. 2007, 2008; Francis et al. 2000; Gelnar et al. 1998; Jang et al. 2013; McGlone et al. 2002; Polonara et al. 1999) whenever the fingers or palm are stimulated.

The relatively high spatial resolution made possible by fMRI has allowed some researchers to identify somatotopic representations within subregions of S1 such as area 3b (Eickoff et al. 2008; Gelnar et al. 1998; Kurth et al. 1998; Martuzzi et al. 2014; McGlone et al. 2002; Moore et al. 2000; Nelson et al. 2008; Overduin and Servos 2004), area 1 (Eickoff et al. 2008; Gelnar et al. 1998; Kurth et al. 1998; Martuzzi et al. 2014; McGlone et al. 2002; Moore et al. 2000; Nelson et al. 2008; Overduin and Servos 2004), area 2 (Eickoff et al. 2008; Gelnar et al 1998; Kurth et al. 1998; Martuzzi et al. 2014; McGlone et al. 2002; Moore et al. 2000; Nelson et al. 2008), and area 3a (Gelnar et al. 1998; McGlone et al. 2002; Moore et al. 2000). Moreover, progress has been made in delineating four possible subregions within S2 namely OP1, OP2, OP3, and OP4 (Burton et al. 2008; Eickoff et al. 2007, 2008, 2010). Three of these subregions, OP1, OP3, and OP4, have somatosensory functions and correspond respectively to macaque S2, VS, and PV whereas OP2 appears to be involved in vestibular processing (Eickoff et al. 2007). The subregions sensitive to somatosensory stimuli appear to differ in their patterns of cortical connectivity. Eickhoff and colleagues examined the cortical connectivity of OP1 and OP4 using diffusion tensor imaging. Relative to OP4, OP1 was found to have stronger connectivity with parietal regions (presumably involved in tactile perception) whereas OP4 had stronger connections to frontal motor areas—regions presumably involved in sensorimotor processing and action control (Eickhoff et al. 2010) (Figure 4).

Parietal regions caudal to S1 have also been implicated (Francis et al. 2000; Gelnar et al. 1998; Jang et al. 2013; Kurth et al. 1998; McGlone et al. 2002). Some work has also demonstrated the role of ipsilateral somatosensory cortex in processing tactile inputs from the fingers or hand. Ipsilateral activations have been observed in Brodmann's areas 1 and 2 as well as S2 (Burton et al. 2008; Eickoff et al. 2008; Kurth et al. 1998).

fMRI work (Maldjian et al. 1999; Martuzzi et al. 2014; McGlone et al. 2002; Nelson et al. 2008; Schweizer et al. 2008) has also confirmed the mediolateral

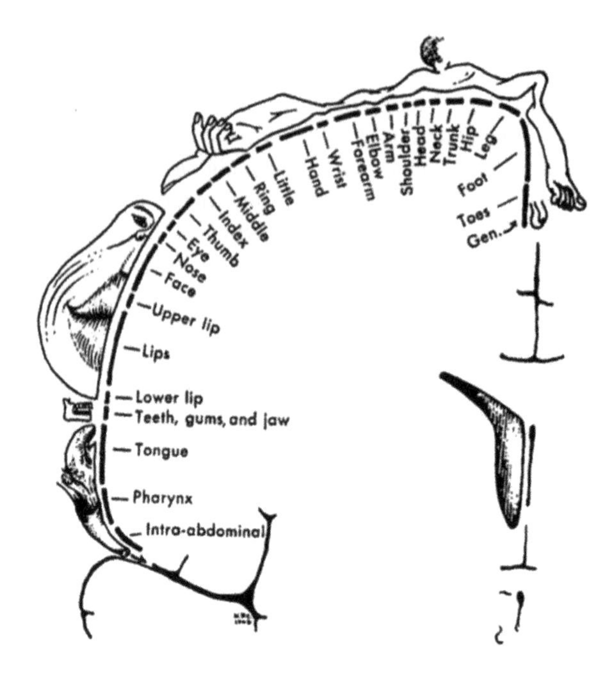

Figure 4 Somatosensory Homunculus as depicted in Figure 17 of the 1950 Penfield and Rasmussen book, The Cerebral Cortex of Man

ordering of the digit representations from little finger to thumb along the postcentral gyrus as originally observed with electrical stimulation mapping (Penfield and Boldrey 1937) and recording from surgically exposed cortex (Woolsey et al. 1979). It should be noted that earlier MEG work also demonstrated this somatotopic ordering of the digits (Baumgartner et al. 1991) [although it is not entirely clear given the limits of the spatial localization of MEG if the activation was specifically localized to the postcentral gyrus] and a more recent optical imaging study as well (Sato et al. 2005).

Phase Paradigm

In most of the above-mentioned studies, the BOLD (blood oxygenation level dependent) signal that occurs as a result of cutaneous stimulation is typically compared to a control condition in which no cutaneous stimulation occurs allowing investigators to delineate a somatosensory organization on the basis of distinct activation 'hotspots' in response to the stimulation. Another approach is to use a sliding window of stimulation and then analyze the phase characteristics of the BOLD signal. This technique was pioneered for retinotopic mapping studies of visual cortex (Engel et al. 1994; Sereno et al. 1995) and has subsequently been applied to the touch system (Besle et al. 2013; Huang et al. 2007; Overduin and Servos 2004, 2008; Sanchez-Panchuelo et al. 2010, 2012; Servos et al. 1998). The basic principle of this technique (using an example in which a single finger is stimulated) is that the phase lag of the BOLD signal should be relatively short for cortical regions representing portions of the finger that were stimulated early on in the stimulus paradigm (e.g., the distal pad of the finger) whereas the phase lag of the BOLD signal should be relatively long for cortical regions representing portions of the finger that were stimulated later in the paradigm (e.g., the more proximal portions of the finger).

One advantage of the phase-delay technique is that it is a much more efficient method to investigate the relative position of the finger within its cortical representation or the relative position of each finger within for example, S1. Such investigations involving on-off paradigms would require an unmanageable number of on/off experiments at many different locations on the fingers.

Early work imaging the arm representation in S1 determined the relative position of the arm within its cortical representation i.e., the relative position of the distal and proximal portions of the arm (Servos et al. 1998). The relative position of the index finger (i.e., its distal and proximal aspects) in areas 3b and 1 has also been imaged using the phase-delay technique (Blankenburg et al. 2003).

The phase mapping technique has also been used to confirm the mediolateral ordering of the digit representations from little finger to thumb in S1 (Overduin and Servos 2004). Later studies have refined these maps (Besle et al. 2013; Huang et al. 2007; Sanchez-Panchuelo et al. 2010).

Some fMRI work involving phase mapping has demonstrated a rostral-caudal gradient in area 3b for the distal-to-proximal portions of the finger and a reversal of this gradient in area 1 (Blankenburg et al. 2003). As promising as these results are they are a bit puzzling because one would have expected the gradient in 3b to be in the caudal-rostral direction given the neurophysiological recordings made in monkey S1 (Kaas et al. 1979). More recent fMRI work has demonstrated the expected caudal-rostral gradient in area 3b for the distal-to-proximal portions of the finger. Moreover, the phase reversals between adjacent regions within S1 that are predicted on the basis of unit recording work in the monkey (Kaas et al. 1979) were also confirmed for areas 3a, 3b, 1, and 2 (Sanchez-Panchuelo et al. 2012).

In the non-human primate somatosensory cells sensitive to cutaneous inputs within the postcentral gyrus are known to project to the precentral gyrus (Strick et al. 1978). Consistent with this, quite a few human neuroimaging studies of somatosensation have observed activation within regions in the precentral gyrus (Besle et al. 2013; Francis et al. 2000; Huang et al. 2007; McGlone et al. 2002; Moore et al. 2000; Overduin and Servos 2008). Some fMRI evidence based on the phase mapping technique involving the finger representations in S1 suggests that there may be a relatively high degree of symmetry between the digit representations in S1 and corresponding regions in primary motor cortex, that is, in the precentral gyrus (Overduin and Servos 2008) (Figure 5).

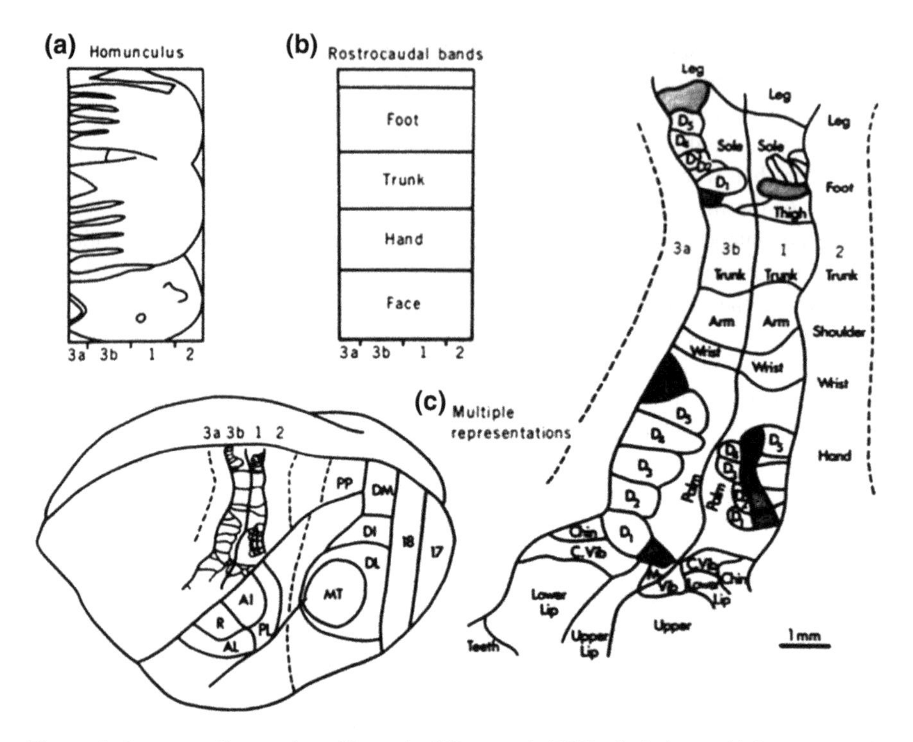

Figure 5 Summary diagram from Figure 1 of Kaas et al. (1979) depicting multiple somatotopic maps in S1 of the owl monkey

Face Somatotopy

Some fMRI work involving the somatotopic organization of the face representation has challenged the original Penfield work depicting the face as being organized in a superior-to-inferior direction along the postcentral gyrus (Penfield and Rasmussen 1950). Servos et al. (1999) provided evidence that the face representation is in an inverted position relative to the postcentral gyrus consistent with the orientation of the face in S1 in the monkey (Kaas et al. 1979). Earlier MEG work also suggests that the face representation is inverted (Yang et al. 1993). However, other research involving human intra-operative optical imaging studies that electrically stimulated different parts of the face (Sato et al. 2005; Schwartz et al. 2004) as well as an fMRI study (Huang and Sereno 2007) have supported to some degree the original Penfield work. Some fMRI studies have suggested other organizational possibilities for the face in S1, such as a representation featuring spatial overlap (Iannetti et al. 2003; Kopietz et al. 2009; Nguyen et al. 2004 (note that this is a MEG study)) or a segmented 'onion-like' organization that follows the three branches of the trigeminal nerve (Moulton et al. 2009).

Non-human primate work is consistent with the latter studies suggesting a relative degree of spatial overlap for the various portions of the face representation in

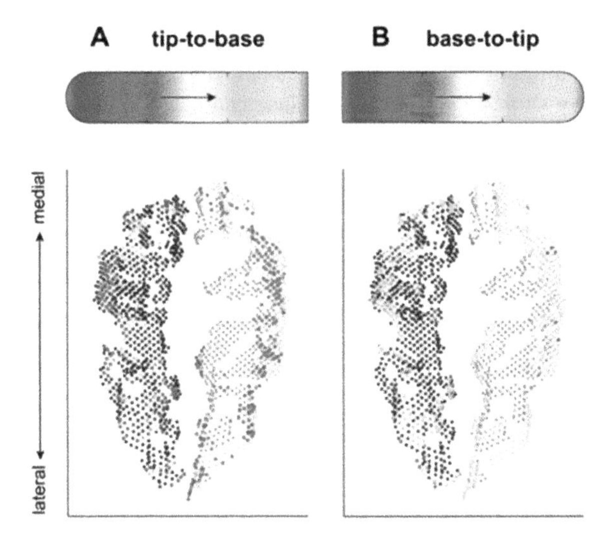

Figure 6 Activation tuning and symmetry of the index finger in flattened representations of pericentral cortex. In each plot the cortex is shown as a 2D surface, oriented in approximate rostrocaudal and lateromedial directions. The *plot* shows functional data, representing phase lag according to the reference bar at *top*, superimposed over M1 (rostral—*left portion* of each plot) and S1 (caudal—*right portion* of each plot). Figure 4 from Overduin and Servos (2008)

human S1. These studies suggest that there is a relatively complex, interdigitated organization of the face representation in S1 not only in the squirrel monkey (Manger et al. 1995) but also in the macaque (Manger et al. 1996). Interestingly, these latter two studies provide some evidence for a posterior to anterior gradient of the face representation with more superior portions of the face corresponding to more posterior regions in area 3b and more inferior portions of the face corresponding to more anterior regions—rather than the superior inferior gradient for respectively the more inferior and superior portions of the face in the owl monkey as described by Kaas et al. (1979) (Figure 6).

Other Factors Affecting Somatotopy

In addition to the standard methods of stimulating the touch system to investigate somatotopic organization (e.g., brush strokes, piezoelectric actuators, air jets, and electrical stimulation) other factors have been investigated namely the interpersonal aspects of the tactile stimulus and the effects of disuse.

Gazzola et al. (2012) observed enhanced activation in S1 when the legs of male subjects were manually caressed by a female relative to when a male did the caressing. Malinen et al. (2014) administered three types of touch to a subject's hand: poking, holding, and a smoothing movement. The latter two stimuli were considered pleasant as compared to the poking stimulus. Responses within area 3b differentiated between the three types of stimuli although the pleasantness rating of the various stimuli did not correlate with the characteristics of the area 3b response.

Even relatively short periods of limb disuse can have an effect on the nature of the somatotopic maps in somatosensory cortex. Lissek et al. (2009) studied individuals who had their limbs immobilized for around 6 weeks (their arms were in a cast as a consequence of bone fracture). Following removal of the cast the subjects' tactile sensitivity of the fingers was lower than control subjects. Moreover, the BOLD activation within the somatotopic maps corresponding to the fingers was also decreased. These effects persisted for 2–3 weeks.

Mapping Studies of Touch-Related Perception and Cognition

Tactile Perception of Shape, Texture, and Hardness

In addition to the above-mentioned PET and fMRI studies examining the cortical representation of the body surface, these techniques have also been used to examine the perceptual and cognitive aspects of touch.

PET work by O'Sullivan et al. (1994) provide some of the earliest work dissociating cortical regions for the processing of shape (in this instance, length) and roughness. Roughness discriminations activated primarily the contralateral postcentral gyrus whereas the length discrimination task not only activated the contralateral postcentral gyrus but also contralateral and ipsilateral regions within the angular and supramarginal gyri. Subsequent work, however, suggested that shape and length discriminations activate a region within the intraparietal sulcus (IPS) whereas roughness discrimination activates the parietal operculum (Eck et al. 2013; Roland et al. 1998; Sathian et al. 2011; Stilla and Sathian 2008). Other work (Bodegård et al. 2001) suggests a hierarchical processing of object features such that tactile motion, length, curvature and roughness are processed within areas 3b and 1, curvature (see also Bodegård et al. 2000) and shape discrimination activate area 2, and the IPS and supramarginal gyrus are activated by shape discrimination (Bodegård et al. 2001). Additional work by Servos et al. (2001) observed contralateral postcentral gyrus activation following haptic texture, hardness and shape discriminations and an additional region within the parietal operculum during hardness discriminations. A more recent study implicates areas 3b, 1, and 2 in roughness perception (Carey et al. 2008). It should be noted that unlike the previously described studies involving discrimination tasks this study involved a simple stimulation task so direct comparisons with these other studies are problematic.

As is evident from the above-cited studies multiple cortical regions appear to be involved in the processing of haptic shape, texture, and hardness. However, currently there is not a clear consensus about exactly which regions are specialized for a given process and which brain regions may play a role in the processing of multiple stimulus attributes.

A recurrent finding in the touch-related perceptual and cognitive imaging literature is that compared to simple tactile stimulation such as that used in most somatotopic mapping studies, when more complicated perceptual and cognitive processes are involved one observes activation not only in S1 and S2 but also in additional cortical regions not classically associated with somatosensation.

For example, haptic shape identification tasks activate the striate visual cortex (Deibert et al. 1999), the lateral occipital complex (LOC) (Deibert et al. 1999; Stilla and Sathian 2008) and inferior parietal cortex (Deibert et al. 1999; Stilla et al. 2007; Stilla and Sathian 2008). Yang et al. (1993) observed IPS and LOC activation when subjects performed a tactile angle discrimination task and visual cortex activity has been observed during tactile texture perception (Eck et al. 2013; Sathian et al. 2011; Stilla and Sathian 2008). Tactile localization and spatial acuity are associated with IPS activation (Sathian et al. 2011; Stilla et al. 2007). A further insight from the Stilla et al. study is that IPS activity predicted individual differences in tactile acuity. Additionally, Kilgour and colleagues observed activation within the fusiform gyrus and parahippocampal region during haptic exploration of face masks (Kilgour et al. 2005) and increased BOLD in the fusiform gyrus when subjects haptically scanned familiar faces as compared to unfamiliar faces (James et al. 2006). Taken together the quite common observation of neural activity within brain regions usually associated with visual processing during tactile perception tasks suggests that subjects may be using visual strategies such as visual imagery to help them solve such tasks.

Tactile Illusions

The tactile motion aftereffect has been investigated with fMRI (Planetta and Servos 2012). During the inducing tactile stimulation (a rotating ridged drum in contact with the volar skin of the hand) contralateral activity is observed in S1 and S2. Interestingly, during the period in which subjects experience the illusion i.e., when the drum is stationary and the subjects report feeling as if the drum is rotating in the opposite direction, the BOLD effect is only observed in the contralateral S1 not the contralateral S2.

The neural substrates of the cutaneous rabbit have also been studied. In this illusion a series of rapid taps are delivered to a particular skin location and then immediately delivered at another location 5-10 cm away. Subjects often feel illusory taps along the skin between the two locations as if a small rabbit were hopping along the skin (Geldard and Sherrick 1972). Blankenburg et al. (2006) used fMRI to identify the somatotopic representation of the forearm in S1 following tactile stimulation to three locations along the forearm. They then examined the neural response to the illusion following stimulation by the two lateral locations along the forearm. Remarkably, during the illusion the central portion of the arm representation was also activated even though tactile stimulation had been restricted to only the two lateral forearm locations.

Work examining tactile-visual illusions such as the rubber hand illusion (Botvinick et al. 1998) have implicated regions posterior to S1 in the IPS that play a role in binding the synchronous visual and tactile stimulation necessary for this illusion and a role of the premotor cortex in establishing body ownership of the rubber hand (Ehrsson et al. 2004). Other work involving the touch-induced visual illusion (Violentyev et al. A et al. 2005) has also implicated a role for the IPS as well as the lingual gyrus in binding the tactile and visual streams (Servos and Boyd 2012).

Tactile-Visual Interactions

The neural bases of tactile-visual interactions have also been investigated. In research involving cross-modal priming of haptically examined novel objects and visual images of those objects, regions previously considered to be largely visual in nature such as the LOC (Amedi et al. 2001; James et al. 2002) as well the striate cortex and lingual gyrus (James et al. 2002) are also active during haptic exploration of objects. Other work has investigated the relative contribution of non-tactile stimuli (in this case visual) to processing in somatosensory cortex. In a tactile-visual integration task Dionne et al. (2010) observed greater BOLD signals in S1 as compared to when the task involved only tactile stimuli. Consonant with this, Eck et al. (2013) presented subjects with either visual or tactile textures and observed a surprisingly high degree of activation overlap for both texture types in primary somatosensory and primary visual areas.

Mapping the Time Course of Touch: Event-Related Potentials (ERP) and Magnetoencephalography (MEG)

Although fMRI has excellent spatial resolution with many fMRI studies of touch acquiring images with in-plane resolutions of 2 mm or less, it is challenging to obtain second to sub-second temporal resolution using this method. Some questions, such as the time course of information flow within and between brain regions are better answered with techniques that have sub-second temporal resolution such as magnetoencephalography (MEG) and event related potentials (ERPs).

The bulk of ERP and MEG investigations of touch have focused on the fingers. Various ERP components (collectively known as somatosensory evoked potentials (SEPs)) or MEG components (collectively known as somatosensory evoked magnetic fields (SEFs)) associated with finger stimulation (either by direct stimulation of peripheral receptors or by median or ulnar nerve stimulation) have been investigated. The human ERP and MEG work on touch complement and parallel each other quite well so will be discussed jointly.

Early work by Stohr and Goldring on SEPs set the stage for subsequent research (Stohr and Goldring 1969). Our current understanding of the neural significance of the various SEPs and SEFs is as follows: after the onset of a tactile stimulus on the fingers several event-related components can be seen such as the N25 (Baumgartner et al. 1991; Hari et al. 1993; Schubert et al. 2006), P50, N80, P100, and N140 (Allison et al. 1989a, b; Eimer and Forster 2003; Forss et al. 1994; Hari et al. 1993). The P50 and N80 components are generated in contralateral S1 and are related to the processing of the physical attributes of the stimulus. About 100 ms after stimulus contact, additional cortical regions are activated such as S2 marked by a parietal P100 and frontal cortices marked by a frontal N140 (Allison et al. 1989b; Eimer and Forster 2003; Forster and Eimer 2004). Such mid-latency components are usually ascribed to processing within a frontoparietal network contributing to conscious perception and attention (Allison et al. 1992; Auksztulewicz et al. 2012; Forster and Eimer 2004; Schubert et al. 2006). Late components such as the P300 are typically related to novelty, change detection, and response inhibition (Auksztulewicz et al. 2013; Nakata et al. 2010) and are observed in contralateral S1, ipsilateral and contralateral S2 as well as frontal regions. Intriguingly, there is evidence suggesting that neural activity in the gamma range (30–70 Hz) may be phase locked between S1 and S2 for the early component SEFs (up to 100 ms post-stimulus onset) but not for the later components (Hagiwara et al. 2010).

SEP and SEF studies have also shed light on the nature of connectivity between S1 and S2. Consistent with neurophysiological work in the monkey time course analyses have demonstrated the flow of sensory signals from S1 to S2 in a serial fashion (Hu et al. 2012; Inui et al. 2004; Schnitzler et al. 1999).

ERP work investigating the neural bases of tactile illusions such as the Aristotle and Reverse illusions has demonstrated that S1 activity is associated with experience of the illusions whereas activity in more posterior parietal regions are associated with the instances in which subjects resist these illusions (Bufalari et al. 2014).

Finally, there is a growing ERP literature related to the active scanning of tactile patterns. Using a simple geometric shape discrimination task, Lucan et al. (2010) observed LOC activation about 150 ms after stimulus presentation (a timeline that is consistent with the time course of visual shape recognition). An ambitious recent ERP study by Adhikari et al.(2014) examined the time course of multiple cortical regions during a tactile acuity task. The earliest cortical signals arose in S1 at 45 ms post-stimulus with subsequent LOC activity at 130 ms, IPS activity at 160 ms, and dorsolateral prefrontal cortex (dlPFC) activity at 175 ms. Strikingly, between 130–175 ms after stimulus onset ERP signals on correct trials differed from those generated during incorrect trials. Adhikari and colleagues analyzed beta band activity (12–30 Hz) and identified a feedforward network involving all four regions.

Additionally, they analyzed gamma band activity (30–100 Hz) and identified a recurrent network involving S1, IPS, and dlPFC. Most interestingly, measures of network activity within both bands was correlated with accuracy of task performance.

Conclusion

With the development of non-invasive imaging tools such as PET and fMRI having a relatively high degree of spatial resolution and tools with high temporal resolution such as ERP and MEG much has been learned about the brain regions involved in human touch. Future work will no doubt delineate further the somatotopic maps involved in touch as well as shedding light on the network of brain regions and their timing relationships involved in higher-order touch processes such as tactile object recognition. New imaging techniques such as optical imaging (Habermehl et al. 2012) complement existing methods and inevitably will also move the field forward.

References

Adhikari, B M; Sathian, K; Epstein, C M; Lamichhane, B and Dhamala, M (2014). Oscillatory activity in neocortical networks during tactile discrimination near the limit of spatial acuity. *NeuroImage* 91: 300-310.

Allison, T; McCarthy, G and Wood, C C (1992). The relationship between human long-latency somatosensory evoked potentials recorded from the cortical surface and from the scalp. *Electroencephalography and Clinical Neurophysiology* 84: 301–314.

Allison, T et al. (1989a). Human cortical potentials evoked by stimulation of the median nerve. I. Cytoarchitectonic areas generating short-latency activity. *Journal of Neurophysiology* 62: 694–710.

Allison, T; McCarthy, G; Wood, C C; Williamson, P D and Spencer, D D (1989b). Human cortical potentials evoked by stimulation of the median nerve. I. Cytoarchitectonic areas generating long-latency activity. *Journal of Neurophysiology* 62: 711–722.

Amedi, A; Malach, R; Hendler, T; Peled, S and Zohary, E (2001). Visuo-haptic object-related activation in the ventral visual pathway. *Nature Neuroscience* 4: 324–330.

Auksztulewicz, R and Blankenburg, F (2013). Subjective rating of weak tactile stimuli is parametrically encoded in event-related potentials. *The Journal of Neuroscience* 33: 11878–11887.

Auksztulewicz, R; Spitzer, B and Blankenburg, F (2012). Recurrent neural processing and somatosensory awareness. *The Journal of Neuroscience* 32: 799–805.

Baumgartner, C et al. (1991). Neuromagnetic investigation of somatotopy of human hand somatosensory cortex. *Experimental Brain Research* 87: 641–648.

Besle, J; Sanchez-Panchuelo, R-M; Bowtell, R; Francis, S and Schluppeck, D (2013). Single-subject fMRI mapping at 7 T of the representation of fingertips in S1: a comparison of event-related and phase-encoding designs. *Journal of Neurophysiology* 109: 2293–2305.

Blankenburg, F; Ruben, J; Meyer, R; Schwiemann, J and Villringer, A (2003). Evidence for a rostral-to-caudal somatotopic organization in human primary somatosensory cortex with mirror-reversal in areas 3b and 1. *Cerebral Cortex* 13: 987–993.
Blankenburg, F; Ruff, C C; Deichmann, R; Rees, G and Driver, J (2006). The cutaneous rabbit illusion affects human primary sensory cortex somatopically. *PLoS Biology* 4: e69.
Bodegård, A; Geyer, S; Grefkes, C; Zilles, K and Roland, P E (2001). Hierarchical processing of tactile shape in the human brain. *Neuron* 31: 317–328.
Bodegård, A et al. (2000). Object shape differences reflected by somatosensory cortical activation in human. *The Journal of Neuroscience* 20(1): RC51, 1–5.
Botvinick, M and Cohen, J (1998). Rubber hands "feel" touch that eyes see. *Nature* 391: 756.
Bufalari, I; Di Russo, F and Aglioti, S M (2014). Illusory and veridical mapping of tactile objects in the primary somatosensory and posterior parietal cortex. *Cerebral Cortex* 24: 1867–1878.
Burton, H; Sinclair, R J; Wingert, J R and Dierker, D L (2008). Multiple parietal operculum subdivisions in humans: tactile activation maps. *Somatosensory & Motor Research* 25: 149–162.
Burton, H; Videen, T O and Raichle, M E (1993). Tactile-vibration-activated foci in insular and parietal-opercular cortex studied with positron emission tomography: mapping the second somatosensory area in humans. *Somatosensory & Motor Research* 10: 297–308.
Carey, L M; Abbott, D F; Egan, G F and Donnan, G A (2008). Reproducible activation in BA2, 1 and 3b associated with texture discrimination in healthy volunteers over time. *NeuroImage* 39: 40–51.
Cushing, H (1909). A note upon the Faradic stimulation of the postcentral gyrus in conscious patients. *Brain* 32: 44–53.
Dana, C T (1888). The cortical localization of cutaneous sensations. *Journal of Nervous and Mental Disease* 15: 650–684.
Deibert, E; Kraut, M; Kremen, S and Hart, J (1999). Neural pathways in tactile object recognition. *Neurology* 52: 1413–1417.
Dionne, J K; Meehan, S K; Legon, W and Staines, W R (2010). Crossmodal influences in somatosensory cortex: interaction of vision and touch. *Human Brain Mapping* 31: 14–25.
Eck, J; Kaas, A L and Goebel, R (2013). Crossmodal interactions of haptic and visual texture information in early sensory cortex. *NeuroImage* 75: 123–135.
Ehrsson, H; Spence, C and Passingham, R E (2004). That's my hand! Activity in premotor cortex reflects feeling of ownership of a limb. *Science* 305: 875–877.
Eickoff, S B; Grefkes, C; Fink, G R and Zilles, K (2008). Functional lateralization of face, hand, and trunk representation in anatomically defined human somatosensory areas. *Cerebral Cortex* 18: 2820–2830.
Eickoff, S B; Grefkes, C; Zilles, K and Fink, G R (2007). The somatotopic organization of cytoarchitectonic areas on the human parietal operculum. *Cerebral Cortex* 17: 1800–1811.
Eickhoff, S B et al. (2010). Anatomical and functional connectivity of cytoarchitectonic areas within the human parietal operculum. *The Journal of Neuroscience* 30: 6409–6421.
Eimer, M and Forster, B (2003). The spatial distribution of attentional selectivity in touch: evidence from somatosensory ERP components. *Clinical Neurophysiology* 114: 1298–1306.
Engel, S A et al. (1994). FMRI of human visual cortex. *Nature* 369: 525.
Forss, N et al. (1994). Activation of the human posterior parietal cortex by median nerve stimulation. *Experimental Brain Research* 99: 309–315.
Forster, B and Eimer, M (2004). The attentional selection of spatial and non-spatial attributes in touch: ERP evidence for parallel and independent processes. *Biological Psychology* 66: 1–20.
Fox, P T; Burton, H and Raichle, M (1987). Mapping human somatosensory cortex with positron emission tomography. *Journal of Neurosurgery* 67: 34–43.
Fox, P T and Raichle, M (1986). Focal physiological uncoupling of cerebral blood flow and oxidative metabolism during somatosensory stimulation in human subjects. *Proceedings of the National Academy of Sciences of the United States of America* 83: 1140–1144.
Francis, S T et al. (2000). fMRI of the responses to vibratory stimulation of digit tips. *NeuroImage* 11: 188–202.

Gazzola, V et al. (2012). Primary somatosensory cortex discriminates affective significance in social touch. *Proceedings of the National Academy of Sciences of the United States of America* 109: E1657-E1666.

Geldard, F A and Sherrick, C E (1972). The cutaneous "rabbit": A perceptual illusion. *Science* 178: 178–179.

Gelnar, P A; Krauss, B R; Szeverenyi, N M and Apkarian, A V (1998). Fingertip representation in the human somatosensory cortex: An fMRI study. *NeuroImage* 7: 261–283.

Geyer, S; Schleicher, A and Zilles, K (1999). Areas 3a, 3b, and 1 of human primary somatosensory cortex 1. Microstructural organization and interindividual variability. *NeuroImage* 10: 63–83.

Geyer, S; Schormann, T; Mohlberg, H and Zilles, K (2000). Areas 3a, 3b, and 1 of human primary somatosensory cortex 2. Spatial normalization to standard anatomical space. *NeuroImage* 11: 684–696.

Ginsberg, M D et al. (1987). Human task-specific somatosensory activation. *Neurology* 37: 1301–1308.

Greenberg, J H et al. (1981). Metabolic mapping of functional activity in human subjects with the [18F] fuorodeoxyglucose technique. *Science* 212: 678–680.

Habermehl, C et al. (2012). Somatosensory activation of two fingers can be discriminated with ultrahigh-density diffuse optical tomography. *NeuroImage* 59: 3201–3211.

Hagen, M C and Pardo, J V (2002). PET studies of somatosensory processing of light touch. *Behavioural Brain Research* 135: 133–140.

Hagiwara, K et al. (2010). Oscillatory gamma synchronization binds the primary and secondary somatosensory areas in humans. *NeuroImage* 51: 412–420.

Hari, R and Karhu, J (1993). Functional organization of the human first and second somatosensory cortices: A neuromagnetic study. *European Journal of Neuroscience* 5: 724–734.

Hu, L; Zhang, Z G and Hu, Y (2012). A time-varying source connectivity approach to reveal human somatosensory information processing. *NeuroImage* 62: 217–228.

Huang, R-S and Sereno, M I (2007). Dodecapus: An MR-compatible system for somatosensory stimulation. *NeuroImage* 34: 1060–1073.

Iannetti, G D et al. (2003). Representation of different trigeminal divisions within the primary and secondary human somatosensory cortex. *NeuroImage* 19: 906–912.

Inui, K; Wang, X; Tamura, Y; Kaneoke, Y and Kakigi, R (2004). Serial processing in the human somatosensory system. *Cerebral Cortex* 14: 851–857.

James, T W et al. (2002). Haptic study of three-dimensional objects activates extrastriate visual areas. *Neuropsychologia* 40: 1706–1714.

James, T W; Servos, P; Huh, E; Kilgour, A R and Lederman, S (2006). The influence of familiarity on brain activation during haptic exploration of facemasks. *Neuroscience Letters* 397: 269–273.

Jang, S; Seo, J; Ahn, S and Lee, M. (2013). Comparison of cortical activation patterns by somatosensory stimulation on the palm and dorsum of the hand. *Somatosensory & Motor Research* 30: 109–113.

Kaas, J H; Nelson, R J; Sur, M; Lin, C-S and Merzenich, M M (1979). Multiple representations of the body within the primary somatosensory cortex of primates. *Science* 204: 521–523.

Kilgour, A R; Kitada, R; Servos, P; James, T W and Lederman, S (2005). Haptic face identification activates ventral occipital and temporal areas: An fMRI study. *Brain and Cognition* 59: 246–257.

Kopietz, R et al. (2009). Activation of primary and secondary somatosensory regions following tactile stimulation of the face. *Clinical Neuroradiology* 19: 135–144.

Kurth, R et al. (1998). FMRI assessment of somatotopy in human Brodmann area 3b by electrical finger stimulation. *NeuroReport* 9: 207–212.

Lissek, S et al. (2009). Immobilization impairs tactile perception and shrinks somatosensory cortical maps. *Current Biology* 19: 837–842.

Lucan, J N; Foxe, J J; Gomez-Ramirez, M; Sathian, K and Molholm, S (2010). Tactile shape discrimination recruits human lateral occipital complex during early perceptual processing. *Human Brain Mapping* 31: 1813–1821.

Maegaki, Y; Yamamoto, T and Takeshita, K (1995). Plasticity of central motor and sensory pathways in a case of unilateral extensive cortical dysplasia: investigation of magnetic resonance imaging, transcranial magnetic stimulation, and short-latency somatosensory evoked potentials. *Neurology* 45: 2255–2261.

Maldjian, J et al. (1996). Functional magnetic resonance imaging of regional brain activity in patients with intracerebral arteriovenous malformations before surgical or endovascular therapy. *Journal of Neurosurgery* 84: 477–483.

Maldjian, J A; Gottschalk, A; Patel, R S; Detre, J A and Alsop, D C (1999). The sensory somatotopic map of the human hand demonstrated at 4 Tesla. *NeuroImage* 10: 55–62.

Malinen, S; Renvall, V and Hari, R (2014). Functional parcellation of the human primary somatosensory cortex to natural touch. *European Journal of Neuroscience* 39: 738–743.

Manger, P R; Woods, T M and Jones, E G (1995). Representation of the face and intraoral structures in area 3b of the squirrel monkey ("Saimiri sciureus") somatosensory cortex with special reference to the ipsilateral representation. *Journal of Comparative Neurology* 363: 597–607.

Manger, P R; Woods, T M and Jones, E G (1996). Representation of face and intra-oral structures in area 3b of Macaque monkey somatosensory cortex. *Journal of Comparative Neurology* 371: 513–521.

Martuzzi, R; van der Zwaag, W; Farthouat, J; Gruetter, R and Blanke, O (2014). Human finger somatotopy in areas 3b, 1, and 2: A 7T fMRI study using a natural stimulus. *Human Brain Mapping* 35: 213–226.

McGlone, F et al. (2002). Functional neuroimaging studies of human somatosensory cortex. *Behavioural Brain Research* 135: 147–158.

Moore, C I et al. (2000). Segregation of somatosensory activation in the human rolandic cortex using fMRI. *Journal of Neurophysiology* 84: 558–569.

Moulton, E A et al. (2009). Segmentally arranged somatotopy within the face representation of human primary somatosensory cortex. *Human Brain Mapping* 30: 757–765.

Nakata, H; Sakamoto, K and Kakigi, R (2010). Characteristics of No-go-P300 component during somatosensory Go/No-go paradigms. *Neuroscience Letters* 478: 124–127.

Nelson, A J and Chen, R (2008). Digit somatotopy within cortical areas of the postcentral gyrus in humans. *Cerebral Cortex* 18: 2341–2351.

Nguyen, B T; Tran, T D; Hoshiyama, M; Inui, K and Kakigi, R (2004). Face representation in the human primary somatosensory cortex. *Neuroscience Research* 50: 227–232.

O'Sullivan, B; Roland, P E and Kawashima, R (1994). A PET study of somatosensory discrimination in man. Microgeometry versus macrogeometry. *European Journal of Neuroscience* 6: 137–148.

Overduin, S A and Servos, P (2004). Distributed digit somatotopy in primary somatosensory cortex. *NeuroImage* 23: 462–472.

Overduin, S A and Servos, P (2008). Symmetric sensorimotor somatotopy. *PLoS ONE* 3: e1505.

Penfield, W (1958). *The Excitable Cortex in Conscious Man*. Liverpool: Liverpool University Press.

Penfield, W and Boldrey, E (1937). Somatic motor and sensory representation in the cerebral cortex of man as studied by electrical stimulation. *Brain* 60: 389–443.

Penfield, W and Rasmussen, T (1950). *The Cerebral Cortex of Man*. New York: MacMillan.

Planetta, P J and Servos, P (2012). The postcentral gyrus shows sustained fMRI activation during the tactile motion aftereffect. *Experimental Brain Research* 216: 535–544.

Polonara, G; Fabri, M; Manzoni, T and Salvolini, U (1999). Localization of the first and second somatosensory areas in the human cerebral cortex with functional MR imaging. *American Journal of Neuroradiology* 20: 199–205.

Puce, A et al. (1995). Functional magnetic resonance imaging of sensory and motor cortex: Comparison with electrophysiology localization. *Journal of Neurosurgery* 83: 262–270.

Reivich, M et al. (1978). Metabolic mapping of functional cerebral activity in man using the 18F-2-fluoro-2-deoxyglucose technique. *Journal of Computer Assisted Tomography* 2: 656.

Reivich, M et al. (1979). The use of the 18F-fluoro-deoxyglucose technique for mapping of functional neural pathways in man. *Acta Neurologica Scandinavica* 60 (Suppl. 72): 198–199.

Roland, P E; O'Sullivan, B and Kawashima, R (1998). Shape and roughness activate different somatosensory areas in the human brain. *Proceedings of the National Academy of Sciences of the United States of America* 95: 3295–3300.
Sakai, K et al. (1995). Functional mapping of the human somatosensory cortex with echo-planar MRI. *Magnetic Resonance in Medicine* 33: 736–743.
Sanchez-Panchuelo, R-M et al. (2012). Within-digit functional parcellation of Brodmann areas of the human primary somatosensory cortex using functional magnetic resonance imaging at 7 tesla. *The Journal of Neuroscience* 32: 15815–15822.
Sanchez-Panchuelo, R-M; Francis, S; Bowtell, R and Schluppeck, D (2010). Mapping human somatosensory cortex in individual subjects with 7T functional MRI. *Journal of Neurophysiology* 103: 2544–2556.
Sathian, K et al. (2011). Dual pathways for haptic and visual perception of spatial and texture information. *NeuroImage* 57: 462–475.
Sato, K et al. (2005). Functional representation of the finger and face in the human somatosensory cortex: Intraoperative intrinsic optical imaging. *NeuroImage* 25: 1292–1301.
Schnitzler, A et al. (1999). Different cortical organization of visceral and somatic sensation in humans. *European Journal of Neuroscience* 11: 305–315.
Schubert, R; Blankenburg, F; Lemm, S; Villringer, A and Curio, G (2006). Now you feel it - now you don't: ERP correlates of somatosensory awareness. *Psychophysiology* 43: 31–40.
Schwartz, T H; Chen, L M; Friedman, R M; Spencer, D D and Roe, A W (2004). Intraoperative optical imaging of human face cortical topography: A case study. *NeuroReport* 15: 1527–1531.
Schweizer, R; Voit, D and Frahm, J (2008). Finger representations in human primary somatosensory cortex as revealed by high-resolution functional MRI of tactile stimulation. *Neuroimage* 42: 28–35.
Sereno, M I et al. (1995). Borders of multiple visual areas in humans revealed by functional magnetic resonance imaging. *Science* 268: 889–893.
Servos, P and Boyd, A (2012). Probing the neural basis of perceptual phenomenology with the touch-induced visual illusion. *PLoS ONE* 7: e47788.
Servos, P; Engel, S A; Gati, J and Menon, R (1999). fMRI evidence for an inverted face representation in human somatosensory cortex. *NeuroReport* 10: 1393–1395.
Servos, P; Lederman, S; Wilson, D and Gati, J (2001). FMRI-derived cortical maps for shape, texture, and hardness. *Cognitive Brain Research* 12: 307–313.
Servos, P; Zacks, J; Rumelhart, D E and Glover, G H (1995). Somatotopy of the arm in humans using fMRI. *Society for Neuroscience Abstracts* 21: 118.
Servos, P; Zacks, J; Rumelhart, D E and Glover, G H (1998). Somatotopy of the human arm using fMRI. *NeuroReport* 9: 605–609.
Stilla, R; Deshpande, G; LaConte, S; Hu, X and Sathian, K (2007). Posteromedial parietal cortical activity and inputs predict tactile spatial acuity. *The Journal of Neuroscience* 27: 11091–11102.
Stilla, R and Sathian, K (2008). Selective visuo-haptic processing of shape and texture. *Human Brain Mapping* 29: 1123–1138.
Stippich, C et al. (1999). Somatotopic mapping of the human primary somatosensory cortex by fully automated tactile stimulation using functional magnetic resonance imaging. *Neuroscience Letters* 277: 25–28.
Stohr, P E and Goldring, S (1969). Origin of somatosensory evoked scalp responses in man. *Journal of Neurosurgery* 31: 117–127.
Strick, P L and Preston, J B (1978). Two representations of the hand in area 4 of a primate. II. Somatosensory input organization. *Journal of Neurophysiology* 48: 150–159.
Violentyev, A; Shimojo, S and Shams, L (2005). Touch-induced visual illusion. *NeuroReport* 16: 1107–1110.
Weiller, C; Ramsay, S C; Wise, R J; Friston, K J and Frackowiak, R S (1993). Individual patterns of functional reorganization in the human cerebral cortex after capsular infarction. *Annals of Neurology* 33: 181–189.

White, L E et al. (1997). Structure of the human sensorimotor system. I: Morphology and cytoarchitecture of the central sulcus. *Cerebral Cortex* 7: 18–30.

Woolsey, C N; Erickson, T C and Gilson, W E (1979). Localization in somatic sensory and motor area of human cerebral cortex as determined by direct recording of evoked potentials and electrical stimulation. *Neurosurgery* 51: 476–506.

Yang, J; Han, H; Chui, D; Shen, Y and Wu, J (2012). Prominent activation of the intraparietal and somatosensory areas during angle discrimination by intra-active touch. *Human Brain Mapping* 33: 2957–2970.

Yang, T T; Gallen, C C; Schwartz, B J and Bloom, F E (1993). Noninvasive somatosensory homunculus mapping in humans by using a large-array biomagnetometer. *Proceedings of the National Academy of Sciences of the United States of America* 90: 3098–3102.

Primate S1 Cortex

Jon H. Kaas

Introduction

All mammals have a region of somatosensory cortex that is called primary somatosensory cortex or S1. The term stands for the cortical representation of touch that was discovered and described first. The name became useful when other cortical representations of touch were discovered, starting with the second representation, S2. Early investigators used Roman numerals for cortical areas, but now Arabic numerals have become common. S1 or SI has also been referred to as SmI by Clinton Woolsey and co-workers to recognize that somatosensory area I also has a small motor (m) component. S1 was first described in cats in 1941 as a single, systematic representation of the contralateral body surface located in parietal cortex of each cerebral hemisphere. The experimental results, obtained by recording the activities of neurons in many locations throughout the representation with penetrating electrodes, indicated that medial recording sites in S1 are activated by touching the hind foot, more lateral sites by the forefoot, and the most lateral sites by the face. The disclosed representation of the contralateral body surface was described as somatotopic because neurons in S1 were activated from tail to tongue in a mediolateral sequence across cortex. From this beginning, S1 has been considered to be a single, roughly somatotopic representation of the contralateral body surface that has a characteristic mediolateral orientation and internal organization across mammal species.

Subsequent studies have associated S1 across species with a strip of cortex that has a distinctive laminar appearance, as cortical layer 4 is densely packed with small neurons that receive activating inputs via axons from the ventroposterior nucleus of the somatosensory thalamus. S1 sends somatosensory information to other areas of somatosensory cortex, including S2. Neurons in S1 (3b) are activated

J.H. Kaas (✉)
Department of Psychology, Vanderbilt University, Nashville, TN, USA

T.J. Prescott et al. (eds.), *Scholarpedia of Touch*, Scholarpedia,
DOI 10.2991/978-94-6239-133-8_39

by light touch on the skin. Each neuron is typically activated by inputs from a small region of skin, such as part of the tip of a single finger, and this activating surface is called the receptive field. Touch outside the activating receptive field usually reduces the response to touch within the receptive field, or has little effect. After damage to S1, small objects may be unrecognized by touch and ignored, and grasping with the hand may be difficult.

The S1 Region of Monkeys and Humans In early studies of parietal cortex in monkeys and humans, S1 was not correctly identified. In brief, the region often called S1 in monkeys, apes and humans mistakenly included four different architectonic fields that are called Brodmann's areas 3a, 3b, 1 and 2 after the early studies of Brodmann over 100 years ago. Each of these areas has its own representation of the body, but only the area 3b representation has the defining characteristics of S1 as described in cats, rats and other mammals. These include having a representation that is almost exclusively responsive to the activation of touch receptors, a characteristic internal somatotopic organization that is coextensive with a cytoarchitectonic field that has a layer 4 that is more densely packed with small neurons than adjoining areas, and more dense inputs from the ventroposterior nucleus to layer 4 than any other area. In monkeys and humans, areas 3a, 1 and 2 contain additional representations of the body surface that parallel S1 in somatotopic organization. While area 3a is dominated by inputs from muscle spindle receptors for proprioception, and area 2 also has major inputs from proprioceptors, neurons in area 1 and at least much of area 2 respond well to light touch. The responsiveness of these other cortical areas to touch was part of the problem of defining S1 in early recording studies of somatosensory cortex in monkeys and humans. The other part of the problem was that most of the recordings in early experiments were done with surface electrodes on the exposed surface of somatosensory cortex, and most of area 3b is hidden on the caudal bank of the deep central sulcus of Old World monkeys, apes and humans. Thus, most of the recordings were from areas 1 and 2, as they were largely on the cortical surface. As areas 1 and 2 paralleled area 3b in their mediolateral somatotopic sequence, recordings from areas 1 and 2 were considered to be from S1. Furthermore, as areas 3b and 3a were also recognized as responsive to touch, all four architectonic strips, each now known to contain a separate representation of the body, were considered to be parts of a single representation, S1, by early investigators. This misnomer persists in many reports today, as S1 is often used to refer broadly to anterior parietal cortex in current studies in humans and monkeys. Since each of the four architectonic fields has its own representation and functions, this careless misuse creates confusion and impedes further progress and understanding.

Here, S1, sometimes called S1 proper, refers only to the area 3b representation. Electrophysiological evidence that area 3b of monkeys contains a complete representation of touch receptors of the contralateral body surface was first obtained in New World owl and squirrel monkeys, as these monkeys had most of area 3b exposed on the surface of the brain. Owl monkeys have only a shallow dimple in cortex rather than a deep central sulcus, and squirrel monkeys have only a short, shallow central sulcus. Thus, it was possible to visually place microelectrodes into hundreds of sites in area 3b of these monkeys, record the responses of neurons at these sites to touch on the body, and determine where on the body touch activated neurons for each site. In this manner, receptive fields were determined for neurons at each site. By outlining the cortical territories of sites where neurons were activated by the same body part, such as digit 1 (thumb), maps of where body parts are represented in area 3b were constructed (Figure 1). The results indicate that the area 3b representation (S1) extends from the medial wall of the cerebral hemisphere lateral to the margin of the lateral sulcus where it curves rostrally to end near the ventral margin of the hemisphere. Part of the posterior leg and tail are represented on cortex of the medial wall, followed in a mediolateral sequence by representation of the toes and foot, anterior leg, genitals and trunk, arm, and hand and digits. In cortex just lateral to the hand representation, parts of the face are represented, followed as the representation curves anteriorly by the teeth, tongue, and finally the ipsilateral teeth and tongue as these important mouth parts are represented in S1 of both cerebral hemispheres. The cortical territories devoted to the teeth are activated by sensitive receptors around the roots of the teeth that respond to when the teeth are touched. Large proportions of area 3b (S1) that respond to touch on the digits of

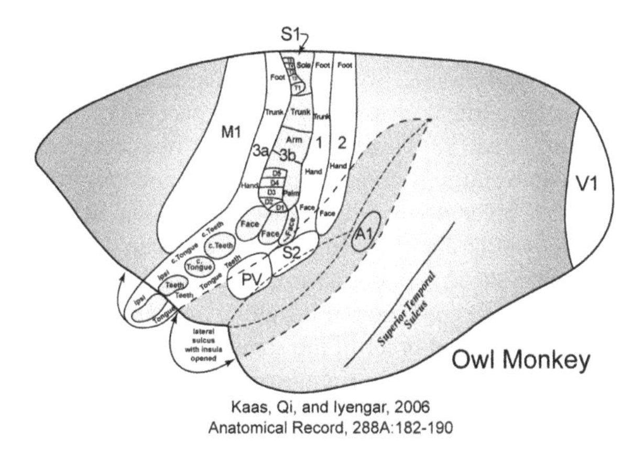

Figure 1 The location and somatotopic organization of S1 (area 3b) in an owl monkey, a small New World monkey. S1 constitutes a mediolateral strip of cortex that extends from cortex on the medial wall of the cerebral hemispheres toward the lateral sulcus, where it curves to the anterior and ends on the ventral surface of frontal cortex. The contralateral tail, leg, and foot are represented medially, with the toes represented along the anterior border of S1 from 1 to 5 (T1–T5) in a lateromedial sequence. As a continuation of this mediolateral sequence, the trunk, arm and hand are represented, with the digits of the hand represented in the anterior half of area 3b from D5 to D1. More laterally, area 3b represents the face followed by the contralateral (c) teeth, contralateral (c) tongue, and finally the ipsilateral (ipsi) teeth and tongue. The most ventral part of S1, devoted to the ipsilateral tongue, has been folded out from the underside of the brain. Parallel representations of the body exist in bordering areas 3a, 1, and 2 (see text). The locations of second somatosensory area (S2) and the parietal ventral area (PV), with inputs from S1, are shown on the upper bank of the opened lateral sulcus. Primary motor cortex (M1), primary auditory cortex (A1) and primary visual cortex (V1) are shown for reference. Modified from Kaas et al. (2006)

the hand, the teeth and the tongue, show how important these parts of the body are in providing touch information to the cortex.

A similar area 3b (S1) representation has been revealed in all studied primates. This representation was examined first in microelectrode mapping studies in other New World monkeys, such as marmosets and titi monkeys, and in prosimian galagos, where a central sulcus is absent, but area 3b has been explored in some detail in Old World macaque monkeys with electrodes that penetrated into the cortex along the posterior bank of the central sulcus. In addition, a similar organization of area 3b has been determined in less detail in humans using functional magnetic imaging.

Area 3b (S1) is bordered along its anterior border in all primates by a narrow strip of cortex, area 3a that is activated mainly by muscle spindle receptors and other proprioceptors that respond while the body moves and provide a sense of the position of body parts. This information is relayed to area 3a from a separate nucleus in the somatosensory thalamus called the ventroposterior superior nucleus (VPS). A similar proprioceptive strip of cortex is found along the anterior border of S1 in at least most mammals. As area 3a receives inputs from area 3b (S1), neurons in area 3a also responds to touch, but such responses are often not apparent in anesthetized animals. Area 3a provides an important source of proprioceptive information to the anteriorly adjoining motor cortex. The posterior border of area 3b (S1) is formed by area 1 in primates. Area 1 receives direct inputs from area 3b, and it appears to be dependent on area 3b for most of its neural responsiveness. Thus, area 1 is a second order cortical area, much like the second visual area, V2, or the second somatosensory area, S2. Nevertheless, area 1 does receive somatosensory inputs directly from the ventroposterior nucleus of the somatosensory thalamus, but to layer 3 rather than layer 4. In most monkeys neurons in area 1 are highly responsive to touch, even in anesthetized animals, and area 1 contains a complete somatotopic representation of the contralateral body that parallels that in area 3b (S1), thus accounting for the confusion of including area 1 in S1 in early studies on monkeys. However, the internal order of the somatotopy in area 1 differs from that in area 3b in that area 1 basically forms a mirror reversal of the somatotopy in area 3b. For example, the digits tips are represented anteriorly in area 3b and posteriorly in area 1. A variously named strip of cortex with inputs from S1 exists along the posterior border of S1 in most mammals, but this cortex is not highly responsive to touch in anesthetized animals. This is also the case in prosimian primates, and even in marmoset monkeys, where the area 1 cortex is unresponsive or poorly responsive to touch in anesthetized animals. Thus, in these primates, and in most mammals, there would be little chance of mistakenly including the area 1 territory in S1 in recording experiments.

In a similar manner, a strip of cortex along the posterior border of area 1 is variably responsive to touch across primates. At least in parts of this area 2 field neurons are highly responsive to touch in macaque monkeys, and this responsive region was included in S1 in early studies in Old World macaque monkeys. Area 2 receives proprioceptive information directly from the ventroposterior superior (VPS) nucleus of the thalamus, and touch information from area 3b (SI) and area 1.

Neurons in area 2 are not very responsive to touch in some New World monkeys, and in prosimian primates, and even in Old World monkeys, neurons in all parts of the architectonically defined field are not highly responsive to touch. Area 2 provides an important source of somatosensory information to posterior parietal cortex.

Suggested Readings

Kaas, J H (1983). What, if anything is SI? Organization of first somatosensory area of cortex. *Physiological Reviews* 63: 206–230.

Qi, H-X; Preuss, T M and Kaas, J H (2007). Somatosensory areas of the cerebral cortex: Architectonic characteristics and modular organization. In: E Gardner and J Kaas (Vol. Eds.), *The Senses: A Comprehensive Reference, Vol. 6 Somatosensation* (pp. 142–169). London: Elsevier.

S1 Laminar Specialization

Jochen F. Staiger

The neocortex is classically considered to be organized into 6 different but "equally well-developed" layers. However, unlike this basic Bauplan of the neocortex, primary somatosensory cortex (S1) possesses a very well-developed layer IV and thus qualifies itself as a typical koniocortical area, as all primary sensory cortical areas. This laminar organization can be easily visualized with many different types of cell stains (e.g., Nissl-staining). It is based on the differential number, size and packing density of neurons in different layers, which show a characteristic pattern when compared with each other, both qualitatively and quantitatively. It will be argued in the main text, however, that when considering more fine-grained morphological properties of the neurons and especially their connections, more than 6 layers should be distinguished. Based on the hypothesis that cortical lamination promotes economic and efficient wiring (Kaas 1997), the focus will be on the cellular composition of excitatory projection neurons and how their assembly into cortical layers organizes intra- but also extracortical connectivity. Here, the 6 layers, from the pial surface to the white matter, will be shortly introduced.

Layer I is a cell-sparse layer that houses a dense connectivity matrix between the apical dendrites of pyramidal cells of virtually all other layers with pathways of many types of origin (e.g. intracortical, thalamic, transmitter-specified from brain stem). Layer II, which is often pooled with layer III because of the lack of a clear cytoarchitectonic border into layer(s) II/III, is mainly composed of medium-sized pyramidal cells which have the potential to receive and issue cortico-cortical projections with the same (associational) and contralateral (callosal) hemisphere but do not project to subcortical sites. Layer IV is characterized by a very cell-dense appearance, mainly consisting of spiny stellate and star pyramidal cells which on the one hand serve as targets of lemniscal thalamic projections and on the other hand efficiently distribute this sensory information to the other layers. Layer V

J.F. Staiger (✉)
Center for Anatomy, Institute for Neuroanatomy, University Medicine Göttingen, Göttingen, Germany
e-mail: jochen.staiger@med.uni-goettingen.de

T.J. Prescott et al. (eds.), *Scholarpedia of Touch*, Scholarpedia,
DOI 10.2991/978-94-6239-133-8_40

houses the largest pyramidal cells and is (together with layer I) the prime target layer for paralemniscal thalamic projections and, in rodents, can issue virtually any of the known local and distant cortical projections. Layer VI is composed of a morphologically highly variable population of excitatory neurons, which seem to subserve two basic functions: (i) to provide specific feedback to the lemniscal thalamus and (ii) to orchestrate local intracolumnar with longer-distance intra-areal activity. In all layers, a complimentary set of GABAergic inhibitory interneurons can be found, that are too diverse in form and function to shortly integrate them into this summary.

What Is S1?

S1 represents a certain nomenclature of cortical areas and stands for primary somatosensory cortex. Somatosensory tells us that here the information coming from exteroceptors (nociceptors, thermoreceptors and, within the scope of this chapter, mainly mechanoreceptors) gains access to the cortex for its conscious perception and fine-grained analysis (Diamond et al. 2008). Primary in this case means that a hierarchy of cortical areas is assumed in which the initial processing of tactile information is continued in secondary (S2) and so forth "higher" somatosensory areas. In rodents as well as in other species, the lemniscal pathway is the classic route by which somatosensory information reaches the cortex (S1) after processing in the thalamus (dorsomedial part of ventroposteromedial nucleus; VPMdm; Figure 1). However, the recently discovered extralemniscal thalamic somatosensory nucleus (ventrolateral part of ventroposteromedial nucleus; VPMvl) and the paralemniscal nucleus (medial portion of the posterior nucleus; POm) directly issue parallel projections to S1 and S2 (Carvell and Simons 1987; Pierret et al. 2000; Diamond et al. 2008). Within the scope of this chapter it should be added that the lemniscal pathway mainly terminates, in a barrel-aligned manner, in layer IV and (still ill-defined) on the layer V/VI border region. The extralemniscal pathway shows no distinct termination tiers (and seemingly avoids layer I) and it should be noted that it is mainly septum-aligned (see below for explanation of barrel versus septum compartments). Finally, the paralemniscal pathway has two nearly exclusive layers of termination, i.e. layer I and layer V(a), which do, however, distribute freely across barrel and septum compartments (see also Figure 3).

One has to concur that at present it is completely unclear how much sequential (hierarchical) or simultaneous (parallel/"democratic") processing takes place in all parts of the somatosensory system, especially when looking at the layers forming a cortical column or the columns communicating with each other across different areas (Hendry and Hsiao 2003).

Rodent S1 is a peculiar kind of cortical area. On the one hand it houses the entire representation of the body (as a "ratunculus" or "musculus", in analogy to the famous "homunculus") and it does so in the form of a typical primary sensory

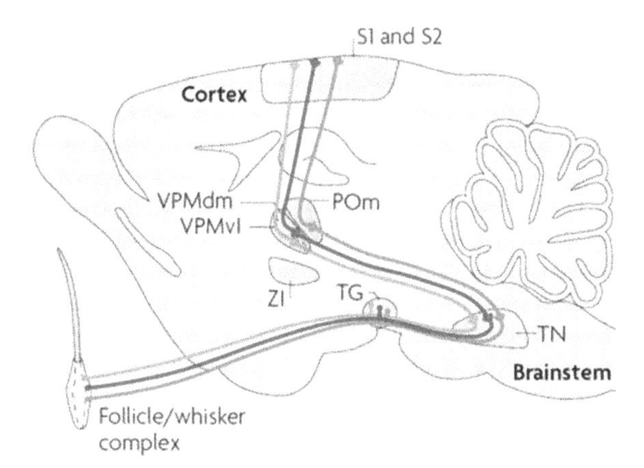

Figure 1 This schematic diagram shows the three parallel pathways transmitting different aspects of the tactile information acquired by the receptors (trigeminal ganglion cells, TG) associated with the large mystacial whiskers to the primary (S1) and secondary somatosensory cortex (S2) (from Diamond et al. 2008; reprinted with permission). The pathways possess subcortical processing stations, some of which are shown for the brainstem (trigeminal nuclei, TN) and the diencephalon (posterior thalamic nucleus, POm; dorsomedial part of the ventroposterior medial thalamic nucleus, VPMdm; ventrolateral part of the ventroposterior medial thalamic nucleus, VPMvl). In red, the "classic" lemniscal pathway, in green, the paralemniscal pathway and in blue, the newly discovered extralemniscal pathway

cortex, i.e. a "koniocortex" showing a very well developed granular cell layer (IV) which leads to the highest gray level index of the entire rat cortex (Palomero-Gallagher and Zilles 2004).

As seen in Figure 2, some parts of the body are mapped in a one-to-one fashion (e.g. each whisker to a corresponding barrel in the posteromedial barrel subfield; S1BF, Paxinos and Watson 1998), whereas other parts are mapped to barrels in a non-one-to-one fashion (e.g. the forepaw digits and palm to the forelimb area barrels; S1FL) and still further body parts are represented in subdivisions of S1 not being modularly organized at all (trunk in S1TR) (Welker and Woolsey 1974; Chapin and Lin 1984). On the other hand, the modules (here: barrels) which represent the granular cortex are embedded in a matrix of dysgranular cortex (to which also the septal compartments belong) with very different connectional and functional properties. Here, a tight relationship to ipsilateral primary motor and contralateral primary somatosensory cortex can be observed (Alloway 2008). The situation is further complicated by the fact that rodent S1 contributes heavily to the subcortical projections reaching the brainstem and the spinal cord, which is reflected by a prominent layer Vb housing large pyramidal neurons so that some researchers have described at least part of the primary somatosensory cortex as a sensory-motor amalgam (Donoghue et al. 1979).

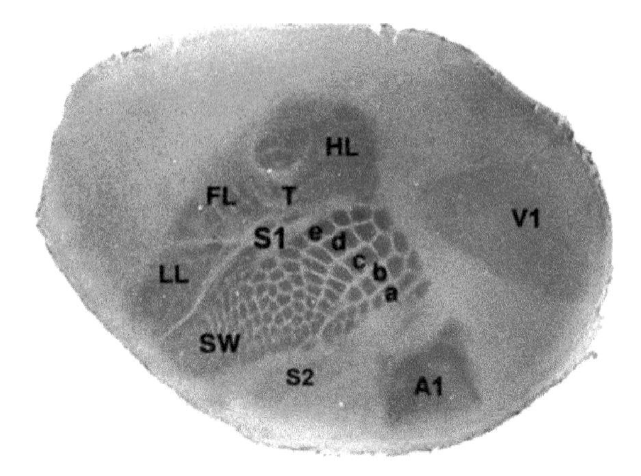

Figure 2 Flattened left hemisphere of GAP-43 wildtype mouse cortex immunostained with serotonin transporter (see Barrels web: http://simonslab.neurobio.pitt.edu/barrels/Figure3.html); rostral is to the left, caudal to the right, medial to the top and lateral to the bottom. The primary somatosensory cortex is represented by a disproportionately large portion of the cortex and consists of the large whisker representation (isomorphic to the five rows on the snout, named A-E), the small whisker representation (SW), the lower lip (LL), forelimb (FL), hindlimb (HL) and trunk (T) representation. The secondary somatosensory cortex (S2) is located laterally adjacent to S1. Also the primary auditory (A1) and the primary visual (V1) cortex get demarcated by the brown reaction product

Species Similarities and Differences

It is well known that the complexity of the primary visual cortex (V1) shows huge differences according to the importance and functional specialization of the sense of vision for a given species (Van Hooser 2007). It should also be stated that V1-circuitry in monkey and cat is well established (Callaway 1998; Binzegger et al. 2004) and would lend itself to direct comparison of rodent S1-circuitry as described below. When one looks for species differences in S1, it becomes clear that they are less striking (Welker and Woolsey 1974; Leclerc et al. 1994; Geyer et al. 1999). This might be explained by the fact that this is a basic sensory modality of similar importance for all mammals (Glassman 1994). An obvious difference is the existence of barrels is some species and the lack of such cellular agglomerations in others. However, no systematic relationship between the appearance of barrels and phylogenetic or functional aspects could be found with certainty (Woolsey et al. 1975). Another obvious difference is that in rodent S1 the infragranular layers take up half of the cortical thickness whereas in feline or primate S1 the supragranular layers (especially layer III) are developed prominently. Layer III in the latter, furthermore, shows very large pyramidal cells whereas they are rare in layer V when compared to the rodent. This is very likely to possess a correlate in the functional connectivity of S1 with different cortical and subcortical areas, which might have a stronger "motor component" in rodents and a more prominent "associative component" in primates (Diamond 1979).

What Are Cortical Layers?

In an earlier review on cortical lamination, Jones (1981) stated at the beginning of p. 201: "The first scheme of cortical lamination was proposed by Meynert in 1867. Despite many years of subsequent work, the significance of lamination is still little understood. To a large extent, this is because the cortex changes its laminar pattern from area to area [...]. Any analysis is further complicated by the fact that pyramidal cell somata situated in one lamina have dendrites that extend through several supra- and subjacent laminae. Finally, there is no agreement regarding the classes and distribution of interneurons [...]." In the present chapter on S1 laminar organization it will become clear that although much has been learned since then on cortical cell types and their physiology, neurochemistry and morphology, the big picture of the functional meaning of cortical lamination has remained obscure.

The 6 (or so, see below) cortical layers are established in an inside-out sequence. They probably are "by-products" of the developmental dynamics of the ventricular and subventricular zones, the germinal neuroepithelium which gives rise to the neuroblasts which differentiate into the different excitatory cells of the neocortex (Mountcastle 1997; Molyneaux et al. 2007).

When all 6 cortical layers have been formed after the first postnatal week (Rice 1995), they can be described as tiers of neurons that share a similar appearance or packing density (classically observed in Nissl staining), which differs qualitatively and/or quantitatively from the neighboring tiers (Figure 3A).

However, with more refined methods massive differences in gross morphology or neurochemistry of the individual neurons forming one cytoarchitectonically coherent layer can be observed (for details see following paragraphs). Let us testify here that delineating 6 layers is a convention, which from the beginning did not remain without alternatives (Jones 1984a). Especially recent molecular studies offer a glimpse into a much richer stratification (Figure 3B; (Lein et al. 2007) which is very likely to reflect also functional differences, as has been shown, for example, when layer Va was compared with Vb (Schubert et al. 2007; Groh et al. 2010).

Thus, below, each layer will shortly be characterized in terms of its classical appearance in Nissl staining. Furthermore, a necessarily eclectic (due to space and time constraints) description of the finer morphological details of the individual cells, their associated functional properties, their neurochemistry and their connectivity will be added. Especially GABAergic interneurons are not featured extensively here, due to their rich diversity of basically all features, except clear-cut laminar specificities (i.e. I would consider them under a conceptual framework different from layers and in a separate chapter). The interested reader is advised to consider seminal original work or some of the recent extensive reviews (as a very

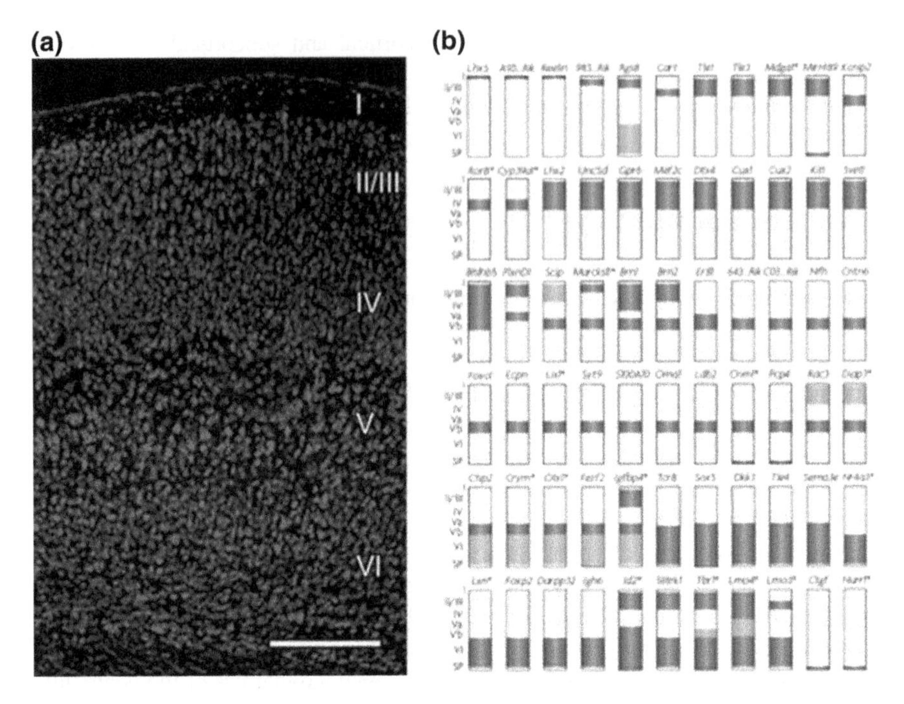

Figure 3 Different views on cortical layers. **a** "Six layers" (Roman numerals) shown by "fluorescent Nissl-staining" (Neurotrace) staining (*scale bar* 250 μm). **b** Layer-specific markers suggest a much richer stratification (Molyneaux et al. 2007)

incomplete suggestion: (Kawaguchi 1995; Cauli et al. 1997; Markram et al. 2004; Ascoli et al. 2008; Helmstaedter et al. 2009a). Most emphasis will be placed on laminar connectivity differences.

Laminar Specificity of Long-Range Input-Output Connectivity

In order to understand the functional capacity of a certain brain area, it is a prerequisite to know its inputs, the local circuitry and the outputs. Here, no written overview will be given on intracortical and thalamo-cortical layer-specific input of rodent S1, which is either badly studied (intracortical) or subject to a different chapter of Scholarpedia. A summary diagram of patterns of cortical inputs can be found in Figure 4. The local circuitry is presented in a very simplified manner as the "canonical microcircuit", a concept which highlights only a selected series of excitatory feedforward projections that can also be found embedded in Figure 6. The remaining of this paragraph is focused on the layer-specific outputs that have been determined for rodent S1 in the recent years. Considering only the excitatory

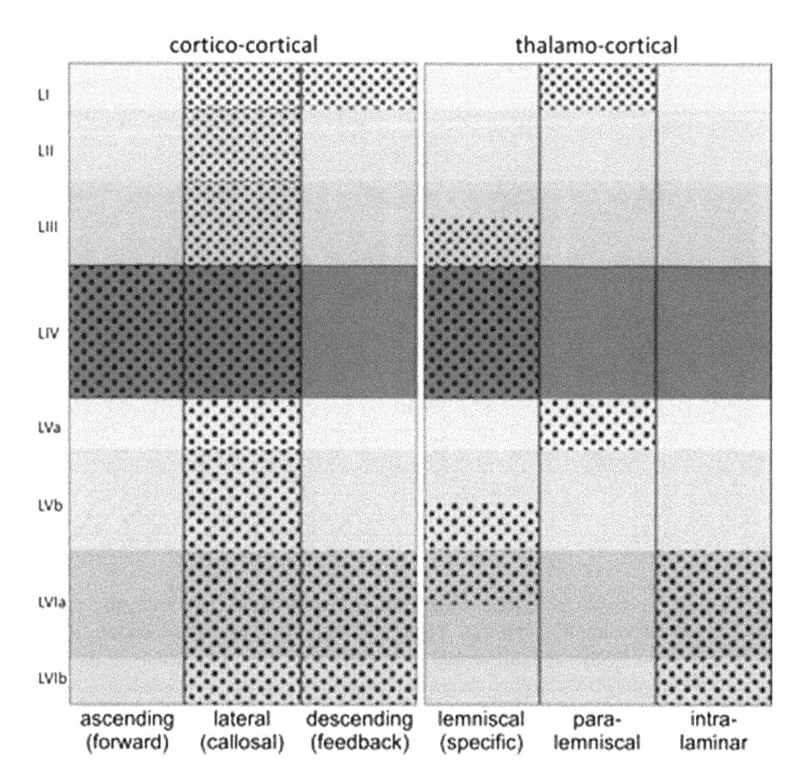

Figure 4 Cortical layer-specific inputs. Cortical layers are targeted in a specific manner by intracortical and thalamic projections but much less so (if at all) by major ascending (transmitter-specified modulatory) pathways (not shown). Due to our sparse knowledge on the cellular origin and the precise laminar termination of these connections in rodent S1, a very schematic and thus oversimplifying (to the extent of a putative faultiness) diagram has been generated. Here, the pattern of cortico-cortical connections is adopted from the concept by Felleman and van Essen (Felleman and van Essen 1991). For thalamo-cortical projections, the concept worked out by Herkenham (Herkenham 1980, 1986) is used

principal cells of the neocortex (i.e. the glutamatergic pyramidal, star pyramidal and spiny stellate neurons), a striking feature of cortical laminar organization is an extensive similarity of projection targets for each individual layer (Figure 5), although this organizing principle is much less clear in the rodent than in the primate (Jones 1984b; White and Keller 1989).

The long-range projections of layer II pyramidal cells (Jones 1984b) are issued over shorter-distance (than layer III, see directly below) for both, associational as well as commissural projections. According to the classical scheme L III pyramidal cells are responsible for the majority of long-range associational and a portion of the callosal projections (Jones 1984b). In S1 this seems, at least partially, to fulfill a sensory-motor integration function since a major associative route of supragranular pyramidal neurons is to the primary motor cortex whereas a major homotypical callosal projection is targeted to septa of the contralateral S1 (Alloway 2008).

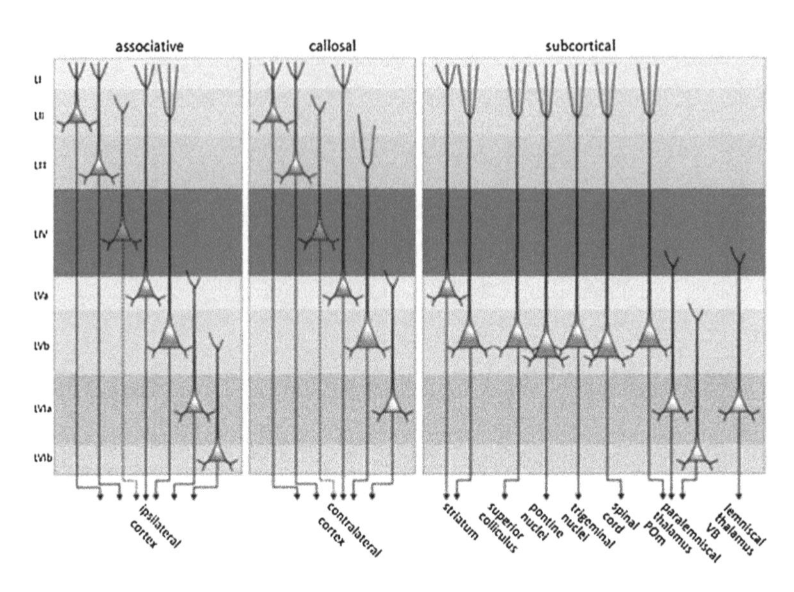

Figure 5 Cortical layers organize long-range (but also short-range; see following paragraphs) projections. Please note that although this scheme has been derived from literature specific for rodent S1, it does not consider the differential origin of some of the projections in terms of their relationship to barrel-related vs. septum-related columns (see references below). It also does not reflect all known morphological details of the neurons of origin, for the sake of simplicity. Stippled layer IV-neurons represent (star-)pyramidal cells which are not considered to be projection neurons in the rodent but can be seen in (Chakrabarti and Alloway 2006) for associative and were described in (Jones 1984b) for callosal projections. Pyramidal cells with an asymmetric apical dendritic tuft either form no tuft at all, or, a very small one. For layer Vb-callosal cells, this was reported e.g. by (Le Be et al. 2007). For the subcortical projections originating in layer Vb a uniform symbol was used because there are only very few reports on the detailed somatodendritic morphology of projection-identified neurons. Such papers suggest that regular-spiking and intrinsically burst-spiking cells do, at least partially, project to different targets (Larsen et al. 2007; Hattox and Nelson 2007). A separate neuron is shown for each of the different subcortical targets since there is no agreement in the field to what extent collateralization of the axon to different targets does exist. Major additional papers consulted: (Wise and Jones 1977; Killackey et al. 1989; Koralek et al. 1990; Mercier et al. 1990; Bourassa et al. 1995; Veinante et al. 2000; Wright et al. 2001; Alloway et al. 2004)

The circuitry of L IV neurons has now been thoroughly studied. In terms of outputs, in contrast to cat and monkey, in the rodent it is usually stated that layer IV-spiny neurons are “excitatory interneurons” without associational, commissural or corticofugal projections. However, upon closer look, a number of studies show retrogradely-labeled neurons in layer IV (mainly septal compartment) after tracer injections in ipsilateral S2 (Chakrabarti and Alloway 2006) or contralateral S1 (Hayama and Ogawa 1997).

In terms of their long-range connections, the bilateral projections of L Va pyramidal cells to primary motor cortex and striatum are especially well characterized (Wright et al. 2001; Alloway 2008). This strongly suggests a participation in the control of motor and sensory aspects of whisking behavior (Derdikman et al. 2006), a notion which is further supported by the organization of the local intracortical circuits into which L Va pyramidal cells are integrated.

The long-range projections of L Vb pyramidal cells in rodent S1 are very diverse. They form part of the neurons-of-origin of associational and callosal projections, just like virtually all other layers except I (and maybe IV). However, they represent the only neurons which are able to project to target areas "below" the thalamus, such as the superior colliculus, pontine and trigeminal nuclei or spinal cord (Jones 1984b; Welker et al. 1988; White and Keller 1989). Thus they are in a position to directly influence circuits mediating behavior as well as controlling the ascending flow of sensory information at various hierarchical levels. Due to their firing rate being the highest of all neurons in a barrel-related column, the large-tufted (IB) pyramidal cells are supposed to be the major contributors to driving subcortical circuits (de Kock et al. 2007; de Kock and Sakmann 2008). In general, however, only little is known about cell type-specific subcortical projection patterns, but promising work is underway to understand this type of specificity by looking at developmental genetic programs (Molyneaux et al. 2007).

The subcortical projections of L VIa are very restricted, as are the intracortical. The projections to the thalamus end exclusively in (i) the lemniscal ventrobasal nucleus and (ii) the reticular nucleus which themselves are reciprocally connected (Bourassa et al. 1995).

Layers and Their Short-Range Connectivity Re-visited

Taking the "canonical microcircuit" in its updated form as a guideline (cf. (Silberberg et al. 2005; Lübke and Feldmeyer 2007; Douglas and Martin 2007; Thomson and Lamy 2007), we will start with the input layer IV, continue with layers III and II to proceed to the major output layers Va, Vb and VIa (Figure 6). This scheme has recently gained important extensions by additional experimental data (Schubert et al. 2007; Lefort et al. 2009). In short, the "canonical microcircuit" is regarded as a series of excitatory feedforward projections, in which the lemniscal thalamus innervates layer IV that projects to layers II/III that innervate layer V, which on the way to their subcortical target sites give off a collateral projection to layer VI that closes the loop with the lemniscal thalamus. We will end with a characterization of layers I and VIb which are by now not reasonably well integrated into this scheme.

Layer IV (Lamina Granularis Interna)

for nomenclature see Box 1.

Cellular composition: This layer shows a high packing density of relatively small-sized neuronal somata providing it in Nissl-stainings with a grain-like

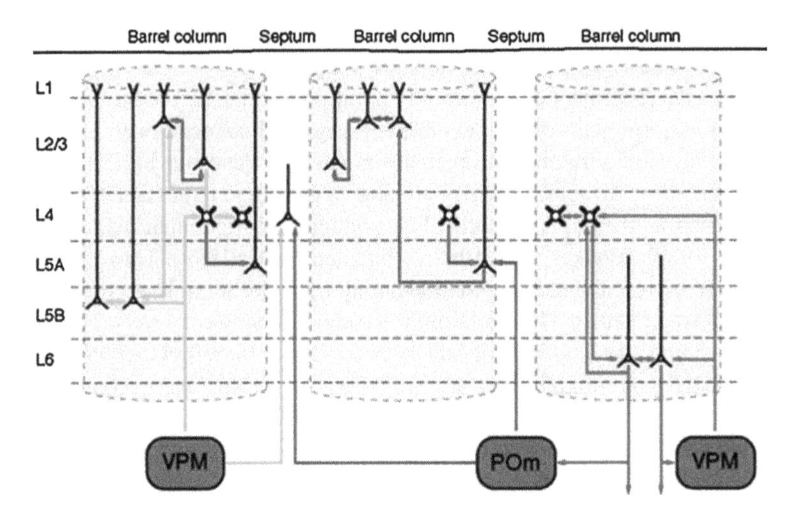

Figure 6 Recent update on canonical microcircuits with a more differentiated view on different intracortical and thalamo-cortico-thalamic interactions (Reproduced with permission from (Lübke and Feldmeyer 2007). These circuits are shown in 3 panels for didactic reasons. In vivo they are all implemented into one module

Box 1 Nomenclature of cortical layers (L)

Layer number	Standard name	Synonym
I	Molecular layer	None
II	External granular layer	Corpuscular layer
III	External pyramidal layer	Pyramidal layer
IV	Internal granular layer	Granular layer
V	Internal pyramidal layer	Ganglionic layer
VI	Multiform layer	Polymorph layer

Please note the right column which makes the widely used terms supragranular (L I-III) and infragranular (L V-VI) layers understandable

appearance. With the advent of Golgi labeling or biocytin filling it has become clear that medium-sized neurons with symmetrical (spherical) or asymmetrical small dendritic trees populate this layer, dendrites which show either (i) high numbers of spines (spiny neurons) or (ii) zero to medium numbers of spines (smooth/aspiny or sparsely spiny stellate cells) (Simons and Woolsey 1984; Lorente de Nó et al. 1992). The spiny stellate neurons are considered to represent excitatory pyramidal neurons which have lost their apical dendrite during development (Vercelli et al. 1992). Indeed, in addition to classical spiny stellate cells pyramidal-like cell types can also be found in this layer, namely symmetrical and asymmetrical star pyramidal neurons as well as classical pyramids (Staiger et al. 2004). These are, however, often not well differentiated in terms of their basal dendritic configuration and the terminal tuft of the apical dendrite which is mostly missing completely. The actual proportion of these cells in L IV is highly area and species specific (Jones 1975; Martin and Whitteridge 1984; Lund 1984; Smith and Populin 2001; Saez and Friedlander 2009). However, precise quantifications of their absolute or proportional numbers are still missing due to the lack of appropriate markers at the population level. In terms of neurochemistry little more than the fact that they express low levels of calbindin d-28 K, like their "relatives" in the supragranular layers, is known (Celio 1990; van Brederode et al. 1991; Alcantara et al. 1993). The spine-poor or spine-free neurons are considered to be GABAergic interneurons which are especially numerous in L IV (Ren et al. 1992).

Connectivity: Until recently it was believed that L IV-neurons possess only two substantial input sources: (i) the lemniscal thalamic projection (Gibson et al. 1999; Bureau et al. 2006) and (ii) local collaterals of neurons residing within the same column (Feldmeyer et al. 1999; Petersen and Sakmann 2000). Although these inputs can still be considered as the two quantitatively most imported ones, evidence has accumulated that, in a cell type-specific manner, intracolumnar feedback projections originating from supra- and infragranular layers as well as transcolumnar projections directly originating in L IV of neighboring columns can also be consistently found (Martin and Whitteridge 1984; Staiger et al. 2000; Schubert et al. 2003; Egger et al. 2008). These horizontal L IV-connections are functional and contribute to the receptive fields of these neurons (Fox et al. 2003) which can be fairly large at the subthreshold level (Moore and Nelson 1998; Brecht and Sakmann 2002). Nevertheless, it can be stated that L IV is likely to process sensory information with the lowest level of context-dependent integration of what is going on in the neighboring sensory periphery as well as in hierarchically higher cortical areas.

Layer III (Lamina Pyramidalis Externa)

In certain species and areas (e.g. rodent S1), there are no cytoarchitectural differences between layers II and III. Thus, these two layers are often pooled into a "layer II/III". However, several recent papers have clearly demonstrated that the input-output function (see respective chapters below) of pyramidal cells located close to the pia is different from those located close to the barrels (Staiger et al. 2015). Therefore, it was chosen to present the two layers here as distinguishable entities, the exact border location of which may become visible with future molecular markers.

Cellular composition: This layer shows, in Nissl-stained sections, a high density of pyramidal-shaped cell bodies which is due to the presence of a morphologically homogeneous population of typical medium-sized pyramidal cells. In Golgi studies they were revealed in more detail, with little variation of their somatodendritic morphological features. In fact, together with the L V (a and b) pyramidal cells they represent the "purest" representatives of this cell type in the brain (Peters and Kara

1985). The apical dendrite always reaches layer I where it forms a well-defined tuft, whereas the basal dendrites mainly originate from the opposite pole of the soma ramifying in layers III and IV. This was later confirmed by biocytin fillings in S1 (Schröder and Luhmann 1997; Lübke et al. 2003) as earlier studies already had shown in cat primary visual cortex (Gilbert and Wiesel 1979; Martin and Whitteridge 1984). The molecular or neurochemical characterization of these neurons is not very well advanced at the identified single-cell level (Andjelic et al. 2009; Karagiannis et al. 2009), although numerous layer-specific traits can be expected (Arion et al. 2007). The other major group of neurons are the non-pyramidal cells: These are very frequent in the supragranular layers III and II (Peters et al. 1985) and probably cover all possible types of inhibitory interneurons which occur in the neocortex (Markram et al. 2004; Helmstaedter et al. 2009a; Helmstaedter et al. 2009b). It is obvious, however, that bipolar and bitufted cells expressing many different neuropeptides are especially numerous here (Cauli et al. 2000).

Connectivity: The intracolumnar circuitry of layer III pyramidal neurons is dominated by two sources: (i) local intralaminar connections (Feldmeyer et al. 2006), which are, however, often outweighed by (ii) L IV translaminar input (Yoshimura et al. 2005; Shepherd and Svoboda 2005; Schubert et al. 2007; Lefort et al. 2009). Recent evidence suggests that this is also true for at least a subpopulation of L III inhibitory cells, i.e. the fast-spiking basket cells (Xu and Callaway 2009). The transcolumnar circuitry of L III pyramidal cells has been more difficult to study in the slice. Although structurally and functionally supragranular transcolumnar pathways have been described (Fox 2002; Brecht et al. 2003; Broser et al. 2008), they are less numerous in L III than in L II (Larsen and Callaway 2006; Bruno et al. 2009; Staiger et al. 2015). This may be one reason for the so far lacking paired recordings of L III pyramidal neurons located in neighboring columns, in vitro and in vivo. Since the connection probability decreases monotonically with distance (Holmgren et al. 2003), new methods to pre-identify connected neurons (Wickersham et al. 2007) have to be further refined (Boldogkoi et al. 2009), in order to study the precise functional and morphological determinants of transcolumnar L III circuits (which is true for all other layers as well). Concerning the output of L III, consistently, L V(b) has been found to be the major intracolumnar target structure which represents one of the backbone feedforward projections of the "canonical microcircuitry" (Martin and Whitteridge 1984; Thomson and Bannister 2003; Kampa et al. 2006; Lefort et al. 2009; Staiger et al. 2015). However, evidence has accumulated that also a functionally weak but anatomically consistent feedback projection to L IV excitatory neurons is formed (Martin and Whitteridge 1984; Schubert et al. 2003; Larsen and Callaway 2006; Lefort et al. 2009).

Concerning their function, the circuits described above can be interpreted to link ongoing tactile information processing in S1 with (i) the related activity of a multitude of afferent and efferent columns and (ii) different functional cortical areas outside S1. Furthermore, reliable experience-dependent plasticity can be found in these cells (Diamond et al. 1994; Huang et al. 1998; Fox 2002; Feldman and

Brecht 2005), and also a memory storage capacity was repeatedly proposed (Goldman-Rakic 1996; Harris et al. 2002) which should be enabled by the strong context- or learning task-related modulation of the above described circuits.

Layer II (Lamina Granularis Externa)

Cellular composition: Due to the close proximity to the pial surface, most pyramidal neurons of this layer do not form a typical apical dendrite, the origin of which must not strictly be located at the somal pole pointing toward the pia. Thus, these cells have a less triangular and more ovoid to round appearance which led to the classification as a granular layer despite the presence of mainly modified (obliquely oriented or tilted) pyramidal cells (Peters and Kara 1985; van Brederode et al. 2000). Concerning the molecular and neurochemical features of L II neurons, they have not been shown to differ significantly from their L III counterparts. This holds true by and large for excitatory as well as inhibitory neurons (see L III-paragraph). It is, however, conceivable, that fine scale molecular differences between layer II and III (Lein et al. 2007; Molyneaux et al. 2007) do translate to neurochemical differences, yet to be identified at the single cell level. So a significant difference, when layer II neurons are compared to layer III, should be found in their connectivity (see below), in order to justify a separate account of these layers instead of just merging them into a "single layer II/III" as is usually done in rodents. However, recent data suggest a continuum that can be roughly parceled in 3 tiers on statistical grounds (Staiger et al. 2015).

Connectivity: Their local circuitry is not simply dominated by the same afferent pathways as described for layer III. In addition to local and layer IV inputs, a reproducible finding was a prominent layer Va input, not only for septum-related (Shepherd and Svoboda 2005) but also for barrel-related neurons (Lefort et al. 2009; Staiger et al. 2015). In the rodent, a synopsis of recent data suggests that in terms of intracolumnar-translaminar and transcolumnar projections the pyramidal cells in layers II and III differ significantly (Shepherd and Svoboda 2005; Larsen and Callaway 2006; Lefort et al. 2009; Staiger et al. 2015). L II pyramidal cells possess denser and more extensive transcolumnar axonal arbors that preferentially target layers II and Va within the home and neighboring columns whereas L III pyramidal cells issue a much sparser intralaminar transcolumnar projection and mainly target L Vb. Furthermore, L III pyramidal cells also have a substantial number of recurrent axonal collaterals in L IV.

In terms of function, it is presently unclear whether L II really differs from L III because no studies have rigorously addressed this issue. This also holds true for other cortical areas (than S1) or species (than rodents) where these layers have been found to be distinguishable on a reliable basis. However, evidence has recently been presented that some key neuronal properties are actually different between pyramidal cells in these two layers which allows to assume a functional segregation as well (Gur and Snodderly 2008).

Layer Va (Lamina Pyramidalis Interna "a")

Cellular composition: This layer shows a low packing density of medium-sized pyramidal cell bodies and serves as a prominent feature to delineate S1 from the neighboring areas (Palomero-Gallagher and Zilles 2004). The pyramidal cells usually form a slender dendritic tree with a little elaborated terminal tuft in layer I (Schubert et al. 2006; Frick et al. 2007). Since we and others have recently shown that their morphology and its correlation with intrinsic and extrinsic physiological parameters differs dramatically from those of layer Vb, we have suggested that these neurons form a genuine layer Va which should not be regarded as a mere "sublayer" of layer V (Ahissar et al. 2001; Manns et al. 2004; Schubert et al. 2006). In other cortical areas different types of specialization of layer Va may occur (Molnar and Cheung 2006). These authors also presented some data on molecular and neurochemical features of layer Va-pyramidal neurons, which might, however, be mainly of developmental importance. The GABAergic interneurons, on the other hand, so far received little attention in layer Va, although they occur rather frequently at the layer IV/Va-border (Ren et al. 1992; Porter et al. 2001). Of course, Martinotti cells have to be mentioned in this context (Fairen et al. 1984), which in their classic form have a soma located in the infragranular layer and an axon that targets layer I (but also often strongly ramifies in layer IV; (Wang et al. 2004). This has recently been associated with "linearization"-effect on different tactile stimulation strengths (Murayama et al. 2009).

Connectivity: Local circuits are dominated by two sources: (i) intra- and transcolumnar inputs from L Va itself have been found to be especially numerous and strong (Schubert et al. 2006; Frick et al. 2008; Lefort et al. 2009) and (ii) apart from L II (Lefort et al. 2009) translaminar input derives mainly from L IV (Feldmeyer et al. 2005; Schubert et al. 2006) making L Va a putative early cortical interface for lemniscal and paralemniscal sensory information. The latter two publications further suggest that the axonal output of L Va-pyramidal cells is targeted to the same layers from which they receive their major inputs.

Layer Vb (Lamina Pyramidalis Interna "b")

Cellular composition: This layer was usually equated with layer V as such and consists of a mixture of differently sized pyramidal cells. In rodent S1 L Vb can be easily distinguished from L Va due a much higher cell density and occurrence of very large pyramidal cells in L Vb. The ratio of medium-sized to large pyramidal neurons has been estimated to be 2,5: 1 in rat visual cortex (Peters and Kara 1985). Interestingly, here the somato-dendritic morphology of the pyramidal cells correlates well with many connectional, physiological and pharmacological features (cf.(White et al. 1994; Kasper et al. 1994; Schubert et al. 2001; Bodor et al. 2005; Christophe et al. 2005; de Kock et al. 2007; Larsen et al. 2007). Upon somatic current injection and dye filling in vitro (see below) but also in vivo (Zhu and Connors 1999), a basic distinction can be made between (i) neurons with medium-sized somata and a sparse to mediocre dendritic tree which display a so-called regular spiking pattern (RS; consisting of single action potentials with various adaptation rates) and (ii) neurons with large somata and a very well elaborated dendritic tree which display a so-called intrinsically burst-spiking firing pattern (IB; consisting of an initial burst of usually two or three but up to five action potential with interspike intervals of less than 10 ms which ride on a depolarizing envelop) (Connors et al. 1982; McCormick et al. 1985; Chagnac-Amitai et al. 1990; Mason and Larkman 1990; Hefti and Smith 2000; Schubert et al. 2001). Other groups, mostly those which used parasagittal slices, did not find this characteristic bursting behavior and usually call their labeled cells "slender" versus "thick-tufted" neurons but it is very likely that both terminologies represent the same types of neurons (e.g. Markram 1997; Angulo et al. 2003). Recently, a third population of pyramidal cells was characterized in L V(b): cortico-callosal pyramidal cells whose dendrites did not reach L I and did not form a terminal tuft (Le Be et al. 2007). GABAergic interneurons, on the other hand, did not receive the same attention in this layer as the principal cells. It is very likely that all types of GABAergic interneurons are present in L Vb in order to function according to their "standard mode" of operation (Thomson et al. 1996; Markram et al. 2004; Silberberg and Markram 2007).

Connectivity: In terms of the local columnar and transcolumnar connectivity, much has been learned in recent years. It appears that a selective pattern of local connections exists between cells differing in their long-distance targets which is not dependent on the firing pattern or the specific minicolumnar identity of the respective neurons (Deuchars et al. 1994; Markram et al. 1997; Kozloski et al. 2001; Morishima and Kawaguchi 2006; Krieger et al. 2007; Brown and Hestrin 2009). Generally it can be stated that RS pyramidal cells have a more home column-focused connectivity whereas IBs are much more transcolumnarly connected. In addition, RS cells have a patchy input pattern with some focal hot-spots, in contrast to IB cells which possess spatially more uniform and dense input with on average lower strength (Schubert et al. 2001). In the latter study we also showed a strong feedback projection from layer VI which was recently corroborated by paired recordings and shown to originate from the subpopulation of cortico-cortical cells (Mercer et al. 2005).

Functionally, this makes IB-pyramidal neurons the prime candidates for multi-whisker integration (Simons 1978; Ito 1992; de Kock et al. 2007) and cortical pacemaker function which has indeed been shown experimentally (Chagnac-Amitai and Connors 1989).

Layer VIa (Lamina Multiformis "a")

Cellular composition: In no other layer than (the entire) L VI such a high shape variability for spiny neurons ranging from classical over oblique to inverted pyramidal cells to all variations of spiny stellate-like neurons has been found

(Tömböl 1984; Kaneko and Mizuno 1996; Zarrinpar and Callaway 2006; Chen et al. 2009; Andjelic et al. 2009). However, the less multiform "sublayer" VIa (of L VI) shows a high packing density of more uniform small somata often possessing a pyramidal to ovoid shape.

Connectivity: Intracolumnarly, cortico-thalamic projection neurons rarely connect to each other but seem to prefer GABAergic interneurons as their targets (West et al. 2006). On the other hand, they are sparsely afferented by local cortico-cortical excitatory, L V and IV neurons (Figure 7) (Mercer et al. 2005; Zarrinpar and Callaway 2006; Lefort et al. 2009).

Their efferents are not only directed to the thalamus but they also possess terminal axons in L IV (Zhang and Deschenes 1997; Kumar and Ohana 2008). It is debated how specific this feedback projection is in terms of targeting GABAergic interneurons but it is very likely that both, excitatory and inhibitory L IV neurons receive synapses (White and Keller 1987; Staiger et al. 1996). In contrast to the thalamic "driver" input to L IV, these recurrent collaterals have been considered as "modulators" (Stratford et al. 1996; Lee and Sherman 2008).

From a functional point of view we are left with a fascinating circuit construction. Layer VIa forms an important input structure for the very same thalamic nucleus to which it directly projects back. At the same time it issues a second feedback projection to the other (primary) input layer of the thalamus: L IV. The above-mentioned input from L Vb can be considered to close the intracortical loop (IV → III/II → Vb → VI). As such L VI holds a "raw" representation of a tactile

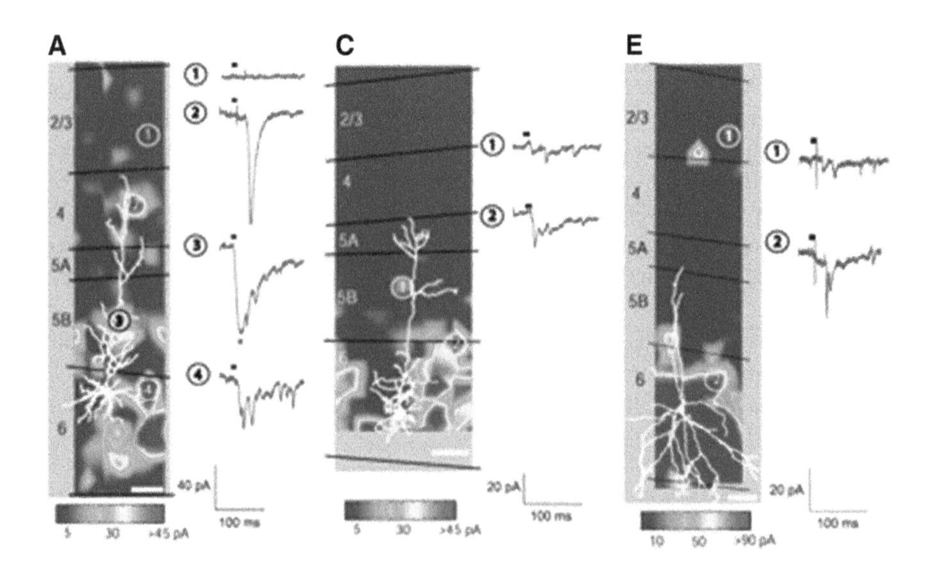

Figure 7 This Figure 4 from (Zarrinpar and Callaway 2006) shows the excitatory (color-coded strength) intracortical translaminar connectivity of three types of cells in L VI of rat visual cortex. It is obvious that only putative cortico-thalamic pyramidal cells (**a**) but not putative cortico-cortical pyramidal cells (**c**) or putative inhibitory interneurons (**e**) receive a multiple "extra-L VI-input". Reproduced with permission

stimulus and is capable of comparing this to the same but cortically processed information. The cortical feedback has a facilitatory effect on the thalamus (Yuan et al. 1986) which is considered, together with the reticular nucleus' effects, to focus attention to salient sensory stimuli (Sillito et al. 1994; Temereanca and Simons 2004).

The following two layers have been studied only infrequently and are not well understood concerning their contribution to cortical circuitry and thus tactile information processing. They will only be briefly touched upon here for the sake of completeness. Interestingly, during cortical development both layers together formed the primordium of the cortex, the so-called preplate, before they were split by the growing cortical plate (Molyneaux et al. 2007).

Layer VIb (Lamina Multiformis "B")

Cellular composition: This layer shows less densely packed neurons than L VIa, already tightly intermingled with many horizontally oriented myelinated fiber pathways. The spiny excitatory neurons in L VIb, which some also call L VII (Reep 2000), are considered to represent the remnants of the subplate but also contain cortical plate-derived neurons, with pyramidal cells being found only very rarely (Friauf et al. 1990; Chen et al. 2009; Andjelic et al. 2009).

Connectivity: The connections of these cells beyond the initial developmental circuits they are involved in (Hanganu et al. 2002) are ill-defined. There seems to be a population of associationally projecting neurons (Zhang and Deschenes 1997) with a strong layer I component (Clancy and Cauller 1999) as well as one projecting to various nuclei of the thalamus, a connection which is likely to be organized reciprocally (Zhang and Deschenes 1998).

Layer I (Lamina Molecularis)

Cellular composition: This layer shows the lowest cell density of all, the rat being cell poorer than the mouse. It is generally accepted that, at the adult stage, virtually all cells located there are GABAergic interneurons (Ren et al. 1992; Beaulieu 1993; Hestrin and Armstrong 1996). A very low number of Cajal-Retzius cells which are now considered as glutamatergic interneurons (Hevner et al. 2003; Ina et al. 2007) were reported in some studies to survive the first two postnatal weeks of development (Derer and Derer 1990; Zhou and Hablitz 1996). However, from this age on, after the completion of the cortical layers, they are of unknown function. It could be hypothesized that in areas in which adult neurogenesis takes place they could still serve the function to guide the migration of these neurons to their final target layer; alternatively, neuronal plasticity may be influenced via interaction of reelin with NMDA receptors (Herz and Chen 2006).

Connectivity: As for the subplate neurons, some of the transient connectivity of the Cajal-Retzius cells has been also inferred by electrical or pharmacological stimulation, the cellular or laminar origin of these connections, however, has remained largely obscure. Altogether, these transient circuits have been proposed to play a role in the formation of an early columnar circuitry (Dupont et al. 2006). For L I-GABAergic interneurons, a plausible hypothesis is that they directly regulate (i) feedforward information-transfer from thalamus (Galazo et al. 2008) and (ii) feedback from "higher" cortical areas which could co-innervate these neurons as well as the terminal tufts of pyramidal cell arborizing in layer I (Vogt 1995; Zhu and Zhu 2004).

Summary and Conclusions

This chapter offers but a S1-biased glimpse on the very rich cortical circuit architecture which is strongly dependent on the individual layers but certainly possesses some cell type-, area- and species-specific traits as well. This circuit analysis is very advanced in the primary somatosensory (barrel) cortex which leads to the expectation that by modeling the basic circuitry contained within one barrel-related column we will begin to understand the basic modes of operation in input, local and output circuits of the cerebral cortex (Markram 2006; Sarid et al. 2007). Based on this in-depth knowledge of cell types and their connections, a reasonable study of experience-dependent plasticity should finally become feasible (Feldman and Brecht 2005; Fox and Wong 2005; Staiger 2006; Clem et al. 2008; Bruno et al. 2009). However, one has to realize that a clear-cut unraveling of the functions of these layer-dependent circuits has still not been achieved. Although recently concepts like decision making by layer Vb-pyramidal neurons (Helmstaedter et al. 2007) have been advanced, most researchers agree that such more complex functions have to be carried out by larger ensembles of neurons distributed over several layers and spatially distant areas in cortical as well as subcortical structures (Krupa et al. 2004; Burke et al. 2005; Diamond et al. 2008).

References

Ahissar, E; Sosnik, R; Bagdasarian, K and Haidarliu, S (2001). Temporal frequency of whisker movement. II. Laminar organization of cortical representations. *Journal of Neurophysiology* 86: 354–367.

Alcantara, S; Ferrer, I and Soriano, E (1993). Postnatal development of parvalbumin and calbindin D28 k immunreactivities in the cerebral cortex of the rat. *Anatomy and Embryology* 188: 63–73.

Alloway, K D (2008). Information processing streams in rodent barrel cortex: the differential functions of barrel and septal circuits. *Cerebral Cortex* 18: 979–989.

Alloway, K D; Zhang, M L and Chakrabarti, S (2004). Septal columns in rodent barrel cortex: Functional circuits for modulating whisking behavior. *Journal of Comparative Neurology* 480: 299–309.

Andjelic, S et al. (2009). Glutamatergic nonpyramidal neurons from neocortical layer VI and their comparison with pyramidal and spiny stellate neurons. *Journal of Neurophysiology* 101: 641–654.
Angulo, M C; Staiger, J F; Rossier, J and Audinat, E (2003). Distinct local circuits between neocortical pyramidal cells and fast-spiking interneurons in young adult rats. *Journal of Neurophysiology* 89: 943–953.
Arion, D; Unger, T; Lewis, D A and Mirnics, K (2007). Molecular markers distinguishing supragranular and infragranular layers in the human prefrontal cortex. *European Journal of Neuroscience* 25: 1843–1854.
Ascoli, G A et al. (2008). Petilla terminology: nomenclature of features of GABAergic interneurons of the cerebral cortex. *Nature Reviews Neuroscience* 9: 557–568.
Beaulieu, C (1993). Numerical data on neocortical neurons in adult rat, with special reference to the GABA population. *Brain Research* 609: 284–292.
Binzegger, T; Douglas, R J and Martin, K A C (2004). A quantitative map of the circuit of cat primary visual cortex. *The Journal of Neuroscience* 24: 8441–8453.
Bodor, A L et al. (2005). Endocannabinoid signaling in rat somatosensory cortex: Laminar differences and involvement of specific interneuron types. *The Journal of Neuroscience* 25: 6845–6856.
Boldogkoi, Z et al. (2009). Genetically timed, activity-sensor and rainbow transsynaptic viral tools. *Nature Methods* 6: 127–130.
Bourassa, J; Pinault, D and Deschenes, M (1995). Corticothalamic projections from the cortical barrel field to the somatosensory thalamus in rats: a single-fibre study using biocytin as an anterograde tracer. *European Journal of Neuroscience* 7: 19–30.
Brecht, M; Roth, A and Sakmann, B (2003). Dynamic receptive fields of reconstructed pyramidal cells in layers 3 and 2 of rat somatosensory barrel cortex. *Journal of Physiology (London)* 553: 243–265.
Brecht, M and Sakmann, B (2002). Dynamic representation of whisker deflection by synaptic potentials in spiny stellate and pyramidal cells in the barrels and septa of layer 4 rat somatosensory cortex. *Journal of Physiology (London)* 543: 49–70.
Broser, P; Grinevich, V; Osten, P; Sakmann, B and Wallace, D J (2008). Critical period plasticity of axonal arbors of layer 2/3 pyramidal neurons in rat somatosensory cortex: Layer-specific reduction of projections into deprived cortical columns. *Cerebral Cortex* 18: 1588–1603.
Brown, S P and Hestrin, S (2009). Intracortical circuits of pyramidal neurons reflect their long-range axonal targets. *Nature* 457: 1133–1U89.
Bruno, R M; Hahn, T T; Wallace, D J; de Kock, C P and Sakmann, B (2009). Sensory experience alters specific branches of individual corticocortical axons during development. *The Journal of Neuroscience* 29: 3172–3181.
Bureau, I; Saint Paul, F and Svoboda, K (2006). Interdigitated paralemniscal and lemniscal pathways in the mouse barrel cortex. *Plos Biology* 4: 2361–2371.
Burke, S N (2005). Differential encoding of behavior and spatial context in deep and superficial layers of the neocortex. *Neuron* 45: 667–674.
Callaway, E M (1998). Local circuits in primary visual cortex of the macaque monkey. *Annual Review of Neuroscience* 21: 47–74.
Carvell, G E and Simons, D J (1987). Thalamic and corticocortical connections of the second somatic sensory area of the mouse. *Journal of Comparative Neurology* 265: 409–427.
Cauli, B et al. (1997). Molecular and physiological diversity of cortical nonpyramidal cells. *The Journal of Neuroscience* 17: 3894–3906.
Cauli, B (2000). Classification of fusiform neocortical interneurons based on unsupervised clustering. *Proceedings of the National Academy of Sciences of the United States of America* 97: 6144–6149.
Celio, M R (1990). Calbindin D-28 k and parvalbumin the rat nervous system. *Neuroscience* 35: 375–475.
Chagnac-Amitai, Y and Connors, B W (1989). Synchronized excitation and inhibition driven by intrinsically bursting neurons in neocortex. *Journal of Neurophysiology* 62: 1149–1162.

Chagnac-Amitai, Y; Luhmann, H J and Prince, D A (1990). Burst generating and regular spiking layer 5 pyramidal neurons of rat neocortex have different morphological features. *Journal of Comparative Neurology* 296: 598–613.

Chakrabarti, S and Alloway, K D (2006). Differential origin of projections from SI barrel cortex to the whisker representations in SII and MI. *Journal of Comparative Neurology* 498: 624–636.

Chapin, J K and Lin, C S (1984). Mapping the body representation in the SI cortex of anesthetized and awake rats. *Journal of Comparative Neurology* 229: 199–213.

Chen, C C; Abrams, S; Pinhas, A and Brumerg, J C (2009). Morphological heterogeneity of layer VI neurons in mouse barrel cortex. *Journal of Comparative Neurology* 512: 726–746.

Christophe, E et al. (2005). Two populations of layer V pyramidal cells of the mouse neocortex: Development and sensitivity to anesthetics. *Journal of Neurophysiology* 94: 3357–3367.

Clancy, B and Cauller, L J (1999). Widespread projections from subgriseal neurons (Layer VII) to layer I in adult rat cortex. *Journal of Comparative Neurology* 407: 275–286.

Clem, R L; Celikel, T and Barth, A L (2008). Ongoing in vivo experience triggers synaptic metaplasticity in the neocortex. *Science* 319: 101–104.

Connors, B W; Gutnick, M J and Prince, D A (1982). Electrophysiological properties of neocortical neurons in vitro. *Journal of Neurophysiology* 48: 1302–1335.

de Kock, C P J; Bruno, R M; Spors, H and Sakmann, B (2007). Layer and cell type specific suprathreshold stimulus representation in primary somatosensory cortex. *Journal of Physiology (London)* 581: 139–154.

de Kock, C P J and Sakmann, B (2008). High frequency action potential bursts (> = 100 Hz) in L2/3 and L5B thick tufted neurons in anaesthetized and awake rat primary somatosensory cortex. *Journal of Physiology (London)* 586: 3353–3364.

Derdikman, D et al. (2006). Layer-specific touch-dependent facilitation and depression in the somatosensory cortex during active whisking. *The Journal of Neuroscience* 26: 9538–9547.

Derer, P and Derer, M (1990). Cajal-Retzius cell ontogenesis and death in mouse brain visualized with horseradish peroxidase and electron microscopy. *Neuroscience* 36: 839–856.

Deuchars, J; West, D C and Thomson, A M (1994). Relationships between morphology and physiology of pyramid-pyramid single axon connections in rat neocortex in vitro. *Journal of Physiology (London)* 478: 423–435.

Diamond, I T (1979). The subdivisions of neocortex: A proposal to revise the traditional view of sensory, motor and association areas. J M Sprague and A N Epstein (Eds.) (pp. 1–43). New York: Academic Press.

Diamond, M E; Huang, W and Ebner, F F (1994). Laminar comparison of somatosensory cortical plasticity. *Science* 265: 1885–1888.

Diamond, M E; Von Heimendahl, M; Knutsen, P M; Kleinfeld, D and Ahissar, E (2008). 'Where' and 'what' in the whisker sensorimotor system. *Nature Reviews Neuroscience* 9: 601–612.

Donoghue, J P; Kerman, K L and Ebner, F F (1979). Evidence for two organizational plans within the somatic sensory-motor cortex of the rat. *Journal of Comparative Neurology* 183: 647–663.

Douglas, R J and Martin, K A C (2007). Mapping the matrix: The ways of neocortex. *Neuron* 56: 226–238.

Dupont, E; Hanganu, I L; Kilb, W; Hirsch, S and Luhmann, H J (2006). Rapid developmental switch in the mechanisms driving early cortical columnar networks. *Nature* 439: 79–83.

Egger, V; Nevian, T and Bruno, R M (2008). Subcolumnar dendritic and axonal organization of spiny stellate and star pyramid neurons within a barrel in rat somatosensory cortex. *Cerebral Cortex*18: 876–889.

Fairen, A; DeFelipe, J and Regidor, J (1984). Nonpyramidal neurons. In: A Peters and E G Jones (Eds.), *Cellular Components of the Cerebral Cortex* (pp. 201–253). New York: Plenum Press.

Feldman, D E and Brecht, M (2005). Map plasticity in somatosensory cortex. *Science* 310: 810–815.

Feldmeyer, D; Egger, V; Lübke, J and Sakmann, B (1999). Reliable synaptic connections between pairs of excitatory layer 4 neurones within a single 'barrel' of developing rat somatosensory cortex. *Journal of Physiology (London)* 521: 169–190.

Feldmeyer, D; Lübke, J and Sakmann, B (2006). Efficacy and connectivity of intracolumnar pairs of layer 2/3 pyramidal cells in the barrel cortex of juvenile rats. *Journal of Physiology (London)* 575: 583–602.

Feldmeyer, D; Roth, A and Sakmann, B (2005). Monosynaptic connections between pairs of spiny stellate cells in layer 4 and pyramidal cells in layer 5A indicate that lemniscal and paralemniscal afferent pathways converge in the infragranular somatosensory cortex. *The Journal of Neuroscience* 25: 3423–3431.

Felleman, D J and van Essen, D C (1991). Distributed hierarchical processing in the primate cerebral cortex. *Cerebral Cortex* 1: 1–47.

Fox, K (2002). Anatomical pathways and molecular mechanisms for plasticity in the barrel cortex. *Neuroscience* 111: 799–814.

Fox, K and Wong, R O L (2005). A comparison of experience-dependent plasticity in the visual and somatosensory systems. *Neuron* 48: 465–477.

Fox, K; Wright, N; Wallace, H and Glazewski, S (2003). The origin of cortical surround receptive fields studied in the barrel cortex. *The Journal of Neuroscience* 23: 8380–8391.

Friauf, E; McConnell, S K and Shatz, C J (1990). Functional synaptic circuits in the subplate during fetal and early postnatal development of cat visual cortex. *The Journal of Neuroscience* 10: 2601–2613.

Frick, A; Feldmeyer, D; Helmstaedter, M and Sakmann, B (2008). Monosynaptic connections between pairs of L5A pyramidal neurons in columns of juvenile rat somatosensory cortex. *Cerebral Cortex* 18: 397–406.

Frick, A; Feldmeyer, D and Sakmann, B (2007). Postnatal development of synaptic transmission in local networks of L5A pyramidal neurons in rat somatosensory cortex. *Journal of Physiology (London)* 585: 103–116.

Galazo, M J; Martinez-Cerdeño, V; Porrero, C and Clasca, F (2008). Embryonic and postnatal development of the layer I-directed ("Matrix") thalamocortical system in the rat. *Cerebral Cortex* 18: 344–363.

Geyer, S; Schleicher, A and Zilles, K (1999). Areas 3a, 3b, and 1 of human primary somatosensory cortex 1. Microstructural organization and interindividual variability. *Neuroimage* 10: 63–83.

Gibson, J R; Beierlein, M and Connors, B W (1999). Two networks of electrically coupled inhibitory neurons in neocortex. *Nature* 402: 75–79.

Gilbert, C D and Wiesel, T N (1979). Morphology and intracortical projections of functionally characterized neurons in the cat visual cortex. *Nature* 280: 120–125.

Glassman, R B (1994). Behavioral specializations of SI and SII cortex: A comparative examination of the neural logic of touch in rats, cats, and other mammals. *Experimental Neurology* 125: 134–141.

Goldman-Rakic, P S (1996). Regional and cellular fractionation of working memory. *Proceedings of the National Academy of Sciences of the United States of America* 93: 13473–13480.

Groh, A et al. (2010). Cell-type specific properties of pyramidal neurons in neocortex underlying a layout that is modifiable depending on the cortical area. *Cerebral Cortex* 20: 826–836.

Gur, M and Snodderly, D M (2008). Physiological differences between neurons in layer 2 and layer 3 of primary visual cortex (V1) of alert macaque monkeys. *Journal of Physiology (London)* 586: 2293–2306.

Hanganu, I L; Kilb, W and Luhmann, H J (2002). Functional synaptic projections onto subplate neurons in neonatal rat somatosensory cortex. *The Journal of Neuroscience* 22: 7165–7176.

Harris, J A; Miniussi, C; Harris, I M and Diamond, M E (2002). Transient storage of a tactile memory trace in primary somatosensory cortex. *The Journal of Neuroscience* 22: 8720–8725.

Hattox, A M and Nelson, S B (2007). Layer V neurons in mouse cortex projecting to different targets have distinct physiological properties. *Journal of Neurophysiology* 98: 3330–3340.

Hayama, T and Ogawa, H (1997). Regional differences of callosal connections in the granular zones of the primary somatosensory cortex in rats. *Brain Research Bulletin* 43: 341–347.

Hefti, B J and Smith, P H (2000). Anatomy, physiology, and synaptic responses of rat layer V auditory cortical cells and effects of intracellular GABA(A) blockade. *Journal of Neurophysiology* 83: 2626–2638.

Helmstaedter, M; de Kock, C P J; Feldmeyer, D; Bruno, R M and Sakmann, B (2007). Reconstruction of an average cortical column in silico. *Brain Research Reviews* 55: 193–203.

Helmstaedter, M; Sakmann, B and Feldmeyer, D (2009a). L2/3 interneuron groups defined by multiparameter analysis of axonal projection, dendritic geometry, and electrical excitability. *Cerebral Cortex* 19: 951–962.

Helmstaedter, M; Sakmann, B and Feldmeyer, D (2009b). Neuronal correlates of local, lateral, and translaminar inhibition with reference to cortical columns. *Cerebral Cortex* 19: 926–937.

Hendry, S H and Hsiao, S S (2003). The somatosensory system. In: L R Squire et al. (Eds.), *Fundamental Neuroscience* (pp. 668–697). San Diego: Academic Press.

Herkenham, M (1980). Laminar organization of thalamic projections to the rat neocortex. *Science* 207: 532–535.

Herkenham, M (1986). New perspectives on the organization and evolution of nonspecific thalamocortical projections. In: E G Jones and A Peters (Eds.), *Cerebral Cortex* (pp. 403–445). New York: Plenum Press.

Herz, J and Chen, Y (2006). Reelin, lipoprotein receptors and synaptic plasticity. *Nature Reviews Neuroscience* 7: 850–859.

Hestrin, S and Armstrong, W E (1996). Morphology and physiology of cortical neurons in layer I. *The Journal of Neuroscience* 16: 5290–5300.

Hevner, R F; Neogi, T; Englund, C; Daza, R A and Fink, A (2003). Cajal-Retzius cells in the mouse: transcription factors, neurotransmitters, and birthdays suggest a pallial origin. *Brain Research. Developmental Brain Research* 141: 39–53.

Holmgren, C; Harkany, T; Svennenfors, B and Zilberter, Y (2003). Pyramidal cell communication within local networks in layer 2/3 of rat neocortex. *Journal of Physiology (London)* 551: 139–153.

Huang, W; Armstrong-James, M A; Rema, V; Diamond, M E and Ebner, F F (1998). Contribution of supragranular layers to sensory processing and plasticity in adult rat barrel cortex. *Journal of Neurophysiology* 80: 3261–3271.

Ina, A et al. (2007). Cajal-Retzius cells and subplate neurons differentially express vesicular glutamate transporters 1 and 2 during development of mouse cortex. *European Journal of Neuroscience* 26: 615–623.

Ito, M (1992). Simultaneous visualization of cortical barrels and horseradish peroxidase-injected layer 5b vibrissa neurones in the rat. *Journal of Physiology (London)* 454: 247–265.

Jones, E G (1975). Varieties and distribution of non-pyramidal cells in the somatic sensory cortex of the squirrel monkey. *Journal of Comparative Neurology* 160: 205–267.

Jones, E G (1981). Anatomy of cerebral cortex: columnar input-output organization. In: F O Schmitt, F G Worden, G Adelmann and S G Dennis (Eds.), *The Organization of the Cerebral Cortex* (pp. 199–235). Cambridge: MIT Press.

Jones, E G (1984a). History of cortical cytology. In: A Peters and E G Jones (Eds.), *Cellular Components of the Cerebral Cortex* (pp. 1–32). New York: Plenum Press.

Jones, E G (1984b). Laminar distributions of cortical efferent cells. In: A Peters and E G Jones (Eds.), *Cellular Components of the Cerebral Cortex* (pp. 521–553). New York: Plenum Press.

Kaas, J H (1997). Topographic maps are fundamental to sensory processing. *Brain Research Bulletin* 44: 107–112.

Kampa, B M; Letzkus, J J and Stuart, G J (2006). Cortical feed-forward networks for binding different streams of sensory information. *Nature Neuroscience* 9: 1472–1473.

Kaneko, T and Mizuno, N (1996). Spiny stellate neurones in layer VI of the rat cerebral cortex. *Neuroreport* 7: 2331–2335.

Karagiannis, A et al. (2009). Classification of NPY-expressing neocortical interneurons. *Journal of Neuroscience* 29: 3642–3659.

Kasper, E M; Larkman, A U; Lübke, J and Blakemore, C (1994). Pyramidal neurons in layer 5 of the rat visual cortex. I. Correlation among cell morphology, intrinsic electrophysiological properties, and axon targets. *Journal of Comparative Neurology* 339: 459–474.

Kawaguchi, Y (1995). Physiological subgroups of nonpyramidal cells with specific morphological characteristics in layer II/III of rat frontal cortex. *The Journal of Neuroscience* 15: 2638–2655.

Killackey, H P; Koralek, K A; Chiaia, N L and Rhoades, R W (1989). Laminar and areal differences in the origin of the subcortical projection neurons of the rat somatosensory cortex. *Journal of Comparative Neurology* 282: 428–445.

Koralek, K A; Olavarria, J and Killackey, H P (1990). Areal and laminar organization of corticocortical projections in the rat somatosensory cortex. *Journal of Comparative Neurology* 299: 133–150.

Kozloski, J; Hamzei-Sichani, F and Yuste, R (2001). Stereotyped position of local synaptic targets in neocortex. *Science* 293: 868–872.

Krieger, P; Kuner, T and Sakmann, B (2007). Synaptic connections between layer 5B pyramidal neurons in mouse somatosensory cortex are independent of apical dendrite bundling. *The Journal of Neuroscience* 27: 11473–11482.

Krupa, D J; Wiest, M C; Shuler, M G; Laubach, M and Nicolelis, M A L (2004). Layer-specific somatosensory cortical activation during active tactile discrimination. *Science* 304: 1989–1992.

Kumar, P and Ohana, O (2008). Inter- and intralaminar subcircuits of excitatory and inhibitory neurons in layer 6a of the rat barrel cortex. *Journal of Neurophysiology* 100: 1909–1922.

Larsen, D D and Callaway, E M (2006). Development of layer-specific axonal arborizations in mouse primary somatosensory cortex. *Journal of Comparative Neurology* 494: 398–414.

Larsen, D D; Wickersham, I R and Callaway, E M (2007). Retrograde tracing with recombinant rabies virus reveals correlations between projection targets and dendritic architecture in layer 5 of mouse barrel cortex. *Frontiers in Neural Circuits* 1: 5.

Le Be, J V; Silberberg, G; Wang, Y and Markram, H (2007). Morphological, electrophysiological, and synaptic properties of corticocallosal pyramidal cells in the neonatal rat neocortex. *Cerebral Cortex* 17: 2204–2213.

Leclerc, S S; Avendano, C; Dykes, R W; Waters, R S and Salimi, I (1994). Reevaluation of area 3b in the cat based on architectonic and electrophysiological studies: Regional variability with functional and anatomical consistencies. *Journal of Comparative Neurology* 341: 357–374.

Lee, C C and Sherman, S M (2008). Synaptic properties of thalamic and intracortical inputs to layer 4 of the first- and higher-order cortical areas in the auditory and somatosensory systems. *Journal of Neurophysiology* 100: 317–326.

Lefort, S; Tomm, C; Sarria, J C F and Petersen, C C H (2009). The excitatory neuronal network of the C2 barrel column in mouse primary somatosensory cortex. *Neuron* 61: 301–316.

Lein, E S et al. (2007). Genome-wide atlas of gene expression in the adult mouse brain. *Nature* 445: 168–176.

Lorente de Nó, R; Fairen, A; Regidor, J and Kruger, L (1992). The cerebral cortex of the mouse (A first contribution - The "Acoustic" Cortex). *Somatosensory & Motor Research* 9: 3–36.

Lübke, J and Feldmeyer, D (2007). Excitatory signal flow and connectivity in a cortical column: focus on barrel cortex. *Brain Structure and Function* 212: 3–17.

Lübke, J; Roth, A; Feldmeyer, D and Sakmann, B (2003). Morphometric analysis of the columnar innervation domain of neurons connecting layer 4 and layer 2/3 of juvenile rat barrel cortex. *Cerebral Cortex* 13: 1051–1063.

Lund, J S (1984). Spiny stellate neurons. In: A Peters and E G Jones (Eds.), *Cellular Components of the Cerebral Cortex* (pp. 255–308). New York: Plenum Press.

Manns, I D; Sakmann, B and Brecht, M (2004). Sub- and suprathreshold receptive field properties of pyramidal neurones in layers 5A and 5B of rat somatosensory barrel cortex. *Journal of Physiology (London)* 556: 601–622.

Markram, H (1997). A network of tufted layer 5 pyramidal neurons. *Cerebral Cortex* 7: 523–533.

Markram, H (2006). The blue brain project. *Nature Reviews Neuroscience* 7: 153–160.

Markram, H; Lübke, J; Frotscher, M; Roth, A and Sakmann, B (1997). Physiology and anatomy of synaptic connections between thick tufted pyramidal neurones in the developing rat neocortex. *Journal of Physiology (London)* 500: 409–440.

Markram, H et al. (2004). Interneurons of the neocortical inhibitory system. *Nature Reviews Neuroscience* 5: 793–807.

Martin, K A and Whitteridge, D (1984). Form, function and intracortical projections of spiny neurones in the striate visual cortex of the cat. *Journal of Physiology (London)* 353: 463–504.

Mason, A and Larkman, A (1990). Correlations between morphology and electrophysiology of pyramidal neurons in slices of rat visual cortex. II Electrophysiology. *The Journal of Neuroscience* 10: 1415–1428.

McCormick, D A; Connors, B W; Lighthall, J W and Prince, D A (1985). Comparative electrophysiology of pyramidal and sparsely spiny stellate neurons of the neocortex. *Journal of Neurophysiology* 54: 782–806.

Mercer, A et al. (2005). Excitatory connections made by presynaptic cortico-cortical pyramidal cells in layer 6 of the neocortex. *Cerebral Cortex* 15: 1485–1496.

Mercier, B E; Legg, C R and Glickstein, M (1990). Basal ganglia and cerebellum receive different somatosensory information in rats. *Proceedings of the National Academy of Sciences of the United States of America* 87: 4388–4392.

Molnar, Z and Cheung, A F P (2006). Towards the classification of subpopulations of layer V pyramidal projection neurons. *Neuroscience Research* 55: 105–115.

Molyneaux, B J; Arlotta, P; Menezes, J R L and Macklis, J D (2007). Neuronal subtype specification in the cerebral cortex. *Nature Reviews Neuroscience* 8: 427–437.

Moore, C I and Nelson, S B (1998). Spatio-temporal subthreshold receptive fields in the vibrissa representation of rat primary somatosensory cortex. *Journal of Neurophysiology* 80: 2882–2892.

Morishima, M and Kawaguchi, Y (2006). Recurrent connection patterns of corticostriatal pyramidal cells in frontal cortex. *The Journal of Neuroscience* 26: 4394–4405.

Mountcastle, V B (1997). The columnar organization of the neocortex. *Brain* 120: 701–722.

Murayama, M et al. (2009). Dendritic encoding of sensory stimuli controlled by deep cortical interneurons. *Nature* 457: 1137–1141.

Palomero-Gallagher, N and Zilles, K (2004). Isocortex. In: G Paxinos (Ed.), *The Rat Nervous System* (pp. 729–757). San Diego: Academic Press.

Paxinos, G and Watson, C (1998). *The Rat Brain in Stereotaxic Coordinates*. San Diego: Academic Press.

Peters, A and Kara, D A (1985). The neuronal composition of area 17 of rat visual cortex. I. The pyramidal cells. *Journal of Comparative Neurology* 234: 218–241.

Peters, A; Kara, D A and Harriman, K M (1985). The neuronal composition of area 17 of rat visual cortex. III. Numerical considerations. *Journal of Comparative Neurology* 238: 263–274.

Petersen, C C H and Sakmann, B (2000). The excitatory neuronal network of rat layer 4 barrel cortex. *The Journal of Neuroscience* 20: 7579–7586.

Pierret, T; Lavallee, P and Deschenes, M (2000). Parallel streams for the relay of vibrissal information through thalamic barreloids. *The Journal of Neuroscience* 20: 7455–7462.

Porter, J T; Johnson, C K and Agmon, A (2001). Diverse types of interneurons generate thalamus-evoked feedforward inhibition in the mouse barrel cortex. *The Journal of Neuroscience* 21: 2699–2710.

Reep, R L (2000). Cortical layer VII and persistent subplate cells in mammalian brains. *Brain Behavior and Evolution* 56: 212–234.

Ren, J Q; Aika, Y; Heizmann, C W and Kosaka, T (1992). Quantitative analysis of neurons and glial cells in the rat somatosensory cortex, with special reference to GABAergic neurons and parvalbumin-containing neurons. *Experimental Brain Research* 92: 1–14.

Rice, F L (1995). Comparative aspects of barrel structure and development. In: E G Jones and I T Diamond (Eds.), *Cerebral Cortex* (pp. 1–75). New York: Plenum Press.

Saez, I and Friedlander, M J (2009). Synaptic output of individual layer 4 neurons in guinea pig visual cortex. *The Journal of Neuroscience* 29: 4930–4944.

Sarid, L; Bruno, R; Sakmann, B; Segev, I and Feldmeyer, D (2007). Modeling a layer 4-to-layer 2/3 module of a single-column in rat neocortex: Interweaving in vitro and in vivo experimental observations. *Proceedings of the National Academy of Sciences of the United States of America* 104: 16353–16358.

Schröder, R and Luhmann, H J (1997). Morphology, electrophysiology and pathophysiology of supragranular neurons in rat primary somatosensory cortex. *European Journal of Neuroscience* 9: 163–176.

Schubert, D; Kötter, R; Luhmann, H J and Staiger, J F (2006). Morphology, electrophysiology and functional input connectivity of pyramidal neurons characterizes a genuine layer Va in the primary somatosensory cortex. *Cerebral Cortex* 16: 223–236.
Schubert, D; Kötter, R and Staiger, J F (2007). Mapping functional connectivity in barrel-related columns reveals layer- and cell type-specific microcircuits. *Brain Structure and Function* 212: 107–119.
Schubert, D; Kötter, R; Zilles, K; Luhmann, H J and Staiger, J F (2003). Cell type-specific circuits of cortical layer IV spiny neurons. *The Journal of Neuroscience* 23: 2961–2970.
Schubert, D et al. (2001). Layer-specific intracolumnar and transcolumnar functional connectivity of layer V pyramidal cells in rat barrel cortex. *The Journal of Neuroscience* 21: 3580–3592.
Shepherd, G M G and Svoboda, K (2005). Laminar and columnar organization of ascending excitatory projections to layer 2/3 pyramidal neurons in rat barrel cortex. *The Journal of Neuroscience* 25: 5670–5679.
Silberberg, G; Grillner, S; LeBeau, F E; Maex, R and Markram, H (2005). Synaptic pathways in neural microcircuits. *Trends in Neurosciences* 28: 541–551.
Silberberg, G and Markram, H (2007). Disynaptic inhibition between neocortical pyramidal cells mediated by Martinotti cells. *Neuron* 53: 735–746.
Sillito, A M; Jones, H E; Gerstein, G L and West, D C (1994). Feature-linked synchronization of thalamic relay cell firing induced by feedback from the visual cortex. *Nature* 369: 479–482.
Simons, D J (1978). Response properties of vibrissa units in rat SI somatosensory neocortex. *Journal of Neurophysiology* 41: 798–820.
Simons, D J and Woolsey, T A (1984). Morphology of Golgi-Cox-impregnated barrel neurons in rat SmI cortex. *Journal of Comparative Neurology* 230: 119–132.
Smith, P H and Populin, L C (2001). Fundamental differences between the thalamocortical recipient layers of the cat auditory and visual cortices. *Journal of Comparative Neurology* 436: 508–519.
Staiger, J F (2006). Immediate-early gene expression in the barrel cortex. *Somatosensory & Motor Research* 23: 135–146.
Staiger, J F; Bojak, I; Miceli, S and Schubert, D (2015). A gradual depth-dependent change in connectivity features of supragranular pyramidal cells in rat barrel cortex. *Brain Structure & Function* 220: 1317–1337.
Staiger, J F et al. (2004). Functional diversity of layer IV spiny neurons in rat somatosensory cortex: Quantitative morphology of electrophysiologically characterized and biocytin labeled cells. *Cerebral Cortex* 14: 690–701.
Staiger, J F; Kötter, R; Zilles, K and Luhmann, H J (2000). Laminar characteristics of functional connectivity in rat barrel cortex revealed by stimulation with caged-glutamate. *Neuroscience Research* 37: 49–58.
Staiger, J F; Zilles, K and Freund, T F (1996). Recurrent axon collaterals of corticothalamic projection neurons in rat primary somatosensory cortex contribute to excitatory and inhibitory feedback-loops. *Anatomy and Embryology* 194: 533–543.
Stratford, K J; Tarczy-Hornoch, K; Martin, K A C; Bannister, N J and Jack, J J B (1996). Excitatory synaptic inputs to spiny stellate cells in cat visual cortex. *Nature* 382: 258–261.
Temereanca, S and Simons, D J (2004). Functional topography of corticothalamic feedback enhances thalamic spatial response tuning in the somatosensory whisker/barrel system. *Neuron* 41: 639–651.
Thomson, A M and Bannister, A P (2003). Interlaminar connections in the neocortex. *Cerebral Cortex* 13: 5–14.
Thomson, A M and Lamy, C (2007). Functional maps of neocortical local circuitry. *Frontiers in Neuroscience* 1: 19–42.
Thomson, A M; West, D C; Hahn, J and Deuchars, J (1996). Single axon IPSPs elicited in pyramidal cells by three classes of interneurones in slices of rat neocortex. *Journal of Physiology (London)* 496: 81–102.
Tömböl, T (1984). Layer VI cells. In: A Peters and E G Jones (Eds.), *Cellular Components of the Cerebral Cortex* (pp. 381–407). New York: Plenum Press.

van Brederode, J F M; Foehring, R C and Spain, W J (2000). Morphological and electrophysiological properties of atypically oriented layer 2 pyramidal cells of the juvenile rat neocortex. *Neuroscience* 101: 851–861.
van Brederode, J F M; Helliesen, M K and Hendrickson, A E (1991). Distribution of the calcium-binding proteins parvalbumin and calbindin-D28 k in the sensorimotor cortex of the rat. *Neuroscience* 44: 157–171.
Van Hooser, S D (2007). Similarity and diversity in visual cortex: is there a unifying theory of cortical computation? *The Neuroscientist* 13: 639–656.
Veinante, P; Lavallee, P and Deschenes, M (2000). Corticothalamic projections from layer 5 of the vibrissal barrel cortex in the rat. *Journal of Comparative Neurology* 424: 197–204.
Vercelli, A; Assal, F and Innocenti, G M (1992). Emergence of callosally projecting neurons with stellate morphology in the visual cortex of the kitten. *Experimental Brain Research* 90: 346–358.
Vogt, B A (1995). The role of layer I in cortical function. In: A Peters and E G Jones (Eds.), *Cerebral Cortex* (pp. 49–80). New York: Plenum Press.
Wang, Y (2004). Anatomical, physiological and molecular properties of Martinotti cells in the somatosensory cortex of the juvenile rat. *Journal of Physiology (London)* 561: 65–90.
Welker, C and Woolsey, T A (1974). Structure of layer IV in the somatosensory neocortex of the rat: Description and comparison with the mouse. *Journal of Comparative Neurology* 158: 437–454.
Welker, E; Hoogland, P V and van der Loos, H (1988). Organization of feedback and feedforward projections of the barrel cortex: A PHA-L study in the mouse. *Experimental Brain Research* 73: 411–435.
West, D C; Mercer, A; Kirchhecker, S; Morris, O T and Thomson, A M (2006). Layer 6 cortico-thalamic pyramidal cells preferentially innervate interneurons and generate facilitating EPSPs. *Cerebral Cortex* 16: 200–211.
White, E L; Amitai, Y and Gutnick, M J (1994). A comparison of synapses onto the somata of intrinsically bursting and regular spiking neurons in layer V of rat SmI cortex. *Journal of Comparative Neurology* 342: 1–14.
White, E L and Keller, A (1987). Intrinsic circuitry involving the local axon collaterals of corticothalamic projection cells in mouse SmI cortex. *Journal of Comparative Neurology* 262: 13–26.
White, E L and Keller, A (1989). *Synaptic Organization of the Cerebral Cortex: Structure, Function, and Theory*. Boston: Birkhäuser.
Wickersham, I R et al. (2007). Monosynaptic restriction of transsynaptic tracing from single, genetically targeted neurons. *Neuron* 53: 639–647.
Wise, S P and Jones, E G (1977). Cells of origin and terminal distribution of descending projections of the rat somatic sensory cortex. *Journal of Comparative Neurology* 175: 129–158.
Woolsey, T A; Welker, C and Schwartz, R H (1975). Comparative anatomical studies of the SmI face cortex with special reference to the occurrence of - barrels - in layer IV. *Journal of Comparative Neurology* 164: 79–94.
Wright, A K; Ramanathan, S and Arbuthnott, G W (2001). Identification of the source of the bilateral projection system from cortex to somatosensory neostriatum and an exploration of its physiological actions. *Neuroscience* 103: 87–96.
Xu, X M and Callaway, E M (2009). Laminar specificity of functional input to distinct types of inhibitory cortical neurons. *The Journal of Neuroscience* 29: 70–85.
Yoshimura, Y; Dantzker, J L M and Callaway, E M (2005). Excitatory cortical neurons form fine-scale functional networks. *Nature* 433: 868–873.
Yuan, B; Morrow, T J and Casey, K L (1986). Corticofugal influences of S1 cortex on ventrobasal thalamic neurons in the awake rat. *The Journal of Neuroscience* 6: 3611–3617.
Zarrinpar, A and Callaway, E M (2006). Local connections to specific types of layer 6 neurons in the rat visual cortex. *Journal of Neurophysiology* 95: 1751–1761.

Zhang, Z W and Deschenes, M (1997). Intracortical axonal projections of lamina VI cells of the primary somatosensory cortex in the rat - A single cell labeling study. *The Journal of Neuroscience* 17: 6365–6379.

Zhang, Z W and Deschenes, M (1998). Projections to layer VI of the posteromedial barrel field in the rat: a reappraisal of the role of corticothalamic pathways. *Cerebral Cortex* 8: 428–436.

Zhou, F M and Hablitz, J J (1996). Morphological properties of intracellularly labeled layer I neurons in rat neocortex. *Journal of Comparative Neurology* 376: 198–213.

Zhu, J J and Connors, B W (1999). Intrinsic firing patterns and whisker-evoked synaptic responses of neurons in the rat barrel cortex. *Journal of Neurophysiology* 81: 1171–1183.

Zhu, Y H and Zhu, J J (2004). Rapid arrival and integration of ascending sensory information in layer 1 nonpyramidal neurons and tuft dendrites of layer 5 pyramidal neurons of the neocortex. *The Journal of Neuroscience* 24: 1272–1279.

S1 Long-Term Plasticity

Daniel E. Shulz and Valerie Ego-Stengel

S1 long-term plasticity refers to persistent modifications in the structure or functioning of the primary somatosensory cortex (S1). These modifications are proposed to underlie learning and memory of tactile information, as well as recovery of function after injury. As in other primary cortical areas, long-term plasticity can arise as a result of peripheral injury (lesion-induced plasticity) or after changes in the spatial or temporal pattern of the sensory input (use-dependent and experience-dependent plasticity).

This article focuses on studies of long-term changes in the barrel cortex, the area of S1 containing a topographic map of the whiskers found on the snout of rodents. It only describes plasticity in the adult brain, as opposed to developmental plasticity in the young during a critical period. Indeed, in the 1960s, it was thought that functional properties in primary sensory cortices were fixed in the adult. In the 1980s, the first experimental evidence for adult plasticity was provided by exploring the interaction between sensory representations and behavioral learning.

Plastic modifications fall in two broad categories. Structural plasticity identifies changes in anatomical properties of neurons and circuits and is relatively restricted in the adult brain. Functional plasticity refers to changes in the response properties of neurons and neural networks, and can be mediated by intrinsic plasticity or synaptic plasticity.

Lesion-Induced Plasticity

Adult plasticity of primary sensory cortices is more readily observed after peripheral injury than after mere alterations of the sensory experience. For the whisker to barrel cortex system, lesion studies have employed either electrolytic or surgical ablation of whisker follicles, or sections of the infraorbital nerve, the branch of the trigeminal nerve that carries sensory information from the whiskers.

D.E. Shulz (✉) · V. Ego-Stengel
Unite de Neurosciences, Information et Complexite, CNRS,
Gif-sur-Yvette, France

T.J. Prescott et al. (eds.), *Scholarpedia of Touch*, Scholarpedia,
DOI 10.2991/978-94-6239-133-8_41

These experimental protocols have two consequences. First, they result in the removal of all sensory evoked activity, and thus cause changes similar to those observed in experience-dependent plasticity. Second, the physical damage to axons or nerve endings triggers additional factors such as the release of neurotrophic factors, which are responsible for modifications at all levels of the system. Peripherally, this is visible by the degeneration and sprouting of nerve endings, and eventually their regeneration (Waite 2001). At the cortical level, lesion-induced plasticity is also more pronounced than experience-dependent plasticity and resembles in some ways developmental plasticity (Fox and Wong 2005), albeit on an attenuated scale and without exhibiting the profound structural changes that can be triggered in neonates.

After nerve transection or ablation of all follicles, electrophysiological recordings reveal a long-term expansion of the representation of body surfaces adjacent to that of the whisker pad into the area initially corresponding to the barrel cortex (Kis et al. 1999; Waite 2001). If peripheral reinnervation is permitted, neurons can regain responsiveness to whiskers or to the scar tissue replacing the whiskers. Otherwise, a central cortical zone stays unresponsive.

Cortical mapping of active regions by 2-deoxyglucose autoradiography after ablation of all follicles in the adult rat led to the same result (Siucinska and Kossut 1994). When some follicles were ablated and others spared, the area activated by the spared vibrissae progressively increased and invaded the neighbouring deprived barrels in the rat (Kossut et al. 1988) and in the mouse (Melzer and Smith 1995, 1998).

Experience-Dependent Plasticity

The characteristics of experience-dependent plasticity in the barrel cortex depend on the nature of the altered experience. Most protocols involve removal of the sensory input from one or several whiskers, either by trimming or by careful whisker plucking sparing the follicle and its innervation. The whiskers are allowed to grow back before testing functional responses. However, the reorganization of cortical networks following a drastic modification of the afferent activity represents an extreme form of plasticity. During the interaction of the animal with its environment, more subtle and specific modifications result from the differential use of peripheral receptors. Modifications in the functional organization of the cortex intervene after sensory discrimination learning or after a classical conditioning procedure.

Plasticity Following Sensory Input Deprivation

Deprivation of input from all whiskers but one for 18 days results in an expansion of the cortical area in which individual neurons respond to the spared whisker

(Glazewski et al. 1996). In the deprived barrels, responses for the spared whisker typically increase by a factor of two, while responses for surround deprived whiskers decrease by about 50 %, and responses to the principal whisker remain constant. More complex deprivation patterns have been used, such as sparing two adjacent whiskers (whisker-pairing), sparing a row of whiskers, removing a single whisker or a single row, or implementing a chessboard pattern. They all conclude that adult cortical plasticity is largely the result of a slow potentiation of responses to the spared peripheral input, on a time course of a few days to several weeks depending on the particular deprivation. Depression for the deprived input, which is dominant and rapid in the developing brain, seems limited in the adult brain (Fox and Wong 2005), although it is sometimes still present (Diamond et al. 1993). Notably, the group of Feldman has shown that the connections from layer IV to layer II/III do exhibit depression in the mature brain after whisker deprivation (Bender et al. 2006). Overall, similar results are obtained by other techniques such as 2-deoxyglucose mapping (Kossut, 1998) and functional magnetic resonance imaging (Alonso et al. 2008).

Plasticity Following Changes in Vibrissal Sensory Experience

Rapid changes in receptive field properties occur when stimulation of an individual whisker is immediately and repeatedly followed by stimulation of a bundle of whiskers in the awake rat. Excitatory responses are enhanced and suppressive responses decreased so that the stimulated whiskers elicit more overall discharge (Delacour et al. 1987).

Aversive or appetitive classical conditioning using a row of vibrissae as the conditioned stimulus also results in an expansion of the cortical area active after stimulation of the trained row compared to the contralateral one (Siucinska and Kossut 1996; Kossut 2001a). This is in agreement with the enhanced responses recorded by evoked potentials in the awake rat during aversive classical conditioning (Wrobel and Kublik 2001).

Alterations of sensory experience can also induce shrinkage of cortical areas, in contrast with the expansions and enhancements more predominantly observed. Passive overstimulation of three whiskers in a row leads to a decrease of their representation measured by autoradiographic deoxyglucose uptake (Welker et al. 1992). Similarly, continued naturalistic experience over weeks leads to a sharpening of the map such that individual whisker representations viewed by intrinsic optical imaging shrink to about half their original size (Polley et al. 2004). Whisker trimming coupled to daily naturalistic experience results in a shrinkage of the cortical representation of the spared input, contrary to the expansion observed in rats remaining in their home cage (Polley et al. 1999).

Mechanisms of Long-Term Plasticity

Any sensory alteration is likely to set into motion multiple plasticity mechanisms operating at multiple sites within a large functionally interconnected circuit (Nelson and Turrigiano 2008). Gross observation can cancel out or mask underlying subtle modifications. Thus, measuring the distributed parallel changes requires a detailed approach that is slowly revealing different components of long-term plasticity.

Cortical Vs Subcortical Sites of Plasticity

Modifications of cortical responses could in principle result from modifications in the subcortical structures that provide the sensory input to the cortex, from changes in the intracortical network itself or from both. At present, most studies have concluded that subcortical plasticity in the adult brain is very limited and cannot explain cortical changes. This is in contrast to plasticity in the neonate, where cortical modifications are considerable and largely reflect cytoarchitectural and functional changes in the brainstem and thalamus (Kossut 2001b).

Arguments in favor of a cortical origin of plasticity in the adult vibrissal system fall in several categories:

- Response changes are absent in the brainstem and thalamus after sensory experience alterations (Glazewski et al. 1998; Wallace and Fox 1999) and much smaller than cortical changes after peripheral injury (Klein et al. 1998; Kis et al. 1999). Furthermore, regions showing adjacent reorganization in the cortex are not necessarily adjacent in the thalamic map, suggesting that the intracortical circuitry is the anatomical substrate (Waite 2001).
- If postsynaptic activity is blocked in the barrel cortex by local muscimol application, the plasticity normally induced by a sensory deprivation does not occur anymore, even though the thalamus responses are intact (Wallace et al. 2001).
- The earliest changes in cortical responses are observed in supragranular layers and, although less prominently, in infragranular layers. In the thalamorecipient layer IV, changes are generally absent (Diamond et al. 1994; Glazewski et al. 1996; Polley et al. 2004). However, several studies have found significant response changes in layer IV barrels (Welker et al. 1992; Siucinska and Kossut 1994; Wallace and Fox 1999). It is likely that these layer IV modifications occur on a longer timescale and as a consequence of supragranular changes (Kossut 2001a; Feldman and Brecht 2005).
- Modifications of the short-latency component of cortical responses, which corresponds to direct thalamocortical excitation, occur very slowly. In contrast, changes in the late component, thought to arise from intracortical connections, are rapid (Armstrong-James et al. 1994). This differential modulation is compatible with the superposition of two plasticity processes, one purely cortical

and rapidly induced, and a second on a longer timescale recruiting the cortico-thalamo-cortical loop.
- The potentiated response to a spared whisker in surround barrels is reduced or abolished by a focal lesion in the spared barrel, or by a cut between the spared and surround barrels, indicating that neural activity takes a cortical route (Fox 1994).
- Structural changes can be directly observed in the cortex (see below).

Structural Changes

In contrast to the profound cytoarchitectural changes that can be observed when manipulations are carried out during development, up to the absence of barrel formation, structural changes are rarely detected in the adult brain and limited to subtle modifications of the circuitry among existing neurons. Maturation of layer IV and its thalamocortical afferents seems to be irreversibly fixed after the end of the critical period, even following peripheral injury. Only one study reported a surprising enlargement of the barrel cortex along the arc axis after long-term naturalistic experience (Polley et al. 2004), which is yet to be explained.

- Axonal remodeling
  Axonal intracortical projections remain susceptible to structural changes in the adult after destruction of afferents. Focal anatomical tracings show that after ablation of all follicles but one, the spared barrel column both sends and receives more elongated and more branching projections to neighbouring barrel columns than deprived ones (Kossut and Juliano 1999). Evidence is sparser in experience-dependent plasticity. In a whisker deprivation study involving whisker trimming, subtle axonal remodeling between neurons already synaptically connected has been shown in adolescent animals (Cheetham et al. 2008). Whisker plucking probably has a stronger effect, inducing axonal restructuring involving both retraction and elongation of inhibitory and excitatory neurons (Marik et al. 2010).
- Axonal endings and bouton dynamics
  Despite the general stability of the overall axon structure, high-resolution two-photon microscopy in vivo has revealed substantial axonal dynamics in the adult brain even in normal conditions (DePaola et al. 2006). This remodeling of branch endings and boutons is cell-type specific. It is highly likely that it also exists, perhaps at an enhanced rate, during long-term plasticity, and thus could participate in the storage of new memories.
- Dendritic growth
  Dendritic branches of excitatory neurons are remarkably stable in the adult barrel cortex in normal conditions, or during experience-dependent plasticity (Trachtenberg et al. 2002). In contrast, as for the axonal structure, changes in

dendritic trees have been observed two months after follicle ablation (Tailby et al. 2005). Interestingly, contrary to excitatory neurons, the dendritic trees of cortical inhibitory interneurons display substantial dynamics even in normal conditions, at least in the visual cortex (Lee et al. 2006).

- Spine turnover and spine morphology changes
  Most dendritic spines are firmly stabilized by the extracellular matrix in the adult brain, so that spine density is roughly fixed after adolescence. Large spines are the postsynaptic substrate of persistent synapses. Based on the measured geometry of axons and dendrites, and on the maximal distance between a presynaptic and a postsynaptic process required for a synapse to form, it has been estimated that about 10 % of all possible synapses actually exist and are persistent in the adult brain (Stepanyants et al. 2002). Spine morphology dynamics and turnover are limited to a small fraction of transient spines constantly trying out new connections. Spine formation and elimination has been shown to be modulated by prolonged altered sensory experience (Zuo et al. 2005, Holtmaat et al. 2006). These studies suggest that during long-term plasticity, a subset of the transient spines eventually forms synapses with presynaptic boutons and stabilizes into large persistent spines (Knott et al. 2006). Concomitantly, some previously-persistent spines are lost (Holtmaat et al. 2006). Both new excitatory and inhibitory synapses may be formed, probably depending on the specific connection considered and the precise nature of the plasticity protocol (Knott et al. 2002, 2006).

Functional Synaptic Changes

Modifications of functional connectivity may occur by changes of synaptic strength. In the adult barrel cortex, because plasticity first appears in layer II/III, focus has been on synapses from layer IV neurons to layer II/III pyramidal neurons. Changes in short-term dynamics and LTD-like depression have been observed after whisker deprivation (Finnerty et al. 1999; Allen et al. 2003). These synapses are indeed susceptible in vitro to spike-timing dependent plasticity (STDP, Feldman 2000; for a general review on STDP see Shulz and Feldman 2011). Remarkably, the timing of inputs to layer II/III neurons changes from one favoring LTP to one favoring LTD during adult cortical map plasticity protocols (Celikel et al. 2004), confirming that plasticity is likely to occur as a result of changes in precise temporal patterns of activity in this system. Although not yet compelling, evidences for the STDP rule in the intact brain, including S1 cortex, have been provided recently (see e.g. Bell et al. 1997; Meliza and Dan 2006; Cassenaer and Laurent 2007; Jacob et al. 2007). From insects to mammals, the presentation of precisely timed sensory inputs drives synaptic and functional plasticity in the intact central nervous system, with similar timing requirements than the in vitro defined STDP rule. Indirect evidence for STDP in the primary somatosensory cortex comes from a combined electrical stimulation of somatosensory afferents and transcranial magnetic stimulation (TMS) of the somatosensory cortex in humans (Wolters et al. 2005). Evoked potentials induced by the TMS were either enhanced or depressed as a function of the order of the paired associative stimulation. More direct evidence of STDP at the synaptic level comes from studies in the primary somatosensory cortex of anesthetized adult rats. Pairing of spontaneous or electrically induced postsynaptic action potentials with afferent excitation elicited by whisker deflections lead to depression of responses to the paired whisker with no significant changes to the unpaired whisker (Jacob et al. 2007; review in Shulz and Jacob 2010).

Synaptic strength changes have also been documented in horizontal connections, notably in the supragranular layers (Lebedev et al. 2000; Cheetham et al. 2007).

Molecular Changes

The initial events in the plasticity processes are stabilized into long-term changes by triggering intracellular protein cascades which eventually lead to modifications in the anatomical and electrical properties of cells and synapses. For example, the presence of the regulatory enzyme alpha-CAMKII and of the transcription factor CREB, both known to underlie various plastic changes, are required for barrel cortex adult plasticity (Fox 2008).

Similarly, there is abundant evidence for changes in the glutamatergic and GABAergic systems, both in lesion-induced and experience-dependent plasticity. AMPA and NMDA glutamate receptors, GABA receptors and local GABA synthesis show modifications whose amplitude and direction depend on the particular protocol employed (Skangiel-Kramska 2001).

General Homeostatic Phenomena

Sudden changes in the afferent sensory drive necessarily result in downstream modifications in the patterns of activity flowing through the sensory system, so that the balance between the different inputs to a postsynaptic neuron may be lost. This can lead to saturation or silencing of the neuron, preventing any form of information processing. Homeostatic mechanisms work to restore the balance of activity and thus stabilize the function of neurons and networks (Nelson and Turrigiano 2008).

- Unmasking of previously ineffective afferents
  After loss of sensory input, a well-known immediate effect in the corresponding cortex is the appearance of a novel response for stimulation of a neighboring input (Wall 1977). This unmasking of previously ineffective input is thought to result from a sudden release from inhibition, triggered by the unbalance of activity of the recorded cell. Evidence has been observed for this short-term mechanism in the barrel cortex after sensory experience manipulation (Kelly et al. 1999; Lebedev

et al. 2000) and after peripheral injury (Waite 2001). In the long-term, it is thought to give way to structural and functional changes that it might contribute to trigger.

- Upregulation of inhibition
  A potentiation of the inhibitory system has been observed, in particular in layer IV and after protocols shown to induce modifications in that layer. For example, an associative fear learning paradigm results in a selective increase in the frequency of spontaneous IPSPs in layer IV excitatory neurons (Tokarski et al. 2007). It has been proposed that the upregulation of inhibition suppresses chronic inputs that are not behaviorally relevant (Welker et al. 1992; Gierdalski et al. 2001), thus leading to a form of sensory habituation.
- Excitability changes
  Conversely, a decrease in the excitability of inhibitory FS interneurons has been observed after 3-week whisker trimming in the mouse, as measured by changes in the intrinsic properties of these cells. These changes were not present in non-FS non-pyramidal cells, suggesting that subnetworks can be regulated independently (Sun 2009).

Influence of the Neuromodulatory Context

The activity within primary sensory cortices is influenced by neuromodulatory afferents, which carry information about the general state of the animal, notably the attention level and cognitive aspects of the ongoing behavior. In addition to their role in sensory perception, neuromodulatory inputs are involved in the cortical plasticity accompanying learning and memory, and in the cellular and synaptic events thought to underlie it (Gu 2002).

Studies of adult plasticity in the barrel cortex have focused in particular on one of the neuromodulators, acetylcholine. Cortical cholinergic afferents originate in the nucleus basalis magnocellularis (NBM) in the basal forebrain (Mesulam et al. 1983). Two main approaches have been used: manipulations of the cholinergic system preventing its action on cortical targets, and direct sensori-cholinergic pairing protocols.

- blockade of cholinergic actions
  Lesions of the basal forebrain cholinergic system generally prevent cortical plasticity from occurring. For example, the plasticity induced by trimming all whiskers but two is reduced if the cortical cholinergic input is selectively lesioned with Ig-saporin (Baskerville et al. 1997; Sachdev et al. 1998). Likewise, systemic or local administration of a cholinergic antagonist prevent the cortical changes in evoked responses observed after pairing two temporally-ordered vibrissal stimuli (Delacour et al. 1990; Maalouf et al. 1998). However, a later study showed that if the animals are engaged in learning a vibrissal task requiring the barrel cortex, plasticity is restored (Sachdev et al.

2000). This suggests that acetylcholine may be necessary for passive experience-dependent plasticity, but that behaviorally-relevant learning triggers other factors that can compensate the cholinergic depletion.

- pairing a sensory event with a cholinergic input
  Two studies have directly tested whether the association between a sensory stimulus and a cholinergic manipulation can induce long-lasting changes in cortical processing. The first one has used pairing of a vibrissal stimulation at a particular frequency with local application of acetylcholine. Response to the paired stimulus was enhanced relative to responses to other frequencies of stimulation. Interestingly, this was only true when acetylcholine was supplied again, suggesting that not only the induction of plasticity but also its expression are dependent on the cholinergic system (Shulz et al. 2000; Ego-Stengel et al. 2001; Shulz et al. 2003). In the second study, an aversive-conditioning protocol was modified by replacing the unconditioned stimulus (classically an electric shock) by an electrical stimulation of the NBM in an anesthetized animal. Cortical changes in evoked responses were very similar to those observed in an awake animal undergoing the normal aversive-conditioning protocol (Wrobel and Kublik 2001).

The role of other neuromodulators, like NA, DA or 5HT in adult barrel cortex plasticity has not been investigated, except for one study suggesting that noradrenaline may also play a permissive role (Levin et al. 1988). In the light of what has been observed in other primary sensory cortices (Bao et al. 2001), it is possible that other neuromodulators than ACh influence the induction and the expression of functional plasticity. Cortical release of noradrenaline for example, produces a reduction of spontaneous and evoked activity in the visual cortex (Ego-Stengel et al. 2002). Through this inhibitory action, the noradrenergic system might provide a reset signal (Dayan and Yu 2006), which is broadcast to the whole cortex, leading to an optimized level of activity for the induction of spike timing dependent plasticity. Other neuromodulators can dynamically regulate timing-based plasticity rules by modifying the biophysical properties of dendrites and the efficacy of spike back propagation (Tsubokawa and Ross 1997; Sandler and Ross 1999). In vitro experiments combining STDP induction protocols concomitant with an increase in neuromodulatory concentrations (Lin et al. 2003; Couey et al. 2007; Seol et al. 2007; Pawlak and Kerr 2008; Zhang et al. 2009) explored how local rules of synaptic plasticity are regulated by global factors acting on several spatial (dendrites, neurons, network) and temporal (milliseconds to minutes) scales.

Summary

S1 long-term plasticity encompasses a wide range of phenomena, underlied by different mechanisms depending on the particular functional modification considered. Many of these mechanisms, if not all, also take place in the plasticity of other sensory cortices or other regions of the brain, and sometimes during development. One challenge of future studies will be to disentangle this complexity and assign identified computational properties in the framework of information processing to each of these modifications.

Internal References

Abraham, W C and Philpot, B (2009). Metaplasticity. *Scholarpedia* 4(5): 4894. http://www.scholarpedia.org/article/Metaplasticity.

Blais, B S and Cooper, L (2008). BCM theory. *Scholarpedia* 3(3): 1570. http://www.scholarpedia.org/article/BCM_theory.

Cudmore, R H and Desai, N S (2008). Intrinsic plasticity. *Scholarpedia* 3(2): 1363. http://www.scholarpedia.org/article/Intrinsic_plasticity.

Shouval, H Z (2007). Models of synaptic plasticity. *Scholarpedia* 2(7): 1605. http://www.scholarpedia.org/article/Models_of_synaptic_plasticity.

Sjöström, J and Gerstner, W (2010). Spike-timing dependent plasticity. *Scholarpedia* 5(2): 1362. http://www.scholarpedia.org/article/Spike-timing_dependent_plasticity.

External References

Allen, C B; Celikel, T and Feldman, D E (2003). Long-term depression induced by sensory deprivation during cortical map plasticity in vivo. *Nature Neuroscience* 6: 291–299.

Alonso, B C; Lowe, A S; Dear, J P; Lee, K C; Williams, S C and Finnerty, G T (2008). Sensory inputs from whisking movements modify cortical whisker maps visualized with functional magnetic resonance imaging. *Cerebral Cortex* 18(6): 1314–1325.

Armstrong-James, M; Diamond, M E and Ebner, F F (1994). An innocuous bias in whisker use in adult rats modifies receptive fields of barrel cortex neurons. *The Journal of Neuroscience* 14: 6978–6991.

Bao, S; Chan, V T and Merzenich, M M (2001). Cortical remodelling induced by activity of ventral tegmental dopamine neurons. *Nature* 412: 79–83.

Baskerville, K A; Schweitzer, J B and Herron, P (1997). Effects of cholinergic depletion on experience-dependent plasticity in the cortex of the rat. *Neuroscience* 80: 1159–1169.

Bell, C C; Han, V Z; Sugawara, Y and Grant, K (1997). Synaptic plasticity in a cerebellum-like structure depends on temporal order. *Nature* 387: 278–281.

Bender, K J; Allen, C B; Bender, V A and Feldman, D E (2006). Synaptic basis for whisker deprivation-induced synaptic depression in rat somatosensory cortex. *The Journal of Neuroscience* 26: 4155–4165.

Cassenaer, S and Laurent, G (2007). Hebbian STDP in mushroom bodies facilitates the synchronous flow of olfactory information in locusts. *Nature* 448: 709–713.

Celikel, T; Szostak, V A and Feldman, D E (2004). Modulation of spike timing by sensory deprivation during induction of cortical map plasticity. *Nature Neuroscience* 7: 534–541.

Cheetham, C E J; Hammond, M S L; Edwards, C E J and Finnerty, G T (2007). Sensory experience alters cortical connectivity and synaptic function site specifically. *The Journal of Neuroscience* 27: 3456–3465.

Cheetham, C E J; Hammond, M S L; McFarlane, R and Finnerty, G T (2008). Altered sensory experience induces targeted rewiring of local excitatory connections in mature neocortex. *The Journal of Neuroscience* 28: 9249–9260.

Couey, J J et al. (2007). Distributed network actions by nicotine increase the threshold for spike-timing dependent plasticity in prefrontal cortex. *Neuron* 54: 73–87.
Dayan, P and Yu, A J (2006). Phasic norepinephrine: A neural interrupt signal for unexpected events. *Network* 17: 335–350.
Delacour, J; Houcine, O and Talbi, B (1987). *Learned* changes in the responses of the rat barrel field neurons. *Neuroscience* 23: 63–71.
Delacour, J; Houcine, O and Costa, J C (1990). Evidence for a cholinergic mechanism of *learned* changes in the responses of barrel field neurons of the awake and undrugged rat. *Neuroscience* 34: 1–8.
DePaola, V et al. (2006). Cell type-specific structural plasticity of axonal branches and boutons in the adult neocortex. *Neuron* 49: 861–875.
Diamond, M E; Armstrong-James, M and Ebner, F F (1993). Experience-dependent plasticity in adult rat barrel cortex. *Proceedings of the National Academy of Sciences of the United States of America* 90: 2082–2086.
Diamond, M E; Huang, W and Ebner, F F (1994). Laminar comparison of somatosensory cortical plasticity. *Science* 265: 1885–1888.
Ego-Stengel, V; Shulz, D E; Haidarliu, S; Sosnik, R and Ahissar, E (2001). Acethylcholine dependent induction and expression of functional plasticity in the barrel cortex of the adult rat. *Journal of Neurophysiology* 86: 422–437.
Ego-Stengel, V; Bringuier, V and Shulz, D E (2002). Noradrenergic modulation of functional selectivity in the cat visual cortex: An in vivo extracellular and intracellular study. *Neuroscience* 111: 275–289.
Feldman, D E (2000). Timing-based LTP and LTD at vertical inputs to layer II/III pyramidal cells in rat barrel cortex. *Neuron* 27: 45–56.
Feldman, D E and Brecht, M (2005). Map plasticity in somatosensory cortex. *Science* 310: 810–815.
Finnerty, G T; Roberts, L S and Connors, B W (1999). Sensory experience modifies the short-term dynamics of neocortical synapses. *Nature* 400: 367–371.
Fox, K (1994). The cortical component of experience-dependent synaptic plasticity in the rat barrel cortex. *The Journal of Neuroscience* 14: 7665–7679.
Fox, K (2008). *Barrel Cortex.* Cambridge University Press.
Fox, K and Wong, R O (2005). A comparison of experience-dependent plasticity in the visual and somatosensory systems. *Neuron* 48: 465–477.
Gierdalski, M et al. (2001). Rapid regulation of GAD67 mRNA and protein level in cortical neurons after sensory learning. *Cerebral Cortex* 11: 806–815.
Glazewski, S; Chen, C M; Silva, A and Fox, K (1996). Requirement for alpha-CaMKII in experience-dependent plasticity of the barrel cortex. *Science* 272: 421–423.
Glazewski, S; McKenna, M; Jacquin, M and Fox, K (1998). Experience-dependent depression of vibrissae responses in adolescent rat barrel cortex. *European Journal of Neuroscience* 10: 2107–2116.
Gu, Q (2002). Neuromodulatory transmitter systems in the cortex and their role in cortical plasticity. *Neuroscience* 111: 815–835.
Hasselmo, M E (1995). Neuromodulation and cortical function: Modeling the physiological basis of behavior. *Behavioural Brain Research* 67: 1–27.
Holtmaat, A; Wilbrecht, L; Knott, W; Welker, E and Svoboda, K (2006). Experience-dependent and cell-type-specific spine growth in the neocortex. *Nature* 441: 979–983.
Jacob, V; Brasier, D J; Erchova, I; Feldman, D and Shulz, D E (2007). Spike timing-dependent synaptic depression in the in vivo barrel cortex of the rat. *Journal of Neuroscience* 27: 1271–1284.
Kelly, M K; Carvell, G E; Kodger, J M and Simons, D J (1999). Sensory loss by selected whisker removal produces immediate disinhibition in the somatosensory cortex of behaving rats. *The Journal of Neuroscience* 19: 9117–9125.

Kis, Z et al. (1999). Comparative study of the neuronal plasticity along the neuraxis of the vibrissal sensory system of adult rat following unilateral infraorbital nerve damage and subsequent regeneration. *Experimental Brain Research* 126: 259–269.

Klein, B G; White, C F and Duffin, J R (1998). Rapid shifts in receptive fields of cells in trigeminal subnucleus interpolaris following infraorbital nerve transection in adult rats. *Brain Research* 779: 136–148.

Knott, G W; Quairiaux, C; Genoud, C and Welker, E (2002). Formation of dendritic spines with GABAergic synapses induced by whisker stimulation in adult mice. *Neuron* 34: 265–273.

Knott, G W; Holtmaat, A; Wilbrecht, L; Welker, E and Svoboda, K (2006). Spine growth precedes synapse formation in the adult neocortex in vivo. *Nature Neuroscience* 9: 1117–1124.

Kossut, M (1998). Experience-dependent changes in function and anatomy of adult barrel cortex. *Experimental Brain Research* 123: 110–116.

Kossut, M (2001a). Imaging functional representations in the barrel cortex - Effects of denervation, deprivation and learning. In: M Kossut (Ed.), *Plasticity of Adult Barrel Cortex* (pp. 175–201). Johnson City, TN: F P Graham Publishing Co.

Kossut, M (2001b). *Plasticity of Adult Barrel Cortex*. Johnson City, TN: F P Graham Publishing Co.

Kossut, M and Juliano, S L (1999). Anatomical correlates of representational map reorganization induced by partial vibrissectomy in the barrel cortex of adult mice. *Neuroscience* 92: 807–817.

Kossut, M; Hand, P J; Greenberg, J and Hand, C L (1988). Single vibrissal cortical column in SI cortex of rat and its alterations in neonatal and adult vibrissa-deafferented animals: A quantitative 2DG study. *Journal of Neurophysiology* 60: 829–852.

Lebedev, M A; Mirabella, G; Erchova, I and Diamond, M E (2000). Experience-dependent plasticity of rat barrel cortex: redistribution of activity across barrel-columns. *Cerebral Cortex* 10: 23–31.

Lee, A et al. (2006). Dynamic remodeling of dendritic arbors in GABAergic interneurons of adult visual cortex. *PLoS Biology* 4(2): e29.

Levin, B E; Craik, R L and Hand, P J (1988). The role of norepinephrine in adult rat somatosensory (Sml) cortical metabolism and plasticity. *Brain Research* 443: 261–271.

Lin, Y W; Min, M Y; Chiu, T H and Yang, H W (2003). Enhancement of associative long-term potentiation by activation of beta-adrenergic receptors at CA1 synapses in rat hippocampal slices. *Journal of Neuroscience* 23: 4173–4181.

Maalouf, M; Miasnikov, A A and Dykes, R W (1998). Blockade of cholinergic receptors in rat barrel cortex prevents long-term changes in the evoked potential during sensory preconditioning. *Journal of Neurophysiology* 80: 529–545.

Marik, S A; Yamahachi, H; McManus, J N J; Szabo, G and Gilbert, C D (2010). Axonal dynamics of excitatory and inhibitory neurons in somatosensory cortex. *PLoS Biology* 8(6): e1000395.

Meliza, C D and Dan, Y (2006). Receptive-field modification in rat visual cortex induced by paired visual stimulation and single-cell spiking. *Neuron* 49: 183–189.

Melzer, P and Smith, C B (1995). Whisker follicle removal affects somatotopy and innervation of other follicles in adult mice. *Cerebral Cortex* 5: 301–306.

Melzer, P and Smith, C B (1998). Plasticity of cerebral metabolic whisker maps in adult mice after whisker follicle removal–I. Modifications in barrel cortex coincide with reorganization of follicular innervation. *Neuroscience* 83: 27–41.

Mesulam, M M; Mufson, E J; Wainer, B H and Levey, A I (1983). Central cholinergic pathways in the rat: an overview based on an alternative nomenclature (Ch1-Ch6). *Neuroscience* 10: 1185–1201.

Nelson, S B and Turrigiano, G G (2008). Strength through diversity. *Neuron* 60: 477–482.

Pawlak, V and Kerr, J N (2008). Dopamine receptor activation is required for corticostriatal spike-timing-dependent plasticity. *Journal of Neuroscience* 28: 2435–2446.

Polley, D B; Chen-Bee, C H and Frostig, R D (1999). Two directions of plasticity in the sensory-deprived adult cortex. *Neuron* 24: 623–637.

Polley, D B; Kvasnák, E and Frostig, R D (2004). Naturalistic experience transforms sensory maps in the adult cortex of caged animals. *Nature* 429: 67–71.

Sachdev, R N; Lu, S M; Wiley, R G and Ebner, F F (1998). Role of the basal forebrain cholinergic projection in somatosensory cortical plasticity. *Journal of Neurophysiology* 79: 3216–3228.

Sachdev, R N; Egli, M; Stonecypher, M; Wiley, R G and Ebner, F F (2000). Enhancement of cortical plasticity by behavioral training in acetylcholine-depleted adult rats. *Journal of Neurophysiology* 84: 1971–1981.

Sandler, V M and Ross, W N (1999). Serotonin modulates spike backpropagation and associated [Ca2+]i changes in the apical dendrites of hippocampal CA1 pyramidal neurons. *Journal of Neurophysiology* 81: 216–224.

Seol, G H et al. (2007). Neuromodulators control the polarity of spike-timing-dependent synaptic plasticity. *Neuron* 55: 919–929.

Shulz, D E; Sosnik, R; Ego, V; Haidarliu, S and Ahissar, E (2000). A neuronal analogue of state-dependent learning. *Nature* 403: 549–553.

Shulz, D E; Ego-Stengel, V and Ahissar, E (2003). Acetylcholine-dependent potentiation of temporal frequency representation in the barrel cortex does not depend on response magnitude during conditioning. *Journal of Physiology Paris* 97: 431–439.

Shulz, D E and Jacob, V (2010). Spike timing dependent plasticity in the intact brain: Counteracting spurious spike coincidences. *Frontiers in Neuroscience* 2: 137.

Shulz, D E and Feldman, D (2011). Spike timing-dependent plasticity. In: P Rakic and J Rubenstein (Eds.), *Comprehensive Developmental Neuroscience*.

Siucinska, E and Kossut, M (1994). Plasticity of mystacial fur representation in SI cortex of adult vibrissectomized rats–A 2DG study. *Neuroreport* 5: 1605–1608.

Siucinska, E and Kossut, M (1996). Short-lasting classical conditioning induces reversible changes of representational maps of vibrissae in mouse SI cortex–a 2DG study. *Cerebral Cortex* 6: 506–513.

Skangiel-Kramska, J (2001). Glutamatergic and GABAergic systems in the adult barrel cortex plasticity. In: M Kossut (Ed.), *Plasticity of Adult Barrel Cortex* (pp. 203–227). Johnson City, TN: F P Graham Publishing Co.

Sun, Q Q (2009). Experience-dependent intrinsic plasticity in interneurons of barrel cortex layer iv. *Journal of Neurophysiology* 102(5): 2955–2973.

Stepanyants, A; Hof, P R and Chklovskii, D B (2002). Geometry and structural plasticity of synaptic connectivity. *Neuron* 34: 275–288.

Tailby, C; Wright, L L; Metha, A B and Calford, M B (2005). Activity-dependent maintenance and growth of dendrites in adult cortex. *Proceedings of the National Academy of Sciences of the United States of America* 102: 4631–4636.

Tokarski, K; Urban-Ciecko, J; Kossut, M and Hess, G (2007). Sensory learning-induced enhancement of inhibitory synaptic transmission in the barrel cortex of the mouse. *European Journal of Neuroscience* 26(1): 134–141.

Trachtenberg, J T et al. (2002). Long-term in vivo imaging of experience-dependent synaptic plasticity in adult cortex. *Nature* 420: 788–794.

Tsubokawa, H and Ross, W N (1997). Muscarinic modulation of spike backpropagation in the apical dendrites of hippocampal CA1 pyramidal neurons. *Journal of Neuroscience* 17: 5782–5791.

Waite, P M (2001). Somatotopic reorganization in the barrel field after peripheral injury. In: M Kossut (Ed.), *Plasticity of Adult Barrel Cortex* (pp. 151–174). Johnson City, TN: F P Graham Publishing Co.

Wall, P D (1977). The presence of ineffective synapses and the circumstances which unmask them. *Philosophical Transactions of the Royal Society B: Biological Sciences* 278: 361–372.

Wallace, H and Fox, K (1999). The effect of vibrissa deprivation pattern on the form of plasticity induced in rat barrel cortex. *Somatosensory & Motor Research* 16: 122–138.

Wallace, H; Glazewski, S; Liming, K and Fox, K (2001). The role of cortical activity in experience-dependent potentiation and depression of sensory responses in rat barrel cortex. *The Journal of Neuroscience* 21: 3881–3894.

Welker, E; Rao, S B; Dörfl, J; Melzer, P and van der Loos, H (1992). Plasticity in the barrel cortex of the adult mouse: effects of chronic stimulation upon deoxyglucose uptake in the behaving animal. *The Journal of Neuroscience* 12: 153–170.

Wolters, A et al. (2005). Timing-dependent plasticity in human primary somatosensory cortex. *Journal of Physiology* 565: 1039–1052.

Wrobel, A and Kublik, E (2001). Modifications of evoked potentials in the rat's barrel cortex induced by conditioning stimuli. In: M Kossut (Ed.), *Plasticity of Adult Barrel Cortex* (pp. 229–239). Johnson City, TN: F P Graham Publishing Co.

Zhang, J C; Lau, P M and Bi, G Q (2009). Gain in sensitivity and loss in temporal contrast of STDP by dopaminergic modulation at hippocampal synapses. *Proceedings of the National Academy of Sciences of the United States of America* 106: 13028–13033.

Zuo, Y; Yang, G; Kwon, E and Gan, W B (2005). Long-term sensory deprivation prevents dendritic spine loss in primary somatosensory cortex. *Nature* 436: 261–265.

Recommended Readings

Feldman, D E and Brecht, M (2005). Map plasticity in somatosensory cortex. *Science* 310: 810–815.

Fox, K (2008). *Barrel Cortex*. Cambridge, UK: Cambridge University Press.

Fox, K and Wong, R O (2005). A comparison of experience-dependent plasticity in the visual and somatosensory systems. *Neuron* 48: 465–477.

Kossut, M (2001). *Plasticity of Adult Barrel Cortex*. Johnson City, TN: F P Graham Publishing Co.

S1 Microcircuits

Dirk Feldmeyer

The whisker-related portion of the primary somatosensory (S1) cortex (the 'barrel cortex') in rodents exhibits a topological arrangement that mirrors that of the peripheral (contralateral) tactile receptors, the vibrissae (whisker hairs) on the rodent's snout. In cortical layer 4, barrel-like cytoarchitectonic units are discernible, each of which represents a single whisker hair (Woolsey and van der Loos 1970). The extension of the borders of layer 4 (L4) barrels throughout all cortical layers has been taken to define a vertical barrel 'column' which is the anatomical correlate of a cortical column (Szentágothai 1975). Because of this well-defined organisation, the barrel cortex has become a model system for investigating synaptic microcircuits and even long-range synaptic connectivity related to the structural representation of sensory receptors (for recent reviews see e.g. Fox 2008; Bosman et al. 2011; Feldmeyer et al. 2013).

This chapter will concentrate on the excitatory microcircuit in the S1 barrel cortex, for which a comprehensive structure and function relationship is now emerging.

Thalamocortical Input to the Barrel Cortex

Sensory information detected by the whisker hairs on the rodent's snout arrives in the barrel cortex via several different 'pathways'. Each of these whisker-to-barrel cortex pathways consists of a four-neuron relay that links the sensory receptor with the S1 barrel cortex: a trigeminal ganglion cell, a trigeminothalamic neuron, a thalamocortical neuron and the intracortical target neurons (see the Scholarpedia chapter by Deschênes 2009 for details). The most prominent parallel pathways are

D. Feldmeyer (✉)
Research Centre Jülich, Institute of Neuroscience and Medicine,
INM-2, Leo-Brandt-Str, 52425 Jülich, Germany

D. Feldmeyer
Department of Psychiatry, Psychotherapy and Psychosomatics,
RWTH Aachen University Hospital, Pauwelsstr. 30, 52074 Aachen, Germany

T.J. Prescott et al. (eds.), *Scholarpedia of Touch*, Scholarpedia,
DOI 10.2991/978-94-6239-133-8_42

the so-called lemniscal, extralemniscal and paralemniscal pathways (Yu et al. 2006; Diamond et al. 2008; Bosman et al. 2011; see also Scholarpedia chapter by Deschênes 2009). These pathways encode different aspects of vibrissae sensory signals.

The lemniscal pathway runs through the principal trigeminal nucleus and projects to the dorsomedial (dm) portion of the ventroposterior medial (VPM) nucleus of the whisker-related thalamus; its cortical targets are described below. Cytoarchitectonic units analogous to the cortical L4 barrels can be found both in the principal trigeminal nucleus and the VPM where they are called barrelettes and barreloids, respectively. The lemniscal pathway combines both whisker motion (whisking) and object location (touch) information and conveys them to the S1 barrel cortex (Yu et al. 2006).

The extralemniscal pathway relays whisker signals through the interpolar spinal trigeminal nucleus. Its trigeminothalamic afferents synapse in the ventrolateral portion of the VPM. Extralemniscal thalamocortical afferents have sparse projections to layers 3, 6 and the septa between the L4 barrels of S1 barrel cortex. However, they densely innervate layers 4 and 6 of the secondary somatosensory (S2) whisker-related cortex (see the Scholarpedia chapter by Deschênes 2009 for details). The extralemniscal pathway conveys only touch signals to the S1 and S2 barrel-related cortex (Yu et al. 2006). Despite its functional relevance, this pathway will not be described in detail in this overview because little is known about their target S1 microcircuits.

The paralemniscal pathway also runs through the interpolar spinal trigeminal nucleus and connects with the posterior medial thalamic nucleus of the thalamus (Pom); the intracortical targets of Pom afferents are described in detail below. In contrast to VPM, the Pom does not show cytoarchitectonic units such as barreloids. The paralemniscal pathway codes only whisker motion information (Yu et al. 2006). All layers of the barrel cortex receive excitatory synaptic input from at least one of these thalamic nuclei (Alloway 2008; Meyer et al. 2010a; Oberlaender et al. 2012; Constantinople and Bruno 2013). In this chapter, the focus is on the intracortical sections of these pathways and their interactions.

Lemniscal Thalamocortical Targets

The *major* thalamorecipient layer in the somatosensory barrel cortex is layer 4 because it contains the highest density of thalamocortical (TC) axons per dendritic length (Jensen and Killackey 1987; Chmielowska et al. 1989; Oberlaender et al. 2012). These TC afferents emanate from VPM and target both excitatory and inhibitory neurons in a L4 barrel (Bruno and Simons 2002). In addition, VPM axons also innervate pyramidal cells and inhibitory GABAergic interneurons in layers 3, 5B and 6A (Meyer et al. 2010b; Meyer et al. 2011; Oberlaender et al. 2012). The majority of their boutons establish synapses onto excitatory neurons because L4 excitatory neurons outnumber L4 interneurons by far (inhibitory/excitatory neuron ratio in layer 4 ~ 8 % vs. 92 %; Lefort et al. 2009; Meyer et al. 2011). Synaptic contacts formed by TC axons comprise only about 10-20 % of the total number of synaptic contacts in layer 4 (White and Rock 1979; Benshalom and White 1986; Schoonover et al. 2014) and are therefore considerably fewer than intracortical synaptic connections. Under in vivo conditions, the average amplitude of unitary VPM-L4 spiny neuron EPSPs is ~1 mV suggesting that this synapse is of very low efficacy (Bruno and Sakmann 2006). However, after whisker stimulation, VPM synaptic input to L4 spiny neurons is highly coincident and synchronous resulting in an efficient TC signal transfer (Jia et al. 2014; Schoonover et al. 2014) and does not require intralaminar amplification (Brumberg et al. 1999; Miller et al. 2001; Bruno and Sakmann 2006).

Paralemniscal Thalamocortical Targets

The major target regions of Pom (paralemniscal) TC input are layers 5A, 1 and 2. In addition, Pom axons also innervate the septa between layer 4 barrels (Alloway 2008; Meyer et al. 2010b; Meyer et al. 2010a; Wimmer et al. 2010; Oberlaender et al. 2012) and layer 3 where they overlap with VPM axons. VPM and Pom inputs have been proposed to be the constituent elements of distinct intracortical columnar pathways, the 'barrel column' and the 'septal column' (Alloway 2008).

In layer 5A, Pom afferents probably establish synaptic contacts with basal dendrites of L5A pyramidal neurons (Petreanu et al. 2009) while in layer 1 and upper layer 2 they may target apical dendritic tufts of L2, L3 and L5 pyramidal neurons as well as L1 and L2 interneurons.

Excitatory S1 Microcircuits

Excitation by the whisker-related thalamus is distributed within the barrel cortex via many distinct microcircuits. These microcircuits can be grouped into different subnetworks or microcircuits; however, these subnetworks are not separate and independent entities but interact at many different levels.

The first prototypical neuronal microcircuit that describes synaptic signalling from the sensory thalamus to and within a primary sensory cortex was the so-called 'canonical microcircuit' by Rodney Douglas and Kevan Martin for the primary visual (V1) cortex (Douglas and Martin 1991; Binzegger et al. 2004; Douglas and Martin 2004; Figure 1). The morphological and synaptic properties of S1 neuronal microcircuits described in this chapter were to a large extent obtained from in vitro

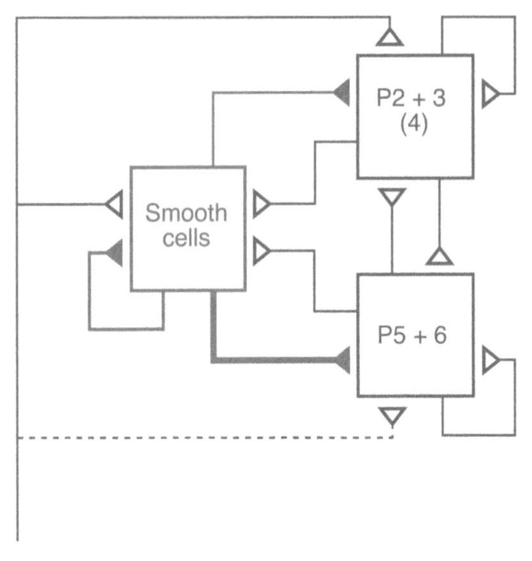

Figure 1 'Canonical neuronal microcircuit' as proposed by Douglas and Martin (1991).Three groups of neurons interact with each other: 'smooth' inhibitory cells, superficial excitatory neurons (P2+3) together with L4 excitatory neurons, and deep layer 5 and 6 neurons. Neurons from each group receive thalamic input, most prominently the superficial neurons. Note the strong recurrent connectivity in this microcircuit model. Dashed and thicker lines indicate weaker or stronger synaptic drive, respectively. *Red*, 'smooth' inhibitory cells; blue, excitatory connections and cells. Modified from Douglas and Martin (1991) with permission of John Wiley and Sons, Inc

paired recording studies. An example of such a correlated structural-functional analysis is shown in Figure 2 for the intralaminar synaptic connection between two neighbouring L2/3 pyramidal cells (Feldmeyer et al. 2006; see also Table 1).

The 'Canonical' S1 Microcircuit

In this 'canonical' S1 microcircuit, afferents from the primary sensory thalamic nucleus target excitatory neurons in the granular and supragranular cortical layers. In addition, inhibitory interneurons receive thalamic input along with pyramidal cells in the infragranular layer, to a lesser extent. Thus, the signal flow in this microcircuit is —generally speaking—from granular and supragranular layers to infragranular layers and subsequently to other cortical areas and subcortical brain regions. However, there is a substantial degree of intra- and translaminar feedback excitation.

In the S1 barrel cortex, a similar but not identical 'canonical' microcircuit exists which can—in approximation—be considered as intracortical part of the lemniscal pathway (Bureau et al. 2006). The main intracortical elements of this subnetwork

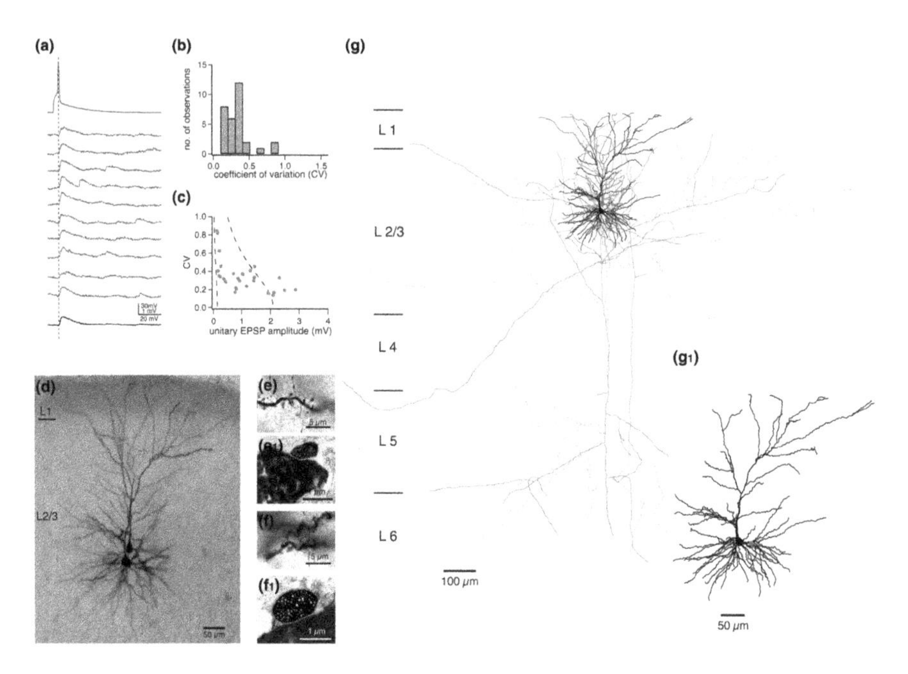

Figure 2 Characterization of a synaptic microcircuit in the S1 barrel cortex. **a** Electrophysiological recordings from synaptically coupled pyramidal cells in layer 2/3. A presynaptic action potential (*red* trace) elicits unitary EPSPs (*grey* traces) in the postsynaptic neuron. The *bottom* trace (black) shows the average time course of a unitary EPSP. **b** Distribution of the coefficient of EPSP amplitude variation (CV) for all recorded L2/3 pyramidal cell pairs. A low CV is indicative of a high reliability of a synaptic connection. **c** The EPSP vs. CV plot shows that even connections in S1 barrel cortex with small mean EPSP amplitudes are relatively reliable. **d** Biocytin-filled, synaptically coupled neuron pair shown in **a**. The connection had two synaptic contacts shown in **e** and **f**; e1 and f1 show electron microscopic verifications of these synaptic contacts; the small structures containing white vesicles are axonal boutons. **g** Reconstruction of the L2/3 pyramidal cell pair shown in **a** and **d**. (g1) Enlarged dendritic domain of the postsynaptic pyramidal cell; the location of the two synaptic contacts is marked by magenta-coloured *circles*. Presynaptic dendrites, *red*; presynaptic axons, blue; postsynaptic dendrites, *black*; axons, *green*. Modified from Feldmeyer et al. 2006 with permission of John Wiley and Sons, Inc

are: L4 excitatory neurons (spiny stellates, star pyramids and pyramidal cells; Staiger et al. 2004; Oberlaender et al. 2012); L2 and L3 pyramidal cells and L5B pyramidal cells (Figure 3). Most studies on identified synaptic connections have concentrated on vertical, feed-forward signalling but a high degree of intralaminar reciprocal synaptic connections have also been identified. In addition to the vertical signal flow in a barrel column, there is mounting evidence for strong horizontal, transcolumnar synaptic signalling in the S1 barrel cortex and to other cortical areas (Zhang and Deschênes 1997; Deschênes et al. 1998; Bruno et al. 2009; Narayanan et al. 2015). However, they will only be mentioned in passing as part of this review.

Table 1 Characteristics of excitatory synaptic connections in a 'canonical' microcircuit' of S1 barrel cortex

Connection type	Apparent connectivity	Mean EPSP amplitude	CV	Failure rate	Species	References
L4 exc. neuron → L4 exc. neuron	0.3	1.6 ± 1.5 mV	0.37 ± 0.16	5.3 ± 7.8	Rat	Feldmeyer et al. (1999)
	0.24	1.0 ± 0.1 mV	n.d.	n.d.	Mouse	Lefort et al. (2009)
L4 exc. neuron → L2/3 PC	~0.1-0.2	0.7 ± 0.6 mV	0.27 ± 0.13[a]	4.9 ± 8.8 %	Rat	Feldmeyer et al. (2002)
	0.12 (L2)	1.0 ± 0.2 mV (L2	n.d	n.d	Mouse	Lefort et al. (2009)
	0.15 (L3)	0.6 ± 0.1 mV (L3)	n.d	n.d		
L2/3 PC → L2/3 PC	0.09 (local)	0.7 ± 0.6 mV	n.d	n.d	Rat	Holmgren et al. (2003)
	~0.1	1.0 ± 0.7 mV	0.33 ± 0.18	3.2 ± 7.8 %	Rat	Feldmeyer et al. (2006)
	0.09 (L2)	0.6 ± 0.1 mV (L2)	n.d	n.d	Mouse	Lefort et al. 2009
	0.19 (L3)	0.8 ± 0.1 mV (L3)	n.d	n.d		
L2/3 PC → tt L5 PC	0.08 (L2)	0.2 ± 0.0 mV (L2)	n.d	n.d	Mouse	Lefort et al. (2009)
	0.12 (L3)	0.2 ± 0.0 mV (L2)	n.d	n.d		
tt L5 PC → tt L5 PC	~0.1	1.3 ± 1.1 mV	0.52 ± 0.41	14 ± 7 %	Rat	Markram et al. (1997)
	−0.07	0.7 ± 0.2 mV	n.d.	n.d.	Mouse	Lefort et al. (2009)

Each synaptic connection is characterised by the apparent connectivity, the mean EPSP amplitude, the coefficient of EPSP amplitude variation (CV) and the failure rate (precentage of APs that did not elicit an EPSP). All synaptic parameters listed here were determined in the presence of 2 mM CaCl2 in the aCSF. L4 exc. neuron: L4 excitatory neuron, i.e. L4 spiny stellate, L4 star pyramid or L4 pyramidal cell. L2/3 PC, pyramidal cells in layer 2 and 3; tt L5 PC, thick-tufted L5 pyramidal cell

[a]for this connection a release probability of 0.8 was determined (Silver et al. 2003)

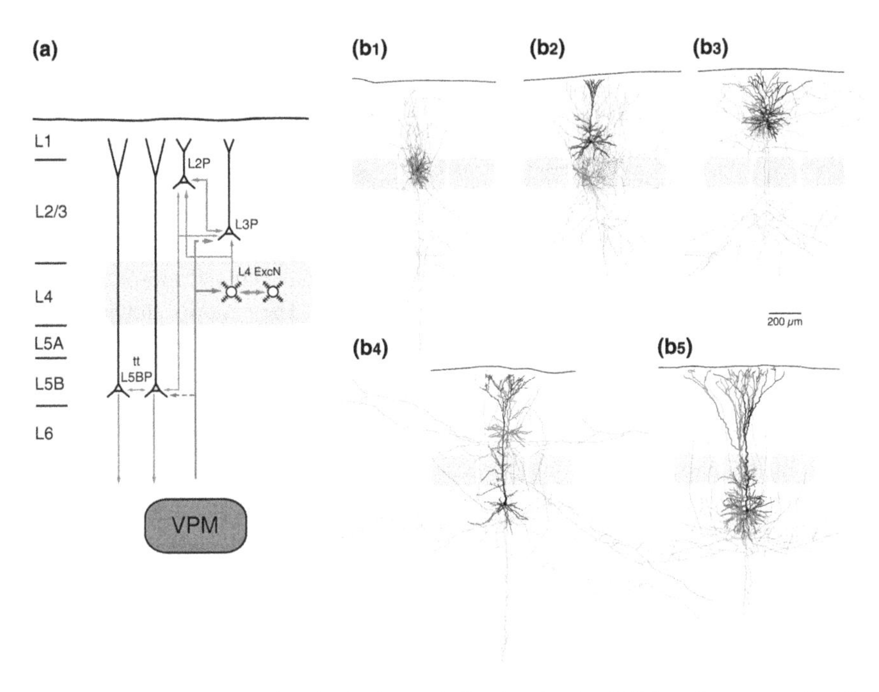

Figure 3 Synaptic connections in the 'canonical' microcircuit of S1 barrel cortex. **a** Simplified schematic drawing of synaptic connections in the synaptic connectivity in the S1 barrel cortex. Only L4 excitatory neurons (L4 ExcN: spiny stellate, star pyramids and pyramidal cells), L2, L3 and the thick-tufted (tt) L5B pyramidal cells are included here. **b** Morphological reconstructions of individual pairs of the excitatory synaptic connection depicted in (**a**): 1, L4-L4 connection, 2, L4-L2/3 connection, 3, L2/3-L2/3 connection, 4, L2-L5B connection and 5, L5B-L5B connection. Reconstructions were obtained from paired recordings and simultaneous biocytin fillings of synaptically coupled neurons in barrel cortex brain slices (b1–b5 modified from Feldmeyer et al. 1999; Silver et al. 2003; Feldmeyer et al. 2006; Reyes et al. 1999; Markram et al. 1997). L4 barrels are depicted in light grey; L2P, L3P, L2 and L3 pyramidal cells; L4ExcN, L4 excitatory neuron, ttL5BP, thick-tufted L5B pyramidal cell. Schematic barrels are depicted in *light grey*. Colour code as for Figure 2g

In simplified terms, excitation arriving from the VPM results in strong recruitment of L4 excitatory neurons. L3 and L5B pyramidal cells are also recruited, but to a lesser degree. Within a L4 barrel, these neurons are recurrently interconnected with a high connectivity ratio of $\sim$ 0.3 (see Table 1). Their translaminar targets are mainly pyramidal cells in supragranular layers (i.e. layers 2 and 3; but see below). In particular spiny stellate cells, the most numerous L4 neuron type, show a largely 'barrel column'-confined, vertical axonal projection and predominantly innervate pyramidal cells in both layer 2 and 3 (Lübke et al. 2000). L2 and L3 pyramidal cells innervate other pyramidal cells in their 'home' layer but provide also substantial synaptic input to thick-tufted pyramidal neurons in infragranular layer 5B. In turn, these L5B pyramidal cells provide output to subcortical target structures such as the thalamus, caudate-putamen, inferior colliculi and cerebellum (e.g. Zingg et al. 2014).

The individual intracortical excitatory synaptic connections in this 'canonical' microcircuit in the S1 barrel cortex have been characterised in detail. These microcircuits comprise the synaptic connections between L4 excitatory neurons and L2/3 pyramidal cells, between L2/3 pyramidal cells and thick-tufted L5B pyramidal cells as well as the intralaminar reciprocal, recurrent connections between L4, L2/3 and L5B neurons (Markram et al. 1997; Feldmeyer et al. 1999; Feldmeyer et al. 2002; Holmgren et al. 2003; Silver et al. 2003; Feldmeyer et al. 2006; Lefort et al. 2009). Figure 3 shows a schematic wiring diagram (Figure 3a) and morphological reconstructions of these synaptic connections (Figure 3b). The majority shows a high release probability, as evidenced by a low EPSP failure rate in response to a presynaptic action potential (AP), a low variation in the EPSP amplitude (low coefficient of EPSP variation, c.v.) and a low paired pulse ratio (see Figure 2 for an example; Table 1). For the L4-L2/3 connections (Figure 3b2), this has been tested directly. In the presence of 2 mM extracellular Ca^{2+}, the release probability of L4-L2/3 synapses was 0.8 (Silver et al. 2003). The number of synaptic contacts for individual connections varied between 2 and 8, the majority of which was located on the basal dendrites.

This data shows that synaptic connections in the 'canonical' microcircuit are reliable, thereby ensuring an efficient sensory signal transfer in the barrel column. However, in vivo labelling of the axons of L2/3 and L5 pyramidal cell also revealed a strong and prominent horizontal projection domain both in layers 2, 3 and/or 5. These horizontal collaterals project across the entire barrel field (Bruno et al. 2009; Oberlaender et al. 2011; Narayanan et al. 2015), thereby integrating whisker-touch induced synaptic excitation in different barrel columns. In addition, there are long-range axon collaterals projecting to cortical areas outside the S1 barrel cortex that serve the interaction between is and other cortical areas. (e.g. the sensorimotor loop; see also Petreanu et al. 2007; Aronoff et al. 2010; Mao et al. 2011; Petreanu et al. 2012; Zingg et al. 2014).

Non-canonical, Intracortical Neuronal Networks

Apart from the excitatory neurons in the 'canonical' microcircuit described above, there are several other synaptic connection types that do not fit this scheme. In general, these synaptic connections are part of the paralemniscal thalamic pathway from the Pom to the neocortex. However, the 'paralemniscal' intracortical microcircuit described here is not a segregated, independent pathway but highly interdigitated with the 'canonical', largely lemniscal pathway (Figure 4a). Constituent elements of the intracortical paralemniscal pathway are L5A pyramidal cells as well as the L2, L3 pyramidal cells and the apical tufts of L5B pyramidal cells, since these neurons are located in the major target regions of Pom afferents (Figure 4a; Bureau et al. 2006 see also Oberlaender et al. 2012). L4 septal neurons are also part of this microcircuit (see above; Wimmer et al. 2010). The major target neurons for

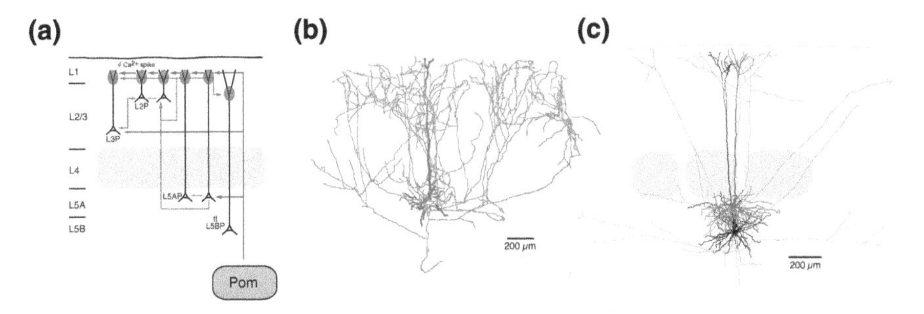

Figure 4 Synaptic connections in the intracortical paralemniscal microcircuit of S1 barrel cortex (**a**) Simplified schematic drawing of synaptic connections in the barrel cortex in the Pom pathway. Note that pyramidal cells in layers 2,3 and 5 receive thalamic input at their apical dendritic tuft where the Ca^{2+} spike initiation zone is located (marked by *light red* ellipse). L2 and L3 pyramidal cell are also likely to receive Pom input onto their basal dendrites. **b** Axonal projection pattern of L5A pyramidal cell filled in vivo revealing an extensive axonal collaterisation at the layer 1/layer 2 border. Modified from Oberlaender et al. 2011. **c** Synaptic connection between two L5A pyramidal cells. L2P, L3P, L5A, L5BP: pyramidal cells in layer 2, 3, 5A and 5B, respectively. Barrels are depicted in *light grey*. Colour code for reconstructions as in Figure 2g. *Panel b* reproduced with permission of the National Academy of Science of the USA

the Pom afferents are the slender-tufted L5A pyramidal cells. These pyramidal cells have an extensive axonal projection in layers 1 and 2 (Figure 4b; Oberlaender et al. 2011). The Pom afferents will establish synaptic contacts not only with other L5A pyramidal cells (Figure 4c) and L2 pyramidal cells, but also the apical dendritic tufts of pyramidal cells in layer 3, 5A and 5B. This is of functional relevance because the dendritic tuft region of pyramidal cells has a high density of Ca^{2+} channels. This site has been shown to be the initiation zone for dendritic Ca^{2+} spikes (Figure 4a; Larkum et al. 1999; Larkum and Zhu 2002; for reviews see Spruston 2008; Larkum 2013) whose activation is involved in coincidence detection mechanisms. Although the L5A-L2/3 pyramidal cell connectivity should be high (given their axodendritic domains; s. Figure 4b), Lefort and coworkers (Lefort et al. 2009) report only low connectivity ratios for supragranular projections (L3: 0.02, L2: 0.04). This is despite the strong synaptic input from these layers (L2: 0.1; L3: 0.06) and the fact that the L2 and L3 pyramidal cell axon density is comparatively lower in layer 5A. These conflicting results are most likely due to slice artifacts (i.e. the truncation of axonal collaterals) but also to the difficulty in detecting EPSPs from synaptic contacts on distal dendritic structures.

The only synaptic connection in the 'paralemniscal' microcircuitry that has been studied in significant detail is the L5A-L5A connection (Figure 4c; Frick et al. 2008) which has relatively high connectivity ratio of $\sim$ 0.2 (Lefort et al. 2009). It exhibits a low EPSP failure rate (2 %) and CV (0.3) indicating that it similarly reliable as most other excitatory synaptic connections in the barrel cortex. Synaptic contacts are found on both basal dendrites and in the apical dendritic tuft. To date, the properties of translaminar connections in this 'paralemniscal' microcircuit are not known.

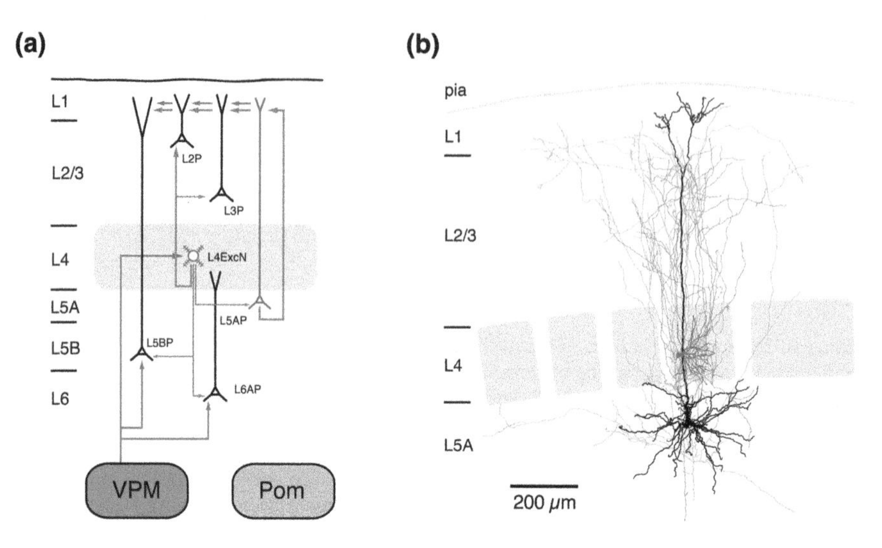

Figure 5 Interdigitation of lemniscal (VPM) and paralemniscal (Pom) intracortical pathways (**a**) Simplified schematic drawing of synaptic connections showing interaction between the two TC pathways. L4 excitatory neurons which receive VPM TC input are connected to excitatory neurons in all cortical layers although the connectivity in granular and supragranular layers is high than in infragranular layers. They interdigitate with L5A pyramidal cells which are the major targets of the Pom TC input. In addition, L5A pyramidal cells innervate also L5B pyramidal cells, which are also target neurons of the VPM TC input on both basal and apical dendrites. Finally, L4 excitatory neurons innervate L2 and L3 pyramidal cells, mostly on their basal dendrites; both neuron types receive either direct or indirect Pom input via Pom afferents and/or L5A pyramidal cell axons. **b** Synaptically coupled pair of a L4 spiny stellate neurons and a slender-tufted L5A pyramidal cell. This synaptic connection represents an early, short latency link between the VPM and Pom TC input into the neocortex. L4 barrels are depicted in *light grey*; L2P, L3P, L2 and L3 pyramidal cells; L4ExcN, L4 excitatory neuron, ttL5BP, thick-tufted L5B pyramidal cell. Colour code for reconstruction as in Figure 2g

Early Convergence of VPM and POm Pathways in the S1 Barrel Cortex

As mentioned earlier, the lemniscal 'canonical' microcircuit and the paralemniscal microcircuit interact at several different stages. First of all, both thick-tufted L5B and L3 pyramidal cells receive monosynaptic and prominent TC input from both VPM (to their basal dendrites) and Pom (to their apical tufts, Figure 5a); this input is near-simultaneous. Further interaction occurs at the level of layer 4 and 5A which are the dominant target regions of VPM and Pom TC afferents, respectively. L4 excitatory neurons are also monosynaptically connected to L5A pyramidal cell with a connectivity ratio of 0.1. This connection shows a high reliability (similar to other excitatory connections in the S1 barrel cortex, see Table 1); synaptic contacts are established on both basal and apical dendrites. This connection provides another short-latency link between the lemniscal and the paralemniscal pathways in the barrel cortex (Figure 5b; Feldmeyer et al. 2005; Bureau et al. 2006; Schubert et al. 2006; Lefort et al. 2009).

The lemniscal and paralemniscal pathways also converge at several other stages of the neuronal network of the barrel column as shown in Figures 4a and 5a, albeit disynaptically via slender-tufted L5A pyramidal cells and/or L4 excitatory neurons. For example, L5A pyramidal cells innervate apical tufts of L2, L3 and thick-tufted L5B pyramidal cells, all of which receive monosynaptic (as mentioned earlier) and/or disynaptic VPM input to their basal dendrites (via L4 excitatory neurons; Lefort et al. 2009; Feldmeyer et al. 2002; Staiger et al. 2014; Qi et al. 2015) (Figure 5a). The near-coincident activation of basal dendrites and the apical dendritic tufts of these pyramidal cells through VPM and Pom synaptic input has been suggested to play a role during sensorimotor behavioural paradigms, such as object location during active whisking (Ahissar et al. 2000; Oberlaender et al. 2011). Alternatively, this 'sensorimotor' coincidence detection may result from simultaneous synaptic input from both the primary motor cortex and the VPM (Xu et al. 2012).

Corticothalamic Projections

The S1 barrel cortex contains also several synaptic microcircuits that provide TC-CT feedback. The central elements of this feedback circuit are the L6 corticothalamic (CT) pyramidal cells that receive strong and depressing synaptic input from VPM (Figure 6a; Beierlein and Connors 2002; Cruikshank et al. 2010). L6A CT axon project back mainly to the same thalamic nucleus; however, projections to Pom as well as to both of these thalamic nuclei have been described (Figure 6a; s. Zhang and Deschênes 1997; Deschênes et al. 1998). In contrast to pyramidal cells from other cortical layers, the majority of L6A pyramidal cells have short apical dendrites with sparse or even no apical tufts terminating predominantly in layers 3 to 5A. L6A pyramidal cells that provide only CT to VPM are predominantly located in the upper half of layer 6. Most L6A CT pyramidal cells have rather short axons—even when in vivo dye fillings are used—with a narrow axonal domain that is almost confined to a 'barrel column'. In marked contrast to L6A CT pyramidal cells, corticocortically (CC) projecting pyramidal cells in layer 6A have extensive, profusely branching axons and send collaterals to other cortical areas such as the S2 cortex or motor cortex but have no obvious subcortical target (Zhang and Deschênes 1997; Kumar and Ohana 2008; Chen et al. 2009; Tanaka et al. 2011; Pichon et al. 2012).

In general, L6 pyramidal cells have been reported to receive synaptic input only from layers 4, 5 and 6; the synaptic connectivity is reportedly low for these pathways, ranging from 0.03 to 0.06 (Lefort et al. 2009). Thus, apart from direct, monosynaptic TC VPM input, L4 excitatory neurons and thick-tufted L5B pyramidal cells provide disynaptic VPM input and thus serve as intracortical elements

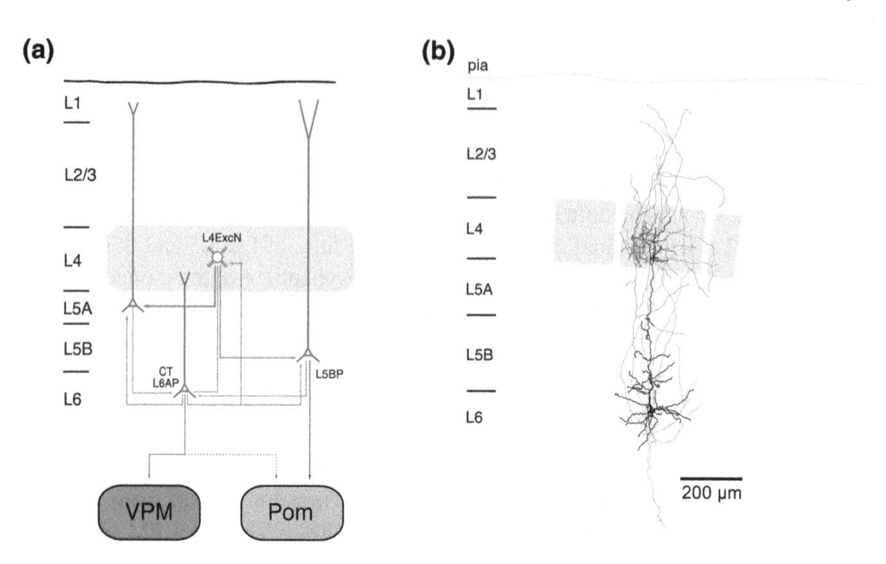

Figure 6 Synaptic microcircuitry in corticothalamic connections. Simplified schematic drawing of synaptic connections that play a role in corticothalamic feedback. VPM and its primary intracortical target neurons are shown in *red*, Pom and its primary cortical target in magenta. Corticothalamic output pathways is coloured in violet. L6A CT pyramidal cells provide synaptic input to the same thalamic nucleus from which they receive synaptic input; in contrast, L5B pyramidal cells receive mainly VPM TC input but project to Pom. L4 excitatory neurons interact with L6A pyramidal cells in an intracortical feedback loop (see text for details). **b** Synaptically coupled pair of a presynaptic L4 spiny stellate neuron and a L6A corticothalamic pyramidal cell. L4 barrels are depicted in *light grey*; L4ExcN, L4 excitatory neuron, L5AP, L5A pyramidal cell; L5BP, thick-tufted L5B pyramidal cell; CT L6AP, L6A CT pyramidal cell. Colour code for reconstruction as in Figure 2g

of a TC-CT feedback loop (Figure 6b). Translaminar synaptic input showed paired pulse EPSP depression indicative of a high release probability (Qi and Feldmeyer 2015; see also Mercer et al. 2005; Thomson 2010). L6A CC and CT pyramidal project also back to layers 5 and 6 (Mercer et al. 2005; Thomson 2010). In addition, L6 projections to L4 excitatory neurons have been identified in the barrel cortex (Qi et al. 2015, cf. Stratford et al. 1996 for V1 cortex). These connections are unique because they show EPSP facilitation (Qi et al. 2015) in contrast to the paired-pulse depression observed for most other TC and CC connections in the S1 barrel cortex. Thus, L4 excitatory neurons and L6 pyramidal cells also provide a secondary, intracortical feedback control with the L4-L6 connection showing a delayed recruitment because of the paired pulse depression.

It has been suggested, that the L4 input to L6 CT pyramidal cells is strong and focussed to its home 'barrel column', indicating that neurons in this layer are involved in shaping the cortical modulation of activity in the somatosensory thalamus (Tanaka et al. 2011; Qi and Feldmeyer 2015). Notably, most synaptic contacts established by L4 excitatory neurons are on apical tufts of L6A pyramidal cells

(Qi and Feldmeyer 2015). Under this condition dendritic filtering is substantial, which may explain the low L4-L6 connectivity ratio found in in vitro experiments (Lefort et al. 2009). In addition to the L4 input, a strong synaptic input from L6 itself is likely to exist because CC projecting L6 pyramidal cells have extensive, profusely branching axons that project to other cortical areas such as the S2 cortex or the motor cortex, without obvious subcortical target structures. The connection probability is comparable to other intracortical excitatory connections (Mercer et al. 2005; Qi and Feldmeyer, unpublished; but see Lefort et al. 2009 for lower connectivity ratios). This is another example of a distinct, apical dendritic tuft targeting and basal dendrite targeting synaptic arrangement (as found in VPM and Pom innervation or L4 and L5A input to basal and apical dendritic tufts, respectively, of L5B pyramidal cells) and may serve in the detection of coincident synaptic stimuli.

Another neuronal population involved in CT signalling is a subpopulation of thick-tufted L5B pyramidal cells that is innervated by VPM afferents but projects back to Pom. In contrast tp L6A CT pyramidal cells, this CT projection does not target the same thalamic nucleus from which it receives TC input. L5B CT pyramidal cells may serve as a feedback control of the VPM over the Pom in a VPM-L5B-Pom loop. They receive VPM input (see above) and have axons that form one or two clusters of large diameter (2-8 μm) presynaptic boutons in Pom (Hoogland et al. 1991; Bourassa et al. 1995; Groh et al. 2008; Groh et al. 2013). L5B-Pom synapses show a high release probability of 0.8 and are therefore very efficacious. Single unitary EPSPs can elicit several APs in the thalamic relay neurons thereby acting as 'drivers' of the Pom (Groh et al. 2008). However, spontaneous activity of the L5B pyramidal cells significantly reduces this 'driving' action through a strong short-term synaptic depression. Because of this it has been suggested that the L5B-Pom synapse works in two functional modes. When the spontaneous activity of this synapse is high, only synchronous activity of several L5B inputs will induce spiking of the thalamic neurons. Therefore, the synapse will act as a coincidence detector. When the spontaneous activity is sparse (which is the case during active whisking or cortical silence) a single L5B input will 'fire' the postsynaptic Pom neuron. Thus, the degree of spontaneous activity determines whether the CT L5B-Pom synapse acts as a detector of synchronous neuronal activity or of cortical silence (Groh et al. 2008).

According to another hypothesis the L5B-Pom connection is part of a feed-forward, trans-thalamic signalling pathway from VPM via L5B pyramidal cells of barrel cortex to Pom and from where it 'drives' higher order cortical areas, e.g. the S2 cortex (see e.g. Killackey and Sherman 2003; Theyel et al. 2010; Sherman and Guillery 2011; for a review see Guillery and Sherman 2011). However, it is likely that the L5B CT pyramidal cells are elements in both the feed-forward and the feedback pathways described above.

Conclusion

It should be noted that this review provides a simplified concept of the neuronal microcircuits in the S1 barrel cortex. The excitatory S1 microcircuits described here are certainly not separate subnetworks but interact at several distinct stages, thereby complicating the understanding of their functional properties. It is likely that with increasing knowledge of the morphological, electrophysiological and synaptic data from both in vivo and in vitro experiments, we will understand more about the structure-function relationship of barrel cortex synaptic microcircuit, both within a 'barrel column' and beyond.

References

Ahissar, E; Sosnik, R and Haidarliu S (2000). Transformation from temporal to rate coding in a somatosensory thalamocortical pathway. *Nature* 406: 302–306.

Alloway, K D (2008). Information processing streams in rodent barrel cortex: The differential functions of barrel and septal circuits. *Cerebral Cortex* 18: 979–989.

Aronoff, R et al. (2010). Long-range connectivity of mouse primary somatosensory barrel cortex. *European Journal of Neuroscience* 31: 2221–2233.

Beierlein, M and Connors, B W (2002). Short-term dynamics of thalamocortical and intracortical synapses onto layer 6 neurons in neocortex. *Journal of Neurophysiology* 88: 1924–1932.

Benshalom, G and White, E L (1986). Quantification of thalamocortical synapses with spiny stellate neurons in layer IV of mouse somatosensory cortex. *Journal of Comparative Neurology* 253: 303–314.

Bosman, L W et al. (2011). Anatomical pathways involved in generating and sensing rhythmic whisker movements. *Frontiers in Integrative Neuroscience* 5: 53.

Bourassa, J; Pinault, D and Deschênes, M (1995). Corticothalamic projections from the cortical barrel field to the somatosensory thalamus in rats: a single-fibre study using biocytin as an anterograde tracer. *European Journal of Neuroscience* 7: 19–30.

Brumberg, J C; Pinto, D J and Simons, D J (1999). Cortical columnar processing in the rat whisker-to-barrel system. *Journal of Neurophysiology* 82: 1808–1817.

Bruno, R M and Sakmann, B (2006). Cortex is driven by weak but synchronously active thalamocortical synapses. *Science* 312: 1622–1627.

Bruno, R M and Simons, D J (2002). Feedforward mechanisms of excitatory and inhibitory cortical receptive fields. *Journal of Neuroscience* 22: 10966–10975.

Bruno, R M; Hahn, T T; Wallace, D J; de Kock, C P and Sakmann, B (2009). Sensory experience alters specific branches of individual corticocortical axons during development. *The Journal of Neuroscience* 29: 3172–3181.

Bureau, I; von Saint Paul, F and Svoboda, K (2006). Interdigitated paralemniscal and lemniscal pathways in the mouse barrel cortex. *PLoS Biology* 4: e382.

Chen, C C; Abrams, S; Pinhas, A and Brumberg, J C (2009). Morphological heterogeneity of layer VI neurons in mouse barrel cortex. *Journal of Comparative Neurology* 512: 726–746.

Chmielowska, J; Carvell, G E and Simons, D J (1989). Spatial organization of thalamocortical and corticothalamic projection systems in the rat SmI barrel cortex. *Journal of Comparative Neurology* 285: 325–338.

Constantinople, C M and Bruno, R M (2013). Deep cortical layers are activated directly by thalamus. *Science* 340: 1591–1594.

Cruikshank, S J; Urabe, H; Nurmikko, A V and Connors, B W (2010). Pathway-specific feedforward circuits between thalamus and neocortex revealed by selective optical stimulation of axons. *Neuron* 65: 230–245.

Deschênes, M (2009). Vibrissal afferents from trigeminus to cortices. *Scholarpedia* 4(5): 7454. http://www.scholarpedia.org/article/Vibrissal_afferents_from_trigeminus_to_cortices. (see also pages 657–672 of this book).

Deschênes, M; Veinante, P and Zhang, Z W (1998). The organization of corticothalamic projections: Reciprocity versus parity. *Brain Research. Brain Research Reviews* 28: 286–308.

Douglas, R J and Martin, K A (1991). A functional microcircuit for cat visual cortex. *Journal of Physiology* 440: 735–769.

Douglas, R J and Martin, K A (2004). Neuronal circuits of the neocortex. *Annual Review of Neuroscience* 27: 419–451.

Feldmeyer, D; Roth, A and Sakmann, B (2005). Monosynaptic connections between pairs of spiny stellate cells in layer 4 and pyramidal cells in layer 5A indicate that lemniscal and paralemniscal afferent pathways converge in the infragranular somatosensory cortex. *The Journal of Neuroscience* 25: 3423–3431.

Feldmeyer, D; Lübke, J and Sakmann, B (2006). Efficacy and connectivity of intracolumnar pairs of layer 2/3 pyramidal cells in the barrel cortex of juvenile rats. *Journal of Physiology* 575: 583–602.

Feldmeyer, D; Egger, V; Lübke, J and Sakmann, B (1999). Reliable synaptic connections between pairs of excitatory layer 4 neurones within a single 'barrel' of developing rat somatosensory cortex. *Journal of Physiology* 521(Pt 1): 169–190.

Feldmeyer, D; Lübke, J; Silver, R A and Sakmann, B (2002). Synaptic connections between layer 4 spiny neurone-layer 2/3 pyramidal cell pairs in juvenile rat barrel cortex: Physiology and anatomy of interlaminar signalling within a cortical column. *The Journal of Physiology* 538: 803–822.

Feldmeyer, D et al. (2013). Barrel cortex function. *Progress in Neurobiology* 103: 3–27.

Fox, K D (2008). *Barrel Cortex*, 1st Edition. Cambridge, UK: Cambridge University Press.

Frick, A; Feldmeyer, D; Helmstaedter, M and Sakmann, B (2008). Monosynaptic connections between pairs of L5A pyramidal neurons in columns of juvenile rat somatosensory cortex. *Cerebral Cortex* 18: 397–406.

Groh, A; de Kock, C P; Wimmer, V C; Sakmann, B and Kuner, T (2008). Driver or coincidence detector: Modal switch of a corticothalamic giant synapse controlled by spontaneous activity and short-term depression. *The Journal of Neuroscience* 28: 9652–9663.

Groh, A et al. (2013). Convergence of cortical and sensory driver inputs on single thalamocortical cells. *Cerebral Cortex.*

Guillery, R W and Sherman, S M (2011). Branched thalamic afferents: What are the messages that they relay to the cortex? *Brain Research Reviews* 66: 205–219.

Holmgren, C; Harkany, T; Svennenfors, B and Zilberter, Y (2003). Pyramidal cell communication within local networks in layer 2/3 of rat neocortex. *Journal of Physiology* 551: 139–153.

Hoogland, P V; Wouterlood, F G; Welker, E and van der Loos, H (1991). Ultrastructure of giant and small thalamic terminals of cortical origin: a study of the projections from the barrel cortex in mice using Phaseolus vulgaris leucoagglutinin (PHA-L). *Experimental Brain Research* 87: 159–172.

Jensen, K F and Killackey, H P (1987). Terminal arbors of axons projecting to the somatosensory cortex of the adult rat. I. The normal morphology of specific thalamocortical afferents. *The Journal of Neuroscience* 7: 3529–3543.

Jia, H; Varga, Z; Sakmann, B and Konnerth, A (2014). Linear integration of spine Ca^{2+} signals in layer 4 cortical neurons in vivo. *Proceedings of the National Academy of Sciences of the United States of America* 111: 9277–9282.

Killackey, H P and Sherman, S M (2003). Corticothalamic projections from the rat primary somatosensory cortex. *The Journal of Neuroscience* 23: 7381–7384.

Kumar, P and Ohana, O (2008). Inter- and intralaminar subcircuits of excitatory and inhibitory neurons in layer 6a of the rat barrel cortex. *Journal of Neurophysiology* 100: 1909–1922.

Larkum, M (2013). A cellular mechanism for cortical associations: An organizing principle for the cerebral cortex. *Trends in Neuroscience* 36: 141–151.

Larkum, M E and Zhu, J J (2002). Signaling of layer 1 and whisker-evoked Ca^{2+} and Na^{+} action potentials in distal and terminal dendrites of rat neocortical pyramidal neurons in vitro and in vivo. *The Journal of Neuroscience* 22: 6991–7005.

Larkum, M E; Zhu, J J and Sakmann, B (1999). A new cellular mechanism for coupling inputs arriving at different cortical layers. *Nature* 398: 338–341.

Lefort, S; Tomm, C; Floyd Sarria, J C and Petersen, C C (2009). The excitatory neuronal network of the C2 barrel column in mouse primary somatosensory cortex. *Neuron* 61: 301–316.

Lübke, J; Egger, V; Sakmann, B and Feldmeyer, D (2000). Columnar organization of dendrites and axons of single and synaptically coupled excitatory spiny neurons in layer 4 of the rat barrel cortex. *The Journal of Neuroscience* 20: 5300–5311.

Mao, T et al. (2011). Long-range neuronal circuits underlying the interaction between sensory and motor cortex. *Neuron* 72: 111–123.

Markram, H; Lübke, J; Frotscher, M; Roth, A and Sakmann, B (1997). Physiology and anatomy of synaptic connections between thick tufted pyramidal neurones in the developing rat neocortex. *Journal of Physiology* 500(Pt 2): 409–440.

Mercer, A et al. (2005). Excitatory connections made by presynaptic cortico-cortical pyramidal cells in layer 6 of the neocortex. *Cerebral Cortex* 15: 1485–1496.

Meyer, H S et al. (2010a). Number and laminar distribution of neurons in a thalamocortical projection column of rat vibrissal cortex. *Cerebral Cortex* 20: 2277–2286.

Meyer, H S et al. (2011). Inhibitory interneurons in a cortical column form hot zones of inhibition in layers 2 and 5A. *Proceedings of the National Academy of Sciences of the United States of America* 108: 16807–16812.

Meyer, H S et al. (2010b). Cell type-specific thalamic innervation in a column of rat vibrissal cortex. *Cerebral Cortex* 20: 2287–2303.

Miller, K D; Pinto, D J and Simons, D J (2001). Processing in layer 4 of the neocortical circuit: New insights from visual and somatosensory cortex. *Current Opinion in Neurobiology* 11: 488–497.

Oberlaender, M et al. (2011). Three-dimensional axon morphologies of individual layer 5 neurons indicate cell type-specific intracortical pathways for whisker motion and touch. *Proceedings of the National Academy of Sciences of the United States of America* 108: 4188–4193.

Oberlaender, M et al. (2012). Cell type-specific three-dimensional structure of thalamocortical circuits in a column of rat vibrissal cortex. *Cerebral Cortex* 22: 2375–2391.

Petreanu, L; Huber, D; Sobczyk, A and Svoboda, K (2007). Channelrhodopsin-2-assisted circuit mapping of long-range callosal projections. *Nature Neuroscience* 10: 663–668.

Petreanu, L et al. (2012). Activity in motor-sensory projections reveals distributed coding in somatosensation. *Nature* 489: 299–303.

Pichon, F; Nikonenko, I; Kraftsik, R and Welker, E (2012). Intracortical connectivity of layer VI pyramidal neurons in the somatosensory cortex of normal and barrelless mice. *European Journal of Neuroscience* 35: 855–869.

Qi, G and Feldmeyer, D (2015). Dendritic target region-specific formation of synapses between excitatory layer 4 neurons and layer 6 pyramidal cells. *Cerebral Cortex* Epub.

Qi, G; Radnikow, G and Feldmeyer, D (2015). Electrophysiological and morphological characterization of neuronal microcircuits in acute brain slices using paired patch-clamp recordings. *Journal of Visualized Experiments* e52358.

Schoonover, C E et al. (2014). Comparative strength and dendritic organization of thalamocortical and corticocortical synapses onto excitatory layer 4 neurons. *The Journal of Neuroscience* 34: 6746–6758.

Schubert, D; Kötter, R; Luhmann, H J and Staiger, J F (2006). Morphology, electrophysiology and functional input connectivity of pyramidal neurons characterizes a genuine layer Va in the primary somatosensory cortex. *Cerebral Cortex* 16: 223–236.

Sherman, S M and Guillery, R W (2011). Distinct functions for direct and transthalamic corticocortical connections. *Journal of Neurophysiology* 106: 1068–1077.

Silver, R A; Lübke, J; Sakmann, B and Feldmeyer, D (2003). High-probability uniquantal transmission at excitatory synapses in barrel cortex. *Science* 302: 1981–1984.
Spruston, N (2008). Pyramidal neurons: Dendritic structure and synaptic integration. *Nature Reviews Neuroscience* 9: 206–221.
Staiger, J F; Bojak, I; Miceli, S and Schubert, D (2015). A gradual depth-dependent change in connectivity features of supragranular pyramidal cells in rat barrel cortex. *Brain Structure & Function* 220(3): 1317–1337.
Staiger, J F et al. (2004). Functional diversity of layer IV spiny neurons in rat somatosensory cortex: quantitative morphology of electrophysiologically characterized and biocytin labeled cells. *Cerebral Cortex* 14: 690–701.
Stratford, K J; Tarczy-Hornoch, K; Martin, K A; Bannister, N J and Jack, J J (1996). Excitatory synaptic inputs to spiny stellate cells in cat visual cortex. *Nature* 382: 258–261.
Szentágothai, J (1975). The 'module-concept' in cerebral cortex architecture. *Brain Research* 95: 475–496.
Tanaka, Y R et al. (2011). Local connections of excitatory neurons to corticothalamic neurons in the rat barrel cortex. *The Journal of Neuroscience* 31: 18223–18236.
Theyel, B B; Llano, D A and Sherman, S M (2010). The corticothalamocortical circuit drives higher-order cortex in the mouse. *Nature Neuroscience* 13: 84–88.
Thomson, A M (2010). Neocortical layer 6, a review. *Frontiers in Neuroanatomy* 4: 13.
White, E L and Rock, M P (1979). Distribution of thalamic input to different dendrites of a spiny stellate cell in mouse sensorimotor cortex. *Neuroscience Letters* 15: 115–119.
Wimmer, V C; Bruno, R M; de Kock, C P; Kuner, T and Sakmann, B (2010). Dimensions of a projection column and architecture of VPM and POm axons in rat vibrissal cortex. *Cerebral Cortex* 20: 2265–2276.
Woolsey, T A and van der Loos, H (1970). The structural organization of layer IV in the somatosensory region (SI) of mouse cerebral cortex. The description of a cortical field composed of discrete cytoarchitectonic units. *Brain Research* 17: 205–242.
Xu, N L et al. (2012). Nonlinear dendritic integration of sensory and motor input during an active sensing task. *Nature* 492: 247–251.
Yu, C; Derdikman, D; Haidarliu, S and Ahissar, E (2006). Parallel thalamic pathways for whisking and touch signals in the rat. *PLoS Biology* 4: e124.
Zhang, Z W and Deschênes, M (1997). Intracortical axonal projections of lamina VI cells of the primary somatosensory cortex in the rat: A single-cell labeling study. *The Journal of Neuroscience* 17: 6365–6379.
Zingg, B et al. (2014). Neural networks of the mouse neocortex. *Cell* 156: 1096–1111.

S1 Somatotopic Maps

Stuart P. Wilson and Chris Moore

S1 somatotopic maps refers to spatial patterns in the functional organization of neuronal responses in the mammalian primary somatosensory cortex (S1/SI). Here the term 'map' refers to a population of neurons that respond selectively to the presence of stimuli that collectively sample from an underlying stimulus space. Maps are referred to as 'somatotopic' when that space is related to locations on the body, such that adjacent neurons in the neural tissue respond selectively to stimuli presented to adjacent locations on the body (Figure 1).

SI refers to a somatosensory neocortical area that responds primarily to tactile stimulation of the skin (and/or hair). Among neocortical areas, neurons in SI representations typically have the smallest receptive fields and receive the shortest-latency input from the sensory periphery. This area is properly defined as containing a single map of the periphery (for example, as found in Brodmann area 3b of the monkey and human), though in some cases several areas in the post-central gyrus of the primate are referred to as SI, in historical deference to early mapping methods that did not recognize the presence of multiple maps in this gyrus as discussed below.

Discovery of Somatotopic Maps

The modern concept of systematic body maps in the brain is most directly attributable to the ideas of British physician John Hughlings Jackson, who in 1886 published the observation that epileptic seizures progress across the body in somatotopic sequence,

---

Dr. Chris Moore accepted the invitation on 11 November 2008 (self-imposed deadline: 11 May 2009).

S.P. Wilson (✉)
Dept. Psychology, The University of Sheffield, Sheffield, UK
e-mail: s.p.wilson@sheffield.ac.uk

C. Moore
Department of Neuroscience, Brown University, Providence, RI, USA

T.J. Prescott et al. (eds.), *Scholarpedia of Touch*, Scholarpedia,
DOI 10.2991/978-94-6239-133-8_43

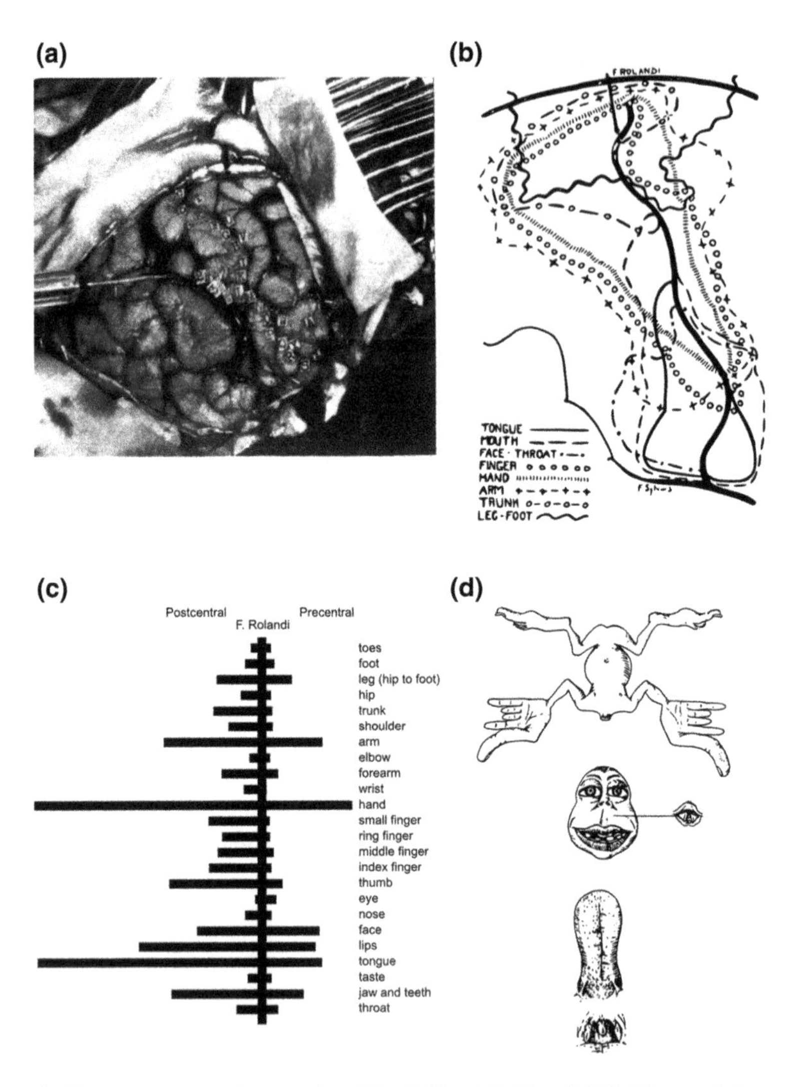

Figure 1 The somatosensory homunculus of Penfield and Boldrey (1937). The panels increase in the level of abstraction with which they depict the human SI somatotopic body map. **a** Photograph of the human cortex during surgery on a patient with epilepsy, with paper tickets used to mark sites of electrical stimulation that elicited reports of localized tactile sensations. **b** Drawing of the cortex with boundaries overlaid to show the zones in which stimulation could elicit tactile sensations at each body part (summarising data from 126 operations; note considerable overlap between the boundaries not apparent in alternative visualisations of the same data). **c** Bar chart indicating the medial-lateral sequence in which body parts are localized to the cortical surface (vertical), as well as the overall size of the representation for each body part, and their distribution caudal or rostral to the fissure of Rolando. **d** First image of the famous sensory and motor homunculus, with the vertical sequence of body parts and relative size depicting their gross somatotopic arrangement in SI and the amount of cortical territory dedicated to each body part. Panels correspond to Figures 4, 25b, 27 and 28 in Penfield and Boldrey (1937); panels B-D were traced from the original text using a vector drawing package, and modified only superficially (e.g., replacing text fonts) for clarity

i.e., with spasms progressing from the hands to the arms to the shoulders, or originating from the feet and progressing up the leg to the trunk, in what is now commonly referred to as a Jacksonian march. Based on these observations, Hughlings Jackson inferred the presence of orderly representations of the body in the brain, and particularly in the cortical hemispheres (see York and Steinberg 2011, for a recent review). In 1870 German medical professors Gustav Fritsch and Eduard Hitzig published a seminal study showing that twitches of specific muscles could be caused by electrical stimulation of specific sites in the frontal cortex of dogs (see Taylor and Gross 2003). In 1874 physician David Ferrier published The Functions of the Brain, a book dedicated to his mentor John Hughlings Jackson, detailing numerous lesion and electrical stimulation experiments on the cortices of birds, cats, guinea-pigs, rabbits, rodents, dogs, jackals, and monkeys. Ferrier used long trains of electrical stimulation, which often yielded more complex and coordinated movements of muscle groups than the local twitches of individual muscles elicited by the shorter electrical pulses delivered by Fritsch and Hitzig (see Graziano 2009, for a review). This book introduced some of the first illustrations of the localization and systematic mapping of motor function to specific cortical sites, as drawn by the brother-in-law of Ferrier, painter Ernest Waterlow (Sandrone and Zanin 2014). However Ferrier was not convinced that maps for touch in the cortex were also topographically organized; "…I am inclined to think that the experimental evidence is against any absolute differentiation of centers of tactile sensation for special regions […] though probably the various motor centres are each anatomically related by associating fibres with corresponding regions of the falciform lobe. This association would form the basis of a musculo-sensory localisation." (Ferrier 1886, pp. 344–345). Some of the earliest images of high-resolution motor maps in primates originate in Leyton and Sherrington (1917) (see Lemon 2008 for a historical perspective).

The first evidence for localization of cortical responses to somatosensory stimuli (rather than motor responses) came from two case studies on epileptic human patients (a 15-year-old boy and a 44-year-old man) reported by Cushing (1909). Cushing found that weaker electrical stimulation of the cortex could elicit the subjective experience of somatosensation localized to specific parts of the hand, without eliciting muscle movements; "…in both of these patients stimulation of the post-central convolution gave definite sensory impressions which were likened in one case to a sensation of numbness, and in the other to definite tactual impulses." (p. 53). These observations paved the way for the discovery and investigation of SI somatosensory maps in the decades to follow.

The Somatosensory Homunculus

The prototypical SI somatotopic map is the human somatosensory 'homunculus', meaning 'little man.' In the context of somatotopic maps, the term homunculus was introduced by Wilder Penfield in what remains one of the most influential articles in neurology, summarizing data from electrical stimulation of the cortex in 126 operations performed on conscious epileptic patients (Penfield and Boldrey 1937).

The somatosensory homunculus is now synonymous with cartoon drawings of the little man commissioned by Penfield to artist Mrs H. P. Cantile, however Penfield and Boldrey (1937) employed a variety of formats to depict their original data, including; (i) detailed dictations of patients' subjective reports of tactile sensation and movements elicited by stimulation at each cortical site; (ii) accompanying photographs of the exposed cortex overlaid with paper tickets labeling the corresponding stimulation sites; (iii) drawings of the cortex marked with the location of stimulation sites corresponding to each body part; (iv) drawings in which stimulation sites reliably correlating with specific body parts are enclosed by (often largely overlapping) boundaries; (v) bar charts for motor and sensory responses indicating the proportion of responses evoked by stimulation of each major body part and drawn in the sequence in which they tend to appear along the cortical surface; and (vi) a cartoon corresponding to the bar charts drawn with familiar outlines of the shape of the body parts to depict the now famous homunculus. In the words of Penfield and Boldrey (1937); "The homunculus gives a visual image of the size and sequence of cortical areas, for the size of the parts of this grotesque creature were determined not so much by the number of responses but by the apparent perpendicular extent of representation of each part" (p. 431).

The accuracy and usefulness of Cantlie's homunculus cartoons, which were further developed in the book of Penfield and Rasmussen (2008), have been questioned by many, notably by Schott (1993) who highlights the mismatch between the sharp topological borders suggested by the clean lines of the drawings, and a considerable overlap of projections from the body surface as depicted in format iv by Penfield and Boldrey (1937). Nonetheless, these images have been highly influential in suggesting two features of the human SI map organization: First, an exaggerated cortical territory dedicated to the somatosensory and motor representation of the hands and face; and, second, somatotopic discontinuities at the junction between the hands and face representations and at the junction between the feet and genital representations.

Soon after the publication of Penfield's homunculus, Woolsey (1952) published an image of the somatosensory homunculus for the primate, depicting a single continuous map spanning across the four cytoarchitectonic bands of the anterior parietal cortex, Brodman areas 3b, 1 (posterior cutaneous field), 2, and 3a (for details see the Scholarpedia article by Kaas 2013). At a finer-scale resolution, Kaas et al. (1979) found that a full representation of the primate body is repeated approximately four times within the post-central gyrus, such that four homunculi lie in parallel to each other. Woolsey (1952) and Kaas et al. (1979) derived their somatotopic maps not by stimulating the cortex, but instead by measuring electrical responses to the delivery of cutaneous stimulation (i.e., touch) to the body surface. Measured in this way, virtually all mammals tested have revealed similar systematic body maps in their respective SI, and a corresponding cartoon and "unculus" suffix has been used to describe many of them. For example, the term 'simculus' was used by Woolsey (1952) to describe the organization in monkeys, the term 'ratunculus' was used by Woolsey and LeMessurier (1948) for the organization in rats (see also Welker 1976), and the terms 'molunculus' for moles and 'platypunculus' for platypus have also been used colloquially. SI somatotopic maps thus appear to be a highly conserved organizational principle across the mammalian lineages (see Krubitzer 1995, 2007).

Homuncular Discontinuities

The somatotopic discontinuities at the feet/genital and hand/face representations of the original homunculi were celebrated in a song by Penfield's colleague Kershman, which includes the verse "The sensory type he was leering/With hand neatly balance on thumb/His happiness founded on things near his toes/That need not always be numb" (Feinsod 2005). Penfield and Boldrey (1937) were more candid in describing the foot/genital discontinuity, noting; "Presumably rectum and genitalia should be placed above feet, that is within the longitudinal fissure, but our evidence is not sufficient for conclusion and they seem to be somewhat posterior to feet." (p. 433). Despite this observation, the co-localization of the feet and genitals in the original homunculus images has persisted as part of neuroscience "folklore" (see Parpia 2011 for a comprehensive review). For example, Ramachandran has speculated that invasion of the genital region of the cortex by adjacent territory otherwise dedicated to representing the feet may explain the mis-localisation of sexual pleasure to the feet reported by some lower limb amputees (e.g., Ramachandran and Hirstein 1998).

A number of more recent fMRI imaging studies have restored the tactile representation of the genitals to their somatotopically consistent position in the map adjacent to the trunk, e.g., see Kell et al. (2005) for males, Michels et al. (2010) for females. A possible explanation for the hands/face discontinuity is given by Farah (1998), who hypothesized that the organization reflect the position of the fetus in the womb with its hands and face (and feet and genitals) often touching at the same time. The coincident activation arising from their spatial proximity was predicted to drive Hebbian plasticity, and therefore adjacency within the map. Computer models later showed how such mapping could emerge, reproducing continuous somatotopic maps in self-organizing neural networks, with selective discontinuities between co-stimulated areas of the simulated body surface (Stafford and Wilson 2007). This model also demonstrated that Hebbian learning mechanisms alone cannot account for the consistent medial-lateral ordering of somatotopic body maps. Models of this nature have been criticized for failing to incorporate subcortical contributions to the discontinuities that may originate from divergent afferent projections at the spinal cord (Parpia 2011).

Somatotopic Re-Mapping

Studying images of the somatosensory homunculi can give the impression that SI somatotopic maps remain fixed, but there is strong evidence that maps are highly plastic in the proportion and ordering of representations on a variety of timescales, ranging from the trajectory of development and maturation of cortical circuitry to the rapid allocation of attention. Florence et al. (1996) showed that cutting and then repairing the nerves of the hand in monkeys, so as to destroy the topology of its projections into the brain, can still yield somatotopic maps for the hand in SI, thus demonstrating the ability of the cortex to rewire itself during development to reflect the functional (rather than strictly anatomical) relationships between adjacent body parts. Braun et al. (2001) found that tactile representations of the fingers were localized to different positions in SI when human subjects were using the hand to write, as compared to when the hand was at rest, thus suggesting that somatotopic maps dynamically re-organize according to specific motoric tasks. Using similar methods Elbert et al. (1995) found that tactile representations of the fingers in the left hand of string players were larger than in controls (and larger than for the right hand), and they report a strong correlation between the age at which musicians began playing and the magnitude of the exaggerated cortical representation, again suggesting a functional remapping of SI due to motor experience. Similar methods have implicated distortions in somatotopic maps of the hand as an underlying factor in dystonia (Bara-Jimenez et al. 1998).

In humans, Weiss et al. (2000) reported that tactile representations of fingers left intact after others of the same hand were amputated became expanded in under ten days post-amputation, as compared to representations in the other hand of the patient or of either hand in controls. SI representations expand in owl monkeys for tactile stimulation of digits receiving tactile input during behavioral tasks in areas 3b (Jenkins et al. 1990) and 3a (Recanzone et al. 1992). In owl monkeys Merzenich et al. (1984) measured a dramatic functional reorganization of the receptive field properties of SI neurons and in the overall somatotopic map of intact digits following digit amputation, supporting a model in which cortical territories dedicated to spared digits invade those no longer receiving input from the missing digits. These data have also been well captured by self-organizing Hebbian models of somatotopic map formation (e.g., Ritter et al. 1992).

The evidence for somatotopic remapping considered in this section was focussed on primate S1, and we direct the reader to important further studies on primate S1 plasticity by Moligner et al. (1993), Moore et al. (2000) and Tegenthoff et al. (2005). The next sections focus on somatotopic maps in other species.

Cortical Magnification of Highly Relevant Tactile Representations

A key feature of SI organization that is preserved across mammalian species is that different areas of the somatosensory surface are magnified depending on the behavioural relevance of the corresponding sensors. As one example, the representation of the face and hands in human SI occupy disproportionately larger cortical territory relative to their dermatopic size than other body representations such as the trunk. Krubitzer et al. (1995) similarly describes an exaggerated representation of the electro-tactile bill of the platypus, thought to help it localize underwater prey (Pettigrew et al. 1998). Similarly Catania and Kaas (1997) describe the SI representation of the tactile appendages on the face of the star-nosed mole, proposing that the particularly magnified representation of the central appendage constitutes a foveal system in much the same way as the high-resolution sampling of the center of the retina is exaggerated in retinotopic maps (see Catania 2012 for a recent review).

A consistent corollary of expanded representation size for a given representation relative to its spatial occupation of the skin surface is a relative decrease in the size of receptive fields of individual neurons within that SI representation (Sur et al. 1980). This principle of cortical magnification and receptive field restriction can be preserved when the elaboration of cortical representation is driven by plastic changes in the adult (Merzenich et al. 1984).

The Barrel Cortex

The vibrissae or 'whiskers' in many species have a particularly highly magnified representation (Figure 2). Since it was first comprehensively characterized by Woolsey, Welker and colleagues, the cortical representation of the vibrissae in rodent S1 has become one of the most important models in modern neuroscience, due to the precise correspondence between the individual sensor (the vibrissa or 'whisker') and the cortical column, called a 'barrel', with which it correlates most strongly (Woolsey and van der Loos 1970; Welker and Woolsey 1974; Welker 1976). In mice

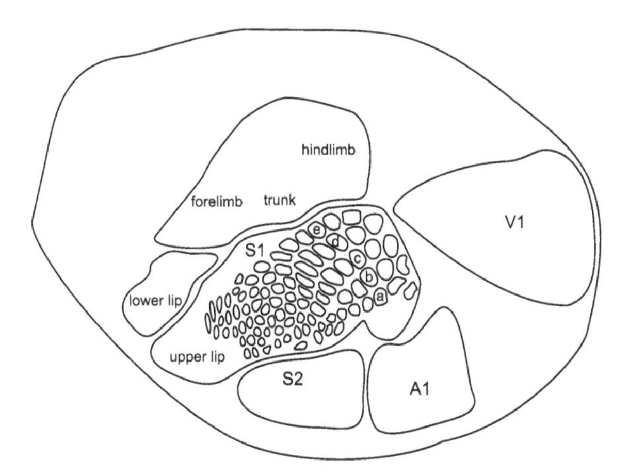

Figure 2 Spatial organisation of the flattened mouse cortical hemisphere. Adapted from Maier et al. (1999), their Figure 2, by tracing the boundaries around all clearly identifiable regions after staining with cytochrome oxidase. Areas marked A1, V1, and S2 correspond to the primary auditory, primary visual, and secondary somatosensory areas, respectively. The barrels are clearly identifiable as a dense array in SI, and are arranged on the cortex so as to create a somatotopic map of the contralateral vibrissae that is somatotopically aligned to an overall 'mouse-unculus' plan

the barrel cortex comprises around 70 % of SI, and 13 % of the entire cortical surface (Lee and Erzurumlu 2005).

The rodent barrel cortex contains a grid of discrete architectonic units, one per vibrissa, which after staining are visible to the naked eye. The pattern of barrels directly corresponds to the layout of the whiskers on the face, such that adjacent whiskers are principal to adjacent barrels. In layer 4 of the mouse, each unit is delineated in the plane tangential to the surface of the brain by a perimeter of cell bodies that is shaped like a barrel. The pattern of delineations between barrels has been described as Dirichlet domains (Senft and Woolsey 1991). Within each barrel, changes in synaptic contact density reveal regular geometric patterns (Land and Erickson 2005; Louderback et al. 2006; see Ermentrout et al. 2009).

When the corresponding vibrissa is deflected, neurons of a given barrel respond with action potential firing at shorter latency and with greater magnitude than the others. Thus, the approximately 10,000 neurons comprising each barrel column (Beaulieu 1993) tend to be mapped primarily to a specific vibrissa in their action potential output, called the principal whisker. Important nuances exist in this assertion, however: Neurons within the barrel column often very large 'sub-threshold' fields driven by vibrissae across the face (reviewed in Moore et al. 1999), and a significant number of individual neurons will respond best to non-aligned vibrissae, particularly when the more global pattern of stimulation is altered (Jacob et al. 2008).

The map for vibrissa identity amongst the barrels of granular layer 4 barrel cortex reflects a similar organisation between nuclei known as barrelettes in the brainstem (Ma and Woolsey 1984), and as barreloids in the thalamus (van der Loos 1976). Upon vibrissa deflection, a rapid pathway for excitation propagates along the neuraxis from barrelette, to barreloid, to barrel, to regions of the supragranular and infragranular cortical layers that are aligned to the corresponding barrel, and laterally into adjacent supra- and infra- granular barrel regions (Armstrong-James et al. 1992; Lefort et al. 2009). It had been suggested that during development the organization of topographic vibrissal maps unfolds in sequence along the pathway, with each vibrissal identity map inheriting the organisation from the antecedent layer (Killackey 1980).

Species with vibrissae that have been found to have barrels include the mouse, rat, hamster, gerbil, muskrat, chipmunk, grey squirrel (although less pronounced), prairie dog, guinea pig, chinchilla, porcupine, mole, rabbit, ferret, wallaby, and the Australian opossum. Species with vibrissae that have been found not to have barrels include the beaver, tree shrew, cat, dog, raccoon, squirrel monkey, rhesus monkey, and the American opossum (summary based on Table 1.1 of Fox 2008; see also Rice 1995).

For a comprehensive review of the anatomy and physiology of the barrel cortex see Fox (2008), see related Scholarpedia article on laminar specialization in barrel cortex by Ahissar and Staiger (2010), and see the recent article by Meyer et al. (2013). For a comprehensive review on plasticity in the barrel cortex see Feldman and Brecht (2005) and see related Scholarpedia article on S1 long-term plasticity by Shulz and Ego-Stengel (2012).

Somatotopic Feature Maps in the Barrels

Within the vibrissal SI representation of rats, additional maps have been discovered that are organized in a somatotopic fashion. Barrel cortex neurons respond selectively depending on the direction in which the corresponding whisker is moved (Simons and Carvell 1989; see also Hellweg et al. 1977). A feature map representing the direction of vibrissal movement is somatotopically aligned within a single barrel column in a pin-wheel type fashion (Andermann and Moore 2006; see also Bruno et al. 2003). Neurons preferentially responsive to the direction of vibrissal deflection towards a given neighboring vibrissa are represented in the somatotopic portion of the barrel column more proximal to that neighboring vibrissa (see also Tsytsarev et al. 2010). Evidence suggests this map is present only in older animals (Kerr et al. 2007), suggesting it may emerge through activity-dependent plasticity (Kremer et al. 2011, see also Wilson et al. 2010). A recent model proposed that the large size of the barrels and relatively slow signal propagation speeds could render neurons at different locations with respect to the barrel centers sensitive to the relative timing of adjacent-whisker deflections, yielding a somatotopic map for the relative timing of tactile stimuli (Wilson et al. 2011; based on evidence from Shimegi et al. 2000 and Jacob et al. 2008). Neimark et al. (2003) have also proposed that the increasing length of the vibrissae from the snout to the ear renders neurons in barrels for the longer vibrissae sensitive to lower resonant frequency vibrations, yielding a somatotopic map of vibration frequencies that spans the barrel cortex in rodent SI, similar to the tonotopy observed in auditory representations.

As the barrel cortex becomes an increasingly important model system in neuroscience, and as efforts become focused on simulating its microcircuitry in exquisite detail (Markram 2006), it is becoming increasingly important to develop complementary models that address how constraints imposed by the overall somatotopic maps in which these circuits are embedded might impact on cortical function.

An important and open question is whether spatial patterning and map organization in the brain is important for neural computation, or whether somatotopic maps, and topological mapping in the brain more generally, are epiphenomena of efficient developmental processes (see Purves et al. 1992, and Wilson and Bednar 2015 for further discussion).

References

Ahissar, E and Staiger, J (2010). S1 laminar specialization. *Scholarpedia* 5(8): 7457. http://www.scholarpedia.org/article/S1_laminar_specialization. (see also pages 505–531 of this book).

Andermann, M L and Moore, C I (2006). A somatotopic map of vibrissa motion direction within a barrel column. *Nature Neuroscience* 9(4): 543–551.

Armstrong-James, M; Fox, K and Das-Gupta, A (1992). Flow of excitation within rat barrel cortex on striking a single vibrissa. *Journal of Neurophysiology* 68(4): 1345–1358.

Bara-Jimenez, W; Catalan, M J; Hallett, M and Gerloff, C (1998). Abnormal somatosensory homunculus in dystonia of the hand. *Annals of Neurology* 44(5): 828–831.

Beaulieu, C (1993). Numerical data on neocortical neurons in adult rat, with special reference to the GABA population. *Brain Research* 609(1–2): 284-292.

Braun, C et al. (2001). Dynamic organization of the somatosensory cortex induced by motor activity. *Brain* 124(11): 2259–2267.

Bruno, R M; Khatri, V; Land, P W and Simons, D J (2003). Thalamocortical angular tuning domains within individual barrels of rat somatosensory cortex. *The Journal of Neuroscience* 23 (29): 9565–9574.

Catania, K C (2012). Evolution of brains and behavior for optimal foraging: A tale of two predators. *Proceedings of the National Academy of Sciences of the United States of America* 109(1): 10701–10708.

Catania, K C and Kaas, J H (1997). Somatosensory fovea in the star-nosed mole: Behavioral use of the star in relation to innervation patterns and cortical representation. *Journal of Comparative Neurology* 387(2): 215–233.

Cushing, H (1909). A note upon the faradic stimulation of the postcentral gyrus in conscious patients. *Brain* 32: 44–53.

Elbert, T; Pantev, C; Wienbruch, C; Rockstroh, B and Taub, E (1995). Increased cortical representation of the fingers of the left hand in string players. *Science* 270(5234): 305–307.

Ermentrout, B; Simons, D J and Land, P W (2009). Subbarrel patterns in somatosensory cortical barrels can emerge from local dynamic instabilities. *PLoS Computational Biology* 5(10): e1000537.

Farah, M J (1998). Why does the somatosensory homunculus have hands next to face and feet next to genitals? A hypothesis. *Neural Computation* 10(8): 1983–1985.

Feinsod, M (2005). Kershman's sad reflections on the homunculus: A historical vignette. *Neurology* 64(3): 524–525.

Feldman, D E and Brecht, M (2005). Map plasticity in somatosensory cortex. *Science* 310(5749): 810–815.

Ferrier, D (1886). "The Functions of the Brain." New York: G P Putman's Sons.

Florence, S L et al. (1996). Central reorganization of sensory pathways following peripheral nerve regeneration in fetal monkeys. *Nature* 381(6577): 69–71.

Fox, K (2008). "Barrel Cortex." Cambridge, UK: Cambridge University Press.

Graziano, M S A (2009). "The Intelligent Movement Machine: An Ethological Perspective on the Primate Motor System." Oxford, UK: Oxford University Press.

Hellweg, F C; Schultz, W and Creutzfeldt, O D (1977). Extracellular and intracellular recordings from cat's cortical whisker projection areas: Thalamocortical response transformation. *Journal of Neurophysiology* 40: 463–479.

Jacob, V; Cam, J L; Ego-Stengel, V and Shulz, D E (2008). Emergent properties of tactile scenes selectively activate barrel cortex neurons. *Neuron* 60(6): 1112–1125.

Jenkins, W M; Merzenich, M M; Ochs, M T; Allard, T and Guc-Robles, E (1990). Functional reorganization of primary somatosensory cortex in adult owl monkeys after behaviorally controlled tactile stimulation. *Journal of Neurophysiology* 63(1): 82–104.

Kaas, J H (2013). Primate S1 cortex. *Scholarpedia* 8(6): 8238. http://www.scholarpedia.org/article/Primate_S1_cortex. (see also pages 499–503 of this book).

Kaas, J H; Nelson, R J; Sur, M; Lin, C S and Merzenich, M M (1979). Multiple representations of the body within the primary somatosensory cortex of primates. *Science* 204(4392): 521–523.

Kell, C A; von Kriegstein, K; Rsler, A; Kleinschmidt, A and Laufs, H (2005). The sensory cortical representation of the human penis: Revisiting somatotopy in the male homunculus. *The Journal of Neuroscience* 25(25): 5984–5987.

Kerr, J N D et al. (2007). Spatial organization of neuronal population responses in layer 2/3 of rat barrel cortex. *The Journal of Neuroscience* 27(48): 13316–13328.

Killackey, H (1980). Pattern formation in the trigeminal system of the rat. *Trends in Neurosciences* 3(12): 303–306.

Kremer, Y; Leger, J-F; Goodman, D; Brette, R and Bourdieu, L (2011). Late emergence of the vibrissa direction selectivity map in the rat barrel cortex. *The Journal of Neuroscience* 31(29): 10689–10700.

Krubitzer, L (1995). The organization of neocortex in mammals: Are species differences really so different? *Trends in Neurosciences* 18(9): 408–417.

Krubitzer, L (2007). The magnificent compromise: Cortical field evolution in mammals. *Neuron* 56(2): 201–208.

Krubitzer, L; Manger, P; Pettigrew, J and Calford, M (1995). Organization of somatosensory cortex in monotremes: In search of the prototypical plan *Journal of Comparative Neurology* 351(2): 261–306.

Land, P W and Erickson, S L (2005). Subbarrel domains in rat somatosensory (S1) cortex. *Journal of Comparative Neurology* 490(4): 414–426.

Lee, L-J and Erzurumlu, R S (2005). Altered parcellation of neocortical somatosensory maps in n-methyl-d-aspartate receptor-deficient mice. *Journal of Comparative Neurology* 485(1): 57–63.

Lefort, S; Tomm, C; Sarria, J-C F and Petersen, C C H (2009). The excitatory neuronal network of the C2 barrel column in mouse primary somatosensory cortex. *Neuron* 61(2): 301–316.

Lemon, R N (2008). An enduring map of the motor cortex. *Experimental Physiology* 93(7): 798–802.

Leyton, A S and Sherrington, C S (1917). Observations on the excitable cortex of the chimpanzee, orangutan, and gorilla. *Quarterly Journal of Experimental Physiology* 11(2): 135–222.

Louderback, K M; Glass, C S; Shamalla-Hannah, L; Erickson, S L and Land, P W (2006). Subbarrel patterns of thalamocortical innervation in rat somatosensory cortical barrels: Organization and postnatal development. *Journal of Comparative Neurology* 497(1): 32–41.

Maier, D L et al. (1999). Disrupted cortical map and absence of cortical barrels in growth-associated protein (gap)-43 knockout mice. *Proceedings of the National Academy of Sciences of the United States of America* 96(16): 9397–9402.

Ma, P M and Woolsey, T A (1984). Cytoarchitectonic correlates of the vibrissae in the medullary trigeminal complex of the mouse. *Brain Research* 306(1–2): 374-379.

Markram, H (2006). The blue brain project. *Nature Reviews Neuroscience* 7(2): 153–160.

Merzenich, M M et al. (1984). Somatosensory cortical map changes following digit amputation in adult monkeys. *Journal of Comparative Neurology* 224(4): 591–605.

Meyer, H S et al. (2013). Cellular organization of cortical barrel columns is whisker-specific. *Proceedings of the National Academy of Sciences of the United States of America.* 110(47): 19113–19118.

Michels, L; Mehnert, U; Boy, S; Shurch, B and Kollias, S (2010). The somatosensory representation of the human clitoris: An fMRI study. *Neuroimage* 49(1): 177–184.

Mogilner, A et al. (1993). Somatosensory cortical plasticity in adult humans revealed by magnetoencephalography. *Proceedings of the National Academy of Sciences of the United States of America* 90: 3593–3597.

Neimark, M A; Andermann, M L; Hopfield, J J and Moore, C I (2003). Vibrissa resonance as a transduction mechanism for tactile encoding. *The Journal of Neuroscience* 23(16): 6499–6509.

Moore, C I; Nelson, S B and Sur, M (1999). Dynamics of neuronal processing in rat somatosensory cortex. *Trends in Neurosciences* 22(11): 513–520.

Moore, C I et al. (2000). Referred phantom sensations and cortical reorganization after spinal cord injury in humans. *Proceedings of the National Academy of Sciences of the United States of America* 97(26): 14703–14708.

Parpia, P (2011). Reappraisal of the somatosensory homunculus and its discontinuities. *Neural Computation* 23(12): 3001–3015.

Penfield, W and Boldrey, E (1937). Somatic motor and sensory representation in the cerebral cortex of man as studied by electrical stimulation. *Brain* 60: 389–443.

Penfield, W and Rasmussen, T (2008). "The Cerebral Cortex of Man." New York: Macmillan.

Pettigrew, J D; Manger, P R and Fine, S L (1998). The sensory world of the platypus. *Philosophical Transactions of the Royal Society B: Biological Sciences* 353(1372): 1199–1210.

Purves, D; Riddle, D R and LaMantia, A S (1992). Iterated patterns of brain circuitry (or how the cortex got its spots). *Trends in Neurosciences* 15(10): 362–368.

Ramachandran, V S and Hirstein, W (1998). The perception of phantom limbs. The D. O. Hebb lecture. *Brain* 121(9): 1603–1630.

Recanzone, G H; Merzenich, M M and Jenkins, W M (1992). Frequency discrimination training engaging a restricted skin surface results in an emergence of a cutaneous response zone in cortical area 3a. *Journal of Neurophysiology* 67(5): 1057–1070.

Rice, F L (1995). Comparative aspects of barrel structure and development. In: E G Jones and I T Diamond (Eds.), "The Barrel Cortex of Rodents," Vol. 11, "Cerebral Cortex" (pp. 1–75). New York: Plenum Press.

Ritter, H; Martinez, T and Schulten, K (1992). "Neural Computation and Self-Organizing Maps: An Introduction." New York: Addison-Wesley.

Sandrone, S and Zanin, E (2014). David Ferrier (1843–1928). *Journal of Neurology* 261: 1247-1248.

Senft, S L and Woolsey, T A (1991). Mouse barrel cortex viewed as Dirichlet domains. *Cerebral Cortex* 1(4): 348–363.

Schott, G D (1993). Penfield's homunculus: A note on cerebral cartography. *Journal of Neurology, Neurosurgery, and Psychiatry* 56(4): 329–333.

Shimegi, S; Akasaki, T; Ichikawa, T and Sato, H (2000). Physiological and anatomical organization of multiwhisker response interactions in the barrel cortex of rats. *The Journal of Neuroscience* 20(16): 6241–6248.

Simons, D J and Carvell, G E (1989). Thalamocortical response transformation in the rat vibrissa/barrel system. *Journal of Neurophysiology* 61(2): 311–330.

Stafford, T and Wilson, S P (2007). Self-organisation can generate the discontinuities in the somatosensory map. *Neurocomputing* 70: 1932–1937.

Shulz, D and Ego-Stengel, V (2012). S1 long-term plasticity. *Scholarpedia* 7(10): 7615. http://www.scholarpedia.org/article/S1_long-term_plasticity. (see also pages 535–547 of this book).

Sur, M; Merzenich, M M and Kaas, J H (1980). Magnification, receptive-field area, and "hypercolumn" size in area 3b and 1 of somatosensory cortex in owl monkeys. *Journal of Neurophysiology* 44: 295–311.

Taylor, C S R and Gross, C G (2003). Twitches versus movements: A story of motor cortex. *Neuroscientist* 9(5): 332–342.

Tegenthoff, M et al. (2005). Improvement of tactile discrimination performance and enlargement of cortical somatosensory maps after 5 Hz rTMS. *PLoS Biology* 3(11): e363.

Tsytsarev, V; Pope, D; Pumbo, E; Yablonskii, A and Hofmann, M (2010). Study of the cortical representation of whisker directional deflection using voltage-sensitive dye optical imaging. *Neuroimage* 53(1): 233–238.

van der Loos, H (1976). Barreloids in mouse somatosensory thalamus. *Neuroscience Letters* 2(1): 1–6.

Weiss, T et al. (2000). Rapid functional plasticity of the somatosensory cortex after finger amputation. *Experimental Brain Research* 134(2): 199–203.

Welker, C and Woolsey, T A (1974). Structure of layer IV in the somatosensory neocortex of the rat: Description and comparison with the mouse. *Journal of Comparative Neurology* 158: 437–453.

Welker, C (1976). Receptive fields of barrels in the somatosensory neocortex of the rat. *Journal of Comparative Neurology* 166(2): 173–189.

Wilson, S P and Bednar, J A (2015). What, if anything, are topological maps for? *Developmental Neurobiology*. doi:10.1002/dneu.22281.

Wilson, S P; Bednar, J A; Prescott, T J and Mitchinson, B (2011). Neural computation via neural geometry: A place code for inter-whisker timing in the barrel cortex? *PLoS Computational Biology* 7(10): e1002188.

Wilson, S P; Law, J S; Mitchinson, B; Prescott, T J and Bednar, J A (2010). Modeling the emergence of whisker direction maps in rat barrel cortex. *PLoS ONE* 5(1): e8778.

Woolsey, C N (1952). Patterns of localization in sensory and motor areas of the cerebral cortex. In: "The Biology of Mental Health and Disease." New York: Hoebner.

Woolsey, C N and LeMessurier, D H (1948). The pattern of cutaneous representation in the rat's cerebral cortex. *Federation Proceedings* 7(1): 137.

Woolsey, T A and van der Loos, H (1970). The structural organization of layer IV in the somatosensory region (SI) of mouse cerebral cortex. The description of a cortical field composed of discrete cytoarchitectonic units. *Brain Research* 17(2): 205–242.

York, G K and Steinberg, D A (2011). Hughlings Jackson's neurological ideas. *Brain* 134(10): 3106–3113.

Balance of Excitation and Inhibition

Michael Okun and Ilan Lampl

In the context of neurophysiology, **balance of excitation and inhibition** (E/I balance) refers to the relative contributions of excitatory and inhibitory synaptic inputs corresponding to some neuronal event, such as oscillation or response evoked by sensory stimulation. In the current literature, owing to the extremely wide range of conditions in which the term is applied, it has several different, albeit related, meanings. As described in more detail below, the precise meaning depends on various considerations, such as averaging across time or population of neurons that is involved; the relevant timescale; whether the synaptic activity is sustained or transient, spontaneous or evoked. In general, excitatory and inhibitory inputs of a neuron are said to be balanced if across a range of conditions of interest the ratio between the two inputs is constant.

In the cortex, interneurons responsible for inhibition comprise just a small fraction of the neurons, yet they have an important function in regulating activity of principal cells. When inhibition is blocked pharmacologically, cortical activity becomes epileptic (Dichter and Ayala 1987), and neurons may lose their selectivity to different stimulus features (Sillito 1975). These and other data indicate that the interplay between excitation and inhibition has an important role in determining the cortical computation. Our understanding of the relationships between these two opposing forces has advanced significantly during the recent years, mainly due to the growing use of in vivo intracellular recording techniques.

Indirect Evidence for E/I Balance

Cortical neurons receive synaptic inputs from thousands of other, mainly excitatory, neurons, most of which evoke only a sub-millivolt response (Bruno and Sakmann 2006; Lefort et al. 2009). If these inputs arrive from neurons that fire at independent

M. Okun (✉)
UCL, London, UK

I. Lampl
Weizmann Institute of Science, Rehovot, Israel

T.J. Prescott et al. (eds.), *Scholarpedia of Touch*, Scholarpedia,
DOI 10.2991/978-94-6239-133-8_44

random times, they are expected to produce an almost constant depolarization leading to a regular firing. However, spike trains extracellularly recorded from single cortical neurons exhibit high variability. For instance, the coefficient of variation of the inter-spike intervals (ISIs) of neurons firing in response to a sensory input for a period of several seconds, is approximately equal to 1, as expected from a Poisson process (Softky and Koch 1993). This apparent paradox between simple probabilistic considerations and the observed statistics of cortical spike trains led to several proposed resolutions.

One early resolution was that excitatory and inhibitory synaptic currents of cortical neurons are approximately balanced in strength, causing the membrane potential to hover somewhat below the spiking threshold, crossing it at random times (Shadlen and Newsome 1994, 1998). Simulations, based on the random walk model of (Gerstein and Mandelbrot 1964) demonstrated that under such a regime of synaptic inputs the ISI variability is in agreement with experimental observations (Shadlen and Newsome 1994, 1998). Furthermore, computational studies of spontaneous activity in neuronal networks showed that E/I balance emerges naturally if the network is sparsely connected (van Vreeswijk and Sompolinsky 1996; Vogels et al. 2005). However, these early theoretical studies were based on crude estimates of the relevant parameters, and therefore cannot be regarded as definitive. In fact, several follow-up studies suggested that other factors, such as synchrony, are required in order to explain the observed ISI statistics, e.g., (Stevens and Zador 1998). Indeed, as described below, it appears that although excitation and inhibition are balanced, the membrane potential of cortical neurons does not necessarily follow the random walk trajectory predicted by these early models (Crochet and Petersen 2006; DeWeese and Zador 2006; Poulet and Petersen 2008; Okun et al. 2010; Polack et al. 2013; Sachidhanandam et al. 2013; Tan et al. 2014).

The possibility of excitation and inhibition having a comparable strength might seem implausible at first, since interneurons comprise only 15–25 % of the population of cortical neurons. However, the synaptic strength and firing rates of inhibitory interneurons are substantially higher than in excitatory neurons, thus inhibitory interneurons have an impact disproportionate to their relatively small number.

Intracellular Measurement of the Excitatory and Inhibitory Synaptic Inputs

In a pioneering study, Borg-Graham and colleagues used intracellular recordings to directly estimate the synaptic conductance changes evoked in cortical neurons by visual stimulation (Borg-Graham et al. 1996, 1998). The average synaptic current evoked by a stimulus is recorded in voltage-clamp mode, using several different clamping voltages. Alternatively, the subthreshold response is recorded in the current-clamp mode at several different clamping currents (Anderson et al. 2000). The behavior of the membrane potential is approximated using a passive, single compartment, conductance-based model of the neuron, described by

$$CdV/dt = -G_{leak}(V(t) - E_{leak}) - G_{ex}(t)(V(t) - E_{ex}) - G_{in}(t)(V(t) - E_{in}) + I_{inj} \quad (1)$$

where E_{leak} is the resting membrane potential of the neuron, C is its capacitance, G_{leak} is the mean conductance in absence of stimulation (the inverse of input resistance), E_{ex} and E_{in} are the reversal potentials of excitation and inhibition, and I_{inj} is the current injected through the recording pipette. By fitting Eq. (1) to the average responses at different holding potentials, the synaptic conductances evoked by the stimulus, $G_{ex}(t)$ and $G_{in}(t)$, can be computed (see Figure 1). For an in-depth review of the method and its caveats an interested reader is referred to (Monier et al. 2008).

Selectivity of Cortical Excitation and Inhibition to Sensory Stimulation

Early models of the visual cortex suggested that the selectivity of cortical cells to sensory stimulation emerges from feedforward inputs. Later models, however, questioned this view by suggesting that cortical inhibition plays a significant role in enhancing the selectivity of cortical response. The best known example for this controversy is the emergence of orientation selectivity in primary visual cortex.

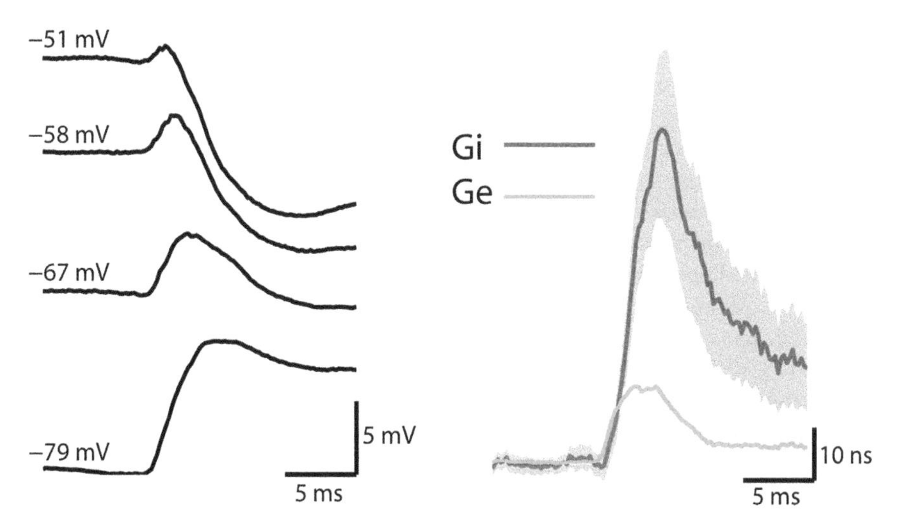

Figure 1 Computation of synaptic conductance evoked by sensory stimulus. The average response to whisker deflection in a spiny stellate neuron in layer IV of the rat primary somatosensory cortex is recorded in current-clamp mode while injecting 4 different currents (*left panel*). In addition, neuron's capacitance and leak conductance are measured (not shown). By fitting the responses to Eq. (1) the average excitatory and inhibitory synaptic conductances evoked by the stimulus are recovered (*right panel*). Adapted from Heiss et al. (2008)

The feedforward model (Hubel and Wiesel 1962) was supported by various studies (Nelson et al. 1994; Alonso and Martinez 1998; Chung and Ferster 1998; Martinez and Alonso 2001), while being challenged by others (Sillito 1975; Volgushev et al. 1996). The feedforward model, however, failed to predict several key experimental findings, and in particular the contrast invariance of orientation tuning (Ferster and Miller 2000). Alternative models proposed that the tuning of inhibitory inputs is wider, so that excitation and inhibition form a 'Mexican hat' interaction pattern which sharpens the selectivity of the cells (Ben-Yishai et al. 1995; Somers et al. 1995; Hansel and Sompolinsky 1996).

In the primary auditory cortex inhibition was similarly suggested to account for the sensory selectivity of the neurons (Calford and Semple 1995; Sutter et al. 1999; Wang et al. 2002).

A breakthrough in the ability to test these models was achieved by the in vivo intracellular conductance measurement methods described above. Over the last 15 years this approach was used in many studies to examine the sensory selectivity of excitatory and inhibitory synaptic inputs in primary sensory areas of several mammalian species. Direct measurements showed that to a first approximation the excitatory and inhibitory inputs are either similarly tuned, or that inhibitory inputs have a somewhat wider tuning. In cat primary visual cortex excitatory and inhibitory synaptic inputs are similarly tuned for orientation (Anderson et al. 2000), as well as for length (Anderson et al. 2001) and the direction of motion (Priebe and Ferster 2005). In the rodent primary auditory cortex inhibition is tuned similarly or somewhat wider than excitation for both frequency and intensity (Wehr and Zador 2003; Wu et al. 2008; Zhou et al. 2014), see Figure 2. Therefore, in these cases the selectivity of the neurons is unlikely to emerge through inhibitory suppression of the response to non-preferred stimuli.

The similar tuning of excitatory and inhibitory inputs to different features of the stimuli space appears to be a rather common organizational principle in the sensory areas, however there are several notable exceptions. The most prominent deviation from co-tuning was observed for orientation selectivity in the mouse primary visual cortex, where the inhibitory input is substantially more broadly tuned than the excitatory input, possibly because rodent primary visual cortex lacks orientation columns (Liu et al. 2011; Atallah et al. 2012; Li et al. 2012; Harris and Mrsic-Flogel 2013). An opposite scenario, where inhibitory inputs have narrower selectivity, was observed for frequency tuning in layer V intrinsically-bursting (but not regular-spiking) neurons of the primary auditory cortex (Sun et al. 2013). Also in the auditory cortex, some intensity-tuned neurons receive excitatory inputs which peak at the preferred intensity, whereas their inhibitory inputs increase monotonically with the stimulus strength (Wu et al. 2006), representing a case where the co-tuning of excitation and inhibition appears to break altogether. Finally, it should be noted that the tuning of inhibitory and excitatory inputs alone is not sufficient to substantiate specific theoretical models for feature selectivity in the cortex, because broad tuning of inhibition may either reflect non-specific convergence of inputs from a population of inhibitory cells that demonstrate highly selective but non-overlapping orientation tuning curves, or simply result from the wide tuning

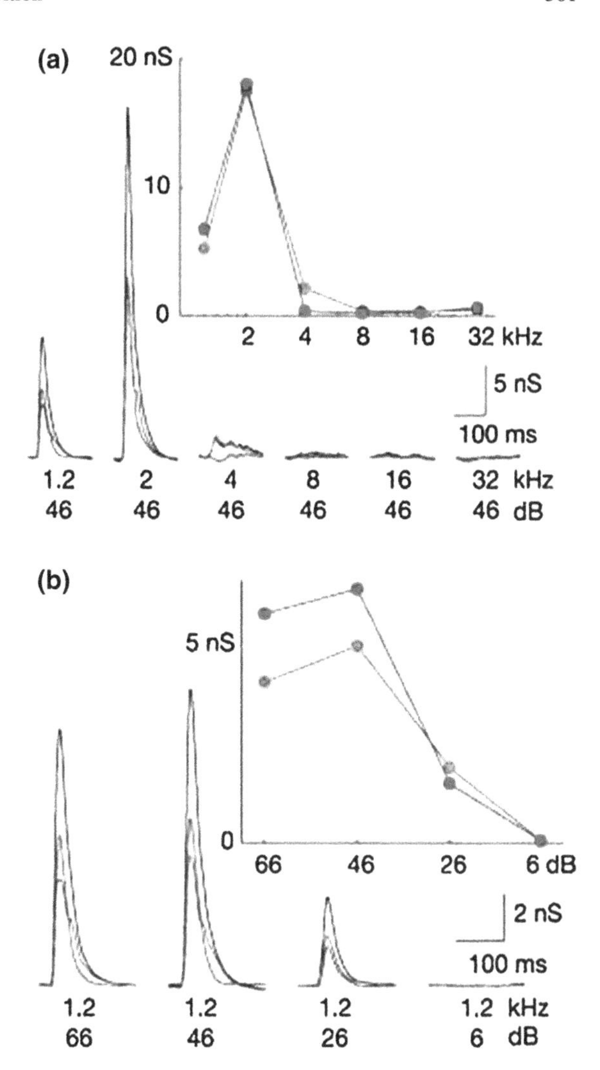

Figure 2 An example of a neuron in the auditory cortex with frequency and intensity co-tuned excitatory and inhibitory inputs. **a** Excitatory and inhibitory synaptic conductances evoked by stimuli of different frequencies and preferred intensity have a similar tuning. The measured conductances are shown at the *bottom* (*green*—excitatory conductance, *red*—inhibitory conductance, *black*—total conductance). **b** The excitatory and inhibitory inputs are also intensity co-tuned, notation as in (**a**). Adapted from Wehr and Zador (2003)

curves of their innervating inhibitory neurons (Shapley and Xing 2013; section "Current Research Directions").

Temporal Structure of Sensory Evoked Excitation and Inhibition

In the auditory and somatosensory cortices sensory stimulation often evokes stereotypic sequence of excitation followed within a few milliseconds by inhibition (Wehr and Zador 2003; Higley and Contreras 2006). Although excitation and inhibition are similarly tuned and hence are said to be balanced, a large imbalance occurs at the fine time scale, as inhibition lags behind excitation by several milliseconds. This lag between excitation and inhibition is likely to determine the integration window for excitation, affecting the number and precise timing of action potentials (Gabernet et al. 2005). In the auditory cortex the lag is independent of the frequency tuning of the cells (Wehr and Zador 2003). In the somatosensory cortex, however, the delay between excitation and inhibition might be related to the stimulus tuning of the neuron, such that at the preferred stimuli the lag between excitation and inhibition is larger than at the non-preferred ones (Wilent and Contreras 2005). Hence, a wider time window is available for integration of excitation for the preferred stimuli, producing more action potentials.

One of the central roles traditionally attributed to inhibition is suppression of neuronal responses during temporal integration of sensory inputs. A widely known example is forward suppression in the auditory cortex, in which the response to a second click presented shortly after the first one is much weaker. Another example is in the barrel cortex, where a response to whisker stimulation is largely suppressed if it is preceded by a stimulation of a neighboring whisker. Such forward suppression was widely believed to be due to inhibition evoked by the first stimuli. However, intracellular conductance measurements found that the duration of inhibitory synaptic input evoked by the first click is too short to account for the duration of forward suppression, so that the above explanation is incomplete at the best (Wehr and Zador 2003, 2005). Similarly, an intracellular recording study in the barrel cortex has shown that cross whisker suppression cannot be fully explained by a postsynaptic inhibitory mechanism (Higley and Contreras 2003). Although inhibition is not the primary cause for forward suppression, in other cases the ratio between the excitatory and inhibitory inputs to a neuron in a primary sensory area does depend not only on the instantaneous properties of the stimulus (its contrast, frequency, intensity, etc.) but also on its history. One particular example is adaptation to repeated stimuli, such as clicks or whisker deflections, which under certain conditions can skew the ratio between excitatory and inhibitory inputs toward excitation (Wehr and Zador 2005; Heiss et al. 2008). Paradoxically, because of a slower recovery of inhibitory inputs from adaptation, neurons become hypersensitive shortly after the termination of the adapting stimulation (Cohen-Kashi Malin a et al. 2013), which might explain why neurons in the barrel cortex respond better to non-periodic stimulation (Lak et al. 2008).

E/I Balance During Spontaneous Activity

Under some anesthesia conditions and during slow wave sleep, the membrane potential of cortical neurons fluctuates between a depolarized state and hyperpolarized state. This behavior is known as Up-Down activity. During the Down phase the neurons receive almost no synaptic inputs, so that the membrane stays near its resting potential. In the Up phase a barrage of synaptic inputs produces a reliable depolarization of 10–20 mV, which occasionally causes spiking (see Figure 1 in Up and down states).

The relation between the average amounts of excitatory and inhibitory synaptic inputs during the Up phase was studied using the conductance measurement method described above. These experiments, conducted both in vitro (Shu et al. 2003) and in vivo (Haider et al. 2006), have shown that excitatory and inhibitory conductances are balanced throughout the Up phase. In the beginning of the Up phase, both the excitatory and the inhibitory synaptic conductances are high and they tend to progressively decrease, but their ratio remains constant and approximately equal to 1.

In awake, drug-free animals the membrane potential dynamics exhibits an entire spectrum of distinct, brain state dependent activity patterns. The highly desynchronized high-conductance state, which is similar to a continuous Up phase (Crochet and Petersen 2006; Destexhe et al. 2007) represents one end of this spectrum. According to an intracellular study in the cortex of awake cats, in this condition the neurons are continuously bombarded by both excitatory and inhibitory inputs, where the total inhibitory conductance is several times higher than the excitatory one (Rudolph et al. 2007), providing a confirmation for the balanced excitation-inhibition hypothesis put forward by (Shadlen and Newsome 1994).

The other end of the spectrum of brain states in awake mammals is the quiet wakefulness condition, which is somewhat similar to light anesthesia, and is characterized by rather short depolarizations ('bumps') and membrane potential distribution that is not bimodal, e.g., (DeWeese and Zador 2006; Poulet and Petersen 2008). In the quiet wakefulness condition and light state of anesthesia there are no stereotypic Up events nor does the activity resemble a single continuous Up phase, therefore the single-electrode conductance measurement method which requires averaging over multiple repeats of some stereotypic event, recorded at different holding potentials, cannot be applied. However, the substantial synchrony of synaptic inputs to closely located neurons (Lampl et al. 1999; Hasenstaub et al. 2005; Okun and Lampl 2008; Poulet and Petersen 2008) which exists in this case allows to continuously monitor both the excitatory and the inhibitory activity in the local network. Toward this end simultaneous recording from a nearby pair of neurons are used, where one cell is hyperpolarized close to the reversal potential of inhibition and the other cell is depolarized sufficiently close to the reversal potential of excitation (Okun and Lampl 2008), Figure 3. This method reveals that in this type of spontaneous activity the excitatory and inhibitory inputs are interlocked in time, with inhibition lagging by several milliseconds behind excitation. Furthermore, the strength of excitatory and inhibitory inputs is (positively) correlated—large bumps typically contain both a strong excitatory and a strong inhibitory components, whereas small bumps are due to weak synaptic inputs, rather than strong inhibition that quenches the excitatory input. These correlations strongly suggest that inhibition plays important role in controlling the excitability of cortical networks at fast time scales.

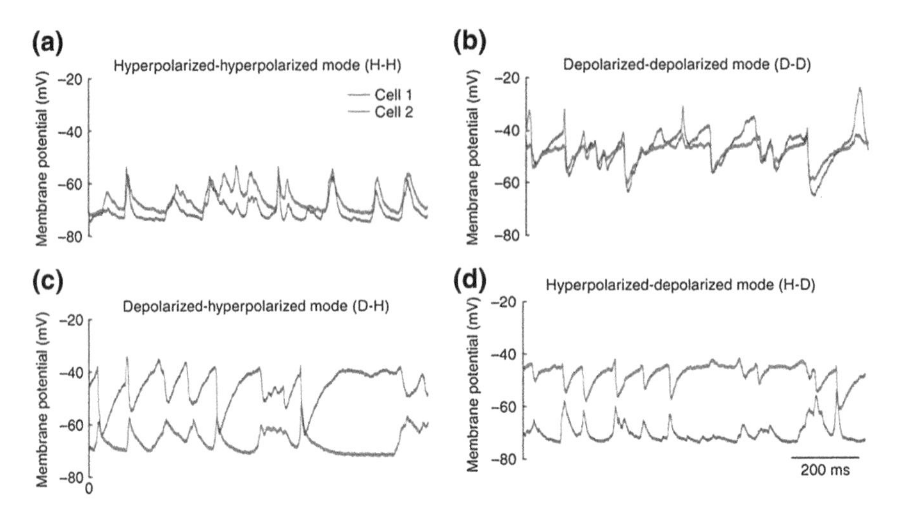

Figure 3 Excitatory and inhibitory inputs are synchronized during spontaneous activity. Two nearby neurons are simultaneously recorded when (**a**) both are at their resting potential, close to the reversal potential of inhibition (hyperpolarized-hyperpolarized mode); **b** both neurons are depolarized close to the reversal potential of excitation (depolarized-depolarized mode); **c–d** one of the neurons is in the hyperpolarized mode while the other is in the depolarized mode. In (**a**) the activity is dominated by excitatory inputs, which are seen to be highly synchronized between the neurons. Similarly, in (**b**) the activity is dominated by inhibitory inputs which are also highly synchronized. Finally, the mixed mode recordings (**c–d**) demonstrate that the excitatory and inhibitory inputs possess a high degree of synchrony. Adapted from Okun and Lampl (2008)

Current Research Directions

In the recent years a whole range of new genetic tools became available, particularly for the mouse (*Mus musculus*) species. In addition, working with awake head-fixed mice is relatively straightforward. These and other recent developments are heavily relied upon in the current research which, in addition to the directions discussed in the previous sections, focuses on new aspects of E/I balance, as described in more detail below.

E/I Balance Across Brain States

To date, only few works investigated how brain state modulation affects E/I balance. A study of primary visual cortex found that in awake mice, when compared to animals under anesthesia, the spatial tuning of inhibitory synaptic inputs is much wider, suggesting that in awake animals the E/I balance is profoundly skewed towards inhibition (Haider et al. 2013). However in the auditory cortex of awake mice excitation and inhibition have similar magnitude and frequency tuning

(Zhou et al. 2014), in agreement with previous results in anesthetized animals. Finally, a study of ongoing activity in the barrel cortex of anesthetized rats found that a switch to lighter anesthesia induces a profound shift toward excitation, probably due to depression of inhibitory synapses in the regime of higher activity under light anesthesia (Taub et al. 2013). At the present time it is not clear whether the differences between these studies are due to differences between brain areas, special connectivity subserving sensory tuning or other factors.

In addition to differences between awake and anesthetized conditions, the effects of transition between quiet wakefulness and locomotion were recently studied. Locomotion was found to have a differential effect on primary visual and auditory cortices, increasing the firing and shifting the balance towards excitation in the former (Bennett et al. 2013), while suppressing firing and equally scaling down both excitation and inhibition in the latter (Zhou et al. 2014). Hence, the impact of locomotion on brain-state and in particular on E/I balance is not uniform across the sensory cortices.

Interneuron Classes and the E/I Balance

In spite of constituting a minority, inhibitory interneurons in the cortex are vastly more diverse than the excitatory cells, with large variety of dendritic and axonal arborization patterns (Cajal 1911; Jones 1975). Histochemical and other methods revealed that GABAergic neurons in the cortex are subdivided into at least 4 almost non-overlapping classes (Kawaguchi and Kubota 1997; Harris and Mrsic-Flogel 2013): Parvalbumin (PV) expressing cells, somatostatin (Sst) expressing cells, vasoactive intestinal peptide (VIP) expressing cells and neurogliaform cells (NGs). Anatomical evidence and recordings in brain-slices suggest that these classes have different roles in the E/I balance and may have different functional roles across cortical layers. Current studies use molecular genetics and imaging methods to understand the role and function of each subtype.

Several converging lines of evidence indicate that PV cells constitute the major source of inhibitory current in principal cells for both spontaneous activity and sensory evoked responses. It follows that the sensory tuning of inhibitory synaptic inputs of pyramidal cells is expected to be the same or wider than the sensory tuning of the individual PV cells. For example, for orientation tuning in the mouse visual cortex, the tuning curves of PV cells were found to be much wider than of the principal cells, explaining the wide tuning of inhibitory inputs of pyramidal neurons (Atallah et al. 2012). In the auditory cortex the PV cells were found to be tuned for frequency, again consistent with inhibitory inputs to pyramidal cells originating in the neighboring PV neurons (Moore and Wehr 2013; Li et al. 2014).

The role of the other classes of inhibitory interneurons is currently investigated in many labs, in particular using the powerful new optogenetic tools. Optogenetic stimulation was recently used to examine the effect of PV and Sst cells on orientation tuning (Atallah et al. 2012; Lee et al. 2012; Wilson et al. 2012). (Atallah et al.

2012) and (Wilson et al. 2012) suggest that PV cells do not alter the tuning of principal cells. Wilson et al. (2012) furthermore attribute to Sst cells the ability to sharpen orientation selectivity of principal cells by a subtraction effect. In contrast, (Lee et al. 2012) report that activation of PV cells was found to sharpen the orientation tuning of principal cells. Whether the contradiction between the studies is real or only at the level of data interpretation is not entirely clear (Lee et al. 2014; Atallah et al. 2014).

Conclusions

The available data, collected under a wide variety of conditions and in distinct cortical areas indicates that co-activation of inhibition and excitation is a basic functional principle underlying various cortical activities (Isaacson and Scanziani 2011). Furthermore, the excitatory and inhibitory synaptic inputs appear to be individually matched in each pyramidal cell (Xue et al. 2014) with a high temporal precision of just a few milliseconds. Yet, whether excitation and inhibition share the same sensory tuning seems to depend on various factors, including animal species, the sensory modality and brain-state.

The E/I balance was studied most extensively in the cortex, however similar principles manifest themselves in many CNS structures, such as the hippocampus (Atallah and Scanziani 2009), superior colliculus (Populin 2005), brain stem (Magnusson et al. 2008), spinal cord (Berg et al. 2007), prefrontal cortex (Yizhar et al. 2011) and others, not covered here in detail. This entry also did not describe E/I balance development and plasticity, e.g., (Froemke et al. 2007; Dorrn et al. 2010; Sun et al. 2010; Li et al. 2012). While the role of the tight coupling between excitation and inhibition is not fully clear, it is most likely to serve as a major gain mechanism that increases the accuracy and speed of neuronal response. By counterbalancing the excitatory drive, inhibitory inputs greatly extend the dynamic range of excitation, allowing a fine and rapid control over the amount of depolarization of the membrane potential. It is apparent that achieving a certain depolarization without a counteracting inhibitory force would have required a much weaker excitatory input, increasing the error and variability of the response.

Internal References

Burke, R E (2008). Spinal cord. *Scholarpedia* 3(4): 1925. http://www.scholarpedia.org/article/Spinal_cord.

Destexhe, A (2007). High-conductance state. *Scholarpedia* 2(11): 1341. http://www.scholarpedia.org/article/High-conductance_state.

Freund, T and Kali, S (2008). Interneurons. *Scholarpedia* 3(9): 4720. http://www.scholarpedia.org/article/Interneurons.

Jonas, P and Buzsaki, G (2007). Neural inhibition. *Scholarpedia* 2(9): 3286. http://www.scholarpedia.org/article/Neural_inhibition.
Llinas, R (2008). Neuron. *Scholarpedia* 3(8): 1490. http://www.scholarpedia.org/article/Neuron.
Meiss, J (2007). Dynamical systems. *Scholarpedia* 2(2): 1629. http://www.scholarpedia.org/article/Dynamical_systems.
Moore, J W (2007). Voltage clamp. *Scholarpedia* 2(9): 3060. http://www.scholarpedia.org/article/Voltage_clamp.
Pikovsky, A and Rosenblum, M (2007). Synchronization. *Scholarpedia* 2(12): 1459. http://www.scholarpedia.org/article/Synchronization.
Skinner, F K (2006). Conductance-based models. *Scholarpedia* 1(11): 1408. http://www.scholarpedia.org/article/Conductance-based_models.
Wilson, C (2008). Up and down states. *Scholarpedia* 3(6): 1410. http://www.scholarpedia.org/article/Up_and_down_states.

External References

Alonso, J M and Martinez, L M (1998). Functional connectivity between simple cells and complex cells in cat striate cortex. *Nature Neuroscience* 1: 395–403.
Anderson, J S; Carandini, M and Ferster, D (2000). Orientation tuning of input conductance, excitation, and inhibition in cat primary visual cortex. *Journal of Neurophysiology* 84: 909–926.
Anderson, J S; Lampl, I; Gillespie, D C and Ferster, D (2001). Membrane potential and conductance changes underlying length tuning of cells in cat primary visual cortex. *The Journal of Neuroscience* 21: 2104–2112.
Atallah, B V and Scanziani, M (2009). Instantaneous modulation of gamma oscillation frequency by balancing excitation with inhibition. *Neuron* 62: 566–577.
Atallah, B V; Bruns, W; Carandini, M and Scanziani, M (2012). Parvalbumin-expressing interneurons linearly transform cortical responses to visual stimuli. *Neuron* 73: 159–170.
Atallah, B V; Scanziani, M and Carandini, M (2014). Atallah et al. reply. *Nature* 508: E3.
Bennett, C; Arroyo, S and Hestrin, S (2013). Subthreshold mechanisms underlying state-dependent modulation of visual responses. *Neuron* 80: 350–357.
Ben-Yishai, R; Bar-Or, R L and Sompolinsky, H (1995). Theory of orientation tuning in visual cortex. *Proceedings of the National Academy of Sciences of the United States of America* 92: 3844–3848.
Berg, R W; Alaburda, A and Hounsgaard, J (2007). Balanced inhibition and excitation drive spike activity in spinal half-centers. *Science* 315: 390–393.
Borg-Graham, L J; Monier, C and Fregnac, Y (1996). Voltage-clamp measurement of visually-evoked conductances with whole-cell patch recordings in primary visual cortex. *Journal of Physiology Paris* 90: 185–188.
Borg-Graham, L J; Monier, C and Fregnac, Y (1998). Visual input evokes transient and strong shunting inhibition in visual cortical neurons. *Nature* 393: 369–373.
Bruno, R M and Sakmann, B (2006). Cortex is driven by weak but synchronously active thalamocortical synapses. *Science* 312: 1622–1627.
Calford, M B and Semple, M N (1995). Monaural inhibition in cat auditory cortex. *Journal of Neurophysiology* 73: 1876–1891.
Chung, S and Ferster, D (1998). Strength and orientation tuning of the thalamic input to simple cells revealed by electrically evoked cortical suppression. *Neuron* 20: 1177–1189.
Cohen-Kashi Malina, K; Jubran, M; Katz, Y and Lampl, I (2013). Imbalance between excitation and inhibition in the somatosensory cortex produces postadaptation facilitation. *The Journal of Neuroscience* 33: 8463–8471.
Crochet, S and Petersen, C C H (2006). Correlating whisker behavior with membrane potential in barrel cortex of awake mice. *Nature Neuroscience* 9: 608–610.

Destexhe, A; Hughes, S W; Rudolph, M and Crunelli, V (2007). Are corticothalamic 'up' states fragments of wakefulness? *Trends in Neurosciences* 30: 334–342.

DeWeese, M R and Zador, A M (2006). Non-Gaussian membrane potential dynamics imply sparse, synchronous activity in auditory cortex. *The Journal of Neuroscience* 26: 12206–12218.

Dichter, M A and Ayala, G F (1987). Cellular mechanisms of epilepsy: A status report. *Science* 237: 157–164.

Dorrn, A L; Yuan, K; Barker, A J; Schreiner, C E and Froemke, R C (2010). Developmental sensory experience balances cortical excitation and inhibition. *Nature* 465: 932–936.

Ferster, D and Miller, K D (2000). Neural mechanisms of orientation selectivity in the visual cortex. *Annual Review of Neuroscience* 23: 441–471.

Froemke, R C; Merzenich, M M and Schreiner, C E (2007). A synaptic memory trace for cortical receptive field plasticity. *Nature* 450: 425–429.

Gabernet, L; Jadhav, S P; Feldman, D E; Carandini, M and Scanziani, M (2005). Somatosensory integration controlled by dynamic thalamocortical feed-forward inhibition. *Neuron* 48: 315–327.

Gerstein, G L and Mandelbrot, B (1964). Random walk models for the spike activity of a single neuron. *Biophysical Journal* 4: 41–68.

Haider, B; Duque, A; Hasenstaub, A R and McCormick, D A (2006). Neocortical network activity in vivo is generated through a dynamic balance of excitation and inhibition. *The Journal of Neuroscience* 26: 4535–4545.

Haider, B; Häusser, M and Carandini, M (2013). Inhibition dominates sensory responses in the awake cortex. *Nature* 493: 97–100.

Hansel, D and Sompolinsky, H (1996). Chaos and synchrony in a model of a hypercolumn in visual cortex. *Journal of Comparative Neuroscience* 3: 7–34.

Harris, K D and Mrsic-Flogel, T D (2013). Cortical connectivity and sensory coding. *Nature* 503: 51–58.

Hasenstaub, A et al. (2005). Inhibitory postsynaptic potentials carry synchronized frequency information in active cortical networks. *Neuron* 47: 423–435.

Heiss, J E; Katz, Y; Ganmor, E and Lampl, I (2008). Shift in the balance between excitation and inhibition during sensory adaptation of S1 neurons. *The Journal of Neuroscience* 28: 13320–13330.

Higley, M J and Contreras, D (2003). Nonlinear integration of sensory responses in the rat barrel cortex: an intracellular study in vivo. *The Journal of Neuroscience* 23: 10190–10200.

Higley, M J and Contreras, D (2006). Balanced excitation and inhibition determine spike timing during frequency adaptation. *The Journal of Neuroscience* 26: 448–457.

Hubel, D H and Wiesel, T N (1962). Receptive fields, binocular interaction and functional architecture in the cat's visual cortex. *Journal of Physiology (London)* 160: 106–154.

Isaacson, J S and Scanziani, M (2011). How inhibition shapes cortical activity. *Neuron* 72: 231–243.

Jones, E G (1975). Varieties and distribution of non-pyramidal cells in the somatic sensory cortex of the squirrel monkey. *Journal of Comparative Neurology* 160: 205–267.

Kawaguchi, Y and Kubota, Y (1997). GABAergic cell subtypes and their synaptic connections in rat frontal cortex. *Cerebral Cortex* 7: 476–486.

Lak, A; Arabzadeh, E and Diamond, M E (2008). Enhanced response of neurons in rat somatosensory cortex to stimuli containing temporal noise. *Cerebral Cortex* 18: 1085–1093.

Lampl, I; Reichova, I and Ferster, D (1999). Synchronous membrane potential fluctuations in neurons of the cat visual cortex. *Neuron* 22: 361–374.

Lee, S H et al. (2012). Activation of specific interneurons improves V1 feature selectivity and visual perception. *Nature* 488: 379–383.

Lee, S H; Kwan, A C and Dan, Y (2014). Interneuron subtypes and orientation tuning. *Nature* 508: E1-E2.

Lefort, S; Tomm, C; Floyd Sarria, J C and Petersen, C C (2009). The excitatory neuronal network of the C2 barrel column in mouse primary somatosensory cortex. *Neuron* 61: 301–316.

Li, Y T; Ma, W P; Pan, C J; Zhang, L I and Tao, H W (2012). Broadening of cortical inhibition mediates developmental sharpening of orientation selectivity. *The Journal of Neuroscience* 32: 3981–3991.

Li, L Y et al. (2014). A feedforward inhibitory circuit mediates lateral refinement of sensory representation in upper layer 2/3 of mouse primary auditory cortex. *The Journal of Neuroscience* 34: 13670–13683.

Liu, B et al. (2011). Broad inhibition sharpens orientation selectivity by expanding input dynamic range in mouse simple cells. *Neuron* 71: 542–554.

Magnusson, A K; Park, T J; Pecka, M; Grothe, B and Koch, U (2008). Retrograde GABA signaling adjusts sound localization by balancing excitation and inhibition in the brainstem. *Neuron* 59: 125–137.

Martinez, L M and Alonso, J M (2001). Construction of complex receptive fields in cat primary visual cortex. *Neuron* 32: 515–525.

Monier, C; Fournier, J and Fregnac, Y (2008). In vitro and in vivo measures of evoked excitatory and inhibitory conductance dynamics in sensory cortices. *Journal of Neuroscience Methods* 169: 323–365.

Moore, A K and Wehr, M (2013). Parvalbumin-expressing inhibitory interneurons in auditory cortex are well-tuned for frequency. *The Journal of Neuroscience* 33: 13713–13723.

Nelson, S; Toth, L; Sheth, B and Sur, M (1994). Orientation selectivity of cortical neurons during intracellular blockade of inhibition. *Science* 265: 774–777.

Okun, M and Lampl, I (2008). Instantaneous correlation of excitation and inhibition during ongoing and sensory-evoked activities. *Nature Neuroscience* 11: 535–537.

Okun, M; Naim, A and Lampl, I (2010). The subthreshold relation between cortical local field potential and neuronal firing unveiled by intracellular recordings in awake rats. *The Journal of Neuroscience* 30: 4440–4448.

Polack, P O; Friedman, J and Golshani, P (2013). Cellular mechanisms of brain state-dependent gain modulation in visual cortex. *Nature Neuroscience* 16: 1331–1339.

Populin, L C (2005). Anesthetics change the excitation/inhibition balance that governs sensory processing in the cat superior colliculus. *The Journal of Neuroscience* 25: 5903–5914.

Poulet, J F and Petersen, C C (2008). Internal brain state regulates membrane potential synchrony in barrel cortex of behaving mice. *Nature* 454: 881–885.

Priebe, N J and Ferster, D (2005). Direction selectivity of excitation and inhibition in simple cells of the cat primary visual cortex. *Neuron* 45: 133–145.

Ramon y Cajal, S (1911). *Histologie du Systeme Nerveux de l'Homme et des Vertebres*. Paris: Maloine.

Rudolph, M; Pospischil, M; Timofeev, I and Destexhe, A (2007). Inhibition determines membrane potential dynamics and controls action potential generation in awake and sleeping cat cortex. *The Journal of Neuroscience* 27: 5280–5290.

Sachidhanandam, S; Sreenivasan, V; Kyriakatos, A; Kremer, Y and Petersen, C C H (2013). Membrane potential correlates of sensory perception in mouse barrel cortex. *Nature Neuroscience* 16: 1671–1677.

Shadlen, M N and Newsome, W T (1994). Noise, neural codes and cortical organization. *Current Opinion in Neurobiology* 4: 569–579.

Shadlen, M N and Newsome, W T (1998). The variable discharge of cortical neurons: Implications for connectivity, computation, and information coding. *The Journal of Neuroscience* 18: 3870–3896.

Shapley, R M and Xing, D (2013). Local circuit inhibition in the cerebral cortex as the source of gain control and untuned suppression. *Neural Networks* 37: 172–181.

Shu, Y; Hasenstaub, A and McCormick, D A (2003). Turning on and off recurrent balanced cortical activity. *Nature* 423: 288–293.

Sillito, A M (1975). The contribution of inhibitory mechanisms to the receptive field properties of neurones in the striate cortex of the cat. *Journal of Physiology* 250: 305–329.

Softky, W R and Koch, C (1993). The highly irregular firing of cortical cells is inconsistent with temporal integration of random EPSPs. *The Journal of Neuroscience* 13: 334–350.

Somers, D C; Nelson, S B and Sur, M (1995). An emergent model of orientation selectivity in cat visual cortical simple cells. *The Journal of Neuroscience* 15: 5448–5465.

Stevens, C F and Zador, A M (1998). Input synchrony and the irregular firing of cortical neurons. *Nature Neuroscience* 1: 210–217.

Sun, Y J et al. (2010). Fine-tuning of pre-balanced excitation and inhibition during auditory cortical development. *Nature* 465: 927–931.

Sun, Y J; Kim, Y J; Ibrahim, L A; Tao, H W and Zhang, L I (2013). Synaptic mechanisms underlying functional dichotomy between intrinsic-bursting and regular-spiking neurons in auditory cortical layer 5. *The Journal of Neuroscience* 33: 5326–5339.

Sutter, M L; Schreiner, C E; McLean, M; O'Connor, K N and Loftus, W C (1999). Organization of inhibitory frequency receptive fields in cat primary auditory cortex. *Journal of Neurophysiology* 82: 2358–2371.

Tan, A Y Y; Chen, Y; Scholl, B; Seidemann, E and Priebe, N J (2014). Sensory stimulation shifts visual cortex from synchronous to asynchronous states. *Nature* 509: 226–229.

Taub, A H; Katz, Y and Lampl, I (2013). Cortical balance of excitation and inhibition is regulated by the rate of synaptic activity. *The Journal of Neuroscience* 33: 14359–14368.

van Vreeswijk, C and Sompolinsky, H (1996). Chaos in neuronal networks with balanced excitatory and inhibitory activity. *Science* 274: 1724–1726.

Vogels, T P; Rajan, K and Abbott, L F (2005). Neural network dynamics. *Annual Review of Neuroscience* 28: 357–376.

Volgushev, M; Vidyasagar, T R and Pei, X (1996). A linear model fails to predict orientation selectivity of cells in the cat visual cortex. *Journal of Physiology (London)* 496: 597–606.

Wang, J; McFadden, S L; Caspary, D and Salvi, R (2002). Gamma-aminobutyric acid circuits shape response properties of auditory cortex neurons. *Brain Research* 944: 219–231.

Wehr, M and Zador, A M (2003). Balanced inhibition underlies tuning and sharpens spike timing in auditory cortex. *Nature* 426: 442–446.

Wehr, M and Zador, A M (2005). Synaptic mechanisms of forward suppression in rat auditory cortex. *Neuron* 47: 437–445.

Wilent, W B and Contreras, D (2005). Dynamics of excitation and inhibition underlying stimulus selectivity in rat somatosensory cortex. *Nature Neuroscience* 8: 1364–1370.

Wilson, N R; Runyan, C A; Wang, F L and Sur, M (2012). Division and subtraction by distinct cortical inhibitory networks in vivo. *Nature* 488: 343–348.

Wu, G K; Li, P; Tao, H W and Zhang, L I (2006). Nonmonotonic synaptic excitation and imbalanced inhibition underlying cortical intensity tuning. *Neuron* 52: 705–715.

Wu, G K; Arbuckle, R; Liu, B H; Tao, H W and Zhang, L I (2008). Lateral sharpening of cortical frequency tuning by approximately balanced inhibition. *Neuron* 58: 132–143.

Xue, M; Atallah, B V and Scanziani, M (2014). Equalizing excitation-inhibition ratios across visual cortical neurons. *Nature* 511: 596–600.

Yizhar, O et al. (2011). Neocortical excitation/inhibition balance in information processing and social dysfunction. *Nature* 477: 171–178.

Zhou, M et al. (2014). Scaling down of balanced excitation and inhibition by active behavioral states in auditory cortex. *Nature Neuroscience* 17: 841–850.

Vibrissa Mechanical Properties

Mitra Hartmann

The vibrissal (whisker) array of the rodent has been an important model for the study of active touch and tactile perception for over a century (Richardson 1909; Vincent 1912). During exploratory behaviors many rodents brush and tap their whiskers against surfaces to tactually extract object features, similar in some ways to how humans use their fingers for tactual exploration.

One of the largest advantages of studying vibrissae is that they are relatively mechanically simple. The whisker can be modeled as a tapered cantilever beam that transmits mechanical information to mechanoreceptors in the follicle at the whisker base.

The relative mechanical simplicity of the whiskers offers a long-term vision in which we can compute the complete set of tactile (mechanical) inputs transmitted by the vibrissae during active tactile exploration. To realize this vision requires careful quantification of whisker mechanics under the full range of behavioral conditions associated with active touch.

The present article reviews whisker mechanics in a manner intended to provide physical intuition for the types of mechanical signals that characterize natural vibrissotactile exploratory behavior.

Introduction: Whisker Geometry and Material Properties

Vibrissal mechanics is an exciting field of study because it allows us to begin to understand how other mammals perceive their world. It is not easy to gain intuition for the vibrissal experience, as humans count as one of very few mammals who do not use vibrissal-based sensing as part of their behavioral repertoire (Muchlinski 2010; Prescott et al. 2011; Grant et al. 2013; Muchlinski et al. 2013). As a start, we can state with confidence that the mechanical behavior of whiskers is governed by their *geometry*, their *density*, and their *elastic moduli*.

M. Hartmann (✉)
Northwestern University, Evanston, USA

T.J. Prescott et al. (eds.), *Scholarpedia of Touch*, Scholarpedia,
DOI 10.2991/978-94-6239-133-8_45

To help gain an intuition for whisker geometry, the schematic of Figure 1 depicts a typical C3 rat whisker whose intrinsic curvature has been removed, and whose base diameter (120 μm) and length (2 cm) have been scaled up by a factor of five. The whisker in the figure has been drawn with a base diameter of 0.6 mm and a length of 10 cm. At this magnified scale it is easier to gain an appreciation for how long the whisker is compared to its diameter. Whiskers are very thin.

The geometric features that characterize a rat whisker are illustrated in Figure 2a. Rat whiskers have an intrinsic curvature well-matched by a quadratic fit (Knutsen et al. 2008; Towal et al. 2011) and they taper approximately linearly from base to tip (Ibrahim and Wright 1975; Williams and Kramer 2010; Hires et al. 2013). The cross-section of a whisker contains three layers: the outermost layer is called the cuticle, followed by the cortex, and then the medulla, which is hollow (Quist et al. 2011; Voges et al. 2012; Adineh et al. 2015). The presence of the medulla changes the cross-sectional geometry of the whisker as well as local average density, but these effects are too detailed for the present article. This article neglects the medulla, and analyzes the whisker as though it were solid and had uniform density. Subsequent sections will show how both whisker taper and intrinsic curvature play important roles in shaping information transmission along the whisker to mechanoreceptors in the follicle at the whisker base.

Whiskers are made of alpha keratin (Fraser and Macrae 1980), with an approximate average density between 1.1 mg/mm^3 and 1.50 mg/mm^3 (Mason 1963; Neimark et al. 2003). With these density values, rat whiskers are expected to range in mass between ~8 micrograms for the smallest, most rostral whiskers (e.g., C6) and ~700 micrograms for the largest most caudal whiskers (e.g., beta, C1) (Hartmann et al. 2003; Neimark et al. 2003). If the whisker depicted at 10× scale in Figure 1 were made of keratin, it would weigh approximately 14 milligrams. This is approximately the same weight as if one were to use heavy-stock paper to construct a whisker of the same size.

The *elastic moduli* of the whisker describe how resistant it is to being deformed in different directions (Beer et al. 2008). All materials, including whiskers, are characterized by many different elastic moduli, but the two most relevant to whisker mechanics are Young's modulus (E) and the shear modulus (G). The shear modulus is important when considering mechanics in three dimensions (3D), discussed only at the end of this article. In contrast, Young's modulus is essential to understanding basic whisker mechanics even in two dimensions (2D), so we describe it here in some detail.

Young's modulus is important because it helps determine the bending stiffness of the whisker, which in turn determines how the whisker will bend when it makes contact with an object.

Figure 1 A schematic of a *straightened* C3 whisker at 5 × scale. The whisker in the figure has been drawn with a base diameter of 0.6 mm and a length of 10 cm. At this scale, the whisker would weigh approximately 14 milligrams

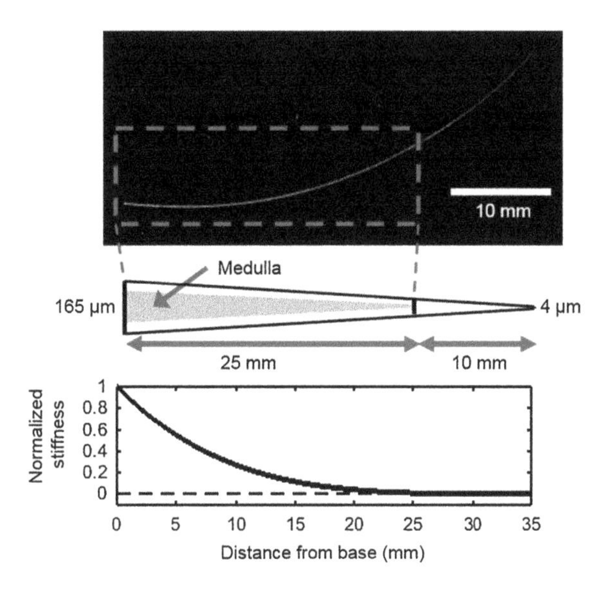

Figure 2 Geometric properties of rat whiskers. *Top* Like all rat whiskers, the C2 whisker, shown here as an image from a flat-bed scan, has an intrinsic curvature that is well-matched by a quadratic fit. The expanded region is a schematic of the whisker with its intrinsic curvature removed, so as better to depict taper (approximately linear) and the presence of a medulla in the proximal region. The medulla dimensions are approximate. *Bottom* Because the whisker tapers, its stiffness decreases rapidly, as the fourth power of the radius. This stiffness calculation was performed neglecting the presence of the medulla

To develop an understanding of how Young's modulus contributes to stiffness, it is useful to perform a thought experiment. Imagine holding a length of stainless steel rod, say 2 cm in diameter, and trying to bend it in the middle. It is clear that you would have to exert considerable force in order for the rod to flex. Now imagine holding a length of nylon rod, also 2 cm in diameter and trying to bend it. Obviously it is much easier to imagine bending the nylon rod than the stainless steel one. Correspondingly, stainless steel has a Young's modulus of approximately 200 gigapascals (GPa) while nylon has a Young's modulus between 2-4 GPa. The stainless steel rod will be about 100 times stiffer than the nylon rod.

But now consider a second thought experiment. Imagine that the stainless steel rod is only 1 mm in diameter. It is now easy to imagine that you could bend the stainless steel rod in the middle. Perhaps your intuition says that a 2 cm nylon rod would be stiffer than a 1 mm stainless steel rod, which is in fact, true. This thought experiment makes it clear that stiffness depends both on the geometry of the rod as well as the material properties of the rod.

It turns out that the bending stiffness of a beam with circular cross section can be written as a product of Young's modulus and the "area moment of inertia" I, defined here as $\boldsymbol{I} = \frac{\pi r^4}{4}$:

$$\textbf{Stiffness} = \boldsymbol{EI} = \frac{\boldsymbol{E\pi r^4}}{\mathbf{4}} \tag{1}$$

An important feature of this equation is that stiffness depends *linearly* on Young's modulus, but depends on the fourth power of the radius (Beer et al. 2008). Notice that stiffness does not depend at all on the density of the whisker.

In the case of a cylindrical rod, the stiffness is equal at all locations along the length of the rod, because the radius does not change. But whiskers are not cylinders—they taper. This means that their stiffness decreases along their length, and it decreases *rapidly*, as the fourth power of the radius. This rapid stiffness drop off is illustrated in the bottom graph of Figure 2.

The rapid drop-off in stiffness explains why it is so difficult to see bending near the base of the whisker during experiments. The distal region of the whisker bends much more than the proximal region because it has a smaller radius—but, as we will see in Sect. 5—this does not mean that the forces and moments are larger in the distal regions.

How stiff are whiskers? Depending on experimental technique, studies typically estimate a range for Young's modulus between 0.33 GPa and 4.92 GPa (Hartmann et al. 2003; Herzog et al. 2005; Birdwell et al. 2007; Quist et al. 2011; Kan et al. 2013). Higher values have been found as outliers (Birdwell et al. 2007) and several studies that rely on bending tests or resonance tests have found significantly higher values, in the range of 7.5 GPa (Neimark et al. 2003; Carl et al. 2012). One recent nanoindentation study (Kan et al. 2013) found evidence that Young's modulus varies considerably from whisker to whisker within a single array, with an overall tendency to increase from caudal to rostral within a row. The most detailed study to date characterized Young's modulus across each of the component layers of the whisker (Adineh et al. 2015). Results showed that the cuticle region had the largest values for Young's modulus, with values gradually decreasing towards the interior, from cortex towards the medulla.

Regardless of the exact value, reporting Young's modulus alone does not provide good physical intuition for whisker stiffness. To improve intuition for whisker stiffness, imagine holding the 10× whisker at its base, recalling that it would weigh approximately as much as if it were constructed of heavy-weight paper. The whisker's own mass will cause it to deflect under the influence of gravity. If the 10× whisker had the same stiffness as a real whisker, its tip would deflect ("droop") under the influence of gravity by no more than approximately the diameter of the whisker base (1.2 mm). The tip of a real rat whisker will thus deflect under its own weight by no more than approximately 250 μm.

In the next three sections, we examine how these properties—geometry, density, and stiffness—influence the mechanical signals that the rat will obtain during whisking behavior.

Non-contact Whisking

Different species of animals exhibit "whisking" behavior to varying degrees (Muchlinski 2010; Prescott et al. 2011; Grant et al. 2013; Muchlinski et al. 2013). Whisking involves an active sweeping motion of the whiskers forwards ("protraction") and backwards ("retraction"), and can occur independent of head movements. The term "non-contact" whisking is used to refer to situations in which the animal actively whisks in free air, that is to say, the whiskers do not touch any object. The term "contact" whisking refers to conditions in which the animal whisks so as to actively brush and tap its whiskers against a surface or object.

The Kinematics of Whisking

The kinematics of whisking is complex (Knutsen 2015). Here we describe those aspects of kinematics essential to understanding the mechanical analysis in Sect. 4. Each of the ***bold, italicized*** terms is defined further in Figure 3.

The most basic description of whisking kinematics is based on the simplifying assumption that each whisker undergoes a rigid two-dimensional (2D) angular rotation in a plane approximately parallel to the ground. Variables are defined within the 2D view that would be obtained from a top-down camera. The very proximal portion of each whisker (usually 0.5–1 cm) is taken to be linear, and the ***position*** of the whisker is then defined as the angle that the linear portion of the whisker makes relative to a reference line, here chosen to be the midline of the rat (Figure 3a). The ***velocity*** of the whisker is taken to be the time derivative of the angular position, and the ***speed*** is the absolute value of the velocity.

Actual whisking motions are considerably more complex than the 2D approximation would suggest. First, there is significant ***translation*** of each whisker's basepoint, in addition to angular rotation. Second, during protraction, each whisker exhibits a small change in ***elevation*** (Bermejo et al. 2002; Knutsen et al. 2008) and each whisker also ***rolls*** (Knutsen et al. 2008) about its own axis (Figure 3b). These effects are small when observed in 2D with a top-down camera, but highly significant when the 3D case is considered (Knutsen et al. 2008; Huet and Hartmann 2014; Huet et al. 2015; Knutsen 2015). The roll and elevation can be written as functions of the protraction angle (Knutsen et al. 2008; Knutsen 2015), so that the entire kinematics of the whisk can be simulated (Huet and Hartmann 2014; Hobbs et al. 2015). Together, these three effects—translation, elevation, and roll—become particularly important when quantifying how much the rodent has deflected its whisker against an object (Sect. 4).

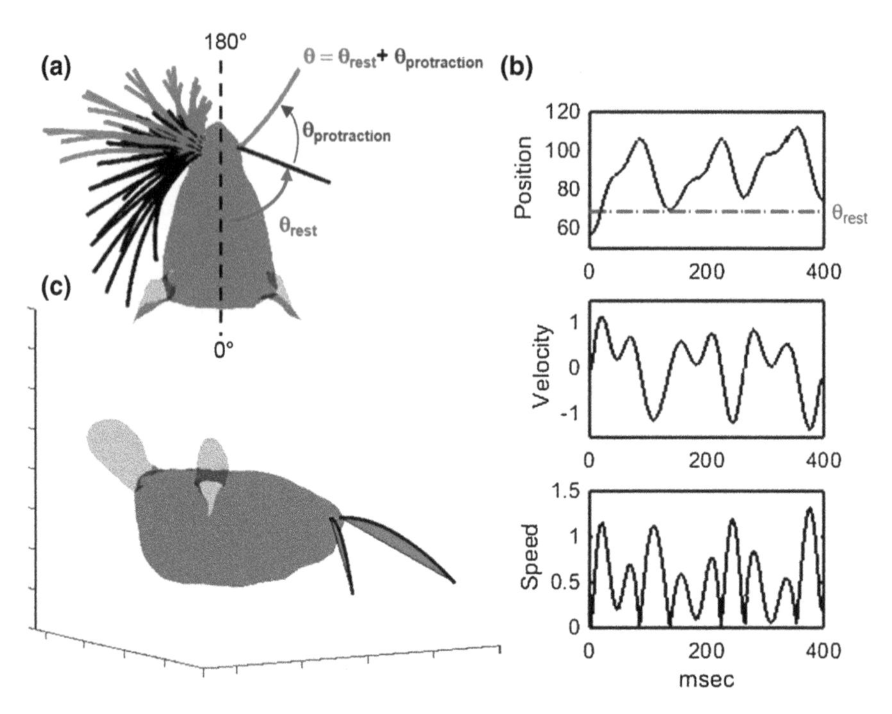

Figure 3 Definitions of a few of the most important kinematic variables. **a** Angular position is defined relative to the rostral-caudal midline of the rat. The values 0° and 180° represent the whisker pointing directly caudal and rostral, respectively. On the right side of the rat the C2 whisker is shown in *black* in its "resting" position, where it would be when none of the vibrissal muscles are contracted. The whisker is shown in *red* after a protraction of 60°. The *left side* of the rat shows how the entire array would look at rest (*black*) and after a 60° protraction (*red*). **b** Position, velocity, and speed of the whisker for three exemplary whisks. Position has units of degrees; velocity and speed both have units of degrees/msec. Note that the whisker can retract further caudal than its resting position. The sign of the velocity indicates the direction in which the whisker is moving (positive is rostral; negative is caudal). The speed is a scalar and is always positive. **c** Although the primary whisker motion is rotation in the horizontal plane, whiskers also exhibit significant translation, roll, and elevation. The schematic shows the C2 whisker at rest and after a 60° protraction. The plane of the whisker has been shaded red to improve visualization. In this schematic, the translation is evident as a shift in the location of the basepoint of the whisker. The roll is evident as a rotation of the whisker about its long axis. The elevation is difficult to observe independent from the roll, but can be seen as a tiny shift in the vertical position of the tip of the whisker

Developing an Intuition for Whisker Kinetics

"Kinetics" is defined by different people in different fields in different ways. Here we use it simply as a broad term to indicate the forces and moments associated with the whisker, as distinct from the "kinematic" variables.

The study of forces and moments can be divided into studying ***quasi-static*** and ***dynamic*** effects. This statement will be more useful if we are able to assign some

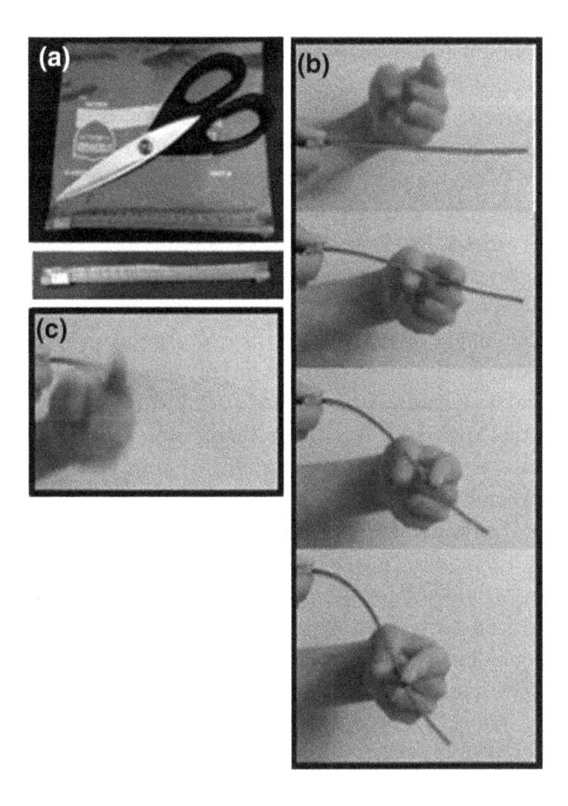

Figure 4 A simple artificial whisker constructed from the zipper portion of a freezer bag (**a**) illustrates the fundamental differences between quasistatic (**b**) and dynamic (**c**) effects

intuitive meaning to these concepts. We will gain intuition by doing some experiments that impose an external force on an artificial whisker. Then we will apply the ideas of quasi-statics and dynamics to understand the forces and moments generated during non-contact whisking, collisions, and bending against an object.

A simple artificial whisker can be constructed by cutting off the zipper portion of a quart-sized freezer bag, as shown in Figure 4a.

For the first experiment, hold the whisker horizontally and use your finger to push down on it very slowly at some point out along its length (Figure 4b). You will see the whisker gradually bend as you deflect it. Notice that all of the bending occurs proximal to your finger. There is no change in the whisker shape distal to your finger.

One of the most important points to notice is that the whisker will bend in the identical way, regardless of whether you push on it at a rate of 1 mm/minute, or 1 mm/hour, or 1 mm/year. The only parameter that matters to the bending is the distance that your finger has deflected the whisker. As long as you push "slowly enough", the rate at which bending occurs does not influence the shape of the bent whisker. This is an example of quasi-static bending. The bending depends only on the stiffness of the whisker (Young's modulus and geometry). The bending does not depend on the whisker's density or mass, and bending does not involve vibrations.

Real whiskers have an intrinsic curvature, so we need to determine how to incorporate the initial curvature into our analysis of bending. As it turns out, the solution is remarkably simple: the final shape of the whisker can be determined simply by adding the whisker's intrinsic curvature to the curvature induced by deflection. Importantly, however, this linear summation works only in the quasi-static regime—that is, if you push on the whisker "slowly enough".

What happens if you don't push "slowly enough"—suppose you push "too fast?" Try deflecting the whisker rapidly. Now you will see that the whisker bends, as before, but it also vibrates (Figure 4c). The vibrations are most visible in the distal region of the whisker, but the proximal region experiences important vibrations as well, even if you can't see them. This is an example of a dynamic effect. The shape of the whisker at every point in time depends on the stiffness of the whisker, the mass of the whisker, how the mass is distributed along the whisker, and how energy is dissipated in the whisker. Whisker dynamics also depend on the exact details of how your finger struck the whisker, and how you happened to choose to decelerate your finger after it struck the whisker. Clearly, whisker dynamics are much more complicated than quasi-statics.

These two experiments get at the heart of the distinction between quasi-statics and dynamics in the context of the vibrissal system. Dynamic effects depend on mass and acceleration (and stiffness, and other parameters); quasi-static effects depend only on stiffness. For completeness, we note that dynamic effects are also sometimes called "inertial effects". These two terms are often used somewhat interchangeably.

An excellent question to ask here is: would it be possible to describe the complete mechanical behavior of the whisker just by adding the dynamic effects on top of the quasistatic effects? This idea is called "linear superposition". The answer is no in general, but yes in the limit of small whisker deformations. This approach recently allowed researchers to correctly capture the behavior of the whisker, at least to within experimental resolution, just after a shock (Boubenec et al. 2012). This method could prove very useful in in other contexts but is expected to become incorrect in configurations in which the whisker experiences large deflections.

We conclude this section with an important final point. All forces and moments—regardless of whether they are generated by quasistatic or dynamic effects—can be computed at every point along the whisker length. In practice, however, it is most useful to compute these quantities at the whisker base, because those are the signals that will directly enter the follicle.

Thus one of the primary reasons that the rat vibrissal system is such an attractive model for the study of active touch is that we can make a very strong statement about the signals sent to the brain by the whisker: ***all*** *mechanical information transmitted by a whisker to the nervous system can be represented by the time series of three forces and three moments at the whisker base.*

The three forces describe how the whisker resists translation in three directions (e.g., x, y, and z). Two of the moments (the two "bending moments") will describe how the whisker tends to bend as it resists rotation. One of the moments (the "twisting moment", or the "torque") will describe how the whisker "twists" about its own axis. We now examine how these forces and moments vary during non-contact whisking.

Kinetics During Non-contact Whisking

To gain intuition for the mechanical effects relevant to non-contact whisking it is useful to perform a few more experiments that highlight the difference between rigid body rotation and flexible body rotation.

During non-contact whisking, the whisker is driven at its base by both intrinsic and extrinsic muscles. To obtain a sense for what whisking would be like if the whisker were a rigid body, try rotating a pen as though it were a whisker, holding it at one end (Figure 5a). Notice that if you use a heavier pen, the mechanical signals you feel at the base during rotation are different than if you use a lighter pen. This is an immediate indication that we are dealing with a dynamic effect.

Next, try rotating the artificial whisker at about the same speed as you did the pen (Figure 5b). The whisker will start to bend during the rotation, and it will be difficult to avoid shaking the whisker at its resonance. As was the case for the pen, the mechanical signals generated during rotation of the artificial whisker are also dynamic effects, and depend on the whisker's mass distribution. You can test the effect of mass distribution by changing the location of the slider on the zipper as shown in Figure 5c. A cone has its center of mass at a distance L/4 from the base, so to gain as realistic impression as possible, place the slider about a quarter of the length out.

The key point here is that in both experiments—rotating either the pen or the artificial whisker—the forces and moments generated at the whisker base are a result of dynamic effects. A dynamic model is needed to compute mechanical signals at the whisker base during non-contact whisking. The open question now is—what does the rodent experience? Does the rodent sense something more similar to what you sensed when you rotated the pen (rigid body), or more similar to when you rotated the artificial whisker (flexible body)?

The experiment that involved rotating the pen (Figure 5a) approximates the whisker as a rigid body. In this case, the moment M at the whisker base can be

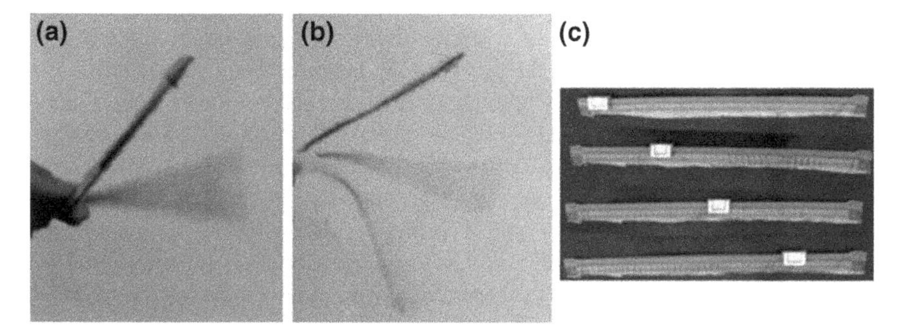

Figure 5 The rotational dynamics of a rigid body are different than those of a flexible body. (**a**) Rotating a pen is an example of rotating a rigid body. (**b**) Rotating an artificial whisker is an example of rotating a flexible body. (**c**) The center of mass of the artificial whisker can be changed by shifting the location of the slider on the zipper

computed directly as $\boldsymbol{M} = \alpha \boldsymbol{I}$, where α is the angular acceleration of the whisker and I is the mass moment of inertia, which depends on the whisker's length, L. Also notice that α is well defined *only* because we are treating the whisker as a rigid body. Once we allow the whisker to bend, the angular acceleration will be different at different points along the whisker, and it is not a particularly meaningful or useful quantity. Importantly, both experimental and modeling studies have found that the first 60–70 %; of the whisker does not appear to deform much during non-contact whisking (Knutsen et al. 2008; Quist et al. 2014), suggesting that the rigid body approximation is likely to be reasonable for much of the whisker.

The experiment that involved rotating the artificial whisker (Figure 5b) approximates whisker mechanics as a resonance phenomenon. Whiskers have first-mode resonance frequencies that range between 25–500 Hz for the fixed-free condition that describes non-contact whisking (Hartmann et al. 2003; Neimark et al. 2003; Boubenec et al. 2012; Yan et al. 2013). A whisker will oscillate with very large amplitude if it is driven near resonance, but the rat typically drives its whisker at much lower frequencies, between 8 and 15 Hz. This means that the whisker will primarily follow the driving frequency, and not much resonance activity will be observed.

With these factors in mind, we might predict that during non-contact whisking the rat would experience mechanical signals that mostly resemble those from rigid body rotation, but also contain some smaller components associated with flexible-body dynamics. A recent simulation study of the dynamics of non-contact whisking confirmed this prediction (Quist et al. 2014). With the important caveat that these simulations were limited to 2D, the study offered the following important conclusions about the mechanical signals generated during non-contact whisking behavior:

- During non-contact whisking, mechanical signals will be dominated by the rigid-body rotation associated with the driving frequency, but there will be small components of the mechanical signals near the resonance frequencies of the whisker.
- Dynamic effects will be larger and more significant for the caudal whiskers, because they are larger than the rostral whiskers and have more mass.
- During noncontact whisking, the time-varying bending moment closely follows the position of the whisker, while the axial force closely follows the whisking speed, which has two maxima per whisk. These findings suggest a basis for the neural coding of angular position (or spatial phase) as well as whisking speed.
- For all whiskers, the mechanical signals generated during non-contact whisking are extremely small. The rat could therefore regulate vibrissal motion by controlling the position of the vibrissa base; no force control is required.
- Because the mechanical signals generated during non-contact whisking are so small, associated neural responses (e.g., in the trigeminal ganglion) are expected to be largely stochastic. This prediction is consistent with the high degree of variability observed experimentally.

We emphasize that adding 3D effects and intrinsic whisker curvature may change some of these results. In addition, an intriguing earlier study (Yan et al. 2013) showed that the natural frequencies of the whisker during both non-contact and contact whisking can be very sensitive to change in the rotational constraint at the base; this could permit the animal to adjust the frequencies of mechanical signals during active behavior.

Overall, however, dynamic effects during non-contact whisking are expected to be well approximated by a rigid body model. It is of questionable value to develop a dynamic model of a flexible whisker just to study non-contact whisking, however, such a model is critical for describing collisions and ensuing vibrations, and for describing the signals the rat will obtain during exploration of a texture (Boubenec et al. 2012; Yan et al. 2013).

Collision

Many readers may recall studying collisions in physics class by performing conservation of momentum calculations. These calculations tended to involve spheres (e.g., billiard balls) that collided and then either stuck together (a "perfectly inelastic" collision) or bounced off each other (an "elastic" collision), with or without some associated loss of energy. If energy losses were zero, the collision was said to be "perfectly elastic".

The reason that these physics problems always involved spheres is that spheres can be treated as rigid point objects. Conservation of momentum calculations become complicated once an object is allowed to contain more than one point, and even more complicated if the object can deform.

Applying conservation of momentum to whiskers is not easy. During and immediately after a collision, each point along the whisker will have different velocities and different accelerations. The energy of impact will be distributed and lost in different ways along the whisker depending on the configuration (shape) of the whisker, which in turn depends on its material properties, including density, stiffness parameters, and damping parameters. At each instant of time, the shape of the whisker changes, which in turn influences how it will lose energy at the next instant of time, which changes local velocities and accelerations, and so on. Numerical models are needed, but very few exist.

One recent study, by Boubenec et al. offered a solution to the collision problem based on decomposing the whisker's deformations into their quasistatic and resonant (dynamic) components, and then performing a linear superposition of these two deformations (Boubenec et al. 2012). This approach yielded a remarkably good fit to experimental data describing the process of whisker contact and detach. This study is also the first to report and quantify the presence of a "shock wave" that travels axially down the whisker. The authors found that the mechanical perturbation induced by the shock changes linearly with velocity measured near the whisker base at the time of collision.

A second recent study, by Quist et al. yielded several additional insights into the dynamics of vibrissal-object collisions (Quist et al. 2014). First, collisions are mostly inelastic, that is, the whiskers will not "bounce" significantly after collision. Thus when the rat explores a complex surface, vibrissae will tend to make contact only once, helping to ensure that tactile signals accurately reflect the object's spatial features.

Second, although most vibrissal-object collisions will generate impact forces that are well above the magnitude of those generated by non-contact whisking, some collisions will not. Specifically, collisions that occur on the distal regions of some of the larger, more caudal whiskers, may not generate signals much larger than those observed during non-contact whisking (Quist et al. 2014). This result may offer a mechanical explanation for "whisking" neurons of the trigeminal ganglion, which respond with equal magnitude to non-contact whisking and light touch (Szwed et al. 2003, 2006; Leiser and Moxon 2007).

The third finding of Quist et al. (2014) was that the mechanical effects of collision depend very little on the velocity of the whisker at the time of impact, and more on how the rat chooses to continue to whisk after the time of collision. This result may initially appear to be in contradiction to results of Boubenec et al. (2012), but the two studies are actually quite reconcilable; they simply emphasize different aspects of the data.

The study of Boubenec et al. (2012) carefully examined the shock wave that travels axially, along the length of the whisker, immediately following a collision. The shock wave is very small, but also very fast, and is likely to be the first indicator to the rat that a collision has occurred. The magnitude of this shock wave depends on the velocity of the whisker at its base at time of impact.

The study of Quist et al. (2014) also observes these shocks immediately following a collision, and notes in passing that their magnitude increases with collision velocity, in agreement with Boubenec et al. (2012). The reported results then focus on quantifying the maximum magnitude of the mechanical signals subsequent to the collision. These maxima are associated with much larger, but slower, components of the signal, and are dominated by the effects of whisker bending and deceleration. It is these larger, slower signals that are mostly independent of the velocity at time of collision.

Summarizing, the two studies agree that although the first shock after collision is small, its magnitude is governed by collision velocity, and its high frequency components suggest that it could be an important cue for the rat. Subsequent changes in mechanical signals (after the shock wave) are governed primarily by far and how fast the rat chooses to push its vibrissa against the object after initial contact. These effects are much larger, but also much slower, than the initial shock.

With these mechanics in mind, several recent studies have shown that rodents respond differently to the first whisk against an object (unexpected) compared to subsequent whisks (expected) (Grant et al. 2009; Deutsch et al. 2012; Grant et al. 2012). Specifically, the first whisk against an object appears to be associated with a fast velocity change immediately after collision. The timing of this large acceleration suggests that it is most likely to result from an involuntary reflex loop

(Deutsch et al. 2012), possibly mediated by the fast-propagating shock wave. In contrast, the rodent's subsequent whisks involve longer latency changes in both ipsi- and contralateral contact profiles (Deutsch et al. 2012). The timing of these changes is most consistent with a voluntary sequence of motor actions, often resulting in a "double pump" against the object (Deutsch et al. 2012).

These voluntary contact-induced signals will have a tremendous effect on the signals received by the rat and are clearly an important area for future investigation. It is not easy to answer a question such as: "how does the C2 whisker respond differently if it collides with an object at 400°/s, compared with a collision that occurs at 700°/s?" After the initial shock wave, the forces and moments at the whisker base will depend almost entirely on the velocity profile that the rat chooses to continue after collision. In addition, if the rat is able to change the stiffness at the whisker base the resonant properties of the whisker may shift considerably (Yan et al. 2013).

Post-collision: Bending of the Whisker Against an Object

After collision, the mechanical signals transmitted by the vibrissa are dominated by quasistatic bending. Quasistatic signals are significantly easier to compute than dynamic signals, though 3D effects make even the quasistatics challenging.

Computing Mechanical Signals During Contact Whisking as the Whisker Bends

As discussed in Sect. 2.1, the stiffness of the whisker in the proximal region is much larger than in the distal region, and it is therefore experimentally challenging to measure bending near the whisker base. Measurement of bending near the base will be dominated by errors in tracking the shape of the whisker. It is therefore essential to develop mechanical models to compute the signals that will enter the follicle.

Using the variable s to indicate the position along the arc length of the whisker, we can write an equation that relates the change (Δ) in bending moment M to the change in curvature κ at each point along the whisker. The equation will use a proportionality constant from Sect. 1, namely the stiffness of the whisker. The bending moment is the product of the stiffness of the whisker (EI) and the curvature of the whisker (κ).

$$\begin{aligned}\Delta M(\mathrm{s}) &= (\text{stiffness})\Delta\kappa(\mathrm{s}) \\ &= EI\Delta\kappa(\mathrm{s}) \\ &= \frac{E\pi}{4}\mathrm{r}(\mathrm{s})^4\Delta\kappa(\mathrm{s}).\end{aligned} \tag{2}$$

Notice that ΔM and $\Delta\kappa$ are both functions of s, as is the radius, r. The bending moment and the curvature are not constant along the length of the whisker. At all locations along the whisker the two variables are related to each other by a constant of proportionality $\frac{E\pi}{4}$ as well as by the fourth power of the radius at that location.

We emphasize that Eq. (2) illustrates why the curvature of a whisker cannot conceptually be simplified into a single value: it will vary along the whisker's length. Because the whisker is much stiffer at its base (by a power of four), changes in curvature will be much more evident distally than proximally. This is why models are essential to accurately determine changes in curvature, and thus the bending moment, at the whisker base, near s = 0.

Multiple studies have solved a two-dimensional version of Eq. (2) to obtain estimates of the signals that the rat will experience during whisker deflection (Kaneko et al. 1998; Ueno et al. 1998; Solomon and Hartmann 2006, 2008; O'Connor et al. 2010; Solomon and Hartmann 2010, 2011; Boubenec et al. 2012; Quist and Hartmann 2012; Hires et al. 2013; Pammer et al. 2013). In two dimensions, these signals consist of the bending moment (M), as defined by Eq. (2), as well as the axial force (Fx), directed along the length of the whisker near the base, and the transverse force (F_y), perpendicular to the axial force. All of these variables (M, F_x, and F_y) change continuously along the length of the whisker, but the signals sent into the follicle are the values of these variables measured at the vibrissal base.

Open source Matlab™ code that implements Eq. (2) and permits an experimentalist to compute forces and moments at the whisker base by tracking the whisker's shape is available here: http://nxr.northwestern.edu/digital-rat.

Mappings Between Vibrissal Mechanics and the Two Dimensional (2D) Location of an Object

One of the longest standing problems in the field of vibrissal research is how the rat might combine various mechanical cues from the vibrissa to determine the location of whisker-object contact. Here we address the most basic 2D version of the mapping problem, which assumes pure rotation of the whisker (no basepoint translation).

Coordinate Systems

The 2D mapping problem can be expressed either in head-centered coordinates, resting-whisker coordinates, or whisker-centered coordinates.

The 2D head-centered coordinate system is illustrated in Figure 6a, with the origin (0, 0) at the tip of the rat's snout. The location of a point object can be expressed either in Cartesian coordinates (x, y), illustrated in blue, or in polar coordinates, (r, θ) illustrated in red. The point object is shaded purple to indicate

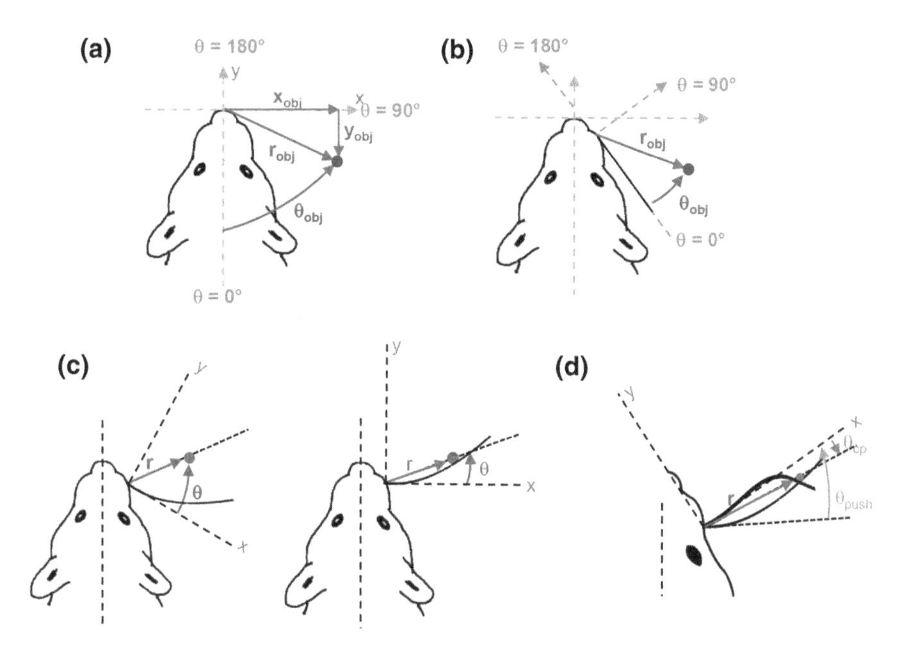

Figure 6 Schematics of the 2D mapping problem. All figures illustrate the most basic version of the 2D problem, which assumes no roll, no elevation, and no basepoint translation. **a Head-centered coordinates**. The 2D location of the object is identified by its position in head-centered coordinates in the horizontal plane in either Cartesian (x_{obj}, y_{obj}) or polar (r_{obj}, θ_{obj}) coordinates. The origin is taken to be the tip of the rat's snout. The angle for θ_{obj} is generally measured relative to the rostral-caudal midline of the head, but in some studies is measured relative to the resting position of an individual whisker. In this particular example, the *y*-coordinate of the object is negative because it is below the *x*-axis, caudal to the rat's snout. **b** Some work has used a "resting-whisker" coordinate system. In this coordinate system the origin is at the whisker base, and the azimuthal angle, θ, is defined uniquely for each whisker relative to its resting position. Each whisker has a unique constant offset relative to the head centered coordinate system shown in (a). **c Whisker-centered coordinates rotate with the whisker**. The 2D location of the point object is again described by its radial distance (r) and horizontal angle (θ), but the angular coordinate changes as the whisker rotates. **d Effect of deflection against an object.** As the whisker rotates against the object, the angle of contact point location (θ_{cp}) changes as the whisker is increasingly pushed against the object (θ_{push}). If the whisker were perfectly straight, θ_{cp} and θ_{push} would be exactly equal and opposite (one would be the negative of the other), but the intrinsic curvature of the whisker means that their relationship depends on the radial distance of contact

that its location is equally well expressed in either coordinate system. The polar coordinates used here differ in one important respect from the standard geometrical definition: by convention, the angle θ is measured relative to the vector pointing caudal along the midline of the rat's head. Note that whiskers play no role in defining the head-centered coordinate system.

Some studies (Szwed et al. 2003, 2006; Bagdasarian et al. 2013) have employed a resting whisker coordinate system very similar to head-centered coordinates. The only two differences are that the origin is placed at the whisker base instead of the rat's snout and the azimuthal angle, θ, is defined relative to the resting angle of the whisker. The resting-whisker coordinate system is illustrated in Figure 6b. In the resting-whisker centered coordinate system, each whisker has a unique, constant offset relative to the head centered coordinate system.

The 2D whisker-centered coordinate system is illustrated in Figure 6c. In this coordinate system, the radial distance of the object is measured from the base of the whisker to the point of contact with the object. Whisker-centered coordinates rotate with the whisker by definition. The whisker base is the origin, and the positive x-axis points along the initial linear portion of the whisker as it emerges from the rat's face (c.f., Figure 3). The horizontal angle θ is measured relative to this x-axis.

An important difference between head-centered and whisker-centered coordinates is that as the rat whisks the location of the object does not change in head-centered coordinates, but it does change in whisker-centered coordinates.

The mapping problem becomes more complicated as we consider what happens as the whisker deflects against the object. As the whisker bends, two more angles become relevant to the mapping problem: θ_{push} and θ_{cp}, identified in Figure 6d. The angle θ_{push}, as its name suggests, is a measure of how far the whisker has pushed against the object since initial contact. The value of θ_{push} is obtained by subtracting the angular position of the whisker at the instant of object contact from the current angular position of the whisker. The angle θ_{cp} is the angle of the contact point (subscript "cp") in standard polar coordinates in the reference frame of the whisker.

If the whisker were perfectly straight, θ_{cp} and θ_{push} would be exactly equal and opposite (one would be the negative of the other). Real whiskers, however, have an intrinsic curvature, so their relationship depends on the radial distance of contact.

Mappings for Object Location in Two Dimensions (2D)

A number of studies have focused on solving the 2D localization problem.

One recent study (Bagdasarian et al. 2013) has addressed how the rat might determine the (r, θ) position of an object, where the azimuthal angle θ is measured in whisker-base coordinates (Figure 6b). The work identifies four morphological variables that can be involved in localization: Global Curvature (G_κ), Angle Absorption (θ_A), Angle Protraction (θ_p), and Base Curvature (B_κ).

In this morphological coding scheme, the parameter θ_p is the angle through which the whisker protracts after making contact with the object, and the base curvature is the curvature near the whisker base. The parameter G_κ is defined as the maximal curvature along a spline fitted to the portion of the whisker between its base and the point of object contact. Angle absorption is defined as the difference between the angle through which the whisker would rotate during non-contact whisking and the angle through which the whisker actually protracted, having been obstructed by an object.

The study experimentally finds that either the pair (G_κ, θ_A) or the pair (θ_p, B_κ) is sufficient to localize the (r, θ) position of an object. In other words, the rat's knowledge of either of the two morphological pairs is sufficient to localize the object in the plane. The work further shows that changes in the angular velocity of the whisker upon contact with an object can code for the object's radial distance.

Some unpublished simulation results from our laboratory lead to significant reservations about the generality of these findings. Although preliminary, the simulations suggest that neither the pair (G_κ, θ_A) nor the pair (θ_p, B_κ) is sufficient to localize the (r, θ) position of an object if the whisker has intrinsic curvature. Results also suggest that—even if the whisker is straight—the azimuthal angle can be determined only if the rat knows the total amplitude of the whisk, and then subtracts off the protraction angle. Finally, results suggest that velocity, in the sense indicated in the paper, is related to the radial distance of contact only if the amplitude of the whisk is assumed to be constant.

Other studies that address the 2D mapping problem have been done with artificial (robotic) whiskers within an engineering context. In 2D, two sets of mechanical variables have been shown to code uniquely for the radial distance of the object in whisker-centered coordinates. The first set is bending moment at the vibrissal base (M) combined with θ_{push}. The second set is bending moment at the vibrissal base (M) and axial force at the vibrissal base (F_x). The combination of M and F_x, both measured at the vibrissal base, has also been shown to code uniquely for the contact point θ_{cp} in whisker centered coordinates.

Kaneko's work in the late 1990's first showed that both radial distance as well as object compliance could be determined by monitoring how the bending moment at the whisker base changed with θ_{push} (Kaneko et al. 1998). The algorithm required a cylindrical "antenna" (whisker) to be rotated so as to ensure a condition of no lateral slip (i.e., the whisker was adjusted in a manner that the problem remained fundamentally two-dimensional).

To generalize this work to rat vibrissae, the mechanical results shown by Kaneko's were extended to include tapered beams with large intrinsic curvature and to include the effect of lateral slip (Solomon and Hartmann 2006). This work enabled development of a robotic whisker array that could push gently against an object to determine the radial distance to each contact point; splining the contact points together then enabled reconstruction of the object surface (Solomon and Hartmann 2006).

The validity of the robotic model for real rat vibrissae was confirmed in a follow-up mechanical study (Birdwell et al. 2007), while on the neurobiological side it was independently demonstrated that neurons of the trigeminal ganglion represented more proximal radial distances with increases in firing rate (Szwed et al. 2003, 2006).

Techniques have now been developed to accommodate for surface friction while determining radial distance (Solomon and Hartmann 2008), and to permit distance extraction as the whisker continuously "sweeps" along the object (in contrast to the discrete pushes of the earlier work) (Solomon and Hartmann 2010). In 2011 Solomon and Hartmann showed that by combining axial force and bending moment the rat could uniquely determine both radial distance as well as the horizontal angle of contact (θ_{cp}) in whisker centered coordinates. These results were shown to hold for a tapered, but not cylindrical, whisker (Solomon and Hartmann 2011).

Experiments on mice subsequently provided behavioral confirmation that axial force and bending moment could be combined to determine radial distance

(Pammer et al. 2013). This study did not specifically demonstrate that results were unique for horizontal angle of contact, but it provided the demonstration that the mouse could distinguish between a compliant object placed at a small radial distance and a non-compliant object placed more distally. Within the limits of a 2D analysis, this result almost completely rules out the possibility that rodents rely exclusively on combining M and θ_{push} to determine radial distance to an object (Pammer et al. 2013). The study suggests it is far more likely that rodents rely on combining information about M and F_x. Thus behavioral experiments come full circle to compare to the original compliance result of Kaneko (Kaneko et al. 1998).

Three painstaking behavioral experiments explored the accuracy with which rodents can determine the horizontal angle of contact with a peg (Knutsen et al. 2006; Mehta et al. 2007; O'Connor et al. 2010). These studies have shown that although rats can localize the relative location of two simultaneously-present poles at hyperacuity (Knutsen et al. 2006), their accuracy drops approximately to the level of inter-whisker spacing when pole location has to be memorized (Mehta et al. 2007). Head-fixed mice can perform these absolute (memorized) estimates of horizontal angle at higher levels of performance, achieving absolute object localizations to better than 0.95 mm in the anterior–posterior dimension ($< 6°$ of azimuthal angle) (O'Connor et al. 2010). One open question left by these studies is whether the rodents perform these tasks based on head-centered cues (e.g., time from whisk start) or whisker-centered cues (e.g., based on differential mechanical signals induced from the collision happening at different phases of the whisk), or a combination of both.

Three Dimensional (3D) Mechanics and Geometry

Neurons throughout the vibrissal-trigeminal system are exquisitely sensitive to deflections of the whisker in all three directions (Simons 1978, 1985; Lichtenstein et al. 1990; Timofeeva et al. 2003; Jones et al. 2004; Furuta et al. 2006; Bellavance et al. 2010; Hemelt et al. 2010). It is therefore evident that 2D mechanical models will fail to capture significant information that the vibrissa transmits to the brain (Knutsen et al. 2008; Quist et al. 2012).

From a 3D kinematic perspective, an important feature of whisking is that the whisker rolls about its own axis, so its orientation changes continuously throughout both protraction and retraction (Knutsen et al. 2008; Knutsen et al. 2015, also see Figure 3). Notice that this orientation change is entirely a 3D effect; in all of the 2D models described above, the orientation of the whisker is assumed not to change during the trajectory.

From a 3D kinetic perspective, recent work has shown that the intrinsic curvature of the whisker will significantly affect the forces and moments at the vibrissal base because the vibrations and bending from collision depend strongly on the whisker's orientation (Boubenec et al. 2012; Quist and Hartmann 2012; Yan et al. 2013; Quist et al. 2014).

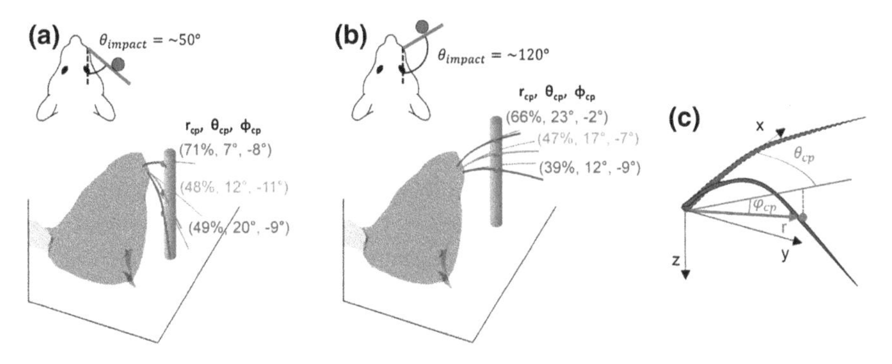

Figure 7 Schematics that illustrate the differences between head-centered and whisker-centered coordinates in 3D. a, b The whiskers are illustrated to collide with a peg at two different locations in head-centered coordinates (θ_{impact} = ~50° and ~120°). Because the whiskers have intrinsic curvature and because they elevate and roll during the whisk (see Figure 3), the location of contact in whisker-centered coordinates is not a simple offset from head-centered coordinates. The 3D location of whisker-peg contact is illustrated for each whisker individually in spherical coordinates (r_{cp}, θ_{cp},) relative to the base of each whisker. The subscript "cp" stands for contact point. *rcp* is the distance from the whisker base to the point of contact. It is measured as a percent distance of the total whisker arc length. **c** Whisker-centered coordinate system. The origin is at the whisker base, and the *x*-axis lies along the linear portion of proximal part of the whisker. The *x*-*y* plane is defined to be the plane of the whisker's intrinsic curvature, and the *z*-axis is perpendicular to the *x*-*y* plane

Putting kinematic and kinetic effects together, it becomes clear that the mechanical effects of a whisker bending against an object will depend strongly on where the whisker contacts the object during its trajectory. In other words, the location of the collision in head-centered coordinates will influence the bending mechanics in whisker-centered coordinates.

This important point is depicted in Figure 7, which illustrates three whiskers colliding with a peg at two different phases of the whisk cycle. In Figure 7a, the three whiskers all collide near 50° relative to the rostral-caudal midline, while in Figure 7b, they collide near 120°. Because the whiskers are in different rows, they also collide at different heights on the peg relative to the rat's snout. Both of these coordinates (θ_{impact} and the height) clearly must be described in a head-centered coordinate system.

But notice also that each whisker makes contact with the peg at a different 3D location relative to its coordinate system. The mechanical signals at each whisker base—the information actually entering each follicle—must be calculated based on how each whisker deflects against the peg in whisker-centered coordinates. The transformation between head and whisker coordinate systems is not simply an offset, because the whisker has intrinsic curvature, and because each whisker translates, elevates, and rolls during both protraction and retraction.

The 3D whisker centered coordinate system is shown in Figure 7c. The origin is at the whisker base, and the *x*-axis lies along the linear portion of proximal part of the whisker. The *x*-*y* plane is defined to be the plane of the whisker's intrinsic curvature, and the z-axis is perpendicular to the x-y plane. This is the coordinate system more relevant to describing the signals experienced by mechanoreceptors within the follicle.

The ideas in this section highlight the critical importance of developing a 3D model of whisker mechanics. Our laboratory recently showed that during active whisking behavior the whisker will often bend out of its plane of rotation, generating sizeable 3D mechanical signals (Huet et al. 2015). In the same work, we developed a model of whisker bending that computes the 3D tactile signals—all six components of force and moment—at the vibrissal base during active whisking behavior of the awake rat (Huet et al. 2015).

A question that remains open is the 3D mapping problem: how might the rat infer the 3D contact point location based only on the tactile signals it receives at the vibrissal base? Our laboratory's preliminary work in this area shows that many different combinations of force and moment components can uniquely determine the point of vibrissal-object contact $(r_{cp}, \theta_{cp}, \phi_{cp})$. Uniqueness of the mapping depends strongly on the exact shape of the whisker (i.e., taper and intrinsic curvature) and is an active topic of investigation.

Mechanics During Electrophysiological Experiments in the Anesthetized Animal

This article has primarily addressed vibrissal mechanics in the awake animal. Two points are important when considering experiments in the anesthetized animal.

The first point is likely to be obvious to many readers, but it is worth stating for emphasis: when studies in the anesthetized animal report that "angular position", "velocity", or "speed" are coded by neurons of the trigeminal system, these variables are completely unrelated to the kinematic variables of the same names used to describe non-contact whisking (Section 2.1 and Figure 3).

In the anesthetized animal stimuli are generated by passively bending the whisker some distance out along its length. This bending most closely resembles the quasi-static bending described in Sect. 2.2 (Figure 4b). The term "position" in this context is fundamentally related to the magnitude and direction of bending, and the term "velocity" reflects the rate of change of bending in a given direction. In the context of active behavior, this type of bending will occur any time that the rat presses its whiskers against an object (Sect. 4.1), and forces and moments at the whisker base may be quite large.

In contrast, during non-contact whisking (Sect. 2.1 and Figure 3) the animal rotates the whisker at its base. The bending of the whisker is very small. The whisker behaves mostly like a rigid body, although it may resonate a bit, as described in Sect. 2.3 and illustrated in Figure 5b. The terms "position" and "velocity" in this context do not depend on bending. The forces and moments at the whisker base will be very small.

Second, if a whisker is trimmed and then fixed (e.g., with glue) to a piezoelectric stimulator (a "piezo") and the piezo is then moved linearly, the whisker will experience forces that tend to pull it out of the follicle. This is a condition that the rat will almost never experience. A more natural experimental paradigm is to let the whisker slip on the piezo a bit, or slip within a capillary tube that serves as an interface between the piezo and the whisker.

Summary

The goal of this article was to provide physical intuition for the types of mechanical signals the rodent will obtain during natural vibrissotactile exploratory behavior. We now summarize some essential points that may help experimentalists identify the mechanical model most appropriate for computing signals during different behaviors.

In the case that the whisker does not make contact with any object (non-contact whisking), then the **relevant question** is to what extent the whisker can be treated as a rigid body versus a flexible body. In many cases, the mechanics of non-contact whisking might be simplified to an analysis of rigid body dynamics, while including resonance components. The rigid-body approximation may not always be appropriate to describe the behavior of the larger, more caudal whiskers, where the flexibility of the whisker will have a larger effect. During non-contact whisking forces and moments will tend to be small relative to those generated during object contact, but larger whiskers and very fast whisking motions will generate larger forces and moments.

In the case that the whisker does make contact with an object (presumed stiffer than the whisker) then the whisker must be treated as a flexible body. The **relevant question** is to what extent whisker mechanics can be described using quasi-static models, which include only stiffness and bending, versus the extent to which dynamic models are necessary. Mechanical descriptions of whisker collisions, vibrations, and interactions with surface texture will require dynamic models. In contrast, quasistatic models will provide good descriptions for most passive-displacement experiments in the anesthetized animal, and for how the rat deflects its whisker against a surface following a collision.

Acknowledgments We thank Lucie Anne Huet for helpful discussions and for running simulations to validate many of the ideas presented here. The work was sponsored by NSF awards CAREER-IOS-0846088; CRCNS-IIS-1208118; EFRI-BioSA-0938007; and IOS-0818414.

References

Adineh, V R; Liu, B; Rajan, R; Yan, W; and Fu, J (2015). Multidimensional characterisation of biomechanical structures by combining Atomic Force Microscopy and Focus Ion Beam: A study of the rat whisker. *Acta Biomaterialia* Epub. doi:10.1016/j.actbio.2015.03.028.

Bagdasarian, K et al. (2013). Pre-neuronal morphological processing of object location by individual whiskers. *Nature Neuroscience* 16(5): 622–631. doi:10.1038/nn.3378.

Beer, F; Johnston, E; DeWolf, J and Mzurek, D (2008). *Mechanics of Materials*, 5th Ed. New York: McGraw-Hill.

Bellavance, M-A; Demers, M and Deschenes, M (2010). Feedforward inhibition determines the angular tuning of vibrissal responses in the principal trigeminal nucleus. *The Journal of Neuroscience* 30(3): 1057–1063. doi:10.1523/jneurosci.4805-09.2010.

Bermejo, R; Vyas, A and Zeigler, H P (2002). Topography of rodent whisking—I. Two-dimensional monitoring of whisker movements. *Somatosensory & Motor Research* 19 (4): 341–346. doi:10.1080/0899022021000037809.

Birdwell, J A et al. (2007). Biomechanical models for radial distance determination by the rat vibrissal system. *Journal of Neurophysiology* 98(4): 2439–2455. doi:10.1152/jn.00707.2006.

Boubenec, Y; Shulz, D E and Debregeas, G (2012). Whisker encoding of mechanical events during active tactile exploration. *Frontiers in Behavioral Neuroscience* 6. doi:10.3389/fnbeh.2012.00074.

Carl, K et al. (2012). Characterization of statical properties of rat's whisker system. *IEEE Sensors Journal* 12(2): 340-349. doi:10.1109/jsen.2011.2114341.

Deutsch, D; Pietr, M; Knutsen, P M; Ahissar, E and Schneidman, E (2012). Fast feedback in active sensing: Touch-induced changes to whisker-object interaction. *Plos ONE* 7(9). doi:10.1371/journal.pone.0044272.

Fraser, R D and Macrae, T P (1980). Molecular structure and mechanical properties of keratins. *Symposia of the Society for Experimental Biology* 34: 211–246.

Furuta, T; Nakamura, K and Deschenes, M (2006). Angular tuning bias of vibrissa-responsive cells in the paralemniscal pathway. *The Journal of Neuroscience* 26(41): 10548–10557. doi:10.1523/jneurosci.1746-06.2006.

Grant, R A; Haidarliu, S; Kennerley, N J and Prescott, T J (2013). The evolution of active vibrissal sensing in mammals: Evidence from vibrissal musculature and function in the marsupial opossum monodelphis domestica. *Journal of Experimental Biology* 216(18): 3483–3494. doi:10.1242/jeb.087452.

Grant, R A; Mitchinson, B; Fox, C W and Prescott, T J (2009). Active touch sensing in the rat: Anticipatory and regulatory control of whisker movements during surface exploration. *Journal of Neurophysiology* 101(2): 862–874. doi:10.1152/jn.90783.2008.

Grant, R A; Sperber, A L and Prescott, T J (2012). The role of orienting in vibrissal touch sensing. *Frontiers in Behavioral Neuroscience* 6. doi:10.3389/fnbeh.2012.00039.

Hartmann, M J; Johnson, N J; Towal, R B and Assad, C (2003). Mechanical characteristics of rat vibrissae: Resonant frequencies and damping in isolated whiskers and in the awake behaving animal. *The Journal of Neuroscience* 23(16): 6510–6519.

Hemelt, M E; Kwegyir-Afful, E E; Bruno, R M; Simons, D J and Keller, A (2010). Consistency of angular tuning in the rat vibrissa system. *Journal of Neurophysiology* 104(6): 3105–3112. doi:10.1152/jn.00697.2009.

Herzog, E K; Bahr, D F; Richards, C D; Richards, R F and Rector, D M (2005). Spatially dependent mechanical properties of rat whiskers for tactile sensing. In: C Viney, et al. (Eds.), *Mechanical Properties of Bioinspired and Biological Materials* (pp. 119–123).

Hires, S A; Pammer, L; Svoboda, K and Golomb, D (2013). Tapered whiskers are required for active tactile sensation. "eLife" 2: e01350. doi:10.7554/eLife.01350.

Hobbs, J; Towal, R and Hartmann, M (2015). Evidence for functional groupings of vibrissae across the rodent mystacial pad. *PLoS Computational Biology* IN PRESS.

Huet, L A; Schroeder, C L and Hartmann, M J Z (2015). Tactile signals transmitted by the vibrissa during active whisking behavior. *Journal of Neurophysiology* Epub. doi:10.1152/jn.00011.2015.

Huet, L A and Hartmann, M J Z (2014). The search space of the rat during whisking behavior. *Journal of Experimental Biology* 217(18): 3365–3376. doi:10.1242/jeb.105338.

Ibrahim, L and Wright, E A (1975). Growth of rats and mice vibrissae under normal and some abnormal conditions. *Journal of Embryology and Experimental Morphology* 33: 831–844.

Jones, L M; Lee, S; Trageser, J C; Simons, D J and Keller, A (2004). Precise temporal responses in whisker trigeminal neurons. *Journal of Neurophysiology* 92(1): 665–668. doi:10.1152/jn.00031.2004.

Kan, Q; Rajan, R; Fu, J; Kang, G and Yan, W (2013). Elastic modulus of rat whiskers-A key biomaterial in the rat whisker sensory system. *Materials Research Bulletin* 48(12): 5026–5032. doi:10.1016/j.materresbull.2013.04.070.

Kaneko, M; Kanayama, N and Tsuji, T (1998). Active antenna for contact sensing. *IEEE Transactions on Robotics and Automation.* 14(2): 278-291. doi:10.1109/70.681246.

Knutsen, P M (2015). Whisking kinematics. *Scholarpedia* 10(3): 7280. doi:10.4249/scholarpedia.7280. http://www.scholarpedia.org/article/Whisking_kinematics. (see also pages 619–629 of this book).

Knutsen, P M; Biess, A and Ahissar, E (2008). Vibrissal kinematics in 3d: Tight coupling of azimuth, elevation, and torsion across different whisking modes. *Neuron* 59(1): 35–42. doi:10.1016/j.neuron.2008.05.013.

Knutsen, P M; Pietr, M and Ahissar, E (2006). Haptic object localization in the vibrissal system: Behavior and performance. *The Journal of Neuroscience* 26(33): 8451–8464. doi:10.1523/jneurosci.1516-06.2006.

Leiser, S C and Moxon, K A (2007). Responses of trigeminal ganglion neurons during natural whisking behaviors in the awake rat. *Neuron* 53(1): 117–133. doi:10.1016/j.neuron.2006.10.036.

Lichtenstein, S H; Carvell, G E and Simons, D J (1990). Responses of rat trigeminal ganglion neurons to movements of vibrissae in different directions. *Somatosensory & Motor Research* 7 (1): 47–65.

Mason, P (1963). Density and structure of alpha-keratin. *Nature* 197(486): 179. doi:10.1038/197179a0.

Mehta, S B; Whitmer, D; Figueroa, R; Williams, B A and Kleinfeld, D (2007). Active spatial perception in the vibrissa scanning sensorimotor system. *PLoS Biology* 5(2): 309-322. doi:10.1371/journal.pbio.0050015.

Muchlinski, M N (2010). A comparative analysis of vibrissa count and infraorbital foramen area in primates and other mammals. *Journal of Human Evolution* 58(6): 447-473. doi:10.1016/j.jhevol.2010.01.012.

Muchlinski, M N; Durham, E L; Smith, T D and Burrows, A M (2013). Comparative histomorphology of intrinsic vibrissa musculature among primates: Implications for the evolution of sensory ecology and 'face touch'. *American Journal of Physical Anthropology* 150(2): 301–312. doi:10.1002/ajpa.22206.

Neimark, M A; Andermann, M L; Hopfield, J J and Moore, C I (2003). Vibrissa resonance as a transduction mechanism for tactile encoding. *The Journal of Neuroscience* 23: 6499–6509.

O'Connor, D H et al. (2010). Vibrissa-based object localization in head-fixed mice. *The Journal of Neuroscience* 30(5): 1947–1967. doi:10.1523/jneurosci.3762-09.2010.

Pammer, L et al. (2013). The mechanical variables underlying object localization along the axis of the whisker. *The Journal of Neuroscience* 33(16): 6726–6741. doi:10.1523/jneurosci.4316-12.2013.

Prescott, T; Mitchinson, B and Grant, R (2011). Vibrissal behavior and function. *Scholarpedia* 6 (10): 6642. http://www.scholarpedia.org/article/Vibrissal_behavior_and_function. (see also pages 103–116 of this book).

Quist, B W; Faruqi, R A and Hartmann, M J Z (2011). Variation in Young's modulus along the length of a rat vibrissa. *Journal of Biomechanics* 44(16): 2775–2781. doi:10.1016/j.jbiomech.2011.08.027.

Quist, B W and Hartmann, M J Z (2012). Mechanical signals at the base of a rat vibrissa: The effect of intrinsic vibrissa curvature and implications for tactile exploration. *Journal of Neurophysiology* 107(9): 2298-2312. doi:10.1152/jn.00372.2011.

Quist, B W; Seghete, V; Huet, L A; Murphey, T D and Hartmann, M J Z (2014). Modeling forces and moments at the base of a rat vibrissa during noncontact whisking and whisking against an object. *The Journal of Neuroscience* 34(30): 9828–9844. doi:10.1523/jneurosci.1707-12.2014.

Richardson, F (1909). A study of sensory control in the rat. In: *Psychology Review Monographs*, Vol. 48 (pp. 1-505). Chicago: University of Chicago.

Simons, D J (1978). Response properties of vibrissa units in rat SI somatosensory neocortex. *Journal of Neurophysiology* 41(3): 798-820.

Simons, D J (1985). Temporal and spatial integration in the rat SI vibrissa cortex. *Journal of Neurophysiology* 54(3): 615-635.

Solomon, J H and Hartmann, M J Z (2006). Robotic whiskers used to sense features. *Nature* 443 (7111): 525–525. doi:10.1038/443525a.

Solomon, J H and Hartmann, M J Z (2008). Artificial whiskers suitable for array implementation: Accounting for lateral slip and surface friction. *IEEE Transactions on Robotics* 24(5): 1157–1167. doi:10.1109/tro.2008.2002562.

Solomon, J H and Hartmann, M J Z (2010). Extracting object contours with the sweep of a robotic whisker using torque information. *International Journal of Robotics Research* 29(9): 1233–1245. doi:10.1177/0278364908104468.

Solomon, J H and Hartmann, M J Z (2011). Radial distance determination in the rat vibrissal system and the effects of weber's law. *Philosophical Transactions of the Royal Society B: Biological Sciences* 366(1581): 3049–3057. doi:10.1098/rstb.2011.0166.

Szwed, M; Bagdasarian, K and Ahissar, E (2003). Encoding of vibrissal active touch. *Neuron* 40(3): 621-630. doi:10.1016/s0896-6273(03)00671-8.

Szwed, M et al. (2006). Responses of trigeminal ganglion neurons to the radial distance of contact during active vibrissal touch. *Journal of Neurophysiology* 95(2): 791–802. doi:10.1152/jn.00571.2005.

Timofeeva, E; Merette, C; Emond, C; Lavallee, P and Deschenes, M (2003). A map of angular tuning preference in thalamic barreloids. *The Journal of Neuroscience* 23(33): 10717–10723.

Towal, R B; Quist, B W; Gopal, V; Solomon, J H and Hartmann, M J Z (2011). The morphology of the rat vibrissal array: A model for quantifying spatiotemporal patterns of whisker-object contact. *PLoS Computational Biology* 7(4). doi:10.1371/journal.pcbi.1001120.

Ueno, N; Svinin, M M and Kaneko, M (1998). Dynamic contact sensing by flexible beam. *IEEE-Asme Transactions on Mechatronics* 3(4): 254–264. doi:10.1109/3516.736160.

Vincent, S (1912). The function of the vibrissae in the behavior of the white rat. In: *Behavior Monographs*, Vol. 1, Series 5 (pp. 1–81). Chicago: University of Chicago.

Voges, D et al. (2012). Structural characterization of the whisker system of the rat. *IEEE Sensors Journal* 12(2): 332–339. doi:10.1109/jsen.2011.2161464.

Williams, C M and Kramer, E M (2010). The advantages of a tapered whisker. *PLoS ONE* 5(1). doi:10.1371/journal.pone.0008806.

Yan, W Y et al. (2013). A truncated conical beam model for analysis of the vibration of rat whiskers. *Journal of Biomechanics* 46(12): 1987-1995. doi:10.1016/j.jbiomech.2013.06.015.

Whisking Kinematics

Per M. Knutsen

Whisking refers to a behavioral process, whereby motile facial vibrissae are repeatedly and rhythmically moved back and forth in order to sample the proximal environment (Figure 1a). The primary functions of whisking are spatial search and tactile exploration of objects and surfaces (see Vibrissal Behavior and Function). Non-rhythmic vibrissae movements also serve many behavioral processes, such as social interactions (Wolfe et al. 2011) and discrimination of lateral gaps (Krupa et al. 2001). Whisking is coordinated with head and body movements, which enables rapid sampling of the proximal environment during spatial exploration. Following contact, motor control of the vibrissae is modulated on several time scales as the animal approaches an object with head, nose and micro-vibrissae. Whisking has been studied primarily in rats, mice and shrews.

Descriptions of Vibrissae Motion

Vibrissae movements are produced by contractions of facial musculature, head movements and locomotion. Vibrissae positions relative to the head are typically measured by high-speed videography (Knutsen et al. 2005), in conjunction with head and body movements (Mitchinson et al. 2007) or while an animal has been partially immobilized (Bermejo et al. 1998).

Vibrissae movements relative to the head are described by three angles of rotation (azimuth, θ; elevation, ϕ; and torsion, ζ) and translation of the vibrissa base (horizontal and vertical; Figure 1b, c). The largest spatial displacement of the vibrissae tips is due to azimuthal rotation, which moves the vibrissae back and forth along the longitudinal axis. This movement co-varies with small changes in elevation. Torsion, or roll, refers to a rotation of a vibrissa about its own axis. The torsional angle is linearly correlated with the azimuth and the ratio $\delta\varsigma/\delta\theta$ changes as a function of the vertical location of the vibrissa follicle on the mystacial pad. Torsion alters the forward facing surface of the vibrissae shaft that contacts the

P.M. Knutsen (✉)
UC San Diego, La Jolla, CA, USA

T.J. Prescott et al. (eds.), *Scholarpedia of Touch*, Scholarpedia,
DOI 10.2991/978-94-6239-133-8_46

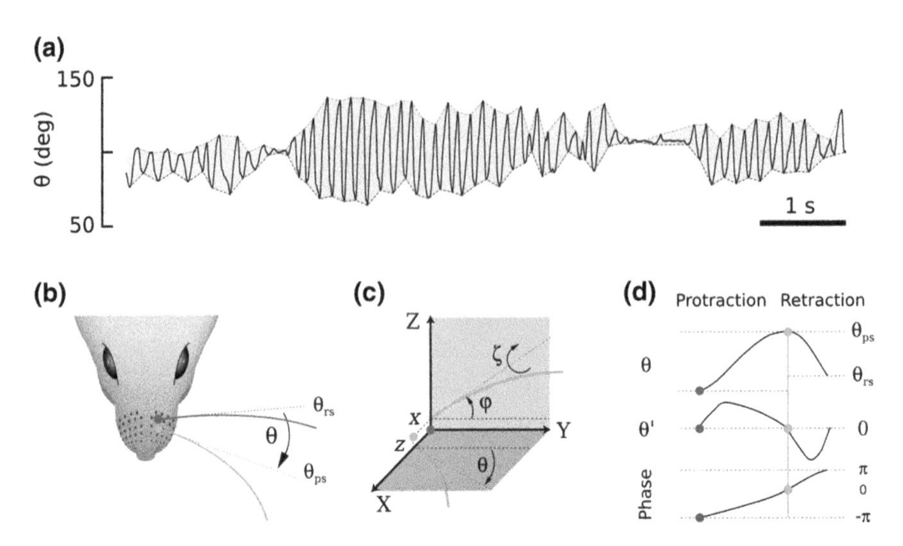

Figure 1 **a** Example of typical whisking behavior. *Solid black line* is the time-varying azimuth angle of the C2 vibrissa relative to the head. The angle envelope (*gray area*) is indicated by the retraction (*blue*) and protraction (*green*) set-points (*dotted lines*) showing that the vibrissa position fluctuates on a slower timescale than the protraction (*up*) and retraction (*down*) whisk cycles. Data was obtained from a head restrained rat, and is adapted from (Pietr et al. 2010) with permission. **b** Representation of the vibrissa azimuth angle (θ), as seen from a *top-down* perspective. *Blue* and *green lines* display the vibrissa at retraction and protraction set-points respectively. *Dotted lines* indicate the vibrissa azimuth, estimated as the tangent at the base of the shaft. **c** Conventions for representing vibrissa position, translation and rotation in head-centered Cartesian coordinates. X indicate direction of nose (anterior), Y points lateral and Z upwards (dorsal). The origin is arbitrarily set to the retraction set-point of the follicle in the previous whisk cycle (*blue dot*). After a protraction, the follicle is displaced a distance x anterior and z dorsal with respect to the skull (*green dot*). The shape and position of the vibrissa at the protraction set-point (*green lines*) is projected *down* and *back* onto the transverse (*light red*) and coronal (*light orange*) planes, respectively. The azimuth (θ), elevation (ϕ) and torsion (ζ) angles are measured with respect to the skull. **d** Single whisk cycle from (**a**). *Top* Vibrissa angle (azimuth, θ). *Middle* Vibrissa angular velocity (θ'). *Bottom* Phase of the whisk cycle. All traces plotted versus time. Conventions as in (**b**)

opposing surface and, because the vibrissae are curved, displaces the whisker tips relative to the head (Knutsen et al. 2008; Huet and Hartmann 2014).

The time-varying three-dimensional position of a vibrissa is described by (1) its shape W, approximated as a quadratic function $W = a + bu + cu^2$ with the intrinsic curvature $\kappa_{(u=0)} = \frac{2b}{(1+c^2)^{1.5}}$, (2) its base position $U(x, y, z)$ which is a single 3D coordinate, and (3) by the angles θ, ϕ and ζ at which the vibrissa protrudes from the mystacial pad, measured with respect to a head-centered coordinate system (Figure 1c). The angles ϕ and ζ are dependent variables, such that $\zeta \approx \alpha\theta + \beta$ for vibrissae rows A, D and E and $\zeta \approx \gamma\theta^2 + \alpha\theta + \beta$ for rows B and C. Since ϕ is typically small and κ is constant, it is therefore in most cases sufficient to describe the kinematics of the vibrissae in terms of θ and translations of the base position U (Knutsen et al. 2008).

Kinematic Profile of the Whisk Cycle

Whisking is a repetition of discrete whisk cycles that is executed and interrupted in discrete epochs (Figure 1a). Vibrissa protraction is initiated by the extrinsic *m. nasalis* muscle which pulls the mystacial pad forward. This phase is followed by contraction of the intrinsic sling muscles associated with individual follicles that rotate the vibrissae further forward. Vibrissa retraction involves the relaxation of protractor muscles and activation of the *m. nasolabialis* and *m. maxillolabialis* muscles that together pull the pad backward (Hill et al. 2008) (see Whisking musculature). This coordinated tri-phasic activation of the mystacial muscles generates stereotyped protraction and retraction trajectories. By definition, the whisk cycle is initiated from the retraction set-point of the previous cycle θ_{rs} and protracts each vibrissa to a new protraction set-point θ_{ps}. Upon reaching θ_{ps}, the vibrissae retracts back to θ_{rs} and the whisk cycle starts over (Figure 1d). The protraction phase is typically slower and lasts longer than the retraction phase. A typical whisk cycle has a single velocity maximum during protraction and retraction. The coordination between the protractor muscles, however, can vary during free-air whisking or contact due to neuromuscular feedback such that two or three velocity maxima can occur during protraction (Towal and Hartmann 2008; Deutsch et al. 2012).

The frequency of whisking is typically described by the peak frequency of vibrissa angle (azimuth). The zeroth harmonic of vibrissa angle typically correspond to the slow fluctuations of the whisking envelope (Figure 1a), and the first harmonic to the coupling of vibrissa movements to basal respiration. Thus, the first harmonic is at a lower frequency in rats (1–2 Hz) compared to mice (1.5–4 Hz) due to the slower respiratory rate of rats. The second harmonic of angle corresponds to the faster whisking, which is also synchronized with respiration (Mitchinson et al. 2011; Moore et al. 2013; Sofroniew et al. 2014). Because velocity is highest during whisking, however, it is more common the estimate the whisking frequency from the angular velocity (θ') which contains a single harmonic typically denoted f_0 (Berg and Kleinfeld 2003). The whisking frequency is both species and context dependent. Rats whisk at rates between 6 and 12 Hz during locomotion, 4−8 Hz during head-restraint and between 15 and 25 Hz during palpation of objects and object localization (Berg and Kleinfeld 2003; Knutsen et al. 2006). Mice whisk at rates between 9 and 16 Hz during locomotion and above 20 Hz during episodes of intense whisking (Jin et al. 2004; Mitchinson et al. 2011; Sofroniew et al. 2014). Etruscan shrews (smallest known mammals by mass) whisk at even higher average rates (12–17 Hz) consistent with their very high basal respiratory rate (10–14 Hz) (Munz et al. 2010). During exploration the whisking frequency can be maintained remarkably stable during each whisking epoch, with frequency changes occurring between whisking epochs (Berg and Kleinfeld 2003).

Whisking can be decomposed into rapidly and slowly varying parameters (Hill et al. 2011). The rapid parameter is the phase which places the whisk position on a normalized time axis from $-\pi$ (protraction onset) to π (retraction set-point;

Figure 1d). Slowly varying protraction and retraction set-points result in fluctuating protraction and retraction amplitudes, sometimes referred to as the whisking envelope (Figure 1a). These slowly varying whisk amplitudes suggest that the control of whisking amplitude is uncoupled from the patterning of whisking frequency. Indeed, direct tests demonstrate that amplitude can be modulated independently of frequency via the cannabinoid type 1 receptor (CB1R). Activation of CB1R, which reduce cortical synchronization in the theta and gamma frequency bands, reduce amplitudes without any effects on whisk timing (Pietr et al. 2010).

Modes of Whisking

It has been proposed that whisking in rats has at least two discrete modes, *foveal* and *exploratory*. During exploratory whisking, the vibrissae are spread further apart, whisks have larger amplitudes and whisking frequencies are in the lower ranges. During foveal whisking, the vibrissae tips are concentrated in a smaller volume while being thrust forward, whisks have small amplitudes and whisking frequencies are in the upper ranges (Berg and Kleinfeld 2003). Thus, different whisking modes result in different instantaneous spatial resolutions of the vibrissae array and varying rates of scanned volumes. While exploratory whisking is well suited for scanning large volumes at low instantaneous resolution, foveal whisking scans a smaller volume but at a higher resolution due to compression of the vibrissae array. Discrete whisking modes have not been described, but are thought to be present, in other species than rat.

Whisking modes can also be described on the basis of kinematic features. During exploratory whisking, retraction set-points are lower, protraction set-points occur later in the whisk cycle and amplitudes ($\theta_{ps} - \theta_{rs}$) are larger (Figures 2a, 3). During foveal whisking, retraction and protraction set-points are higher, amplitudes smaller and angular velocities larger (Figures 2a, b and 3). In the head-fixed rat, the transition between exploratory and foveal whisking is sharply delineated such that practically all foveal whisking occurs above 12 Hz with amplitudes smaller than 20° (Figure 3). High-frequency, foveal whisking is driven almost entirely by the intrinsic sling muscles (Berg and Kleinfeld 2003), which are composed primarily of type 2B muscle fibers thought to provide for the high maximum contraction velocities (Jin et al. 2004).

Comparisons with Saccadic Eye Movements

Vibrissae movements have been compared to saccadic eye-movements. Saccades are ballistic movements, with stereotyped trajectories. Whisks are not ballistic and differ from saccades both in terms of motor generation and sensory processing. The trajectories of saccades are smooth with peak velocities that correlate strongly with

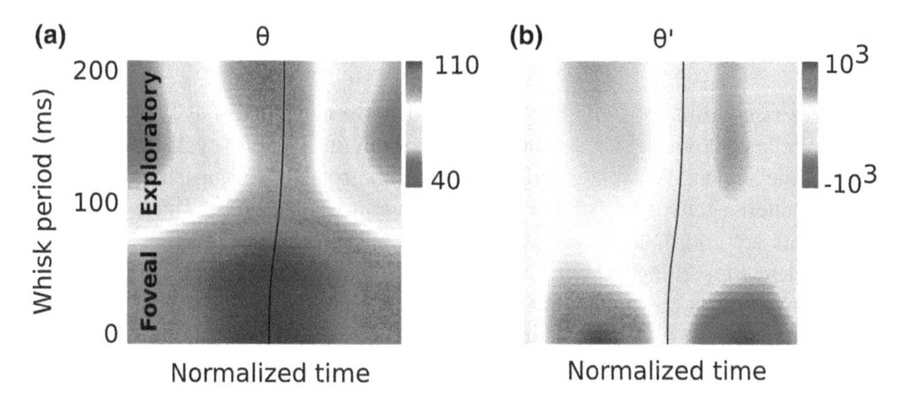

Figure 2 Average azimuth angle (θ) and velocity (θ') during whisking, as a function of whisk period. Time has been normalized so that the abscissa represents the duration of one whisk cycle. The average position of the protraction set-point is indicated by *solid, black lines*. During foveal whisking (periods below 100 ms), both retraction and protraction set-points are on average higher. Data obtained from head-fixed rat, in absence of contact (Knutsen et al. 2008)

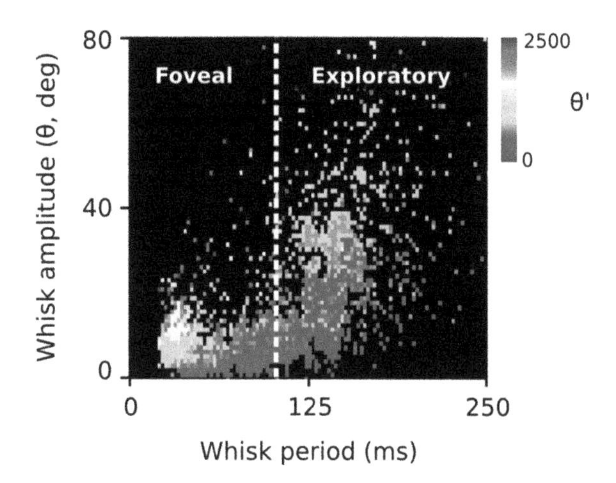

Figure 3 Peak velocities during whisk protraction as a function of whisk amplitude and period (or, instantaneous frequency). Velocity (θ') is represented by color. *Dots* are individual whisk cycles. Same data as in Figure 2

amplitudes, a correlation known as the saccadic ‘main sequence’. Similarly, peak angular velocities during whisking can exhibit strong linear correlations with whisk amplitudes during limited sample windows (Bermejo et al. 1998). Across modes of whisking, however, there is no consistent linear correlation between velocity and amplitude (Figure 3), and the movements of individual whisk cycles can be complex with variable velocity trajectories (Towal and Hartmann 2008; Mitchinson et al. 2011). During saccadic eye movements, visual sensory processing is selectively blocked. During whisking, tactile information can be processed throughout the whisk cycle (O’Connor et al. 2013). Additionally, whisking trajectories are modulated by fast neuromuscular feedback during contact (Nguyen and Kleinfeld 2005; Deutsch et al. 2012).

Coordination with Other Behaviors

Whisking is coordinated with head, nose and body movements as well as respiration (Welker 1964), These correlations occur both on cycle-by-cycle basis and on slower time scales. Whisking derives its rhythm from the brainstem respiratory central pattern generator (see Whisking Pattern Generation). Thus, vibrissae movements are coordinated with respiratory events, both during basal respiration and accelerated sniffing. During accelerated sniffing, above 5 Hz in rat, each inspiratory event is coordinated with individual whisk cycles. During basal respiration, below 3–4 Hz in rat, each inspiration is precisely synchronized to vibrissae protraction followed by multiple intervening whisks with decreasing amplitudes until the next breath occurs (Moore et al. 2013).

During development, the emergence of exploratory whisking occurs alongside improved locomotor abilities. Thus, vibrissae movements early in development are limited to unilateral retractions during head and body turns (Grant et al. 2012a). This behavior is also reflected in adult behavior. In both adult mice and rats, the difference in set-point between left and right side is modulated by running direction, such that right-side vibrissae are relatively retracted compared to left-side vibrissae during turning towards the right (Towal and Hartmann 2006). Additionally, vibrissae are more protracted during faster running and individual strides phase-lock with whisking. Thus, during locomotion the vibrissae are oriented and scan along and ahead of the running trajectory (Mitchinson et al. 2011; Arkley et al. 2014; Sofroniew et al. 2014). This behavior is consistent with the observation that active vibrissae movements are required for localization along the longitudinal (forward direction), but not the transverse (lateral direction), axis (Krupa et al. 2001; Knutsen et al. 2006).

Vibrissae movements are also coordinated with finer head movements during tactile scanning of objects. Following contact and detection of an object with the vibrissae, rats approach objects while making small ($\sim$5 mm) vertical head movements that repeatedly press the microvibrissae against the object at rates up to 8 Hz (Grant et al. 2012b).

Whisking in Spatial Exploration

The vibrissae diverge from a small grid of sinus hair follicles arranged in rows (A through E) and columns (1 through 6 for the largest vibrissae; Figure 4). When the vibrissae do not move, the spatial sampling is discrete and resolution low since the space between vibrissae is not sampled. During exploratory whisking, the entire vibrissae array is swept back and forth and thus samples the covered space continuously along the horizontal plane. The vertical space between vibrissae is sampled by two mechanisms. First, individual vibrissae elevate and lower by a small angle during protraction and retraction, thus tracing the contours of an

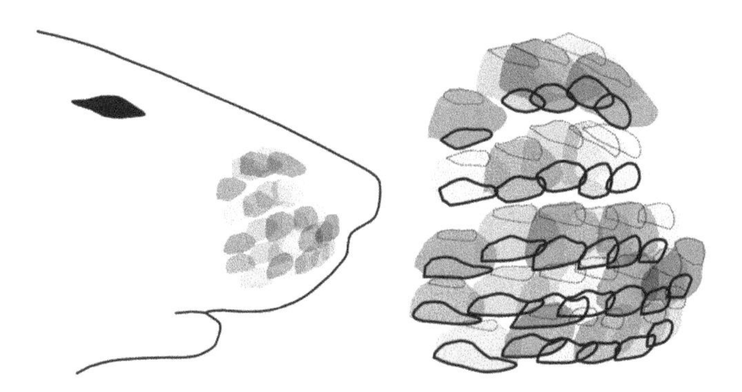

Figure 4 The vibrissal array is displaced by translational shifts in position of the mystacial follicle array. In this example, the array of follicles was tracked by high-speed video. Each *colored* patch represents the extent of translation by a single follicle across a 30-min session. The two *black outlines* inside each patch represent the extent of follicle translation during two different whisking epochs (5–10 s). It can be seen that across the two whisking epochs, a large static vertical translation of the entire follicle array occurred (unpublished results)

approximately curved, elliptical cone in each whisk cycle (Bermejo et al. 2002; Knutsen et al. 2008). Second, the entire vibrissae array can be vertically shifted by the *transversus* muscle which pulls the pad upwards when contracted and lowers it when relaxed (Hill et al. 2008) (Figure 4). Thus, over time, the total volumes scanned by individual vibrissae during exploratory whisking overlap and comprise a complete coverage of the spatial, proximal volume out to the vibrissae tips. The total volume explored per unit time is a function of whisking frequency and the size of the search space. The search space is determined by vibrissae lengths, whisking amplitudes and the angular spread of the rostral and caudal most vibrissae (Huet and Hartmann 2014). The positional envelope of the vibrissa during exploratory whisking is continuously modulated by running speed (Arkley et al. 2014; Sofroniew et al. 2014) and environmental contacts (Mitchinson et al. 2007; Munz et al. 2010).

Kinematics During Contact

Upon contact with objects and surfaces, whisking rodents rapidly modify their behaviors on multiple time scales. Environmental contact activates a fast within-cycle, disynaptic excitatory brainstem loop which results in increased protraction of all vibrissae on the contacting side. This fast neuromuscular modulation occurs without a change in vibrissae movement synchrony or whisk cycle duration (Sachdev et al. 2003; Nguyen and Kleinfeld 2005; Deutsch et al. 2012; Matthews et al. 2014). Contact also results in orienting responses of the head and body which subsequently results in asymmetric modulation of vibrissae set-points on either side of the head (Mitchinson et al. 2007; Sofroniew et al. 2014) (see **Coordination with other behaviors**).

The vibrissae contact objects with their tips or shafts, resulting in bending and distortions of the intrinsic shape. These distortions result in axial forces that activate mechanoreceptors in the follicles. The range of forces and moments at the base of a vibrissa vary as a function of intrinsic curvature, radial contact location as well as the torsional orientation of the shaft (Quist and Hartmann 2012). Simulations suggest that the magnitude of angular velocity has little effect on these contact signals, although the degree of continued protraction after contact does (Quist et al. 2014). Thus, whisking kinematics is a major determinant of contact related signals (see also Vibrissa Mechanical Properties).

There is evidence to suggest that both rats and mice adjust their whisking kinematics during, or in expectation of, contact. During exploration, rats rapidly modify the protraction set-point in response to contact (Mitchinson et al. 2007). In tasks where animals are trained to localize targets along the horizontal (rostro-caudal) axis, mice maximize contact with targets (O'Connor et al. 2010) and rats shift the angular position of their vibrissae to that of targets (Knutsen et al. 2003). Features of whisking kinematics also correlate with object position. The curvature of the vibrissa shaft correlate with the radial distance of a wall during locomotion (Sofroniew et al. 2014) and with the angular position of an object during object localization (Bagdasarian et al. 2013). Performance in an object localization task can additionally be predicted on the basis of whisking spectral power (Knutsen et al. 2003).

Neural Representations of Whisking Kinematics

Whisking is represented in the neural activity of multiple brain regions receiving vibrissal sensory inputs. A subset of neurons in the trigeminal (Gasserian) ganglion and in the posterior medial nucleus (POm) of the thalamus exhibit position dependence during artificial vibrissae movements in anesthetized rats (Szwed et al. 2003; Yu et al. 2006). In awake, freely-moving rats neurons have been found in the primary vibrissa somatosensory cortex (vS1) that exhibit phase dependence during whisking but do not report slowly varying whisking parameters such as amplitude (Fee et al. 1997; Curtis and Kleinfeld 2009). The envelope of whisking is instead reported by efferent signaling in both primary vibrissa motor cortex (vMC) and vS1 (inherited from vMC; see Whisking control by motor cortex), suggesting that both afferent and efferent signals are required for cortical circuits to compute the absolute position, or angle, of the vibrissae and contacted objects (Ahrens and Kleinfeld 2004; Hill et al. 2011).

Evolution of Whisking

Although species in several mammalian orders exhibit motile facial vibrissae, only a few members of the *Muridae* family of the *Rodentia* order (such as rats, mice and shrews), as well as some members of the marsupial *Didelphimorphia* order (opossums), whisk their vibrissae. Thus, the ancestral whisking species is likely to predate the earliest known placental mammal ancestor at about 65 Ma (O'Leary et al. 2013). Whisking species are often prolific climbers and nocturnal. Thus, whisking may have evolved as a strategy to navigate uneven, elevated surfaces in reduced lighting conditions. The evolution of the facial musculature that controls vibrissae movements have been traced to homologies in the gill-arch musculature involved in respiration in teleostome fishes (Huber 1930). Thus, the coordination between whisking and respiration seen in rodents likely represents an early evolutionary re-purposing of facial musculature used for respiration into an active sensing organ. Interestingly, similarly specialized facial active sensory organs have evolved within other classes and orders of animals, such as the fleshy appendages of the star-nosed mole (*Soricomorpha: Condylura cristata*) and catfishes (*Siluriformes*). Whether movements of these organs are similarly coordinated with respiration is unknown.

References

Ahrens, K F and Kleinfeld, D (2004). Current flow in vibrissa motor cortex can phase-lock with exploratory rhythmic whisking in rat. *Journal of Neurophysiology* 92: 1700–1707.

Arkley, K; Grant, R A; Mitchinson, B and Prescott, T J (2014). Strategy change in vibrissal active sensing during rat locomotion. *Current Biology* 24: 1507–1512.

Bagdasarian, K et al. (2013). Pre-neuronal morphological processing of object location by individual whiskers. *Nature Neuroscience* 16(5): 622–631.

Berg, R W and Kleinfeld, D (2003). Rhythmic whisking by rat: Retraction as well as protraction of the vibrissae is under active muscular control. *Journal of Neurophysiology* 89: 104–117.

Bermejo, R; Houben, D and Zeigler, H P (1998). Optoelectronic monitoring of individual whisker movements in rats. *Journal of Neuroscience Methods* 83(2): 89–96.

Bermejo, R; Vyas, A and Zeigler, H P (2002). Topography of rodent whisking-I. Two-dimensional monitoring of whisker movements. *Somatosensory & Motor Research* 19: 341–346.

Curtis, J C and Kleinfeld, D (2009). Phase-to-rate transformations encode touch in cortical neurons of a scanning sensorimotor system. *Nature Neuroscience* 12: 492–501.

Deutsch, D; Pietr, M; Knutsen, P M; Ahissar, E and Schneidman, E (2012). Fast feedback in active sensing: Touch-induced changes to whisker-object interaction. *PLoS One* 7: e44272.

Fee, M S; Mitra, P P and Kleinfeld, D (1997). Central versus peripheral determinants of patterned spike activity in rat vibrissa cortex during whisking. *Journal of Neurophysiology* 78: 1144–1149.

Grant, R A; Mitchinson, B and Prescott, T J (2012a). The development of whisker control in rats in relation to locomotion. *Developmental Psychobiology* 54: 151–168.

Grant, R A; Sperber, A L and Prescott, T J (2012b). The role of orienting in vibrissal touch sensing. *Frontiers in Behavioral Neuroscience* 6: 39.

Hill, D N; Bermejo, R; Zeigler, H P and Kleinfeld, D (2008). Biomechanics of the vibrissa motor plant in rat: rhythmic whisking consists of triphasic neuromuscular activity. *The Journal of Neuroscience* 28: 3438–3455.

Hill, D N; Curtis, J C; Moore, J and Kleinfeld, D (2011). Primary motor cortex reports efferent control of vibrissa motion on multiple timescales. *Neuron* 72: 344–356.

Huber, E (1930). Evolution of facial musculature and cutaneous field of trigeminus: Part I. *The Quarterly Review of Biology* 5: 133–188.

Huet, L A and Hartmann, M J Z (2014). The search space of the rat during whisking behavior. *The Journal of Experimental Biology* 217: 3365–3376.

Jin, T E; Witzemann, V and Brecht, M (2004). Fiber types of the intrinsic whisker muscle and whisking behavior. *The Journal of Neuroscience* 24: 3386–3393.

Knutsen, P M; Biess, A and Ahissar, E (2008) Vibrissal kinematics in 3D: Tight coupling of azimuth, elevation, and torsion across different whisking modes. *Neuron* 59: 35–42.

Knutsen, P M; Derdikman, D and Ahissar, E (2005). Tracking whisker and head movements in unrestrained behaving rodents. *Journal of Neurophysiology* 93: 2294–2301.

Knutsen, P M; Pietr, M D and Ahissar, E (2006). Haptic object localization in the vibrissal system: behavior and performance. *The Journal of Neuroscience* 26: 8451–8464.

Knutsen, P M; Pietr, M D; Derdikman, D and Ahissar, E (2003). Whisking behavior of freely-moving rats in an object localization task. 10: 209.

Krupa, D J; Matell, M S; Brisben, A J; Oliveira, L M and Nicolelis, M A (2001). Behavioral properties of the trigeminal somatosensory system in rats performing whisker-dependent tactile discriminations. *The Journal of Neuroscience* 21: 5752.

Matthews, D W et al. (2014). Feedback in the brainstem: An excitatory disynaptic pathway for control of whisking. *Journal of Comparative Neurology* 523(6): 921–942. doi:10.1002/cne.23724.

Mitchinson, B et al. (2011). Active vibrissal sensing in rodents and marsupials. *Philosophical Transactions of the Royal Society of London B: Biological Sciences* 366: 3037–3048.

Mitchinson, B; Martin, C J; Grant, R A and Prescott, T J (2007). Feedback control in active sensing: rat exploratory whisking is modulated by environmental contact. *Proceedings of the Royal Society of London B* 274: 1035–1041.

Moore, J et al.(2013). Hierarchy of orofacial rhythms revealed through whisking and breathing. *Nature*497: 205–210. doi:10.1038/nature12076.

Munz, M; Brecht, M and Wolfe, J (2010). Active touch during shrew prey capture. *Frontiers in Behavioral Neuroscience* 4: 191.

Nguyen, Q T and Kleinfeld, D (2005). Positive feedback in a brainstem tactile sensorimotor loop. *Neuron* 45: 447–457.

O'Connor, D H et al. (2010). Vibrissa-based object localization in head-fixed mice. *The Journal of Neuroscience* 30: 1947–1967.

O'Connor, D H et al.(2013). Neural coding during active somatosensation revealed using illusory touch. *Nature Neuroscience*16: 958–965.

O'Leary, M et al. (2013). The placental mammal ancestor and the post–K-Pg radiation of placentals. *Science* 339(6120): 662–667.

Pietr, M D; Knutsen, P M; Shore, D I; Ahissar E and Vogel Z (2010). Cannabinoids reveal separate controls for whisking amplitude and timing in rats. *Journal of Neurophysiology* 104: 2532–2542.

Quist, B W and Hartmann, M J Z (2012). Mechanical signals at the base of a rat vibrissa: The effect of intrinsic vibrissa curvature and implications for tactile exploration. *Journal of Neurophysiology*107(9): 2298–2312.

Quist, B W; Seghete, V; Huet, L A; Murphey, T D and Hartmann, M J Z (2014). Modeling forces and moments at the base of a rat vibrissa during noncontact whisking and whisking against an object. *The Journal of Neuroscience* 34: 9828–9844.

Sachdev, R N; Berg, R W; Champney, G; Ebner, F and Kleinfeld, D (2003). Unilateral vibrissa contact: Changes in amplitude but not timing of rhythmic whisking. *Somatosensory & Motor Research* 20: 163–169.

Sofroniew, N J; Cohen, J D; Lee, A K and Svoboda, K (2014). Natural whisker-guided behavior by head-fixed mice in tactile virtual reality. *The Journal of Neuroscience* 34: 9537–9550.
Szwed, M; Bagdasarian, K and Ahissar, E (2003). Encoding of vibrissal active touch. *Neuron* 40: 621–630.
Towal, R B and Hartmann, M J Z (2008). Variability in velocity profiles during free-air whisking behavior of unrestrained rats. *Journal of Neurophysiology* 100: 740–752.
Towal, R B and Hartmann, M J Z (2006). Right-left asymmetries in the whisking behavior of rats anticipate head movements. *The Journal of Neuroscience* 26: 8838–8846.
Welker, W I (1964). Analysis of sniffing of the albino rat. *Behaviour* 22: 223–244.
Wolfe J; Mende, C and Brecht, M (2011). Social facial touch in rats. *Behavioral Neuroscience* 125: 900–910.
Yu, C; Derdikman, D; Haidarliu, S and Ahissar, E (2006). Parallel thalamic pathways for whisking and touch signals in the rat. *PLoS Biology* 4: 819–825.

Whisking Musculature

Sebastian Haidarliu

Definition

Whisking musculature is represented by a group of facial striated muscles that have their insertion sites within the mystacial pad and control vibrissa movements in whisking mammals. The role of the vibrissa movements in active rats for equilibration, determining nearness or position of edges or corners, as well as discrimination of inequalities of surface as a compensation for a poor vision was described for the first time by Vincent (1912). The ability to move vibrissae rhythmically during tactile exploration of the environment (whisking) was then observed in many other rodents, such as mice, hamsters, gerbils, squirrels and porcupines (Welker 1964; Woolsey et al. 1975; Rice et al. 1986; Munz et al. 2010), in insectivores, such as Etruscan shrew (Anjum et al. 2006) and greater hedgehog tenrec (Mitchinson et al. 2011), and in marsupials, such as Brazilian short-tailed and Virginia opossums (Rice et al. 1986; Mitchinson et al. 2011). The whisking musculature moves vibrissa-sinus complexes in such a way that vibrissae can scan the whole space around animal snout for precise detecting eventual objects. Whisking muscles move vibrissae with the aid of connective tissue (collagenous skeleton) and are the principal mover of the vibrissae in the "vibrissal motor plant" described by Hill et al. (2008).

Classification

Whisking musculature is represented by voluntary striated muscles. According to the location of the muscle origins, whisking muscles were grouped into two categories: intrinsic and extrinsic (Dörfl 1982). Intrinsic muscles were first described by

S. Haidarliu (✉)
Department of Neurobiology, Weizmann Institute of Science, Rehovot, Israel

T.J. Prescott et al. (eds.), *Scholarpedia of Touch*, Scholarpedia,
DOI 10.2991/978-94-6239-133-8_47

Vincent (1913) under the name of "follicle muscles". They originate and insert within the mystacial pad. Extrinsic muscles originate outside, and insert within the mystacial pad. According to the direction in which extrinsic muscles move the vibrissae, they can be divided into three groups: protractors, retractors, and vertical vibrissa deflectors. Based on the shape and the orientation of the muscle fibers, whisking muscles can be attributed to parallel, convergent (fan-shaped), and pennate types.

Anatomy

Methodology

The first anatomical schemes of the whisking muscle arrangement within the mystacial pad were obtained by using methods of dissection. Using these methods, majority of whisking muscles, including their origins and insertion sites, were described (Huber 1930a, b; Meinertz 1944; Rinker 1954; Klingener 1964; Ryan 1989). However, a detailed complete map of whisking muscle arrangement was obtained after visualizing muscle fibers in the slices of the mystacial pad in situ using histoenzymatic methods of muscle staining. By these methods, additional muscles, such as Pars interna profunda and pseudointrinsic slips of the Pars interna of the M. nasolabialis profundus, which were not observed by using traditional methods of dissection, were revealed (Haidarliu et al. 2010).

Intrinsic Muscles

Intrinsic muscles connect adjacent vibrissa follicles within the rows of the mystacial pad. Each intrinsic muscle is represented by two extremities (dorsal and ventral). The extremities originate from the rostral surface of the proximal end of the rostrally located vibrissal follicle. Both extremities insert into the distal end of the neighboring, caudally located follicle, and into the contiguous corium. In mice and rats, intrinsic muscles connect adjacent vibrissal follicles only in the same row and look similar in all the vibrissal rows (Dörfl 1982; Haidarliu et al. 2010, 2015) (Figure 1). However, in some species, the arrangement of intrinsic muscles is different. For example, in big-clawed shrews, the arrangement of intrinsic muscles in the dorsal two rows (nasal compartment of the mystacial pad) is similar to that described in mice and rats, but in the ventral rows (maxillary compartment of the mystacial pad), their arrangement is hexagonal, and muscle extremities that originate from one follicle insert into two other caudally located follicles that belong to different rows (Yohro 1977). In the marsupial Monodelphis domestica, in addition to regular intrinsic muscles within the two dorsal-most rows, there are additional oblique extremities that connect follicles within the rows A and B, and may cause

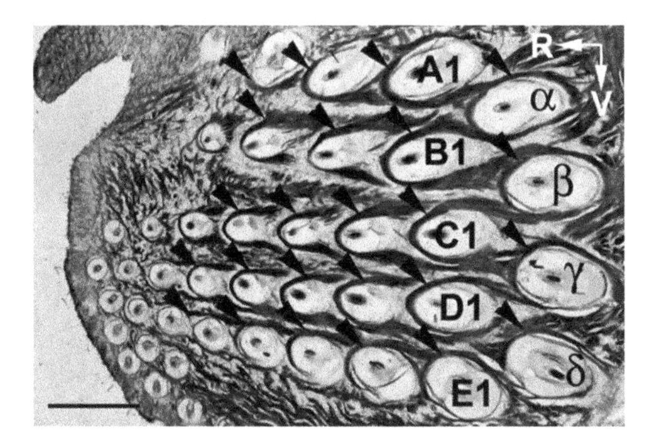

Figure 1 Light microscopy of a tangential slice of the mystacial pad of an adult mouse. Intrinsic muscles are revealed by staining for cytochrome oxidase activity and marked with *arrow heads*. (α–δ) Straddlers; (A1–E1) *first arc* of the five vibrissal rows; (R) rostral; (V) ventral. *Scale bar* = 1 mm

rotational whisker movements (Grant et al. 2013). Intrinsic muscles of the most caudally located follicles (vibrissal arc that is composed of straddlers) insert into the corium caudal to the mystacial pad.

Extrinsic Muscles

Extrinsic muscles take their origin from the bones, cartilages or aponeuroses outside the mystacial pad, and insert into the corium or subcapsular fibrous mat within the mystacial pad. In mice and rats, *extrinsic vibrissa protractors* are represented by four subunits of the M. nasolabialis profundus: Partes media superior et inferior, and two subunits of the Pars interna. The first of them originates from the rostral end of the premaxilla, the second, from the intermuscular septum, and the last two, from the lateral wall of the nasal cartilage. All four extrinsic vibrissa protractors insert into the corium of the mystacial pad, and the last of them (pseudointrinsic) inserts also into the distal ends of the vibrissa follicles of the rows A and B (Haidarliu et al. 2010).

Extrinsic vibrissa retractors are represented by two superficial and three deep muscles. *Superficial extrinsic retractors* (Mm. nasolabialis et maxillolabialis) are similarly represented also in other whisking species, such as hamsters (Wineski 1985) and marsupials (Grant et al. 2013), originate from the skull, caudal to the mystacial pad, and are inserted into the corium of the mystacial pad between the rows of vibrissae (Figure 2). *Deep extrinsic retractors* are represented by three parts of the M. nasolabialis profundus (Pars interna profunda, and Partes maxillares superficialis et profunda) that originate from the nasal cartilage and insert into the subcapsular fibrous mat (Figure 3a). These muscles have a typical bipennate

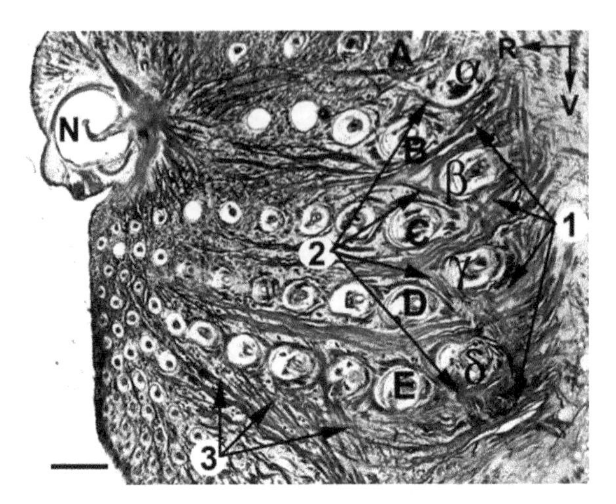

Figure 2 Light microscopy of a superficial tangential slice of the rat mystacial pad. Staining for cytochrome oxidase activity. (α–δ) Straddlers; (A–E) five rows of vibrissae; (N) nostril. (1) M. nasolabialis; (2) M. maxillolabialis; (3) Pars orbicularis oris of the M. buccinatorius. (R) Rostral; (V) ventral. *Scale bar* = 1 mm

architecture (Figure 3b) and were described also in hamsters (Wineski 1985) and marsupials (Grant et al. 2013).

Vertical vibrissa deflectors are represented by two muscles. One of them (M. transversus nasi) originates from the dorsal nasal aponeurosis and forms also myomyous origins along the midline (Figure 4). It inserts into the corium of the nasal compartment of the mystacial pad. The other (Pars orbicularis oris of the M. buccinatorius) originates from the skin of the lower lip, and from the muscle

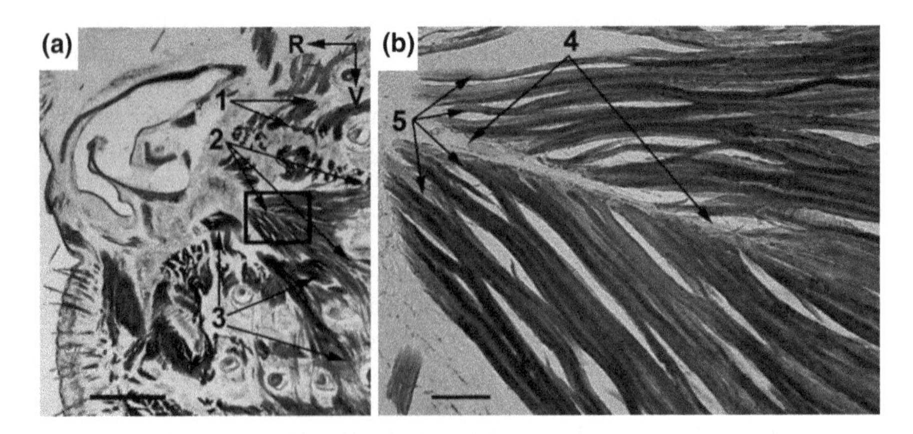

Figure 3 Light microscopy of a deep tangential slice (**a**) of the mouse mystacial pad. (**b**) Enlarged boxed area in (**a**). Staining for cytochrome oxidase activity. (1) Pars interna profunda; (2) and (3), Partes maxillares superficialis et profunda, respectively, of the M. nasolabialis profundus; (4) tendon; (5) muscle fibers. (R) Rostral; (V) ventral. *Scale bars* = 1 mm (**a**) and 0.1 mm (**b**)

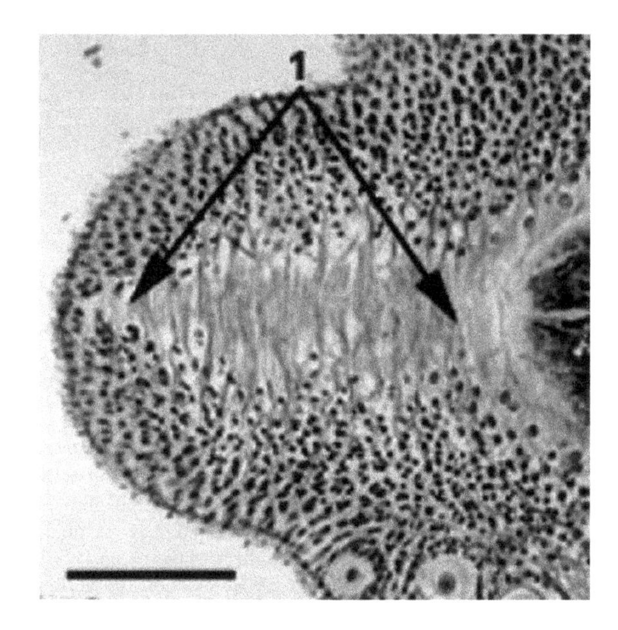

Figure 4 Light microscopy of a horizontal slice of the snout of a young rat. (1) M. transversus nasi. *Scale bar* = 1 mm

bundles of the M. buccinatorius, and inserts into the corium of the maxillary compartment of the mystacial pad (Figure 2). Extrinsic muscles of the mystacial pad and their attachment sites (entheses) are shown schematically in Figure 5.

Muscle Fiber Types

Striated muscles are composed of muscle fibers characterized by different metabolic properties that determine their functional abilities: speed of shortening, duration of high activity, fatigability. Based on the morphological and functional data, striated muscles were divided into red and white types that correspond to slow and fast muscles, respectively (McComas 1996). When the typing is based on the muscle fiber staining, obtained results do not always agree (Staron 1997; Scott et al. 2001). Using histochemical methods of staining for ATPase activity, three types of muscle fibers (red, white, and intermediate) were described in the diaphragm of different mammals (Padykula and Gauthier 1963). After staining for cytochrome oxidase activity that addresses oxidative capacity of the muscle fibers, similar three fiber types were revealed randomly distributed in the M. nasolabialis of the rat mystacial pad (White and Vaughan 1991) and in the Mm. nasolabialis and maxillolabialis in mice (Grant et al. 2014). Similar mosaic pattern of fiber type distribution was found in other extrinsic muscles of the rat mystacial pad (Figure 6) (Haidarliu et al. 2010).

Microscopic appearance of muscle fibers depends on the method that was used for their visualization so that attempts of muscle typing that are based on the comparison of images obtained using different methods can lead to confusing

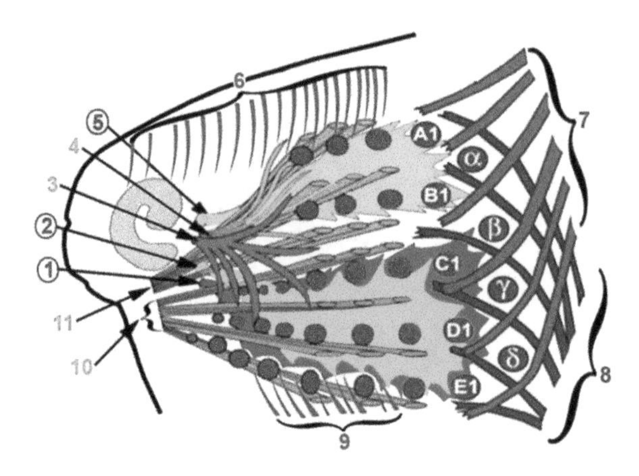

Figure 5 Schematic drawing that represents arrangement of the extrinsic musculature of a rodent mystacial pad. (α–δ) Straddlers; A1–E1, *first arc* of the vibrissal rows; 1–5, 10 and 11 are subunits of the M. nasolabialis profundus: (1) Pars maxillaris profunda; (2) Pars maxillaris superficialis; (3) Posterior slips, and (4) Pseudointrinsic slips of the Pars interna; (5) Pars interna profunda; (10) Pars media inferior; (11) Pars media superior. (6) M. transversus nasi; (7) M. nasolabialis; (8) M. maxillolabialis; (9) Pars orbicularis oris of the M. buccinatorius. Encircled numbers designate extrinsic muscles that are inserted into the deep fibrous mat, the rest of muscles are inserted into the corium. *Green* numbers show extrinsic vibrissa protractors, *red*, vibrissa retractors, and *blue* numbers, vertical vibrissa deflectors

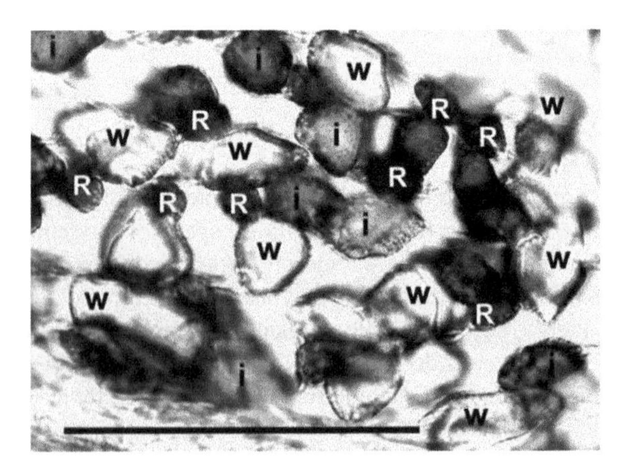

Figure 6 Muscle fiber types in a muscle fascicle of the Pars media inferior of the M. nasolabialis profundus in an adult rat. Staining for cytochrome oxidase activity. (i) Intermediate, (R) *red*, and (W) *white* muscle fibers. *Scale bar* = 0.1 mm

results (Edgerton and Simpson 1969). In addition, metabolic profile and patterns of the distribution of muscle fibers may vary in different compartments of the same muscle (Katsura et al. 1982), can be altered by functional demands and aging (Guth and Yellin 1971; Tomanek et al. 1973; Pette and Staron 1997), and depend also on the body size of the animals (Gauthier and Padykula 1966). The first typing of intrinsic muscle fibers was conducted by Yohro (1977) who stained snout serial slices of the big-clawed shrew with hematoxylin and eosin, and found that intrinsic muscles are composed of typical red fibers. When intrinsic muscles were typed by revealing reactivity of cytochrome oxidase, they were classified as containing mainly white fibers, with a few red and intermediate fibers observable along muscle extremities (Haidarliu et al. 2010). Immunohistochemical staining for myosin heavy chain types has shown that in both whisking (mice and rats) and non-whisking (guinea pigs) species, intrinsic muscles are composed predominantly of type 2B muscle fibers that provide fast vibrissa movements (Jin et al. 2004). Immunohistochemical method is one of the most precise methods in muscle typing, but extrinsic muscles of the mystacial pad were not yet satisfactory revealed by such analysis, and their typing is traditionally performed mostly according to their histochemical staining for ATPase or oxidative enzymes activities which reveal a mixture of various fiber types.

Muscle Fiber Organization

Muscle fiber arrangement within the muscles is one of the factors that contribute to muscle force and speed. Whisking muscles possess different patterns of their fiber arrangement relative to the axis of force generated by a muscle. Intrinsic muscles are characterized by parallel muscle fiber arrangement, with a curved segment at the rostral site of their attachment (muscle origin) to the follicles. Extrinsic vibrissa protractors (Partes mediae superior et inferior, and pseudointrinsic and posterior subunits of the Pars interna of the M. nasolabialis profundus), as well as superficial vibrissa retractors (Mm. nasolabialis et maxillolabialis), and ventral vibrissa deflector (Pars orbicularis oris of the M. buccinatorius) have a divergent (fan-shaped) fiber arrangement that permits to affect large areas of the corium in the entire mystacial pad or at least in one of its two compartments (Figure 2). Deep vibrissa retractors (Partes maxillares superficialis et profunda, and Pars interna profunda) that pull rostrally the deep fibrous mat, together with the proximal ends of vibrissa follicles, have a leaf-shaped profile and bipennate fiber arrangement (Figure 3). Physiological cross sectional area of these bipennate muscles exceeds their anatomical cross section area, that leads to a proportional increase of the total force during contraction (Gans 1982), and simultaneous decrease of the velocity of contraction and of the target (insertion site) excursion (Lieber and Fridén 2002). M. transversus nasi is composed of parallel muscle fibers that are inserted into the corium of the nasal compartment of the mystacial pad (Figure 4).

Function

Whisking muscles are the principal providers of the vibrissal movements, and the distributors of the loads developed by their contraction in order to obtain a relevant whisking pattern. Vibrissa protraction is controlled by a combination of activity of intrinsic and extrinsic muscles, whereas retraction is brought about by extrinsic muscles and elastic forces of the mystacial pad (Hill et al. 2008). Intrinsic muscles can cause also vibrissa torsional rotation that may result from an asymmetric motor innervation of the two intrinsic muscle extremities (Dörfl 1982; Knutsen et al. 2008), or, in addition, from the contraction of oblique intrinsic muscle extremities that were observed in marsupials (Grant et al. 2013). Extrinsic vibrissa protractors pull the corium of the mystacial pad rostrally together with the distal ends of vibrissae. They may also lead to a reduction of the rostrocaudal whisker field as well as of the vertical spread of the vibrissae that has been described during "foveal" whisking (Berg and Kleinfeld 2003) and surface investigation by vibrissae (Grant et al. 2009, 2012). Superficial vibrissa retractors pull the corium of the mystacial pad caudally, together with the distal ends of vibrissae. Deep vibrissa retractors pull rostrally the subcapsular fibrous mat, together with the proximal ends of the vibrissa follicles in all whisking species, even if these muscles differ morphologically: in rodents, they run the full length of the pad (Wineski 1985; Haidarliu et al. 2010, 2015), whereas in opossum, they are much shorter and actuate the pad via collagen bundles (Grant et al. 2013). Simultaneous contraction of the superficial and deep vibrissa retractors results in an enhanced retraction of the vibrissae of the entire mystacial pad. Additional contraction of the extrinsic vibrissa protractors may lead to a rostral translation of the entire mystacial pad. Each extrinsic muscle of the mystacial pad that moves vibrissae in the rostrocaudal direction, can be in synergistic or antagonistic relationships with other extrinsic muscles, and with intrinsic muscles.

Dorsal vertical vibrissa deflector (M. transversus nasi) pulls the corium together with the distal ends of the follicles of the vibrissal rows A and B dorsomedially, while the ventral vibrissa deflector (Pars orbicularis oris of the M. buccinatorius) pulls the corium, together with the distal ends of the follicles of the vibrissal rows C–E, ventrocaudally. Contraction of these two muscles leads to an increase of the vertical vibrissal spread.

Exploratory whisking in air is characterized by simple, rhythmic, synchronous and symmetric movements of the vibrissae of the both mystacial pads (Semba and Egger 1986). If the vibrissae touch or actively palpate an object, their movements become complex and sophisticated (Brecht et al. 2006). Vibrissa movements can be asynchronous and asymmetric (Erzurumlu and Killackey 1979; Kleinfeld et al. 1999; Sachdev et al. 2002; Brecht et al. 2006; Ahissar and Knutsen 2008; Deutsch et al. 2012; Sherman et al. 2013), with simultaneous torsional rotation (Knutsen et al. 2008; Grant et al. 2013). Such movements can be performed voluntarily or be triggered by head turning (Towal and Hartmann 2006; Mitchinson et al. 2011).

Innervation

Mystacial pad is innervated by two cranial nerves: (i) the trigeminal nerve as a sensory nerve, and (ii) the facial nerve as a motor nerve controlling facial musculature. General scheme of the facial muscle innervation in different rodents is apparently similar, though there are interspecies differences in the distribution of small nerves and in their nomenclature. In mice, intrinsic muscles and ventral vibrissa retractor (M. maxillolabialis) receive motor innervation from the branches of the fused together rami buccolabiales superior et inferior of the facial nerve, whereas the dorsal vibrissa retractor (M. nasolabialis) receives branches from the ramus zygomatico-orbitalis after it fuses with the ramus temporalis of the facial nerve (Dörfl 1985). In rats, vibrissal movements are controlled by the buccal branch and the upper division of the marginal mandibular branch of the facial nerve (Semba and Egger 1986; Rice et al. 1993). In hamsters, rami buccolabiales superior et inferior of the facial nerve join together to form a buccal plexus at the anterior edge of the M. masseter and ventral margin of the M. maxillolabialis (Wineski 1985). Buccal plexus innervates majority of facial striated muscles, and it was described in many other muroid rodents (Huber and Hughson 1926; Rinker 1954). There is also evidence that small hypoglossal neurons also project to the extrinsic musculature of the mystacial pad and compose a part of the hypoglossal-trigeminal loop that participates in sensory-motor control of the vibrissal system (Mameli et al. 2008).

Blood Supply

In rodents, external carotid artery is the main source of arterial supply to the face (Priddy and Brodie 1948). It gives rise to external maxillary artery that passes in the rostral direction and gives branches to Pars orbicularis oris of the M. buccinatorius, then it turns dorsorostral and gives rise to its three terminal branches of which two are feeding the muscles of the mystacial pad: angular artery gives numerous branches to the Mm. nasolabialis et maxillolabialis, whereas superior labial artery passes rostrally and gives branches to the subunits of the M. nasolabialis profundus. Intrinsic muscles are supplied with blood from the arterioles that take their origin from the branches of the superior labial artery and pass within the core of mystacial pad giving rise to moderately dense networks of capillaries oriented longitudinally regarding fibers of each intrinsic muscle (Rice 1993). These vessels receive predominantly peptidergic innervation, and during exploratory whisking behavior, they can meet enhanced metabolic demands of intrinsic muscles by increasing the overall blood flow (Fundin et al. 1997).

Development

The muscles of the head arise from somites that develop from the paraxial mesoderm, as well as from more rostral nonsomitic paraxial and prechordal head mesoderm (Kablar and Rudnicki, 2000). Two of these muscles (Mm. platysma myoides and sphincter colli) give rise to the muscles that control vibrissa movement (Huber 1930a, b). Facial muscles that are associated with the rows A and B of vibrissae develop in the mouse embryos on the 12th developmental day, and can be revealed within the lateral nasal prominence, while those that are concerned with rows C–E, within the maxillary prominence (Yamakado and Yohro 1979). Row-wise arrangement of the musculature related to the nasal and maxillary compartments of the mystacial pad suggests separate, though coordinated, development of both muscle groups within entire mystacial pad. Emergence of whisker movement is coming through two consecutive stages: twitching and whisking. Twitching was observed in sleeping newborn rats and was characterized by self-generated sleep-related twitches of single whiskers, as well as adjacent and non-adjacent whiskers, and complex movements comprising various subsets of whiskers moving in various directions (Tiriac et al. 2012). The movements dominate in the rostrocaudal direction, but are observed in other directions as well. The patterns of movements were consistent with the known anatomy of whisking musculature. Twitches correlated with the recorded electromyographic activity of the Mm. nasolabialis and maxillolabialis. Whisking emerges at the end of the second postnatal week (Welker 1964; Landers and Zeigler 2006; Grant et al. 2012) and may result from the maturation of the motor-sensory-motor loops (Saig et al. 2012).

Muscle-Based Models of Whisking

For better understanding of the mechanisms of vibrissa movements, few attempts to create biomechanical models of the whisking musculature were undertaken. Hill et al. (2008) developed a biomechanical model of vibrissa motor plant and replicated the experimental observation that whisking results from three phases of extrinsic and intrinsic muscle activity. The model proposed by Simony et al. (2010) demonstrates direct translation from motoneuron spikes to vibrissa movements, and can serve as a building block in closed-loop motor-sensory models of active touch. Analysis of the model shows that contraction of a single intrinsic muscle results in movement of its two attached whiskers with different amplitudes; the relative amplitudes depend on the resting angles. Towal et al. (2011) quantified vibrissal array morphology and constructed a descriptive model of complete 3D morphology of the rat vibrissal array that was used to determine the constrains conditioned by muscle biomechanics. Combinations of these models can be used to simulate the contact patterns that would be generated as a rat uses its whiskers to tactually explore objects with varying curvatures.

References

Ahissar, E and Knutsen, P M (2008). Object localization with whiskers. *Biological Cybernetics* 98: 449–458.

Anjum F; Turni, H; Mulder, P G H; van der Burg, J and Brecht, M (2006). Tactile guidance of pray capture in Etruscan shrews. *Proceedings of the National Academy of Sciences of the United States of America* 103: 16544–16549.

Berg, R W and Kleinfeld, D (2003). Rhythmic whisking by rat: Retraction as well as protraction of the vibrissae is under active muscular control. *Journal of Neurophysiology* 89: 104–117.

Brecht, M; Grinevich, V; Jin, T E; Margrie, T and Osten, P (2006). Cellular mechanisms of motor control in the vibrissal system. *Pflügers Archiv European Journal of Physiology* 453: 269–281.

Deutsch, D; Pietr, M; Knutsen, P M; Ahissar, E and Schneidman, E (2012). Fast feedback in active sensing: Touch-induced changes to whisker-object interaction. *PLoS ONE* 7: e44272.

Dörfl, J (1982). The musculature of the mystacial vibrissae of the white mouse. *Journal of Anatomy* 135: 147–154.

Dörfl, J (1985). The innervation of the mystacial region of the white mouse: A topographical study. *Journal of Anatomy* 142: 173–184.

Edgerton, V R and Simpson, D R (1969). The intermediate muscle fiber of rats and guinea pigs. *Journal of Histochemistry & Cytochemistry* 17: 828–838.

Erzurumlu, R S and Killackey, H P (1979). Efferent connections of the brainstem trigeminal complex with the facial nucleus of the rat. *Journal of Comparative Neurology* 188: 75–86.

Fundin, B T; Pfaller, K and Rice F L (1997). Different distributions of the sensory and autonomic innervation among the microvasculature of the rat mystacial pad. *Journal of Comparative Neurology* 389: 545–568.

Gans, C (1982). Fiber architecture and muscle function. *Exercise and Sport Science Reviews* 10: 160–207.

Gauthier, G F and Padykula, H A (1966). Cytological studies of fiber types in skeletal muscle. A comparative study of the mammalian diaphragm. *The Journal of Cell Biology* 28: 333–354.

Grant, R A; Mitchinson, B; Fox, C W and Prescott, T J (2009). Active touch sensing in the rat: anticipatory and regulatory control of whisker movements during surface exploration. *Journal of Neurophysiology* 101: 862–874.

Grant, R A; Haidarliu, S; Kennerley, N J and Prescott, T J (2013). The evolution of active vibrissal sensing in mammals: Evidence from vibrissal musculature and function in the marsupial opossum *Monodelphis domestica*. *Journal of Experimental Biology* 216: 3483–3494.

Grant, R A; Mitchinson, B and Prescott, T J (2012). The development of whisker control in rats in relation to locomotion. *Developmental Psychobiology* 54: 151–168.

Grant, R A et al. (2014). Abnormalities in whisking behavior are associated with lesions in brain stem nuclei in a mouse model of amyotrophic lateral sclerosis. *Behavioural Brain Research* 259: 274–283.

Guth, L and Yellin, H (1971). The dynamic nature of the so-called "fiber types" of mammalian skeletal muscle. *Experimental Neurology* 31: 227–300.

Haidarliu, S; Kleinfeld, D; Deschenes, M and Ahissar, E (2015). The musculature that drives active touch by vibrissae and nose in mice. *The Anatomical Record* 298: 1347–1358. In press.

Haidarliu, S; Simony, E; Golomb, D and Ahissar, E (2010). Muscle architecture in the mystacial pad of the rat. *The Anatomical Record* 293: 1192–1206.

Hill, D N; Bermejo, R; Zeigler, H P and Kleinfeld, D (2008). Biomechanics of the vibrissa motor plant in rat: Rhythmic whisking consists of triphasic neuromuscular activity. *The Journal of Neuroscience* 28: 3438–3455.

Huber, E (1930). Evolution of the facial musculature and cutaneous field of the trigeminus: Part I. *The Quarterly Review of Biology* 5: 133–188.

Huber, E (1930). Evolution of the facial musculature and cutaneous field of the trigeminus: Part II. *The Quarterly Review of Biology* 5: 389–437.

Huber, E and Hughson, W (1926). Experimental studies on the voluntary motor innervation of the facial musculature. *Journal of Comparative Neurology* 42: 113–163.

Jin, T E; Witzemann, V and Brecht, M (2004). Fiber types of the intrinsic whisker muscle and whisking behavior. *The Journal of Neuroscience* 245: 3386–3393.

Kablar, B and Rudnicki, M A (2000). Skeletal muscle development in the mouse embryo. *Histology and Histopathology* 15: 649–656.

Katsura, S; Ishizuka, H; Matsumoto, H and Nakae, Y (1982). Histochemical studies on the architecture of rat masseter muscle. *Acta Histochemica et Cytochemica* 15: 527–536.

Kleinfeld, D; Berg, R W and O'Connor, S M (1999). Anatomical loops and their electrical dynamics in relation to whisking by rat. *Somatosensory & Motor Research* 16: 69–88.

Klingener, D (1964). The comparative myology of four dipodoid rodents (Genera *Zapus*, *Napeozapus*, *Sicista*, and *Jaculus*). *Miscellaneous Publications Museum of Zoology University of Michigan* 124: 1–100.

Knutsen, P M and Ahissar, E (2008). Orthogonal coding of object location. *Trends in Neurosciences* 32: 101–109. doi:1016/j.tins.2008. 10.002.

Landers, M and Zeigler, H P (2006). Development of rodent whisking: Trigeminal input and central pattern generation. *Somatosensory & Motor Research* 23: 1–10.

Lieber, R L and Fridén, J (2002). Functional and clinical significance of skeletal muscle architecture. *Muscle Nerve* 23: 1647–1666.

Mameli, O et al. (2008). Hypoglossal nuclei participation in rat mystacial pad control. *Pflügers Archiv European Journal of Physiology* 456: 1189–1198. doi:10.1007/s00424-008-0472-y.

McComas, A J (1996). *Skeletal Muscle: Form and Function*. Champaign, Illinois, USA: Human Kinetics. ISBN 10: 0873227808 / ISBN 13: 9780873227803.

Meinertz, T (1944). Das superfizielle Facialisgebiet der Nager. VII. Die hystricomorphen Nager. *Zeitschr Anat Entwicklungsgesch* 113: 1–38.

Mitchinson, B et al. (2011). Active vibrissal sensing in rodents and marsupials. *Philosophical Transactions of the Royal Society B* 366: 3037–3048. doi:10.1098/rstb.2011.0156.

Munz, M; Brecht, M and Wolfe, J (2010). Active touch during shrew prey capture. *Frontiers in Behavioral Neuroscience* 4: 191. doi:10.3389/fnbeh.2010.00191. eCollection 2010.

Padykula, H A and Gauthier, G F (1963). Cytochemical studies of adenosine triphosphatases in skeletal muscle fibers. *The Journal of Cell Biology* 48: 87–107.

Pette, D and Staron, R S (1997). Mammalian skeletal muscle fiber type transitions. *International Review of Cytology* 170: 143–223.

Priddy, R B and Brodie A F (1948). Facial musculature, nerves and blood vessels of the hamster in relation to the cheek pouch. *Journal of Morphology* 83: 149–180.

Rice, F L (1993). Structure, vascularization, and innervation of the mystacial pad of the rat as revealed by the lectin *Griffonia simplicifolia*. *Journal of Comparative Neurology* 337: 386–399.

Rice, F L; Kinnman, E; Aldskogius, H; Johansson, O and Arvidsson, J (1993). The innervation of the mystacial pad of the rat as revealed by PGP 9.5 immunifluorescence. *Journal of Comparative Neurology* 337: 366–385.

Rice, F L; Mance, A and Munger B L (1986). A comparative light microscopic analysis of the sensory innervation of the mystacial pad. I. Innervation of vibrissal follicle-sinus complexes. *Journal of Comparative Neurology* 252: 154–174.

Rinker, G C (1954). The comparative myology of the mammalian genera *Sigmodon, Oryzomys, Neotoma*, and *Peromyscus* (Cricetinae), with remarks on their intergeneric relationships. *Miscellaneous Publications Museum of Zoology University of Michigan* 83: 1–125.

Ryan, J M (1989). Comparative myology and polygenetic systematics of the Heteromyidae (Mammalia, Rodentia). *Miscellaneous Publications Museum of Zoology University of Michigan* 176: 1–103.

Sachdev, R N S; Sato, T and Ebner, F F (2002). Divergent movement of adjacent whiskers. *Journal of Neurophysiology* 87: 1440–1448. doi:10.1152/jn.00539.2001.

Saig, A; Gordon, G; Assa, E; Ariely A and Ahissar, E (2012). Motor-sensory confluence in tactile perception. *The Journal of Neuroscience* 32: 14022–14032.

Scott, W; Stevens, J and Binder-Macleod S A (2001). Human skeletal muscle fiber type classifications. *Physical Therapy* 81: 1810–1816.

Semba, K and Egger, M D (1986). The facial "motor" nerve of the rat: Control of vibrissal movement and examination of motor and sensory components. *Journal of Comparative Neurology* 247: 144–158.

Sherman, D et al. (2013). Tactile modulation of whisking via the brainstem loop: Statechart modeling and experimental validation. *PLoS ONE* 8: e79831.

Simony, E et al. (2010). Temporal and spatial characteristics of vibrissa responses to motor commands. *The Journal of Neuroscience* 30: 8935–8952.

Staron, R S (1997). Human skeletal muscle fiber types: Delineation, development, and distribution. *Canadian Journal of Applied Physiology* 22: 307–327.

Tiriac, A; Uitermarkt, B D; Fanning, A S; Sokoloff, G and Blumberg, M S (2012). Rapid whisker movements in sleeping newborn rats. *Current Biology* 22: 2075–2080.

Tomanek, R J; Asmundson, C R; Cooper, R R and Barnard, R J (1973). Fine structure of fast-twitch and slow-twitch guinea pig muscle fibers. *Journal of Morphology* 139: 47–66.

Towal, R B and Hartmann, M J (2006). Right-left asymmetries in the whisking behavior of rats anticipate head movements. *The Journal of Neuroscience* 26: 8838–8846.

Towal R B; Quist, B W; Gopal, V; Solomon, J H and Hartmann, M J Z (2011). The morphology of the rat vibrissal array: A model for quantifying spatiotemporal patterns of whisker-object contact. *PLoS Computational Biology* 7: e1001120.

Vincent, S B (1912). The function of the vibrissae in the behavior of the white rat. *Behav Monog* 1: 1–81.

Vincent, S B (1913). The tactile hair of the white rat. *Journal of Comparative Neurology* 23: 1–34.

Welker, W I. (1964). Analysis of sniffing of the albino rat. *Behavior* 22: 223–244.

White, K K and Vaughan, D W (1991). The effects of age on atrophy and recovery in denervated fiber types of the rat nasolabialis muscle. *The Anatomical Record* 229: 149–158.

Wineski, L E (1985). Facial morphology and vibrissal movement in the golden hamster. *Journal of Morphology* 183: 199–217.

Woolsey, T A; Welker, C and Schwartz R H (1975). Comparative anatomical studies of the Sml face cortex with special reference to the occurrence of "barrels" in layer IV. *Journal of Comparative Neurology* 164: 79–94.

Yamakado, M and Yohro, T (1979). Subdivision of mouse vibrissae on an embryological basis, with descriptions of variations in the number and arrangement of sinus hairs and cortical barrels in BALB/c (nu/+ ; nude, nu/nu) and hairless (hr/hr) strains. *American Journal of Anatomy* 155: 153–174.

Yohro, T (1977). Arrangement and structure of sinus hair muscles in the big-clawed shrew, *Sorex unguiculatus*. *Journal of Morphology* 153: 317–331.

Whisking Pattern Generation

Phil Zeigler and Asaf Keller

Introduction

The delineation of central mechanisms underlying the generation and modulation of rhythmic movement patterns in vertebrates [Central Pattern Generators: CPGs]—including respiration and locomotion, swallowing, chewing and licking—has been an important goal of systems neuroscience. Common to many of these functions is the operation of small ensembles of premotor neurons that generate patterned drive to motoneurons, producing relatively simple repetitive patterns of movement even after the removal of any identifiable phasic sensory inputs. However, while the generation of these patterns in the absence of phasic peripheral influences is one defining criterion for CPGs, another is their susceptibility to modulation by both sensory inputs and descending control mechanisms. An additional commonality is the ubiquity of chemical modulation, often by serotonergic mechanisms. These three sets of factors, endogenous rhythm generation, descending (e.g. cortical and neuromodulatory) influences and peripheral sensory inputs, interact to provide adaptive control of the rhythmic behavior pattern.

Rodent whisking behavior has many features that make it an excellent model for the study of such interactions. It has a relatively simple motor plant, its peripheral innervation is well-described, as are its central sensory and motor pathways, and the system is not complicated by a proprioceptive loop. Although the system lacks proprioceptors, reflex arcs are formed by sensorimotor loops (Nguyen and Kleinfeld 2005; Deutsch et al. 2012; Matthews et al. 2015). In recent years, behavioral studies of whisking in both head-fixed and freely moving rodents have been facilitated by

P. Zeigler (✉)
Psychology Department, Hunter College of the City University of New York, New York, USA

A. Keller
Department of Anatomy and Neurobiology, University of Maryland School of Medicine, Baltimore, USA

T.J. Prescott et al. (eds.), *Scholarpedia of Touch*, Scholarpedia,
DOI 10.2991/978-94-6239-133-8_48

the development of optoelectronic and videographic methods (Bermejo et al. 1998; Knutsen et al. 2005). Together, the relatively simple mechanics and advanced monitoring techniques available for this system have facilitated investigations of the underlying control circuitry.

Here we review recent progress in the analysis of central pattern generation mechanisms in rodents and suggest that its mediating mechanisms, though similar in many respects to those of the more classic models, have a number of novel features of interest, including the active participation of vibrissae motoneurons in the generation of the whisking rhythm.

Whisking Musculature

The mystacial vibrissae are sinus hairs, each emerging from a follicle that is embedded in the mystacial pad. Protraction of the vibrissae is an active process, effected by contraction of the intrinsic muscles. These are small sling-like muscles that wrap around the base of each follicle and attach to the pad surrounding the next caudal vibrissa. Retraction of the vibrissae can occur passively, through the elastic properties of the tissue. Retraction is also aided by active contraction of a set of extrinsic muscles that are involved with movement of not only the vibrissae, but also of several parts of the face, such as the lips and nares. Kleinfeld, Zeigler and colleagues (Hill et al. 2008) have developed a model to describe the interactions between these intrinsic and extrinsic musculature in the rat. This model proposes that, during exploratory whisking, the periodic motion of the vibrissae and mystacial pad results from three phases of muscle activity. First, the vibrissae are thrust forward as the rostral extrinsic muscle, musculus (m.) nasalis, contracts to pull the pad and initiate protraction. Second, late in protraction, the intrinsic muscles pivot the vibrissae farther forward. Third, retraction involves the cessation of m. nasalis and intrinsic muscle activity and the contraction of the caudal extrinsic muscles m. nasolabialis and m. maxillolabialis to pull the pad and the vibrissae backward. The model suggests that the combination of extrinsic and intrinsic muscle activity leads to a more extended range of vibrissa motion than would be available from the intrinsic muscles alone.

A similar organization is found in the mouse vibrissae pad (Haidarliu et al. 2014), where, in the rostral part of the mouse snout, there are both protractors and retractors of the vibrissae.

The patterns of muscle activation described above characterize whisking behavior during exploration, when whisking is synchronous with sniffing. During other behavioral states, when breathing rate is slow, the rate of whisking exceed that of breathing. Thus, the activity of the intrinsic muscles leads protraction for both sniffing and slow respiration, and the extrinsic muscle—which is active for every whisk during sniffing—is only active for inspiratory whisks during basal breathing (Moore et al. 2013). Thus, whisking dynamics change during different behavioral states.

The follicular muscles are structurally homogeneous: In addition, essentially all follicular muscles are of the fast-twitch type and lack proprioceptors. Analyses by Brecht and collaborators (Jin et al. 2004) demonstrate that >90 % of the muscle fibers are of type 2B, which have high levels of anaerobic glycolytic enzymes providing a rapid source of ATP and high maximum velocity of contraction but are less fatigue resistant than other muscle fiber types. The high percentage of type 2B fibers distinguishes the intrinsic vibrissa musculature from skeletal muscles and may have evolved for fast scanning of the sensory environment. A comparable analysis of the extrinsic muscles has not been reported, but they are presumed to contain a mix of slow and fast-twitch fibers, as well as proprioceptors (Lazarov 2007; Burrows et al. 2014).

Whisking Motoneurons

Both the extrinsic and follicular muscles are innervated by motoneurons whose parent somata reside in the facial nucleus ipsilateral to the innervated vibrissa pad. Relatively little is known about the motoneurons innervating the extrinsic muscles, but the follicular muscles are innervated by motoneurons ("whisking motoneurons") in the lateral and intermediate subdivision of the facial nucleus (Klein and Rhoades 1985; Hattox et al. 2002; Nguyen and Kleinfeld 2005; Matthews et al. 2015). These motoneurons are arranged, roughly, in a somatotopic manner corresponding to the arrangement of the vibrissae. Available evidence suggests that, with rare exceptions, each motoneuron innervates only one sling muscle, and that the motoneurons have no axon collaterals within the facial nucleus or in any other structure. Further, there are no known interneurons in the lateral facial nucleus, which appears to be composed exclusively of motoneurons and glial cells.

Although the facial nucleus contains a large number of gap junctions, these appear to primarily involve glial cells. Attempts to identify gap junctions, or electrical coupling among facial motoneurons have so far been unsuccessful.

Some rhythmic motor functions, including locomotion, breathing and chewing, involve motoneurons that have intrinsic membrane properties that allow them to function as intrinsic or conditional bursters (Lee and Heckman 1996; Del Negro et al. 1999). For example, they may express plateau potentials, which generate prolonged firing in response to brief current injections, and a hyperpolarization-activated cationic current (Ih) active at or near resting membrane potential. These properties do not characterize whisking motoneurons (Hattox et al. 2003) (although Ih is expressed in facial motoneurons of young rodents (Larkman and Kelly 1998, 2001)), suggesting that these motoneurons are not intrinsically bursting and that they require rhythmic synaptic inputs, or a neuromodulatory drive, to generate rhythmic firing. Below we consider potential sources of these synaptic inputs, and hypothetical mechanisms through which they might generate rhythmic whisking.

Afferents to Whisking Motoneurons

In an attempt to identify potential contributors to rhythm generation in this system Hattox et al. (2002) systematically labeled and identified the origin of afferents to whisking motoneurons. A very large number of brainstem, mesencephalic and midbrain nuclei were found to provide unilateral, and sometimes bilateral, innervation to the lateral facial nucleus. Similar cortical targets were recently identified Petersen and collaborators (Sreenivasan et al. 2015). In addition, Brecht and collaborators (Grinevich et al. 2005), identified a pathway providing a direct, though sparse, connection between the motor cortex and whisking motoneurons.

The plethora of regions innervating the whisking motoneurons suggests that rhythm generation and the regulation of whisking parameters in this system is subject to control by a number of centers that might be active during different behavioral states. Of particular interest were the findings that some of these regions also received direct inputs from the vibrissa representation in the motor cortex, suggesting that these hypothetical rhythm generators could be controlled voluntarily. Subsequent studies have attempted to narrow down this list by seeking to identify causal relationships between activity in these afferents and whisking behaviors.

Brainstem Reticular Formation

The brainstem reticular formation originates a particularly dense projection to the facial nucleus. Several lines of anatomical and physiological evidence implicate it in a whisking CPG. Motoneurons in this region are involved in a number of rhythmic motor acts in mammals, such as licking, mastication, and locomotion (for review see: Buttner-Ennever and Holstege 1986; Moore et al. 2014b). Similarly, the reticular formation in birds contains the CPG for rhythmic acts such as pecking and jaw movements (Berkhoudt et al. 1982; Wild et al. 1985). In rodents, microstimulation of neurons in this region evokes rhythmic whisking, suggesting that these neurons may control whisking behaviors by driving the motoneurons. In support of this idea, electrical stimulation of neurons in the reticular formation evokes monosynaptic EPSPs in facial motoneurons in the cat. Below we discuss the potential role of an important subset of these reticular formation neurons: the serotonergic neurons of the raphe and the lateral paragigantocellularis nucleus.

Recent work by Deschênes, Kleinfeld and collaborators (Moore et al. 2013) identified a region within the intermediate band of the reticular formation (IRt) containing neurons that project directly to whisking motoneurons, and that fire in phase with whisking. Further, lesions in this region abolish whisking behaviors. Thus, these neurons appear to function as a premotor pattern generator for whisking (see also below).

A number of other brainstem nuclei have been implicated in regulating and coordinating various facial rhythmic activities, including whisking, sniffing, licking and breathing (Hattox et al. 2002; Cao et al. 2012; Moore et al. 2014b). Of particular interest is the pre-Bötzinger complex, long implicated in respiratory rhythm generation (Ramirez et al. 2012), and recently shown to be involved also in controlling rhythmic whisking (see below, and: Moore et al. 2013).

Trigeminal Nuclei

The trigeminal nerve carries sensory inputs from the vibrissae and innervates the principal trigeminal nucleus as well as the three spinal trigeminal nuclei (oralis, interpolaris and caudalis). Injection of retrograde markers in facial nucleus produced labeling in the spinal trigeminal nuclei which was sparse, exclusively ipsilateral, and observed predominantly in the nucleus caudalis. Nevertheless, as demonstrated by Nguyen and Kleinfeld (2005) trigeminal inputs to facial motoneurons can evoke a rapidly depressing reflex that might provide a positive sensory feedback to the vibrissa musculature during whisking behaviors.

The spinal trigeminal nucleus pars muralis displays anatomical substrates suggesting that it plays a key role in controlling whisking, and, specifically, in sensorimotor reflex arcs (Matthews et al. 2015). This recently defined nucleus is interspersed between spinal trigeminal nuclei caudal and interpolaris. Glutamatergic projection neurons in this nucleus both receive inputs from sensory afferent fibers and send monosynaptic connections to whisking motoneurons in the facial nucleus. These interactions provide for a disynaptic positive feedback for motor output that drives whisking (Matthews et al. 2015).

Red Nucleus

The dorsal regions of the red nucleus project to the contralateral facial nucleus, suggesting that this structure is involved in relaying inputs from the olivocerebellar system to whisking motoneurons. This is of interest because the characteristic frequency of rhythmic activity in the olivocerebellar system is similar to the frequency of rhythmic whisking (see, e.g., Lang et al. 1997). However, inactivation of the inferior olive does not affect exploratory vibrissa movements (Semba and Komisaruk 1984), so this rhythmic activity may not be causally related to these movements. Furthermore, stimulation of the red nucleus does not reliably evoke vibrissa movements (Isokawa-Akesson and Komisaruk 1987). Thus, the role of the red nucleus in modulating vibrissa movements is at present unclear.

Cholinergic Activating System

The pedunculopontine tegmental nucleus (PPTg) is part of the brainstem cholinergic activating system, critical for controlling arousal and states of vigilance. It sends dense, presumably cholinergic inputs to the lateral facial nucleus.

Superior Colliculus

A particularly dense projection to lateral facial motoneurons arises from both (contralateral and ipsilateral) superior colliculi (SC) and electrical stimulation of SC elicits contralateral vibrissa movements. SC also receives dense projections from the vibrissa representation of the motor cortex suggesting its role in rhythmic whisking can be voluntarily regulated. Because SC also forms reciprocal connections with trigeminal nuclei relaying vibrissal information it is likely to be involved in integrating inputs from motor behaviors with inputs from somatosensory, visual, and auditory sensory modalities. For these reasons the superior colliculus was the object of a series of studies to be discussed in a later section.

Whisking Behavior

We argued above that whisking behavior cannot emerge from the intrinsic properties of the pertinent motoneurons, or from interactions among them. Rather, rhythmic whisking must be governed by inputs these motoneurons receive from some or all the myriad of nuclei that project to these motoneurons. Formulating testable hypotheses regarding the nature of these rhythm generators requires a comprehensive description of the whisking behaviors, which we summarize below. (For a more complete description see other chapters in this edition.)

Whisking: Development, Kinematics and Bilateral Coordination

In rat pups, small, uncoordinated movements of the vibrissae are evident as early as days P10-14, a few days before eye opening and before the initial appearance of reliable motor maps to stimulation of the cortical vibrissal motor area (vMcx; AK, unpublished observations). During the next 2 weeks the movements gradually increase in both amplitude and frequency, maturing at the characteristic modal frequency for whisking in air (5–9 Hz) by the end of the first month (Welker 1964;

Landers and Zeigler 2006). Over the same period, there is a parallel increase in the bilateral coordination of whisking on the two sides.

The emergence of whisking behavior parallels the development of bilateral excitatory inputs to whisking motoneurons from LPGi, and the development of descending axons from the vMCx to LPGi (Takatoh et al. 2013).

In adult rats, the rhythmic vibrissa movements used in "active sensing" (whisking and palpation) exhibit a range of frequencies from 1–20 Hz, with dominant frequencies of 5–9 Hz in both head-fixed and freely moving animals (Carvell and Simons 1990; Gao et al. 2001; Hill et al. 2008). [Note that the whisking parameters reported here were computed from observations on the behavior of the laboratory rat. No comparable data are available for rats observed under more natural conditions and mice are reported to whisk at significantly higher frequencies (Jin et al. 2004)]. However, as with other rhythmic movements, whisking patterns are strongly influenced by signals from peripheral receptors. Higher frequencies (15–25 Hz) have been reported during palpation of objects ("foveation") (Berg and Kleinfeld 2003a) and, during texture discriminations, modulation of movement parameters (amplitude, frequency and bandwidth) is correlated with discriminated properties (Carvell and Simons 1995; Harvey et al. 2001). Whisking rates may also be brought under voluntary control using behavioral contingencies such as operant reinforcement schedules (Gao et al. 2003b). The brain mechanisms thought to mediate such voluntary control are discussed below.

Observations of whisking in air (without vibrissal contacts) convey a strong impression of bilateral synchrony, but even under head fixed conditions the activity of bilaterally homologous vibrissae is not always in phase or identical in amplitude (see Figure 2 in Gao et al. 2001). Indeed, under natural conditions rats may exhibit considerable bilateral asynchrony, with persistent whisking on one side and no movements on the other (Wineski 1985; Towal and Hartmann 2006; Mitchinson et al. 2007). In contrast, during whisking in air, vibrissae movements on the same side of the face are synchronous, with similar protraction amplitudes and topographies (Bermejo et al. 2005; Hill et al. 2011) (but see Sachdev et al. 2002). The mechanisms responsible for either bilateral or unilateral synchrony is presently unknown (see below).

Whisking Behavior Patterns: Effects of Deafferentation

The observation that rhythmic movements persist after sensory denervation (Welker 1964), decerebration (Lovick 1972) and cortical ablations (Semba and Komisaruk 1984) suggested the operation of a central pattern generating mechanism. Subsequently, a detailed analysis of whisking in air in head-fixed animals, using high-resolution optoelectronic monitoring methods for kinematic analysis (Gao et al. 2001) demonstrated that deafferentation (infraorbital nerve section: IOx)—when carried out in a single-stage procedure—did not affect the generation, spectral properties, kinematics, or bilateral coordination of the normal rhythmic whisking pattern (although Berg and Kleinfeld (2003a)) report that IOx produced a slight decrease in whisking frequency). After unilateral section, there was an immediate and significant increase in whisking frequency on both sides of the face that was abolished by subsequent section of the contralateral sensory nerve. Taken in conjunction with the observations of bilateral whisking asynchrony in normal rats, these data suggest, first, the existence of distinct right and left rhythm generators with separate outputs to homolateral motoneurons and, second, some degree of coupling of the right and left rhythm generators. It is possible that respiratory nuclei, that have bilateral influences (see below), are involved in bilateral coordination of whisking (Moore et al. 2014a).

Additional support for these conclusions comes from a deafferentation study carried out in developing rat pups (Landers and Zeigler 2006). Unilateral IO section at P7 (before the emergence of vibrissae movements) has no effect on whisking behavior, while transection at P12 significantly delays the emergence of the normal whisking rhythm, but only on the treated side. Whisking rhythms on the untreated side emerged at the normal time, but with a slightly, but significantly increased frequency. Bilateral IOx delayed the emergence of normal whisking until almost the end of the first postnatal month. Once normal whisking had emerged, re-sectioning of the sensory nerve had no effect on the re-emergence of vibrissae movements. In pups in which unilateral sensory denervation is combined with contralateral motor denervation—thus reducing the afference generated by active whisking—not only is the initial emergence of whisking significantly delayed but whisking frequency remains significantly reduced 2 months postnatally.

Taken together, the deafferentation data from adults and pups help to delineate some of the functional properties of central pattern generation mechanisms for whisking, including independent, but closely coupled rhythm generators on the two sides, and a sensitive period during, but not after which, trigeminal afference is critical for the normal development of rhythmic movement patterns. Given that the circuitry for such complex motor patterns as locomotion and suckling is constructed during embryonic development (Nishimaru and Kudo 2000; Kozlov et al. 2003), trigeminal afference during development seems to contribute primarily to the shaping of pre-existing pattern-generating circuitry.

Hypotheses Concerning the Neural Substrate for the Whisking CPG

As of this writing, four potential rhythm generators for whisking have been studied in detail: the respiratory-whisking ventral medulla region, the serotonergic brainstem nuclei, the motor cortex, and the superior colliculus. Indeed, it is entirely feasible that whisking is not governed by a conventional CPG, but, rather, that whisking is controlled by one or more rhythm generators that do not have an intrinsic propensity to generate output at a fixed rhythm.

The search for rhythm generators often focuses on identifying putative rhythmic pre-motoneurons, neurons that provide rhythmic inputs to motoneurons responsible for producing the rhythmic movement. Significant progress towards this goal was made by Moore et al. (2013) who identified a small population of neurons in the brainstem region they defined as the vibrissal zone of IRT (intermediate band of the reticular formation; vIRT) whose oscillatory activity co-varies in phase with vibrissae contractions. Some neurons in this region project to whisking motoneurons in the facial nucleus that control intrinsic muscles, and most of them contain either GABA or glycine. Chemical activation of neurons in this region evokes whisking, whereas lesions in this area abolish whisking. Moore et al. (2013) conclude that this vibrissal zone of vIRT functions as a premotor pattern generator for rhythmic whisking (see also Moore et al. 2014b). They also demonstrated that this pattern generator participates in a larger circuit involved in respiration (see Ramirez et al. 2012). Specifically, (Moore et al. 2013; Moore et al. 2014b) posit that neurons in the pre-Bötzinger complex—critically involved in respiratory rhythm generation—reset the phase of vIRT neurons and thus coordinate whisking and breathing. Pre-Bötzinger neurons interact reciprocally with neurons in the ventral respiratory group, and the latter drive the rhythmic activity of vibrissae extrinsic muscles, to control motion of the mystical pad.

Serotonin and Rhythmic Whisking

Cranial and spinal motor nuclei, including whisking motoneurons in the lateral facial nucleus, receive some of the densest serotonergic inputs in the brain (Li et al. 1993; Hattox et al. 2003). These inputs arise from several brainstem nuclei known to contain serotonergic neurons, including the raphe magnus and the lateral paragigantocellularis nucleus (LPGi). These serotonergic nuclei also receive dense inputs from the vibrissa representation of the motor cortex (Hattox et al. 2003).

Serotonin is an important modulator of many rhythmic motor acts, including locomotion, respiration, chewing, suckling and licking (Das and Fowler 1995). Like other spinal and cranial motoneurons, facial motoneurons respond to serotonin both in vivo (VanderMaelen and Aghajanian 1980) and in vitro (Larkman et al. 1989) with an increase in excitability mediated by a membrane depolarization, an increase in input resistance, and a decrease in their firing threshold. These observations provide an anatomical, behavioral and physiological rationale for implicating serotonin in the regulation of rhythmic whisking.

A series of studies from the Keller laboratory has provided direct evidence for a role of serotonin in initiating and regulating whisking (Hattox et al. 2002, 2003; Cramer and Keller 2006; Friedman et al. 2006; Cramer et al. 2007). In brief, infusion of serotonin receptor antagonists into the facial nucleus (in vivo) suppresses voluntary whisking, whereas stimulation (electrical or chemical) of LPGi evokes vibrissa movements. In addition, rhythmic whisking evoked by intracortical microstimulation of the rhythmic protraction region of motor cortex is suppressed by serotonin receptor antagonists. In vitro, serotonin, or its receptor agonists, drives facial motoneurons to fire at whisking frequencies, by facilitating a persistent inward current in these motoneurons; The magnitude of this persistent current is positively correlated with the motoneurons' firing rate.

The Motor Cortex and Rhythmic Whisking

Although it is widely assumed that the vibrissal representation of motor cortex (vMCx) participates in voluntary whisking, the mechanisms by which this occurs remain to be established. One line of evidence suggests that whisking might be controlled by the vMCx on a cycle-by-cycle basis. Berg and Kleinfeld (2003b) reported that stimulation of the vMCx at whisking frequencies evokes vibrissae movements entrained to the stimulation frequency. This result, coupled with the recent findings that vibrissa motoneurons receive direct, albeit sparse, projections from vMCx (Grinevich et al. 2005) suggest that vMCx can, in principle, control whisking on a cycle-by-cycle basis.

In contrast, the data described above suggest that rhythmic whisking is generated by a subcortical CPG or rhythm generator, under modulatory control of the vMCx. Whisking persists after decerebration (Lovick 1972), or cortical ablation (Semba and Komisaruk 1984; Gao et al. 2003a), indicating that vMCx is not necessary for rhythmic whisking. Furthermore, recordings of cortical activity during voluntary whisking suggest vMCx does not directly generate whisking. Findings that vMCx activity precedes the onset of voluntary whisking and that rhythmic whisking outlasts vMCx activity are also consistent with activation of a whisking rhythm generator by vMCx (Friedman et al. 2006). Indeed, stimulation of the rhythmic subregion of vMCx evokes whisking epochs that are preceded by relatively long onset latencies, occur at frequencies distinct from the stimulation frequency, and can outlast the stimulus (Haiss and Schwarz 2005; Cramer and Keller 2006). Sreenivasan et al. (2015) recently demonstrated that high frequency optogenetic stimulation of vMCx results in near-immediate onset of rhythmic whisking, but conclude that this results from indirect activation of whisking motoneurons, through pre-motoneurons in the brainstem. Taken together, these observations are consistent with the hypothesis that vMCx does not directly generate whisking but instead acts through a subcortical whisking CPG that contains an essential serotonergic component.

Friedman et al. (2012) revealed significant coherence between the frequency of units in the rhythmic subregion of motor cortex and vibrissae movements during free-air whisking, but not when animals were using their vibrissae to contact an object. Spike rate in vMCx was most frequently correlated with the amplitude of vibrissa movements, whereas correlations with movement frequency did not exceed chance levels. Similarly, Hill et al. (2011) report that most vMCx units are modulated by slow variations in whisking envelope, and few units report rapid changes in whisker position. These findings suggest that the specific parameter under cortical control may be the amplitude of whisker movements.

Obviously, the two control strategies are not mutually exclusive. Indeed, Brecht et al. (2004) found that stimulation of layer V vMCx neurons evokes vibrissae movements entrained to the stimulation frequency, whereas stimulation of layer VI neurons produces bouts of whisking that are out of phase across trials. Thus vMCx might use different control strategies to produce or modulate rhythmic whisking.

Recording from the same, rhythmic subregion, Gerdjikov et al. (2013) report that single units may encode two aspects of whisker movement: (1) whisker position; (2) speed, intensity, and frequency. Information theory analysis suggested that these firing patterns contain information mostly about position and frequency, while intensity and speed are less well represented. These investigators found no evidence for phase locking, movement anticipation, or contact related responses. Gerdjikov et al. (2013) conclude that vMCx neither programs nor initiates vibrissae trajectories, nor does it process contact information. They suggest that vMX has an indirect role in whisking, and that it may be related to movement monitoring, perhaps using feedback from a whisking CPG. In contrast, Kleinfeld et al. (2002) find that firing of vMCx units is modulated as a sinusoid at the repetition rate of the stimulus for whisking frequencies (5–15 Hz).

The Superior Colliculus and Rhythmic Whisking

The superior colliculus sends dense and direct projections to the facial nucleus, where the vibrissa motor neurons are located (Hattox et al. 2002; Miyashita and Mori 1995), and stimulation of the superior colliculus produces movements of the vibrissae (McHaffie and Stein 1982). These findings suggest that the superior colliculus may also have a role in controlling whisking kinematics. Indeed the superior colliculus may have a unique role in whisking behavior by functioning as a sensorimotor loop. Collicular neurons reliably respond to vibrissae contacts with short-latency spikes reflecting their direct and potent inputs from trigeminal nuclei (Hemelt and Keller 2007; Drager and Hubel 1976). This, coupled with its direct projections to the facial nucleus, implies that the superior colliculus functions as part of a closed loop (Kleinfeld et al. 1999) through which vibrissae contacts reliably evoke vibrissae movements.

Consistent with this hypothesis, Hemelt and Keller (2008) found that in anesthetized rats, microstimulation of the colliculus evokes a sustained vibrissa protraction. This suggests that the superior colliculus plays a pivotal role in vibrissa movement—regulating vibrissa set point and whisk amplitude. This result contrasts with the effects of stimulation of vMCx, which produces rhythmic protractions. Movements generated by the superior colliculus are independent of motor cortex and can be evoked at lower thresholds and shorter latencies than those generated by the motor cortex. Thus, with the motor cortex regulating the whisking frequency (through subcortical targets, perhaps vIRT), the superior colliculus control of set point and amplitude would account for the main parameters of voluntary whisking.

The Generation of Whisking Rhythm

As detailed in the sections above, at least four systems are directly involved in generating rhythmic whisking: The vIRT that functions as a premotor pattern generator (Moore et al. 2013); parafacial respiratory pre-motoneurons that may regulate rhythmic movements of the whisker pad (Moore et al. 2013, 2014b); the superior colliculus that may control vibrissae set point and protraction amplitude (Hemelt and Keller 2008); brainstem serotonergic nuclei that may initiate whisking and regulate its frequency (Hattox et al. 2003; Cramer and Keller 2006; Friedman et al. 2006; Cramer et al. 2007).

The challenge, of course, is to determine how these diverse systems interact to perform behaviorally relevant vibrissae movements. For example, Kleinfeld, Deschênes and collaborators (Kleinfeld et al. 2014) posit that activity in vIRT, and in related respiratory rhythm generators, fully account for rhythmic whisking, whereas serotonergic inputs trigger whisking and modulate whisking amplitudes. This conclusion is supported by computational work of Golomb (2014), who shows that moderate periodic inputs from the vIRT and Bötzinger nuclei control whisking frequency, whereas serotonergic neuromodulation controls whisking amplitude.

Rhythm Generation in the Vibrissa System: Some Unanswered Questions

These relate primarily to mechanisms of synchronization and coordination of neuronal activity at several levels of the vibrissa sensorimotor system, and the role of vMCx in whisking. These questions include:

- The simultaneous generation (in unison) of the whisking rhythm by all facial motoneurons on one side of the animal, possibly by actions of the vIRT. Because the facial nucleus is thought not to contain interneurons, and its neurons do not have axon collaterals, Cramer, et al. (2007) suggested that unilateral synchronization of whisking might result from coordinated discharge of electrically coupled vFMNs, a hypothesis consistent with the presence of gap junction proteins in the facial nucleus (Rohlmann et al. 1993). However, unpublished results (Y. Li and A. Keller) have thus far failed to identify gap junctions or electrical coupling among vFMNs.
- The coupling of whisking activity on the two sides of the face. While bilateral synchronization could be mediated by one of the numerous pre-motoneuron groups identified by Hattox et al. (2002), its mechanism remains to be characterized. As noted above, Kleinfeld and collaborators (Moore et al. 2014a) provided intriguing support for the hypothesis that this bilateral coordination is mediated by respiratory pre-motoneurons.

- The role of vMCx—and other descending motor centers, such as the red nucleus —in regulating whisking. As detailed above, there exist conflicting reports regarding covariation of vMCx activity and whisking parameters. These discrepancies may be due, at least in part, to the use of different whisking behaviors to study these correlations, or to recording from different cell types in different vMCx subregions.

Answers to these questions, and to the question of whether there are multiple circuits for multiple forms of whisking, will require further neurobehavioral experiments in awake, behaving rodents, analogous to those common with primates, combining unit recording, experimental control of the whisking response and high resolution, "online" monitoring of the vibrissae movements.

References

Berg, R W and Kleinfeld, D (2003a). Rhythmic whisking by rat: Retraction as well as protraction of the vibrissae is under active muscular control. *Journal of Neurophysiology* 89: 104–117.

Berg, R W and Kleinfeld, D (2003b). Vibrissa movement elicited by rhythmic electrical microstimulation to motor cortex in the aroused rat mimics exploratory whisking. *Journal of Neurophysiology* 90: 2950–2963.

Berkhoudt, H; Klein, B G and Zeigler, H P (1982). Afferents to the trigeminal and facial motor nuclei in pigeon (*Columbia livia L.*): Central connections of jaw motoneurons. *Journal of Comparative Neurology* 209: 301–312.

Bermejo, R; Friedman, W and Zeigler, H P (2005). Topography of whisking II: Interaction of whisker and pad. *Somatosensory & Motor Research* 22: 213–220.

Bermejo, R; Houben, D and Zeigler, H P (1998). Optoelectronic monitoring of individual whisker movements in rats. *Journal of Neuroscience Methods* 83: 89–96.

Brecht, M; Schneider, M; Sakmann, B and Margrie, T W (2004). Whisker movements evoked by stimulation of single pyramidal cells in rat motor cortex. *Nature* 427: 704–710.

Burrows, A M; Durham, E L; Matthews, L C; Smith, T D and Parr, L A (2014). Of mice, monkeys, and men: Physiological and morphological evidence for evolutionary divergence of function in mimetic musculature. *Anatomical record (Hoboken, N.J.)* 297: 1250–1261.

Buttner-Ennever, J and Holstege, G (1986). Anatomy of premotor centers in the reticular formation controlling oculomotor, skeletomotor and autonomic motor systems. *Progress in Brain Research* 64: 89–98.

Cao, Y; Roy, S; Sachdev, R N and Heck, D H (2012). Dynamic correlation between whisking and breathing rhythms in mice. *The Journal of Neuroscience* 32: 1653–1659.

Carvell, G and Simons, D J (1990). Biometric analyses of vibrissal tactile discrimination in the rat. *The Journal of Neuroscience* 10: 2638–2648.

Carvell, G E and Simons, D J (1995). Task- and subject-related differences in sensorimotor behavior during active touch. *Somatosensory & Motor Research* 12: 1–9.

Cramer, N P and Keller, A (2006). Cortical control of a whisking central pattern generator. *Journal of Neurophysiology* 96: 209–217.

Cramer, N P; Li, Y and Keller, A (2007). The whisking rhythm generator: A novel mammalian network for the generation of movement. *Journal of Neurophysiology* 97: 2148–2158.

Das, S and Fowler, S C (1995). Acute and subchronic effects of clozapine on licking in rats: Tolerance to disruptive effects on number of licks, but no tolerance to rhythm slowing. *Psychopharmacology (Berl)* 120: 249–255.

Del Negro, C A; Hsiao, C-F and Chandler, S H (1999). Outward current influencing bursting dynamics in guinea pig trigeminal motoneurons. *Journal of Neurophysiology* 81: 1478–1485.

Deutsch, D; Pietr, M; Knutsen, P M; Ahissar, E and Schneidman, E (2012). Fast feedback in active sensing: Touch-induced changes to whisker-object interaction. *PLoS ONE* 7: e44272.

Drager, U C and Hubel, D H (1976). Topography of visual and somatosensory projections to mouse superior colliculus. *Journal of Neurophysiology* 39: 91–101.

Friedman, W A et al. (2006). Anticipatory activity of motor cortex in relation to rhythmic whisking. *Journal of Neurophysiology* 95: 1274–1277.

Friedman, W A; Zeigler, H P and Keller, A (2012). Vibrissae motor cortex unit activity during whisking. *Journal of Neurophysiology* 107: 551–563.

Gao, P; Bermejo, R and Zeigler, H P (2001). Whisker deafferentation and rodent whisking patterns: Behavioral evidence for a central pattern generator. *The Journal of Neuroscience* 21: 5374–5380.

Gao, P; Hattox, A M; Jones, L M; Keller, A and Zeigler, H P (2003a). Whisker motor cortex ablation and whisker movement patterns. *Somatosensory & Motor Research* 20: 191–198.

Gao, P; Ploog, B O and Zeigler, H P (2003b). Whisking as a “voluntary” response: Operant control of whisking parameters and effects of whisker denervation. *Somatosensory & Motor Research* 20: 179–189.

Gerdjikov, T V; Haiss, F; Rodriguez-Sierra, O E and Schwarz, C (2013). Rhythmic whisking area (RW) in rat primary motor cortex: an internal monitor of movement-related signals? *The Journal of Neuroscience* 33: 14193–14204.

Golomb, D (2014). Mechanism and function of mixed-mode oscillations in vibrissa motoneurons. *PLoS ONE* 9: e109205.

Grinevich, V; Brecht, M and Osten, P (2005). Monosynaptic pathway from rat vibrissa motor cortex to facial motor neurons revealed by lentivirus-based axonal tracing. *The Journal of Neuroscience* 25: 8250–8258.

Haidarliu, S; Kleinfeld, D; Deschenes, M and Ahissar, E (2014). The musculature that drives active touch by vibrissae and nose in mice. *Anatomical Record (Hoboken, N.J.).*

Haiss, F and Schwarz, C (2005). Spatial segregation of different modes of movement control in the whisker representation of rat primary motor cortex. *The Journal of Neuroscience* 25: 1579–1587.

Harvey, M A; Bermejo, R and Zeigler, H P (2001). Discriminative whisking in the head-fixed rat: Optoelectronic monitoring during tactile detection and discrimination tasks. *Somatosensory & Motor Research* 18: 211–222.

Hattox, A; Li, Y; and Keller, A (2003). Serotonin regulates rhythmic whisking. *Neuron* 39: 343–352.

Hattox, A M; Priest, C A and Keller, A (2002). Functional circuitry involved in the regulation of whisker movements. *Journal of Comparative Neurology* 442: 266–276.

Hemelt, M E and Keller, A (2007). Superior sensation: superior colliculus participation in rat vibrissa system. *BMC Neuroscience* 8: 12.

Hemelt, M E and Keller, A (2008). Superior colliculus control of vibrissa movements. *Journal of Neurophysiology* 100: 1245–1254.

Hill, D N; Bermejo, R; Zeigler, H P and Kleinfeld, D (2008). Biomechanics of the vibrissa motor plant in rat: rhythmic whisking consists of triphasic neuromuscular activity. *The Journal of Neuroscience* 28: 3438–3455.

Hill, D N; Curtis, J C; Moore, J D and Kleinfeld, D (2011). Primary motor cortex reports efferent control of vibrissa motion on multiple timescales. *Neuron* 72: 344–356.

Isokawa-Akesson, M and Komisaruk, B R (1987). Difference in projections to the lateral and medial facial nucleus: Anatomically separate pathways for rhythmical vibrissa movement in rats. *Experimental Brain Research* 65: 385–398.

Jin, T E; Witzemann, V and Brecht, M (2004). Fiber types of the intrinsic whisker muscle and whisking behavior. *The Journal of Neuroscience* 24: 3386–3393.

Klein, B G and Rhoades, R W (1985). Representation of whisker follicle intrinsic musculature in the facial motor nucleus of the rat. *Journal of Comparative Neurology* 232: 55–69.

Kleinfeld, D; Berg, R W and O'Connor, S M (1999). Anatomical loops and their electrical dynamics in relation to whisking by rat. *Somatosensory & Motor Research* 16: 69–88.
Kleinfeld, D; Deschenes, M; Wang, F and Moore, J D (2014). More than a rhythm of life: Breathing as a binder of orofacial sensation. *Nature Neuroscience* 17: 647–651.
Kleinfeld, D; Sachdev, R N; Merchant, L M; Jarvis, M R and Ebner, F F (2002). Adaptive filtering of vibrissa input in motor cortex of rat. *Neuron* 34: 1021–1034.
Knutsen, P M; Derdikman, D and Ahissar, E (2005). Tracking whisker and head movements in unrestrained behaving rodents. *Journal of Neurophysiology* 93: 2294–2301.
Kozlov, A P; Petrov, E S; Kashinsky, W; Nizhnikov, M E and Spear, N E (2003). Oral compression activity on a surrogate nipple in the newborn rat: Nutritive and nonnutritive sucking. *Developmental Psychobiology* 43: 290–303.
Landers, M and Philip Zeigler, H (2006). Development of rodent whisking: Trigeminal input and central pattern generation. *Somatosensory & Motor Research* 23: 1–10.
Lang, E J; Sugihara, I and Llinas, R (1997). Differential roles of apamin- and charybdotoxin-sensitive K + conductances in the generation of inferior olive rhythmicity in vivo. *The Journal of Neuroscience* 17: 2825–2838.
Larkman, P M; Penington, N J and Kelly, J S (1989). Electrophysiology of adult rat facial motoneurones: the effects of serotonin (5-HT) in a novel in vitro brainstem slice. *Journal of Neuroscience Methods* 28: 133–146.
Larkman, P M and Kelly, J S (1998). Characterization of 5-HT-sensitive potassium conductances in neonatal rat facial motoneurones in vitro. *Journal of Physiology* 508: 67–81.
Larkman, P M and Kelly, J S (2001). Modulation of the hyperpolarisation-activated current, Ih, in rat facial motoneurones in vitro by ZD-7288. *Neuropharmacology* 40: 1058–1072.
Lazarov, N E (2007). Neurobiology of orofacial proprioception. *Brain Research Reviews* 56: 362–383.
Lee, R H and Heckman, C J (1996). Influence of voltage-sensitive dendritic conductances on bistable firing and effective synaptic current in cat spinal motoneurons in vivo. *Journal of Neurophysiology* 76: 2107–2110.
Li, Y; Takada, M and Mizuno, N (1993). The sites of origin of serotoninergic afferent fibers in the trigeminal motor, facial, and hypoglossal nuclei in the rat. *Neuroscience Research* 17: 307–313.
Lovick, T A (1972). The behavioral repertoire of precollicular decerebrate rats. *Journal of Physiology (London)* 224: 4–6.
Matthews, D W et al. (2015). Feedback in the brainstem: An excitatory disynaptic pathway for control of whisking. *Journal of Comparative Neurology* 523: 921–942.
McHaffie, J G and Stein, B E (1982). Eye movements evoked by electrical stimulation in the superior colliculus of rats and hamsters. *Brain Research* 247: 243–253.
Mitchinson, B; Martin, C J; Grant, R A and Prescott, T J (2007). Feedback control in active sensing: Rat exploratory whisking is modulated by environmental contact. *Proceedings of the Royal Society B: Biological Sciences* 274: 1035–1041.
Miyashita, E and Mori, S (1995). The superior colliculus relays signals descending from the vibrissal motor cortex to the facial nerve nucleus in the rat. *Neuroscience Letters* 195: 69–71.
Moore, J D et al. (2013). Hierarchy of orofacial rhythms revealed through whisking and breathing. *Nature* 497: 205–210.
Moore, J D; Deschenes, M; Kurnikova, A and Kleinfeld, D (2014a). Activation and measurement of free whisking in the lightly anesthetized rodent. *Nature Protocols* 9: 1792–1802.
Moore, J D; Kleinfeld, D and Wang, F (2014b). How the brainstem controls orofacial behaviors comprised of rhythmic actions. *Trends in Neurosciences* 37: 370–380.
Nguyen, Q T and Kleinfeld, D (2005). Positive feedback in a brainstem tactile sensorimotor loop. *Neuron* 45: 447–457.
Nishimaru, H and Kudo, N (2000). Formation of the central pattern generator for locomotion in the rat and mouse. *Brain Research Bulletin* 53: 661–669.
Ramirez, J M et al. (2012). The cellular building blocks of breathing. *Comprehensive Physiology* 2: 2683–2731.

Rohlmann, A et al. (1993). Facial nerve lesions lead to increased immunostaining of the astrocytic gap junction protein (connexin 43) in the corresponding facial nucleus of rats. *Neuroscience Letters* 154: 206–208.
Sachdev, R N S; Sato, T and Ebner, F F (2002). Divergent movement of adjacent whiskers. *Journal of Neurophysiology* 87: 1440–1448.
Semba, K and Komisaruk, B R (1984). Neural substrates of two different rhythmical vibrissal movements in the rat. *Neuroscience* 12: 761–774.
Sreenivasan, V; Karmakar, K; Rijli, F M and Petersen, C C (2015). Parallel pathways from motor and somatosensory cortex for controlling whisker movements in mice. *European Journal of Neuroscience* 41: 354–367.
Takatoh, J et al. (2013). New modules are added to vibrissal premotor circuitry with the emergence of exploratory whisking. *Neuron* 77: 346–360.
Towal, R B and Hartmann, M J (2006). Right-left asymmetries in the whisking behavior of rats anticipate head movements. *The Journal of Neuroscience* 26: 8838–8846.
VanderMaelen, C P and Aghajanian, G K (1980). Intracellular studies showing modulation of facial motoneurone excitability by serotonin. *Nature* 287: 346–347.
Welker, W I (1964). Analysis of sniffing of the albino rat. *Behaviour* 22: 223–244.
Wild, J M; Arends, J M and Zeigler, H P (1985). Telencephalic connections of the trigeminal system in the pigeon Columbia livia: A trigeminal sensorimotor circuit. *Journal of Comparative Neurology* 234: 441–464.
Wineski, L E (1985). Facial morphology and vibrissal movement in the golden hamster. *Journal of Morphology* 183: 199–217.

Vibrissal Afferents from Trigeminus to Cortices

Martin Deschenes and Nadia Urbain

On each side of the rat's snout there are five horizontal rows of vibrissae that form an orderly of low-threshold mechanoreceptors. Each peripheral fiber innervating these mechanoreceptors responds to only one vibrissa and, centrally, the arrangement of the vibrissal pad is mapped into homotopic s of cellular aggregates. In layer 4 of the mouse somatosensory cortex in which they were first observed, aggregates consist of small stellate cells surrounding a 'hollow' core filled with dendrites, axons and glial cells. Thence the term barrel was used to describe their structure (Woolsey and Van der Loos 1970). Homotopic cellular aggregates were later observed in the ventral posterior medial nucleus (VPM) of the thalamus and in trigeminal brainstem nuclei (Van der Loos 1976; Ma and Woolsey 1984). They were called barreloids and barrelettes respectively. So, to each vibrissa correspond a trigeminal barrelette, a thalamic barreloid, and a cortical barrel. Because of this morphologically demonstrable, homologous arrangement of each of its major component parts, the vibrissal system of rodents has become one of the most valuable models for research in sensory physiology, developmental neuroscience, and in studies of experience-dependent synaptic plasticity. The advent of transgenic mice and the development of new imaging techniques in vivo have further contributed to promote the popularity of this sensory system.

By the mid-nineties we knew of only two pathways of information processing in the vibrissal system of rodents; (1) a lemniscal pathway that arises from the principal trigeminal nucleus (PrV), transits through the VPM, and terminates in layer 4 of the barrel cortex; (2) a paralemniscal pathway whose exact origin was uncertain, that transits through the posterior thalamic nuclear group (Po) and terminates in cortical regions surrounding the barrels. Yet, brainstem neurons that gave rise to these pathways were only partially identified, and it was not clear to what degree vibrissal inputs to the VPM and Po arose from separate populations of trigeminothalamic cells. Since then, tract tracing studies and studies that combined electrophysiological recording with single cell labelling have clarified these issues

M. Deschenes (✉)
Université Laval Robert-Giffard, Québec, Canada

N. Urbain
EPFL, Lausanne, Switzerland

T.J. Prescott et al. (eds.), *Scholarpedia of Touch*, Scholarpedia,
DOI 10.2991/978-94-6239-133-8_49

and led to the discovery of additional pathways. So far, four ascending pathways of vibrissal information have been identified: three pathways that relay information through different sectors of the VPM, and another one through the medial part of Po (Pom). The multiplicity of pathways thus necessitates a revision of the current nomenclature to avoid confusion in the research community. The Text Box below describes the updated nomenclature that will be used in the present review.

Nomenclature of thalamic nuclei and pathways

Here we shall refer to the somatosensory thalamic nuclei according to the nomenclature of the Stereotaxic Atlas of the Rat Brain by Paxinos and Watson (1998). Thus, when intranuclear subdivisions do not matter, the ventral posterior medial nucleus and the posterior thalamic nuclear group will be abbreviated as VPM and Po, respectively. However, tract tracing and electrophysiological studies have now identified three pathways that pass through different sectors of the VPM (see Figure 1b). To keep on with names used in prior studies, we shall distinguish two VPM regions: the dorsal medial region (VPMdm) and the ventral lateral region (VPMvl).

VPMdm relays input from the **lemniscal** pathway. Within VPMdm we shall further distinguish the head and core of the barreloids, which will be referred to as VPMh and VPMc, respectively.

VPMvl (i.e., the tail of barreloids) relays input from the **extralemniscal** pathway.

The whisker-responsive part of Po relays input from the **paralemniscal** pathway. This region is often referred to as Pom (the medial part of the posterior thalamic nuclear group). This designation is somewhat confusing since the whisker-related part of Po actually consists of a shell-like region that borders the dorsomedial aspect of VPM (i.e., the lateralmost part of Po). However, to keep on with the tradition we will refer to that region as Pom.

We define a pathway as a 3-neuron chain that links the vibrissae to the cerebral cortex: it comprises a trigeminal ganglion cell, a trigeminothalamic neuron, and a thalamocortical cell. There also exist collateral pathways that process vibrissal input through the cerebellum and superior colliculus before sending information to cortex. The organization of the latter pathways is less well documented, and will not be considered in this review. Pathways are to be distinguished from sensory channels, which are subsystems within a pathway that sample different ranges of whisker deflection (rapidly and slowly adapting fibers, low- and high-threshold afferents; Stüttgen et al. 2006). Parallel channels arise from different populations of ganglion cells that are likely associated with different types of nerve endings in whisker follicle.

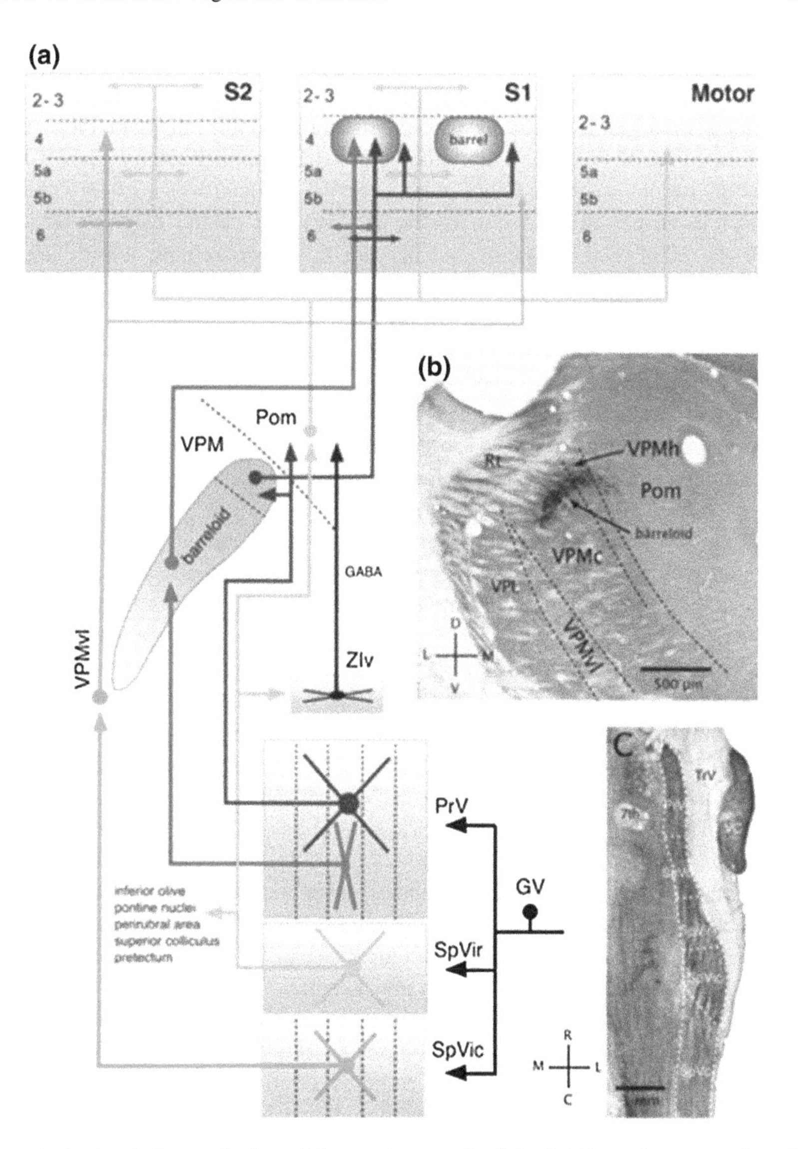

Figure 1 Anatomical organization of four pathways of vibrissal information processing. The wiring diagram in (**a**) summarizes the principal features of these pathways: individual trigeminal ganglion cells (GV) that innervate a vibrissa project to each of the trigeminal subnuclei; each pathway arises from a different cell type, transits through a different thalamic region and projects to different cortical areas or different layers in the same cortical area. In the photomicrograph (**b**) thalamic regions that serve as relay stations for each ascending pathway are delineated by dashed lines. This cytochrome oxidase-stained section also displays a barreloid that was labeled by Fluorogold injection into barrel C2; D, V, M, L stand for dorsal, ventral, medial, and lateral, respectively. Trigeminal subnuclei that give rise to the ascending pathways are outlined in the horizontal section of the brainstem (C; cytochrome oxidase staining); R, C, M, L, stand for rostral, caudal, medial, and lateral, respectively. Abbreviations: 7th, tract of the facial nucleus; TrV, spinal trigeminal tract; VC, ventral cochlear nucleus; VPL, ventral posterior lateral nucleus

Vibrissal Input in Trigeminal Nuclei

The brainstem trigeminal complex is the first processing stage in the vibrissal system; it comprises the PrV and the spinal trigeminal nucleus (SpV), which consists of the oralis (SpVo), interpolaris (SpVi) and caudalis (SpVc) subnuclei. The SpVi is further divided into rostral (SpVir) and caudal (SpVic) territories, which correspond to the magno- and parvocellular cytoarchitectonic divisions of Phelan and Falls (1989), respectively (Figure 1). In cytochrome oxidase stained coronal sections of the brainstem, the PrV, SpVic and caudal part of the SpVc display honeycomb-like patches, termed barrelettes, in which primary vibrissa afferents terminate. In each of these subnuclei barrelettes form rostrocaudally-oriented rods, about 1 mm long and 60 μm wide, whose orderly arrangement replicates that of the vibrissae on the mystacial pad (Ma and Woolsey 1984; Henderson and Jacquin 1995). Barrelettes are not discernible in the SpVo and SpVir.

Regardless of how they respond to whisker deflection, large caliber vibrissa afferents (Aβ) form ladder-like projection patterns in the brainstem, consisting of several puffs of terminations distributed at regular interval (150–200 μm) in each of the trigeminal nuclei (Hayashi 1980; Henderson and Jacquin 1995). The discontinuous arbors from each of the fibers innervating a single whisker interdigitate to produce a rostrocaudally continuous column that is coextensive with the barrelette corresponding to the same vibrissa. Collateral distribution bears no obvious relationship to the functional properties of the axons (Shortland et al. 1996), which indicates that second-order neurons in each subnucleus receive the same sensory messages. Therefore, the parallel pathways that arise from these subnuclei do not relay inputs encoding different features of an object. They likely use the same sensory inputs to inform the brain about whisker motion, texture and shape, and object location in the whisking space (Yu et al. 2006), or again different pathways may operate in different behavioral contexts (e.g., the exploratory and object recognition modes discussed by Curtis and Kleinfeld 2006).

With regard to the innervation of trigeminal nuclei by vibrissal afferents, the caudalis subnucleus deserves special comment. Like the other subnuclei, the SpVc receives profuse vibrissal input but contains relatively few trigeminothalamic neurons. The bulk of caudalis projections target the other trigeminal nuclei. Therefore, what is conveyed to the thalamus by each of the other subnuclei is already a synthesis of peripheral and caudalis inputs.

Parallel Pathways of Vibrissal Information

Figure 1 shows a wiring diagram that summarizes the anatomical organization of the four ascending pathways of vibrissal information described below.

The Lemniscal Pathway (1)

The VPM has long been recognized as the thalamic relay station of the lemniscal pathway. It contains a single type of neurons, the relay cells, which are clustered in whisker-related structures called barreloids. Barreloids form curved, obliquely oriented tapering rods that extend from the border of Pom towards the ventral posterior lateral nucleus (Figure 1a; Land et al. 1995; Haidarliu and Ahissar 2001; Varga et al. 2002). In the ventral lateral part of the VPM (VPMvl), barreloids fade out and become undistinguishable. As described below, that part of the VPM serves as a relay station for a specific population of trigeminothalamic cells. Thus, although the VPMvl does not yet figure in any atlas of the rat's brain, we shall consider it here as a separate entity (for a topographic description of VPMvl see Haidarliu et al. 2008).

Thalamic barreloids receive vibrissal input principally from small-sized PrV neurons (soma diameter < 20 μm) whose dendrites are confined within the limit of their home barrelette (Henderson and Jacquin 1995; Lo et al. 1999). These cells have receptive field dominated by a single whisker, and account for about 75 % of the projection cells in PrV (Minnery and Simons 2003). They project only to the contralateral VPMdm, where they give off small bushy terminal fields (~80 μm in diameter) in the homologous barreloid. Together, cells within a given PrV barrelette innervate the whole of a barreloid, and their projections show little convergence (on average, a VPMdm relay cell receives input from 1–2 PrV neurons; Castro-Alamancos 2002; Deschênes et al. 2003; Arsenault and Zhang 2006). Sensory transmission in this fine-grained map of vibrissa representation is mediated by large-sized perisomatic synapses that ensure a fast and secure relay of information (Spacek and Lieberman 1974; Williams et al. 1994). It is commonly thought that the high spatiotemporal resolution of the lemniscal pathway is ideally suited for texture discrimination.

Like PrV cells, barreloid cells display vigorous, short-latency responses (4–6 ms) to deflection of the principal whisker (Minnery et al. 2003). Weaker responses to 1–5 surrounding whiskers are also recorded in lightly anesthetized animals, but later responses are eliminated by deepening anesthesia (Friedberg et al. 1999; Aguilar and Castro-Alamancos 2005), or by lesion of the intersubnuclear projections from the SpV (Timofeeva et al. 2004). Axons of barreloid cells do not branch locally, but give off collaterals in the reticular thalamic nucleus (Rt; Harris 1987) as they head towards the primary somatosensory cortex (S1), where they innervate profusely layers 3–4 of a barrel column, and more sparsely the upper layer 6 of the same column (Figure 2a; Jensen and Killackey 1987; Pierret et al. 2000; Arnold et al. 2001). Relay cells in barreloids receive excitatory corticothalamic input from upper lamina 6 cells of a barrel column that is distributed principally over the distal dendrites (Hoogland et al. 1987; Bourassa et al. 1995; Mineff and Weinberg 2000), and an inhibitory input from Rt cells that targets the whole of the dendritic tree (Peschanski et al. 1983; Ohara and Lieberman 1993; Varga et al. 2002). At a unitary level, corticothalamic and Rt projections are composed of axons

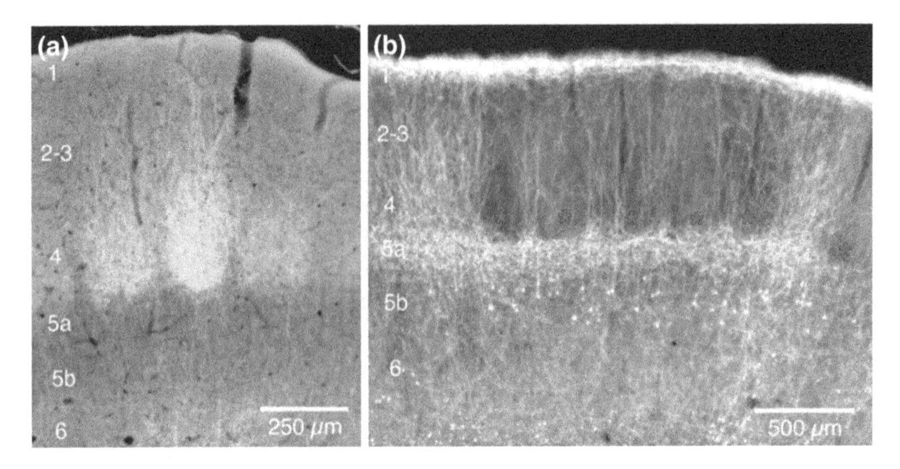

Figure 2 Terminal fields of VPMc and Pom neurons in barrel cortex. Separate injections of biotinylated dextran in VPMc and Pom reveal complementary projection patterns. While VPMc cells target barrels (**a**), Pom cells innervate the surrounding regions with a preferential distribution of terminals in layers 5a and 1. Barrels in (**b**) appear as nest-like regions (*asterisks*) devoid of terminals. The original color of both photomicrographs was inverted to enhance contrast

with terminal field topographically restricted to the barreloid representing the principal whisker of their receptive field (Bourassa et al. 1995; Désîlets-Roy et al. 2002). It is worth reminding that the thalamus of rodents (except for the lateral geniculate nucleus) is devoid of local circuit cells; thus, the Rt represents the sole source of inhibitory inputs to all subdivisions of the VPM (Barbaresi et al. 1986).

The Lemniscal Pathway (2)

At the dorsomedial margin of the VPM, near Pom, there is a stratum of barreloid cells ($\sim$150 μm thick) that receive vibrissal input from an additional population of PrV cells with large multiwhisker receptive field (Veinante and Deschênes 1999). Thus, in contrast with VPMc cells whose receptive field is dominated by a single vibrissa, those situated in the head of the barreloids (i.e., in VPMh) respond equally well to multiple vibrissae (Urbain and Deschênes 2007a). Because of their proximity to Pom, VPMh cells extend dendrites across the VPM/Pom border, which leads to their modulation by corticothalamic fibers from the vibrissa motor cortex that project to Pom.

If VPMc and VPMh relay cells form separate pathways of vibrissal information, one would expect both populations of neurons to differ in the way they innervate the barrel cortex. A recent study indeed demonstrated that VPMh cells project principally to septal regions, whereas VPMc cells innervate mainly the homologous barrel (Furuta et al. 2009). Moreover, septal cells were shown to maintain their multiwhisker receptive field after lesion of the SpV, but that lesion of the PrV abolishes nearly completely vibrissal responses throughout the barrel cortex. These results thus provide conclusive evidence that the multiwhisker receptive field of septal cells derives primarily from their innervation by VPMh cells.

Corticothalamic projections also display specificity with respect to VPMc and VPMh subdivisions. While upper lamina 6 cells of a barrel column establish a one to one reciprocal relationship with the corresponding barreloid, lower lamina 6 cells of a barrel column and lamina 6 cells of septal columns project to Pom, and form rostrocaudally oriented bands in VPMh (Bourassa et al. 1995; Deschênes et al. 1998). Moreover, VPMh cells, but not VPMc cells, receive corticothalamic input from lamina 6 of the vibrissa motor cortex (Urbain and Deschênes 2007a). A similar topographic specificity characterizes Rt projections to the VPM. Some axons only innervate VPMh, while others give off terminations throughout the whole expanse of a barreloid (Désîlets-Roy et al. 2002).

In sum, anatomical and electrophysiological data point to a functional specialization of cells in the head of the barreloids, which suggests that the two subdivisions of the lemniscal pathway that pass through VPMc and VPMh convey different types of vibrissal information.

The Extralemniscal Pathway

The extralemniscal pathway differs from the lemniscal pathway in that thalamic cells exhibit large multiwhisker receptive fields that are independent of input from the PrV (Bokor et al. 2008). This pathway arises from medium-sized SpVic cells (soma size, 25–30 μm) that respond to multiple whiskers (Veinante et al. 2000a). There is no evidence that these cells innervate other regions of the brainstem or thalamus. In contrast with the small size of the terminal field of PrV axons in a barreloid (diameter ~ 80 μm; Veinante and Deschênes 1999), individual SpVic axons form larger, rostrocaudally oriented terminal fields in VPMvl (size ~ 100 μm × 250 μm; Veinante et al. 2000a), suggesting a higher degree of input convergence in this nucleus.

The VPMvl is a crescent-shaped region that approximately corresponds to the lower tier of the VPM. It is thicker caudally and thins out rostrolaterally. It was initially identified as a separate relay station after anterograde tracer injection delineated a ventral lateral region in the VPM that receives input from the SpVic (Pierret et al. 2000). There exist no cytoarchitectonic feature or immunohistochemical stain that permits to clearly delineate its boundary with the rest of the VPM (but see Haidarliu et al. 2008). In cytochrome oxidase stained material, VPMvl does not display whisker-like arrangement (i.e., barreloids).

VPMvl relay cells project to the second somatosensory area (S2) and to the dysgranular zone of S1 by means of axon collaterals (Pierret et al. 2000). Projection foci are dense in layers 4 and 6 of S2, and moderate in layers 3, 4, and 6 of S1. VPMvl receives back projections from specific populations of corticothalamic neurons that reside in lamina 6 of S2 and S1, and from a distinct population of multiwhisker Rt cells (Bokor et al. 2008).

Neurons in S2 and in the septal columns of S1 have large receptive fields (Brumberg et al. 1999; Kwegyir-Afful and Keller 2004). There is no doubt that in anesthetized animals S2 cells derive their response properties directly from the VPMvl, but the actual impact of VPMvl cells on septal neurons remains unknown.

The Paralemniscal Pathway

The paralemniscal pathway arises from large-sized, multiwhisker cells located in the SpVir (Williams et al. 1994; Veinante et al. 2000a). On their way to the thalamus, SpVir axons send branches into a number of brainstem and diencephalic regions, which include the inferior olive, the pontine nuclei, the perirubral region, the superior colliculus, the anterior pretectal nucleus, and the zona incerta (ZI). In Pom, axons terminate principally in a shell-like region that covers the dorsomedial aspect of VPM, where they make large synaptic contacts with the proximal dendrites of relay neurons (Lavallée et al. 2005).

The labelling of single axons issuing from different parts of Po revealed a heterogeneous population of fibers which, collectively, project across all the somatomotor regions of the neocortex: S1, S2, perirhinal, insular and motor cortices (Deschênes et al. 1998). The great majority of these axons divide in the white matter at their exit from the striatum and send branches in different cortical areas. The laminar distribution of terminal fields varies across areas, but layers 5a and 1 are usually the most densely innervated (Figure 2b). As a rule, cells labelled at the same injection site in Po project to the same cortical regions. Thus, Po appears as a collection of discrete neuronal assemblies that have in common a multiareal projection pattern.

Like most higher-order thalamic nuclei, Po receives a dual corticothalamic input: one that arises from layer 6 cells in cortical areas innervated by Po axons, and another one from layer 5b cells that are exclusively located in the granular and dysgranular zones of the barrel field (Bourassa et al. 1995; Veinante et al. 2000b). Latter projection consists of collaterals of long-range axons that project to the tectum, ZI, and other brainstem regions. These collaterals do not supply a branch in the Rt or in the VPM, but establish large synaptic contacts with the proximal dendrites of Pom neurons (Hoogland et al. 1987, 1991; Bourassa et al. 1995; Groh et al. 2008). These large terminals are morphologically similar to those of trigeminothalamic axons, suggesting a strong excitatory cortical drive.

Inhibitory inputs to Pom arise from three sources: the Rt, the ventral division of ZI (ZIv), and the anterior pretectal nucleus (Barthó et al. 2002; Bokor et al. 2005). While single Rt axons innervate extensive regions of Pom and mostly target distal dendrites (Pinault et al. 1995), incertal and pretectal axons make large GABAergic synaptic contacts on the soma and proximal dendrites, next to the SpVir terminals, which suggests that extrathalamic inhibitory inputs can exert a strong control over sensory transmission in this nucleus. It is not yet known whether incertal and pretectal axons contact different subpopulations of Pom cells.

Control of Sensory Transmission in Pom

Electrophysiological studies of Pom neurons were probably those that produced the most puzzling results. While anatomical studies reported that Pom received monosynaptic input from the trigeminal nuclei, Pom cells were found to respond weakly to whisker deflection. When present, responses occurred at long latencies (16–20 ms), and were of much lower magnitude than those observed in the VPM (Diamond et al. 1992a; Sosnik et al. 2001). Moreover, sensory responses in Pom were suppressed by silencing the barrel cortex, which led to the proposal that sensory transmission in this nucleus depends on cortical feedback (Diamond et al. 1992b). Yet, the reason for which Pom cells responded so weakly and tardily to whisker deflection remained intriguing. This issue was finally resolved after it was shown that Pom receives GABAergic input from the ZIv (Barthó et al. 2002), and that most of the trigeminal axons that innervate Pom also project to the ZIv (Veinante et al. 2000a, b). Thereafter, electrophysiological studies convincingly demonstrated that ZIv cells take part in a feedforward inhibitory circuit that gates vibrissal inputs, and that silencing ZIv reinstates short-latency sensory transmission through Pom (Trageser and Keller 2004; Lavallée et al. 2005). It was thus proposed that the relay of vibrissal inputs through this nucleus relies on a mechanism of disinhibition (i.e., inhibition of the inhibitory incerto-thalamic pathway). This possibility received support from a recent study in which it was shown that corticofugal messages from the vibrissa motor cortex suppress vibrissal responses in ZIv, and that suppression is mediated by an intra-incertal GABAergic circuit (Figure 3; Urbain and Deschênes 2007b). Thus, these results suggest that sensory transmission in Pom involves a top-down disinhibitory mechanism that is contingent on motor instructions.

Thalamic Projections from the SpVo and SpVc

Thalamic projection from the SpVo arises from large cells with multiwhisker receptive field (Jacquin and Rhoades 1990; Veinante et al. 2000a). It is the least abundant trigeminothalamic projection, and also the least studied. It terminates in the most posterior part of the VPM and Po, right in front of the pretectum, and also in a caudal thalamic region intercalated between the pretectal and medial geniculate nuclei. These thalamic regions are known to receive multisensory inputs (somatic, visceral, nociceptive, auditory), and to project to the perirhinal cortex, striatum, and

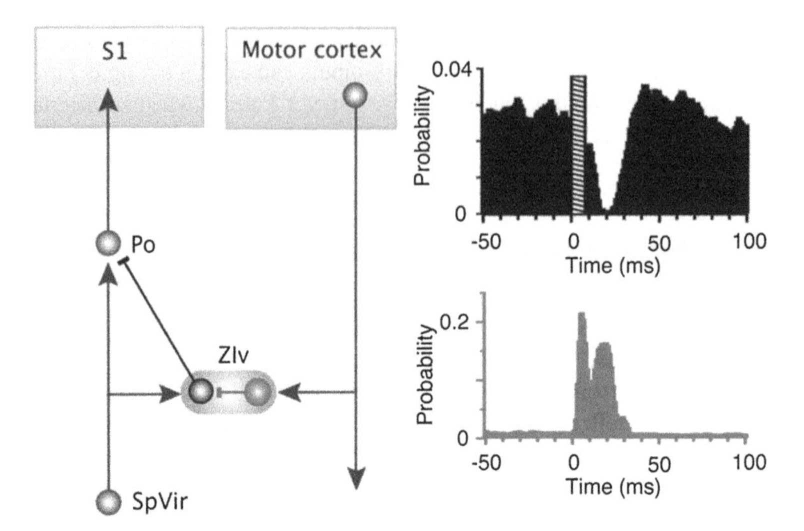

Figure 3 A feedforward inhibitory circuit impedes sensory transmission in Pom. When a whisker is deflected, ascending messages from the SpVir reach both Pom and ZIv. ZIv is a network of interconnected GABAergic cells. Because conduction velocity in SpVir axons slows down passed the branching point, inhibition in Pom occurs before vibrissal excitation, thus blocking sensory transmission through Pom. The excitability of whisker-sensitive incertal cells (*black cell*) is however depressed by motor cortex stimulation. Depression involves a mechanism of lateral inhibition mediated by whisker-insensitive cells (*red cell*). The population peristimulus histograms show the close time relationship between inhibition induced by motor cortex in vibrissa-sensitive cells (*black histogram*), and the excitatory responses of vibrissa-insensitive cells (*red histogram*). Background discharges in whisker-sensitive cells were driven by juxtacellular current injection. These results support the proposal that sensory transmission in Pom operates via a top-down disinhibitory mechanism that is contingent on motor activity

amygdala (Groenewegen and Witter 2004). Oralis cells also provide a substantial projection to the superior colliculus. Although electrophysiological data are not yet available, there is little doubt that the oralis projection constitutes a 'fifth pathway' that might be involved in the association of multiple sensory inputs, and the translation of this information via the amygdala and temporal cortices in behavioral and emotional reactions.

The SpVc contains both mono- and multiwhisker responsive cells (Renehan et al. 1986), but so far there exist no clear evidence that these cells project to the thalamus. However the SpVc projects abundantly to the other trigeminal subnuclei (Jacquin et al. 1990a). As far as we know, no physiological study has yet examined vibrissal response properties in this subnucleus.

Anatomical Basis for Crosstalk Between the Vibrissal Pathways

In the brainstem, the parallel streams of vibrissal information processing are not totally isolated from each other, in that each trigeminal subnucleus that gives rise to an ascending pathway receives projections from the other subnuclei (Figure 4;

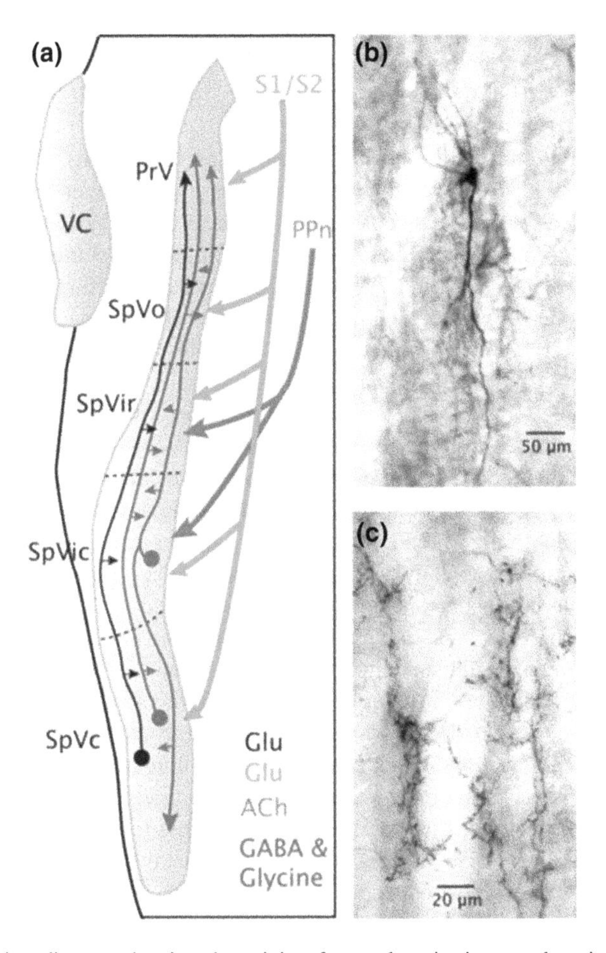

Figure 4 Wiring diagram showing the origin of central projections to the trigeminal nuclei. **a** When retrograde tracers are injected into any of the trigeminal subnuclei, the vast majority of retrogradely labeled cells in the brain are found in the other trigeminal subnuclei, in the somatosensory cortical areas, and in cholinergic neurons located in the pedunculopontine nucleus (PPn). How these inputs modulate intersubnuclear-projecting cells is currently unknown. **b** intersubnuclear-projecting cell labeled with Neurobiotin in the SpVic. **c** Axonal arborization of intersubnuclear-projecting SpVic cells in the PrV. Both photomicrographs were taken from horizontal sections of the brainstem (rostral is up). Note the preferential alignment of dendrites and terminal fields along the rostrocaudal axis (i.e., along the long axis of barrelettes)

Jacquin et al. 1990a; Voisin et al. 2002). Although most subnuclei are reciprocally connected, the most abundant intersubnuclear projections are those of the SpVc to the SpVi and PrV, and those from the SpVic to the PrV. The PrV, for instance, receives inhibitory GABAergic projection from the SpVic, and excitatory glutamatergic projection from the SpVc (Furuta et al. 2008). The inhibitory projection is particularly significant since 86 % of the SpVic cells that project to the PrV express the transcript for VIAAT, a vesicular inhibitory amino acid transporter that is expressed in both GABAergic and glycinergic neurons. That the SpVic exerts a strong inhibitory control over sensory transmission in the PrV is further supported by the near complete elimination of surround-whisker inhibition in the PrV after lesion of the SpVi (Furuta et al. 2008; see also Lee et al. (2008) for additional evidence obtained in behaving rats).

When retrograde tracers are injected into the SpVi of rodents, the vast majority of retrogradely labeled cells in the brain are found in the other trigeminal nuclei (Jacquin et al. 1990a), in the somatosensory cortical areas (Wise and Jones 1977; Wise et al. 1979; Killackey et al. 1989), and in cholinergic neurons located in the pedunculopontine nucleus (Timofeeva et al. 2005). The projection from S1 is heavy and topographically organized, connecting barrels to homotopic barrelettes in the brainstem (Welker et al. 1988; Jacquin et al. 1990b). The actual impact of these inputs on the activity of intersubnuclear-projecting cells is currently unknown, but available evidence suggest that they can exert a decisive influence on the way trigeminothalamic cells respond to vibrissal inputs (Woolston et al. 1983; Jacquin et al. 1990b; Hallas and Jacquin 1990; Timofeeva et al. 2005). This raises the possibility that, by controlling the activity of intersubnuclear-projecting cells, brain regions that project to the trigeminal nuclei may take an active part in selecting the type of information that is conveyed through each of the vibrissal pathways.

A central issue in sensory physiology is to understand how an animal endowed with highly sensitive sensory organs, and exploring the environment, can control the unceasing stream of sensory inputs it receives, and select those that are most relevant to an adaptive behavior. Clearly, there should exist multilevel, state-dependent and context-dependent gating mechanisms that filter out irrelevant sensory inputs. For example, when rats whisk to explore a new environment, they are likely little interested in object texture, no more than we are when we stretch out our arms to locate obstacles in a dark room. Processing texture information in this context appears behaviorally irrelevant; therefore pathways that process texture information might be depressed in this condition. Thus, sensory signals associated with different modes of tactual information processing (exploration, object recognition, whisking in air) might be differentially gated in brainstem trigeminal nuclei by inhibitory intersubnuclear projections.

Internal References

Alonso, J-M and Chen, Y (2009). Receptive field. *Scholarpedia* 4(1): 5393. http://www.scholarpedia.org/article/Receptive_field.
Braitenberg, V (2007). Brain. *Scholarpedia* 2(11): 2918. http://www.scholarpedia.org/article/Brain.
Jonas, P and Buzsaki, G (2007). Neural inhibition. *Scholarpedia* 2(9): 3286. http://www.scholarpedia.org/article/Neural_inhibition.
LeDoux, J E (2008). Amygdala. *Scholarpedia* 3(4): 2698. http://www.scholarpedia.org/article/Amygdala.
Llinás, R (2008). Neuron. *Scholarpedia* 3(8): 1490. http://www.scholarpedia.org/article/Neuron.
Redgrave, P (2007). Basal ganglia. *Scholarpedia* 2(6): 1825. http://www.scholarpedia.org/article/Basal_ganglia.
Roberts, E (2007). Gamma-aminobutyric acid. *Scholarpedia* 2(10): 3356. http://www.scholarpedia.org/article/Gamma-aminobutyric_acid.
Sherman, S M (2006). Thalamus. *Scholarpedia* 1(9): 1583. http://www.scholarpedia.org/article/Thalamus.

External References

Aguilar, J R and Castro-Alamancos, M A (2005). Spatiotemporal gating of sensory inputs in thalamus during quiescent and activated states. *The Journal of Neuroscience* 25: 10990–11002.
Arnold, P B; Li, C X and Waters, R S (2001). Thalamocortical arbors extend beyond single cortical barrels: An in vivo intracellular tracing study in rat. *Experimental Brain Research* 136: 152–168.
Arsenault, D and Zhang, Z W (2006). Developmental remodelling of the lemniscal synapse in the ventral basal thalamus of the mouse. *Journal of Physiology* 573: 121–132.
Barbaresi, P; Spreafico, R; Frassoni, C and Rustioni, A (1986). GABAergic neurons are present in the dorsal column nuclei but not in the ventroposterior complex of rats. *Brain Research* 382: 305–326.
Barthó, P; Freund, T F and Acsády, L (2002). Selective GABAergic innervation of thalamic nuclei from zona incerta. *European Journal of Neuroscience* 16: 999–1014.
Bokor, H; Acsády, L and Deschênes, M (2008). Vibrissal responses of thalamic cells that project to the septal columns of the barrel cortex and to the second somatosensory area. *The Journal of Neuroscience* 28(20): 5169–5177.
Bokor, H et al. (2005). Selective GABAergic control of higher-order thalamic relays. *Neuron* 45: 929–940.
Bourassa, J; Pinault, D and Deschênes, M (1995). Corticothalamic projections from the cortical barrel field to the somatosensory thalamus in rats: a single-fibre study using biocytin as an anterograde tracer. *European Journal of Neuroscience* 7: 19–30.
Brumberg, J C; Pinto, D J and Simons, D J (1999). Cortical columnar processing in the rat whisker-to-barrel system. *Journal of Neurophysiology* 82: 1808–1817.
Castro-Alamancos, M A (2002). Properties of primary sensory (lemniscal) synapses in the ventrobasal thalamus and the relay of high-frequency sensory inputs. *Journal of Neurophysiology* 87: 946–953.
Curtis, J C and Kleinfeld, D (2006). Seeing what the mouse sees with its vibrissae: A matter of behavioral state. *Neuron* 50: 524–526.
Deschênes, M; Timofeeva, E and Lavallée, P (2003). The relay of high-frequency signals in the whisker-to-barrel pathway. *The Journal of Neuroscience* 23: 6778–6787.
Deschênes, M; Veinante, P and Zhang, Z W (1998). The organization of corticothalamic projections: Reciprocity versus parity. *Brain Research Reviews* 28: 286–308.

Désîlets-Roy, B; Varga, C; Lavallée, P and Deschênes, M (2002). Substrate for cross-talk inhibition between thalamic barreloids. *The Journal of Neuroscience* 22: RC218 (1–4).
Diamond, M E; Armstrong-James, M and Ebner, F F (1992a). Somatic sensory responses in the rostral sector of the posterior group (POm) and in the ventral posterior medial nucleus (VPM) of the rat thalamus. *Journal of Comparative Neurology* 318: 462–476.
Diamond, M E; Armstrong-James, M; Budway, M J and Ebner, F F (1992b). Somatic sensory responses in the rostral sector of the posterior group (POm) and in the ventral posterior medial nucleus (VPM) of the rat thalamus: Dependence on the barrel field cortex. *Journal of Comparative Neurology* 319: 66–84.
Friedberg, M H; Lee, S M and Ebner, F F (1999). Modulation of receptive field properties of thalamic somatosensory neurons by the depth of anesthesia. *Journal of Neurophysiology* 81: 2243–2252.
Furuta, T et al. (2008). Inhibitory gating of vibrissal inputs in the brainstem. *The Journal of Neuroscience* 28: 1789–1797.
Furuta, T; Kaneko, T and Deschênes, M (2009). Septal neurons in the barrel cortex derive their receptive field input from the lemniscal pathway. *The Journal of Neuroscience* 29(13): 4089–4095.
Groenewegen, H J and Witter, M P (2004). Thalamus. In: G Paxinos (Ed.), *The Rat Nervous System, Third Edition* (pp. 407–453). Sidney: Academic Press.
Groh, A; de Kock, C P; Wimmer, V C; Sakmann, B and Kuner, T (2008). Driver or coincidence detector: Modal switch of a corticothalamic giant synapse controlled by spontaneous activity and short-term depression. *The Journal of Neuroscience* 28: 9652–9663.
Haidarliu, S and Ahissar, E (2001). Size gradients of barreloids in the rat thalamus. *Journal of Comparative Neurology* 429: 372–387.
Haidarliu, S; Yu, C; Rubin, N and Ahissar, E (2008). Lemniscal and extralemniscal compartments in the VPM of the rat. *Frontiers in Neuroanatomy* 4: Epub.
Hallas, B H and Jacquin, M F (1990). Structure-function relationships in rat brain stem subnucleus interpolaris. IX. Inputs from subnucleus caudalis. *Journal of Neurophysiology* 64: 28–45.
Harris, R M (1987). Axon collaterals in the thalamic reticular nucleus from thalamocortical neurons of the rat ventrobasal thalamus. *Journal of Comparative Neurology* 258: 397–406.
Hayashi, H (1980). Distributions of vibrissae afferent fibers collaterals in the trigeminal nuclei as revealed by intra-axonal injection of horseradish peroxidase. *Brain Research* 183: 442–446.
Henderson, T A and Jacquin, M F (1995). What makes subcortical barrels? In: E G Jones and I T Diamond (Eds.), *Cerebral Cortex, the Barrel Cortex of Rodents*, Vol. 11 (pp. 123–187). New York: Plenum.
Hoogland, P V; Welker, E and Van der Loos, H (1987). Organization of the projections from barrel cortex to thalamus in mice studied with Phaseolus vulgaris-leucoagglutinin and HRP. *Experimental Brain Research* 68: 73–87.
Hoogland, P V; Wouterlood, F G; Welker, E and Van der Loos, H (1991). Ultrastructure of giant and small thalamic terminals of cortical origin: A study of the projections from the barrel cortex in mice using Phaseolus vulgaris leuco-agglutinin (PHA-L). *Experimental Brain Research* 87: 159–172.
Jacquin, M F; Chiaia, N L; Haring, J H and Rhoades, R W (1990a). Intersubnuclear connections within the rat trigeminal brainstem complex. *Somatosensory & Motor Research* 7: 399–420.
Jacquin, M F and Rhoades, R W (1990). Cell structure and response properties in the trigeminal subnucleus oralis. *Somatosensory & Motor Research* 7: 265–288.
Jacquin, M F; Wiegand, M R and Renehan, W E (1990b). Structure-function relationships in rat brain stem subnucleus interpolaris. VIII. Cortical inputs. *Journal of Neurophysiology* 64: 3–27.
Jensen, K F and Killackey, H P (1987). Terminal arbors of axons projecting to the somatosensory cortex of the adult rat. I. The normal morphology of specific thalamocortical afferents. *The Journal of Neuroscience* 7: 3529–3543.
Killackey, H P; Koralek, K A; Chiaia, N L and Rhoades, R W (1989). Laminar and areal differences in the origin of the subcortical projection neurons of the rat somatosensory cortex. *Journal of Comparative Neurology* 282: 428–445.

Kwegyir-Afful, E E and Keller, A (2004). Response properties of whisker-related neurons in rat second somatosensory cortex. *Journal of Neurophysiology* 92: 2083–2092.

Land, P W; Buffer, S A and Yaskosky, J D (1995). Barreloids in adult rat thalamus: Three-dimensional architecture and relationship to somatosensory cortical barrels. *Journal of Comparative Neurology* 355: 573–588.

Lavallée, P et al. (2005). Feedforward inhibitory control of sensory information in higher-order thalamic nuclei. *The Journal of Neuroscience* 25: 7489–7498.

Lee, S; Carvell, G E and Simons, D J (2008). Motor modulation of afferent somatosensory circuits. *Nature Neuroscience* 12: 1430–1438.

Lo, F-S; Guigo, W and Erzurumlu, R S (1999). Electrophysiological properties and synaptic responses of cells in the trigeminal principal sensory nucleus of postnatal rats. *Journal of Neurophysiology* 82: 2765–2775.

Ma, P M and Woolsey, T A (1984). Cytoarchitectonic correlates of the vibrissae in the medullary trigeminal complex of the mouse. *Brain Research* 306: 374–379.

Mineff, E M and Weinberg, R J (2000). Differential synaptic distribution of AMPA receptor subunits in the ventral posterior and reticular thalamic nuclei of the rat. *Neuroscience* 101: 969–982.

Minnery, B S; Bruno, R M and Simons, D J (2003). Response transformation and receptive-field synthesis in the lemniscal trigeminothalamic circuit. *Journal of Neurophysiology* 90: 1556–1570.

Minnery, B S and Simons, D J (2003). Response properties of whisker-associated trigeminothalamic neurons in rat nucleus principalis. *Journal of Neurophysiology* 89: 40–56.

Ohara, P T and Lieberman, A R (1993). Some aspects of the synaptic circuitry underlying inhibition in the ventrobasal thalamus. *Journal of Neurocytology* 22: 815–825.

Peschanski, M; Ralston, H J and Roudier, F (1983). Reticularis thalami afferents to the ventrobasal complex of the rat thalamus: An electron microscope study. *Brain Research* 270: 325–329.

Phelan, K D and Falls, W M (1989). An analysis of the cyto- and myeloarchitectonic organization of trigeminal nucleus interpolaris in the rat. *Somatosensory & Motor Research* 6: 333–366.

Pierret, T; Lavallée, P and Deschênes, M (2000). Parallel streams for the relay of vibrissal information through thalamic barreloids. *The Journal of Neuroscience* 20: 7455–7462.

Pinault, D; Bourassa, J and Deschênes, M (1995). The axonal arborization of single thalamic reticular neurons in the somatosensory thalamus of the rat. *European Journal of Neuroscience* 7: 31–40.

Renehan, W E; Jacquin, M F; Mooney, R D and Rhoades, R W (1986). Structure-function relationships in rat medullary and cervical dorsal horns. II. Medullary dorsal horn cells. *Journal of Neurophysiology* 55: 1187–1201.

Shortland, P J; Demaro, J A; Shang, F; Waite, P M and Jacquin, M F (1996). Peripheral and central predictors of whisker afferent morphology in the rat brainstem. *Journal of Comparative Neurology* 375: 481–501.

Sosnik, R; Haidarliu, S and Ahissar, E (2001). Temporal frequency of whisker movement. I. Representations in brain stem and thalamus. *Journal of Neurophysiology* 86: 339–353.

Spácek, J and Lieberman, A R (1974). Ultrastructure and three-dimensional organization of synaptic glomeruli in rat somatosensory thalamus. *Journal of Anatomy* 117: 487–516.

Stüttgen, M C; Rüter, J and Schwarz, C (2006). Two psychophysical channels of whisker deflection in rats align with two neuronal classes of primary afferents. *The Journal of Neuroscience* 26: 7933–7941.

Timofeeva, E; Dufresne, C; Sík, A; Zhang, Z W and Deschênes, M (2005). Cholinergic modulation of vibrissal receptive fields in trigeminal nuclei. *The Journal of Neuroscience* 25: 9135–9143.

Timofeeva, E; Lavallée, P; Arsenault, D and Deschênes, M (2004). Synthesis of multiwhisker receptive fields in subcortical stations of the vibrissa system. *Journal of Neurophysiology* 91: 1510–1515.

Trageser, J C and Keller, A (2004). Reducing the uncertainty: Gating of peripheral inputs by zona incerta. *The Journal of Neuroscience* 24: 8911–8915.

Urbain, N and Deschênes, M (2007a). A new pathway of vibrissal information processing modulated by the motor cortex. *The Journal of Neuroscience* 27: 12407–12412.

Urbain, N and Deschênes, M (2007b). Motor cortex gates vibrissal responses in a thalamocortical projection pathway. *Neuron* 56: 714–725.

Van der Loos, H (1976). Barreloids in the mouse somatosensory. *Neuroscience Letters* 2: 1–6.

Varga, C; Sík, A; Lavallée, P and Deschênes, M (2002). Dendroarchitecture of relay cells in thalamic barreloids: A substrate for cross-whisker modulation. *The Journal of Neuroscience* 22: 6186–6194.

Veinante, P and Deschênes, M (1999). Single- and multi-whisker channels in the ascending projections from the principal trigeminal nucleus in the rat. *The Journal of Neuroscience* 19: 5085–5095.

Veinante, P; Jacquin, M and Deschênes, M (2000a). Thalamic projections from the whisker sensitive regions of the spinal trigeminal complex in the rat. *Journal of Comparative Neurology* 420: 233–240.

Veinante, P; Lavallée, P and Deschênes, M (2000b). Corticothalamic projections from layer 5 of the vibrissal barrel cortex in the rat. *Journal of Comparative Neurology* 424: 197–204.

Voisin, D L; Domejean-Orliaguet, S; Chalus, M; Dallel, R and Woda, A (2002). Ascending connections from the caudal part to the oral part of the spinal trigeminal nucleus in the rat. *Neuroscience* 109: 183–193.

Welker, E; Hoogland, P V and Van der Loos, H (1988). Organization of feedback and feedforward projections of the barrel cortex: A PHA-L study in the mouse. *Experimental Brain Research* 73: 411–435.

Williams, M N; Zahm, D S and Jacquin, M F (1994). Differential foci and synaptic organization of the principal and spinal trigeminal projections to the thalamus in the rat. *European Journal of Neuroscience* 6: 429–453.

Wise, S P and Jones, E G (1977). Cells of origin and terminal distribution of descending projections of the rat somatic sensory cortex. *Journal of Comparative Neurology* 175: 129–158.

Wise, S P; Murray, E A and Coulter, J D (1979). Somatotopic organization of corticospinal and cortico-trigeminal neurons in the rat. *Neuroscience* 4: 65–78.

Woolsey, T A and Van der Loos, H (1970). The structural organization of layer IV in the somatosensory region (SI) of mouse cerebral cortex. The description of a cortical field composed of discrete cytoarchitectonic units. *Brain Research* 17: 205–242.

Woolston, D C; La Londe, J R and Gibson, J M (1983). Corticofugal influences in the rat on responses of neurons in the trigeminal nucleus interpolaris to mechanical stimulation. *Neuroscience Letters* 36: 43–48.

Yu, C; Derdikman, D; Haidarliu, S and Ahissar, E (2006). Parallel thalamic pathways for whisking and touch signals in the rat. *PLOS Biology* 4: 819–825.

Vibrissal Basal Ganglia Circuits

Kevin Alloway

In species such as rats and mice, the vibrissal basal ganglia circuit is a subcomponent of the sensorimotor channel in the basal ganglia network. As such, it consists of all regions in the basal ganglia that process vibrissa-related information received from the cortex and thalamus. While most of this information comes from the vibrissal regions in the primary somatosensory (SI) and motor (MI) cortical areas, some whisker information is transmitted much more rapidly to the basal ganglia by intralaminar and other thalamic nuclei. A series of interconnected nuclei in the basal ganglia transform these whisker-related inputs, and the processed output is then sent to other thalamic nuclei that project to several cortical areas. Although the vibrissal circuits in the basal ganglia are poorly understood, many findings support the view that these circuits are involved in regulating the movements of the head, neck, and whiskers during a wide range of behaviors.

Many mammals have vibrissae, but only two marsupials (Virginia opossum, Brazilian short-tailed opossum) and a few rodents (e.g., mouse, rat, gerbil, hamster, chinchilla) actively move their whiskers to acquire tactile information about the spatial features of external objects (Rice 1995). In other species with vibrissae (e.g., rabbit, cat, dog, squirrel, chipmunk), tactile information is acquired from passive whisker stimulation that occurs as the animal moves through space or as objects move past the animal's head. Although the basal ganglia process tactile information produced by passive whisker stimulation, scientific interest in using the whisker system to understand the functional mechanisms of the basal ganglia has focused on the active whisking system of rats and mice.

Vibrissal Basal Ganglia Connections

The vibrissal circuit extends across the same set of interconnected nuclei as in other basal ganglia channels. Consistent with this scheme, the striatum receives inputs from sensorimotor cortical areas that are specialized for processing sensory infor-

K. Alloway (✉)
College of Medicine, Pennsylvania State University, State College, USA

T.J. Prescott et al. (eds.), *Scholarpedia of Touch*, Scholarpedia,
DOI 10.2991/978-94-6239-133-8_50

mation received from the peripheral whiskers. Corticostriatal projections originate from cortical layers III and V (Reiner et al. 2003), use glutamate as an excitatory neurotransmitter (Ottersen and Storm-Mathisen 1984; Gundersen et al. 1996), and terminate largely on medium spiny neurons in the striatum, which use GABA as an inhibitory neurotransmitter (Kincaid et al. 1998). Medium spiny neurons represent the source of all efferent projections from the striatum, and axons from these neurons project to the entopeduncular nucleus, the pars reticulata of the substantia nigra, and to the lateral globus pallidus. All of these striatal nuclei use GABA as an inhibitory neurotransmitter, and the projections from the entopeduncular nucleus and the substantia nigra pars reticulata project to motor- related nuclei in the thalamus, especially the ventromedial and ventrolateral nuclei, which project to sensorimotor cortex.

Vibrissal Cortical Areas

In rats and mice, whisker-related sensory information is transmitted to SI cortex by two parallel pathways that originate in the brainstem and proceed to the contralateral thalamus. The lemniscal trigeminal pathway originates from the principal sensory trigeminal (PrV) nucleus and terminates in the ventromedial (VPM) thalamus. The paralemniscal trigeminal pathway originates from the interpolaris division of the spinal trigeminal (SPVi) nucleus and terminates in the medial part of the posterior (POm) thalamus. The thalamic nuclei for both of these pathways convey information to the SI barrel field, which represents the main cortical area for processing vibrissal information (Woolsey and van der Loos 1970; Welker 1976). The lemniscal pathway terminates in the layer IV barrels, which contain high concentrations of cytochrome oxidase and, collectively, form an isomorphic map of the peripheral whisker pad. By comparison, the paralemniscal pathway sends dense projections to the layer IV septa that separate individual barrels from each other.

As seen in Figure 1, several cortical areas receive vibrissal-related inputs from the thalamus and SI barrel cortex. The secondary somatosensory (SII) cortex, for example, receives projections from VPM, POm, and SI barrel cortex (Cavell and Simons 1987; Spreafico et al. 1987; Fabri and Burton 1991). In addition, the

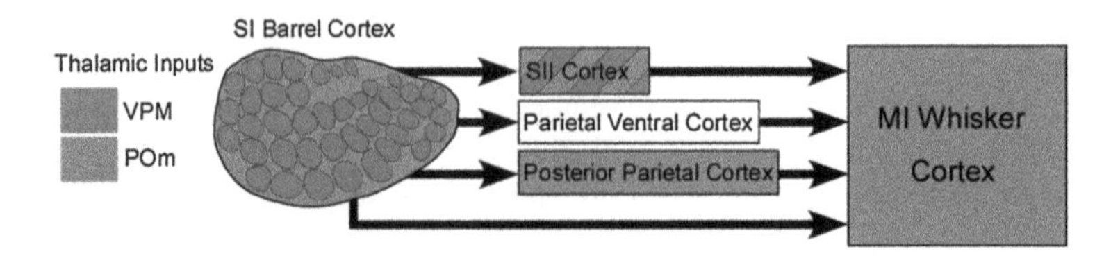

Figure 1 Feedforward projections that convey vibrissal information to the cortical area that projects to the basal ganglia. Cortical regions innervated by VPM and POm are color coded *red* and *green*, respectively

parietal ventral cortex (PVC) receives dense projections from SI barrel cortex (Fabri and Burton 1991), and the posterior parietal cortex (PPC) receives projections from SI barrel cortex and POm, but not from VPM (Reep et al. 1994; Lee et al. 2011). All of these cortical areas, including SI barrel cortex, convey vibrissal information to the basal ganglia.

The MI whisker region also processes vibrissal-related information and conveys it to the striatum. While the MI whisker region is operationally defined by sites in which intracranial microstimulation evokes whisker twitches (Gioanni and Lamarche 1985; Brecht et al. 2004), extracellular recordings demonstrate that mechanical deflections of the whiskers can activate neurons in the deep layers of MI, but not if somatosensory cortex has been inactivated (Chakrabarti et al. 2008). Therefore, even though POm sends some projections directly to MI (Aldes 1988; Alloway et al. 2004; Colechio and Alloway 2009), the functional impact of whisker stimulation on MI is mediated mainly by its cortical inputs from SI and SII.

Vibrissal Corticostriatal Projections

The vibrissal circuit undoubtedly extends through all basal ganglia nuclei, but most studies of this sensorimotor channel have focused on the striatum, especially its dorsolateral region. Like other striatal regions, the dorsolateral part receives dopaminergic inputs from the substantia nigra pars compacta, as well as glutamatergic inputs from the thalamic intralaminar nuclei. Whisker-related regions in the dorsolateral striatum also receive dense thalamic inputs from POm (Smith et al. 2012), which represents part of the ascending paralemniscal trigeminal pathway.

As shown in Figure 2, the dorsolateral striatum receives whisker related sensory inputs from multiple cortical areas including SI barrel cortex, SII, and PPC (Levesque et al. 1996b; Brown et al. 1998; Alloway et al. 2006; Smith et al. 2012). The MI whisker region, which is located in medial agranular cortex (Brecht et al. 2004), projects to the striatum and innervates the dorsolateral and dorsocentral regions (Hoffer and Alloway 2001; Reep et al. 2003). Quantitative analysis of the projections from SI, MI, and other cortical areas has revealed several principles of corticostriatal organization in the vibrissal and related sensorimotor circuits:

- **Somatotopic organization**: Projections from the head, limb, and trunk representations in SI cortex terminate in distinct, yet overlapping, parts of the striatum. Consequently, the striatum contains a crude somatotopic map in which the forepaw region is located ventral to the hindpaw region, and both limb representations are medial and rostral to the main part of the vibrissal representation (Carelli and West 1991; Brown 1992; Hoover et al. 2003). Compared to the forelimb and hindlimb projections, the vibrissal- related projections are more numerous and innervate larger parts of the striatum. This is consistent with the fact that the whisker representation occupies the largest portion of the SI somatotopic map.

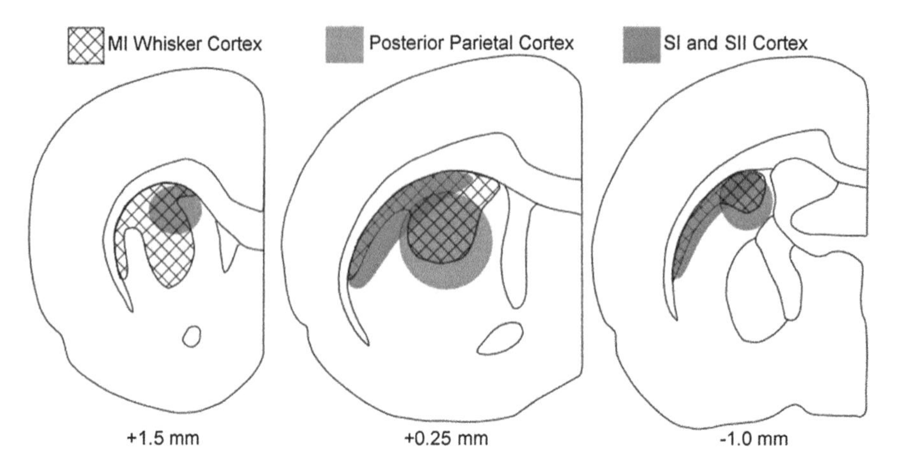

Figure 2 Schematic diagrams indicating the stratal regions that receive the densest projections from whisker-related areas in MI, SI, SII and the posterior parietal cortex (PPC). The dorsolateral striatum receives most of its whisker-related inputs from SI, SII and MI. By comparison, the dorsocentral striatum receives whisker-related inputs from MI and the PPC. Distance from bregma indicated below each section

- **Divergence**: In both rodents and primates, corticostriatal projections in the sensorimotor channel have a one-to-many projection pattern (Flaherty and Graybiel 1991). In the vibrissal system, projections from individual whisker-barrel columns in SI terminate in multiple, discontinuous patches in the dorsolateral striatum (Alloway et al. 1998). Divergent corticostriatal projections represent a mechanism for distributing the same sensorimotor information to multiple processing zones in the striatum.
- **Convergence**: Complementing this one-to-many projection pattern, each striatal region receives overlapping inputs from several cortical sites. An early primate study indicates that interconnected cortical regions project to overlapping parts of the striatum (Yeterian and Van Hoesen 1978). Although this pattern has exceptions (Selemon and Goldman-Rakic 1985), corticostriatal projections from interconnected vibrissal regions are characterized by large amounts of corticostriatal overlap (Alloway et al. 1999, 2000; Hoffer and Alloway 2001).
- **Combinatorial maps**: Experiments using 2-deoxyglucose to reveal striatal regions that are maximal activated during tactile stimulation indicate that elements of different body parts are represented in multiple locations that are juxtaposed in unique combinations (Brown 1992). Consistent with both divergent and convergent projections, these complex patterns suggest that different striatal zones represent the substrate for integrating specific combinations of somatotopic inputs, presumably to mediate specific behaviors or sequences of movements.
- **Collateralization**: Few neurons in rat sensorimotor cortex project exclusively to the striatum. Instead, corticostriatal projections represent collaterals of axons that project to the thalamus, globus pallidus, subthalamic nucleus, superior

colliculus, and certain cortical regions (Levesque et al. 1996a). This indicates that sensorimotor information sent to the striatum is also sent to other brain regions, but the exact function of these projections has not been identified.

Principles of Corticostriatal Convergence

Several principles of corticostriatal convergence have been identified by injecting different anterograde tracers into separate cortical sites of the same animal and then quantifying the amount of tracer overlap in the striatum as a function of the features associated with each pair of injections:

- **Homology**: Homologous functional representations in different cortical areas project to overlapping parts of the striatum. Like primates, in which SI and MI hand representations project to overlapping parts of the putamen (Flaherty and Graybiel 1993), the vibrissal regions in SI, SII, and MI project to adjacent and overlapping parts of the striatum as shown in Figure 2 (Alloway et al. 2000; Hoffer and Alloway 2001). Consistent with these projection patterns, electron microscopy confirms that individual striatal neurons often receive convergent synaptic inputs from whisker regions in both SI and MI (Ramanathan et al. 2002). The whisker regions in SI, SII, and MI are interconnected (Alloway et al. 2004; Chakrabarti and Alloway 2006; Colechio and Alloway 2009), and this adds further support to the view that interconnected cortical areas often project to overlapping parts of the basal ganglia (Yeterian and Van Hoesen 1978).
- **Somatotopic continuity**: Adjacent SI cortical regions representing contiguous somatic areas (e.g., forepaw and wrist) project to overlapping parts of the striatum (Hoover et al. 2003). By comparison, projections from SI regions that represent non-contiguous areas (e.g., forepaw and whisker pad) send few overlapping projections to the striatum. These distinctions are significant because contiguous body parts must cooperate with each other during behavioral movements. By contrast, non-contiguous body parts such as the head and arm can move independently of each other. Hence, corticostriatal projections have a topographic organization that enables integration of inputs from cortical regions that cooperate with each other during behavioral movements.
- **Anisotropic organization**: As shown by Figure 3, the whisker region in the dorsolateral striatum has a row- based organization in which corticostriatal projections from each SI barrel row innervate a curved, lamellar- shaped region along the dorsolateral edge of the striatum (Brown et al. 1998). Corticostriatal projections from whisker barrel row “A” terminate most laterally while those from barrel row “E” terminate most medially (Wright et al. 1999; Alloway et al. 1999).

This anisotropic pattern is noteworthy because exploratory whisking is characterized by whisker motion along the rostrocaudal axis, not along the dorsoventral axis. Furthermore, barrels in the same row have more reciprocal interconnections

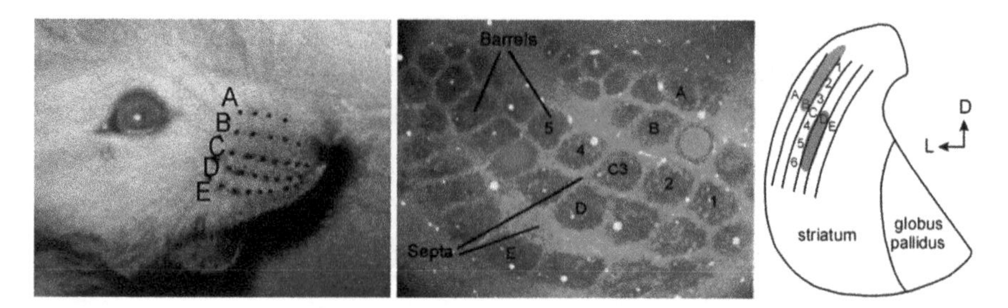

Figure 3 Topography of afferent whisker projections to SI barrel cortex and the dorsolateral striatum. Each mystacial whisker is identified by its row (A,B,C,D,E) and arc (1,2,3,4…) position (*left panel*). When processed for cytochrome oxidase, tangential sections through layer IV of SI barrel cortex reveal an isomorphic representation of the peripheral whiskers (*middle panel*). Following injections of anterograde tracers into the D5 (*red spot*) and B2 (*blue spot*) barrel columns, coronal brain sections reveal a row-based somatotopic organization in the dorsolateral striatum (*right panel*)

than barrels in different rows. Consistent with this row-based organization, corticostriatal projections from the same row overlap more than projections from different barrel rows (Alloway et al. 1999). Hence, the row-like organization of corticostriatal overlap enables greater integration of inputs from the SI barrel columns that are most likely to interact and be coordinated during whisking behavior.

Bilateral Corticostriatal Projections

The striatum receives MI inputs from both hemispheres (Wilson 1987; Reiner et al. 2003), but most of these bilateral projections originate from the MI whisker region, not the MI forepaw region (Alloway et al. 2009). This is significant because most exploratory whisking consists of synchronous whisker movements that are bilaterally symmetric (Mitchison et al. 2007). By comparison, the forepaws are much more likely to move independently. These facts suggest that interhemispheric corticostriatal projections from the MI whisker region represent an important part of the neuroanatomical substrate for coordinating bilateral whisker movements.

Bilateral corticostriatal projections from the MI whisker region are complemented by interhemispheric projections from other vibrissal-related cortices. As indicated by Figure 4, a small deposit of a retrograde tracer in the dorsolateral striatum produces neuronal labeling in MI, SI, and SII of both hemispheres (Alloway et al. 2006). Large numbers of neurons also appear bilaterally in the PVC and PPC regions. Cortical labeling is densest ipsilaterally, and the labeling patterns form mirror-image distributions in the sensorimotor regions of both hemispheres. These data demonstrate that the dorsolateral striatum processes whisker-related information from multiple regions in both hemispheres, and this supports the view that the vibrissal basal ganglia circuits help coordinate bilateral whisker movements.

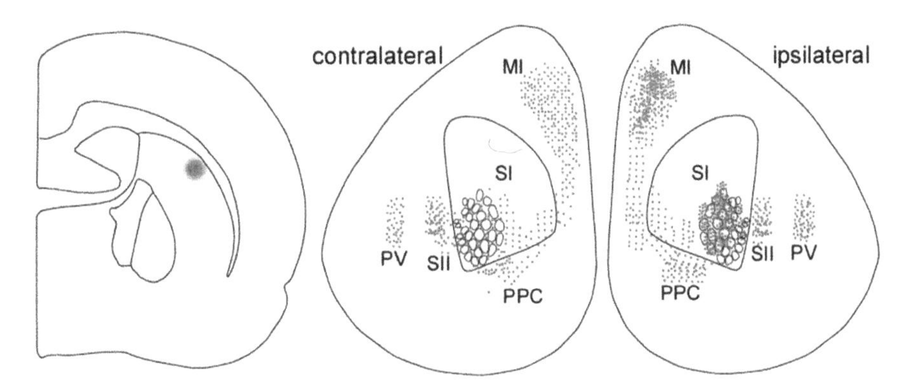

Figure 4 Bilateral distribution of corticostriatal projections to the dorsolateral neostriatum. The schematic diagram in the *left panel* represents experiments in which a small deposit of a retrograde tracer (*red spot*) is been placed in the dorsolateral striatum. Tangential sections (*right panels*) through both cortical hemispheres reveal the bilateral distribution of labeled neurons in SI barrel cortex, SII, PV, PPC and the MI whisker region. Although labeling density is greatest ipsilaterally, the labeling patters form mirror images across the two hemispheres (Alloway et al. 2006)

Dorsocentral Striatum

The MI whisker region sends dense projections to the dorsolateral striatum, but most of its corticostriatal projections terminate in the dorsocentral striatum (Alloway et al. 2009), which receives dense projections from the PPC (Reep et al. 2003). Like many other cortical regions that project to overlapping parts of the striatum, the MI whisker region and PPC are interconnected (Reep et al. 1994; Colechio and Alloway 2009). The PPC receives vibrissal information from both POm and the adjacent SI barrel cortex, as well as inputs from the auditory and visual cortical areas. These connections suggest that the PPC integrates vibrissal inputs with other sensory modalities that contain information about salient stimuli near the animal's head. This view is supported by behavioral lesion studies showing that PPC is critical for guiding head movements during directed attention (Kesner et al. 1989; Crowne et al. 1992; Tees 1999). Collectively, these findings suggest that the dorsocentral striatum uses whisker-related inputs from MI and the PPC to coordinate whisking with head movements and other orienting behaviors that subserve directed attention (Reep and Corwin 2009).

Striatal Computations

Corticostriatal convergence is consistent with prevailing views about the computational functions of the striatum. Medium spiny neurons have strong rectifying potassium currents that shunt small excitatory inputs, but they shift to a depolarized state when they receive strong excitation. Corticostriatal axons traverse the striatal neuropil in a relatively straight path and individual axons contribute only a few synaptic inputs to each striatal neuron that is contacted (Kincaid et al. 1998). Consequently, convergent corticostriatal terminals must discharge simultaneously to drive a striatal neuronal target to its discharge threshold (Wilson 1995). Hence, whisker-sensitive striatal neurons probably signal when multiple regions in cortex are synchronously active during whisking behavior. Interconnections between the whisker representations in SI, MI, and other sensorimotor cortical areas may increase the synchronization of these regions, thereby increasing the probability that convergent corticostriatal projections will depolarize medium spiny neurons sufficiently to elicit striatal activity during whisking behavior.

Striatal Behavioral Responses

Substantial evidence indicates that the dorsolateral striatum is needed to execute sensorimotor habits (Yin et al. 2004, 2006; Redgrave et al. 2010). Such behaviors are highly repetitive, are mediated by stimulus-response (S-R) associations, and are expressed even in the absence of reinforcement. In rats, focal lesions in the dorsolateral striatum disrupt the normal sequence of repetitive, stereotyped grooming behaviors (Cromwell and Berridge 1996). Although the normal sequence of grooming behavior is clearly disrupted, the capacity to emit individual grooming movements is not affected. Consistent with this distinction, neurons in the dorsolateral striatum appear to encode the serial order of sequential grooming movements (Aldridge and Berridge 1998). Furthermore, the striatal sites associated with stereotyped grooming behaviors are located in regions that receive corticostriatal projections from the forepaw and, to a lesser extent, the whisker representations in SI cortex (Hoover et al. 2003).

In rats, exploratory whisking is a stereotyped behavior that is characterized by a series of short (1–2 s) whisking bouts or epochs in which the frequency of whisker motion is relatively constant in each epoch, but shifts to another frequency in the next epoch. When whiskers contact external stimuli during exploratory behavior, the whisking behavior is characterized by stereotyped changes in the bilateral pattern of the rhythmic movements (Mitchinson et al. 2007). These stimulus-induced changes appear to reflect the motor expression of an S-R association. These observations suggest that vibrissal processing in the dorsolateral striatum is responsible for coordinating the sequences of whisker movements that accompany exploratory behaviors, including changes in patterns that are evoked by external stimuli. According to this view, corticostriatal projections to the dorsolateral striatum convey sensorimotor signals that accompany the execution of well-learned, habitual behaviors that are performed automatically.

Rodent whisking behavior has many hallmarks of a sensorimotor habit, but very few studies have characterized how dorsolateral striatal neurons respond to vibrissal inputs. Electrical stimulation of SI barrel cortex evokes neuronal discharges in the

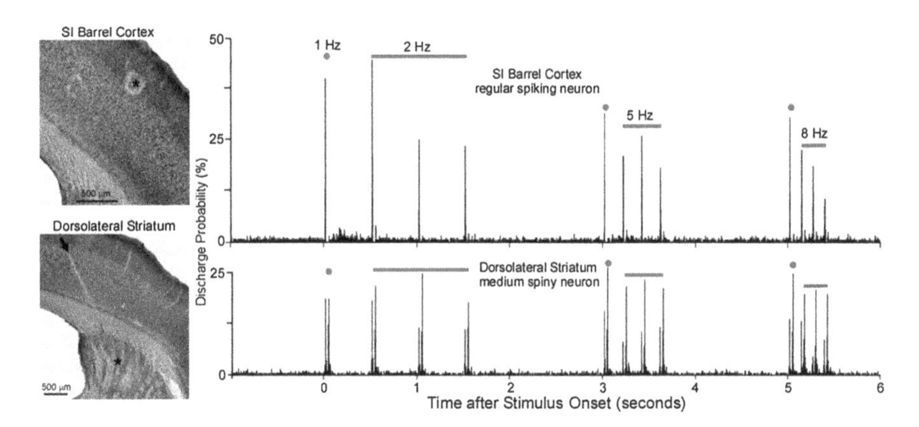

Figure 5 Neuronal responses to whisker deflections recorded simultaneously in SI barrel cortex (*top*) and dorsolateral striatum (*bottom*). The photomicrographs illustrate the recording sites (asterisks) of the SI and striatal neurons. As indicated by the peristimulus-timed histograms (PSTHs), the SI neuron adapted to repetitive whisker deflections, but the neuron in the dorsolateral striatum did not. PSTHs based on 200 trials; binwidths, 2 ms (Mowery et al. 2011)

dorsolateral striatum of anesthetized rats (Wright et al. 2001). In awake rats, neurons in the dorsolateral striatum are excited by passive whisker deflections and discharge rhythmically when rats are actively whisking (Carelli and West 1991). Whisker deflections rarely evoke striatal discharges in deeply anesthetized rat preparations (West 1998; Pidoux et al. 2011), but whisker stimulation can reliably evoke neuronal responses in the dorsolateral striatum of rats that are in a lightly anesthetized state (Mowery et al. 2011). As seen in Figure 5, neurons in the dorsolateral striatum display very little adaptation when the whiskers are repetitively deflected at frequencies up to 8 Hz. When neurons in SI barrel cortex and the dorsolateral striatum are recorded simultaneously, SI neurons display quickly adapt to repetitive whisker movements and usually discharge only after neurons in the dorsolateral striatum have already responded.

These facts suggest that striatal responses to passive or external whisker stimulation do not depend on SI inputs. Furthermore, the relative lack of neuronal adaptation in the striatum suggests that invariant neuronal responses may represent the neural mechanism by which the dorsolateral striatum encodes S-R associations that mediate sensorimotor habits. Consistent with this view, whisker-sensitive regions in the dorsolateral striatum receive inputs from several thalamic nuclei that respond to whisker stimulation including the POm and parafascicular (Pf) nuclei (Smith et al. 2012). Although Pf does not receive whisker- related inputs directly from the trigeminal nuclei in the brainstem, it does receive inputs from whisker- sensitive regions in the intermediate layers of the superior colliculus (Smith and Alloway, unpublished observations). In addition, tracer injections into whisker-sensitive regions in the Pf nucleus have revealed projections to both the dorsocentral and dorsolateral striatum.

The superior colliculus is well known for processing multimodal inputs to enable orientation to salient sensory stimuli (Hemelt and Keller 2007; Schulz et al. 2009).

By receiving direct inputs from the superior colliculus, the Pf nucleus and its projections to the striatum could represent a rapid pre-attentive mechanism that enables highly salient signals to initiate an abrupt change in sensorimotor behavior. Together, the Pf and POm are likely to transmit whisker-sensitive and other somesthetic inputs to the striatum that play an important role in mediating the S-R associations that subserve sensorimotor habits.

References

Aldes, L D (1988). Thalamic connectivity of rat somatic motor cortex. *Brain Research Bulletin* 20: 333–348.

Aldridge, J W and Berridge, K C (1998). Coding of serial order by neostriatal neurons: A "natural action" approach to movement sequence. *The Journal of Neuroscience* 18: 2777–2787.

Alloway, K D; Crist, J; Mutic, J J and Roy, S A (1999). Corticostriatal projections from rat barrel cortex have an anisotropic organization that correlates with vibrissal whisking movements. *The Journal of Neuroscience* 19: 10908–10922.

Alloway, K D; Lou, L; Nwabueze-Ogbo, F and Chakrabarti, S (2006). Topography of cortical projections to the dorsolateral neostriatum in rats: Multiple overlapping sensorimotor pathways. *Journal of Comparative Neurology* 499: 33–48.

Alloway, K D; Mutic, J J; Hoffer, Z S and Hoover, J E (2000). Overlapping corticostriatal projections from the rodent vibrissal representations in primary and secondary somatosensory cortex. *Journal of Comparative Neurology* 426: 51–67.

Alloway, K D; Mutic, J J and Hoover, J E (1998). Divergent corticostriatal projections from a single cortical column in the somatosensory cortex of rats. *Brain Research* 785: 341–346.

Alloway, K D; Smith, J B; Beauchemin, K J and Olson, M L (2009). Bilateral projections from rat MI whisker cortex to the neostriatum, thalamus, and claustrum: Forebrain circuits for modulating whisking behavior. *Journal of Comparative Neurology* 515: 548–564.

Alloway, K D; Zhang, M and Chakrabarti, S (2004). Septal columns in rodent barrel cortex: Functional circuits for modulating whisking behavior. *Journal of Comparative Neurology* 480: 299–309.

Brecht, M et al. (2004). Organization of rat vibrissa motor cortex and adjacent areas according to cytoarchitectonics, microstimulation, and intracellular stimulation of identified cells. *Journal of Comparative Neurology* 479: 360–373.

Brown, L L (1992). Somatotopic organization in rat striatum: Evidence for a combinatorial map. *Proceedings of the National Academy of Sciences of the United States of America* 89: 7403–7407.

Brown, L L; Smith, D M and Goldbloom, L M (1998). Organizing principles of cortical integration in the rat neostriatum: Corticostriate map of the body surface is an ordered lattice of curved laminae and radial points. *Journal of Comparative Neurology* 392: 468–488.

Carelli, R M and West, M O (1991). Representation of the body by single neurons in the dorsolateral striatum of the awake, unrestrained rat. *Journal of Comparative Neurology* 309: 231–249.

Carvell, G E and Simons, D J (1987). Thalamic and corticocortical connections of the second somatic sensory area of the mouse. *Journal of Comparative Neurology* 265: 409–427.

Chakrabarti, S and Alloway, K D (2006). Differential origin of projections from SI barrel cortex to the whisker representations in SII and MI. *Journal of Comparative Neurology* 498: 624–636.

Chakrabarti, S; Zhang, M and Alloway, K D (2008). MI neuronal responses to peripheral whisker stimulation: Relationship to neuronal activity in SI barrels and septa. *Journal of Neurophysiology* 100: 50–63.

Colechio, E M and Alloway, K D (in press). Bilateral topography of the cortical projections to the whisker and forepaw regions in rat motor cortex. *Brain Structure & Function.*

Cromwell, H C and Berridge, K C (1996). Implementation of action sequences by a neostriatal site: A lesion mapping study of grooming syntax. *The Journal of Neuroscience* 16: 3444–3458.
Crowne, D P; Novony, M F; Maier, S E and Vitols, R (1992). Effects of unilateral parietal lesions on spatial localization in the rat. *Behavioral Neuroscience* 106: 808–819.
Fabri, M and Burton, H (1991). Ipsilateral cortical connections of primary somatic sensory cortex in rats. *Journal of Comparative Neurology* 311: 405–424.
Flaherty, A W and Graybiel, A M (1991). Corticostriatal transformations in the primate somatosensory system. Projections from physiologically mapped body-part representations. *Journal of Neurophysiology* 66: 1249–1263.
Flaherty, A W and Graybiel, A M (1993). Two input systems for body representations in the primate striatal matrix: experimental evidence in the squirrel monkey. *The Journal of Neuroscience* 13: 1120–1137.
Gioanni, Y and Lamarche, M (1985). A reappraisal of rat motor cortex organization by intracortical microstimulation. *Brain Research* 344: 49–61.
Gundersen, V; Ottersen, O P and Storm-Mathisen, J (1996). Selective excitatory amino acid uptake in glutamatergic nerve terminals and in glia in the rat striatum: Quantitative electron microscopic immunocytochemistry of exogenous D-aspartate and endogenous glutamate and GABA. *European Journal of Neuroscience* 8: 758–765.
Hemelt, M W and Keller, A (2007). Superior sensation: superior colliculus participation in rat vibrissa system. *BMC Neuroscience* 8: 12.
Hoffer, Z S and Alloway, K D (2001). Organization of corticostriatal projections from the vibrissal representations in the primary motor and somatosensory cortical areas in rodents. *Journal of Comparative Neurology* 439: 87–103.
Hoover, J E; Hoffer, Z S and Alloway, K D (2003). Projections from primary somatosensory cortex to the neostriatum: The role of somatotopic continuity in corticostriatal convergence. *Journal of Neurophysiology* 89: 1576–1587.
Kesner, R P; Farnsworth, G and DiMattia, B V (1989). Double dissociation of egocentric and allocentric space following medial prefrontal and parietal cortex lesions in the rat. *Behavioral Neuroscience* 103: 956–961.
Kincaid, A E; Zheng, T and Wilson, C J (1998). Connectivity and convergence of single corticostriatal axons. *The Journal of Neuroscience* 18: 4722–4731.
Lee, T H; Alloway, K D and Kim, U (2011). Interconnected cortical networks between SI columns and posterior parietal cortex in rat. *Journal of Comparative Neurology* 519: 405–419.
Levesque, M; Charara, A; Gagnon, S; Parent, A and Deschenes, M (1996a). Corticostriatal projections from layer V cells are collaterals of long-range corticofugal axons. *Brain Research* 709: 311–315.
Levesque, M; Gagnon, S; Parent, A and Deschenes, M (1996b). Axonal arborizations of corticostriatal and corticothalamic fibers arising from the second somatosensory area in the rat. *Cerebral Cortex* 6: 759–770.
Mitchison, B; Martin, C J; Grant, R A and Prescott, T J (2007). Feedback control in active sensing: rat exploratory whisking is modulated by environmental contact. *Proceedings of the Royal Society of London B* 274: 1035–1041.
Mowery, T; Harrold, J and Alloway, K D (2011). Repeated whisker stimulation evokes invariant neuronal responses in the dorsolateral striatum of anesthetized rats: A potential correlate of sensorimotor habits. *Journal of Neurophysiology* 105: 2225–2238.
Ottersen, O P and Storm-Mathisen, J (1984). Glutamate- and GABA-containing neurons in the rat and mouse brain, as demonstrated with a new immunocytochemical technique. *Journal of Comparative Neurology* 229: 374–392.
Pidoux, M; Mahon, S; Deniau, J M and Charpier, S (2011). Integration and propagation of somatosensory responses in the corticostriatal pathway: An intracellular study in vivo. *Journal of Physiology* 589: 263–281.
Ramanathan, S; Hanley, J J; Deniau, J M and Bolam, J P (2002). Synaptic convergence of motor and somatosensory cortical afferents onto GABAergic interneurons in the rat striatum. *The Journal of Neuroscience* 22: 8158–8169.

Redgrave, P et al. (2010). Goal-directed and habitual control in the basal ganglia: Implications for Parkinson's disease. *Nature* 11: 760–772.

Reep, R L; Chandler, H C; King, V and Corwin, J V (1994). Rat posterior parietal cortex: Topography of corticocortical and thalamic connections. *Experimental Brain Research* 100: 67–84.

Reep, R L; Cheatwood, J L and Corwin, J V (2003). The associative striatum: organization of cortical projections to the dorsocentral striatum in rats. *Journal of Comparative Neurology* 467: 271–292.

Reep, R L and Corwin, J V (2009). Posterior parietal cortex as part of a neural network for directed attention in rats. *Neurobiology of Learning and Memory* 91: 104–113.

Reiner, A; Jiao, Y; Del Mar, N; Laverghetta, A V and Lei, W L (2003). Differential morphology of pyramidal tract-type and intratelencephalically projecting-type corticostriatal neurons and their intrastriatal terminals in rats. *Journal of Comparative Neurology* 457: 420–440.

Rice, F L (1995). Comparative aspects of barrel structure and development. In: E G Jones and I T Diamond (Eds.), *Cerebral Cortex, Vol. 11, The Barrel Cortex of Rodents* (pp. 1–75). New York: Plenum Press.

Schulz, J M et al. (2009). Short-latency activation of striatal spiny neurons via subcortical visual pathways. *The Journal of Neuroscience* 29: 6336–6347.

Selemon, L D and Goldman-Rakic, P S (1985). Longitudinal topography and interdigitation of corticostriatal projections in the rhesus monkey. *The Journal of Neuroscience* 5: 776–794.

Smith, J B; Mowery, T M and Alloway, K D (2012). Thalamic POm projections to the dorsolateral striatum of rats: Potential pathway for mediating stimulus-response associations for sensorimotor habits. *Journal of Neurophysiology* 108: 160–174.

Spreafico, R; Barbaresi, P; Weinberg, R J and Rustioni, A (1987). SII-projecting neurons in the rat thalamus: a single and double retrograde tracing study. *Somatosensory Research* 4: 359–375.

Tees, R C (1999). The effects of posterior parietal and posterior temporal cortical lesions on multimodal spatial and nonspatial competencies in rats. *Behavioural Brain Research* 106: 55–73.

Welker, C (1976). Receptive fields of barrels in the somatosensory cortex of the rat. *Journal of Comparative Neurology* 199: 205–219.

West, M O (1998). Anesthetics eliminate somatosensory-evoked discharges of neurons in the somatotopically organized sensorimotor striatum of the rat. *The Journal of Neuroscience* 18: 9055–9068.

Wilson, C J (1987). Morphology and synaptic connections of crossed corticostriatal neurons in the rat. *Journal of Comparative Neurology* 263: 567–580.

Wilson, C J (1995). The contribution of cortical neurons to the firing patterns of striatal spiny neurons. In: J C Houk, J L Davis and D G Beiser (Eds.), *Models of Information Processing in the Basal Ganglia* (pp. 29–50). Cambridge: MIT Press.

Woolsey, T A and van der Loos, H (1970). The structural organization of layer IV in the somatosensory region, SI, of mouse cerebral cortex. *Brain Research* 17: 205–242.

Wright, A K; Ramanathan, S and Arbuthnott, G W (2001). Identification of the source of the bilateral projection system from cortex to somatosensory neostriatum and an exploration of its physiological actions. *Neuroscience* 103: 87–96.

Wright, A K; Norrie, L; Ingham, C A; Hutton, E A M and Arbuthnott, G W (1999). Double anterograde tracing of outputs from adjacent "barrel columns" of rat somatosensory cortex. Neostriatal projection patterns and terminal ultrastructure. *Neuroscience* 88: 119–133.

Yeterian, E H and Van Hoesen, G W (1978). Cortico-striate projections in the rhesus monkey: The organization of certain cortico-caudate connections. *Brain Research* 139: 43–63.

Yin, H H; Knowlton, B J and Balleine, B W (2004). Lesions of the dorsolateral striatum preserve outcome expectancy but disrupt habit formation in instrumental learning. *European Journal of Neuroscience* 19: 181–189.

Yin, H H; Knowlton, B J and Balleine, B W (2006). Inactivation of dorsolateral striatum enhances sensitivity to changes in the action-outcome contingency in instrumental conditioning. *Behavioural Brain Research* 166: 189–196.

Vibrissal Midbrain Loops

Manuel Castro-Alamancos and Asaf Keller

Components of the Vibrissa Midbrain Network: Superior Colliculus

The midbrain consists of the tectum and the tegmentum. The tectum comprises the inferior colliculus and superior colliculus. The tegmentum contains several areas including the periaqueductal gray, mesencephalic reticular formation (aka, deep mesencephalic nucleus), cuneiform nucleus (also commonly referred in rodents as deep mesencephalic nucleus), laterodorsal tegmental nucleus (LDT), pedunculopontine tegmental nucleus (PPT), red nucleus, substantia nigra and ventral tegmental area. The midbrain also contains ascending and descending fiber bundles such as the cerebral peduncle, the medial lemniscus and the lateral lemniscus. Among these areas, the superior colliculus is well known to receive vibrissal sensory inputs from the trigeminal complex and to project to vibrissa motoneurons in the facial nucleus. Thus, we will focus on the superior colliculus below. Apart from the superior colliculus, it is worth mentioning that the lateral column of the periaqueductal gray (Beitz 1982) and the mesencephalic reticular formation (Veazey and Severin 1982) receive some trigeminal afferents. Also, the red nucleus, mesencephalic reticular formation and periaqueductal gray oculomotor nucleus project to vibrissa motoneurons (Hattox et al. 2002). Vibrissa motor cortex projects directly to vibrissa motorneurons (Grinevich et al. 2005) and also indirectly by innervating each of the midbrain regions that project to vibrissa motorneurons (Hattox et al. 2002).

M. Castro-Alamancos (✉)
Department of Neurobiology, Drexel University College of Medicine,
Philadelphia, PA, USA

A. Keller
Department of Anatomy and Neurobiology, University of Maryland
School of Medicine, Baltimore, USA

T.J. Prescott et al. (eds.), *Scholarpedia of Touch*, Scholarpedia,
DOI 10.2991/978-94-6239-133-8_51

The rat superior colliculus is divided into superficial, intermediate and deep layers. There are three superficial layers (I–III); the zonal layer (stratum zonale), the superficial gray layer (stratum griseum superficiale) and the optic layer (stratum opticum). There are two intermediate layers (IV–V); the intermediate gray layer (stratum griseum intermediale) and the intermediate white layer (stratum album intermediale). There are two deep layers (VI–VII); deep gray layer (stratum griseum profundum) and deep white layer (stratum album profundum). The superficial layers receive dense visual inputs from retina, primary visual cortex and cholinergic inputs from parabigeminal nucleus (PB), and project to deeper layers in the superior colliculus and to the thalamus. The intermediate layers receive direct somatosensory and auditory inputs, as well as inhibitory inputs from the substantia nigra pars reticulata and cholinergic inputs from the PPT and LDT. These sensory modalities become integrated with visual information arriving from the superficial layers, so that visual, auditory and somatosensory maps lie in spatial register across the collicular layers (Drager and Hubel 1975b; Stein et al. 1975).

Since this article is about vibrissa, we will focus on the intermediate and deep layers (aka, deeper layers). For reviews of the cell types, circuitry and functions of cells in the superficial layers of the superior colliculus the reader is referred to the following references (Stein 1981; May 2005). The deeper layers contain mostly multipolar cells of various types located in all layers and horizontal cells, which are GABAergic interneurons. Large multipolar cells whose dendrites cross several layers give rise to predorsal bundle fibers, a main output of the superior colliculus. Intrinsic collicular circuitry distributes excitatory as well as inhibitory collicular activity within and across layers and across major collicular subdivisions (Zhu and Lo 2000; Meredith and King 2004; Tardif et al. 2005).

Vibrissal Superior Colliculus Inputs

There are two sources of vibrissa inputs that reach the superior colliculus: a direct input from the trigeminal complex (trigeminotectal) and an indirect input from the barrel cortex (corticotectal). In addition, there are other pathways that may regulate vibrissa inputs (see Figure 51.1).

- **Trigeminotectal pathway**: The trigeminal complex projects to the deeper layers of the superior colliculus, particularly the intermediate layers. Trigeminal complex is divided into principal (PrV) and spinal (SpV) nuclear divisions. The spinal nucleus is further divided into three subnuclei: subnucleus interpolaris (SpVi), oralis (SpVo) and caudalis (SpVc). Between 10 and 30 % of PrV cells, which have large multipolar somata with expansive dendritic trees and multi-whisker receptive fields, project to the superior colliculus, and also to POm and zona incerta (ZI) (Killackey and Erzurumlu 1981; Huerta et al. 1983; Bruce et al. 1987; Rhoades et al. 1989; Veinante and Deschenes 1999). Likewise, cells in SpVi and SpVo, which have multiwhisker receptive fields, project to the

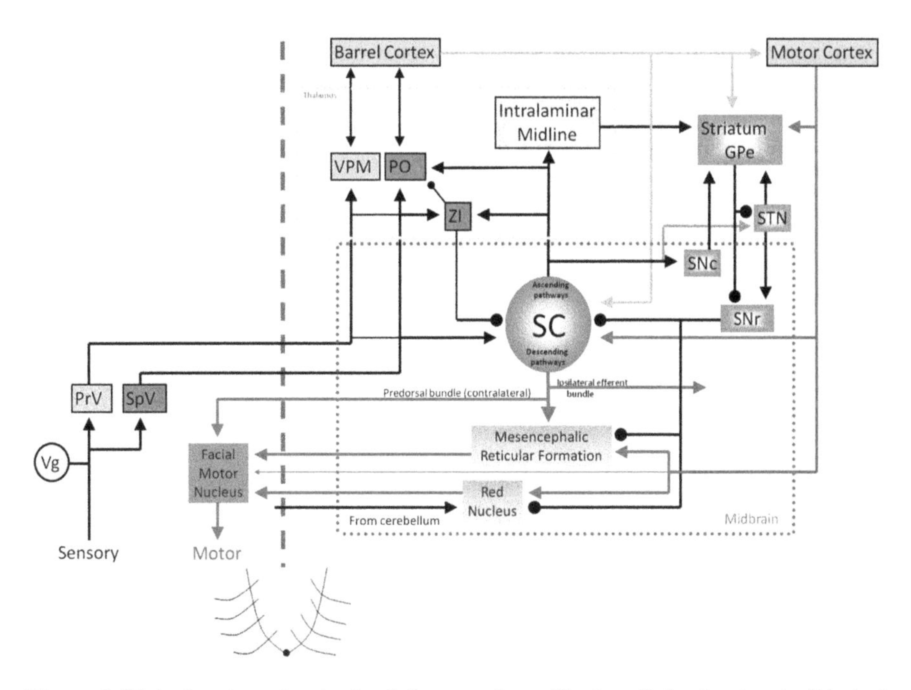

Figure 1 Main inputs and outputs of the superior colliculus vibrissal network. Principal (PrV) and spinal (SpV) trigeminal complex nuclei form pathways that project to the thalamus (VPM/PO) and superior colliculus (SC). Moreover, the SC projects via ascending pathways to several thalamic nuclei, and thalamic activity is relayed back to the SC via the cortex and the basal ganglia. The SC forms two major descending pathways. The predorsal bundle targets the contralateral facial nucleus directly, while the ipsilateral efferent bundle targets the facial nucleus indirectly via the mesencephalic reticular formation. STN, subthalamic nucleus; SNr, SNc, substantia nigra pars reticulata and compacta; GPe, globus pallidus external

superior colliculus; these cells project also to POm and zona incerta. In contrast, SpVc has few projections to the superior colliculus. Thus, the trigeminotectal pathway is formed by cells with multiwhisker receptive fields located in PrV, SpVi and SpVo.

- **Corticotectal pathways**: Layer V cells in both the vibrissa motor cortex (Miyashita et al. 1994) and the barrel cortex (Wise and Jones 1977) project to the deeper layers of the superior colliculus.
- **Nigrotectal pathway**: GABAergic cells in the ipsilateral substantia nigra pars reticulata project to the deeper layers of the superior colliculus where they directly contact cells that form the predorsal bundle (Harting et al. 1988; Redgrave et al. 1992; Kaneda et al. 2008). There is also a smaller projection from the contralateral substantia nigra pars reticulata, and sensory (visual) inputs have opposing effects on crossed and uncrossed nigrotectal cells (Jiang et al. 2003).
- **Incertotectal pathway**: GABAergic cells from the zona incerta project to the superior colliculus (Kolmac et al. 1998). Interestingly, zona incerta cells are

contacted by trigeminal afferents that provide vibrissa information to the superior colliculus, and posterior thalamus (Lavallee et al. 2005). The superior colliculus also projects to the zona incerta (see below).

- **Neuromodulatory pathways**: Cholinergic cells from PPT and LDT (Beninato and Spencer 1986; Krauthamer et al. 1995; Billet et al. 1999), noradrenergic cells from the locus coeruleus (Swanson and Hartman 1975), and serotoninergic cells from the dorsal raphe nucleus (Steinbusch 1981; Beitz et al. 1986) project to the superior colliculus.

Vibrissal Superior Colliculus Outputs

The main efferents of the superior colliculus can be divided into ascending and descending projections (see Figure 51.1).

Ascending Output Pathways

- **Tectothalamic pathways**: A large number of thalamic nuclei are contacted by the superior colliculus, including nuclei in the posterior, lateral, geniculate, intralaminar and midline groups (Roger and Cadusseau 1984; Yamasaki et al. 1986; Linke et al. 1999; Krout et al. 2001). The rostral sector of the posterior nucleus of the thalamus (POm), which is driven by whisker stimulation, also receives afferents from the superior colliculus (Roger and Cadusseau 1984). In turn, POm projects to the barrel cortex providing a loop back to the superior colliculus via corticotectal pathways. The other major superior colliculus projections to the thalamus reach the intralaminar nuclei (i.e. particularly the parafascicularis nucleus, followed by central lateral, paracentral and central medial) and midline nuclei (Yamasaki et al. 1986; Grunwerg and Krauthamer 1990; Krout et al. 2001). In turn, cells in the intralaminar nuclei project to the striatum, giving rise to a loop through the basal ganglia back to the superior colliculus (McHaffie et al. 2005).
- **Tectoincertal pathway**: The superior colliculus projects to the zona incerta (Kolmac et al. 1998) and dorsal aspect of the nucleus reticularis of the thalamus (nRt) (Kolmac and Mitrofanis 1998). The tectoincertal pathway targets a different part of the zona incerta (dorsal) than the origin of the incertotectal pathway (ventral). The GABAergic cells in nRt and zona incerta are well known to modulate vibrissal responses in dorsal thalamus (Trageser and Keller 2004; Hirata et al. 2006).
- **Tectonigral pathway**: The superior colliculus projects to dopaminergic cells in the substantia nigra pars compacta (Coizet et al. 2003, 2007; Comoli et al. 2003). Interestingly, through this pathway sensory stimuli may be identified as

unpredicted and salient (Redgrave and Gurney 2006). The superior colliculus also projects to the subthalamic nucleus in the basal ganglia (Coizet et al. 2009).

- **Pretectal pathways**: The superior colliculus provides an extensive terminal field to the nucleus of the optic tract and the posterior pretectal nucleus, mainly from the superficial layers. In turn, the pretectum provides feedback connections back to the superior colliculus (Taylor et al. 1986).
- **Tectotectal commissural pathway**: This pathway originates in the deeper layers and connects the two superior colliculi on both sides of the brain (Sahibzada et al. 1987). These fibers may mediate mutually suppressive effects on the output of the contralateral colliculus to control competing responses (Sprague 1966).

Descending Output Pathways

The main descending output pathways that originate in separate subregions of the deeper layers and mediate approach and escape responses (Redgrave et al. 1987a, b; Dean et al. 1989; Westby et al. 1990; Redgrave et al. 1993) are:

- **Predorsal bundle pathways**: This is a crossed descending projection that contacts targets in the contralateral brainstem, including precerebellar nuclei (e.g. reticularis tegmenti pontis, inferior olive), and spinal cord. Approach responses toward salient stimuli are mediated by this pathway. In addition, many of the cells that give rise to this pathway (>50 %) are driven by whisker stimulation (Westby et al. 1990).
- **Ipsilateral efferent bundle pathways**: This is an ipsilateral descending projection with terminations in the periaqueductal gray, cuneiform nucleus, lateral pons and ventral pontine/medullary reticular formation. Escape behaviors rely on this ipsilateral efferent pathway. Few cells ($\sim$14 %) that give rise to this pathway respond to whisker stimulation (Westby et al. 1990).

Superior Colliculus Activity Driven by Passive Touch

Several studies have described whisker evoked responses in the superior colliculus of the rat (Stein and Dixon 1979; Fujikado et al. 1981; McHaffie et al. 1989; Grunwerg and Krauthamer 1990; Cohen and Castro-Alamancos 2007; Hemelt and Keller 2007; Cohen et al. 2008), mouse (Drager and Hubel 1975a) and hamster (Chalupa and Rhoades 1977; Finlay et al. 1978; Stein and Dixon 1979; Rhoades et al. 1983, 1987; Larson et al. 1987). Together these studies indicate that whisker-sensitive superior colliculus cells are rapidly adapting cells with large receptive fields and display angular selectivity.

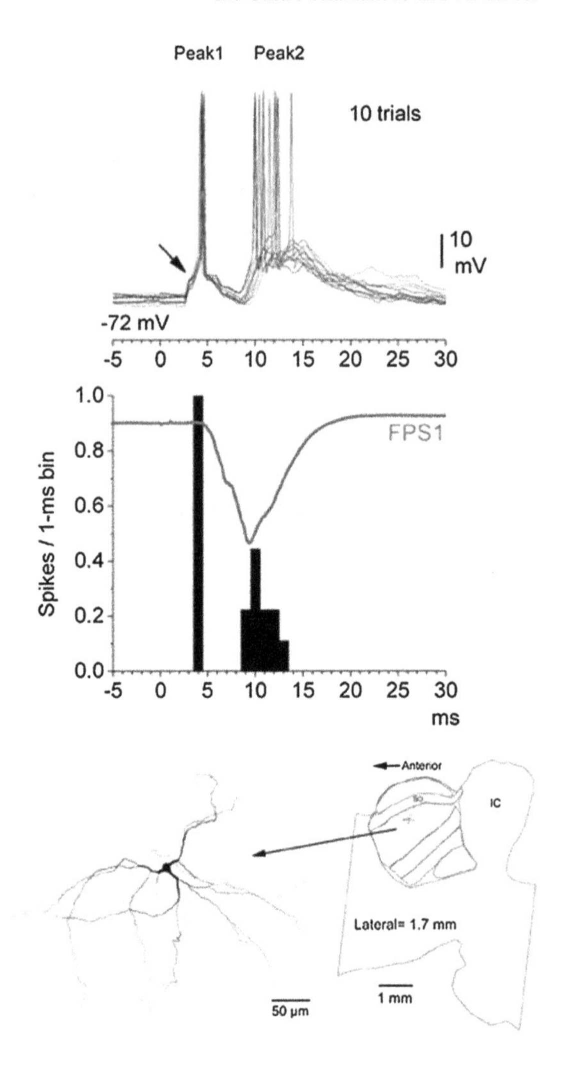

Figure 2 Intracellular recording from a superior colliculus cell driven by passive multiwhisker stimulation. Note the peak1 and peak2 responses. Also shown is an average FP response recorded simultaneously from the barrel cortex, and a PSTH of the SC spikes. The *lower panel* shows reconstruction of a whisker-sensitive multipolar cell after recording [see (Cohen et al. 2008) for details]

Although superior colliculus cells respond relatively effectively to single whiskers, including the principal whisker (PW) and several adjacent whiskers (AWs), cells respond much more robustly to simultaneous, or nearly simultaneous, wide-field (multiwhisker) stimuli (Cohen et al. 2008) (see Figure 51.2). The enhanced multiwhisker response is temporally stereotyped, consisting of two short latency excitatory peaks (peak1 and peak2) separated by ~10 ms that have different characteristics and origins. These properties make superior colliculus cells highly sensitive to the degree of temporal dispersion of stimulated whiskers during multiwhisker stimulation. The spikes evoked during peak1 show very little jitter and are driven by direct trigeminotectal excitatory postsynaptic potentials (EPSPs) from different single whiskers that sum to produce a robust multiwhisker response. The spikes evoked during peak2 are much more dispersed and are driven by EPSPs returning to the superior colliculus from the barrel cortex that ride on top of an evoked inhibitory postsynaptic potential (IPSP). Consistent with their cortical origin, peak2 responses are highly dependent on the level of forebrain activation. These properties make superior colliculus cells highly sensitive to the order and temporal dispersion of multiwhisker stimulation. Cells are most responsive when the PW is stimulated first and AWs follow at short intervals between 0 and 10 ms. However, when the AWs are stimulated first and with an interval above 2 ms, the cells do not respond at all to the PW and this suppression starts to recover at intervals above 50 ms (Cohen et al. 2008).

Populations of superior colliculus cells are tuned to respond robustly when their PW contacts an object first and other whiskers follow within <10 ms, and to remain silent when their PW contacts an object >2 ms after other whiskers. These response characteristics are likely useful to signal contact with an object as rats navigate the environment because a selective population of superior colliculus cells representing the first contacted whiskers will discharge. This may well serve as a signal to orient toward the contact location, which (as discussed below) is a putative function of the superior colliculus.

Superior Colliculus Activity Driven by Whisking Movement, Active Touch and Texture

Rats sense the environment through rhythmic vibrissa protractions, called active whisking, which can be simulated in anesthetized rats by electrically stimulating the facial motor nerve (Zucker and Welker 1969; Szwed et al. 2003). Simultaneous recordings from the barrel cortex and superior colliculus have shown that, similar to passive touch, whisking movement is signaled during the onset of the whisker protraction by short-latency responses in barrel cortex that drive corticotectal responses in superior colliculus, and all these responses show robust adaptation with increases in whisking frequency (Bezdudnaya and Castro-Alamancos 2011). Active touch and texture are signaled by longer latency responses, first in superior colliculus during the rising phase of the protraction, likely driven by trigeminotectal inputs, and later in barrel cortex by the falling phase of the protraction. Thus, superior colliculus can decode whisking movement, active touch and texture.

Vibrissa Movements Driven by Superior Colliculus

The superior colliculus sends dense and direct projections to the rat facial nucleus, where the vibrissa motor neurons are located (Miyashita and Mori 1995; Hattox et al. 2002), and stimulation of the superior colliculus produces movements of the vibrissae (McHaffie and Stein 1982; Hemelt and Keller 2008). Microstimulation throughout the deeper layers of the superior colliculus produces sustained ipsilateral (ventral sites), contralateral (dorsal sites) or bilateral (intermediate sites) vibrissa protractions, frequently outlasting the stimulus duration (see movie clip at http://jn.physiology.org/content/100/3/1245/suppl/DC1). Thus, the deeper layers appear to preferentially target the ipsilateral efferent bundle drive, while more superficial layers target the contralateral predorsal bundle. In addition, tecto-facial neurons rarely (9 %) receive direct trigeminal inputs, so that the colliculus does not seem to act as a simple closed sensorimotor loop (Hemelt and Keller 2008).

The movements elicited by stimulating the superior colliculus are very different from the rhythmic movements evoked by identical stimulation in the motor cortex (Hemelt and Keller 2008). For instance, the motor cortex evokes rhythmic protractions that are much smaller than the sustained protractions produced by superior colliculus. In addition, superior colliculus evoked movements have much shorter latencies than those evoked from the motor cortex (8 ms vs. >30 ms). Motor cortex appears to regulate whisking frequency, acting through a brain stem central patter generator, while the superior colliculus may be controlling the amplitude and set point of whisking.

Behavioral State and Neuromodulator Influences in Superior Colliculus

The spontaneous firing of vibrissa-sensitive cells in the superior colliculus is dependent on the behavioral state of the animal (see Figure 51.3). Firing is much higher during active exploration, which includes active whisking, and paradoxical or rapid eye movement sleep (REM) compared to awake immobility and slow wave sleep. Thus, according to firing rate, the superior colliculus is activated during active exploration and REM sleep, and deactivated during awake immobility and slow wave sleep (Cohen and Castro-Alamancos 2010b). The superior colliculus appears to come online during active exploratory states and rapidly goes offline during awake immobile periods. Particularly interesting are periods during awake immobility when superior colliculus firing ceases completely (Cohen and Castro-Alamancos 2010b).

Firing in superior colliculus is not eliminated during sleep. During slow wave sleep, firing seems to be driven by slow oscillations in neocortex via the corticotectal pathway. This is expected because corticotectal influence is strongest during the deactivated mode (Cohen et al. 2008). The driver for REM related activity is not known, but cannot be movement or sensory input such as during active exploration. One possibility is that superior colliculus during REM is driven by the actions of specific neuromodulator systems. For example, activity in cholinergic brainstem nuclei is significant during REM sleep (el Mansari et al. 1989), these nuclei innervate the superior colliculus (Beninato and Spencer 1986; Krauthamer et al. 1995; Billet et al. 1999), and acetylcholine activates some superior colliculus

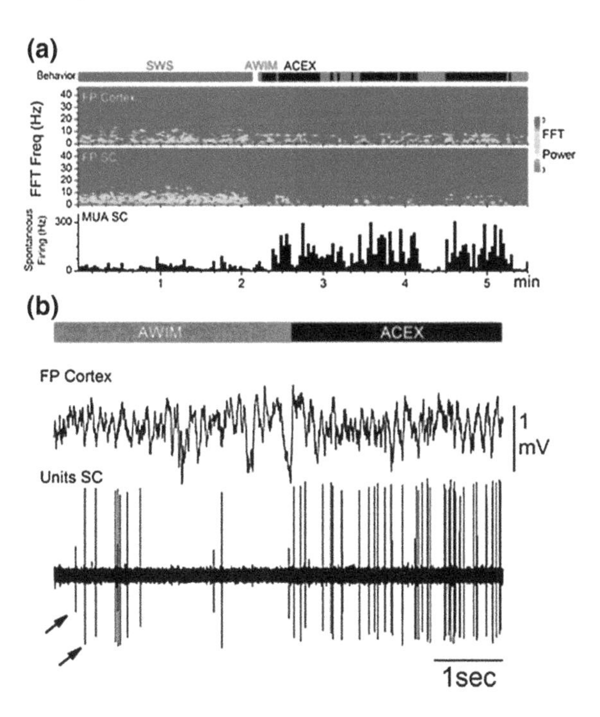

Figure 3 Neural activity recorded in barrel cortex (FP) and superior colliculus (FP and single-units) from freely behaving rats during different behavioral states. a Shows continuous recordings (5.5 min) as the animal transitions between slow wave sleep (SWS), active exploration (ACEX), and awake immobility (AWIM) states. The *color contour plots* depict FFT power spectrums of the spontaneous FP activity. Also shown is the spontaneous firing (Hz) measured in the superior colliculus. Note the large increase in firing during ACEX. **b** Shows typical single-unit raw traces from another animal during a transition between AWIM and ACEX (two well discernible units are marked by *arrows*). [see (Cohen and Castro-Alamancos 2010b) for details]

neurons in vitro (Li et al. 2004; Sooksawate and Isa 2006). Thus, cholinergic activation likely causes enhanced spontaneous firing during REM sleep.

Sensory responses in superior colliculus are also dependent on behavioral state. In particular, corticotectal responses (peak2) evoked by whisker stimuli are larger during the deactivated mode than during the activated mode, just like the cortical responses on which they depend (Cohen et al. 2008; Cohen and Castro-Alamancos 2010b). However, longer latency responses (peak3; of unknown origin) are typically larger during the activated mode. Sensory responses that are driven by different neural circuits are regulated differently by behavioral state. The stronger corticotectal responses of whisker-sensitive cells in the intermediate layers of the superior colliculus during deactivated modes may be useful as a powerful alerting stimulus in an animal that is sleeping, drowsy, or inattentive, and an unknown moving object or animal makes contact with its whiskers. Interestingly, superior colliculus responses evoked by air-puff stimuli are largest when the animal is quiescent and orients to the stimulus (Cohen and Castro-Alamancos 2010b).

As already mentioned, neuromodulators are likely to drive the spontaneous firing and sensory response changes related to specific states. Although, cholinergic, noradrenergic and 5-HT neurons project to the superior colliculus, the effects of these neuromodulators on vibrissa-sensitive superior colliculus cells are unknown.

Functional Role of Superior Colliculus Vibrissa Networks

The superior colliculus is an early sensory hub well suited to mediate sensory detection of stimuli that require immediate action (Sprague and Meikle 1965; Schneider 1969; Sparks 1986; Dean et al. 1989; Westby et al. 1990; Redgrave et al. 1993; Stein and Meredith 1993; McHaffie et al. 2005; Redgrave and Gurney 2006; Cohen and Castro-Alamancos 2007). Considering an evolution perspective, in lower vertebrates without significant cortical organization (e.g. amphibians) the superior colliculus acts like the highest brain center controlling escape and approach. With the evolution of neocortex these and more complex functions incorporated the neocortex but the superior colliculus still retains many of these ancient functions. Below we discuss how the superior colliculus basic functions may be relevant to the vibrissal system.

Detect Sensory Stimuli

As discussed above, cells in the superior colliculus are highly sensitive to passive and active touch. In behaving animals, the trigeminotectal sensory pathway is capable of independently (i.e. in the absence of thalamocortical networks) detecting relevant (i.e. that signal impending danger) sensory stimuli applied to the whisker pad as long as the stimulus is sufficiently salient (Cohen and Castro-Alamancos 2007). However, the trigeminotectal pathway must work in synergy with thalamocortical networks to detect stimuli of low saliency (Cohen and Castro-Alamancos 2010a). Low intensity stimuli may not be detectable by the superior colliculus alone because the neural responses they evoke are weak and dispersed, and this sparse code is further suppressed when corticotectal inputs driven by the stimulus are absent (i.e. during thalamocortical inactivation) (Cohen et al. 2008). Therefore, the superior colliculus may require corticotectal inputs driven by the sensory stimulus to enhance direct trigeminotectal sensory responses for successful detection of low salience stimuli.

Whisker sensitive cells in the intermediate layers of the superior colliculus respond most effectively to multiwhisker stimuli when the PW is stimulated first and other whiskers follow shortly within 10 ms (Cohen et al. 2008). This means that only the cells representing the first stimulated whiskers will fire and the cells representing the succeeding whiskers will be inhibited. Such an arrangement suggests a population code signaling the whiskers that first made contact. This is likely very useful during navigation and exploration in order to detect the presence of objects.

Gate Orienting Responses to Detected Sensory Stimuli Depending on the State of the Animal and the Significance of the Stimulus

After detection of initial contact with a salient stimulus, a primary function of the whisker sensitive cells in the intermediate layers of the superior colliculus may be to gate the induction of appropriate orienting responses depending on the level of alertness of the animal. If the animal is already alert, an initial contact does not usually elicit a strong orienting response. However, if the animal is quiescent or inattentive, an initial contact with a salient stimulus usually elicits strong orienting. The superior colliculus is well known to be involved in orienting responses to innocuous and painful stimuli from a wide range of modalities, including somatosensory, auditory and visual (Sprague and Meikle 1965; Wurtz and Albano 1980; Meredith and Stein 1985; Sparks 1986; Dean et al. 1989; Stein and Meredith 1993; Stein 1998).

As already mentioned, whisker sensitive cells in the intermediate layers elicit highly stereotyped peak1 and peak2 responses in close succession when stimulated by salient stimuli, such as near-simultaneous multiwhisker stimulation. The occurrence of successive spikes separated by short intervals has strong impact on the target neurons due to temporal synaptic summation. The interesting aspect of the superior colliculus response is that the first and successive spikes are independently regulated. In particular, peak2 depends on cortical feedback which is strongly regulated by behavioral state (Castro-Alamancos 2004a, b). The strongest output of whisker sensitive cells in the intermediate layers of the superior colliculus occurs for nearly simultaneous multiwhisker contacts during quiescent states (Cohen et al. 2008; Cohen and Castro-Alamancos 2010b). This is likely useful as a powerful alerting stimulus in an animal that is sleeping, drowsy or inattentive, and an unknown moving object or animal makes contact with its whiskers. Since the target of these cells in deeper layers and in the brainstem drive orienting responses (Redgrave et al. 1987a; Dean et al. 1989; Westby et al. 1990), such a powerful alerting output makes good functional sense. The **two main types of behaviors gated by the superior colliculus** are to:

- **Approach (move toward) sensory stimuli**: Orienting motor behaviors elicited by the superior colliculus occur via the crossed descending projections from the deeper layers (Redgrave et al. 1987a; Dean et al. 1989; Westby et al. 1990; Redgrave et al. 1993). Moreover, the ability of the superior colliculus to control the direction and speed of eye (McHaffie and Stein 1982), head (Dean et al. 1986), and whisker movements (Hemelt and Keller 2008) is particularly germane in this orienting role. Whisker-responsive cells in the intermediate layers of the superior colliculus are associated with the contralaterally projecting predorsal bundle (Westby et al. 1990), which mediates approach movements towards novel stimuli (i.e. orienting responses) (Dean et al. 1989). Enhancement of sensory responsiveness through this pathway during orienting responses makes functional sense; the stronger neural activity during overt orienting responses may well reflect the neural drive for the orienting responses (Cohen and Castro-Alamancos 2010b).

- **Escape (move away) sensory stimuli**: Stimulation of the superior colliculus produces defensive behaviors, such as escape responses (Bandler et al. 1985; Dean et al. 1988, 1989; Brandao et al. 1994, 2003). Cells in deeper layers also respond to noxious stimuli (Stein and Dixon 1979; McHaffie et al. 1989; Redgrave et al. 1996a) and certain nocifensive reactions depend on the integrity of the superior colliculus (Redgrave et al. 1996b; Wang and Redgrave 1997; McHaffie et al. 2002). Interestingly, whisker and nociceptive face inputs converge in the superior colliculus and may control face orientation and withdrawal responses during active whisking exploration (McHaffie et al. 1989), which is critical for navigation in rodents. Escape responses elicited by fear also involve the superior colliculus (Cohen and Castro-Alamancos 2007, 2010a, c); rats use the superior colliculus to detect a whisker pad stimulus that elicits fear because of its association with an aversive event. In essence, the superior colliculus is a site where sensory inputs can directly control behavior, directing the animal towards objects of interest and away from objects that might pose a threat.

Activate (Wake-up) the Forebrain in Response to Sensory Stimuli

The superior colliculus responses driven by sensory stimulation may also serve to trigger forebrain activation in quiescent animals by impacting on neuromodulatory systems in the midbrain and brainstem that cause cortical activation (Castro-Alamancos 2004b). These neuromodulatory nuclei are well known targets of superior colliculus cells (Redgrave et al. 1987b; Dean et al. 1989; Redgrave et al. 1993).

In summary, whisker-sensitive cells in the intermediate layers are excellent detectors of initial whisker contact, and their output (i.e. the ability to drive target cells) is independently regulated by converging trigeminal synaptic inputs and cortical feedback. Because spikes driven by cortical feedback are gated by behavioral state, they provide a powerful mechanism for driving target cells depending on the level of arousal. This mechanism may serve to gate orienting responses and forebrain activation in quiescent animals. The main role of the superior colliculus may be to detect novel or salient sensory stimuli and to elicit an appropriate response to it (approach, ignore or escape). The superior colliculus can serve as an early relay station for rapid detection of sensory signals that have gained behavioral significance through learning and that call for immediate action.

References

Bandler, R; Depaulis, A and Vergnes, M (1985). Identification of midbrain neurones mediating defensive behaviour in the rat by microinjections of excitatory amino acids. *Behavioural Brain Research* 15: 107–119.

Beitz, A J (1982). The organization of afferent projections to the midbrain periaqueductal gray of the rat. *Neuroscience* 7: 133–159.

Beitz, A J; Clements, J R; Mullett, M A and Ecklund, L J (1986). Differential origin of brainstem serotoninergic projections to the midbrain periaqueductal gray and superior colliculus of the rat. *Journal of Comparative Neurology* 250: 498–509.

Beninato, M and Spencer, R F (1986). A cholinergic projection to the rat superior colliculus demonstrated by retrograde transport of horseradish peroxidase and choline acetyltransferase immunohistochemistry. *Journal of Comparative Neurology* 253: 525–538.

Bezdudnaya, T and Castro-Alamancos, M A (2011). Active touch and texture sensitive cells in the superior colliculus during whisking. *Journal of Neurophysiology* (in press).

Billet, S; Cant, N B and Hall, W C (1999). Cholinergic projections to the visual thalamus and superior colliculus. *Brain Research* 847: 121–123.

Brandao, M L; Troncoso, A C; Souza Silva, M A and Huston, J P (2003). The relevance of neuronal substrates of defense in the midbrain tectum to anxiety and stress: Empirical and conceptual considerations. *European Journal of Pharmacology* 463: 225–233.

Brandao, M L; Cardoso, S H; Melo, L L; Motta, V and Coimbra, N C (1994). Neural substrate of defensive behavior in the midbrain tectum. *Neuroscience and Biobehavioral Reviews* 18: 339–346.

Bruce, L L; McHaffie, J G and Stein, B E (1987). The organization of trigeminotectal and trigeminothalamic neurons in rodents: A double-labeling study with fluorescent dyes. *Journal of Comparative Neurology* 262: 315–330.

Castro-Alamancos, M A (2004). Absence of rapid sensory adaptation in neocortex during information processing states. *Neuron* 41: 455–464.

Castro-Alamancos, M A (2004). Dynamics of sensory thalamocortical synaptic networks during information processing states. *Progress in Neurobiology* 74: 213–247.

Chalupa, L M and Rhoades R W (1977). Responses of visual, somatosensory, and auditory neurones in the golden hamster's superior colliculus. *Journal of Physiology* 270: 595–626.

Cohen, J D and Castro-Alamancos, M A (2007). Early sensory pathways for detection of fearful conditioned stimuli: Tectal and thalamic relays. *The Journal of Neuroscience* 27: 7762–7776.

Cohen, J D and Castro-Alamancos, M A (2010). Detection of low salience whisker stimuli requires synergy of tectal and thalamic sensory relays. *The Journal of Neuroscience* 30: 2245–2256.

Cohen, J D and Castro-Alamancos, M A (2010). Behavioral state dependency of neural activity and sensory (whisker) responses in superior colliculus. *Journal of Neurophysiology* 104: 1661–1672.

Cohen, J D and Castro-Alamancos, M A (2010). Neural correlates of active avoidance behavior in superior colliculus. *The Journal of Neuroscience* 30: 8502–8511.

Cohen, J D; Hirata, A and Castro-Alamancos, M A (2008). Vibrissa sensation in superior colliculus: wide-field sensitivity and state-dependent cortical feedback. *The Journal of Neuroscience* 28: 11205–11220.

Coizet, V; Overton, P G and Redgrave, P (2007). Collateralization of the tectonigral projection with other major output pathways of superior colliculus in the rat. *Journal of Comparative Neurology* 500: 1034–1049.

Coizet, V; Comoli, E; Westby, G W and Redgrave, P (2003). Phasic activation of substantia nigra and the ventral tegmental area by chemical stimulation of the superior colliculus: An electrophysiological investigation in the rat. *European Journal of Neuroscience* 17: 28–40.

Coizet, V et al. (2009). Short-latency visual input to the subthalamic nucleus is provided by the midbrain superior colliculus. *The Journal of Neuroscience* 29: 5701–5709.

Comoli, E et al.(2003). A direct projection from superior colliculus to substantia nigra for detecting salient visual events. *Nature Neuroscience* 6: 974–980.

Dean, P; Mitchell, I J and Redgrave, P (1988). Responses resembling defensive behaviour produced by microinjection of glutamate into superior colliculus of rats. *Neuroscience* 24: 501–510.

Dean, P; Redgrave, P and Westby, G W (1989). Event or emergency? Two response systems in the mammalian superior colliculus. *Trends in Neurosciences* 12: 137–147.

Dean, P; Redgrave, P; Sahibzada, N and Tsuji, K (1986). Head and body movements produced by electrical stimulation of superior colliculus in rats: Effects of interruption of crossed tectoreticulospinal pathway. *Neuroscience* 19: 367–380.

Drager, U C and Hubel, D H (1975). Responses to visual stimulation and relationship between visual, auditory, and somatosensory inputs in mouse superior colliculus. *Journal of Neurophysiology* 38: 690–713.

Drager, U C and Hubel, D H (1975). Physiology of visual cells in mouse superior colliculus and correlation with somatosensory and auditory input. *Nature* 253: 203–204.

el Mansari, M; Sakai, K and Jouvet, M (1989). Unitary characteristics of presumptive cholinergic tegmental neurons during the sleep-waking cycle in freely moving cats. *Experimental Brain Research* 76: 519–529.

Finlay, B L; Schneps, S E; Wilson, K G and Schneider, G E (1978). Topography of visual and somatosensory projections to the superior colliculus of the golden hamster. *Brain Research* 142: 223–235.

Fujikado, T; Fukuda, Y and Iwama, K (1981). Two pathways from the facial skin to the superior colliculus in the rat. *Brain Research* 212: 131–135.

Grinevich, V; Brecht, M and Osten, P (2005). Monosynaptic pathway from rat vibrissa motor cortex to facial motor neurons revealed by lentivirus-based axonal tracing. *The Journal of Neuroscience* 25: 8250–8258.

Grunwerg, B S and Krauthamer, G M (1990). Vibrissa-responsive neurons of the superior colliculus that project to the intralaminar thalamus of the rat. *Neuroscience Letters* 111: 23–27.

Harting, J K; Huerta, M F; Hashikawa, T; Weber, J T and Van Lieshout, D P (1988). Neuroanatomical studies of the nigrotectal projection in the cat. *Journal of Comparative Neurology* 278: 615–631.

Hattox, A M; Priest, C A and Keller, A (2002). Functional circuitry involved in the regulation of whisker movements. *Journal of Comparative Neurology* 442: 266–276.

Hemelt, M E and Keller, A (2007). Superior sensation: superior colliculus participation in rat vibrissa system. *BMC Neuroscience* 8: 12.

Hemelt, M E and Keller, A (2008). Superior colliculus control of vibrissa movements. *Journal of Neurophysiology* 100: 1245–1254.

Hirata, A; Aguilar, J and Castro-Alamancos, M A (2006). Noradrenergic activation amplifies bottom-up and top-down signal-to-noise ratios in sensory thalamus. *The Journal of Neuroscience* 26: 4426–4436.

Huerta, M F; Frankfurter, A and Harting, J K (1983). Studies of the principal sensory and spinal trigeminal nuclei of the rat: Projections to the superior colliculus, inferior olive, and cerebellum. *Journal of Comparative Neurology* 220: 147–167.

Jiang, H; Stein, B E and McHaffie, J G (2003). Opposing basal ganglia processes shape midbrain visuomotor activity bilaterally. *Nature* 423: 982–986.

Kaneda, K; Isa, K; Yanagawa, Y and Isa, T (2008). Nigral inhibition of GABAergic neurons in mouse superior colliculus. *The Journal of Neuroscience* 28: 11071–11078.
Killackey, H P and Erzurumlu, R S (1981). Trigeminal projections to the superior colliculus of the rat. *Journal of Comparative Neurology* 201: 221–242.
Kolmac, C I and Mitrofanis, J (1998). Patterns of brainstem projection to the thalamic reticular nucleus. *Journal of Comparative Neurology* 396: 531–543.
Kolmac, C I; Power, B D and Mitrofanis, J (1998). Patterns of connections between zona incerta and brainstem in rats. *Journal of Comparative Neurology* 396: 544–555.
Krauthamer, G M; Grunwerg, B S and Krein, H (1995). Putative cholinergic neurons of the pedunculopontine tegmental nucleus projecting to the superior colliculus consist of sensory responsive and unresponsive populations which are functionally distinct from other mesopontine neurons. *Neuroscience* 69: 507–517.
Krout, K E; Loewy, A D; Westby, G W and Redgrave, P (2001). Superior colliculus projections to midline and intralaminar thalamic nuclei of the rat. *Journal of Comparative Neurology* 431: 198–216.
Larson, M A; McHaffie, J G and Stein, B E (1987). Response properties of nociceptive and low-threshold mechanoreceptive neurons in the hamster superior colliculus. *The Journal of Neuroscience* 7: 547–564.
Lavallee, P et al. (2005). Feedforward inhibitory control of sensory information in higher-order thalamic nuclei. *The Journal of Neuroscience* 25: 7489–7498.
Li, F; Endo, T and Isa, T (2004). Presynaptic muscarinic acetylcholine receptors suppress GABAergic synaptic transmission in the intermediate grey layer of mouse superior colliculus. *European Journal of Neuroscience* 20: 2079–2088.
Linke, R; De Lima, A D; Schwegler, H and Pape, H C (1999). Direct synaptic connections of axons from superior colliculus with identified thalamo-amygdaloid projection neurons in the rat: Possible substrates of a subcortical visual pathway to the amygdala. *Journal of Comparative Neurology* 403: 158–170.
May, P J (2005). The mammalian superior colliculus: laminar structure and connections. *Progress in Brain Research* 151: 321–378.
McHaffie, J G and Stein, B E (1982). Eye movements evoked by electrical stimulation in the superior colliculus of rats and hamsters. *Brain Research* 247: 243–253.
McHaffie, J G; Kao, C Q and Stein, B E (1989). Nociceptive neurons in rat superior colliculus: Response properties, topography, and functional implications. *Journal of Neurophysiology* 62: 510–525.
McHaffie, J G; Wang, S; Walton, N; Stein, B E and Redgrave, P (2002). Covariant maturation of nocifensive oral behaviour and c-fos expression in rat superior colliculus. *Neuroscience* 109: 597–607.
McHaffie, J G; Stanford, T R; Stein, B E; Coizet, V and Redgrave, P (2005). Subcortical loops through the basal ganglia. *Trends in Neurosciences* 28: 401–407.
Meredith, M A and Stein, B E (1985). Descending efferents from the superior colliculus relay integrated multisensory information. *Science* 227: 657–659.
Meredith, M A and King, A J (2004). Spatial distribution of functional superficial-deep connections in the adult ferret superior colliculus. *Neuroscience* 128: 861–870.
Miyashita, E and Mori, S (1995). The superior colliculus relays signals descending from the vibrissal motor cortex to the facial nerve nucleus in the rat. *Neuroscience Letters* 195: 69–71.
Miyashita, E; Keller, A and Asanuma, H (1994). Input-output organization of the rat vibrissal motor cortex. *Experimental Brain Research* 99: 223–232.
Redgrave, P and Gurney, K (2006). The short-latency dopamine signal: a role in discovering novel actions? *Nature Reviews Neuroscience* 7: 967–975.
Redgrave, P; Mitchell, I J and Dean, P (1987). Further evidence for segregated output channels from superior colliculus in rat: Ipsilateral tecto-pontine and tecto-cuneiform projections have different cells of origin. *Brain Research* 413: 170–174.

Redgrave, P; Mitchell, I J and Dean, P (1987). Descending projections from the superior colliculus in rat: A study using orthograde transport of wheat germ-agglutinin conjugated horseradish peroxidase. *Experimental Brain Research* 68: 147–167.
Redgrave, P; Marrow, L and Dean, P (1992). Topographical organization of the nigrotectal projection in rat: evidence for segregated channels. *Neuroscience* 50: 571–595.
Redgrave, P; Westby, G W and Dean, P (1993). Functional architecture of rodent superior colliculus: Relevance of multiple output channels. *Progress in Brain Research* 95: 69–77.
Redgrave, P; McHaffie, J G and Stein, B E (1996). Nociceptive neurones in rat superior colliculus. I. Antidromic activation from the contralateral predorsal bundle. *Experimental Brain Research* 109: 185–196.
Redgrave, P; Simkins, M; McHaffie, J G and Stein, B E (1996). Nociceptive neurones in rat superior colliculus. II. Effects of lesions to the contralateral descending output pathway on nocifensive behaviours. *Experimental Brain Research* 109: 197–208.
Rhoades, R W; Mooney, R D and Jacquin, M F (1983). Complex somatosensory receptive fields of cells in the deep laminae of the hamster's superior colliculus. *The Journal of Neuroscience* 3: 1342–1354.
Rhoades, R W; Fish, S E; Chiaia, N L; Bennett-Clarke, C and Mooney, R D (1989). Organization of the projections from the trigeminal brainstem complex to the superior colliculus in the rat and hamster: Anterograde tracing with *Phaseolus vulgaris* leucoagglutinin and intra-axonal injection. *Journal of Comparative Neurology* 289: 641–656.
Rhoades, R W et al.(1987). The structural and functional characteristics of tectospinal neurons in the golden hamster. *Journal of Comparative Neurology* 255: 451–465.
Roger, M and Cadusseau, J (1984). Afferent connections of the nucleus posterior thalami in the rat, with some evolutionary and functional considerations. *Journal fur Hirnforschung* 25: 473–485.
Sahibzada, N; Yamasaki, D and Rhoades, R W (1987). The spinal and commissural projections from the superior colliculus in rat and hamster arise from distinct neuronal populations. *Brain Research* 415: 242–256.
Schneider, G E (1969). Two visual systems. *Science* 163: 895–902.
Sooksawate, T and Isa, T (2006). Properties of cholinergic responses in neurons in the intermediate grey layer of rat superior colliculus. *European Journal of Neuroscience* 24: 3096–3108.
Sparks, D L (1986). Translation of sensory signals into commands for control of saccadic eye movements: role of primate superior colliculus. *Physiological Reviews* 66: 118–171.
Sprague, J M (1966). Interaction of cortex and superior colliculus in mediation of visually guided behavior in the cat. *Science* 153: 1544–1547.
Sprague, J M and Meikle, T H, Jr. (1965). The role of the superior colliculus in visually guided behavior. *Experimental Neurology* 11: 115–146.
Stein, B E (1981). Organization of the rodent superior colliculus: Some comparisons with other mammals. *Behavioural Brain Research* 3: 175–188.
Stein, B E (1998). Neural mechanisms for synthesizing sensory information and producing adaptive behaviors. *Experimental Brain Research* 123: 124–135.
Stein, B E and Dixon, J P (1979). Properties of superior colliculus neurons in the golden hamster. *Journal of Comparative Neurology* 183: 269–284.
Stein, B E and Meredith, M A (1993). *The Merging of the Senses*. Cambridge, MA: MIT Press.
Stein, B E; Magalhaes-Castro, B and Kruger, L (1975). Superior colliculus: Visuotopic-somatotopic overlap. *Science* 189: 224–226.
Steinbusch, H W (1981). Distribution of serotonin-immunoreactivity in the central nervous system of the rat-cell bodies and terminals. *Neuroscience* 6: 557–618.
Swanson, L W and Hartman, B K (1975). The central adrenergic system. An immunofluorescence study of the location of cell bodies and their efferent connections in the rat utilizing dopamine-beta-hydroxylase as a marker. *Journal of Comparative Neurology* 163: 467–505.
Szwed, M; Bagdasarian, K and Ahissar, E (2003). Encoding of vibrissal active touch. *Neuron* 40: 621–630.
Tardif, E; Delacuisine, B; Probst, A and Clarke, S (2005). Intrinsic connectivity of human superior colliculus. *Experimental Brain Research* 166: 316–324.

Taylor, A M; Jeffery, G and Lieberman, A R (1986). Subcortical afferent and efferent connections of the superior colliculus in the rat and comparisons between albino and pigmented strains. *Experimental Brain Research* 62: 131–142.

Trageser, J C and Keller, A (2004). Reducing the uncertainty: Gating of peripheral inputs by zona incerta. *The Journal of Neuroscience* 24: 8911–8915.

Veazey, R B and Severin, C M (1982). Afferent projections to the deep mesencephalic nucleus in the rat. *Journal of Comparative Neurology* 204: 134–150.

Veinante, P and Deschenes, M (1999) Single- and multi-whisker channels in the ascending projections from the principal trigeminal nucleus in the rat. *The Journal of Neuroscience* 19: 5085–5095.

Wang, S and Redgrave, P (1997). Microinjections of muscimol into lateral superior colliculus disrupt orienting and oral movements in the formalin model of pain. *Neuroscience* 81: 967–988.

Westby, G W; Keay, K A; Redgrave, P; Dean, P and Bannister, M (1990). Output pathways from the rat superior colliculus mediating approach and avoidance have different sensory properties. *Experimental Brain Research* 81: 626–638.

Wise, S P and Jones, E G (1977). Somatotopic and columnar organization in the corticotectal projection of the rat somatic sensory cortex. *Brain Research* 133: 223–235.

Wurtz, R H and Albano, J E (1980). Visual-motor function of the primate superior colliculus. *Annual Review of Neuroscience* 3: 189–226.

Yamasaki, D S; Krauthamer, G M and Rhoades, R W (1986). Superior collicular projection to intralaminar thalamus in rat. *Brain Research* 378: 223–233.

Zhu, J J and Lo, F S (2000). Recurrent inhibitory circuitry in the deep layers of the rabbit superior colliculus. *Journal of Physiology* 523(Pt. 3): 731–740.

Zucker, E and Welker, W I (1969). Coding of somatic sensory input by vibrissae neurons in the rat's trigeminal ganglion. *Brain Research* 12: 138–156.

External Link

http://www.drexelmed.edu/Home/AboutOurFaculty/ManuelCastroAlamancos.aspx Author's website

Vibrissal Thalamic Modes

Manuel Castro-Alamancos

A thalamic mode (or thalamocortical mode) refers to a particular arrangement of the response properties of the thalamocortical network that gives rise to a distinct input-output function. Since the thalamus controls the flow of information to the neocortex, a thalamic mode adjusts how thalamocortical cells relay sensory and corticothalamic information. Thalamic modes can change on a moment-to-moment basis due to the actions of neuromodulators and/or excitatory and inhibitory inputs. In the vibrissal system, thalamic modes lead to robust changes on how sensory information from the vibrissa is processed temporally, integrated spatially and relayed to the neocortex.

Components of the Vibrissa Thalamocortical Network in the VPM Thalamus

The vibrissa thalamocortical network consists of two distinct nuclei. The main thalamocortical nucleus is the ventroposterior medial thalamus (VPM). In addition, cells in the medial sector of the posterior complex (POm) also project to the cortex. Here we will focus on VPM. Thalamocortical cells are at the center of a neuronal network that involves sensory, cortical and modulatory inputs. In VPM, thalamocortical cells form clusters, called barreloids, that project to clusters of cells in layer 4 of somatosensory cortex, called barrels. Thalamocortical cells in VPM receive signals from four main sources (see Figure 1):

- **Lemniscal sensory fibers** originating from clusters of cells, called barrelettes, in the principal trigeminal nucleus (PR5) provide sensory signals.
- **Corticothalamic fibers** originating in layer 6 of barrel cortex provide cortical feedback and top-down influences.

M. Castro-Alamancos (✉)
Department of Neurobiology, Drexel University College of Medicine, Philadelphia, PA, USA

T.J. Prescott et al. (eds.), *Scholarpedia of Touch*, Scholarpedia,
DOI 10.2991/978-94-6239-133-8_52

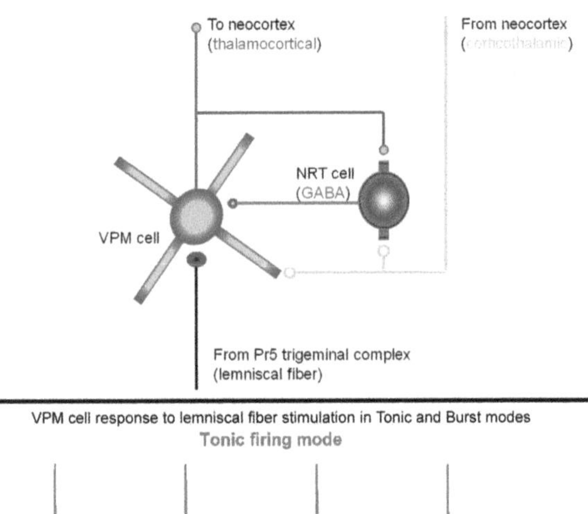

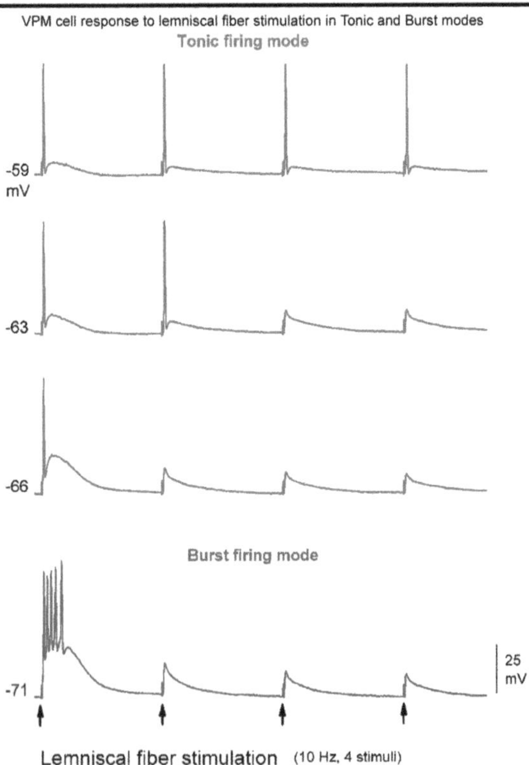

Figure 1 Main components of the thalamocortical network and intracellular responses of a VPM thalamocortical cell to medial lemniscus (lemniscal) fiber stimulation in a slice preparation. The cell is held at different membrane potentials within the tonic firing and burst modes. Electrical stimulation of the lemniscal fibers consists of 4 pulses at 10 Hz. Each train of pulses is delivered every 10 s, so the first pulse arrives at low frequency. Because lemniscal EPSPs depress at high frequencies, the stimulus only evokes spikes to the high frequency pulses when the cell is well depolarized within the tonic mode. (see Castro-Alamancos 2002a for details)

- Fibers from **nucleus reticularis thalamic** (NRT) cells provide the main inhibitory control of thalamocortical cells because there are no inhibitory interneurons in VPM.
- Finally, a variety of fibers that originate mainly in several brainstem nuclei provide **neuromodulator** inputs. These influences can directly affect thalamocortical cells or they can affect the other sources of inputs (NRT, corticothalamic, and Pr5 cells), which will indirectly affect thalamocortical cells.

As discussed later, vibrissa thalamic modes are defined by the properties of these network inputs, which are modified on a moment to moment basis by neuromodulators acting locally or in afferent structures (Castro-Alamancos 2004a). Considering their role in setting network modes, we first discuss neuromodulators.

Neuromodulators Set Network Modes

Neurotransmitters often act through ionotropic receptors (ligand-gated channels), while neuromodulators act through metabotropic receptors (G-protein coupled). The effects of neurotransmitters acting on ionotropic receptors are usually phasic, lasting only 10s of milliseconds. The effects of neuromodulators acting on metabotropic receptors are usually slower and longer lasting, in the range of 100s of milliseconds to seconds or more. However, the distinction between a neurotransmitter and a neuromodulator can be rather arbitrary and most neuroactive substances can function as both. For instance, a substance acting as a neuromodulator can alter the properties of ion channels that are activated by the same substance acting as a neurotransmitter (e.g. by affecting channel opening probabilities, receptor desensitization, release). Substances acting on ionotropic receptors may also (appear to) act as neuromodulators if the presynaptic neuron fires continuously in a sustained manner.

A number of substances are well-known neuromodulators and some of these have significant actions in the thalamocortical network:

- **Glutamate** acts on ionotropic receptors (AMPA, NMDA, kainate) and on metabotropic receptors (mGluR1-8). Glutamate is released by lemniscal and corticothalamic synapses. It appears that metabotropic receptors are selectively activated by corticothalamic synapses innervating NRT and thalamocortical cells (McCormick and von Krosigk 1992), and not by lemniscal synapses (Castro-Alamancos 2002a).
- **GABA** acts on ionotropic receptors ($GABA_A$) and metabotropic receptors ($GABA_B$). GABA is released by NRT cells.
- **Norepinephrine** is a catecholamine acting on α and β type metabotropic receptors. Noradrenergic neurons are found in the locus coeruleus in the brainstem reticular formation, from where they project throughout the brain, including the thalamus. Noradrenergic neurons discharge robustly during high levels of vigilance and attention, reduce their firing during slow-wave sleep and stop firing during paradoxical (also called, "rapid eye movement"; REM) sleep (Foote et al. 1980).
- **Dopamine** is a catecholamine acting on D1 and D2 type metabotropic receptors. So far, there is little evidence of any role of dopamine in the vibrissa thalamocortical network.
- **Histamine** acts on H_1-H_4 type metabotropic receptors. Histamine neurons are found in the posterior hypothalamus, in the tuberomammillary complex, from

where they project throughout the brain, including the thalamus. Histaminergic neurons discharge robustly during wakefulness (Brown et al. 2001).

- **Serotonin** acts on ionotropic (5-HT_3) and metabotropic (5-HT_1, 5-HT_2 5-HT_4, 5-HT_5, 5-HT_6, 5-HT_7) receptors. Serotonin neurons are found in the raphe nuclei in the brainstem reticular formation, from where they project throughout the brain, including the thalamus. Similar to noradrenergic neurons, 5-HT neurons fire tonically during wakefulness, decrease their activity in slow-wave sleep, and are nearly quiet during paradoxical sleep (McGinty and Harper 1976; Trulson and Jacobs 1979).
- **Acetylcholine** acts on ionotropic (nicotinic) and metabotropic (muscarinic) receptors. Acetylcholine neurons projecting to the thalamus are found in the pedunculopontine nuclei (PPT) and in the dorsolateral tegmental nuclei (LDT) in the brainstem. Acetylcholine neurons are also found in the basal forebrain from where they project to the neocortex. Cholinergic neurons in the LDT/PPT complex discharge vigorously during paradoxical sleep and also during wakefulness (el Mansari et al. 1989), and the levels of acetylcholine increase in the thalamus during those states (Williams et al. 1994).
- **Neuropeptides** usually act at metabotropic receptors. Neurons very often make both a conventional neurotransmitter (glutamate, GABA) and one or more neuropeptides. Examples include opioids (endorphins, enkephalins, dynorphins), substance P, etc.
- **Hormones** are chemicals released by cells that affect cells in other parts of the organism generally through the bloodstream. For example, epinephrine (adrenaline) is a catecholamine that is released by the adrenal gland.
- Other **intrinsic neuroactive substances** released within the thalamus may include adenosine, cannabinoids, growth factors, cytokines, etc.
- **Extrinsic neuroactive substances** that reach the thalamus may also affect thalamic modes. For example, nicotine from tobacco, caffeine from coffee, etc.

Trigeminal Complex Cells Provide Sensory Inputs and Sensory (Lemniscal) Synapses Act as Low-Pass Filters

Trigeminal complex cells projecting to VPM give rise to lemniscal synapses that form specialized glomeruli (for a review see Castro-Alamancos 2004a). Lemniscal synapses release the neurotransmitter glutamate and trigger EPSPs in thalamocortical cells by activating primarily AMPA receptors and may also activate NMDA receptors to some extent, but they do not seem to activate mGLUR receptors. Lemniscal EPSPs are highly specialized for the effective transmission of information. They have short latencies, fast rise times, large amplitudes, are highly reliable, and depress at frequencies above 2 Hz (Castro-Alamancos 2002a). The short latencies and fast rise times reflect the thickness and myelination of lemniscal fibers, and the fact that the synapses they form are electrotonically close to the soma. The large amplitude and security at low frequencies and the depression at high frequencies reflect a large number of release sites at synaptic glomeruli, and a high release probability (i.e. chance that a release site will fuse a vesicle and excrete neurotransmitter to the synaptic cleft) at those sites during low frequency inputs. Release probability is likely reduced during high frequency presynaptic activity. The depression of lemniscal synapses acts as a low-pass filter, enabling the relay of low-frequency sensory inputs under most circumstances, while the relay of high-frequency sensory inputs is subject to the membrane potential (Vm) of thalamocortical cells (Castro-Alamancos 2002a, b) (see Figure 1). The efficacy (synaptic strength) of lemniscal synapses per se is not affected by certain neuromodulators, such as acetylcholine and norepinephrine (Castro-Alamancos 2002a). But these neuromodulators can impact the transmission of lemniscal inputs by affecting the membrane potential of thalamocortical cells postsynaptically.

Layer 6 Cells Provide Corticothalamic Feedback and Corticothalamic Synapses Act as High-Pass Filters

Upper layer 6 cells located in barrel cortex project to VPM and leave a collateral fiber in NRT. Thalamocortical cells in VPM and layer 6 corticothalamic cells form closed-loops for the flow of information between a thalamic barreloid and a cortical barrel column (Bourassa et al. 1995). For each thalamocortical fiber ascending to barrel cortex there are many more corticothalamic fibers coming back to thalamus. Corticothalamic fibers form corticothalamic synapses that release glutamate and trigger EPSPs in thalamocortical and NRT cells by activating AMPA, NMDA and mGLUR receptors (Golshani et al. 2001). Corticothalamic EPSPs mediated by ionotropic receptors are very different compared to lemniscal EPSPs. They have long latencies, slow rise times, small amplitudes, are unreliable, and facilitate at frequencies above 2 Hz (for a review see Castro-Alamancos 2004a). The long latencies and slow rise times reflect the thinness and sparse myelination of corticothalamic fibers, and the fact that the synapses they form are located in distal dendrites; electrotonically far from the soma. The small amplitude and low security at low frequencies and the facilitation at high frequencies reflect a small number of release sites per synapse (estimated to be 1), and a low release probability at those sites during low frequency inputs that sharply increases during high frequencies inputs. Corticothalamic synapses display a robust form of LTP when stimulated repetitively at relatively high frequencies (10 Hz and above), while LTD is also induced when repetitive stimulation occurs at low frequencies (1 Hz), thereby providing mechanisms for bidirectional changes in synaptic efficacy (Castro-Alamancos and Calcagnotto 1999). Corticothalamic EPSPs mediated by mGLUR receptors are triggered by high-frequency stimulation and produce a long-lasting slow depolarization (McCormick and von Krosigk 1992). The efficacy of corticothalamic synapses is suppressed by some neuromodulators, such as acetylcholine and norepinephrine (Castro-Alamancos and Calcagnotto 2001). The amplitude of EPSPs evoked in NRT neurons by stimulating single corticothalamic fibers is several times larger than those evoked in thalamocortical neurons, and the number of GluR4-receptor subunits at these synapses may provide a basis for the differential synaptic strength (Golshani et al. 2001). The stronger corticothalamic EPSPs on NRT cells assures that low-frequency corticothalamic activity drives NRT cells and triggers robust feedforward inhibition in VPM thalamocortical cells.

NRT Cells Provide Inhibitory Inputs

NRT cells project to thalamocortical cells in VPM and are also thought to influence each other through inhibitory collateral fibers (Shu and McCormick 2002; Sohal and Huguenard 2003) and gap junctions (Landisman et al. 2002). NRT synapses are inhibitory, release GABA and trigger IPSPs in thalamocortical cells by activating $GABA_A$ and $GABA_B$ receptors. $GABA_A$ receptors are activated by the amount of GABA released by a single action potential in an NRT fiber, while $GABA_B$ receptor activation appears to require more GABA, usually released by bursts of action potentials (Kim and McCormick 1998). Similar to thalamocortical cells, NRT cells have two intrinsic firing modes: burst and tonic firing (Steriade et al. 1997; Hartings et al. 2003). The relay of sensory information to the neocortex by thalamocortical cells is strongly influenced by the TRN. Thus, TRN is probably the main gatekeeper of the neocortex (Crick 1984). The reciprocal synaptic connectivity between TRN and thalamocortical cells in VPM is critical for generating normal and abnormal rhythmic activities, such as spindle oscillations (see http://www.scholarpedia.org) and absence seizures (Beenhakker and Huguenard 2009).

Thalamocortical Cells Provide the Output to Neocortex via Thalamocortical Synapses

Thalamocortical cells have two intrinsic firing modes: tonic and burst firing modes. Bursts are due to the Ca2 + conductance of T type channels, which gives rise to the low-threshold calcium current (I_T) (Steriade et al. 1997). Because of its voltage-dependence, this conductance in thalamocortical cells is controlled by inhibitory inputs from NRT cells. In particular, when an NRT neuron fires a burst of action potentials, it activates both $GABA_A$ and $GABA_B$ receptors in thalamocortical cells, which hyperpolarizes these neurons and deinactivates low-threshold T-type Ca2+ channels, enabling thalamocortical neurons to produce regenerative calcium spikes that can trigger bursts of action potentials. In response to NRT inhibition, thalamocortical cells fire with a delay on a postinhibitory rebound, and the stronger the inhibition the stronger the rebound excitation. Even without NRT inhibition, if thalamocortical cells are sufficiently hyperpolarized, resulting in the deinactivation of the low-threshold calcium current, excitatory inputs (corticothalamic or lemniscal) can directly drive regenerative calcium spikes that can trigger bursts of action potentials. Moreover, when thalamocortical cells are more depolarized, resulting in the inactivation of the low-threshold calcium current, thalamocortical cells enter the tonic firing mode and produce single action potentials, instead of bursts, in response to excitatory inputs.

The output of the thalamocortical cells is transmitted by thalamocortical synapses that reach the barrel cortex and terminate in layer 4. As thalamocortical fibers ascend to layer 4 they leave fiber collaterals in NRT and in layer 6. In layer 4, thalamocortical fibers produce thalamocortical synapses that release glutamate and trigger EPSPs on cortical neurons.

What Determines a Thalamocortical Mode?

A thalamocortical mode is a particular arrangement of the properties of the thalamocortical network components that gives rise to a distinct input-output function. The properties that are most commonly affected to determine a thalamocortical mode include the membrane potential (Vm), intrinsic firing mode, intrinsic excitability and the strength of synaptic inputs. Neuromodulators act directly on thalamocortical cells and on afferent synapses within the thalamus to change these properties. Neuromodulators may also indirectly influence thalamocortical cells by affecting the activity of the main excitatory and inhibitory inputs, within the trigeminal complex, layer 6 and NRT, respectively. Activity in these afferents changes their strength via short-term synaptic plasticity but may also affect the Vm, firing mode, and intrinsic excitability of thalamocortical cells. Thus, thalamocortical modes are set in a complex way, through direct and indirect effects of neuromodulators affecting several variables. The main variables determining a thalamocortical mode are:

- **Vm of thalamocortical cells**: This critical variable is highly dynamic because it is affected by most, if not all, neurotransmitters and neuromodulators present in the thalamocortical network.
- **Intrinsic excitability**: Neurons express a number of voltage-dependent conductances that endow them with different response properties. For example, thalamocortical cells are characterized by strong hyperpolarization-activated cation currents (I_H) and low-threshold calcium currents (I_T). These currents are not only affected by Vm but can be directly affected by many neuromodulators. Excitatory and inhibitory inputs can affect the intrinsic excitability of thalamocortical cells by changing the Vm and engaging voltage-dependent currents. In addition, when the resting Vm of thalamocortical cells is at the reversal of incoming synaptic inputs, the increased conductance produced by the synaptic

inputs can affect the integrative properties of the cell, without changing the Vm, by shunting the membrane (e.g. shunting inhibition).

- **Thalamocortical firing mode and rate**: Thalamocortical cells are characterized by two distinct firing modes: bursting and tonic. These firing modes are set primarily by the Vm of thalamocortical cells. Thus, factors that influence Vm also determine firing mode. During bursting, thalamocortical cells produce a cluster of action potentials (usually between 3-6 action potentials) at very high frequencies (>100 Hz) riding on the low-threshold calcium spike. However, thalamocortical cells are limited by how fast they can produce bursts because of the dependence of bursts on the low-threshold calcium current, which must be deinactivated by hyperpolarization. Thus, cells can usually burst at <15 Hz. In contrast, during tonic firing, cells can produce action potentials at much higher constant firing rates.
- **Activity and strength of excitatory and inhibitory afferents**: Thalamocortical cells in VPM receive excitatory (glutamatergic) and inhibitory (GABAergic) afferent inputs from three main sources: lemniscal synapses from trigeminal nucleus (mostly Pr5), corticothalamic synapses from layer 6 of barrel cortex and inhibitory synapses from NRT. Activity in the afferent cells drives their synapses and can change the Vm of thalamocortical cells, which can result in changes in intrinsic excitability and intrinsic firing mode. In addition, some of these afferents can activate metabotropic receptors leading to a modulator action on thalamocortical cells. For example, glutamate released from corticothalamic synapses can activate mGLUR, and GABA released from inhibitory synapses can activate $GABA_B$ receptors. The frequency and pattern of activity in the synaptic afferents also sets the strength of these synapses by affecting short-term synaptic plasticity and temporal integration. For example, high-frequency activity in corticothalamic synapses will increase release probability at these synapses and enhance the strength of this pathway. In contrast, activity in Pr5 cells will depress lemniscal synapses and decrease the strength of this pathway. Moreover, neuromodulators released in the thalamus can directly affect the efficacy of excitatory and inhibitory neurotransmission rather selectively. For example, acetylcholine and norepinephrine depresses corticothalamic but not lemniscal synaptic strength (Castro-Alamancos and Calcagnotto 2001; Castro-Alamancos 2002a).

The Main Thalamocortical Modes

Slow Oscillation Mode During Slow-Wave Sleep and Anesthesia

- **When does it happen**: The slow oscillation or quiescent mode is considered here as a broad baseline state during which active sensory processing per se does not occur because animals are either sleeping, inattentive/drowsy or anesthetized.

While there may well be different modes within these states, we encompass them here within a single mode for simplicity. In this mode, slow synchronous oscillations are common, particularly when animals are sleeping in non-REM sleep. In addition, this mode can be induced by surgical anesthesia, which is when most electrophysiological studies take place. During non-REM sleep, slow wave oscillations occur often but are most prevalent in the deeper stage(s) (typically referred as stage 3 or 3/4). In less deep stages of sleep, slow oscillations can occur interposed with other rhythms, such as spindle oscillations. Similar to non-REM sleep stages, there are also stages of anesthesia. During the surgical anesthesia stage, the slow oscillation mode is evident but can vary significantly depending on the level or plane of surgical anesthesia. For example, the frequency of slow oscillations can vary quite significantly within this mode. Moreover, some anesthetics will tend to produce more rhythmic activity of a certain frequency range than others. Thus, the depth of anesthesia and the specific effects of the anesthetic used are critical at setting the particular characteristics of this mode.

- **Spontaneous activity**: During slow-wave sleep and surgical anesthesia thalamocortical cells fire at low frequencies producing either bursts or single spikes. This slow activity in thalamocortical cells can be driven by ongoing slow oscillations generated intrinsically in the neocortex (also known as Up and Down states) (Steriade et al. 1997; Rigas and Castro-Alamancos 2007). But slow activity can also be driven by intrinsic currents in thalamocortical cells and persist in the absence of corticothalamic activity (Hughes et al. 2002; Rigas and Castro-Alamancos 2007). Thus, full expression of slow oscillations in thalamocortical cells appears to require both thalamic and cortical oscillators (Crunelli and Hughes 2010). During the slow oscillation mode, thalamocortical cells are usually fairly hyperpolarized close to the reversal potential of K^+ (Down), and they may transition for short periods of time to more depolarized states due to synaptic bombardment, usually produced by spontaneous corticothalamic and NRT activity (Up) (Steriade et al. 1997). In this situation, nRt cells can burst and drive strong IPSPs in thalamocortical cells. The hyperpolarization caused by the IPSPs deinactivates I_T and activates I_H in thalamocortical cells. This sets up thalamocortical cells so that at the outset of the IPSP a rebound depolarization occurs caused by activation of I_T. The rebound triggers a burst of action potentials in thalamocortical cells that feedback to NRT and cortex. Such an interplay between NRT and thalamocortical cells repeated in a sequence at 5–12 Hz is responsible for the generation of spindle oscillations that recur every few seconds (McCormick and Bal 1997). Spindles are waxing and waning rhythms with dominant frequencies of 7–14 Hz, grouped in sequences that last 1–3 s and recur periodically at 0.1–0.2 Hz (see http://www.scholarpedia.org). Spindle oscillations are common during the slow oscillation mode and are prominent at sleep onset, during loss of awareness, and are prevalent during barbiturate anesthesia, which enhances inhibitory efficacy. Apart from the occasional spindle oscillations, thalamocortical activity in VPM during this state is of low frequency (<1 Hz) (Castro-Alamancos 2002b; Aguilar and Castro-Alamancos 2005; Hirata et al. 2006).

- **Frequency-dependent sensory responses (rapid sensory adaptation)**: Sensory responses driven by whisker stimulation during the slow oscillation mode are of high probability as long as the stimulus is delivered at low frequencies. As soon as the vibrissa stimulus augments in frequency, the thalamocortical response is strongly depressed (Castro-Alamancos 2002b). Thus, thalamocortical neurons follow high-frequency whisker stimulation with great difficulty in the slow oscillation mode; thalamocortical sensory responses are low-pass filtered. Intracellular recordings in urethane anesthetized rats during the slow oscillation mode show that whisker stimulation evokes EPSP–IPSP sequences in thalamocortical neurons (see http://www.scholarpedia.org), and both the EPSPs and IPSPs depress with repetitive whisker stimulation at frequencies above 2 Hz (Castro-Alamancos 2002b). The underlying cause of this low-pass filter is the frequency-dependent depression of lemniscal synapses (Castro-Alamancos 2002a).Thus, sensory inputs at frequencies above 2 Hz reduce the efficacy of lemniscal synapses, which drives the lemniscal EPSP away from the discharge threshold of the cell resulting in a low probability of firing for thalamocortical cells. However, as described below, a major impact of activated states is to change this low-pass filtering. In addition, feedback inhibition from NRT, driven by sensory inputs, also contributes to the low-pass filtering of sensory inputs during the slow oscillation mode (Castro-Alamancos 2002b). This effect is most notable at the beginning of a high-frequency sensory stimulus train, when IPSPs are more robust and produce a stronger hyperpolarization. However, less effect of feedback inhibition is observed for sensory responses occurring at the end of a long high-frequency train (Castro-Alamancos 2002b; Hirata et al. 2009). Those responses are mostly depressed by lemniscal synaptic depression with little contribution of synaptic inhibition from NRT.
- **Spatial dependent sensory responses (receptive fields and selectivity)**: Excitatory receptive fields of VPM cells consist of an excitatory center, the principal whisker (PW), and an excitatory surround, the adjacent whiskers (AWs). For low-frequency sensory inputs, during the slow oscillation mode, the response to the PW (receptive field excitatory center) is much stronger and faster than the response to AWs (receptive field excitatory surround) (Aguilar and Castro-Alamancos 2005). As mentioned above, for high-frequency sensory inputs, both PW and AW responses are depressed because of the low-pass filtering at the lemniscal pathway. Simultaneous stimulation of the PW and several AWs (i.e. multiwhisker stimulation) produces a response in thalamocortical cells that matches the PW, as if the AWs had not been stimulated (Aguilar and Castro-Alamancos 2005; Hirata et al. 2006). Interestingly, simultaneous multiwhisker responses are distinguishable from PW responses in the next stage of processing, the barrel cortex (Hirata and Castro-Alamancos 2008). The size of the receptive field measured as the number of whiskers that evoke a response depends on the level of anesthesia within this mode (Friedberg et al. 1999; Aguilar and Castro-Alamancos 2005).
- **Corticothalamic feedback**: The amplitude of corticothalamic EPSPs is relatively small during low frequency corticothalamic activity because corticothalamic synapses have a low release probability and occur at distal portions of the

dendritic tree. However, during high frequency corticothalamic activity (> 5 Hz) the probability of release at these synapses sharply increases due to synaptic facilitation producing large amplitude EPSPs that can be as powerful than those produced by lemniscal sensory afferents. Thus, the corticothalamic pathway is an activity dependent driver of thalamocortical activity, which can be demonstrated by using electrical stimulation to stimulate corticothalamic fibers (Castro-Alamancos 2004a). However, it is not clear when corticothalamic cells discharge at high-frequencies to engage the activity dependent facilitation of corticothalamic synapses.

In Vivo Intracellular Recording of a VPM Cell During the Slow Oscillation Mode

Spontaneous and whisker-evoked intracellular activity of a VPM thalamocortical cell in a urethane-anesthetized rat. The 20 s of continuous recording shown in the panel below displays spontaneous spikes and large amplitude lemniscal EPSPs, and two instances of whisker stimulation that evoke the lemniscal EPSPs and feedback IPSPs from NRT. The insets that appear during whisker stimulation show a close-up of responses evoked by the first and tenth stimulus in a 10 Hz train. During the recording the thalamus is in the slow oscillation mode and the VPM cell only evokes spikes (sometimes) to the first stimulus in the whisker train. Note also the large amplitude IPSP to the first stimulus in a train and the depression of both the IPSP and EPSP at 10 Hz. (See Castro-Alamancos 2002b for details) http://www.scholarpedia.org.

Activated Modes During Arousal and BRF Stimulation

- **When does it happen**: The activated mode is typical when animals are awake during arousal, and it is most robust when animals are in a state of vigilance during attentive processing, such as during performance in a behavioral task (Castro-Alamancos 2004b). A somewhat similar activated mode to that observed during waking occurs when animals enter REM or paradoxical sleep. The activated mode can be induced in anesthetized animals that are in a slow oscillation mode by electrically stimulating the brainstem reticular formation (BRF), and this is a useful method to determine the impact of the activated mode on sensory thalamocortical responses because it allows to compare slow oscillation and activated sensory responses in the same neurons (Moruzzi and Magoun 1949; Castro-Alamancos 2002b; Castro-Alamancos and Oldford 2002; Aguilar and Castro-Alamancos 2005).
- **Spontaneous activity**: Spontaneous thalamocortical activity can vary from nil to high frequency tonic firing. During BRF stimulation, thalamocortical activity

consists of high frequency tonic firing, which outlasts the stimulation by several seconds (Castro-Alamancos 2002b; Castro-Alamancos and Oldford 2002). Bursts in this state are uncommon although they may occur as thalamocortical cells transition from the slow oscillation mode to the activated mode.

- **Frequency-dependent sensory responses (rapid sensory adaptation)**: In contrast to the slow oscillation mode, during the activated mode, low-frequency sensory responses are a bit stronger (increase slightly in probability) and become faster (evoked spikes display shorter latencies) (Castro-Alamancos 2002b; Aguilar and Castro-Alamancos 2005). But the most robust change occurs at the level of high-frequency sensory processing. Thus, during the activated mode thalamocortical cells robustly enhance their responses to high-frequency sensory signals, virtually eliminating the low-pass filtering typical of the slow oscillation mode (Castro-Alamancos 2002b). These effects are similar to those produced by cholinergic activation of the thalamus, as discussed below.
- **Spatial dependent sensory responses (receptive fields and selectivity)**: Excitatory receptive fields of VPM cells consist of an excitatory center, the PW, and an excitatory surround, the AWs. Activation produced by BRF stimulation in anesthetized animals enlarges the excitatory surround of VPM cells (Aguilar and Castro-Alamancos 2005). Thus, for low-frequency sensory inputs, during quiescent states, the response to the PW (receptive field excitatory center) is much stronger than the response to most AWs (receptive field excitatory surround), but during activation, there is an enhancement of the response to AWs, which can reach response levels similar to the PW. For high-frequency sensory inputs, during quiescent states, both PW and AW responses are depressed because of the low-pass filtering at the lemniscal pathway. However, during the activated mode, there is a significant increase in both PW and AW responses so that they become similar, but PW responses are generally stronger than AW responses at high frequencies (Aguilar and Castro-Alamancos 2005).
- **Corticothalamic feedback**: During the activated mode, produced by BRF stimulation in anesthetized animals, corticothalamic responses are further high-pass filtered (i.e. only high-frequency corticothalamic activity is allowed) (Castro-Alamancos and Calcagnotto 2001). This effect is mimicked by specific neuromodulators (i.e. norepinephrine), as described below.

Activated Modes Produced by Specific Neuromodulators

Application of specific neuromodulators into the thalamus leads to characteristic activated modes. Neuromodulators have highly selective effects that set different modes of thalamocortical and corticothalamic information processing. Natural behavioral states are likely set by a combination of neuromodulators acting in synergy.

Cholinergic Mode

- **When does it happen**: Cholinergic neurons in the LDT/PPT complex discharge vigorously during paradoxical sleep and also during wakefulness (el Mansari et al. 1989), and the levels of acetylcholine increase in the thalamus during those states (Williams et al. 1994). Thus, the cholinergic mode is expected to occur during both REM/paradoxical sleep and during states of vigilance in awake animals.
- **Spontaneous activity**: Cholinergic activation leads to a sharp increase of spontaneous thalamocortical tonic firing in VPM (see Figure 2), which reduces signal to noise ratios (Aguilar and Castro-Alamancos 2005; Hirata et al. 2006). Typically, cells increase their spontaneous firing by more than 10-fold compared to the slow oscillation mode. The effect of cholinergic activation on spontaneous firing is explained by both a direct depolarization of VPM cells and a suppression of NRT cell firing. The depolarizing effect of acetylcholine on thalamocortical cells is mediated by muscarinic receptors, which block a resting K^+ conductance, and the hyperpolarizing effect of acetylcholine on NRT cells is produced by activation of a K^+ conductance (McCormick, 1992).
- **Frequency-dependent sensory responses (rapid sensory adaptation)**: During the cholinergic activated mode, low-frequency sensory responses are a bit stronger (increase slightly in probability) and become faster (evoked spikes display shorter latencies) compared to the slow oscillation mode but signal to noise ratios are sharply reduced (Castro-Alamancos 2002b; Aguilar and Castro-Alamancos 2005; Hirata et al. 2006). Another robust change occurs at the level of high-frequency sensory processing. Thus, during cholinergic activation thalamocortical cells robustly enhance their responses to high-frequency sensory signals, virtually eliminating the low-pass filtering of sensory signals in the sensory thalamus (Castro-Alamancos 2002b). The postsynaptic depolarization of thalamocortical neurons produced by cholinergic activation is sufficient to eliminate the effect of lemniscal synaptic depression on the relay of high frequency inputs (Castro-Alamancos 2002a, b). Cholinergic activation also reduces the effects of inhibition from NRT by hyperpolarizing NRT cells and suppressing IPSPs in thalamocortical cells (Castro-Alamancos 2002a, b).
- **Spatial dependent sensory responses (receptive fields and selectivity)**: Cholinergic activation enlarges the excitatory surround of VPM cells just like BRF stimulation does. Thus, for low-frequency sensory inputs during the slow oscillation mode, the response to the PW (receptive field excitatory center) is stronger than the response to AWs (receptive field excitatory surround), but during activation, there is an enhancement of the response to AWs, which can reach response levels similar to the PW. For high-frequency sensory inputs during the slow oscillation mode, both PW and AW responses are depressed because of the low-pass filtering at the lemniscal pathway. However, during cholinergic activation, there is a significant increase in both PW and AW responses so that they become similar, but PW responses are generally stronger than AW responses at high frequencies.

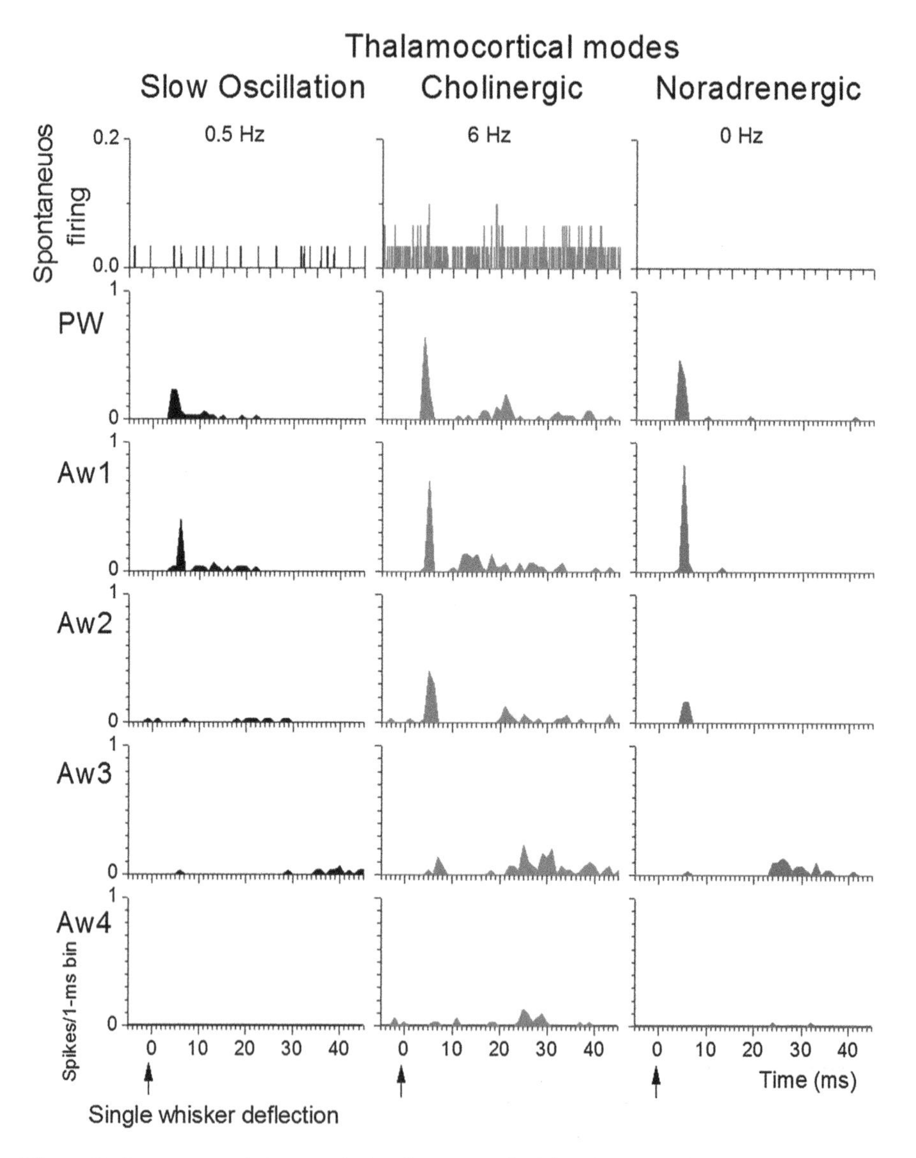

Figure 2 Spontaneous thalamocortical spikes or evoked by single-whisker deflections of the principal whisker (PW) or 4 adjacent whiskers (AWs) in a urethane anesthetized rat during the slow oscillation, cholinergic and noradrenergic modes. The cholinergic mode was induced by application of a cholinergic agonist into the thalamus and the noradrenergic mode was induced by subsequent application of a noradrenergic agonist into the thalamus. (see Hirata et al. 2006 for details)

- **Corticothalamic feedback**: Corticothalamic EPSPs are suppressed by acetylcholine, an effect that is independent of the postsynaptic actions of acetylcholine (Castro-Alamancos and Calcagnotto 2001). However, cholinergic activation augments low-frequency corticothalamic responses, which reduces the amount of facilitation in corticothalamic responses, making thalamocortical cells responsive to a wide frequency band of cortical signals (Hirata et al. 2006). Hence, during cholinergic activation, the selectivity of VPM cells for high-frequency corticothalamic signals (high-pass filtering) is lost. This may cause a major problem for thalamocortical sensory processing, because it allows low-frequency cortical signals to become effective driver of thalamocortical cells. Such an effect seems undesirable during sensory processing, because thalamocortical cells may not be able to distinguish sensory and cortical inputs. One possibility is that the enhanced responsiveness to low-frequency cortical signals during cholinergic activation is related to sensory experiences that are driven by internal, top-down, representations during paradoxical sleep (when cholinergic activation is strong). During paradoxical sleep, cortical cells may be strong drivers of thalamocortical neurons, which could serve to feed top-down representations to upper layers of primary sensory cortex via the thalamus, perhaps related to sensory experiences during this phase of sleep.

In conclusion, cholinergic activation has the following effects: increases the spontaneous tonic firing of thalamocortical cells; strongly eliminates the low-pass filtering of sensory inputs, allowing high-frequency sensory relays; broadens thalamocortical receptive fields; allows low-frequency corticothalamic signals, which are usually blocked by the high-pass filtering at this connection; reduces signal-to-noise ratios. Possibly, the larger receptive fields, lower signal-to-noise ratios, and broad-frequency spectrum corticothalamic responses characteristic of cholinergic activation may be related to activated modes during paradoxical sleep and non-attentive wakefulness.

Noradrenergic Mode

- **When does it happen**: Noradrenergic neurons discharge robustly during high levels of vigilance and attention, reduce their firing during slow-wave sleep, and stop firing during paradoxical sleep (Hobson et al. 1975; Foote et al. 1980; Aston-Jones and Bloom 1981). Thus, the noradrenergic mode is expected to occur during states of vigilance in attentive animals.
- **Spontaneous activity**: Noradrenergic activation leads to a reduction of thalamocortical cell firing so that they have basically nil spontaneous firing (see Figure 2). The effect of noradrenergic activation on spontaneous thalamocortical firing is completely mediated by the NRT because during thalamic disinhibition (block of GABA receptors) norepinephrine no longer suppresses thalamocortical cells (Hirata et al. 2006). In fact, during disinhibition, thalamocortical cells in VPM are excited by norepinephrine. Thus, noradrenergic activation strongly excites NRT cells, which inhibit thalamocortical cells in VPM. The depolarizing effect of norepinephrine on NRT cells is mediated by α-adrenergic receptors, and attributable to a decrease of a resting K^+ conductance (McCormick, 1992).

- **Frequency-dependent sensory responses (rapid sensory adaptation)**: The effects of norepinephrine on sensory responses are similar to those produced by cholinergic activation but without the increase in spontaneous firing (Hirata et al. 2006). For sensory signals, noradrenergic activation sets sensory processing to a focused and noise-free excitatory receptive field, which contrasts with the broad and noisy excitatory receptive field characteristic of cholinergic activation. Norepinephrine also facilitates the high-frequency responses to whisker stimulation, albeit less effectively than cholinergic activation.
- **Spatial dependent sensory responses (receptive fields and selectivity)**: Noradrenergic activation enhances AW responses but only for one whisker and for low-frequency responses. Whereas cholinergic activation enhances high-frequency responses for several AWs, norepinephrine only enhances high-frequency responses for the PW. This indicates that high-frequency sensory inputs are highly focused to the center of the receptive field during noradrenergic activation. Consequently, VPM receptive fields are more focused during noradrenergic activation than during cholinergic activation.
- **Corticothalamic feedback**: Corticothalamic EPSPs are suppressed by norepinephrine, an effect that is independent of the postsynaptic actions of norepinephrine (Castro-Alamancos and Calcagnotto 2001). Moreover, noradrenergic activation further high-pass filters corticothalamic responses (Hirata et al. 2006). The high-pass filtering ensures that thalamocortical cells are not driven by cortical signals unless those signals arrive at high frequencies. This effect is similar to that observed after BRF stimulation in anesthetized animals. Thus, for corticothalamic signals, noradrenergic activation sets corticothalamic processing to a noise-free high-frequency signal detection mode.

In conclusion, noradrenergic activation may provide a dynamic mechanism to (1) focus thalamocortical receptive fields, (2) high-pass filter corticothalamic signals, and (3) enhance signal-to-noise ratios. Possibly, the more focused receptive fields and higher signal-to-noise ratios during noradrenergic activation reflect a more appropriate, information processing mode for spatial discrimination of sensory inputs.

Active Whisking Mode

- **When does it happen**: Rodents use their vibrissae to navigate the environment by performing fast rhythmic vibrissa movements. During active exploration, whisking consists in ellipsoid movements (which are characterized by vibrissa protractions) through the air and over objects at between 4 and 15 Hz.
- **Spontaneous activity**: During active whisking in air, thalamocortical cell activity in VPM increases compared to non-whisking (Fanselow and Nicolelis 1999; Lee et al. 2008). Spontaneous activity also increases for most cells in

VPM during artificial whisking in air, which is induced by electrical stimulation of motor nerves in a pattern resembling active whisking (Yu et al. 2006).

- **Frequency-dependent sensory responses (rapid sensory adaptation)**: Sensory responses evoked by stimuli delivered during active whisking are usually suppressed compared to during non-whisking. For example, whisker follicle or infraorbital nerve stimulation evokes a smaller field potential and/or fewer spikes in VPM during active whisking periods than during non-whisking (Fanselow and Nicolelis 1999; Lee et al. 2008). Also, paired-pulse ratios (amplitude of the response to the second stimulus divided by the amplitude of the first) are significantly smaller during non-whisking, indicating stronger paired pulse suppression. Thus, just like during activated modes (Castro-Alamancos 2002b), thalamocortical cells appear to follow high-frequency stimuli much better during active whisking. During artificial whisking, most VPM cells enhance their response when the whiskers contact an object compared to the response during whisking in air, while other cells suppress their responses (Yu et al. 2006); cells in the ventrolateral portion of VPM appear to convey a pure touch signal because they mostly fire when a whisker contacts an object but not during whisking in air.
- **Spatial dependent sensory responses (receptive fields and selectivity)**: Because whiskers are moving during active whisking it is difficult to map receptive fields during this mode.
- **Corticothalamic feedback**: Corticothalamic responses have not been monitored during active whisking.

Pathological Modes During Epilepsy

- **When does it happen**: Epilepsy has many different causes and there are a number of rodent genetic models that produce spontaneous seizures involving the thalamocortical network. A well known example is the Genetic Absence Epilepsy Rat from Strasbourg (GAERS). Apart from genetic models, the simplest way to generate seizures in the brain is to impair the control that GABA-mediated inhibition has on excitation. This can be accomplished by blocking GABA receptors (disinhibition) using specific antagonists. Disinhibition may occur naturally in the brain due to a variety of mechanisms including withdrawal of inhibitory synapses or death of inhibitory cells caused by various insults, developmental disorders and/or activity-dependent mechanisms.
- **Spontaneous activity**: In the vibrissa system, block of thalamic $GABA_A$ receptors in vivo leads to $\sim$3 Hz activity in thalamocortical cells that is translated into $\sim$3 Hz spike-wave discharges in the neocortex, and these discharges are abolished by subsequent block of thalamic $GABA_B$ receptors (Castro-Alamancos 1999). Work in vitro has shown that when thalamic $GABA_A$ receptors are blocked, $GABA_B$-mediated responses are observed in thalamocortical cells due to longer and higher frequency bursts in NRT neurons caused

by a reduction of intra-NRT inhibition (Huntsman et al. 1999). The longer time constants of $GABA_B$-mediated hyperpolarization drive the slower ~3 Hz activity, which is then logically abolished by blocking $GABA_B$ receptors. This ~3 Hz activity resembles the activity observed in children during absence seizures, and has been proposed as a laboratory model of this disorder (for a recent review see Beenhakker and Huguenard 2009).

- **Frequency-dependent sensory responses (rapid sensory adaptation)**: Block of thalamic GABA receptors has robust consequences on thalamocortical sensory responses (Hirata et al. 2009). During high-frequency (10 Hz) whisker stimulation, thalamic disinhibition enhances short-latency multiwhisker (PW and AWs) and PW responses but only of "transition stimuli", which are those stimuli in between the first stimulus and the last of a 10 Hz train (see Hirata et al. 2009). Thalamic disinhibition also enhances long-latency multiwhisker and PW responses evoked by all stimuli in a train regardless of their frequency and position within a train.
- **Spatial dependent sensory responses (receptive fields and selectivity)**: Thalamic disinhibition slightly enhances the short-latency response of the strongest whisker in the surround during low-frequency stimulation. In addition, thalamic disinhibition enhances the long-latency response of most of the whiskers in the surround during low-frequency stimulation.
- **Corticothalamic feedback**: During thalamic disinhibition, there are two major effects on corticothalamic responses. First, low-frequency responses are strongly enhanced. Thus responses to all 10 stimuli in a train at 2 and 5 Hz are significantly enhanced by thalamic disinhibition. Second, there are complex effects of thalamic disinhibition on frequency-dependent facilitation evoked by corticothalamic stimulation. Steady-state facilitated responses (i.e., last 5 stimuli in a 10 stimulus train), evoked at 5 and 10 Hz, are further enhanced by disinhibition. However, the last five stimuli in 20 and 40 Hz trains do not reach a steady facilitated state; instead these responses depress after reaching peak facilitation. This depression phenomenon appears to be related to the ability of high-frequency corticothalamic stimulation (facilitation) to trigger epileptic-like discharges (leading to post-discharge depression). These discharges are not evoked during thalamic disinhibition when high-frequency whisker stimulation is used. Thus, it appears that during thalamic disinhibition thalamocortical cells are sensitive to high-frequency corticothalamic activity, which can trigger epileptic-like seizure activity.

Thalamocortical Modes Set Neocortex Modes

The considerable differences in spontaneous firing of thalamocortical cells during different Thalamic modes may lead to different modes in the barrel neocortex. For example, thalamic noradrenergic and cholinergic activation produce two very

Table1. Effects of thalamocortical modes on:

Thalamocortical modes	Spontaneous thalamocortical Firing	Relay of sensory inputs	Sensory response receptive field	Corticothalamic feedback
Slow oscillation	Low tonic and bursts	Relay of low frequency inputs (low-pass filter)	Focused	Frequency-dependent facilitation (high-pass filter)
Activated	Noisy tonic	Relay of low and high frequency inputs	Broader	Strong high-pass filter
Cholinergic	Noisy tonic	Relay of low and high frequency inputs	Broader	Removal of high-pass filter by low frequency response enhancement
Noradrenergic	Quiet tonic	Relay of Low and high frequency Inputs	Highly focused	Strong high-pass filter
Epileptic ($GABA_A$ block)	Rhythmic ~3 Hz bursts	Relay of low frequency inputs; long-latency response enhancement	Broader (long-latency)	Removal of high-pass filter by low-frequency response enhancement

Figure 3 Table listing the main effects of different thalamocortical modes on spontaneous thalamocortical firing, the relay of sensory inputs to the neocortex, sensory response receptive fields, and corticothalamic feedback responses

distinct modes of thalamocortical firing, and it is possible that this has consequences on cortical activity. Indeed, recent work has shown that the distinct thalamic modes set by these thalamic neuromodulators produces different cortical modes (Hirata and Castro-Alamancos 2010). Thalamic cholinergic activation of the thalamus makes thalamocortical cells very responsive to whisker stimuli but also increases their spontaneous tonic firing and this leads to cortical activation in the barrel cortex. In contrast, thalamic noradrenergic activation also makes thalamocortical cells very responsive to sensory stimuli but abolishes their spontaneous firing, and this leads to cortical deactivation or slow oscillations in the barrel cortex. Thus, cholinergic and noradrenergic thalamic activation lead to two well differentiated thalamocortical modes that directly set two distinct cortical modes. The cholinergic thalamocortical mode has abundant presynaptic thalamocortical activity (relay cell noise) but little postsynaptic activity (cortical cell spontaneous activity or noise), while the noradrenergic mode has nil presynaptic activity (no thalamocortical relay cell noise) but plenty of postsynaptic activity (cortical noise) (Figure 3).

Internal References

Alonso, J-M and Chen, Y (2009). Receptive field. *Scholarpedia* 4(1): 5393. http://www.scholarpedia.org/article/Receptive_field.
Bouret, S and Sara, S J (2010). Locus coeruleus. *Scholarpedia* 5(3): 2845. http://www.scholarpedia.org/article/Locus_coeruleus.
Braitenberg, V (2007). Brain. *Scholarpedia* 2(11): 2918. http://www.scholarpedia.org/article/Brain.
Freund, T and Kali, S (2008). Interneurons. *Scholarpedia* 3(9): 4720. http://www.scholarpedia.org/article/Interneurons.
Hille, B (2008). Ion channels. *Scholarpedia* 3(10): 6051. http://www.scholarpedia.org/article/Ion_channels.
Izhikevich, E M (2006). Bursting. *Scholarpedia* 1(3): 1300. http://www.scholarpedia.org/article/Bursting.
Johnson, D H (2006). Signal-to-noise ratio. *Scholarpedia* 1(12): 2088. http://www.scholarpedia.org/article/Signal-to-noise_ratio.
Jonas, P and Buzsaki, G (2007). Neural inhibition. *Scholarpedia* 2(9): 3286. http://www.scholarpedia.org/article/Neural_inhibition.
Llinas, R (2008). Neuron. *Scholarpedia* 3(8): 1490. http://www.scholarpedia.org/article/Neuron.
Moehlis, J; Josic, K and Shea-Brown, E T (2006). Periodic orbit. *Scholarpedia* 1(7): 1358. http://www.scholarpedia.org/article/Periodic_orbit.
Pikovsky, A and Rosenblum, M (2007). Synchronization. *Scholarpedia* 2(12): 1459. http://www.scholarpedia.org/article/Synchronization.
Rayner, K and Castelhano, M (2007). Eye movements. *Scholarpedia* 2(10): 3649. http://www.scholarpedia.org/article/Eye_movements.
Roberts, E (2007). Gamma-aminobutyric acid. *Scholarpedia* 2(10): 3356. http://www.scholarpedia.org/article/Gamma-aminobutyric_acid.
Saper, C (2009). Hypothalamus. *Scholarpedia* 4(1): 2791. http://www.scholarpedia.org/article/Hypothalamus.
Sherman, S M (2006). Thalamus. *Scholarpedia* 1(9): 1583. http://www.scholarpedia.org/article/Thalamus.
Ward, L M (2008). Attention. *Scholarpedia* 3(10): 1538. http://www.scholarpedia.org/article/Attention.
Wilson, C (2008). Up and down states. *Scholarpedia* 3(6): 1410. http://www.scholarpedia.org/article/Up_and_down_states.

External References

Aguilar, J R and Castro-Alamancos, M A (2005). Spatiotemporal gating of sensory inputs in thalamus during quiescent and activated states. *The Journal of Neuroscience* 25: 10990–11002.
Aston-Jones, G and Bloom, F E (1981). Norepinephrine-containing locus coeruleus neurons in behaving rats exhibit pronounced responses to non-noxious environmental stimuli. *The Journal of Neuroscience* 1: 887–900.
Beenhakker, M P and Huguenard, J R (2009). Neurons that fire together also conspire together: is normal sleep circuitry hijacked to generate epilepsy? *Neuron* 62: 612–632.
Bourassa, J; Pinault, D and Deschenes, M (1995). Corticothalamic projections from the cortical barrel field to the somatosensory thalamus in rats: A single-fibre study using biocytin as an anterograde tracer. *European Journal of Neuroscience* 7: 19–30.
Brown, R E; Stevens, D R and Haas, H L (2001). The physiology of brain histamine. *Progress in Neurobiology* 63: 637–672.
Castro-Alamancos, M A (1999). Neocortical synchronized oscillations induced by thalamic disinhibition in vivo. *The Journal of Neuroscience* 19: RC27.

Castro-Alamancos, M A (2002a). Properties of primary sensory (lemniscal) synapses in the ventrobasal thalamus and the relay of high-frequency sensory inputs. *Journal of Neurophysiology* 87: 946–953.

Castro-Alamancos, M A (2002b). Different temporal processing of sensory inputs in the rat thalamus during quiescent and information processing states in vivo. *Journal of Physiology* 539: 567–578.

Castro-Alamancos, M A (2004a). Dynamics of sensory thalamocortical synaptic networks during information processing states. *Progress in Neurobiology* 74: 213–247.

Castro-Alamancos, M A (2004b). Absence of rapid sensory adaptation in neocortex during information processing states. *Neuron* 41: 455–464.

Castro-Alamancos, M A and Calcagnotto, M E (1999). Presynaptic long-term potentiation in corticothalamic synapses. *The Journal of Neuroscience* 19: 9090–9097.

Castro-Alamancos, M A and Calcagnotto, M E (2001). High-pass filtering of corticothalamic activity by neuromodulators released in the thalamus during arousal: in vitro and in vivo. *Journal of Neurophysiology* 85: 1489–1497.

Castro-Alamancos, M A and Oldford, E (2002). Cortical sensory suppression during arousal is due to the activity- dependent depression of thalamocortical synapses. *Journal of Physiology* 541: 319–331.

Crick, F (1984). Function of the thalamic reticular complex: The searchlight hypothesis. *Proceedings of the National Academy of Sciences of the United States of America* 81: 4586–4590.

Crunelli, V and Hughes, S W (2010). The slow (< 1 Hz) rhythm of non-REM sleep: a dialogue between three cardinal oscillators. *Nature Neuroscience* 13: 9–17.

el Mansari, M; Sakai, K and Jouvet, M (1989). Unitary characteristics of presumptive cholinergic tegmental neurons during the sleep-waking cycle in freely moving cats. *Experimental Brain Research* 76: 519–529.

Fanselow, E E and Nicolelis, M A (1999). Behavioral modulation of tactile responses in the rat somatosensory system. *The Journal of Neuroscience* 19: 7603–7616.

Foote, S L; Aston-Jones, G and Bloom, F E (1980). Impulse activity of locus coeruleus neurons in awake rats and monkeys is a function of sensory stimulation and arousal. *Proceedings of the National Academy of Sciences of the United States of America* 77: 3033–3037.

Friedberg, M H; Lee, S M and Ebner, F F (1999). Modulation of receptive field properties of thalamic somatosensory neurons by the depth of anesthesia. *Journal of Neurophysiology* 81: 2243–2252.

Golshani, P; Liu, X B and Jones, E G (2001). Differences in quantal amplitude reflect GluR4-subunit number at corticothalamic synapses on two populations of thalamic neurons. *Proceedings of the National Academy of Sciences of the United States of America* 98: 4172–4177.

Hartings, J A; Temereanca, S and Simons, D J (2003). State-dependent processing of sensory stimuli by thalamic reticular neurons. *The Journal of Neuroscience* 23: 5264–5271.

Hirata, A and Castro-Alamancos, M A (2008). Cortical transformation of wide-field (multi-whisker) sensory responses. *Journal of Neurophysiology* 100: 358–370.

Hirata, A and Castro-Alamancos, M A (2010). Neocortex network activation and deactivation states controlled by the thalamus. *Journal of Neurophysiology* 103: 1147–1157.

Hirata, A; Aguilar, J and Castro-Alamancos, M A (2006). Noradrenergic activation amplifies bottom-up and top-down signal-to-noise ratios in sensory thalamus. *The Journal of Neuroscience* 26: 4426–4436.

Hirata, A; Aguilar, J and Castro-Alamancos, M A (2009). Influence of subcortical inhibition on barrel cortex receptive fields. *Journal of Neurophysiology* 102: 437–450.

Hobson, J A; McCarley, R W and Wyzinski, P W (1975). Sleep cycle oscillation: Reciprocal discharge by two brainstem neuronal groups. *Science* 189: 55–58.

Hughes, S W; Cope, D W; Blethyn, K L and Crunelli, V (2002). Cellular mechanisms of the slow (< 1 Hz) oscillation in thalamocortical neurons in vitro. *Neuron* 33: 947–958.

Huntsman, M M; Porcello, D M; Homanics, G E; DeLorey, T M and Huguenard, J R (1999). Reciprocal inhibitory connections and network synchrony in the mammalian thalamus. *Science* 283: 541–543.

Kim, U and McCormick, D A (1998). The functional influence of burst and tonic firing mode on synaptic interactions in the thalamus. *The Journal of Neuroscience* 18: 9500–9516.

Landisman, C E et al. (2002). Electrical synapses in the thalamic reticular nucleus. *The Journal of Neuroscience* 22: 1002–1009.

Lee, S; Carvell, G E and Simons, D J (2008). Motor modulation of afferent somatosensory circuits. *Nature Neuroscience* 11: 1430–1438.

McCormick, D A (1992). Neurotransmitter actions in the thalamus and cerebral cortex and their role in neuromodulation of thalamocortical activity. *Progress in Neurobiology* 39: 337–388.

McCormick, D A and von Krosigk, M (1992). Corticothalamic activation modulates thalamic firing through glutamate "metabotropic" receptors. *Proceedings of the National Academy of Sciences of the United States of America* 89: 2774–2778.

McCormick, D A and Bal, T (1997). Sleep and arousal: Thalamocortical mechanisms. *Annual Review of Neuroscience* 20: 185–215.

McGinty, D J and Harper, R M (1976). Dorsal raphe neurons: Depression of firing during sleep in cats. *Brain Research* 101: 569–575.

Moruzzi, G and Magoun, H W (1949). Brain stem reticular formation and activation of the EEG. *Electroencephalography and Clinical Neurophysiology* 1: 455–473.

Rigas, P and Castro-Alamancos, M A (2007). Thalamocortical up states: Differential effects of intrinsic and extrinsic cortical inputs on persistent activity. *The Journal of Neuroscience* 27: 4261–4272.

Shu, Y and McCormick, D A (2002). Inhibitory interactions between ferret thalamic reticular neurons. *Journal of Neurophysiology* 87: 2571–2576.

Sohal, V S and Huguenard, J R (2003). Inhibitory interconnections control burst pattern and emergent network synchrony in reticular thalamus. *The Journal of Neuroscience* 23: 8978–8988.

Steriade, M; Jones, E G and McCormick, D A (1997). *Thalamus*. New York: Elsevier.

Trulson, M E and Jacobs, B L (1979). Raphe unit activity in freely moving cats: Correlation with level of behavioral arousal. *Brain Research* 163: 135–150.

Williams, J A; Comisarow, J; Day, J; Fibiger, H C and Reiner, P B (1994). State-dependent release of acetylcholine in rat thalamus measured by in vivo microdialysis. *The Journal of Neuroscience* 14: 5236–5242.

Yu, C; Derdikman, D; Haidarliu, S and Ahissar, E (2006). Parallel thalamic pathways for whisking and touch signals in the rat. *PLoS Biology* 4: e124.

External Link

http://www.drexelmed.edu/Home/AboutOurFaculty/ManuelCastroAlamancos.aspx Author's website

Vibrissal Location Coding

Ehud Ahissar and Per M. Knutsen

Vibrissal location coding refers to the ways by which the location of external objects is coded (represented) in the vibrissal system of rodents. The vibrissal system contains the vibrissae (whiskers) and the follicles, neurons and muscles associated with them. Coding is traditionally sub-categorized to encoding, i.e., coding at the whisker-object interaction phase, and recoding, i.e., coding at processing stages that are remote from this direct interaction. The vibrissal system is an active-sensing system—the system acquires information about objects in its environment by moving its whiskers ("whisking", see Vibrissal behavior and function, Whisking kinematics) and interpreting the resulting sensations (Figure 1). In recent years, this system has attracted interest from researchers that study the emergence of perception from motor-sensory interactions and from engineers who regard whisking as a useful model system for developing robotic touch platforms, such as whiskered robots. This article reviews recent progress and our current understanding of vibrissal object location coding in rodents.

Coordinates of Object Location

Object location can be specified in Cartesian head-centered coordinates by three spatially-orthogonal axes; rostro-caudal, dorso-ventral and medio-lateral. These axes are often also referred to as the horizontal, vertical and radial axes, respectively. Spherical object coordinates are specified in terms of azimuth, elevation and radial distance relative to a plane intersecting the eyes and nose (Figure 2).

E. Ahissar (✉)
Department of Neurobiology, Weizmann Institute of Science, Rehovot, Israel

P.M. Knutsen
UC San Diego, La Jolla, CA, USA

T.J. Prescott et al. (eds.), *Scholarpedia of Touch*, Scholarpedia,
DOI 10.2991/978-94-6239-133-8_53

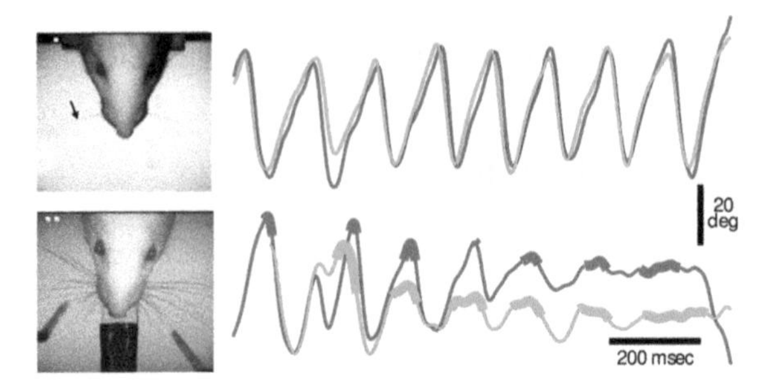

Figure 1 Whisking behavior of freely-moving rats during two different tasks. Left and right C2 whiskers denoted in red and green, respectively. Top: Rhythmic whisking during whisking in air. Bottom: Whisking during an object localization task. Contact periods denoted by thick lines

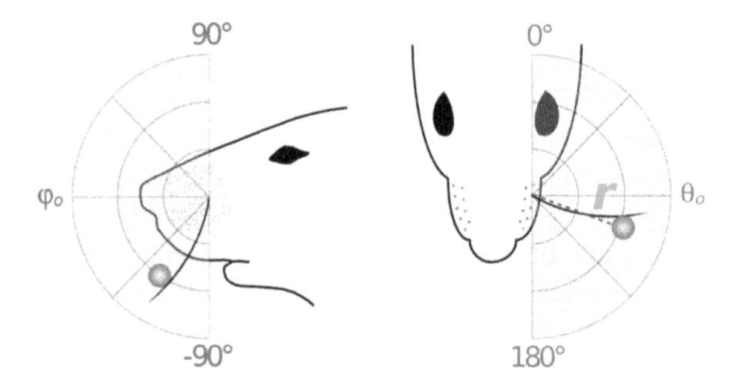

Figure 2 Object location in spherical coordinates. Elevation, ϕ; azimuth, θ; radial distance, r. Blue circle is object

Possible Coding Dimensions

The location of an object could in principle be specified by information contained in four different encoding domains:

Space: Spatial encoding refers to the spatial distribution of activated sensory neurons. For instance, an encoding scheme whereby individual neurons are exclusively activated when objects are located at specific locations would constitute coding by a spatial parameter (neuron identity). An example of such encoding is retinotopic representation of object location. In the vibrissal system, individual primary sensory neurons have single-whisker receptive fields. Thus, these neurons encode the identity of contacting whiskers. The identities of contacting whiskers are, however, ambiguous with respect to the location of an object. Figure 3 illustrates how an object positioned at different coordinates may result in the same combination of contacting whiskers. By increasing the number of contacting whiskers the location of the object can be ascertained with higher accuracy.

Intensity: In addition to being activated by contact, primary sensory neurons are also tuned to contact parameters such as force, velocity and direction. During active

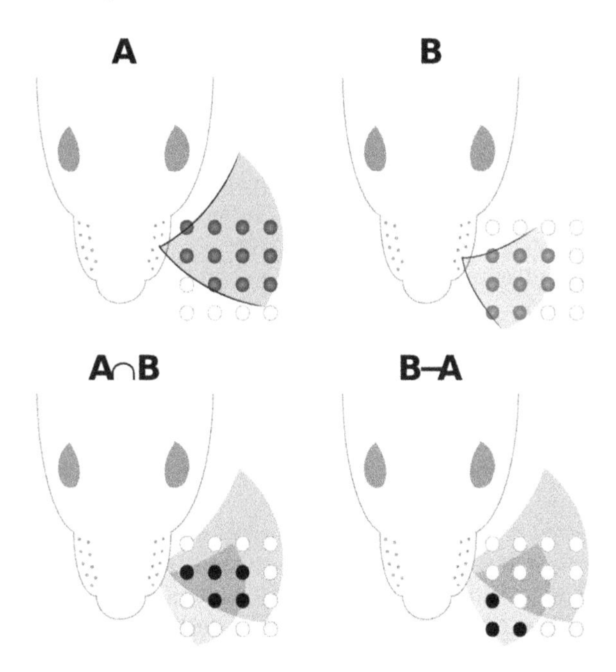

Figure 3 Spatial coding. Two whiskers (*A*, *B*) scan regions of space that are determined by their positions on the whisker pad, lengths and movement patterns. An object positioned at different coordinates (*circles*) can be contacted (*filled circles*) if it is inside the field scanned by a whisker. Individual whiskers may contact an object at many locations and contact by itself does not encode location. Combining contact across whiskers (*bottom panels*) improves accuracy. The intersect A + B outlines the approximate location of an object if both whiskers contact. The difference B–A outlines the location if whisker B, but not A, contacts

sensing, the intensity of receptor activation depends on the location of the contacted object, through the dependency on object location of the rotational and translational forces applied on the follicle (Birdwell et al. 2007). The exact mapping of spatial coordinates onto different levels of neural activity depends on the details of the interactions between the whisker and the object.

Time: The whiskers provide a very sparse ($\sim$1/1000) snapshot of the environment. This is because the whiskers are thin (microns) and relatively widely spaced apart (millimeters), thus introducing 'blind spots' between whiskers that limit the resolution of spatial and intensity coding. During whisking, however, the entire spatial continuum reached by the fanned-out array of whiskers is scanned through time. Thus, space and location can additionally be encoded by the timing of contact events. Sensory neurons along the trigeminal pathway are sensitive to the timing of whisker deflections, accurately encoding events with sub-millisecond precision (Arabzadeh et al. 2005; Jones et al. 2004). During whisking, temporally encoded tactile events can therefore be encoded with higher angular precision than spatially encoded events.

Posture: Crucially, temporal encoding requires temporal reference signals, such as proprioceptive or other motor signals derived from corollary discharges or re-afference. Proprioceptive signals can also code object location directly when contact is controlled; for example, the angle of the whisker encodes the azimuth of the object in head-related coordinates when contact is controlled to be of minimal impingement. The postural dimension also includes morphological coding by which object location is coded by relationships between morphological variables of the whisker at any given moment of interaction with the object.

Neural Encoding of Object Location by Primary Afferents

In awake animals, the behaviors generating sensory inputs can be highly dynamic, reflecting continuous updating of motor output in response to sensory inputs. This makes systematic investigation of sensory signals difficult, prompting the development of anesthetized preparations where naturalistic whisking patterns can be mimicked by electrical stimulation of central motor regions or nerves supplying the mystacial musculature (referred to as *electrical*, *artificial* or *fictive* whisking).

During artificial whisking, primary afferents of the trigeminal ganglion (TG; about 200–400 neurons per whisker) encode tactile events and object location (Szwed et al. 2003; Szwed et al. 2006). Based on the type(s) of event responded to, four principal classes of primary afferents have been characterized (Figure 4). "**Contact**" cells respond briefly and with short latency when a whisker contacts an object. "**Pressure**" cells respond throughout contact. "**Detach**" cells respond when a whisker detaches from an object. "**Whisking**" cells respond only to whisker movement. "Contact", "Pressure" and "Detach" cells are referred to as "**Touch**" cells. Additionally, a class of cells referred to as "**Whisking/Touch**" respond during whisking and increase their firing upon and during touch.

These experiments demonstrate that together, neurons of all these classes encode the 3D location of an object during whisking. This coding exhibits an orthogonal scheme, in which each spatial dimension is coded by an independent neuronal variable. "Contact" cells use a temporal code to encode the **horizontal coordinate** (azimuth) of an object. These cells are activated upon contact, and therefore encode the protraction angle at which the object was encountered by the timing of the first evoked spike. In order for this latency code to be correctly decoded, read-out circuits must compare the time of contact with a reference signal indicating whisker position. This reference signal is provided by "Whisking" cells, whose activity is locked to specific phases of the movement cycle. The **radial coordinate** is primarily encoded by an intensity code. Upon contact, "Touch" cells respond with spiking rates of varying intensity that depends on the radial location of the object. Typically, evoked responses drop in intensity with increasing radial distance. In some cases, this intensity coding is reduced to a binary code with some "Touch" cells responding only when an object is positioned very close to the whisker pad. Not every "Touch" cell, however, encodes radial position in a monotonic manner

Figure 4 Classes of trigeminal ganglion (TG) cells that respond to movement and contact events

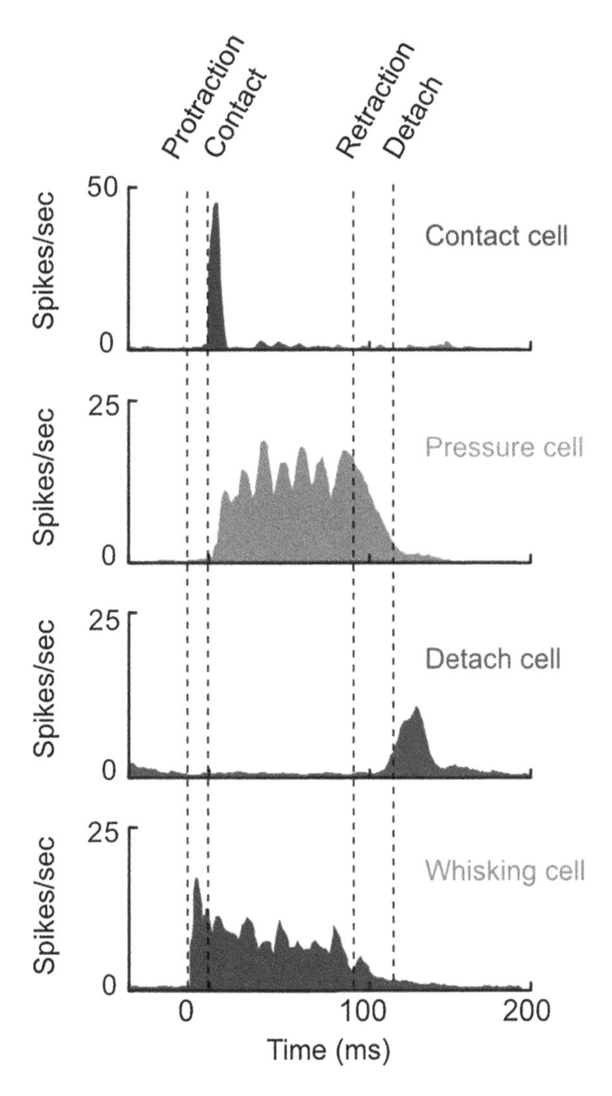

(Szwed et al. 2006). Thus, reliable decoding by read-out circuits should pool the activity of many "Touch" cells. Encoding of vertical object location is determined by anatomy. Because primary afferents have single-whisker receptive fields and because whiskers move along the axis of rows (Bermejo et al. 2002; Knutsen et al. 2008), any afferent is therefore activated only when an object is present at the elevation of its receptive whisker. Thus, the **vertical coordinate** of contact is encoded by neuron identity.

Behavioral Aspects of Object Localization

Behavioral studies of object localization in rats have confirmed the contribution of temporal, intensity and spatial information in object localization, and are consistent with the orthogonal scheme revealed using artificial whisking (Ahissar and Knutsen 2008). Behavioral studies have isolated the optimal behavior for each spatial dimension by maintaining object location constant in two and varying location only along one dimension. The selective contribution of spatial information has been probed by removing sub-sets of whiskers, and the contribution of temporal cues facilitated by training animals to make relative comparisons that exceed the acuity limits of spatial coding (Figure 5; Krupa et al. 2001; Shuler et al. 2002; Knutsen et al. 2006; Mehta et al. 2007).

The ability of rats to localize objects along the horizontal direction depends on both prior training conditions and kinematics of whisker movements. Two independent studies of horizontal object localization in rats (Knutsen et al. 2006; Mehta et al. 2007) agree on the following; (1) Rats can accurately localize with a single whisker on each side of the face, provided they first learn the same task using many whiskers, (2) the identity and number of whiskers contacting the objects does not determine horizontal acuity, (3) whisker movements are required to localize, (4) the energy of whisking correlates with acuity, and (5) relative localization (between co-existing objects) is more accurate then absolute (memory-guided) localization. Head-fixed studies have also confirmed that mice can localize along the horizontal dimension with a single whisker (O'Connor et al. 2010). Single-whisker localization precludes any dependency on spatial encoding of horizontal object location.

Horizontal acuity during relative localization (as fine as $\sim 1°$ when comparing positions of two co-existing objects) is an order of magnitude better than the limit imposed by the spacing of adjacent same-row whiskers ($\sim 20°$), a performance level referred to as vibrissal hyperacuity (Knutsen et al. 2006). During absolute (memory-guided) localization the behavioral resolution ($\sim 15°$) is closer to the whisker spacing limit. Vibrissal hyperacuity is also achieved by rats with just one whisker remaining on each mystacial pad (i.e. one whisker contacting each object).

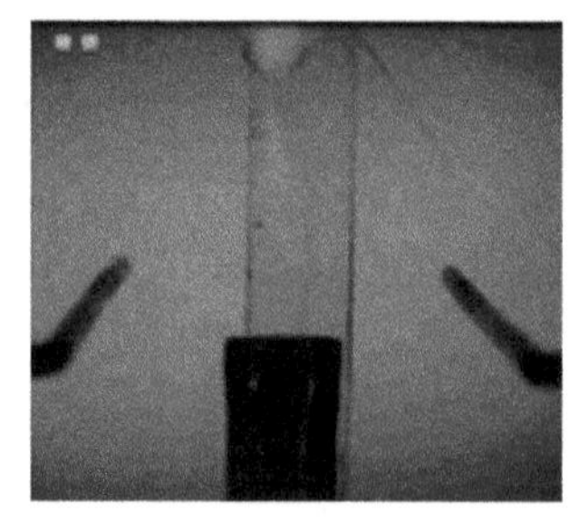

Figure 5 Head and whisker movements during an object localization task (described in Knutsen et al., 2006). This rat was trained to detect which of the two vertical poles was closer to the home cage, using a single whisker on each side

This manipulation excludes the possibility that the identity of contacting whiskers is compared (see Figure 3). Instead, other kinematic variables afforded by whisker movements, such as the relative angles or contact-times, may be important for horizontal encoding. Horizontal location may also be encoded in part by a labeled-line code (e.g. by a set of neurons, each of which fires if and only if contact occurs at a certain horizontal coordinate) due to interactions between torsional whisker rotation and directional tuning of afferents (Knutsen et al. 2008).

Behavior during localization of objects along the radial axis differs from that during horizontal localization. Whisking is not always required during radial localization. Instead, the whiskers can be brought in contact with objects through slow adjustments in whisker set-point and head/body movements. In one paradigm of radial localization, radial acuity correlates with the number (but not the identities) of intact whiskers available to the rat, and has not been shown to improve with the presence of a reference object (Krupa et al. 2001; Shuler et al. 2002). The following conclusions about radial encoding have been inferred from these available observations. In freely moving rodents a simple labeled-line code can probably be ruled out since whisker identities used to localize do not determine performance (note that head fixed rodents may exhibit a different strategy). Suppression of whisking in freely moving rodents suggests that temporal cues are not important for radial localization. Rather, the observation that radial acuity falls proportionally to the number of removed whiskers, suggests that radial object location is encoded by a sensory cue accumulated across all whiskers, such as the firing intensity of populations of Touch neurons (see above).

Thus far, no behavioral test of vertical object localization has been reported.

Orthogonal Coding of Object Location

Behavioral studies of object localization agree with the observations that primary sensory afferents encode object coordinates by orthogonal neuronal codes. Primary afferents have been shown to be highly temporally tuned and encode horizontal object location on an individual basis, consistent with behavioral observations that the highest acuity can be achieved with a single whisker contacting an object. Radial location is poorly encoded by individual afferents and requires the pooled activity of multiple afferents to reach the precision exhibited by behaving animals. Consistent with this, behaving animals also need multiple whiskers to localize objects along the radial axis. These are consistent with a temporal and a population rate code for the horizontal and radial coordinates, respectively. No behavioral support has so far been found for the labeled-line spatial code of vertical coordinates (Figure 6).

Orthogonal coding appears to be an efficient scheme for object localization as during natural behavior object location is encoded for all spatial dimensions simultaneously (Knutsen and Ahissar 2009). By relying on an independent neuronal variable for each dimension, 3D location can be read out from the same afferents using orthogonal decoding circuits in parallel.

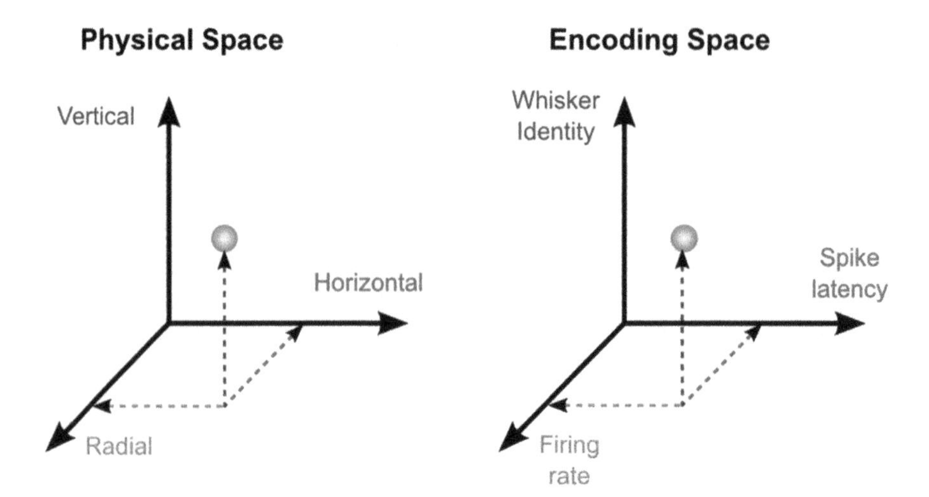

Figure 6 Proposed orthogonal encoding scheme of object location. During exploration, whisker movements are mainly along the horizontal plane. Upon contact with an object (*blue dot*) the timing of the contact response (latency to spikes) encodes the horizontal dimension (*red*). The vertical dimension (*blue*) is encoded by the identity of contacting whiskers. The radial dimension (*green*) is encoded by the intensity of activation (e.g. rate of evoked spikes) due to bending and mechanical forces acting upon whisker shaft. These three codes for location are orthogonal and the spatial dimensions can thus be encoded independently of each other

Morphological Coding of Object Location

Object location in the horizontal plane, spanned by the azimuthal and radial coordinates, is also encoded by phase planes of whisker-related morphological variables (Bagdasarian et al. 2013). Phase planes spanned by angular and curvature variables encode object location reliably already 15 ms after object contact onset, and reliability increases as long as contact continues. Coding is based on motor-sensory contingencies rather on sensory cues alone (Figure 7). Whisker rigidity allows direct transformation of morphological coding to mechanical coding within the follicle (Bagdasarian et al. 2013; Quist and Hartmann 2012).

Recoding

Eventually, proposed encoding schemes of object location must also take into account realistic read-out circuits that could possibly recode each variable in some common internal language. Such recoding is sometimes termed "decoding", although unlike engineered devices, no restoration of the original encoding signals is attempted in the brain. Recoding of temporal information requires an additional reference signal that signals whisker position. Such a reference signal can be generated internally as a corollary discharge, by proprioceptors or by

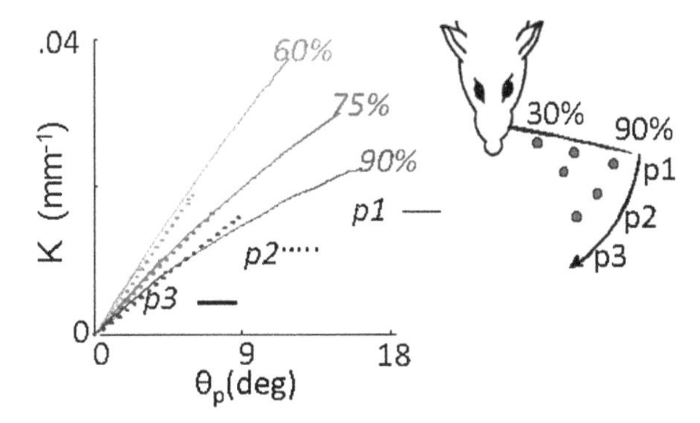

Figure 7 Morphological coding by whisker angle and curvature. Trajectories of whiskers in the θp—K phase plain (θp: push angle, maximal change in whisker angle from contact onset during contact; K: curvature at whisker's base) are shown. Neither of the two variables provide unambiguous coding of object location by itself; for example, K around .02 mm-1 (at max protraction) codes for both ~[p2, 60 %] and ~[p1, 90 %]. In contrast, a motor-sensory contingency, between the motor (θp) and sensory (K) parameters, provides unique coding of both azimuthal and radial coordinates (Bagdasarian et al. 2013)

mechanoreceptors (re-afference) responsive to whisker motion (e.g. as signaled by "Whisking" cells, Szwed et al. 2003). The comparison between Touch and reference signals could be implemented by phase-locked loops or phase detectors. Re-afferent signaling of whisker motion has been observed in both the somatosensory thalamus (Yu et al. 2006) and cortex (Fee et al. 1997; Brecht et al. 2006; Crochet and Petersen 2006; Derdikman et al. 2006; Curtis and Kleinfeld 2009). Evidence for recoding the horizontal coordinate of object location in firing rates, via phase detection in the thalamocortical network, was recently demonstrated (Curtis and Kleinfeld 2009; Yu et al. 2013). Intensity coding of radial position is likely based on interpolating firing intensity across multiple whiskers. This could be implemented by neural integrators, peak detectors, attractor neural networks or synfire chains. The vertical coordinate, encoded by labeled-lines, can be read out by threshold detectors. Recoding may involve iterative processes within thalamocortical loops (Yu et al. 2013; Edelman 1993) as well as across motor-sensory-motor loops involving multiple object contacts (Knutsen et al. 2006; Saig et al. 2012). During such iterative processes, various coding schemes may be used in parallel or in sequence (Horev et al. 2011).

Role of Motor Strategies

Motor control of head, pad and whiskers is crucial for the consistency, reliability and resolution of each of the localization codes: pad orientation, whisking velocity and contact force will directly affect the mapping of the spatial coordinates into neuronal variables. The closed-loop architecture of the vibrissal system allows efficient adaptive control of vibrissal touch, though it remains to be seen to what extent adaptive behavior is indeed manifested during perceptual tasks, which motor variables are controlled in which contexts, and how they are controlled.

A perceptual coding scheme, unlike a sensory coding scheme, must include motor variables in its definition. This is made clear by the fact that the sensory cues that serve perceptual comparisons are not necessarily the same cues that convey primary sensory data, where the latter refers to the manner of receptor activation. The former depend on motor strategies, which involve multiple levels of sensory-motor control loops encompassing body, head and whisker movements, while the latter depend only on the peripheral level of vibrissal interaction with the external world. For example, one rat may whisk synchronously on both sides and use the time difference between left and right contacts as a primary cue for discrimination (Knutsen et al. 2006). Another rat may instead aim to contact (rather than move) synchronously and use the bilateral angular difference as a primary cue. In these sensory-motor processes, both whisker angle and contact times (the sensory data) are crucial, but each serves as a perceptual cue in only one of the strategies, and as a sensory-motor coordinator in the other.

References

Ahissar, E and Knutsen, P M (2008). Object localization with whiskers. *Biological Cybernetics* 98 (6): 449–458. doi:10.1007/s00422-008-0214-4.

Arabzadeh, E; Zorzin, E and Diamond, M E (2005). Neuronal encoding of texture in the whisker sensory pathway. *PLoS Biology* 3(1): 155–165. doi:10.1371/journal.pbio.0030017.

Bagdasarian, K et al. (2013). Pre-neuronal morphological processing of object location by individual whiskers. *Nature Neuroscience* 16: 622–631. doi:10.1038/nn.3378.

Bermejo, R; Vyas, A and Zeigler, H P (2002). Two-dimensional monitoring of whisker movements. *Somatosensory & Motor Research* 19(4): 341–346. doi:10.1080/0899022021000037809.

Birdwell, J A et al. (2007). Biomechanical models for radial distance determination by the rat vibrissal system. *Journal of Neurophysiology* 98(4): 2439–2455. doi:10.1152/jn.00707.2006.

Brecht, M; Grinevich, T E; Jin, T E; Margrie, T and Osten, P (2006). Cellular mechanisms of motor control in the vibrissal system. *European Journal of Neuroscience* 453: 269–281. doi:10.1007/s00424-006-0101-6.

Curtis, J C and Kleinfeld, D (2009). Phase-to-rate transformations encode touch in cortical neurons of a scanning sensorimotor system. *Nature Neuroscience* 12: 492–501. doi:10.1038/nn.2283.

Crochet, S and Petersen, C C (2006). Correlating whisker behavior with membrane potential in barrel cortex of awake mice. *Nature Neuroscience* 9(5): 608–610. doi:10.1038/nn1690.

Derdikman, D; Haidarliu, S; Bagdasarian, K; Arieli, A and Ahissar, E (2006). Layer-specific touch-dependent facilitation and depression in the somatosensory cortex during active whisking. *The Journal of Neuroscience* 26(37): 9538–9547. doi:10.1523/jneurosci.0918-06.2006.

Edelman, G M (1993). Neural Darwinism: Selection and reentrant signaling in higher brain function. *Neuron* 10: 115–125.

Fee, M A; Mitra, P P and Kleinfeld, D (1997). Central versus peripheral determinants of patterned spike activity in rat vibrissa cortex during whisking. *Journal of Neurophysiology* 78(2): 1144–1149.

Horev, G et al. (2011). Motor-sensory convergence in object localization: A comparative study in rats and humans. *Philosophical Transactions of the Royal Society of London Series B, Biological Sciences* 366(1581): 3070–3076. doi:10.1523/JNEUROSCI.2432-12.2012.

Jones, L M; Depireux, D A; Simons, D J and Keller, A (2004). Robust temporal coding in the trigeminal system. *Science* 304: 1986–1989. doi:10.1126/science.1097779.

Knutsen, P M and Ahissar, E (2009). Orthogonal coding of object location. *Trends in Neuroscience* 32: 101–109.

Knutsen, P M; Pietr, M and Ahissar, E (2006). Haptic object localization in the vibrissal system: behavior and performance. *The Journal of Neuroscience* 26(33): 8451–8464. doi:10.1523/jneurosci.1516-06.2006.

Knutsen, P M; Biess, A and Ahissar, E (2008). Vibrissal kinematics in 3D: Tight coupling of azimuth, elevation, and torsion across different whisking modes. *Neuron* 59(1): 35–42. doi:10.1016/j.neuron.2008.05.013.

Krupa, D; Matell, M S; Brisben, A J; Oliveira, L M and Nicolelis, M A L (2001). Behavioral properties of the trigeminal somatosensory system in rats performing whisker-dependent tactile discriminations. *The Journal of Neuroscience* 21(15): 5752–5763.

Mehta, S M; Whitmer, D; Figueroa, R; Williams, B A and Kleinfeld, D (2007). Active spatial perception in the vibrissa scanning sensorimotor system. *PLoS Biology* 5(2): 309–322. doi:10.1371/journal.pbio.0050015.

O'Connor, D et al. (2010). Vibrissa-based object localization in head-fixed mice. *The Journal of Neuroscience* 30: 1947–1967. doi:10.1523/jneurosci.3762-09.2010.

Quist, B W and Hartmann, M J (2012). Mechanical signals at the base of a rat vibrissa: The effect of intrinsic vibrissa curvature and implications for tactile exploration. *Journal of Neurophysiology* 107: 2298–2312. doi:10.1152/jn.00372.2011.

Saig, A; Gordon, G; Assa, E; Arieli, E and Ahissar, E (2012). Motor-sensory confluence in tactile perception. *The Journal of Neuroscience* 32(40): 14022–14032. doi:10.1523/JNEUROSCI.2432-12.2012.

Shuler, M G; Krupa, D and Nicolelis, M A L (2002). Integration of bilateral whisker stimuli in rats: Role of the whisker barrels cortices. *Cerebral Cortex* 12(1): 86–97. doi:10.1093/cercor/12.1.86.

Szwed, M et al. (2006). Responses of trigeminal ganglion neurons to the radial distance of contact during active vibrissal touch. *Journal of Neurophysiology* 95(2): 791–802. doi:10.1152/jn.00571.2005.

Szwed, M; Bagdasarian, K and Ahissar, E (2003). Encoding of vibrissal active touch. *Neuron* 40(3): 621–630. doi:10.1016/s0896-6273(03)00671-8.

Yu, C; Derdikman, D; Haidarliu, S and Ahissar, E (2006). Parallel thalamic pathways for whisking and touch signals in the rat. *PLoS Biology* 4: 819–825. doi:10.1371/journal.pbio.0040124.

Yu, C et al. (2013). Coding of object location in the vibrissal thalamocortical system. *Cerebral Cortex* Epub: bht241. doi:10.1093/cercor/bht241.

Vibrissal Texture Decoding

Ehsan Arabzadeh, Moritz von Heimendahl and Mathew Diamond

Texture is a central component of touch. To learn how contact with a surface gives rise to a sensation of texture, many laboratories have examined the vibrissae system of rodents—a highly efficient sensory system with well-studied structural organization (Kleinfeld et al. 2006). **Vibrissal texture decoding** summarizes current knowledge about how whisking on surfaces leads to texture sensation. The vibrissae system of rats presents a unique opportunity for investigating how sensory receptors generate signals through their interaction with the environment, and how the brain reads and interprets the afferent signals.

Anatomy and Physiology

Rodents have a set of 30-some long whiskers on each side of the snout, together with short hairs packed more densely around the nose and mouth (Figure 1). The long whiskers, also called *mystacial vibrissae*, are the focus of this chapter. Several hundred primary afferent fibers innervate specialized receptors on each whisker shaft (Ebara et al. 2002), and these are excited by whisker movement. Signals travel along the sensory nerve, past the cell body in the trigeminal ganglion, and form synapses in the brain stem. The axons of second-order neurons cross the brain midline and travel to the thalamic somatosensory nuclei, where the second synapse is located. Thalamic neurons project to the primary somatosensory cortex, conveying information to layer IV cell populations as well as target populations in other layers (see Kleinfeld

E. Arabzadeh (✉)
Eccles Institute of Neuroscience, John Curtin School of Medical Research, the Australian National University, Canberra, ACT, Australia

M.v. Heimendahl
Bernstein Center for Computational Neuroscience, Humboldt Universität zu Berlin, Berlin, Germany

M. Diamond
Tactile Perception and Learning Lab, International School for Advanced Studies, Trieste, Italy

T.J. Prescott et al. (eds.), *Scholarpedia of Touch*, Scholarpedia,
DOI 10.2991/978-94-6239-133-8_54

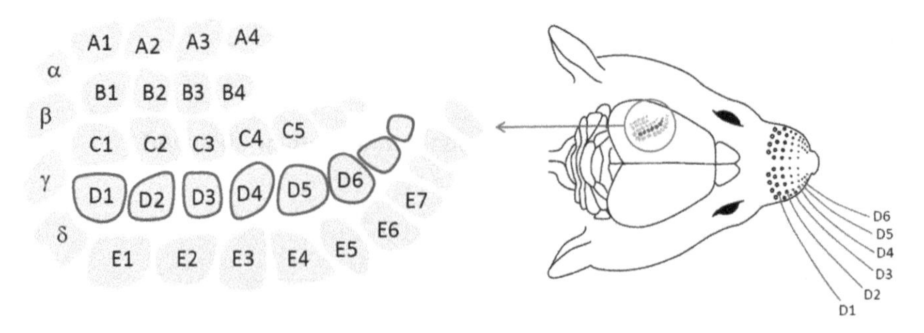

Figure 1 The arrangement of vibrissae on the snout and their topographic cortical representation

et al. 2006). Here, the whisker area—also known as barrel cortex—is arranged as a topographic map where neurons in a given *barrel* and its associated column respond most strongly to the corresponding whisker (Woolsey and van der Loos 1970). Thus, sensory signals arising from individual whiskers are channeled through restricted population of neurons that can be identified and sampled by recording electrodes. The fact that the functional map, revealed by neuronal receptive fields, matches the readily visible anatomical map (the barrel field), makes this sensory system particularly attractive to neuroscientists.

Behavior

Along with olfaction, whisker touch represents the major channel through which rodents collect information from the nearby environment. The signals arise through an active process called whisking, a sweeping motion of the mystacial vibrissae forward and backward to encounter objects and palpate them (Berg and Kleinfeld 2003; Hill et al. 2008), often in conjunction with movement of the head (Mitchinson et al. 2007; Towal and Hartmann 2006). Characterizing the way by which rats actively engage with a texture to extract its identity is an essential step in understanding vibrissal texture decoding.

Several laboratories have trained rats to distinguish between rough and smooth textures (Guic-Robles et al. 1989; Carvell and Simons 1990; Prigg et al. 2002; von Heimendahl et al. 2007; Ritt et al. 2008). When probed for their psychophysical discrimination threshold, the animals were able to discriminate a smooth surface from one with grooves that were 50 μ deep and spaced at 90 μ. They failed, though, for 15 μ deep, 50 μ spaced grooves (Carvell and Simons 1990). In a similar texture discrimination task (von Heimendahl et al. 2007 and Figure 2) using P100 sandpaper as a rough surface, the rats' stereotypical discrimination behavior was filmed with a high-speed camera to quantify whisker use: Intervals of whisker contact with the texture were brief; in a typical trial, the rat made 1–3 touches per whisker of 24–62 ms duration each. The rat then began to withdraw its head, which the

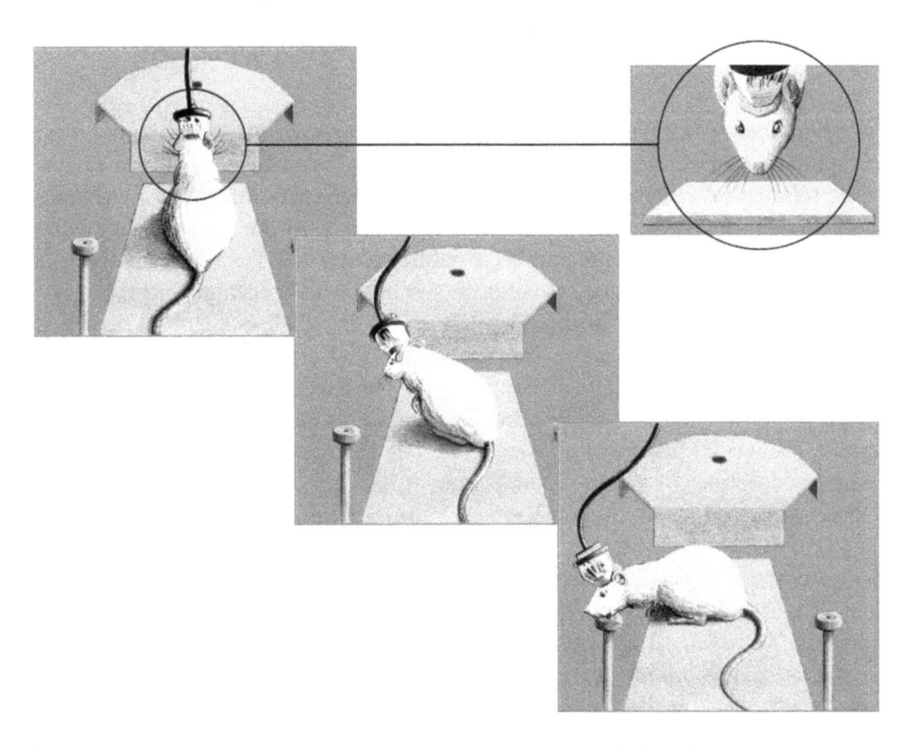

Figure 2 Behavioral paradigm from von Heimendahl et al. (2007). On each trial, the rat approached the texture plate (Rough or Smooth) which signaled the reward spout (*Left* or *Right*). After touching the texture (*upper panel* and the *inset*) the rat made a behavioral choice by turning towards one of the spouts (*middle panel*) and collected the water reward if the choice was correct (*lower panel*)

investigators took to be the first visible sign of its choice. The time from first contact to this moment of choice was 98–330 ms (interquartile ranges). Yet another similar study found that, when rats palpate a surface before their snout, typically only the more rostral (frontal) whiskers tend to reach the surface, and less often the (longer) more caudal ones (Ritt et al. 2008).

Whisker Kinetics

When a whisker's tip or shaft makes contact with a texture, its movement changes; whisker motion signals report to the brain what the whiskers have contacted. How do whiskers interact with a textured object to prepare a meaningful message for the brain? This is an important question because the capacity of the behaving animal to discriminate between textures must be based upon (and can never exceed) the information contained in the movement signals.

Hypotheses

Two hypotheses compete to explain which features of whisker motion vary according to texture. According to the "resonance hypothesis", a given texture drives mechanical resonance specifically in those whiskers that possess the resonance frequencies best matching the input frequency to the whisker. Input frequency is the product of the texture's spatial frequency with the whisker's speed of translation across the texture. On the rat's snout, whisker length increases systematically from the front to the back (Brecht et al. 1997; Neimark et al. 2003) generating a spatial gradient in frequency tuning (Neimark et al. 2003). In an analogy to the cochlea—a frequency analyzer par excellence—the map-like projection from vibrissae to cortex causes each texture to excite a specific subset of barrels. The spatial pattern of activity in the barrel cortex thus encodes the spatial frequency spectrum of the contacted texture.

The "kinetic signature hypothesis" views resonance as an unavoidable consequence of the whisker structure (a tapered elastic beam), but not necessarily central to the sensation of texture. Instead, this view stresses the conversion of surface shape into trains of discrete motion events, sometimes called stick/slip events, by individual whiskers (Arabzadeh et al. 2005). As a whisker contacts a textured surface, the grains produce irregularities in the movement trajectory. Among these irregularities, those with high-velocity (and high-acceleration) are hypothesized to encode textures by their occurrence, number, and possibly timing. Sensory receptors as well as barrel cortex neurons are tuned to the key features of the signature—the high velocity jumps over texture grains (Arabzadeh et al. 2005).

Whisker Tracking Experiments

To select between these hypotheses, whisker motion has been studied in the following preparations:

- In contact with a rotating textured cylinder
  Whisker motion was monitored when an anchored whisker (fixed boundary condition) was held in contact with a rotating cylinder (Neimark et al. 2003; Andermann et al. 2004; Moore and Andermann 2005; Ritt et al. 2008). The data were consistent with the resonance hypothesis: whiskers, according to their length, exhibited vibrations of maximal amplitude in contact with specific textures. Measurements, however, were made not during brief touches but rather when whiskers had reached a steady state vibration.
- During head sweep
  Rats were trained to touch a plate containing rough and smooth regions (Ritt et al. 2008). During a brief initial approach the rat identified the contacted texture, and subsequently made a head turn towards the reward port. High frame-rate videos showed, for the first time, the presence of high-velocity slip

events in awake animals performing a discrimination task, with high velocity, high amplitude events occurring preferentially on the rough surface. The slip events were followed by microvibrations when mechanical energy transferred to the whisker by hitting, or being released from, the surface generated ringing at the resonance frequency (about 100–300 Hz) of the whisker. However, given the design of the experiment, the rats' head trajectory towards a reward location should be taken as the expression of a completed sensory decision, occurring only *after* critical texture information provided by the whisker signal has been integrated. The whisker movement analysis was likely dominated by the post-choice data, when displacements are greater and speeds faster. Therefore, the study provided interesting observations concerning whisker motion during high-speed translation across a surface, but did not speak to the question of which features of whisker motion inform the rat about texture. To find the behaviorally relevant features, it would have been essential to focus on the whiskers exactly when the rat did so—during the initial contact phase, when it performed the discrimination (Diamond et al. 2008a, b c). In response to this observation, Ritt et al. reanalysed their data excluding the post-choice phase and reported the presence of whisker-specific micromotions during the pre-decision palpation phase (Ritt et al. 2008). However, beyond being a characteristic of whisker, it was not specified to what extent these early micromotions were informative about the contacted texture.

- Electrical whisking

  In this paradigm (Arabzadeh et al. 2005), rats were anesthetized and "electrical whisking" was induced at 8 Hz by direct stimulation of the facial nerve, the motor bundle innervating the muscles of the whisker pad. Figure 3 shows the movement profile as the whisker moved either freely in the air or on sandpapers of different grades. Under free whisking condition, the trajectory was a smooth ellipsoid. In contrast, whisking across grainy surfaces produced irregularities in the trajectory; each texture was associated with a distinct trajectory, giving rise to the abovementioned hypothesis of "kinetic signatures". Kinetic signatures are best visualized in the velocity profile—that is, the temporal sequence of velocity features across the course of a whisk. When whiskers were not touching a texture but freely moving in air, their motion was continuous and fluid. But moving along the texture, the whisker tip tended to get fixed in place before bending and springing loose. This pattern of movement gave each texture a characteristic velocity profile—the number of high-velocity events and their temporal pattern differed across textures. The same findings have been reported in an additional study based on electrical whisking (Lottem and Azouz 2008). Although these experiments show clear texture-dependent differences in whisker motion, they leave open the possibility that electrical whisking might differ from the animal's own self-generated motion in some important way.

- Whisking in awake animal

  Additional experiments were carried out (Wolfe et al. 2008) to determine whether the encoding of texture under awake behaving conditions is more

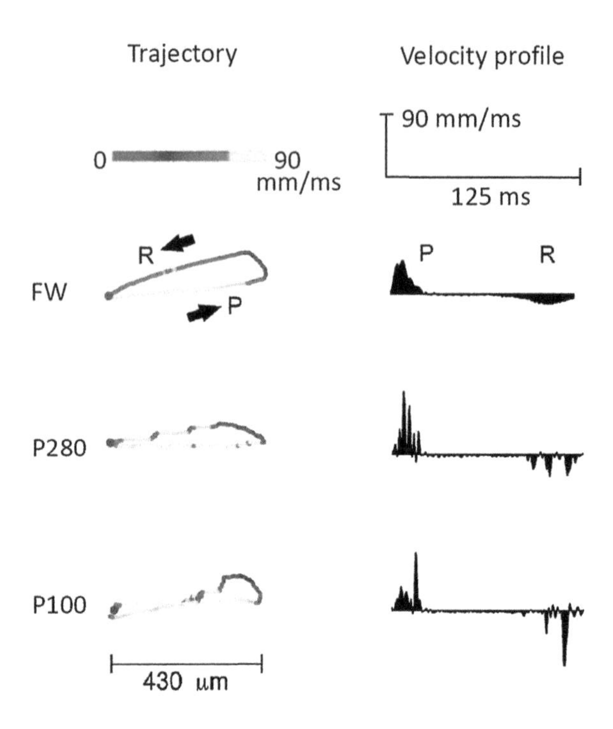

Figure 3 Whisker trajectory and velocity profile measured when the whiskers moved through the air with no texture present (*upper trace*), and when the whiskers contacted texture P280 (*middle trace*) and texture P100 (*lower trace*). Adapted from (Arabzadeh et al. 2005)

consistent with the resonance hypothesis or the kinetic signature hypothesis. Wolfe et al. trained rats to whisk against sandpapers and whisker motion was recorded by an optic sensor. Similar to the electrical whisking data, moving along the texture, the whiskers' trajectory was characterized by an irregular, skipping motion: the whisker tip tended to get fixed in place ("stick"), before bending and springing loose ("slip") only to get stuck again, reminiscent of the motion seen during electrical whisking. On progressively coarser textures there were progressively more high speed and high acceleration stick-slip events; on progressively smoother textures there were progressively more low speed and low acceleration stick-slip events. So the ratio of the number of high to low magnitude events gave a remarkably fine kinetic signature of the contacted texture. It is also significant that one candidate encoding mechanism could be ruled out: ringing at the resonance frequency carried no information about texture in the experimental conditions of Wolfe.

Outcome

Although original experiments with the rotating cylinder were consistent with the resonance hypothesis (Neimark et al. 2003; Andermann et al. 2004; Moore and Andermann 2005;), further experiments (Ritt et al. 2008; Wolfe et al. 2008) showed that texture-specific ringing does not generalize to an actively whisking animal. In natural settings when the whisker palpates the texture (instead of the texture moving against an end-fixed whisker), after detaching from the texture the whisker sometimes vibrates at its resonance frequency but this "ringing" is characteristic of the whisker, not of the texture.

Any argument for the kinetic signature hypothesis would be bolstered by the demonstration that kinetic signatures can robustly encode textures under different behavioral conditions. Because whisking is an actively controlled sensory-motor behavior, its parameters vary from moment to moment. Moreover, a rat may encounter the textured surface to the side of its snout (Wolfe et al. 2008), in front of the snout (Carvell and Simons 1990; von Heimendahl et al. 2007; Ritt et al. 2008), or on the ground. Can a single encoding mechanism work under all these conditions? To answer this it is useful to compare the whisker motion obtained under two very different settings: electrical whisking (Arabzadeh et al. 2005) versus Wolfe et al.'s behavioral experiment (Wolfe et al. 2008). In the case of electrical whisking, textures were positioned 7 mm from the snout (about 20 mm in Wolfe) and whiskers moved in a plane parallel to the textured surface (orthogonal in Wolfe). Despite the diversity of conditions, the same encoding mechanism was found in both studies: each texture generated a kinetic signature, a distinct motion pattern folded into the whisker trajectory. Observing side-by-side traces of whisker velocity obtained from the two studies reveals a striking similarity (Figure 4). Large

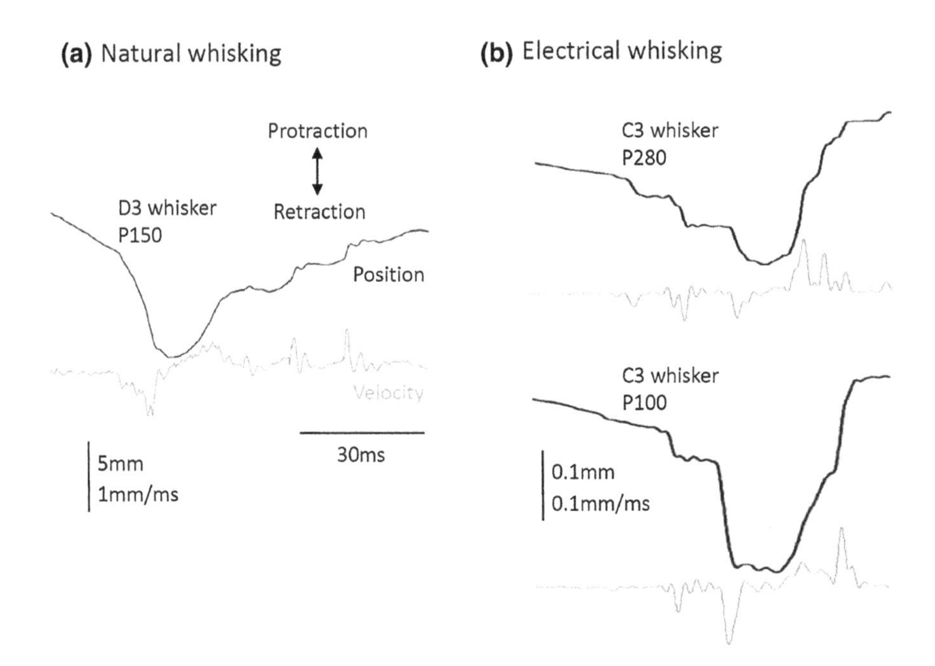

Figure 4 Comparison of whisker motion profiles collected under different conditions. (**a**) natural whisking and (**b**) electrical whisking. Adapted from Diamond et al. (2008a)

amplitude slip/stick events were also observed in the behavioral paradigm used by Ritt et al. (2008), as rats turned their head towards a reward port and whiskers swept over a surface in front of the rat. More recently, it has been found that kinetic signatures obtained during electric whisking are highly informative about texture identity even when varying key parameters such as whisking frequency and distance of the surface from the follicle (Lottem and Azouz 2008).

Though currently there is no evidence to support the hypothesis that the full set of whiskers encodes texture the way that the basilar membrane encodes acoustic frequency (Neimark et al. 2003; Andermann et al. 2004), there remain several possible interpretations for resonance-related microvibrations, among them are the following three examples. First, they may play a role when there is a sustained, driving input with constant relative translation between the whisker and the surface. An example might be a rat running down a tunnel with textured walls—concrete versus metal. These could produce conditions that resemble the rotating cylinder. Second, microvibrations may have some function unrelated to the perception of texture. For example, they may amplify contact signals to enhance edge detection, as suggested by Hartmann et al. (2003); or they may serve to maintain high-velocity input during prolonged contact, so that neuronal responsiveness in cortex does not diminish through adaptation (Arabzadeh et al. 2003; Maravall et al. 2007). Third, they may have no perceptual significance whatsoever. Whiskers undergo high frequency vibrations because they are tapered elastic beams and their resonance follows from mechanical principles. It cannot be excluded that resonance is an unhelpful but unavoidable consequence of the physical properties of whiskers. Indeed, at moments when resonance would add noise to the afferent signal, rats may whisk in such a way as to suppress resonance—for example by increasing the damping (vibration absorption) in the follicle.

Kinetic signatures thus seem the most plausible texture encoding mechanism. But to specify in a definitive way which features of whisker kinetics are relevant to texture perception, more evidence is needed. Any candidate feature must occur in the short interval during which the animal forms its percept of texture; it must vary according to texture during this critical interval; it must evoke neuronal activity that carries information about texture; those neuronal response features must influence the animal's percept.

Neuronal Encoding of Texture

How is the texture-specific motion signal transformed by the nervous system into spiking activity? In anesthetized animals, neurons at all stages of the sensory pathway, from the trigeminal ganglion to barrel cortex, are effectively driven by high-speed and high-acceleration kinetic events (Shoykhet et al. 2006; Arabzadeh et al. 2003; Jones et al. 2004; Arabzadeh et al. 2005). Thus, texture-specific kinetic signatures obtained through electrical whisking are represented by differences in the overall rate of neuronal firing, which follow from the number and size of kinetic

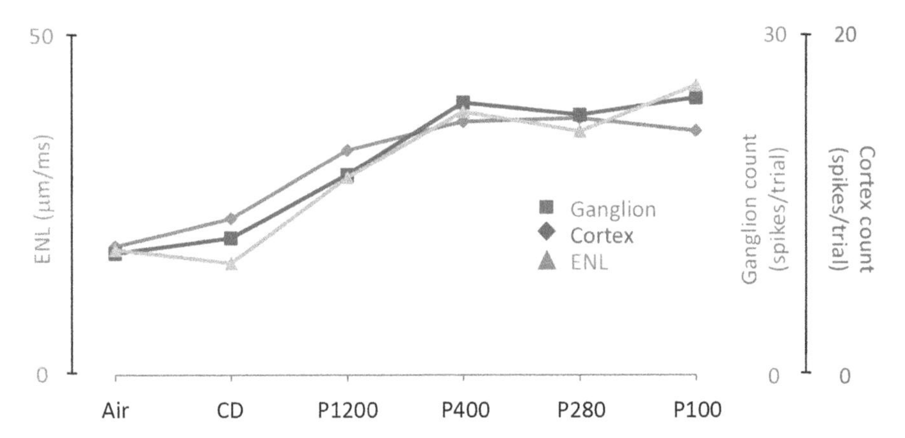

Figure 5 Electrical stimulation of the facial nerve induced whisking in the air, on the smooth surface of a compact disk (CD) and on sandpapers of four different grades: P1200, P400, P280, P100 (from fine-grained to coarse-grained). Progressively coarser textures evoked kinetic signatures with higher Equivalent Noise Levels (*green triangles*). Higher ENLs led to higher firing rates in the trigeminal ganglion (*blue squares*) and barrel cortex (*red diamonds*). Adapted from (Arabzadeh et al. 2005)

events. They are also represented by differences in the temporal pattern of neuronal firing, caused by the temporal pattern of kinetic events (Arabzadeh et al. 2005, 2006). The firing-rate coding mechanism is illustrated in Figure 5, where spike count per trial, measured in the trigeminal ganglion and the barrel cortex (blue squares and red diamonds, respectively) are given for the corresponding set of textures (Arabzadeh et al. 2005). It is evident that the value of ENL (Equivalent Noise Level is a measure of energy related to the number and size of kinetic events) for a given texture's motion profile was accurately translated to firing rate. The temporal pattern coding mechanism is illustrated in Figure 5, where electrical whisking on textures P400, P280 and P100 evoked distinctive spike sequences in a ganglion cell. More studies are needed in order to quantify the relation between whisker kinetics and neuronal firing in alert, behaving animals.

Decoding of the Texture Signal

A third problem in texture sensation is to understand how spike trains are "read out" to allow a behaving animal to discriminate between textures. The two coding mechanisms described above suggest two corresponding read out mechanisms, *temporal integration* and *temporal pattern*. According to the first hypothesis, the brain identifies texture by extracting a single integrated number of spikes (total number of spikes, or else spikes per unit of time) accumulated across a contact interval. In short, when the texture-specific kinetic signature causes high energy movements to reach the receptors in the follicle, high firing rates are evoked and the

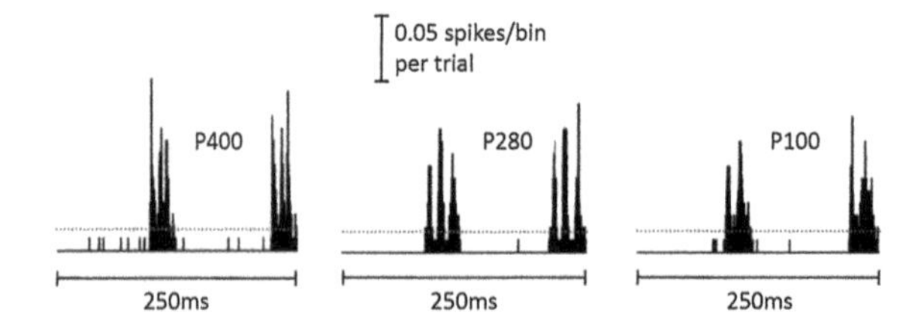

Figure 6 Texture-specific spike patterns. Peristimulus time histogram (2 ms bins; 100 trials) of a ganglion cell for two whisks on textures P400 (*left*), P280 (*middle*) and P100 (*right*). Mean firing rate (*dashed red lines*) were similar, suggesting temporal firing pattern as a critical coding mechanism. Adapted from Arabzadeh et al. (2005)

contacted texture is decoded as rough; lower energy movement and lower firing rate is decoded as smoother. According to the second hypothesis, the brain identifies texture by extracting from each contact interval the temporal sequence of high energy events within the kinetic signature. For example, one texture may evoke a kinetic signature with regularly timed stick-slip events, and a second texture may evoke a signature with alternating long and short intervals. The temporal pattern of kinetic events would be captured in the neuronal spiking sequence (e.g. Figure 6), and if the readout mechanism can decode firing patterns, then the animal would possess a much higher capacity for representing textures than if it used only the firing rate decoding mechanism (Arabzadeh et al. 2006).

Existing evidence supports temporal integration as a plausible readout mechanism. In a rough versus smooth texture discrimination task, contacts with the rough texture evoked significantly higher firing rates in barrel cortex than did contact with the smooth texture (von Heimendahl et al. 2007). On trials when the rat correctly identified the stimulus, the firing rate of neurons in barrel cortex was higher for rough than for smooth during a temporal window immediately preceding the instant of choice. This firing-rate code was reversed on error trials (lower for rough than for smooth) suggesting that the rat made its decision based upon the magnitude of whisker-evoked activity in barrel cortex. But temporal firing patterns may provide supplementary information in other texture discrimination tasks; if a pair of textures evokes nearly the same firing rate, differences in spiking sequences could be crucial (Arabzadeh et al. 2006). Just as rats shift their whisking strategy according to the textures they must discriminate (Carvell and Simons 1995), so might they adapt their strategy for decoding neuronal activity.

Sensorimotor Integration

In the sense of touch, it is the motion of the sensory receptors themselves that leads to an afferent signal—whether these receptors are in our fingertips sliding along a surface (Gamzu and Ahissar 2001) or a rat's whiskers palpating an object. Thus, tactile exploration entails the interplay between motor output and sensory input (reviewed in Kleinfeld et al. 2006; Diamond et al. 2008b; also see Kleinfeld et al. 2006). Just as we would not be able to estimate the weight of an object we are lifting without taking into account the motor signals that produce muscle contraction, nor can the afferent signal from a whisker be optimally decoded without information about the movement that generated the tactile signal to begin with. Anatomical and physiological evidence (Gioanni and Lamarche 1985; Kleinfeld et al. 2002) indicate that barrel cortex has access to motor signals as it is a direct participant in the motor network. The technical term for a signal in which motor areas inform sensory areas about outgoing motor signals—or, similarly, about their expected sensory consequences—is "corollary discharge" (Crapse and Sommer 2008). While this has been studied in detail in many species, including song birds (Troyer and Doupe 2000) and bats (Ulanovsky and Moss 2008), in rats we know neither the signal's neuronal substrate, nor its coding properties nor how and where it is integrated during sensory processing. Understanding these aspects of sensorimotor integration will be the focus of future research in the field.

Internal References

Alonso, J-M and Chen, Y (2009). Receptive field. *Scholarpedia* 4(1): 5393. http://www.scholarpedia.org/article/Receptive_field.

Braitenberg, V (2007). Brain. *Scholarpedia* 2(11): 2918. http://www.scholarpedia.org/article/Brain.

Llinas, R (2008). Neuron. *Scholarpedia* 3(8): 1490. http://www.scholarpedia.org/article/Neuron.

Meiss, J (2007). Dynamical systems. *Scholarpedia* 2(2): 1629. http://www.scholarpedia.org/article/Dynamical_systems.

Prescott, T J (2008). Vibrissal behavior and function. *Scholarpedia* 6(10): 6642. http://www.scholarpedia.org/article/Vibrissal_behavior_and_function. (see also pages 737–749 of this book).

Schultz, W (2007). Reward. *Scholarpedia* 2(3): 1652. http://www.scholarpedia.org/article/Reward.

Sherman, S M (2006). Thalamus. *Scholarpedia* 1(9): 1583. http://www.scholarpedia.org/article/Thalamus.

External References

Andermann, M L; Ritt, J; Neimark, M A and Moore, C I (2004). Neural correlates of vibrissa resonance; band-pass and somatotopic representation of high-frequency stimuli. *Neuron* 42(3): 451–463.

Arabzadeh, E; Panzeri, S and Diamond, M E (2006). Deciphering the spike train of a sensory neuron: counts and temporal patterns in the rat whisker pathway. *The Journal of Neuroscience* 26(36): 9216–9226.

Arabzadeh, E; Petersen, R S and Diamond, M E (2003). Encoding of whisker vibration by rat barrel cortex neurons: Implications for texture discrimination. *The Journal of Neuroscience* 23(27): 9146–9154.

Arabzadeh, E; Zorzin, E and Diamond, M E (2005). Neuronal encoding of texture in the whisker sensory pathway. *PLoS Biology* 3(1): e17.

Berg, R W and Kleinfeld, D (2003). Rhythmic whisking by rat: Retraction as well as protraction of the vibrissae is under active muscular control. *Journal of Neurophysiology* 89(1): 104–117.

Brecht, M; Preilowski, B and Merzenich, M M (1997). Functional architecture of the mystacial vibrissae. *Behavioural Brain Research* 84(1–2): 81-97.

Carvell, G E and Simons, D J (1990). Biometric analyses of vibrissal tactile discrimination in the rat. *The Journal of Neuroscience* 10(8): 2638–2648.

Carvell, G E and Simons, D J (1995). Task- and subject-related differences in sensorimotor behavior during active touch. *Somatosensory & Motor Research* 12(1): 1–9.

Crapse, T B and Sommer, M A (2008). Corollary discharge across the animal kingdom. *Nature Reviews Neuroscience* 9(8): 587–600.

Diamond, M E; von Heimendahl, M and Arabzadeh, E (2008a). Whisker-mediated texture discrimination. *PLoS Biology* 6(8): e220.

Diamond, M E; von Heimendahl, M; Knutsen, P M; Kleinfeld, D and Ahissar, E (2008b). Where and what in the whisker sensorimotor system. *Nature Reviews Neuroscience* 9(8): 601–612.

Diamond, M E; von Heimendahl, M; Itskov, P and Arabzadeh, E (2008c). Response to: Ritt et al., Embodied information processing: Vibrissa mechanics and texture features shape micromotions in actively sensing rats. *Neuron* 60(5): 743–744.

Ebara, S; Kumamoto, K; Matsuura, T; Mazurkiewicz, J E and Rice, F L (2002). Similarities and differences in the innervation of mystacial vibrissal follicle-sinus complexes in the rat and cat: A confocal microscopic study. *Journal of Comparative Neurology* 449(2): 103–119.

Gamzu, E and Ahissar, E (2001). Importance of temporal cues for tactile spatial-frequency discrimination. *The Journal of Neuroscience* 21(18): 7416–7427.

Gioanni, Y and Lamarche, M (1985). A reappraisal of rat motor cortex organization by intracortical microstimulation. *Brain Research* 344(1): 49–61.

Guic-Robles, E; Valdivieso, C and Guajardo, G (1989). Rats can learn a roughness discrimination using only their vibrissal system. *Behavioural Brain Research* 31(3): 285–289.

Hartmann, M J; Johnson, N J; Towal, R B and Assad, C (2003). Mechanical characteristics of rat vibrissae: Resonant frequencies and damping in isolated whiskers and in the awake behaving animal. *The Journal of Neuroscience* 23(16): 6510–6519.

Hill, D N; Bermejo, R; Zeigler, H P and Kleinfeld, D (2008). Biomechanics of the vibrissa motor plant in rat: rhythmic whisking consists of triphasic neuromuscular activity. *The Journal of Neuroscience* 28(13): 3438–3455.

Jones, L M; Depireux, D A; Simons, D J and Keller, A (2004). Robust temporal coding in the trigeminal system. *Science* 304(5679): 1986–1989.

Kleinfeld, D; Ahissar, E and Diamond, M E (2006). Active sensation: Insights from the rodent vibrissa sensorimotor system. *Current Opinion in Neurobiology* 16(4): 435–444.

Kleinfeld, D; Sachdev, R N S; Merchant, L M; Jarvis, M R and Ebner, F F (2002). Adaptive filtering of vibrissa input in motor cortex of rat. *Neuron* 34(6): 1021–1034.

Lottem, E and Azouz, R (2008). Dynamic translation of surface coarseness into whisker vibrations. *Journal of Neurophysiology* 100(5): 2852–2865.

Maravall, M; Petersen, R S; Fairhall, A L; Arabzadeh, E and Diamond, M E (2007). Shifts in coding properties and maintenance of information transmission during adaptation in barrel cortex. *PLoS Biology* 5(2): e19.

Mitchinson, B; Martin, C J; Grant, R A and Prescott, T J (2007). Feedback control in active sensing: rat exploratory whisking is modulated by environmental contact. *Proceedings of the Royal Society B: Biological Sciences* 274(1613): 1035–1041.

Moore, C and Andermann, M (2005). The vibrissa resonance hypothesis. In: F F Ebner (Ed.), *Somatosensory Plasticity* (chapter 2, pp. 21–60). CRC Press.

Neimark, M A; Andermann, M L; Hopfield, J J and Moore, C I (2003). Vibrissa resonance as a transduction mechanism for tactile encoding. *The Journal of Neuroscience* 23(16): 6499–6509.

Prigg, T; Goldreich, D; Carvell, G E and Simons, D J (2002). Texture discrimination and unit recordings in the rat whisker/barrel system. *Physiology & Behavior* 77(4–5): 671-675.

Ritt, J T; Andermann, M L and Moore, C I (2008). Embodied information processing: Vibrissa mechanics and texture features shape micromotions in actively sensing rats. *Neuron* 57(4): 599–613.

Shoykhet, M; Doherty, D and Simons, D J (2000). Coding of deflection velocity and amplitude by whisker primary afferent neurons: Implications for higher level processing. *Somatosensory & Motor Research* 17(2): 171–180.

Towal, R B and Hartmann, M J (2006). Right-left asymmetries in the whisking behavior of rats anticipate head movements. *The Journal of Neuroscience* 26(34): 8838–8846.

Troyer, T W and Doupe, A J (2000). An associational model of birdsong sensorimotor learning I. Efference copy and the learning of song syllables. *Journal of Neurophysiology* 84(3): 1204–1223.

Ulanovsky, N and Moss, C F (2008). What the bat's voice tells the bat's brain. *Proceedings of the National Academy of Sciences of the United States of America* 105(25): 8491–8498.

von Heimendahl, M; Itskov, P M; Arabzadeh, E and Diamond, M E (2007). Neuronal activity in rat barrel cortex underlying texture discrimination. *PLoS Biology* 5(11): e305.

Wolfe, J et al. (2008). Texture coding in the rat whisker system: Slip-stick versus differential resonance. *PloS Biology* 6(8): e215.

Woolsey, T A and van der Loos, H (1970). The structural organization of layer IV in the somatosensory region (SI) of mouse cerebral cortex. the description of a cortical field composed of discrete cytoarchitectonic units. *Brain Research* 17(2): 205–242.

External Links

http://www.sissa.it/cns/tactile/index.php.
http://jcsmr.anu.edu.au/neural-coding.

Whisking Control by Motor Cortex

Cornelius Schwarz and Shubhodeep Chakrabarti

Nomenclature of vM1 modules

Various authors have used various nomenclature systems to refer to the different modules within vM1. Some of them were based on cytoarchitectonically defined subregions observed in histological examination of tissue whereas others were derived from functional assessment of cortical function. We use the following nomenclature in this article.

AGm: AGm is the medial agranular cortex, defined cytoarchitectonically. Functionally it corresponds to the vM1 representation. Part of AGm overlaps with the functionally defined area, rhythmic whisking region or RW.

TZ: The transitional zone between AGm and AGl, previously classified as part of the retraction face (RF) region, was first called TZ by Alloway and colleagues, again using cytoarchitectonic criteria. TZ has previously been shown to be interconnected with both AGm and AGl (Weiss and Keller 1994), and to be the main recipient of vS1 projections to vM1 (Smith and Alloway 2013).

AGl: AGl is the lateral agranular cortex, defined using cytoarchitectonic boundaries and mainly corresponds, functionally, to the forelimb region of M1.

PMPF: The prefrontal and premotor subregion of M1, defined using functional criteria lies towards the frontal pole of M1 bordering pre-frontal cortex. Part of it was also previously classified as belonging to the so-called RF. For a graphical representation of the above areas, see Figure 1.

C. Schwarz (✉)
Hertie Institute for Clinical Brain Research, University of Tübingen, Tübingen
Germany

S. Chakrabarti
Werner Reichardt Center for Integrative Neuroscience; Hertie Institute for Clinical Brain Research, Tübingen, Germany

T.J. Prescott et al. (eds.), *Scholarpedia of Touch*, Scholarpedia,
DOI 10.2991/978-94-6239-133-8_55

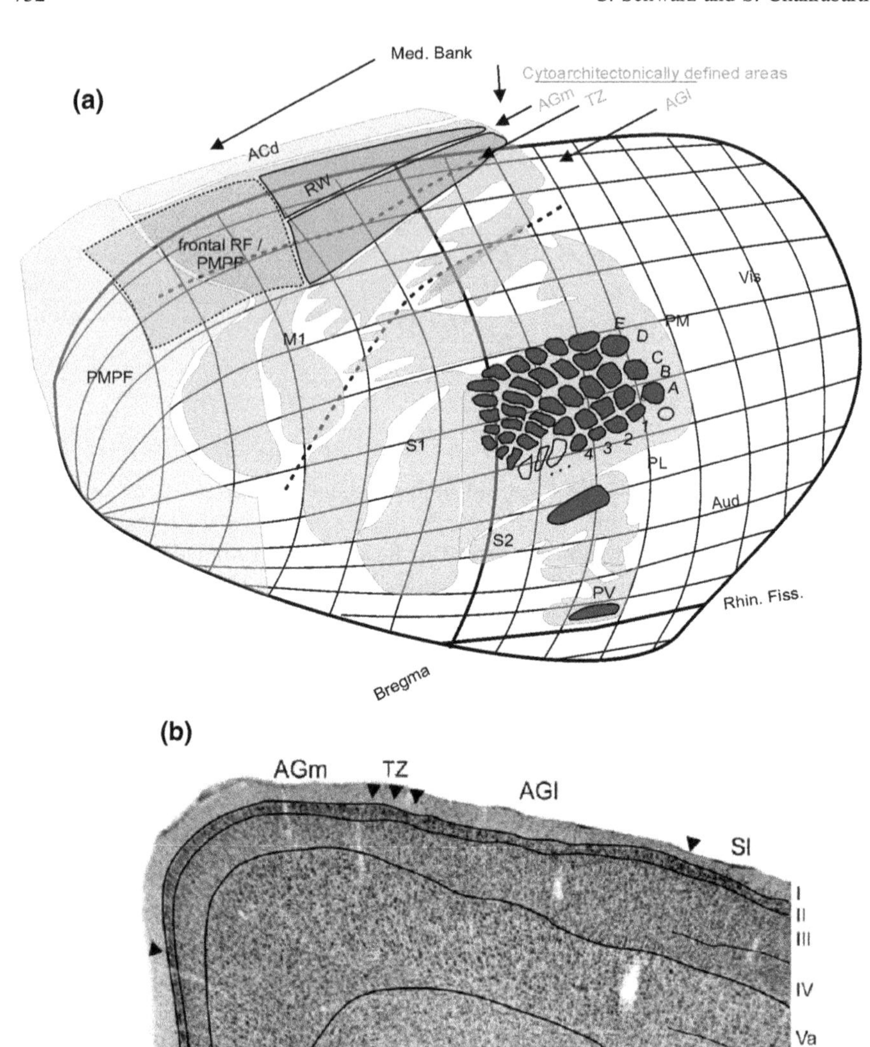
(a)
Med. Bank
Cytoarchitectonically defined areas
AGm
TZ
AGl
ACd
RW
frontal RF / PMPF
Vis
PM
E
D
C
B
A
M1
PMPF
S1
4
3
2
1
PL
Aud
S2
PV
Rhin. Fiss.
Bregma
(b)
AGm
TZ
AGl
SI
I
II
III
IV
Va
Vb
VI
Cg

◀ **Figure 1** The modular nature of vM1. A. A surface map of the rat sensorimotor cortex. The primary somatosensory cortex (S1) and tactile, partly multimodal, association areas are depicted in *light red*. Motor areas are in *light green* and premotor/prefrontal areas in *light blue*. The strong colors indicate whisker representations. The modularity of the primary motor cortex (M1) whisker representation (vM1) is indicated. The rhythmic whisker area (RW) reaches the dorsal surface of the neocortex but likely extends well into the medial bank. The barrel cortex recipient zone, also known as the TZ (transitional zone) from cytoarchitectonic markers, is located on the dorsal surface of the neocortex, a new module defined which was part of the previously described retraction face (RF) module. Frontal RF and whisker representations in the premotor/prefrontal cortex (PMPF) are little investigated. Their delineation is unclear and within PMPF the detailed topography of limb and head representations is not consistent in the data available today. To indicate this uncertainty, these modules have been paled in *color* and limits are depicted by *broken lines*. The *blue* text labels mark cytoarchitectonically defined areas whereas the *black* text labels denote functionally defined areas. ACd dorsal anterior cingulate cortex, AGm medial agranular cortex, AGl lateral agranular cortex. *Thick broken lines* indicate borders between AGm and AGl and between AGl and S1. S2 secondary somatosensory cortex, AGm houses the head and whisker representations while AGl houses trunk and limb representations (indicated by *arrows*), PV, PL, PM posterior ventral, lateral and medial cortex, Aud auditory cortex, Vis visual cortex. Med. Bank medial bank of the hemisphere (the parts of the map extending into the medial bank are folded up for clarity. Rhin. Fiss. rhinal fissure. The *coronal section line* corresponding to anterior–posterior coordinate 0 is labeled Bregma. B. A coronal section through vM1, stained for Nissl material illustrating the locations of AGm, AGl and TZ. Reproduced from Smith and Alloway 2013 with permission

Introduction

The rodent whisker-related sensorimotor system is outstanding as a model system because these animals use their mobile whiskers to "actively scan" their environment (Carvell and Simons 1990). Active scanning means that rodents deploy energy to objects (via whisker movements) and gain information by sensing the object's reflections, in the form of fine object-dependent whisker vibrations (Hentschke et al. 2006; Ferezou et al. 2007; Wolfe et al. 2008). This is akin to active scanning using echo-location or electro-sensation of bats, cetaceans, and fish, and is performed by humans in very similar ways using their finger tips (Gamzu and Ahissar 2001). It is comprehensible that the active scanning property will emphasize sensorimotor interplay even more as known from other motor systems, as any change in movement strategy fundamentally changes the character of incoming sensory sensation and vice versa. One critical constraint of the tactile system is that it is a 'near' sense, i.e. whiskers need to be held in touch of the object of interest, and while in touch, movements need to be optimized to serve the purpose of fine discrimination of location, texture, and shape. Despite the critical dependence on tactile inputs, scanning movements are voluntarily initiated and maintained—even without tactile input—and have a stereotyped, relatively simple rhythmic basis, onto which the modulation by tactile signals is superimposed (Gao et al. 2001; Landers and Zeigler 2006).

The study of active scanning whisker movements and their modulation by tactile inputs is in its early stages. However, considering the above-mentioned sensorimotor interplay, and other functions of the whisker system like animal navigation, one should not be surprised to find a functionally parceled system serving different functions with sensorimotor interconnections on each level of the neuronal hierarchy. In fact, active ongoing research has revealed a highly modulatory system serving either more basic motor functions (rhythmic scanning) or sensorimotor feedback of different types and complexities.

The Organization of the Rat Vibrissal Motor Cortex (VM1)

Spatial Extent and Motor Map

The first functional mapping of rat vM1 was performed using electrical stimulation on the cortical surface (Settlage et al. 1949) and later refined using intracortical microstimulation (ICMS) (Hall and Lindholm 1974; Hicks and D'Amato 1977; Sapienza et al. 1981; Neafsey and Sievert 1982; Gioanni and Lamarche 1985; Neafsey et al. 1986). These studies together delineated a motor cortical area that occupied the frontal and dorsomedial aspects of the rat neocortex and contained a movement map of body parts arranged topographically. Cytoarchitectonically, M1 can be subdivided into three fields which occupy the area between the midline and the primary somatosensory cortex (Figure 1). The two more medial fields together constitute vibrissal M1, and are related mainly to whisker and face movements (medial agranular and transitional zone; AGm and TZ) while the lateral one, adjoining the somatosensory cortex, is related to trunk and limb movements (lateral agranular zone AGl) (Donoghue and Parham 1983; Brecht et al. 2004a; Smith and Alloway 2013). As schematized in Figure 1, AGm lies along the midline extending into the medial bank down to the cingulate cortex and is characterized by a broad layer V and a thin layer III whereas AGl has a thick layer III and a reduced layer V. TZ is characterized by intermediate layer III and V thickness (Zilles et al. 1980; Donoghue and Wise 1982; Brecht et al. 2004a; Smith and Alloway 2013). Whereas AGm and AGl neurons primarily have dense interconnections within their respective cytoarchitectonic boundaries, TZ neurons are interconnected with both AGm and AGl (Weiss and Keller 1994). Corresponding to their body representations, one target of subcortical projections from AGm and TZ is the superior colliculus, while AGl connects to the spinal cord instead (Neafsey et al. 1986). The exact reaches of M1 toward the frontal pole and its borders with the premotor and prefrontal cortices is poorly understood (Neafsey et al. 1986; Conde et al. 1990, 1995; Uylings et al. 2003) and therefore we call this anterior aspect the premotor and prefrontal cortices (PMPF). On its extreme medial aspect, M1 contains a special representation of movements of the eye (saccades) and eyelids (Hall and Lindholm 1974; Neafsey et al. 1986; Brecht et al. 2004a).

The vM1 occupies a large portion of the motor cortex and there is considerable debate about the exact nature of the topographic map of the vibrissal pad. Some authors have described single whisker responses using ICMS although others have shown that the number of whiskers showing evoked movement varies with the level of anesthesia used (Brecht et al. 2004a; Haiss and Schwarz 2005). Single cell microstimulation in vivo has consistently yielded multi-whisker movements (Brecht et al. 2004b) supporting the hypothesis that muscle synergies or movements might be represented in vM1 rather than individual muscles.

Connectivity

Cortical Connections

Vibrissal M1 is densely and reciprocally connected with almost all other vibrissal cortical areas including the primary and secondary somatosensory cortices (S1 and S2) and vM1 in the contralateral hemisphere (Figure 2). The most important cortical projection to vM1 is the one originating from ipsilateral S1 barrel cortex (vS1). These projections arise predominantly from neurons in the supra- and infragranular layers of vS1 that are aligned with the layer IV inter-barrel septa, although layer IV itself does not project to M1 (Alloway et al. 2004; Chakrabarti and Alloway 2006). This projection from the vS1 is most likely the main source for tactile input to vM1

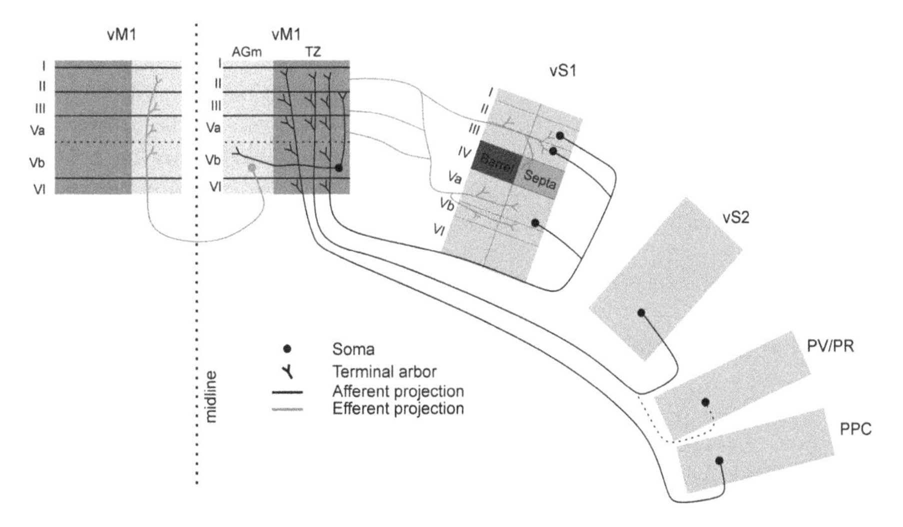

Figure 2 Cortical connectivity of vM1. The detailed laminar connectivity of AGm (*blue*), and transitional zone (TZ, *red*) with the contralateral vM1 and ipsilateral sensory cortices have been shown. Layer IV barrels have been indicated as well as inter-barrel septa and the barrel and septal columns indicated using a *vertical line*. The *midline* is indicated by a *vertical dashed line*. Afferent projections to vM1 have been indicated using *black* and efferent projections from vM1 using *green*. The projection from PV/PR to vM1 is poorly documented and hence has been shown using a *dashed line*

as tactile responses in vM1 are dependent on a viable vS1 (Farkas et al. 1999; Chakrabarti et al. 2008; Aronoff et al. 2010). Furthermore, the projection from vS1 to vM1 is anisotropic with septal regions located along rows showing significantly greater convergence in vM1 than the ones along the whisker arcs (Hoffer et al. 2003). The projection from vS1 to vM1 is limited to a millimeter wide area straddling TZ in the medio lateral direction, an area from which tactile responses can be readily recorded in vM1 (Smith and Alloway 2013). More medial areas in vM1 are devoid of tactile responses (Gerdjikov et al. 2013; Smith and Alloway 2013). The connection between vS1 and vM1 is reciprocal with the vM1-vS1 projection arising mainly from layers II/III and Va and targeting preferentially the deeper layers but also layer I as well as the septal regions in Layer IV in vS1 (Sato and Svoboda 2010; Mao et al. 2011; Petreanu et al. 2012; Zagha et al. 2013; Kinnischtzke et al. 2014). The vM1 projections to vS1 primarily target VIP expressing interneurons in vS1 resulting in whisking related modulation of vS1 activity (Lee et al. 2013).

The S2 vibrissal region (vS2) also projects to vM1 with the terminals intermingled with those arising from S1 (Reep et al. 1990; Colechio and Alloway 2009; Smith and Alloway 2013). Projections from the posterior parietal cortex (PPC) terminate in AGm adjoining the TZ (Fabri and Burton 1991; Reep et al. 1994; Colechio and Alloway 2009; Smith and Alloway 2013). Projections to vibrissal M1 connections from other somatosensory cortical regions lateral to S2 such as the parietal ventral cortex (PV) and perirhinal cortex (PR) have also been reported but the exact region of termination within vM1 remains unclear (Krubitzer et al. 1986; Reep et al. 1990; McIntyre et al. 1996; Kyuhou and Gemba 2002; Colechio and Alloway 2009).

The intrinsic connections of the different vM1 sub-divisions are strikingly different. Using anterograde tracer deposits in AGm (vM1) and the TZ, Weiss and Keller showed that whereas the majority of the axons labeled following a AGm tracer deposit were restricted in the same compartment, TZ injections produced axonal labeling in both AGm, TZ as well as AGl hinting at different functional connectivities of these modules (Weiss and Keller 1994). Further, the intrinsic connectivity has a distinct laminar organization with Layer V cells projecting horizontally to Layers V and III whereas the Layer III collaterals tend to be restricted to the superficial layers (Aroniadou and Keller 1993).

Finally both vM1's in the two hemispheres are strongly interconnected with each other (Porter and White 1983; Miyashita et al. 1994), this interconnection being significantly stronger than the one connecting the two M1 forelimb representations (Colechio and Alloway 2009). This might have implications for the bilateral co-ordination of whisking as reported in many behavioral studies (Gao et al. 2003; Towal and Hartmann 2006; Mitchinson et al. 2007).

Thalamic Connections

Vibrissal M1 also has both afferent and efferent connectivity with various ipsilateral thalamic nuclei, viz., the mediodorsal group of nuclei (MD), the centrolateral group (CL) and the medial aspect of the posterior nucleus (POm) (Cicirata et al. 1986; Rouiller et al. 1991; Miyashita et al. 1994; Alloway et al. 2008a, 2009). Figure 3 provides a complete schematic of thalamic and subcortical connectivities of vM1, both ipsi- and contralaterally. Further, the interanteromedial group (IAM), the anteromedial group (AM) and the ventrolateral (VL) and ventromedial (VM) groups of thalamic nuclei receive projections from vM1 of both hemispheres and may reflect the thalamic counterpart of interhemispheric whisking coordination pathways (Cicirata et al. 1986; Rouiller et al. 1991; Miyashita et al. 1994; Alloway et al. 2008b, 2009; Hooks et al. 2013). It has been suggested that the reciprocal connections between vM1 and POm play a vital role in the motor gating of ascending sensory information via the paralemniscal pathway. One possible hypothesis holds that bi-synaptic disinhibition triggered by vM1 connections to the zona incerta switches POm neurons from burst to regular firing during whisking activity (Lavallee et al. 2005; Trageser et al. 2006; Urbain and Deschenes 2007) although zona incerta has a highly complex projection pattern which could modulate POm activity in a variety of ways.

Other Sub-cortical Connections

In addition, vM1 also projects to a number of other subcortical structures such as the pontocerebellum (Schwarz and Möck 2001), dorsolateral neostriatum (Alloway et al. 2006, 2009), intermediate and deep layers of the superior colliculus (Miyashita et al. 1994; Alloway et al. 2010) as well as bilaterally to the claustrum (Smith and Alloway 2010; Smith et al. 2012). The main subcortical recipients of vM1 cortical motor output are summarized schematically in Figure 3. However, arguably the most critical for vibrissal movement are the putative pathways from vM1 that convey motor commands to vibrissal musculature. Direct projections from vM1 to vibrissal motoneurons in the facial nucleus have been difficult to demonstrate with anterograde tracing methods (Miyashita et al. 1994; Hattox et al. 2002; Grinevich et al. 2005; Alloway et al. 2010). However, monosynaptic tracing, a method that uses injection of a deficient rabies virus in the muscles with targeted expression of the deficient protein in motoneurons gave rise to an important breakthrough—clearly revealing the presence of direct CM projections (Sreenivasan et al. 2014). This feature is remarkable because it likens the vibrissal system to that of the primate hand, which also is characterized by direct projections of M1 to motoneurons (Rathelot and Strick 2006). Despite the presence of direct connections, both motor systems are strongly dependent on subcortical circuitry to generate normal movements, and the function of CM connections remains a mystery. Especially, the role of M1 in computing motor commands needed to realize detailed patterns of muscle activity remains unknown. Hand spasticity

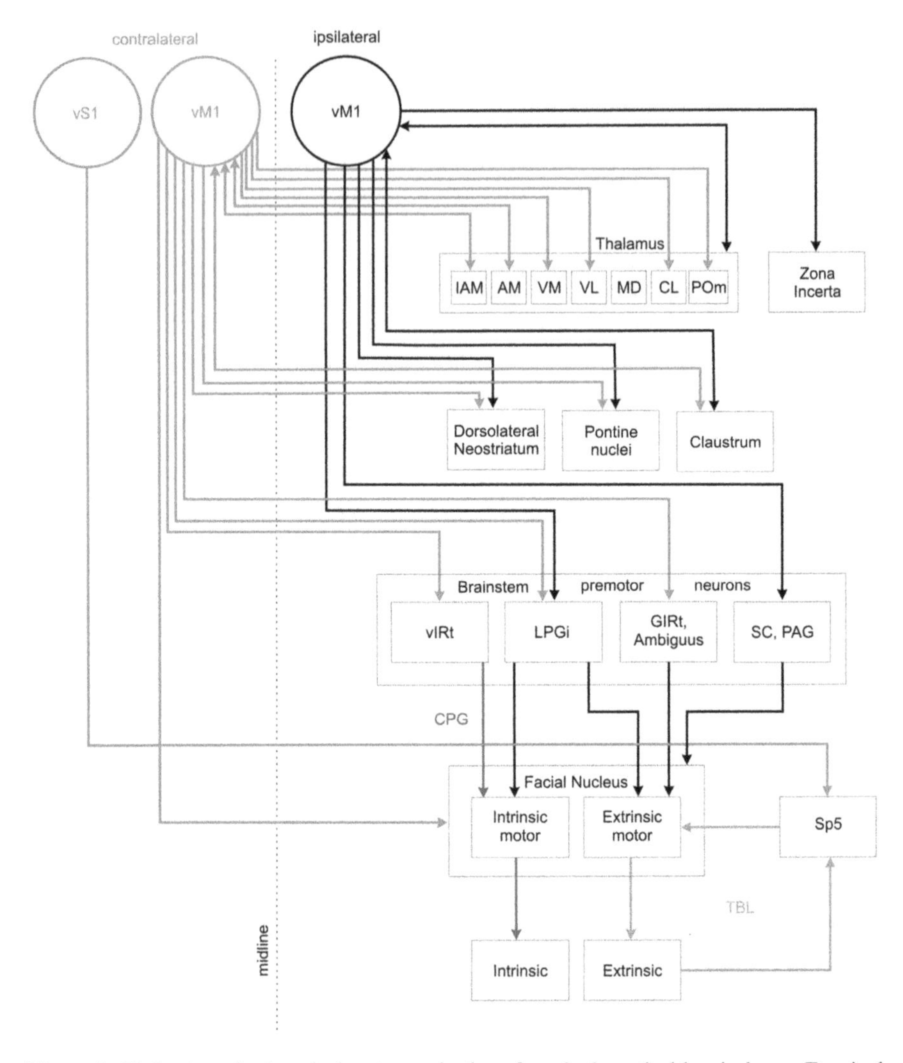

Figure 3 Thalamic and subcortical motor projections from both cortical hemispheres. Terminal projections are marked by *arrows*, contralateral projections are shown in *green* and ipsilateral projections in *black*. The connections of the central pattern generator (CPG) are shown in *red* whereas those subserving the trigeminal brainstem loop (TBL), the putative whisker reflex arc are shown in *blue*. IAM interanteromedial nucleus, AM anteromedial nucleus, VM ventromedial nucleus, VL ventrolateral nucleus, MD mediodorsal nucleus, CL centrolateral nucleus, POm medial posterior nucleus, vIRT ventral intermediate reticular formation, LPGi lateral paragigantocellularis, GIRt gigantocellular reticular formation, Ambiguus nucleus ambiguus, SC superior colliculus, PAG periaqueductal gray, Sp5 spinal trigeminal nuclei

developing in primates after M1 lesions clearly points to an important functional role of remaining subcortical inputs for patterning hand muscle activity. The analogous situation in the whisker motor system allows an even clearer interpretation: whisking movements after a vM1 lesion recover to almost normal kinematic profiles (Gao et al. 2003), suggesting that vM1 is not needed at all to compute normal spatiotemporal muscle activities. The enigmatic CM connection appears to have more modulatory functions—likely on a slower timescale than that used to convey patterned drive to sets of muscles.

Which subcortical centers are responsible for detailed pattern generation? Several brainstem regions receive motor cortex projections and in turn project to the facial motoneurons and are therefore candidates for an oligosynaptic motor pathway controlling whisking. Amongst these are the reticular formation, superior colliculus, nucleus ambiguus, the deep mesencephalic nucleus, the periaqueductal gray, the interstitial nucleus of the medial longitudinal fasciculus and the red nucleus (Reep et al. 1987; Miyashita et al. 1994; Hattox et al. 2002; Smith and Alloway 2010). Particularly promising candidates are the intermediate reticular formation (IRt) containing a central pattern generator (CPG) (Moore et al. 2013) which generates rhythmic whisking movements and receives projections from the contralateral vM1 (Alloway et al. 2010; Sreenivasan et al. 2014) as well as the lateral paragigantocellularis (LPGi) which also receives projections from both ipsi- and contralateral vM1 and projects in turn to the facial motoneurons (Takatoh et al. 2013).

In summary, the subcortical motor projections can be organized into four basic groups as shown in Figure 3. First, there are the vM1 projections to thalamus and zona incerta. Second there are vM1 projections to the neostriatum, pontine nuclei and claustrum which form cortico-subcortical loops for higher order functions which are not yet clearly understood. Third, there are vM1 projections to various groups of brainstem nuclei, the second order motoneurons which in turn project to the intrinsic/extrinsic premotor neurons in the facial nucleus. Finally there are the direct CM projections to the facial nucleus.

Functional Modules of VM1

There is mounting evidence that the motor control of whisking is achieved through the concerted action of different motor cortical modules with different functional specializations. As discussed above they are connected either directly to the vibrissal motoneurons in the facial nucleus or indirectly via brainstem centers such as CPGs, shown in red arrows in Figure 3 or the TBL, depicted in blue (Chakrabarti and Schwarz 2014a, b). Presently, three modules have been described in more detail, and can be delineated using functional assessment (ICMS and tactile responses/connectivity). A region located medio-caudally evokes rhythmic whisking upon prolonged ICMS, and was thus named the rhythmic whisking region or RW (Haiss and Schwarz 2005). In contrast a fronto-lateral area evokes whisker retraction accompanied by face and body movements and was called the retraction face region (RF) (Haiss and Schwarz 2005). The presence of RW and RF modules have been also confirmed in the mouse vM1 (Ferezou et al. 2007; Matyas et al.

2010). Electrophysiological monitoring and tract tracing has led to the redefinition of RF into two distinct regions—a so-called transitional zone or TZ located in the cytoarchitectonically defined transitional region between AGm and AGl and receiving tactile input from vS1 (Smith and Alloway 2013) and a rostral area devoid of such tactile input which we name PMPF because of its putative premotor and prefrontal functions (Neafsey and Sievert 1982; Uylings et al. 2003). It is quite possible that with future explorations in this area, of which very little is known, further modules will be discovered. Further, vS1 has been shown to access premotor neurons in the spinal trigeminal nuclei (Sreenivasan et al. 2014) that are part of the TBL, the brainstem sensorimotor reflex arc (Nguyen and Kleinfeld 2005). Via these connections it evokes whisker retraction movements upon ICMS (Matyas et al. 2010). As will be apparent from the following sections, even the functional roles of RW and TZ, the best investigated parts of vM1, are presently far from clear. What is missing (and therefore should be an immediate target of research) is the elucidation of the differential projection patterns of each of these modules to the brainstem premotor circuits, which hopefully will help to clarify their functions.

Rhythmic Whisking Region (RW)

As shown in Figure 4a, prolonged ICMS in RW in awake rats elicits rhythmic whisking which is virtually indistinguishable from natural whisking in rats (Haiss and Schwarz 2005). Under different types of anesthesia, these rhythmic whisking movements have been found to be either strongly suppressed (Cramer and Keller 2006) or reduced to monophasic protraction movements (Sanderson et al. 1984; Haiss and Schwarz 2005). Electrophysiological recordings in RW in awake rats (Figure 4b), trained to generate explorative whisking, showed no coherence of spiking with whisking on the frequency scale of a whisker stroke, ~10 Hz (Friedman et al. 2006, 2012; Gerdjikov et al. 2013). RW, therefore, does not seem to be involved in the programming of whisker trajectories in a stroke by stroke fashion. Interestingly, RW cell firing rates first encode either decrements or increments of these movement parameters, and second, follow rather than anticipate abrupt whisking onset (Gerdjikov et al. 2013). RW, thus, shows a loose relation to movement with part of the activity signaling whisker quiescence rather than movement and little involvement in movement initiation. Further, RW does not receive any tactile signals about whisker contact with an object (Gerdjikov et al. 2013).

The rather abstract RW motor signals may constitute input signals to the CPG, depicted in Figure 4c, which has been recently found in the ventral part of the intermediate band of the reticular formation (vIRt) near the Bötzinger complex (Moore et al. 2013). This CPG receives projections from vM1, as shown in Figure 4d (Sreenivasan et al. 2014). In contrast to RW (Gao et al. 2003), the blockade of vIRt entirely prevents whisking and its activation generates continuous rhythmic whisking (Moore et al. 2013; Figure 4c). As activity anticipating whisker

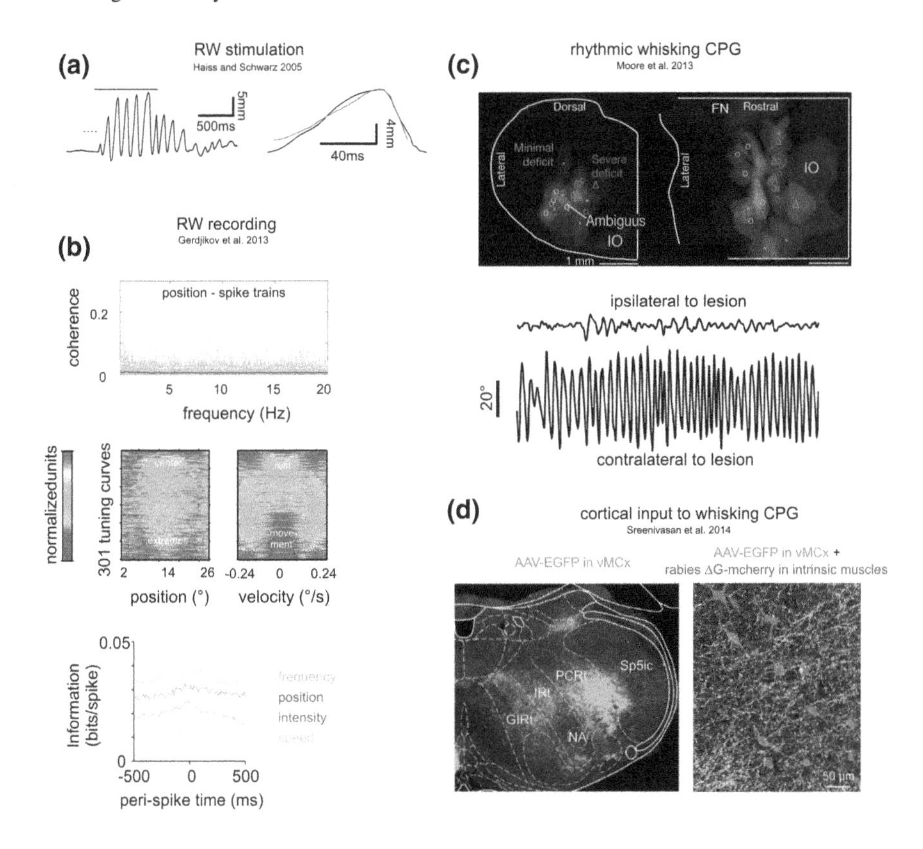

movement has not been found in RW, speaking against a classical motor function, an alternative speculation is that RW may be an internal monitor of CPG function (Gerdjikov et al. 2013).

Premotor and Prefrontal Region (PMPF)

The earlier described Retraction Face region (RF) was defined as the area of vM1 where ICMS in the awake rat evokes monophasic whisker retraction in concert with face and body movements. In the anesthetized rat, face and body movements are typically absent (Haiss and Schwarz 2005). In the light of new evidence, we classify the region previously described as RF into two distinct functional zones—the transitional zone (TZ) and the frontal Prefrontal and Premotor region (PMPF), the latter devoid of any sensory input (Smith and Alloway 2013). Not much is known about the functionality of PMPF in comparison to its extensively studied neighbor TZ, except that ICMS here evokes whisker and face retraction as in TZ. A possible hypothesis that whisker retraction seen in both PMPF and TZ could be involved in orientation responses is discussed below in next section but a delineation of the functional properties of this region needs systematic future exploration.

◀ **Figure 4** Functional organization of RW and rhythmic whisking CPG. **a** Rhythmic whisking evoked by long ICMS in RW. The *line* above the whisker trace on the *left* indicates the duration of 60 Hz ICMS. *Right* Individual strokes, one evoked by ICMS (*thin line*) and another voluntarily generated by the rat (*thick line*). Note the close similarity between the two. Modified from Haiss and Schwarz (2005) with permission. **b** Unitary recordings from RW in awake head-fixed animals engaged in a whisking task. *Top* Coherence between the spike train and the whisker position trace. The coherence function of all RW units is low and flat, excluding any significant stroke-by-stroke coding in RW (line colors: *gray* individual single (n = 301) and multi units (n = 261); *red* median of distribution; *yellow* 90 % percentile). *Center* Color coded tuning curves for position (*left*) and velocity (*right*) calculated from spike trains of 301 single units. The tuning strength (rainbow color code *violet-blue-green-yellow-red*) is scaled in normalized units. Note that the neurons' tuning curves were ordered according to the coefficient of the first principal component obtained from the sample of tuning curves to reveal different types of tuning (i.e. *lines* in the two panels do not correspond to the same cell). *Bottom* Average Shannon information carried by a single RW spike about the whisker trajectory at a certain latency. Information transferred from different whisking variables is shown. A bootstrap procedure using scrambled spike trains indicated that the majority of RW neurons convey significant information about the whisker trajectory. Importantly, information about a large interval around the spike (time 0) is present, making a pure causal role of RW for whisker movement unlikely. Modified from Gerdjikov et al. (2013) with permission. **c** The rhythmic whisking CPG. *Top* Effective lesion (*red symbols*) sites in the medulla as seen in the frontal (*left*) and horizontal plane (*right*). The location of the rhythmic whisking CPG is in the ventral intermediate band of the reticular formation (vIRt). FN facial nucleus, IO inferior olive. Ambiguus: Nucleus ambiguous. *Bottom* Two whisking traces ipsilateral and contralateral of the electrolytic lesion in the medulla are shown. Rhythmic whisking requires intactness of the lesioned site in the medulla. Modified from Moore et al. (2013) with permission. **d** Cortical input to the putative CPG. *Left* Anterograde terminal labeling in the medulla following a AAV-EGFP injection in the contralateral vM1. *Right* Similar anterograde injection in the vM1 along with injection of rabies ΔG-mcherry in the intrinsic muscles and labeling of premotor neurons in the medulla using expression of the deficient protein. The *green* terminals are projections from contralateral vM1, whereas the *red* retrogradely labeled neurons project to the intrinsic motor neurons in the facial nucleus which in turn project to the intrinsic muscles responsible for whisker protraction. GIRt gigantocellular reticular formation, IRt intermediate reticular formation, PCRt parvocellular reticular formation, NA nucleus ambiguus, Sp5ic caudal portion of the trigeminal nucleus pars interpolaris. Modified from Sreenivasan et al. 2014 with permission

Transitional Zone (TZ)

The best studied part of the erstwhile RF is the more caudal part sandwiched between RW and the body and limb representations which has been shown to receive direct projections from vS1 (Krubitzer et al. 1986; Koralek et al. 1990; Reep et al. 1990; Miyashita et al. 1994; Alloway et al. 2004; Chakrabarti and Alloway 2009; Colechio and Alloway 2009; Aronoff et al. 2010; Tennant et al. 2011; Mao et al. 2011; Smith and Alloway 2013). Judging from published coordinates of electrode placement, and reports of tactile inputs many previous studies on rat vM1 neuronal activity have focused on TZ (Figure 5a). Sensory responses to whisker deflections in the TZ are dependent on the intactness of vS1 (Farkas et al. 1999; Chakrabarti et al. 2008), as shown in Figure 5b and have been observed by several studies (Farkas et al. 1999; Kleinfeld et al. 2002; Chakrabarti et al. 2008; Aronoff et al. 2010; Petreanu et al. 2012). Experiments in anesthetized animals have shown

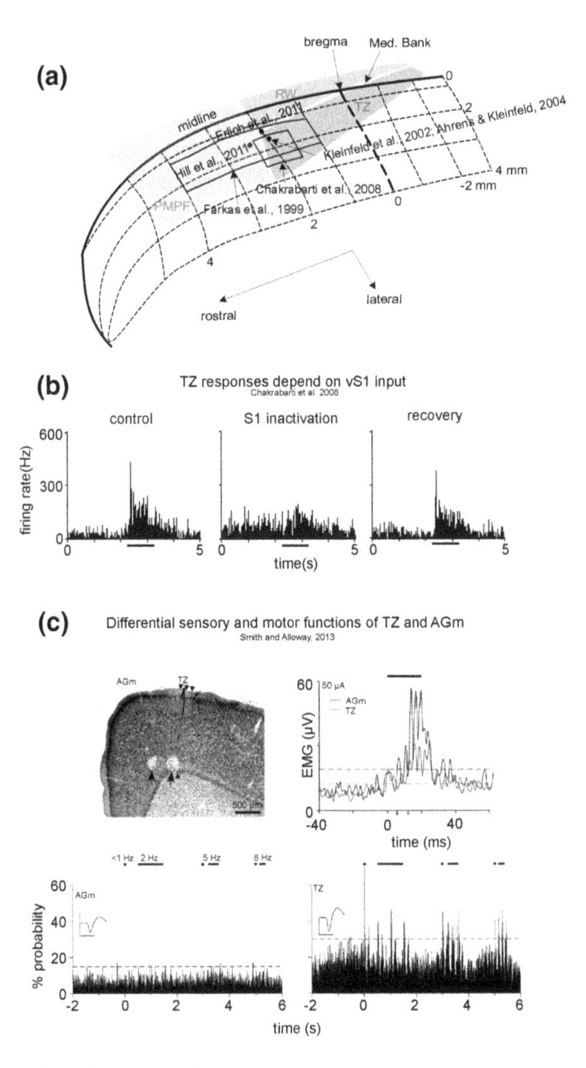

Figure 5 Organization of module TZ with sensory evoked responses. **a** A map of the stereotaxic co-ordinates used by various studies overlaid on the surface map of cortex from Figure 1a showing the barrel cortex recipient tactile transitional zone (TZ) in *red*, the prefrontal and premotor cortex (PMPF) in *blue* and RW in *green*. **b** Sensory evoked responses in vM1 are abolished followed inactivation of S1 barrel cortex. Peri-stimulus time histograms or PSTHs showing vM1 responses to whisker deflection using a moving airjet under control conditions (*left*), during S1 barrel cortex inactivation (*middle*) and during recovery phase (*right*). The *horizontal black bars* below the PSTHs indicate periods during which the stimulus was on. Modified from Chakrabarti et al. 2008 with permission. **c** The TZ module has stronger evoked responses but higher ICMS thresholds. *Top left* Recording/stimulation sites in the AGm and TZ marked by electrolytic lesions (*arrowheads*) in a coronal section through vM1 stained for Nissl material. *Top right* Electromyograph (EMG) recordings from the whisker pad showing muscle responses to ICMS applied using 50 μA of cathodal current in either AGm (*black trace*) or TZ (*red trace*). Horizontal dashed lines show maximum pre-stimulus activity. *Bottom left* PSTH depicting firing probability of a single neuron recorded from AGm during multiwhisker stimulation using a window screen at frequencies of <1(*dot*), 2, 5 and 8 Hz (*horizontal bars*). *Bottom right* Identical PSTH for a single neuron recorded from TZ. Mean waveforms shown in insets, scales 200 μV, 1 ms. *Horizontal dashed lines* indicate 99 % confidence intervals based on pre-stimulus firing rates. Modified from Smith and Alloway 2013 with permission

that when moving the electrode from TZ toward the medial bank (likely corresponding to RW), the strength of ICMS evoked movements becomes stronger while tactile responses vanish as illustrated in Figure 5c (Smith and Alloway 2013). It is important to mention here that part of what has been cytoarchitectonically defined as AGm has been functionally classified as RW (Haiss and Schwarz 2005). In line with the above evidence from the Alloway lab, our recordings from RW also failed to show any sensory responses in the awake animal upon contact with a real object (Gerdjikov et al. 2013). Sensory projections from vS1 to TZ have also been shown to activate inhibitory TZ neurons with shorter latencies and larger magnitudes than their excitatory neighbors, thus selectively activating a feedforward inhibitory network in vM1 and possibly allowing sensory input to dynamically recruit different motor cortical modules (Murray and Keller 2011). Further, clipping of whiskers during early development has been shown to cause a reduction in the size of TZ as measured using ICMS raising the possibility of a critical phase for a dependency of TZ on tactile inputs during development (Keller et al. 1996).

Reports on movement coding of TZ neurons are diverse. LFP recordings, most likely collected in the TZ, were reported to reflect rhythmic whisking (Ahrens and Kleinfeld 2004) although unit recordings failed to confirm the same. There are some rare cells in a wide region overlapping PMPF and TZ that are modulated by whisking rhythm (Hill et al. 2011), but whether a specific readout is formed from them to generate rhythmic whisking is an open question. Calcium transients in axonal terminals of TZ neurons projecting to layer 1 in vS1, recorded in mice performing an object localization task, were reported to carry a host of behavioral signals on a slow time scale, which encompass whisking activity and touch (Petreanu et al. 2012).

Another possible functional interpretation of the ICMS-evoked whisker retraction is that it takes part in guiding navigation. Neurons presumably located in TZ were found to encode direction of orientation responses, including whole body orientation movements accompanied by concomitant whisker retraction (Erlich et al. 2011). Inactivation of vM1 impaired such orientation movements which argues in favor of TZ being involved in the co-ordination of head, body and whisker movements and therefore having possible connections to a far wider variety motor centers of brainstem (face) and spinal cord (body) than RW.

Still another possible functional explanation comes from the observation that ICMS-evoked retraction movements give way to rhythmic whisker movements after lesions of vS1 (Matyas et al. 2010). According to this view TZ is the putative vM1 module that is perhaps involved in adapting rhythmic whisking patterns according to tactile inputs.

Barrel Cortex (VS1)

From the findings of Matyas et al. (2010) which showed that vM1 mediated whisker retraction is critically dependent on the intactness of vS1 and vS1 by itself evokes retraction movements upon ICMS (Matyas et al. 2010), it could well be concluded that vS1 itself is a motor structure. These investigators further showed that vM1 and vS1 project to partially overlapping subgroups of premotor neurons in the brainstem. The biggest difference was found for regions in the intermediate and parvocellular portions of the reticular formation and the facial nucleus which receive predominant vM1 input whereas vS1 predominantly targeted the trigeminal nuclei pars oralis, interpolaris and to a lesser extent caudalis (Sreenivasan et al. 2014).

Whisking Control—the Extent of Cortical Involvement

In summary the understanding of control of whisking movements by sensorimotor cortex is complicated by the existence of different functional motor modules in vM1 and contributions of vS1, and by their differential connection to a variety of brainstem centers including only partially known CPGs and reflex arcs—not to speak of interconnections on all hierarchical levels of sensorimotor integration that we ignored in this review. Despite the complexity, the functional organization of whisker motor control is beginning to emerge. One basic property of the system is that sensorimotor cortex is not likely to contain rhythm generating functions itself, despite the existence of CM cells in vM1 and activity modulated by whisker phase in vM1 and vS1. This will allow the future characterization of CM connections, the function of which is ignored in the primate fingertip system as well. The second basic property is that brainstem premotor circuits are differentially controlled by vM1 and vS1 and fall apart in at least two functional domains, one motor (the CPG(s)), and one sensorimotor (the TBL(s)). Sorting out the premotor neuronal elements contributing to these brainstem networks and their cortical control presents the main challenge for immediate future research.

References

Ahrens, K F and Kleinfeld, D (2004). Current flow in vibrissa motor cortex can phase-lock with exploratory rhythmic whisking in rat. *Journal of Neurophysiology* 92: 1700-1707.

Alloway, K D; Lou, L; Nwabueze-Ogbo, F and Chakrabarti, S (2006). Topography of cortical projections to the dorsolateral neostriatum in rats: Multiple overlapping sensorimotor pathways. *Journal of Comparative Neurology* 499: 33–48.

Alloway, K D; Olson, M L and Smith J B (2008). Contralateral corticothalamic projections from MI whisker cortex: Potential route for modulating hemispheric interactions. *Journal of Comparative Neurology* 510: 100–116.

Alloway, K D; Olson, M L and Smith J B (2008). Contralateral corticothalamic projections from MI whisker cortex: potential route for modulating hemispheric interactions. *Journal of Comparative Neurology* 510: 100–116.

Alloway, K D; Smith, J B and Beauchemin, K J (2010). Quantitative analysis of the bilateral brainstem projections from the whisker and forepaw regions in rat primary motor cortex. *Journal of Comparative Neurology* 518: 4546–4566.

Alloway, K D; Smith, J B; Beauchemin, K J and Olson, M L (2009). Bilateral projections from rat MI whisker cortex to the neostriatum, thalamus, and claustrum: Forebrain circuits for modulating whisking behavior. *Journal of Comparative Neurology* 515: 548–564.

Alloway, K D; Zhang, M and Chakrabarti, S (2004). Septal columns in rodent barrel cortex: Functional circuits for modulating whisking behavior. *Journal of Comparative Neurology* 480: 299–309.

Aroniadou, V A and Keller, A (1993). The patterns and synaptic properties of horizontal intracortical connections in the rat motor cortex. *Journal of Neurophysiology* 70: 1553–1569.

Aronoff, R et al.(2010). Long-range connectivity of mouse primary somatosensory barrel cortex. *European Journal of Neuroscience* 31: 2221–2233.

Brecht, M et al.(2004a). Organization of rat vibrissa motor cortex and adjacent areas according to cytoarchitectonics, microstimulation, and intracellular stimulation of identified cells. *Journal of Comparative Neurology* 479: 360–373.

Brecht, M; Schneider, M; Sakmann, B and Margrie, T W (2004b). Whisker movements evoked by stimulation of single pyramidal cells in rat motor cortex. *Nature* 427: 704–710.

Carvell, G E and Simons, D J (1990). Biometric analyses of vibrissal tactile discrimination in the rat. *The Journal of Neuroscience* 10: 2638–2648.

Chakrabarti, S and Alloway, K D (2006). Differential origin of projections from SI barrel cortex to the whisker representations in SII and MI. *Journal of Comparative Neurology* 498: 624–636.

Chakrabarti, S and Alloway, K D (2009). Differential response patterns in the si barrel and septal compartments during mechanical whisker stimulation. *Journal of Neurophysiology* 102: 1632–1646.

Chakrabarti, S and Schwarz, C (2014). Studying motor cortex function using the rodent vibrissal system. *e-Neuroforum* 5: 20–27.

Chakrabarti, S and Schwarz, C (2014). The rodent vibrissal system as a model to study motor cortex function. In: Groh and Krieger (Eds.), *Sensorimotor Integration in the Whisker System.* Springer.

Chakrabarti, S; Zhang, M and Alloway, K D (2008). MI neuronal responses to peripheral whisker stimulation: Relationship to neuronal activity in si barrels and septa. *Journal of Neurophysiology* 100: 50–63.

Cicirata, F; Angaut, P; Cioni, M; Serapide, M F and Papale, A (1986). Functional organization of thalamic projections to the motor cortex. An anatomical and electrophysiological study in the rat. *Neuroscience* 19: 81–99.

Colechio, E M and Alloway, K D (2009). Differential topography of the bilateral cortical projections to the whisker and forepaw regions in rat motor cortex. *Brain Structure & Function* 213: 423–439.

Conde, F; Audinat, E; Maire-Lepoivre, E and Crepel, F (1990). Afferent connections of the medial frontal cortex of the rat. A study using retrograde transport of fluorescent dyes. I. Thalamic afferents. *Brain Research Bulletin* 24: 341–354.

Conde, F; Maire-Lepoivre, E; Audinat, E and Crepel, F (1995). Afferent connections of the medial frontal cortex of the rat. II. Cortical and subcortical afferents. *Journal of Comparative Neurology* 352: 567–593.

Cramer, N P and Keller, A (2006). Cortical control of a whisking central pattern generator. *Journal of Neurophysiology* 96: 209–217.

Donoghue, J P and Parham, C (1983). Afferent connections of the lateral agranular field of the rat motor cortex. *Journal of Comparative Neurology* 217: 390–404.

Donoghue, J P and Wise, S P (1982). The motor cortex of the rat: Cytoarchitecture and microstimulation mapping. *Journal of Comparative Neurology* 212: 76–88.

Erlich, J C; Bialek, M and Brody, C D (2011). A cortical substrate for memory-guided orienting in the rat. *Neuron* 72: 330–343.

Fabri, M and Burton, H (1991). Ipsilateral cortical connections of primary somatic sensory cortex in rats. *Journal of Comparative Neurology* 311: 405–424.

Farkas, T; Kis, Z; Toldi, J and Wolff, J R (1999). Activation of the primary motor cortex by somatosensory stimulation in adult rats is mediated mainly by associational connections from the somatosensory cortex. *Neuroscience* 90: 353–361.

Ferezou, I et al. (2007). Spatiotemporal dynamics of cortical sensorimotor integration in behaving mice. *Neuron* 56: 907–923.

Friedman, W A et al. (2006). Anticipatory activity of motor cortex in relation to rhythmic whisking. *Journal of Neurophysiology* 95: 1274–1277.

Friedman, W A; Zeigler, H P and Keller, A (2012). Vibrissae motor cortex unit activity during whisking. *Journal of Neurophysiology* 107: 551–563.

Gamzu, E and Ahissar, E (2001). Importance of temporal cues for tactile spatial-frequency discrimination. *The Journal of Neuroscience* 21: 7416–7427.

Gao, P; Bermejo, R and Zeigler, H P (2001). Whisker deafferentation and rodent whisking patterns: Behavioral evidence for a central pattern generator. *The Journal of Neuroscience* 21: 5374–5380.

Gao, P; Hattox, A M; Jones, L M; Keller, A and Zeigler, H P (2003). Whisker motor cortex ablation and whisker movement patterns. *Somatosensory & Motor Research* 20: 191–198.

Gerdjikov, T V; Haiss, F; Rodriguez-Sierra, O E and Schwarz, C (2013). Rhythmic whisking area (RW) in rat primary motor cortex: An internal monitor of movement-related signals? *The Journal of Neuroscience* 33: 14193–14204.

Gioanni, Y and Lamarche, M (1985). A reappraisal of rat motor cortex organization by intracortical microstimulation. *Brain Research* 344: 49–61.

Grinevich, V; Brecht, M and Osten, P (2005). Monosynaptic pathway from rat vibrissa motor cortex to facial motor neurons revealed by lentivirus-based axonal tracing. *The Journal of Neuroscience* 25: 8250–8258.

Haiss, F and Schwarz, C (2005). Spatial segregation of different modes of movement control in the whisker representation of rat primary motor cortex. *The Journal of Neuroscience* 25: 1579–1587.

Hall, R D and Lindholm, E P (1974). Organization of motor and somatosensory neocortex in the albino rat. *Brain Research* 66: 23–38.

Hattox, A M; Priest, C A and Keller, A (2002). Functional circuitry involved in the regulation of whisker movements. *Journal of Comparative Neurology* 442: 266–276.

Hentschke, H; Haiss, F and Schwarz, C (2006). Central signals rapidly switch tactile processing in rat barrel cortex during whisker movements. *Cerebral Cortex* 16: 1142–1156.

Hicks, S P and D'Amato, C J (1977). Locating corticospinal neurons by retrograde axonal transport of horseradish peroxidase. *Experimental Neurology* 56: 410–420.

Hill, D N; Curtis, J C; Moore, J D and Kleinfeld, D (2011). Primary motor cortex reports efferent control of vibrissa motion on multiple timescales. *Neuron* 72: 344–356.

Hoffer, Z S; Hoover, J E and Alloway, K D (2003). Sensorimotor corticocortical projections from rat barrel cortex have an anisotropic organization that facilitates integration of inputs from whiskers in the same row. *Journal of Comparative Neurology* 466: 525–544.

Hooks, B M et al.(2013). Organization of cortical and thalamic input to pyramidal neurons in mouse motor cortex. *The Journal of Neuroscience* 33: 748–760.

Keller, A; Weintraub, N D and Miyashita, E (1996). Tactile experience determines the organization of movement representations in rat motor cortex. *Neuroreport* 7: 2373–2378.

Kinnischtzke, A K; Simons, D J and Fanselow, E E (2014). Motor cortex broadly engages excitatory and inhibitory neurons in somatosensory barrel cortex. *Cerebral Cortex* 24: 2237–2248.

Kleinfeld, D; Sachdev, R N; Merchant, LM; Jarvis, M R and Ebner, F F (2002). Adaptive filtering of vibrissa input in motor cortex of rat. *Neuron* 34: 1021–1034.

Koralek, K A; Olavarria, J and Killackey, H P (1990). Areal and laminar organization of corticocortical projections in the rat somatosensory cortex. *Journal of Comparative Neurology* 299: 133–150.

Krubitzer, L A; Sesma, M A and Kaas, J H (1986). Microelectrode maps, myeloarchitecture, and cortical connections of three somatotopically organized representations of the body surface in the parietal cortex of squirrels. *Journal of Comparative Neurology* 250: 403–430.

Kyuhou, S and Gemba, H (2002). Projection from the perirhinal cortex to the frontal motor cortex in the rat. *Brain Research* 929: 101–104.

Landers, M and Zeigler, H P (2006). Development of rodent whisking: trigeminal input and central pattern generation. *Somatosensory & Motor Research* 23: 1–10.

Lavallee, P et al. (2005). Feedforward inhibitory control of sensory information in higher-order thalamic nuclei. *The Journal of Neuroscience* 25: 7489–7498.

Lee, S; Kruglikov, I; Huang, Z J; Fishell, G and Rudy, B (2013). A disinhibitory circuit mediates motor integration in the somatosensory cortex. *Nature Neuroscience* 16: 1662–1670.

Mao, T et al. (2011). Long-range neuronal circuits underlying the interaction between sensory and motor cortex. *Neuron* 72: 111–123.

Matyas, F et al. (2010). Motor control by sensory cortex. *Science* 330(6008): 1240–1243.

McIntyre, D C; Kelly, M E and Staines, W A (1996). Efferent projections of the anterior perirhinal cortex in the rat. *Journal of Comparative Neurology* 369: 302–318.

Mitchinson, B; Martin, C J; Grant, R A and Prescott, T J (2007). Feedback control in active sensing: rat exploratory whisking is modulated by environmental contact. *Proceedings of the Royal Society B: Biological Sciences* 274: 1035–1041.

Miyashita, E; Keller, A and Asanuma, H (1994). Input-output organization of the rat vibrissal motor cortex. *Experimental Brain Research* 99: 223–232.

Moore, J D et al. (2013). Hierarchy of orofacial rhythms revealed through whisking and breathing. *Nature* 497: 205–210.

Murray, P D and Keller, A (2011). Somatosensory response properties of excitatory and inhibitory neurons in rat motor cortex. *Journal of Neurophysiology* 106: 1355–1362.

Neafsey, E J et al. (1986). The organization of the rat motor cortex: A microstimulation mapping study. *Brain Research* 396: 77–96.

Neafsey, E J and Sievert, C (1982). A second forelimb motor area exists in rat frontal cortex. *Brain Research* 232: 151–156.

Nguyen, Q-T and Kleinfeld, D (2005). Positive feedback in a brainstem tactile sensorimotor loop. *Neuron* 45: 447–457.

Petreanu, L et al. (2012). Activity in motor-sensory projections reveals distributed coding in somatosensation. *Nature* 489: 299–303.

Porter, L L and White, E L (1983). Afferent and efferent pathways of the vibrissal region of primary motor cortex in the mouse. *Journal of Comparative Neurology* 214: 279–289.

Rathelot, J and Strick, P (2006). Muscle representation in the macaque motor cortex: An anatomical perspective. *Proceedings of the National Academy of Sciences of the United States of America* 103: 8257–8262.

Reep, R L; Chandler, H C; King, V and Corwin, J V (1994). Rat posterior parietal cortex: Topography of corticocortical and thalamic connections. *Experimental Brain Research* 100: 67–84.

Reep, R L; Corwin, J V; Hashimoto, A and Watson, R T (1987). Efferent connections of the rostral portion of medial agranular cortex in rats. *Brain Research Bulleti*n 19: 203–221.

Reep, R L; Goodwin, G S and Corwin, J V (1990). Topographic organization in the corticocortical connections of medial agranular cortex in rats. *Journal of Comparative Neurology* 294: 262–280.

Rouiller, E M; Liang, F Y; Moret, V and Wiesendanger, M (1991). Patterns of corticothalamic terminations following injection of *Phaseolus vulgaris* leucoagglutinin (PHA-L) in the sensorimotor cortex of the rat. *Neuroscience Letters* 125: 93–97.

Sanderson, K J; Welker, W and Shambes, G M (1984). Reevaluation of motor cortex and of sensorimotor overlap in cerebral cortex of albino rats. *Brain Research* 292: 251–260.

Sapienza, S; Talbi, B; Jacquemin, J and Albe-Fessard, D (1981). Relationship between input and output of cells in motor and somatosensory cortices of the chronic awake rat. A study using glass micropipettes. *Experimental Brain Research* 43: 47–56.

Sato, T R and Svoboda, K (2010). The functional properties of barrel cortex neurons projecting to the primary motor cortex. *The Journal of Neuroscience* 30: 4256–4260.

Schwarz, C and Möck, M (2001). Spatial arrangement of cerebro-pontine terminals. *Journal of Comparative Neurology* 435: 418–432.

Settlage, P H; Bingham, W G; Suckle, H M; Borge, A F and Woolsey, C N (1949). The pattern of localization in the motor cortex of the rat. *Federation Proceedings* 8: 144.

Smith, J B and Alloway, K D (2010). Functional specificity of claustrum connections in the rat: Interhemispheric communication between specific parts of motor cortex. *The Journal of Neuroscience* 30: 16832–16844.

Smith, J B and Alloway, K D (2013). Rat whisker motor cortex is subdivided into sensory-input and motor-output areas. *Frontiers in Neural Circuits* 7: 4.

Smith, J B; Radhakrishnan, H and Alloway, K D (2012). Rat claustrum coordinates but does not integrate somatosensory and motor cortical information. *The Journal of Neuroscience* 32: 8583–8588.

Sreenivasan, V; Karmakar, K; Rijli, F M and Petersen, C C H (2014). Parallel pathways from motor and somatosensory cortex for controlling whisker movements in mice. *European Journal of Neuroscience* 41(3): 354–367.

Takatoh, J et al. (2013). New modules are added to vibrissal premotor circuitry with the emergence of exploratory whisking. *Neuron* 77: 346–360.

Tennant, K A at al.(2011). The organization of the forelimb representation of the C57BL/6 mouse motor cortex as defined by intracortical microstimulation and cytoarchitecture. *Cerebral Cortex* 21: 865–876.

Towal, R B and Hartmann, M J (2006). Right-left asymmetries in the whisking behavior of rats anticipate head movements. *The Journal of Neuroscience* 26: 8838–8846.

Trageser, J C et al.(2006). State-dependent gating of sensory inputs by zona incerta. *Journal of Neurophysiology* 96: 1456–1463.

Urbain, N and Deschenes, M (2007). Motor cortex gates vibrissal responses in a thalamocortical projection pathway. *Neuron* 56: 714–725.

Uylings, H B; Groenewegen, H J and Kolb, B (2003). Do rats have a prefrontal cortex? *Behavioural Brain Research* 146: 3–17.

Weiss, D S and Keller, A (1994). Specific patterns of intrinsic connections between representation zones in the rat motor cortex. *Cerebral Cortex* 4: 205–214.

Wolfe, J et al.(2008). Texture coding in the rat whisker system: slip-stick versus differential resonance. *PLoS Biology* 6: e215.

Zagha, E; Casale, A E; Sachdev, R N; McGinley, M J and McCormick, D A (2013). Motor cortex feedback influences sensory processing by modulating network state. *Neuron* 79(3): 567–578.

Zilles, K; Zilles, B and Schleicher, A (1980). A quantitative approach to cytoarchitectonics VI - The areal pattern of the cortex of the albino rat. *Anatomy and Embryology* 159: 335–360.

Tactile Attention in the Vibrissal System

Ben Mitchinson

Attention is a rich research area concerning itself with many aspects of brain function and animal behaviour (even as far as consciousness). This article focuses on the most prosaic aspect of attention, however, that of overt orienting of sensory receptors to the location of an 'attended stimulus' in space. Whilst overt orienting of the eyes has been studied in great detail, orienting of the vibrissal array of rodents has only recently begun to be investigated, despite the importance of this organ as an experimental model to a wide variety of investigations. This article reviews current understanding of orienting in this organ and identifies major open questions.

The Expression of Attention

Attention as a faculty of neural systems need not, in general, have a direct behavioural correlate (or, in principle, any behavioural correlate at all). Purported system variables associated with 'covert' attention management can be measured behaviourally only indirectly through the impact they have on subsequent stimulus-response or object search times or efficiencies (Ward 2008). Models of attentional mechanisms can be complex and multi-faceted in attempting to account for the effects of immediate sensory stimuli, prior knowledge, motivation, instruction, and so on, on the variables visible through these relatively small apertures (Frintrop et al. 2010). These models have been developed using data from visual system studies (in models such as cat and primate); attention is often studied using human participants since the required manipulations are often not invasive (Posner 1980).

Despite the complexity of the evolution of these internal states, however, the 'overt' orienting of the sensory organs, the head, and the body, to focus sensory apparatus onto a selected region of space (the 'orienting reflex') is both a rather more straightforward expression of attention switching and absolutely central to

B. Mitchinson (✉)
Adaptive Behaviour Research Group, The University of Sheffield,
Sheffield, South Yorkshire, UK

T.J. Prescott et al. (eds.), *Scholarpedia of Touch*, Scholarpedia,
DOI 10.2991/978-94-6239-133-8_56

everyday behaviour (Itti and Koch 2001). In primates, orienting is typically led by the visual system's fovea and is modelled as a series of discrete 'saccades' of the visual fixation point from one spatial location to the next (Itti et al. 1998). This behaviour is predictable, and has been characterised in a range of tasks including, for instance, scene search (Yarbus 1967), reading (Rayner 1998) and locomotion (Hollands et al. 1995). The contemporary ease of measuring eye movements in a human subject facilitates the use of this model (Duchowski 2007).

Rodent Models of Attention

Whilst observation of primate visual systems provides relatively controlled access to attentional mechanisms, and particularly to spatial variables, the use of rodent models may be desirable for some studies, for example if the goal is drug discovery or if the manipulation is otherwise invasive. Techniques for observing attentional mechanisms in rodents are less comprehensive, but a variety of choice/reward protocols have been employed which make use of the spatial component of the apparatus only for the indication of the choice (Sagvolden et al. 2005). One well-established paradigm is the 5-choice serial reaction time task (5CSRTT) (Robbins 2002) which comprises the measurement of the procedure and timing of nose-poking into apertures signalled by (paradigmatically, visual) stimuli. Results from the 5CSRTT are described as indicating characteristics such as 'impulsivity' and 'compulsivity' (Robbins 2002; Sanchez-Roige et al. 2012). The potential relevance of this paradigm to research into contemporary medical problems such as attention-deficit disorders is clear (in fact, the task was originally designed with just such investigations in mind).

The sensory modality upon which rodents rely the most is, under many conditions, the tactile whisker (vibrissa) system rather than vision (see Vibrissal behavior and function). One consequence of that in this context follows from the observation that 'orienting' of the tactile whisker sensory organ to a spatial location implies the bringing of the organ *to*, rather than just the pointing of the organ *towards*, the target of attention, a result of the fundamentally limited range of tactile sensing (Mitchinson and Prescott 2013). Thus, the animal's morphology dictates a convergence between sensory orienting and 'taking action' (which might include, for example, the orienting of motor resources such as the mouth and locomotion to a location from where action can be taken). Investigations using Whiskered robots have demonstrated that linking a changing focus of spatial attention to orienting of the snout generates plausible sequences of gross behaviour (Mitchinson et al. 2012; Mitchinson and Prescott 2013; Prescott et al. in press). This convergence has previously been described using the expression *orienting is acting* (Mitchinson in press). Note that this convergence is present in all systems to an extent even though it is much less marked when it is remote sensors that are being oriented.

Meanwhile, whilst many animals can move their eyes within their heads, so many whiskered animals—certainly, rodents—appear able to reposition their whiskers about their heads (Sofroniew and Svoboda 2015). Most models to date of the control of these movements have been based fairly directly on measured relationships between whisker movements and immediate sensory and motor variables rather than on central mechanisms (Towal and Hartmann 2006; Mitchinson et al. 2007; Grant et al. 2009; Deutsch et al. 2012; Arkley et al. 2014), but an analogy between these movements and saccadic movements of the eyes (which are defined in physical space) has been drawn more than once (Towal and Hartmann 2006; Pearson et al. 2007; Mitchinson et al. 2014; Sofroniew and Svoboda 2015). Taken together, these observations suggest an identification between the control of the whiskers and snout and the control of the eyes and head, both driven by the animal's changing focus of spatial attention.

A Model of Spatial Attention Mediated by Vibrissae

This identification is supported, also, by the nature of observed whisker movements, and one model of whisker movement control has been proposed that makes this identification explicit, having at its core a state representing the region of space attended by the animal (Mitchinson and Prescott 2013). The model provides a succinct explanation for several observations of whisker movement and makes predictions as to how direct manipulation of the animal's spatial attention will be reflected through those movements. Experimental implementation of the model expresses rodent-like exploratory behaviour as a 'series of orients' analogous to the 'series of saccades' model of visual orienting behaviour, underlining both this analogous relationship and also the connection in the case of the rodent model with gross behaviour.

Description of the Model

In brief, the model dictates that whiskers will be moved so as to bring them to bear on attended objects. Thus, its central variable is a representation of the immediate region of space attended by an animal, which we can refer to as the locus of spatial attention or the 'attended region' (Figure 1). This region may be compact or extensive, though 'objects' are not explicitly delineated (a distinction is usually drawn between location-based and object-based attention).

Behavioural studies have indicated that the movement of all of the whiskers on each side of the face can be well-summarised by only two variables (Grant et al. 2009); however, such a summary does not account for the shaping of the 2D surface formed by the whisker tips (Huet and Hartmann 2014). Accordingly, a fixed transform, which may be arbitrarily non-linear, maps from the representation of the attended region to the maximum protraction angle of each of the whiskers individually. This transform is chosen so that each whisker tends to be driven forward

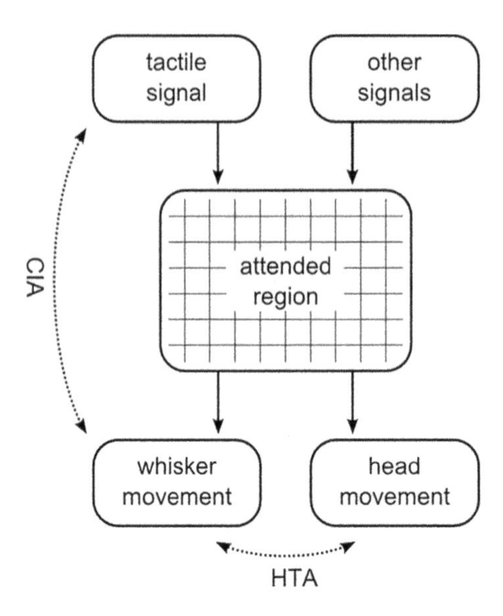

Figure 1 Summary of model of expression of rodent spatial attention. The locus of spatial attention is driven by sensory signals and, in turn, drives movement of the whiskers and the head. Specific observations that have been reported (CIA, HTA, see text) can be understood as correlations between states of this model. Taken from Mitchinson and Prescott (2013)

into the attended region by a small amount. A separate model component generates the rhythmic 'whisking' motion characteristic of this system (see Whisking kinematics and Whisking pattern generation) so that the instantaneous protraction angle of each whisker is derived by combining the outputs of these two systems, and whiskers are palpated in and out of the attended region.

Meanwhile, the movement of the snout of the animal (specifically, the positioning of a generalised sensory 'fovea' around the mouth) is driven by the same representation on a longer timescale. In this model, whisker and head movements are both, therefore, generated as overt expressions of this attended region. Interactions between the animal and its environment then modulate the locus of spatial attention through the sensory systems; in particular, contacts between the whiskers and objects in the environment excite attention directly.

Key Results

The model has been used to simulate a range of behavioural experimental paradigms and has generated results in each case that are qualitatively and quantitatively similar to those reported in animals. First, 'Head-Turning Asymmetry' (HTA) is a relationship that has been observed in rat, mouse, and opossum (Towal and Hartmann 2006; Mitchinson et al. 2011). Turning of the head is preceded by

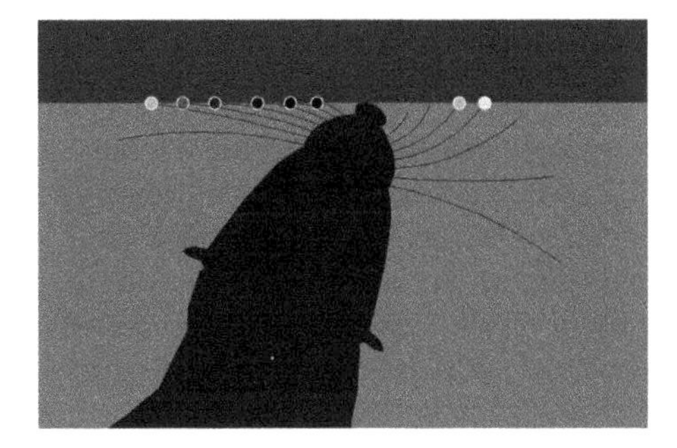

Figure 2 Still from video of simulated rat behaviour comprising plane model of rat snout, one row of whiskers, and stationary flat obstacle. Contact-induced asymmetry is seen, as the whiskers on the *right* are protracted more than those on the *left*. Taken from Mitchinson and Prescott (2013)

asymmetry in the protraction angles of the whiskers on each side of the head such that the whiskers appear to 'lead' the head movement. Second, 'Contact-Induced Asymmetry' (CIA, Figure 2) is a relationship between whisker-environment contact and protraction angle asymmetry (Mitchinson et al. 2007, 2011; Deutsch et al. 2012). Contact with the environment on one side only leads to less/more protraction on the ipsilateral/contralateral side. Third, 'spread reduction' (SR) denotes the finding that the spread between the protraction angles of whiskers on each side is reduced during environmental contact (Grant et al. 2009; Huet and Hartmann 2014).

Discussion

Each of the observations listed can be alternatively interpreted as indicating a single direct reactive mechanism. However, the model based on spatial attention is sufficient to produce all of them from a single, albeit more complex, mechanism. Moreover, the attention-based model can explain anticipatory modulations of whisker movement, which reactive mechanisms cannot, and there are now many observations of these (Sachdev et al. 2003; Berg and Kleinfeld 2003; Mitchinson et al. 2007; Grant et al. 2009; Arkley et al. 2014; Voigts et al. 2015).

Direct manipulation of the animal's attention will be required to make a direct test of the model, and this experiment has not yet been performed. Given the potential for confound with direct influences on whisker movement, manipulating attention using a non-tactile modality may be the preferred experimental approach (audio or olfactory stimuli might be suitable, for instance).

Methodological Potential

Methodologically, the measurement of whisker motions has become far easier over recent years primarily owing to the increased availability of high-speed and high-sensitivity videography (Diamond and Arabzadeh 2013) (coincidentally, eye movement tracking has become more accessible over much the same period, though for different reasons). As a result, measurement of whisker motion has become a potentially-useful general tool for accessing the animal's state, apart from its use in the analysis of whisker operation per se (for instance, see (Wolfe et al. 2011)). Whilst whisker motions can now be measured even in freely-behaving preparations (Arkley et al. 2014), they are available in the head-fixed preparation also (which would stand in the way of choice protocols such as 5CSRTT), and measurement can be largely automated in both cases (see Clack et al. 2012 and references therein).

Potentially, then, whisker motion measurement provides a window onto attentional mechanisms in conditions that are highly flexible and amenable to direct neural measurements and interventions. The establishment of direct connections between whisker movements and hidden attentional state variables would offer a powerful new tool for the study of attention in a rat model. Such a tool would provide access to states not available through established paradigms such as the 5CSRTT and, since whisker movements are known to respond to sensory events on very short timescales (10–15 ms, Mitchinson et al. 2007), with good temporal as well as spatial resolution. The degree to which vibrissal attention would share mechanisms and resources in rodents with other modalities would be an open question, but cross-modal effects on attention are very well established, at least, in humans (Driver and Spence 1998).

In Context

If a connection between attention and whisker movement can be validated, it will have two impacts. First, as discussed, by providing a minimally-invasive tool for accessing attentional states in the behaving animal. Second, by providing a more complete understanding of how the whiskers are used by the animal to serve the purpose of Active tactile perception. Through the relationship with gross behaviour, it is perhaps particularly clear in animals that rely so heavily on their whiskers how closely these two concepts—attention and active sensing—are linked.

Given the importance of this system to many areas of neuroscience, a complete understanding of how the whiskers are moved in response to exogenous and endogenous influences would be very valuable. Indeed, experimental paradigms whereby whisker stimulation occurs as a result of whisker movement, rather than of stimulator movement, are necessary if the decoding of sensory signals is to be understood (see Vibrissal location coding).

Any commonalities identified between these two popular models (visual and vibrissal) would, per se, be a great step forward in terms of synthesis. Lines of enquiry in this regard could include investigations of whisker sensory and motor characteristics of brain areas involved in the management of spatial attention and the generation of spatially-targeted behaviours including visual orienting. An obvious example is superior colliculus, which is a well-studied area in general (May 2006; Gandhi and Katnani 2011) and has generated a slew of promising results in studies of the vibrissae both historically and more recently (see Bezdudnaya and Castro-Alamancos 2014 for a recent contribution, also see references in Mitchinson and Prescott 2013).

Intriguingly, though, it may be the differences between these systems that has the potential to be most revealing (Mitchinson and Prescott 2013). There is both neurophysiological (Benedetti 1991) and morphological (Huet and Hartmann 2014) evidence that whisker sensory signals are brought into register with visual signals in rat colliculus, a result which is itself unsurprising (Stein and Stanford 2008). Bringing into register sensory signals from modalities with mobile sensory organs requires dynamic remapping, however. In the case of the whiskers, which constantly move back and forth through space during active sensing, this remapping must occur at multiple cycles each second, a fact which may help to highlight the underlying mechanism. Meanwhile, orienting of the eyes—a remote sensor—requires the control only of azimuth and elevation, and representations of space in colliculus are exclusively described as being two-dimensional (Itti and Koch 2001) (very likely, it seems, correspondingly). In contrast, the control of just the snout (let alone the whiskers) requires a richer representation of space; on the face of it, all three dimensions must be represented, since the snout must be directed *to* not just *towards* the target of attention (Mitchinson and Prescott 2013). Understanding how these two systems, primate and rodent, correspond in these regards is bound, it seems, to be instructive.

References

Arkley, K; Grant, R A; Mitchinson, B and Prescott, T J (2014). Strategy change in vibrissal active sensing during rat locomotion. *Current Biology* 24(13): 1507–1512.

Benedetti, F (1991). The postnatal emergence of a functional somatosensory representation in the superior colliculus of the mouse. *Developmental Brain Research* 60(1): 51–57.

Berg, R W and Kleinfeld, D (2003). Rhythmic whisking by rat: retraction as well as protraction of the vibrissae is under active muscular control. *Journal of Neurophysiology* 89(1): 104–117.

Bezdudnaya, T and Castro-Alamancos, M A (2014). Neuromodulation of whisking related neural activity in superior colliculus. *The Journal of Neuroscience* 34(22): 7683–7695.

Clack, N G et al. (2012). Automated tracking of whiskers in videos of head fixed rodents. *PLoS Computational Biology* 8(7): e1002591.

Deutsch, D; Pietr, M; Knutsen, P M; Ahissar, E and Schneidman, E (2012). Fast feedback in active sensing: touch-induced changes to whisker-object interaction. *PloS ONE* 7(9): e44272.

Diamond, M E and Arabzadeh, E (2013). Whisker sensory system-from receptor to decision. *Progress in Neurobiology* 103: 28–40.

Driver, J and Spence, C (1998). Cross-modal links in spatial attention. *Philosophical Transactions of the Royal Society of London B: Biological Sciences* 353(1373): 1319–1331.

Duchowski, A (2007). *Eye Tracking Methodology: Theory and Practice*, Vol. 373. Springer Science & Business Media.

Frintrop, S; Rome, E and Christensen, H I (2010). Computational visual attention systems and their cognitive foundations: A survey. *ACM Transactions on Applied Perception (TAP)* 7(1): 6.

Gandhi, N J and Katnani, H A (2011). Motor functions of the superior colliculus. *Annual Review of Neuroscience* 34: 205.

Grant, R A; Mitchinson, B; Fox, C W and Prescott, T J (2009). Active touch sensing in the rat: anticipatory and regulatory control of whisker movements during surface exploration. *Journal of Neurophysiology* 101(2): 862–874.

Hollands, M A; Marple-Horvat, D E; Henkes, S and Rowan, A K (1995). Human eye movements during visually guided stepping. *Journal of Motor Behavior* 27(2): 155–163.

Huet, L A and Hartmann, M J (2014). The search space of the rat during whisking behavior. *The Journal of Experimental Biology* 217(18): 3365–3376.

Itti, L and Koch, C (2001). Computational modelling of visual attention. *Nature Reviews Neuroscience* 2(3): 194–203.

Itti, L; Koch, C and Niebur, E. (1998). A model of saliency-based visual attention for rapid scene analysis. *IEEE Transactions on Pattern Analysis and Machine Intelligence* 20(11): 1254–1259.

May, P J (2006). The mammalian superior colliculus: laminar structure and connections. *Progress in Brain Research* 151: 321–378.

Mitchinson, B (in press). Attention and orienting. In: T J Prescott and P F M J Verschure (Eds.), *Living Machines: A Handbook of Research in Biomimetic and Biohybrid Systems.* Oxford University Press.

Mitchinson, B et al. (2011). Active vibrissal sensing in rodents and marsupials. *Philosophical Transactions of the Royal Society B: Biological Sciences* 366(1581): 3037–3048.

Mitchinson, B; Martin, C J; Grant, R A and Prescott, T J (2007). Feedback control in active sensing: rat exploratory whisking is modulated by environmental contact. *Proceedings of the Royal Society B: Biological Sciences* 274(1613): 1035–1041.

Mitchinson, B; Pearson, M J; Pipe, A G and Prescott, T J (2014). Biomimetic tactile target acquisition, tracking and capture. *Robotics and Autonomous Systems* 62(3): 366–375.

Mitchinson, B; Pearson, M; Pipe, A and Prescott, T. (2012). The emergence of action sequences from spatial attention: Insight from rodent-like robots. In: *Biomimetic and Biohybrid Systems*, Vol. 7375 of *Lecture Notes in Computer Science* (pp. 168–179). Springer.

Mitchinson, B and Prescott, T J (2013). Whisker movements reveal spatial attention: a unified computational model of active sensing control in the rat. *PLoS Computational Biology* 9(9): e1003236.

Pearson, M J; Pipe, A G; Melhuish, C; Mitchinson, B and Prescott, T J (2007). Whiskerbot: A robotic active touch system modeled on the rat whisker sensory system. *Adaptive Behavior* 15 (3): 223–240.

Posner, M I (1980). Orienting of attention. *Quarterly Journal of Experimental Psychology* 32(1): 3–25.

Prescott, T J et al. (in press). The robot vibrissal system: Understanding mammalian sensorimotor co-ordination through biomimetics. In: P Krieger and A A Groh (Eds.), *Sensorimotor Integration in the Whisker System.* Springer.

Rayner, K (1998). Eye movements in reading and information processing: 20 years of research. *Psychological Bulletin* 124(3): 372.

Robbins, T (2002). The 5-choice serial reaction time task: Behavioural pharmacology and functional neurochemistry. *Psychopharmacology* 163(3–4): 362–380.

Sachdev, R N; Berg, R W; Champney, G; Kleinfeld, D and Ebner, F F (2003). Unilateral vibrissa contact: changes in amplitude but not timing of rhythmic whisking. *Somatosensory & Motor Research* 20(2): 163–169.

Sagvolden, T; Russell, V A; Aase, H; Johansen, E B and Farshbaf, M (2005). Rodent models of attention-deficit/hyperactivity disorder. *Biological Psychiatry* 57(11): 1239–1247.

Sanchez-Roige, S; Pena-Oliver, Y and Stephens, D N (2012). Measuring impulsivity in mice: the five-choice serial reaction time task. *Psychopharmacology* 219(2): 253–270.
Sofroniew, N J and Svoboda, K (2015). Whisking. *Current Biology* 25(4): R137–R140.
Stein, B E and Stanford, T R (2008). Multisensory integration: current issues from the perspective of the single neuron. *Nature Reviews Neuroscience* 9(4): 255–266.
Towal, R B and Hartmann, M J (2006). Right-left asymmetries in the whisking behavior of rats anticipate head movements. *The Journal of Neuroscience* 26(34): 8838–8846.
Voigts, J; Herman, D H and Celikel, T (2015). Tactile object localization by anticipatory whisker motion. *Journal of Neurophysiology* 113(2): 620–632.
Ward, L M (2008). Attention. *Scholarpedia* 3(10): 1538. http://www.scholarpedia.org/article/Attention.
Wolfe, J; Mende, C and Brecht, M (2011). Social facial touch in rats. *Behavioral Neuroscience* 125(6): 900.
Yarbus, A L (1967). *Eye Movements and Vision.* New York: Plenum Press.

Part IV
Synthetic Touch

Tactile Sensors

Uriel Martinez-Hernandez

Tactile sensors are data acquisition devices, or transducers, that are designed to sense a diversity of properties via direct physical contact (Nicholls and Lee 1989). Tactile sensor designs are based around a range of different technologies some of which are directly inspired by research on biological touch. The growth of robotic applications in healthcare, agriculture, social assistance, autonomous systems and unstructured environments has created a pressing need for effective tactile sensors. Their deployment plays an important role permitting the detection, measurement and conversion of information, acquired by physical interaction with objects, into an appropriate form to be processed and analysed by higher level modules within an intelligent system (Najarian et al. 2009). Although, in recent decades, tactile sensor technology has shown great advances in design and capability, tactile sensing systems are still relatively undeveloped compared to the sophisticated technology accomplished in vision (Lee and Nicholls 1999). The relatively slow development attained thus far is possibly related to the inherent complexity of the sense of touch (Siciliano and Khatib 2008). Another limiting factor is that, by their very nature, tactile sensors require direct contact to be made on surfaces and objects, and are therefore subject to wear and risk of damage than some other sensor types.

The Human Sense of Touch and Tactile Sensors

Vision is sometimes asserted to be the most important human sensory modality perhaps underestimating the role of the sense of touch. Certainly, losing the capabilities offered by touch can cause catastrophic impairments to posture, locomotion and control of limbs, extraction of object properties and in general to any physical interaction with the environment (Robles-De-La-Torre 2006).

Psychophysical studies have shown that the human **haptic touch** is rich in information for interaction, exploration, manipulation and extraction of object properties such as texture, shape, hardness and temperature (Lederman and Klatzky 1987).

U. Martinez-Hernandez (✉)
University of Sheffield, Sheffield, UK

T.J. Prescott et al. (eds.), *Scholarpedia of Touch*, Scholarpedia,
DOI 10.2991/978-94-6239-133-8_57

This information is registered by various types of receptors, e.g. mechanoreceptors (pressure and vibration), thermoreceptors (temperature), and nociceptors (pain and damage) distributed all over the body with variable density and located in the various layers of the skin (Johansson and Westling 1984). Human hands have a particularly high density of **mechanoreceptors**, being one of the parts of the body most specialised to provide accurate tactile feedback (Vallbo et al. 1984).

Tactile sensing was relatively neglected in the early years of robotics, with only a handful of devices developed by the end of the 1970s and with relatively limited integration of these systems into robots. In contrast, the 1980s saw substantial advances in tactile sensor technology, accompanied by a reduction in manufacturing costs. Progress was made in sensor materials, design and fabrication technologies, and in transduction methods for integration in various robotic platforms (Lee 2000). The main tactile sensing technologies developed by this time were capacitive, piezoresistive, piezoelectric, magnetic, inductive, optical and strain gauges, allowing the successful development of accurate devices for detection of object shape, size, texture, force and temperature (De Rossi 1991).

Criteria for Sensor Design

The human hand has a wide range of sensor types that support several different forms of touch. Whilst it would be desirable to create a robotic device with similar sensing capability, particularly for applications in prosthetics, to do so would involve addressing a large list of design specifications (Castelli 2002). Artificial touch sensing has therefore largely been focused towards less ambitious targets. The first design criteria for tactile sensors were proposed by Harmon (1982) and were motivated by the design requirements for industrial robots in the 1980s. As the technology has evolved so have the criteria. A key target for contemporary systems is humanoid robotics and the capacity to emulate human in-hand manipulation. To achieve this goal, Yousef et al. (2011) proposed the following list of functional requirements:

- Contact detection and release of an object.
- Lifting and replacement of an object.
- Detection of shape and force distribution for object recognition.
- Detection of dynamic and static forces.
- Tracking of contact points during manipulation.
- Estimation and detection of grip forces for manipulation.
- Detection of motion and direction during manipulation.
- Detection of tangential forces to prevent slip.

Beginning with the desirable features for in-hand manipulation, a set of general design guidelines for tactile sensors was presented by (Dargahi and Najarian 2004), considering also the limitations and possibilities of sensors. The suggested

Table 1 Design guidelines for tactile sensors in robotics

Parameter	Guidelines
Force direction	Normal and tangential
Temporal variation	Dynamic and static
Spatial resolution	1 mm in fingertips to 5 mm in palm of hand
Time response	1 ms
Force sensitivity	0.01–10 N
Linearity/hysteresis	Stable, repeatable and monotonic with low hysteresis
Robustness	Resistant to the application and the environment
Tactile cross-talk	Minimal cross-talk
Shielding	Electronic and/or magnetic shielding
Integration and fabrication	Simple mechanical integration. Minimal wiring with low power consumption and cost

guidelines, shown in Table 1, draw inspiration from the sensing capacities of the human hand (Dahiya et al. 2010).

Beyond the guidelines presented in Table 1, temperature tolerance, size, weight, power consumption and durability are some additional important criteria (Najarian et al. 2009).

Multi-purpose sensors that address all of the above criteria remain a significant technological challenge. For this reason, a more limited set of constraints will be identified when designing sensors for specific applications, reducing cost and complexity.

Tactile Sensor Technologies

Human tactile sensory system is composed by sensors classified by the type of measurements registered and the way they are obtained. For example, tactile sensors can be sensitive to either *static* or *dynamic* forces, and can be employed for *proprioception* or for *exteroception* (Fraden 2004). Proprioceptive sensors are responsible for measuring the internal state of the system, e.g. joint angles, limbs positions, velocity and motor torque. Exteroceptive sensors measuring the characteristics of the physical contact when touching objects in the environment, e.g. sensor surface deformation, contact area, contact pattern and pressure measurements (Siciliano and Khatib 2008). The remainder of this review focuses on artificial touch sensors that are appropriate for applications in exteroception.

Tactile sensor technologies are classified by the transduction method employed to convert stimuli from external environment in a proper form for an intelligent system (Martinez-Hernandez 2014). The most widely used tactile sensor technologies in robotics are based on capacitive, piezoresistive, optical, magnetic, binary and piezoelectric transduction methods are described in the following sections.

Capacitive Sensors

Tactile sensors based on capacitive transduction operate by measuring the variations of capacitance from an applied load over a parallel plate capacitor. The capacitance is related to the separation and area of the parallel plate capacitor, which uses an elastomeric separator to provide compliance. Although Harmon has noted that capacitive sensors are susceptible to external fields (Harmon 1982), this sensor technology has become popular in robotics for the development of "taxels" that mimic aspects of mechanoreception in human fingers (Schmidt et al. 2006; Muhammad et al. 2011). Capacitive sensors can be fabricated in very small sizes, permitting their construction and integration into dense arrays in reduced spaces, e.g. palms and fingertips (Schmitz et al. 2011). This technology also presents various advantages in terms of high sensitivity, long-term drift stability, low temperature sensitivity, low power consumption and sensing of normal or tangential forces (Lee and Wise 1982). Limitations include significant hysteresis.

Piezoresistive Sensors

This transduction method measures changes in the resistance of a contact when force is applied. Piezoresistive sensors are generally fabricated in conductive rubber or made with piezoresistive ink and stamped with a pattern. A maximum resistance value is generated when no contact or stress is applied to the sensor. Conversely, the resistance decreases with increasing pressure or stress to the contact (Webster 1988). The benefits of this transduction method for integration into sensor arrays was initially demonstrated by Snyder and St Clair (1978), Briot et al. (1979), and Russell (1987). The advantages offered by this technology include its wide dynamic range, durability, good overload tolerance, low cost and ability for fabrication in very small sizes. Disadvantages include limited spatial resolution, the challenge of individually wiring multiple sensor elements, susceptibility to drift and hysteresis. The investigation and development of fabric-based piezoresistive sensors has offered an alternative material to improve the durability and reduce hysteresis (Siciliano and Khatib 2008). Piezoresistive tactile sensors have been used in many robotic applications, particularly where high accuracy is not a design criteria (Beebe et al. 1995; Kerpa et al. 2003; Weiss and Woern 2004).

Optical Sensors

Optical sensors operate by transducing mechanical contact, pressure, or directional movement, into changes in light intensity or refractive index, which are then detected using state-of-the-art vision sensors. A drawback is the need to include light emitters and detectors (e.g. CCD arrays), leading to increased bulk. However, optical sensors are attractive due to their potential for high-spatial resolution, robustness to electrical interference, light weight, and their potential to resolve the wiring complexity problem presented by other sensor types such as capacitive and piezoresistive (Nicholls and Lee 1989; Yousef et al. 2011). This has led to the integration of optical tactile sensors into various robotic systems. Begej (1988) described a robotic system for investigating dexterous object manipulation that integrated two 32 × 32 planar sensor arrays built with optical fibres. In Yamada et al. (2005) sub-millimetre resolution for object contact and location detection is described using an optical fingertip that operated by measuring the intensity and direction of reflected light. Fabrication of optical taxels that are capable of measuring normal forces has been described in Heo et al. (2006). This optical system used an LED emitter together with a CCD array to measure force transduced into changes in light intensity. As a final example, Hsiao et al. (2009) describe an optical device for robust detection of object contact and grasping in a three-fingered robot hand.

Magnetic Sensors

This technology operates by detecting changes in magnetic flux, induced by an applied force, through the use of Hall effect, magnetoresistive or magnetoelastic sensors. Hall effect sensors operate by measuring variations in the voltage that is generated by an electric current passing through a conductive material immersed in a magnetic field (Najarian et al. 2009). In Kinoshita et al. (1983) robot gripper using this sensing technology was integrated with twenty Hall effect sensors permitting the robot to perform an object tracking experiment. Contact detection and fingertip deformation were investigated with a 4 × 4 array of Hall effect sensors mounted on a rigid base by Nowlin (1991). Hall effect sensors have also proved to be an effective way of detecting multi-directional deflections of an **artificial whisker** (Pearson et al. 2007; Sullivan et al. 2012). Magnetoresistive and magnetoelastic sensors detect variations in magnetic fields generated by the application of mechanical stress. In the 1970s a robot tactile sensor using this magnetic approach was developed for object classification based on their contours (Pugh 1986). Despite the relatively large size of magnetic sensing elements, Jayawant (1989) achieved recognition of 2D images using a sensor array fabricated with 256 magnetic elements. Advantages of magnetic sensor technologies include high sensitivity, wide dynamic range, very low hysteresis, linear response and general robustness. However, they are susceptible to magnetic interference and noise. Applications are limited by the physical size of the sensing device, and by the need to operate in nonmagnetic environments (Dahiya et al. 2010).

Binary Sensors

Contact switches permit the detection of discrete on/off events brought about by mechanical contact (Webster 1988). The ease of designing and building this type of sensor has permitted its integration into a wide variety of robotic systems. A five-fingered prosthetic robotic hand developed by Edin et al. (2006) used on/off sensors embedded in the fingertips to support a grasping procedure. It is possible to design contact devices that go beyond a simple binary code. For example, Tajima et al. (2002) describe a sensor capable of encoding variations in pressure in a discrete multi-state code. Lack of resolution is the primary disadvantage to this sensor technology limiting applications to problems such as contact or collision detection.

Piezoelectric Sensors

Piezoelectric sensors produce an electric charge proportional to an applied force, pressure or deformation. Limitation to dynamic measurements and susceptibility to temperature are the main drawbacks of this sensing technology. However, they are suitable for measurement of vibrations and widely used due to their sensitivity, high frequency response and availability in various forms, e.g. plastics, crystals, ceramics and polyvinylidene fluoride (PVDF) (Lee and Nicholls 1999; Schmidt et al. 2006). Grahn and Astle (1986) achieved robust object detection using a tactile sensor covered with a piezoelectric material based on a layer of silicon rubber. Here, the electric charge used for object detection was generated by contacting and deforming the silicon layer. Yamada and Cutkosky (1994) used piezoelectric technology in an artificial skin technology designed to be sensitive to force, vibration and slip. From the repertoire of piezoelectric materials, PVDF is the most commonly preferred for fabrication of tactile sensors given its flexibility, workability and chemical stability (Dahiya et al. 2010). Dario and De Rossi (1985) have described the use of PVDF for building and integration of tactile sensors in a robotic gripper.

Hydraulic Sensors

Hydraulic technology uses a type of actuator that converts fluid pressure into mechanical motion. Recent industrial and medical applications require microscopic servomechanisms, known as microactuators, to detect pressure and measure force based on hydraulics (De Volder and Reynaerts 2010). Micro-hydraulics structures developed in (Sadeghi et al. 2011) allowed the fabrication of a low-power, accurate and robust flow sensor. This sensor, composed of a biomimetic hair-like structure, allows to translate flow into hydraulic pressure offering a large measurement range and high sensitivity. Force sensor arrays, similar to the human fingertip size, were able to achieve high sensitivity based on the micro-hydraulics sensing technology (Sadeghi et al. 2013). These low cost force sensors, fabricated with a stereo-lithography technique, provided robust tactile data and high spatial resolution, making them suitable for skin-like sensing applications.

Applications

Since the 1980s, when robotics was defined as the science for studying perception, action and their intelligent interconnection (Siciliano and Khatib 2008), the integration of tactile sensors has played an important role in the development of robust, flexible and adaptable robots capable of exploring their environments and interacting safely with humans.

A variety of open robotic platforms that employ different tactile sensor technologies in their hands, fingertips, arms, forearms and torso have been developed for the study of embodied cognition, exploration, perception, recognition, learning and interaction (Schmitz et al. 2011; Weiss and Woern 2004; Edin et al. 2006; Metta et al. 2008; Brooks et al. 1999). For example, Figure 1 shows the iCub humanoid robot (Metta et al. 2008) equipped tactile sensors in its torso, arms, forearms, palms and fingertips.

Robotic fingertips equipped with piezoelectric sensing elements have been used to recognise various object properties such as texture, hardness and shape by performing different exploratory procedures, e.g. sliding, squeezing, pushing and tapping over various objects (Hosoda et al. 2006; Takamuku et al. 2007). Various robotics hands integrated with capacitive tactile sensors have also been developed for exploration and recognition (see Son et al. 1996; Schneider et al. 2009). Pressure and force sensors in robotic fingertips have allowed the achievement of reliable control during object detection and manipulation (Dang et al. 2011; Chen et al. 1995b). Design and integration of tactile sensors in prosthetic hands permitted to mimic the natural motion and contact detection observed in humans (Carrozza et al. 2003). Investigations of tactile exploration, perception and interaction have been successfully achieved using the tactile sensory system of the iCub humanoid robot (Martinez-Hernandez 2014; Lepora et al. 2013).

Object shape exploration has been investigated using different perception and control approaches implemented in a variety of tactile robotic platforms.

For instance, a PUMA robot integrated with a planar tactile sensor array was used to extract object edge and orientation based on tactile images and geometrical moments (Muthukrishnan et al. 1987; Chen et al. 1995a). A similar approach based on geometrical moments, but using a KUKA arm with a planar tactile sensors, was able to explore and recognise the shape of various objects (Li et al. 2013). An adaptive threshold approach applied to tactile images obtained from a CMU DD Arm II allowed the recognition of object orientation (Berger and Khosla 1988).

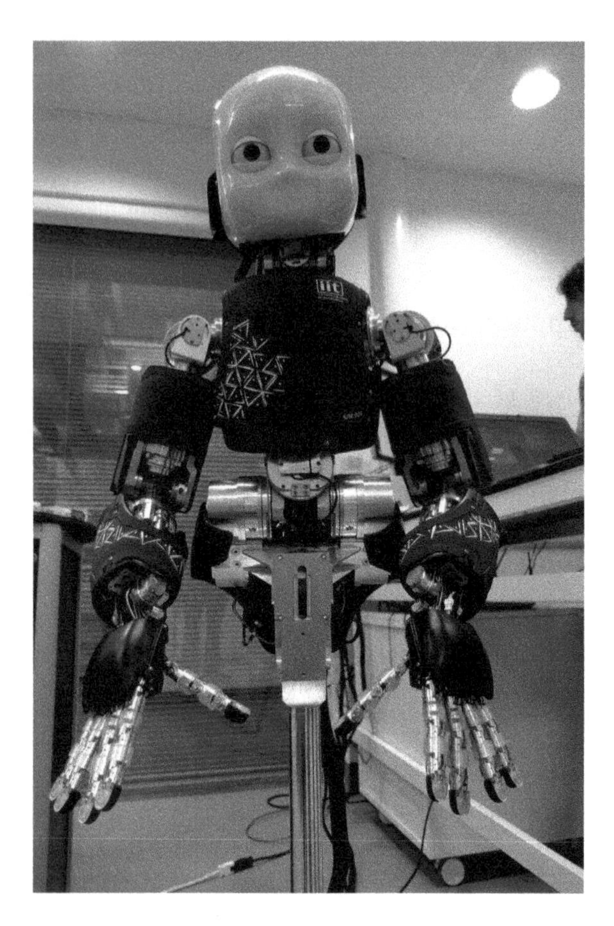

Figure 1 Contemporary humanoids such as the iCub are beginning to be equipped with tactile sensors on body surfaces suh as the hands/fingers, arms, and torso

The use of the iCub fingertip sensor with a probabilistic approach presented an intelligent and robust object shape exploration and extraction (Martinez-Hernandez et al. 2013). Figure 2 shows a robotic hand integrated with tactile fingertip sensors that has been mounted on a KUKA arm and programmed to perform robust exploratory procedures.

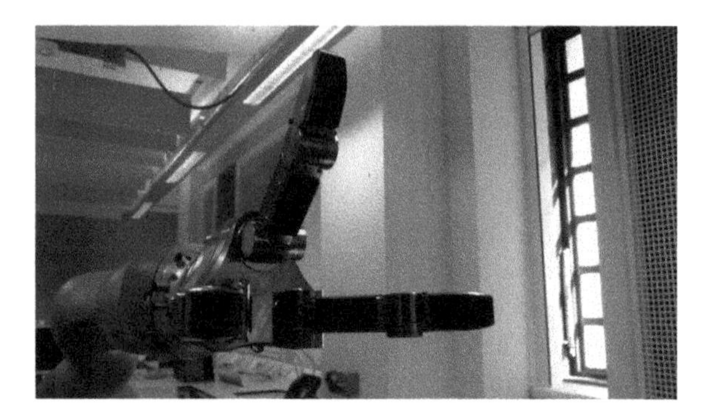

Figure 2 Schunk hand mounted on a KUKA arm for manipulations tasks

A rolling and enclosing exploration procedure was implemented for robust object recognition based on the kurtosis observed on each tactile fingertip of a five-fingered robotic hand (Nakamoto et al. 2008). A shadow robotic platform performing an enclosing procedure was able to classify a variety of objects using tactile information and a Self-Organising Map (SOM) (Ratnasingam and McGinnity 2011). The iCub humanoid robot using tactile information from its hands and fingers during an enclosure procedure was able to achieve high accuracy for an object recognition task. These results were obtained by the implementation of a biologically inspired method based on a probabilistic and temporal approach (Soh et al. 2012). The robotic hand from Barrett Hand Inc. covered with tactile sensors in its palm and fingertips has been widely used for dexterous manipulation and perception with the sense of touch (Figure 3).

Tactile sensors have been integrated into a number of biomimetic robots both as a means to understand tactile sensing in animals and as a path towards the development of useful robotic technologies. Perception of stimulus attributes, such as texture, distance to contact, and speed and direction of moving stimuli, has been demonstrated in a number of whiskered robots equipped with artificial vibrissae (Pearson et al. 2007; Prescott et al. 2009; Sullivan et al. 2012; Lungarella et al. 2002). Tactile sensing based on artificial whiskers has also been demonstrated to be able to support complex surface detection and reconstruction (Solomon and Hartmann 2006), tactile simultaneous localisation and mapping (Pearson et al. 2013), and moving object detection and tracking (Mitchinson et al. 2014). The development of artificial antennae fabricated with pressure and force sensors has allowed modelling of the contact detection and exploration behaviour of insects such as ants and **cockroaches** (Kaneko 1994; Ueno and Kaneko 1994; Kaneko et al. 1998). Research and integration of tactile sensors has also reached the field of underwater robotics in the form of artificial whiskers, modelled on the remarkable perceptual capabilities of **seals**, that are able to measure speed and direction of fluid motion, angle and wake detection (Eberhardt et al. 2011; Beem et al. 2013).

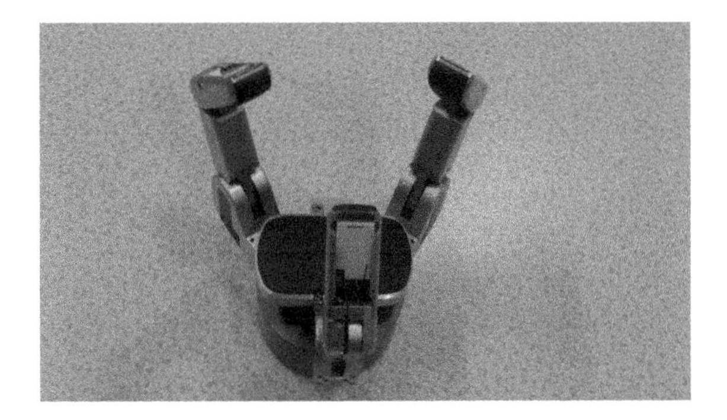

Figure 3 Robotic hand from Barrett Hand Inc. used for dexterous manipulation and exploration based on active tactile perception

Conclusions

Inspired by the repertoire of capabilities and benefits that the sense of touch offers to the animal kingdom, different tactile sensor technologies have been developed, since the early days of robotics, that can enhance the performance and functionality of robotic systems allowing them to touch, feel and explore their environments. A range of different sensor technologies have been developed each with their own advantages and limitations, though we have yet to come close to emulating the versatility and richness of the human sense of touch.

Internal References

Hanke, W and Dehnhardt, G (2015). Vibrissal touch in pinnipeds. Scholarpedia 10(3): 6828. http://www.scholarpedia.org/article/Vibrissal_touch_in_pinnipeds. (see also pages 125–139 of this book)
Klatzky, R and Reed, C L (2009). Haptic exploration. Scholarpedia 4(8): 7941. http://www.scholarpedia.org/article/Haptic_exploration. (see also pages 177–183 of this book)
Moayedi, Y et al. (2015). Mammalian mechanoreception. Scholarpedia 10(3): 7265. http://www.scholarpedia.org/article/Mammalian_mechanoreception. (see also pages 423–435 of this book)
Okada, J (2009). Cockroach antennae. Scholarpedia 4(10): 6842. http://www.scholarpedia.org/article/Cockroach_antennae. (see also pages 31–43 of this book)
Pipe, T and Pearson, M J (2015). Whiskered robots. Scholarpedia 10(3): 6641. http://www.scholarpedia.org/article/Whiskered_robots. (see also pages 809–815 of this book)

External References

Beebe, D J; Hsieh, A S; Denton, D D and Radwin, R G (1995). A silicon force sensor for robotics and medicine. *Sensors and Actuators A: Physical* 50(1): 55–65.
Beem, H; Hildner, M and Triantafyllou, M (2013). Calibration and validation of a harbor seal whisker-inspired flow sensor. *Smart Materials and Structures* 22(1): 014012.
Begej, S (1988). Planar and finger-shaped optical tactile sensors for robotic applications. *IEEE Journal of Robotics and Automation* 4(5): 472–484.
Berger, A D and Khosla, P K (1988). Edge detection for tactile sensing. In: *Proceedings of SPIEs Cambridge Symposium on Optical and Optoelectronic Engineering: Advances in Intelligent Robotics Systems.*
Briot, M et al. (1979). The utilization of an artificial skin sensor for the identification of solid objects. In: *9th International Symposium on Industrial Robots* (pp. 13–15).
Brooks, R A; Breazeal, C; Marjanović, M; Scassellati, B and Williamson, M M (1999). The cog project: Building a humanoid robot. In: *Computation for Metaphors, Analogy, and Agents* (pp. 52–87). Springer.
Carrozza, M et al. (2003). The cyberhand: on the design of a cybernetic prosthetic hand intended to be interfaced to the peripheral nervous system. In: *2003 IEEE/RSJ International Conference on Intelligent Robots and Systems (IROS 2003). Proceedings,* Vol. 3 (pp. 2642–2647).

Castelli, F (2002). An integrated tactile-thermal robot sensor with capacitive tactile array. *IEEE Transactions on Industry Applications* 38(1): 85–90.

Chen, N; Rink, R and Zhang, H (1995a). Efficient edge detection from tactile data. In: *1995 IEEE/RSJ International Conference on Intelligent Robots and Systems 95. 'Human Robot Interaction and Cooperative Robots', Proceedings,* Vol. 3 (pp. 386–391).

Chen, N., Zhang, H., and Rink, R. (1995b). Edge tracking using tactile servo. In: *1995 IEEE/RSJ International Conference on Intelligent Robots and Systems 95. 'Human Robot Interaction and Cooperative Robots.' Proceedings,* Vol. 2 (pp. 84–89).

Dahiya, R S; Metta, G; Valle, M and Sandini, G (2010). Tactile sensing from humans to humanoids. *IEEE Transactions on Robotics* 26(1): 1–20.

Dang, H; Weisz, J and Allen, P K (2011). Blind grasping: Stable robotic grasping using tactile feedback and hand kinematics. In: *2011 IEEE International Conference on Robotics and Automation (ICRA)* (pp. 5917–5922).

Dargahi, J and Najarian, S (2004). Human tactile perception as a standard for artificial tactile sensing a review. *The International Journal of Medical Robotics and Computer Assisted Surgery* 1(1): 23–35.

Dario, P and De Rossi, D (1985). Tactile sensors and the gripping challenge: Increasing the performance of sensors over a wide range of force is a first step toward robotry that can hold and manipulate objects as humans do. *IEEE Spectrum* 22(8): 46–53.

De Rossi, D (1991). Artificial tactile sensing and haptic perception. *Measurement Science and Technology* 2(11): 1003.

De Volder, M and Reynaerts, D (2010). Pneumatic and hydraulic microactuators: a review. *Journal of Micromechanics and Microengineering* 20(4): 043001.

Eberhardt, W; Shakhsheer, Y; Calhoun, B; Paulus, J and Appleby, M (2011). A bio-inspired artificial whisker for fluid motion sensing with increased sensitivity and reliability. In: *2011 IEEE Sensors* (pp. 982–985).

Edin, B B et al. (2006). Bio-inspired approach for the design and characterization of a tactile sensory system for a cybernetic prosthetic hand. In: *Proceedings of the IEEE International Conference on Robotics and Automation, 2006. ICRA 2006* (pp. 1354–1358).

Fraden, J (2004). *Handbook of Modern Sensors: Physics, Designs, and Applications.* Springer.

Grahn, A R and Astle, L (1986). Robotic ultrasonic force sensor arrays. *Robot Sensors* 2: 297–315.

Harmon, L D (1982). Automated tactile sensing. *The International Journal of Robotics Research* 1(2): 3–32.

Heo, J-S; Chung, J-H and Lee, J-J (2006). Tactile sensor arrays using fiber bragg grating sensors. *Sensors and Actuators A: Physical* 126(2): 312–327.

Hosoda, K; Tada, Y and Asada, M (2006). Anthropomorphic robotic soft fingertip with randomly distributed receptors. *Robotics and Autonomous Systems* 54(2): 104–109.

Hsiao, K; Nangeroni, P; Huber, M; Saxena, A and Ng, A Y (2009). Reactive grasping using optical proximity sensors. In: *IEEE International Conference on Robotics and Automation, 2009. ICRA'09* (pp. 2098–2105).

Jayawant, B (1989). Tactile sensing in robotics. *Journal of Physics E: Scientific Instruments* 22(9): 684.

Johansson, R and Westling, G (1984). Roles of glabrous skin receptors and sensorimotor memory in automatic control of precision grip when lifting rougher or more slippery objects. *Experimental Brain Research* 56(3): 550–564.

Kaneko, M (1994). Active antenna. In: *1994 IEEE International Conference on Robotics and Automation. Proceedings* (pp. 2665–2671).

Kaneko, M; Kanayama, N and Tsuji, T (1998). Active antenna for contact sensing. *IEEE Transactions on Robotics and Automation* 14(2): 278–291.

Kerpa, O; Weiss, K and Worn, H (2003). Development of a flexible tactile sensor system for a humanoid robot. In: *2003 IEEE/RSJ International Conference on Intelligent Robots and Systems. (IROS 2003). Proceedings,* Vol. 1 (pp. 1–6).

Kinoshita, G; Hajika, T and Hattori, K (1983). Multifunctional tactile sensors with multi-elements for fingers. In: *Proceedings of the International Conference on Advanced Robotics* (pp. 195–202).

Lederman, S J and Klatzky, R L (1987). Hand movements: A window into haptic object recognition. *Cognitive Psychology* 19(3): 342–368.

Lee, M H (2000). Tactile sensing: new directions, new challenges. *The International Journal of Robotics Research* 19(7):636–643.

Lee, M H and Nicholls, H R (1999). Review article tactile sensing for mechatronics a state of the art survey. *Mechatronics* 9(1): 1–31.

Lee, Y S and Wise, K D (1982). A batch-fabricated silicon capacitive pressure transducer with low temperature sensitivity. *IEEE Transactions on Electron Devices* 29(1): 42–48.

Lepora, N F; Martinez-Hernandez, U and Prescott, T J (2013). A solid case for active bayesian perception in robot touch. In: *Biomimetic and Biohybrid Systems* (pp. 154–166). Springer.

Li, Q; Schurmann, C; Haschke, R and Ritter, H (2013). A control framework for tactile servoing. Presented at the 2013 Robotics: Science and Systems Conference.

Lungarella, M; Hafner, V V; Pfeifer, R and Yokoi, H (2002). An artificial whisker sensor for robotics. In: *IEEE/RSJ International Conference on Intelligent Robots and Systems, 2002,* Vol. 3 (pp. 2931–2936).

Martinez-Hernandez, U (2014). Autonomous active exploration for tactile sensing in robotics. PhD thesis, The University of Sheffield, Sheffield, UK.

Martinez-Hernandez, U; Dodd, T; Prescott, T J and Lepora, N F (2013). Active bayesian perception for angle and position discrimination with a biomimetic fingertip. In: *2013 IEEE/RSJ International Conference on Intelligent Robots and Systems (IROS)* (pp. 5968–5973).

Metta, G; Sandini, G; Vernon, D; Natale, L and Nori, F (2008). The icub humanoid robot: An open platform for research in embodied cognition. In: *Proceedings of the 8th Workshop on Performance Metrics for Intelligent Systems* (pp. 50–56). ACM.

Mitchinson, B; Pearson, M J; Pipe, A G and Prescott, T J (2014). Biomimetic tactile target acquisition, tracking and capture. *Robotics and Autonomous Systems* 62(3): 366–375.

Muhammad, H et al. (2011). Development of a bioinspired MEMS based capacitive tactile sensor for a robotic finger. *Sensors and Actuators A: Physical* 165(2): 221–229.

Muthukrishnan, C; Smith, D; Myers, D; Rebman, J and Koivo, A (1987). Edge detection in tactile images. In: *1987 IEEE International Conference on Robotics and Automation. Proceedings,* Vol. 4 (pp. 1500–1505).

Najarian, S; Dargahi, J and Mehrizi, A (2009). *Artificial Tactile Sensing in Biomedical Engineering.* McGraw Hill Professional.

Nakamoto, H; Kobayashi, F; Imamura, N; Shirasawa, H and Kojima, F (2008). Shape classification in rotation manipulation by universal robot hand. In: *IEEE/RSJ International Conference on Intelligent Robots and Systems, 2008. IROS 2008* (pp. 53–58).

Nicholls, H R and Lee, M H (1989). A survey of robot tactile sensing technology. *The International Journal of Robotics Research* 8(3): 3–30.

Nowlin, W C (1991). Experimental results on bayesian algorithms for interpreting compliant tactile sensing data. In: *IEEE International Conference on Robotics and Automation, 1991. Proceedings,* (pp. 378–383).

Pearson, M J; Pipe, A G; Melhuish, C; Mitchinson, B and Prescott, T J (2007). Whiskerbot: A robotic active touch system modeled on the rat whisker sensory system. *Adaptive Behavior* 15 (3): 223–240.

Pearson, M et al. (2013). Simultaneous localisation and mapping on a multi-degree of freedom biomimetic whiskered robot. *IEEE International Conference on Robotics and Automation (ICRA).*

Prescott, T J; Pearson, M; Mitchinson, B; Sullivan, J C W and Pipe, A G (2009). Whisking with robots: From rat vibrissae to biomimetic technology for active touch. *IEEE Robotics and Automation Magazine* 16(3): 42–50.
Pugh, A (1986). *Robot Sensors: Tactile and Non-Vision,* Vol. 2. Springer.
Ratnasingam, S and McGinnity, T (2011). Object recognition based on tactile form perception. In: *2011 IEEE Workshop on Robotic Intelligence In Informationally Structured Space (RiiSS)* (pp. 26–31).
Robles-De-La-Torre, G (2006). The importance of the sense of touch in virtual and real environments. *IEEE MultiMedia* 13(3): 24–30.
Russell, R A (1987). Compliant-skin tactile sensor. In: *1987 IEEE International Conference on Robotics and Automation. Proceedings,* Vol. 4 (pp. 1645–1648).
Sadeghi, M M; Peterson, R L and Najafi, K (2011). Micro-hydraulic structure for high performance bio-mimetic air flow sensor arrays. In: *2011 IEEE International Electron Devices Meeting (IEDM)* (pp. 29–4).
Sadeghi, M M; Peterson, R L and Najafi, K (2013). High-speed electrostatic micro-hydraulics for sensing and actuation. In: *2013 IEEE 26th International Conference on Micro Electro Mechanical Systems (MEMS)* (pp. 1191–1194).
Schmidt, P A; Maël, E and Wurtz, R P (2006). A sensor for dynamic tactile information with applications in human–robot interaction and object exploration. *Robotics and Autonomous Systems* 54(12): 1005–1014.
Schmitz, A et al. (2011). Methods and technologies for the implementation of large-scale robot tactile sensors. *IEEE Transactions on Robotics* 27(3): 389–400.
Schneider, A et al. (2009). Object identification with tactile sensors using bag-of-features. In: *IEEE/RSJ International Conference on Intelligent Robots and Systems, 2009. IROS 2009* (pp. 243–248).
Siciliano, B and Khatib, O (2008). *Springer Handbook of Robotics.* Springer.
Snyder, W E and St Clair, J (1978). Conductive elastomers as sensor for industrial parts handling equipment. *IEEE Transactions on Instrumentation and Measurement* 27(1): 94–99.
Soh, H; Su, Y and Demiris, Y (2012). Online spatio-temporal gaussian process experts with application to tactile classification. In: *2012 IEEE/RSJ International Conference on Intelligent Robots and Systems (IROS)* (pp. 4489–4496).
Solomon, J H and Hartmann, M J (2006). Biomechanics: Robotic whiskers used to sense features. *Nature* 443(7111): 525.
Son, J S; Howe, R D; Wang, J and Hager, G D (1996). Preliminary results on grasping with vision and touch. In: *Proceedings of the 1996 IEEE/RSJ International Conference on Intelligent Robots and Systems '96, IROS 96,* Vol. 3 (pp. 1068–1075).
Sullivan, J et al. (2012). Tactile discrimination using active whisker sensors. *IEEE Sensors Journal* 12(2): 350–362.
Tajima, R; Kagami, S; Inaba, M and Inoue, H (2002). Development of soft and distributed tactile sensors and the application to a humanoid robot. *Advanced Robotics* 16(4): 381–397.
Takamuku, S; Gomez, G; Hosoda, K and Pfeifer, R (2007). Haptic discrimination of material properties by a robotic hand. In: *IEEE 6th International Conference on Development and Learning, 2007. ICDL 2007* (pp. 1–6).
Ueno, N and Kaneko, M (1994). Dynamic active antenna-a principle of dynamic sensing. In: *IEEE International Conference on Robotics and Automation, 1994. Proceedings* (pp. 1784–1790).
Vallbo, A B et al. (1984). Properties of cutaneous mechanoreceptors in the human hand related to touch sensation. *Human Neurobiology* 3(1): 3–14.
Webster, J G (1988). *Tactile Sensors for Robotics and Medicine.* John Wiley & Sons, Inc.
Weiss, K and Woern, H (2004). Tactile sensor system for an anthropomorphic robotic hand. In: *IEEE International Conference on Manipulation and Grasping IMG.*
Yamada, Y and Cutkosky, M R (1994). Tactile sensor with 3-axis force and vibration sensing functions and its application to detect rotational slip. In: *IEEE International Conference on Robotics and Automation, 1994. Proceedings,* (pp. 3550–3557).

Yamada, Y; Morizono, T; Umetani, Y and Takahashi, H (2005). Highly soft viscoelastic robot skin with a contact object-location-sensing capability. *IEEE Transactions on Industrial Electronics* 52(4): 960–968.

Yousef, H; Boukallel, M; and Althoefer, K (2011). Tactile sensing for dexterous in-hand manipulation in robotics a review. *Sensors and Actuators A: Physical* 167(2): 171–187.

Models of Tactile Perception and Development

Goren Gordon

Models of tactile perceptions are mathematical constructs that attempt to explain the process with which the tactile sense accumulates information about objects and agents in the environment. Since touch is an active sense, i.e., the sensor organ is moved during the process of sensation, these models often describe the motion strategies that optimize the perceptual outcome.

Models of tactile development attempt to explain the emergence of perception and the accompanying motor strategies from more basic principles. These models often involve learning of exploration strategies and are aimed at explaining ontogenetic development of behavior.

These models are used in two complementary ways. The first is in an attempt to explain and predict animal and human behaviors. For this purpose, the vibrissae system of rodents is often used as it is a well-studied system in neuroscience. Vibrissae behaviors, i.e., movement strategies of rodent's facial hairs, during different perceptual tasks are modeled in an attempt to uncover the underlying common principles, as well as the neuronal mechanism of tactile perception and development. The same models are also used in artificial constructs, e.g., robots, in an attempt to both validate the emergence of tactile sensorimotor strategies, as well as to try and optimize tactile perception in novel robotic platforms.

Introduction

Tactile perception means the information gathered on tactile objects in the environment. This information can be the position, shape, material or surface texture of the object. Models of tactile perception are thus aimed at explaining how this information is accumulated, integrated and used in tactile tasks, such as discrimination and localization.

Touch is an active sense, i.e., the sensory organ is usually moved in order to perceive the environment. Hence modeling tactile perception involves modeling the

G. Gordon (✉)
Curiosity Lab, Department of Industrial Engineering, Tel-Aviv University, Tel-Aviv, Israel

T.J. Prescott et al. (eds.), *Scholarpedia of Touch*, Scholarpedia,
DOI 10.2991/978-94-6239-133-8_58

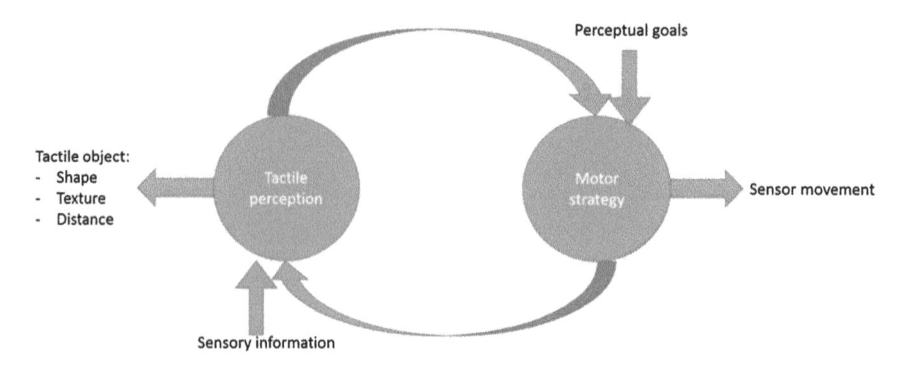

Figure 1 Active tactile perception models architecture

sensorimotor strategy that results in the accumulation of tactile information. In other words, these models describe the behavior, or motion, of the sensory organ as it interacts with the tactile object. Models attempt to either *describe* observed tactile-oriented behaviors in animals and humans or *derive* optimal perceptual strategies and then compare them to observed behaviors (Figure 1).

Since touch, as opposed to vision, audition and smell, is a proximal sense, i.e., the sensory organ must be in contact with the object in order to perceive, locomotion is often part of the description of the tactile strategy. In nocturnal animals, such as many rodents, the vibrissae system, an array of moveable facial hairs, is used to perceive the environment in darkness. Navigation and object recognition is hence done mainly by the tactile sense. Several models of tactile-guided locomotion have been developed to address this cross-modality integration.

As any other sense, tactile perception changes during ontological development, based on the agent's experience and interaction with the environment. Part of this change is the emergence of sensorimotor tactile strategies that explore tactile objects. For example, pups' vibrissae have been shown to move in different ways as they mature to adulthood (Grant et al. 2012). Developmental models try to describe this emergence of exploration behavior using basic principles of sensory-guided motor learning and intrinsic motivation exploration.

Model Types

Tactile perception modeling is usually composed of two main components, namely, perception and action. The perception component attempts to describe the integration of tactile information into a cohesive percept. The action component attempts to describe the motor strategies used in order to move the sensory organ so that it can acquire this information.

Tactile perception is usually modeled by either artificial neural networks or Bayesian inference. Artificial neural networks (ANN) are used in order to describe the learning process during the perceptual task. They are more closely related to the biological neural system and there are many computationally efficient tools to implement them. ANN are usually used in a supervised learning fashion, where the aim is to learn either tactile discrimination via labeled training sets, or continuous-variable forward models that capture the entire sensorimotor agent-environment interaction. Bayesian inference models capture the optimal integration of new observed information into a single framework of perceptual updates. Each new evidence from the possibly noisy environment is used in an optimal way to update the tactile perception in the current task. These models have fewer free parameters to tune and have been shown in recent years to describe many perceptual tasks in humans and animals very well.

Motor strategies for tactile perception are usually modeled either by optimal control theory or reinforcement learning. Optimal control theory is a mathematical formalism wherein one defines a cost function and then uses known mathematical techniques to find the optimal trajectory or policy that minimizes the cost. In tactile perceptual tasks, the cost function is usually a combination of perceptual errors, e.g., discrimination ambiguity, and energy costs of moving the sensory organ. Thus an optimal control solution can give the policy, or the optimal behavior, that maximizes perception while minimizing energy costs. Reinforcement learning is a computational paradigm that attempts to find the policy or behavior that maximizes future accumulated rewards. This is a gradual learning process where repeated interactions with the environment result in convergence to an optimal policy. In tactile perceptual tasks the reward is the completion of the task and the model results in a converged sensorimotor tactile strategy. A major difference between optimal control and reinforcement learning is that the former is solved *off-line*, while the latter is a learning algorithm that takes into account the interaction with the environment. While both result in an optimal strategy or policy, their formalism and mathematical techniques are different.

Model Application

Tactile perception and developmental models can be used in several ways. The first one is to attempt and describe, explain and predict animal and human tactile behavior. In each tactile task, the observed behavior is recorded and analyzed. Models are then constructed to attempt and re-capture the same behavior and then produce prediction of behaviors in novel tasks. The models are then validated in these new predicted tasks.

The second application for tactile models is the understanding of the underlying neuronal mechanism. For example, the vibrissae system of rodents have been studied for decades and have produced a deep understanding of the underlying neuronal networks that result in tactile perception. Linking model components that describe tactile perception to specific brain areas or functions can increase understanding of these areas and may attempt to explain abnormal behaviors in model and neurological terms.

Another application for tactile models is their implementation in artificial agents such as robots. Robotic platforms that have tactile senses are inspired by new understanding of biological tactile perception models. Integrating motors into the sensory organ, e.g., artificial whisker robots or tactile-sensor covered robotic fingers, enables new capabilities of object perception. However, controlling these robotic platforms become non-trivial as known motor-oriented control strategies fail in these perception-oriented domains. Implementing biologically-inspired sensorimotor models results in better performing robots.

Active Sensing

Biological Application

In an attempt to properly understand the tactile sensorimotor strategy rodents employ during a well-known perceptual task called pole-localization, humans were used as models for rodents (Saig et al. 2012). Subjects were equipped with artificial whiskers at their fingertips and were asked to localize a vertical pole, i.e., determine which pole was more posterior, using only information they got from their whiskers, since their vision and audition were blocked. Force and position sensors were placed on the finger-whisker connection, which enabled full access to the information into the *system*, i.e., the human subject. It was shown that humans spontaneously employed similar strategies as rodents, i.e., they *whisked* with their artificial whiskers by moving their hands synchronously and perceiving temporal differences according to pole location. In other words, they determined which pole was more posterior by moving their hands together and detecting which hand touched a pole first. While there were other possible non-active strategies to solve the task, e.g. by positioning their hands over the pole and sensing the angular difference between the hands, participants chose to employ an active sensing strategy.

In order to model this behavior, a Bayesian inference approach was selected for the tactile perception, whereas an optimal control theory approach was selected for the motor strategy analysis. The task was then described as a simple binary discrimination task, i.e., which pole is more posterior, and a Bayes update rule was modeled by integrating perceived temporal differences between the two hands. A Gaussian noise model was assumed for the perceived temporal difference, introducing a parameter of the temporal noise, i.e., how close in time can two stimuli be to still be perceived as distinct. Another important parameter introduced in the Bayesian inference model was the confidence probability above which subjects decided to report their perceived answer. In other words, after repeated contacts with the poles, the probability of one pole being perceived as more posterior increases;

above which threshold does the subject stop the interaction and report the perceived result?

The selected Bayesian inference model of this tactile perception task resulted in only two parameters, temporal noise and confidence probability, and allowed their estimation based on fitting to experimental results. The number of contacts prior to reporting was shown to increase with task difficulty, measured by decreased distance between the poles, as was predicted by the Bayesian model. Fitting the model prediction to the experimental results enabled estimation of the parameters: the temporal noise was assessed to be 312 ms and the confidence probability 84 %. The temporal noise was somewhat higher than previously reported purely *tactile* temporal discrimination thresholds, due to the fact that this experimental setup was an *active sensing* setup which introduced also motor noise. The confidence probability was comparable to many other psychological experiments, within which subjects had to report their perceived result after accumulating information. Hence, the Bayesian inference model of tactile perception eloquently described the accumulation and integration of tactile information.

The motor strategies employed by the subjects were also structured and exhibited initial longer, larger amplitude movements followed by decreasingly shorter and smaller amplitude ones. To model this behavior, an optimal control theory approach was taken, where a cost function was defined, followed by optimization techniques that resulted in an optimal policy that minimized costs. The cost function had three components: a perceptual error term representing the task; an energy cost term representing penalty for laborious actions; and a perceptual cost term, symmetrically identical to the energy term representing a cost to too much information. The model captured the behavior exhibited by the subjects and resulted in a simple principle governing it, namely, maintaining a constant information flow. In other words, the optimal control model *distilled* the complex tactile-perceptual driven behavior to a single guiding principle.

Robotic Application

Inspired by the rodents vibrissae system, a robotic platform was constructed that had fully controlled moving artificial whiskers (Sullivan et al. 2012). The robot was used in similar tasks to those studied in rodents, namely, surface distance and texture estimation. In other words, the robot moved its whiskers in biologically-inspired motor strategies and collected information about the surfaces via sensors located at the base of the whiskers. The robot employed models of both tactile perception and motor strategies designed based on understanding of the biological vibrissae system.

Tactile perception was modeled using a naive Bayes approach, where during training the robot collected sensory information on each type of surface and each distance from the surface, constructing labeled probability distributions for each. Then, during validation, the robot whisked upon a surface, collected information and classified the texture and distance according to the most probable class, based on the trained distributions.

The motor strategy employed an observed behavior in rodents, namely, Rapid Cessation of Protraction (RCP), which means that rodents whisk with smaller amplitude after an initial contact with an object. This strategy results in *light touch* upon the second whisk and onward with the surface. The same behavior was modeled and executed in the robotic rodent, whereupon the amplitude of the whisking was decreased after an initial perceived contact with the surface. The goal of the task and the specific models was to ascertain the potential benefits rodents might have for employing such a strategy.

The results of the study showed that the robot performed much more efficient and accurate classification of both texture and distance of the surface when employing the Rapid Cessation of Protraction (RCP) strategy, compared to unmodulated whisking. Further analysis of the results showed that using RCP resulted in less noisier sensory information which in turn resulted in improved classification. This model thus suggests that rodents employ the RCP strategy not only to keep their whisker intact, but also to improve signal-to-noise ratio and tactile perception. It also enables the development of more robust and more accurate artificial agents with moving tactile sensors.

Tactile Navigation

Biological Application

Since touch is a proximal sense, direct contact with the objects in the environment are mandatory for tactile perception (Gordon et al. 2014a, b). In order to understand exploration behavior of rodents, a model was constructed that attempted to capture the complexity and structure of their exploration patterns. When rodents are allowed to explore a new round dark arena on their own, they move around the arena and sense its walls using their whiskers. They exhibit a complex exploration pattern in which they first explore the entrance to the arena, only then walk along the circumference walls of the arena and only then explore the open space in the center of the arena. Their exploration is composed of excursions made up of an outbound exploratory part and a fast retreat part in which they return to their home cage.

This tactile-driven exploration strategy was modeled using a novelty-based approach which combined tactile-perceptual representation of the arena and a motor strategy that balances between exploration motor primitives and retreats. For tactile perception of the arena a Bayesian inference approach was taken to represent the forward model of locomotion. In other words, the arena was represented as the prediction of the sensory information in a given location and orientation, e.g., a wall is represented as *in location x and orientation o, the left whisker is predicted to experience touch*. This representation is updated whenever the animal perceives a new tactile perception in any location using Bayes rule and assuming sensory noise, i.e., the perceived tactile sensation is not necessarily the correct one.

The exploration motor strategy was taken to consist of a balance between exploration motor primitives and retreats, where novelty was used as the thresholding factor. Exploration motor primitives are policies that determine the locomotive behavior of the rodent based on its perceived tactile sensation, e.g., wall-following primitive is the policy *if left whisker senses a wall, go forward*, whereas wall-avoidance primitive is the policy *if right whisker senses a wall turn left*. Three motor primitives were modeled, namely, circle-in-place, wall-following and wall-avoidance. Another *retreat primitive* was modeled as, given the current estimation of the arena, take the shortest path from the current location to the home cage.

The balance between these motor primitives was dictated based on novelty, measured as the information gain in each time step that the arena model was updated. In other words, whenever the tactile forward model of the arena was updated, the number of bits that were updated, quantified by the Kullback-Leibler divergence between the prior and posterior distributions, represented the novelty. Whenever novelty was higher than a certain threshold, the retreat primitive was employed. Whenever novelty was lower than a certain threshold for a certain amount of time, the next exploration motor primitive was employed. This generative model captured many of the observed behaviors in tactile-driven exploring rodents and showed that the basic principle of novelty management can be used to model complex and structured exploration behaviors (Figure 2).

Robotic Application

A robotic platform with actuated artificial whiskers was used to study a tactile-based Simultaneous Localization And Mapping (tSLAM) model (Pearson et al. 2013). In this setup, the perceptual task is dual, i.e., the robot needs to both localize itself in space and also map out the objects in the environment. As opposed to many other SLAM models, this model used only odometry and tactile sensation from the whisker-array as its input, i.e., it had no vision.

The tactile-driven exploration of the environment consisted of an occupancy map particle filter-based model of tactile perception and an attention-based *orient* motor strategy. The tactile perception model was composed of an occupancy map in which each cell in the modeled environmental grid had a probability of being occupied by an object. Each whisk of the artificial whisker on the robot updated this occupancy map in the estimated location of the robot, i.e., if a whisker made contact with an object, the probably of occupancy in that cell was increased. To optimize the simultaneous estimation of location and mapping, a particle filter algorithm was used, where each particle had its own occupancy map that was updated according to the *flow of information* from the whiskers. For estimation, the particle with the highest posterior probability was taken.

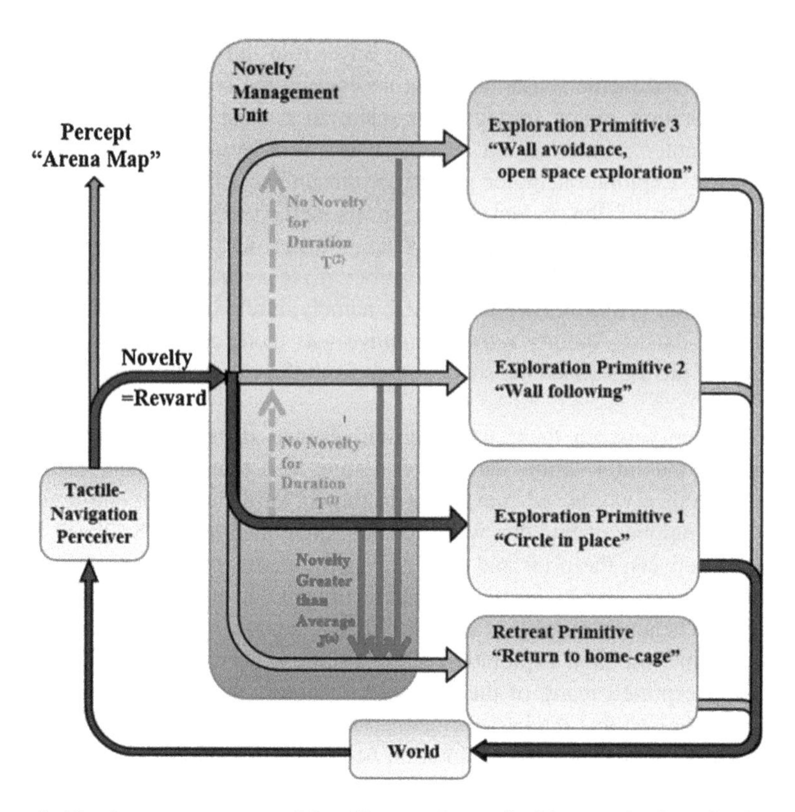

Figure 2 Novelty management model architecture for tactile-driven navigation (Gordon et al. 2014b)

The motor strategy employed governed the motion of the moveable whisker array and was based on an attention model that executed an orienting behavior. In other words, a salience-based attention map was constructed based on the perceived whisker information, resulting in an orienting behavior of the entire *head* of the robot towards the salient tactile object. Thus, once contact was made with an object in the environment, the robot explored that object in greater detail. This increased the information collection required for the tSLAM algorithm.

The results of the study showed that the robot, which made several exploratory bouts in an arena with several geometric shapes, had performed a simultaneous localization and mapping of its environment, with an impressive agreement with the ground truth, as measured by an overhead camera. This model shows how known and well established models from other senses can be adapted to the unique properties of the tactile domain and inform of possible perceptual characteristics of exploring rodents, as well as improve the performance of tactile-based robotic platforms.

Development of Tactile Perception

Biological Application

Developmental models attempt to explain the emergence of tactile perception and their accompanying motor strategies from more basic principles (Gordon and Ahissar 2012). The latter assume repeated interaction between the agent and its environment, thus accumulating statistical representations of the underlying mechanism of sensory perception. Furthermore, the optimal sensorimotor strategies that maximize the perceptual confidence are learned in these developmental models, and not assumed or pre-designed.

One framework of developmental models is artificial curiosity, wherein a reinforcement learning paradigm is used to learn the optimal policy, yet the reward function is intrinsic and is proportional to the learning progress of sensory perception. In one instantiation of this framework in the tactile domain, an artificial neural network was used to model the tactile forward model, i.e., the network predicted the next sensory state based on the current state and the action performed. More specifically, the network was implemented on the vibrissae system, where the sensory states were composed of whisker angle and binary contact information and the action was protraction (increased whisker angle) or retraction (decreased whisker angle). Thus, the ANN learned to map objects in the whisker field, e.g., given the current whisker angle and no contact, if the whisker protracts will it induce contact (there is an object) or not. By moving the whisker, the tactile perceptual model learned about the environment (Figure 3).

The question the developmental model tries to answer is, what is the optimal way to move the whisker so as to maximize the efficiency of mapping the environment? For this purpose, an intrinsic reward reinforcement learning was used, where the reward was proportional to the prediction error of the perceptual ANN. Thus, the more prediction errors made, the higher the reward, exemplifying the concept of *you learn by making mistakes*. The policy converged to moving the whisker to the more unknown places.

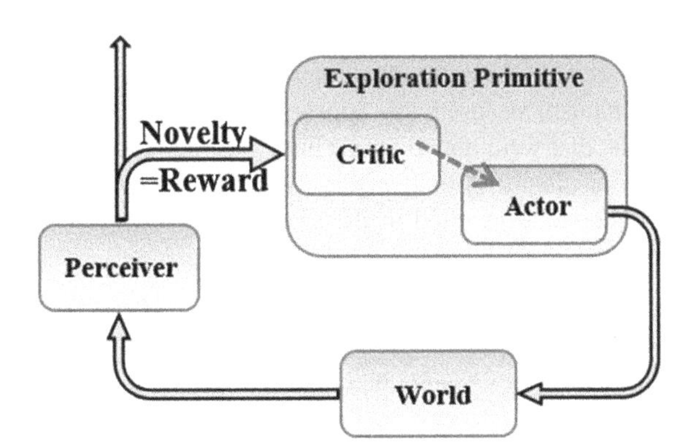

Figure 3 Intrinsic reward reinforcement learning model architecture (Gordon et al. 2014b)

The results of this developmental model showed the convergence of whisking behaviors, starting from random motion and ending up with behaviors observed in adult rodents, e.g., periodic whisking for learning free-space and touch-induced pumps (Deutsch et al. 2012) for localizing tactile objects in the whisker field. The model suggests that these behaviors are learned during development and are not innate in the rodent brain. Furthermore, the model suggests developmental-specific brain connectivity, between the perceptual-learning brain areas, e.g., barrel cortex, and the reward system, e.g., basal ganglia, such that the former supplied the reward signal to the latter.

Robotic Application

A study of artificial curiosity principles on a finger robotic platform with tactile sensors was also performed (Pape et al. 2012). The goal was to study the emergence of tactile-oriented finger movements, that optimize tactile perception of surface textures. For the robotic platform, a robotic finger with two tendon-based actuators and a 2×2 array of 3D Micro-Electro-Mechanical System (MEMS) tactile sensors at its tip was used. The finger was able to flex in order to touch a surface with changing textures.

For tactile perception, a clustering algorithm was used to distinguish between the resulting frequency spectra of the MEMS recordings during 0.33s. This unsupervised learning model represented the abstraction of tactile sensory information into discrete tactile perceptions. However, the clustering was performed only on recent observations and was thus dependent on the movement of the finger, e.g., free movements without contact resulted in different spectra than tapping on the surface. The question asked in this study was *Which skills will be learned by intrinsically motivating the robotic finger to learn about different tactile perceptions?*.

For this purpose, a rewarding mechanism was developed such that intrinsic rewards were given to various aspects of exploration: reward was high for unexplored states of the finger position encouraging exploration; reward was given for ending up in a tactile perceptual state, thus driving sensation towards a specific tactile perception, embodying active sensing principles; reward was given for skills that are still changing, thus focusing on stabilization of skills. This complex reward mechanism ensured the appearance of several intrinsically motivated stabilized skills, that were aimed at reaching specific tactile perceptions. Each developed skill thus resulted in a unique perception in a repeatable manner.

The study resulted in the emergence of several specific intrinsically motivated skills:

1. free movements that avoided the surface that resulted in free-air tactile perception;
2. tapping movements that resulted in unique spectra of the surface and;
3. sliding movements that resulted in a texture-specific spectra.

These well-known and documented tactile strategies of human finger-driven tactile perceptions emerged from intrinsic motivation and were not pre-designed. Thus, the developmental model resulted in learned tactile skills that were associated with unique tactile perceptions.

Internal References

Johnson, D H (2006). Signal-to-noise ratio. *Scholarpedia* 1(12): 2088. http://www.scholarpedia.org/article/Signal-to-noise_ratio.

Prescott, T J; Mitchinson, B and Grant, R A (2011). Vibrissal behavior and function. *Scholarpedia* 6(10): 6642. http://www.scholarpedia.org/article/Vibrissal_behavior_and_function. (see also pages 103–116 of this book).

Schultz, W (2007). Reward signals. *Scholarpedia* 2(6): 2184. http://www.scholarpedia.org/article/Reward_signals.

Sporns, O (2007). Brain connectivity. *Scholarpedia* 2(10): 4695. http://www.scholarpedia.org/article/Brain_connectivity.

Woergoetter, F and Porr, B (2008). Reinforcement learning. *Scholarpedia* 3(3): 1448. http://www.scholarpedia.org/article/Reinforcement_learning.

External References

Deutsch, D; Pietr, M; Knutsen, P M; Ahissar, E and Schneidman, E (2012). Fast feedback in active sensing: Touch-induced changes to whisker-object interaction. *PLoS One* 7(9): e44272.

Gordon, G and Ahissar, E (2012). Hierarchical curiosity loops and active sensing. *Neural Networks* 32: 119–129.

Gordon, G; Fonio, E and Ahissar, E (2014a). Emergent exploration via novelty management. *The Journal of Neuroscience* 34(38): 12646–12661.

Gordon, G; Fonio, E and Ahissar, E (2014b). Learning and control of exploration primitives. *Journal of Computational Neuroscience* 37(2): 259–280.

Grant, R A; Mitchinson, B and Prescott, T J (2012). The development of whisker control in rats in relation to locomotion. *Developmental Psychobiology* 54(2): 151–168.

Pape, L et al. (2012). Learning tactile skills through curious exploration. *Frontiers in Neurorobotics* 6: 6.

Pearson, M J et al. (2013). Simultaneous localisation and mapping on a multi-degree of freedom biomimetic whiskered robot. In: *2013 IEEE International Conference on Robotics and Automation* (*ICRA*) (pp. 586–592).

Saig, A; Gordon, G; Assa, E; Arieli, A and Ahissar E (2012). Motor-sensory confluence in tactile perception. *The Journal of Neuroscience* 32(40): 14022–14032.

Sullivan, J C et al. (2012). Tactile discrimination using active whisker sensors. *IEEE Sensors Journal* 12(2): 350–362.

Further Readings

Gordon, G; Fonio, E and Ahissar, E (2014). Emergent exploration via novelty management. *The Journal of Neuroscience* 34(38): 12646–12661.

Gordon, G; Fonio, E and Ahissar, E (2014). Learning and control of exploration primitives. *Journal of Computational Neuroscience* 37(2): 259–280.

External Link

http://gorengordon.com/ Goren Gordon's home page.

Whiskered Robots

Tony Pipe and Martin J. Pearson

Whiskered robots. This article summarises the motivation and history of whisker inspired tactile sensors for robotics and the contribution that this field of research has made in our understanding of the biological analogue (Figure 1).

Overview

Touch is currently an underutilized sensory mode in robotics, with vision remaining the preferred method of spatial exploration. There are many examples in the animal kingdom of creatures that live in confined and visually occluded environments where a developed sense of touch rather than vision is advantageous for survival. Such environments are challenging operational domains for mobile robots as conventional proximity sensors and artificial vision systems do not perform well. Perhaps more persuasively, there are many situations where a developed sense of whisker-like touch on a robot would serve as a beneficial complement, rather than replacement, for vision. Of particular interest is that the flexible shaft of a whisker provides an intrinsically compliant interface between the robot and a surface under inspection thereby minimising the risk of damage to both. It follows that this reduces the required precision for motion planning whilst not compromising the efficacy of the tactile inspection.

In this article the use of whiskers as a method of endowing robots with a sense of whisker touch is summarised through a brief history of whiskered robots. This is followed by a discussion on the utility in adopting a biomimetic approach to whiskered robot development and how this has resulted in a 2-way exchange of ideas between engineers and biologists.

T. Pipe (✉) · M.J. Pearson
Bristol Robotics Laboratory, Bristol, UK

T.J. Prescott et al. (eds.), *Scholarpedia of Touch*, Scholarpedia,
DOI 10.2991/978-94-6239-133-8_59

Figure 1 The tactile whisker array of the Shrewbot platform was inspired by the form and function of the whiskers of small mammals

History

Whisker based sensors have been deployed on mobile robots since the mid 1980s. Initially they were used as simple binary proximity sensors to assist in navigation and obstacle avoidance (Russell 1984; Brooks 1989; Hirose et al. 1990).

Examples of whiskers used on robots for more detailed spatial exploration can be divided into either active or passive touch approaches. Active touch whiskers measure the bending torque of the whisker as it makes contact with objects during a controlled movement, analogous to the whisking behaviour of rats (Ueno and Kaneko 1994), (Kaneko et al. 1998) (Top left panel of Figure 2). Passive touch whiskers measure the torque of whisker bend in response to contacts made with objects as the platform on which the whiskers are mounted moves passed them, i.e., the whiskers are not 'directly' actuated. For example spring loaded potentiometers have been extensively used to measure this torque (Russell 1992), (Jung and Zelinsky 1996), (Wijaya and Russell 2002). More recent examples of active whisker sensors for surface inspection use load cells to measure the whisker deflection torque (Clements and Rahn 2006), whilst others use resonating piezo-electric stimulators (Muraoka 2005) or strain gauges (Solomon and Hartmann 2006) (Lower left panel of Figure 2). There are also a number of examples of more biologically plausible whisker sensory arrays installed on mobile robots to conduct biomimetic study of the whisker sensory system. The aMOUSE project integrated a bilateral array of real rat whiskers glued to the diaphragm of electrohet microphones and mounted onto a small mobile robot platform to investigate texture classification of surfaces (Lungarella et al. 2002) (Top right panel of Figure 2). The ability to extract surface shape was also demonstrated using an actuated array of flexible artificial whiskers instrumented using Hall effect sensors at their base (Kim and Moller 2004) (Lower right panel of Figure 2).

The Whiskerbot platform (built as part of EPSRC grant no. GR/S19639/01) had a small array of glass-fibre moulded whiskers instrumented at their base using micro-strain gauges to measure 2-dimensional deflections of the shaft. This analogue information was converted into a series of empirically based spike train

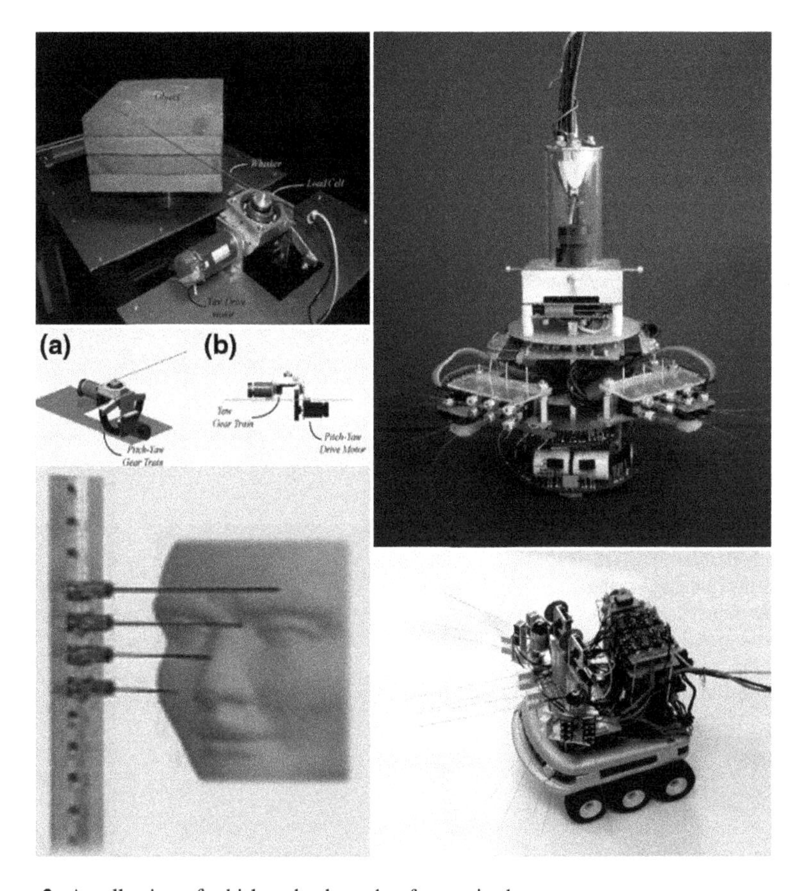

Figure 2 A collection of whiskered robots that feature in the text

models of the rat vibrissal primary afferent (Mitchinson et al. 2004) and passed to an embedded real-time spike based model of the rodent trigeminal complex and superior colliculus (Pearson et al. 2007) (Figures 3 and 4).

The SCRATCHbot (Spatial Cognition and Representation through Active TouCH bot) developed as part of the ICEA project (Integrated Cognition, Emotion and energy Autonomy, EU FP6) had a much larger number of whiskers and motile degrees of freedom to position the array within the environment. Hall effect sensors were used for measuring the deflections in the whisker shafts and standard DC motors used to actuate the "whisking" behaviour observed in rats. This platform was used to reproduce different implementations of whisking pattern generation and coordination to allow a quantitative comparison of each hypothesis with regards to the quality of sensory information derived. It was also used to demonstrate a simple model of tactile attention that could be used to direct the exploration of the robot through its environment. Both of these research questions were developed further using the Shrewbot platform (Shown in Figure 1) which was built as part of the

Figure 3 The Whiskerbot platform had an array of 6 glass-fibre whiskers and a spike based model of the trigeminal ganglion and rat brain stem to process tactile information and generate motor behaviour

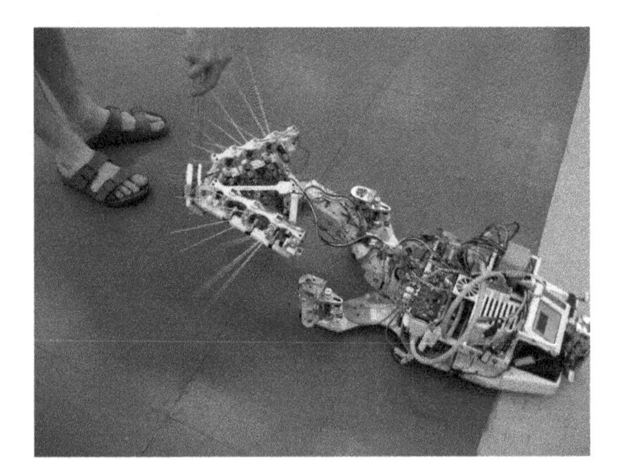

Figure 4 The SCRATCHbot platform had an array of 18 whiskers instrumented using Hall effect sensors and used a biomimetic model of attention to direct the array toward points of tactile salience

BIOTACT project (Biomimetic technology for vibrissal active touch, EU FP7) along with the G2 Sensor which was used to develop algorithms to extract texture and object form from an active tactile whisker array (Sullivan et al. 2012) (Shown in Figure 5). Both of these platforms used miniaturised modular whisker sensorimotor assemblies to form their array.

Biomimetic Whiskers

To design an artificial whisker sensory system that can derive useful tactile information from the environment gives rise to a number of lines of inquiry that are of equal importance to both engineers and biologists. From an engineering perspective, one issue is to identify any advantages of using whiskers for tactile exploration

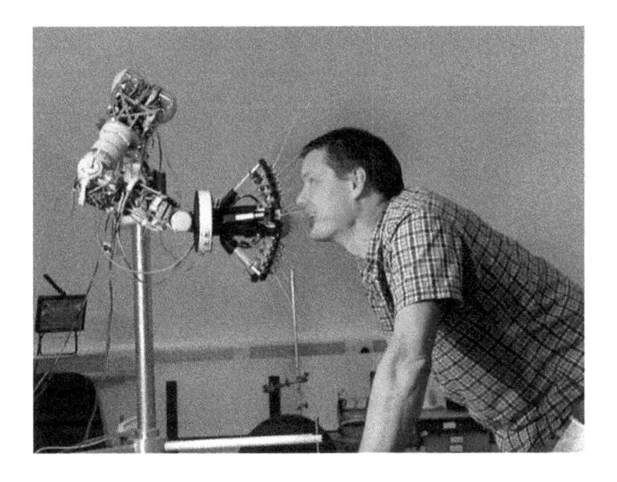

Figure 5 The G2 sensor developed in the BIOTACT project to demonstrate biomimetic tactile object and texture classification algorithms

over other approaches, for example, based more directly on human touch (cutaneous touch). Through the observation of whisker development (Sullivan et al. 2003), array morphology (Brecht et al. 1997), and how animals position and move their whiskers during natural exploratory behaviour (Grant et al. 2009), we can intuit advantages such as robustness, speed of response and size of sensory field. Other behavioural studies have found that whiskered animals, such as rat, can discriminate intricate surface features with the same acuity as human touch (Carvell and Simons 1990). Specific questions that are generated through the development of an engineering specification have also inspired further biological investigation. For example, to determine how best to move an artificial array of whiskers to gather information from the environment, video footage of rat whisking has been analysed to develop posits for possible control strategies (Mitchinson et al. 2007). Intriguingly, these model control strategies were then physically tested using the robotic artifact (Pearson et al. 2007) which inspired further investigation of the original biological analogue, such as, how to deal with "noise" generated by whisker self-motion (Anderson et al. 2010).

A biomimetic design approach that encapsulates a two-way exchange of ideas and skills, can be applied to many areas of the design of a useful whisker based sensory system. In addition to modeling the physical mechanics of whiskers to develop a sensory tool for robotics, whiskers also act as extremely powerful "probes" with which to observe the function of the brain. The neural components of the whisker sensory system of the rat maintain an exquisite degree of topological preservation from whisker follicle through to cortex (Kleinfeld et al. 2006). This allows for controlled observations of the neural response to stimuli applied to the whisker shaft at multiple levels of this neuraxis. Model systems can be constructed from such observations which are then reinforced, or at least validated, by behavioural observations of the animal, an approach known as neuroethology (Camhi 1941). Using robotic artifacts to test such models (computational neuroethology) introduces a degree of experimental flexibility to allow experiments that may be either impossible or unethical to achieve through animal testing. It also provides roboticists the opportunity to work with neuroscientists to evaluate novel brain-based control approaches for future autonomous robotic systems.

References

Anderson, S R et al. (2010). Adaptive cancelation of self-generated sensory signals in a whisking robot. *IEEE Transactions on Robotics* 26(6): 1065–1076.

Brecht, M; Preilowski, B and Merzenich, M M (1997). Functional architecture of the mystacial vibrissae. *Behavioural Brain Research* 84: 81–97.

Brooks, R A (1989). A robot that walks: Emergent behaviours from a carefully evolved network. *Neural Computation* 1: 253–262.

Camhi, J M (1941). *Neuroethology: Nerve Cells and the Natural Behavior of Animals.* Sunderland, MA: Sinauer Associates, Inc.

Carvell, G A and Simons, D J (1990). Biometric analyses of vibrissal tactile discrimination in the rat. *The Journal of Neuroscience* 10(8): 2638–2648.

Clements, T N and Rahn, C D (2006). Three-dimensional contact imaging with an actuated whisker. *IEEE Transactions on Robotics* 22(4): 844–848.

Grant, R A; Mitchinson, B; Fox, C W and Prescott, T J (2009). Active touch sensing in the rat: Anticipatory and regulatory control of whisker movements during surface exploration. *Journal of Neurophysiology* 101(2): 862–874.

Hirose, S; Inoue, S and Yoneda, K (1990). The whisker sensor and the transmission of multiple sensor signals. *Advanced Robotics* 4(2): 105–117.

Jung, D and Zelinsky, A (1996). Whisker based mobile robot navigation. In: *International Conference on Intelligent Robots and Systems*, Vol. 2 (pp. 497–504).

Kaneko, M; Kanayama, N and Tsuji, T (1998). Active antenna for contact sensing. *IEEE Transactions on Robotics and Automation* 14(2): 278–291.

Kim, D and Moller, R M (2004). A biomimetic whisker for texture discrimination and distance estimation. *International Conference on the Simulation of Adaptive Behaviour.*

Kleinfeld, D; Ahissar, E and Diamond, M E (2006). Active sensation: Insights from the rodent vibrissa sensorimotor system. *Current Opinion in Neurobiology* 16: 435–444.

Lungarella, M; Hafner, V V; Pfeifer, R and Yokoi, H (2002). Whisking: An unexplored sensory modality. In: *International Conference on Simulation of Adaptive Behaviour* (pp. 58–69).

Mitchinson, B et al. (2004). Empirically inspired simulated electro-mechanical model of the rat mystacial follicle-sinus complex. *Proceedings of the Royal Society of London B* 271: 2509–2516.

Mitchinson, B; Martin, C J; Grant, R A and Prescott, T J (2007). Feedback control in active sensing: Rat exploratory whisking is modulated by environmental contact. *Proceedings of the Royal Society of London B* 274: 1035–1041.

Muraoka, S (2005). Environmental recognition using artificial active antenna system with quartz resonator force sensor. *Measurement* 37: 157–165.

Pearson, M J; Pipe, A G; Melhuish, C; Mitchinson, B and Prescott, T J (2007). Whiskerbot: A robotic active touch system modelled on the rat whisker sensory system. *Adaptive Behavior* 15 (3): 223–240.

Russell, R A (1984). Closing the sensor-computer-robot control loop. *Robotics Age* 15–20.

Russell, R A (1992). Using tactile whiskers to measure surface contours. In: *International Conference on Robotics and Automation* (pp. 1295–1299).

Solomon, J H and Hartmann, M J (2006). Robotic whiskers used to sense features. *Nature* 443: 525.

Sullivan, J C et al. (2012). Tactile discrimination using active whisker sensors. *IEEE Sensors* 12 (2): 350–362.

Sullivan, R M; Landers, M S; Flemming, J; Young, T A and Polan, H J (2003). Characterizing the functional significance of the neonatal rat vibrissae prior to the onset of whisking. *Somatosensory & Motor Research* 20(2): 157–162.

Ueno, N and Kaneko, M (1994). Dynamic active antenna: A principle of dynamic sensing. In: *IEEE International Conference on Robotics and Automation* (pp. 1784–1790).

Wijaya, J A and Russell, R A (2002). Object exploration using whisker sensors. *Australasian Conference on Robotics and Automation* (pp. 180–185).

External Link

http://www.brl.ac.uk/bnr Bristol Robotics Laboratory.

Haptic Displays

Yeongmi Kim and Matthias Harders

Display devices are essential for the computer-based rendering, i.e. presentation and output of data and information. From binary pixels to photorealistic three-dimensional images, visual data can, for example, be rendered via visual displays (e.g. monitors or screens of computers, smart phones, or e-book readers). Similarly, auditory displays render corresponding output, ranging from simple binary signals to multi-channel surround sound. In this context, the term **Haptic Displays** refers to interfaces delivering haptic feedback, typically by stimulating somatic receptors to generate a sensation of touch. The rendered feedback is generally categorized into cutaneous as well as kinesthetic or proprioceptive feedback, respectively. The former concerns sensations via skin receptors (mechanoreceptors), while the latter deals with stimulation of receptors in muscles, tendons and joints (proprioceptors).

Early Development of Haptic Displays

Some of the earliest displays capable of generating haptic signals have been developed even before the emergence of digital electronic computers. Already in the 1920s, Gault described a device to transform speech into vibrational stimuli applied to the skin. The method enabled subjects to distinguish colloquial sentences and certain vowels. His "teletactor", a multi-vibration unit delivering stimuli to five fingers, allowed subjects with auditory disability to perceive speech, music, and other sounds (Gault 1927). This is a typical example of sensory substitution, where one sensory stimuli is replaced with another, e.g. sound to tactile substitution. In general, the development of devices and mechanisms to assist persons with visual or hearing impairment has been a driving force in the emergence of tactile devices in the 20th century. In this context, devices to display Braille—a widely employed tactile writing system—via a refreshable output have played an important role.

Y. Kim (✉)
The Department of Mechatronics, MCI, Innsbruck, Austria

M. Harders
University of Innsbruck, Innsbruck, Austria

T.J. Prescott et al. (eds.), *Scholarpedia of Touch*, Scholarpedia,
DOI 10.2991/978-94-6239-133-8_60

Related to this, in the 1960s the Optacon (Optical to Tactile Converter) was developed to enable visually impaired users to access tactile representations of black-and-white images captured by the system (Linvill and Bliss 1966). The device comprised of a photoelectric sensor and a 24 × 6 array of piezo-electrically driven tactile pins. While the Optacon presented tactile images of small areas, the Tactile Vision Sensory Substitution system developed by Bach-Y-Rita et al. provided a wider range of tactile images. For these, solenoid stimulators were employed that provided stimuli according to the visual image of a video camera (Bach-Y-Rita et al. 1969).

Also the first devices akin to kinesthetic displays operated without computers. In the 1950s, Goertz developed at the Argonne National Lab the first master/slave remote-operation manipulator designed to safely handle radioactive material (Goertz and Thompson 1954). The mechanism was capable of providing force feedback from the remote site to an operator. The first notable initiatives employing computer-based haptic rendering have been the series of GROPE projects, which started in the 1960s and extended to the 1980s (Brooks et al. 1990). In fact, one of the devices employed for **haptic display** was the above mentioned Argonne Remote Manipulator. The projects focused on the rendering of force fields, for instance in the context of molecular docking. Feedback was initially provided two, and ultimately in six degrees of freedom.

Another event with a notable influence on the recent considerable growth of the field of haptics, has been the emergence of commercially available haptic displays in the 1990s. Examples are the PHANTOM device (Massie and Salisbury 1994), providing three degrees of freedom of force display, the Impulse Engine, which was optimized for haptic feedback in medical simulations (Jackson and Rosenberg 1995), as well as the CyberTouch and CyberGrasp gloves (CyberGlove Systems LLC 2009); the former providing vibration feedback to the fingertips, and the latter employing exoskeletal actuation to display forces on the fingers. The wider availability of these and similar haptic displays led to a growth in haptics research addressing questions of perception, control, rendering, and applications in haptics. In turn, new results also led to an acceleration of the development of new haptic displays. More in-depth coverage of the historical steps of haptic displays can be found, for instance, in (Burdea 1999) and (Stone 2001).

Categories of Haptic Displays

Haptic displays can be categorized according to different characteristics and metrics. As already utilized in the previous section, an oft-employed discrimination is by feedback type, i.e. kinesthetic vs. tactile displays (often also complemented with a hybrid, i.e. both kinesthetic and tactile category). This is often broken down further by considering the device attachment (body- vs. ground-mounted) (see e.g. Kurfess 2005) or the device portability. A similar classification is given in (Vu and Proctor 2011), who consider input vs. output, tactile vs. force feedback, as well as location of the feedback (e.g. hand, arm, body). Recently, Gonzalez et al. (2013) proposed a “Hand Interaction Tree”, based on which haptic displays can also be categorized. They focused on examining the complexity and efficiency of spatial and temporal interaction on the hand. Similar notions concerning the study of hand grasps and manipulation tasks were also discussed in (Dollar 2014). In the following a number of key characteristics, and thus associated categories, of haptic displays will be outlined.

Kinesthetic Versus Tactile Displays

Haptic displays can be categorized by the type of stimuli/output they generate, and correspondingly by the type of sensory receptors that are stimulated. kinesthetic displays is concerned with the rendering of forces. The latter are generated based on computational models, remote interaction, recordings or data-driven approaches. Displays in this category can be further subdivided according to the type of actuation (i.e. hardware) employed for force generation, such as electric motors, pneumatic/hydraulic actuators, etc. A large number of kinesthetic displays are actuated using electric motors (e.g. see Massie and Salisbury 1994; Campion et al. 2005; Lee et al. 2010). This is due to the latter being robust and compact, while offering a high bandwidth, as well as being easy to control. In general they exhibit high torque-to-power as well as torque-to-inertia ratios. Nevertheless, the provided torques are often smaller than those obtainable with actuation based on circulating fluids. The latter comprise hydraulic and pneumatic actuators. Hydraulic actuation has for instance been employed to build haptic displays capable of generating high forces (see e.g. Frey et al. 2008; Lee and Ryu 2008). Along these lines, also pneumatic actuation has been used, for instance to drive pistons in an exoskeletal glove displaying forces to the fingers (Bouzit et al. 2002). Another alternative is the use of magnetic levitation for haptic display (Berkelman 2007). Related to the notion of actuation principle is also the applied approach of transmission for generating feedback, as well as the kinematic design (parallel vs. serial). This is for instance addressed in (Massie and Salisbury 1994; Liu et al. 2014).

The force feedback rendered on kinesthetic displays often tries to represent objects or physical phenomena (e.g. force fields). Concerning the former, the output comprises at least of the object shape as well as often its compliance. Related to this is the notion of the degrees of freedom for in- and output. For instance, three degrees of freedom of force output allow the display of forces in three dimensions arising from point-contacts on the surface of a spatial object representation. In such a setting, feedback is usually generated according to active user input, generated during active spatial exploration of a virtual object. However, display of haptic feedback can also have other purposes, such as the rendering of magnetic force fields or the display of abstract data.

A special subcategory of kinesthetic displays is concerned with proprioception. Typical applications are dealing with rehabilitation or skill transfer (see e.g.

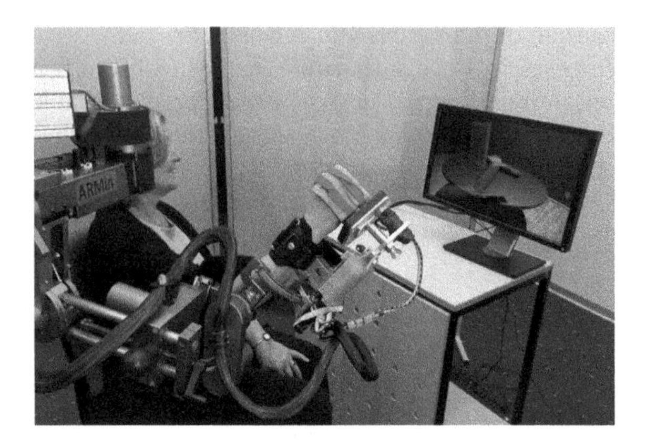

Figure 1 An exoskeletal interface. With kind permission of Dietmar Heinz

Lambercy et al. 2011; Bluteau et al. 2008). The main idea is to provide forces guiding or assisting a user. The employed devices often take the form of an exoskeletal setup(see Figure 1) or an end-effector of a haptic mechanism attached to a limb (Nef et al. 2009; Ozkul et al. 2012).

Kinesthetic displays provide stimuli mainly perceived through receptors located inside of muscles, tendons, and joints (proprioceptors). In contrast to this, tactile displays focus on stimulation of the mechanoreceptors found inside the skin. The different types of cutaneous (skin) receptors are sensitive to vibration, stretch, pressure, deformation, etc. and thus, for instance, stimulated by surface textures. This is complemented by further receptors sensitive to temperature as well as pain stimuli. Tactile displays are commonly categorized either according to the tactile sensation they provide or to the type of employed actuation. A well-known example of the use of tactile displays is the vibration motors found in most mobile phones. Various technological solutions for providing vibration sensations to the skin have been realized, such as eccentric rotating mass actuation, linear resonant actuation, piezoelectric actuators, shape-memory alloys, or electro-active polymers. Other stimuli, such as skin stretch or local deformation can be generated, for instance, by air-jet displays, pneumatic balloons (see e.g. Santos-Carreras et al. 2012), pin array displays (e.g. actuated by solenoids or linear motors, see Kyung and Lee 2009) as shown in Figure 2 (Furrer 2011) or rheological fluids (see e.g. Lee and Jang 2011). Furthermore, temperature stimuli are typically realized with Peltier-elements (Jones and Ho 2008).

Finally, also the notion of haptic illusions has to be addressed in this context. Similar to optical illusions, the former relate to stimuli that are perceived differently than actually provided physical stimuli. For instance, low frequency vibrations at about 80 Hz applied to a tendon result in the illusory percept of muscle elongation (Roll et al. 1989). Various other tactile illusions have been reported (e.g. cutaneous rabbit illusion, tau effect, kappa effect, etc., see Helson 1930; Geldard and Sherrick 1972; Bekesy 1958; Sherrick and Rogers 1966) and others are an active field of current research.

Figure 2 A braille display driven by Tiny Ultrasonic Linear Actuators

Contact Versus Non-contact Displays

Unlike visual and auditory stimuli, most haptic signals do not travel through air and require direct contact with a user's body to stimulate the haptic sensory receptors (mechanoreceptors and proprioceptors). Thus, the majority of haptic displays are contact-type interfaces. The contact with a user's body can differ in size, location, attachment, etc. It can be realized via tools, held in a hand, that are attached to a haptic mechanism, exoskeletal mechanisms that are attached to body parts, or direct application of actuators to the skin.

Overall peripheral force feedback interfaces with a joystick, pen-type/sphere type end-effector are sorted as a tool-based haptic display (Massie and Salisbury 1994 see Figure 3; Jackson and Rosenberg 1995; Grange et al. 2001). The tool-based display can be effective when generating haptic feedback of analogous circumstance simulations such as needle insertion, dental training or surgical simulation of minimally invasive surgery. Whereas various kinesthetic devices are based on tool-based interface, there are few tool-based tactile displays to deliver tactile stimuli mainly through a tool instead of direct contact with actuators. McMahan and Kuchenbecker developed a stylus comprised of a voice coil actuator (i.e. Haptuator) on top of the handle (McMahan and Kuchenbecker 2014).

Exoskeleton type haptic displays are often wearable system (Frisoli et al. 2009; Fontana et al. 2013). Thus it is possible to afford wider workspaces and elaborated haptic feedback during dexterous manipulation of a virtual object (e.g. CyberGrasp™). However, the complexity of donning is a drawback compared to the tool-based haptic displays.

Majority of tactile displays directly stimulate the skin. Wearable tactile displays designed to a headband, wrist band, arm band, glove, vest, glasses, or belt enable to receive haptic stimuli passively (Van Erp et al. 2005; Kim et al. 2009; Kajimoto et al. 2006; Jones et al. 2006). Handheld displays of direct skin stimulation are often contact localized area such as fingertips (Pasquero et al. 2007).

Figure 3 A tool-based haptic display

Nevertheless, while most haptic displays rely on direct contact with body parts to stimulate the receptors, some haptic actuation principles allow for the generation of non-contact haptic stimulation. Recently, there has been an increased interest in these approaches. For instance, air-jets are a comparatively simple technical solution to generate non-contact haptic feedback (Tsalamlal et al. 2013; Kim et al. 2008 see Figure 4). However, it is difficult to create complex haptic sensations, and the

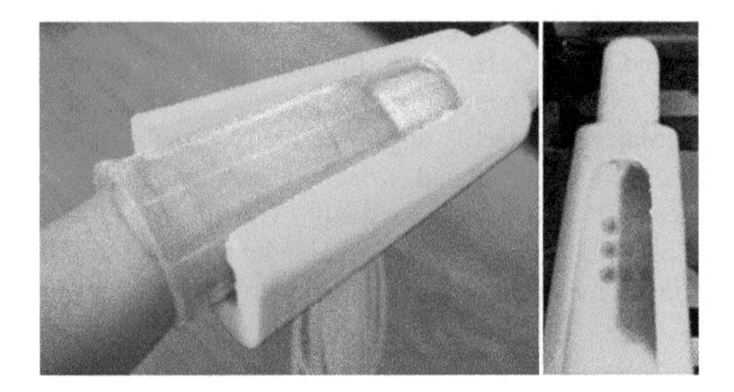

Figure 4 5 × 5 and 3 × 3 arrays of fingertip air-jet displays

range of haptic interaction is limited due to dissipation effects. A more advanced strategy has recently been reported in (Gupta et al. 2013; Sodhi et al. 2013). They employ pneumatics to create air vortices, based on which haptic feedback is generated over a distance. Another non-contact approach is based on focused ultrasound waves (see e.g. Hoshi et al. 2010; Carter et al. 2013). The key idea is to employ acoustic radiation pressure to stimulate the skin at a distance. Solutions employing a 2D grid of ultrasound transducers have been employed, however, recently reported developments even tried to extend this to a volumetric interaction space fully enclosed by transducers. These non-contact haptic displays provide stimuli directly to the body over distance, and have the potential for new applications of haptic display and interaction.

Recent Trends in Haptic Display Development

Currently, the arguably most widely used haptic displays are mobile phones. In the first phone generations, simple vibratory stimuli perceivable through the entire case were employed to display alerts. With the advent of the current phone generation employing touch screens, there has been an increased interest in providing richer and more localized haptic feedback, possibly also tightly coupled with the interaction of a user (Freeman et al. 2014). In this regard, for instance tactile pattern design tools have been developed allowing for the creation of more complex vibration patterns that exhibit complex temporal/intensity variations, thus going beyond simple On/Off signals.

Fueled by the large prevalence of tablets and smart phones, a focus of recent research and development has been on providing more sophisticated tactile feedback during active exploration on a touch surface with one or more fingers. In this regard, Winfield et al. (2007) introduced the TPad; in this device a piezoelectric disk is employed to create ultrasonic vibration waves, which allow to modulate the coefficient of friction according to the squeeze film effect. Various haptic sensations (e.g. bumps, textures, geometry, and edges) can be generated via this approach, however, active user movement is required for the sensations to appear. The technology has recently also been extended to provide feedback on commercially available tablets (see TPad Tablet Project (Neuroscience and Robotics Laboratory, Northwestern University 2015)). Related to this, another strategy is the use of electrostatic vibration. As implemented by Bau et al. (2010), tactile information can be displayed on a surface by controlling friction through the electro-vibration principle. By applying a periodic voltage on an electrode covered by an insulator, an electrostatic attractive force is generated between a sliding finger and the insulator plate. The aforementioned approaches of providing tactile signals on flat displays is often also referred to as "surface haptics". In the past few years, the first commercial devices (e.g. Senseg FeelScreen™, Esterline and Pacinian—HapticTouch™) employing the previously described techniques have started to emerge.

Other strategies of providing haptic feedback on touch screens have also recently been reported. For instance, Rantala et al. (2009) employed a piezoelectric actuator placed underneath a touchscreen of a commercial tablet, to render six braille dots on the device. This was used to make the touch screen device accessible to visually impaired persons. Further, also the use of microfluidics in a deformable tactile display has been suggested, in order to create physically embossed buttons on a screen (e.g. Tactus Tactile Layer™).

Conclusion

Over the last few decades, more and more haptic displays have appeared, in synergy with the development of new actuation and sensing technologies, mechanical designs, and control strategies in parallel. The use of haptic displays has been suggested in a wide range of diverse applications, such as education, entertainment, simulation and training, rehabilitation, tele-operation, assistive technology, prototyping, etc. However, except for the widely prevalent basic vibration actuators in mobile phones, more sophisticated haptic displays have not yet spread into everyday use. This can often be attributed to involved costs and considerations of robustness as well as usability. Haptic feedback is often more compelling in combination with matching auditory and visual stimuli, for instance applied to enhance interaction in virtual or augmented realities. Yet, there is currently no multi-purpose haptic display available that is capable of generating highly realistic sensations of touch. In this regard, the technology is still lagging behind visual and auditory displays. Nevertheless, in some clearly-defined cases plausible haptic feedback can be displayed. In comparison to the other sensory channels, receptors of haptic stimuli are diverse and spread over the whole body. However, haptic exploration of the environment is typically carried out with the hand; and is associated with both tactile and kinesthetic sensations. Therefore, the combination of kinesthetic with tactile displays has been suggested in the past as a possible approach to enhance overall realism. In this context, the notion of indirect tool-based vs. direct hand-based interaction has to be considered. Convincing future haptic displays will have to go beyond the tool-mediated rendering of forces and allow direct manual contact. In this regard, the recent developments of surface haptics as well as the direction of non-contact displays may pave the way to more natural and intuitive haptic feedback. Still, the development of haptic displays has only just left its infancy, and considerable further developments still remain to achieve highly realistic haptic rendering.

References

Bach-Y-Rita, P; Collins, C C; Saunders, F A; White, B and Scadden, L (1969). Vision substitution by tactile image projection. *Nature* 221(5184): 963–964.

Bau, O; Poupyrev, I; Israr, A and Harrison, C (2010). TeslaTouch: Electrovibration for touch surfaces. In: *Proceedings of the 23nd Annual ACM Symposium on User Interface Software and Technology* (pp. 283–292).

Berkelman, P (2007). A novel coil configuration to extend the motion range of lorentz force magnetic levitation devices for haptic interaction. In: *IEEE/RSJ International Conference on Intelligent Robots and Systems, 2007. IROS 2007.* (pp. 2107–2112).
Bekesy, G V (1958). Funneling in the nervous system and its role in loudness and sensation intensity on the skin. *The Journal of the Acoustical Society of America* 30(5): 399–412.
Bluteau, J; Coquillart, S; Payan, Y and Gentaz, E (2008). Haptic guidance improves the visuo-manual tracking of trajectories. *PLoS One* 3(3): e1775.
Burdea, G C (1999). Keynote address: Haptics feedback for virtual reality. In: *Proceedings of International Workshop on Virtual Prototyping,* Laval, France (pp. 87–96).
Brooks, F P, Jr.; Ouh-Young, M; Batter, J J and Kilpatrick, P J (1990). Project GROPEHaptic displays for scientific visualization. In: *ACM SIGGraph Computer Graphics,* Vol. 24, No. 4 (pp. 177–185).
Bouzit, M; Popescu, G; Burdea, G and Boian, R (2002). The rutgers master ii-nd force feedback glove. In: *Proceedings of the 10th Symposium on Haptic Interfaces for Virtual Environment and Teleoperator Systems, 2002. HAPTICS 2002* (pp. 145–152).
Campion, G; Wang, Q and Hayward, V (2005). The pantograph Mk-II: A haptic instrument. In: *IEEE/RSJ International Conference on Intelligent Robots and Systems, 2005. (IROS 2005)* (pp. 193–198).
Carter, T; Seah, S A; Long, B; Drinkwater, B and Subramanian, S (2013). Ultrahaptics: Multi-point mid-air haptic feedback for touch surfaces. In: *Proceedings of the 26th Annual ACM Symposium on User Interface Software and Technology* (pp. 505–514).
CyberGlove Systems LLC (2009). CyberTouch^{TM}: Tactile feedback for the CyberGlove System. http://www.cyberglovesystems.com/sites/default/files/CyberTouch_Brochure_2009.pdf.
Dollar, A M (2014). Classifying human hand use and the activities of daily living. In: *The Human Hand as an Inspiration for Robot Hand Development* (pp. 201–216). Springer International Publishing.
Frey, M; Johnson, D E and Hollerbach, J (2008). Full-arm haptics in an accessibility task. In: *Symposium on Haptic Interfaces for Virtual Environment and Teleoperator Systems, 2008. HAPTICS 2008* (pp. 405–412). IEEE.
Fontana, M; Fabio, S; Marcheschi, S and Bergamasco, M (2013). Haptic hand exoskeleton for precision grasp simulation. *Journal of Mechanisms and Robotics* 5(4): 041014.
Freeman, E; Brewster, S and Lantz, V (2014). Tactile feedback for above-device gesture interfaces: Adding touch to touchless interactions. In: *Proceedings of the 16th International Conference on Multimodal Interaction* (pp. 419–426). ACM.
Frisoli, A; Salsedo, F; Bergamasco, M; Rossi, B and Carboncini, M C (2009). A force-feedback exoskeleton for upper-limb rehabilitation in virtual reality. *Applied Bionics and Biomechanics* 6(2): 115–126.
Furrer, J (2011). Augmented white cane II: Towards an effective electronic mobility aid for the blind. Master Thesis, Rehabilitation Engineering Lab, ETH Zurich & ZHAW, Switzerland.
Gault, R H (1927). "Hearing" through the sense organs of touch and vibration. *Journal of the Franklin Institute* 204(3): 329–358.
Geldard, F A and Sherrick, C E (1972). The cutaneous "rabbit": A perceptual illusion. *Science* 178 (4057): 178–179.
Goertz, R C and Thompson, W M (1954). Electronically controlled manipulator. *Nucleonics* (pp. 46–47).
Gonzalez, F; Gosselin, F and Bachta, W (2013). A framework for the classification of dexterous haptic interfaces based on the identification of the most frequently used hand contact areas. In: *World Haptics Conference (WHC), 2013* (pp. 461–466). IEEE.
Grange, S; Conti, F; Rouiller, P; Helmer, P and Baur, C (2001). Overview of the Delta Haptic Device. In: *Proceedings of EuroHaptics '01.*
Gupta, S; Morris, D; Patel, S N and Tan, D (2013). Airwave: Non-contact haptic feedback using air vortex rings. In: *Proceedings of the 2013 ACM International Joint Conference on Pervasive and Ubiquitous Computing* (pp. 419–428).

Helson, H (1930). The tau effect—an example of psychological relativity. *Science* 71(1847): 536–537.

Hoshi, T; Takahashi, M; Iwamoto, T and Shinoda, H (2010). Noncontact tactile display based on radiation pressure of airborne ultrasound. *IEEE Transactions on Haptics* 3(3): 155–165.

Jackson, B and Rosenberg, L (1995). *Force Feedback and Medical Simulation. Interactive Technology and the New Paradigm for Healthcare* (pp. 147–151). Amsterdam: IOS Press.

Jones, L A; Lockyer, B and Piateski, E (2006). Tactile display and vibrotactile pattern recognition on the torso. *Advanced Robotics* 20(12): 1359–1374.

Jones, L A and Ho, H N (2008). Warm or cool, large or small? The challenge of thermal displays. *IEEE Transactions on Haptics* 1(1): 53–70.

Kajimoto, H; Kanno, Y and Tachi, S (2006). Forehead electro-tactile display for vision substitution. In: *Proceedings of EuroHaptics.*

Kim, Y; Cha, J; Oakley, I and Ryu, J (2009). Exploring tactile movies: An initial tactile glove design and concept evaluation. *MultiMedia, IEEE* PP(99): 1.

Kim, Y; Oakley, I and Ryu, J (2008). Human perception of pneumatic tactile cues. *Advanced Robotics* 22(8): 807–828.

Kurfess, T R (Ed.). (2005). *Robotics and Automation Handbook.* CRC Press.

Kyung, K U and Lee, J Y (2009). Ubi-Pen: A haptic interface with texture and vibrotactile display. *IEEE Computer Graphics and Applications* (1): 56–64.

Lambercy, O; Robles, A J; Kim, Y and Gassert, R (2011). Design of a robotic device for assessment and rehabilitation of hand sensory function. In: *2011 IEEE International Conference on Rehabilitation Robotics (ICORR)* (pp. 1–6).

Linvill, J G and Bliss, J C (1966). A direct translation reading aid for the blind. *Proceedings of the IEEE* 54(1): 40–51.

Liu, L; Miyake, S; Maruyama, N; Akahane, K and Sato, M (2014). Development of two-handed multi-finger haptic interface SPIDAR-10. In: *Haptics: Neuroscience, Devices, Modeling, and Applications* (pp. 176–183). Berlin Heidelberg: Springer.

Lee, L F; Narayanan, M S; Mendel, F; Krovi, V N and Karam, P (2010). Kinematics analysis of in-parallel 5 dof haptic device. In: *2010 IEEE/ASME International Conference on Advanced Intelligent Mechatronics (AIM)* (pp. 237–241).

Lee, C H and Jang, M G (2011). Virtual surface characteristics of a tactile display using magneto-rheological fluids. *Sensors* 11(3): 2845–2856.

Lee, Y and Ryu, D (2008). Wearable haptic glove using micro hydraulic system for control of construction robot system with VR environment. In: *IEEE International Conference on Multisensor Fusion and Integration for Intelligent Systems, 2008. MFI 2008* (pp. 638–643).

McMahan, W and Kuchenbecker, K J (2014). Dynamic modeling and control of voice-coil actuators for high-fidelity display of haptic vibrations. In: *2014 IEEE Haptics Symposium (HAPTICS)* (pp. 115–122).

Massie, T H and Salisbury, J K (1994). The phantom haptic interface: A device for probing virtual objects. In: *Proceedings of the ASME Winter Annual Meeting, Symposium on Haptic Interfaces for Virtual Environment and Teleoperator Systems,* Vol. 55, No. 1 (pp. 295–300).

Nef, T; Guidali, M and Riener, R (2009). ARMin III–arm therapy exoskeleton with an ergonomic shoulder actuation. *Applied Bionics and Biomechanics* 6(2): 127–142.

Neuroscience and Robotics Laboratory, Northwestern University (2015). TPad tablet. http://www.nxr.northwestern.edu/tpad-tablet.

Ozkul, F; Barkana, D E; Demirbas, S B and Inal, S (2012). Evaluation of elbow joint proprioception with RehabRoby: A pilot study. *Acta Orthopaedica et Traumatologica Turcica* 46(5): 332–338.

Pasquero, J et al. (2007). Haptically enabled handheld information display with distributed tactile transducer. *IEEE Transactions on Multimedia* 9(4): 746–753.

Rantala, J et al. (2009). Methods for presenting braille characters on a mobile device with a touchscreen and tactile feedback. *IEEE Transactions on Haptics* 2(1): 28–39.

Roll, J P; Vedel, J P and Ribot, E (1989). Alteration of proprioceptive messages induced by tendon vibration in man: A microneurographic study. *Experimental Brain Research* 76(1): 213–222.

Santos-Carreras, L; Leuenberger, K; Samur, E; Gassert, R and Bleuler, H (2012). Tactile feedback improves performance in a palpation task: Results in a VR-based testbed. *Presence: Teleoperators and Virtual Environments* 21(4): 435–451.

Sherrick, C E and Rogers, R (1966). Apparent haptic movement. *Perception & Psychophysics* 1 (3): 175–180.

Sodhi, R; Poupyrev, I; Glisson, M and Israr, A (2013). AIREAL: Interactive tactile experiences in free air. *ACM Transactions on Graphics (TOG)* 32(4): 134.

Stone, R J (2001). Haptic feedback: A brief history from telepresence to virtual reality. In: *Haptic Human-Computer Interaction* (pp. 1–16). Berlin Heidelberg: Springer.

Tsalamlal, M Y; Ouarti, N and Ammi, M (2013). Psychophysical study of air jet based tactile stimulation. In: *World Haptics Conference (WHC), 2013* (pp. 639–644). IEEE.

Van Erp, J B; Van Veen, H A; Jansen, C and Dobbins, T (2005). Waypoint navigation with a vibrotactile waist belt. *ACM Transactions on Applied Perception (TAP)* 2(2): 106–117.

Vu, K P L and Proctor, R W (Eds.) (2011). *Handbook of Human Factors in Web Design.* Boca Raton, FL: CRC Press.

Winfield, L; Glassmire, J; Colgate, J E and Peshkin, M (2007). T-PaD: Tactile pattern display through variable friction reduction. In: *EuroHaptics Conference, 2007 and Symposium on Haptic Interfaces for Virtual Environment and Teleoperator Systems. World Haptics 2007. Second Joint* (pp. 421–426). IEEE.

Tactile Substitution for Vision

Yael Zilbershtain-Kra, Amos Arieli and Ehud Ahissar

Sensory Substitution (SenSub) is an approach that allows perceiving environmental information that is normally received via one sense (e.g., vision) via another sense (e.g., touch or audition). A typical SenSub system includes three major components: (a) a sensor that senses information typically received by the substituted modality (e.g., visual), (b) a coupling system that can process the sensor's output and drive the actuator, and (c) an actuator that activates receptors of the substituting modality (e.g., skin mechanoreceptors or auditory hair cells) (Bach-y-Rita 2002; Bach-y-Rita and Kercel 2003; Lenay et al. 2003; Renier and De Volder 2005; Ziat et al. 2005). Vision, the predominant sense in sighted humans, is typically the substituted modality in SenSub. The substituting modalities are touch and hearing (Renier and De Volder 2005). This chapter focuses on visual-to-touch SenSub (VTSenSub).

VTSenSub Feasibility

The ability to use a novel sensory modality, such as that offered by a SenSub, builds on the plasticity and flexibility of the perceptual system (Bach-y-Rita 2004). Yet, as the range of plastic changes is limited, not every SenSub system is feasible. There is an advantage to SenSub that bridges between modalities that operate on similar low-level principles and share high-level classifications. Vision and touch share several major perceptual principles. Both modalities are based on two-dimensional (2D) arrays of receptors that actively scan the environment (Bach-y-Rita 1972; Geldard 1960), encoding its spatial aspects via spatial and temporal cues (Collins and Saunders 1970). In comparison, acoustic signals activate a one-dimensional

Y. Zilbershtain-Kra (✉)
Department of Neurobiology, Weizmann Institute of Science, Rehovot, Israel

A. Arieli
Department of Neurobiology, Weizmann Institute of Science, Rehovot, Israel

E. Ahissar
Department of Neurobiology, Weizmann Institute of Science, Rehovot, Israel

T.J. Prescott et al. (eds.), *Scholarpedia of Touch*, Scholarpedia,
DOI 10.2991/978-94-6239-133-8_61

array of cochlear receptors in a relatively passive ear, coding spectral and temporal cues. Furthermore, vision and touch exhibit similar strategies of active sensing (Ahissar and Arieli 2001; Bach-y-Rita 2002). These similarities between vision and touch enable exploitation of existing natural mechanisms in tactile substitutions for vision, an advantage that cannot be applied to hearing. This, together with other disadvantages of hearing as a substituting sense for vision (Hanneton et al. 2010; Kim and Zatorre 2008; Tang and Beebe 1998; Visell 2009), make VTSenSub a natural choice. Indeed, there is evidence for perception in visual terms (e.g., shadow or luminance) while using VTSenSub systems (Bach-y-Rita 1969, 1972, 1987, 1995; Bach-y-Rita et al. 1998; White et al. 1970). Interestingly, attempts to use combined tactile-hearing SenSub did not succeed, possibly due to a cognitive overload (Chekhchoukh et al. 2011; Jansson and Pedersen 2005).

In spite of these similarities between vision and touch, the nature of the qualia perceived via tactile substitution is not yet clear: is it visual-like (Hurley and Noë 2003; Noë 2004;O'Regan and Noë 2001), tactile-like (Block 2003; Prinz 2006) or representing a completely new modality (Lenay et al. 2003). The latter is consistent with the limited resolution of VTSenSub (Bach-y-Rita 2002; Lenay et al. 2003). Yet, VTSenSub experiences had been reported to be rich enough to evoke emotional excitements, for example when being able to watch a moving candle flame (Guarniero 1974) or finding a toy for a blind child (Bach-y-Rita 2002).

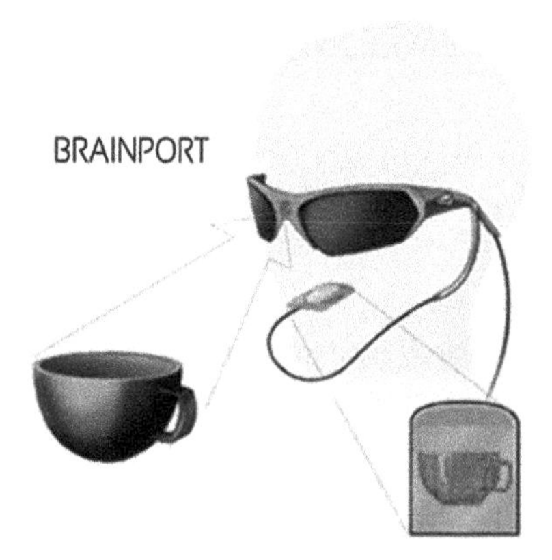

Fundamental Limitations of VTSenSub

Most existing VTSenSub devices convert visual space to tactile space and visual luminance to tactile vibrations (Bach-y-Rita et al. 1969; Krishna et al. 2010; Ziat et al. 2005). The idea of conveying information through vibrating array of pins on the skin is known from the 1920s and the first attempt to convey visual information via touch was done in the 60s by Starkiewicz and later by Bach-y-Rita who coined the term sensory substitution (Starkiewicz et al. 1971; Bach-y-Rita 1969). With such active conversion, conveyed to a passive sensory organ, the crucial factor determining perceptual resolution is the resolution of the actuators array. This poses a significant limitation on VTSenSub as actuators arrays usually contain only tens of actuators, with the densest device containing 1000 actuators (Visell 2009). The amount of information that can be conveyed via such arrays is several orders of magnitude lower than what can be conveyed by an intact retina, which severely limits the functioning of VTSenSub as visual substitute.

Another fundamental limitation of VTSenSub is its competition with other essential functions of the blind person. In order to maximize perceptual resolution a VTSenSub system should be attached to a sensitive skin surface, such as that of the fingertips or the tongue. This of course comes with a cost of losing co-functioning of these organs in other essential functions (Bach-y-Rita 1972). Finally, energy consumption of VTSenSub systems is typically high (Auvray et al. 2007; Lenay et al. 2003).

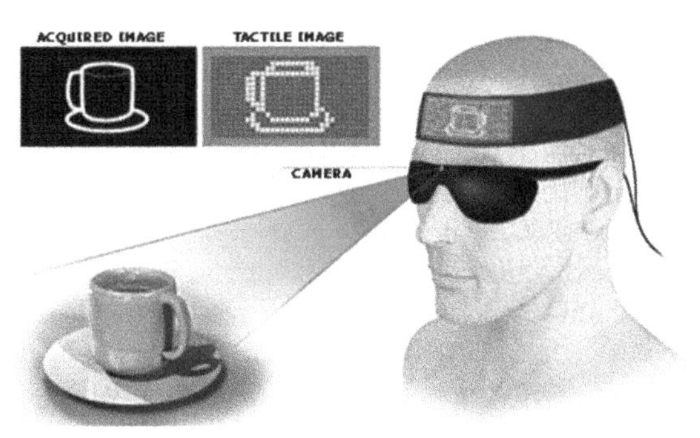

Active SenSub (ASenSub)

Visual and tactile sensory organs are attached to muscles whose activation enables information acquisition. Although muscle-driven active sensing can be bypassed by flashing stimuli on passive sensory organs, active sensing typically outperforms passive sensing (Gamzu and Ahissar 2001; Heller 1980; Heller and Myers 1983; LaMotte and Whitehouse 1986; Lederman and Klatzky 1993; Loomis and Lederman 1986; Saida and Wake 1982; Saig et al. 2012; Yanagida et al. 2004). With active sensing, motor-sensory relations, and not sensory signals per se, are the relevant cues for the perception of external objects (Held and Hein 1973; O'Regan and Noë 2001; Ahissar and Vaadia 1990, Ahissar and Arieli 2001, Ahissar and Assa 2014; Gibson 1962; Katz 1989; Saig 2012; Bagdasarian et al. 2013).

Active haptic exploration via a SenSub system enables the development of a unique scanning strategy for each participant (Tang and Beebe 1998), which is instrumental for accurate perception (Jansson 1998; Rovira et al. 2010). Sensor movement can facilitate perception even when dissociated from the sensory organ. For example, with head-attached video camera and fingers-attached actuators, recognition and learning improve when participants are allowed to move their heads

(Siegle and Warren 2010). This improvement is often associated with "externalization" of the sensed object, i.e., feeling it in a remote location rather than on the skin, a phenomenon crucially dependent on active control over sensor motion (Harman 1990; Lenay et al. 1999; Siegle and Warren 2010).

Bach-y-Rita and others demonstrated that in order to achieve externalization, the user must be trained, the sensor must be placed on one of the user's motor systems and a motor-sensory control should be obtained (Bach-y-Rita 2002, 2005, Loomis et al. 1992; White et al. 1970; Epstein et al. 1986). Bach-y-Rita also suggested that the sensor movement can be replaced with a virtual movement, which can also result in externalization and assumed that there is no importance to the position of the sensor and actuator (Bach-y-Rita and Kercel 2003). Consistent with this, no significant difference was found between having the actuator on the same hand that moved the sensor or at the other hand, although some participants claimed for disruption at the split-hand condition (Pietrzak et al. 2009). This assumption, however, appears to be contrasted by experiments in humans and animals demonstrating the dependency of active sensing on natural sensory-motor loops (Visell 2009; Saig et al. 2010; Ahissar and Knutsen 2008). Indeed, SenSub devices in which the sensor and actuator are attached to the same organ, show improved performance (Chan et al. 2007; Zilbershtain-Kra et al. 2014).

Having the sensor and actuator on the same organ is not sufficient for driving active sensing. As naturally a perceptual system acquires its sensations by movements, the SenSub device must not induce any active actuation by itself—actuation should result only from sensor motion. Indeed, devices in which no tactile vibrations are used, and actuation depends solely on sensor motion, show improved performance (Chan et al. 2007; Zilbershtain-Kra et al. 2014). Such devices, in which the sensor and actuator are attached to the same organ and sensations are generated only via sensor motion, are termed Active (ASenSub) devices.

Evaluating the Perceptual Power of VTSenSub Devices

Evaluation of the relative perceptual power of SenSub systems can be done by comparing behavioral performance in different tasks (Visell 2009). Tasks that has been used for this purpose include: distance measurement (Bach-y-Rita et al. 1969), line orientation detection (Bach-y-Rita et al. 1969; Chekhchoukh et al. 2011; Ziat et al. 2005), shape identification (Bach-y-Rita et al. 1969; Chan et al. 2007; Shinohara et al. 1998; Tang and Beebe 1998; Ziat et al. 2005; Zilbershtain-Kra et al. 2014), object identification (Bach-y-Rita et al. 1969; Shinohara et al. 1998; Zilbershtain-Kra et al. 2014), face recognition (Bach-y-Rita et al. 1969), letter reading (Bliss et al. 1970; Chan et al. 2007; Linvill and Bliss 1966; Loomis 1974; Ziat et al. 2005), movement detection (Chekhchoukh et al. 2011), body movement recognition (Miletic 1988; Jansson 1983; Mandik 1999), hand—"eye" coordination (Guarniero 1974; Miletic 1988), navigation (Segond et al. 2005) and assembling tasks (Bach-y-Rita 1995).

Comparisons across tasks and research groups still lack a common acceptable metrics. Importantly, in most cases performance depends crucially on learning and learning time and effort should be taken into account when comparing different approaches.

Exploration Strategy with VTSenSub

Differences in performance levels between subjects in the same task can often be accounted for by differences in motion strategies (Gamzu and Ahissar 2001; Rovira et al. 2010). To date, information about these dependencies is sparse. Existing data are based on both subjective reports and objective measurements of sensor motion. Strategies reported so far include: contour following, orthogonal (horizontal and vertical) scanning, random scanning and feature oriented scanning (Chan et al. 2007; Guarniero 1974; Hsu et al. 2013; Rovira et al. 2010; Ziat et al. 2005; Zilbershtain-Kra et al. 2014).

Challenges for VTSenSub

Although the first study with VTSenSub system was done five decades ago, those systems are still not used by the blind community. It seems that the two major technical challenges of these systems remain the limited resolution of "foveal" sensation and the lack of "peripheral" sensation. Unlike in vision, with currently available VTSenSub devices subjects not only receive impoverished "foveal" information, but are also in the dark in respect to the rest of the visual field. While foveal acuity can be facilitated in ASenSub devices, using appropriate motor-sensory strategies, peripheral sensation appears to require specific design of the actuators array and of its coupling with the sensor.

Additional challenges include financial and ergonomics considerations. The system should better be comfortable and easy to use, enabling free movement, as well as esthetical, while keeping the costs affordable to the end customer (Lenay et al. 2003).

Research Interests

The basic practical aim of SenSub is to provide a tool that can help the visually impaired in everyday tasks such as navigation, object localization and identification, and communication. In addition, SenSub serves as a research tool for neuroscientists addressing perceptual mechanisms, learning and plasticity (Bach-y-Rita 1995; Hanneton et al. 2010; Lenay et al. 2003; Sampaio et al. 2001; Visell 2009).

Importantly, SenSub gives a unique possibility to study the emergence of perception via a novel modality, addressing the development of specific perceptual aspects such as distal awareness (externalization), environmental structure and object familiarity (Loomis et al. 2012; Siegle and Warren 2010).

References

Ahissar, E and Arieli, A (2001). Figuring space by time. *Neuron* 32: 185–201.

Ahissar, E and Assa, E (2014). Theories of mammalian perception: Open and closed loop modes of brain-world interactions. COSYNE Workshop, Snowbird, Utah.

Ahissar, E and Knutsen, P M (2008). Object localization with whiskers. *Biological Cybernetics* 98: 449–458.

Ahissar E and Vaadia, E (1990). Oscillatory activity of single units in a somatosensory cortex of an awake monkey and their possible role in texture analysis. *Proceedings of the National Academy of Sciences of the United States of America* 87: 8935–8939.

Auvray, M; Hanneton, S and O'Regan, J K (2007). Learning to perceive with a visuo-auditory substitution system: Localisation and object recognition with 'The vOICe'. *Perception* 36: 416–430.

Bach-y-Rita, P (1972). *Brain Mechanisms in Sensory Substitution.* New York: Academic Press.

Bach-y-Rita, P (1987). Brain plasticity as a basis of sensory substitution. *Neurorehabilitation and Neural Repair* 1: 67–71.

Bach-y-Rita, P (1995). *Nonsynaptic Diffusion Neurotransmission and Late Brain Reorganization* New York: Demos.

Bach-y-Rita, P (2002). Sensory substitution and qualia. In: A Noë and E Thompson, *Vision and Mind: Selected Readings in the Philosophy of Perception* (pp. 497–514).

Bach-y-Rita, P (2004). Tactile sensory substitution studies. *Annals of the New York Academy of Sciences* 1013: 83–91.

Bach-y-Rita, P (2005). Emerging concepts of brain function. *Journal of Integrative Neuroscience* 4: 183–205.

Bach-y-Rita, P; Collins, C C; Saunders, F A; White, B and Scadden, L (1969). Vision substitution by tactile image projection. *Nature* 221: 963–964.

Bach-y-Rita, P and Kercel, S (2003). Sensory substitution and the human–machine interface. *Trends in Cognitive Sciences* 7: 541–546.

Bach-y-Rita, P; Kaczmarek, K A; Tyler, M E and Garcia-Lara, J (1998). Form perception with a 49-point electrotactile stimulus array on the tongue: A technical note. *Journal of Rehabilitation Research and Development* 35: 427–430.

Bagdasarian, K et al. (2013). Pre-neuronal morphological processing of object location by individual whiskers. *Nature Neuroscience* 16: 622–631.

Bliss, J C; Katcher, M H; Rogers, C H and Shepard, R P (1970). Optical-to-tactile image conversion for the blind. *IEEE Transactions on Man-Machine Systems* 11: 58–65.

Block, N (2003). Tactile sensation via spatial perception. *Trends in Cognitive Sciences* 7: 285.

Chan, J S et al. (2007). The virtual haptic display: A device for exploring 2-D virtual shapes in the tactile modality. *Behavior Research Methods* 39: 802–810.

Chekhchoukh, A; Vuillerme, N and Glade, N (2011). Vision substitution and moving objects tracking in 2 and 3 dimensions via vectorial electro-stimulation of the tongue. In: *Actes de ASSISTH 2011, 2ème Conférence internationale sur l'Accessibilitè et les Systèmes de Supplèance aux personnes en situaTions de Handicaps.*

Collins, C and Saunders, F (1970). Pictorial display by direct electrical stimulation of the skin. *Journal of Biomedical Syst* 1: 3–16.

Epstein, W; Hughes, B; Schneider, S and Bach-y-Rita, P (1986). Is there anything out there? A study of distal attribution in response to vibrotactile stimulation. *Perception* 15: 275–284.
Gamzu, E and Ahissar, E (2001). Importance of temporal cues for tactile spatial-frequency discrimination. *The Journal of Neuroscience* 21: 7416–7427.
Geldard, F A (1960). Some neglected possibilities of communication. *Science* 131(3413): 1583–1588.
Gibson, J J (1962). Observations on active touch. *Psychological Review* 69: 477–491.
Guarniero, G (1974). Experience of tactile vision. *Perception* 3: 101–104.
Hanneton, S; Auvray, M and Durette, B (2010). The Vibe: A versatile vision-to-audition sensory substitution device. *Applied Bionics and Biomechanics* 7: 269–276.
Harman, G (1990). The intrinsic quality of experience. *Philosophical Perspectives* 4: 31–52.
Held, R and Hein, A (1973). Movement-produced stimulation in the development of visually guided behavior. *Perception: An Adaptive Process* 56: 182.
Heller, M A (1980). Reproduction of tactually perceived forms. *Perceptual and Motor Skills* 50: 943–946.
Heller, M A and Myers, D S (1983). Active and passive tactual recognition of form. *The Journal of General Psychology* 108: 225–229.
Hsu, B et al. (2013). A tactile vision substitution system for the study of active sensing. In: *Engineering in Medicine and Biology Society (EMBC), 2013 35th Annual International Conference of the IEEE (IEEE)* (pp. 3206–3209).
Hurley, S and Noë, A (2003). Neural plasticity and consciousness. *Biology and Philosophy* 18: 131–168.
Jansson, G (1983). Tactile guidance of movement. *International Journal of Neuroscience* 19: 37–46.
Jansson, G (1998). Haptic perception of outline 2D shape: The contributions of information via the skin, the joints and the muscles. In: B Bril, A Ledebt, G Dietrich and A Roby-Brami (Eds.), *Advances in Perception-Action Coupling* (pp. 25–30). Paris: Editions EDK.
Jansson, G and Pedersen, P (2005). Obtaining geographical information from a virtual map with a haptic mouse. In: *XXII International Cartographic Conference (ICC2005).*
Katz, D (1989). *The World of Touch.* Hillsdale, NJ: Erlbaum.
Kim, J-K and Zatorre, R J (2008). Generalized learning of visual-to-auditory substitution in sighted individuals. *Brain Research* 1242: 263–275.
Krishna, S; Bala, S; McDaniel, T; McGuire, S and Panchanathan, S (2010). VibroGlove: An assistive technology aid for conveying facial expressions. *In CHI'10 Extended Abstracts on Human Factors in Computing Systems (ACM)* (pp. 3637–3642).
LaMotte, R H and Whitehouse, J (1986). Tactile detection of a dot on a smooth surface: Peripheral neural events. *Journal of Neurophysiology* 56: 1109–1128.
Lederman, S J and Klatzky, R L (1993). Extracting object properties through haptic exploration. *Acta Psychologica* 84: 29–40.
Lenay, C; Gapenne, O; Hanneton, S; Marque, C and Genouëlle, C (2003). Sensory substitution: Limits and perspectives. In: Y Hatwell, A Streri and E Gentaz, *Touching for Knowing: Cognitive Psychology of Haptic Manual Perception* (pp. 275–292). Amsterdam: John Benjamins Publishing.
Lenay, C; Gapenne, O; Hanneton, S and Stewart, J (1999). Perception et couplage sensori-moteur: Expériences et discussion épistémologique. *Intelligence Artificielle Située* 99: 71–86.
Linvill, J G and Bliss, J C (1966). A direct translation reading aid for the blind. *Proceedings of the IEEE* 54: 40–51.
Loomis, J M (1974). Tactile letter recognition under different modes of stimulus presentation. *Perception & Psychophysics* 16: 401–408.
Loomis, J M; Da Silva, J A; Fujita, N and Fukusima, S S (1992). Visual space perception and visually directed action. *Journal of Experimental Psychology: Human Perception and Performance* 18: 906.

Loomis, J M; Klatzky, R L and Giudice, N A (2012). Sensory substitution of vision: Importance of perceptual and cognitive processing. In: R Manduchi and S Kurniawan (Eds.), *Assistive Technology for Blindness and Low Vision* (pp. 161–193). Boca Raton: CRC Press.

Loomis, J M and Lederman, S J (1986). Tactual perception. In: K R Boff, L Kaufman and J P Thomas (Eds.), *Handbook of Perception and Human Performances, Vol. 2* (pp. 31/1–31/41). New York: Wiley.

Mandik, P (1999). Qualia, space, and control. *Philosophical Psychology* 12: 47–60.

Noë, A (2004). *Action in perception* Cambridge, MA: The MIT Press.

O'Regan, J K and Noë, A (2001). A sensorimotor account of vision and visual consciousness. *Behavioral and Brain Sciences* 24: 939–972; discussion 973–1031.

Pietrzak, T; Crossan, A; Brewster, S A; Martin, B and Pecci, I (2009). Exploring geometric shapes with touch. In: *Human-Computer Interaction–INTERACT 2009* (pp. 145–148). Springer.

Prinz, J (2006). Putting the brakes on enactive perception. *Psyche* 12: 1–19.

Renier, L and De Volder, A G (2005). Cognitive and brain mechanisms in sensory substitution of vision: A contribution to the study of human perception. *Journal of Integrative Neuroscience* 4: 489–503.

Rovira, K; Gapenne, O and Ammar, A A (2010). Learning to recognize shapes with a sensory substitution system: A longitudinal study with 4 non-sighted adolescents. In: *Development and Learning (ICDL), 2010 IEEE 9th International Conference on (IEEE)* (pp. 1–6).

Saida, S and Wake, T (1982). Computer-controlled TVSS and some characteristics of vibrotactile letter recognition. *Perceptual and Motor Skills* 55: 651–653.

Saig, A; Gordon, G; Assa, E; Arieli, A and Ahissar, E (2012). Motor-sensory confluence in tactile perception. *The Journal of Neuroscience* 32: 14022–14032.

Saig, A; Shore, D I; Arieli, A and Ahissar, E (2010). Action-perception coupling: Importance of actuator-sensor relative placing. In: *FENS*, vol. 5 (p. 084.012).

Sampaio, E; Maris, S and Bach-y-Rita, P (2001). Brain plasticity: "Visual" acuity of blind persons via the tongue. *Brain Research* 908: 204–207.

Segond, H; Weiss, D and Sampaio, E (2005). Human spatial navigation via a visuo-tactile sensory substitution system. *Perception* 34: 1231–1249.

Shinohara, M; Shimizu, Y and Mochizuki, A (1998). Three-dimensional tactile display for the blind. *IEEE Transactions on Rehabilitation Engineering* 6: 249–256.

Siegle, J H and Warren, W H (2010). Distal attribution and distance perception in sensory substitution. *Perception* 39: 208.

Starkiewicz, W; Kuprianowicz, W and Petruczenko, F (1971). 60-channel elektroftalm with CdSO4 photoresistors and forehead tactile elements: Visual prosthesis: The interdisciplinary dialogue. New York: Academic Press.

Tang, H and Beebe, D J (1998). A microfabricated electrostatic haptic display for persons with visual impairments. *IEEE Transactions on Rehabilitation Engineering* 6: 241–248.

Visell, Y (2009). Tactile sensory substitution: Models for enaction in HCI. *Interacting with Computers* 21: 38–53.

White, B W; Saunders, F A; Scadden, L; Bach-Y-Rita, P and Collins, C C (1970). Seeing with the skin. *Perception & Psychophysics* 7: 23–27.

Yanagida, Y; Kakita, M; Lindeman, R W; Kume, Y and Tetsutani, N (2004). Vibrotactile letter reading using a low-resolution tactor array. In: *Proceedings of the 12th International Symposium on Haptic Interfaces for Virtual Environment and Teleoperator Systems, HAPTICS '04 (IEEE)* (pp. 400–406).

Ziat, M; Gapenne, O; Stewart, J and Lenay, C (2005). A comparison of two methods of scaling on form perception via a haptic interface. *ACM ICMI* 236–243.

Zilbershtain-Kra, Y; Ahissar, E and Arieli, A (2014). Speeded performance with active- sensing based vision- to- touch substitution. FENS Abstract D036.

Virtual Touch

Adrian David Cheok and Gilang Andi Pradana

Introduction

As we move further into the digital age we are growing physically further apart from family and friends. The rapid development of society brings about a vicious cycle that can result in feelings of isolation, loneliness and a lack of sense of value (Slater 1990). Research in Social Presence Theory states that less rich computer mediated communication environments inhibit communicating emotional expression, while in much richer environments in which non-verbal cues like touch are available, a full range of emotional information can be communicated due to greater social presence (Short et al. 1976). While the proliferation of computers and the Internet enables us to exchange information and perform certain tasks in a quicker and more efficient manner, we are isolating ourselves from the real world where actual physical touch is very important as a communication means.

The Importance of Touch

Human touch has a long history in healing and medical therapy. In ancient Greece, Hippocrates (ca 460–370 BC), the father of Western medicine, hailed "rubbing" as an important physician's skill (Kline 2004). Modern empirical work has also shown that human touch in massage reduces stress hormones and increases the circulation of chemicals that counteract physiological arousal (Morhenn et al. 2011). Simple touches, such as holding hands or tapping another's forearm, can dispel the threat and promote calm. (Feldman et al. 2010).

A.D. Cheok (✉) · G.A. Pradana
Imagineering Institute, City University, London, UK

T.J. Prescott et al. (eds.), *Scholarpedia of Touch*, Scholarpedia,
DOI 10.2991/978-94-6239-133-8_62

Touch is able to signal deeper meanings than words alone. It enables us to communicate on a social platform at deeper affectual level compared to mere words, better signaling affiliation and support. For example, while the exchange of words may vary in the greeting and farewell rituals of family members, friends and even political representatives, these rituals consistently involve tactile exchanges such as hugging, kissing or shaking hands (Heslin and Boss 1980). Likewise, interpersonal touch as seen in team sport accompanies or replaces verbal communication during exciting moments in the game. Touch is also important in smaller groups such as dyads, when one individual shares positive or negative news or seeks support and confirmation (Henley 1977).

This problem is more pronounced for parents with young children. Children of these young ages need high care, guidance and love (Falicov 1995). Parents are generally able to reach their children by telephone or video phone, but communication purely by voice or video lacks the physical interaction which has been shown in previous research to be vital in effective communication (Bakeman and Brown 1980). Younger children might have difficulties understanding the true meaning of words spoken by their parents. As a consequence, we require a more effective way of remote communication between parents and young children. While it may not always be possible for parents to decline work commitments (such as long office hours and business trips), remote haptic interaction may be a feasible alternative when the parent must be away from the home. Love, closeness, and intimacy are important for people's psychological well-being (Thoits 1985). If we look at another case, in couples for example, they often live apart nowadays. Accordingly, there has been a growing and flourishing interest in designing technologies that mediate a feeling of relatedness when being separated, beyond the explicit verbal communication and simple emoticons available technologies offer (Hassenzahl et al. 2012).

In this chapter, we present two novel virtual touch systems for family, friends, and lovers. The first one is Huggy Pajama, a mobile and wearable human-computer-human interaction system that allows users to send and receive touch and hug interactions. Huggy Pajama consists of two physical entities. On one end, a novel hugging interface takes the form of a small, mobile doll with embedded touch and pressure sensing circuits. It is connected via the Internet to a haptic wearable pajama with embedded air pockets, heating elements and color changing fabric. Advances in mobile wearable technology have made us believe that there is a need to implement the idea into a compact wearable device. The second system is RingU, a ring-shaped wearable system aimed at promoting emotional communications in remote communication between people using the vibro-tactile and color lighting expressions. A ring is one of the fashion accessories between couples to represent their relationships. Using this metaphor, we believe that a ring is a perfect symbol of something emotionally close and connected, which fits really well with RingU aim to create a communication system that makes users feel even more connected and emotionally close. The RingU system consists of a wearable ring-shaped device and a smart phone. When a user squeezes the ring, a signal will be sent via bluetooth low energy to his/her smartphone, and then through the internet to his/her partner's system, and it allows a virtual mini-hug and color to be sent to a paired partner's ring. For that very instant, they will feel each other's warm presence. The result of our experiment has shown that these augmented cues can help to stimulate a better assessment to emotional states in a computer mediated communication environment.

Touch in Human-Computer Interaction

Touch is really important in social interaction and essential in forming bonds and building trust (Henley 1977), and these characteristics are also expected in virtual touch. Providing touch based interaction in Human-Computer Interaction (HCI) will increase the effectiveness in communicating emotions, building trust, and achieving behavioural changes. Humans can communicate distinct emotions through touch, and this is also possible through mediated virtual touch (Van Erp and Toet 2013). Touch based interaction in HCI has grown rapidly over the last few years. A range of new applications has become possible now that touch can be used as an interaction technique. One example of a system that uses tactile information is 'inTouch' (Brave and Dahley 1997) first described in 1997. It introduced the method of applying haptic feedback to interpersonal communication providing a physical haptic link between users separated by distance. Another related field called Affective haptics focuses on the study and design of systems that can enhance the emotional state of a human by means of the sense of touch. Rehman and Liu from Digital Media Lab in Sweden have proposed iFeeling interface, where they implemented vibro-tactile rendering of human emotions on mobile phones, and has shown a potential to enrich mobile phones communication among the users through the touch channel (Liu 2010). Another paper discovered about our readiness to empathize with and support that person by being touched by another person influences, and suggests that touch seems to be a special sensory signal that influences recipients in the absence of conscious reflection and that promotes pro social behavior (Schirmer et al. 2011).

In a more closely related project, there is some work such as the CuteCircuit's Hug Shirt (http://www.cutecircuit.com/now/projects/wearables/fr-hugs), which has detachable pads containing sensors which senses touch pressure, heart beat and warmth, and actuators which reproduces them. This system also utilises vibration actuators to generate the hug. Poultry Internet, (Lee et al. 2006) in 2005 presented a remote human pet interaction system using a jacket specially designed with vibrotactile actuators embedded for pets. These projects indicate the attempts in the past to achieve remote haptics in close relation to hugging. Another system called HugMe (Cha et al. 2008) enables touch interaction with Haptic Jacket that is synchronized with audio/visual information. The 'Hug Shirt', Poultry Internet, and HugMe use vibration to provide a sense of a remote hug which does not correspond to the feeling of natural human touch. However, through the Huggy Pajama we attempt to recreate a main property of a hug, the pressure. 'Hug over a distance'

(Mueller et al. 2005) uses a koala teddy to sense a hug and send it wirelessly to the air inflatable jacket to recreate a hugging feeling. The Koala teddy has a PDA in it of which the screen is touched by the user to send a hug. The PDA on the inflatable jacket upon receiving the hug activates serial controller to simulate the hug. Shifting attention to sensing technologies, we realize the importance of accurately measuring the haptic properties such as the force of a hug or touch before transmission. Even though the 'Hug over a distance' work closely relates to our concept, this system simply does not sense the force of a hug thus it simply transmits a command on an execution basis in a binary mode of "on" or "off". However, 'Huggy Pajama' is more concerned with the importance of regenerating the hugging feeling with accurate pressure levels corresponding to the input force exerted by the sender. Huggy Pajama uses the QTC sensor (http://www.peratech.com/qtc-touch-processing-unit.html) for accurate pressure measurement. The easy usage and manipulation of the sensor eased through the design and integration phases of the input pressure-sensing module, but the key motivation factor is the accurate pressure measurement suitable for accurate tactile sensing.

Regarding haptics in clothing, we set out to identify and compare key concepts and technologies relevant to the design space of the Huggy Pajama. Touch Sensitive by MIT Media Lab (Vaucelle and Abbas 2007) lists down four different methods they explored for haptic apparel for massage on the move. In one prototype, thermally responsive metallic wires embedded in the apparel caused it to shrink mechanically when a current is passed through. In other prototypes, silicon buttons, vinyl inflatable air pockets and vinyl pockets filled with liquids that diffuse around a wooden ball during a massage were used. In (Haans et al. 2007) the researchers used a neoprene vest with two arm straps to produce mediated touch. They also used vibrotactile actuators to enable haptic communication and have conducted a study to evaluate the effect of mediated touch. In another project (Lindeman et al. 2004) the authors again use vibrotactile units to develop a haptic feedback vest to deliver haptic cues for immersive virtual environments through garments worn on the body. However many of these systems use vibrotactile actuation to enable haptics. In this project, Huggy Pajama, we stress the accurate reproduction of the pressure in a remote hugging system. As mentioned above, even though some of the systems focused on remote haptic communication, most of them focus on just the context of remote touch whereas in our project we try to recreate in high fidelity, each hug giving attention to the pressure, which is an essential property of the touch/hug. Even though most of the aforementioned works relate to remote touch, a high number of them use the vibrotactile actuation as a solution to the output haptics generation. Many of them justify this by claiming that it makes the wearable system lighter and more long lasting in terms of the battery life. However with Huggy Pajama, we employ novel techniques and equipment and use an air actuating system embedded in a jacket to exert exact amounts of pressure on the wearer simulating a realistic hug. We believe that, even though right now it may not be comparable to commercial systems in terms of the usability, this research will open up avenues for more precise communication using haptics in the future, thus enabling more effective communication of feelings.

Proposed Interface

Huggy Pajama

Huggy Pajama is a novel wearable system aimed at promoting physical interaction in remote communication between parent and child. This system enables parents and children to hug one another through a hugging interface device and a wearable, hug reproducing pajama connected through the Internet. The hug input device is a small, mobile doll with an embedded pressure sensing circuit that is able to accurately sense varying levels of pressure along the range of human touch produced from natural touch. This device sends hug signals to a haptic jacket that simulates the feeling of being hugged to the wearer. It features air pocket actuators that reproduce hug sensations, heating elements to produce warmth that accompanies hugs, and a color changing pattern and accessory to indicate distance of separation and communicate expressions.

A general overview of the system is shown in Figure 1. On the left of the figure, an input device acts as a cute interface that allows parents to hug their child and send mood related cheerful expressions to them. On the right side of the figure, connected through the Internet, an air actuating module and color-changing clothing reproduces the hug sensation and connects the parent and child.

This Pajama is able to simulate hugs to the wearer in the form of pressure that is accurately reproduced according to the inputs from the hugging interface accompanied by the generation of warmth, color changes of the fabric according to distance of separation between parent and child, as well as displaying emoticons. Our system provides a semantically meaningful interface that can be easily understood by children and parents as a reproduction of hugging. Furthermore, the hug sensation is produced in a calm and relaxing way through gentle air actuators rather than through vibration or other mechanical means. We aim to have an

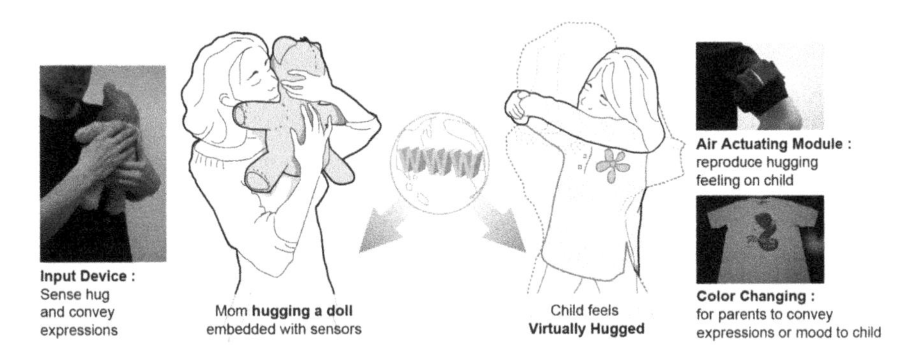

Figure 1 Overview of Huggy Pajama

"impedance matching" between the input (a soft, cute, calm touch sensing interface) and output (ambient, calm, cute, hugging output).

Although never intended to replace real physical hugging, we believe this system would be of great benefit for times when the parent and child cannot be at the same physical place. Related research provides scientific evidence which showed that infant monkeys grew up to be more healthy when artificial contact comfort was given even in the total absence of their real mothers (although it would be unethical to carry out the same tests to deprive human infants artificially) (Harlow 1958).

The interaction between parent and child can be bi-directional. The parent and child each can wear the Pajama. Each interacts with the other through a mobile hugging interface. The bi-directionality is left as an option for the users, because at this stage, such a wearable device is not suitable to be worn at work for parents. The interactive modes are summarized in Table below.

Interactive modes	Description
Remote touch and hug	Transmit human touch and hug on doll to wearer of haptic pajama
Haptic pajama	Reproduce hug sensation and warmth on wearer
Distance and emotion indication	Color changing clothes and accessories to give indication of separation distance between parent and child, and emotion data

Huggy Pajama focuses on developing a mobile hug communication system for parent and child, and provides a realistic soft touch sensation. We enable users to hug or touch different areas on the hug sensing interface, and then map this to actuate different parts of the haptic Pajama. Besides that, the hug sensing doll senses varying levels of force acting on it in an accurate and analog manner. The output air actuating pockets applies different levels of pressure to the human body according to the input force. Also, we experimented with color changing cloths to give an indication of distance of separation, and display emotion data of the parent and child.

In addition, our pilot study using psycho-physiological methods shows that there are no significant differences in certain aspects in comparing effects of mediated touch using our system versus real physical touch from a friend to a human participant. This is encouraging as it shows that we are able to elicit the same response from the human in remote mediated touch as compared to real physical touch.

The main functionality of the touch device is sensing the location and strength of the touch as input, and then encapsulating and transmitting this data over the internet and reproducing the force at the receiving end. Thus, it has several electronic hardware modules for each feature.

Input touch sensing module:	is used to sense the touch levels and the area of touch of the intended recipient. The pressure variation is sensed by this module, digitized, and this information is transmitted via the Internet
Output touch actuation module:	is used to reproduce the touch levels and positions related to the received digitized data from the input touch system. This module consists of a pneumatic system controlled electronically with air pouch actuators
Fabric display module:	consists of a fabric coated with thermochromic ink and a temperature control system made from Peltier (p-n junction) cooling technology. The thermo-chromic ink color is changed according to the different levels of temperature applied

The overall system is a wearable remote hugging jacket which includes all three modules mentioned above. A block diagram of the mediated touch system is shown in Figure 2. Touch sensing, touch reproduction and a color changing module with thermal control modules are connected via the Internet to reproduce real-time touch sensation and affective communication.

This system presents the flexibility for either one-way or two-way communication between the sender and the receiver. For example the sender (parent) sends a hug to their child (receiver) and if parents are at work in a business meeting, it might not be suitable for them to put on the Pajama. They can easily hug their child by using only the input device. However in the case of being in office, hotel, or airport, the parent could wear the pajama and have two-way hugging with the child.

In Huggy Pajama system, there are 12 unique input sensors which corresponds to 12 unique output modules. These 12 output modules are integrated into a pajama for the children. Currently, as the result is a prototype with the need to fit all modules, we integrate them into a more sturdy soft jacket-like construction. Figure 3 shows

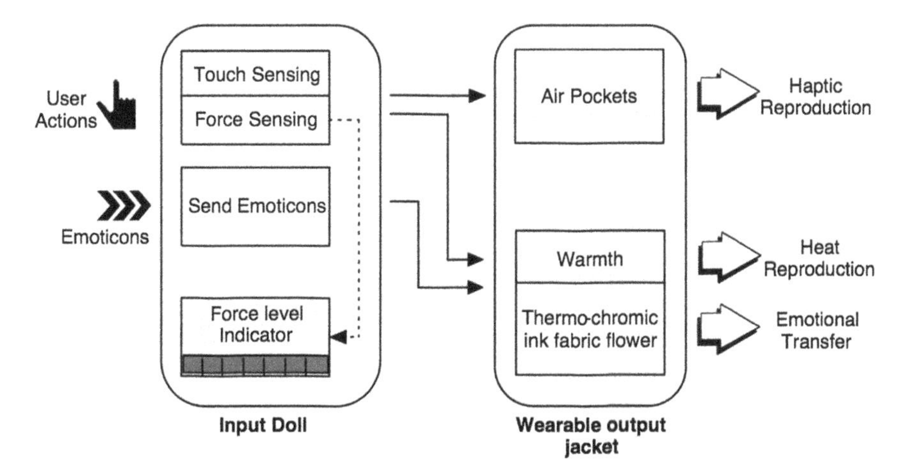

Figure 2 Overall block diagram showing different modules of system

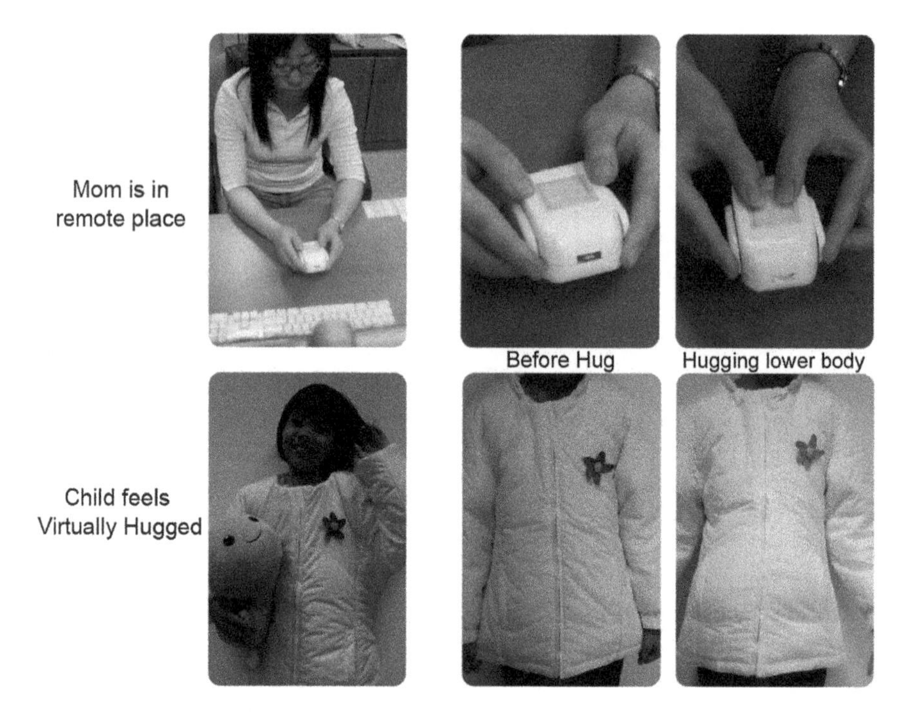

Figure 3 Huggy Pajama system overview

the Huggy Pajama actual prototype in action. QTC (http://www.peratech.com/qtc-touch-processing-unit.html) based material was used to construct the input sensing module. We chose the QTC Sheet form factor as it allows flexibility in terms of size and shape. This allowed considerable freedom in designing the sensing component of the input interface. Figure 4 shows the touch and hug sensing and actuating

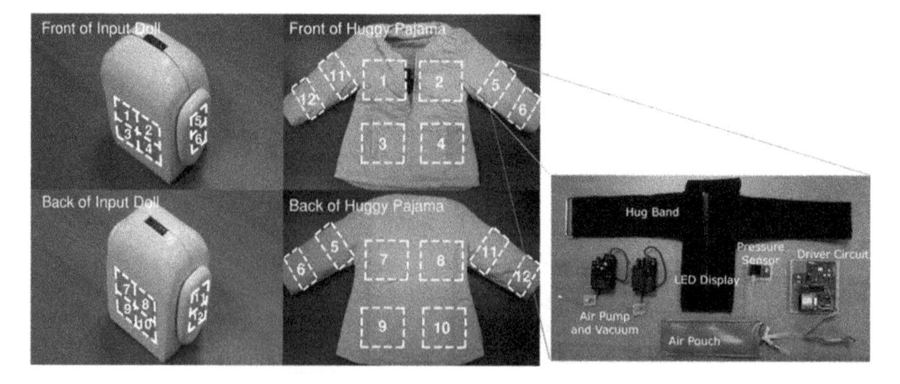

Figure 4 Mapping of input sensors to output actuators, with a zoomed in view of single module of air actuator system

device, with a zoomed in view of the individual air pouch with its controlling air actuation module. The aesthetic design of the input device is shaped like a small doll with a body and arms along the sides to correspond to the human body where we want to reproduce the haptic sensation. The appearance of the input device is important to consider as this directly affects the user experience in affective communication and the ability to intuitively understand the system image. the system mappings and affordance. There are 12 QTC sensing areas, which cover the front and back of the doll body, as well as both the side arms. The output touch actuation system consists of air bags which inflate and deflate according to the remote input while maintaining the pressure in a constant level. Overall remote hugging jacket consists of 12 individual air pouches which corresponds to each of the 12 sensors on the input doll.

RingU

Advances in mobile technology have enabled us to effectively interact with mobile devices to do our task, mainly communication, regardless of the place. Recent technology has also made it possible to package a communication device into a range of wearable devices. We present RingU, a ring-shaped wearable system aimed at promoting emotional communications in remote communication between people using the vibro-tactile and color lighting expressions. Traditionally, a ring has been used as a symbolic present to deliver a message from the sender (the one who gives present), to the receiver. A ring is one of the fashion accessories between couples to represent their relationships. A ring is an unbroken circle, which many cultures understand as the representative of eternity, which symbolizes the eternized promise between them on their engagement and wedding. The ring can acts as a reminder and an outward symbol to others that a person is currently on an eternal commitment.

Not only for couples, ring also has a meaning as a source of unity. People wear rings to join others and symbolise that they are in the same cause. We can see some examples of the use of a purity ring, or when a group of supporters wear rings after the victory of their team. Using this metaphor, we believe that a ring is a perfect symbol of something emotionally close and connected, which fits really well with RingU aim to create a communication system that makes users feel even more connected and emotionally close.

The RingU system consists of a wearable ring-shaped device and a smart phone. When a user squeezes the ring, a signal will be sent via bluetooth low energy to his/her smartphone, and then through the internet to his/her partner's system, and it allows a virtual mini-hug and color to be sent to a paired partner's ring. For that very instant, they will feel each other's warm presence, by interpreting partner's emotion from the vibrotactile and color lighting feedback. Figure 5 shows the general concept of this type of interaction in two paired rings.

Figure 5 Squeezing the ring to send a lighting and a vibro-tactile signal to the paired partner

Ambient multi-color glow has adopted in some research to create a certain mood and emotional feeling (Chang et al. 2001). The same approach was implemented into the wearable ring, along with the vibro-tactile, which acts as a non-verbal cue in the communication channel. This system can also be integrated with social network system like Facebook, Twitter, and Google+, allowing not only one-to-one communication, but also one-to-many communication by sending a group hug to many friends in the social network at once. The scheme of RingU system is shown in Figure 6.

The user can send 14 different combinations of lighting and 3 levels of vibration intensity through the setting of RGB color control (e.g. red, green, blue, red + green, red + blue, green + blue, red + green + blue) and vibration control (controlling the pattern with Pulse Width Modulation) using the smart phone app in current RingU system. Once the user sets the value of lighting and vibration

Figure 6 RingU system scheme

Figure 7 Squeezing the ring to send a lighting and a vibration signal to the paired partner through the internet using smart phone app

expression, then he/she can send it to the partner through pressing the button from the smartphone app or squeezing the ring. By implementing force sensor on the surface of the ring, where user can do a squeezing interaction, we can measure the pressure applied by the user. The pressure exerted determines the intensity of the color and the vibration. Figure 7 shows how users can interact with the ring and control it from the smartphone app.

Several prototypes have been developed as an implementation of RingU. The prototype shown as prototype 1 in Figure 8 is the first prototype of RingU system designed for the proof of concept and it consists of a processor, a push button, an RGB LED, a vibration motor, a XBee module, and a battery. In this prototype, the push button, the LED, and the vibration motor were built into the ring and a separated box that is connected with a wire with the ring has been designed to contain comparatively bigger modules such as the processor, the Xbee, and the battery. We continued to develop RingU by implementing force sensor into the surface of the ring as an input, and making the form factor smaller, wireless, and

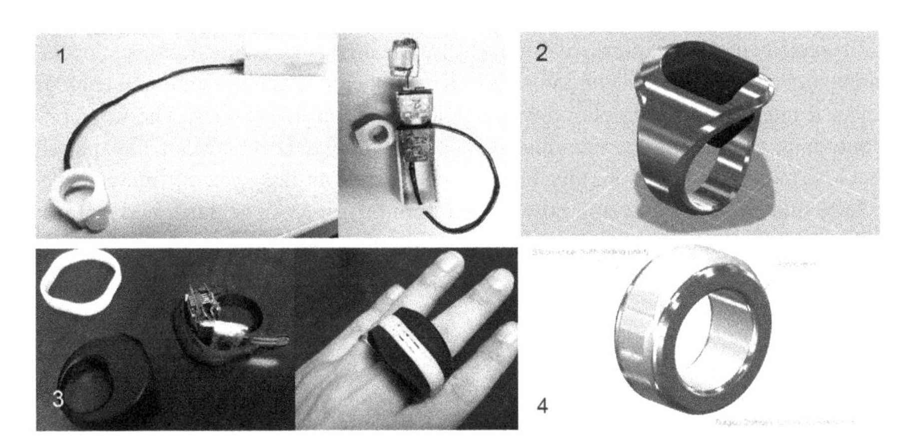

Figure 8 RingU prototype iteration. **1.** First prototype of RingU designed for the proof of concept **2.** RingU Prototype 1.1 **3.** RingU Prototype 1.2 **4.** Future RingU Prototype Design (in development)

more comfortable to wear, shown as prototype 2 and 3 in Figure 8. Currently we are developing a new design of the ring and we are rebuilding the hardware and ring design toward the aesthetic shown as prototype 4 in Figure 8.

Based on the first prototype of the RingU system, we researched how the RingU system can convey the people's intimacy and emotional communication messages through the subtle lightings and tactile. We explored how non-verbal stimuli implemented in the ring can prime the emotion of a text message, by assigning participants to rate their emotional responses corresponding to the message, vibro-tactile stimuli, and color lighting stimuli they received during the experiment using an emotion wheel evaluation system (Feldman Barrett and Russel 1998). For the emotional interaction, we are continuing our study about the correlation of the emotional feedback and the various lightings and tactile expressions from the RingU system. We also are researching how tactile feedback from the ring can accompany a more meaningful transfer of emotion compared to a mobile phone.

Conclusion

We believe that computer mediated touch is an important form of human communication and will allow a major improvement in achieving meaningful remote presence. To further this goal, we have developed two remote touch systems which believe can provide great benefits in remote mediated human communication. Huggy Pajama is a novel wearable system that promotes physical interaction in remote communication between parents and children by enabling each to hug one another through a hugging interface device and a wearable, hug reproducing jacket connected through the Internet. One major contribution is the design of a remote communication system with the ability to sense and reproduce touch and hugs between two people. An additional mode of communication is provided by the incorporated cute and expressive interfaces for the conveyance of emotions between parent and children. Our second interface, RingU, is a wearable system that enables physical interaction in remote communication between loved ones. The key contribution is the ability to reproduce a mini-hug experience between two paired people remotely in a natural, physical manner. Based on this system, future works include studying and analyzing how natural mini-hugging can be generated and if it helps people's intimate interactions while being physically separated.

References

Bakeman, R and Brown, J V (1980). Early interaction: Consequences for social and mental development at three years. *Child Development* 51(2): 437–447.

Brave, S and Dahley, A (1997). InTouch: A medium for haptic interpersonal communication. *Proceedings of CHI* (pp. 363–364).

Cha, J; Eid, M; Rahal, L and Saddik, A E (2008). HugMe: An interpersonal haptic communication system. *Haptic, Audio and Visual Environments and Games, 2008* (pp. 99–102).
Chang, A et al. (2001). Lumitouch: anemotional communication device. *Conference on Human Factors in Computing Systems: CHI'01 Extended Abstracts on Human Factors in Computing Systems* 31: 313–314.
Falikov, C J (1995). Training to think culturally: A multidimensional comparative framework. *Family Process* 34(4): 373–388.
Feldman, R; Singer, M and Zagoory, O (2010). Touch attenuates infants' physiological reactivity to stress. *Developmental Science* 13(2): 271–278.
Feldman Barrett, L and Russel, J A (1998). Independence and bipolarity in the structure of current affect. *Journal of Personality and Social Psychology* 74(4): 967.
Haans, A; de Nood, C and Ijsselsteijn, W A (2007). Investigating response similarities between real and mediated social touch: A first test. *CHI '07: CHI '07 Extended Abstracts on Human Factors in Computing Systems* (pp. 2405–2410).
Harlow, H F (1958). The nature of love. First published in *American Psychologist* 13: 673–685. http://psychclassics.yorku.ca/Harlow/love.htm.
Hassenzahl, M et al. (2012). All you need is love: Current strategies of mediating intimate relationships through technology. *ACM Transactions on Computer-Human Interaction* 19(4): 30:1–30:19.
Henley, N (1977). *Body Politics: Power, Sex and Nonverbal Communication.* Prentice Hall.
Heslin, R and Boss, D (1980). Nonverbal intimacy in airport arrival and departure. *Personality and Social Psychology Bulletin* 6: 248–252.
Kline, G A (2004). The discovery, elucidation, philosophical testing and formal proof of various exceptions to medical sayings and rules. *Canadian Medical Association Journal* 171(12): 1491–1492.
Lee, P et al. (2006). A mobile pet wearable computer and mixed reality system for human & poultry interaction through the internet. *Personal Ubiquitous Computing* 10(5): 301–317.
Lindeman, R W; Page, R; Yanagida, Y and Sibert, J L (2004). Towards full-body haptic feedback: the design and deployment of a spatialized vibrotactile feedback system. *VRST '04: Proceedings of the ACM Symposium on Virtual Reality Software and Technology* (pp. 146–149).
Liu, L (2010). iFeeling: Vibrotactile rendering of human emotions on mobile phones. *The First International Workshop on Mobile Multimedia Processing, 2008,* and *Lecture Notes in Computer Science 5960* (pp. 1–20).
Morhenn, V et al. (2011). Massage increases oxytocin and reduces adrenocorticotropin hormone in humans. *Alternative Therapies in Health and Medicine* 18(6): 11–18.
Mueller, F et al. (2005). Hug over a distance. *CHI '05: CHI '05 Extended Abstracts on Human Factors in Computing Systems* (pp. 1673–1676).
Schirmer, A et al. (2011). Squeeze me, but don't tease me: Human and mechanical touch enhance visual attention and emotion discrimination. *Social Neuroscience* 6(3): 219–230.
Short, J; Williams, E and Christie, B (1976). *The Social Psychology of Telecommunications.* New York: Wiley.
Slater, P (1990). *The Pursuit of Loneliness.* Boston: Beacon Press.
Thoits, P A (1985). Social support and psychological well-being: Theoretical possibilities. In: I G Sarason and B R Sarason (Eds.), *Social Support: Theory, Research and Applications* (pp. 51–72). Boston, MA: Martinus Nijhoff.
Van Erp, J B F and Toet, A (2013). How to touch humans: Guidelines for social agents and robots that can touch. *2013 Humaine Association Conference on Affective Computing and Intelligent Interaction (ACII)* (pp. 780–785).
Vaucelle, C and Abbas, Y (2007). Touch: Sensitive apparel. *CHI '07: CHI '07 Extended Abstracts on Human Factors in Computing Systems* (pp. 2723–2728).

Index

T.J. Prescott et al. (eds.), *Scholarpedia of Touch*, Scholarpedia,
DOI 10.2991/978-94-6239-133-8

Lightning Source UK Ltd.
Milton Keynes UK
UKHW02n0245200818
327436UK00001B/3/P